7th International Workshop DICE2014 Spacetime – Matter – Quantum Mechanics

Journal of Physics: Conference Series Volume 626

Castiglioncello, Italy
15-19 September 2014

ISBN: 978-1-5108-0801-0
ISSN: 1742-6588

Printed from e-media with permission by:

Curran Associates, Inc.
57 Morehouse Lane
Red Hook, NY 12571

Some format issues inherent in the e-media version may also appear in this print version.

Printed by Curran Associates, Inc. (2015)

For permission requests, please contact the Institute of Physics at the address below.

Institute of Physics
Dirac House, Temple Back
Bristol BS1 6BE UK

Phone: 44 1 17 929 7481
Fax: 44 1 17 920 0979

techtracking@iop.org

Additional copies of this publication are available from:

Curran Associates, Inc.
57 Morehouse Lane
Red Hook, NY 12571 USA
Phone: 845-758-0400
Fax: 845-758-2634
Email: curran@proceedings.com
Web: www.proceedings.com

Table of contents

Volume 626

7th International Workshop DICE2014 Spacetime – Matter – Quantum Mechanics
15–19 September 2014, Castiglioncello, Italy

Accepted papers received: 2 June 2015
Published online: 3 July 2015

Preface

Papers

Invited Lectures

Contributed Papers

Journal of Physics: Conference Series **626** (2015) 011001 doi:10.1088/1742-6596/626/1/011001

DICE 2014 : Spacetime Matter Quantum Mechanics ... news on missing links

Presented in this volume are the Invited Lectures and the Contributed Papers of the *Seventh International Workshop on Decoherence, Information, Complexity and Entropy – DICE 2014*, held at **Castello Pasquini, Castiglioncello (Tuscany), September 15-19, 2014**.

These proceedings are intended to reflect the lively exchange of ideas during the meeting for the interested public and the wider scientific community, as well as to provide a document of the scientific works presented. The number of participants has continued to grow, which may correspond to an increasing attraction, if not need, of such conference: Our very intention has always been to bring together leading researchers, advanced students, and renowned scholars from various areas, in order to stimulate new ideas and their exchange across the borders of specialization.

In this way, the series of meetings successfully continued from the beginning with DICE 2002,[1] followed by DICE 2004,[2] DICE 2006,[3] DICE 2008,[4] DICE 2010,[5] and DICE 2012.[6] This time, DICE 2014 brought together more than 120 participants representing more than 30 countries.

It has been a great honour and inspiration that we had with us Nobel Prize laureate Gerard 't Hooft (Utrecht – Keynote Lecture *"The Cellular Automaton Interpretation and Bell's Theorem"*), Fields Medal winner Alain Connes (Paris – Keynote Lecture *"Quanta of geometry"*), Professor Avshalom Elitzur (Rehovot – Keynote Lecture *"Voices of silence, novelties of noise: on some quantum hairsplitting methods with nontrivial consequences"*, in this volume) and Professor Mario Rasetti (Torino – Keynote Lecture *"The topological field theory of data: a possible new venue for data mining"*, in this volume). The opening Keynote Lecture *"History of electroweak symmetry breaking"* was presented by Sir Tom Kibble (London), co-discoverer of the Higgs mechanism, Sakurai Prize laureate and winner of, i.a., Dirac and Einstein Medals.

[1] *Decoherence and Entropy in Complex Systems* ed H-T Elze *Lecture Notes in Physics* **633** (Berlin: Springer, 2004)

[2] *Proceedings of the Second International Workshop on Decoherence, Information, Complexity and Entropy – DICE 2004* ed H-T Elze *Braz. Journ. Phys.* **35**, 2A & 2B (2005) pp 205–529; free access at: **www.sbfisica.org.br/bjp**

[3] *Proceedings of the Third International Workshop on Decoherence, Information, Complexity and Entropy – DICE 2006* eds H-T Elze, L Diósi and G Vitiello *Journal of Physics: Conference Series* **67** (2007); free access at: **www.iop.org/EJ/toc/1742-6596/67/1**

[4] *Proceedings of the Fourth International Workshop on Decoherence, Information, Complexity and Entropy – DICE 2008* eds H-T Elze, L Diósi, L Fronzoni, J J Halliwell and G Vitiello *Journal of Physics: Conference Series* **174** (2009); free access at: **http://www.iop.org/EJ/toc/1742-6596/174/1**

[5] *Proceedings of the Fifth International Workshop on Decoherence, Information, Complexity and Entropy – DICE 2010* eds H-T Elze, L Diósi, L Fronzoni, J J Halliwell, E Prati, G Vitiello and J Yearsley *Journal of Physics: Conference Series* **306** (2011); free access at: **http://iopscience.iop.org/1742-6596/306/1**

[6] *Proceedings of the Sixth International Workshop on Decoherence, Information, Complexity and Entropy – DICE 2012* eds H-T Elze, L Diósi, L Fronzoni, J J Halliwell, E Prati, G Vitiello and J Yearsley *Journal of Physics: Conference Series* **442** (2013); free access at: **http://iopscience.iop.org/1742-6596/442/1**

Journal of Physics: Conference Series **626** (2015) 011001 doi:10.1088/1742-6596/626/1/011001

Professor Kibble gave this lecture also at DISCRETE 2014 – Fourth Symposium on Prospects in the Physics of Discrete Symmetries, King's College London, 2-6 December 2014 – see: https://indico.cern.ch/event/328215/ – and we are greatly honoured and glad that it is included in the present volume.

Entitled *"Spacetime - Matter - Quantum Mechanics ... news on missing links"*, the conference took place in the very pleasant and inspiring atmosphere of Castello Pasquini – in beautiful surroundings, overlooking a piece of Tuscany's coast. The 5-day program addressed aspects of the following topics:

- Decoherence, quantum-to-classical transition, measurement problem
- Quantum-classical hybrids
- Quantum coherent processes all the way ... to biology
- Key experiments
- Emergent spacetime, matter, and symmetries
- Quantum gravity / cosmology and high-energy physics

A Roundtable Discussion dedicated to the public under the theme *"Progresso scientifico: speranze e paure"* formed an integral part of the program. With participation of V Domenici (Comune di Rosignano), H-T Elze (Università di Pisa), M Rasetti (Politecnico di Torino, Fondazione ISI), and G Vitiello (Università di Salerno), it attracted a large audience involved in the extended lively discussions.

The workshop has been organized by L Diósi (Budapest), H-T Elze (Pisa, chair), L Fronzoni (Pisa), J J Halliwell (London), C Kiefer (Cologne), E Prati (Milano) and G Vitiello (Salerno), with most essential help from our conference secretaries L Fratino (London), D Gigli (Pisa), A Lampo (Barcelona), and I Pozzana (Pisa).

Several institutions and sponsors supported the workshop and their representatives and, in particular, the citizens of Rosignano / Castiglioncello are deeply thanked for the generous help and kind hospitality:

- *Comune di Rosignano* – A Franchi (Sindaco di Rosignano), with V Domenici, D Donati, L Montagnani, V Moretti, P Nocchi and M Pia (Giunta Comunale di Rosignano).
- *Associazione Armunia* – A Fumarola and F Masi (Direttore), G Mannari (Programmazione), C Perna, F Bellini, M Nannerini, P Bruni and L Meucci (Tecnici). Special thanks go to G Mannari and her collaborators for advice and great help in all the matters that had to be dealt with, in order to run the meeting at Castello Pasquini smoothly.
- *Solvay Italia SA, Rosignano* – A De Lorenzo (Site Communication & Total Quality Manager).

Funds made available by Università di Pisa, Domus Galilaeana (Pisa), Centro Interdisciplinare per lo Studio dei Sistemi Complessi – CISSC (Pisa), Dipartimento di Fisica (Università di Salerno), Istituto Italiano per gli Studi Filosofici – IISF (Napoli), Institute of Physics Publishing – IOP (Bristol) and Springer Verlag (Heidelberg) are gratefully acknowledged.

We especially thank Laura Pesce (Vitrium Galleria – http://www.laurapesce.it) for the exposition of her artwork *"Riflessioni della Fisica"* at Castello Pasquini during the conference.

The proceedings have been edited by L Diósi, H-T Elze, L Fronzoni, J J Halliwell, C Kiefer, E Prati, and G Vitiello and the accepted papers follow essentially the order of presentation during the conference, separating, however, invited lectures and contributed papers.[7] In the name of all

[7] We regret that a number of invited lectures, cf. list and program at **mail.df.unipi.it/ elze/DICE2014.html**, could not be reproduced here, partly for copyright reasons.

Journal of Physics: Conference Series **626** (2015) 011001 doi:10.1088/1742-6596/626/1/011001

participants, we would like to thank S Toms with her collaborators at IOP Publishing (Bristol) for friendly advice and most valuable immediate help during the editing process and, especially, for their continuing efforts to make the Journal of Physics: Conference Series available to all.

Budapest, Pisa, London, Cologne, Milano, Salerno, May 2015

Lajos Diósi Hans-Thomas Elze Leone Fronzoni Jonathan Halliwell
Claus Kiefer Enrico Prati Giuseppe Vitiello

Peer review statement

All papers published in this volume of *Journal of Physics: Conference Series* have been peer reviewed through processes administered by the proceedings Editors. Reviews were conducted by expert referees to the professional and scientific standards expected of a proceedings journal published by IOP Publishing.

History of electroweak symmetry breaking

TWB Kibble
Blackett Laboratory, Imperial College, London SW7 2AZ
E-mail: t.kibble@imperial.ac.uk

Abstract. In this talk, I recall the history of the development of the unified electroweak theory, incorporating the symmetry-breaking Higgs mechanism, as I saw it from my standpoint as a member of Abdus Salam's group at Imperial College. I start by describing the state of physics in the years after the Second World War, explain how the goal of a unified gauge theory of weak and electromagnetic interactions emerged, the obstacles encountered, in particular the Goldstone theorem, and how they were overcome, followed by a brief account of more recent history, culminating in the historic discovery of the Higgs boson in 2012.

1. Introduction

I have always believed that it was a stroke of great good fortune that I was able to join the theoretical physics group at Imperial College less than three years after it was founded by Abdus Salam. Salam already had a top-rank international reputation for his work on renormalization theory. In 1959, the year I joined the group, he was elected as the youngest Fellow of the Royal Society, at the age of 33. He was a man of immense vivacity and charisma, and a most inspiring leader of the group.

It was a wonderful place for a young postdoctoral fellow setting out on a research career. Imperial was a very lively place, with numerous visitors from around the world, including Murray Gell-Mann, Stanley Mandelstam, Steven Weinberg, Ken Johnson and many others.

2. Physics in the 1950s

So what were we doing? The great success of the immediate post-war period had been the development of renormalization theory. This banished the infinities that had plagued quantum electrodynamics (QED), allowing calculations beyond the lowest order of approximation. With experiments measuring the magnetic moment of the electron and the Lamb shift in hydrogen, QED rapidly became the most accurately verified theory in the history of physics.

Following this triumph, the natural next goal was to find similarly successful theories of the other fundamental interactions, the strong nuclear force that binds the constituents in atomic nuclei and the weak nuclear interactions responsible for radioactive beta decay, which also play a key role in the energy generation mechanism in the Sun. Even better, one might hope for a unified theory of all of these; that is a goal that is still ahead of us. We hoped too at some stage to incorporate the fourth fundamental force, gravity, but that of course is so weak as to be negligible on particle-physics scales. That too is still a work in progress.

Most interest initially fell on the strong interactions, and there were candidate quantum field theories that were widely discussed, in particular Yukawa's meson theory, in which the π-mesons or pions played the role of the force carriers. The problem there, however, was that it was hard to do any meaningful calculations, because the only known calculational technique, perturbation theory, does not work if the 'small' dimensionless parameter, the analogue of the fine structure constant of QED, is of order unity, as appeared to be the case.

As a result, quantum field theory fell out of favour. Many physicists abandoned it in favour of S-matrix theory, based on a study of the complex analytic properties of scattering amplitudes, working on concepts like Regge poles and dispersion relations. There were, however, places, like Imperial College, where the flag of field theory was kept flying — another was Harvard. Certainly Salam never gave up his belief that a unified field theory of all the interactions would eventually be found.

However, the difficulties of calculating with strong interactions did lead some people to speculate that weak interactions might be a more promising initial target.

3. Approximate symmetries

The experimenters had been busy too. Using cloud chambers and later bubble chambers, they discovered a huge array of new 'elementary' particles. Over a few years the number went from a handful to a hundred or more.

To try to make sense of this zoo of particles, theorists looked for approximate symmetries, trying to find patterns among the particles, much as chemists had done in developing the periodic table in the nineteenth century.

The first of these symmetries was *isospin*, originally proposed by Heisenberg [1]. He noted that protons and neutrons are in many ways very similar; they have the same spin, the same strong interactions and nearly the same mass. The name 'isospin' does not imply any physical relation to spin angular momentum, but stems from a mathematical similarity with the description of electron spin. Heisenberg suggested that the proton and neutron might be regarded as two different states of a single entity, the *nucleon*, and that there is a symmetry under which one can rotate the proton into a neutron or a combination of the two, much as one can rotate the up-spin state of an electron into the down-spin state or something in between. In fact isospin and spin are both described by SU(2) groups.

This is of course an *approximate* symmetry: the proton and the neutron are not identical;. The proton has an electric charge, the neutron does not. The symmetry is *broken* by the electromagnetic interaction. However, on a nuclear scale, the electromagnetic interaction is relatively weak compared to the strong nuclear force, so isospin is still a fairly good approximation. It proved very useful for example in classifying low-lying energy levels of light nuclei.

Later, Kemmer [2] showed that the symmetry could be extended to the other known strongly-interacting particles, the pions, which form an isospin triplet of charge states (π^+, π^0, π^-), analogous to the spin states of a particle of spin $S = 1$ (in units of $\hbar$), and that one could write down a version of Yukawa's theory that was invariant under the isospin transformations. The isospin symmetry is now seen as fundamentally a symmetry between the two lightest *quarks*, the *up* and *down* quarks (u and d), the constituents of nucleons and pions.

The next step was taken in 1961, independently by Gell-Mann [3] and by Yuval Ne'eman [4], who was at the time a student of Salam's at Imperial College. They proposed a larger, more approximate, SU(3) symmetry, and showed that the newly discovered *hadrons*, or strongly interacting particles, could be grouped into two-dimensional patters of octets and decuplets, forming multiplets related by SU(3) symmetry transformations. Gell-Mann called this scheme the *eightfold way*. It is now seen as a symmetry of the *three* lightest quarks, up, down and *strange* (u, d, s).

4. Gauge theories

Quantum electrodynamics is a *gauge theory*. This means it has a special kind of symmetry, a *local* symmetry. In the quantum mechanics of a single electron, the physical situation is unchanged if the electron wave function is multiplied by a phase factor; we have invariance under the transformation

$$\psi(x) \to \psi(x)e^{i\alpha} \tag{1}$$

for any constant α. But in QED we have a more general invariance for the electron field, in which α becomes a function of space and time, *provided* that a corresponding transformation is applied to the electromagnetic potentials $A_\mu(x)$; specifically

$$\psi(x) \to \psi(x)e^{ie\alpha(x)}, \qquad A_\mu(x) \to A_\mu(x) - \partial_\mu\alpha(x). \tag{2}$$

This is a vitally important feature: the success of the renormalization procedure relies on its existence.

For this reason, many people, including Salam, believed that viable theories of the other interactions should also be gauge theories. The first gauge theory beyond QED was the Yang-Mills theory [5], proposed in 1954, a gauge theory based on the isospin SU(2) symmetry. The same theory was in fact written down in the same year by a student of Salam's, Ronald Shaw, though he never published it except as a Cambridge University PhD thesis [6]. Ultimately this did not prove to be the correct theory of strong interactions — that is a very different kind of gauge theory, *quantum chromodynamics* (QCD) — but it is nevertheless the foundation for all later work on gauge theories.

Because isospin is an approximate symmetry, it must be broken in some way. But that immediately led to problems, because simply adding symmetry-breaking terms destroys many of the nice properties of gauge theories. So people began to ask the question: could this be some sort of *spontaneous* symmetry breaking, avoiding the need to add explicit symmetry-breaking terms? This is a subject I shall return to shortly.

5. The goal of unification

Because of the difficulties already mentioned of calculating with a strong-interaction theory, people began to think that perhaps the weak interactions would be a better immediate target, especially after the development in 1958 of the $V - A$ theory by Feynman and Gell-Mann [7] and by Sudarshan and Marshak [8], which showed that they could be seen as proceeding via the exchange of spin-1 $W^\pm$ bosons, just as electromagnetic interactions proceed via the exchange of spin-1 photons. This led Schwinger [9] to suggest a gauge theory of weak interactions mediated by W^+ and $W-$ exchange. He even asked: could there be a unified theory of weak and electromagnetic interactions, involving three gauge bosons, W^+, W^- and the photon γ?

But this idea immediately ran into difficulty. If there is in fact a symmetry between these three gauge bosons, it clearly must be severely broken, because there are major differences between them. The fact that the weak interactions are short-range implies that the bosons that carry this interaction must be very massive, whereas the photon, mediating the long-range electromagnetic force, is massless. It was clear that if the interaction strengths are comparable, the $W^\pm$ must have masses of the order of 100 GeV.

There was another key difference: it was known that the weak interactions do not conserve parity — they violate mirror symmetry — whereas the electromagnetic interactions are parity-conserving. So how could there be a symmetry between the two?

This latter problem was solved in 1961 by Glashow [10], who proposed an extended model with a larger symmetry group, SU(2)×U(1), and a fourth gauge boson Z^0. He showed that by an intriguing mixing mechanism between the two neutral gauge bosons, one could end up with one boson (γ) with parity-conserving interactions and three that violate parity, W^+, W^- and Z^0.

In 1964, Salam and his long-term collaborator John Ward, apparently unaware of Glashow's work, proposed a very similar model also based on SU(2)×U(1) [11].

But in all these models, the symmetry breaking, giving the W and Z bosons masses, had to be inserted by hand, and theories of spin-1 bosons with explicit masses were well known to be non-renormalizable and thus unphysical. The big question therefore was: could this be a case of spontaneous symmetry breaking?

6. Spontaneous symmetry breaking

Spontaneous symmetry breaking occurs when the ground state or vacuum, or equilibrium state of a system does not share the underlying symmetries of the theory. It is ubiquitous in condensed matter physics, associated with phase transitions. Often, there is a high-temperature symmetric phase and a critical temperature below which the symmetry breaks spontaneously. A simple example is crystalization. If we place a round bowl of water on a table, it looks the same from every direction, but when it freezes the ice crystals form in specific orientations, breaking the full rotational symmetry. The breaking is spontaneous in the sense that, unless we have extra information, we cannot predict in which directions the crystals will line up. A similar thing happens when a ferromagnet is cooled through its Curie point; it may acquire a magnetization in one direction or the reverse, but we cannot say in advance which it will choose.

In 1960, Nambu [12] pointed out that gauge symmetry is broken in a superconductor when it goes through the transition from normal to superconducting, and that this gives a mass to the plasmon, although this view was still quite controversial in the superconductivity community (see also Anderson [13]). Nambu suggested that a similar mechanism might give masses to elementary particles. The next year, with Jona-Lasinio [14], he proposed a specific model, though not a gauge theory, a four-fermion interaction based on the interaction Lagrangian

$$\mathcal{L}_{\text{int}} = g[(\bar{\psi}\psi)^2 - (\bar{\psi}\gamma_5\psi)^2]. \tag{3}$$

Here the phase symmetry $\psi \to e^{i\alpha}\psi$ is unbroken, but the chiral symmetry $\psi \to e^{\alpha\gamma_5}\psi$ is spontaneously broken because $\langle\bar{\psi}\psi\rangle \neq 0$. This means that there is a non-zero mass m_ψ.

The model has a significant feature, a massless pseudoscalar particle, which Nambu and Jona-Lasinio tentatively identified with the pion. To account for the non-zero (though small) pion mass, they suggested that the chiral symmetry was not quite exact even before the spontaneous symmetry breaking.

Attempts to apply this idea to symmetry breaking of fundamental gauge theories however ran into a severe obstacle, the *Goldstone theorem*.

7. Nambu–Goldstone bosons

As in the Nambu–Jona-Lasinio model, the spontaneous breaking of a continuous symmetry often leads to the appearance of massless spin-0 particles. The simplest model that illustrates this is the *Goldstone model* [15], a model of a complex scalar field with the Lagrangian density

$$\mathcal{L} = \partial_\mu\phi^*\partial^\mu\phi - \mathcal{V}, \tag{4}$$

where $\mathcal{V}$ is the *sombrero potential*,

$$\mathcal{V} = \frac{1}{2}\lambda\left(\phi^*\phi - \frac{1}{2}\eta^2\right)^2, \tag{5}$$

where λ and η are positive constants. This Lagrangian is clearly invariant under global phase rotations, $\phi(x) \to \phi(x)e^{i\alpha}$.

The key feature of the potential is that it has a *maximum* at $\phi = 0$. Consequently, in the vacuum state, the expectation value of ϕ is not zero, but lies somewhere on the circle of minima:

$$\langle 0|\phi(x)|0\rangle = \frac{1}{\sqrt{2}}\eta e^{i\alpha}. \tag{6}$$

This breaks the symmetry. We have a continuous infinity of degenerate vacuum states, distinguished by the value of α, though the underlying symmetry is exact; all are in fact physically equivalent.

Suppose for example we choose α to be real, and set

$$\phi(x) = \frac{1}{\sqrt{2}}[\eta + \varphi_1(x) + i\varphi_2(x)], \tag{7}$$

where $\varphi_{\{1,2\}}$ are real. Then, expanding (5), we find

$$\mathcal{V} = \frac{1}{2}\lambda\eta^2\varphi_1^2 + \text{cubic and quartic terms.} \tag{8}$$

Thus φ_1 has a mass but φ_2 is massless. Looking at the figure, it is easy to see why this is. The φ_1 mass is determined by the curvature of $\mathcal{V}$ in the radial direction, but the φ_2 mass vanishes because in the transverse direction $\mathcal{V}$ is flat. The φ_2 excitations correspond to space-time dependent variations in the angle α defining the direction of symmetry breaking.

The appearance of these massless spin-zero *Nambu–Goldstone bosons* was believed to be an inevitable consequence of spontaneous symmetry breaking in a relativistic theory; this is the content of the *Goldstone theorem.* That is a problem because such massless particles, if they had any reasonable interaction strength, should have been easy to see, but none had been seen.

8. The Goldstone theorem

This problem was obviously of great concern to all those who were trying to build a viable gauge theory of weak interactions. When Steven Weinberg came to spend a sabbatical at Imperial College in 1961, he and Salam spent a great deal of time discussing the obstacles. They developed a proof of the Goldstone theorem, published jointly with Goldstone [16].

The argument is straightforward. We start with two assumptions. First, the symmetry in question corresponds to a current $j^\mu(x)$ which is conserved, i.e,

$$\partial_\mu j^\mu = 0, \tag{9}$$

and which generates the transformations in the sense that the variation of any field $\phi(x)$ is given by

$$\delta\phi(0) = i\epsilon \int d^3\mathbf{x}\, [\phi(0), j^0(0,\mathbf{x})], \tag{10}$$

where ϵ is an infinitesimal parameter. Second, there is a scalar field ϕ whose vacuum expectation value is not invariant under this transformation:

$$\langle 0|\delta\phi(0)|0\rangle \neq 0. \tag{11}$$

Now the continuity equation (9) would seem to imply the existence of a conserved charge operator:

$$\frac{dQ}{dt} = 0, \qquad Q(t) = \int d^3\mathbf{x}\, j^0(t,\mathbf{x}). \tag{12}$$

Then the broken symmetry condition (10) can be written

$$i\langle 0|[\phi(0), Q]|0\rangle = \eta \neq 0. \tag{13}$$

But if Q is time-independent, then the only intermediate states that can contribute in the commutator are *zero-energy* states, which can only appear if there are massless particles in the theory.

This argument seemed conclusive, with no escape. Spontaneous symmetry breaking implied massless spin-zero bosons, which should have been easy to see but had not been seen. On the other hand adding explicit symmetry-breaking terms led to non-remnormalizable theories predicting infinite results. Weinberg commented 'Nothing will come of nothing; speak again', a quotation from *King Lear*. Fortunately, however, our community was able to speak again.

9. Resolution of the impasse

In 1964 a young American postdoctoral fellow arrived at Imperial College, Gerald Guralnik. He had been a student of Walter Gilbert at Harvard, while Gilbert had earlier been a student of Salam in Cambridge. Guralnik had been studying the problem of giving masses to gauge bosons, and had already published some ideas about it [17, 18]. We began collaborating, with another American visitor, Carl Richard Hagen, to find a way round the obstacle posed by the Goldstone theorem. We, and of course others, eventually succeeded.

The argument fails in the case of a gauge theory, for quite subtle reasons to which I will return. The proof is valid, but there is a hidden assumption which, though seemingly natural, is violated by gauge theories. This was discovered independently by three groups, first Englert and Brout from Brussels [19], then Higgs from Edinburgh [20, 21] and finally Guralnik, Hagen and myself from Imperial College [22]. All three groups published papers in Physical Review Letters during the summer and autumn of 1964, papers that were selected by the journal, during its 50th anniversary celebrations as among the most significant of that year. (Higgs also wrote a paper for Physics Letters.)

The simplest model that illustrates this mechanism is the *Abelian Higgs model*, which is just a gauged version of the Goldstone model, with the Lagrangian

$$\mathcal{L} = D_\mu\phi^* D^\mu\phi - \frac{1}{4}F_{\mu\nu}F^{\mu\nu} - \mathcal{V}, \tag{14}$$

where

$$D_\mu\phi = \partial_\mu\phi + ieA_\mu\phi, \qquad F_{\mu\nu} = \partial_\mu A_\nu - \partial_\nu A_\mu, \tag{15}$$

and $\mathcal{V}$ as before is the sombrero potential (5). Once again, ϕ acquires a non-zero vacuum expectation value (vev), lying somewhere on the circle of minima (6).

Let us again choose $\alpha = 0$, and write ϕ in terms of vev plus real and imaginary parts. It is convenient also to define a new vector field

$$B_\mu = A_\mu - \frac{1}{e\eta}\partial_\mu\varphi_2. \tag{16}$$

Note that because this has the form of a gauge transformation, we have

$$F_{\mu\nu} = \partial_\mu B_\nu - \partial_\nu B_\mu. \tag{17}$$

Substituting into (14), we find

$$\mathcal{L} = \frac{1}{2}\partial_\mu\varphi_1\partial^\mu\varphi_1 - \frac{1}{4}F_{\mu\nu}F^{\mu\nu} - \frac{1}{2}\lambda\eta^2\varphi_1^2 + \frac{1}{2}e^2\eta^2 B_\mu B^\mu + \text{cubic terms} + \dots, \tag{18}$$

Thus we have arrived at a model with no massless particles, only a massive scalar φ_1 and a massive vector, B_μ. Seemingly almost by magic, the massless gauge and Goldstone bosons have combined to give a massive gauge boson.

But there is more to the story than this. We have to ask about the original fields ϕ and A_μ, The field equations, in particular the Maxwell equation

$$\partial_\mu F^{\mu\nu} = j^\nu = -e^2\eta^2 B^\nu + \dots \tag{19}$$

are satisfied by any arbitrary value of φ_2 so long as B_μ, given by (16), vanishes, or is some other solution. This is simply a statement of the gauge invariance of the original model.

To tie down not only B_μ but also the original fields A_μ and φ_2 we need to impose a gauge condition. There are two obvious choices. We could select the Coulomb gauge condition $\partial_k A^k = 0$. Then $B_\mu = 0$ requires that φ_2 also vanishes, or at most is a constant.

Alternatively, we may choose the Lorentz gauge condition $\partial_\mu A^\mu = 0$. In that case $B_\mu = 0$ only requires that φ_2 satisfy the wave equation $\partial_\mu \partial^\mu \varphi_2 = 0$. In this manifestly covariant gauge, the Goldstone theorem does apply; φ_2 is a massless scalar field. But the Goldstone boson is a pure gauge mode; φ_2 has vanishing matrix elements between physical states.

10. Avoiding the Goldstone theroem

How then did we manage to avoid the Goldstone theorem? Where did the argument go wrong?

In the proof, I said that the continuity equation (9) implied the existence of a conserved charge, (12). But this is only true if we can drop a surface integral at infinity:

$$\frac{dQ(t)}{dt} = \int d^3\mathbf{x}\, \partial_0 j^0(t,\mathbf{x}) = -\int d^3\mathbf{x}\, \partial_k j^k(t,\mathbf{x}) = -\int dS_k\, j^k(t,\mathbf{x}). \tag{20}$$

Dropping this is legitimate in a manifestly Lorentz-invariant theory (e.g. Lorentz-gauge QED), because what we are really interested in are commutators, and for such a theory the commutators vanish outside the light cone. But it is *not* legitimate in Coulomb-gauge QED where commutators fall off quite slowly with distance.

In fact, as shown in our paper [22], it turns out that when the symmetry is spontaneously broken, the integral defined in (12) is not merely time-varying, but actually *does not exist* as a self-adjoint operator. For example, in the Abelian Higgs model, it has the form

$$Q = -e^2\eta^2 \int d^3\mathbf{x}\, B^0(t,\mathbf{x}) + \dots \tag{21}$$

and this integral is clearly divergent.

Related to the non-existence of Q is the fact that no unitary transformation connecting the distinct degenerate vacuum states can be constructed from the field operators. The distinct degenerate vacua belong to distinct orthogonal Hilbert spaces bearing *unitarily inequivalent representations* of the canonical commutation relations. This is now recognized as a defining feature of spontaneous symmetry breaking.

11. Electroweak unification

The 1964 papers from the three groups attracted very little attention at the time. Talks on the subject were often greeted with scepticism. By the end of that year, the mechanism was known, and Glashow's (and Salam and Ward's) SU(2) × U(1) model was known. But, surprisingly perhaps, it still took three more years for anyone to put the two together. This may have been in part at least because many of us were still thinking primarily of a gauge theory of strong interactions, not weak.

In early 1967, I did some further work on the detailed application of the mechanism to models with larger symmetries than U(1), in particular on how the symmetry breaking pattern determines the numbers of massive and massless particles [23]. I had some lengthy discussions with Salam on this subject, which I believed helped to renew his interest in the subject.

A unified gauge theory of weak and electromagnetic interactions of leptons was first proposed by Weinberg later that year [24]. Essentially the same model was presented independently by Salam in lectures he gave at Imperial College in the autumn of 1967 — he called it the *electroweak theory*. (I was not present because I was in the United States, but I have had accounts from others who were.) Salam did not publish his ideas until the following year, when he spoke at a Nobel Symposium [25], largely perhaps because his attention was concentrated on the development in its crucial early years of his International Centre for Theoretical Physics in Trieste.

Weinberg and Salam both speculated that their theory was renormalizable, but they could not prove it. An important step was the working out by Faddeev and Popov of a technique for applying Feynman diagrams to gauge theories [26]. Renormalizability was finally proved by a young student, Gerard 't Hooft [27], in 1971, a real *tour de force* using methods developed by his supervisor, Martinus Veltman, especially the computer algebra programme *Schoonship*.

In 1973, the key prediction of the electroweak theory, the existence of the *neutral current* interactions — those mediated by Z^0 — was confirmed at CERN [28]. That discovery led to the award of Nobel Prizes to Glashow, Salam and Weinberg in 1979. However Ward was left out, even though he had been a co-author on almost all of Salam's papers on the subject, possibly because of the 'rule of three', the restriction that the Prize cannot be split among more than three people. But 't Hooft and Veltman had to wait until 1999 for their Nobel Prizes.

The next major step was the discovery of the W and Z particles at CERN in 1983 [29, 30], which led to Nobel Prizes for the experimenters, Carlo Rubbia and Simon van der Meer.

Meanwhile, during the 1970s and 1980s there had been separate, parallel development of a gauge theory of strong interactions, *quantum chromodynamics* (QCD). This is a gauged SU(3) theory, which solves the problem of making the interaction short range by a completely different mechanism, not involving spontaneous symmetry breaking. So we now have the SU(3) $\times$ SU(2) $\times$ U(1) *standard model*, whose predictions have been verified with increasingly impressive accuracy.

12. The Higgs boson

In 1964, or 1967, the existence of a massive scalar boson had been a rather minor and unimportant feature. The important thing was the *mechanism* for giving masses to gauge bosons and avoiding the appearance of massless Nambu–Goldstone bosons.

But after 1983, the Higgs boson began to assume a key importance as the *only* remaining undiscovered piece of the standard-model jigsaw — apart that is from the last of the six quarks, the *top quark*. The standard model worked so well that the Higgs boson, or something else doing the same job, more or less *had* to be present. Finding the boson was one of the main motivations for building the Large Hadron Collider (LHC) at CERN. Over a period of more than twenty years, the two great collaborations, ATLAS and CMS, have designed, built and operated their two huge and massively impressive detectors.

As is by now well known, their efforts were rewarded in 2012 by the unambiguous discovery of the Higgs boson by each of the two detectors [31, 32]. This led to Nobel Prizes for Englert and Higgs in 2013. Nambu had earlier in 2008 received a Prize for his role in introducing the idea of spontaneous symmetry breaking into particle physics.

This is the end of a chapter, the completion of the standard model. But it is certainly not the end of the story. The model has been amazingly successful; almost all the experimental tests have proved positive. But the model cannot be the last word. It is in some ways a mess, with around twenty arbitrary parameters such as mass ratios that we cannot predict. It is not truly a unified model, because the symmetry group has three factors, each with its own coupling strength, though there have of course been efforts to combine the three pieces into a *grand unified theory* with an even larger symmetry that shows up at energies orders of magnitude above the scale accessible by present-day experiments. This might require the existence of *supersymmetry*, relating fermions to bosons and vice versa, but disappointingly there is as yet no real empirical evidence for its existence.

Moreover, there are quite a few things that the standard model does not explain. We do not know what constitutes the dark matter revealed by rotation curves of galaxies, or the dark energy responsible for its recently accelerating expansion. We do not know why the elementary particles come in three families with very similar structure but wildly differing masses. Nor do we know why the neutrinos have non-zero, though tiny, masses. And finally of course, we cannot yet find a place for gravity in the standard model. Perhaps for that we need string theory or M-theory — or who knows what?

Acknowledgments

I am grateful to the organizers of DICE2014, held at Castiglioncello, 15–19 September 2014 for inviting me to give this talk, and also to the organizers of DISCRETE 2014, at King's College London, 2–6 December 2014, who asked me to repeat it.

I wish especially to acknowledge the great debt I owe to two people who are sadly no longer with us. Abdus Salam had a huge influence on the direction of my research. He was a very supportive and inspiring leader of our group, but sadly died prematurely in 1996. And earlier this year, I lost my erstwhile collaborator and lifelong close friend Gerry Guralnik, who died of a sudden heart attack just after giving a lecture.

References

[1] Heisenberg W 1932 Über den Bau der Atomkerne I *Z. Phys.* **77** 1–11
[2] Kemmer N 1938 The charge-dependence of nuclear forces *Proc. Camb. Phil. Soc.* **34** 354–364
[3] Gell-Mann M 1961 *The Eightfold Way: A theory of strong interaction symmetry*, California Institute of Technology Synchrotron Laboratory Report, CTSL-20, unpublished
[4] Ne'eman Y 1961 Derivation of strong interactions from a gauge invariance *Nucl. Phys.* **26** 222–229
[5] Yang CN and Mills RL 1954 Conservation of isotopic spin and isotopic gauge invariance *Phys. Rev.* **96** 191–195
[6] Shaw R 1955 *Invariance under general isotopic gauge transformations* Cambridge University PhD thesis, part II, chapter III, unpublished.
[7] Feynman RP and Gell-Mann M 1958 Theory of the Fermi interaction *Phys. Rev.* **109** 193–197
[8] Sudarshan ECG and Marshak RE. 1958 Chirality invariance and the universal Fermi interaction *Phys. Rev.* **109** 1860–1862
[9] Schwinger JS 1957 A theory of the fundamental interactions *Annals Phys.* **2** 407–434
[10] Glashow SL 1961 Partial symmetries of weak interactions *Nucl. Phys.* **22** 579–588
[11] Salam A and Ward JC 1964 Electromagnetic and weak interactions *Physics Letters* **13** 168–171
[12] Nambu Y 1960 Quasi-particles and gauge invariance in the theory of superconductivity *Phys. Rev.* **117** 648–663
[13] Anderson PW 1963 Plasmons, gauge Invariance, and mass *Phys. Rev.* **130** 439–442
[14] Nambu Y and Jona-Lasinio G 1961 Dynamical model of elementary particles based on an analogy with superconductivity. I *Phys. Rev.* **122** 345–358
[15] Goldstone J 1961 Field theories with superconductor solutions *Nuovo Cim.* **19** 154–164
[16] Goldstone J, Salam A and Weinberg S 1962 Broken symmetries *Phys. Rev.* **127** 965–970
[17] Guralnik GS 1964 Photon as a symmetry-breaking solution to field theory. I *Phys. Rev.* **136** B1404–1416
[18] Guralnik GS 1964 Photon as a symmetry-breaking solution to field theory. II *Phys. Rev.* **136** B1417–1422
[19] Englert F and Brout R 1964 Broken symmetry and the mass of gauge vector bosons *Phys. Rev. Lett.* **13** 321–323
[20] Higgs PW 1964 Broken symmetries, massless particles and gauge fields *Phys. Lett.* **12** 132–133
[21] Higgs PW 1964 Broken symmetries and the masses of gauge bosons *Phys. Rev. Lett.* **13** 508–509
[22] Guralnik GS, Hagen CR and Kibble TWB. 1964 Global conservation laws and massless particles *Phys. Rev. Lett.* **13** 585–587
[23] Kibble TWB 1967 Symmetry breaking in non-Abelian gauge theories *Phys. Rev.* **155** 1554–1561

Journal of Physics: Conference Series **626** (2015) 012001 doi:10.1088/1742-6596/626/1/012001

[24] Weinberg S 1967 A model of leptons *Phys. Rev. Lett.* **19** 1264–1266
[25] Salam A. 1968 Weak and electromagnetic interactions *Elementary particle theory, Proceedings of the Nobel Symposium held in 1968 at Lerum, Sweden* ed N Svartholm (Stockholm: Almqvist & Wiksell) pp 367–377
[26] Faddeev LD and Popov VN 1967 Feynman diagrams for the Yang–Mills field Phys. Lett. B **25** (1967) 29.
[27] 't Hooft G 1971 Renormalizable Lagrangians for massive Yang-Mills fields Nucl. Phys. B **35** (1971) 167–188
[28] Hasert FJ *et al* [Gargamelle Neutrino Collaboration] 1973 Observation of neutrino like interactions without muon or electron in the Gargamelle neutrino experiment *Phys. Lett. B* **46** 138–140
[29] Arnison G *et al* [UA1 Collaboration] 1983 Experimental observation of isolated large transverse energy electrons with associated missing energy at $s^{1/2} = 540$ GeV *Phys. Lett. B* **122** (1983) 103–116
[30] Banner M *et al* [UA2 Collaboration] 1983 Observation of single isolated electrons of high transverse momentum in events with missing transverse energy at the CERN anti-p p collider *Phys. Lett. B* **122** 476–485
[31] Aad G *et al* [ATLAS Collaboration] 2012 Observation of a new particle in the search for the standard model Higgs boson with the ATLAS detector at the LHC *Physics Letters B* **716** 1–29
[32] Chatrchyan S *et al* [CMS Collaboration] 2012 Observation of a new boson at a mass of 125 GeV with the CMS experiment at the LHC *Physics Letters B* **716** 30–61

Inflation from gravitino condensates

Nick E Mavromatos

King's College London, Department of Physics, Theoretical Particle Physics and Cosmology Group, Strand London WC2R 2LS, UK;
Currently also at: CERN, Theory division, CH 1211 Geneva 23, Switzerland.

E-mail: Nikolaos.Mavromatos@kcl.ac.uk

Abstract. We review work on the formation of gravitino condensates via the super-Higgs effect in the early Universe. This is a scenario for both inflating the early universe and breaking local supersymmetry (supergravity), entirely independent of any coupling to external matter. The goldstino mode associated with the breaking of (global) supersymmetry is "eaten" by the gravitino field, which becomes massive (via its own vacuum condensation) and breaks supergravity dynamically. The most natural association of gravitino condensates with inflation proceeds in an indirect way, via a Starobinsky-type inflation, in the massive gravitino phase. This inflationary phase is associated with scalar modes hidden in the higher order curvature corrections of the effective action arising from integrating out massive gravitino degrees of freedom. The scenario is in agreement with Planck data phenomenology in a natural and phenomenologically-relevant range of parameters, namely Grand-Unified-Theory values for the supersymmetry breaking energy scale and dynamically-induced gravitino mass. A hill-top inflation, on the other hand, which could also occur in the model, whereby the rôle of the inflaton field is played by the gravitino condensate itself, would require significant fine tuning in the inflaton's wave function renormalisation and thus may be discarded on naturalness grounds.

1. Introduction and summary

The inflationary paradigm offers an elegant solution to the horizon and flatness problems of the standard Big Bang cosmology, whilst simultaneously seeds both the large-scale structure of the universe and temperature anisotropies of the CMB via quantum fluctuations occurring during the inflationary epoch. The precise microscopic mechanism of inflation is however unknown at present.

The data favour, or - from a rather more conservative viewpoint - are in agreement with, a scalar field or fields with canonical kinetic terms slowly rolling down an almost flat potential in the context of Einstein gravity, generating in the process 50 - 60 e-folds of inflation, along with adiabatic, nearly scale invariant primordial density perturbations [1, 2]. From the best fit value of the running spectral index $n_s \sim 0.96$ for the gravitational perturbations in the slow-roll parametrisation, found by Planck [1], and the usual relations among the slow-roll inflationary parameters [2] $n_s = 1 - 6\epsilon + 2\eta\,, r = 16\epsilon$, one may deduce $r \le 0.11$, given the non-observation of primordial gravitational wave-like (transverse and traceless) perturbations by Planck or WMAP collaborations (that is the absence of B-mode polarisations). This observational fact implies that the energy scale E_I of inflation is much smaller than the Planck scale m_P, lying in the ballpark of the Grand Unified Theory (GUT) scale [1, 2] $E_I = \left(3\,H_I^2 M_{\rm Pl}^2\right)^{1/4} \simeq 2.1 \times 10^{16} \times \left(\frac{r}{0.20}\right)^{1/4}$ GeV, with $M_{\rm Pl} = 2.4 \times 10^{18}$ GeV the reduced Planck mass, H_I the Hubble scale during inflation and

Journal of Physics: Conference Series **626** (2015) 012002 doi:10.1088/1742-6596/626/1/012002

r the tensor-to-scalar perturbation ratio [2].The upper bound on $r < 0.11$ placed by the Planck Collaboration [1] implies

$$H_I = 1.05 \left(\frac{r}{0.20}\right)^{1/2} \times 10^{14}\ \text{GeV} \leq 0.78 \times 10^{14}\ \text{GeV}\ , \tag{1}$$

that is an upper bound four orders of magnitudes smaller than the reduced Planck mass.

An important issue at present is the extent to which this inflationary process is tied to physics at the Grand Unification (GUT) scale, and in particular, to a possible supersymmetric phase transition occurring in the early universe. Links of supersymmetry to inflation may be arguably expected from the fact that supersymmetry provides a rather natural reason [3] for the smallness (compared to Planck scale) value of the inflationary Hubble scale (1). If supersymmetry is realised in nature however, it is certainly broken. It is known that simple realisations of global supersymmetry (SUSY) breaking, such as in the Wess-Zumino model [4], can provide, when embedded in gravitational environments, slow-roll models for inflation consistent with the Planck data [1]. Rigorous embeddings of global SUSY to local supersymmetry (SUGRA) have also been considered and explored in the literature over the years in connection with various scenarios for inflation [5], such as hybrid [6], chaotic [7], no-scale SUGRA [8], Jordan-frame-SUGRA [9] and others [10]. In the case of *some* no-scale SUGRA models, the inflationary potential can be made similar to that of the so-called Starobinsky model of inflation [11], by a specific choice of parameters. Unlike the SUGRA case, however, in the original Starobinsky model, the inflaton scalar mode represents collectively the effects of the quadratic-order scalar curvature terms (R^2) in the effective gravitational action, obtained by integrating out (conformal) matter in the early Universe [11]. Starobinsky-type models seem to be preferred by the Planck data [1] [1].

In this talk we shall present a rather minimal inflationary scenario which is associated indirectly with a Starobinsky type inflation in its original sense [11], namely based on scalar modes associated with higher curvature terms of the (one-loop) effective action of simple (3+1)-dimensional supergravity models, obtained by integrating out massive gravitino degrees of freedom in the broken supergavity phase. The approach is documented in a series of previous publications [15, 16, 17], and is based on the possibility of dynamically breaking SUGRA solely by means of exploiting the four-gravitino interactions that characterise (any) supergravity action, via the fermionic torsion parts of the spin connection. The primary example, where the calculations of the effective potential were detailed, was that of $\mathcal{N} = 1$, $D = 4$ simple SUGRA without matter [18, 19]. The dynamical breaking process may be concretely realised by means of a phase transition from the supersymmetric phase where the bilinear $\langle \overline{\psi}_\mu \psi^\mu \rangle$ representing the effective scalar degree of freedom has zero vacuum expectation value, to one where $\sigma \equiv \langle \overline{\psi}_\mu \psi^\mu \rangle \neq 0$. The quantum excitations about this condensate vacuum are then identified with a gravitino condensate scalar field. Since this must be an energetically favourable process to occur, it then follows that the effective potential experienced by the gravitino condensate must be locally concave about the origin.

The corresponding one-loop effective potential of the gravitino condensate scalar field, obtained after integrating out fermionic (gravitino) and bosonic (graviton) degrees of freedom therefore has the characteristic form of a Coleman-Weinberg double well potential, offering the possibility of hilltop-type inflation, with the condensate field playing the role of the inflaton [15, 16], which would be the simplest scenario. However, in order to guarantee a slow-roll inflation one needs unnaturally large values of the gravitino-condensate wave function

[1] The most recent Planck data, released in February 2015 [1], did not confirm previous claims by the BICEP2 Collaboration [12] on the observation of a large $r \sim 0.16$ due to fundamental primordial graviton fluctuations. These most recent data confirmed earlier indications by the Planck collaboration [13] towards significant foreground pollution in the BICEP2 region that would prevent the interpretation of the BICEP2 B-mode polarization data as being due to fundamental physics (see also ref. [14]).

renormalisation. This prompted us to discuss a second, rather indirect way, by means of which the gravitino condensate is associated with inflation [17]. This is realised via the higher (in particular quadratic-order) curvature corrections of the (quantum) effective action of the gravitino condensate field, obtained after integration of graviton and gravitino degrees of freedom in the massive gravitino phase. These corrections induce a Starobinsky-type inflation [11], which occurs for quite natural values of the parameters of the $\mathcal{N}=1$ SUGRA model and its variants, as we shall review below.

The structure of the talk is as follows: In Section 2 we review the formalism and physical concepts underlying dynamical breaking of SUGRA and the associated super-Higgs effect, within the context of simple four dimensional $\mathcal{N}=1$ models, including superconformal extensions thereof (with broken conformal symmetry) that are necessitated for phenomenological reasons, as explained in the text. In Section 3 we discuss the simplest possible scenario for hilltop inflation, where gravitino condensate fields near the origin of the effective potential play the role of the inflation field. Unfortunately, for (phenomenologically desirable) supersymemtry breaking scales that are near or below the GUT scale, the model is compatible with slow roll for very large (unnatural) values of the condensate wave function renormalisation. This prompts us to discuss in Section 4 alternative scenarios for inflation of Starobinsky type that may occur in the massive gravitino phase, near the non-trivial minimum of the effective potential. In such scenarios, which are compatible with the Planck results, the rôle of the inflaton field is played by the scalar mode that describes the effects of scalar-curvature-square terms that characterise the gravitational sector of the effective action in the broken SUGRA phase, after integrating out the massive gravitinos. Finally, conclusions and outlook are presented in section 5.

2. Super-Higgs effect and dynamical breaking of $\mathcal{N}=1$ SUGRA

Our starting point is the $\mathcal{N}=1$ $D=4$ (on-shell) action for 'minimal' Poincaré supergravity in the second order formalism (with explicit factors of the (dimensionful) gravitational constant $\kappa^2 = 8\pi \mathrm{G} = 1/M_{\mathrm{Pl}}^2$, in units $\hbar = c = 1$, where M_{Pl} the reduced Planck mass in four space-time dimensions) [18, 19] :

$$S_{\mathrm{SG}} = \int d^4x\, e\left(\frac{1}{2\kappa^2}R\,(e) - \overline{\psi}_\mu \gamma^{\mu\nu\rho} D_\nu \psi_\rho + \mathcal{L}_{\mathrm{torsion}}\right),\ \gamma^{\mu\nu\rho} = \frac{1}{2}\left\{\gamma^\mu, \gamma^{\nu\rho}\right\},\ \gamma^{\nu\rho} = \frac{1}{2}\left[\gamma^\nu, \gamma^\rho\right], \quad (2)$$

where e is the vierbein determinant and $R(e)$ and $D_\nu\psi_\rho \equiv \partial_\nu\psi_\rho + \frac{1}{4}\omega_{\nu ab}\,(e)\,\gamma^{ab}\psi_\rho$ are defined via the torsion-free connection. Given the gauge condition

$$\gamma_\mu\, \psi^\mu = 0\ , \quad (3)$$

one can write the term $\mathcal{L}_{\mathrm{torsion}}$ as:

$$\mathcal{L}_{\mathrm{torsion}} = -\frac{1}{8}\,\kappa^2\left(\left(\overline{\psi}^\rho \gamma^\mu \psi^\nu\right)\left(\overline{\psi}_\rho \gamma_\mu \psi_\nu + 2\overline{\psi}_\rho \gamma_\nu \psi_\mu\right)\right). \quad (4)$$

This term arises from the fermionic torsion parts of the spin connection [2].

[2] Note that four-fermion interactions are characteristic of any Einstein-Cartan theory of fermions in curved space-time [20], with non-zero fermionic torsion. In fact, in a standard spin-1/2 fermion-gravity theory, the torsion-induced four fermion interactions assume a repulsive axial (pseudovector)-current-current form $-\left(\overline{\psi}\gamma^\mu\gamma^5\psi\right)\left(\overline{\psi}\gamma_\mu\gamma^5\psi\right)$. As we demonstrate in the second article of ref. [16], a corresponding repulsive axial-current-current term for the gravitino torsion terms can also be obtained by appropriately utilising Fierz identities in analogy with the Einstein-Cartan theory. One thus would naively conclude that dynamical breaking of supergravity may not be possible. However, in all such theories, the Fierz identities among the fermions, including the case of gravitinos, make the actual coefficient of such terms ambiguous. Only the non-perturbative physics can settle the value of the four-fermion terms [21], and hence dynamical breaking of symmetry via scalar gravitino condensates is a realistic possibility.

Extending the action off-shell requires the addition of auxiliary fields to balance the graviton and gravitino degrees of freedom. These fields however are non-propagating and may only contribute to topic at hand through the development of scalar vacuum expectation values, which would ultimately be resummed into the cosmological constant. Making further use of the gauge condition (3) in concert with the Fierz identities (as detailed in the second reference in [16]), we may write

$$\mathcal{L}_{\text{torsion}} = \lambda_{\text{S}} \left(\overline{\psi}^{\rho} \psi_{\rho}\right)^2 + \lambda_{\text{PS}} \left(\overline{\psi}^{\rho} \gamma^5 \psi_{\rho}\right)^2 + \lambda_{\text{PV}} \left(\overline{\psi}^{\rho} \gamma^5 \gamma_{\mu} \psi_{\rho}\right)^2 \tag{5}$$

where the couplings λ_{S}, λ_{PS} and λ_{PV} express the freedom we have to rewrite each quadrilinear in terms of the others via Fierz transformation. They satisfy the constraint: $\lambda_{\text{S}} - \lambda_{\text{PS}} + 4\lambda_{\text{PV}} = -3\kappa^2/4$. This freedom in turn leads to a known ambiguity in the context of mean field theory [21], which we addressed in the second reference in [16], where we refer the reader for details.

Following the original ideas of dynamical symmetry breaking by Nambu and Jona-Lasinio [22], we wish to linearise these four-fermion interactions via suitable auxiliary fields, e.g.

$$\frac{1}{4}\left(\overline{\psi}^{\rho} \psi_{\rho}\right)^2 \sim \sigma \left(\overline{\psi}^{\rho} \psi_{\rho}\right) - \sigma^2 , \tag{6}$$

where the equivalence (at the level of the action) follows as a consequence of the subsequent Euler-Lagrange equation for the auxiliary scalar σ. Our task is then to look for a non-zero vacuum expectation value $\langle\sigma\rangle$ which would serve as an effective mass for the gravitino. To induce the super-Higgs effect [23] we also couple in the Goldstino associated to global supersymmetry breaking via the addition of

$$\mathcal{L}_{\lambda} = f^2 \det \left(\delta_{\mu\nu} + \frac{i}{2f^2} \overline{\lambda} \gamma_{\mu} \partial_{\nu} \lambda\right)\Big|_{\gamma\cdot\psi=0} = f^2 + \dots , \tag{7}$$

where λ is the Goldstino, $\sqrt{f}$ expresses the scale of global supersymmetry breaking, and $\dots$ represents higher order terms which may be neglected in our weak-field expansion of the determinant. It is worth emphasising at this point the universality of (7); any model containing a Goldstino may be related to $\mathcal{L}_{\lambda}$ via a non-linear transformation [24], and thus the generality of our approach is preserved.

Upon the specific gauge choice (3) for the gravitino field and an appropriate redefinition, one may eliminate any presence of the Goldstino field from the final effective action describing the dynamical breaking of local supersymmetry, except the cosmological constant term f^2 in (7), which serves as a reminder of the pertinent scale of supersymmetry breaking. The non-trivial energy scale this introduces, along with the disappearance (through field redefinitions) of the Goldstino field from the physical spectrum and the concomitant development of a gravitino mass, characterises the super-Higgs effect.

We may then identify in the broken phase an effective action

$$S = \frac{1}{2\kappa^2} \int d^4x\, e \left(\left(R\left(e\right) - 2\Lambda\right) - \overline{\psi}_{\mu} \gamma^{\mu\nu\rho} D_{\nu} \psi_{\rho} + m_{\text{dyn}} \left(\overline{\psi}_{\mu} \psi^{\mu}\right)\right) , \tag{8}$$

where Λ is the renormalised cosmological constant, to be contrasted with the (negative) tree level cosmological constant

$$\Lambda_0 \equiv \kappa^2 \left(\sigma^2 - f^2\right) , \tag{9}$$

and $m_{\text{dyn}} \propto \langle\sigma\rangle$ is a dynamically generated gravitino mass, the origin of which will be explained presently. It is worth stressing at this point that Λ_0 must be negative due to the incompatibility of supergravity with de Sitter vacua; if SUGRA is broken at tree level, then of course no further dynamical breaking may take place.

For phenomenological reasons which have been outlined in detail in refs. [15, 16], and we shall discuss below, we adopt an extension of $\mathcal{N} = 1$ SUGRA which incorporates local supersymmetry in the Jordan frame, enabled by an associated dilaton superfield [9]. The scalar component φ of the latter can be either a fundamental space-time scalar mode of the gravitational multiplet, i.e. the trace of the graviton (as happens, for instance, in supergravity models that appear in the low-energy limit of string theories), or a composite scalar field constructed out of matter multiplets. In the latter case these could include the standard model fields and their superpartners that characterise the Next-to-Minimal Supersymmetric Standard Model, which can be consistently incorporated in such Jordan frame extensions of SUGRA [9]. Upon appropriate breaking of conformal symmetry, induced by specific dilaton potentials (which we do not discuss here), one may assume that the dilaton field acquires a non-trivial vacuum expectation value $\langle\varphi\rangle \neq 0$. One consequence of this is then that in the broken conformal symmetry phase, the resulting supergravity sector, upon passing (via appropriate field redefinitions) to the Einstein frame is described by an action of the form (2), but with the coupling of the gravitino four-fermion interaction terms being replaced by

$$\tilde{\kappa} \equiv e^{-\langle\varphi\rangle}\kappa \ , \tag{10}$$

while the Einstein term in the action carries the standard gravitational coupling $1/2\kappa^2$.

Expanding the graviton field about a de Sitter background [25] (under the assumption that it is a solution of the one-loop effective equations) with renormalised cosmological constant $\Lambda > 0$, and integrating out both bosonic and fermionic quantum fluctuations to one loop yields the following effective potential for the gravitino condensate field σ in the flat space-time limit $\Lambda \to 0$, as detailed in ref. [16],

$$V_{\text{eff}} = V_B^{(0)} + V_B^{(1)} + V_F^{(1)} = -\frac{\Lambda_0}{\kappa^2} + V_B^{(1)} + V_F^{(1)} \ , \tag{11}$$

where Λ_0 is given by (9) and

$$V_B^{(1)} = \tfrac{45\kappa^4}{512\pi^2}\left(f^2-\sigma^2\right)^2\left(3 - 2\ln\left(\tfrac{3\kappa^2(f^2-\sigma^2)}{2\mu^2}\right)\right) \ ,$$
$$V_F^{(1)} = \left(\tfrac{\tilde{\kappa}}{\kappa}\right)^4 \tfrac{\kappa^4\sigma^4}{30976\pi^2}\left(30578\ln\left(\left(\tfrac{\tilde{\kappa}}{\kappa}\right)^2\tfrac{\kappa^2\sigma^2}{3\mu^2}\right) - 45867 + 29282\ln\left(\tfrac{33}{2}\right) + 1296\ln\left(\tfrac{54}{11}\right)\right) \ , \tag{12}$$

indicate the contributions to the effective potential from bosonic and fermionic fields respectively, and μ is an inverse renormalisation group (RG) scale. The one-loop effective potential (11) is depicted in fig. 1 and it has the characteistic double-well shape of dynamical breaking of a symmetry.

We may firstly note that as we flow from UV to IR (*i.e.* in the direction of increasing μ), we obtain the correct double-well shape required for the super-Higgs effect, and secondly that tuning f allows us to shift V_{eff} and thus attain the correct vacuum structure (*i.e.* non-trivial minima σ_c such that $V_{\text{eff}}(\sigma_c) = 0$). Moreover, the shape of the effective potential changes, as one varies the (renormalisation) scale μ from ultraviolet to infrared values (*i.e.* flowing in the direction of increasing μ), in such a way that the broken symmetry phase (double-well shaped potential) is reached in the IR. This indicates that the dynamical generation of a gravitino mass is actually an IR phenomenon, in accordance with the rather general features of dynamical mass in field theory.

In the broken phase, the mass of the gravitino condensate is then given by

$$m_\sigma^2 \equiv V''_{\text{eff}}(\sigma_c) \ , \tag{13}$$

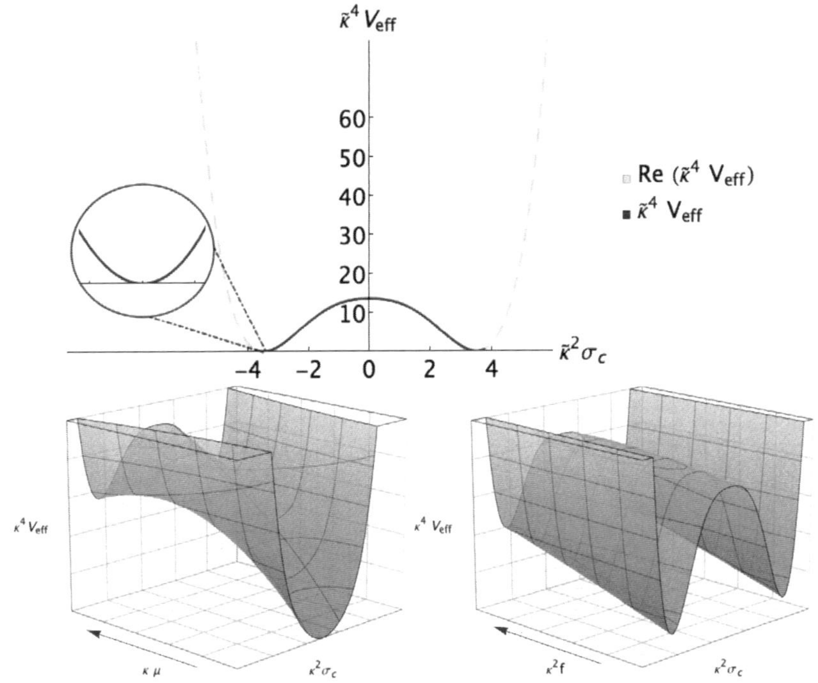

Figure 1. *Upper panel*: The effective potential (11), expressed in units of the coupling $\tilde{\kappa}$ (10). *Lower panel*: As above, but showing schematically the effect of tuning the RG scale μ and the supersymmetry breaking scale f, whilst holding, respectively, f and μ fixed. The arrows in the respective axes correspond to the direction of increasing μ and f.

where σ_c is the minimum of $V_{\rm eff}$ and a prime denotes a functional derivative with respect to the gravitino-condensate field. As observed from (12), the bosonic contributions to the effective potential contain logarithmic terms which would contribute imaginary terms, leading to instabilities, unless

$$\sigma_c^2 < f^2 \,. \tag{14}$$

From (11) it is straightforward to see that this condition is equivalent to the negativity of the tree-level cosmological constant Λ_0, which is entirely sensible; if $\Lambda_0 > 0$ then SUGRA is broken at tree level (given the incompatibility of supersymmetry with de Sitter vacua) and there can be no dynamical breaking. As such, we must then tune f for a given value of μ to find self consistent minima σ_c satisfying (14), thereby ensuring a real $V_{\rm eff}$. In fact, here lies the importance of the super-Higgs effect, and thus of a non-zero positive $f^2 > \sigma_c^2 > 0$, in allowing dynamical breaking of local supersymmetry [3].

As discussed in refs. [15, 16], phenomenologically realistic situations, where one avoids transplanckian gravitino masses, for supersymmetry breaking scales $\sqrt{f}$ at most of order of the Grand Unification (GUT) scale 10^{15-16} GeV, as expected from arguments related to the

[3] It should be mentioned at this point that in refs. [26], the importance of the super-Higgs effect was ignored, which led to the incorrect conclusion that imaginary parts exist necessarily in the one-loop effective potential (in the same class of gauges as the one considered in ref. [16] and here) and hence dynamical breaking of SUGRA was not possible. As we have seen above, such imaginary parts are absent when the condition (14) is satisfied, and thus dynamical breaking of SUGRA occurs.

stability of the electroweak vacuum, can occur only for large $\tilde{\kappa}$ couplings, typically of order $\tilde{\kappa} \sim \left(10^3 - 10^4\right)\kappa$. Given the relation (10) this corresponds to dilaton vev of $\mathcal{O}(-10)$, where the negative sign may be familiar in the context of dilaton-influenced cosmological scenarios [27].

If we consider for concreteness the case $\tilde{\kappa} = 10^3\kappa$, which is a value dictated by the inflationary phenomenology of the model [15], we may find solutions with a vanishing one-loop effective potential at the non-trivial minima corresponding to:

$$\tilde{\kappa}^2\,\sigma_c \simeq 3.5\ , \quad \tilde{\kappa}^2\,f \simeq 3.7\ , \quad \tilde{\kappa}\,\mu \simeq 4.0\ , \tag{15}$$

which leads to a global supersymmetry breaking scale

$$\sqrt{f} \simeq 4.7 \times 10^{15}\ \text{GeV}\ , \tag{16}$$

and dynamical gravitino mass

$$m_{\rm dyn} \simeq 2.0 \times 10^{16}\ \text{GeV}\ . \tag{17}$$

At the non-trivial minima we find $\tilde{\kappa}^4 V_F^{(1)} \simeq -1.4$, $\tilde{\kappa}^4 V_B^{(1)} \simeq 5.9 \times 10^{-13}$, with tree-level cosmological constant $\tilde{\kappa}^2\Lambda_0 \simeq -1.4$. We thus observe that fermion contributions to the effective potential are much stronger than the corresponding bosonic contributions for the cases of large couplings $\tilde{\kappa} \gg \kappa$. These values are phenomenologically realistic, thereby pointing towards the viability (from the point of view of producing realistic results of relevance to phenomenology) of the scenarios of dynamical breaking of local supersymmetry in conformal supergravity models [4].

On the other hand, in standard SUGRA scenarios, where $\tilde{\kappa} = \kappa$, one finds, as already mentioned, transplanckian values for the dynamically generated gravitino mass [16]: $m_{\rm dyn} \simeq 2.0 \times 10^{19}$ GeV, and a global supersymmetry breaking scale $\sqrt{f} \simeq 4.7 \times 10^{18}$ GeV, far too high to make phenomenological sense.

3. Slow-roll hill-top inflation

In order to discuss the possible connection with inflation, we need to calculate one more important ingredient; the wave-function renormalisation. In principle, this should be calculated in a curved de Sitter space-time, which characterises the (unbroken) phase of SUGRA, when the condensate field is near the trivial maximum of the effective potential (11). This is a complicated task. However, it turns out that, since, according to the data [1, 2], the de Sitter phase Hubble parameter in phenomenologically relevant inflationary models is expected to be several orders of magnitude smaller than the Planck scale, m_P (1), the space-time curvature during inflation

[4] A comment concerning SUGRA models in the Jordan frame with such large values for their frame functions is in order here. In our approach, the dilaton φ could be a genuine (dimensionless) dilation scalar field $\varphi = 2\phi$ arising in the gravitational multiplet of string theory, whose low-energy limit may be identified with some form of SUGRA action. In our normalization the string coupling would be $g_s \equiv e^{\phi} = \tilde{\kappa}^{-1/2}$. In such a case, a value of $\tilde{\kappa} = e^{-\langle\varphi\rangle}\,\kappa = \mathcal{O}(10^{3-4})$ would imply a large negative v.e.v. of the (four-dimensional) dilaton field of order $\langle\phi\rangle = -\mathcal{O}(5) < 0$, and thus a weak string coupling squared $g_s = \mathcal{O}(10^{-2})$, which may not be far from values attained in phenomenologically realistic string cosmologies [27]. On the other hand, in the Jordan-frame SUGRA models of [9], the frame function reads $\Phi \equiv e^{-\varphi} = 1 - \frac{1}{3}\left(S\overline{S} + \sum_{u,d} H_i H_i^\dagger\right) - \frac{1}{2}\chi\left(-H_u^0\,H_d^0 + H_u^+\,H_d^- + \text{h.c.}\right)$, in the notation of [28] for the various matter super fields of the next-to-minimal supersymmetric standard model that can be embedded in such supergravities. The quantity χ is a constant parameter. At energy scales much lower than GUT, it is expected that the various fields take on subplanckian values, in which case the frame function is almost one, and hence $\tilde{\kappa} \simeq \kappa$ for such models today. To ensure $\tilde{\kappa} \gg \kappa$, and thus large values of the frame function, $\Phi \gg 1$, as required in our analysis, one needs to invoke trasnplanckian values for some of the fields, $H^0_{u,d}$, and large values of χ, which may indeed characterize the inflationary phase of such theories. A similar situation occurs for the values of the higgs field (playing the role of the inflaton) in the non-supersymmetric Higgs inflation models [29].

Journal of Physics: Conference Series **626** (2015) 012002 doi:10.1088/1742-6596/626/1/012002

is not too large, and thus a flat space-time estimate of the wave function renormalisation may suffice.

The effective Lagrangian describing the gravitino bound state, with a non-trivial wave-function renormalization, is

$$\mathcal{L}_{\text{eff}} = \frac{Z\kappa^2}{2}\partial_\mu\sigma\partial^\mu\sigma - V_{\text{eff}}(\sigma) \ , \tag{18}$$

where the rescaling $\sigma = \tilde{\sigma}/\kappa\sqrt{Z}$ leads to the canonically normalised Lagrangian

$$\tilde{\mathcal{L}}_{\text{eff}} = \frac{1}{2}\partial_\mu\tilde{\sigma}\partial^\mu\tilde{\sigma} - \tilde{V}_{\text{eff}}(\tilde{\sigma}) \ , \tag{19}$$

and the coupling constants in the potential $\tilde{V}_{\text{eff}}$ are defined as

$$\tilde{V}_{\text{eff}}^{(n)}(0) \equiv \frac{V_{\text{eff}}^{(n)}(0)}{Z^{n/2}} \ . \tag{20}$$

The latter normalisations ultimately yield the slow roll parameters

$$\epsilon = \frac{1}{Z}\frac{M_{\text{Pl}}^2}{2}\left(\frac{V'_{\text{eff}}}{V_{\text{eff}}}\right)^2 \ , \quad \eta = \frac{1}{Z}M_{\text{Pl}}^2\frac{V''_{\text{eff}}}{V_{\text{eff}}} \ , \quad \xi = \frac{1}{Z^2}M_{\text{Pl}}^4\frac{V'_{\text{eff}}V'''_{\text{eff}}}{V_{\text{eff}}^2} \ . \tag{21}$$

That large values of $Z \gg 1$ are necessarily linked to slow-roll hilltop inflation in this case is to be expected from the fact that the effective potential (11) can be approximated near the origin (*i.e.* for small field values of the condensate $\tilde{\sigma} \to 0$) as:

$$V_{\text{eff}} \simeq f^2 - (Z\kappa^2)^{-1}\tilde{\sigma}^2 \ , \quad \tilde{\sigma} \to 0 \ , \tag{22}$$

for a canonically normalised condensate field $\tilde{\sigma}$. To ensure that the slow-roll parameter $|\eta| < 1$ (21), then, we must have

$$Z \gg \frac{M_{\text{Pl}}^4}{f^2} \ . \tag{23}$$

Since the (observed) running spectra index is of order $\eta_s \simeq 0.96$ [1], we must further impose that $|\eta| < 10^{-2}$.

As already mentioned, we assume that we can use the flat space-time results for the wave-function renormalisation obtained in [16], to obtain a correct order of magnitude estimate that is valid in the curved space-times during the inflationary period [15]. We first note, that in the broken phase, with the phenomenologically acceptable values of the gravitino mass m_{dyn} and supersymmetry breaking scales $\sqrt{f}$ the function Z is of order one, which is consistent with the exit from the slow-roll inflationary phase. Near the origin of the potential (11), the wave-function renornmalization is [16]

$$Z \simeq -\frac{1}{2\pi^2}\ln\left(\omega_{\tilde{\mu}}^2\right) \ , \quad \omega_{\tilde{\mu}} \equiv \tilde{\mu}/C_{\text{off}}, \quad g \equiv \lambda_S C_{\text{off}}^2/2\pi^2 \ . \tag{24}$$

where $\tilde{\mu}$ is a transmutation mass scale and C_{off} is a flat-space-time cutoff, which may be taken to be the Planck scale, for low energy theories. For appropriate values of g this corresponds to the limit of large $Z \gg 1$ and small $\omega_{\tilde{\mu}}$, both of which are phenomenologically desirable [5].

[5] Note that larger than one values of the wave-function renormalisation for the composite gravitino condensate fields do not contradict unitarity. A similar situation is encountered in composite Higgs symmetry breaking models in field theory [30].

Phenomenologically realistic models of broken SUSY have $\sqrt{f} < 10^{16}$ GeV $= 10^{-2}\, M_{\rm Pl}$ (*cf.* (16)), hence we must have $Z \gg 10^{10}$, implying very small, practically vanishing, transmutation mass scales. A typical case, compatible with the phenomenologically acceptable values (16) and (17) is given in figure 2, from which we observe that agreement with Planck results is achieved for values of the wave function renormalisation of order $Z \sim \mathcal{O}\,(10^{16})$ for the phenomenologically relevant values of the couplings $\tilde{\kappa}/\kappa \sim 10^3$. This corresponds to practically zero transmutation mass scales of $\tilde{\mu} \to 0$. It may be interesting to notice that increasing $\tilde{\kappa}/\kappa$ higher still has the effect of scaling $V_{\rm eff}$ whilst leaving the shape of the potential qualitatively unchanged, allowing smaller and smaller values of $\sqrt{f}$ and $m_{\rm dyn}$. Whilst this decrease in f tends to naturally increase the slow-roll parameter η, by virtue of (23) this scenario may still be rendered compatible with slow-roll inflation if Z is scaled accordingly to counteract this. As such, Planck compatible inflation as demonstrated in figure 2 can be achieved for any value of $\tilde{\kappa}/\kappa$. The Planck-compatible result is $(0.959, 0.04) \leq \{n_s, r\} \leq (0.964, 0.03)$ for 50 and 60 e-folds, respectively, corresponding to $\sqrt{f} \sim 5\, e^{\langle\varphi\rangle} \times 10^{18}$ GeV. This is the case for any value of the (negative) dilaton vev $\langle\varphi\rangle$, however, as already mentioned, for realistic supersymmetry breaking phenomenology one should really fix $\sqrt{f}$ around or below the GUT scale.

One way to interpret this result is the following. Near the origin of the potential one is in the unbroken phase, and hence the gravitino condensate has not yet fully formed, or rather is beginning to form, corresponding to a very small value of the gravitino mass. This small value grows in actual time, until the condensate sits in the minimum of the potential after rolling downhill, at which point the gravitino mass is stabilised at phenomenologically acceptable value, *e.g.* of order the GUT scale. The duration of the whole process is that of the slow-roll inflation period, and exit from this phase occurs near the non-trivial minimum of the potential (11).

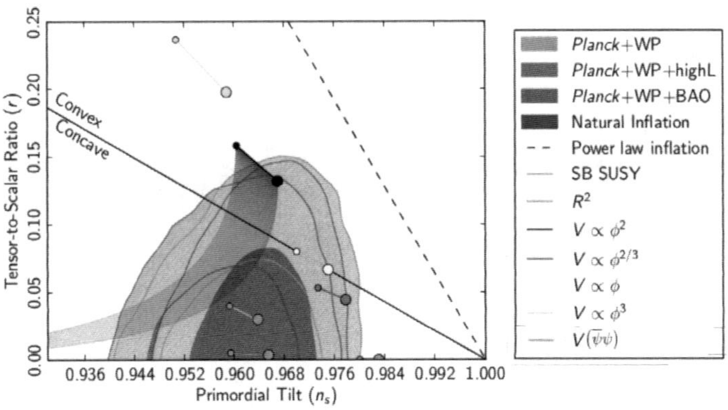

Figure 2. Planck data [1] for n_s and r with the gravitino-condensate hill-top inflation indicated explicitly (dark green). The latter model leads to higher r than Starobinsky-type R^2 inflation (orange), although requires a very high value of the gravitino-condensate wave-function renormalisation, of order larger than $O(10^{16})$.

One may object to the huge value of the wave function renormalisation (23) during the slow-roll inflationary phase as unnatural [6]. There are alternative scenarios of slow-roll inflation

[6] One may be tempted to discuss, within the context of our minimal model, an alternative scenario, according to which global SUSY breaks at a transplanckian scale $\sqrt{f} \gg 1$ (in Planck units). In this case, the SUSY matter partners would completely decouple from the low-energy spectrum, and hence there would be no experimental evidence for SUSY. On the other hand, local SUSY (SUGRA) would ensure inflation via the gravitino condensation

Journal of Physics: Conference Series **626** (2015) 012002 doi:10.1088/1742-6596/626/1/012002

linked to this model which do not require such large Z, which we shall now come to discuss. These are associated with another type of inflation that may occur in the broken SUGRA phase, where, in contrast to the hill-top inflationary scenario discussed so far, the gravitino condensate field lies near its value that minimises the potential (11). In this scenario, the inflaton field is not the gravitino condensate, but it is linked to the scalar mode that parametrises a R^2-Starobinsky-like [11] inflation that is associated with the effective gravitational action obtained after integrating out the massive gravitino-condensate degrees of freedom. This scenario was discussed in detail in ref. [17], and we now proceed to review it briefly.

4. Starobinsky-type inflation in the broken SUGRA phase

Starobinsky inflation is a model for obtaining a de Sitter (inflationary) cosmological solution to gravitational equations arising from a (four space-time-dimensional) action that includes higher curvature terms. Specifically, an action of the type in which the quadratic curvature corrections consist only of scalar curvature terms [11]

$$\mathcal{S} = \frac{1}{2\,\kappa^2}\int d^4x\sqrt{-g}\left(R + \beta\,R^2\right)\;,\;\beta = \frac{8\,\pi}{3\,\mathcal{M}^2}\;, \tag{25}$$

where $\kappa^2 = 8\pi G$, and $G = 1/m_P^2$ is Newton's (gravitational) constant in four space-time dimensions, with m_P the Planck mass, and $\mathcal{M}$ is a constant of mass dimension one, characteristic of the model.

The important feature of this model is that inflationary dynamics is driven purely by the gravitational sector, through the R^2 terms, and that the scale of inflation is linked to $\mathcal{M}$. From a microscopic point of view, the scalar curvature-squared terms in (25) are viewed as the result of quantum fluctuations (at one-loop level) of conformal (massless or high energy) matter fields of various spins, which have been integrated out in the relevant path integral in a curved background space-time [31]. The quantum mechanics of this model, proceeding by means of tunnelling of the Universe from a state of "nothing" to the inflationary phase of ref. [11] has been discussed in detail in ref. [32]. The above considerations necessitate truncation to one-loop quantum order and to curvature-square (four-derivative) terms, which implies that there must be a region of validity for curvature invariants such that $\mathcal{O}(R^2/m_P^4) \ll 1$. This is of course a condition satisfied in phenomenologically realistic scenarios of inflation [1, 2], for which the inflationary Hubble scale H_I satisfies (1) (the reader should recall that $R = 12H_I^2$ in the inflationary phase).

Although the inflation in this model is not driven by rolling scalar fields, nevertheless the model (25) (and for that matter, any other model where the Einstein-Hilbert space-time Lagrangian density is replaced by an arbitrary function $f(R)$ of the scalar curvature) is conformally equivalent to that of an ordinary Einstein-gravity coupled to a scalar field with a potential that drives inflation [33]. To see this, one firstly linearises the R^2 terms in (25) by means of an auxiliary (Lagrange-multiplier) field $\tilde{\alpha}(x)$, before rescaling the metric by a conformal transformation and redefining the scalar field (so that the final theory acquires canonically-normalised Einstein and scalar-field terms):

$$g_{\mu\nu} \to g^E_{\mu\nu} = \left(1 + 2\,\beta\,\tilde{\alpha}(x)\right)\,g_{\mu\nu}\;,\quad \tilde{\alpha}\,(x) \to \rho(x) \equiv \sqrt{\frac{3}{2}}\ln\,\left(1 + 2\,\beta\,\tilde{\alpha}\,(x)\right)\;. \tag{26}$$

mechanism described in this work, while the induced transplanckian dynamical mass for the gravitino, would remove any possibility of observing it as well. From (23) we can then conclude that slow-roll inflation could be achieved for natural values of the wave-function renormalisation $Z < \mathcal{O}(10)$, but in this case the stability of the electroweak vacuum would be delinked from any SUSY arguments. One could also try to relax the slow-roll assumption but this opens up a whole new game, where comparison with data may be complicated, and we do not consider it here.

Journal of Physics: Conference Series **626** (2015) 012002 doi:10.1088/1742-6596/626/1/012002

These steps may be understood schematically via

$$\int d^4x\sqrt{-g}\,\left(R+\beta\,R^2\right)\hookrightarrow\int d^4x\sqrt{-g^E}\,\left(R^E+g^{E\,\mu\nu}\,\partial_\mu\,\rho\,\partial_\nu\,\rho-V\big(\rho\big)\right)\,, \tag{27}$$

where the arrows have the meaning that the corresponding actions appear in the appropriate path integrals, with the potential $V(\rho)$ given by:

$$V(\rho)=\frac{\left(1-e^{-\sqrt{\frac{2}{3}}\,\rho}\right)^2}{4\,\beta}=\frac{3\mathcal{M}^2\left(1-e^{-\sqrt{\frac{2}{3}}\,\rho}\right)^2}{32\,\pi}\,. \tag{28}$$

This potential is sufficiently flat for large values of ρ (compared to the Planck scale) to produce phenomenologically acceptable inflation, with the scalar field ρ playing the role of the inflaton. In fact, the Starobinsky model, with its characteristically low value of the tensor-to-scalar ratio r, provides an excellent fit to the recent Planck data on inflation [1].

The agreement of the model of ref. [11] with the Planck data has triggered an enormous interest in the current literature in revisiting the model from various points of view, such as its connection with no-scale supergravity [8] and (super)conformal versions of supergravity and related areas [7, 10]. In the latter works however the Starobinsky scalar field is fundamental, arising from the appropriate scalar component of some chiral superfield that appears in the superpotentials of the model. Although of great value, illuminating a connection between supergravity models and inflationary physics, and especially for explaining the low-scale of inflation compared to the Planck scale, it can be argued that these works contradict the original spirit of the Starobinsky model (25). Therein, higher curvature corrections are viewed as arising from quantum fluctuations of matter fields in a curved space-time background, such that inflation is driven by the pure gravity sector in the absence of fundamental scalars.

In this section we consider an extension of the analysis of ref. [16] to the case where the de Sitter parameter Λ is perturbatively small compared to m_P^2, but not zero, so that truncation of the series to order Λ^2 suffices. This is in the spirit of the original Starobinsky model [11], with the rôle of matter fulfilled by the now-massive gravitino field. Specifically, we are interested in the behaviour of the effective potential near the non-trivial minimum, where σ_c is a non-zero constant.

It is important to notice at this point that, in contrast to the original Starobinsky model [11], where the crucial for inflation R^2 terms have been argued to arise from the *conformal anomaly* in the path integral of massless (conformal) matter in a de Sitter background, and thus their coefficient was arbitrary, in our scenario, such terms arise in the one-loop effective action of the gravitino condensate field, evaluated in a de Sitter background, after integrating out massive gravitino fields, whose mass was generated dynamically. The order of the de Sitter cosmological constant, $\Lambda > 0$ that breaks supersymmetry, and the gravitino mass are all evaluated dynamically (self-consistently) in our approach from the minimization of the effective potential. Thus, the resulting R^2 coefficient, which determines the phenomenology of the inflationary phase, is calculable [17]. Moreover, in our analysis, unlike Starobinsky's original work, we will keep the contributions from both graviton (spin-two) and gravitino quantum fluctuations. Specifically, we are interested in the behaviour of the effective potential near the non-trivial minimum, where σ_c is a non-zero constant (*cf* fig. 3).

The one-loop effective potential, obtained by integrating out gravitons and (massive) gravitino fields in the scalar channel (after appropriate euclideanisation), may be expressed as a power series in Λ:

$$\Gamma\simeq S_{\rm cl}-\frac{24\pi^2}{\Lambda^2}\left(\alpha_0^F+\alpha_0^B+\left(\alpha_1^F+\alpha_1^B\right)\Lambda+\left(\alpha_2^F+\alpha_2^B\right)\Lambda^2+\dots\right)\,, \tag{29}$$

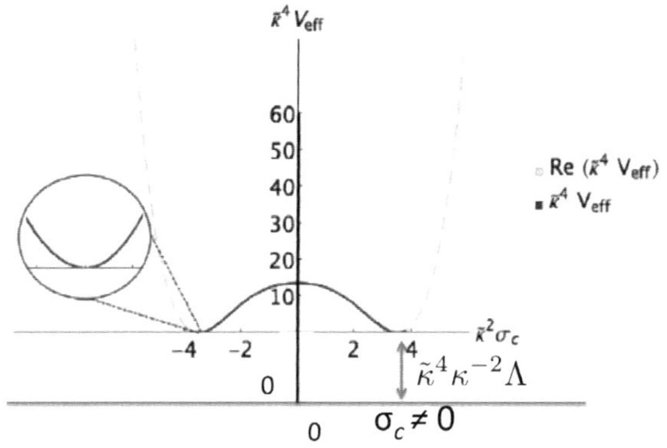

Figure 3. Generic shape of the one-loop effective potential for the gravitino condensate field σ_c in dynamically broken (conformal) Supergravity models in the presence of a non-trivial de Sitter background with cosmological constant $\Lambda > 0$ [17]. The Starobinsky inflationary phase is associated with fluctuations of the condensate and gravitational field modes near the non-trivial minimum of the potential, where the condensate $\sigma_c \neq 0$, and the potential assumes the value $\Lambda > 0$, consistent with supersymmetry breaking. The dashed green lines denote "forbidden" areas of the condensate field values, violating the condition (14), for which imaginary parts appear in the effective potential, thereby destabilising the broken symmetry phase.

where $S_{\rm cl} = -\frac{1}{2\kappa^2}\int d^4x\sqrt{g}\left(\widehat{R} - 2\Lambda_0\right)$ is the classical action with Λ_0 given by (9) and $\widehat{R}$ denoting the fixed S^4 background we expand around ($\widehat{R} = 4\Lambda$, Volume $= 24\pi^2/\Lambda^2$); the $\alpha_i^{B(F)}$, $i = 1, 2$ in (29) indicate the bosonic (B) and fermionic (F) quantum corrections at a given order ($i = 1, 2$) in Λ, respectively.

The leading order term in Λ is then the effective action found in [16] in the limit $\Lambda \to 0$,

$$\Gamma_{\Lambda\to 0} \simeq -\frac{24\pi^2}{\Lambda^2}\left(-\frac{\Lambda_0}{\kappa^2} + \alpha_0^F + \alpha_0^B\right) \equiv \frac{24\pi^2}{\Lambda^2}\frac{\Lambda_1}{\kappa^2}, \tag{30}$$

and the remaining quantum corrections , proportional to Λ and Λ^2, may be identified respectively with Einstein-Hilbert R-type and Starobinsky R^2-type terms in an effective action (31) of the form

$$\Gamma \simeq -\frac{1}{2\kappa^2}\int d^4x\sqrt{g}\left(\left(\widehat{R} - 2\Lambda_1\right) + \alpha_1\,\widehat{R} + \alpha_2\,\widehat{R}^2\right), \tag{31}$$

where we have combined terms of order Λ^2 into curvature scalar square terms. For general backgrounds such terms would correspond to invariants of the form $\widehat{R}_{\mu\nu\rho\sigma}\,\widehat{R}^{\mu\nu\rho\sigma}$, $\widehat{R}_{\mu\nu}\,\widehat{R}^{\mu\nu}$ and $\widehat{R}^2$, which for a de Sitter background all combine to yield $\widehat{R}^2$ terms [7]. The coefficients α_1 and

[7] In general, however, the metric tensor fluctuations of such terms may differ from the ones due to the R^2 term. We note that, in four-dimensions, in the absence of dilatons, one may actually deal with just one more independent invariant, $R_{\mu\nu}R^{\mu\nu}$, in addition to R^2, since the Gauss-Bonnet combination is a total derivative. This is not the case, however, for Jordan-frame SUGRA, where dilaton factors accompany the higher-order curvature terms.

α_2 absorb the non-polynomial (logarithmic) in Λ contributions, so that we may then identify (31) with (29) via

$$\alpha_1 = \frac{\kappa^2}{2}\left(\alpha_1^F + \alpha_1^B\right) , \quad \alpha_2 = \frac{\kappa^2}{8}\left(\alpha_2^F + \alpha_2^B\right) . \tag{32}$$

To identify the conditions for phenomenologically acceptable Starobinsky inflation around the non-trivial minima of the broken SUGRA phase of our model, we impose first the *cancellation* of the "classical" Einstein-Hilbert space term $\widehat{R}$ by the "cosmological constant" term Λ_1, *i.e.* that $\widehat{R} = 4\,\Lambda = 2\,\Lambda_1$. This condition should be understood as a necessary one characterising our background in order to produce phenomenologically-acceptable Starobinsky inflation in the broken SUGRA phase following the first inflationary stage, as discussed in ref. [15]. This may naturally be understood as a generalisation of the relation $\widehat{R} = 2\Lambda_1 = 0$, imposed in ref. [16] as a self-consistency condition for the dynamical generation of a gravitino mass. From thid it follows that the cosmological constant Λ satisfies the four-dimensional Einstein equations in the non-trivial minimum, and in fact coincides with the value of the one-loop effective potential of the gravitino condensate at this minimum. As we discussed in [16], this non-vanishing positive value of the effective potential is consistent with the generic features of dynamical breaking of supersymmetry [34]. In terms of the Starobinsky inflationary potential (28), the value $\Lambda > 0$ corresponds to the approximately constant value of this potential in the high φ-field regime, where Starobinsky-type inflation takes place. Thus we may set $\Lambda \sim 3\,H_I^2$, where H_I the (approximately) constant Hubble scale during inflation, which is constrained by the current data to satisfy (1).

The effective Newton's constant in (31) is then $\kappa_{\rm eff}^2 = \kappa^2/\alpha_1$, and from this, we can express the effective Starobinsky scale (25) in terms of $\kappa_{\rm eff}$ as $\beta_{\rm eff} \equiv \alpha_2/\alpha_1$. This condition thus makes a direct link between the action (29) with a Starobinsky type action (25). Comparing with (25), we can then identify the Starobinsky inflationary scale in this case as

$$\mathcal{M} = \sqrt{\frac{8\pi}{3}\frac{\alpha_1}{\alpha_2}} . \tag{33}$$

We may then determine the coefficients α_1 and α_2 in order to evaluate the scale $1/\beta$ of the effective Starobinsky potential in this case, and thus the scale of the inflationary phase. To this end, we use the results of ref. [16], derived via an asymptotic expansion as explained in the appendix therein, to obtain the following forms for the coefficients

$$\begin{aligned}
\alpha_1^F &= 0.067\,\tilde{\kappa}^2\sigma_c^2 - 0.021\,\tilde{\kappa}^2\sigma_c^2 \ln\left(\frac{\Lambda}{\mu^2}\right) + 0.073\,\tilde{\kappa}^2\sigma_c^2 \ln\left(\frac{\tilde{\kappa}^2\sigma_c^2}{\mu^2}\right) , \\
\alpha_2^F &= 0.029 + 0.014 \ln\left(\frac{\tilde{\kappa}^2\sigma_c^2}{\mu^2}\right) - 0.029 \ln\left(\frac{\Lambda}{\mu^2}\right) , \\
\alpha_1^B &= -0.083\Lambda_0 + 0.018\,\Lambda_0 \ln\left(\frac{\Lambda}{3\mu^2}\right) + 0.049\,\Lambda_0 \ln\left(-\frac{3\Lambda_0}{\mu^2}\right) , \\
\alpha_2^B &= 0.020 + 0.021 \ln\left(\frac{\Lambda}{3\mu^2}\right) - 0.014 \ln\left(-\frac{6\Lambda_0}{\mu^2}\right) ,
\end{aligned} \tag{34}$$

where σ_c denotes the gravitino scalar condensate at the non-trivial minimum of the one-loop effective potential (*cf.* fig. 3), and $\tilde{\kappa} = e^{-\langle\varphi\rangle}\,\kappa$ is the conformally-rescaled gravitational constant in the model of ref. [9], defined previously via (10).

To check the compatibility of the Starobinsky inflationary phase with the Planck data [1] we search numerically [17] for points in the parameter space such that the effective equations

$\frac{\partial \Gamma}{\partial \Lambda} = 0$, $\frac{\partial \Gamma}{\partial \sigma} = 0$, are satisfied; Λ is small and positive ($0 < \Lambda < 10^{-5} M_{\rm Pl}^2$, to ensure the validity of our expansion in Λ) and $10^{-6} < \mathcal{M}/M_{\rm Pl} < 10^{-4}$, to match with known phenomenology of Starobinsky inflation [1]. For $\tilde{\kappa} = \kappa$ (*i.e.* for non-conformal supergravity), we were unable to find any solutions satisfying these constraints, which was to be expected, given the previously demonstrated non-phenomenological suitability of this simple model [16]. If we consider $\tilde{\kappa} >> \kappa$, however, we find that we are able to satisfy the above constraints for a range of values [17], with gravitino mass and global SUSY breaking scale in the ball-park of GUT scale, and the Starobinsky scale of order $\mathcal{M} \sim 10^{-5} M_{\rm Pl}$, which leads to phenomenologically acceptable inflation in the massive gravitino phase, consistent with the Planck-satellite data [1].

Exit from the inflationary phase is a complicated issue which we shall not discuss here, aside from the observation that it can be achieved by coherent oscillations of the gravitino condensate field around its minima, or tunnelling processes à la Vilenkin [32]. This is still an open issue, which may be addressed via construction of more detailed supersymmetric models, including coupling of the matter sector to gravity.

5. Conclusions

In this talk we considered a minimal inflationary scenario, by means of which a gravitino condensate in supergravity models is held responsible for breaking local supersymmetry dynamically and inducing inflation in an indirect way by means of a Starobinsky-type inflation in the massive gravitino phase. Although Inflation of hilltop type via gravitino condensate, where the inflaton is the gravitino condensate field itself, appears at first sight the simplest scenario, nevertheless to ensure slow roll in such a case would require unnaturally high values of the condensate wave-function renormalization, unless the supersymmetry-breaking scale assumed transplanckian values. It is in this sense that the Starobinsky-type scenario for inflation, which is associated with the scalar mode that collectively parametrizes the effects of the quadratic-curvature contributions to the effective action of the gravitino condensate, after integrating out graviton and massive gravitino degrees of freedom, appears quite natural. It involves parameters that assume values of a natural and phenomenologically relevant order of magnitude, specifically global supersymmetry scale and gravitino masses of the order of GUT mass scales or less.

Such a scenario is a truly minimal scenario for natural inflation, in the sense that it involves two scalar primordial composite modes to achieve dynamical breaking of a gauge symmetry (supergravity) and inflation. From our analysis above, it seems that in order to ensure phenomenologically relevant (for the stability of the electroweak vacuum) supersymmetry breaking scales and gravitino masses, one needs to apply the above ideas, not to the minimal SUGRA model, but to Jordan-frame extensions thereof, involving a third scalar field (dilaton). In the context of the next to minimal supersymmetric standard model, which Jordan-frame SUGRA models can incorporate, such dilatons may be composite of appropriate matter superfields, involving Higgs (supermultiplets). Details of the microscopic matter model are important to ensure the correct cosmological evolution, in particular satisfaction of the Big-Bang-Nucleosynthesis constraints. A GUT scale gravitino can be made to decay fast enough so as not to disturb the BBN, but this depends on the details of the matter sector of the theory. This is left for the future.

Nevertheless, we believe that the above scenario for dynamical breaking of supergravity and its links with Starobinsky inflation (which is thus in agreement with the most recent Planck data that seem to exclude fundamental interpretations of the BICEP2 data) deserve further study, and have a serious chance of leading to realistic phenomenological scenarios compatible with the cosmological and particle physics data.

Acknowledgements
N.E.M. would like to thank the organisers of the DICE 2014 conference for their kind invitation to give a plenary talk. This work is supported in part by the London Centre for Terauniverse Studies (LCTS), using funding from the European Research Council via the Advanced Investigator Grant 267352 and by STFC (UK) under the research grant ST/L000326/1.

References

[1] Ade P A R *et al* [Planck Collaboration] 2015 *Preprint* arXiv:1502.02114
[2] Martin J, Ringeval C and Vennin V 2013 *Preprint* arXiv:1303.3787
[3] Ellis J R, Nanopoulos D V, Olive K A and Tamvakis K 1982 *Phys. Lett.* B **118** 335
[4] Croon D, Ellis J and Mavromatos N E 2013 *Phys. Lett.* B **724** 165 (*Preprint* arXiv:1303.6253); Ellis J, Mavromatos N E and Mulryne D J 2014 *JCAP* **1405** 012 (*Preprint* arXiv:1401.6078)
[5] Yamaguchi M 2011 *Class. Quant. Grav.* **28** 103001 (*Preprint* arXiv:1101.2488)
[6] Linde A D and Riotto A 1997 *Phys. Rev.* D **56** 1841 (*Preprint* arXiv:hep-ph/9703209); Halyo E 1996 *Phys. Lett.* B **387** 43 (*Preprint* arXiv:hep-ph/9606423)
[7] Kawasaki M, Yamaguchi M and Yanagida T 2000 *Phys. Rev. Lett.* **85** 3572 (*Preprint* arXiv:hep-ph/0004243); Ellis J, Garcia M A G, Nanopoulos D V and Olive K A 2014 *Preprint* arXiv:1403.7518; Kallosh R, Linde A and Westphal A 2014 *Preprint* arXiv:1405.0270
[8] Ellis J, Nanopoulos D V and Olive K A 2013 *JCAP* **1310** 009 (*Preprint* arXiv:1307.3537); Buchmuller W, Domcke V and Kamada K 2013 *Phys. Lett.* B **726** 467 (*Preprint* arXiv:1306.3471); Farakos F, Kehagias A and Riotto A 2013 *Nucl. Phys.* B **876** 187 (*Preprint* arXiv:1307.1137); Ketov S V and Terada T 2013 *JHEP* **1312** 040 (*Preprint* arXiv:1309.7494); Pallis C 2014 *JCAP* **1404** 024 (*Preprint* arXiv:1312.3623; Ferrara S, Kallosh R and Van Proeyen A 2013 *JHEP* **1311** 134 (*Preprint* arXiv:1309.4052)
[9] Ferrara S, Kallosh R, Linde A, Marrani A and Van Proeyen A 2010 *Phys. Rev.* D **82** 045003 (*Preprint* arXiv:1004.0712); *do.* 2011 *Phys. Rev.* D **83** 025008 (*Preprint* arXiv:1008.2942)
[10] Garcia M A G and Olive K A 2013 *Preprint* arXiv:1306.6119; Roest D, Scalisi M and Zavala I 2013 *Preprint* arXiv:1307.4343; Ferrara S, Kallosh R, Linde A and Porrati M 2013 *Preprint* arXiv:1307.7696
[11] Starobinsky A A 1980 *Phys. Lett.* B **91** 99
[12] Ade P A R *et al* [BICEP2 Collaboration] 2014 *Phys. Rev. Lett.* **112** 241101 (*Preprint* arXiv:1403.3985)
[13] Adam R *et al* [Planck Collaboration], *Preprint* arXiv:1409.5738
[14] Mortonson M J and Seljak U 2014 *Preprint* arXiv:1405.5857; Flauger R, Hill J C and Spergel D N 2014 *Preprint* arXiv:1405.7351; Liu H, Mertsch P and Sarkar S 2014 *Preprint* arXiv:1404.1899
[15] Ellis J and Mavromatos N E 2013 *Phys. Rev.* D **88** 085029 (*Preprint* arXiv:1308.1906)
[16] Alexandre J, Houston N and Mavromatos N E 2013 *Phys. Rev.* D **88** 125017 (*Preprint* arXiv:1310.4122); *do.* 2014 *Preprint* arXiv:1409.3183, *Int. J. Mod. Phys.* D, to appear [doi: 10.1142/S0218271815410047]
[17] Alexandre J, Houston N and Mavromatos N E 2014 *Phys. Rev.* D **89** 027703 (*Preprint* arXiv:1312.5197)
[18] Freedman D Z and Van Proeyen A 2012 *Supergravity* (Cambridge, UK: Cambridge Univ. Press)
[19] Van Nieuwenhuizen P 1981 *Phys. Rept.* **68** 189
[20] Mercuri S 2006 *Phys. Rev.* D **73** 084016 (*Preprint* arXiv:gr-qc/0601013)
[21] Jaeckel J and Wetterich C 2003 *Phys. Rev.* D **68** 025020 (*Preprint* arXiv:hep-ph/0207094)
[22] Nambu Y and Jona-Lasinio G 1961 *Phys. Rev.* **122** 345
[23] Deser S and Zumino B 1977 *Phys. Rev. Lett.* **38** 1433
[24] Komargodski Z and Seiberg N 2009 *JHEP* **0909** 066 (*Preprint* arXiv:0907.2441)
[25] Fradkin E S and Tseytlin A A 1984 *Nucl. Phys.* B **234** 509
[26] Buchbinder I L and Odintsov S D 1989 *Class. Quant. Grav.* **6** 1955; see also Odintsov S D 1988 *Phys. Lett.* B **213** 7
[27] Antoniadis I, Bachas C, Ellis J R and Nanopoulos D V 1988 *Phys. Lett.* B **211** 393
[28] Einhorn M B and Jones D R T 2010 *JHEP* **1003** 026 (*Preprint* arXiv:0912.2718)
[29] Bezrukov F L and Shaposhnikov M 2008 *Phys. Lett.* B **659** 703 (*Preprint* arXiv:0710.3755); Bezrukov F, Magnin A, Shaposhnikov M and Sibiryakov S 2011 *JHEP* **1101** 016 (*Preprint* arXiv:1008.5157)
[30] Bardeen W A, Hill C T and Lindner M 1990 *Phys. Rev.* D **41** 1647
[31] Davies P C W, Fulling S A, Christensen S M and Bunch T S 1977 *Annals Phys.* **109** 108; Bunch T S and Davies P C W 1978 *Proc. Roy. Soc. Lond.* A **360** 117; Birrell N and Davies P 1982 *Quantum Fields in Curved Space* (Cambridge Monogr. Math. Phys.)
[32] Vilenkin A 1985 *Phys. Rev.* D **32** 2511
[33] Stelle K S 1978 *Gen. Rel. Grav.* **9** 353; Whitt B 1984 *Phys. Lett.* B **145** 176
[34] Witten E 1981 *Nucl. Phys.* B **188** 513

What can quantum cosmology say about the inflationary universe?

Gianluca Calcagni [1], Claus Kiefer [2] and Christian F Steinwachs [3]

[1] Instituto de Estructura de la Materia, CSIC, calle Serrano 121, 28006 Madrid, Spain
[2] Institut für Theoretische Physik, Universität zu Köln, Zülpicher Strasse 77, 50937 Köln, Germany
[3] Physikalisches Institut, Albert-Ludwigs-Universität Freiburg, Hermann-Herder-Strasse 3, 79104 Freiburg, Germany

Abstract. We propose a method to extract predictions from quantum cosmology for inflation that can be confronted with observations. Employing the tunneling boundary condition in quantum geometrodynamics, we derive a probability distribution for the inflaton field. A sharp peak in this distribution can be interpreted as setting the initial conditions for the subsequent phase of inflation. In this way, the peak sets the energy scale at which the inflationary phase has started. This energy scale must be consistent with the energy scale found from the inflationary potential and with the scale found from a potential observation of primordial gravitational waves. Demanding a consistent history of the universe from its quantum origin to its present state, which includes decoherence, we derive a condition that allows one to constrain the parameter space of the underlying model of inflation. We demonstrate our method by applying it to two models: Higgs inflation and natural inflation.

1. Introduction

It is generally assumed that the Universe underwent a period of quasi-exponential expansion very early in its evolution. This phase is called *inflation* and has the advantage of giving a causal explanation for structure formation (see e.g. [1] for a textbook introduction). But while the kinematic features of inflation are well understood, its precise dynamical origin remains unclear. There exist plenty of different models, mostly phenomenologically devised, and one must at present resort to observations in order to constrain the class of allowed models [2].

How can one select inflationary models from a theoretical point of view? The ideal situation would be to have an established fundamental theory at one's disposal from which one can derive the dynamics of inflation, for example in the form of an inflaton field φ and its potential $V(\varphi)$. Unfortunately, we do not have a theory of this kind. This is partly related to the open question of constructing a consistent and empirically correct quantum theory of gravity [3]. While the energy scale of inflation is most likely separated from the Planck scale by some orders of magnitude, its origin can probably not be entirely understood without reference to quantum gravity.

A conservative approach to quantum gravity, which should give reliable results at least somewhat away from the Planck scale, is the direct quantization of general relativity in standard metric variables [3]. This approach is called quantum geometrodynamics and has the important feature that it leads back to general relativity in the semiclassical limit. One might then use this framework to derive the desired constraints on inflationary models. But how is this possible?

Journal of Physics: Conference Series **626** (2015) 012003 doi:10.1088/1742-6596/626/1/012003

If applied to the cosmic regime, quantum geometrodynamics leads to the framework of quantum cosmology, with the wave function of the universe as its central concept. This wave function obeys the Wheeler–DeWitt equation [3]. One could thus attempt to derive constraints on inflation from this equation. But this is not easy, because the interpretation of the wave function in quantum cosmology is far from being straightforward. In our contribution to these Proceedings, we shall review one popular method to obtaining predictions in quantum cosmology and apply it to two models of inflation.

In the following Sec. 2, we shall briefly discuss how one can envisage predictions in quantum cosmology and how one can apply them to inflation. In Sec. 3, we present the energy scales relevant for inflation. In slow-roll models, the relevant energy scale is derived from the corresponding potential. If primordial gravitational waves can be observed, one can directly extract the inflationary energy scale, which must then be consistent with the scale derived from the potential. A third scale is the one obtained from quantum cosmology, which must be consistent with the two other scales if we adopt the criterion of selection presented in Sec. 2. In Sec. 4, we apply these considerations to the models of non-minimal Higgs inflation and natural inflation. In this, we follow closely our earlier papers [4] and [5], where more details and references can be found. We end with a brief conclusion and outlook.

2. Predictions in quantum cosmology

At the most fundamental level, quantum gravity — and therefore also quantum cosmology — is timeless [3, 6]. This is a direct consequence of the fact that general relativity does not contain any absolute notion of time; after quantization, the dynamical spacetime vanishes in the same way as the classical trajectories do in quantum mechanics. In the Wheeler–DeWitt equation, there is thus no t, and the notions of probability and probability conservation (unitarity) seem to lose their usual meaning.

How, then, can one extract predictions from quantum cosmology? So far, only heuristic ideas are available. It has been suggested that a minimal scheme is to look for peaks in the wave function and to interpret them as predictions, see, for example, [7, 8]. If the wave function vanishes in a certain region of configuration space, this means that the corresponding values will never occur; this property is important for the discussion of singularity avoidance.

Inflation is a (semi)classical concept, so a prerequisite for obtaining inflation from quantum cosmology is an efficient quantum-to-classical transition. This is achieved by decoherence, a process that is well understood and experimentally explored in quantum mechanics [9, 10]. It has been shown that decoherence is efficient at the "onset of inflation", which thus justifies the use of robust semiclassical components of the universal wave function and the neglect of interference terms, see, for example, [3, 11] and the references therein. It has been suggested that one should interpret the wave function only in the semiclassical limit [12], but we leave this as an open question. For the purpose of this contribution, we adopt the heuristic proposal that a strong peak in the wave function is interpreted as a prediction, while we will not attempt to infer anything from it in the general case. In the semiclassical limit, from the wave function after decoherence one can get also a restriction on the allowed classical trajectories, that is, one obtains a selection criterion for trajectories.

In the general case, one could envisage a derivation of probabilities from quantum entanglement in the manner attempted in [13], see also [10], but this is still an open issue.

The idea to get a probability for inflation from the wave function in this way was entertained already in [14]. A more precise formulation was obtained in [15] and later papers (see the references in [4]) by emphasizing in particular the need to go to the one-loop level in quantum field theory in order to obtain a normalizable wave function. At the tree level, the slow-roll approximation does, in general, not lead to a peak because of the small field derivatives.

Even in the semiclassical limit, the form of the wave function will depend on the employed

Journal of Physics: Conference Series **626** (2015) 012003 doi:10.1088/1742-6596/626/1/012003

boundary condition. The two most popular boundary conditions are the 'no-boundary' and the 'tunneling' conditions; see, for example, [3, 8] for detailed reviews. In general, the no-boundary condition will not predict the occurrence of inflation. This is why attention is concentrated on the tunneling proposal. It should be emphasized that 'tunneling' is meant here only as a metaphor because tunneling has no meaning in a timeless context, except in a limited sense in the semiclassical approximation; see [11, 16, 17] for a more detailed discussion. We shall also use the tunneling boundary condition in our contribution and shall see in which sense one can get a prediction for inflation from it. Instead of using the Wheeler–DeWitt equation, we shall employ the equivalent path-integral formulation and its semiclassical limit.

The issues of probability and probability measure are even more subtle and contrived in the recently discussed 'multiverse' context (consult, for example, [18] and the references therein), but this will not be addressed here.

3. Energy scales of inflation

In this section, we will summarize how to extract in detail the energy scale of inflation from inflationary models themselves, from observation, and from quantum cosmological considerations.

3.1. *Slow-roll predictions for the energy scale of inflation*

For all inflationary models, the main observables are the power spectra of primordial scalar and tensor perturbations which are generated during inflation on top of an already existing homogeneous and isotropic Friedmann-Robertson-Walker background space-time,

$$\mathcal{P}_{\mathrm{t}} := A_{\mathrm{t}} \left(\frac{k}{k_*}\right)^{n_{\mathrm{t}}}, \qquad \mathcal{P}_{\mathrm{s}} := A_{\mathrm{s}} \left(\frac{k}{k_*}\right)^{n_{\mathrm{s}}-1}. \tag{1}$$

The mode k_* corresponds to a pivot scale k_*^{-1} (to be chosen according to the observational window of the experiment) when the mode k_* first crosses the Hubble scale, $k_* = a_* H_*$. Here, $H(t) = \dot{a}(t)/a(t)$ denotes the Hubble parameter. Within the slow-roll approximation, deviations from a perfect de Sitter stage can be parametrized to first order in terms of the two slow-roll parameters

$$\epsilon_{\mathrm{V}} := \frac{M_{\mathrm{P}}^2}{2}\left(\frac{V'}{V}\right)^2, \qquad \eta_{\mathrm{V}} := M_{\mathrm{P}}^2 \frac{V''}{V}, \tag{2}$$

where M_{P} denotes the (reduced) Planck mass. The amplitudes of the power spectra are given by A_{t} and A_{s}. The tensor and scalar spectral indices n_{t} and n_{s} encode the scale dependence of the power spectra (its slope). These parameters can be entirely expressed in terms of V, ϵ_{V}, and η_{V}

$$A_{\mathrm{t}} = \frac{2\,V}{3\,\pi^2\,M_{\mathrm{P}}^4}, \qquad A_{\mathrm{s}} = \frac{V}{24\,\pi^2\,M_{\mathrm{P}}^4\,\epsilon_{\mathrm{V}}}, \qquad n_t = -2\,\epsilon_{\mathrm{V}}, \qquad n_{\mathrm{s}} = 1 + 2\,\eta_{\mathrm{V}} - 6\,\epsilon_{\mathrm{V}}. \tag{3}$$

All quantities in (1)–(2) must be evaluated for the value of the inflaton field at Hubble-scale exit, φ_*, which, in turn, can be expressed in terms of the number of e-folds N_* by integrating and inverting the relation

$$N_* = \int_{t_*}^{t_{\mathrm{end}}} \mathrm{d}t\, H \simeq \frac{1}{M_P^2}\int_{\varphi_*}^{\varphi_{end}} \mathrm{d}\varphi\, \frac{V}{V'}. \tag{4}$$

The value of $\varphi_{\rm end}$, where inflation ends, is defined by the breakdown of the slow-roll approximation, when $\epsilon_{\rm v} \simeq \mathcal{O}(1)$, which leads to the convention

$$\epsilon_{\rm v}(\varphi_{\rm end}) := 1\,. \tag{5}$$

The energy scale predicted by inflationary slow-roll models is then given by

$$E_{\rm infl}^{\rm model} := V_*^{1/4} := [V(\varphi_*)]^{1/4}\,. \tag{6}$$

3.2. Observational constraints for the energy scale of inflation

The observational energy scale of inflation $E_{\rm inf}^{\rm obs}$ is unknown and so far there only exists an upper bound. Observations of primordial gravitational waves that leave their imprint in the B-polarization spectrum of the cosmic microwave radiation would allow to determine $E_{\rm inf}^{\rm obs}$ in a model independent way. But increasing precision will lead to stronger bounds and eventually even to a detection that would allow to uniquely fix $E_{\rm inf}^{\rm obs}$. In the following, we will derive how $E_{\rm infl}^{\rm obs}$ can be expressed in terms of observable quantities.

The amplitude of the scalar power spectrum $A_{\rm s}$ is fixed by the measured temperature anisotropies of the CMB, $A_{\rm s} \propto (\Delta T/T)^2$. For the pivot scale $k_* = 0.05\ {\rm Mpc}^{-1}$, the best PLANCK fit by the ΛCDM model for the scalar amplitude in the absence of tensor modes and with lensing and polarization data is [20]

$$A_{{\rm s}*} = (2.139 \pm 0.063) \times 10^{-9} \tag{7}$$

at the 68% confidence level.

The tensor-to-scalar ratio — to first order in the slow roll approximation — is defined as

$$r := \frac{A_{\rm t}}{A_{\rm s}} = 16\,\epsilon_{\rm v} = -8\,n_{\rm t}\,. \tag{8}$$

The B-polarization spectrum of the CMB is produced only by tensorial perturbations. A detection of B-modes would allow one to fix $A_{{\rm t}*}$ and with (7) also r_*. Finally, this would allow us to determine the energy scale of inflation in a model-independent way from observations,

$$(E_{\rm infl}^{\rm obs})^4 := \frac{3\,\pi^2\,M_{\rm P}^4}{2}\,A_{{\rm t}*} = \frac{3\,\pi^2\,M_{\rm P}^4}{2}\,A_{{\rm s}*}\,r_*. \tag{9}$$

So far, observations managed to obtain only an upper bound on r_*. The current bound from [19] is $r_* < 0.12$ at 95% confidence level at $k_* = 0.05\,{\rm Mpc}^{-1}$. Taking the central values $A_{{\rm s}*} = 2.2 \times 10^{-9}$ and saturating the bound for r, one obtains an upper bound for the energy scale

$$E_{\rm infl}^{\rm obs} < 1.9 \times 10^{16}\,{\rm GeV}\,. \tag{10}$$

Obviously, all cosmological models have to satisfy the condition

$$E_{\rm infl}^{\rm model} \approx E_{\rm infl}^{\rm obs} \tag{11}$$

in order not to be in conflict with observational data.

Journal of Physics: Conference Series **626** (2015) 012003 doi:10.1088/1742-6596/626/1/012003

3.3. Quantum cosmological energy scale of inflation

As discussed in Sec. 2, we use a heuristic approach and interpret peaks in the tunneling probability distributions as setting the initial conditions for inflation. The tunneling distribution in the semiclassical limit is found to be [4, 5]

$$\mathcal{T}(\varphi) := e^{-\Gamma(\varphi)} = \exp\left[-\frac{24\,\pi^2\,M_{\rm P}^4}{V(\varphi)}\right]. \tag{12}$$

A peak corresponds to a maximum of (12). Finding this peak is equivalent to finding the maxima of the potential $V_{\rm max} := V_{\rm eff}(\varphi_{\rm max})$. This leads to the simple conditions

$$\frac{\mathrm{d}\,V(\varphi)}{\mathrm{d}\varphi}\bigg|_{\varphi=\varphi_{\rm max}} = 0, \qquad \frac{\mathrm{d}^2\,V(\varphi)}{\mathrm{d}\varphi^2}\bigg|_{\varphi=\varphi_{\rm max}} < 0\,. \tag{13}$$

The peak $\varphi_{\rm max}$ in (12) corresponds to the value of φ that selects the most probable value of $\Lambda_{\rm eff} = V(\varphi_{\rm max})/M_{\rm P}^2$ for which the universe starts after tunneling. In this way, the quantum scale of inflation was obtained in [21, 22, 23, 24, 25]. The predictability of the tunneling distribution (12) can be quantified by the sharpness of the peak at $\varphi_{\rm max}$,

$$\mathcal{S} := \frac{(\Delta\varphi)^2}{E_{\rm infl}^{\rm QC}}. \tag{14}$$

Here, the variance $(\Delta\varphi)^2$ is a measure of the width of the peak, while $E_{\rm infl}^{\rm QC}$ is a measure for the height of the peak. We can get a rough estimate of the variance by fitting a normal distribution around the peak $\varphi_{\rm max}$. Taking $\varphi_{\rm max}$ as the mean value and expanding Γ to second order around $\varphi_{\rm max}$, we obtain

$$(\Delta\varphi)^2 := \left[\Gamma''(\varphi_{\rm max})\right]^{-1}. \tag{15}$$

In the inflationary slow-roll regime, $\varphi \approx$ const and the energy density is completely dominated by $V_{\rm max} := V(\varphi_{\rm max})$. Therefore, the peak value $\varphi_{\rm max}$ allows one to determine the energy scale of inflation as

$$E_{\rm infl}^{\rm QC} := V_{\rm max}^{1/4}\,. \tag{16}$$

Demanding a consistent quantum cosmological history of the universe, beginning with the quantum creation via tunneling, we extend the consistency condition (11) and require

$$E_{\rm infl}^{\rm QC} \approx E_{\rm infl}^{\rm model} \approx E_{\rm infl}^{\rm obs}\,. \tag{17}$$

This implies that the energy scale of the inflationary model must not only agree with present observations but must also be of the same order as the prediction from quantum cosmology.

Two points deserve further discussion. First, in the presence of multiple maxima of the effective potential, there might be several (possibly degenerated) peaks in the probability distribution. In such a case, the environment of these peaks, in particular the neighboring minima, has to be investigated. Second, it should be noted that in the context of eternal inflation and the landscape picture [26, 27], such a strong condition as $E_{\rm infl}^{\rm QC} \approx E_{\rm infl}^{\rm obs}$ might not hold and one has to resort to the somewhat weaker condition $E_{\rm infl}^{\rm QC} \geq E_{\rm infl}^{\rm obs}$. Even if there was only one global maximum of the effective potential, corresponding to the unique single highest peak in the tunneling distribution, one must be careful in case the effective potential also features several metastable minima with different energy densities. Then, starting from the global maximum, inflation could happen in a cascade-like process decaying from higher vacua to lower ones step by step (in the following labeled by $\varphi_{\min,n}$ where higher n mean lower values of $V(\varphi_{\min,n})$) or decaying directly in some lower metastable (or eventually even stable) vacuum. We would not be sure whether 'our inflation,' which produced our exponentially blown patch of the universe we can observe today, was due to the initial inflationary phase that started at the global maximum of the effective potential or due to another inflationary period that started in a lower metastable minimum of the effective potential. In other words, it is logically possible that, depending on the structure of the effective potential, the phase of inflation triggered by the quantum creation of the universe leads to a phase of eternal inflation with $V(\varphi_{\max})$. Then, in this eternally inflating universe at some moment in time in some region of space, the inflaton field could decay in one of the metastable vacua $\varphi_{\min,n}$, starting another phase of inflation, with a different initial condition set by the local minimum of the effective potential $V(\varphi_{\min,n}) < V(\varphi_{\max})$ and this can happen several times. We cannot really say whether the energy scale of 'our' inflation is $V(\varphi_{\max})$ or $V(\varphi_{\min,n})$. Therefore, we can realistically only demand $E_{\rm infl}^{\rm QC} \geq E_{\rm infl}^{\rm obs}$.

In this context, it might be interesting to note that according to [28] inflation can exist only eternally to the future direction, but not to the past, assuming a universe that is on average always expanding (not necessarily accelerated). This supports the assumption of an initial moment of creation, in contrast to an eternally existing inflationary universe with no beginning.

4. Special models

In the following, we will apply the general method presented in the previous sections to two particular models of inflation: non-minimal Higgs inflation and natural inflation; these two scenarios are among the class of inflationary models currently favored by observational data [29, 30]. In the following, we quote the results for the 2013 data release of PLANCK.

While natural inflation already admits a quantum cosmological analysis at the tree level, quantum corrections are essential in non-minimal Higgs inflation as they lead to the formation of a strict maximum in the potential and therefore to a sharp peak in (12).

4.1. Non-minimal Higgs inflation

In the non-minimal Higgs inflation model [31, 32, 33, 34, 35, 36, 37, 38, 39], the Standard Model Higgs boson and the cosmological inflaton are identified to be one and the same scalar field φ – the Higgs inflaton. The other essential assumption of this model is a strong non-minimal coupling $\xi \sim 10^4 - 10^5$ of the Higgs inflaton to gravity.[1] The interactions relevant for cosmology can be summarized by the graviton-Higgs sector of the model,

$$S[g,\varphi] = \int \mathrm{d}^4x\, g^{1/2}\left[U(\varphi)R - \frac{1}{2}\partial_\mu\varphi\partial^\mu - V(\varphi)\right]. \tag{18}$$

The coupling to the Ricci scalar R and the Higgs potential are given by

$$U(\varphi) = \frac{1}{2}\left(M_{\rm P}^2 + \xi\varphi^2\right), \qquad V(\varphi) = \frac{\lambda}{4}\left(\varphi^2 - \nu^2\right)^2. \tag{19}$$

Here, ξ is the non-minimal coupling constant, λ the quartic self-coupling, and $\nu \simeq 246$ GeV the symmetry breaking scale. The matter sector is given by the Standard Model interaction

[1] We note that, for an interacting scalar field φ, a non-mininal coupling of the form $\propto \varphi^2 R$ will be unavoidably induced already at the one-loop order. Even within an effective field theoretical approach, where higher order terms are supposed to be sufficiently suppressed, consistency of the renormalization procedure would require to include such a term already from the very beginning. Regarding the strength of ξ, we note that in view of the rather small mass of the discovered Higgs boson $M_{\rm H} \simeq 126$ GeV, the condition of a large non-minimal coupling can be relaxed somewhat; see the discussion at the end of this section and, e.g., [39, 40].

Lagrangian

$$\mathcal{L}_{\rm int} = -\sum_{\chi} \frac{1}{2} \lambda_\chi \chi^2 \varphi^2 - \sum_{A} \frac{1}{2} g_A^2 A_\mu^2 \varphi^2 - \sum_{\Psi} y_\Psi \varphi \bar{\Psi}\Psi. \tag{20}$$

The sum extends over scalar fields χ, vector gauge bosons A_μ and fermions Ψ. The matter content can be restricted to the dominant contributions that come from the heavy $W^\pm$ and Z bosons and the Yukawa top quark $q_{\rm t}$. Their masses follow from the relations

$$m_{W^\pm}^2 = \frac{1}{4} g\, \varphi^2, \quad m_{\rm Z}^2 = \frac{1}{4} (g^2 + g'^2)\, \varphi^2, \quad m_{\rm t}^2 = \frac{1}{2} y_{\rm t}^2 \varphi^2, \quad m_{\rm H}^2 = \lambda\, (3\varphi^2 - \nu^2), \tag{21}$$

with the electroweak and strong gauge couplings g, g' and $g_{\rm s}$ as well as the Yukawa top quark coupling $y_{\rm t}$. This matter content results in essential quantum contributions to the effective potential —the quantity that encodes the relevant information for the cosmological analysis. Since the energy scales of the electroweak vacuum and inflation are separated by many orders of magnitude, one also needs to take into account the dependence of the coupling constants on the energy scale. In order to evaluate the coupling constants at the high energy scale of inflation, one needs to calculate the renormalization group flow that connects the electroweak scale with the energy scale of inflation [33, 34, 35]. The renormalization group running of the couplings, in turn, is encoded in the beta functions which give rise to a system of coupled ordinary differential equations that has to be solved numerically,

$$\frac{dg_i(t)}{dt} = \beta_{g_i}, \qquad g_i = (\lambda, \xi, g, g', g_{\rm s}, y_{\rm t}), \qquad \frac{dZ(t)}{dt} = \gamma Z. \tag{22}$$

Here, $t = \ln\varphi/\mu$ is the logarithmic running scale and μ is an arbitrary renormalization point. The wave function renormalization $Z(t)$ is determined by the anomalous dimension γ of the Higgs field.

In order to make use of the standard slow-roll formalism for the cosmological analysis, it is convenient to transform to the Einstein frame by a conformal transformation of the metric field and a redefinition of the Higgs inflaton,

$$\hat{g}_{\mu\nu} = \frac{2U(\varphi)}{M_{\rm P}^2} g_{\mu\nu}, \quad \left(\frac{d\hat{\varphi}}{d\varphi}\right)^2 = \frac{M_{\rm P}^2}{2} \frac{(U + 3U'^2)}{U^2}, \quad \hat{V} = \frac{M_{\rm P}^2}{4} \frac{V}{U^2}\bigg|_{\varphi=\hat{\varphi}}. \tag{23}$$

The one-loop renormalization group improved effective potential in the Einstein frame reads [35]

$$\hat{V} \simeq \frac{M_{\rm P}^4}{4} \frac{\lambda}{\xi^2} \left[1 - \frac{2M_{\rm P}^2}{\xi\varphi^2} + \frac{\boldsymbol{A_I}}{16\pi^2} \ln\left(\frac{\varphi}{\mu}\right)\right], \tag{24}$$

where $\boldsymbol{A_I}$ represents the inflationary anomalous scaling [35]

$$\boldsymbol{A_I}(t) := \frac{3}{8\lambda(t)} \left[2g^4(t) + \left(g^2(t) + g'^2(t)\right)^2 - 16y_{\rm t}^4(t)\right] - 6\lambda(t). \tag{25}$$

4.1.1. Slow-roll predictions Using the slow-roll formulas of Sec. 3.1 for the Einstein-frame renormalization-group-improved effective potential, and taking the derivatives with respect to the Einstein frame scalar field $\hat{\varphi}$, one can express the the slow-roll parameters in terms of the original Jordan frame variables [35]

$$\hat{\varepsilon} = \frac{M_P^2}{2} \left(\frac{1}{\hat{V}} \frac{d\hat{V}}{d\hat{\varphi}}\right)^2 = \frac{4}{3} \left(\frac{M_P^2}{\xi\, \varphi^2} + \frac{\boldsymbol{A_I}}{64\pi^2}\right)^2, \tag{26}$$

$$\hat{\eta} = \frac{M_P^2}{\hat{V}} \frac{d^2\hat{V}}{d\hat{\varphi}^2} = -\frac{4M_P^2}{3\xi\varphi^2}. \tag{27}$$

Journal of Physics: Conference Series **626** (2015) 012003 doi:10.1088/1742-6596/626/1/012003

For the expressions of the remaining cosmological parameters, it is convenient to introduce the abbreviation

$$x := \frac{N\boldsymbol{A_I}}{48\pi^2}, \tag{28}$$

which can be interpreted as a measure of the strength of quantum corrections, resulting from $\boldsymbol{A_I}$. In this way, for the scalar amplitude one finds [35]

$$\hat{A}_s = \frac{\lambda}{96\pi^2\xi^2\hat{\varepsilon}} = \frac{N^2}{72\pi^2}\frac{\lambda}{\xi^2}\left(\frac{e^x-1}{x\,e^x}\right)^2. \tag{29}$$

The scalar spectral index and the tensor-to-scalar ratio are then found to be [35]

$$n_s = 1 - \frac{2}{N}\frac{x}{e^x-1}, \tag{30}$$

$$r = \frac{12}{N^2}\left(\frac{xe^x}{e^x-1}\right)^2. \tag{31}$$

All these quantities have to be evaluated at the energy scale of of inflation, the moment of first horizon crossing, when the inflaton value is φ_*. This, in turn, means that one has to numerically integrate the system of renormalization-group equations (22) from the electroweak vacuum $t_{\rm ew} \simeq 0$ up to the scale t_*, corresponding to φ_*, and then evaluate all running couplings at t_*.

Fixing the arbitrary renormalization point at the top mass scale $\mu = M_{\rm t}$ and assuming a modified convention for the condition of the end of inflation $\hat{\varepsilon}|_{t=t_{\rm end}} := 3/4$ (instead of the convention (5)), the times t_* and $t_{\rm end}$ can be determined via the relation $\varphi_{*/{\rm end}} = M_t \exp(t_{*/{\rm end}})$ and the estimate for the number of e-folds [32],

$$N_* \simeq \frac{3}{4}\frac{\xi(t_*)}{M_{\rm P}^2}\left(\varphi_*^2 - \varphi_{\rm end}^2\right). \tag{32}$$

In the large ξ approximation, these times read [35]

$$t_* = \ln\frac{M_{\rm P}}{M_{\rm t}} + \frac{1}{2}\ln\frac{4N}{3\xi_*} + \frac{1}{2}\ln\frac{\exp x_* - 1}{x_*}, \tag{33}$$

$$t_{\rm end} = \ln\frac{M_{\rm P}}{M_{\rm t}} + \frac{1}{2}\ln\frac{4}{3\xi_{\rm end}}. \tag{34}$$

While we have taken into account a running ξ, numerically the running is very slow, i.e. $\xi(t_{\rm ew}) \simeq \xi(t_*)$. Nevertheless, in view of the fact that there is no initial condition for $\xi(t_{\rm ew})$, one has to impose a 'final condition' $\xi(t_*)$, determined by the correct normalization of the scalar amplitude (29) evaluated at t_*, i.e. $A_{s*} \propto \lambda_*/\xi_*^2 \propto (\Delta T/T)^2 \sim 10^{-10}$. Note that for $N = 50 \div 60$, the duration of inflation in terms of the logarithmic scale t is numerically very short $t_* - t_{\rm end} \simeq 2$ compared to the post-inflationary running $t_{\rm end} - t_{\rm ew} \simeq 35$ [35].

The numerical predictions for (30) and (31) depend on the details of the renormalization-group flow and are, in particular, very sensitive to the initial conditions at the electroweak scale. A more precise analysis including beta functions up to two and three loops has become available, see e.g. [36, 37] and, since the discovery of the Higgs boson, also the initial conditions at the electroweak scale (in particular the top mass $M_{\rm t}$) are known to a higher precision. Another aspect is connected to the rather light value of the measured Higgs mass $M_{\rm H} \simeq 126\,{\rm GeV}$. As a consequence, $\lambda(t_{\rm in})$ can be very small by itself at the energy scale of inflation and therefore

Journal of Physics: Conference Series **626** (2015) 012003 doi:10.1088/1742-6596/626/1/012003

allows for much smaller non-minimal couplings $\xi(t_{\rm in})$; see e.g. a discussion in [39, 40]. For certain initial values the renormalization group flow can even drive λ to negative values and therefore lead to an unstable (or meta-stable) vacuum, see e.g. [41]. In general, the precise inflationary predictions of this model are very sensible to small changes in initial values for the Standard Model masses.

4.1.2. Quantum cosmological predictions Following the general method presented in Sec. 3.3, the tunneling amplitude for the non-minimal Higgs inflation model is given by [4]

$$\Gamma(\varphi) = 24\pi^2 \frac{M_{\rm P}^4}{\hat{V}(\varphi)} \simeq 96\pi^2 \frac{\xi^2}{\lambda}\left(1 + \frac{2M_{\rm P}^2}{\xi Z^2 \varphi^2}\right). \tag{35}$$

The peak position $\varphi_{\rm max}$ is determined by the extrema

$$\varphi\frac{d\Gamma}{d\varphi} = \frac{d\Gamma}{dt} = -\frac{6\xi^2}{\lambda}\left(\boldsymbol{A_I} + \frac{64\pi^2 M_{\rm P}^2}{\xi Z^2\varphi^2}\right) = 0, \tag{36}$$

The solution of this condition in terms of the probability peak reads [4]

$$\varphi_{\rm max}^2 = -\left.\frac{64\pi^2 M_{\rm P}^2}{\xi \boldsymbol{A_I} Z^2}\right|_{t=t_{\rm max}}. \tag{37}$$

The peak is very narrow, as can be estimated by the sharpness

$$\mathcal{S} = \frac{(\Delta\varphi)^2}{E_{\rm inf}^{\rm QC}} \simeq \left.\frac{\frac{d^2\Gamma(t)}{dt^2}}{\hat{V}(t)}\right|_{t=t_{\rm max}} = -\left.\frac{\lambda}{12\xi^2}\frac{1}{\boldsymbol{A_I}}\right|_{t=t_{\rm max}} \sim 10^{-10}. \tag{38}$$

In view of $\boldsymbol{A_I}(t_{\rm max}) \simeq \boldsymbol{A_I}(t_{\rm end})$, the point of the horizon crossing φ_* for the pivot scale k, chosen to correspond to $N = 60$, is very close to the point of quantum creation $\varphi_{\rm max}$. Their ratio for different modes, corresponding to different N, therefore takes the form [4]

$$\frac{\varphi_*^2}{\varphi_{\rm max}^2} = 1 - \exp\left[-N\frac{|\boldsymbol{A_I}(t_{\rm end})|}{48\pi^2}\right]. \tag{39}$$

Thus, (39) indicates that, for wavelengths longer than the pivotal one, the instant of horizon crossing approaches the moment of 'creation' of the Universe, but it is always posterior to it ($\varphi_{\rm max} > \varphi_*$), as required for a consistent quantum cosmological history of the universe.

4.2. Natural inflation

Another inflationary model which is in agreement with the Planck data is that of natural inflation [42]. The inflaton potential for natural inflation reads

$$V = \Lambda^4\left[1 + \cos\left(\varphi/f\right)\right]. \tag{40}$$

In this model, φ is supposed to be a pseudo Nambu–Goldstone boson taking values on a circle with radius f and angle $\varphi/f \in [0, 2\pi)$ [42]. The two constants Λ and f determine the height and the slope of the potential and have physical dimension of mass in natural units. The interpretation of φ as a pseudo Nambu–Goldstone field suggests that $f = O(M_{\rm P})$ and $\Lambda \approx M_{\rm GUT} \sim 10^{16}$ GeV.

4.2.1. Slow-roll predictions The cosmological parameters in the inflationary slow-roll analysis can again be derived from the general expressions in Sec. 3.1. From (2), the first two slow-roll parameters read

$$\epsilon_{\rm V} = \frac{M_{\rm P}^2}{2\,f^2}\tan^2\left[\varphi/(2\,f)\right], \qquad \eta_{\rm V} = -\frac{M_{\rm P}^2\cos(\varphi/f)}{f^2\left[1+\cos(\varphi/f)\right]}. \tag{41}$$

The scalar spectral index and the tensor-to-scalar ratio then take the form

$$n_{\rm s} = -\frac{M_{\rm P}^2}{f^2}\frac{3-\cos(\varphi/f)}{1+\cos(\varphi/f)}, \qquad r = \frac{8\,M_{\rm P}^2}{f^2}\tan^2\left[\varphi/(2\,f)\right]. \tag{42}$$

All cosmological observables must again be evaluated at φ_*, the field value that corresponds to the moment where the pivot mode k_* first crosses the Hubble scale. The number of e-folds N_* connecting the end of inflation $\varphi_{\rm end}$ with the value φ_* is

$$N_* = \frac{2\,f^2}{M_{\rm P}^2}\ln\left[\frac{\sin\left(\frac{\varphi_{\rm end}}{2\,f}\right)}{\sin\left(\frac{\varphi_*}{2\,f}\right)}\right]. \tag{43}$$

The value $\varphi_{\rm end}$ that determines the upper integration bound in (43) is determined to be

$$\varphi_{\rm end} = 2\,f\arctan(\sqrt{2}\,f/M_{\rm P}). \tag{44}$$

Inserting (44) in (43), solving for φ_* and parametrizing f in units of $M_{\rm P}$, we find

$$\varphi_* = 2M_{\rm P}\,\alpha\arcsin\left(\frac{\alpha\,e^{-N_*/2\alpha^2}}{\sqrt{1/2+\alpha^2}}\right), \tag{45}$$

where $\alpha := f/M_{\rm P}$. Evaluating the potential (40) at φ_* yields

$$V(\varphi_*) = 2\,\Lambda^4\left[1-\delta_V(\alpha,N_*)\right], \tag{46}$$

where, following [5], we have defined

$$\delta_V(N_*,\alpha) := \frac{2\,e^{-N_*/\alpha^2}\,\alpha^2}{1+2\,\alpha^2}. \tag{47}$$

The expressions for $n_{\rm s}$ and r evaluated at φ_* can be expressed in terms of δ_V and α:

$$n_{{\rm s}*} = 1+\frac{1}{\alpha^2}\frac{\delta_V(N_*,\alpha)+1}{\delta_V(N_*,\alpha)-1}, \qquad r_* = \frac{8}{\alpha^2}\frac{\delta_V(N_*,\alpha)}{1-\delta_V(N_*,\alpha)}. \tag{48}$$

For $N_* = 60$, Planck 2013 data [29] constrain α to lie in the interval [43]

$$5.1 < \alpha < 7.9 \quad (68\%\,{\rm CL}). \tag{49}$$

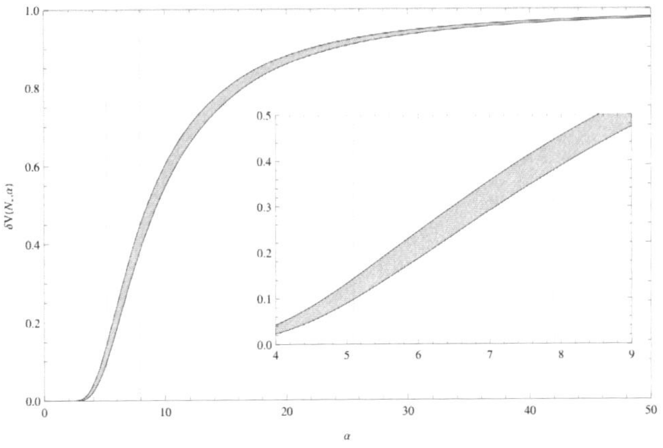

Figure 1. The function $\delta_V(N_*, \alpha)$ as a function of α for values of $N_* \in [50, 60]$, taken from [5]. The upper line corresponds to $N_* = 50$, the lower line to $N_* = 60$. The inset shows the region with α in the 68% CL range $5.1 < \alpha < 7.9$ (see (49)).

4.2.2. Quantum cosmological predictions Following again the general algorithm of Sec. 3.3, we first have to calculate the extrema of (40),

$$\left.\frac{\mathrm{d}V}{\mathrm{d}\varphi}\right|_{\varphi=\varphi_{\mathrm{ext}}} = -\frac{\Lambda^4}{f}\sin\left(\varphi_{\mathrm{ext}}/f\right) = 0\,. \tag{50}$$

If φ_{ext} is a maximum, peak values correspond to

$$\varphi_{\mathrm{max}} := 2\,\pi n f\,. \tag{51}$$

The potential at φ_{max} has the value

$$V_{\mathrm{max}} = 2\,\Lambda^4\,. \tag{52}$$

With the width $\Delta\varphi$ of the distribution given by

$$(\Delta\varphi)^2 = \left.\frac{1}{\Gamma''}\right|_{\varphi=\varphi_{\mathrm{max}}} = \frac{1}{6\,\pi^2}\frac{f^2\,\Lambda^4}{M_{\mathrm{P}}^4}\,, \tag{53}$$

the sharpness of the peak φ_{max} is estimated as

$$\mathcal{S} = \frac{(\Delta\varphi)^2}{V_{\mathrm{max}}^{1/2}} \approx \frac{1}{6\,\pi^2}\frac{f^2\,\Lambda^2}{M_{\mathrm{P}}^4} \sim \frac{\Lambda^2}{M_{\mathrm{P}}^2} \sim 10^{-4}\,, \tag{54}$$

where we have used $f \sim M_{\mathrm{P}}$ and $E_{\mathrm{inf}}^{\mathrm{QC}} \sim \Lambda$, according to (16) and (52).

As can be inferred from Fig. 2, it was shown in [5] that the requirement of a deviation from the approximate consistency requirement (17) of not more than one order of magnitude leads to the following constraint for the parameter α:

$$\alpha \lesssim 710 \text{ for } N_* = 50 \qquad \text{and} \qquad \alpha \lesssim 780 \text{ for } N_* = 60. \tag{55}$$

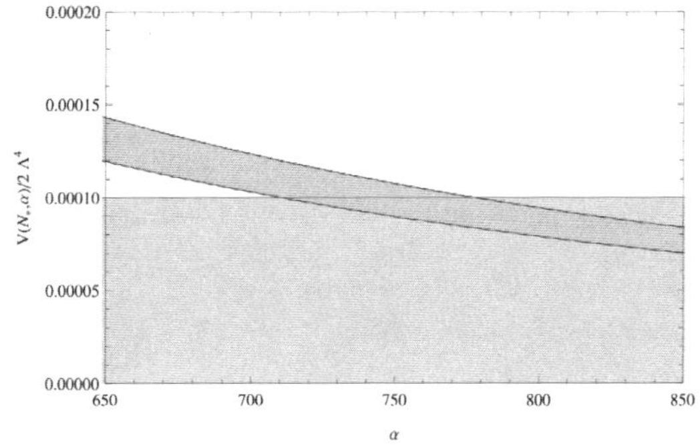

Figure 2. A zoomed-in region of the function $V(N_*, \alpha)/(2\,\Lambda^4) = 1 - \delta_V$ as a function of α for values of $N_* \in [50, 60]$, taken from [5]. The upper purple line corresponds to $N_* = 60$, the lower purple line to $N_* = 50$. The lower area, colored in light red (in black-and-white printing: light gray), corresponds to the region where $E_{\rm inf}^{\rm model} < 10^{-1}\, E_{\rm inf}^{\rm QC}$.

Although a quantum cosmological bound on α derived in this way depends on the allowed tolerance for a violation of (17), this bound is clearly not as restrictive as the constraints on α coming from the comparison with the observational constraints of the spectral index and the tensor-to-scalar ratio by PLANCK. As shown in Fig. 1, in this range $\delta_V \approx 0.1 \div 0.5$ is still small enough to respect the approximate condition (17) to the tolerated accuracy.

Thus, consistency of classical inflationary predictions with observational data (49) result in a much sharper bound $\alpha \sim O(10)$ far below the threshold $\alpha \approx 700$. We can therefore conclude that to a good approximation no conflict with the a quantum origin or our universe does arise in natural inflation, since the consistency condition is satisfied for all experimentally allowed values of α.

5. Conclusions and outlook

In this contribution, we have presented a general method that allows one to derive predictions from quantum cosmology by assuming a consistent history of our universe from its initial quantum creation up to its present state. We have restricted our analysis here to the tunneling scenario, but the method can also be extended to other quantum initial conditions such as the no-boundary condition, although this condition does not lead naturally to inflationary initial conditions.

We have in detail investigated two particular models of inflation: non-minimal Higgs inflation [31, 32, 33, 34, 35, 36, 37, 38, 39] and natural inflation [42]. We have found that both models allow for a consistent cosmic history starting from a quantum tunneling process.

In principle, all inflationary single-field models favored by recent PLANCK data can be summarized by the general class of scalar-tensor theories with the action

$$S = \int \mathrm{d}^4x\,\sqrt{g}\left[U(\varphi)\,R - \frac{G(\varphi)}{2}\,(\nabla\varphi)^2 - V(\varphi)\right], \tag{56}$$

and the method presented here could, in principle, be applied also to this general class parametrized in terms of the arbitrary functions $U(\varphi)$, $G(\varphi)$ and $V(\varphi)$. As has been discussed in the context of non-minimal Higgs inflation, quantum corrections can become important and modify the shape and the location of the extrema of the effective potential. The one-loop divergences for the general action (56), necessary for renormalization, were obtained in [44] in a closed form for an even more general setup of a symmetric $O(N)$ invariant multiple of scalar fields.

Finally, a note regarding the parametrization dependence of these quantum corrections is in order. While in the transition from the Jordan to the Einstein frame parametrizations leads to equivalent formulations at tree level, in the usual quantum field theoretical formalism such a field transformation will in general induce an off-shell parametrization dependence of the effective action [45, 46, 47]. In [45], a geometric approach to the effective action was suggested to overcome the problem with non-covariance (with respect to the configuration space of field) of the ordinary formalism. Recently, this idea has been adopted in [48] in the context of non-minimal Higgs inflation.

References

[1] Mukhanov V 2012 *Physical Foundations of Cosmology* third edition (Cambridge: Cambridge University Press)
[2] Martin J 2015 The observational status of cosmic inflation after Planck *Preprint* astro-ph.CO/1502.05733
[3] Kiefer C 2012 *Quantum Gravity* third edition (Oxford: Oxford University Press)
[4] Barvinsky A O, Kamenshchik A Yu, Kiefer C and Steinwachs C F 2010 Tunneling cosmological state revisited: Origin of inflation with a non-minimally coupled Standard Model Higgs inflaton *Phys. Rev.* D **81** 043530 (*Preprint* hep-th/0911.1408)
[5] Calcagni G, Kiefer C and Steinwachs C F 2014 Quantum cosmological consistency condition for inflation *J. Cosmol. Astropart. Phys.* **10** 026
[6] Kiefer C 2013 Conceptual problems in quantum gravity and quantum cosmology *ISRN Math. Phys.* **2013** 509316 (*Preprint* gr-qc/1401.3578)
[7] Hartle J B 1987 *Prediction in quantum cosmology* (Gravitation in astrophysics) ed B Carter and J B Hartle (New York: Plenum Press) pp 329–60
[8] Halliwell J J 1991 *Introductory lectures on quantum cosmology* (Quantum cosmology and baby universes) ed S Coleman, J B Hartle, T Piran and S Weinberg (Singapore: World Scientific) pp 159–243
[9] Joos E, Zeh H D, Kiefer C, Giulini D, Kupsch J and Stamatescu I O 2003 *Decoherence and the Appearance of a Classical World in Quantum Theory* second edition (Berlin: Springer)
[10] Schlosshauer M 2007 *Decoherence and the quantum-to-classical transition* (Berlin: Springer)
[11] Zeh H D 2007 *The Physical Basis of the Direction of Time* fifth edition (Berlin: Springer)
[12] Vilenkin A 1989 Interpretation of the wave function of the universe *Phys. Rev.* D **39** 1116
[13] Zurek W H 2005 Probabilities from entanglement, Born's rule $p_k = |\psi_k|^2$ from envariance *Physical Review* A **71** 052105
[14] Hawking S W and Page D N 1986 Operator ordering and the flatness of the Universe *Nucl. Phys.* B **264** 185
[15] Barvinsky A O and Kamenshchik A Yu 1990 One loop quantum cosmology: the normalizability of the Hartle–Hawking wave function and the probability of inflation *Class. Quantum Grav.* **7** L181
[16] Conradi H D 1998 Tunneling of macroscopic universes *Int. J. Mod. Phys.* D **7** 189
[17] Mithani A T and Vilenkin A 2015 Tunneling decay rate in quantum cosmology *Preprint* hep-th/1503.00400
[18] Page D N 2014 Spacetime average density (SAD) cosmological measures *J. Cosmol. Astropart. Phys.* **11** 038
[19] Ade P A R *et al.* 2015 A joint analysis of BICEP2/Keck Array and Planck data *Preprint* astro-ph.CO/1502.00612
[20] Ade P A R *et al.* 2015 Planck 2015 results. XIII. Cosmological parameters *Preprint* astro-ph.CO/1502.01589
[21] Barvinsky A O and Kamenshchik A Yu 1996 Quantum origin of the early universe and the energy scale of inflation *Int. J. Mod. Phys.* D **5** 825-844 (*Preprint* gr-qc/9510032)
[22] Barvinsky A O, Kamenshchik A Yu and Mishakov I V 1997 Quantum origin of the early inflationary universe *Nucl. Phys.* B **491** 387-426 (*Preprint* gr-qc/9612004)
[23] Barvinsky A O 1999 Open inflation from quantum cosmology with a strong nonminimal coupling *Nucl. Phys.* B **561** 159-187 (*Preprint* gr-qc/9812058)
[24] Barvinsky A O and Kamenshchik A Yu 1998 Effective equations of motion and initial conditions for inflation in quantum cosmology *Nucl. Phys.* B **532** 339-360 (*Preprint* hep-th/9803052)
[25] Barvinsky A O, Kamenshchik A Yu and Kiefer C 1990 Origin of the inflationary universe *Mod. Phys. Lett.* A **14** 1083 (*Preprint* gr-qc/9905098)
[26] Carr B (ed) 2007 *Universe or Multiverse?* (Cambridge: Cambridge University Press)
[27] Douglas M R 2003 The statistics of string/M theory vacua *J. High Energy Phys.* **05** 046
[28] Borde A, Guth A H and Vilenkin A 2003 Inflationary space-times are incomplete in past directions *Phys. Rev. Lett.* **90** 151301 (*Preprint* gr-qc/0110012)

[29] Ade P A R *et al.* Planck Collaboration 2013 Planck 2013 results. XXII. Constraints on inflation *Preprint* astro-ph.CO/1303.5082

[30] Ade P A R *et al.* Planck Collaboration 2015 Planck 2015. XX. Constraints on inflation *Preprint* astro-ph.CO/1502.02114

[31] Bezrukov F L and Shaposhnikov M 2008 The Standard Model Higgs boson as the inflaton *Phys. Lett. B* **659** 703-706 (*Preprint* hep-th/0710.3755)

[32] Barvinsky A O, Kamenshchik A Yu and Starobinsky A A 2008 Inflation scenario via the Standard Model Higgs boson and LHC *J. Cosmol. Astropart. Phys* **11** 021 (*Preprint* hep-ph/0809.2104)

[33] Bezrukov F L, Magnin A and Shaposhnikov M 2009 Standard Model Higgs boson mass from inflation *Phys. Lett. B* **675** 88-92 (*Preprint* hep-ph/0812.4950)

[34] De Simone A, Hertzberg M P and Wilczek F 2009 Running inflation in the Standard Model *Phys. Lett. B* **678** 1-8 (*Preprint* hep-ph/0812.4946)

[35] Barvinsky A O, Kamenshchik A Yu, Kiefer C, Starobinsky A A and Steinwachs C 2009 Asymptotic freedom in inflationary cosmology with a non-minimally coupled Higgs field *J. Cosmol. Astropart. Phys* **12** 003 (*Preprint* hep-ph/0904.1698)

[36] Bezrukov F and Shaposhnikov M 2009 Standard Model Higgs boson mass from inflation: two loop analysis *J. High Energy Phys.* **07** 089 (*Preprint* hep-ph/0904.1537)

[37] Bezrukov F, Magnin A, Shaposhnikov M and Sibiryakov S 2011 Higgs inflation: consistency and generalisations *J. High Energy Phys.* **01** 016 (*Preprint* hep-ph/1008.5157)

[38] Barvinsky A O, Kamenshchik A Yu, Kiefer C, Starobinsky A A and Steinwachs C F 2012 Higgs boson, renormalization group, and naturalness in cosmology *Eur. Phys. J. C* **72** 2219 (*Preprint* hep-ph/0910.1041)

[39] Allison K 2014 Higgs xi-inflation for the 125-126 GeV Higgs: a two-loop analysis *J. High Energy Phys.* **1402** 040 (*Preprint* hep-ph/1306.6931)

[40] Bezrukov F, Rubio J and Shaposhnikov M 2014 Living beyond the edge: Higgs inflation and vacuum metastability *Preprint* hep-ph/1412.3811

[41] Buttazzo D, Degrassi G, Giardino P P, Giudice G F, Sala F, Salvio A and Strumia A 2013 Investigating the near-criticality of the Higgs boson *J. High Energy Phys.* **1312** 089 (*Preprint* hep-ph/1307.3536)

[42] Freese L, Frieman J A, Olinto A V 1990 Natural inflation with pseudo-Nambu–Goldstone bosons *Phys. Rev. Lett.* **65** 3233-3236

[43] Tsujikawa S, Ohashi J, Kuroyanagi S and De Felice A 2013 Planck constraints on single-field inflation *Phys. Rev.* D **88** 023529 (*Preprint* astro-ph.CO/1305.3044)

[44] Steinwachs C F and Kamenshchik A Yu 2011 One-loop divergences for gravity non-minimally coupled to a multiplet of scalar fields: calculation in the Jordan frame. I. The main results *Phys. Rev.* D **84** 024026 (*Preprint* gr-qc/1101.5047)

[45] Vilkovisky G A 1984 The unique effective action in quantum field theory *Nucl.Phys.* B **234** 125-137

[46] Steinwachs C F and Kamenshchik A Yu 2012 Non-minimal Higgs inflation and frame dependence in cosmology *AIP Conf.Proc.* **1514** 161-164 (*Preprint* gr-qc/1301.5543)

[47] Kamenshchik A Yu and Steinwachs C F 2014 Frame dependence of quantum corrections in cosmology (*Preprint* gr-qc/1408.5769)

[48] Moss I G 2014 Covariant one-loop quantum gravity and Higgs inflation *Preprint* hep-th/1409.2108

Journal of Physics: Conference Series **626** (2015) 012004 doi:10.1088/1742-6596/626/1/012004

Isochronous spacetimes and cosmologies

Fabio Briscese [1,2] **and Francesco Calogero** [3,4]

[1] Departamento de Física, CCEN, Universidade Federal da Paraíba, Cidade Universitária, 58051-970 João Pessoa, PB, Brazil
[2] Istituto Nazionale di Alta Matematica Francesco Severi, Gruppo Nazionale di Fisica Matematica, Città Universitaria, p. Aldo Moro 5, I-00185 Rome, Italy
[3] Physics Department, University of Rome "La Sapienza", p. Aldo Moro, I-00185 Rome, Italy
[4] Istituto Nazionale di Fisica Nucleare, Sezione di Roma

E-mail: `briscese.phys@gmail.com`, `francesco.calogero@uniroma1.it`, `francesco.calogero@roma1.infn.it`

Abstract. It is shown, in the context of general relativity, how to modify any cosmological model so that its local (in time) properties are unchanged, but its global time evolution is *isochronous*, i. e. completely periodic with a fixed period independent of the initial data. In this manner the Big Bang singularity might be avoided, even if this singularity is featured by the original model.

1. Introduction

The possibility has been recently demonstrated [1] to identify (nonrelativistic, Hamiltonian) many-body problems which feature an isochronous time evolution with an arbitrarily assigned period T yet mimic with an arbitrarily good approximation, or even exactly, the time evolution of any given many-body problem (belonging to a very large, quite realistic, class: arbitrary dimensions of ambient space, arbitrary number of particles with arbitrary masses, arbitrary interparticle forces) over a time $\tilde{T}$ which may also be arbitrarily large (but of course such that $\tilde{T} < T$). This finding is valid both in a classical and quantal context; hence—rather disturbingly—it applies to most of nonrelativistic physics. Purpose and scope of this presentation is to report the possibility to extend it to a general relativity context (classical, non quantal) as described in [2, 3]. In particular we show how to modify any cosmological model so that its local (in time) properties are unchanged, but its global time evolution is *isochronous* (and might avoid, in the past and future, the Big Bang, even if this singularity is featured by the original model).

2. Synthetic recall of previous results

The Hamiltonian characterizing the standard, overall translation-invariant, nonrelativistic N-body problem reads as follows (here, merely for notational simplicity, we consider unit-mass particles in a one-dimensional space environment, but this is not a significant limitation, see below):

$$H(\underline{p},\underline{q}) = \frac{1}{2}\sum_{n=1}^{N} p_n^2 + V(\underline{q}), \qquad V(\underline{q}+a) = V(\underline{q}). \qquad (1)$$

Here of course $\underline{p}$ respectively $\underline{q}$ denote the N-vectors the components of which, p_n respectively q_n, are the standard canonical coordinates.

Let us now review some standard related developments, trivial as they are. We hereafter denote with P the total momentum, and with Q the (canonically-conjugate) centre-of-mass coordinate:

$$P \equiv \sum_{n=1}^{N} p_n, \qquad Q \equiv \frac{1}{N} \sum_{n=1}^{N} q_n \ . \tag{2}$$

Thanks to the translation invariance of the Hamiltonian the total momentum Poisson-commutes with the Hamiltonian, $[H, P] = 0$, where—here and hereafter—the Poisson bracket of two generic functions of the canonical variables is defined as usual,

$$[F, G] \equiv \sum_{n=1}^{N} \left[\frac{\partial F}{\partial p_n} \frac{\partial G}{\partial q_n} - \frac{\partial F}{\partial q_n} \frac{\partial G}{\partial p_n} \right] \ ; \tag{3}$$

and let us recall that the evolution of any function $F(p, q)$ of the canonical coordinates is determined by the equation $F' = [F, H]$, where the appended prime denotes differentiation with respect to the "timelike" variable corresponding to the evolution induced by the Hamiltonian H.

It is now convenient to introduce the "relative coordinates" and the "relative momenta" via the standard definitions

$$x_n \equiv q_n - Q; \qquad y_n \equiv p_n - \frac{P}{N} \ . \tag{4}$$

Note that these are not canonically conjugated quantities, since $[x_n, y_n] = \delta_{nm} - 1/N$, and they are not independent since obviously their sum vanishes: $\sum_{n=1}^{N} x_n = 0$, $\sum_{n=1}^{N} y_n = 0$.

It is moreover convenient to introduce the "relative-motion" Hamiltonian $h(\underline{y}, \underline{x})$ via the formula

$$h(y, x) \equiv \frac{1}{2} \sum_{n=1}^{N} y_n^2 + V(\underline{x}) = \frac{1}{4N} \sum_{n,m=1}^{N} (p_n - p_m)^2 + V(\underline{q}) \ , \tag{5}$$

implying

$$H(\underline{p}, \underline{q}) = \frac{P^2}{2\,N} + h(\underline{y}, \underline{x}) \ . \tag{6}$$

Note that this definition of the relative-motion Hamiltonian $h(\underline{y}, \underline{x})$ entails that it Poisson-commutes with both P and Q: $[P, h] = [Q, h] = 0$.

For completeness and future reference let us also display the equations of motion implied by the original Hamiltonian $H(p, q)$:

$$q_n' = p_n \ ; \quad p_n' = -\frac{\partial V(q)}{\partial q_n} = q_n'' \ , \tag{7}$$

where (for reasons that will be clear below) we denote as τ the independent variable corresponding to this Hamiltonian flow and with appended primes the differentiations with respect to this variable:

$$q_n \equiv q_n(\tau); \qquad p_n \equiv p_n(\tau); \qquad q_n' = \frac{dq_n}{d\tau}; \qquad p_n' = \frac{dp_n}{d\tau} \ . \tag{8}$$

Hence $Q' = P/N$ and $P' = 0$, yielding

$$Q(\tau) = Q(0) + \frac{P(0)}{N}\tau; \qquad P(\tau) = P(0) \ , \tag{9}$$

Journal of Physics: Conference Series **626** (2015) 012004 doi:10.1088/1742-6596/626/1/012004

as well as

$$x'_n = y_n = \frac{\partial h(y,x)}{\partial y_n}, \qquad y'_n = -\frac{\partial V(x)}{\partial x_n} = -\frac{\partial h(y,x)}{\partial x_n} . \tag{10}$$

Note that these equations, (10), have the standard Hamiltonian form even though, as mentioned above, the coordinates y_n and x_n are not quite canonically conjugated variables.

This ends our review of quite standard results for the classical nonrelativistic N-body problem. Let us also emphasize that, above and below, the restriction to unit-mass particles, and to one-dimensional space, is merely for simplicity: generalizations—also of the following results—to the more general case with different masses and arbitrary space dimensions is quite elementary, essentially trivial. As for the number of particles N, it is an arbitrary positive integer: interesting cases are $N = 2$ (the generally solvable two-body problem), $N \approx 10^{25}$ (the standard macroscopic system, where statistical mechanics and thermodynamics rules), $N \approx 10^{85}$ (maybe the entire Universe, the topic of cosmology; which however should be treated in a general relativity context, see below).

3. An isochronous Hamiltonian

A "Ω-modified" isochronous Hamiltonian $\tilde{H}(p, q; \Omega)$ is now defined [1] by the formula

$$\tilde{H}(\underline{p}, \underline{q}; \Omega) \equiv \frac{1}{2}\left[\left(P + \frac{h(\underline{y}, \underline{x})}{d}\right)^2 + \Omega^2 \, Q^2\right] , \tag{11}$$

where d is an irrelevant arbitrary constant (introduced for dimensional reasons: it has the dimensions of a momentum[1]) and Ω is hereafter a positive constant the choice of which will be our privilege. Let us emphasize that hereafter the evolution of the various quantities is that caused by this new Ω-modified Hamiltonian; the corresponding independent variable is hereafter denoted as t and interpreted as "time", and differentiations with respect to this variable will be denoted, as usual, by superimposed dots; and of course the time evolution of any function $F \equiv F(\underline{p}, \underline{q})$ of the canonical variables will be determined by the standard equation $\dot{F} = [\tilde{H}, F]$.

The solution of the time-evolution yielded by the isochronous Hamiltonian $\tilde{H}(p, q; \Omega)$ is [1]

$$Q(t) = Q(0)\cos(\Omega t) + \dot{Q}(0)\frac{\sin(\Omega t)}{\Omega} = C\, d\,\frac{\sin\left(\Omega(t - \bar{t})\right)}{\Omega} , \tag{12a}$$

$$P(t) = P(0)\cos(\Omega t) + \dot{P}(0)\frac{\sin(\Omega t)}{\Omega} + \frac{h(y(0), x(0))}{d}\left[\cos(\Omega t) - 1\right] , \tag{12b}$$

where $C = \sqrt{2\tilde{H}(\underline{p}, \underline{q}; \omega)}/d$ is a dimensionless constant, and

$$\tilde{x}_n(t) = x_n(\tau); \qquad \tilde{y}_n(t) = y_n(\tau) , \tag{12c}$$

where (changing for convenience notation) we now denote as $\tilde{x}_n \equiv \tilde{x}_n(t)$ and $\tilde{y}_n \equiv \tilde{y}_n(t)$ the dependent variables whose time-evolution is determined by the Ω-modified Hamiltonian $\tilde{H}(\underline{p}, \underline{q}; \Omega)$ and as $x_n \equiv x_n(\tau)$ and $y_n \equiv y_n(\tau)$ the dependent variables whose τ-evolution is determined by the original, unmodified Hamiltonian $H(\underline{p}, \underline{q})$. And here (most importantly)

$$\tau \equiv \tau(t) = \frac{A\,\sin(\Omega t) + B\,\left[1 - \cos(\Omega t)\right]}{\Omega} = \frac{Q(t)}{d} , \tag{12d}$$

where the constants A and B are given by simple explicit formulas in terms of the initial position $Q(0)$ and velocity $P(0)$ of the centre-of-mass of the system and the value of the Hamiltonian

[1] Note that (5) implies that $h(y, x)$ has the dimension of a squared momentum.

$\tilde{H}(\underline{p}, \underline{q}; \Omega)$ (which is of course a constant of motion) [1]. The crucial observation is that $\tau(t)$—hence the entire solution, see (12)—is a periodic function of t with period $T = 2\pi/\Omega$.

The solution formulas displayed above (see in particular (12c) and (12d)) show that the dynamical evolution implied by the *isochronous* Hamiltonian (11) features exactly the *same* trajectories in the (phase) space of the coordinates x_n and y_n as those yielded by the standard N-body Hamiltonian (1); but in the case of the standard N-body Hamiltonian (1) the corresponding evolution corresponds to a uniform travel along those trajectories; while in the case of the *isochronous* Hamiltonian (11) the evolution of the system corresponds to an oscillatory travel—periodic with the fixed period T—along those same trajectories, only a portion of which gets therefore explored. It also entails that the dynamics yielded by the *isochronous* Hamiltonian (11) does not differ—on a time scale much shorter than the period T—from that yielded by the original Hamiltonian (1) (up to a *constant* shift and rescaling of time). Indeed clearly $\tau \equiv \tau(t)$ on a sufficiently short time scale varies linearly in t, since in the neighborhood of any time $\bar{t}$—except when $\dot{\tau}(\bar{t})$ vanishes—clearly

$$\tau(t) = \bar{C} + \bar{D}\ (t - \bar{t}) + O\left[\left(\frac{t-\bar{t}}{T}\right)^2\right] , \tag{13a}$$

$$\bar{C} = \frac{A\ \sin(\Omega\bar{t}) + B\ [1 - \cos(\Omega\bar{t})]}{\Omega} , \tag{13b}$$

$$\bar{D} = A\ \cos(\Omega\bar{t}) + B\ \sin(\Omega\bar{t})\ . \tag{13c}$$

Remark 3.1. A different modified Hamiltonian can be introduced which also yields an *isochronous* evolution with the arbitrary period T but reproduces the dynamics of the original, standard N-body Hamiltonian (1) over an arbitrary time $\tilde{T}$ (of course such that $\tilde{T} < T$) *exactly* (rather than with arbitrary accuracy). And the property of *isochrony* remains true in a *quantal* context. [1] □

4. Extension to general relativity

The following presentation is terse, focusing on the main ideas.

The standard approach. The following space-time metric is generally introduced as appropriate to describe a homogeneous and isotropic Universe:

$$ds^2 = dt^2 - a(t)^2\ d\vec{x}^2\ . \tag{14}$$

Our approach. We generalize the class of metrics to be considered, writing instead

$$ds^2 = b(t)^2\ dt^2 - a(t)^2\ d\vec{x}^2\ , \tag{15}$$

where $b(t)^2$ is commonly termed the "lapse" function.

Standard objection in the general relativity context. This new metric is physically equivalent to that written above, since it can be reduced to it by a change of the time variable from t to τ with

$$d\tau = b(t)dt; \qquad \tau(t) = \int^t dt' b(t')\ , \tag{16}$$

whereby the metric (15) becomes

$$ds^2 = d\tau^2 - \alpha(\tau)^2\ d\vec{x}^2\ , \tag{17}$$

of course with $a(t) = \alpha(\tau) = \alpha(\tau(t))$. Indeed, such a redefinition of the time-variable has no physical meaning in general relativity, where a change of coordinates (in this case, from t to τ)

is *physically irrelevant*: what is relevant is the geometry of space-time, not the variables used to describe it.

Our rejoinder. Yes, in the general relativity context this is indeed the case, but only if $b(t)$ is a one-to-one invertible function (in which case the two metrics (14) and (15) are related to each other by a diffeomorphism). However, this is certainly *not* the case if $b(t)$ is a periodic function (with arbitrarily assigned period T and vanishing mean value):

$$b(t+T) = b(t) \; ; \qquad B(t) \equiv \int_0^t b(t') \, dt' \, , \quad B(t+T) = B(t) \; . \tag{18}$$

Then the two metrics (14) and (15) are indeed *locally equivalent* (in time), but they are *not globally equivalent* (in time). And clearly with such an assignment one arrives at an *isochronous* cosmology.

This is essentially our argument: given an arbitrary cosmology, one can construct another one (indeed an infinity of other ones) which is *locally equivalent* (in time) but *not globally equivalent* (in time) to the previous one: for instance one, but not the other, might feature a Bing Bang. Clearly the corresponding space-times would be physically different; but they would not be distinguishable by any (local in time) measurement... And the new cosmology might be *isochronous*... i. e., entail a time evolution which is periodic with an arbitrary, *a priori* assigned, period T.

Let us now be a bit more specific. As usual in cosmological investigations based on the Einstein equations, we assume that the energy-momentum tensor of the Universe is that of an ideal fluid with a 4-velocity given by $U_\mu = |b(t)| \; \delta_{\mu 0}$, so that the ideal fluid has zero spatial velocity and the frame of reference is co-moving, while the zero component of the 4-velocity is always nonnegative (as it should be). The Einstein equations for the metric (15) are then drastically simplified and read

$$3 \left[\frac{\dot{a}(t)}{a(t)} \right]^2 = k \; b(t)^2 \; \rho(t) \; , \tag{19a}$$

$$\dot{\rho}(t) + 3 \left[\frac{\dot{a}(t)}{a(t)} \right] \; [\rho(t) + P(t)] = 0 \; . \tag{19b}$$

Here and hereafter superimposed dots denote differentiations with respect to the time t; ρ and P are respectively the energy density and pressure of the universe; $k = 8\pi G/c^4$; G is the gravitational constant; c the speed of light.

This system of two ODEs must be complemented by an equation of state relating the pressure P of the universe to its energy density ρ: we shall here use for simplicity the simple ideal fluid relation $P = \omega \rho$, with constant ω. But note that, after this assignment, this system of two ODEs is not quite closed: it is still possible to assign, essentially arbitrarily, the function $b(t)$. Of course this arbitrariness corresponds to the freedom in general relativity to reparametrize time; but—as discussed above—it goes beyond this if $b(t)$ is *not* invertible, hence if the relevant reparametrization is *not* a global diffeomorphism.

If we consider a perfect fluid with constant equation of state parameter $\omega > -1$, it follows from the system (19) that the energy density of the fluid is

$$\rho(t) = \rho(0) \left[\frac{a(0)}{a(t)} \right]^{3(\omega+1)} \; , \quad \text{if} \quad \omega > -1 \; , \tag{20}$$

while for $\omega = -1$ one has a constant energy density,

$$\rho = \Lambda \quad \text{if} \quad \omega = -1 \; . \tag{21}$$

Journal of Physics: Conference Series **626** (2015) 012004 doi:10.1088/1742-6596/626/1/012004

It is now convenient [2, 3] to introduce a new variable τ by setting (see (18))

$$\tau(t) \equiv B(t) \ , \tag{22}$$

and new auxiliary functions $\alpha(\tau)$, $r(\tau)$, $p(\tau)$ by setting

$$a(t) \equiv \alpha(\tau(t)) \ , \quad \rho(t) \equiv r(\tau(t)) \ , \quad P(t) \equiv p(\tau(t)) \ . \tag{23}$$

Then the functions $a(t)$ and $\rho(t)$ are solutions of (19) provided $\alpha(\tau)$ and $r(\tau)$ are solutions of the following system:

$$3\left[\frac{\alpha'(\tau)}{\alpha(\tau)}\right]^2 = k\ r(\tau) \ , \tag{24a}$$

$$r'(\tau) + 3\left[\frac{\alpha'(\tau)}{\alpha(\tau)}\right][r(\tau) + p(\tau)] = 0 \ . \tag{24b}$$

Here the prime denotes differentiation with respect to τ, and $p(\tau)$ and $r(\tau)$ satisfy the same equation of state $p(\tau) = \omega\, r(\tau)$ as $P(t)$ and $\rho(t)$. And since the definition of $\tau(t)$ (see (22) with (18)) implies that this function is periodic with period T,

$$\tau(t+T) = \tau(t) \ , \tag{25}$$

the same property is inherited via the definitions (23) by the physical quantities $a(t)$, $\rho(t)$ and $P(t)$:

$$a(t+T) = a(t) \ , \quad \rho(t+T) = \rho(t) \ , \quad P(t+T) = P(t) \ . \tag{26}$$

This property holds for the solutions of the system (19) characterized by any initial conditions: the physical quantities $a(t)$, $\rho(t)$ and $P(t)$ are therefore *isochronous*. Moreover, since $b(t)$ can be assigned arbitrarily, the period T of these *isochronous* solutions is a parameter which can be freely assigned.

It is immediate to identify the system (24) with the standard Friedman-Robertson-Walker (FRW) system characterized by the metric (17) with the energy momentum tensor of a perfect fluid in co-moving coordinates (see [4] for a review on FRW equations), while the system (19) is a generalized FRW system corresponding to the metric (15). But let us re-emphasize that the new variable τ cannot be given globally the significance of "time", hence the fact that it might change sign—indeed, it certainly does so, since it depends periodically on the time t—has no unphysical connotation.

Let us now show two explicit examples of *isochronous* cosmological solutions of Einstein's equations obtained in this manner. In both cases we assign for simplicity the function $b(t)$ and $B(t)$ as follows:

$$b(t) \equiv \cos(\Omega t) \ , \quad B(t) = \frac{\sin(\Omega t)}{\Omega} \ , \quad \Omega = \frac{2\pi}{T} \ . \tag{27}$$

As first example, consider a universe filled with a dark energy fluid with constant density $\rho = \Lambda$ and equation of state parameter $\omega = -1$. Assuming that the metric tensor is given by (15), the system (19) has the following solution:

$$a(t) = a_0 \exp\left[\sqrt{\frac{k\Lambda}{3}}\frac{\sin(\Omega t)}{\Omega}\right] \ , \quad a_0 \equiv a(0) \ . \tag{28}$$

Note that at the times $t_n = (1/2+n)\pi/\Omega$ with n integer the scale factor $a(t)$ reaches its maximum (for n even) and minimum (for n odd) values $a_\pm \equiv a_0 \exp\left[\pm\sqrt{k\Lambda/3\Omega^2}\right]$. Also note that, even if the metric (15) is degenerate since $b(t_n) = 0$, all physical quantities such as the energy density

Journal of Physics: Conference Series **626** (2015) 012004 doi:10.1088/1742-6596/626/1/012004

and pressure, the Ricci scalar curvature R, etc., are *not* singular since they can be expressed as functions of the scale factor $a(t)$ (e. g. $R = k[\rho(a) - 3P(a)]$), which is finite for all time, see (28)). Therefore t_n is a fictitious singularity corresponding to the time when the universe passes from an expanding to a contracting epoch and viceversa.

As second example, consider the case of a perfect fluid with equation of state parameter $\omega > -1$. Then the scale factor is

$$a(t) = a_0 \left[1 + (1+\omega)\sqrt{\frac{3k\rho_0}{4}}\frac{\sin(\Omega t)}{\Omega}\right]^{\frac{2}{3(1+\omega)}} , \quad a_0 \equiv a(0) \; . \tag{29}$$

The maxima and minima of the scale factor are now again reached at $t_n = (1/2 + n)\pi/\Omega$ and have the values $a_\pm = a_0\left[1 \pm \sqrt{3k\rho_0}\frac{1+\omega}{2\Omega}\right]^{\frac{2}{3(1+\omega)}}$. Also in this case t_n corresponds to the transition from the expanding to the contracting phases and viceversa and to a fictitious singularity of the metric, since $b(t_n) = 0$ but the scale factor $a(t)$ and the relevant physical quantities may remain finite for all time. Indeed for any assignment of the parameter Ω such that

$$\Omega > \frac{\sqrt{3k\rho_0}}{2}\,(1+\omega) \; , \tag{30}$$

the scale factor $a(t)$ is positive for all time (see (29)) and the big bang singularity of the FRW model is avoided. Hereafter we assume that this restriction, (30), is always enforced. Of course with a different assignment of $b(t)$ one obtains different solutions, all of which yield the same dynamics locally (in time); but possibly quite different dynamics globally (in time).

Furthermore, it has been shown that, by choosing $b(t)$ so that $a(t) = \alpha(\tau(t)) > 0$ for any t, such *isochronous* metrics can be manufactured to be geodesically complete [2] and therefore singularity-free [2], so that the geodesic motion as well as all physical quantities described by scalar invariants are always well defined [2] and the Big Bang singularity may be avoided. This implies that these *isochronous* metrics may describe a *singularity-free* universe which—while featuring a time evolution which reproduces identically (up to diffeomorphic time reparameterization; *locally* in time, except at a discrete set of instants t_n) that of the FRW metric (24)—features an expansion which stops at some instant, to be followed by a period of contraction, until this phase of evolution stops and is again followed by an expanding phase, this pattern being repeated *ad infinitum*.

Let us also note that this class of metrics realize *de facto* the reversal of time's arrow, as discussed for instance in [5].

It is also evident that these *isochronous* metrics must be degenerate at a discrete set of instants t_n when the time reversals occur, say from expansion to contraction and viceversa; a phenomenology whose inclusion in general relativity might be considered problematic, because at these instants t_n—when the metric is degenerate—the "equivalence principle" corresponding to the requirement that the metric tensor have a Minkowskian signature is violated. On the other hand a *physically unobservable* violation of a "principle" can be hardly considered *physically relevant*; and the *isochronous* metrics are solutions of a version of general relativity whose dynamics, while still governed by Einstein's equations[3], does allow the violation of the equivalence principle at a discrete set of hypersurfaces.

Let us note that the consideration in the context of general relativity of degenerate metrics is not new. For instance, they were already introduced in [7, 8] and subsequently investigated in a series of papers [9, 10, 11, 12, 13, 14, 15, 16, 17, 18, 19, 20, 21, 22, 23, 24, 25], in order to

[2] A spacetime is singularity-free if it is geodesically complete, i.e. if its geodesics can be always past- and future-extended [4].

[3] Einstein's equations by itself do not determine the signature of spacetime, see the discussion in [6, 7].

study signature-changing spacetimes which are a classical realization of the change of signature in quantum cosmology conjectured by Hartle and Hawking [26] (see also [27, 28, 29]), which has philosophical implications on the origin of the universe [30].

Let us also mention that in [3] a different realization of *isochronous* cosmologies has been considered: via *nondegenerate* metrics featuring a jump in their first derivatives at the inversion times t_n, which then implies a distributional contribution in the stress-energy tensor at t_n. These metrics are diffeomorphic to the continuous and degenerate *isochronous* metrics everywhere except on the inversion hypersurfaces $t = t_n$; hence they describe the same physics except at the infinite set of discrete times t_n; and clearly they may also be manufactured so as to agree with all cosmological observations (*locally* in time). Since these two realizations of *isochronous* cosmologies are *locally* but *not globally* diffeomorphic, they correspond to different spacetimes and may be considered to emerge from different theories: the *nondegenerate* ones are generalized (in the sense of distributions) solutions of Einsteinian general relativity (including universal validity of the equivalence principle, but allowing jumps—whose physical significance and justification is moot—at the discrete times t_n); the *degenerate* ones feature no mathematical or physical pathologies but fail to satisfy the "equivalence principle" at the discrete times t_n.

Finally let us emphasize that the argument based on the transition in the reduced Einstein equations from the time t to the auxiliary variable τ—illustrated above when the right-hand side is characterized by the simple ideal fluid relation $P = \omega\rho$—remains valid in the more general case in which the system (24) corresponds to a universe filled with a mixture of ultra-relativistic and nonrelativistic perfect fluids plus a cosmological constant and an inflaton scalar field (the currently most popular candidate to describe our Universe); then the solution of the (so-modified) system (24) gives, for any time evolution over a limited time period such that $|t - \bar{t}| \ll T$, exactly the "right" sequence of inflation / radiation domination / matter domination / late time acceleration epochs characterizing the Λ-CDM universe. Note that it shall then also display any phenomenology interpretable as observable consequence of a past Big Bang singularity implied by this evolution; in spite of the fact that, as discussed above, a suitable assignment of Ω hence of the isochrony period T entails that the evolution of the so-modified system (24) would never feature such an event.

5. Conclusions

We have shown that for any given FRW cosmology described by the metric (14) it is possible to find an infinite number of *isochronous cosmologies* described by the periodic metric (15) with (18), and that one can always assign the arbitrary function $b(t)$ so that the new metric (15) is *singularity-free* and yields an *isochronous* time evolution which coincides—locally in time—with that yielded by the FRW metric (14). One might wonder how general this finding is. To answer this question, we note that the mechanism to manufacture (theoretically) these *isochronous cosmologies* can be applied to any synchronous metric, i.e. to any metric with $g_{00} = 1$ and $g_{0j} = 0$. Therefore, as long as one can perform a coordinate transformation that makes a solution of the Einstein equations *synchronous*, one can also find an *isochronous* solution of the relevant field equations which is locally—but only locally (in time)—physically identical to the original solution up to a diffeomorphic reparametrization of the time variable. Since it is possible to write most metrics in *synchronous* form by a diffeomorphic change of coordinates, this makes our finding quite general.

Let us also note (see *Remark 3.1* above) that, while the results that are valid for the quite general (nonrelativistic) many-body problems mentioned at the beginning of this presentation can be extended from a classical to a quantal context in a (theoretically and mathematically) rigorous manner, an analogous quantal extension of the approach described above is unfeasible as long as there is no mathematically rigorous theory encompassing general relativity and quantization. And let us conclude by emphasizing that we are *not* arguing that our Universe evolves periodically indeed *isochronously*, nor do we wish to enter the related *philosophical* issues (not our cup of tea!); we merely point out a (disturbing indeed unpleasant!) fact: in the context of theoretical and mathematical physics such a possibility does not seem to be excluded.

Acknowledgments
F.B. thanks CNPq for grant 304494/2014-3.

References
[1] Calogero F and Leyvraz F 2007 General technique to produce isochronous Hamiltonians *J. Phys. A: Math. Theor.* **40** 12931; Calogero F and Leyvraz F 2009 How to extend any dynamical system so that it becomes isochronous, asymptotically isochronous or multi-periodic *J. Nonlin. Math. Phys.* **16** 311; Calogero F and Leyvraz F 2011 Isochronous systems, the arrow of time, and the definition of deterministic chaos *Lett. Math. Phys.* **96** 37; Calogero F 2008 *Isochronous systems* (Oxford Univ. Press) – marginally updated paperback edition, 2012; Calogero F 2011 Isochronous dynamical systems *Phil. Trans. R. Soc.* A **369** 1118
[2] Briscese F and Calogero F 2014 Isochronous cosmologies *Int. J. Geom. Meth. Mod. Phys.* **11** 1450054 (*Preprint* arXiv:1402.0704 [gr-qc])
[3] Briscese F and Calogero F 2014 Isochronous spacetimes *Acta Appl. Math.* in press (*Preprint* arXiv:1406.7156 [gr-qc])
[4] Misner C W, Thorne KS and Wheeler J A 1973 *Gravitation* (Freeman & Co.)
[5] Sakharov A D 1980 Cosmological models of the Universe with reversal of time's arrow *Zh. Eksp. Teor. Fiz.* **79** 689 [1980 Sov. Phys. JETP **52** 349]
[6] Teitelboim C 1980 Hamiltonian structure of spacetimes, in *General Relativity and Gravitation: One Hundred Years after the Birth of Albert Einstein* ed A Held (New York: Plenum) p 23
[7] Ellis G, Sumerukt A, Coulet D and Hellabyt C 1992 Change of signature in classical relativity *Class. Qu. Grav.* **9** 1535
[8] Ellis G 1992 Covariant Change of Signature in Classical Relativity *Gen. Rel. Grav.* **24** No. 10
[9] Carfora M and Ellis G 1995 The Geometry of Classical Change of Signature *Int. J. Mod. Phys.* D **4** 175
[10] Dray T, Ellis G and Hellaby C 2001 Note on Signature Change and Colombeau Theory *Gen. Rel. Grav.* **33** 1041
[11] Mansouri R and Nozari K 2000 A New Distributional Approach to Signature Change *Gen. Rel. Grav.* **32** 253
[12] Dray T, Manogue C A and Tucker R W 1991 Particle Production from Signature Change *Gen. Rel. Grav.* **23** 967
[13] Dray T, Manogue C A and Tucker R W 1993 The Scalar Field Equation in the Presence of Signature Chang *Phys. Rev.* D **48** 2587
[14] Hellaby C and Dray T 1994 Failure of standard conservation laws at a classical change of signature *Phys. Rev.* D **49** 5096
[15] Hayward S A 1994 Weak Solutions Across a Change of Signature *Class. Qu. Grav.* **11** L87
[16] Dray T, Manogue C A and Tucker R W 1995 Boundary Conditions for the Scalar Field in the Presence of Signature Change *Class. Qu. Grav.* **12** 2767
[17] Hayward S A 1995 Comment on "Failure of Standard Conservation Laws at a Classical Change of Signature" *Phys. Rev.* D **52** 7331
[18] Hellaby C and Dray T 1995 Reply Comment: Comparison of Approaches to Classical Signature Change *Phys. Rev.* D **52** 7333
[19] Hayward S A 1992 Signature Change in General Relativity *Class. Qu. Grav.* **9** 1851; *erratum:* 1992 *Class. Qu. Grav.* **9** 2543
[20] Kossowski M and Kriele M 1993 Smooth and Discontinuous Signature Type Change in General Relativity *Class. Qu. Grav.* **10** 2363
[21] Dray T 1996 Einstein's Equations in the Presence of Signature Change *J. Math. Phys.* **37** 5627
[22] Dray T, Ellis G, Hellaby C and Manogue C A 1997 Gravity and Signature Change *Gen. Rel. Grav.* **29** 591
[23] Kamleh W 2000 Signature Changing Space-times and the New Generalised Functions *Preprint* arXiv:gr-qc/0004057
[24] Dray T and Hellaby C 1996 Comment on 'Smooth and Discontinuous Signature Type Change in General Relativity' *Gen. Rel. Grav.* **28** 1401
[25] Hartley D, Tucker R W, Tuckey PA and Dray T 2000 Tensor distributions on signature changing space-times *Gen. Rel. Grav.* **32** 491 (*Preprint* arXiv:gr-qc/9701046)
[26] Hartle J B and Hawking S W 1983 Wave function of the universe *Phys. Rev.* D **28** 2960
[27] De Witt B S 1967 Quantum Theory of Gravity. I. The Canonical Theory *Phys. Rev.* **160** 1113

[28] Hawking S W 1982 in *Astrophysical Cosmology* ed H A Bruck, G V Coyne and M S Longair (Vatican City: Pontifica Academia Scientarium) p 563
[29] Hawking S W 1984 The Quantum State of the Universe *Nucl. Phys.* B **239** 257
[30] Hawking S W 1989 *A Brief History of Time* (London: Bantam)

Journal of Physics: Conference Series **626** (2015) 012005 doi:10.1088/1742-6596/626/1/012005

The Topological Field Theory of Data: a program towards a novel strategy for data mining through data language

M Rasetti [1] **and E Merelli** [2]

[1]ISI Foundation, Via Alassio 11-C, 10126 Torino (Italy)
[2] School of Science and Technology, University of Camerino, Via del Bastione 1, 62032 Camerino (Italy)

E-mail: rasetti@isi.it ; emanuela.merelli@unicam.it

Abstract. This paper aims to challenge the current thinking in IT for the 'Big Data' question, proposing – almost verbatim, with no formulas – a program aiming to construct an innovative methodology to perform data analytics in a way that returns an automaton as a recognizer of the data language: a *Field Theory of Data*. We suggest to build, directly out of probing data space, a theoretical framework enabling us to extract the manifold hidden relations (patterns) that exist among data, as correlations depending on the semantics generated by the mining context. The program, that is grounded in the recent innovative ways of integrating data into a topological setting, proposes the realization of a Topological Field Theory of Data, transferring and generalizing to the space of data notions inspired by physical (topological) field theories and harnesses the theory of formal languages to define the potential semantics necessary to understand the emerging patterns.

1. The landscape

Complex Systems are ubiquitous; complex, multi-level, multi-scale systems are everywhere: in Nature, but also in the Internet, the brain, the climate, the spread of pandemics, in economy and finance; in other words in Society. A deep, intriguing question that has been raised about complex systems is: can we envisage the construction of a *bona fide* Complexity Science Theory? In other words, does it make sense to think of a conceptual construct playing for complex systems the same role that Statistical Mechanics played for Thermodynamics?

The challenge is indeed enormous. In statistical mechanics a number of basic restrictive assumptions play a crucial role: *ergodicity* ensures that all accessible states of a system are reached with equal probability; the *thermodynamic limit*, $N \to \infty$, induces the number of particles into play, measured in terms of the Avogadro's number, to be essentially infinite; *particles* are *identical* and *indistinguishable*, constraint that is not even mentioned when studying the features of collections of particles, but it is there – particles of the same species are identical and interact with each other pairwise all in the same way, that is, with the same interaction law – in the quantum case they are indistinguishable; *analytical structure* can be defined for the *underlying dynamics*, that is, regular equations of motion exist at the micro-scale – analyticity breaking and singularities only appear as signal of the macro-phenomenon of phase transitions; *'experiment-based' phenomenology* is repeatable, as in reductionist science, under the same

initial and boundary conditions the same experiment should give the same outcome.
Only under these conditions statistical mechanics is able to provide a reliable representation of most of the facts of thermodynamics.

On the contrary, typically complex systems, in particular those representing societal phenomena, have the following hallmarks: they are *NOT* ergodic; their number of *agents*, N, is ordinarily finite, even though it can be large on a social scale; their agents are *NOT* identical – they are distinguishable complex systems themselves, with their strategies and autonomous behaviors; they are *NEVER* representable by analytic, perhaps in certain cases not even by recursive, functions; above all, they are *DATA-based*, usually *NO* repeatable experiment is possible under external control.

Today our world is no doubt complex and data-based: about 4 billion people own a cell phone (which makes this the first device in human history owned by more than a half of the world population), every day over 300 billion e-mails and 25 billion SMSs are exchanged, 500 million pictures are uploaded on Facebook, etc. The information created and exchanged in a year added up in 2013 to 4 zettabyte (1 zettabyte = 10^{21} bytes) and every year it grows 40% (in 4 years it will reach a yottabyte, 10^{24}, a number larger than Avogadro's number!). For this reason we concentrate on the last item of the list first: data; indeed Big Data. The challenge is to extract the huge amount of information, as a norm hidden, flowing in and around complex systems.

Big Data have a variety of diverse features; they are present in science (see, e.g., the Hubble and Genoma projects, the CERN data archive, etc.): typically well organized in high quality data-bases; but now also in society, where they might for the first time allow for a true societal tomography making possible predictions not envisioned before (see, e.g., the H1N1 pandemics of 2009); or for unprecedented targets and strategies: the Big Data hardware challenge (high performance computing); the Big Data manipulation challenge: both in computer science (new computing paradigm; interaction-based computing; beyond Turing machine) and data analytics (new approach for data mining; non-linear causal inference). Also, the ever-more blurred boundaries between the digital and physical worlds that characterize our digitalized global world are bound to progressively fade away, as ICT becomes an integral part of the fabric of nature and society.

Goal of the game is to endow ICT with new, more and more efficient tools to play its role in the difficult process of turning *data* into *information*; information into *knowledge*; and eventually knowledge into *wisdom*. In other word, to contribute for a new computing paradigm able to manage the dialectic relations between structural and functional properties in the same way as the human brain interacts with information and behaves as a set of embodied computers.

Aim of this manuscript is to explore the possibility of taming Big Data with Topology (the geometry of ′shapes′), bearing on a fundamental notion from computer science when dealing with data; the concept of ′space of data′. It is the latter that provides the structure (geometrically represented) in which information is encoded; the frame for algorithmic (digital) thinking; the lode where to perform data mining, that is, to extract patterns of information. It generates the goal as well: finding new ways of mining data spaces for information resorting to geometrical – indeed topological and combinatorial – methods.

First of all we claim – extending ideas proposed by Carlsson, Edelsbrunner and others to deal with the problem of image reconstruction – that geometry and topology are the natural tools to handle large, high-dimensional, complex spaces of data. *Why?* Because: *global, though qualitative information* is relevant: the data user aims to obtain knowledge, i.e., to understand how data is organized on large scale. *Metrics are not theoretically warranted*: in physics, most phenomena naturally support elegant, clear-cut theories which imply exactly what metrics to use; in the life or social sciences this is much less cogent. *Coordinates are not natural*: data are typically conveyed and received in the form of vector-like object (strings of symbols, typically numbers in some field), but the ′components′ of these vectors have no meaning as such and their

Journal of Physics: Conference Series **626** (2015) 012005 doi:10.1088/1742-6596/626/1/012005

linear combinations are not objects in data space. In other words, the space of data is not a vector space and those properties of data space that depend on a specific choice of coordinates cannot be considered relevant. *Summaries are most valuable*: the conventional method of handling data is to build a graph (*network*) whose vertex set is the collection of points in data space (each point possibly itself a collection of data) where two vertices are connected by an edge if their 'proximity measure' is, say, less than some given η; then try and determine the optimal choice of η. The complete diagram that illustrates the arrangement produced by hierarchical clustering is however much more informative, as it captures at once the summary of all relevant features under all possible values of η. The problem here is to try and find a way to know how the global features of data space vary varying η.

For all these reasons the methods to be adopted should be inspired by topology, because: *Topology* is the branch of mathematics that deals with both local and global qualitative geometric information in a space, namely, connectivity, classification of loops and higher dimensional manifolds, and invariants, that are, properties which are preserved under homeomorphisms of the ambient space. *Topology* studies geometric properties in a way less sensitive to metrics than geometry, it ignores the value of distance functions and replaces it just with a notion of proximity ($\eta \approx$ connective 'nearness'). *Topology* deals with those properties of geometric objects that do not depend on coordinates but only on intrinsic geometric features; it is *coordinate-free*.

Moreover, in topology, relationships involve maps between objects; thus they are a manifestation of *functoriality* and, what is more subtle and deeper, the invariants are related not just to objects, but to maps between objects as well. Functoriality reflects then a *categorical* structure, allowing for computation of global invariants from local information.

Last, typically full information about topological spaces is faithfully contained in their *simplicial representation*, a piece-wise linear (PL), combinatorially complete, discrete realization of functoriality. As already recalled, the conventional way to convert a collection of points in data space into a global object is to use the vertex set of a network, whose edges are determined by proximity. Such graph, however, whilst it captures well data connectivity (local), it ignores a wealth of higher order (global) features, well discerned instead thinking of it as the scaffold (1-skeleton) of a higher-dimensional object: the *simplicial complex*. The latter is a PL space built from simple pieces (simplices) identified combinatorially along their faces, obtained by completion of the graph.

2. The challenge

Current thinking in Information Technology is at the crossroad of different evolution pathways. On the one hand, is what is now universally referred to as the 'Big Data' question [1], which urges fully innovative methodologies to approach data analytics – in particular data mining, to be able to extract information from data with the required efficiency. On a different side is the ever increasing number of practical instances, in science as well as in the analysis of societal issues, of problems that require going computationally 'beyond Turing' [2], which touch on decidability and computability, as well as on embodied computation. These bear on the recently Cerf's raised question of whether or not there is any real 'science' in computer science [3], namely if all the well posed questions can be approached by a truly scientific methodology, universal and self-contained.

Following Cerf's view, we assume modeling of computational processes as the most inspiring candidate for the construction of a true 'theory', able to sustain predictions about complex processes through the analysis of large data sets. We acknowledge also Pearl's diagrams [4] – surprisingly analogous to Feynman's representation of interactions in quantum field theory, with cause-effect relations replacing time flow direction – aiming to construct analytic equations that not only fully characterize a problem of this sort, but make its solution accessible. Main point at issue is indeed to eliminate unimportant details while revealing the underlying structure: a method well known to statistical mechanics (renormalization group), dealing with fluctuations and noise induced by interactions, and to chaos theory (dynamical effects of nonlinearity), where patterns emerge despite the apparent randomness of the processes.

Present paper intends to describe a program designed to suggest a possible novel way to face some of the challenges posed above, in particular, the issue of sustaining predictions about the dynamics of complex processes through the analysis of big data sets, by paving the way for the creation of high-level query languages that allow insignificant details to be suppressed and information to emerge as 'mined out' correlations. Principal goal of the paper is the definition of a theoretical framework, described essentially as a non-linear topological field theory, as possible alternative to conventional machine learning or other artificial intelligence data mining techniques, allowing for an efficient analysis of and extraction of information from large sets of data: *information as element of a language.*

The approach proposed differs from all others for its deep roots in the inference of globally rather than locally coded data properties. Its focus is on the integration of constructive elements of topological data analysis (i.e., facts as forms) into a topological field theory for the data space (the logical space of forms), relying on the structural and syntactical features generated by the formal language whereby the transformation properties of the space of data are described. A sort of *language of forms* recognized by a proper automaton closely related to the field theory.

Incidentally, but not unexpectedly, this has a profound, far reaching philosophical meaning: in the words of Wittgenstein [5], " The world is the totality of facts, not things. ... The facts in logical space are the world. ... The logical picture of the facts is the thought." and " To imagine a language means to imagine a form of life. ... The meaning of a word may be defined by how the word can be used as an element of a language game."

Outcome of the approach proposed will be a way to discover directly from the space of observations (the collection of data) those relations that encode (by mens of a language) emergent features in the complex systems; *patterns* that data describe as correlations among events at the global level, result of interactions among systemic components at local level. Complex systems global properties are hard to represent and even harder to predict, just because – contrary to the information conventional science deals with – complex systems knowledge is typically based not on repeatable experiments and shared phenomenology that, incidentally, provide information about the statistical features of the system, but on *data* or on virtual artificial representations of real systems built out of data. Emergent features of complex systems arise out of these interactions; whilst the global patterns they create react back on those low lying levels, with no need of developing tools to sustain a learning process.

Tree bodies of knowledge, that are the three pillars our scheme rest on, need to operate synergically: i) *Singular Homology Methods*, tools for the efficient (re-)construction of the (simplicial) topological structures which encode patterns in the space of data; it enables to make *Topological Data Analysis* – homology driven – resting on the global topological, algebraic and combinatorial architectural features of the data space, equipped with an appropriate 'measure'; ii) *Topological Field Theory*, a construct mimicking physical field theories, to extract the necessary characteristic information about such patterns in a way that – in view of the field non-linearity and self-interaction – might generate as well, as feedback, the reorganization of the data set itself; it supports the construction of *Statistical/Topological Field Theory of Data Space*, as generated by the simplicial structure underlying data space, an 'action', a suitable gauge group and a corresponding fibre (block) bundle; iii) *Formal Language Theory*, a way to study the syntactical aspects of languages - the inner structure of patterns - and to reason and understand how they behave; it allows to map the semantics of the transformations implied by the non-linear field dynamics into automated self-organized learning processes. The three pillars interlaced in such a way as to allow us to identify structural patterns in large data sets and efficiently perform there data mining. The outcome is a new *Pattern Discovery* method, based on extracting information from field correlations, that produces an automaton as a recognizer of the data language.

3. Topological data analysis

Main pillar of the whole construction is the notion of *data space*, the crucial feature of which is that it is neither a metric space nor a vector space[1], but is a topological space. This is at the very root of all crucial questions at the basis of the scheme proposed: whether the higher dimensional, global structures encoding relevant information can be efficiently inferred from lower dimensional, local representations; whether the necessary reduction process (filtration; the progressive finer and finer simplicial complex representation of the data space) may be implemented in such a way as to preserve maximal information about the space global structure; whether the process can be carried over in a truly metric-free way [7]; whether such global topological information can be utilized to extract *knowledge* as well as correlated information, in the form of patterns in data space.

The basic principles of the approach stem out of the seminal work of a number of authors: G. Carlsson [8], H. Edelsbrunner and J. Harer [9], A.J. Zomorodian [10], and others. Fundamental goal of it is to overcome the conventional method of simply converting the collection of points in data space into a combinatorial graph $\mathcal{G}$ (a *network*); an object encompassing all relevant local topological features, whose edges are determined by some given notion of 'proximity', characterized by a parameter η somehow fixing a coordinate-free metric for 'distance'. Indeed, while $\mathcal{G}$ captures pretty well *local* connectivity data, it ignores an abundance of higher order features, most of which have *global* nature. Such features can instead be accurately perceived and captured by focusing on a different object than $\mathcal{G}$, say $\mathcal{S}$. $\mathcal{S}$ is a higher-dimensional, discrete object, of which $\mathcal{G}$ is the 1-skeleton, generated by combinatorially completing the graph $\mathcal{G}$ to a *simplicial complex*. As such, $\mathcal{S}$ is built from higher and higher dimensional simple pieces (simplices) identified (combinatorially) along their faces. It is this operation that makes the subtlest features of the data set, seen eventually as a topological space $X \sim \mathcal{S}$, manifest and accessible.

In this representation X has an hypergraph structure whose hyperedges generate, for a given η, the set of relations induced by η itself as a measure of proximity. In other words, each hyperedge is a *relational* simplex, i.e., a simplicial complex built by gluing together lower-dimensional relational simplices that satisfy the η property. This makes η effectively metric independent: in fact an n-relation here is nothing but a subset of n related data points satisfying the property represented by η. Dealing with the simplicial complex representation of X by the methods of algebraic topology, specifically the theory of persistent homology that explores it at various proximity levels by varying η, namely filtering relations by their robustness with respect to η, allows for the construction of a parameterized ensemble of inequivalent representations of X. The filtration process identifies those topological features which persist over a significant parameter range, qualifying them as candidates to be considered as *signal*, while those that have short-lived features can be assumed to characterize *noise*. Moreover, it defines the notion of an η-driven semigroup connecting spaces in the ensemble.

Key ingredients of this analysis are the homology groups, $H_i(X)$, $i = 0, 1, \ldots$, of X and in particular the associated *Betti numbers*, a basic set of topological invariants of X, the i-th Betti number, $b_i = b_i(X)$, being the rank of $H_i(X)$. Intuitively, homology groups are a functional algebraic tool easy to deal with (as they are abelian) to pick up the qualitative features of a topological space represented by a simplicial complex connected with the existence of *i-holes*

[1] This is unfortunately still acritically assumed by even the most distinguished authors (see, e.g., the book by Hopcroft and Kannan [6]).

(holes in i dimensions) in X. Holes simply mean cycles which don't arise as boundaries of higher-dimensional objects. This is why the number of i-dimensional holes, b_i, is realized by the dimension of $H_i(X)$, quotient vector space of the group of i-cycles with the group of i-boundaries. In the torsion-free cases knowing the b_i's is equivalent to knowing the full space homology. Efficient algorithms are known for the computation of homology groups [11]. Persistent homology is generated recursively, beginning with a specific complex characterized by a given $\eta = \eta_0$ and constructing from it the succession of chain complexes and chain maps for an increasing sequence of values of η, say $\eta_0 \leq \eta \leq \eta_0 + \Lambda$, for some Λ. Complexes grow with η, thus such chain maps can be naturally identified with a sequence of successive inclusions.

Whilst most invariants in algebraic topology are difficult to compute efficiently, homology is somewhat exceptional not only because its invariants arise simply as quotients of finite-dimensional spaces but also because some of its features can be possibly derived from 'physical' models. Indeed, whereas in standard topology invariants were *constructed* out of geometric properties to distinguish between manifest global, homeomorphically invariant objects, it was shown in physics that other invariants could instead be *discovered*, based, e.g., on topological quantum field theory technology [12]. Such invariants provide information about purely topological properties one cannot detect, not even hint, based on purely geometric representation.

It is this latter perspective that is adopted here, namely the idea of constructing a reliable 'physical' scenario for data spaces, where no structure is visible a priori. 'Physical' should of course be interpreted metaphorically. We want to figure out a coherent formal framework in an abstract space of data, where no equation is available, to describe through its topology the hidden correlation patterns coding data into information, in a way analogous to that one follows, say, in general relativity, where a given distribution of masses returns the full geometry of space-time. Here we expect that information hidden in data should return the full topology of data space. Of course we don't have a priori equations to rely on, yet we argue that a topological, nonlinear field theory can be designed over data space whereby global, topology-related pattern structures can indeed be reconstructed, providing a key to the information they encode.

This touches of course on how patterns are to be interpreted, as it bears rather on pattern discovery than on pattern recognition. In logic there are approaches that, drawing on abstract algebra and on the theory of relations in formal languages – as opposed to others that deal with patterns via the theory of algorithms and effective constructive procedures – define a pattern as exactly that kind of structural regularity, namely organization of configurations or symmetry, that one identifies with the notion of *correlations* in (statistical) physics [13]. These logical paradigms will guide our strategy.

A delicate issue is here that simplicial complex $\mathcal{S}$ (typically but not automatically a finite CW complex whose cellular chain complex is endowed with Poincaré duality) is not necessarily a manifold; it is only if the links of all vertices are simplicial spheres. The difficulty resides in the feature that n-spheres are straightforwardly identifiable only for $n = 1, 2$. The problem is tractable for $n = 3$ and possibly 4 only with exponential resources and it is undecidable for $n \geq 5$ [14]. However, given a singular chain complex $\mathcal{S}$, a normal map endows it with the homotopy-theoretic global structure of a closed manifold. Sergei P. Novikov proved that for dim $\mathcal{S} \geq 5$ there is only the surgery obstruction to $\mathcal{S}$ being homotopy equivalent to a closed manifold, and if $\mathcal{S}$ is homotopy equivalent to a manifold then the complex behaves as the base space of a unique Spivak normal fibration[2]. This entails that all finite simplicial complexes have

[2] A (smooth) manifold has a unique tangent bundle and a unique stable normal bundle; but a finite Poincaré complex does not possess such a unique bundle. Nevertheless, it possesses a substitute, in some sense spherical, fibration that is unique: the Spivak normal fibration. This has a property that if $\mathcal{S}$ is homotopy equivalent to a manifold then the spherical fibration associated to the pullback of the normal bundle of that manifold is isomorphic to the Spivak normal fibration, whose fibre is homotopically equivalent to a sphere.

the homotopy type of manifolds with boundary.

An additional observation is that available algorithms to compute persistent homology groups all focus on the notion of filtered simplicial complex, consisting of pairs built out of the simplex obtained at each given step in the recursive construction and the order-number of the step, the time-like discrete ordering parameter (label) at which that simplex appears in the filtration; a picture that can be naturally interpreted as the representation of a *process* endowed with a proper inherent dynamics, similar to a discrete-time renormalization group flow [15]. One may then expect, in analogy with dynamical triangulations of simplicial gravity, that the combinatorially different ways in which one may realize the sampling of *inequivalent* structures in the persistence construction process, varying the simplex structure, give raise to a 'natural' probability measure. The measure thus generated is consistent with the data space invariants and transformation properties.

4. From topology to field theory

Besides the customary filtrations due to Vietoris-Rips [16], whose k-simplices are the unordered $(k+1)$-tuples of points pairwise within distance η, and to Čech [17], where k-simplices are instead unordered $(k+1)$-tuples of points whose $\frac{1}{2}\eta$-ball neighborhoods intersect, or other complexes such as the witness complex [18], which provide natural settings in which to implement persistence, another filtration needs to be considered, entering here naturally into play, Morse filtration. In the case of simplicial complexes that are manifolds, this is a filtration by excursion sets, in terms of what for differentiable manifolds would be curvature-like data; here it is a non-smooth, discretized, intrinsic, metric-free version thereof, appropriate for the wild simplicial complex that is data space, that can be obtained by the simplicial, combinatorial analog of the Hodge construction.

Even though apparently dealing with metric-dependent features Morse filtration is indeed purely topological. Morse theory generates a set of inequalities for alternating sums of Betti numbers in terms of a corresponding alternating sum of the numbers of critical points of the Morse function for each given index. Also, simplicial Morse theory generates intrinsic, discrete notions of gradient vector field and gradient flow associated to any given Morse function f_M. The latter, particularly significant if interpreted in the framework of discrete differential calculus, had applications in classical field theory over arbitrary discrete sets [19], described as well in a non-commutative geometry setting [20].

The crucial feature here is that a Morse complex, built out of the critical points of a Morse function with support on the vertices of $\mathcal{S}$, has the same homology as this underlying structure. This assumes particular significance, because the induced Morse stratification, [21], is essentially the same as the Harder-Narasimhan [22] stratification of algebraic geometry, thus one can construct the analog of local 'co-ordinates' around the Morse strata and represent the negative normal bundle to the critical sets. What relates Morse with homology is the property that the number ν_i of critical points of index i of a given function f_M is equal to the number of i cells in the simplicial complex structure obtained 'climbing' f_M, that bears on b_i. Noticeably, Morse homology, which can be defined using a generic f_M and an induced, local, topologically invariant metric (i.e., one independent of the function and on the metric itself), is isomorphic to the singular homology: in other words, Morse and Betti numbers encode the same information, yet Morse numbers allow us to think of and underlying 'manifold'.

Inspired by what happens in the context of gravity, Gromov-Hausdorff (GH) [23, 24] topology is selected to construct a self-consistent measure over $\mathcal{S}$. Gromov's spaces of bounded geometries provide in fact the natural framework for addressing the questions posed by high-dimensional simplicial geometry; in particular for establishing entropy estimates that characterize the distribution of combinatorially inequivalent simplicial configurations (a problem solved in gravity theory [25], where however it is much easier, as one deals there with an underlying metric

Journal of Physics: Conference Series **626** (2015) 012005 doi:10.1088/1742-6596/626/1/012005

vector space that gives rise, under triangulation, to a simplicial complex which is a Lorentz manifold). This choice leads naturally to the construction of a statistical field theory of data, as the statistical features of GH topology are fully determined by the *homotopy* types of data space [26]. Both complexity and randomness of the emerging structure can be quite large in the case of big data, since the number of 'coverings' of a simplicial complex of bounded geometry is expected to grow exponentially with the volume. A "thermodynamic limit" needs then to be built over the growing filtrations of simplicial complexes, that look more and more random. As a well defined statistical field theory requires dealing with the extension of the statistical notion of Gibbs field to the case where the substrate is not simply a graph but a simplicial complex, this amounts to including the property that the substrate underlying the Gibbs field may itself be in some way random. One needs therefore to resort to Gibbs 'families' [27], so that the ensuing ensemble of geometric systems – a sort of phase space endowed with a natural measure – behaves as a statistical mechanics object. Notice that this may lead to critical behavior, as diversified phase structures emerge – entailing a sort of phase transition when the system passes from an homotopy type to another. A scenario emerges: the deep connection between the simplicial complex structure of data space and the information that such space hides, encoded at its deepest levels, resides in the property that data can be partitioned in a variety of equivalence classes classified by their homotopy type, all elements of each of which encode similar information. In other words, in X information behaves as a sort of 'order parameter'.

5. A Field Theory of Data

A single mathematical object may be assumed to encompass most of the information about the global topological structure of the data space: $\mathcal{P}(z)$, the Hilbert-Poincaré series (indeed a polynomial, in some indeterminate z); generating function for the Betti numbers of the related simplicial complex. $\mathcal{P}(z) = \sum_{i\geq 0} b_i\, z^i$, can be generated through a field theory, as it turns out to be nothing but one of the functors of the theory itself for an appropriate choice of the field action.

The best known 'template' to refer to for this formal setup – naturally keeping in mind not only the analogies but mostly the deep structural differences: continuous vs. discrete, tame vs. wild, finite vs. infinite gauge group – is Yang-Mills' field theory (YMFT) [28]. In YMFT the variables are the connection field over a manifold M (a Riemann surface), and the gauge group G, under which the Chern-Simons' (CS) action (i.e., the ($2k$-1)-form defined in such a way that its exterior derivative equals the trace of the k-th power of the curvature) is invariant, is $SU(N)$.

To begin with, let us recall – paraphrasing Terry Tao [29] – that a *gauge* is simply a global 'coordinate system' that varies depending on one's location with respect to the reference space. A gauge transformation is a change of coordinates consistently applied to *each* such location, and a gauge theory is the model for a system to which gauge transformations can be applied letting dynamics unchanged. A global coordinate system is nothing but an isomorphism between some geometric or combinatorial objects in a given class and a standard reference object in that same class, as opposed to a local coordinate system, in which the analogous isomorphism is between local pieces of objects. In a coordinate-invariant perspective all geometric quantities must be converted to the values they assume in that specific representation, so that every statement is invariant under coordinate changes. If this can actually be done, the theory can be cast into coordinate-free form. Given the coordinate system and an isomorphism of the standard object, a new coordinate system is simply obtained by composing the global coordinate system and the standard object isomorphism. Every coordinate system arises in this manner. Thus, the space of coordinate systems can be identified with the isomorphism group G of the standard object. This group is the *gauge group* for the class of objects considered. Such general definition will allow us to still think of 'coordinates'; however, not as one does for vector spaces but simply as an intrinsic way to identify mutual relations between objects.

Let us continue the analogy, yet stressing out the differences. The arena for YMFT is a generic, smooth manifold, M, over which the connection field is well defined and allows for a consistent definition of the action (the curvature is simply the exterior derivative of the connection plus the wedge product of the connection by itself). Field equations are then nothing but a *variational* 'machine' which takes as input a symmetry constraint, expressed as invariance with respect to G, and produces as output a field satisfying that constraint. In YMFT connections allow us to do calculus with the appropriate type of field attaching to each point p on M a vector space, i.e., a *fiber* over that point: the field at p is simply an element of such fiber. The resulting collection of objects (manifold M plus fibers at every point $p \in M$) is a *vector bundle*. In the presence of a gauge symmetry, the fiber must be the copy of a representation of the corresponding group G. In other words, the field structure is that of a G-bundle. Atiyah and Bott [30], via an infinite-dimensional Morse theory with the CS action functional as Morse function, and Harder and Narasimhan [22], via a purely combinatorial approach, have established a powerful formula expressing the Hilbert-Poincaré series (global) in terms of those corresponding to all Levi subgroups of G as a functor of the YMFT, in a form that is reminiscent of the relation between grand-canonical and canonical partition functions in statistical mechanics.

For the space of data the picture is much more complex, because of the more complex underlying structure. Let us however recall that vector bundles, proper to the differential category, have a PL category analogue, referred to as *block bundles* [31]. These allow us to reduce geometric and transformation problems over manifolds to homotopy theory for the various groups and complexes involved. This provides a natural way to reconstruct the moduli space of G-bundles in a discretized setting. The construction needs to be extended to a simplicial complex that, as already noticed, in general is not a manifold. Fortunately, Novikov's lesson is that this can be done in homotopy terms [14]. Since the homotopy class of a map fully determines its homology class, the simplicial block-bundle construction clearly furnishes all necessary tools to compute the Poincaré series. Also, in spite of its topological complexity, data space offers a natural, simple choice for the action. Indeed an obvious candidate to start with the exponentiated action is the Heat Kernel $\mathcal{K}$, because the Heat Kernel's trace gives just the Poincaré series [32]. $\mathcal{K}$ can be obtained by constructing over the simplicial complex an intrinsic (metric-free) combinatorial Laplacian [33]. This is done by the *ad hoc* construction of the Hodge decomposition over $\mathcal{S}$ and the related Dirac operator.

Regarding the group G, notice that the space of data has a deep, far reaching property: it is fully characterized only by its topological properties, neither metric nor geometric, thus – as the objects of the theory have no internal degrees of freedom, but are constrained by the manipulation processes they can be submitted to – there is only one natural symmetry it needs to satisfy: invariance under all those transformations of data that don't change its topology and are consistent with the constraints. This implies that the gauge group be the semidirect product $\mathcal{G} \wedge \mathfrak{G}_{\mathfrak{MC}}$ of the group $\mathcal{G}$ associated with the characteristic process algebra of the data set and the (simplicial analog) $\mathfrak{G}_{\mathfrak{MC}}$ of the *mapping class group* [34] for the space of data.

Recall that 'process algebra' refers to the *behavior* of a *system* [35]. A system is indeed anything showing behavior, which is the total of events or actions that it can perform, together with the order in which they are executed and other aspects of this execution, such as timing or probabilities, that define the process. The term algebra refers instead simply to the fact that the approach taken to represent behavior is algebraic and axiomatic. That is, operations on processes are defined, and their effects are represented formally in terms of methods and techniques of universal algebra.

In analogy with the definition of a group as any mathematical structure consisting of a single universe of elements, with operators on this universe of elements that satisfy the group axioms, a *process algebra* is any mathematical structure satisfying the axioms given for its operators, and a process is then an element of the universe of this process algebra. The axioms allow calculations with processes. Process algebra has thus its roots in universal algebra, however it often goes beyond the strict bounds of universal algebra: the restriction to a single universe of elements can be relaxed, e.g., different types of elements can be used, and sometimes binding operators are considered. It supports mathematical reasoning about behavioral equivalences that, independent of the specific approach followed for its definition, must be a congruence with respect to behavioral operators.

In automata theory, on the other hand, a process is modeled as an automaton. An automaton has a number of *states* and of *transitions*, namely, ways of going from a state to its 'neighbor' states through the execution of elementary actions, the basic units of behavior. Besides, an automaton has an initial state (possibly, more than one) and a number of final states. A behavior is an execution path of a number of elementary actions that leads from the initial state to a final state. Given this basic behavioral abstraction, an important issue is to decide when two automata can be considered equal, expressed by a notion of equivalence, the *semantic equivalence*. In automata, the basic semantic equivalence is language equivalence: an automaton is characterized by its set of execution paths, and two automata are equal when they have the same set of execution paths. In this context, an algebra that allows reasoning about automata is the algebra of regular expressions [36].

In the automata model the notion of 'interaction' is missing, yet during the execution from initial state to final state, a system may interact with another system. When dealing with models representing interacting systems, 'concurrency theory' is used: the theory of interacting, parallel, distributed or reactive systems. Process algebra is an algebraic approach to concurrency theory, and as such it may have parallel composition among its basic operators. In this case, automata are 'transition systems' for which the notion of equivalence is not necessarily restricted to language equivalence. Prominent among the equivalences is *bisimilarity*, which considers two transition systems equal if and only if they can mimic each other's behavior in any state they may reach.

Finally, any algebra with a finite number of generators and a finite number of relations can be written as a quiver with relations (though not necessarily in a unique way) thinking of the set of execution paths of the automaton's actions as the basis of a *k-path algebra* with composition law induced by the structure of the combinatorial data of a suitable k-Quiver (kQ). Moreover, for a given quiver kQ, a relation is simply a k-linear combination of paths in kQ. Given a finite number of relations, one can form their two sided ideal $\mathcal{R}$ in the path algebra, and thus define the algebra $\mathcal{A} \sim kQ/\mathcal{R}$ as a 'quiver with relations'. Process algebras can always be assumed to be representable by a quiver with relations. $\mathcal{G}$ is the group associated with $\mathcal{A}$.

As for $\mathfrak{G}_{\mathfrak{MC}}$, the actions of (extended) mapping class groups on spaces of different sorts encoding geometric and topological objects, like homotopy classes, foliations, conformal structures have been extensively studied [37]. These actions are all induced from the actions of homeomorphisms of the base space on the corresponding objects. Moreover, the spaces on which the mapping class groups act can be equipped with various structures, e.g., groups, simplicial complexes, or manifolds, and the mapping class groups are embedded accordingly into groups of algebraic isomorphisms, simplicial automorphisms, isometries of the related metrics – if any. For most of these actions, the natural homomorphism from the mapping class group to the automorphism group of the given structure is an isomorphism. Among these, particularly interesting in present context are actions by simplicial automorphisms on the different abstract simplicial complexes associated to X; namely, actions by piecewise linear automorphisms of the associated measured foliations space, equipped, for example, with the train-track piecewise linear structure introduced by Thurston [38] or with the set of self-preserving intersection functions.

The latter structure is related with the braid group, whose central extension – not unexpectedly – is the group of permutations. $\mathfrak{G}_{\mathfrak{MC}}$ is finite and finitely presented; its presentation, as well as its representations, can be completely constructed once one knows the full homotopy of the simplicial complex. Recently, a complete representation of $\mathfrak{G}_{\mathfrak{MC}}$ realized in terms of the group $SU(1,1)$ of hyperbolic rotations has been obtained by the authors (and is reported in [39]).

We claim that, in spite of the formal difficulties, mimicking the block bundle approach for the appropriate simplicial complex structure and given G, the data space topological invariants (among which Betti numbers) can be computed in the context of the proposed field theory through the (recursively computable) subsets of symmetries of $\mathcal{G} \wedge \mathfrak{G}_{\mathfrak{MC}}$. The benefit is twofold, on the one hand the cosets of $\mathcal{G} \wedge \mathfrak{G}_{\mathfrak{MC}}$ order data in equivalence classes with respect to isotopy, leading to a canonical system in the related process algebras. On the other hand, one can make a unique choice among the several possible theories – the multiplicity being related with the plurality of topological structures due to the passage through Morse numbers (Morse and Betti numbers are related through inequalities, <u>not</u> equalities) – in the following way. One begins by constructing, for all manifolds in the family generated by the collection of Morse numbers, the 'free' field theories whose exponentiated action is simply the Heat Kernel, for which the partition function is the generating function of the manifold Betti numbers. By self-consistency, i.e., simply comparing the coefficients of $\mathcal{P}(z)$ with the Betti numbers outcome of the 'phenomenological' persistent homology one identifies which is the effective data manifold.

In this way, not only we fully recover through the construction proposed the whole data space topology (for example the set of Betti numbers) the space of data, but we are able to continue construct an autonomous, self-consistent Topological Data Field Theory (TDFT) on the space of data: once more 'the fascination of unexpected links in mathematics' [40].

As a final remark, notice that the resulting picture comprises a surprising amount of information on the associated moduli spaces as well; markedly the quiver representation for the path algebra $\mathcal{A}$ (see also [41, 42]), basic tools for the description of processes involving maps and transformations of data sets.

6. The Formal Language Theory facet

The construction outlined so far naturally brings to light a new facet, as much unexpected as elegant in its form: a Formal Language Theory (FLT) dimension.

A preliminary question to raise is then whether the topological landscape adopted is inherently coherent with the structure of FLT. As we know a central issue in the theory of computation is to determine classes of languages whose representation has finite specification [36]. A formal language defined over a finite alphabet $\mathfrak{A}$ of symbols is a subset of the set $\mathfrak{A}^*$ of all strings of any length that can be represented by that alphabet. As a consequence, the number of possible representations is countably infinite and the set of all possible languages over a given alphabet $\mathfrak{A}$ is uncountably infinite. Under these conditions we are obviously unable to represent all languages. Coupled with this issue there is the limit posed by well-known Gold's theorem for which the 'minimum automaton identification from given data is NP-Complete' [43]. In the TDFT context, the challenge is to construct a finite representation of the language defined over the alphabet whose symbols are the generators of gauge group G, whose cosets partition the data space X in equivalence classes of finitely presented objects. Such languages can be finite or infinite; what is interesting here is that their presentation can always be finitely given in $\mathcal{G} \wedge \mathfrak{G}_{\mathfrak{MC}}$. In other words, such languages are each a collection of discrete spaces containing a finite number of homeomorphic objects. In other words, by the TDFT we construct indeed a language of data, the language proper to topological shape $\mathcal{S}$.

Interpreting the gauge group G as topological shape language requires to resort to a notion of duality somehow similar to that entering the construction of Langland's dual group, yet designed to represent the relationship between structure and function of a behavior: a 'mirror' symmetry that allows each to affect the other in the same way. As a consequence, we can characterize the data language as the process algebra whose processes are well-behaved with respect to 'modulo bisimulation' [44], by attributing them the same, unique (bi-)algebra induced by the gauge group $\mathcal{G} \wedge \mathfrak{G}_{\mathfrak{M}\mathfrak{C}}$, with $\mathcal{G}$, as mentioned, the group of $\mathcal{A}$.

The role of $\mathfrak{G}_{\mathfrak{M}\mathfrak{C}}$ in the discrete case can be naturally traced back to *Automatic Groups* [45], i.e., finitely generated groups equipped with several finite-state automata able to distinguish whether or not a given word, representation of a group element, is in 'canonical form', and hence if two elements in canonical form differ and by which generators. It may be worth recalling that automatic groups were actually originally introduced in connection with topology, in particular with the study of the fundamental group, and hence of the homotopy, of 3-manifolds, because the class of automatic groups can be extended to include the fundamental group of every compact 3-manifold satisfying Thurston's geometrization [38]. In the topological structure we are dealing with here – where we consider collections of *relational* simplexes, built by combinatorially gluing together relational simplices – the task is much more complex. However, as the basic structure is fully controlled by homotopy types, turning the generation of a family of parametrized simplicial complexes into a classification problem in formal language theory is natural and straightforward in its statement, if not in its solution. One should be aware, however, that issues of uncontrollable algorithmic complexity or even of undecidability may possibly arise.

Moreover, the syntax of a language in FLT is traditionally described by using of the notion of grammar, defined by the relations necessary to build correct syntax constructs from atomic entities (symbols). This is what allows us to describe the syntax of a formal language universally, in spite of the representation of its texts. In addition, the syntax constructs are typically described resorting to the notion of syntax diagram, $\mathbb{D}$, that is the connected multigraph with nodes labeled in terms of the formal languages alphabet $\mathfrak{A}$ and connections – in our representation not only edges or links, but also higher dimensional simplices – that represent the syntax relations. The multigraph of a syntax diagram may be directed or not, and in view of its combinatorial structure, inherited from the simplicial structure of data space and accounted for in the FTL vision, it is itself to all effects a simplicial complex. It is possible to select specific syntax diagrams (referred to as 'correct', as defined below) out of the set of all syntax diagrams on $\mathfrak{A}$ to construct different grammars. The formalism used to do this requires a fundamental notion: that of neighbor grammars [46], whose meaning is the following. Define first, for each $\mathbb{D}$, the collection of subdiagrams labelled by the set of pairs $(\mathbb{D}', \mathfrak{s})$ where $\mathbb{D}' \subseteq \mathbb{D}$ is another syntax diagram and $\mathfrak{s}$ is the inclusion map of $\mathbb{D}'$ into $\mathbb{D}$. The neighborhood of a symbol of $\mathfrak{A}$ is a syntax diagram that contains the node singled out by this symbol. The neighbor grammar of the given grammar consists of the finite family of neighborhoods defined for each symbol of $\mathfrak{A}$. A given syntax diagram is said to be 'correct' if for each of its nodes, labelled by some symbol of $\mathfrak{A}$, it includes some neighborhood of this symbol. Such neighborhood should contain all simplices adjoining to its center. There is therefore at least one cover consisting of neighborhoods for each correct syntax diagram in the given neighbor grammar. Such cover is the *syntax*. Furthermore, the category $\mathfrak{D}$ of syntax diagrams over the given alphabet can be introduced, based on the neighboring grammar. It is known [46] that the category of correct syntax diagrams, defined as $\mathfrak{D}$ but limited to correct syntax diagrams, admits a Grothendieck topology [47] (the Grothendieck topology is a structure on a category that makes its objects act like the open sets of a topological space). It is the formal language generated by the field theory through its gauge group that makes the TDFT consistent with a formal language architecture.

7. Language, structure and behavior, automata

On a different front, the notions the TDFT construct outlined have crucial consequences in terms of information theory. In particular, the three basic identifications: the *architectural structure* as a *fiber bundle*, consisting of a base space, here the space of data X, viewed as a topological space, and a fibre attached to each point of it; the *fibers*, as each a representation of the *gauge*

Journal of Physics: Conference Series **626** (2015) 012005 doi:10.1088/1742-6596/626/1/012005

group $G = \mathcal{G} \wedge \mathfrak{G}_{\mathfrak{M}\mathfrak{C}}$; the *field* as an element of the *fiber* at each point of the data space; an *action* – here, in the simplest non-interacting case, the combinatorial Laplacian – able to describe the variations of the global topological landscape, have a far reaching interpretation.

Such architecture is indeed what allows us to touch the final goal: the definition of a methodology whereby, starting from the exploration of (large) sets of data one could extract a 'language' capable of describing them as a unified system of structure and behavior. This new object can be interpreted as a *true* (effective, extended) *data space* including, besides the topological features inherent in the data set, the set transformations allowed among them, generated by the group of all its possible topology-preserving transformations and reflected in the ensuing equivalence classes. In such perspective, the system is itself a *self-organizing* 'program', whose identifiers are the interactions that the field action implies. These interactions correlate parts of potential processes (embedded programs) of real life applications: a characteristic feature caught in the $S[B]$ paradigm [48].

The principle of *self-organization* has long entered as fundamental feature the theory of nonlinear (possibly discrete [49]) dynamical systems. It provides the clue to represent the diverse pictures of the relation between lower level elements and higher order structures in a complex system. Its basic idea is that the interactions among low-level elements, in which each element adjusts to the others, is local, because it does not make reference to patterns, that are global. It is however this latter feature that leads to the emergence of highly coherent structures and complex behavior over the system as a whole. Such structures, in turn, are able to provide correlations for the lower level elements with no need of higher order agents [50, 51] to induce their emergence. In other words, rather than being imposed from above or from outside, the higher order structures emerge from the interactions internal to the system or between the system and its environment.

In an algebraic prospective the language signature becomes a measure of the interactions that generate the environment associated with the data set. This is exactly what is done in the $S[B]$ model when one establishes which states connected to $'B'$ (i.e., which *behavior*) satisfy the constraints imposed by the set of states $'S'$ (i.e., the states defining the system's *structure*). That is equivalent in TFTD to determine which is the fiber bundle associated to a given element of a fiber attached to a given point of the topological space.

The structure of $S[B]$ can be naturally identified as a fiber bundle: $S[B] = \big(B, S, \pi, \mathcal{B} \doteq \{B_j \,|\, j \in \mathcal{J}\}\big)$, with total space B, base space S, projection map $\pi : B \to S$ and fiber set $\mathcal{B}$. $\mathcal{J}$ is a label set tagging points $s_j \in S$, $j \in \mathcal{J}$. In $\mathcal{B}$ each single fiber B_j specifies the global topological constraints conditioning all the correlations of s_j. It should be recalled here that S is a higher dimensional 'standard' object that provides the frame for the data space X. This defines, for any subset of constraints corresponding to a given choice of the global invariants, the inner homeomorphisms among equivalence classes on the fiber.

An important, though simpler, analog of this construction comes from conformal field theory (CFT), that it somehow mimics. In statistical field theory a *conformal theory* is fully determined by its correlation functions, exactly like it happens in TDFT and in $S[B]$. In CFT correlation functions are bilinear combinations of conformal blocks (sets of correlators that implement the identities and constraints that follow from the global gauge symmetries of the theory), and a monodromy for conformal blocks arises that is encoded into a category $\mathfrak{T}$ (the so called Modular Tensor Category). Given conformal blocks with monodromy described by $\mathfrak{T}$, specifying the correlation functions is equivalent to selecting another category, the 'module category' $\mathfrak{M}$ over $\mathfrak{T}$. Also, in a CFT conformal blocks are controlled by a 'vertex' algebra $\mathcal{V}$ [52]. A deep theorem [53] states that for $\mathfrak{M}$ indecomposable over the representations of $\mathcal{V}$ one can combine conformal blocks of $\mathcal{V}$ into a globally consistent system of correlation functions.

In this complex, articulated construction a crucial notion emerges: that of (asynchronously) $\mathfrak{L}$-*combable* group [54], namely a group to each element of which can be associated a word in some free group in an arbitrary, abstract family of languages $\mathfrak{L}$. The nature of $\mathfrak{L}$ is rather flexible: it can be the family of regular languages, context-free languages, or indexed languages. Words representing group elements in some of these languages [55] describe transformations (flows) of the data set. The class of combable regular languages consists of precisely those groups that are asynchronously automatic. Recalling Atiyah-Bott and Harder-Narashiman results for manifolds, it is relevant to try and classify the (normal) sub-groups of G, and this can be done in the group and language theoretical setting. In the algebraic theory of languages, a regular language is fully represented by its syntactic monoid (meaning that the properties of that language, e.g., the expressive power of its first-order logic, are totally contained in the structure of the monoid), which is typically finite. In this context regular languages are referred to as languages of $'$data words$'$. A rigorous, but simple construction of data words consists in identifying first alphabet $\mathfrak{A}$, and focusing then the attention on words and languages over $\mathfrak{A}$ and on the algebraic theory they generate. The field theoretical construction of the Betti number generating function for data sets is an instance of representation of the complex language of data words associated with the simplicial realization of $\mathfrak{G}_{\mathfrak{MC}}$, which is known to be combable (though in some cases possibly not-automatic) [56].

Automata models can be developed for such languages, whose basic feature is that they provide a trade-off among three important properties: *expressivity*, good *closure* properties and decidable (or efficiently decidable) *emptiness*; striking an acceptable balance in this trade-off. Logics have been developed to establish the properties of data words: in particular a language of data words is definable in first-order logic iff its syntactic monoid is aperiodic; a statement that links the feature of definability in first-order logic to a property that in our framework is dynamical.

The relevant emerging relationships naturally involve, in the topological setting, invariants are related not only to objects but maps between pairs of objects as well. This is an explicit manifestation of *functoriality*. Functoriality is central in algebraic topology as it permits computation of global homological invariants from local information, a feature that reflects the inherent, natural categorical structure of this tool. The theory of automata and formal languages merge with the field theoretical picture just in this way, because the field theory generates sequences of symbols, that enter into play in the simplicial construction of the G-bundle associated with with the gauge group G, and relations among them. This bears on the enumerative combinatorics content of the theory, because G is reduced essentially to homotopy braids, that provide the language recognized by the automata. Moreover, combinatorics on words pertains to the wide set of natural operations on languages, in particular to the property – crucial for the final step of pattern discovery in data space – that the orbit of any language in $\mathfrak{L}$ under the *monoid* generated by such set is finite and bounded, independently of $\mathfrak{L}$.

The use of formal languages leads as well to the recognition of automatically generated domain-specific languages. The latter are languages appropriate to single out specific topological objects (concepts) and their mutual relations, hidden in the noisy landscape of the very large data space, and to manage, query and reason over those concepts so as to infer new knowledge. This is remindful of Codd's theory of database management with its basic tool, *relational algebra*, derived from the algebra of sets and first-order logic when dealing with finite relations closed under specific operations. Yet, while Codd's approach tackles the problem top-down, first defining the conceptual model, then classifying data through relations, and finally manipulating such relations through their schemas, the approach based on the topology of data space, on the contrary, tackles the problem bottom-up. The two approaches can thus be associated to two different yet complementary ways of thinking; the former based on the assumption that the agent knows a-priori, at least in part, the properties of data (characteristic, e.g., of artificial intelligence approaches to data mining, such as machine learning), the latter aimed at inferring new knowledge for the agent, extracting from data (*ontological emergence*) those relations that define the hidden structural aspects, with no a-priori information on what data is about.

The dialectical question about the nature of patterns, grounded in the antithesis between pattern recognition and pattern discovery, has guided us naturally – in the field theoretical context – to search for a way to describe patterns at once algebraic, computational, intrinsically probabilistic, yet causal. In TDFT, patterns can be collected in ensembles resorting to equivalence classes of histories, or of sets of states. The strength of such patterns (e.g., their predictive, i.e. information retrieval, capability) and their statistical complexity (via state entropy, or the amount of information retained) provide, for each particular process, a measure of the forecasting ability of the theory over the entire data space.

8. Patterns

The conclusive step, that results by merging all the above ingredients into a unique, consistent field-theoretical picture, in a representation of the space of data equivariant with respect to the transformation properties induced by the simplicial topological scheme itself and by the processes the system may undergo, bears on the individuation of characteristic patterns within the data set via the field correlation functions. The weights depend on the notion of proximity adopted, on the formal language on which the theory is based, on the field action functional selected and on the Morse stratification corresponding to it as well as on the set of transformations of the data space into itself that preserve its topology. The choice of correlations to represent patterns is crucial though delicate. Use of the characteristic patterns of the data system enables us to make predictions without violating the unavoidable restriction (a mixture of the second law of thermodynamics with the principle of relativity) that predictions can only be based on the process's past, not on any outside source of information except the data in X. In such a perspective, patterns belong to the intrinsic structure of the process, not to the rest of the universe; aggregated pieces of information that share a common structure, saying little about what that pattern is: and this is what correlations are about.

Patterns as represented by field correlations are: *robust*, because they are derived from persistent homology (mediated, if necessary, by the statistical mechanics manipulation process, smoothing out the role of very high order topological invariants) and hence free, to any desired accuracy, of irrelevant noisy components; *global*, as they describe deep lying correlations dictated by the non-local features of the space topology inherited by the field; *optimal*, based as they are on the variational principle proper to the field theory; *flexible*, due to the vast diversity inherent in their language theoretic structure. This is why they provide strategic directions as how to search data space. Whilst several details of the theory remain to be exhaustively worked out, its grand design – just presented – at least programmatically does not, except perhaps in some of its subtlest technicalities, a number of applications have started to confirm its potential reach and validity. Among these we mention in particular two: the formulation of a novel 'many body' approach to the construction of an effective immune system model [48], and the analysis of the nature of altered consciousness in the 'psychedelic state' based on functional magnetic resonance imaging data [7, 57].

9. Conclusions

A few final remarks. We have outlined the construction of a topological gauge field theory able to act as a machine whose inputs are a space of data and the symmetry group generated by its simplicial complex approximation as resulting from persistent homology, while the output is pattern sets in the form of field correlations as generated by the field equations. Such correlation functions fully encode information about patterns in data space, where the relevant information about the system which the data refer to is encoded. The field theory is self-consistent. It is topological because the data space features it resorts to are topological invariants and because the gauge group embodies the most general transformations of data space which leave such global topological features unchanged. Finally, the field evolution – due to the simplicial nature of the construct – has a natural implementation in terms of finite state automata, which maps both the emergence of patterns and the identification of correlations into well-defined formal language theoretical questions.

Acknowledgments
The financial support is acknowledged of the Future and Emerging Technologies (FET) programme within the Seventh Framework Programme (FP7) for Research of the European Commission, under the FET-Proactive grant agreement TOPDRIM, number FP7-ICT-318121.

References

[1] Various Authors 2011 Special Section: Dealing with Data *Science* **331**
[2] Barry Cooper S 2012 Incomputability after Alan Turing *Notices AMS* **59** (6) 776–784
[3] Cerf V 2012 Where is the science in computer science? *Commun. ACM* **55** (10) 5
[4] Pearl J 2009 *Causality: Models, Reasoning, and Inference* (Cambridge: Cambridge University Press)
[5] Wittgenstein L 1921 Logisch-Philosophische Abhandlung *Annalen der Natur und Kulturphilosophie, Leipzig* **14** 184–262
[6] Hopcroft J and Kanna R 2013 *Foundations of Data Science* (http://blogs.siam.org/the-future-of-computer-science/)
[7] Petri G, Scolamiero M, Donato I and Vaccarino F 2013 Topological Strata of Weighted Complex Networks *PLoS ONE* **8** (6): e66506. DOI: 10.1371/journal.pone.0066506
[8] Carlsson G 2009 Topology and data *Bull. AMS* **46** (2) 255–308
[9] Edelsbrunner H and Harer J 2010 *Computational Topology, an Introduction* (Providence: Amer. Math. Soc.)
[10] Zomorodian AJ 2009 *Topology of Computing* (Cambridge: Cambridge University Press)
[11] Basu S, Pollack R and Roy M-F 2006 *Algorithms in Real Algebraic Geometry* (New York: Springer-Verlag)
[12] Witten E 1989 Quantum field theory and the Jones polynomial *Commun. Math. Phys.* **121** (3) 351–399
[13] Shalizi CR and Crutchfield JP 2001 Computational Mechanics: Pattern and Prediction, Structure and Simplicity *J. Stat. Phys.* (3-4) **104** 816–879
[14] Novikov SP 1996 On manifolds with free abelian fundamental group and applications *Izv. Akad. Nauk SSSR* ser. mat. **30** (1) 208–246 – English translation: 1967 *A.M.S. Transl.* **67** (2) 1–42
[15] Zinn-Justin J 2002 *Quantum field theory and critical phenomena* (Oxford: Clarendon Press)
[16] Vietoris L 1927 Über den höeren Zusammenhang kompakter Räume und eine Klasse von zusammenhangstreuen Abbildungen *Math. Ann.* **97** 454–472
[17] Čech E 1932 Théorie générale de l'homologie dans un espace quelconque *Fund. Math.* **19** 149–183
[18] de Silva V 2003 A weak definition of Delaunay triangulation (*Preprint* arXiv cs.CG/0310031 v1)
[19] Auchmann B abd Kurz S 2006 A geometrically defined discrete Hodge operator on simplicial cells *IEEE Trans.* MAG-**42** (4), 643–646
[20] Battaglia D and Rasetti M 2003 Quantum-like Diffusion over Discrete Sets *Phys. Lett.* A **313** 8–15
[21] Harada M and Wilkin G 2011 Morse Theory of the Moment Map for Representations of Quivers *Geom. Dedicata* **150** 307-353 (*Preprint* arXiv math.DG 0807.4734v3)
[22] Harder G and Narasimhan MS 1974/75 On the cohomology groups of moduli spaces of vector bundles on curves *Math. Ann* **212** 215–248
[23] Gromov M 1981 *Structures métriques pour les variétés Riemanniennes* (Paris: Conception Edition, Diffusion Information Communication, Nathan)
[24] Fukaya K 1990 Hausdorff convergence of riemannian manifolds and its applications 1990 *Advanced Studies Pure Math.* **18** (1)
[25] Ambjørn J, Carfora M and Marzuoli A 1997 *The Geometry of Dynamical Triangulations* (Lecture Notes in Physics Springer-Verlag)
[26] Wilkin G 2009 Homotopy groups of moduli spaces of stable quiver representations, *Int. J. Math.* in press (*Preprint* arXiv 0901.4156)
[27] Diaconis P, Khare K and Saloff-Coste L 2008 Gibbs Sampling, Exponential Families and Orthogonal Polynomials *Statistical Science* **23** (2) 151–178
[28] Yang CN and Mills R 1954 Conservation of Isotopic Spin and Isotopic Gauge Invariance *Phys. Rev.* **96** (1) 191–195
[29] Tao T 2008 (http://terrytao.wordpress.com/2008/09/27/what-is-a-gauge/)
[30] Atiyah MF and Bott R 1983 The Yang-Mills equations over Riemann surfaces *Philos. Trans. Roy. Soc. London* Ser. A **308** (1505) 523–615

[31] Rourke CP and Sanderson BJ 1968 Block Bundles: I, II, III *Annals Math.* **87** (1) 1–28, (2) 256–278, (3) 431–483

[32] Knill O 2013 The Dirac operator of a graph (*Preprint* arXiv math.CO 1306.2166v1)

[33] Hodge WVD 1941 *The Theory and Applications of Harmonic Integrals* (Cambridge: Cambridge University Press)

[34] Farb B and Margalit D 2011 *A primer on Mapping Class Group* (Princeton: Princeton University Press)

[35] Baeten J 2005 *The history of process algebra* **335**(2-3), 131-146

[36] Lewis HH and Papadimitriou CH 1998 *Elements of the theory of computation* (Prentice-Hall)

[37] McCarthy JD and Papadopoulos A 2012 Simplicial actions of mapping class groups, in Papadopoulos A (Ed.) *Handbook of Teichmüller theory* Vol. III (European Mathematical Society Publishing House, Zürich) 297–423

[38] Thurston WP 1997 *Three-Dimensional Geometry and Topology* (Princeton: Princeton University Press)

[39] Rasetti M 2014 Is quantum simulation of turbulence within reach? *Int. J. Quant. Inf.* (In press DOI: 10.1142/ S0219749915600084)

[40] Asok A, Doran B and Kirwan F 2008 Yang-Mills theory and Tamagawa numbers: The fascination of unexpected links in mathematics *Bull. London Math. Soc.* **40** (4), 533–567

[41] Crawley-Boevey W 2001 Geometry of the moment map for representations of quivers *Compositio Math.* **126** (3), 257–293

[42] Zagier D 1996 Elementary aspects of the Verlinde Formula and of the Harder-Narasimhan-Atiyah-Bott Formula *Israel Mathematical Conference Proceedings* 445–462

[43] Gold E. M. 1978 Complexity of Automaton Identification from Given Data *Information and Control* **37** 302-320

[44] Baeten J, Corradini F and Grabmayer CA 2007 A characterization of regular expression under bisimulation *Jou. of ACM* **54** (2)

[45] Mosher L 1995 Mapping Class Groups are Automatic *Ann. Math.* **142** (2) 303–384

[46] Lapshin V The topology of syntax relations of a formal language *Preprint* arXiv math.CT 0802.4181v1

[47] Artin M, Grothendieck A and Verdier J-L (Eds.) 1972 Théorie des topos et cohomologie étale des schémas *Séminaire de Géométrie Algébrique du Bois Marie* 1963-64 (SGA 4) Vol. 1 Lecture notes in mathematics (in French) **269** Berlin: Springer-Verlag, xix+525

[48] Merelli E, Pettini M and Rasetti M 2014 Topology driven modeling – the IS metaphor *Natural Computing* (In press DOI: 10.1007/s11047-014-9436-7)

[49] Barrett C, Hunt III HB, Marathe MV, Ravi SS, Rosenkrantz DJ, Stearns RE and Thakur M 2007 Predecessor existence problems for finite discrete dynamical systems *Theor. Computer Sci.* **386** 3–37

[50] Haken H 2010 *Information and Self-Organization: A Macroscopic Approach to Complex Systems* (New York: Springer-Verlag, Series in Synergetics)

[51] Kelso JAS 1995 *Dynamic patterns – The self-organization of brain and behavior* (Cambridge: The MIT Press)

[52] Frenkel E 2005 Lectures on the Langlands Program and Conformal Field Theory (*Preprint* arXiv hep-th math.AG math.QA /0512172v1)

[53] Runkel I, Fjelstad J, Fuchs J and Schweigert C 2007 Topological and conformal field theory as Frobenius algebras *Contemp. Math.* **431** 225–248

[54] Rees S 1998 Hairdressing in groups: a survey of combings and formal languages in: *Geometry & Topology Monographs* Vol. 1: The Epstein birthday schrift 493–509

[55] Epstein DBA, Cannon JW, Holt DF, Levy SVF, Paterson MS and Thurston WP 1992 *Word processing in groups* (Boston: Jones and Bartlett)

[56] Bridson MR and Gilman RH 1996 Formal Language Theory and the Geometry of 3-Manifolds *Comment. Math. Helv.* **71** 525–555

[57] Petri G, Expert P, Turkheimer F, Carhart-Harris R, Nutt D, Hellyer PJ and Vaccarino F. 2014 Homological scaffolds of brain functional networks *J. Roy. Soc. Interface* **11** 20140873 DOI: 10.1098/rsif.2014.0873

Journal of Physics: Conference Series **626** (2015) 012006 doi:10.1088/1742-6596/626/1/012006

Proposing new experiments to test the quantum-to-classical transition

M Bahrami and A Bassi

Department of Physics, University of Trieste, Strada Costiera 11, 34014 Trieste, Italy
Istituto Nazionale di Fisica Nucleare, Sezione di Trieste, Via Valerio 2, 34127 Trieste

E-mail: Mohammad.Bahrami@ts.infn.it; Bassi@.ts.infn.it

Abstract. An open problem in modern physics is why microscopic quantum objects can be at two places at once (i.e. a superposed quantum state) while macroscpoic classical object never show such a behaviour. Collapse models provides a quantitative answer for this problem and explain how macroscopic classical world emerges out of microscopic quantum world. A universal noise field is postulated in collapse models, inducing appropriate Brownian-motion corrections to standard quantum dynamics. The strength of collapse-driven Brownian fluctuations depend on: (i) the parameters characterizing the system (e.g., mass, size, density), and (ii) two phenomenological parameters defining the statistical properties of the collapse noise. The collapse-driven Brownian motion works such that microscopic systems behave quantum mechanically, while macroscopic objects are classical. At the intermediate mesocopic scale, collapse models predict deviations from standard quantum predictions. This issue has been subject of experimental tests. All experiments to date have been at the scales where collapse effects are negligible for all practical purposes. However, recent experimental progress in revealing quantum features of larger objects, increases the hope for testing at unprecedented scales where collapse models can be falsified. Current experiments are mainly focused on the preparation of macroscopic systems in a spatial quantum superposition state. The collapse effects would then manifest as loss of visibility in the observed inference pattern. However, one needs a quantum interference with single particles of mass $\sim 10^{10}$amu for a decisive test of collapse models. Creating such massive superpositions is quite challenging, and beyond currect state-of-the-art. Quite recently, an alternative approach has been proposed where the collapse manifests in the fluctuating properties of light interacting with the quantum system. The great advantage of this new approach is that here there is no need for the preparation of a quantum superposed state. It has been discussed that promising results can be revealed in the spectrum of light interacting with a radiation pressure-driven mechanical oscillator in a cavity optomechanics setting. Here, we review the theoretical modelling of the above optomechenical proposal. We discuss how collapse-driven Brownian motion modifies the spectrum. We quantify the collapse effect and explain how it depends on the parameters of the mechanical oscillator (e.g., mass, density, geometry).

1. Introduction

Physics at the quantum-classical border is one of the crucial fields of today's research, both theoretical and experimental [1, 3, 2, 4, 5, 6, 7]. One of the challenging questions is to understand if, and under which conditions, quantum linearity fails when the size and complexity of the system increases. The exploration of this question has substantial consequences because if the quantum superposition principle fails beyond a certain scale (e.g., a mass scale), then it necessitates the modification of quantum theory [8, 9]. This might result in a correct scenario for unifying quantum theory with gravity [10].

Collapse models are one possible way to modify the standard quantum theory in a fully-consistent way [8, 9, 11, 12, 13, 14, 15, 16]. Collapse model assumes a universal noise field that, when acting on matter, introduces non-linear effects on the dynamics, which explain the collapse of the wave function. In other words, one can derive the random localization of the wave function at the end of a measurement, of course with the correct quantum probabilities. The strength of the collapse process scales with the size of the system, thus the wave function of microscopic systems can be superimposed, while macroscopic objects are always well-localized. To summarize, the collapse noise breaks the validity of the quantum superposition principle beyond a certain scale, and explains how the classical macroscopic world emerges from the quantum microscopic one.

Recently, there has been rapid experimental progress in revealing quantum features such as particle-wave duality for large objects with tiny de Broglie wavelength of only a few hundred femtometer [17]. Such objects were successfully decoupled from environmental noises, thus overcoming the technological limit and thereby extending the realm of quantum theory to new regimes. This progress provides the possibility to search for test of collapse models, as well. In this regard, quite a few new experimental schemes have been proposed [9]. All these proposals are based on the natural idea of creating a macroscopic quantum superposition in space, in order to test the superposition principle. Creating macroscopic superpositions is very difficult, source of formidable technological challenges.

Quite recently, an alternative approach has been proposed, based on measuring light emitted by radiative transitions of excited states of matter [18]. Here, the collapse manifests as an extra broadening and shift in lineshapes of the spectral density. The most important advantage of this new approach is that here there is no need for the preparation of a quantum superposed state. Later, it was shown that more promising results can be revealed in the spectrum of light interacting with a radiation pressure-driven mechanical oscillator in a cavity optomechanical setting [19]. The collapse results in an increase of the intensity of the noise spectrum of the light driving the mechanical oscillator. A Similar result has been obtained in [20].

In this paper, we will review the theoretical modelling of the above optomechenical proposal. The structure of this paper is as follows. We first derive the collapse dynamical equation for the center-of-mass of a complex system. We use the mass-proportional Continuous Spontaneous Localization (CSL) model [15], which is the most-studied collapse model in the literature. Then in the narrow wave function limit, we discuss how statistical effects of the CSL collapse noise can be reproduced by adding a linear random potential into the Schrödinger equation. We discuss how this random potential modifies the dynamic of an optomechanical oscillator. In our setting, the oscillator is the moving mirror of a Fabry-Perot cavity, that couples to an external laser field and is immersed in a finite-temperature bath. Therefore, the noise sources are: the thermal-driven Brownian motion of the oscillator interacting with the bath, the input laser noise and the CSL collapse noise. We use the Langevin formalism to account for the dynamics of the oscillator, where the above noises are introduced by using suitable noise operators. Any observable quantity is obtained after averaging over the noises. We deduce the thermal-driven Brownian noise- and the input laser noise-operators by using standard techniques of open quantum systems theory. In this way we derive an expression for the spectrum of the output light, which shows modifications due to all noises, in particular from the CSL collapse noise. We finally discuss how the collapse frequency scales with the size and the shape of the mechanical oscillator.

2. The collapse effect on the center-of-mass motion

In this section, we show how the motion of center-of-mass is modified due to the collapse noise that interacts with a many-body system, as prescribed in collapse models. We shall focus on

Journal of Physics: Conference Series **626** (2015) 012006 doi:10.1088/1742-6596/626/1/012006

the mass proportional Continuous Spontaneous Localization (CSL) collapse model, which is the most studied collapse model in the literature. The CSL evolution of the wave function reads [15]:

$$\begin{aligned} \mathrm{d}|\Psi_t\rangle = \Big[-\frac{i}{\hbar}\hat{H}_0\,\mathrm{d}t + \sqrt{\gamma}\int \mathrm{d}^3\mathbf{x}\,\Big(\hat{L}(\mathbf{x}) - \langle\hat{L}(\mathbf{x})\rangle\Big)\,\mathrm{d}W(t,\mathbf{x}) \\ -\frac{\gamma}{2}\int \mathrm{d}^3\mathbf{x}\,\Big(\hat{L}(\mathbf{x}) - \langle\hat{L}(\mathbf{x})\rangle\Big)^2\,\mathrm{d}t\Big]\,|\Psi_t\rangle, \end{aligned} \tag{1}$$

which is a non-linear stochastic differential equation in Itō form. Here $\gamma \simeq 10^{-28}\,\mathrm{m}^3\cdot\mathrm{s}^{-1}$ is the localization parameter which is a new phenomenological parameter of collapse models, $W(t,\mathbf{x})$ is standard Wiener process giving a noise $\xi(t,\mathbf{x}) = \mathrm{d}W(t,\mathbf{x})/\mathrm{d}t$, which is white both in space and time (i.e. $\mathbb{E}(\xi(t,\mathbf{x})) = 0$ and $\mathbb{E}(\xi(t_1,\mathbf{x})\xi(t_2,\mathbf{y})) = \delta(t_1-t_2)\,\delta(\mathbf{x}-\mathbf{y})$, with $\mathbb{E}(\cdots)$ the stochastic average), $\langle\hat{L}(\mathbf{x})\rangle = \langle\Psi_t|\hat{L}(\mathbf{x})|\Psi_t\rangle$ which induces the nonlinearity in the dynamics, and $\hat{L}(\mathbf{x})$ is the coarse-grained mass density operator:

$$\hat{L}(\mathbf{x}) = \int \mathrm{d}^3\mathbf{y}\, g(\mathbf{x}-\mathbf{y})\sum_{\mathrm{j}}\frac{m_{\mathrm{j}}}{m_0}\sum_{s}\hat{a}_{\mathrm{j}}^{\dagger}(s,\mathbf{y})\hat{a}_{\mathrm{j}}(s,\mathbf{y}), \tag{2}$$

where $m_0 = 1\,\mathrm{amu}$, $\hat{a}_{\mathrm{j}}(s,\mathbf{y})$ is the annihilation operator of particle of type-j with mass m_{j} and the spin s at position $\mathbf{y}$; and

$$g(\mathbf{r}) = \exp(-\mathbf{r}^2/2r_C^2)/(\sqrt{2\pi}r_C)^3, \tag{3}$$

with $r_C \simeq 10^{-7}\,\mathrm{m}$ the correlation length, which is the second phenomenological parameter of collapse models. In the CSL model a system is well-localized when its position spread is smaller than r_C.

Taking the system as a rigid body and averaging over the relative coordinates, we can derive the CSL motion for the center-of-mass, where the spread of center-of-mass's position is smaller than the CSL correlation length $r_C = 10^{-7}\,\mathrm{m}$. Since we work in the non-relativistic regime, in Eq. (2) the index j runs over electrons and nucleons where the number of particles is also a constant of motion. Since masses are divided by the mass of a nucleon m_0 (see Eq.(2)), we can also neglect electrons. Accordingly, in the subspace with a fixed number of particles, we can write: $\hat{L}(\mathbf{x}) \approx \sum_{k=1}^{N} A_k\, g(\mathbf{x}-\hat{\mathbf{x}}_k)$, where N is the number of atomic nuclei, A_k is the atomic mass number, and $\mathbf{x}_k$ is the nuclear position. We assume that the system is a rigid body where nuclei have a fixed relative distance from the center-of-mass. We also assume that the spatial fluctuating around the equilibrium relative coordinates are much smaller than r_C, implying that relative coordinates are classical variables with their equilibrium values. Accordingly, the Lindblad operator reads as:

$$\hat{L}(\mathbf{x}) \approx \hat{L}_{\mathrm{com}}(\mathbf{x}) = \sum_{k=1}^{N} A_k\, g(\mathbf{x}-\mathbf{r}_k-\hat{\mathbf{q}}) = \int \mathrm{d}^3\mathbf{r}\,\varrho(\mathbf{r})\,g(\mathbf{x}-\mathbf{r}-\hat{\mathbf{q}}), \tag{4}$$

with $\hat{\mathbf{q}}$ the center-of-mass position operator, $\mathbf{r}_k$ the equilibrium position of k-th nucleus, and $\varrho(\mathbf{r})$ the number density of the system where the origin is the center-of-mass: $\varrho(\mathbf{r}) = \sum_{k=1}^{N}(m_k/m_0)\,\delta(\mathbf{r}-\mathbf{r}_k)$. If we assume that the system is in a separable state $|\Psi\rangle = |\psi\rangle\otimes|\phi_{\mathrm{rel}}\rangle$ with $|\psi\rangle$ the center-of-mass state and $|\phi_{\mathrm{rel}}\rangle$ the relative coordinates state, then the CSL dynamics for the center-of-mass reads as:

$$\begin{aligned} \mathrm{d}|\psi_t\rangle = \Big[-\frac{i}{\hbar}\hat{H}_{\mathrm{com}}\,\mathrm{d}t + \sqrt{\gamma}\int \mathrm{d}^3\mathbf{x}\,\Big(\hat{L}_{\mathrm{com}}(\mathbf{x}) - \langle\hat{L}_{\mathrm{com}}(\mathbf{x})\rangle\Big)\,\mathrm{d}W(t,\mathbf{x}) \\ -\frac{\gamma}{2}\int \mathrm{d}^3\mathbf{x}\,\Big(\hat{L}_{\mathrm{com}}(\mathbf{x}) - \langle\hat{L}_{\mathrm{com}}(\mathbf{x})\rangle\Big)^2\,\mathrm{d}t\Big]\,|\psi_t\rangle, \end{aligned} \tag{5}$$

where $\hat{H}_{\rm com}$ contain only the center-of-mass degrees of freedom. In fact, we have: $\hat{H} = \hat{H}_{\rm com} + \hat{H}_{\rm rel}$ where $\hat{H}_{\rm rel}$ is the Hamiltonian of relative coordinates and $[\hat{H}_{\rm com}, \hat{H}_{\rm rel}] = 0$. In other words, the CSL collapse field only acts on the center-of-mass and the relative coordinates follow standard quantum dynamics determined by the Hamiltonian $\hat{H}_{\rm rel}$. For cases where the center-of-mass is distributed around a time-dependent mean value $\langle \hat{\mathbf{q}} \rangle = \langle \psi(t)|\hat{\mathbf{q}}|\psi(t)\rangle$ with the spread much smaller than r_C, one can Taylor-expand $g(\mathbf{x} - \mathbf{r} - \hat{\mathbf{q}})$ to the first order around $\langle \hat{\mathbf{q}} \rangle$. Accordingly, $\hat{L}_{\rm com}(\mathbf{x})$ can be approximated as follows:

$$\hat{L}_{\rm com}(\mathbf{x}) \approx \int \mathrm{d}^3\mathbf{r}\, \varrho(\mathbf{r})\, g(\mathbf{x} - \mathbf{r} - \langle \hat{\mathbf{q}} \rangle) \left(1 + \frac{1}{r_C^2} (\mathbf{x} - \mathbf{r} - \langle \hat{\mathbf{q}} \rangle) \cdot (\hat{\mathbf{q}} - \langle \hat{\mathbf{q}} \rangle)\right). \tag{6}$$

Putting above equation into Eq.(5) and doing some lengthy calculations, one eventually obtain:

$$\frac{\mathrm{d}}{\mathrm{d}t}|\psi_t\rangle = \left[-\frac{i}{\hbar}\hat{H}_{\rm com} + \sqrt{\gamma}\sum_{k=1}^{3} (\hat{q}_k - \langle \hat{q}_k \rangle)\, \mathrm{d}W_k(t) - \frac{\gamma}{2}\sum_{k,l=1}^{3} \eta_{k,l}\, (\hat{q}_k - \langle \hat{q}_k \rangle)\, (\hat{q}_l - \langle \hat{q}_l \rangle) \right] |\psi_t\rangle \tag{7}$$

where $\hat{H}_{\mathbf{q}}$ is the standard quantum Hamiltonian of the center-of-mass,

$$\eta_{k,l} = \iint \mathrm{d}^3\mathbf{r}\, \mathrm{d}^3\mathbf{r}'\, \frac{\exp\left(-\frac{|\mathbf{r}-\mathbf{r}'|^2}{4r_C^2}\right)}{(2\sqrt{\pi}\, r_C)^3}\, \frac{\partial\varrho(\mathbf{r})}{\partial r_k}\, \frac{\partial\varrho(\mathbf{r}')}{\partial r'_l}, \tag{8}$$

and $w_k(t) = \mathrm{d}W_k(t)/\mathrm{d}t$ are white noises:

$$w_k(t) = \int \mathrm{d}^3\mathbf{x}\, \xi(t,\mathbf{x}) \int \mathrm{d}^3\mathbf{r}\, g(\mathbf{x} - \mathbf{r} - \langle \hat{\mathbf{q}} \rangle_{\Psi_t})\, \frac{\partial\varrho(\mathbf{r})}{\partial r_k}, \tag{9}$$

with a zero mean (i.e. $\mathbb{E}(w_k(t)) = 0$) and correlation: $\mathbb{E}(w_k(t)\, w_l(s)) = \delta(t-s)\, \eta_{k,l}$. For an object with cuboid, or disc, or spherical geometry, one finds $\eta_{k,l} = \delta_{kl}\, \eta_{k,k}$ which implies that white noises $w_k(t)$ are also independent. Henceforth, we will consider objects which are cuboid, or disc, or spherical.

Instead of working with the stochastic *nonlinear* dynamics given in Eq.(7), we work with the Schrödinger equation with a stochastic potential as follows:

$$i\hbar\frac{\mathrm{d}}{\mathrm{d}t}|\psi(t)\rangle = (\hat{H}_{\rm com} + \hat{V}(t))|\psi(t)\rangle, \qquad \hat{V}(t) = -\hbar\sqrt{\gamma}\sum_{k=1} \hat{q}_k\, w_k(t), \tag{10}$$

which is a linear stochastic differential equation in Stratonovich form. As discussed several times in the literature [22, 23, 24, 25, 26], the effects of nonlinear terms in Eq.(7), at the statistical level, can be mimicked also by linear random potentials. For individual realizations of the noise, the effects are very different (those of a linear dynamics vs. those of a nonlinear one), while at the statistical level they coincide, if the potential is suitably chosen.

3. Quantum Langevin equations for an optomechanical oscillator

We now discuss how the collapse mechanism modifies the dynamical equation for an optomechanical system. The approximations we used in this section are all standard approximations in optomechanics (e.g., see Ref. [27]). In fact, the final quantum Langevin equations are the standard ones in optomechanics and are only modified by a noise term induced by the collapse noise. As discussed before, to the extent that one is only concerned with the statistical observable effects, the collapse can be included as a random linear potential $V(t)$

Journal of Physics: Conference Series **626** (2015) 012006 doi:10.1088/1742-6596/626/1/012006

as given in Eq.(10). We shall denote the position operator along the cavity axis by $\hat{q}$, and its conjugate momentum by $\hat{p}$. We will consider the cavity axis along the x-axis. We shall denote the components of the Hamiltonian and of the noise term along the cavity axis by the subscript q (see Eqs.(11,12). We also assume that the Hamiltonian of the optomechanical system, in the rotating frame of the laser frequency, is given by [27]:

$$\hat{H}_q = \hbar(\omega_c - \omega_l)\hat{a}^\dagger\hat{a} - \hbar\,\chi\,\hat{a}^\dagger\hat{a}\,\hat{q} + \frac{1}{2}m\omega_m^2\,\hat{q}^2 + \frac{1}{2m}\hat{p}^2 + i\hbar\varepsilon(\hat{a}^\dagger - \hat{a}), \tag{11}$$

where m is the effective mass of the moving mirror (i.e., the mass of the mechanical oscillator), ω_l is the frequency of the external laser, ω_c is the frequency of the cavity mode driven by the laser, ω_m is the harmonic frequency of the mechanical oscillator, $\chi = \omega_c/L$ is the optomechanical coupling constant between the cavity and the mechanical oscillator with L the length of the cavity, and $\varepsilon = \sqrt{2\kappa\mathcal{P}/\hbar\omega_l}$ with $\mathcal{P}$ the laser power and κ the cavity photon decay rate. Along the cavity axis, the random potential is given by:

$$V_q(t) = -\hbar\sqrt{\eta}\,w(t)\,\hat{q}, \tag{12}$$

where $\eta = \gamma\eta_{1,1}$ and $w(t)$ is a white noise with zero mean (i.e. $\mathbb{E}(w(t)) = 0$) and correlation $\mathbb{E}(w(t)w(s)) = \delta(t-s)$. Notice that one can separate the motion along different axes, only for the case where white noises in Eq.(9) are independent (e.g, for a cuboid object).

For convenience, we will use the dimensionless operators $\hat{Q} = \hat{q}\sqrt{m\omega_m/\hbar}$ and $\hat{P} = \hat{p}/\sqrt{m\hbar\omega_m}$. Accordingly, the Heisenberg equations of the motion are:

$$\frac{\mathrm{d}}{\mathrm{d}t}\hat{Q} = \omega_m\,\hat{P} \tag{13}$$

$$\frac{\mathrm{d}}{\mathrm{d}t}\hat{P} = \omega_m\bar{\chi}\,\hat{a}^\dagger\hat{a} - \omega_m\hat{Q} + \sqrt{\lambda}\,w(t) \tag{14}$$

$$\frac{\mathrm{d}}{\mathrm{d}t}\hat{a} = -i(\omega_c - \omega_l - \omega_m\bar{\chi}\,\hat{Q})\hat{a} - \kappa\,\hat{a} + \varepsilon, \tag{15}$$

where:

$$\bar{\chi} = \frac{\chi}{\omega_m}\sqrt{\frac{\hbar}{m\omega_m}}; \qquad \lambda = \frac{\hbar\eta}{m\omega_m} = \frac{\hbar}{m\omega_m}\gamma\iint \mathrm{d}^3\mathbf{r}\,\mathrm{d}^3\mathbf{r}'\,\frac{\exp\left[-\frac{|\mathbf{r}-\mathbf{r}'|^2}{4r_C^2}\right]}{(2\sqrt{\pi}\,r_C)^3}\,\frac{\partial\varrho(\mathbf{r})}{\partial\mathbf{r}_1}\,\frac{\partial\varrho(\mathbf{r}')}{\partial\mathbf{r}'_1}. \tag{16}$$

One should remember that above stochastic differential equations should be understood in the Stratonovich sense. Using standard procedures [28], one can obtain the corresponding Itō form, which coincides with the Stratonovich form in this special case. Now we can add noise terms: the thermal Brownian motion induced by a bath, and the input noise of the laser field. Therefore the quantum Langevin equations are:

$$\frac{\mathrm{d}}{\mathrm{d}t}\hat{Q} = \omega_m\,\hat{P} \tag{17}$$

$$\frac{\mathrm{d}}{\mathrm{d}t}\hat{P} = \omega_m\bar{\chi}\,\hat{a}^\dagger\hat{a} - \omega_m\hat{Q} + \sqrt{\lambda}\,w(t) - \gamma_m\,\hat{P} + \hat{W}(t) \tag{18}$$

$$\frac{\mathrm{d}}{\mathrm{d}t}\hat{a} = -i(\omega_c - \omega_l - \omega_m\bar{\chi}\,\hat{Q})\hat{a} - \kappa\,\hat{a} + \varepsilon + \hat{a}_{\mathrm{in}}\sqrt{2\kappa} \tag{19}$$

$$\frac{\mathrm{d}}{\mathrm{d}t}\hat{a}^\dagger = i(\omega_c - \omega_l - \omega_m\bar{\chi}\,\hat{Q})\hat{a}^\dagger - \kappa\,\hat{a}^\dagger + \varepsilon + \hat{a}_{\mathrm{in}}^\dagger\sqrt{2\kappa} \tag{20}$$

where $\hat{a}_{\text{in}}(t)$ is the input noise with zero mean the correlation function[1]

$$\langle\, \delta\hat{a}_{\text{in}}(t)\, \delta\hat{a}^{\dagger}_{\text{in}}(s)\rangle = \delta(t-s), \tag{21}$$

and $\hat{W}(t)$ is the quantum thermal Brownian motion operator with zero mean the correlation function:

$$\langle\, \hat{W}(t)\, \hat{W}(s)\rangle = \frac{\gamma_m}{2\pi\omega_m}\int_{-\infty}^{+\infty} \mathrm{d}\omega\, \omega\, e^{-i\omega(t-s)}\left[1+\coth\left(\frac{\hbar\omega}{2k_BT}\right)\right] \tag{22}$$

Since here the noise operators $\hat{W}$ and $\hat{a}_{\text{in}}$ are of quantum origin, we represent their stochastic average by $\langle\cdots\rangle$, meaning the trace over the corresponding quantum mechanical degrees of the freedom.

We now linearise the first term on the right-side of Eqs.(19,20). We write each operator as follows: $\hat{A} = A_s + \delta\hat{A}$, where A_s is the classical steady-state value of the operator $\hat{A}$, and $\delta\hat{A}$ is a small quantum fluctuation around this steady-state value. Doing so, and also assuming $a_s \gg 1$, the linearised quantum Langevin equations take the following form:

$$\frac{\mathrm{d}}{\mathrm{d}t}\delta\hat{Q} = \omega_m\, \delta\hat{P} \tag{23}$$

$$\frac{\mathrm{d}}{\mathrm{d}t}\delta\hat{P} = \omega_m\bar{\chi}\,(a_s\,\delta\hat{a}^{\dagger} + a_s^*\,\delta\hat{a}) - \omega_m\,\delta\hat{Q} + \sqrt{\lambda}\,w(t) - \gamma_m\,\delta\hat{P} + \hat{W}(t) \tag{24}$$

$$\frac{\mathrm{d}}{\mathrm{d}t}\delta\hat{a} = -i(\omega_c - \omega_l - \omega_m\bar{\chi}\,Q_s)\delta\hat{a} + i\omega_m\bar{\chi}\,a_s\,\delta\hat{Q} - \kappa\,\delta\hat{a} + \delta\hat{a}_{\text{in}}\sqrt{2\kappa} \tag{25}$$

$$\frac{\mathrm{d}}{\mathrm{d}t}\delta\hat{a}^{\dagger} = i(\omega_c - \omega_l - \omega_m\bar{\chi}\,Q_s)\delta\hat{a}^{\dagger} - i\omega_m\bar{\chi}\,a_s^*\,\delta\hat{Q} - \kappa\,\delta\hat{a}^{\dagger} + \delta\hat{a}^{\dagger}_{\text{in}}\sqrt{2\kappa}. \tag{26}$$

For the dimensionless position operator $\delta\hat{Q}$, the solution of the Fourier transform of above equations is given by:

$$\delta\hat{Q}(\omega) = -\frac{\omega_m}{d(\omega)}\Big[\Big(\hat{W}(\omega) + \sqrt{\lambda}w(\omega)\Big)\left(\Delta^2 + (\kappa + i\omega)^2\right) \\ -i\omega_m\sqrt{2\kappa}\,\bar{\chi}\left(a_s^*(\kappa + i(\omega-\Delta))\,\delta\hat{a}_{\text{in}}(\omega) + a_s(\kappa + i(\omega+\Delta))\,\delta\hat{a}^{\dagger}_{\text{in}}(-\omega)\right)\Big], \tag{27}$$

where $d(\omega) = (\Delta^2 + (\kappa + i\omega)^2)(\omega(\omega - i\gamma_m) - \omega_m^2) + 2|a_s|^2\bar{\chi}^2\omega_m^3\Delta$, and $\Delta = \omega_c - \omega_l - \omega_m\bar{\chi}Q_s$. Taking the Fourier transforms, the CSL noise and also the thermal Brownian noise operator yield:

$$\mathbb{E}(w^*(\omega)w(\omega')) = \frac{1}{2\pi}\delta(\omega-\omega'); \qquad \langle\,\delta\hat{a}_{\text{in}}(\omega)\,\delta\hat{a}^{\dagger}_{\text{in}}(\omega')\rangle = \frac{1}{2\pi}\delta(\omega-\omega') \tag{28}$$

$$\langle\hat{W}^{\dagger}(\omega)\hat{W}(\omega')\rangle = \frac{\gamma_m\,\omega}{2\pi\omega_m}\left[1+\coth\left(\frac{\hbar\omega}{2k_BT}\right)\right]\delta(\omega-\omega') \tag{29}$$

From the Wiener-Khinchin theorem [29], we also have:

$$\mathbb{E}(\langle\delta\hat{Q}^{\dagger}(\omega)\delta\hat{Q}(\omega')\rangle) = \mathcal{S}(\omega)\,\delta(\omega-\omega'), \tag{30}$$

where $\mathcal{S}(\omega)$ is the corresponding spectral density of the dimensionless position operator $\delta\hat{Q}$. Thus, one finds:

$$\mathcal{S}(\omega) = \frac{1}{2}\int_{-\infty}^{+\infty}\mathrm{d}\omega'\left(\mathbb{E}(\langle\delta\hat{Q}^{\dagger}(\omega)\delta\hat{Q}(\omega')\rangle) + \mathbb{E}(\langle\delta\hat{Q}^{\dagger}(\omega')\delta\hat{Q}(\omega)\rangle)\right). \tag{31}$$

[1] See the following paragraphs for the precise definition of $\delta\hat{a}_{\text{in}}(t)$.

Introducing Eq.(27) into Eq.(30) yields:

$$\mathbb{E}(\langle\delta\hat{Q}^\dagger(\omega)\delta\hat{Q}(\omega')\rangle) = \delta(\omega-\omega')\frac{\omega_m^2}{2\pi|d(\omega)|^2}\times \tag{32}$$

$$\left\{4\kappa\bar{\chi}^2|a_s|^2\omega_m^2(\kappa^2+\omega^2+\Delta^2)+((\Delta^2+\kappa^2-\omega^2)^2+4\kappa^2\omega^2)\left(\frac{\gamma_m\,\omega}{\omega_m}\left[1+\coth\left(\frac{\hbar\omega}{2k_BT}\right)\right]+\lambda\right)\right\}$$

Accordingly, for the spectral density, we find:

$$\mathcal{S}(\omega) == \frac{\omega_m^2\left\{2\kappa\bar{\chi}^2|a_s|^2\omega_m^2(\kappa^2+\omega^2+\Delta^2)+\left[(\Delta^2+\kappa^2-\omega^2)^2+4\kappa^2\omega^2\right]\left(\frac{\gamma_m\,\omega\,\coth(\hbar\beta\omega)}{\omega_m}+\lambda\right)\right\}}{\pi|d(\omega)|^2} \tag{33}$$

with $\beta = 1/2k_BT$. Here the term $1/|d(\omega)|^2$ determines the shape of the spectrum, while other terms in the nominator contribute to the intensity of the spectrum. Notice that the spectral density of the intracavity field $\mathcal{S}_{\text{cav}}(\omega)$ is related to $\mathcal{S}(\omega)$ via the simple relation: $\mathcal{S}_{\text{cav}}(\omega) = |f(\Delta)|^2\,\mathcal{S}_{\text{cav}}(\omega)$+additional terms where $f(\Delta) = 2\bar{\chi}a_s\frac{\kappa-i\omega}{\Delta^2+(\kappa-i\omega)^2}$ (see Eq.(15) in [30]). So the following arguments for $\mathcal{S}(\omega)$ can be directly also applied for $\mathcal{S}_{\text{cav}}(\omega)$. The contribution of the CSL noise is shown in Fig.1, which manifests in the intensity of the spectrum by increasing the area under $\mathcal{S}(\omega)$. The relative increase of the area under the spectral density can be considered as a measure for the CSL effect on the spectrum [19]:

$$\mathcal{A} = \frac{\int_{-\infty}^{+\infty}\mathrm{d}\omega\,\mathcal{S}(\omega) - \lim_{\lambda\to0}\int_{-\infty}^{+\infty}\mathrm{d}\omega\,\mathcal{S}(\omega)}{\lim_{\lambda\to0}\int_{-\infty}^{+\infty}\mathrm{d}\omega\,\mathcal{S}(\omega)}. \tag{34}$$

Here $\mathcal{A}$ gives us a quantitative estimate of the relative increase of the area under the spectral density when the CSL effects are included. As one can see in Fig.1, the spectrum is well-localized around $\omega\sim\omega_m$. Accordingly, with a very good approximation the spectral density can be written as follows:

$$\mathcal{S}(\omega) \approx \frac{\omega_m^2\left\{2\kappa\bar{\chi}^2|a_s|^2\omega_m^2(\kappa^2+\omega_m^2+\Delta^2)+\left[(\Delta^2+\kappa^2-\omega_m^2)^2+4\kappa^2\omega_m^2\right](\gamma_m\coth(\hbar\beta\omega_m)+\lambda)\right\}}{\pi|d(\omega)|^2}. \tag{35}$$

Introducing Eq.(35) into Eq.(34), one finds:

$$\mathcal{A} \approx \lambda\frac{(\Delta^2+\kappa^2-\omega_m^2)^2+4\kappa^2\omega_m^2}{2\kappa\bar{\chi}^2|a_s|^2\omega_m^2(\kappa^2+\omega_m^2+\Delta^2)+\left[(\Delta^2+\kappa^2-\omega_m^2)^2+4\kappa^2\omega_m^2\right](\gamma_m\coth(\hbar\beta\omega_m))} \tag{36}$$

As one can see, $\mathcal{A}$ scales linearly with the collapse frequency λ. Accordingly, the larger λ, the stronger the CSL effects and thus the better for experimental searches. For example, considering the numerical values introduced in the caption of Fig.1, one finds: $\mathcal{A} \approx 10^{-10}(\lambda/1\text{Hz})$. A very important question is that how the collapse frequency λ depends on parameters characterizing the mechanical oscillator (e.g., mass, size, density and etc.). As given in Eq.(16), the collapse frequency is $\lambda = \eta\,x_0^2$. It is similar to the famous Joos/Zeh decoherence rate (see Eq.(3.59) in [31]), however here $x_0 = \sqrt{\hbar/m\omega_m}$ and depends on the mass. The term η is usually called as the localization strength (or the localization rate), and it depends on the mass and the geometry of the oscillator (see Eq.(8)). Accordingly, λ depends on the interplay between x_0 and η; for example the delocalization distance x_0 decreases by increasing the mass while the localization strength η increases by increasing the mass. The detailed behaviour of λ with the mass and geometry of the mechanical oscillator is the topic of our future research.

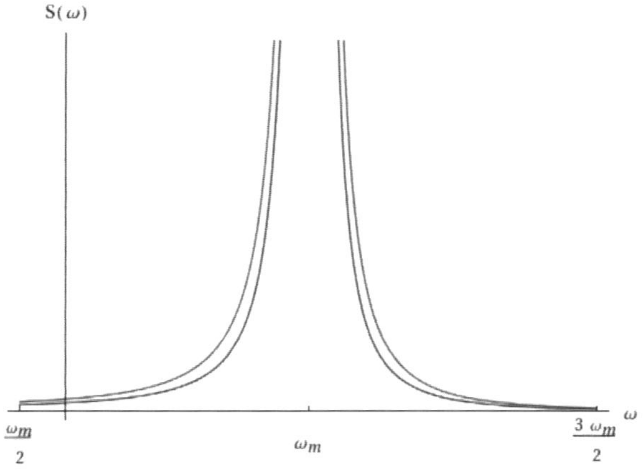

Figure 1. The spectral density as given in Eq.(33). For the blue line, we set $\lambda = 0$ (no CSL effect) and for the red line, we set: $\lambda = 10^{10}\,\mathrm{Hz}$ (with CSL effect), which is a quite unrealistic value that we consider only for the sake of illustration. The parameters are set as follows: $m = 150\,\mathrm{ng}$, $\omega_m/2\pi = 2.75 \times 10^5\,\mathrm{Hz}$, $\gamma_m/2\pi = 10^{-5}\omega_m$, $\Delta = 4\kappa$, $\mathcal{P} = 4\,\mathrm{mW}$, $\kappa = 5\times10^7\,\mathrm{Hz}$, $T = 1\,\mathrm{mK}$, $\omega_l/2\pi = 2.78 \times 10^{15}\,\mathrm{Hz}$, $L = 25\,\mathrm{mm}$.

4. Concluding remarks

Collapse models provide a quantitative explanation on how macroscopic classical world emerges from the microscopic quantum world. They assume a universal noise field inducing appropriate Brownian motion corrections to standard Schrödinger dynamics. Accordingly, one *derives* the random localization of the wave function at the end of a measurement, of course with the correct probabilities given by the Born rule. The strength of the collapse process grows with the size of the system. Therefore, microscopic systems behave quantum mechanically, while macroscopic objects are classical. At the intermediate mesocopic scale, collapse models predict deviations from standard quantum predictions. These deviations depend on the statistical properties of the collapse noise and also on parameters characterizing the system (e.g., mass, size, density and etc.). In the CSL model, statistical properties of the collapse noise (i.e., its mean value and its autocorrelation) are fixed by two phenomenological parameters.

All experiments to date have been at scales where the collapse effects are negligible for all practical purposes. However, recent experimental progress in revealing quantum features of larger objects, opens new path to search for testing at unprecedented scales where collapse models can be falsified. The most serious challenge in this regard is creating macroscopic superpositions, because all experimental schemes proposed so far are based on the natural idea of creating a macroscopic quantum superposition in space.

Quite recently, an alternative approach has been proposed, based on measuring the fluctuating properties of light (e.g., the spectral density) which interacts with a quantum system. The collapse manifests as a change (e.g., an extra broadening) in lineshapes of the light spectrum. The great advantageous of this new approach is that here there is no need for the preparation of a quantum superposed state. It has been discussed that promising results can be revealed in the spectrum of light interacting with a radiation pressure-driven mechanical oscillator in a cavity optomechanics setting. Proper measures have been introduced to quantify the collapse effect on the spectrum. These measures depend on the collapse frequency. Therefore, it is very important to quantify the collapse frequency, e.g., to explain how it depends on the parameters of the mechanical oscillator (e.g., it size, mass, density and ect.). Here, the collapse frequency depends linearly on the localization strength η and quadratically on the delocalization distance x_0 (i.e. $\lambda = \eta x_0^2$). The localization strength increases with the mass, as expected; while the delocalization distance decreases with the mass. This is the very particular thing that distinguishes our example from other cases studied so far in the literature. In previous cases, only the behaviour of the localization strength with the mass has been investigated, because the delocalization distance was independent from the mass; while in our case it depends also on the mass, and thus one needs to carefully study the interplay of the mass on the localization strength and on the delocalization distance. Accordingly, the precise behaviour of the collapse frequency depends on this dual role of the mass, and for different geometries it may produce different behaviours.

Acknowledgments

We would like to thank Prof. H.Ulbricht (Southampton Univ., UK) and Prof. M.Paternostro (Queen's Univ., Belfast) for helpful comments. We also thank Prof. K.Hornberger (Univ. of Duisburg, Germany) for his comments on our early work on this topic (Ref. [19]). We gratefully acknowledge funding and support through the EU project NANOQUESTFIT, the COST Action MP1006 "Fundamental Problems in Quantum Physics", the John Templeton foundation (grant 39530), FRA 2013 (UniTS) and the Istituto Nazionale di Fisica Nucleare (INFN). We gratefully thank the organizers of the Seventh International Workshop "DICE2014: Spacetime - Matter - Quantum Mechanics", Italy.

References

[1] Leggett A J 2002 *J. Phys.: Condens. Matter* **14** R415
[2] Adler S L and Bassi A 2009 *Science* **325** 275
[3] Adler S L 2004 *Quantum Theory as an Emergent Phenomenon* (Cambridge: Cambridge Univ. Press)
[4] Weinberg S 2012 *Phys. Rev.* A **85** 062116
[5] Joos E *et al* 2003 *Decoherence and the Appearance of a Classical World in Quantum Theory* 2nd edition (New York: Springer)
[6] Breuer H P and Petruccione F 2002 *The Theory of Open Quantum Systems* (Oxford: Oxford Univ. Press)
[7] Adler S L 2003 *Stud. Hist. Philos. Mod. Phys.* **34** 135; Vacchini B 2007 *J. Phys. A: Math. Theor.* **40** 2463; Adler S L 2014 Gravitation and the noise needed in objective reduction models *Preprint* arXiv:1401.0353v4; Arndt M and Hornberger K 2014 *Nat. Phys.* **10** 271
[8] Bassi A and Ghirardi G C 2003 *Phys. Rep.* **379** 257
[9] Bassi A, Lochan K, Satin S, Singh T P and Ulbricht H 2013 *Rev. Mod. Phys.* **85** 471
[10] Penrose P 1996 *Gen. Rel. Grav.* **28** 581; *do.* 2014 *Found. Phys.* **44** 557
[11] Pearle P 1976 *Phys. Rev.* D **13** 857; *do.* 1989 *Phys. Rev.* A **39** 2277
[12] Ghirardi G C, Rimini A and Weber T 1986 *Phys. Rev.* D **34** 470
[13] Diósi L 1987 *Phys. Lett.* A **120** 377; *do.* 1989 *Phys. Rev.* A **40** 1165
[14] Ghirardi G C, Pearle P and Rimini A 1990 *Phys. Rev.* A **42** 78
[15] Ghirardi G C, Grassi R and Benatti F 1995 *Found. Phys.* **25** 5
[16] Adler S L 2007 *J. Phys.* A **40** 2935
[17] Marquardt F and Girvin S M 1993 *Physics* **2** 40; Aspelmeyer M *et al* 2010 *J. Opt. Soc. Am.* B **27** A189; Aspelmeyer M, Kippenberg T J and Marquardt F 2014 *Rev. Mod. Phys.* **86** 1391; Romero-Isart O 2011 *Phys. Rev.* A **84** 052121; Hornberger K, Gerlich S, Haslinger P, Nimmrichter S and Arndt M 2012 *Rev. Mod. Phys.* **84** 157; Juffmann T, Ulbricht H and Arndt M 2013 *Rep. Prog. Phys.* **76** 086402
[18] Bahrami M, Bassi A and Ulbricht H 2014 *Phys. Rev.* A **89** 032127
[19] Bahrami M, Paternostro M, Bassi A and Ulbricht H 2014 *Phys. Rev. Lett.* **112** 210404
[20] Nimmrichter S, Hornberger K and Hammerer K 2014 *Phys. Rev. Lett.* **113** 020405
[21] For a more detailed calculation, see Section II of Bedingham D J 2013 *Phys. Rev.* D **88** 045032
[22] Gisin N 1989 *Hel. Phys. Acta* **62** 363; *do.* 1990 *Phys. Lett.* A **143** 1
[23] Gisin N and Rigo M 1995 *J. Phys.* A **28** 7375
[24] Polchinski J 1991 *Phys. Rev. Lett.* **66** 397
[25] Adler S L 2000 *J. Math. Phys.* **41** 2485; *do.* Statistical Dynamics of Global Unitary Invariant Matrix Models as Pre-Quantum Mechanics 2002 *Preprint* arXiv:hep-th/0206120, Section 5F; *do.* 2003 *Phys. Rev.* D **67** 025007, added note
[26] Diósi L 1988 *Phys. Lett.* A **129** 419
[27] Agarwal G S 2012 *Quantum Optics* (NY: Cambridge Univ. Press) Chapter 20
[28] Arnold L 1974 *Stochastic differential equations: Theory and applications* (New York: Wiley) Section 10.2.
[29] Mandel L and Wolf E 1995 *Optical Coherence and Quantum Optics* (Cambridge: Cambridge Univ. Press) Section 2.4.
[30] Paternostro M, Gigan S, Kim M S, Blaser F, Böhm H R and Aspelmeyer M 2006 *New J. Phys.* **8** 107
[31] Joos E and Zeh H D 1985 *Z. Phys.* B **59** 223

Journal of Physics: Conference Series **626** (2015) 012007 doi:10.1088/1742-6596/626/1/012007

Time symmetry in wave function collapse models

Daniel Bedingham

Faculty of Philosophy, University of Oxford, OX2 6GG, United Kingdom.

E-mail: daniel.bedingham@philosophy.ox.ac.uk

Abstract. A framework for wave function collapse models that is symmetric under time reversal is presented. Within this framework there are equivalent pictures of collapsing wave functions evolving in both time directions. The backwards-in-time Born rule can be broken by an initial condition on the Universe resulting in asymmetric behaviour. Similarly the forwards-in-time Born rule can in principle be broken by a final condition on the Universe.

1. Introduction

1.1. A time-asymmetric example

Consider the experimental situation depicted in figure 1[1] (taken from reference [1]). A photon is emitted from a source S. It encounters a beam splitter B from which it is either transmitted then detected by a detector D, or it is reflected up towards the ceiling C with no detection made at D. The Born rule tells us that a single photon arriving at the beam splitter has a probability of 0.5 for each possibility. If we perform this experiment many times we can confirm this by comparing the number of detection events with the number of times the source emitted a photon. Both of these types of events are presumably observable to us. The source could be a reliable source of single photons emitting at a fixed rate in which case we do not directly observe individual emission events but we can be fairly sure of the number of emissions in a fixed time period. Perhaps when the detector registers there is a click.

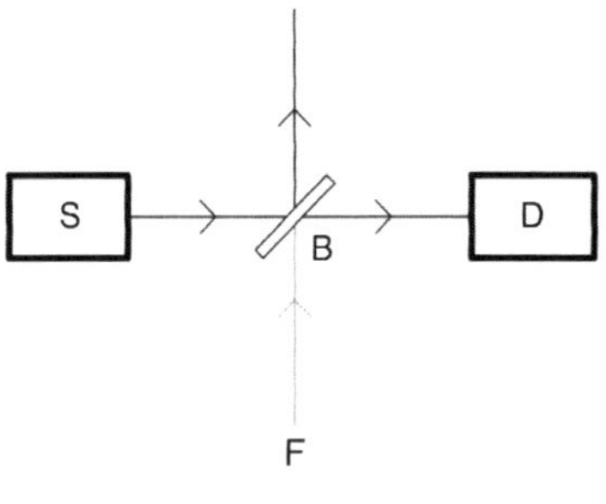

Figure 1. Beam splitter thought experiment.

[1] Thanks to Avshalom Elitzur for suggesting this example in a question at this talk.

Now consider this experiment backwards in time. That is not to say that we propose to actually go backwards in time or to try to build a source of backwards-in-time photons. We simply consider the events of the experiment which involve a photon interacting either with the source S, or with the beam splitter B, or with the detector D, in the backwards-in-time order. The story starts with a click at D as the photon heads backwards in time towards B where then with certainty it is transmitted through the beam splitter arriving back at the source (since all photons detected at D originated at the source S).

Now if we apply the Born rule to the photon going backwards in time, then as it interacts with the beam splitter there would be a probability amplitude to two possible trajectories. We would predict that only half the photons should arrive back at the source whilst the other half are sent to the floor F. In reference [1] Penrose has argued that this example indicates a time asymmetry in wave function collapse. We might imagine that the backwards-in-time photon, once it encounters the beam splitter enters a superposition state of wave packets at the two locations S and F. However, the collapse of the wave function, were it to happen in the backwards-in-time direction would always choose the history SBD. There are no photons with history FBD (let us assume that there is some cold opaque object at F to ensure this). The Born rule simply doesn't work for the photon going backwards in time.

In drawing this conclusion we are stripping down the physical system to consider only the behaviour of the photon. It is one of the strengths of quantum mechanics that we can do this. We only need to know the amplitudes for the various options and not all the details and circumstances of other things going on around the experiment. Indeed there will be records in the world that the source is emitting photons at the rate that it is. Perhaps there are other signatures such as a depletion of energy in the source. If the source functions by optical pumping of a small number of atoms then the transparency of the atoms might be dependent on the emission rate. There may be records of the cold opaque object at F being kept at absolute zero such that there is simply no possibility for a photon to originate from it. These signatures are part of the extended experiment and they are correlated to the specific behaviour of the photons but the Born rule applied to the photon alone shouldn't care about all this. There is no obligation for the wave function of the photon to collapse to those particular outcomes that result in correlations with the wider world.

1.2. Inherent time asymmetry or the effect of an initial condition?

There seems to be no doubt that the Born rule doesn't work in the reverse-time direction in the example of figure 1. There is then the possibility that the Born rule has an inherent time asymmetry—that it only works in the forwards-in-time direction. But there is also the possibility that the time asymmetry is due to an initial boundary condition which serves as a constraint. Furthermore, this initial constraint could be the (low-entropy) initial condition of the Universe. This view is in line with the standard explanation of time asymmetric behaviour in the world. For the photon travelling backwards in time, having its wave packet split into a superposition by the beam splitter, it is not simply left with a 50-50 choice of collapse to F or collapse to S, it must also weigh this up against the need of the Universe to achieve its low entropy initial state. By collapsing the wave packet at S many correlations are established with the extended experiment and entropy is reduced.

We might also consider the possibility of some far-off final condition of the Universe which would ultimately cause a breakdown of forwards-in-time Born-rule behaviour. Note that in this picture the boundary conditions are responsible for modifying the probabilistic behaviour of the world in order to fulfil the condition. This is characteristically different from a deterministic world where the state at any given moment must be such that the boundary conditions are satisfied when we evolve in time.

It is therefore by no means certain that collapse of the wave function is inherently time asymmetric. However, if we are to treat collapse as a genuine physical process we face further issues to do with time symmetry.

2. Collapse models

In models of wave function collapse we regard the wave function as physically real and the collapse of the wave function as a real physical process. This requires a deviation from the standard Schrödinger dynamics. The general idea is to modify the Schrödinger equation in such a way that certain types of large-scale superposition states are unstable and will collapse. Yet the modified equation should be universally applicable and so when applied to micro systems it should be well approximated by the standard Schrödinger equation. These modifications need to be both stochastic (since collapse is random) and non linear (since the chance of a collapse outcome depends on the state itself). The most well known of these are the GRW model [2] and the CSL model [3, 4]. There are also relativistic versions of these models [5, 6, 7].

2.1. The GRW model

Let us briefly summarise the GRW model for the sake of a definite example. The GRW model concerns a set of N distinguishable particles with wave function $\psi_t(x_1, \ldots, x_2)$. The wave function usually satisfies the usual Schrödinger equation but at random times chosen with fixed probability per unit time per particle i, the wave function makes a jump of the form

$$\psi_t \to \psi_{t+} = j(z - x_i)\psi_t, \tag{1}$$

with z a random variable. This happens for every particle i independently. Each particle has its own random sequence of jumps occurring at a different set of random times. The jump operator j takes the form

$$j(x) = \frac{1}{(\pi a^2)^{1/4}} \exp\left(\frac{-x^2}{2a^2}\right), \tag{2}$$

where a is a fixed parameter with units of length chosen by GRW to be 10^{-7}m. The action of the jump operator is that of a quasi projection of the ith particle about position z. The complete set of operators $\{j^2(z - x_i)|z \in \mathbb{R}\}$ form a POVM. All the collapses occur in this smeared position state basis. This is the preferred basis of this model. It is a remarkable feature that even with collapses only occurring in the position basis, models of this type are able to account for collapse of the wave function in any conceived measurement situation.

Finally, the random collapse centres z are drawn from a probability distribution of the form

$$\mathbb{P}_t\left(z\right) = \frac{\int dx_1 \cdots dx_N |j(z - x_i)\psi_t|^2}{\int dx_1 \cdots dx_N |\psi_t|^2}. \tag{3}$$

This is nothing but the Born rule probability distribution for a quasi projection of the form (1). This makes the model non linear.

The rate of collapse per particle is chosen to be sufficiently small that individual particles very rarely undergo jumps (GRW choose a rate of 10^{-16}s^{-1}, this is about once every lifetime of the Universe). The physics of a few particles is therefore left unaffected. However, for bulk matter with of order 10^{24} particles, there are many jumps each second causing, for example, pointers to commit to definite readings etc. The theory does the job of making a very minor modification to the micro physics of individual particles which leads to radical effects at the macro level. Similar principles are used in other collapse models.

Journal of Physics: Conference Series **626** (2015) 012007 doi:10.1088/1742-6596/626/1/012007

2.2. Time reversal symmetry in collapse models

Now let us return to the question of time reversal symmetry. When a single jump occurs, the wave function of a single particle goes from being dispersed in space to suddenly being localised about the random position z. This is clearly a time-asymmetric process. Viewed backwards in time what we have is a localised wave function spontaneously becoming dispersed in space. When the jump happens it affects only the future state of the particle and this introduces a preferred time direction.

Note that often solutions to the Schrödinger equation seem to display time directed behaviour such as quantum dispersion. However, it is straightforward to show the that reverse-time process, quantum focussing let's call it, is also a solution to the Schrödinger equation[2]. Quantum dispersion is more common than quantum focussing simply because the initial conditions for quantum focussing are not prevalent in the Universe.

But the situation with the GRW jumps is different. We can't make the time reverse of a jump look like a jump. The dynamical rule which is composed of the Schrödinger equation with jumps does not apply to a reverse time sequence of wave functions. Faced with a sequence of wave functions and asked to identify the forward direction of time, it can be done if a jump occurs (but not if one doesn't).

One way to restore time reversal symmetry to this picture is to introduce a separate backwards-in-time wave function (we can call it $\bar{\psi}$). This new wave function might look quite similar in form to ψ but it has the jump rule

$$\bar{\psi}_t \to \bar{\psi}_{t-} = j(z - x_i)\bar{\psi}_t. \tag{4}$$

The jumps affect the past state of $\bar{\psi}$ rather than the future state. This is the same jump rule as for ψ but acting in the reverse-time direction. To be clear, the jump rule is equivalent to a pair of left and right boundary conditions on $\bar{\psi}$ at time t: on the $t+$ side the state is $\bar{\psi}_t$; on the $t-$ side the state is $j(z - x_i)\bar{\psi}_t$. At all other times the wave function satisfies the (time-reversal-symmetric) Schrödinger equation.

What ties the two wave functions ψ and $\bar{\psi}$ together? At a given moment they might look broadly similar and make almost identical predictions. However, what guarantee is there that this will be the case at other times? And if they are not the same which one should we trust to tell us about the state of the world?

We can in fact ensure that the two wave functions are empirically equivalent at all times by having their jumps occur at identical locations in space and time. To understand this consider a single particle which suffers a sequence of jumps at times $t_1 < t_2 < \cdots < t_{n-1} < t_n$ and at locations $z_1, z_2, \ldots, z_{n-1}, z_n$. The forwards-in-time wave function ψ undergoes first a jump about z_1 at t_1, then a jump about z_2 at t_2, etc. By contrast the evolution of the backwards-in-time wave function $\bar{\psi}$ involves first a jump about z_n at time t_n, then a jump about z_{n-1} at time t_{n-1}, etc. The history of the particle is the same in each case as it passes through the same series of locations at the same times. This is despite the fact that the detailed behaviour of the two wave functions will be quite different from one another. In particular the cycle of sudden collapse followed by dispersion will be occurring in opposite time directions.

The same idea works when we have more particles. We have consistency between ψ and $\bar{\psi}$ if they share the same set of collapse centres $\{z\}$. It has even been argued that the ontology of collapse models should be just this set of collapse centres [8]. As Bell puts it, the collapse centres *"...are the mathematical counterparts in the theory to real events at definite places and times in the real world. [] A piece of matter then is a galaxy of such events."*[8]. The collapse centres are the part of the theory that exist in ordinary space and time (the wave function exists

[2] provided that we remember to take the complex conjugate of the wave function

is an infinite dimensional Hilbert space). It therefore makes sense to treat the collapse centres as the basis for local beables in the theory.

Further developing this picture it has been shown in reference [9] for a lattice model of wave function collapse that, given a set of collapses generated by some specific wave function, a generic wave function (subject to some constraints such as consistent particle number) will, after undergoing this same set of collapses, tend towards the original wave function. This implies that the information contained in the wave function at any given point is also contained in the collapse data from a sufficient period of history. We might therefore be able to abandon the wave function altogether [10]. Of course these considerations will apply to both the forwards and backwards in time wave functions.

In order to demonstrate that this picture is indeed time symmetric it must be the case that the probability rule (3) is also true for $\bar{\psi}$, i.e.

$$\bar{\mathbb{P}}_t\left(z\right) = \frac{\int dx_1 \cdots dx_N |j(z - x_i)\bar{\psi}_t|^2}{\int dx_1 \cdots dx_N |\bar{\psi}_t|^2}. \tag{5}$$

This has been shown in reference [11] for two different models of wave function collapse. (Specifically these models are the lattice model of collapse of references [9, 12, 13] and the QMUPL model of reference [14] in the localised particle limit.) The demonstration involves a statistical test. A forwards-in-time wave function is used to generate a set of collapses according to rules of the type outlined in section 2.1. A test wave function is then used as a backwards-in-time wave function and evolved backwards using (4) with the same set of collapses in the reverse order. The likelihood that these collapses could have instead been generated by the backwards-in-time probability rule (5) is then tested by calculating a statistic designed to indicate this (along the lines of the Kolmogorov-Smirnov test).

It is found that to within statistical error the probability rule (5) (which is essentially the Born rule) works perfectly well backwards in time. What this means is that if you were given the set of collapses $\{z\}$ along with the two wave functions ψ and $\bar{\psi}$ (of course not being told which was which), there would be no way for you to determine by analysing all this information, which was the forward direction of time and which was the backward direction of time. The dynamical collapse rules work identically in both directions. The collapse model is time reversal symmetric.

2.3. The effect of an initial condition

As we have seen in section 1.1, this conclusion cannot be universally valid. We can envisage examples such as the beam splitter experiment of figure 1 where the Born rule is not expected to hold backwards in time. However, we have argued that this does not necessarily imply an inherent time asymmetry in the collapse of the wave function. A possible explanation is that it is the effect of a low entropy initial condition. Indeed neither of the examples in reference [11] involved large numbers of particles with highly ordered initial conditions. Moreover, the precise details of the initial wave function are explicitly washed away after a sufficient period of time in the lattice model of reference [9]. It is conceivable that in these cases the effect of the initial condition is innocuous or that the system is in a state of equilibrium. The idea that asymmetries result from boundary conditions at least allows for an explanation for the observed time symmetry in these examples which would be harder to do otherwise.

We can propose a modification of the backwards-in-time probability rule valid in the presence of a fixed initial condition on the state $\bar{\psi}_0$,

$$\bar{\mathbb{P}}_t\left(z\right) \rightarrow \bar{\mathbb{P}}_t\left(z|\bar{\psi}_0\right) = \frac{\bar{\mathbb{P}}_t\left(z \cap \bar{\psi}_0\right)}{\bar{\mathbb{P}}_t\left(\bar{\psi}_0\right)}. \tag{6}$$

Journal of Physics: Conference Series **626** (2015) 012007 doi:10.1088/1742-6596/626/1/012007

Here $\bar{\mathbb{P}}_t(\bar{\psi}_0)$ means the probability that the state $\bar{\psi}$ will take the form $\bar{\psi}_0$ at the start of the Universe as a result of the backwards-in-time collapse dynamics, given that it takes the form $\bar{\psi}_t$ at time t. Typically this would be very complicated to determine based on all possible collapse events happening between time t and the start of the Universe.

Similarly a final time boundary condition on the state ψ_f would modify the forwards-in-time probability rule

$$\mathbb{P}_t(z) \to \mathbb{P}_t(z|\psi_f) = \frac{\mathbb{P}_t(z \cap \psi_f)}{\mathbb{P}_t(\psi_f)}. \tag{7}$$

These modified probability rules potentially account for the time asymmetric behaviour of the Born rule identified in the Introduction in a way which retains the time symmetry of the collapse dynamics. In terms of the beam splitter example, the probability rule (6) would weight the probabilities in favour of the backwards-in-time photon being transmitted rather than reflected. (Of course the photon itself would not be expected to undergo a GRW collapse, the collapse would happen when the photon has become entangled with the displacement of a sufficient number of atoms in one of the devices of the experiment.)

We finally note that collapse models are typically known to result in a gradual increase in energy. If this happens to the future it cannot also happen to the past. But again a simple reason for this behaviour is that energies in the past are constrained by the initial condition.

3. Summary

We have argued that wave function collapse models can be understood in a way in which the dynamical rules are symmetric under time reversal symmetry. This implies that the physical process of the collapse of the wave function is symmetric in time. We have argued that certain examples where the Born rule does not hold in the reverse time direction can be attributed to the effect of an initial condition of the Universe. This is the standard explanation of time asymmetric behaviour resulting from time symmetric laws.

Acknowledgments

This work was supported by a grant from the Templeton World Charity Foundation.

References

[1] Penrose R 2004 *The Road to Reality* (London: Jonathan Cape) section 30.3 pp 819-823
[2] Ghirardi GC, Rimini A and Weber T (1986) Phys. Rev. D **34** 470
[3] Pearle P 1989 Phys. Rev. A **39** 2277
[4] Ghirardi GC, Pearle P and Rimini A 1990 Phys. Rev. A **42** 78
[5] Tumulka R 2006 J. Stat. Phys. **125** 821
[6] Bedingham D 2011 Found. Phys. **41** 686
[7] Bedingham D 2011 J. Phys.: Conf. Ser. **306** 012034
[8] Bell J 2004 *Speakable and Unspeakable in Quantum Mechanics* (Cambridge) chapter 22 pp 201-212
[9] Dowker F and Herbauts I 2005 Found. Phys. Lett. **18** 499
[10] Kent A 1989 Mod. Phys. Lett. A **4** 1839
[11] Bedingham D and Maroney O 2015 Time reversal symmetry and collapse models *Preprint* quant-ph/1502.06830
[12] Dowker F and Henson J 2004 J. Stat. Phys. **115** 1349
[13] Dowker F and Herbauts I 2004 Class. Quant. Grav. **21** 2963
[14] Diósi L 1988 Phys. Lett. A **132** 233

Journal of Physics: Conference Series **626** (2015) 012008 doi:10.1088/1742-6596/626/1/012008

Is spontaneous wave function collapse testable at all?

Lajos Diósi

Wigner Research Centre for Physics, H-1525 Budapest 114, P.O.Box 49, Hungary

E-mail: diosi.lajos@wigner.mta.hu

Abstract. Mainstream literature on spontaneous wave function collapse never reflects on or profits from the formal coincidence and conceptual relationship with standard collapse under time-continuous quantum measurement (monitoring). I propose some easy lessons of standard monitoring theory which would make spontaneous collapse models revise some of their claims. In particular, the objective detection of spontaneous collapse remains impossible as long as the correct identification of what corresponds to the signal in standard monitoring is missing from spontaneous collapse models, the physical detectability of the "signal" is not stated explicitly and, finally, the principles of physical detection are not revealed.

1. Introduction

Spontaneous wave function models [1, 2, 3, 4, 5], reviewed by [6, 7], dynamically violate the superposition principle of quantum mechanics, assuming tiny spontaneous time-continuous collapse of the wave function. For massive degrees of freedom spontaneous collapse gets amplified and will result in classical behaviour in the objective way. The toolkit of standard quantum measurements is no more requested, it is always replaced by spontaneous collapses. Since collapse and classicality only appear at the level of the mathematical formalism, additional considerations are used to identify which mathematical object of the given spontaneous collapse model should represent the emerged classical entities.

I complain about mainstream literature on spontaneous collapse for it ignores the lessons of standard collapse. Lessons of time-continuous measurement (monitoring) theory [8, 9] are obligatory and instructive for spontaneous collapse, even if one would not implement them all.

The present analysis shows that in current spontaneous collapse models the proposed collapse is illusory because it is not testable by objective detection. The sole testable effect is spontaneous decoherence, i.e., the degradation of certain interference terms. Mathematical apparatus of spontaneous collapse models is redundant: the stochastic Schrödinger equation (SSE) is untestable. The master equation (ME), governing the density matrix, does encode all possible objectively testable effects.

2. Continuous measurement and collapse in standard quantum mechanics

Considering the continuous measurement (monitoring) of the position $\hat{q}$ of a quantized particle, I published two Ito stochastic equations in 1988 [10, 11]. A plausible expression yields the measurement outcome (also called signal) q_t:

$$q_t dt = \langle\hat{q}\rangle_t dt + \frac{1}{2\sqrt{D}} dW_t, \qquad (1)$$

where $\langle\hat{q}\rangle_t = \langle\psi_t|\hat{q}|\psi_t\rangle$ is the expectation value of $\hat{q}$ in the current quantum state ψ_t while W_t is the standard Wiener process (cf.: white-noise). The power of the noise is inverse proportional to the parameter D which, as we see below, controls the precision of monitoring. The evolution of the state vector ψ_t under coordinate monitoring is governed by the SSE:

$$d\psi_t = -\frac{i}{\hbar}\hat{H}\psi_t dt - \frac{D}{2\hbar^2}(\hat{q} - \langle\hat{q}\rangle_t)^2\psi_t dt + \frac{\sqrt{D}}{\hbar}(\hat{q} - \langle\hat{q}\rangle_t)\psi_t dW_t. \tag{2}$$

It causes dynamical collapse (localization) of the wave function which is the natural consequence of position measurements. The theory of q-monitoring contains *two* equations: (1) and (2).

If the signal q_t is not accessible or just not recorded, the SSE becomes redundant. It implies the following ME for the ensemble average (density matrix) $\hat{\rho}_t = \langle\psi_t\psi_t^\dagger\rangle_{st}$ of the stochastic ψ_t:

$$\frac{d\hat{\rho}_t}{dt} = -\frac{i}{\hbar}[\hat{H}, \hat{\rho}_t] - \frac{D}{2\hbar^2}[\hat{q}, [\hat{q}, \hat{\rho}_t]]. \tag{3}$$

This ME predicts decoherence only; it does not predict collapse.

It turned out soon that the above equations of continuous measurement (monitoring) do follow from standard quantum theory of collapse if we interpret monitoring in terms of infinite frequent repeated standard measurements. Quantum monitoring has since 1988 become a precise discipline of standard quantum theory, an unavoidable model of modern experiments, see monographs [8, 9].

3. Spontaneous collapse in modified quantum mechanics

When the above dynamical theory of *real* monitoring and *standard* collapse emerged, about the same years, various dynamical models of hypothetical *spontaneous* collapse emerged (see Sec. 1). My proposal [3] mimics as if hidden devices would be monitoring the whole Universe. In the introductory model, hidden monitoring concerns all particle positions. More physical is the gravity-related model, where hidden monitoring concerns the (non-relativistic) mass density $\hat{f}(\mathbf{r})$ at each point $\mathbf{r}$ of the Universe. Full analogy with dynamical theory of real monitoring (Sec. 2) is exploited by [12] in path integral formalism equivalent to Ito's.

However, the rest of the spontaneous collapse models and the mainstream works [6, 7] never reflect on the mathematical coincidence and conceptual relationship with standard quantum monitoring. Hence they do not profit from the *lessons* of standard quantum monitoring. Wearing such blindfold against these lessons might be an innocent stance but it is not.

4. Continuous Spontaneous Localization

For concreteness, we consider the Continuous Spontaneous Localization (CSL) model [4, 5] which is similar to the gravity-related model [3] in that it localizes the mass density $\hat{f}(\mathbf{r})$ instead of particle positions. The definitive equation of CSL coincides with the trivial generalization of the SSE (2) for $\hat{f}(\mathbf{r})$—more precisely: for its coarse graining $\hat{f}^\sigma(\mathbf{r})$—in place of $\hat{q}$ [13]:

$$d\psi_t = -\frac{i}{\hbar}\hat{H}\psi_t dt - \frac{\gamma}{2m_0^2}\int d\mathbf{r}[\hat{f}^\sigma(\mathbf{r}) - \langle\hat{f}^\sigma(\mathbf{r})\rangle_t]^2\psi_t dt + \frac{\sqrt{\gamma}}{m_0}\int d\mathbf{r}[\hat{f}^\sigma(\mathbf{r}) - \langle\hat{f}^\sigma(\mathbf{r})\rangle_t]\psi_t dW_t(\mathbf{r}) \tag{4}$$

where $\sigma = 10^{-5}$cm is the standard width of Gaussian course-graining, γ is a second CSL-specific parameter and m_0 is the atomic mass unit. The noise $W_t(\mathbf{r})$ is a field of spatially uncorrelated standard Wiener processes.

The definitive equation of CSL, when used to evolve the density matrix, yields the following closed linear ME:

$$\frac{d\hat{\rho}_t}{dt} = -\frac{i}{\hbar}[\hat{H}, \hat{\rho}_t] - \frac{\gamma}{2m_0^2}\int d\mathbf{r}[\hat{f}^\sigma(\mathbf{r}), [\hat{f}^\sigma(\mathbf{r}), \hat{\rho}_t]]. \tag{5}$$

There is a tendency to attribute physical status to the noise field $W_t(\mathbf{r})$ but the mainstream judgement admits that "at this stage this is only speculation. [...] one has [yet] to justify the non-Hermitian coupling and the nonlinear character of the collapse equations" [7]. CSL is a little perplexed by the missing interpretation of the noise.

Four helpful lessons of standard monitoring, relevant for CSL, follow.

4.1. Lesson I: Noise is measurement signal noise

Standard quantum mechanics offers the following unique interpretation for the CSL noise. CSL corresponds to standard monitoring (1-3)—by hidden devices this time—of the mass density $\hat{f}^\sigma(\mathbf{r})$ at all locations $\mathbf{r}$. The signal equation (1) would then read

$$\mathcal{S}(\mathbf{r},t) = \langle\hat{f}^\sigma(\mathbf{r})\rangle_t + \frac{m_0}{2\sqrt{\gamma}}w_t(\mathbf{r}), \tag{6}$$

were the Ito differentials $dW_t(\mathbf{r})$ have formally been replaced by $w_t(\mathbf{r})dt$ and $w_t(\mathbf{r})$ are independent standard white-noises for all $\mathbf{r}$. The natural interpretation of CSL noise $W_t(\mathbf{r})$ is obvious: it is signal noise where the signal $\mathcal{S}(\mathbf{r},t)$ is the classical outcome of monitoring $\hat{f}^\sigma(\mathbf{r})$. However, CSL turns blind eyes toward $\mathcal{S}(\mathbf{r},t)$ and its equation (6).

4.2. Lesson II: Signal is the only tangible variable

It is correctly felt in the literature [6, 7] that a complete CSL model should specify what unique classical configurations the basic SSE (4) is to describe. The preferred choice [6] is that it is the quantum average of (σ-smeared) mass density:

$$\mathcal{M}(\mathbf{r},t) = \langle\hat{f}^\sigma(\mathbf{r})\rangle_t = \langle\psi_t|\hat{f}^\sigma(\mathbf{r})|\psi_t\rangle. \tag{7}$$

It is thought, in particular, that this mean-field "is accessible at the macrolevel" and "behaves in a classical way" [6]. Unfortunately, this is not really so.

If we compare $\mathcal{M}(\mathbf{r},t)$ to the signal (6), we observe that the signal contains a noise term as well:

$$\mathcal{S}(\mathbf{r},t) = \mathcal{M}(\mathbf{r},t) + \frac{m_0}{2\sqrt{\gamma}}w_t. \tag{8}$$

The role of the noise term is crucial if we desire that the mass distribution in question be common sense classical field. I proposed the term *tangible* for such variables (fields) because they can be coupled at will to other fields, including that they can be used at will to control any feed-back on the quantum system itself. This is why I talked about "Free Will Test" (FWT) of tangibility [14]. Since the signal $\mathcal{S}(\mathbf{r},t)$ is nothing else just a sequence of standard quantum measurement outcomes, it passes the FWT. Let us, for instance, modify the Hamiltonian by a simple feed-back term:

$$\hat{H} + g\int d\mathbf{r}\hat{f}^\sigma(\mathbf{r})\mathcal{S}(\mathbf{r},t). \tag{9}$$

Substituting (6) for $\mathcal{S}(\mathbf{r},t)$ and the above Hamiltonian for $\hat{H}$ in the SSE (4), we get a modified SSE such that for $\hat{\rho}_t$ the ME (5) does survive with modified Hamiltonian and modified decoherence coefficient, resp.:

$$\hat{H} \Rightarrow \hat{H} + \tfrac{1}{2}g\int d\mathbf{r}[\hat{f}^\sigma(\mathbf{r})]^2, \quad \frac{\gamma}{m_0^2} \Rightarrow \frac{\gamma}{m_0^2} + \frac{g^2}{4\hbar^2}\frac{m_0^2}{\gamma}. \tag{10}$$

(To reconstruct the derivation, works [12],[15] or [9] may be studied.) If, however, we complete the SSE (4) by a feed-back controlled by $\mathcal{M}(\mathbf{r},t)$ instead of $\mathcal{S}(\mathbf{r},t)$:

$$\hat{H} + g\int d\mathbf{r}\hat{f}^\sigma(\mathbf{r})\mathcal{M}(\mathbf{r},t), \tag{11}$$

Journal of Physics: Conference Series **626** (2015) 012008 doi:10.1088/1742-6596/626/1/012008

then the modified SSE will no more allow for any closed linear evolution of $\hat{\rho}_t$. Any feed-back control variable, different from the signal (or functional of it) will jeopardize the autonomous linear equation for $\hat{\rho}_t$ whose loss means loss of consitency [16, 17, 18, 19]. That makes the feed-back (11) illegitimate, hence mean-field mass density $\mathcal{M}(\mathbf{r}, t)$ is not tangible, it does not behave in a "classical way".

Of course the mean-field $\mathcal{M}(\mathbf{r}, t)$ will approximate the signal $\mathcal{S}(\mathbf{r}, t)$ if the noise on the right-hand-side of (6) is averaged out. Therefore $\mathcal{M}(\mathbf{r}, t)$ might approximate the predicted classical configuration if we suitably impose a time-average on it.

4.3. Lesson III: Bell chooses tangible variables

The GRW jump model of spontaneous collapse [1], whose diffusive mass-proportional limit is CSL, could also be identified as position monitoring of constituents by randomly fired standard (though hidden) von Neumann detectors [20]. When in 1987 John Bell casts the GRW model into its ultimate form, he also asks for the "mathematical counterparts in the theory to real events at definite places and times in the real world" [2]. His choice is those space-time points, later called flashes (cf., e.g. [21] and references therein), where GRW jumps are being centred. These flashes correspond exactly what the said (hidden) von Neumann detectors would record as measurement outcomes, hence flashes are perfect classical (tangible) entities, they pass the FWT. Although Bell does not mention the resemblance of GRW to standard monitoring (by hidden detectors) his intuition works perfectly. Subsequent works on CSL ignore the fact that the signal (6) would be nothing else than the diffusive limit of flashes of a GRW-like jump model (where random jumps localize on the field $\hat{f}^\sigma(\mathbf{r})$ instead of positions). The contrary is believed: mean-field matter density (7) "must be taken because it [CSL] does not work with flashes" [22]. I think CSL does work with flashes which are just the signal (6).

4.4. Lesson IV: SSE is empirically redundant

In standard theory of monitoring and collapse (Sec. 2) it is crucial for the empirical testability of the state vector ψ_t under monitoring that the signal q_t be accessible and recorded. If it is not accessible for some technical reasons, or it is just not recorded, then the behaviour of the state vector ψ_t is irrelevant because it is not testable/tested empirically. All possible testable predictions are encoded in the average state, i.e., in the density matrix $\hat{\rho}_t$. In this case the SSE (2) becomes redundant, the ME (3) yields all testable/tested effects which consist of decoherence, as we already said.

The above lesson from standard monitoring, if applied to CSL, tells us the following. Since CSL does not interpret the signal at all, it remains empirically unaccessible, it can of course not be recorded either. Therefore the CSL stochastic wave function ψ_t is empirically irrelevant: all testable predictions of CSL are encoded in the density matrix $\hat{\rho}_t$ and its ME (5). These testable predictions consist of decoherence. Spontaneous collapse is not testable at all, the SSE (4) is physically redundant.

5. Closing remarks

It is obvious that my criticising the choice of mean density $\mathcal{M}(\mathbf{r}, t)$ for classical predictions is irrelevant as long as its detection is meant by *subjective perception*. That $\mathcal{M}(\mathbf{r}, t)$ is not tangible (fails FWT) becomes relevant only when the goal is *objective detection*, typically by *coupling* $\mathcal{M}(\mathbf{r}, t)$ to a device or elsewhere.

It is also obvious that spontaneous collapse models differ from standard quantum monitoring, they may well differ in much more than "hiddennes" of the fictitious monitoring devices, especially because we assume a lot of freedom in how we wish to interpret spontaneous collapse. My work warns that such freedom may not be that much as believed.

The statements that the SSE is redundant, the ME is sufficient, spontaneous decoherence is testable, spontaneous collapse is not, appeared already in 1989 [3]. These statements hold for all models of spontaneous collapse. Detailed arguments were given for CSL, just for concreteness. Independently of the power (even of validity) of my arguments, let me formulate the central claim in the practical way. Apparently, any *objective* detection proposed so far turns out to test decoherence effects fully calculable via the corresponding master equation. Therefore, apparently, the promise of spontaneous collapse theories to objectify collapse does not fulfils yet.

Acknowledgments
Support by the Hungarian Scientific Research Fund under Grant No. 103917 and by the EU COST Actions MP1006, MP1209 are acknowledged.

References

[1] Ghirardi G C, Rimini A and Weber T 1986 *Phys. Rev.* D **34** 470
[2] Bell J S 1987 Are there quantum jumps? In *Schrödinger Centenary of a polymath* (Cambridge: Cambridge University Press)
[3] Diósi L 1989 *Phys. Rev.* A **40** 1165
[4] Ghirardi G C, Pearle P and Rimini A 1990 *Phys. Rev.* A **42** 78
[5] Ghirardi G C, Grassi R and Benatti F 1995 *Found. Phys.* **25** 5
[6] Bassi A and Ghirardi G C 2003 *Phys. Rep.* **379** 257
[7] Bassi A, Lochan K, Satin S, Singh T P and Ulbricht H 2013 *Rev. Mod. Phys.* **85** 471
[8] Carmichael H 1993 *An open system approach to quantum optics* (Berlin: Springer-Verlag)
[9] Wiseman H M and Milburn G J 2010 *Quantum measurement and control* (Cambridge: Cambridge University Press)
[10] Diósi L 1988 *Phys. Lett.* **129A** 419
[11] Diósi L 1988 *Phys. Lett.* **132A** 233
[12] Diósi L 1990 *Phys. Rev.* A **42** 5086
[13] My notation $\hat{f}^\sigma(\mathbf{r})$ and σ correspond to $M(\mathbf{r})$ and r_C, respectively, in [6, 7].
[14] Diósi L 2012 *J. Phys.: Conf. Ser.* **361** 012028
[15] Diósi L and Gisin N 1995 *Phys. Rev. Lett.* **72** 4053
[16] Diósi L 1988 *J. Phys.* A **21** 2885
[17] Gisin N 1989 *Hel. Phys. Acta* **62** 363
[18] Gisin N 1990 *Phys. Lett.* **143A** 1
[19] Gisin N and Rigo M 1995 *Journ. Phys.* A **28** 7375
[20] Diósi L 2000 *Lect. Notes Phys.* **538** 243
[21] Allori V, Goldstein S, Tumulka R and Zanghi N 2008 *Brit. J. Philos. Sci.* **59** 353
[22] Feldmann W and Tumulka R 2012 *J. Phys. A: Math. Theor.* **45** 065304

Possible role of gravity in collapse of the wave-function: a brief survey of some ideas

Tejinder P Singh

Tata Institute of Fundamental Research, Homi Bhabha Road, Mumbai 400005, India

E-mail: tpsingh@tifr.res.in

Abstract. This article is a brief survey of some approaches to implementing the suggestion that collapse of the wave function is mediated by gravity. These approaches include: a possible connection between the problem of time and problem of quantum measurement, decoherence models based on space-time uncertainty, the Schrödinger-Newton equation, attempts to introduce gravity into collapse models such as CSL, ideas based on the black hole - elementary particle complementarity, and the possible role of a complex space-time metric.

1. Introduction

No experiment to date disagrees with quantum theory. However, it is often not appreciated that the theory has not been tested in all parts of the parameter space (as defined by the number of degrees of freedom in the system being tested). The largest objects for which the theory has been tested have a mass of about 10^5 amu, and the smallest objects for which classical mechanics is known to hold have a mass of about 10^{-6} grams [about 10^{18} amu]. This leaves a mesoscopic desert of some thirteen orders of magnitude over which we do not know whether the centre of mass motion of an object obeys quantum mechanics, or classical mechanics, or some new mechanics, to which the classical and quantum are approximations. Nor do we know for sure as to exactly where in this untested parameter space the quantum-classical transition takes place, and what the mechanism for this transition is [1–3].

[This is perhaps a suitable place to remark that while macroscopic quantum behaviour is indeed observed in superconductors, superfluids, and Bose-Einstein condensates, this by itself does not preclude the need to test quantum theory in the desert. The macroscopic quantum states in these systems are collective internal states akin to one-particle states, and what we are referring to above is the centre of mass motion of the object as a whole. One should also note that such macroscopic states put very weak bounds on collapse models such as Continuous Spontaneous Localization (CSL) and do not rule out these models [4]].

We could ask if we have any reason to suspect that quantum theory might break down in the desert. The quantum measurement problem which has been with us for nearly a century now, does give rise to the belief in a possible breakdown. It might well be that the problem will eventually go away in a suitable re-interpretation of the theory (many-worlds, consistent histories, ...) or in a suitable reformulation (Bohmian mechanics, ...). On the other hand, there remains open the possibility that the theory has to be modified to solve the measurement problem. It is precisely in the untested mesoscopic desert where the modified theories make experimental predictions different from quantum theory, making it all the more relevant that this regime be experimentally tested. If no departures from the standard theory are found, we can be sure that the answer to the foundational problems of quantum theory will lie in a re-interpretation [3].

Mention must also be made of the curious status of the Born probability rule, a rule which has to be invoked in order to explain the random probabilistic outcomes of measurements. The status is curious because even though the Schrödinger equation is perfectly deterministic, and the initial conditions unambiguously specified by prescribing the initial wave-function, the interaction of the quantum system with a classical apparatus produces unpredictable outcomes. The only meaningful way to remedy this unphysical situation is to introduce randomness in the initial conditions (Bohmian mechanics) or in the dynamics (collapse models) or to reinterpret measurements in such a way that collapse is only apparent, but not for real. Apart from the need to understand the Born rule, another shortcoming of the standard (Copenhagen) explanation is the vaguely defined classical apparatus - quantum theory has to depend on its own limit for properly understanding outcomes of experiments. This indeed is not satisfactory [5, 6].

In our opinion, the most significant incompleteness in our present understanding of quantum theory is the problem of time, even though this problem gets little attention in the literature. The time which describes evolution in quantum theory is part of a classical space-time whose geometry is produced by classical bodies, which in turn are a limiting case of quantum theory. Once again, the theory has to depend on its own limit for its formulation, and there ought to exist a more complete equivalent formulation which does not refer to classical time. As we briefly note in the next section, maybe when we have such a reformulation we will also understand how to solve the measurement problem [7].

In summary, we may list the four frontiers along which we need to do better with quantum theory: (i) experimental tests of the theory in the yet unexplored part of the parameter space (ii) obtaining a time-independent reformulation of the theory (iii) finding a universally acceptable resolution of the measurement problem (iv) understanding the origin of the Born probability rule in a deterministic theory.

Models for quantum theory without classical time have not received much attention, and are briefly discussed in the next section. Models which seek to modify quantum theory so as to solve the measurement problem, while taking classical time as given, are themselves severely constrained on theoretical and experimental grounds. The modified theory should be stochastic nonlinear, and must include anti-Hermitean modifications to the Hamiltonian [8–13]. The most successful and well-studied such phenomenological model is Continuous Spontaneous Localization (CSL) [14, 15]. CSL bridges the microscopic quantum theory and macroscopic classical mechanics, acting as an underlying generalisation to which the former two are approximations, and a generalisation which solves the measurement problem (collapse is dynamical) and explains the Born rule (strictly speaking, the theory is constructed in such a way that the Born rule follows, and a fundamental explanation of the Born rule still alludes us).

Since CSL is a phenomenological theory, eventually one would like to look for a deeper explanation for the CSL mechanism, if it proves to be correct. Possible underlying physics could be that the stochastic field which drives the CSL effect is cosmological in origin [maybe having something to do with the cosmic microwave background] or it is a relic thermodynamic effect, as in Adler's theory of Trace Dynamics [16–18]. Alternatively, the origin of the CSL effect could lie in the universal interaction of gravitation. There are intrinsic quantum fluctuations in the gravitational field, because the objects which produce it are not exactly classical but possess an intrinsic uncertainty, and these fluctuations might feed back to decohere the quantum wave function and possible induce collapse. Such an idea forms the basis for investigating the possible role of gravity in collapse of the wave function., and a possible connection with CSL. Alternatively, maybe it is not the quantum fluctuations of gravity, but some other feature of gravity, not yet appreciated, which might account for collapse.

There is unfortunately no concrete theory to this effect as yet, although there have been a series of investigations. In the subsequent sections of the present modest review, we briefly survey some of the gravity related ideas that have been proposed in the context of quantum collapse. These include: a possible connection between the problem of time and problem of quantum measurement, decoherence models based on space-time uncertainty, the Scrödinger-Newton equation and stochastic gravity, attempts to introduce gravity into collapse models such as CSL, ideas based on the black hole - elementary particle complementarity, and the possible role of a complex space-time metric. Earlier reviews of gravity based models include [19–21].

2. The problem of time in quantum theory

The time that is used to describe evolution in quantum theory is classical time. It is a part of classical space-time on which there overlies a space-time geometry which is determined by classical matter fields, in accordance with the laws of general relativity. The Einstein hole argument implies that in order to give operational meaning to points on the space-time manifold, one needs to assign a metric at each space-time point. It evidently follows that classical matter fields are essential for this operational definition, and since such fields are a limiting case of quantum fields, we are forced to conclude that in the definition of the time that quantum theory employs, the theory depends on its own limit (Fig. 1).

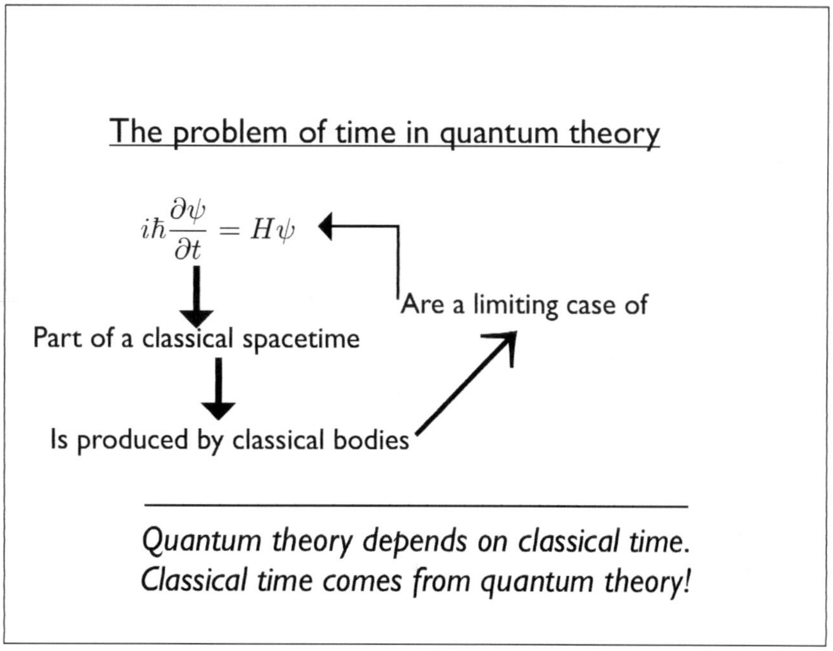

Figure 1. The circularity of time in quantum theory. From [7]

A more complete formulation would be one where such dependence could be avoided, and we call such a formulation 'quantum theory without classical time' [22]. For brevity, we venture to name this as pure quantum theory. It is hoped that once such a pure formulation has been developed, standard quantum theory, with its classical time, will emerge as an approximation to the more general formulation.

Elsewhere we have suggested that pure quantum theory should be formulated by starting from a noncommutative generalisation of special relativity described by the line-element [23]

$$ds^2 = Trd\hat{s}^2 \equiv Tr[d\hat{t}^2 - d\hat{x}^2 - d\hat{y}^2 - d\hat{z}^2] \tag{1}$$

Space-time coordinates and the matter degrees of freedom that reside on them acquire the status of matrices (equivalently operators). The statistical equilibrium thermodynamics of this noncommutative theory, in parallel with Trace Dynamics, is the sought for pure quantum theory [24]. In the transition from here to a classical world, the role of gravity in causing collapse of the wave-function is implicated. This is expected to come about because CSL type Brownian fluctuations around equilibrium cause collapse of the wave function describing the quantum space-time in pure quantum theory, as well as of the wave function describing the matter degrees of freedom.

A detailed discussion of this ongoing program can be found in [7], and an overview in the context of collapse theories is given in [3].

3. Spacetime uncertainty and decoherence

In the absence of a concrete formulation for pure quantum theory, it is useful to build heuristic models by asking how quantum fluctuations in space-time geometry can affect the propagation of the wave-function obeying the Schrödinger equation. In order to make progress, one is compelled to start from some sort of description of space-time uncertainty, model it by a classical stochastic potential, and ask how the Schrödinger propagation is affected by such a stochastic potential. Decoherence can be expected because of the stochasticity, and it maybe attributed to gravity. One of the earliest such studies was by Karolyhazy, as early as 1966 [25] and then followed by Karolyhazy and collaborators in several papers [26–35]. It was first shown that because of the intrinsic quantum nature of objects which produce gravity, there is a minimal length uncertainty Δs in a geodesic of length Δs, which is given by the following universal relation:

$$\Delta s^3 \sim l_p^2 s \tag{2}$$

where l_p is Planck length. One may imagine that Minkowski space-time has a perturbing stochastic gravitational field $\gamma(\mathbf{x}, t)$ whose two point correlation is such that it is consistent with the above universal uncertainty. For a matter distribution described by the mass density operator $f(\mathbf{x}, t)$ one then writes the stochastic Schrödinger equation:

$$i\hbar\frac{\partial\psi(\mathbf{x},t)}{\partial t} = \left(H + \int d^3x' f(\mathbf{x}')\gamma(\mathbf{x}',t)\right)\psi(\mathbf{x},t) \tag{3}$$

The fluctuating gravity causes distant parts of the wave-function to decohere, and by considering the evolution of a spherical object of mass m and size R one obtains the following criteria for deciding whether the object is microscopic (quantum dominates over gravity) or macroscopic (gravity dominates over quantum, and the object is classical), and expressions for the critical separation length a_c beyond which decoherence occurs:

$$a_c \gg R \quad \rightarrow \quad \frac{\hbar^2}{G} \gg m^3 R \qquad : micro - region, \tag{4}$$

$$a_c \approx R \quad \rightarrow \quad \frac{\hbar^2}{G} \approx m^3 R \qquad : transition - region, \tag{5}$$

$$a_c \ll R \quad \rightarrow \quad \frac{\hbar^2}{G} \ll m^3 R \qquad : macro - region. \tag{6}$$

There arise two limiting cases: one for $R \gg a_c$ and another for $R \ll a_c$. For a micro-object of linear size $R \ll a_c$, the expression for the coherence length reduces to that for an elementary particle of mass m and the coherence length a_c over which decoherence effects become relevant is given by

$$a_c \approx \frac{\hbar^2}{G}\frac{1}{m^3} = \left(\frac{L}{l_p}\right)^2 L; \qquad L = \frac{\hbar}{mc}. \tag{7}$$

For $R \gg a_c$, the critical length can be expressed as,

$$a_c \approx \left(\frac{\hbar^2}{G}\right)^{1/3} \frac{R^{2/3}}{m} = \left(\frac{R}{l_p}\right)^{2/3} L. \tag{8}$$

In a different treatment, Diósi [36–39] considered a somewhat different starting point for understanding the intrinsic space-time uncertainty, by asking how much the uncertainty in a measured Newtonian gravitational field would be, if the intrinsic quantum uncertainty of the measuring probe is taken into account. It is shown that if a gravitational field $\mathbf{g}$ is probed over a volume V for a time interval T, then the uncertainty $\delta\tilde{\mathbf{g}}$ in the averaged field $\tilde{\mathbf{g}}$ is given by

$$(\delta\tilde{\mathbf{g}})^2 \geq G\hbar/VT\,. \tag{9}$$

Once again, this is modelled by a classical stochastic perturbing gravitational potential ϕ whose two-point correlation is consistent with the above uncertainty bound, and the impact of such a potential on the stochastic Schrödinger equation is given by

$$i\hbar\dot{\psi}(t) = \left(H + \int \phi(\mathbf{r},t) f(\mathbf{r}) d^3r\right)\psi(t) \tag{10}$$

The conditions for decoherence and the micro-macro divide are the same as in the Karolyhazy model; however the actual expressions for decoherence length are different:

$$\begin{aligned} a_c &\sim (\hbar^2/Gm^3)^{1/4} R^{3/4}, \quad \text{if} \quad Rm^3 \gg \hbar^2/G, \\ &\sim (\hbar^2/Gm^3)^{1/2} R^{1/2}, \quad \text{if} \quad Rm^3 \ll \hbar^2/G\,. \end{aligned} \tag{11}$$

The reasons for this difference were first studied by Diósi and Lukacs [39], and have also been recently studied in [40] and are reported by Bera et al. in the present volume.

These models based on modelling of space-time uncertainty can account for decoherence, but do not actually produce collapse to one of the many possible outcomes. This is because the modified Schrödinger equation lacks an anti-Hermitean component. This lacking plagues essentially all treatments of collapse models: the anti-Hermitean component has to be put in by hand, for we know what outcome we desire. The fundamental origin for such a component remains to be understood.

4. Schrödinger-Newton equation and stochastic gravity

While on the one hand one takes into account the impact on the Schrödinger equation of space-time uncertainty produced by other objects, one should also consider how the self-gravity of a quantum object influences its Schrödinger evolution. This self-gravity itself has an inbuilt quantum uncertainty and even though we do not really know at present how to describe the self-gravity of a quantum object, we can attempt to model it. One possible approach is the so-called Schrödinger-Newton [SN] equation. It is proposed that that the self-gravitational potential Φ produced by a quantum source satisfies a semiclassical Poisson equation

$$\nabla^2\Phi = 4\pi G m|\psi|^2 \tag{12}$$

whose solution is fed into the potential dependent Schrödinger equation

$$i\hbar\frac{\partial\psi}{\partial t} = -\frac{\hbar^2}{2m}\nabla^2\psi + m\Phi\psi \tag{13}$$

so as to obtain the Schrödinger-Newton equation [41–44]

$$i\hbar\frac{\partial\psi(\mathbf{r},t)}{\partial t} = -\frac{\hbar^2}{2m}\nabla^2\psi - Gm^2\int\frac{|\psi(\mathbf{r}',t)|^2}{|\mathbf{r}-\mathbf{r}'|}d\mathbf{r}'\psi(\mathbf{r},t) \tag{14}$$

This equation has been discussed extensively in the literature [45–55] and also in the present volume by Giulini. The equation is deterministic nonlinear, but not stochastic. This makes it different in character from the Karolyhazy and Diósi equation, where by virtue of the stochastic potential the models predict decoherence at the level of the master equation. The SN equation does not predict decoherence but is instead suggested as an equation whose stationary solutions are ones to which the system will proceed upon decoherence / collapse, once a gravity-based decoherence / collapse mechanism can be incorporated in this system. What has been shown is that there is a gravitationally induced inhibition of dispersion of an expanding wave-packet, at the critical length $a_c \sim \hbar^2/Gm^3$.

The reason decoherence is not observed is easy to see: one is only dealing with the mean potential in the semiclassical Poisson equation, whereas the Karolyhazy and Diósi models incorporate stochastic fluctuations by way of two point correlation of the stochastic potential, which is patterned after the space-time geodesic uncertainty. One needs to bring in consideration of the stochastic fluctuations, apart from the mean, into the SN equation. One possible way to do this is to employ the theory of stochastic gravity [56], which takes into account corrections to the semiclassical Einstein equations

$$R_{\mu\nu} - \frac{1}{2}g_{\mu\nu}R = \frac{8\pi G}{c^4}\langle\Psi|\hat{T}_{\mu\nu}|\Psi\rangle \tag{15}$$

by considering the role of the two-point fluctuations of the energy-momentum tensor [going from semiclassical gravity to the Einstein-Langevin equation].

In order to take into account the effect of stress-tensor fluctuations on gravity, a classical stochastic perturbation is introduced in the metric:

$$g_{\mu\nu} \to g_{\mu\nu} + h_{\mu\nu} \tag{16}$$

and a Gaussian stochastic tensor field $\xi_{\mu\nu}$ is defined though its two-point correlation as follows

$$\langle\xi_{\mu\nu}(x)\xi_{\alpha\beta}(y)\rangle_c = N_{\alpha\beta\mu\nu} = \langle\{\hat{t}_{\alpha\beta}(x)\hat{t}_{\mu\nu}(y)\}\rangle \tag{17}$$

where

$$\hat{t}_{\alpha\beta} = \hat{T}_{\alpha\beta} - \langle\hat{T}_{\alpha\beta}\rangle \tag{18}$$

The Einstein-Langevin equation for the perturbed metric is given by

$$R_{\mu\nu} - \frac{1}{2}g_{\mu\nu}R = \frac{8\pi G}{c^4}\left(\langle\Psi|\hat{T}_{\mu\nu}|\Psi\rangle + 2\xi_{\mu\nu}\right) \tag{19}$$

The Newtonian approximation to this equation is given by the modified Poisson equation

$$\nabla^2\Phi = \frac{4\pi G}{c^2}\left(\langle\Psi|\hat{T}_{00}|\Psi\rangle + \xi_{00}\right) \tag{20}$$

It is hoped that since now the gravitational potential Φ is stochastic, its incorporation in the SN equation can induce decoherence. This is at present under investigation [57]. A stochastic modification of the Schrödinger equation has also been considered by [58].

Journal of Physics: Conference Series **626** (2015) 012009 doi:10.1088/1742-6596/626/1/012009

5. Collapse models and gravity

A collapse model modifies the Schrödinger equation to include a stochastic element, which can dynamically explain the random outcomes of quantum measurements obeying the Born probability rule. The most successful example of a collapse model is [mass-proportional] CSL, described by the equation

$$d\psi_t = \left[-\frac{i}{\hbar}H dt + \frac{\sqrt{\gamma}}{m_0}\int d\mathbf{x}(M(\mathbf{x}) - \langle M(\mathbf{x})\rangle_t)dW_t(\mathbf{x}) \right.$$
$$\left. - \frac{\gamma}{2m_0^2}\int d\mathbf{x}\, d\mathbf{y}\, g(\mathbf{x}-\mathbf{y})(M(\mathbf{x}) - \langle M(\mathbf{x})\rangle_t)(M(\mathbf{y}) - \langle M(\mathbf{y})\rangle_t)dt\right]\psi_t \tag{21}$$

H is the standard quantum Hamiltonian of the system and the other two terms induce the collapse of the wave function in space. The mass m_0 is a reference mass, which is usually taken to be nucleon mass. The parameter γ is a positive coupling constant which sets the strength of the collapse process, while $M(\mathbf{x})$ is a smeared *mass density* operator:

$$M(\mathbf{x}) = \sum m_j \delta(\mathbf{x} - \mathbf{r_j}), \tag{22}$$

The smearing function $g(\mathbf{x})$ which encodes the noise correlation is taken equal to

$$g(\mathbf{x}) \;=\; \frac{1}{\left(\sqrt{2\pi}r_c\right)^3}\, e^{-\mathbf{x}^2/2r_C^2}, \tag{23}$$

r_c is the second new phenomenological constant of the model. $W_t(\mathbf{x})$ is an ensemble of independent Wiener processes, one for each point in space. Collapse models are discussed in detail in this volume by Bassi.

It would be significant progress if the origin of the CSL mechanism was shown to lie in some universal phenomenon such as gravity. This would imply a gravitational origin for the two new fundamental constants. However, to date such an understanding has not been achieved. In order to incorporate gravity into collapse models Diósi [38] proposed to replace the CSL noise correlation by the Diósi white noise correlation of his model discussed above, and a CSL-like collapse equation

$$d\psi_t = \left[-\frac{i}{\hbar}H dt + \int d\mathbf{x}(M(\mathbf{x}) - \langle M(\mathbf{x})\rangle_t)dW_t(\mathbf{x}) \right.$$
$$\left. - \int d\mathbf{x}\, d\mathbf{y}\, g(\mathbf{x}-\mathbf{y})(M(\mathbf{x}) - \langle M(\mathbf{x})\rangle_t)(M(\mathbf{y}) - \langle M(\mathbf{y})\rangle_t)dt\right]\psi_t \tag{24}$$

The noise correlation is now of gravitational origin, and is given by

$$g(\mathbf{x}-\mathbf{y}) = \frac{G}{\hbar}\frac{1}{|\mathbf{x}-\mathbf{y}|} \tag{25}$$

The corresponding master equation has divergences; to regularise these a cut-off was proposed, at the natural choice of a nucleon scale [38]. Unfortunately a cut-off at this scale produces excessive stochastic heating inconsistent with observations. It was then proposed to raise this cut-off to the much higher value r_c which is the new length scale in the CSL model [59]. This solves the heating problem but it is difficult to physically justify the inclusion of r_c in the Diósi collapse model, one of whose motivations was to achieve a parameter free description of collapse. In an interesting recent discussion [60] Bahrami et al. compare the Diósi model with CSL in some detail, and also consider the possibility of avoiding overheating by introducing dissipation in the dynamics, instead of by going to a high cut-off.

6. The black-hole elementary particle complementarity

Considering that black holes and a large class of elementary particles are both fundamental entities characterized by mass, charge and intrinsic angular momentum, it is suggestive, almost cliched, to consider a possible correspondence between the two. While the task of searching for such a correspondence is made difficult by the fact that black holes are classical whereas elementary particles are quantum [and an $\hbar$ of course enters in the description of a quantised spin] here we discuss some ongoing work on developing such a correspondence. This work also opens up a possible way in which one might understand involvement of gravity in collapse of the wave-function.

We begin by returning to Eqn. (8), which is Karolyhazy's expression for the coherence length of an extended object. It can be shown that this relation continues to hold for a micro-object as well, and hence it can be treated as a general relation. It can be written in the following useful form

$$\frac{a_c}{R} \approx \left(\frac{L}{R_S}\right)^{2/3} \left(\frac{R_S}{R}\right)^{1/3} \tag{26}$$

where $R_S \sim Gm/c^2$ is of the order of the Schwarzschild radius of the object. If we specialise to the case $R = R_S$ we would be dealing only with black holes [assuming they can be described classically] and this relation becomes

$$\frac{a_c}{R} = \left(\frac{L}{R_S}\right)^{2/3} \tag{27}$$

In the macroscopic limit $a_c \ll R$ the Schwarzschild radius exceeds the Compton wavelength, and this also implies that $m \gg m_{pl}$, $R_S \gg l_p$ and $L \ll l_p$. The macroscopic black hole limit is obtained for large masses such that Schwarzschild radius is much larger than Planck length, and Compton wavelength is much less than Planck length. Conversely the quantum microscopic limit $a_c \gg R$ is obtained if Compton wavelength exceeds Schwarzschild radius, i.e. $m \ll m_{pl}$, $R_S \ll l_p$ and $L \gg l_p$. The quantum limit is obtained for small masses such that Schwarzschild radius is much less than Planck length, and Compton wavelength is much larger than Planck length: the concept of black hole is not defined in this limit, and we are dealing with a quantum object characterized by its Compton wavelength. It is reassuring that Karolyhazy's analysis for the role of gravity in decoherence produces expected results for the quantum-classical transition from classical black holes to quantum objects.

The macroscopic limit is described accurately by general relativity, accompanied by classical equations of motion for the particle. The microscopic limit is described by the Dirac equation for a relativistic particle of mass m. (For now, it is easier to consider the relativistic case, as opposed to the non-relativistic Schrödinger equation). In both these cases [general relativity and the Dirac equation] the mass m acts as a source term. One could enquire as to how the particle 'decides' whether it should obey Einstein equations or the Dirac equation? There is no information either in Einstein equations or in the Dirac equation which would provide an answer to this question, because neither of the equations involve Planck mass. It is only through experience that we know that micro-objects should be described by the Dirac equation and macro-objects [such as an astrophysical rotating black hole] should be described by the Kerr solution of Einstein equations. And according to the argument presented above, for gravity induced quantum-classical transition, the behaviour of an object is quantum or classical, depending on whether or not the associated Compton wavelength exceeds the Schwarzschild radius. It would be significant if there was a system of equations describing the dynamics of a mass m, irrespective of whether or not this mass satisfies $m \ll m_{pl}$ or $m \gg m_{pl}$, and to which system the Einstein equations and the Dirac equation were approximations. Such a system of equations would significantly help in understanding the transition region $m \sim m_{pl}$ as well as provide a more fundamental understanding of gravity induced quantum collapse.

Journal of Physics: Conference Series **626** (2015) 012009 doi:10.1088/1742-6596/626/1/012009

In order to make some progress towards describing Einstein equations and Dirac equations as part of a common structure, we first look for a common language for describing gravity and the Dirac field. Gravity described via the metric and the Dirac field described by spinors look very different. On the other hand, it is promising to consider Einstein gravity in the tetrad formalism; in particular through the spin coefficients which are employed for writing the Riemann tensor in the Newman-Penrose formalism, via the so-called Ricci identities. The Dirac equation too can be written for a curved space in the Newman-Penrose formalism, and then the set of four Dirac equations look strikingly similar to the Riemann tensor equations [the Ricci identities] written in terms of spin coefficients in the N-P language [61].

The N-P formalism is a tetrad formalism in which the basis vectors are null vectors $\mathbf{l}, \mathbf{n}, \mathbf{m}, \overline{\mathbf{m}}$ where $\mathbf{l}$ and $\mathbf{n}$ are real, and $\mathbf{m}$ and $\overline{\mathbf{m}}$ are complex conjugates of each other. These basis vectors are regarded as directional derivatives and are denoted by the symbols

$$D = \mathbf{l}, \quad \Delta = \mathbf{n}, \quad \delta = \mathbf{m}, \quad \delta^* = \overline{\mathbf{m}} \tag{28}$$

Ricci rotation coefficients (i.e. spin coefficients) arise in the definition of the covariant derivatives of these four null vectors. There are twelve such spin coefficients, conventionally denoted by the standard symbols

$$\kappa, \sigma, \lambda, \nu, \rho, \mu, \tau, \pi, \epsilon, \gamma, \alpha, \beta \tag{29}$$

The ten independent components of the Weyl tensor are expressed by five complex Weyl scalars, denoted as

$$\Psi_0, \Psi_1, \Psi_2, \Psi_3, \Psi_4 \tag{30}$$

The ten components of the Ricci tensor are defined in terms of four real scalars and three complex scalars

$$\Phi_{00}, \Phi_{22}, \Phi_{02}, \Phi_{20}, \Phi_{11}, \Phi_{01}, \Phi_{10}, \Lambda, \Phi_{12}, \Phi_{21} \tag{31}$$

The Riemann tensor can be expressed in terms of Weyl scalars and Ricci scalars, and directional derivatives of the spin coefficients. There are eighteen complex equations to this effect, known as Ricci identities. A typical Ricci identity has the form

$$D\rho - \delta^*\kappa = (\rho^2 + \sigma\sigma^*) + \rho(\epsilon + \epsilon^*) - \kappa^*\tau - \kappa(3\alpha + \beta^* - \pi) + \Phi_{00} \tag{32}$$

The Ricci components are determined from the Einstein equations, and the eighteen complex Ricci identities obey sixteen real constraints known as eliminant conditions, since there are only twenty independent Riemann components [61].

The four Dirac equations for the four spinor components F_1, F_2, G_1, G_2 can also be written in the N-P language, and a typical equation has the form

$$(D + \epsilon - \rho)F_1 + (\delta^* + \pi - \alpha)F_2 = i\mu_* G_1 \tag{33}$$

where $\mu_* = mc/\sqrt{2}\hbar$. The Dirac equation has a striking similarity with the Ricci identity, with both having a pair of derivatives of spin-coefficients / Dirac spinors. This similarity can be converted into an actual correspondence, and the four Dirac equations can be recovered as special cases of the Ricci identities, provided eight of the spin coefficients are set to zero

$$\rho = \mu = \tau = \pi = \epsilon = \gamma = \alpha = \beta = 0 \tag{34}$$

and the following fundamental identification is *assumed* between the four Dirac spinors and the remaining four non-zero spin-coefficients:

$$F_1 = \frac{1}{\sqrt{l_p}}\,\lambda, \quad F_2 = -\frac{1}{\sqrt{l_p}}\,\sigma, \quad G_1 = \frac{1}{\sqrt{l_p}}\,\kappa^*, \quad G_2 = \frac{1}{\sqrt{l_p}}\,\nu^* \tag{35}$$

The Dirac equations follow provided one assumes relations between the Riemann components and the Dirac mass, a typical example being

$$\Phi_{20} + \Phi_{01} = (i\mu_* + \nu)\kappa^* \tag{36}$$

Unfortunately however, the sixteen constraints [the eliminant conditions alluded to above] on the Ricci identities lead to undesirable constraints on the Dirac equation, and this particular approach to the Einstein-Dirac correspondence has to be abandoned.

One then looks for possible circumstances when there are no constraints on the thirty-six real Ricci identities. This can happen when the Riemann tensor has thirty-six independent components, instead of twenty. Indeed there is such a situation possible, and that is when the space-time admits torsion, and hence a Riemann-Cartan geometry, torsion being the anti-symmetric part of the connection:

$$\Gamma_{\mu\nu}{}^{\lambda} = \left\{{}_{\mu\nu}{}^{\lambda}\right\} - K_{\mu\nu}{}^{\lambda} \tag{37}$$

Here $\left\{{}_{\mu\nu}{}^{\lambda}\right\}$ is the Christoffel symbol of the second kind, and $K_{\mu\nu}{}^{\lambda}$ is the contortion tensor which describes the presence of torsion. Now the Ricci tensor is no longer necessarily symmetric, and there are six additional components in the Ricci tensor, described by the three complex quantities $(\Phi_0, \Phi_1$ and $\Phi_2)$. The Weyl tensor has ten additional components, described by the real quantities $(\Theta_{00}, \Theta_{11}, \Theta_{22}, \chi)$ and the complex quantities $(\Theta_{01}, \Theta_{02}, \Theta_{12})$. For definition of these notations the reader is referred to [62]. Thus the Riemann tensor now has exactly thirty-six components, for which there are thirty-six real Ricci identities. There are no eliminant conditions.

The spin coefficients now have an additional term due to torsion, which is denoted by γ_1^{nlm}. Thus, the spin coefficients can now be written as $\gamma_{lmn} = \gamma^{\circ}_{lmn} + K_{nlm}$, where the first term corresponds to the torsion free part and the second term corresponds to the torsion component of the spin coefficients. We further use the following notation to represent the spin-coefficients (see [62] for details):

$$\kappa = \kappa^{\circ} + \kappa_1, \qquad \rho = \rho^{\circ} + \rho_1 \tag{38}$$

etc.

A few remarks on the fundamental significance of torsion are in order. In special relativity, the symmetry group of the Minkowski space-time is the Poincaré group, which includes both Lorentz transformations and translations. In quantum theory, elementary particles are irreducible, unitary representations of the Poincaré group (not the Lorentz group), and these are labelled by mass as well as spin. On the other hand, in general relativity, which describes a curved space-time, the structure group acting on the tangent space is the Lorentz group, *not* the Poincaré group. Translations are not included. By introducing torsion and relating it to intrinsic angular momentum, translations are included, and the structure group of the space-time becomes the Poincaré group, and is no longer the Lorentz group [Riemann-Cartan geometry]. Curvature is related to Lorentz transformations in the same way that torsion is related to translations [63,64]. Thus, since both the Dirac equation and Einstein gravity with torsion have the Poincaré group as their symmetry group, it seems that correspondence between Einstein equations and Dirac equations might be more easily established if torsion is included. Of course we have to respect the fact that to date there is no observational evidence for torsion in the astrophysical world, and in empty space outside of matter, torsion vanishes. This can be achieved in the construction we now describe.

One first writes down the eighteen complex Ricci identities, now with torsion included in the spin-coefficients [62]. A typical example is

$$\begin{aligned} D\rho - \delta^*\kappa = \quad & \rho(\rho + \epsilon + \epsilon^*) + \sigma\sigma^* - \tau\kappa^* - \kappa(3\alpha + \beta^* - \pi) + \Phi_{00} \\ & -\rho(\rho_1 - \epsilon_1 + \epsilon_1^*) - \sigma\sigma_1^* + \tau\kappa_1^* + \kappa(\alpha_1 + \beta_1^* - \pi_1) + i\Theta_{00} \end{aligned} \tag{39}$$

Journal of Physics: Conference Series **626** (2015) 012009 doi:10.1088/1742-6596/626/1/012009

We consider two limits. In the first limit, the torsion part of the spin-coefficients is set to zero. In this limit the Ricci identities reduce to the ones discussed previously, and if the source for the Ricci tensor is taken as the energy-momentum tensor, one recovers Einstein gravity. In the other limit, we set the torsion free part of the spin-coefficients to zero, and retain only the torsion part. We now assume that eight of these torsion dominated spin-coefficients are zero, precisely as in (34) above, and the remaining four non-vanishing spin-coefficients are proportional to the Dirac spinors, as in (35). The Dirac equations then follow from the Ricci identities provided the Riemann tensor obeys the following conditions:

$$\Phi_{20} + i\Theta_{20} + \Phi_{01} + i\Theta_{01} - \Psi_1 - \Phi_0 = i\mu_*\kappa^* \tag{40}$$

$$\Phi_{21} + i\Theta_{21} + \Phi_2 - \Psi_3 + \Phi_{02} + i\Theta_{02} = i\mu_*\nu^* \tag{41}$$

$$i\Theta_{12} - \Phi_{12} + i\Theta_{00} - \Phi_{00} + \Phi_2^* - \Psi_3^* = i\mu_*\sigma \tag{42}$$

$$i\Theta_{10} - \Phi_{10} - \Phi_0^* - \Psi_1^* + i\Theta_{22} - \Phi_{22} = i\mu_*\lambda \tag{43}$$

Thus we have the torsion-dominated limit, which are the Dirac equations, and we have the gravity dominated limit, which are the Einstein equations. In the former case, gravity is absent (Minkowski space-time) and the material behaviour is quantum. In the latter case matter behaviour is classical, and gravity dominates. Thus there must be a more general underlying theory in which the torsion-free and the torsion part of the spin-connection are both present, and to which general relativity and quantum theory are both approximations. We conjecture that GR is the $m \gg m_{pl}$ approximation, and Dirac theory is the $m \ll m_{pl}$ approximation, and the more general underlying theory, which is valid for $m \sim m_{pl}$ has the schematic structure

$$\frac{c^3}{G} G_{\mu\nu} - \frac{i\hbar}{l^2} [Riem]_{\mu\nu}^{torsion} = \frac{mc}{l^2} f_{\mu\nu} \tag{44}$$

Although the dynamics of this theory remains to be understood, some general remarks could be made. The length scale l should be determined by the solution of the field equations. In the first term on the left, $G_{\mu\nu}$ is the Einstein tensor. If this term dominates over the second term, the right hand side reduces to the symmetric stress-energy tensor. The function $f_{\mu\nu}$ then defines an inverse length scale $1/l$, in which case the right hand side reduces to a mass density times speed of light. It can be inferred that when the second term on the left is negligible, the length scale l is the Schwarzschild radius.

The second term on the left side is the new idea: the Dirac field could be identified with a complex torsion in a Riemann-Cartan spacetime. If this term dominates over the first term on the left hand side, then by the quantities $[Riem]_{\mu\nu}^{torsion}$ we mean the physics in the left hand side of the four equations (40) to (43) above. Now the quantities $f_{\mu\nu}$ on the right hand should be linearly proportional to the corresponding spin coefficients, so that the Dirac equations maybe recovered. If the spin coefficient is of the order $1/l$, then the scale l is found to be of the order of Compton wavelength L, as expected. If the first term dominates we have $R_S \gg L$ and $m \gg m_{pl}$ and if the second term dominates the inequalities are reversed [65, 66].

When $m \sim m_{pl}$ we expect the two terms on the left side to be comparable, and this is the key point of our argument: to propose the Dirac equations as consequence of a modified Einstein gravity, with the modification being caused by inclusion of a complex torsion on a Riemann-Cartan space-time, via the Newman-Penrose equations. We can think of the first term, which contains gravity, as introducing self-gravity into the formalism, which is the modification we seek. Since the non-relativistic limit of the Dirac equation is the Schrödinger equation, if we could construct the non-relativistic limit of the gravity-Dirac equation (44) we expect to obtain a non-linear Schrödinger equation which incorporates the effect of self-gravity, and which could help understand role of gravity in collapse of the wave-function. This work is under progress.

7. Complex space-time metric, and other ideas

CSL sets out to construct a model for dynamical collapse based on a stochastic non-linear modification of the Schrödinger equation. The structure of the CSL model is uniquely determined by imposing two requirements, whose eventual origin would have to be explained by a more fundamental theory underlying CSL. These requirements are: preservation of the norm of the state-vector, and no faster than light signalling. An anti-Hermitean linear term is added for obtaining wave-vector reduction, and a supplementary quadratic term is enforced by the requirement of norm preservation. Stochastic noise is introduced to achieve random outcomes of measurements. In order to avoid conflict with experiments [67] the noise is assumed to be mass-proportional, and couples to the local mass density operator. This is the preferred form of the CSL model, described in Eqn. (21) above. To complete the specification of the model, the form of the noise correlation must be prescribed.

How does one introduce gravity into the CSL model? One idea, as we saw above, is due to Diósi, where the noise correlation of the collapse model is assumed to be the same as in the decoherence model based on gravity. However, this cannot really be called a fundamental approach, because the stochastic part of the Schrödinger equation has already been introduced, without ascribing any connection with gravity. In fact, as Adler points out [68], and as we also noted above, real valued metric fluctuations lead to unitary evolution, and can at best induce decoherence, but not collapse of the wave-function.

An interesting way to relate gravity to actual collapse is to propose that the classical gravitational metric has, apart from the conventional real space-time metric $\overline{g}_{\mu\nu}$, a complex fluctuating term $\phi_{\mu\nu}$, so that [68]

$$g_{\mu\nu} = \overline{g}_{\mu\nu} + \phi_{\mu\nu} \tag{45}$$

and a noise correlation is prescribed for the fluctuating part. From the standard definition of the energy-momentum tensor in terms of variation of the metric, it can be shown that the imaginary part of ϕ_{00} gives rise to a real noise coupling in the CSL equation of the type (21), thus providing a gravitational origin for the CSL noise. It should be noted that the fluctuation term is of an entirely classical nature, and in fact it is assumed that gravity itself does not have to be quantised. Although complex valued classical metrics have been considered earlier in the literature, and in some sense could be considered more natural than real metrics, it remains to understand why the complex part should be stochastic.

The issue of whether or not gravity should be quantised deserves some more discussion perhaps. If by 'quantisation' is meant the application of the rules of quantum field theory to a classical theory of gravity, then the answer is likely to be no. There are various conceptual reasons to believe this, including the one that rules of quantisation are written down assuming a classical space-time background, and it does not appear logically sound to apply those rules to the very background whose pre-existence was assumed for writing down those rules. What then is the status of gravity? Can it be classical, and be described by say semiclassical Einstein equations: a quantum matter source for a classical gravity? Maybe. But then it has been argued that when fluctuations in the quantum energy-momentum tensor operator become significant, the semiclassical description will break down, and will have to be replaced at least by an Einstein-Langevin equation, with the metric acquiring a stochastic component. To counter such arguments, Dyson has noted that objections against coupling quantum matter to classical gravity apply only to the semiclassical Einstein equations model, and further that the Bohr-Rosenfeld type argument in favour of quantising the electromagnetic field does not apply in the case of gravity [69]. Still, we do not believe that a universe made of only quantum matter fields can coexist with a classical space-time manifold: the quantum theory requires a classical time to describe evolution, and classical time is produced by classical bodies: there is no self-consistent demonstration that quantum matter can produce a classical space-time background in which the quantum fields evolve. Perhaps a universe made of only quantum fields requires the space-time to be noncommutative, which in itself is a 'nonclassical' feature, which might act as an effective source of stochasticity and gravity induced collapse.

We briefly mention three other ideas where gravity has been invoked to explain decoherence or collapse of the wave-function. Pearle and Squires [70] proposed to relate the CSL noise to fluctuations in the Newtonian potential or the curvature scalar. The theory of Primary State Diffusion due to Percival [71, 72], proposes non-differentiable fluctuations in space-time on the Planck scale to introduce stochastic terms into the evolution equation for a quantum state. The theory bears resemblance to the work of Karolyhazy described earlier. Ellis, Mohanty and Nanopoulos [73] discussed how interaction of a microscopic system with space-time wormholes can induce a non-unitary modification in the Hamiltonian evolution.

Another theoretical testing ground for CSL and gravity induced collapse is the very early universe, where these models have been applied to understand the origin of quantum to classical transition of inflationary density perturbations. While theoretical understanding is still in early stages, many investigations have been carried out, and this promises to be a useful line of research where it might be possible to put constraints on gravity induced collapse from observations of the cosmic microwave background [74–79]. Constraints on CSL from the bounds of $\mu-$ and $y-$ distortions of the CMB have been studied in [80].

It has also been suggested that a modification in quantum theory of the CSL type (or gravity induced modification) can help resolve the black hole information loss paradox [81]. Thus it seems that time is ripe to consider applications of collapse models in diverse theoretical, astrophysical and cosmological situations, apart from solving the measurement problem or understanding the quantum-classical transition in laboratory systems. Doing so will on the one hand constrain collapse models, and on the other hand might even assist in solving long-standing theoretical puzzles.

8. Possible experimental tests

The most direct experimental tests of departure from quantum theory in the mesoscopic regime are being carried out through molecular interferometry, and optomechanics [3, 82–86]. These tests look for a possible breakdown of the quantum superposition principle. No breakdown has been observed to date. Even if a breakdown were to be discovered, it will signal that quantum theory is approximate, but the source of the departure cannot be pinned down to gravity. It is curious that we do not have a gravity specific test, even at the gedanken level, using which we can say that if a violation is discovered, it is necessarily due to gravity. One reason for this is that predictions of gravity induced decoherence / collapse are not sufficiently accurate quantitatively - for instance the quantitative predictions of the Karolyhazy model and the Diósi model differ substantially. Theoretical studios have to improve considerably in their precision and reliability, before a gravity specific test can be proposed. In the same vein, bounds on CSL models coming from laboratory and astrophysical data are also bounds on gravity models [3].

Another promising avenue, apart from interferometry and optomechanics, is a laboratory test for anomalous random walk of a quantum object induced by CSL or by gravity. It is known that the CSL effect induces a tiny stochastic heating of the collapsed particle, leading to a very small energy-momentum violation, too small to contradict any of the known data. However, the momentum violation induces an anomalous random walk which can be in principle be detected at low temperatures and extremely low pressures. Such an experiment has been proposed, but not yet planned or executed. The CSL induced random walk competes with two other standard effects, these latter two being thermal Brownian motion of the particle under consideration, coming from its interaction with ambient radiation (scattering, absorption and emission). The second standard effect is collision with molecules of the ambient gas (normal Brownian motion). Both these effects can be reduced sufficiently by going to pressures of about a pico Torr (achievable) and a temperature of 100 Kelvin or lower. It also turns out that the

Journal of Physics: Conference Series **626** (2015) 012009 doi:10.1088/1742-6596/626/1/012009

anomalous effect is easier to detect via the rotational displacement of a disc, as opposed to the translational displacement of a sphere, by levitating the disc, its ideal linear size being about 10^{-5} cm [87, 88]. Gravity induced collapse / decoherence also predicts such a random walk, though the effect is typically a few orders of magnitude smaller than in the case of CSL [87, 88]. Another promising approach which has recently gained prominence is investigation of spectral line broadening due to stochastic heating [89, 90]. While the focus has been on CSL, it would be worthwhile looking into gravity induced effects as well. It seems particularly relevant and important to look for an experimental effect which could discriminate between a gravity induced collapse and a non-gravitational CSL mechanism.

9. Concluding Remarks

There are indeed very many ideas which have been tried to investigate if decoherence and collapse of the wave-function can be caused by a gravitational effect. Maybe finding a formulation of quantum theory which does not refer to classical time will throw light on the nature and origin of the quantum-classical transition. The models of Karolyhazy, Diósi and Percival, though built on heuristic principles, give very suggestive results. Similarly, including stress tensor fluctuations in the Schrödinger-Newton equation indicates gravity induced decoherence. The idea of a complex metric with a stochastic imaginary part is also very attractive, as a neat source for the CSL noise. The black hole - elementary particle complementarity suggests that a gravity theory with torsion bridges general relativity and relativistic quantum mechanics contained in the Dirac equation, and the underlying theory incorporates quantum effects as well as self-gravity. None of the models proposed thus far provide a fundamental explanation for the Born probability rule, nor definitively conclude whether or not special relativity needs to be modified when gravity and quantum mechanics are combined. It is difficult at this stage of our understanding to say whether gravity should be 'quantised' in the sense of there being a description of gravity which necessarily has quantum fluctuations as opposed to classical stochastic fluctuations. Nevertheless, diverse investigations of this nature, where the gravity-quantum interface is studied to address unresolved issues in quantum theory, have fascinating conceptual facets, and should be pursued vigorously. These should be accompanied by experimental efforts to ascertain the domain of validity of quantum theory.

"...I would like to suggest that it is possible that quantum mechanics fails at large distances and for large objects. Now, mind you, I do not say that quantum mechanics does fail at large distances, I only say that it is not inconsistent with what we know. If this failure is connected with gravity, we might speculatively expect this to happen such that $GM^2/hc = 1$, or M near 10^{-5} gm....If there was some mechanism by which phase evolution had a little bit of smearing in it, so it was not absolutely precise, then our amplitudes would become probabilities for very complex objects. But surely, if the phases did have this built in smearing, there might be some consequences to be associated with this smearing. If one such consequence were to be the existence of gravitateion itself, then there would be no quantum theory of gravitation, which would be a terrifying idea..."

- Feynman (1961)

Acknowledgements: I would like to thank the organisers of DICE2014 for giving me an opportunity to speak at the conference. It is a pleasure to thank my collaborators Angelo Bassi, Shreya Banerjee, Sayantani Bera, Suratna Das, Sandro Donadi, Kinjalk Lochan, Satyabrata Sahu, Seema Satin, Anushrut Sharma, and Hendrik Ulbricht for interaction and discussions. This work is supported by a grant from the John Templeton Foundation (#39530).

References

[1] Adler S L and Bassi A 2009 *Science* **325** 275
[2] Bassi A and Ghirardi G C 2003 *Phys. Rep.* **379** 257
[3] Bassi A, Lochan K, Satin S, Singh T P and Ulbricht H 2013 *Rev. Mod. Phys.* **85** 471
[4] Adler S L 2007 *J. Phys. A* **40** 2935
[5] Bell J S 1990 *Physics World* 8, 33
[6] Landau L D and Lifshitz E M 1965 *Quantum Mechanics* (Oxford : Pergamon Press)
[7] Singh T P 2015 The problem of time and the problem of quantum measurement *in: Re-thinking time at the interface of physics and philosophy: the forgotten present* ed Filk T, von Muller A and Filk T (Berlin-Heidelberg : Springer)
[8] Pearle P 1976 *Phys. Rev. D* **13** 857
[9] Gisin N 1981 *J. Phys. A* **14** 2259
[10] Gisin N 1984 *Phys. Rev. Lett.* **52** 1657
[11] Gisin N 1989 *Helv. Phys. Acta* **62** 363
[12] Gisin N and Rigo M 1995 *J. Phys. A* **28** 7375
[13] Weinberg S 2012 *Phys. Rev. A* **85** 062116
[14] Pearle P 1989 *Phys. Rev. A* **39** 2277
[15] Ghirardi G C, Pearle P and Rimini A 1990 *Phys. Rev. A* **42** 78
[16] Adler S L 2004 *Quantum theory as an emergent phenomenon* (Cambridge: Cambridge University Press)
[17] Adler S L 1994 *Nucl. Phys. B* **415** 195
[18] Adler S L and Millard A C 1996 *Nucl. Phys. B* **473** 199
[19] Diósi L 2005 *Braz. J. Phys.* **35** 260
[20] Gao S 2013 *Stud. Hist. Philos. Mod. Phys.* **44** 148
[21] Diósi L 2013 *J. Phys. Conf. Ser.* **442** 012001
[22] Singh T P 2006 *Bulg. J. Phys.* **33** 217
[23] Lochan K and Singh T P 2011 *Phys. Lett. A* **375** 3747
[24] Lochan K, Satin S and Singh T P 2012 *Found. Phys.* **42** 1556
[25] Karolyhazy F 1966 *Nuovo Cimento A* **42** 390
[26] Karolyhazy F, Frenkel A and Lukács B 1986 On the possible role of gravity in the reduction of the wave function *Quantum concepts in space and time* ed Penrose R and Isham C J (Oxford: Clarendon)
[27] Karolyhazy F 1974 *Magy. Fiz. Foly.* **22** 23
[28] Karolyhazy F 1990 *Sixty-two years of uncertainty* ed Miller A (New York: Plenum)
[29] Karolyhazy F 1995 *Fundamental problems of quantum physics* ed Ferrero M and van der Merwe A (Netherlands: Kluwer Acad. Publ.)
[30] Karolyhazy F, Frenkel A and Lukacs B 1982 *Physics as natural philosophy* ed Shimony A and Feshbach H (Cambridge: MIT Press)
[31] Frenkel A 1977 *Quantum mechanics - a half century later* ed Lopes J L and Paty M (Dordrecht: Reidel)
[32] Frenkel A 1990 *Found. Phys.* **20** 159
[33] Frenkel A 1995 *Fundamental problems of quantum physics* ed Ferrero M and van der Merwe A (Netherlands: Kluwer Acad. Publ.)
[34] Frenkel A 2002 *Found. Phys.* **32** 751
[35] Frenkel A 1997 *Experimental Metaphysics* vol 1 ed Shimony A, Cohen R S, Horne M and Stachel J (Boston: Kluwer Acad. Publ.) p 39
[36] Diósi L 1987 *Physics Letters A* **120** 377
[37] Diósi L 2007 *J. Phys. A* **40** 2989
[38] Diósi L 1989 *Phys. Rev. A* **40** 1165
[39] Diósi L and Lukacs B 1987 *Annalen der Physik* **44** 488
[40] Bera S, Donadi S, Lochan K and Singh T P 2014 A comparison between models of gravity induced decoherence *Preprint* aXiv:1408.1194
[41] Penrose R 1996 *Gen. Rel. Grav.* **28** 581
[42] Penrose R 1998 *Phil. Trans. R. Soc. London A* **356** 1927

[43] Penrose R 2000 Wavefunction collapse as a real gravitational effect *Mathematical physics 2000* ed Fokas A, Grigoryan A, Kibble T and Zegarlinski B (London: Imperial College Press) p 266
[44] Diósi L 1984 *Phys. Lett. A* **105** 199
[45] Bernstein D H, Giladi E and Jones K W R 1998 *Mod. Phy. Letts. A* **13** 2327
[46] Giulini D and Großardt A 2011 *Class. Quant. Grav.* **28** 195026
[47] Harrison R, Moroz I and Tod K P 2003 *Nonlinearity* **16** 101
[48] Harrison R, Moroz I and Tod K P 1998 *Class. Quant. Grav.* **15** 2733
[49] Ruffini R and Bonazzola S 1969 *Phys. Rev.* **187** 1767
[50] Giulini D and Grossardt A 2012 *Class. Quant. Grav.* **29** 215010
[51] Giulini D and Grossardt A 2013 *Class. Quant. Grav.* **30** 155018
[52] Hu B L 2014 *J. Phys. Conf. Ser.* **504** 012021
[53] Anastapoulos C and Hu B L 2014 *New J. Phys.* **16** 085007
[54] Bahrami M, Grossardt A, Donadi S and Bassi A 2014 *New J. Phys.* **16** 11
[55] Collin S, Durt T and Willox R 2014 *Class. Quant. Grav.* **31** 24503
[56] Hu B L and Verdaguer E 2004 *Living Rev. Relativity* **7** 3
[57] Mohan R, Bera S and Singh T P 2015 *in preparation*
[58] Nimmrichter S and Hornberger K 2015 *Phys. Rev. D* **91** 024016
[59] Ghirardi G, Grassi R and Rimini A 1990 *Phys. Rev. A* **42** 1057
[60] Bahrami M, Smirne A and Bassi A 2014 *Phys. Rev. A* **90** 062105
[61] Chandrasekhar S 1983 *The Mathematical Theory of Black Holes* (Oxford: Clarendon)
[62] Jogia S and Griffiths J B 1980 *Gen. Rel. Grav.* **12** 597
[63] Hehl F W, von der Hyde P, Kerlick G D and Nester J M 1976 *Rev. Mod. Phys.* **48** 393
[64] Trautmann A 2006 Einstein-cartan theory *Encyclopedia of Mathematical Physics* ed Francoise J P, Naber G L and Tsou S (Oxford: Elsevier) p 189
[65] Sharma A and Singh T P 2014 *Gen. Rel. Grav.* **46** 1821
[66] Sharma A and Singh T P 2014 *Int. J. Mod. Phys. D* **23** 1442007
[67] Pearle P and Squires E 1994 *Phys. Rev. Lett.* **73** 1
[68] Adler S L 2014 Gravitation and the noise needed in objective reduction models *Preprint* arXiv:1401.0353
[69] Dyson F 2012 *Int. J. Mod. Phys. A* **28** 1330041
[70] Pearle P and Squires E 1996 *Found. Phys.* **26** 291
[71] Percival I C 1995 *Proc. Roy. Soc. London Ser. A* **451** 503
[72] Percival I 2005 *Quantum State Diffusion* (Cambridge: Cambridge University Press)
[73] Ellis J, Nanopoulos D and Mohanty S 1984 *Nucl. Phys. B* **241** 381
[74] Das S, Sahu S, Banerjee S and Singh T P 2014 *Phys. Rev. D* **90** 043503
[75] Das S, Lochan K, Sahu S and Singh T P 2013 *Phys. Rev. D* **88** 085020
[76] Martin J, Vennin V and Peter P 2012 *Phys. Rev. D* **86** 103524
[77] Leon G and Bengochea G R 2015 Emergence of inflationary perturbations in the CSL model *Preprint* arXiv:1502.04907
[78] De Unanue A and Sudarsky D 2008 *Phys. Rev. D* **78** 043510
[79] Castagnino M, Fortin S, Laura R and Sudarsky D 2014 Interpretations of quantum theory in the light of modern cosmology *Preprint* arXiv:1412.7576
[80] Lochan K, Das S and Bassi A 2012 *Phys. Rev. D* **86** 065016
[81] Okon E and Sudarsky D 2014 The black hole information paradox and the collapse of the wave function *Preprint* arXiv:1406.2011
[82] Kleckner D, Pikovski I, Jeffrey E, Ament L, Eliel E, van den Brink J and Bouwmeester D 2008 *New J. Phys.* **10** 095020
[83] Marshall W, Simon C, Penrose R and Bouwmeester D 2003 *Phys. Rev. Lett.* **91** 130401
[84] Pikovski I, Vanner M, Aspelmeyer M, Kim M and Brukner C 2012 *Nat. Phys.* **8** 393
[85] Nimmrichter S, Hornberger K, Haslinger P and Arndt M 2011 *Phys. Rev. A* **83** 043621
[86] Hornberger K, Gerlich S, Haslinger P, Nimmrichter S and Arndt M 2011 *Rev. Mod. Phys.* **84** 157
[87] Bera S, Motwani B, Singh T P and Ulbricht H 2015 *Scientific Reports* **5** 7664
[88] Collett B and Pearle P 2003 *Found. Phys.* **33** 1495
[89] Bahrami M, Bassi A and Ulbricht H 2014 *Phys. Rev. A* **89** 032127
[90] Bahrami M, Paternostro M, Bassi A and Ulbricht H 2014 *Phys. Rev. Lett.* **112** 210404

Some not-so-common ideas about gravity

Raúl Carballo-Rubio[1], Carlos Barceló[1] and Luis J Garay[2,3]

[1] Instituto de Astrofísica de Andalucía (IAA-CSIC), Glorieta de la Astronomía, 18008 Granada, Spain

[2] Departamento de Física Teórica II, Universidad Complutense de Madrid, 28040 Madrid, Spain

[3] Instituto de Estructura de la Materia (IEM-CSIC), Serrano 121, 28006 Madrid, Spain

E-mail: `raulc@iaa.es`

Abstract. Most of the approaches to the construction of a theory of quantum gravity share some principles which do not have specific experimental support up to date. Two of these principles are relevant for our discussion: (i) the gravitational field should have a quantum description in certain regime, and (ii) any theory of gravity containing general relativity should be relational. We study in general terms the possible implications of assuming deviations from these principles, their compatibility with current experimental knowledge, and how can they affect future experiments.

1. Introduction

The acceptance of certain principles or axioms is a necessary scientific practice. In natural sciences, these principles should have a strong experimental corroboration in the best of the cases or, at least, do not contradict any experimental fact. When a theoretical paradigm shows signs of needing a revision, either because of internal or external (when compared with experiments) inconsistencies, these principles which lack of any direct experimental support are usually the first suspects in order to be replaced.

There is accumulative evidence that general relativity must be completed at high energies by a yet unknown theory (see, e.g., [1] and references therein). This observation follows from purely theoretical considerations, as there is no experiment up to date which is not consistent with the theory [2]. Theoretical investigations of the high-energy properties of the gravitational interaction routinely assume certain principles as valid (see, e.g., [3]). Of particular interest for us are the following two: (i) the gravitational field should have a quantum description in certain regime, and (ii) any theory of gravity containing general relativity should be relational. In this contribution we emphasize that there exist alternatives to these principles which are compatible with current experimental knowledge. Not only this but, if these ideas are taken seriously, they can lead to specific experimental signatures. Personally, I think that diversity is beneficial (not only for science) but, moreover, that it is worth it to understand the possible implications of deviations from these principles in order to build empirical confidence on them.

It is our aim to explore two ideas, in principle independent, although they could present interesting interconnections. The first idea is that some properties or elements of the gravitational interaction could be essentially classical in nature. This idea has been proposed by many people at different times; here we just focus in some general conceptual issues which appear to be model-independent. The second idea is that a convenient theory of gravity do not need to be relational, in the sense that it could contain non-dynamical structures such as a preferred volume element or a trivial (i.e., with no horizons) non-dynamical causal structure.

Our goal is to understand the possible consequences of these ideas. We will see that, if gravity has to be essentially classical, then one should expect deviations from standard quantum mechanics in mesoscopic experiments involving gravity and matter fields. On the other hand, if one has a non-relational theory of gravity, one should expect deviations from the usual gravitational phenomenology. In particular, a fixed volume element may permit to avoid the cosmological constant problem, while having a trivial non-dynamical underlying causality makes possible the construction of totally different stories for the gravitational collapse process.

2. Hybrid classical-quantum theories

In the first part we are going to study a specific model for a general classical-quantum hybrid system. The possibility of formulating a consistent hybrid classical-quantum theory has been approached in many different ways. In all these approaches one starts with initially separated purely classical and quantum sectors and then makes them interact in order to analyze the outcome. Without pretending to be exhaustive, we can classify these approaches in the following categories: (1) approaches that try to maintain the use of quantum states (or density matrices) to describe the quantum sector and trajectories for the classical sector [4, 5], (2) those that first formulate the classical sector as a quantum theory [6, 7, 8] and then work with a formally completely quantum system [9, 10, 11, 12, 13, 14, 15, 16, 17], (3) conversely, those that first formulate the quantum sector as a classical theory [18] and then work with a formally completely classical system [19, 20, 21], and (4) approaches that take the quantum and the classical sectors to a common language and then extend it to a single framework in the presence of interactions, for instance, using Hamilton-Jacobi statistical theory for the classical sector and Madelung representation for the quantum sector [22, 23, 24] or modeling classical and quantum dynamics starting from Ehrenfest equations [25]. This classification is not sharp and in some cases it is subject to interpretation, but it may be useful as a way to organize the possible procedures and conceptual viewpoints in the enterprise of constructing a hybrid theory. For more details on issues with these formulations, see [26, 27] and references therein.

Approaches in category 2 deal with a perfectly defined quantum theory based on Koopman-von Neumann-Sudarshan [6, 7, 8] translation of classical mechanics into the language of Hilbert spaces. Within this approach, the work of Peres-Terno is often mentioned as an important objection to the consistency of hybrid configurations. They proved that, in the presence of interaction, Heisenberg's equations of motion of the canonical-variable operators (Heisenberg picture) are necessarily different from what would have been obtained by assuming the two sectors to be quantum. From this they conclude (as we will see, somewhat precipitately) that the classical (Ehrenfest) limit of the classical-quantum (CQ) system will be necessarily different from that obtained directly from the corresponding quantum-quantum (QQ) system. On the other hand, they argue that these two limits should coincide, that is, that any hybrid theory should comply with this *definitive benchmark*, as they call it. Then, they finally conclude that hybrid systems, at least of this kind, are unphysical.

Here, first of all, we want to stress that this benchmark only makes sense if one tacitly assumes that a hybrid system should be just some suitable limit of a purely quantum system. If this is not assumed, we can invert the logic and see the failure to satisfy this condition as a clear sign of new physics in the CQ interplay. Second, as we will see the Peres-Terno model has many interesting features, being more complex than just exposed. When looking at the quadratic Peres-Terno model in more detail, one realizes that *it can* be made to comply with the above benchmark by just adding some constraints to the system. In fact, to add constraints to the system is more than reasonable because its very formulation incorporates more degrees

Journal of Physics: Conference Series **626** (2015) 012010 doi:10.1088/1742-6596/626/1/012010

of freedom than the QQ theory with which it is compared. Nonetheless, in a second twist, we have shown that this hybrid system still exhibits some sort of refined Peres-Terno no-go result: One cannot require that the dynamical equations for the second-order momenta (i.e., the dispersions) be equal to those in the QQ system (which for these system are equal to the CC ones). Going beyond quadratic interactions we also show that there is no way in general to fulfill the Peres-Terno benchmark by adding constraints. Overall, we are led to conclude that either the evolution of the expectation values or the evolution of the dispersions will exhibit some new physics beyond both QQ and CC systems.

The starting point of this particular hybrid scheme is the Koopman-von Neumann-Sudarshan formulation of classical mechanics [6, 7, 8, 28]. This formulation translates the usual description of classical mechanics in terms of symplectic manifolds to a quantum-mechanical language by associating to each physical observable a self-adjoint operator in a suitable Hilbert space and by implementing the time evolution as a unitary operator. Once the classical sector is treated formally as quantum, one can describe a CQ interaction by using the tensor product of Hilbert spaces, as in a pure quantum-mechanical theory [29]. It is in developing this program that one has to face a number of difficulties.

To make the discussion self-contained let us briefly summarize the Koopman-von Neumann-Sudarshan formalism. For the sake of simplicity we shall deal with one-dimensional systems although the discussion can be easily generalized to an arbitrary number of degrees of freedom. The quantum sector will be represented by a Hilbert space $\mathcal{H}_{\mathrm{Q}}$ with the standard position and momentum operators,

$$[\hat{x}, \hat{k}] = i\hbar \ . \tag{1}$$

In the classical sector we will also have position and momentum operators defined over a Hilbert space $\mathcal{H}_{\mathrm{C}}$, but in this case they commute,

$$[\hat{q}, \hat{p}] = 0 \ . \tag{2}$$

The elements of the pre-Hilbert space $\overline{\mathcal{H}}_{\mathrm{C}}$ are taken to be the “classical” wave functions $\psi(q,p)$, whose square equal the classical distribution functions:

$$|\psi|^2 = \rho_{\mathrm{C}}(q,p) \ . \tag{3}$$

The action of the classical operators (2) is then multiplicative,

$$\hat{q}\,\psi(q,p) = q\psi(q,p) \ , \qquad \hat{p}\,\psi(q,p) = p\psi(q,p) \ . \tag{4}$$

As is well known from classical statistical mechanics, the evolution of a classical probability distribution $\rho_{\mathrm{C}}(q,p)$ is given by the Liouville equation $\partial_t\rho_{\mathrm{C}} = \hat{L}\rho_{\mathrm{C}}$ [29]. Since the Liouville operator $\hat{L}$ is linear in the derivatives, the evolution equation for ψ is the same as that for ρ_{C}:

$$i\hbar\partial_t\psi = \hat{H}_{\mathrm{CL}}\psi \ , \qquad \hat{H}_{\mathrm{CL}} := i\hbar\hat{L} = \widehat{\partial_p H_{\mathrm{C}}}\ \hat{p}_q + \widehat{\partial_q H_{\mathrm{C}}}\ \hat{q}_p \ . \tag{5}$$

Here $H_{\mathrm{C}}(q,p)$ is the classical Hamiltonian and we have defined the operators $(\hat{p}_q, \hat{q}_p)$ to be canonically conjugate to $(\hat{q}, \hat{p})$:

$$[\hat{q}, \hat{p}_q] = i\hbar = [\hat{q}_p, \hat{p}] \ , \qquad [\hat{q}, \hat{q}_p] = [\hat{p}, \hat{p}_q] = 0 \ ; \tag{6}$$

which are correspondingly represented by

$$\hat{p}_q = -i\hbar\partial_q \ , \qquad \hat{q}_p = i\hbar\partial_p \ . \tag{7}$$

In the following these variables will be called *unobservable* variables.

Moreover, under reasonable assumptions about the classical Hamiltonian, the Liouville Hamiltonian $\hat{H}_{\rm CL}$ is essentially self-adjoint in the inner product,

$$(\psi, \phi) := \int \mathrm{d}q\mathrm{d}p\, \psi^* \phi \;, \qquad \psi, \phi \in \overline{\mathcal{H}}_{\rm C} \;, \tag{8}$$

so it generates a unitary evolution [30]. The completion of $\overline{\mathcal{H}}_{\rm C}$ in the inner product (8) constitutes the classical Hilbert space $\mathcal{H}_{\rm C}$.

Hereafter let us work for convenience in the Heisenberg picture in which the operators (2) carry the time dependence. With the previous definitions it is direct to check that the Heisenberg equations for the operators $(\hat{q}, \hat{p})$ have the same form as the classical Hamilton equations. The unobservable variables $(\hat{q}_p, \hat{p}_q)$ have their own evolution without influencing the physical sector $(\hat{q}, \hat{p})$. So, at least at this level, their evolution is irrelevant. Let us point out that, concerning the classical observable sector and directly related to this decoupling, there exists a symmetry under phase transformations of the form

$$\psi(q, p) \longrightarrow e^{i\varphi(q,p)} \psi(q, p) \;. \tag{9}$$

As we will see, this ambiguity disappears when developing a CQ interaction scheme. To end this summary, let us mention that it can be rigorously proved [6, 7] that the Koopman-von Neumann-Sudarshan formulation of classical mechanics is equivalent to the standard one in terms of symplectic manifolds and permits the use of operator techniques to treat classical problems. In particular, it is quite useful to study some aspects of statistical mechanics and ergodic theory [30].

Let us now consider an interacting CQ system [8, 12]. The space of states is the tensor product $\mathcal{H}_{\rm C} \otimes \mathcal{H}_{\rm Q}$, and the total Hamiltonian operator is

$$\hat{H}_{\rm T} = \hat{H}_{\rm CL} + \hat{H}_{\rm Q} + \hat{H}_{\rm I} \;, \tag{10}$$

where $\hat{H}_{\rm Q}$ stands for the Hamiltonian of the quantum system and $\hat{H}_{\rm I}$ provides the CQ interaction. At this point, the first question with a non-straightforward answer appears: How do we determine the form of the operator $\hat{H}_{\rm I}$ from its classical counterpart $H_{\rm I}$?

Given any interaction term, one can easily check that the classical Hamiltonian equations of motion (considering the system completely classical) and quantum Heisenberg equations are formally equivalent. The proposal of Peres and Terno [12] as the definite benchmark for an acceptable classical-quantum hybrid was that any hybridization of this system should respect this equivalence. Then, the result which they proved is that even for this simple system there is no term $\hat{H}_{\rm I} = \hat{H}_{\rm I}(\hat{q}, \hat{p}, \hat{x}, \hat{k}, \hat{q}_p, \hat{p}_q)$ which fulfills this condition, so they rejected this kind of hybrid CQ dynamics for not being physically meaningful.

In brief, the problem emerges as follows. If one wants the operators of the quantum sector $(\hat{x}, \hat{k})$ to appear in the equations of motion of the classical variables $(\hat{q}, \hat{p})$, one has to introduce the unobservable variables $(\hat{q}_p, \hat{p}_q)$ in the interaction term. But, by doing that, the decoupling of the unobservable sector no longer holds. That is, the equations of motion of the variables in the physical sector will contain the unobservable operators explicitly. As the unobservable operators do not appear in the QQ theory, the equations of motion of the CQ and QQ theories are different. Notice that it is the description of the effect of the quantum variables on the classical variables which leads to these features, which goes further than strict semiclassical physics, i.e., quantum mechanics in presence of classical fields. At this point it is interesting to have in mind that experiments have explored so far the semiclassical realm, so that we have no empirical clue about the behavior of physical systems near the classical-quantum interface.

With the risk of overinterpreting Peres and Terno's logic, in our view they took this condition as a definite benchmark because they identified formally having equal Heisenberg equations with obtaining an appropriate correspondence principle. However these are, in principle, logically distinct issues, as we have shown [26]. Moreover, this benchmark is fully appropriate if one assumes that the hybrid theory is just a particular approximation of the straight QQ theory (here by straight we mean the standard quantization one would have performed to the classical Hamiltonian for two interacting harmonic oscillators). However, if one is looking at hybrid systems as examples of new physics, then the violation of this benchmark could be interpreted positively. The precise prescription of an interaction term $\hat{H}_I$ should take into account additional physical insights coming from the detailed characteristics of the variable that is being regarded as classical, and how it is affected by the interaction with quantum systems.

All in all, our analysis shows that one can formulate consistent hybrid theories, but these theories incorporate new physics. They are not suitable limits of a direct quantization of the entire system. We should distinguish this straight quantum theory from the possibility of using a quantum formalism to describe all the physics of the hybrid system, as happens in the Koopman-von Neumann-Sudarshan formalism. Among other things, the new physics appears to bring about new degrees of freedom, which should be associated with the additional level of description of the quasi-classical (macroscopic) variables one would need to prescribe the precise form of the quantum backreaction to these variables. In this respect, the measurement process as presented by the Copenhagen interpretation could be seen as an extreme case of interaction between a classical and a quantum system in which there is no transfer of fluctuations into the classical variables. The measurement theory of Sudarshan [8, 9, 10] goes one step further, allowing some transfer of fluctuations while they do not compromise the classical integrity of the variables. Another interesting lesson is that in a hybrid context pure notions such as quantum or classical are only limiting notions, which are very useful in isolated situations but are, strictly speaking, nonexistent; in a hybrid setting everything might have a certain degree of "quantumness" (classicality).

At the end of the day, only new experiments in the mesoscopic realm can help us decide whether in the behavior of a complete system there is something beyond quantum dynamics of microscopic quantum constituents. Proposals such that those of Diósi and Penrose [31, 32, 33] or Guirardi et al. [34] of new physics at the classical-quantum cut find additional justification in light of these hybrid models. In particular, the proposal that gravity, with its stubborn resistance to quantization, might play a role in going beyond standard quantum mechanics is a powerful idea that we should not forsake.

3. Non-geometric gravity

The geometric vision of gravity embodied in general relativity has been highly successful up to date in the explanation of gravitational phenomena. However, as discussed above, general relativity should be replaced at high energies. While the majority of approaches to construct a theory of quantum gravity consider geometry as a fundamental tool to describe the high-energy regime, emergent gravity approaches show that non-relational structures could play an important role. It is important to notice that, within this context, these non-relational properties are not regarded as fundamental, but rather as being also emergent.

3.1. Volume element

One of the constructive processes which can be used to reach the theory of general relativity is the self-coupling procedure of gravitons (see [35, 36] for a thorough discussion of the history of the subject and the different assumptions needed to obtain the result). It is commonly said that general relativity is the only solution to the self-coupling problem [37]. However, it has been stated recently [36, 38] that there exists another solution which is not structurally equivalent to general relativity, but it is described by the structure of Weyl transverse gravity [39, 40, 41, 42].

Journal of Physics: Conference Series **626** (2015) 012010 doi:10.1088/1742-6596/626/1/012010

Most importantly for our purposes, the resulting nonlinear theory displays a background, non-dynamical volume element.

The nonlinear action of Weyl transverse gravity takes the following form in the framework of the self-coupling problem:

$$\mathscr{A} := \frac{1}{\lambda^2}\int \mathrm{d}\mathscr{V} R(\hat{g}), \tag{11}$$

where $R(\hat{g})$ has the same functional form as the Ricci scalar of a metric $\hat{g}_{ab}$ whose determinant is constrained by the condition $\det(\hat{g}) = \det(\eta)$, and $\mathrm{d}\mathscr{V}$ is the flat volume element, which is non-dynamical. For us $\hat{g}_{ab}$ is just a tensor field which lives in a flat background. To make this explicit, let us use the well-known fact [43] that one can express the Ricci scalar in terms of the covariant derivatives associated with the flat metric η_{ab}, denoted by ∇, and integrate by parts to put the action in the form:

$$\mathscr{A} = \frac{1}{4\lambda^2}\int \mathrm{d}\mathscr{V}\left[2\hat{g}_{bj}\delta^a_k\delta^i_c - \hat{g}^{ai}\hat{g}_{bj}\hat{g}_{ck}\right]\nabla_a\hat{g}^{bc}\nabla_i\hat{g}^{jk}. \tag{12}$$

Written in this form, it is clear that it describes a special relativistic field theory, albeit nonlinear. This form of the action is the natural one in the framework of the self-coupling problem. Starting with (11) is just convenient to make contact with our knowledge of general relativity but, indeed, one first obtains the action (12), which can be optionally supplemented with an irrelevant surface term to give (11); see [36] for an analogue discussion in the case of general relativity. The covariant notation we are using makes clear that this theory is invariant under general coordinate transformations. Moreover, by construction this action is invariant under transverse diffeomorphisms, whose infinitesimal form is

$$\delta_\xi \hat{g}^{ab} := \mathcal{L}_\xi \hat{g}^{ab}, \qquad \nabla_a\xi^a = 0. \tag{13}$$

The composite object $\hat{g}_{ab}$ is defined as

$$\hat{g}_{ab} := \kappa^{-1/4} g_{ab}. \tag{14}$$

Here $\kappa := \det(g)/\det(\eta)$. This definition makes the action automatically invariant under Weyl transformations, whose infinitesimal version is

$$\delta_\omega g_{ab} := \omega\, g_{ab}. \tag{15}$$

Written in terms of g_{ab} the action is given by:

$$\mathscr{A} = \frac{1}{4\lambda^2}\int \mathrm{d}\mathscr{V}\,\kappa^{1/4}\Big\{[2g_{bj}\delta^a_k\delta^i_c - g^{ai}g_{bj}g_{ck}]\nabla_a g^{bc}\nabla_i g^{jk} + \\ + \frac{1}{\kappa}\delta^a_b\delta^i_c\nabla_a g^{bc}\nabla_i\kappa - \frac{1}{2\kappa}g^{ai}g_{bc}\nabla_a g^{bc}\nabla_i\kappa - \frac{1}{8\kappa^2}g^{ai}\nabla_a\kappa\nabla_i\kappa\Big\}. \tag{16}$$

One can see that the transformations (13) and (15) combine in a way that makes this action invariant under transverse diffeomorphisms (now acting on the field g_{ab}) as well as Weyl transformations. The action (16) is thus the most general nonlinear covariant action quadratic in the derivatives of the field g_{ab} and satisfying these invariance requisites.

To complete the definition of the theory we need to decide how to include matter. The gravity-matter coupling is not unique, but we can appeal to two principles to resolve its non-uniqueness. The first one is the quantum version of the equivalence principle shown as a consequence of Poincaré invariance in [44]: the coupling of the field which describes gravity to matter and to itself must be governed by the same coupling constant λ. The second principle rests in the observation that conformal invariance is not a symmetry of matter in our world. A safe way to proceed is then to consider that matter fields are not affected by the Weyl transformations (15) (in the language of [41], matter fields are inert under Weyl transformations). Both principles imply that the resulting action of matter is obtained by replacing η_{ab} with the composite field $\hat{g}_{ab}$. Non-minimal couplings to $\hat{g}_{ab}$ are allowed by construction. Notice that no constraints are imposed on the matter sector.

When expressed in terms of the composite variable $\hat{g}_{ab}$, the nonlinear theory of gravity we are considering here has the form of general relativity plus matter, but it is a theory formulated in flat spacetime and the determinant of the metric is fixed by $\det(\hat{g}) = \det(\eta)$. On these grounds it can be understood as an extension of unimodular gravity which is unconstrained and invariant under general coordinate transformations. By the condition on the determinant, the corresponding (regulated) radiative corrections to the cosmological constant sector (vacuum zero-point energies) in the effective action would have the form

$$\int \mathrm{d}\mathscr{V} E_{\text{vac}}. \tag{17}$$

These contributions act as a mere constant shift in the effective action, as we are used to in any special relativistic quantum field theory which does not contain gravity. This does not happen when considering general relativity (as an effective quantum field theory [45, 46]) instead, leading to the cosmological constant problem [47, 48].

Radiative corrections are intrinsically a feature of the semiclassical theory in which matter fields are quantized, so that the form of the matter action is fundamental to address any question concerning them. In Weyl transverse gravity, Weyl invariance dictates a fairly different coupling between gravity and matter fields with respect to the situation in general relativity. Although it should be clear that any argument trying to read the form of these contributions from the form of the classical equations of motion alone is misleading, there still exist some confusion in the community in this respect. A well-known procedure to obtain the form of these radiative corrections leads to a definite answer which settles down the issue of radiative stability unambiguously, as dictated by the form of the symmetries of the gravitational action [49].

One could expect that deviating from the usual solution to the self-coupling problem would incur in disagreement with experimental facts. On the contrary, the theory constructed here indeed describes gravity in a way which is compatible with all the known experiments in gravitation. The field equations are, by construction, traceless. In particular, in the gauge $\det(g) = \det(\eta)$ they reduce to the trace-free Einstein equations:

$$R_{ab} - \frac{1}{4} R g_{ab} = \lambda^2 \left(T_{ab} - \frac{1}{4} T g_{ab} \right). \tag{18}$$

As a consequence, any solution of the Einstein field equations for arbitrary values of the cosmological constant is imported to this theory. In principle one could expect additional solutions in which the Ricci scalar and the trace of the stress-energy momentum tensor are not tied up through a constant but a general function, but this is not the case.

This can be seen by means of the following argument. Given the fact that the canonical stress-energy tensor of gravity and matter is conserved under solutions [43], we recover the Einstein field equations with a phenomenological integration constant, unrelated to zero-point energies of matter, playing the role of a cosmological constant; this is shown in Sec. VI-A of [36]. The arguments above show that this effective cosmological constant is not renormalized by radiative corrections: its value is protected by local symmetries. The resulting field equations include potential energies in the matter sector as, even if gravity is not directly coupled to these terms by construction, they inevitably appear in the definition of the canonical stress-energy tensor. Notice that the conservation of the canonical stress-energy tensor is not an additional equation on its own, but a consequence of the translational symmetry of the theory and the field equations (18). This is in contrast with the situation in usual formulations of unimodular gravity (see the corresponding discussion in [36] and references therein).

In summary, Weyl transverse gravity arises as a promising candidate for reconciling the classical and quantum-mechanical aspects of the cosmological constant problem. Any theory of emergent gravity which is described by means of this theory in the infrared regime would surpass this problem. On the one hand, all the necessary input to obtain this is deceptively simple: that gravity is mediated by gravitons; on the other hand, the cosmological constant problem may be avoided by the occurrence of a non-dynamical volume element.

3.2. Causal structure

One of the most important differences between Newtonian gravity and general relativity is that, in the latter, the causal structure of spacetime is dynamical with its evolution being dictated by the gravitational field equations. It is this departure the one behind the qualitative deviations in the gravitational collapse process [50].

In Newtonian gravity one can perfectly imagine arbitrarily small balls with arbitrarily large masses. One can design non-collision scattering processes with arbitrarily small impact parameter. These scatterings are characterized by an acceleration phase (falling into the gravitational potential) followed by a deceleration phase (climbing the gravitational potential). The entire scattering is perfectly time symmetric. By considering the balls as hard spheres subject to elastic collisions or non-interacting with each other when coming into superposition (i.e., transparent to each other), one could even make a precise head-on collision. This process will also consist of the previous two phases.

Now, general-relativistic situations in which a lump of matter is climbing a gravitational potential can easily be found: one can just think of throwing a stone upwards over our heads. One can also think of the following idealized thought experiment. Two equal balls with sufficiently small mass-to-radius ratio are thrown directly towards each other. Imagine that they are transparent to each other (for simplicity one could imagine them to be internally rigid to a first approximation and forget about changes in their form). In the case in which no horizon forms (i.e., sufficiently low kinetic energy in the collision), if one neglects the gravitational wave emission due to the encounter, the resulting trajectories will be perfectly time symmetric, again with acceleration and deceleration phases. Within standard classical general relativity the situation radically changes if a horizon forms. In a collision with sufficiently high kinetic energy the two balls in the previous example will form a horizon. The time-symmetric deceleration phase will disappear and be substituted by a black hole remnant. Somewhat surprisingly, although white-hole solutions exist mathematically, in supposedly realistic situations one never encounters potential-climbing cases beyond the Schwarzschild radius.

The fading into oblivion of the white-hole district is arguably the strongest departure of general relativity from Newtonian gravity. There seems to be a prejudice against exploring the possibility and implications of really having genuine bouncing solutions. As we will see, there are indeed good reasons for this prejudice to exist, but also for taking the bold leap of exploring beyond it. Our analysis shows that this can be done consistently only under the following assumption: it should exist an underlying causality that is explored only when Planck energies are at stake. This background causality should be non-dynamical and trivial in the sense of containing no horizons whatsoever, the simplest example one can think of being a Minkowskian structure. Otherwise general-relativistic light cones could not suffer such a dramatic turn. The only possibility apparently left would be that the light cones did not quickly reverse its tilting but only slowly recover their unbent positions before collapse. The geometry would be the one described, for instance, in [51], or more recently in [52].

Our assumption can be motivated both from particle physics and condensed matter physics. On the one hand, our proposal connects naturally with Rosen's reformulation of general relativity as a nonlinear theory on a flat Minkowski background [43]. This reformulation indeed goes further than the standard formulation of general relativity in the sense that it is a convenient effective framework to describe the switching-off of gravity at high energies. Rosen's reformulation can be understood as the long-wavelength limit of a nonlinear theory of gravitons (see [36] and references therein). It is still an open possibility that an ultraviolet completion of such a theory would exhibit asymptotic freedom (as its QCD cousin). On the other hand, similar ideas also appear when thinking of gravity as an emergent notion in a condensed matter framework [53] (see also [54]). The nonlinear theory of gravity describes in that case the behavior of collective degrees of freedom. There, it is reasonable to think that the first quantum gravitational effect is that, above some Planckian energy scale, the collective degrees of freedom corresponding to gravity are diluted, leaving a Minkowskian background for the matter excitations.

Let us detail how the switching-off of gravity can be described. The action of a theory of this sort should be expressible in terms of a composite field $g_{ab} = \eta_{ab} + \lambda h_{ab}$, where η_{ab} is the flat background metric, h_{ab} the graviton field and λ the coupling constant which controls the nonlinearities of the gravitational sector as well as the coupling to matter (take Rosen's formulation as an example, see [36]). For any nonzero value of λ the field equations of this theory are equivalent to the Einstein field equations for a metric g_{ab}. Thus in this regime one recovers general relativity, with a (generally curved) metric g_{ab} controlling the effective causality of the spacetime. However, the structure of the theory permits us to consider the limit $\lambda \to 0$, in which the nonlinearities disappear and matter is effectively decoupled from the graviton field, which evolves separately as a free field. In this limit the causality of the spacetime, given by η_{ab}, is no longer dynamical. The two conceptual frameworks above suggest that when high-energy phenomena are involved, this limit is indeed reached so that the underlying causality in the system, which is Minkowskian with no horizons whatsoever, is unveiled.

This view prioritizes the role of matter with respect to the dynamical causal structure contained in g_{ab}, and brings the general-relativistic and Newtonian descriptions of matter scattering in the presence of gravity a step closer to each other. Let us discuss the qualitative picture of our proposal, thinking about what happens in a local region around the distribution of matter undergoing gravitational collapse and making use of the figure 1. The mathematical details of the solution can be found in [55, 50]. When a lump of matter undergoes the gravitational collapse process, at some point it will enter the regime in which the local causal structure is Minkowskian and there is no trace of gravity. After a scattering process which takes place in the absence of gravity and which can be idealized as a first approximation as dissipationless, the lump of matter will effectively bounce back, now expanding in time. Notice that it is not needed to consider that some kind of repulsive force acts in this regime, and that its existence would only lead to quantitative changes in this picture. If we keep following the expanding distribution of matter we will exit the high-energy regime in which the causal structure is not dynamical and the usual general-relativistic picture with a gravitational field g_{ab} will be restored. However, this dynamical causal structure is a secondary character in this story, so that it will naturally adapt itself to the distribution of matter in spacetime, with the corresponding light cones pointing outwards.

The gravitational switching-off and -on process we have described would have an interpretation in general relativity in terms of an effective energy content that violates certain energy conditions in some regions. We advance here the following interpretation of the ideal dissipationless bouncing process. When the collapsing structure reaches Planckian densities it acquires a "quantum modification" leading to an effective density $\rho - \rho_{\rm q}$. At some point this effective density becomes negative and thus repulsive. This negative density is compensated

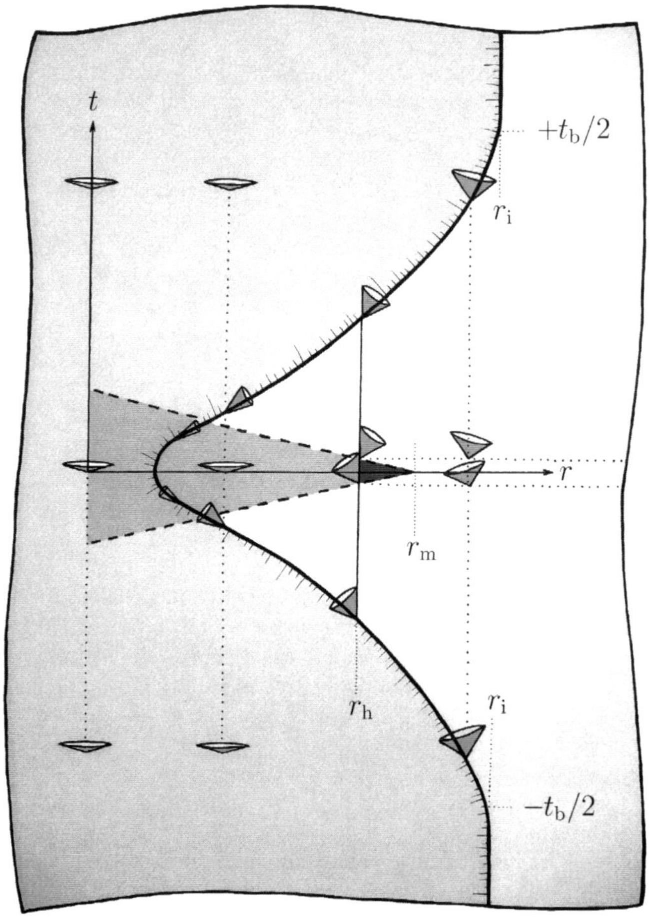

Figure 1. The figure represents the collapse and time-symmetric bounce of a stellar object in our proposal (the thick line). The past thick dashed line from $r = 0$ to r_m marks the boundary where the non-standard gravitational effects start to occur. In all the external white region the metric is Schwarzschild. In the region between the two thick dashed lines (which extends outside the stellar matter itself) the metric is not Schwarzschild, including the small dark gray triangle outside the Schwarzschild radius r_h. The drawing tries to capture the general features of any interpolating geometry. The slope of the almost Minkowskian cones close to the origin has been taken larger than the usual 45 degrees to cope with a convenient and explicit time-symmetric drawing.

by a burst of positive energy that is expelled out of the structure through the underlying causality (this is what the bottom thick dashed line in Fig. 1 intends to represent). This positive energy burst reaches some point r_m outside the gravitational radius. The time-reversed process should not be understood as a sort of collapse of this energy burst, but rather as describing the attenuation of the effect of its propagation through the dynamical geometry.

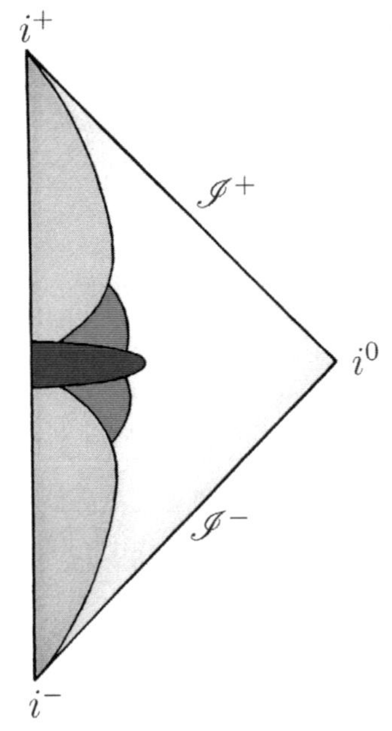

Figure 2. The figure represents the Penrose diagram of the proposed geometry. Globally it has the same causality than Minkowski spacetime. Locally it has some peculiarities. The dark gray region represents a non-standard gravitational field, while the down and up gray regions are respectively regions with outer and inner trapped surfaces. The light gray regions on the left-hand side are those filled by matter. There is no long-lived trapping horizons of any sort.

In more formal terms, we have shown that there exists an entire family of geometries, with functional and parametric degrees of freedom which at the moment cannot be fixed without knowing additional features of the underlying theory. However, the most relevant details are robust with respect to these variations, and can be summarized in two points:

- The time lapse associated with the collapse is very short (of the order of milliseconds for neutron-star-like initial configurations) for geometries which simply regularize the behavior of the collapse near the singularity. This is equally true both for observers attached to the structure as well as for distant stationary observers.
- It is mandatory that a certain open region outside the Schwarzschild radius deviates from the usual static and spherically symmetric solution. This deviation as well as the matter bounces will lead to characteristic imprints in the transient phase. While a detailed study must be done in order to find experimental signatures of these phenomena, the short timescales associated with all these processes are encouraging.

The plausibility of the time-symmetric bounce process rests on the assumption that the general-relativistic description of gravity is not fundamental, so that modifications to the dynamical behavior of the spatiotemporal causal structure at high energies are conceivable. The standard view is that "quantum corrections" to a general-relativistic geometry could only occur in high curvature regions. Our discussion points out that this assertion contains at least one assumption that is typically unstated: The non-existence of a deeper causality of Minkowskian character (that is, with no horizons). If this causality exists, it is not difficult to imagine that the echo of the high-energy collision produced in the location where the classical singularity was supposed to appear would be transmitted outside through the deeper (high-energy) causality, and modify the effective light-cone structure (i.e., the light-cone structure of general relativity) even in places with very small effective curvature. One also has to take into account that the non-general-relativistic modification of the effective light-cone structure we speak about occurs just in a brief transient region, being the manifestation of the rapid process of switching off and on of gravity. The existence of this deeper causality is indeed a necessary and sufficient assumption, thus making a tight connection between certain properties of the high-energy behavior of the gravitational interaction and low-energy solutions which do not present the lifetime problem.

This rapid bounce alternative radically changes the discussion of the possible endpoints of gravitational collapse. In particular, it makes plausible that the final object is a compact object with no horizons whatsoever. As with other compact objects in nature, the structure of the transient would be given by "dirty" (non-geometrical) physics whose details still need to be filled out. If this view is indeed realized in nature, black holes could just be an idealized approximation to the ultimate stationary objects.

The transient phase should leave some traces, for instance, in the physics of gamma ray bursts (GRBs) and its coincident gravitational-wave emission. One would expect some signatures associated with a reverberant collapse. In the collapsar model of GRBs however (see ,e.g., [56]) the emission zone is supposed to be very far from the collapsed core. This means that the connection between the processes at the core and those at the external wind shells could be very far from direct. The signatures should be clearer however in the spectrum of gravitational waves. The last part of the gravitational-wave signal should clearly distinguish between the standard relaxation towards a black hole and the presence of bouncing processes.

4. Conclusions

There are sensible alternatives to the philosophical assumptions behind the predominant research programs in quantum gravity. In particular, we have considered what we may regard as two of the most rooted principles: that a quantum description of gravity is needed in some regime, and that any gravitational theory containing general relativity should be relational. We have stressed that specific deviations from these alternatives can be compatible with current experimental knowledge, while leading to characteristic experimental signatures which could be possibly tested in future experiments, ranging from mesoscopic physics to astrophysical observations. While work remains to be done in order to understand in detail the possible implications, we have displayed some arguments that point out to generic properties which should be model-independent. Only further research could determine whether these assumptions may stop being philosophical, to become physical.

Acknowledgments

Financial support was provided by the Spanish MICINN through Projects No. FIS2011-30145-C03-01 and FIS2011-30145-C03-02 (with FEDER contribution), and by the Junta de Andalucía through Project No. FQM219. R. C-R. acknowledges support from CSIC through the JAE-predoc program, cofunded by FSE.

Journal of Physics: Conference Series **626** (2015) 012010 doi:10.1088/1742-6596/626/1/012010

[1] Wald R M 2010 *General Relativity* (University of Chicago Press)
[2] Will C M 2006 *Living Reviews in Relativity* **9** 3
[3] Rovelli C 2004 *Quantum Gravity* (Cambridge University Press)
[4] Boucher W and Traschen J H 1988 *Phys. Rev.* **D37** 3522
[5] Anderson A 1995 *Phys. Rev. Lett.* **74**(5) 621–625
[6] Koopman B O 1931 *Proceedings of the National Academy of Sciences* **17** 315–318
[7] von Neumann J 1932 *Ann. Math.* **33** 587
[8] Sudarshan E C G 1976 *Pramana* **6**(3) 117–126
[9] Sherry T N and Sudarshan E C G 1978 *Phys. Rev. D* **18**(12) 4580–4589
[10] Sherry T N and Sudarshan E C G 1979 *Phys. Rev. D* **20**(4) 857–868
[11] Gautam S R, Sherry T N and Sudarshan E C G 1979 *Phys. Rev. D* **20**(12) 3081–3094
[12] Peres A and Terno D R 2001 *Phys. Rev. A* **63**(2) 022101
[13] Terno D R 2006 *Found. Phys.* **36** 102
[14] Radonjić M, Prvanović S and Burić N 2012 *Phys. Rev. A* **85**(6) 064101
[15] Burić N, Popović D B, Radonjić M and Prvanović S 2013 *Phys. Rev. A* **87** 054101
[16] Burić N, Popović D B, Radonjić M and Prvanović S 2013 *Foundations of Physics* **43** 1459–1477
[17] Radonjić M, Popović D B, Prvanović S and Burić N 2014 *Phys. Rev. A* **89** 024104
[18] Heslot A 1985 *Phys. Rev. D* **31**(6) 1341–1348
[19] Elze H T 2012 *Phys. Rev.* **A85** 052109
[20] Alonso J L, Castro A, Clemente-Gallardo J, Cuchí J C, Echenique P and Falceto F 2011 *Journal of Physics A Mathematical General* **44** 5004 ; *do.* 2010 *Preprint* quant-ph/1010.1494
[21] Alonso J L, Clemente-Gallardo J, Cuchí J C, Echenique P and Falceto F 2012 *J. Chem. Phys.* **137** 054106
[22] Hall M J W and Reginatto M l 2005 *Phys. Rev.* **A72** 062109
[23] Hall M J W 2008 *Phys. Rev. A* **78**(4) 042104
[24] Chua A J K, Hall M J W and Savage C M 2012 *Phys. Rev. A* **85**(2) 022110
[25] Bondar D I, Cabrera R, Lompay R R, Ivanov M Y and Rabitz H A 2012 *Phys. Rev. Lett.* **109** 190403
[26] Barceló C, Carballo-Rubio R, Garay L J and Gómez-Escalante R 2012 *Phys. Rev. A* **86** 042120
[27] Elze H T 2013 *Journal of Physics Conference Series* **442** 012007
[28] Mauro D 2003 *Preprint* quant-ph/0301172
[29] Ballentine L E 1998 *Quantum Mechanics: A Modern Development* (World Scientific Publishing Co.)
[30] Reed M and Simon B 1981 *Methods of Modern Mathematical Physics: Functional Analysis, Volume 1* (Academic Press)
[31] Diósi L 1987 *Physics Letters A* **120** 377 – 381
[32] Penrose R 1996 *Gen. Rel. Grav.* **28** 581–600
[33] Marshall W, Simon C, Penrose R and Bouwmeester D 2003 *Phys. Rev. Lett.* **91** 130401
[34] Ghirardi G C, Rimini A and Weber T 1986 *Phys. Rev.* **D34** 470
[35] Ortín T 2007 *Gravity and Strings* (Cambridge University Press)
[36] Barceló C, Carballo-Rubio R and Garay L 2014 *Phys. Rev.* **D89** 124019
[37] Deser S 1970 *General Relativity and Gravitation* **1** 9–18
[38] Barceló C, Carballo-Rubio R and Garay L J 2014 *Preprint* gr-qc/1406.7713
[39] Álvarez E, Blas D, Garriga J and Verdaguer E 2006 *Nuclear Physics B* **756** 148–170
[40] Blas D 2007 *Journal of Physics A Mathematical General* **40** 6965–6971
[41] Alvarez E and Vidal R 2010 *Phys.Rev.* **D81** 084057
[42] Alvarez E 2012 *JCAP* **1207** 002
[43] Rosen N 1940 *Phys. Rev.* **57**(2) 147–150
[44] Weinberg S 1964 *Physical Review* **135** 1049–1056
[45] Donoghue J F 1994 *Phys. Rev. D* **50** 3874–3888
[46] Burgess C P 2004 *Living Reviews in Relativity* **7** 5
[47] Burgess C P 2013 *Preprint* hep-th/1309.4133
[48] Martin J 2012 *Comptes Rendus Physique* **13** 566–665
[49] Carballo-Rubio R 2015 *Preprint* gr-qc/1502.05278
[50] Barceló C, Carballo-Rubio R, Garay L J and Jannes G 2015 *Classical and Quantum Gravity* **32** 035012
[51] Roman T A and Bergmann P G 1983 *Phys. Rev.* **D28** 1265–1277
[52] Rovelli C and Vidotto F 2014 *International Journal of Modern Physics D* **23** 1442026
[53] Barceló C, Garay L J and Jannes G 2011 *Found. Phys.* **41** 1532–1541
[54] Jannes G 2015 *Foundations of Physics* 1–16
[55] Barceló C, Carballo-Rubio R and Garay L J 2014 *International Journal of Modern Physics D* **23** 1442022
[56] MacFadyen A and Woosley S E 1999 *Astrophys. J.* **524** 262

Towards room temperature solid state quantum devices at the edge of quantum chaos for long-living quantum states

Enrico Prati

Istituto di Fotonica e Nanotecnologie, Consiglio Nazionale delle Ricerche, Piazza Leonardo da Vinci 32, I-20133 Milano, Italy

E-mail: enrico.prati@cnr.it

Abstract. Long living coherent quantum states have been observed in biological systems up to room temperature. Light harvesting in chromophores is realized by excitonic systems living at the edge of quantum chaos, where energy level distribution becomes semi-Poissonian. On the other hand, artificial materials suffer the loss of coherence of quantum states in quantum information processing, but semiconductor materials are known to exhibit quantum chaotic conditions, so the exploitation of similar conditions are to be considered. The advancements of nanofabrication, together with the control of implantation of individual atoms at nanometric precision, may open the experimental study of such special regime at the edge of the phase transitions for the electronic systems obtained by implanting impurity atoms in a silicon transistor. Here I review the recent advancements made in the field of theoretical description of the light harvesting in biological system in its connection with phase transitions at the few atoms scale and how it would be possible to achieve transition point to quantum chaotic regime. Such mechanism may thus preserve quantum coherent states at room temperature in solid state devices, to be exploited for quantum information processing as well as dissipation-free quantum electronics.

1. Introduction

After the seminal observations of quantum coherence in biological systems in the last few years[1], a model has been developed for explaining such unpredicted phenomenon due to Vattay, Kauffman and Niiranen [2]. The protection of the coherence of quantum states in disordered systems is granted by the multifractality of wavefunction when delocalization starts to take place between insulating and disordered conductive regime. Doped silicon where impurity disorder occurs represents the ideal environment to create such conditions artificially.[3] The study of the Anderson-Mott transition from delocalized to localized electrons in semiconductors triggered by the doping concentration is a metal-insulator transition (MIT)[4, 5, 6, 7]. Such MIT has represented a hard problem to solve theoretically and experimentally, even in the case of the most simple and well known case of doped silicon such as Si:As and Si:P.[7] The possibilities opened by nanofabrication of semiconductor devices [8, 9] consist in the control of the position of impurity atoms at nanometric precision by a one-by-one method of implantation called single ion implantation (SII). Taking advantage of this method [10, 11, 12], it has been possible to create chains of individually implanted ions in silicon [13, 14], and to investigate the effects caused by their proximity and the residual disorder affecting their position and ground state energy distribution [15]. The transition from a single particle behavior to collective phenomena is directly observed. In the following, I review the recent advancements in the light harvesting biological systems, the physics observed in single atom transistors including the band formation in arrays of donors, and the key features of disordered doped semiconductors in the Anderson-Mott transition to emulate light harvesting systems where long-living quantum states may exist. A semi-Poisson edge of quantum chaos regime is predicted to exist in suitably deterministically doped [16] group IV semiconductor devices.

2. Anderson-Mott transition and disorder in doped semiconductors

In this section the metal-to-insulator transition generated by impurity doping in semiconductors such as silicon are considered. MIT phase transitions where the electron states evolve from delocalized to localized wavefunctions, are generated by means of two possible mechanisms: a structural change due to a change in the ionic lattice, or purely electronic. In this second case, two limiting cases are classified: Mott phase transitions, triggered by electron correlation and Anderson phase transitions, triggered by disorder. Transitions in which the competition of the two latter mechanisms concur are referred as Anderson-Mott transition. To start, let's first consider the purely electronic phase transitions. The Mott transition is captured by the Hubbard Hamiltonian:

$$H = t\sum_{\langle i,j\rangle}\left(a^{+}_{i\uparrow}a_{j\uparrow} + a^{+}_{i\downarrow}a_{j\downarrow}\right) + U\sum_{i} n_{i\uparrow}n_{i\downarrow} \tag{1}$$

where

$$n_{i\sigma} = a^{+}_{i\sigma}a_{j\sigma}, \tag{2}$$

t is the hopping matrix element and U is the on-site repulsion energy ($U > 0$). Though, impurity doping in semiconductors induces disorder, which in turn is captured by the Anderson transition governed by the Hamiltonian

$$H = t\sum_{\langle i,j\rangle} a^{+}_{i}a_{j} + \sum_{i}\epsilon_i a^{+}_{i}a_{i} \tag{3}$$

which is a non-interacting electrons model, so spin here only provides a trivial factor of two which is omitted. The energies ϵ_i are distributed within a bandwidth W: Anderson predicted for W/t large but finite a transition from delocalized states to insulating regime. Delocalization is achieved by either decreasing the density of scatterers, or, for soft scattering potential, increasing energy of scattered particles. Pure Anderson transition is not expected to be realized in nature, as Coulomb interaction between electrons is always present. To describe the effects of the Anderson transition in semiconductors such as silicon in which disorder is introduced by the addition of donor atoms, such as P and As, the Mott transition, which would be inadequate if considered alone, has to be taken into account. Neither Anderson's nor Mott's picture itself are sufficient separately to understand the observed MIT in semiconductors[7]. The resulting quantum phase transition, which carries aspects of both types of transitions, is called an Anderson-Mott transition.

3. Quantum chaos in disordered semiconductors

Manifestation of quantum chaos has been investigated in the nineties [3], by revealing a rich variety of different regimes. Let us first recall that discrimination among integrable and chaotic system is reflected in the level spacing distribution $P(s)$ where the spacings s of neighboring energy levels differs depending on the regime of the system: the energy levels of integrable systems are not correlated, and not prohibited from crossing, so their distribution is Poissonian (P):

$$P_{\mathtt{P}}(s) = e^{-s} \tag{4}$$

while in chaotic systems the eigenvalues become correlated and crossings are avoided, which determines level repulsion, so their distribution is described by the Wigner-Dyson (WD) distribution,

$$P_{\mathtt{WD}}(s) = \frac{\pi s}{2} e^{-\pi s^2/4} \tag{5}$$

as predicted by random matrix theory. The form of the Wigner-Dyson distribution depends on the symmetry properties of the Hamiltonian. Time reversal invariance imply a Gaussian orthogonal ensemble (GOE). The edge between the two regimes where the transition between the two regimes occur and where the wavefunction of the system becomes multifractal is described by the semi-Poisson distribution (SP):

$$P_{\mathtt{SP}}(s) = 4se^{-2s} \tag{6}$$

The latter is the regime where long-living quantum states may exist. Band structure calculations reveal that crystalline materials unambiguously manifest quantum chaos. Silicon can be seen as a quantum chaotic system with a particularly simple diatomic unit cell and the false time-reversal violation carries GOE distribution in the entire Brilluoin zone away from symmetry points, described by

$$P_{\mathtt{GOE}}(s) \propto se^{-\pi s^2/4} \tag{7}$$

On the contrary, alloy III-V semiconductors such as AlGaAs lack of inversion symmetry of the unit cell to T is broken outside the Γ point and the ensemble is unitary (GUE). The latter is expressed as:

$$P_{\mathtt{GUE}}(s) \propto s^2 e^{-4s^2/\pi} \tag{8}$$

being generated by unitary instead of orthogonal random matrix ensemble. Simulation show that in the proximity of the symmetry points the band distribution is Poissonian, while away from such points is GOE in silicon.

4. Experiment with few atom transistors

Single electron phenomena in defects have been originally observed in silicon/silicon oxide interfaces from random telegraph signal in small devices [17, 18, 19, 20, 21, 22]. Later, single electron effects have been observed in donor at the silicon/silicon oxide interface.[23] Individual atoms centered between the source and the drain of a small device provide the energy levels for sequential tunneling, in which an impurity was randomly diffused from the doping of one of the contacts. The presence of the individual atom is observed by looking at the quantum transport of the device at cryogenic temperature [24, 25, 26]. The condition necessary to observe the quantum transport is that the overlap integral between the electron wavefunction and the conduction electron wavefunction exponentially decaying out of the doped region is sufficiently large so that the probability of creating sequential tunneling from the contact to the donor site and from the donor site to the second contact is measurable. When the single donor is positioned along the channel, it provides a set of localized eigenstates of electrons corresponding to permitted energy levels below the conduction band edge. In Ref. [15] transistors fabricated on 100 silicon-on-insulator (SOI) substrates with 125-nm-thick buried oxide (BOX), which acts as a back-gate oxide, are considered. The nominal channel length is 200 nm, the width and thickness of the channel are 100 nm and 90 nm, respectively. The arsenic and phosphorous ions implanted two by two at 60 keV into the channel through the surface oxide with 10nm thick are expected to distribute around the depth of 53 nm from the interface on the back side.

Transistors with only two atoms implanted along the channel, for which their average distance is of the order of 20-40 nm, are treated as isolated particles. This is experimentally confirmed by looking at the quantum transport below the threshold voltage V_G which probes the silicon band gap, which shows Coulomb blockade at cryogenic temperature. When one moves from such an extremely dilute impurity concentration, which is conventionally defined as $n < 1 \cdot 10^{16}$ cm^{-3} in bulk silicon, such that r_C >46.4 nm where $r_C = n^{-1/3}$.[27], to higher concentrations, one first achieves a linear density which corresponds to a cubic lattice density n between $1 \cdot 10^{16}$ cm^{-3} and 10^{17} cm^{-3} ($r_C = 21.5$ nm), called semidilute regime, characterized in bulk silicon by the formation of pairs and complexes because of the randomness of the donor distribution. At higher linear density, which corresponds to approximately 10^{17} cm^{-3}, Hubbard bands are formed (intermediate regime) with activation energies of ϵ_2 and ϵ_3 for the upper and lower bands, respectively, triggered by temperature. The Hubbard impurity bands are observed also for arrays of few atoms as flat bands below the conduction band. The number of atoms needed to create such bands is very small. Four and six atoms are sufficient to observe band formation. By increasing the temperature from base temperature, the combination of the correlation and of the thermal activation determines the rise of the two partially overlapped Hubbard bands. As the frozen electron system is created at 4 K, we may call it a Wigner-like phase (each electron occupies a site of the potential landscape created by the 4 donors). By raising the temperature, the electron system is described as a Fermi glass.[5]

5. Discussion: application of deterministic doping towards the edge of quantum chaos in silicon quantum devices

The combination of the concepts reviewed in the previous sections opens novel scenarios of capital interest for solid state quantum information. The discovery of the mechanism enabling nature to preserve the quantum coherence at longer timescales thanks to the special properties of multifractal systems suggest a method for extending quantum coherence in quantum devices where fast decoherence proved to be a major issue. Such issue is technologically so severe that currently, after almost 20 years of development of solid state quantum information devices, tuning four qubits for circuital quantum intormation represents the maximum success, while at least 1000-10000 qubits are needed to implement quantum error correction codes. The special properties in the multifractal regime at the edge of quantum chaos consist in the simultaneous realization of sufficiently extended delocalization of wavefunction, together with power law decay of the coherence instead of exponentially fast loss of coherence of delocalized systems. In order to exploit the lesson learned from nature in the light harvesting systems, which display long-living quantum states up to room temperature thanks to a sort of lucky low-dissipation energy state distribution, and to apply to quantum information solid state devices, one may exploit a controlled quantum transition in semiconductor energy bands. Group IV semiconductors such as silicon and germanium are considered suitable for the long coherence times permitted by the lack of spin-orbit interaction, which on the contrary affects III-V semiconductor quantum devices. Group IV semiconductor bands engineering from undoped to impurity doping condition naturally carries the transition across the edge of quantum chaos. Indeed, if one consider high symmetry points such as Γ point for holes in undoped silicon and both electrons and holes in germanium, as well as X point for electrons in silicon, the energy spacing distribution is Poisson-like.[3] At the same time, doping is able to move the system to a GUE condition *with no need* to move away from such symmetry point, by gradually increasing the degree of disorder with the employment of deterministic doping method. Such an aspect represents the key fact of tuning the transition from Poisson to semi-Poisson condition, i.e. the chance to maintain the carriers within the same point in reciprocal space where conduction already takes place. According to what observed in biological light harvesting systems, where exciton states live for unexpectedly long times even at room temperature, we may expect that a similar behaviour may arise in quantum states living in suitably doped group IV semiconductors. Single ion implantation can thereofre play a major role to implement the designed doping in realistic solid state quantum devices created to coherently transfer quantum information.

6. Conclusion

To conclude, recent experiments have shown that the long-living coherent quantum states may exist up to room temperature in biological systems thanks to quantum chaos. Such phenomenon may in principle be swapped in doped semiconductors, where a high degree of control down to single atom control at nanometric precision has been proved. The combination of quantum chaos with doped semiconductor physics may lead to long-living quantum states to be exploited for quantum information processing as well as dissipationless quantum electronics.

References

[1] Panitchayangkoona G et al 2010 Long-lived quantum coherence in photosynthetic complexes at physiological temperature *PNAS* **107** 29 1276612770
[2] Vattay G, Kauffman S and Niiranen S 2014 Quantum biology on the edge of quantum chaos. *PloS One*, **9** 3 e89017
[3] Mucciolo ER, Capaz RB, Altshuler BL and Joannopoulos JD 1994 Manifestation of quantum chaos in electronic band structures *Phys. Rev. B* **50** 12 8245-8251
[4] Mott NF and Twose WD 1961 The theory of impurity conduction *Adv. in Phys.* **10** 38 107-163
[5] Anderson PW 1958 Absence of diffusion in certain random lattices *Phys. Rev.* **109** 1492-1505
[6] Hubbard J 1963 Electron Correlations in Narrow Energy Bands *Proc. R. Soc. Lond.* **A 276** 1365 238-257
[7] Belitz D and Kirkpatrick TR 1994 The Anderson-Mott transition *Rev. Mod. Phys.* **66** 261-380
[8] Prati E et al.: Few Electron Limit of n-type Metal Oxide Semiconductor Single Electron Transistors *Nanotechnology* **23** 215204 (2012)
[9] De Michielis M, Prati E, Fanciulli M, Fiori G, Iannaccone G 2012 Geometrical Effects on Valley-Orbital Filling Patterns in Silicon Quantum Dots for Robust Qubit Implementation *Appl. Phys. Expr.* **5** 124001
[10] Shinada T, Okamoto S, Kobayashi T and Ohdomari I 2005 Enhancing semiconductor device performance using ordered dopant arrays *Nature* **437** 1128-1131
[11] Tan K Y et al 2010 Transport spectroscopy of single phosphorus donors in a silicon nanoscale transistor *Nano Lett.* **10** 11-15
[12] Schenkel T, Persaud A, Park SJ, Nilsson J, Bokor J, Liddle JA, Keller R, Schneider DH, Cheng DW, Humphries DE 2003 Solid state quantum computer development in silicon with single ion implantation *J. Appl. Phys.* **94** 7017-7024
[13] Shinada T, Hori M, Guagliardo F, Ferrari G, Komatubara A, Kumagai K, Tanii T, Endo T, Ono Y, Prati E 2011 Quantum transport in deterministically implanted single-donors in Si FETs. 2011 *IEEE International Electron Devices Meeting (IEDM)* 5-7 Dec. 2011 30.4.1-30.4.4
[14] Hori M, Shinada T, Guagliardo F, Ferrari G and Prati E 2012 Quantum Transport Property in FETs with Deterministically Implanted Single-Arsenic Ions Using Single-ion Implantation *Silicon Nanoelectronic Workshop (SNW), 2012 IEEE*, Honolulu, pp. 1-2
[15] Prati E, Hori M, Guagliardo F, Ferrari G, Shinada T 2012 Anderson-Mott transition in arrays of a few dopant atoms in a silicon transistor *Nature Nanotechnology* **7** 443-447
[16] Shimizu Y, Takamizawa H, Inoue K, Yano F, Nagai Y, Lamagna L, Perego M, Mazzeo G and Prati E 2014Behavior of phosphorous and contaminants from molecular doping combined with a conventional spike annealing method *Nanoscale* **6** 2 706-710
[17] Prati E, Fanciulli M, Calderoni A, Ferrari G and Sampietro M 2007 Microwave irradiation effects on random telegraph signal in a MOSFET *Phys. Lett.* **A 370** 491-493
[18] Prati E, Fanciulli M 2008 Manipulation of localized charge states in n-MOSFETs with microwave irradiation *Phys. Lett.* **A 372** 3102-3104
[19] Prati E, Fanciulli M, Ferrari G, Sampietro M 2008 Giant Random Telegraph Signal Generated by Single Charge Trapping in sub-micron n-MOSFETs *J. Appl. Phys.* **103** 123707
[20] Prati E, Fanciulli M, Calderoni A, Ferrari G, Sampietro M 2008 Effects of microwave irradiation on the emission and capture dynamics in silicon metal oxide semicondutcor field effect transistors *J. Appl. Phys.* **103** 104503
[21] Prati E et al 2010 Measuring the Temperature of a Mesoscopic Electron System by means of Single Electron Statistics *Applied Physics Letters* **96** 113109

[22] Prati E 2010 The finite quantum grand canonical ensemble and temperature from single-electron statistics for a mesoscopic device *J. Stat. Mech.* **1** P01030

[23] Prati E, Latempa R and Fanciulli M 2009 Microwave Assisted Transport in a Single Donor Silicon Quantum Dot, *Phys. Rev. B* **80** 165331 (2009)

[24] Prati E et al 2011 Adiabatic charge control in Silicon single donor quantum dots *Appl. Phys. Lett.* **98** 053109

[25] Mazzeo G et al. 2012 Charge dynamics of a single donor coupled to a few electrons quantum dot in silicon *Appl. Phys. Lett.* **100** 213107

[26] Leti G, Prati E, Belli M, Petretto G, Fanciulli M, Wacquez R, Vinet M, Sanquer M 2011 Switching quantum transport in a three donors silicon Fin-field effect transistor *Appl. Phys. Lett.* **99** 242102

[27] Thomas GA, Capizzi M, DeRosa F, Bhatt RN, Rice MT 1981 Optical study of interacting donors in semiconductors *Phys. Rev. B* **23** 10 5471-4394

An instability of unitary quantum dynamics

Jasper van Wezel

Institute for Theoretical Physics, IoP, University of Amsterdam, 1090 GL Amsterdam, The Netherlands

E-mail: vanwezel@uva.nl

Abstract. Instabilities of equilibrium quantum mechanics are common and well-understood. They are manifested for example in phase transitions, where a quantum system becomes so sensitive to perturbations that a symmetry can be spontaneously broken. Here, we consider the possibility that the time evolution governing quantum dynamics may be similarly subject to an instability, at which its unitarity spontaneously breaks down owing to an extreme sensitivity towards perturbations. We find that indeed such an instability exists, and we explore its immediate consequences. Interpretations of the results both in terms of extreme sensitivity to the influence of environmental degrees of freedom, and in terms of a possible fundamental violation of unitarity are discussed.

1. Singular limits

Instabilities in theories of physics are often signalled by the presence of a singular limit [1], which describes the extreme sensitivity of a physical situation to perturbations. The stability of a ball balanced on the top of a hill, the turbulent flow of a zero-viscosity fluid, and the chaotic motion of Saturn's moon Hyperion, are all examples of situations in which a singular limit heralds the fact that one physical concept has reached the bounds of its usefulness, and is surpassed by another [1]. In many cases, the culprit is the thermodynamic limit, implementing Anderson's celebrated expression that 'More is Different' [2].

In all of these cases, the singularity of the involved limit, and the associated sensitivity to perturbations can be expressed mathematically by the presence of a set of non-commuting limits. As an elementary example, consider the function $y = \arctan(zx)$, where z is a parameter (see figure 1). For any finite value of z the function is perfectly smooth, and the limit of $y(x)$ as x goes to zero is simply zero. However, upon taking the limit $z \to \infty$, the arctangent turns into a step function, and at that point the limit of $y(x)$ as x approaches zero from above, is one. We thus find that the order in which the limits are taken matters:

$$\begin{aligned} &\lim_{z\to\infty} \lim_{x\downarrow 0} \arctan(zx) = 0 \\ &\lim_{x\downarrow 0} \lim_{z\to\infty} \arctan(zx) = 1. \end{aligned} \qquad (1)$$

The limit $z \to \infty$ is said to be singular, because it is qualitatively different from the situation with any finite z, no matter how large z gets. At the same time, the value of $y(0)$ in the singular limit $z \to \infty$ is very sensitive to perturbations. In fact, the value of $y(\epsilon)$ differs qualitatively from $y(0)$, even for infinitesimally small ϵ. Such an extreme sensitivity to even infinitesimally small perturbations is characteristic of singular limits throughout physics.

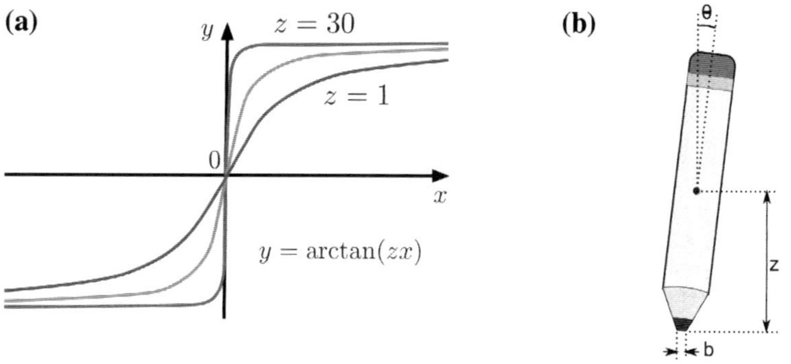

Figure 1. (a) The function $y = \arctan(zx)$ for different values of the parameter z. The value of the function at $x = 0$ is infinitely sensitive to perturbations in the limit $z \to \infty$. **(b)** A pencil balanced on its tip. The pencil is not perfectly sharp, having a blunt area with diameter b, and it is not perfectly balanced, dipping by an angle θ. The fate of the center of mass height z depends on the order in which the limits $b \to 0$ and $\theta \to 0$ are taken, signalling an extreme sensitivity to perturbations of θ in the limit of infinite sharpness.

A physical example of this idea can be seen when we consider a sharp pencil being balanced on its tip (see figure 1). The state of the pencil is encoded in the height z of its center of mass. After sharpening the pencil such that it only has a blunt area of diameter b left at its tip, and balancing it to within an angle θ of being perfectly upright, the fate of the pencil is determined by a singular limit:

$$\lim_{b\to 0}\lim_{\theta\to 0} z = 1 \tag{2}$$

$$\lim_{\theta\to 0}\lim_{b\to 0} z = 0. \tag{3}$$

That is, although it may be possible to perfectly balance a somewhat blunt pencil, starting with a sufficiently sharp pencil ($b \to 0$) makes it infinitely sensitive to perturbations so that even an infinitesimal perturbation away from $\theta = 0$ causes the pencil to tip over. In practice, it is therefore impossible to balance a sharp pencil on its tip. Notice that we can draw this conclusion about real pencils, even if neither of the limits is ever actually realised. The practical implication of the singular nature of the limit is that there is an extreme sensitivity to perturbations, and this sensitivity is manifested already for very (but not infinitely) sharp, well (but not infinitely-well) balanced pencils.

2. Instabilities of equilibrium quantum mechanics

The most common occurrence of singular limits in quantum mechanics, is in the spontaneous breakdown of symmetries. In this case, the thermodynamic limit presents a boundary to the applicability of the usual equilibrium description, and the state of the system is instead determined by infinitesimally weak perturbations. To see how this instability arises, it is instructive to consider the example of a harmonic crystal, described by the Hamiltonian:

$$\hat{H} = \sum_j \frac{\hat{P}_j^2}{2m} + \frac{1}{2} m\omega^2 \left(\hat{X}_j - \hat{X}_{j+1}\right)^2. \tag{4}$$

Journal of Physics: Conference Series **626** (2015) 012012 doi:10.1088/1742-6596/626/1/012012

Here, $\hat{X}_j$ and $\hat{P}_j$ are the position and momentum of particle j, which has mass m and is connected to neighbouring particles by a harmonic potential with natural frequency ω. It is easy to check that this Hamiltonian commutes with the operator for total momentum $\hat{P}_{\text{tot}} = \sum_j \hat{P}_j$. In fact, this remains true even in the presence of anharmonic potentials, proportional to higher powers of the distance between neighbouring particles. The fact that the Hamiltonian for a crystal and the operator for total momentum commute, implies that they share a common set of eigenstates. The eigenstates of total momentum are all plane waves, with an even distribution of the centre of mass position over all of space. According to the usual rules of quantum mechanics, the state of the crystal in thermal equilibrium is given by a thermally weighted mixture of eigenstates of the Hamiltonian. It thus follows that for any temperature whatsoever, the equilibrium centre of mass position of a crystal is undefined, and its fluctuations diverge: $\langle \hat{X}^2_{\text{c.o.m.}} \rangle = \infty$. In other words, under equilibrium conditions, any crystal ought to be spread out over all of space.

Of course, this situation is a direct consequence of the presence of translational symmetry, and the reason that localised objects do exist in our everyday world is that translational symmetry can be spontaneously broken in the thermodynamic limit. In that limit, the ground state of the Hamiltonian becomes extremely sensitive to perturbations of the form:

$$\hat{H}' = \hat{H} + \frac{1}{2} N m \Omega^2 \left(\hat{X}_{\text{c.o.m.}} - X_0 \right)^2, \tag{5}$$

where Ω is the natural frequency of a harmonic potential which tends to localize the centre of mass coordinate $\hat{X}_{\text{c.o.m.}}$ of the crystal with total mass Nm at the position X_0. The strength of the perturbation is given by the steepness of the potential, encoded in the natural frequency Ω, which is considered to be infinitesimally small. In the limit of a crystal containing infinitely many particles, the ground state in the presence of any perturbation is qualitatively different from the ground state with Ω strictly zero. In that thermodynamic limit, even an infinitesimally small perturbation suffices to completely localize the centre of mass position of the crystal, and entirely suppress its fluctuations [3]:

$$\lim_{N\to\infty} \lim_{\Omega\to 0} \langle \hat{X}^2_{\text{c.o.m.}} \rangle = \infty \tag{6}$$

$$\lim_{\Omega\to 0} \lim_{N\to\infty} \langle \hat{X}^2_{\text{c.o.m.}} \rangle = 0. \tag{7}$$

As before, the harmonic crystal is only strictly localised in the presence of an infinitesimally small perturbation, if it is truly infinitely large. For a crystal of finite size, the non-commuting limits instead signal an extreme sensitivity to small, but non-zero, perturbations. Since the strength of the perturbation required to localise an object consisting of N particles scales as $1/N$, objects as large as tables and chairs are susceptible to even the smallest perturbation in their environment, and in practice can never avoid its symmetry breaking influence.

Notice that any symmetric state can be written as a linear combination or wave packet of localised states. In the case of the harmonic crystal, the symmetric ground state of $\hat{H}$ can be formed from a superposition of localised states which are related to each other by rigid translations of the centre of mass position. These rigid translations are the symmetry operations corresponding to the translational symmetry of the Hamiltonian, and are generated by the total momentum operator. They do therefore not affect the internal degrees of freedom of the crystal, and form only a very small subset of all available states. Seen from the opposite point of view, one could equally well say that the localised ground state of the harmonic crystal in the presence of a perturbation, contains only a small subset of all eigenstates of the symmetric Hamiltonian. That is, the localised wave packet only contains states with differ in their values of the total momentum, but which have exactly equal quantum numbers for all internal degrees of freedom (phonon excitations). Moreover, precisely these total momentum eigenstates which make up the localised wave packet become degenerate in the thermodynamic limit, enabling the formation of a superposition with only an infinitesimal cost in energy, provided by an infinitesimally weak perturbation. The sparsity of the states contributing to this process as well as the collapsing nature of this tower of states in the thermodynamic limit, has earned them the designation of 'thin spectrum' states [2–6].

The extreme sensitivity to perturbations signalled by the presence of a singular limit, and the central role played by a collapsing tower of thin spectrum states are features which are present whenever a continuous global symmetry is spontaneously broken. A description precisely mirroring the above analysis can be given for the symmetry breaking observed in antiferromagnets, Bose-Einstein condensates, superconductors, and any other system with a spontaneously broken continuous symmetry [2–11].

3. An instability of quantum dynamics

The ubiquity of the singular thermodynamic limit in equilibrium quantum mechanics and the resulting extreme sensitivity of ground states to perturbations, suggests that the time evolution operator describing the dynamics of quantum objects in the thermodynamic limit may be similarly susceptible to infinitesimal disturbances. The defining symmetry of quantum dynamics is its unitary nature. The question we would thus like to pose is:

"*Is the unitary time evolution implied by Schrödinger's equation* stable *against infinitesimally small perturbations?*"

Or equivalently:

"*Can unitarity be spontaneously broken?*"

To answer this question, we follow the same procedure as that for the usual, static symmetries: an infinitesimally small perturbation is added to the Hamiltonian, and if this qualitatively affects the dynamics in the thermodynamic limit, the corresponding non-commuting set of limits signals the presence of an instability [12–14].

For concreteness, consider a quantum Hamiltonian $\hat{H}$ and the time evolution generated by it: $i\hbar\partial/\partial t|\psi\rangle = \hat{H}|\psi\rangle$. When exploring infinitesimal additions to $\hat{H}$, it is clear that perturbations of the form $V\hat{O}$ with V real and $\hat{O}$ a Hermitian operator will never break the unitarity of time evolution. To obtain non-unitary evolution, one needs to add an explicitly non-Hermitian perturbation. The simplest way to do this is to consider a perturbation of the form $iV\hat{O}$, with V real and $\hat{O}$ Hermitian. Of course such perturbations are not allowed in the quantum mechanical description of any closed system. The study of spontaneous unitarity violations in closed quantum systems is therefore by definition a study of the stability of quantum dynamics towards perturbations *beyond* quantum theory [12–14]. An alternative interpretation could be to treat the perturbed dynamics as describing the perfectly quantum mechanical, but effective dynamics of only part of a full system. It is well-known that the influence of the ignored environmental degrees of freedom can on average be modeled by a non-Hermitian term in the dynamics of the open quantum system [15–17]. We will comment on the implications of both of these interpretations after first considering the dynamics resulting from the presence of an infinitesimal unitarity breaking term.

As we are looking for a perturbation which will have an effect in the infinite particle limit $N \to \infty$, even for infinitesimally small strength of the perturbation $V \to 0$, it is clear that we need to consider perturbations which couple to an observable $\hat{O}$ that scales with the number of particles. Only in that way can the product of the two remain finite if the correct order of limits is taken. A natural candidate for such an extensive observable is the order parameter in a system which breaks a continuous symmetry. Choosing the perturbation $\hat{O}$ to be an order parameter moreover has the additional advantage that even the non-unitary time evolution will conserve energy in the thermodynamic limit. The reason for this, is that the order parameter only has non-zero matrix elements between states within the thin spectrum, which all collapse onto the ground state energy in the limit $N \to \infty$. In other words, a non-unitary term proportional to the order parameter field of a symmetry-broken system will affect the global value or orientation of the order parameter, but it cannot excite any of the internal degrees of freedom.

For concreteness, consider again the harmonic crystal of equation (4). The unitarity breaking field in that case couples to the centre of mass position (the order parameter) of the crystal [12–14]:

$$i\hbar\frac{\partial}{\partial t}|\psi\rangle = \left[\frac{\hat{P}^2_{\text{tot}}}{2Nm} + i\frac{1}{2}Nm\Omega^2\left(X_{\text{c.o.m.}} - X_0\right)^2\right]|\psi\rangle. \tag{8}$$

Notice that the unitarity breaking field in this expression is an imaginary version of the symmetry breaking field in equation (5), and that in the unitary part we focus on the collective motion of the crystal as a whole only, since the order parameter field will not affect the internal dynamics.

The effect of the non-unitary dynamics defined by equation (8) depends not only on the time evolution operator, but also on the initial state it is applied to. To characterise the possible effects, consider the three different initial conditions displayed on the left side of figure 3: (a) a uniformly spread, zero-momentum wave packet with equal amplitude everywhere in space, (b) a sharply localised wave packet at a position different from $x = X_0$, and (c) a superposition of two sharply localised wave packets with different amplitudes and different distances from $x = X_0$. The first two initial states can be seen as the extreme cases representing the ground states of the symmetric Hamiltonian and a symmetry broken, localised crystal respectively. The third initial state is an interpolation between these extremes. The initial states can be straightforwardly numerically integrated forward in time [12], resulting in the final states at late times displayed on the right of figure 3. Notice that during the time evolution defined by equation (8), the norm of the total wave function is not conserved. This is not a problem, since in quantum mechanics only relative amplitudes lead to observable predictions. All the usual interpretations of the laws of quantum theory remain consistent and well-defined, as long as expectation values are taken relative to the instantaneous norm of the wave function, so that:

$$\langle\hat{O}\rangle(t) \equiv \frac{\langle\psi(t)|\hat{O}|\psi(t)\rangle}{\langle\psi(t)|\psi(t)\rangle}. \tag{9}$$

The eventual fate of the wave function at infinite time $t \to \infty$ is found to be a localised state at $x = X_0$ regardless of the initial state. This can be easily understood, as the non-Hermitian term in the evolution of equation (8) suppresses any component of the wave function away from $x = X_0$. For the completely delocalised initial state, which is an eigenstate of the total momentum, the suppression of weight by the non-unitary term is the dominant process, and the time scale on which the wave packet is localised can be estimated by considering the evolution in the absence of any Hermitian term. A straightforward calculation then shows that the spread of an initially delocalised wave packet will be reduced to a single lattice spacing, a, within the time $t_{\text{loc}} = \hbar/(2Nm\Omega^2a^2)$. Direct numerical integration shows that a non-zero kinetic energy term will affect the prefactor of this localisation time, but not its proportionality to $\hbar/(Nm\Omega^2a^2)$ [12]. Notice that this proportionality immediately implies that the thermodynamic limit and the limit of vanishing perturbation do not commute:

$$\lim_{N\to\infty}\lim_{\Omega\to 0} t_{\text{loc}} = \infty \tag{10}$$

$$\lim_{\Omega\to 0}\lim_{N\to\infty} t_{\text{loc}} = 0. \tag{11}$$

Journal of Physics: Conference Series **626** (2015) 012012 doi:10.1088/1742-6596/626/1/012012

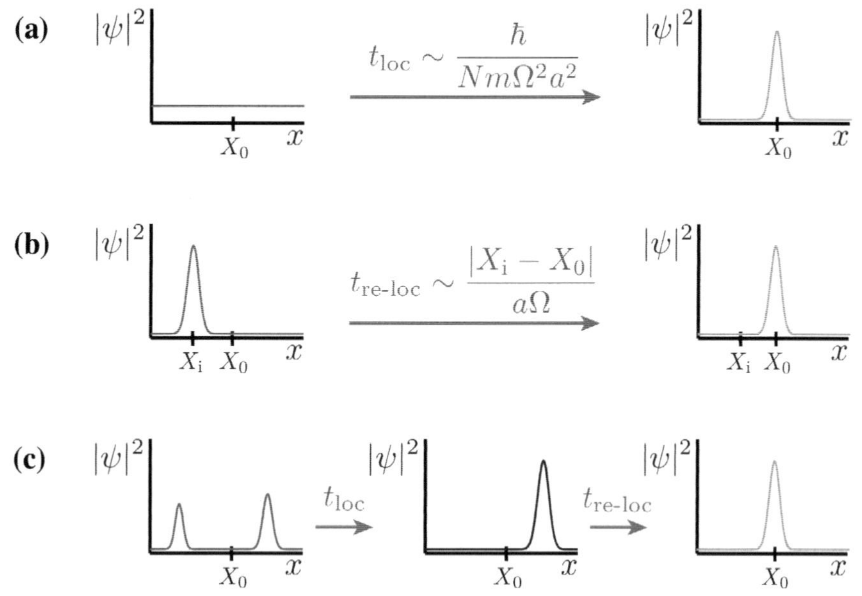

Figure 2. Sketch of the dynamics resulting from the non-unitary time evolution defined by equation (8) for different initial conditions. **(a)** Starting from a completely delocalised initial state, the wave function will be localised by the non-unitary perturbation within the time $t_{\text{loc}} \sim \hbar/(Nm\Omega^2a^2)$. **(b)** Starting from an already localised initial state at position $x = X_{\text{i}}$, the wavefunction will be re-located to position $x = X_0$ within the time $t_{\text{re-loc}} \sim |X_{\text{i}} - X_0|/(a\Omega)$. **(c)** A superposed initial state combines the two types of dynamics, and will first be reduced to a localised wave function at a single position within time t_{loc}, after which it will re-locate to position $x = X_0$ within the time $t_{\text{re-loc}}$.

We thus find that infinitely large quantum systems are sensitive to even infinitesimally small non-unitary perturbations, which will instantaneously localise any delocalised state. In other words, the unitary quantum dynamics of a delocalised state is *unstable* against non-unitary perturbations. For finite sized objects this means that the dynamics of large delocalised quantum systems will be dominated entirely by the influence of arbitrarily small non-unitary terms, if any exist.

Starting from an already localised initial state, the dynamics resulting from equation (8) behaves differently. In this case, the initial wave packet has zero amplitude at position $x = X_0$, and the non-unitary term can therefore not amplify the relative weight of that component. The initial time evolution is instead dominated by the Hermitian kinetic energy, which tends to spread a localised state. Disregarding the non-unitary term entirely, it is straightforward to see that the time it takes for an initially localised wave packet to be spread to a width $|X_{\text{i}} - X_0|$, is given by $t_{\text{spread}} = 2Nm|X_{\text{i}} - X_0|^2/\hbar$. Once the wave function has been spread over this distance, it develops a non-zero amplitude at the position $x = X_0$, and the localising effect of the non-unitary term takes over once again. The combined time scale needed for the wave packet to first spread in space and then localise again, can be estimated by the geometric mean of the individual contributions. This gives a re-location time for the initially localised state of $t_{\text{re-loc}} = \sqrt{t_{\text{spread}}t_{\text{loc}}} = |X_{\text{i}} - X_0|/(a\Omega)$. Once again, the direct numerical evaluation

Journal of Physics: Conference Series **626** (2015) 012012 doi:10.1088/1742-6596/626/1/012012

of the dynamics confirms that the proportionality of the re-location time to $|X_i - X_0|/(a\Omega)$ is maintained also if both terms in the evolution operator are simultaneously present [12]. Notice that in the case of the re-location time, the thermodynamic limit does commute with the limit of vanishing perturbation:

$$\lim_{N\to\infty}\lim_{\Omega\to 0} t_{\text{re-loc}} = \infty \tag{12}$$

$$\lim_{\Omega\to 0}\lim_{N\to\infty} t_{\text{re-loc}} = \infty. \tag{13}$$

Even in the presence of a non-unitary perturbation, the dynamics of a large, localised quantum object is therefore *stable*. The non-unitary perturbation will not have an effect on a localised macroscopic state within any measurable time.

Finally, consider an initial state superposed with different weights over two positions, at different distances to $x = X_0$. In this case, the dynamics consists of both a localisation and a re-location. At first, the component closest to $x = X_0$ is amplified by the non-unitary term relative to the further component. As before, this results in a localised wave packet, but this time the centre of mass is determined by the position of the amplified component of the initial wave function. The time scale over which the localisation takes place is again given by t_{loc}. Once the wave function is localised, the slower re-location process takes over, and the wave packet is re-located to $x = X_0$ within the time $t_{\text{re-loc}}$. Taking the thermodynamic limit and the limit of vanishing perturbation, we then find that the dynamics of an initially superposed state is *unstable* agains non-unitary perturbations. Even an infinitesimally small non-unitary perturbation will cause a spatially superposed macroscopic quantum state to collapse onto just one of its components, which then is stable agains any further effects of non-unitary terms in the dynamics.

4. Interpretation and speculation

So far, all presented results are exact. It is a mathematical fact that the time evolution generated by the quantum mechanical Hamiltonian is unstable against even infinitesimally small non-unitary perturbations in the thermodynamic limit. To see if this mathematical observation also has any physical consequences, we need to consider whether an arbitrarily small, but non-zero, unitarity breaking perturbation could exist even in principle. If we consider closed systems only, and if we additionally assume quantum mechanics to hold universally, then the answer is clearly that such perturbations cannot exist, and the mathematical instability described above is irrelevant to physics. Relaxing either one of the constraints however, non-unitary perturbations are no longer strictly forbidden, and we can speculate about possible origins of non-unitary dynamics.

If we consider open quantum systems for example, it is well known that the influence of an unobserved environment on the system of interest can in many cases be effectively described by a unitarity violating term in its dynamics [15–17]. If the environment tends to localise a macroscopic object (which is typically the case), the non-unitary evolution will be similar to that of equation (8). One should keep in mind however that the environment is generically unknown, unpredictable, and evolving uncontrollably in time. The position X_0 at which it tends to localise the object of interest is thus an unknown parameter, whose value moreover changes unpredictably with time. This effective behaviour can be modelled by taking $X_0(t)$ to be a randomly fluctuating variable, and considering the time evolution generated by:

$$i\hbar\frac{\partial}{\partial t}|\psi\rangle = \left[\frac{\hat{P}^2_{\text{tot}}}{2Nm} + iN\gamma\left(X_{\text{c.o.m.}} - X_0(t)\right)^2\right]|\psi\rangle. \tag{14}$$

Here γ depends on the coupling of the system to the environmental degrees of freedom, and the scaling of the interaction with N has been made explicit. Assuming the environment to be infinitely large, the randomly fluctuating variable $X_0(t)$ may be modeled by white noise over the entire range of available positions. In this interpretation of the origin of non-unitary evolution, the implication of the singular thermodynamic limit is that it is impossible even in principle to isolate macroscopic objects from their environments. Large objects are sensitive to infinitesimally small interactions with the environment, which will instantaneously affect their dynamics.

An alternative way of arriving at a similar description of time evolution for macroscopic quantum objects, is to consider closed systems only, but allow for the possibility that quantum theory may not be an exact description of nature. It has been proposed for example, that the diffeomorphism invariance underlying general relativity is incompatible with the unitary nature of quantum dynamics, and that this must result in either one or both of these principles being violated at the borderline between gravity and quantum physics [18]. In one particular proposal, dimensional analysis is used to argue that this borderline may lie close to the length scales at which we cease to be able tot observe quantum behaviour in experiments [18, 19]. The non-unitary evolution discussed here, could then be interpreted as arising from a non-unitary modification of the laws of quantum mechanics due to gravity [12, 13, 20]. The strength of the non-unitary perturbation in that case should be similar to the cross-over energy scale identified as the border between the two theories. Moreover, the principle of diffeomorphism invariance can be argued to cause any measure of a distance between objects in different components of a superposed wave function (in the position basis) to be inherently ill-defined [18]. As a poor man's approach, one could model such an ill-defined distance by adding a randomly fluctuating correction to the definition of the position operator: $\hat{X} \to \hat{X} - X_0(t)$. The fluctuations $X_0(t)$ do not correspond to any physical quantum field, but instead represent the fact that when the effects of diffeomorphism invariance start to play a role, distances between different components of the wave function cannot be predicted with certainty anymore [12, 13]. The dynamics arising from such considerations of a macroscopic object in the region where quantum mechanics just begins to be affected by the principles underlying the theory of gravity, can then be written as [12, 13]:

$$i\hbar\frac{\partial}{\partial t}|\psi\rangle = \left[\frac{\hat{P}_{\text{tot}}^2}{2Nm} + i\frac{1}{2}NmG\rho\left(X_{\text{c.o.m.}} - X_0(t)\right)^2\right]|\psi\rangle. \tag{15}$$

Here G is Newton's gravitational constant, ρ is the density of the object, and Nm its total mass. $X_0(t)$ is again a random variable that can be modeled by white noise over the entire range of positions. If the non-unitary term is fundamental in the sense that it comes from physics outside of quantum mechanics and applies to the evolution of single-shot experiments (as opposed to the ensemble averaged description implicit in the case of an open quantum system), then the implication of the singular thermodynamic limit must be that the dynamics of macroscopic quantum objects is unstable by itself. Large enough objects can thus not be forçed to evolve quantum mechanically under any circumstances, because they always probe the influence of physics beyond Schrödinger's equation.

The overall effect of the perturbed evolution in both equation (14) and (15) is the same as that of equation (8) which we considered before. That is, the unitary dynamics of an initially localised object is stable agains perturbations, while the dynamics of delocalised or superposed initial states are unstable. This time however, the infinite sensitivity to non-unitary perturbations does not lead to the deterministic outcome of a localised state at a predictable position. The reason is that the wave function component whose amplitude is effectively amplified by the non-unitary term in the dynamics, is the component closest to $x = X_0(t)$, which now randomly fluctuates in time. Starting for example from a superposed wave function over two initial locations, it may be that first one component is closer to $X_0(t)$, and hence is amplified, while an instant later the other component is selected, and so on. In this way, a randomly switching sequence of amplifications is applied to the two components, which terminates only when one of the components is suppressed to zero amplitude. The dynamics thus becomes a realisation of the "Gambler's Ruin" game of probability theory [21, 22]. Using the principle of 'envariance', it is possible to show that the only possible average outcome of a series of such evolutions, is the emergence of Born's rule [12, 14, 22]. That is, in each individual realisation of the dynamics of equation (14) or (15) applied to the initial state $|\psi(0)\rangle = \alpha|x = X_1\rangle + \beta|x = X_2\rangle$, the outcome is unpredictable, but will be either $|x = X_1\rangle$ or $|x = X_2\rangle$. Repeating the same evolution many times, the proportion of times that the state $|x = X_1\rangle$ is realised will be given by $|\alpha|^2$, and the proportion of outcomes at $|x = X_2\rangle$ will be $|\beta|^2$. This conclusion is confirmed by direct numerical integration of equations (14) and (15) [12, 13].

We thus find that spatial superpositions of macroscopic objects spontaneously collapse into the position basis according to the prescription of Born's rule. This collapse process is a direct consequence of the instability of their unitary quantum dynamics to even infinitesimally small non-unitary perturbations. The non-unitary term can be provided either by a modification to quantum mechanics originating from gravity, or from an effective description of the environment in an open quantum system.

5. Experimental consequences

The final result of the sensitivity of macroscopic objects to non-unitary perturbations is that such objects cannot retain a state of spatial superposition. The absence of macroscopic superpositions in the world around us is indeed easily verified. If an interaction between a microscopic quantum particle in a spatially superposed state and a localised macroscopic object is instantaneously turned on, this may instantaneously cause the macroscopic object to become entangled with the microscopic particle, and to enter into a superposed state. The subsequent very fast collapse to just one of the components, in accordance with Born's rule can then be interpreted as a measurement of the particle's position by the macroscopic object. Indeed, this process satisfies all of the usual requirements for a description of wavefunction collapse during a quantum measurement. To determine whether an evolution like the one imposed by equation (14) or (15) is responsible for these observations however, requires the experimental observation of the actual dynamical evolution, rather than only its eventual outcome.

To observe the dynamics imposed by a non-unitary perturbation within experimentally accessible time scales, the object to be studied should not be too macroscopic, so that the effect of the perturbation is not too strong and hence t_{loc} is large enough to be observable. On the other hand, it should also not be too microscopic, as otherwise the perturbation has no effect at all, and t_{loc} becomes too large to be observed. The region in between, where collapse times can be expected to be of observable sizes, falls precisely in the region that is hard to access experimentally, and which is only very recently starting to be explored by modern experiments involving macroscopic superpositions [13, 18, 23].

If such experiments succeed in observing collapse dynamics, the scaling of the time scales involved may give some hint as to the origin of the non-unitary perturbation that underlies it. A more pertinent question however, will be whether such experiments can also distinguish between perturbations coming from the effective action of an unobserved environment, and perturbation coming from a fundamental modification of Schrödinger's equation. The expected dynamics resulting from these sources is practically indistinguishable, as shown by the similarity of equations (14) and (15). There is however an essential difference between the effective dynamics of an open system, and the possibly fundamentally non-unitary dynamics of a closed system [13]: in the former case, the effective dynamics has to emerge from the neglect (tracing

Journal of Physics: Conference Series **626** (2015) 012012 doi:10.1088/1742-6596/626/1/012012

out) of some environmental degrees of freedom which are beyond the control of the experimental setup. The environment may be external or internal and involve spin, charge, massive, massless, or any other kinds of degrees of freedom, but it has to become entangled with the observed properties of the macroscopic object for the dynamics to become effectively non-unitary. Since the experiment by definition has no control over the environmental degrees of freedom, the entanglement between observed properties and the state of the environment cannot be undone once it has been created, and the collapse dynamics cannot be turned off once it has started. If on the other hand the non-unitary term is fundamental, no entanglement with environmental degrees of freedom is required. In that case, one can create a macroscopic superposition, let it begin its collapse dynamics, and then bring back both superposed components onto the initial position before the collapse is completed. Since there is no entanglement with the environment, the situation is just as it was before creating the macroscopic superposition, and this procedure can be repeated an arbitrary number of times. When entanglement with environmental degrees of freedom is present however, bringing the components of the superposed object back to their initial positions will not undo the entanglement, and hence a subsequent run of the experiment will be influenced by the entanglement left behind from previous runs. This will enhance the effectively non-unitary dynamics, which thus occurs on a shorter and shorter time scale as the experiment is repeated again and again. In this way, sequences of collapse experiments can be used to distinguish between effectively non-unitary dynamics in open systems, and truly non-unitary dynamics resulting from a fundamental modification of quantum theory [13].

6. Conclusions

We have found that the unitarity of quantum mechanical time evolution can be spontaneously broken, in precise analogy with the usual breaking of continuous symmetries under equilibrium conditions. The loss of unitary dynamics arises from an extreme sensitivity in the evolution of large quantum objects to even infinitesimally small non-unitary perturbations. If such perturbations can (effectively) exist, then regardless of their origin, it immediately follows that truly macroscopic objects can never be observed to evolve unitarily, as even the weakest unitarity violating interaction will instantaneously dominate its dynamics. The effect of this instability is that delocalised or spatially superposed macroscopic objects are instantaneously reduced to a single location. Objects which are already localised on the other hand, are insensitive to the non-unitarity, and will remain localised at the same position.

Two possible sources of unitarity violating perturbations to quantum dynamics are the influence of an unseen environment, or a fundamental modification of Schrödinger's equation. In the former case, the dynamics of an open system entangled with environmental degrees of freedom can be effectively described by a non-unitary evolution after tracing over the environmental degrees of freedom. The effective dynamics generated in this way generically tends to localise large objects. For the latter case, it has been suggested that the conflict between general relativity and quantum physics originates in the fundamentally non-unitary nature of the diffeomorphism invariant theory of gravity. It may then be expected that the laws of quantum mechanics break down at an energy scale where gravity starts to appear, and that the first effect of this breakdown can be modelled by a non-unitary modification of quantum dynamics. In either case, the scaling of the non-unitary term with system size and the randomly fluctuating nature of the location at which it tends to localise large objects, is the same. The result of a non-unitary perturbation which includes random fluctuations, is the emergence of probabilistic dynamics for large objects. That is, the instability of unitary quantum dynamics in this case implies that a superposition of a macroscopic object over multiple spatial locations is instantaneously reduced to just one of these positions. The selection of which place the wave function collapses to, is a probabilistic process. The probability for a particular component of the initial state to be selected according to the dynamics generated by an infinitesimal, fluctuating, non-unitary perturbation, is given by Born's rule.

Experimentally capturing the predicted non-unitary dynamics of macroscopic objects requires studying the dynamics of objects which are large enough to be influenced by the non-unitarity, but small enough not to be instantaneously reduced to a localised state. Such mesoscopic quantum superpostions have very recently begun to become accessible in state of the art experimental setups. As further technological development allows these experiments to proceed to spatial superposition of large enough objects for the instability to non-unitary perturbations to become noticeable, it will be necessary to develop protocols which can distinguish between the effectively non-unitary, but strictly quantum mechanical, evolution induced by environmental degrees of freedom, and fundamental modifications of Schrödinger's equation. The presence of entanglement with environmental degrees of freedom in the former case, but not the latter, is an unavoidable difference that may be employed to tell them apart.

Regardless of the source of non-unitary contributions to the (effective) dynamics, the fact that spatial superpositions of sufficiently large quantum objects are inherently unstable against arbitrarily small perturbations, will affect experiments accessing quantum states at these scales within the near future.

Acknowledgments
This work was supported by a VIDI grant financed by the Netherlands Organisation for Scientific Research (NWO).

References
[1] Berry M V 2002 *Physics Today* **55** 10–11
[2] Anderson P W 1972 *Science* **177** 393–396
[3] van Wezel J and van den Brink J 2007 *Am. J. Phys.* **75** 635–638
[4] Anderson P W 1952 *Phys. Rev.* **86** 694–701
[5] van Wezel J, van den Brink J and Zaanen J 2005 *Phys. Rev. Lett.* **94** 230401
[6] van Wezel J, Zaanen J and van den Brink J 2006 *Phys. Rev. B* **74** 094430
[7] Kaiser C and Peschel I 1989 *J. Phys. A* **22** 4257–4261
[8] Kaplan T A, von der Linden W and Horsch P 1990 *Phys. Rev. B* **42** 4663–4669
[9] van Wezel J and van den Brink J 2008 *Phys. Rev. B* **77** 064523
[10] van Wezel J and van den Brink J 2008 *Physica B* **403** 3206–3210
[11] Birol T, Dereli T, Müstecaplıoğlu Ö E and You L 2007 *Phys. Rev. A* **76** 43616
[12] van Wezel J 2010 *Symmetry* **2** 582–608
[13] van Wezel J and Oosterkamp T H 2011 *Proc. R. Soc. A* **468** 35–56
[14] van Wezel J 2008 *Phys. Rev. B* **78** 054301
[15] Caldeira A O and Leggett A J 1983 *Physica A* **121** 587–616
[16] Caldeira A O and Leggett A J 1985 *Phys. Rev. A* **31** 1059–1066
[17] Unruh W G and Zurek W H 1989 *Phys. Rev. D* **40** 1071–1094
[18] Penrose R 1996 *Gen. Rel. Grav.* **28** 581–600
[19] Diósi L 1987 *Phys. Lett. A* **120** 377–381
[20] van Wezel J and van den Brink J 2008 *Phil, Mag.* **88** 1659–1671
[21] Pearle P 1976 *Phys. Rev. D* **13** 857–868
[22] Zurek W H 2005 *Phys. Rev. A* **71** 052105
[23] Marshall W, Simon C, Penrose R and Bouwmeester D 2003 *Phys. Rev. Lett.* **91** 159903

Voices of Silence, Novelties of Noise: Oblivion and Hesitation as origins of quantum mysteries

E Cohen[1] and A C Elitzur[2]

[1]School of Physics and Astronomy, Tel Aviv University, Ramat-Aviv, Tel Aviv 6997801, Israel
[2]Iyar, The Israeli Institute for Advanced Research, Rehovot, Israel

E-mail: eliahuco@post.tau.ac.il

Abstract. Among the (in)famous differences between classical and quantum mechanics, quantum counterfactuals seem to be the most intriguing. At the same time, they seem to underlie many quantum oddities. In this article, we propose a simple explanation for counterfactuals, on two levels. Quantum Oblivion (QO) is a fundamental type of quantum interaction that we prove to be the origin of quantum counterfactuals. It also turns out to underlie several well-known quantum effects. This phenomenon is discussed in the first part of the article, yielding some novel predictions. In the second part, a hypothesis is offered regarding the unique spacetime evolution underlying QO, termed Quantum Hesitation (QH). The hypothesis invokes advanced actions and interfering weak values, as derived first by the Two-State-Vector Formalism (TSVF). Here too, weak values are argued to underlie the familiar "strong" quantum values. With these, an event that appears to have never occurred can exert causal effects and then succumb to QO by another time-evolution involving negative weak values that eliminate them. We conclude with briefly discussing the implications of these ideas on the nature of time.

1. Introduction

Seeking a suitable thank to the organizers of this unique conference and volume, we chose to present a nascent work in which rigor and speculation are hopefully well balanced. Accordingly, it has two parts:

i) Quantum Oblivion (QO) refers to a very simple yet hitherto unnoticed type of quantum interaction, where momentum appears not to be conserved. The problem's resolution lies within a brief critical interval, during which more than one interaction takes place. Rapid self-cancellation, however, leaves only one interaction completed. While the paradox's resolution is novel, the interaction itself turns out to underlie several well-known quantum peculiarities. This gives a new, realistic twist to one of QM's most uncanny hallmark, namely, the *counterfactual*. Although rigorously derived from standard QM, Oblivion offers some new experimental predictions, as well as new insights into other quantum oddities.

ii) Quantum Hesitation (QH) proceeds further to theorizing, as to *how* QO takes place. It reformulates Oblivion within time-symmetric interpretations of QM, mainly Aharonov's Two-State-Vector Formalism (TSVF). We assume that, beneath several momentary interactions, out of which only one is completed, there were several possible *histories*, which have left only one finalized. From the TSVF we then take one of its most exotic features, namely, weak values unbounded by the eigenvalue spectrum. This allows too-large/too-small/negative weak values appearing under special pre- and post-selections. Such "unphysical" values are assumed to evolve along both time directions, over the same spacetime trajectory, eventually making some interactions "unhappen" while prompting a single one to "complete its happening," until all conservation laws are satisfied over the entire spacetime region.

Naturally, the QO phenomenon and the QH hypothesis should be considered separately. Their combination, however, strives to make quantum mechanics more comprehensible and realistic, at the price of admitting that spacetime has some aspects still unaccounted for in current physical theory.

2. "If grandma had wheels, she would be a wagon": Counterfactual as the essential difference between classical and quantum mechanics

It is the quantum effects themselves, regardless of any interpretation, which are best illustrated by the cynical Yiddish aphorism quoted in this section's title. In classical physics, grandmother's wheels, which she never had, trivially play no role in her dynamics. A quantum grandmother, in contrast, manages somehow to employ them, even to the point of outrunning vehicles. This is the quantum counterfactual [1], illustrated by the following examples:

i) In the Elitzur-Vaidman experiment [2], a bomb is prepared and positioned such that, if struck by an appropriately prepared photon, it will explode. Surprisingly, even when no explosion happens, its mere *potentiality* suffices to disturb the photons interference: a non-explosive bomb leaves it undisturbed.
ii) Hence, in any interaction of a particle with more than one possible absorber, each absorber's capability of absorbing the particle takes part in determining its final position ("collapse"). Consider, e.g., a position measurement of a photon whose wave-function spreads from the source towards a circle of remote detectors: Every detector's *non-clicking* incrementally contribute to the photon's eventual position, just like the single detector that does click.
iii) Such counterfactuals underlie also quantum non-locality. Bell's theorem [3] implies the following counterfactual: Had Bob chosen to measure his particle's spin along an axis other than the one he actually chose, then Alice's spin measurement would give the opposite value.

This peculiar status of counterfactuals in QM stems from the uncertainty principle. Consider the simple setting in Fig. 1. A beam-splitter (BS) is positioned between four equidistant mirrors, Left/Right/Up/Down. We know only that, within the device, a single photon crosses the BS back and forth. This situation, with four possible positions and momenta of the photon, encapsulates the uncertainty $\Delta x \Delta p \geq \hbar/2$, well-known from the standard Mach-Zehnder interferometer (MZI).

To see that, loosen mirror L_u to enable it to measure the photon's possible interaction with it. In 50% of the cases, the detector will remain silent. This means, with certainty, that the photon hits only L_d and then reflected back to BS, where it splits towards *both* R_u and R_d, then reflected back, reunited by BS and returns *only* to L_d. We can therefore leave the detector in L_u and be sure that, despite the photon's endless oscillations, it will *never* hit this mirror. What has happened here? Apparently nothing. A possible event did not occur. Yet the fact that it *could* occur endows it a physical say. For if we turn also mirror R_u into a detector, then again in 50% cases it will not click, but now, sooner or later, L_u is bound to click! The second non-event gave rise to a real event, which until that moment, was impossible.

These four possible positions and momenta are typical quantum counterfactuals. As such,

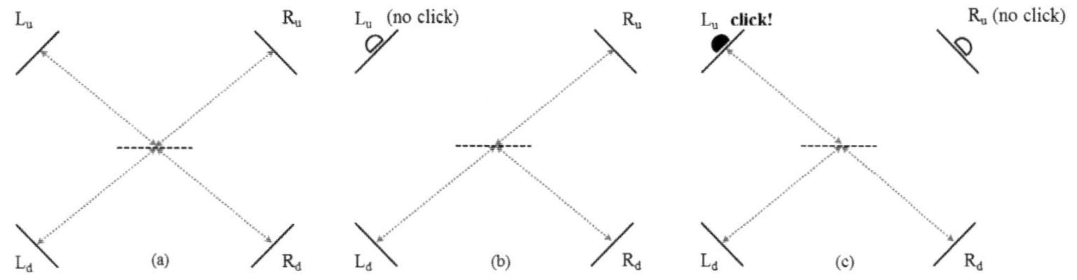

Figure 1. (a) A symmetric interferometer containing a single photon of unknown position and momentum. (b) A non-click on one corner permanently banishes the photon from that corner. (c) That this banishment was an objective effect can be proved by the fact that a second non-click causes its detection in the excluded location.

they present an acute duality:
i) As long as they are not verified, they take an essential causal part in the system's dynamics.
ii) Once, however, one of them is verified, all others vanish without a trace.

How, then, can non-events play part in the quantum process? The challenge's acuity is reflected in the various radical moves it has elicited from the reveal interpretations of QM, of which two famous schools have desperately resort to the extreme opposites:
Abandonment of ontology: Copenhagen. Since counterfactuals are facts of our knowledge, just like actual facts, let us define QM as dealing only with knowledge, information, etc., rather than with objective reality.
Excess ontology: Many worlds. The counterfactual does occur, but in a different world, split from ours at the instant of measurement.

In what follows we offer an explanation to quantum counterfactuals which is purely physical, derived from quantum theory alone: At the sub-quantum level, grandmother Nature utilizes her proverbial wheels by taking advantage of an otherwise-unhappy consequence of her age: *Amnesia*.

3. When an event is forgotten by Nature herself: Quantum Oblivion

We begin with a very simple yet surprising quantum interaction where one particle emerges from it visibly affected, while the other seems to "remember" nothing about it.

3.1. A non-reciprocal interaction

Let an electron and a positron, with spin states $|\uparrow_z\rangle = \frac{1}{\sqrt{2}}(|\uparrow_x\rangle + |\downarrow_x\rangle)$ and momenta $\langle(p_x)_{e^-}\rangle < \langle(p_x)_{e^+}\rangle$, $\langle(p_y)_{e^-}\rangle = \langle(p_y)_{e^+}\rangle$, enter two Stern-Gerlach magnets (SGMs) (drawn for simplicity as beam-splitters) positioned at (t_0, x_{e^-}, y_0) and (t_0, x_{e^+}, y_0) respectively (see Fig. 2). The SGMs split the particles' paths according to their spins in the x-direction:

$$|\psi_{e^-}\rangle = \frac{1}{\sqrt{2}}(|1'_{e^-}\rangle + |1''_{e^-}\rangle), \tag{1}$$

and

$$|\psi_{e^+}\rangle = \frac{1}{\sqrt{2}}(|2'_{e^+}\rangle + |2''_{e^+}\rangle). \tag{2}$$

Let technical care be taken such that, if the particles turn out to reside in the intersecting paths, they would meet, at t_1 or t_2, ending up in annihilation. Two nearby detectors $|READY_1\rangle$, $|READY_2\rangle$ are set to measure the photons emitted upon pair annihilation, thereupon they would change their states to $|CLICK_1\rangle$ or $|CLICK_2\rangle$.

Let us follow the time evolution of these particles. Initially, at t_0, the total wave-function is the separable state:

$$|\psi\rangle = \frac{1}{2}(|1'_{e^-}\rangle + |1''_{e^-}\rangle)(|2'_{e^+}\rangle + |2''_{e^+}\rangle)|READY_1\rangle|READY_2\rangle. \tag{3}$$

Depending on the two particles' positions at times t_1 or t_2, they may (not) annihilate and consequently (not) release photons, which would in turn (not) trigger one of the detectors. At $t_0 \leq t < t_1$, then, the superposition is still unchanged, as in 3. But at $t_1 < t < t_2$, either photons are emitted, indicating that the system ended up in $|1''_{e^-}\rangle)|2'_{e^+}\rangle|CLICK_1\rangle|READY_2\rangle$ *or not*, and then

$$|\psi\rangle = \frac{1}{\sqrt{3}}[(|1'_{e^-}\rangle + |1''_{e^-}\rangle)|2''_{e^+}\rangle + |1'_{e^-}\rangle|2'_{e^+}\rangle]|READY_1\rangle|READY_2\rangle, \tag{4}$$

which is entangled: one component of it is a definite state, while the other is a superposition in itself.

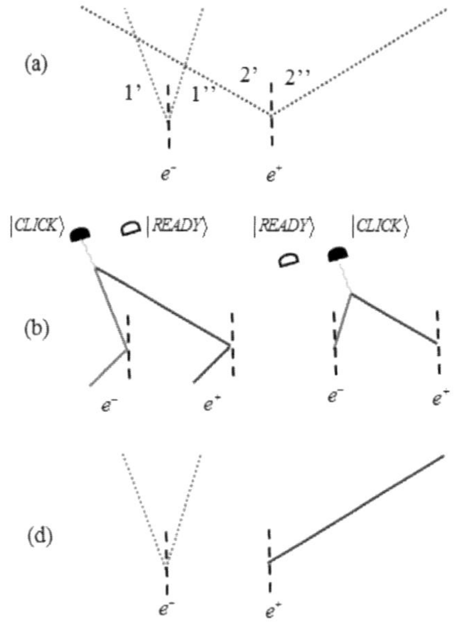

Figure 2. An electron-positron interaction and its possible outcomes. (a) The setting. (b-c) Annihilation. (d) Oblivion.

Similarly at $t > t_2$: If a photon pair is emitted, we know that the particles ended up in paths $1'$ and $2'$, i.e. $|1'_{e^-}\rangle|2'_{e^+}\rangle|READY_1\rangle|CLICK_2\rangle$, otherwise we find the non-entangled state

$$|\psi\rangle = \frac{1}{\sqrt{2}}(|1'_{e^-}\rangle + |1''_{e^-}\rangle)|2''_{e^+}\rangle|READY_1\rangle|READY_2\rangle, \tag{5}$$

Journal of Physics: Conference Series **626** (2015) 012013 doi:10.1088/1742-6596/626/1/012013

which is peculiar. The positron is observably affected: If we time-reverse its splitting, it may fail to return to its source. Its momentum has thus changed. Not so with the electron: It remains superposed, hence its time-reversibility remains intact (Fig. 3).

Summarizing, one party of the interaction "remembers" it through momentum change, while the other remains "oblivious," apparently violating momentum conservation. This is Quantum Oblivion (QO).

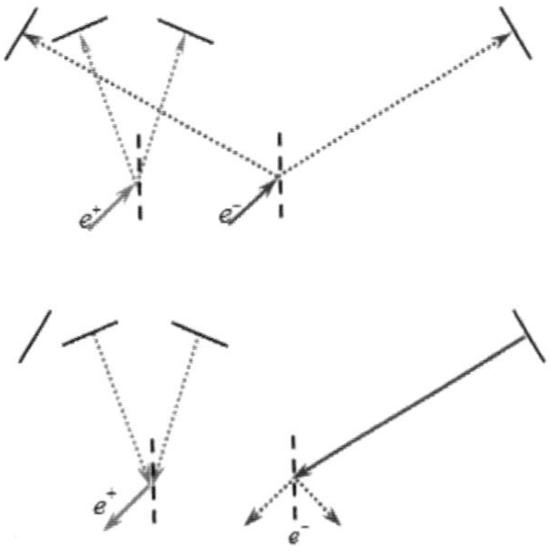

Figure 3. The peculiar momentum change due to outcome (d) in Fig. 2: The positrons interference is disturbed while that of the electron remains intact.

3.2. How is momentum conservation restored? The Critical Interval

It is obviously the intermediate time-interval $t_1 < t < t_2$ that conceals the momentum conservation in QO. The details, however, are no less interesting.

The two particles, during this interval, become entangled in their positions and momenta. Suppose, e.g. that we reunite the two halves of each particle's wave-function through the original BS, to see whether they return to their source. Either one of the particles may fail to do that, on which case the other must remain intact. Similarly for their positions. This is entanglement, identical to that of the electron-positron pair in Hardy's experiment [4], $|\psi\rangle = \frac{1}{2}(|1'_{e^-}\rangle + |1''_{e^-}\rangle)(|2'_{e^+}\rangle + |2''_{e^+}\rangle)$.

It is during this interval that the positions and momenta of the two particles become, in contrast with ordinary quantum measurement, first entangled and then non-entangled (for greater detail of this effect and its implications see [5]). It remains to be shown that something similar occurs also to the two macroscopic detectors that finalize a experiment. This is done next.

3.3. Oblivion's ubiquity: Every quantum detector's pointer must be superposed in the conjugate variable

Rather than a curious effect of a specific interaction, Oblivion is present in every routine quantum measurement. Its elucidation can therefore shed new light on the nature of measurement, further enabling some novel varieties thereof.

Ordinary quantum measurement requires a basic preparation often considered trivial. Consider e.g., a particle undergoing simple detection (as routinely employed during spin measurements [6]). The detector's pointer, positioned at a specific location, reveals the particle's arrival by receiving momentum from it. This, by definition, requires the pointer to have considerable momentum certainty (usually 0). In return, however, the pointer must have *position uncertainty*. Let this tradeoff be illustrated with a slight modification of our first experiment. In the original version (Fig. 2), the two possible interactions were annihilations, which were mutually exclusive. For the present purpose, however, let us replace annihilation by mere (elastic) collision (Fig. 4). In other words, two superposed atoms $A1$ and $A2$ interact like the electron and positron in Fig. 2, but instead of annihilating, they just collide. This can now happen on *both* possible occasions at t_1 and t_2, namely at the two locations where $A1$ can reside. With annihilations thus dropped, the outcome is even more interesting. Suppose that the detector on path $2''$ remains silent (Fig. 4b): We are now certain that the two atoms have collided, but remain oblivious about this collision's location. What we thus measure is an *ordinary momentum exchange*: Both atoms' momenta have been reversed along the horizontal axis. Here, oblivion is small, affecting only the two atoms' positions at the time of the collision. Yet, because $A2$ has vanished from the distant location on $2''$, the final outcome is a coarse position measurement of $A2$.

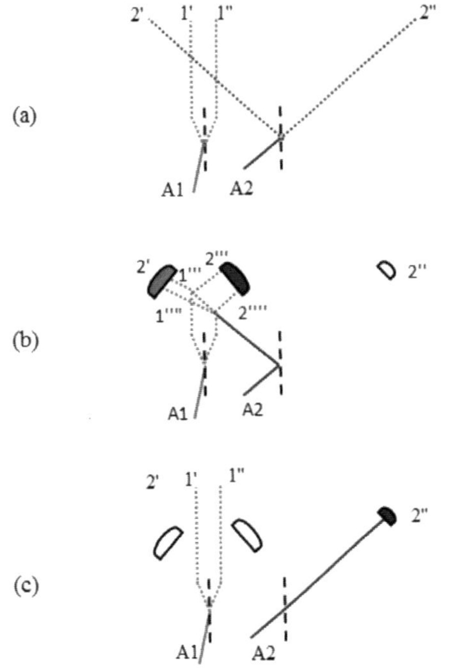

Figure 4. Same interaction as in Fig. 2 but with two atoms that do not annihilate upon interaction, hence (a) merely form elastic collision. In this version, the critical interval is the detector's long exposure time which does not allow the precise detection time. (b) Measurement ending up with collision, where the nearby detectors' widths signify that A1's and A2's positions remain unmeasured. (c) IFM of $A2$'s position.

The situation is much more surprising when we *fail* to detect $A1$ and $A2$ in the paths to which they would be diverted in case of collision, namely $1'''$, $1''''$, $2'''$, $2''''$. We are now certain that $A2$'s superposition is reduced to path $2''$ while $A1$ has returned to its initial superposition over $1'$ and $1''$ (Fig. 4c). In other words, $A2$ undergoes momentum oblivion, which is again an ordinary position measurement of $A1$, but this time the measurement is interaction-free, offering a special opportunity to observe how a counterfactual takes an integral part in the quantum evolution.

To summarize: We have studied an asymmetric interaction between two atoms, where two halves of $A1$'s wave-function interact with one half of $A2$. Two momentum exchanges between the atoms can occur: Either
i) $A2$ turns out to have collided with $A1$. This amounts to $A2$ undergoing position measurement. The price exacted by the uncertainty principle is a minor position oblivion of $A1$ and $A2$.
Or
ii) $A2$ turns out to have not collided with $A1$. This again amounts to $A2$ undergoing position measurement- collapsing it to the remote $2''$ path. Hence its momentum (interference) is visibly disturbed. This time, however, if $A1$ serves as a detector's pointer, its oblivion is amplified to a macroscopic scale, making the measurement interaction-free.

As both (i) and (ii) occur under unusually high space- and time-resolution, they enable a novel study of the Critical Interval. During this interval, entanglement between the two atoms has ensued, as they have assumed new possible locations

$$|\psi\rangle = \frac{1}{2}[|2'''\rangle(|1'''\rangle + |1''''\rangle) + |2''\rangle(|1'\rangle + |1''\rangle)], \tag{6}$$

which have remained undistinguished until the macroscopic detectors indicated that no annihilation has occurred. This has broken the symmetry between the two wave-functions, finalizing the interaction and sealing the oblivion.

The generalization is therefore natural: During every quantum measurement, the detector's pointer interacts with the particle in the same asymmetric manner as atoms 1 and 2 above: Part of the particle's wave-function with the whole wave-function of the pointer. To make the analogy complete, recall that in reality the pointer's superposition is continuous rather than discrete. As the pointer thus resides over a wide array of locations, momentum measurement becomes much more precise.

3.4. From IFM to AB: Several quantum phenomena based on QO

Not only is Quantum Oblivion essential for every quantum measurement, it is also present beneath several well-known variants thereof. In [5] we have demonstrated QO's underlying IFM [2], Hardy's paradox [4], weak measurements [7, 8], partial measurement [9, 10], the Aharonov-Bohm [12] and the quantum Zeno [13] effects. This passage from discrete to continuous superposition also opens the door for several interesting interventions, studied in [5].

3.5. Summary: Momentum measurement as an example of Oblivion

Let us summarize the Oblivion effect with the commonest and most basic example. Every time a detector's pointer is set to measure the momentum of a particle that hits it, the following happens.
i) The detector's pointer, prepared with (almost) precise zero momentum, is consequently superposed in space.
ii) Therefore, not only one particle-pointer interaction takes place; rather, several interactions occur, one after another, in all the pointer's superposed locations, as the particle proceeds along them during the Critical Interval.
iii) In all these interactions, the photon and the pointer momentarily "measure" each other's state,
iv) And each of these mutual "measurements" mixes position and momentum, hence being very inaccurate.

v) Yet none of them is amplified to a full measurement. A complex superposition of correlated states thus builds up during the Critical Interval.
vi) Only then does the pointer's state undergo amplification into a full measurement, as follows: All its possible positions fall prey to Oblivion, while all possible momenta add up to a precise value.
vii) All the above equally holds for IFM, where the momentum measured is zero.
Our counterfactuals, then, become demystified. During the CI, grandmother's wheels do appear and vanish time and again beneath the quantum noise. In the present case, these wheels are the pointer's possible locations. It is only because we choose to measure the pointer's momentum that we give up the options to extract these positions, that have now turned into counterfactuals. Otherwise, as shown in the previous section, had we chosen to measure them instead of the momentum, other quantum surprises would emerge.

4. Time-symmetric quantum causality: TSVF and weak measurements

Like two tunnels dug underneath towards one another until merging into one, the above line of investigation turns out to complement another research program, much older and renowned for its surprising predictions, and also verified time and again in laboratories. In what follows we introduce the Two-State Vector Formalism and its offshoot, weak measurement, before proceeding to describe how the two tunnels meet.

4.1. Why two-state-vectors?

TSVF originates from the work of Aharonov, Bergman and Lebowitz [14]. It asserts that every quantum system is determined by two wave-functions: One (also known as the pre-selected wave-function) evolves forward in time while the other (post-selected) evolves backward. The forward- and backward-evolving wave-functions, $|\psi_i\rangle$ and $\langle\psi_f|$ respectively, define the so called two-state vector $\langle\psi_f|\ |\psi_i\rangle$. The two wave-functions are equally important for describing the present of the quantum system via the weak value of any operator A defined by:

$$\langle A\rangle_w = \frac{\langle\psi_f|A|\psi_i\rangle}{\langle\psi_f|\psi_i\rangle} \tag{7}$$

Here a logical catch ensues: *"The state between two measurements" cannot be revealed by measurement.* Weak measurement [7, 8] was conceived in order to bypass this obstacle as well as to test other TSVF's predictions. This led to numerous intriguing works, both theoretical and experimental.

4.2. Anomalous momentum exchanges predicted under special pre- and post-selections

Apart from its time-symmetry, TSVF reveals even subtler symmetries with respect to values usually considered inherently positive. Consider the surprising values for mass and momentum in the following two experiments
i) The Three Boxes Paradox [15] is a by-now familiar surprise yielded by TSVF, of which the underlying logic can serve as an introduction to the idea of QH. A particle is prepared with equal probability to reside in one out of three boxes:

$$|\psi_i\rangle = \frac{1}{\sqrt{3}}(|1\rangle + |2\rangle + |3\rangle). \tag{8}$$

Later it is post-selected in the state

$$|\psi_f\rangle = \frac{1}{\sqrt{3}}(|1\rangle + |2\rangle - |3\rangle). \tag{9}$$

Journal of Physics: Conference Series **626** (2015) 012013 doi:10.1088/1742-6596/626/1/012013

What is the particle's state between these two strong measurements? By definition, *projective measurement is unsuitable for answering this question*, as it reveals the state upon the intermediate measurement. It is *weak* measurement, again, that comes to help. TSVF predicts the following weak values of the projection operators, $P_i \equiv |i\rangle\langle i|$ for $i = 1, 2, 3$:

$$\begin{aligned}\langle P_1\rangle_w &= 1\\ \langle P_2\rangle_w &= 1\\ \langle P_3\rangle_w &= -1.\end{aligned} \tag{10}$$

Therefore, the total number of particles is 1, as it should be, but it is a sum of two ordinary particles plus one odd.

The last equation denotes negative weak value for the very exitance in the third box. In order to fully grasp the paradoxical nature of this term, let us consider within the context of its standard versions:

If "probability 1" means "The particle certainly resides within this box";

and "probability 0" means "The particle has never resided within this box";

then "probability −1" means "The particle certainly *un*resides within this box."

As absurd as third expression may sound, this is its simplest non-mathematical meaning, the alternative being dismissing it as meaningless. The choice in [15] to trust the mathematics, has led to assigning a negative sign to every interaction involving the third box, as long as it is weak enough [16]. Obviously, this cannot be related to the particle's charge, as it played no role from the beginning. The remaining choice is mass. The simplest way to prove this prediction is through the particle's momentum: A collision with another particle must give the latter a "pull" rather than a "push," even though their initial velocities were opposite.

In passing, it is worth comparing this step of the TSVF to Dirac's choice to trust the mathematics upon encountering the negative value for the electron's charge, following the dual solution for his famous equation. That choice has later led to the discovery of the positron. This may be the case with Aharonov's present choice as well.

Can this prediction be put to test? A preliminary version has been carried out by Steinberg's group [17]

ii) The importance of negative weak values becomes highly visible through Hardy's experiment [4, 18, 19]. Two MZIs overlap in one corner (See Fig. 5). They are tuned such that electron entering the first MZI will always arrive at detector C_-, while a positron entering the second MZI will always arrive at detector C_+. Therefore, when the electron and positron simultaneously traverse the setup, they might annihilate at the intersection or make their partner reach the "forbidden" detector D_-/D_+ for the electron/positrion respectively. In case no annihilation was detected we can exclude the case they booth took the overlapping path O and their state becomes

$$|\psi_i\rangle = \frac{1}{\sqrt{3}}(|O\rangle_+|NO\rangle_- + |NO\rangle_+|O\rangle_- + |NO\rangle_+|NO\rangle_-), \tag{11}$$

i.e., at least one of the particles took the non-overlapping (NO) state. The interferometers were tuned such that C_- clicks for the electron constructive interference state $\frac{1}{\sqrt{2}}(|O\rangle_- + |NO\rangle_-)$, D_- clicks for the position constructive interference state $\frac{1}{\sqrt{2}}(|O\rangle_- - |NO\rangle_-)$ state, and similarly for C_+ and D_+. Therefore, observing clicks at D_- and D_+ is equivalent to post-selection of the state

$$|\psi_f\rangle = \frac{1}{2}(|O\rangle_+ - |NO\rangle_+)(|O\rangle_- - |NO\rangle_-). \quad (12)$$

This post-selection is rather peculiar since a click at D_- naively tells us that the positron took its overlapping path, while click at D_+ naively tells us that the electron took its overlapping path. This scenario, however, is impossible, because we know annihilation did not occur.

The paradox can be resolved within the TSVF. When we calculate the weak values of the various projection operators we find out that

$$\langle \Pi_O^- \Pi_O^+ \rangle_w = 0 \quad (13)$$

and

$$\langle \Pi_{NO}^- \Pi_O^+ \rangle_w = \langle \Pi_O^- \Pi_{NO}^+ \rangle_w = +1, \quad (14)$$

while

$$\langle \Pi_{NO}^- \Pi_{NO}^+ \rangle_w = -1. \quad (15)$$

This leads us to conclude that although the number of pairs is 1, we have two "positive" pairs and one "negative" pair- a pair of particles with opposite properties. The pair in the "NO-NO" path creates a negative "weak potential" [?], that is, when weakly interacting with any other particle in the intermediate time, its effect will have a negative sign. Moreover, we can see now the cancelation of positive and negative weak values by considering projections on the non-overlapping paths:

$$\langle \Pi_{NO}^- \rangle_w = \langle \Pi_{NO}^- \Pi_O^+ \rangle_w + \langle \Pi_{NO}^- \Pi_{NO}^+ \rangle_w = 1 - 1 = 0 \quad (16)$$

$$\langle \Pi_{NO}^+ \rangle_w = \langle \Pi_O^- \Pi_{NO}^+ \rangle_w + \langle \Pi_{NO}^- \Pi_{NO}^+ \rangle_w = 1 - 1 = 0 \quad (17)$$

iii) We can now derive from these experiments precisely the conclusion needed for our thesis: *The negative values that they reveal exist also in ordinary measurements, only perfectly hidden.* Consider then, a photon split by a BS, into 1/3 and 2/3 beams, then split the 2/3 beam further into two 1/3 beams (this setup was recently suggested by Vaidman to answer the question "where have the photons been?" [20, 21, 22]. Perform weak measurements on all three thirds of the wave-function. Now delay all three parts in separate boxes. This delay enables you to choose between two options: Either
a) Reunite the three beams for interference, and post-select destructive interference with negative sign of the third. You will get the results of in the above three boxes experiment, namely, two photon and one negative-mass photon.
Or
b) Reunite only the two split 1/3s back into the original 2/3 beam, then perform strong position measurements on each beam. You will get a photon on one side and nothing on the other. No post-selection can reveal nothing unusual in the earlier weak measurements. What was so far a trivial "nothing" thus turns out to be something much more profound, namely, a perfect mutual cancellation of positive and negative masses. Similarly for the real photon detected on the other half: It turns out to be the sum of two odd weak values.

5. The synthesis: How does grandma dispose of the wheels she denies she ever had?

So far, our discussion went strictly within standard quantum formalism. Oblivion straightforwardly follows under a finer time-resolution of the quantum interaction. Encouraged by TSVF, we now take a step further asking: How does oblivion evolve in space and time? This

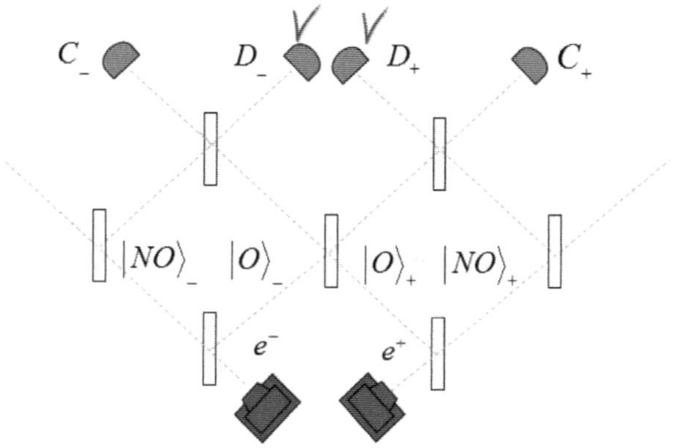

Figure 5. Hardy's setup.

is admittedly a dissembling question, pretending not to know that quantum measurement itself still poses unresolved issues; that "collapse of the wave-function" is hotly debated, etc. We thus enter the daunting minefield of interpretations of QM, where each choice carries its penalty. Equally, however, these difficulties may serve as positive incentives: To the extent that QO and TSVF illuminate the nature of quantum measurement, it is worth rephrasing older questions in the terms of these frameworks.

5.1. Grandma's wheels revisited: How do counterfactuals vanish without a trace?

Our proposal focuses on aspect (ii) of the quantum counterfactuals' dual nature (Section 2), namely, their perfect self-annihilation. Recall that any possibility to observe a counterfactual after the measurement, such as retrieving information about a particle's momentum after its position was measured, entails straightforward violation of the uncertainty principle. The challenge, therefore, is to explain not only how the counterfactual exerts such an essential role, but also how, having done that, it manages to disappear without a trace.

5.2. Clue 1: Counterfactuals momentarily show up prior to Oblivion

The first clue is already in our possession: QO has shown that counterfactuals can be observed, although indirectly beneath great quantum noise, during the Critical Interval (Sec. 3.2).

5.3. Clue 2: Time-symmetric causality may be the source of quantum oddities

We now return to Aharonov's Two State-Vector formalism described in Sec. 4.1 above, where causality is highly time-symmetric at the quantum level. Elsewhere [8, 23] we have explained our endorsement of this approach. First, it is appealingly simple. Take for example the EPR experiment: One experimenter's choice of measurement obviously affects the outcome of another measurement chosen at that moment, far away, by the other experimenter. This appears disturbing, however, only as long as one assumes the familiar causation, going from the past emission of the particles to the later measurements. If, in contrast, one allows causal relations to somehow go back and forth between the past event and its successors, much of the mystery dissolves. *What appears to be nonlocal in space, becomes perfectly local in spacetime.* Similar explanations, equally elegant, await many other quantum paradoxes, from Schrödinger's cat to the Quantum Liar paradox [11].

To be sure, the above "somehow", by which causality runs along both time directions, still needs much clarification. The model also takes the price of admitting that there is something to spacetime that still lies beyond physics' comprehension. These issues will be discussed in consecutive works. Yet we find the advantages of time-symmetric quantum causality compelling enough to pursue them, even before they mature into a fully-fledged physical theory. Our second clue is therefore: *Quantum peculiarities can be better explained by allowing causality to be time-symmetric at the quantum level.*

5.4. Clue 3: Time-symmetric quantum causality entails unusual physical values

While time-symmetric causality has been conceived by several quantum physicists (most notably Cramer's Transactional Interpretation [24]), surprising experimental derivation of it are solely due to the TSVF. These are the odd weak values described in Section 4.2 above. Although they could be derived from standard quantum theory, the fact is that they never crossed anybody's mind unless guided by TSVF. Paradoxically, it is these exotic features that go beyond interpretation, having won several laboratory verifications worldwide. Here then is our third clue: *Unusual quantum values that appear to be mere noise due to the pointer's uncertainty, turn out to be physically real, revealed by weak measurements. They may be unusually large, unusually small, and negative, even in the case of the mass, which in classical physics is known to be only positive.*

5.5. Synthesis: Quantum Hesitation

We have what we need. The established Oblivion effect (Section 3), plus the hypothesis of time-symmetric quantum causality (Section 4.1), plus the latter's more specific discovery of odd values (Section 4.2), merge into the following coherent picture: Oblivion is an evolution not only of states, but of histories. "Collapse," whereby quantum counterfactuals first play an essential role and then succumb to oblivion, occurs by (i) retarded and advanced actions complementing one another along the same spacetime trajectory, and (ii) negative values taking place alongside with positive ones. To prove this assertion, we have to show that its most exotic ingredient, namely the odd values, rather than being fringe phenomena, are in fact omnipresent. As paradoxical as this may sound, the mathematical proof is fairly straightforward.

5.6. Ubiquity: Weak values affect all quantum values

As in the case of oblivion, we now show that weak values, rather than an exotic curiosity, are part and parcel of every quantum value. The proof relies on two ranges of continuum. Suppose that we gradually move (i) from weak measurements to strong ones, and/or (ii) from special pre- and post-selections to non-special ones. Would odd weak values vanish? This indeed appears to be the case, but in reality they become stronger, only counterbalanced by opposite weak values so as to give the familiar quantum values. Moreover, as one of us has shown in [27] there exists a continuous mapping between weak and projective measurement, allowing to decompose the latter into a long sequence of weak measurements. Performing a sequence of weak measurements on a single particle, is amount to a biased random walk with incremental. Finally, the measured state is randomly driven into a definite vector out of the measurement basis.

Ergo, Upon gradually moving from weak to strong measurements, odd values do not diminish, but rather increase. Its only through their addition with other weak values that they add up or cancel out, either completing or canceling each other to give the familiar quantum values, including 0.

5.7. The missing link: Weak values play an essential role in quantum interference and contextuality

The above idea seem to be independently emerging in the very current literature on weak values. The main argument of this work nothing short of astonishing, but perfectly converges with ours: Weak values take part in the formulation of the most salient features of the quantum realm.
i) Contextuality: When you measure a particle's spin along a certain direction and get the value "up," you can be sure that, had the SGM been oriented to the opposite direction, the results would have been "down." Similarly for other directions: The outcome should deviate from your "up" by the angle difference between the directions. This is common logic. Quantum contextuality, however, poses an odd restriction on these counterfactual outcomes: They have no objective existence, because they depend on the very orientation you chose first. In other words, there cannot be some objective "up/down" along a certain spin direction which has existed prior to your choice, such that the angle you choose would be related to it. On the contrary, it is the direction that you chose for whatever reason that serves as the basis for all other possible choices. Although contextuality and non-locality of quantum phenomena are dominating strong values, the work of Pusey [25], as well as our previous works [23, 19] have recently shown respectively that interfering weak values can well account for the contextuality and non-locality of QM.
ii) Interference: Classical waves interfere in strict accordance with everyday intuition. Single-particle interference, in contrast, is paradoxical in that the location of a single particle after passing the interference device is strictly determined by counterfactual paths. Here too, Mori and Tsutumi [26] have recently argued that weak values take part in the formation of this quantum phenomenon. Their "weak trajectories," which the single particle is supposed not to have traversed, are shown on one hand to be detectable by weak measurement and on the other hand to add up to give the wave-function's familiar undulatory motion.

6. Summary: Silence and noise as the progenitors of the quantum signal

This article recounts a search for a better understanding of the unique causal efficacy of quantum events that failed to occur. We have revisited two opposite characteristics unique to quantum physical values. In terms of information theory, classical values constitute, by definition, signals. Quantum values, in contrast, are sometimes equivalent to lack of signals, namely, silence, and sometimes to a surplus of signals, namely, noise. Perhaps, then, instead of ignoring the former and trying to filter out the latter, it is time to take them as complementary to signals, even as their constituents? We have therefore studied two such quantum phenomena.
Interaction-Free Measurement indicates that a detector's silence is as causally potent as its click. Seeking to understand this potency, we pointed out Quantum Oblivion as the basic mechanism underlying it. Upon further study, QO turned out to underlie every standard quantum measurement.
Weak Values appear amid enormous quantum noise inflicting the quantum weak measurement. TSVF, however, has extracted from them a great deal of novel information about the quantum state. They reveal, e.g. negative weak values emerging in many quantum states. Recently, it has been argued that these weak values take part also in more common quantum phenomena, such as contextuality and interference.

Our synthesis of these two lines of research has led to the model of Quantum Hesitation: *The weak values, taking part in the time-symmetric evolution of the quantum state, are responsible for the very phenomenon of quantum measurement which we take as real.* Following the TSVF, we suggest that the measurement's outcome, randomly coming out of several possible ones, is the sum of the two state vectors going back and forth in time. This outcome may therefore emerge from weak values, even odd ones, brought by these two state vectors. Similarly for the even more intriguing disappearance of the other possible outcomes: It may be negative weak values, contributed by one state vector, which precisely cancel the positive weak value of the other state vector. Hence, perhaps, the apparently innocuous "0," which nevertheless endows quantum grandmother such intriguing mobility.

Can this account boast greater rigor than just one more interpretation of QM? In conclusion we want to point out what we consider to be the greatest difficulty hindering such an advance. The above account invokes two distinct causal chains contacting between past and future events, *traversing the same spacetime trajectory*. Such a far-reaching idea may invoke more new questions than answers for the old ones. The most acute problem concerns the nature of time: Is there more to physical time then the geometrical characteristics assigned to it by special and general relativity? Or does it allow some yet unknown "becoming" [28] that involves a privileged "now" moving from past to future?

The two major time-symmetric interpretations of quantum mechanics, namely TSVF and the Transactional Interpretation, openly admit that these issues are sorely obscure. Recently both schools produced novel accounts that deal with time's nature in radical ways [29, 30]. As our own contribution to these issues has been only to stress their acuity [31], we can only hope for this kind of theorizing to be even more radical in the future.

Acknowledgments
We thank Tomer Landsberger and Tomer Shushi for very helpful comments and discussions. E.C. was supported in part by Israel Science Foundation Grant No. 1311/14.

[1] Vaidman L 2009 Counterfactuals in quantum mechanics, in *Compendium of Quantum Physics: Concepts, Experiments, History and Philosophy* ed D Greenberger, K Hentschel and F Weinert (Berlin: Springer-Verlag) p 132 (*Preprint* arXiv:0709.0340)
[2] Elitzur A C and Vaidman L 1993 *Found. Phys.* **23** 987
[3] Bell J S 1964 *Physics* **1** 195
[4] Hardy L 1992 *Phys. Rev. Lett.* **68** 298
[5] Elitzur A C and Cohen E 2015 *Int. J. Quantum Inf.* **12** 1560024
[6] Wan K K and McLean R G 1991 *J. Phys. A: Math. Gen.* **24** L425
[7] Aharonov Y, Albert D and Vaidman L 1988 *Phys. Rev. Lett.* **60** 1351
[8] Aharonov Y, Cohen E and Elitzur A C 2014 *Phys. Rev.* A **89** 052105
[9] Elitzur A C and Dolev S 2001 *Phys. Rev.* A **63** 062109
[10] Elitzur A C and Cohen E 2011 *AIP Conf. Proc.* **1408** 120
[11] Elitzur A C and Dolev S 2006 *AIP Conf. Proc.* **863** 27
[12] Aharonov Y and Bohm D 1959 *Phys. Rev.* **115** 485
[13] Sudarshan E C G and Misra B 1977 *J. Math. Phys.* **18** 756
[14] Aharonov Y, Bergman P G and Lebowitz J L 1964 *Phys. Rev.* **134** 1410
[15] Aharonov Y and Vaidman L 1991 *J. Phys. A: Math. Gen.* **24**, 2315
[16] Aharonov Y, Cohen E and Ben-Moshe S 2014 *EPJ Web Conf.* **70** 00053
[17] Resch K J, Lundeen J S and Steinberg A M 2004 *Phys. Lett.* A **324** 125
[18] Aharonov Y, Botero A, Popescu S, Reznik B and Tollaksen J 2002 *Phys. Lett.* A **301** 130
[19] Aharonov Y and Cohen E 2015 Weak values and quantum nonlocality, to be published in *Quantum Nonlocality and Reality* ed M Bell and G Shan (Cambridge: Cambridge Univ. Press) (*preprint* http://www.ijqf.org/wps/wp-content/uploads/2015/01/AharonovCohen.pdf)
[20] Vaidman L 2013 *Phys. Rev.* A **87** 052104
[21] Danan A, Farfurnik D, Bar-Ad S and Vaidman L 2013 *Phys. Rev. Lett.* **111** 240402
[22] Vaidman L 2014 *Phys. Rev.* A **89** 024102
[23] Aharonov Y, Cohen E and Elitzur A C (2015) *Ann. Phys.* **355** 258
[24] Cramer J G 1986 *Rev. Mod. Phys.* **58** 647
[25] Pusey M F 2014 *Phys. Rev. Lett.* **113** 200401

[26] Mori T and Tsutsui I 2014 Weak value and the wave-particle duality *Preprint* arXiv:1410.0787
[27] Tamir B, Cohen E and Priel A 2015 *Quantum Stud.: Math. Found.* **3**
[28] Davies P 1996 *About time: Einstein's unfinished revolution* (New-York: Simon and Schuster)
[29] Kastner R E 2012 *The transactional interpretation of quantum mechanics: the reality of possibility* (Cambridge: Cambridge Univ. Press)
[30] Aharonov Y, Popescu S and Tollaksen J 2014 Each instant of time new universe, in *Quantum Theory: A Two-Time Success Story* ed D C Struppa and J M Tollaksen (Milan: Springer) pp 21-36
[31] Cohen E and Elitzur A C 2014 *EPJ Web Conf.* **71** 00028

Journal of Physics: Conference Series **626** (2015) 012014 doi:10.1088/1742-6596/626/1/012014

Noncommutative spectral geometry, Bogoliubov transformations and neutrino oscillations

Maria Vittoria Gargiulo[1], **Mairi Sakellariadou**[2] **and Giuseppe Vitiello**[1]

[1] Dipartimento di Fisica "E.R.Caianiello", I.N.F.N., Università di Salerno, I-84100 Salerno, Italy

[2] Department of Physics, King's College, University of London, Strand WC2R 2LS, London, U.K.

E-mail: vitiello@sa.infn.it

Abstract. In this report we show that neutrino mixing is intrinsically contained in Connes' noncommutative spectral geometry construction, thanks to the introduction of the doubling of algebra, which is connected to the Bogoliubov transformation. It is known indeed that these transformations are responsible for the mixing, turning the mass vacuum state into the flavor vacuum state, in such a way that mass and flavor vacuum states are not unitary equivalent. There is thus a red thread that binds the doubling of algebra of Connes' model to the neutrino mixing.

1. Introduction

In the noncommutative spectral geometry (NCSG) construction [1]-[17], introduced by Alain Connes, non commutative geometry is combined with a spectral triples so to obtain the full Lagrangian of the Standard Model (SM), minimally coupled with gravity and compatible with neutrino mixing. The coupling with gravity is obtained thanks to the fact that this construction uses a group of symmetry which encodes both, the diffeomorphism which control general relativity, and the local gauge invariance which rules the gauge theories the SM is based on.

The key point in Connes' construction is the doubling of the space and of the algebra. It has been shown [18] that this doubling is related to the gauge field structure, to dissipation and that the NCSG classical construction carries in itself the seeds of quantization. Closely following ref. [19], in this report we show that the doubling of the algebra is also the main mathematical tool on which neutrino mixing is based.

A brief summary of the NCSG construction is presented in Section 2. The doubling of the algebra and the Bogoliubov transformations are discussed in Sections 3 and 4 and the neutrino mixing is studied in Section 5. Section 6 is devoted to the conclusions.

2. A brief summary of NCSG construction

The skeleton of the NCSG construction for the Standard Model is presented below.

Considering the action functional S at low energy scale, $S = S_{\mathcal{E}-\mathcal{H}} + S_{\mathcal{SM}}$, which is the sum of the Einstein-Hilbert action ($S_{\mathcal{E}-\mathcal{H}}$) and SM action ($S_{\mathcal{SM}}$), it can be noticed that the two parts do not share the same symmetries. The former is ruled by outer

Journal of Physics: Conference Series **626** (2015) 012014 doi:10.1088/1742-6596/626/1/012014

automorphism invariance (diffeomorphism) the latter by inner automorphism (local gauge transformation).Connes considers then a model of a two-sheeted space, realized as the product of a four dimensional smooth compact Reimannian manifold $\mathcal{M}$ with a fixed spin structure by a discrete noncommutative space $\mathcal{F}$ composed by only two points. The geometric space is thus defined as the tensor product of the continuous geometry of $\mathcal{M}$ by an internal geometry $\mathcal{F}$.

In this approach, the SM is seen as a phenomenological model with the geometry of space-time such that the Maxwell-Dirac action functional leads to the SM action. The noncommutative nature of the discrete space $\mathcal{F}$ is expressed by the real spectral triple $\mathcal{F} = (\mathcal{A}_{\mathcal{F}}, \mathcal{H}_{\mathcal{F}}, D_{\mathcal{F}})$, where $\mathcal{A}_{\mathcal{F}}$ is an involution of operators on the finite-dimensional Hilbert space $\mathcal{H}_{\mathcal{F}}$ of Euclidean fermions, and $\mathcal{D}_{\mathcal{F}}$ is a self-adjoint unbounded operator in $\mathcal{H}_{\mathcal{F}}$. The spectral nature of the triple implies that the Dirac operator $D_{\mathcal{F}}$ of the internal space is the fermionic mass matrix. The Dirac operator is the inverse of the Euclidean propagator of fermions. The information carried by the metric are contained in the algebra $\mathcal{A}_{\mathcal{F}}$. The spectral nature approach is intrinsic to the the noncommutative spectral geometry. The $\mathcal{M} \times \mathcal{F}$ product geometry is specified by the rules:

$$\mathcal{A} = C^{\infty}(\mathcal{M}) \otimes \mathcal{A}_{\mathcal{F}} = C^{\infty}(\mathcal{M}, \mathcal{A}_{\mathcal{F}}), \tag{1}$$
$$\mathcal{H} = \mathcal{L}^2(\mathcal{M}, S) \otimes \mathcal{H}_{\mathcal{F}} = \mathcal{L}^2(\mathcal{M}, S \otimes \mathcal{H}_{\mathcal{F}}) \tag{2}$$
$$D = \partial\!\!\!/_{\mathcal{M}} \otimes 1 + \gamma_5 \otimes D_{\mathcal{F}}, \tag{3}$$

where $\mathcal{L}^2(\mathcal{M}, S)$ is the space of square integrable Dirac spinors over $\mathcal{M}$, the Dirac operator on $\mathcal{M}$ is denoted by $\partial\!\!\!/_{\mathcal{M}} = i\gamma^{\mu}\nabla^s_{\mu}$ (with ∇^s_{μ} is the spin connection $\nabla^s_{\mu} = \partial_{\mu} + \frac{1}{2}\omega^{ab}_{\mu}\gamma_{ab}$), γ_5 is the chirality operator in the four-dimensional case, $C^{\infty}(\mathcal{M})$ is the algebra of smooth functions on $\mathcal{M}$ acting on $\mathcal{H}_{\mathcal{F}}$ as simple multiplication operators. Within the noncommutative spectral geometry D plays the role of the inverse of the line element ds.

The main results of the model are obtained by using the spectral action principle, which is of the form

$$Tr(f(\frac{\mathcal{D}}{\Lambda})), \tag{4}$$

where $\mathcal{D}$ denotes the inverse of the Dirac operator, Λ fixes the energy scale and f is a cut-off positive even function of the real variable; it falls to zero for large values of its argument. The function f only plays a role through its momenta f_0, f_2 and f_4, where $f_k = \int_0^{\infty} f(u)u^{k-1}\, du$ for $k > 0$ and $f_0 = f(0)$, which are three parameters of the model and are related to the coupling constant at unification, the gravitational constant and the cosmological constant, respectively. Its asymptotic expansion is:

$$Tr(f(\frac{\mathcal{D}}{\Lambda})) \sim 2\Lambda^4 f_4 a_0 + 2\Lambda^2 f_2 a_2 + f_0 a_4. \tag{5}$$

The action functional, applied to inner fluctuations, only accounts for the bosonic part of the model. The term in Λ^4 gives the cosmological term, the term in Λ^2 gives the Einstein-Hilbert action functional, and the Λ-indipendent term yields the Yang-Mills action for the gauge fields. The coupling with fermions is obtained by adding the term $\frac{1}{2} < J\psi, \mathcal{D}\psi >$, where J is a real structure on the spectral triple and ψ is a spinor in the Hilbert space $\mathcal{H}_{\mathcal{F}}$ of the quarks and leptons. The computation of the spectral action functional gives the full Lagrangian for the Standard Model minimally coupled with gravity, with neutrino mixing and Majorana mass terms. The vacuum expectation value of the Higgs field is related to the non commutative distance between the two sheets. In the model neutrino mixing is obtained in analogy with the quarks case.

3. Doubling the Algebra and the deformed Hopf algebra coproduct

In this Section we show that Bogoliubov transformations are "built in" in the algebra doubling in the NCSG. This implies in turn that the the NCSG construction insists on a collection of unitarily inequivalent Hilbert spaces which are related among themselves (phase transitions) by Bogoliubov transformations. Since these also characterize the mixing transformations of neutrinos, the NCSG construction also contains implicitly the neutrino mixing phenomenon.

Consider the Dirac spectral triple $(\mathcal{A}_1, \mathcal{H}_1, \mathcal{D}_1) = (C^\infty(\mathcal{M}, \mathcal{L}^2(\mathcal{M}, S), \partial\!\!\!/_{\mathcal{M}})$ and its product with the finite geometry $(\mathcal{A}_2, \mathcal{H}_2, \mathcal{D}_2) = (\mathcal{A}_\mathcal{F}, \mathcal{H}_\mathcal{F}, \mathcal{D}_\mathcal{F})$. The product geometry $\mathcal{M} \times \mathcal{F}$ is given by

$$\begin{aligned} \mathcal{A} &= \mathcal{A}_1 \otimes \mathcal{A}_2 \ , \ \mathcal{H} = \mathcal{H}_1 \otimes \mathcal{H}_2 \ , \\ \mathcal{D} &= \mathcal{D}_1 \otimes 1 + \gamma_1 \otimes \mathcal{D}_2 \ , \\ \gamma &= \gamma_1 \otimes \gamma_2 J = J_1 \otimes J_2 \ , \end{aligned} \tag{6}$$

with

$$J^2 = -1, \quad [J, \mathcal{D}] = 0, \quad [J_1, \gamma_1] = 0, \quad \{J, \gamma\} = 0, \tag{7}$$

where J is an antilinear isometry and commutators and anticommutators are denoted by square and curl brackets, respectively. In the formalism of the algebra doubling an important rôle is played by the noncommutative q-deformed Hopf algebra, pointing to a deep physical meaning of the noncommutativity in this construction. Indeed, the map $\mathcal{A} \to \mathcal{A}_1 \otimes \mathcal{A}_2$ is just the Hopf coproduct map $\mathcal{A} \to \mathcal{A} \otimes \mathbf{1} + \mathbf{1} \otimes \mathcal{A}$. Actually for a noncommutative algebra the coproduct is defined by

$$\Delta a_q = a_q \otimes q^H + q^{-H} \otimes a_q \tag{8}$$

$$\Delta a_q^\dagger = a_q^\dagger \otimes q^H + q^{-H} \otimes a_q^\dagger. \tag{9}$$

The fermionic algebra $h(1|1)$ is generated by a set of operators $\{a, a^\dagger, H, N\}$ with anticommutation relations:

$$\{a, a^\dagger\} = 2H, \quad [N, a] = -a, \quad [N, a^\dagger] = a^\dagger, \quad [H, \bullet] = 0. \tag{10}$$

The deformed algebra $h_q(1 \mid 1)$ relations are

$$\{a_q, a_q^\dagger\} = [2H]_q, \quad [N, a_q] = -a_q, \quad [N, a_q^\dagger] = a_q^\dagger, \quad [H, \bullet] = 0, \tag{11}$$

where $N_q \equiv N$ and $H_q \equiv H$. The Casimir operator $\mathcal{C}_q$ is given by $\mathcal{C}_q = N[2H]_q - a_q^\dagger a_q$, where

$$[x]_q = \frac{q^x - q^{-x}}{q - q^{-1}}.$$

and together with the coproduct defined in Eqs.(8) we have

$$\Delta H = H \otimes \mathbf{1} + \mathbf{1} \otimes H, \tag{12}$$

$$\Delta N = N \otimes \mathbf{1} + \mathbf{1} \otimes N. \tag{13}$$

Requiring that a and $a^\dagger$, and a_q and $a_q^\dagger$ are adjoint operator implies that q can only be real or of modulus one; for fermion, in fact, $q \sim e^{i\theta}$. Considering a two-mode Fock space $F_2 = F_1 \otimes F_1$ it is possible to write

$$\Delta a_q = a_1 q^{\frac{1}{2}} + q^{-\frac{1}{2}} a_2, \tag{14}$$

$$\Delta a_q^\dagger = a_1^\dagger q^{\frac{1}{2}} + q^{-\frac{1}{2}} a_2^\dagger, \tag{15}$$

$$\Delta N = N_1 + N_2, \quad \Delta H = 1. \tag{16}$$

One key point is that the noncommutative Hopf coproduct turns out to be strictly related to the Bogoliubov transformations [20]. It is indeed possible to identify $a_1 \equiv a$, $a_1^\dagger \equiv a^\dagger$, and $a_2 \equiv \tilde{a}$, $a_2^\dagger \equiv \tilde{a}^\dagger$, where $\tilde{O}$, with O a generic operator, denotes the "tilde-coniugation" operation defined by the "tilde-coniugation" rules [21]:

$$\widetilde{(OO')} = \tilde{O}\tilde{O}', \tag{17}$$

$$\widetilde{(\alpha O + \beta O')} = \alpha^* \tilde{O} + \beta^* \tilde{O}', \tag{18}$$

$$\widetilde{(O^\dagger)} = \tilde{O}^\dagger, \tag{19}$$

$$\widetilde{(\tilde{O})} = O, \tag{20}$$

$$\{O, \tilde{O}'\} = \{O, \tilde{O}'^\dagger\} = 0, \tag{21}$$

$$\widetilde{|O\rangle} = |O\rangle. \tag{22}$$

It is now possible to define the new operators A_q and B_q as follow

$$A_q \equiv \frac{\Delta a_q}{\sqrt{[2]_q}} = \frac{1}{\sqrt{[2]_q}}(e^{i\theta} a + e^{-i\theta}\tilde{a}), \tag{23}$$

$$B_q \equiv \frac{1}{i\sqrt{[2]_q}} \frac{\delta}{\delta\theta} \Delta a_q = \frac{2q}{\sqrt{[2]_q}} \frac{\delta}{\delta q} \Delta a_q = \frac{1}{\sqrt{[2]_q}}(e^{i\theta} a - e^{-i\theta}\tilde{a}). \tag{24}$$

and h.c., with $q = q(\theta) \equiv e^{i2\theta}$. The anti-commutation relation are:

$$\{A_q, A_q^\dagger\} = 1, \quad \{B_q, B_q^\dagger\} = 1, \quad \{A_q, B_q\} = 0, \tag{25}$$

$$\{A_q, B_q^\dagger\} = -i \tanh i2\theta, \tag{26}$$

whereas all other anti-commutators are equal to zero. A set of operators A and B with canonical commutation relations and commuting among themselves is given by

$$A(\theta) \equiv \frac{\sqrt{[2]_q}}{2\sqrt{2}} \left[A_{q(\theta)} + A_{q(-\theta)} + A^\dagger_{q(\theta)} - A^\dagger_{q(-\theta)}\right], \tag{27}$$

$$B(\theta) \equiv \frac{\sqrt{[2]_q}}{2\sqrt{2}} \left[B_{q(\theta)} + B_{q(-\theta)} - B^\dagger_{q(\theta)} + B^\dagger_{q(-\theta)}\right]. \tag{28}$$

and h.c., so that

$$\{A(\theta), A^\dagger(\theta)\} = 1, \quad \{B(\theta), B^\dagger(\theta)\} = 1, \quad \{A(\theta), B^\dagger(\theta)\} = 0. \tag{29}$$

and all other anti-commutators equal to zero. It's also possible to write:

$$A(\theta) = \frac{1}{\sqrt{2}}\big(a(\theta) + \tilde{a}(\theta)\big), \qquad B(\theta) = \frac{1}{\sqrt{2}}\big(a(\theta) - \tilde{a}(\theta)\big). \tag{30}$$

with

$$a(\theta) = \frac{1}{\sqrt{2}}\big(A(\theta) + B(\theta)\big) = a \cosh i\theta - \tilde{a}^\dagger \sinh i\theta, \tag{31}$$

$$\tilde{a}(\theta) = \frac{1}{\sqrt{2}}\big(A(\theta) - B(\theta)\big) = \tilde{a} \cosh i\theta - i a^\dagger \sinh i\theta, \tag{32}$$

and

$$\{a(\theta), a^\dagger(\theta)\} = 1, \quad \{\tilde{a}(\theta), \tilde{a}^\dagger(\theta)\} = 1. \tag{33}$$

All others anti-commutators are equal to zero and $a(\theta)$ and $\tilde{a}(\theta)$ anti-commute among themselves. Eqs. (31) and (32) are nothing but the Bogoliubov transformations of the pair of creation and annihilation operators $(a, \tilde{a})$ into a new pair $(a(\theta), \tilde{a}(\theta))$. In other words, Eqs. (31)-(33) show that the Bogoliubov-transformed operators $a(\theta)$ and $\tilde{a}(\theta)$ are linear combinations of the coproduct operators defined in terms of the deformation parameter $q(\theta)$ and their θ-derivatives.

4. Neutrino mixing

In the context on NCSG, neutrinos appear as Majorana spinors. Therefore, we refer below to Majorana neutrinos; however, provided that convenient changes are introduced, our formalism can be readily extended to Dirac neutrinos and to other particle mixing.

Majorana fermions are self-conjugate particles. The charge-conjugation operator C satisfies the relations

$$C^{-1}\gamma_\mu C = -\gamma_\mu^T, \qquad C^\dagger = C^{-1}, \qquad C^T = -C. \tag{34}$$

The charge conjugate ψ^c of ψ is defined by

$$\psi^c(x) \equiv \gamma_0 C \psi^*(x). \tag{35}$$

A Majorana fermion is a field that satisfies both, the Dirac equation and the self-conjugation relation:

$$(i\partial\!\!\!/ - m)\psi = 0, \tag{36a}$$

$$\psi = \psi^c. \tag{36b}$$

In NCSG, the neutrino mass terms in the Lagrangian is written as

$$\frac{1}{2}\sum_{\lambda\kappa} \bar{\psi}_{\lambda L}\mathcal{S}_{\lambda\kappa}\hat{\psi}_{\kappa R} + \frac{1}{2}\sum_{\lambda\kappa} \overline{\bar{\psi}_{\lambda L}\mathcal{S}_{\lambda\kappa}\hat{\psi}_{\kappa R}}\,. \tag{37}$$

The subscripts $_{L,R}$ denote left-handed and right-handed states, respectively. The off-diagonal elements of the symmetric matrix $\mathcal{S}_{\lambda\kappa}$ are the Dirac mass terms, the diagonal ones are the Majorana mass terms.

¿From the equations of motion one has that the largest eigenvalue of the Majorana mass matrix M_R is of the order of the unification scale, while the Dirac mass M_ν is of the order of the Fermi energy and thus much smaller. Neutrino mixing and the seesaw mechanism are thus built in in the NCSG construction. We show below that neutrino mixing is implied by the doubling of the algebra, which is the core of Connes construction.

In order to proceed in our discussion, we first summarize briefly the QFT formalism for the neutrino mixing [22]-[26] We present a number of relations which we need in order to show the role played in such a phenomenon by the Bogoliubov transformations. We will closely follow ref. [19], where a more detailed presentation is reported.

We start by introducing the Lagrangian:

$$L(x) = \bar{\psi}_f(x)(i\partial\!\!\!/ - M)\psi_f(x) = \bar{\psi}_m(x)(i\partial\!\!\!/ - M_d)\psi_m(x), \tag{38}$$

with $\psi_f^T = (\nu_e, \nu_\mu)$ being the flavor fields and $M = \begin{pmatrix} m_e & m_{e\mu} \\ m_{e\mu} & m_\mu \end{pmatrix}$.

The flavor fields are connected to the free fields $\psi_m^T = (\nu_1, \nu_2)$, with $M_d = diag(m_1, m_2)$, by the the Pontecorvo mixing transformation

$$\nu_e(x) = \nu_1(x)\cos\theta + \nu_2(x)\sin\theta, \tag{39}$$

$$\nu_\mu(x) = -\nu_1(x)\sin\theta + \nu_2(x)\cos\theta. \tag{40}$$

The free fields are given by

$$\nu_i(x) = \sum_{r=1,2}\int \frac{d^3\mathbf{k}}{(2\pi)^{\frac{3}{2}}} e^{i\mathbf{k}\cdot\mathbf{x}}\left[u^r_{\mathbf{k},\mathbf{i}}(t)\alpha^r_{\mathbf{k},\mathbf{i}} + v^r_{-\mathbf{k},\mathbf{i}}(t)\alpha^{r\dagger}_{-\mathbf{k},\mathbf{i}}\right] \quad , \quad i = 1,2 \tag{41}$$

where $u^r_{\mathbf{k},\mathbf{i}}(t) = e^{-i\omega_{\mathbf{k},\mathbf{i}}t}u^r_{\mathbf{k},\mathbf{i}}$, $v^r_{\mathbf{k},\mathbf{i}}(t) = e^{i\omega_{\mathbf{k},\mathbf{i}}t}v^r_{\mathbf{k},\mathbf{i}}$ and $\omega_{\mathbf{k},\mathbf{i}} = \sqrt{\mathbf{k^2}+\mathbf{m_i^2}}$. In order for the Majorana condition (36b) to be satisfied, the four spinors must also satisfy the following condition:

$$u^s_{\mathbf{k},\mathbf{i}} = \gamma_0 C(v^s_{\mathbf{k},\mathbf{i}})^*, \qquad v^s_{\mathbf{k},\mathbf{i}} = \gamma_0\, C(u^s_{\mathbf{k},\mathbf{i}})^*. \tag{42}$$

The equal-time anti-commutation relations are

$$\{\nu^\alpha_i(x), \nu^{\beta\dagger}_j(y)\}_{t=t'} = \delta^3(\mathbf{x}-\mathbf{y})\delta_{\alpha\beta}\delta_{\mathbf{ij}}, \tag{43}$$

$$\{\nu^\alpha_i(x), \nu^{\beta}_j(y)\}_{t=t'} = \delta^3(\mathbf{x}-\mathbf{y})(\gamma_\mathbf{0}\mathbf{C})_{\alpha\beta}\delta_{\mathbf{ij}}, \tag{44}$$

with $\alpha, \beta = 1, \ldots, 4$. And

$$\{\alpha^r_{-\mathbf{k},\mathbf{i}}, \alpha^{s\dagger}_{-\mathbf{q},\mathbf{i}}\}_{t=t'} = \delta^3(\mathbf{k_q})\delta_{\beta\mathbf{rs}}\delta_{\mathbf{ij}}, \qquad \mathbf{i} = \mathbf{1}, \mathbf{2}. \tag{45}$$

All the other anti-commutators are zero. The ortonormality and completeness relations are :

$$u^{r\dagger}_{\mathbf{k},i}u^s_{\mathbf{k},i} = v^{r\dagger}_{\mathbf{k},i}v^s_{\mathbf{k},i} = \delta rs, \qquad u^{r\dagger}_{\mathbf{k},i}v^s_{-\mathbf{k},i} = v^{r\dagger}_{-\mathbf{k},i}u^s_{\mathbf{k},i} = 0, \tag{46}$$

$$\sum r = 1, 2u^r_{\mathbf{k},i}u^{r\dagger}_{\mathbf{k},i} + v^r_{\mathbf{k},i}v^{r\dagger}_{\mathbf{k},i} = \mathbf{1}. \tag{47}$$

Eqs. (39) and (40) can be recast in the form :

$$\nu^\alpha_e(x) = G^{-1}_\theta(t)\nu^\alpha_1(x)G_\alpha(t), \tag{48}$$

$$\nu^\alpha_\mu(x) = G^{-1}_\theta(t)\nu^\alpha_2(x)G_\alpha(t), \tag{49}$$

where $G_\theta(t)$ is given by

$$G_\theta(t) = \exp\left[\frac{\theta}{2}\int d^3\mathbf{x}\big(\nu^\dagger_\mathbf{1}(\mathbf{x})\nu_\mathbf{2}(\mathbf{x}) - \nu^\dagger_\mathbf{2}(\mathbf{x})\nu_\mathbf{1}(\mathbf{x})\big)\right]. \tag{50}$$

One has $G_\theta(t) = \prod_\mathbf{k} G^\mathbf{k}_\theta(t)$. For a given $\mathbf{k}$, in the reference frame where $\mathbf{k} = (\mathbf{0}, \mathbf{0}, |\mathbf{k}|)$, the spins decuple and one has $G^\mathbf{k}_\theta(t) = \prod_r G^{\mathbf{k},\mathbf{r}}_\theta(t)$ with

$$G^\mathbf{k}_\theta(t) = \exp\left\{\theta\left[U^*_\mathbf{k}(t)\alpha^{r\dagger}_{\mathbf{k},\mathbf{1}}\alpha^r_{\mathbf{k},\mathbf{2}} - U_\mathbf{k}(t)\alpha^{r\dagger}_{-\mathbf{k},\mathbf{2}}\alpha^r_{-\mathbf{k},\mathbf{1}} - \epsilon^r V^*_\mathbf{k}(t)\alpha^r_{-\mathbf{k},\mathbf{1}}\alpha^r_{\mathbf{k},\mathbf{2}} + \epsilon^r V_\mathbf{k}(t)\alpha^{r\dagger}_{\mathbf{k},\mathbf{1}}\alpha^{r\dagger}_{-\mathbf{k},\mathbf{2}}\right]\right\}, \tag{51}$$

where $U_\mathbf{k}(t)$ e $V_\mathbf{k}(t)$ are Bogoliubov coefficients given by

$$U_\mathbf{k}(t) \equiv |U_\mathbf{k}|e^{i(\omega_{\mathbf{k},\mathbf{2}}-\omega_{\mathbf{k},\mathbf{1}})t} \qquad , \qquad V_\mathbf{k}(t) \equiv |V_\mathbf{k}|e^{i(\omega_{\mathbf{k},\mathbf{2}}+\omega_{\mathbf{k},\mathbf{1}})t} \qquad , \tag{52}$$

$$|U_\mathbf{k}| \equiv \left(\frac{\omega_{\mathbf{k},\mathbf{1}}+m_1}{2\omega_{\mathbf{k},\mathbf{1}}}\right)^{\frac{1}{2}}\left(\frac{\omega_{\mathbf{k},\mathbf{2}}+m_2}{2\omega_{\mathbf{k},\mathbf{2}}}\right)^{\frac{1}{2}}\left(1 + \frac{|\mathbf{k}|^2}{(\omega_{\mathbf{k},\mathbf{1}}+m_1)(\omega_{\mathbf{k},\mathbf{2}}+m_2)}\right), \tag{53}$$

$$|V_\mathbf{k}| \equiv \left(\frac{\omega_{\mathbf{k},\mathbf{1}}+m_1}{2\omega_{\mathbf{k},\mathbf{1}}}\right)^{\frac{1}{2}}\left(\frac{\omega_{\mathbf{k},\mathbf{2}}+m_2}{2\omega_{\mathbf{k},\mathbf{2}}}\right)^{\frac{1}{2}}\left(\frac{|\mathbf{k}|}{(\omega_{\mathbf{k},\mathbf{2}}+m_2)} - \frac{|\mathbf{k}|}{(\omega_{\mathbf{k},\mathbf{1}}+m_1)}\right), \tag{54}$$

$$|U_\mathbf{k}|^2 + |V_\mathbf{k}|^2 = 1. \tag{55}$$

The flavor fields can be thus expanded as :

$$\nu_\sigma(x) = \sum_{r=1,2}\int \frac{d^3\mathbf{k}}{(2\pi)^{\frac{3}{2}}}e^{i\mathbf{k}\cdot\mathbf{x}}\left[u^r_{\mathbf{k},\mathbf{j}}(t)\alpha^r_{\mathbf{k},\sigma} + v^r_{-\mathbf{k},\mathbf{j}}(t)\alpha^{r\dagger}_{-\mathbf{k},\sigma}\right] \quad , \tag{56}$$

with $\sigma, j = (e, 1), (\mu, 2)$ and the flavor annihilation operators given by (for $\mathbf{k} = (\mathbf{0}, \mathbf{0}, |\mathbf{k}|)$):

$$\alpha^r_{\mathbf{k},\mathbf{e}} \equiv G^{-1}_\theta(t)\, \alpha^r_{-\mathbf{k},\mathbf{1}} G_\theta(t) = \cos\theta\, \alpha^r_{\mathbf{k},\mathbf{1}} + \sin\theta\big(U^*_{\mathbf{k}}(t)\, \alpha^r_{\mathbf{k},\mathbf{2}} + \epsilon^r V^*_{\mathbf{k}}(t)\, \alpha^{r\dagger}_{-\mathbf{k},\mathbf{2}}\big), \tag{57}$$

$$\alpha^r_{\mathbf{k},\mu} \equiv G^{-1}_\theta(t)\, \alpha^r_{-\mathbf{k},\mathbf{2}} G_\theta(t) = \cos\theta\, \alpha^r_{\mathbf{k},\mathbf{2}} - \sin\theta\big(U_{\mathbf{k}}(t)\, \alpha^r_{\mathbf{k},\mathbf{1}} + \epsilon^r V_{\mathbf{k}}(t)\, \alpha^{r\dagger}_{-\mathbf{k},\mathbf{1}}\big). \tag{58}$$

Notice the presence of the Bogoliubov coefficients $U_{\mathbf{k}}(t)$ and $V_{\mathbf{k}}(t)$ in Eqs. (57) and (58). *These relations clearly show how Bogoliubov transformations enter the neutrino mixing transformations.*

Next, we consider the action of the mixing generator on the vacuum $|0\rangle_{1,2}$. The flavor vacuum is defined as:

$$|0(\theta, t)\rangle_{e,\mu} \equiv G^{-1}_\theta(t)|0\rangle_{1,2} \tag{59}$$

The state for a mixed particle with definite flavor, spin and momentum is given by:

$$|\alpha^r_{\mathbf{k},\mathbf{e}}\rangle \equiv \alpha^{r\dagger}_{\mathbf{k},\mathbf{e}}|0(\theta, t)\rangle_{e,\mu} = G^{-1}_\theta(t)\alpha^{r\dagger}_{\mathbf{k},\mathbf{1}}|0\rangle_{1,2} \quad . \tag{60}$$

The anti-commutators of the flavor ladder operators at different times are then computed and the following quantity is found to be constant in time:

$$|\{\alpha^r_{\mathbf{k},\mathbf{e}}(t), \alpha^{r\dagger}_{\mathbf{k},\mathbf{e}}(t')\}|^2 + |\{\alpha^{r\dagger}_{-\mathbf{k},\mathbf{e}}(t), \alpha^{r\dagger}_{\mathbf{k},\mathbf{e}}(t')\}|^2 + |\{\alpha^r_{\mathbf{k},\mu}(t), \alpha^{r\dagger}_{\mathbf{k},\mathbf{e}}(t')\}|^2 + |\{\alpha^{r\dagger}_{-\mathbf{k},\mu}(t), \alpha^{r\dagger}_{\mathbf{k},\mathbf{e}}(t')\}|^2 = 1 \tag{61}$$

The energy-momentum tensor for the fermion field is defined by $\mathcal{J}^{\mu\nu} \equiv i\bar{\psi}\gamma^\nu\partial_\mu\psi$ and the momentum operator is given by $P^j \equiv \int d^3\mathbf{x} \quad \mathcal{J}^{\mathbf{0j}}(\mathbf{x})$. For the free fields we have:

$$\mathbf{P_i} = \int \mathbf{d^3x}\psi^\dagger_{\mathbf{i}}(\mathbf{x})(-\mathbf{i}\nabla)\psi_{\mathbf{i}}(\mathbf{x}) = \int \mathbf{d^3k} \sum_{\mathbf{r=1,2}} \mathbf{k}\Big(\alpha^{\mathbf{r}\dagger}_{\mathbf{k},\mathbf{i}}\alpha^{\mathbf{r}}_{\mathbf{k},\mathbf{i}} - \alpha^{\mathbf{r}\dagger}_{-\mathbf{k},\mathbf{i}}\alpha^{\mathbf{r}}_{-\mathbf{k},\mathbf{i}}\Big), \tag{62}$$

for $i = 1, 2$. The momentum operator for mixed fields is given by:

$$\mathbf{P}_\sigma(\mathbf{t}) = \int \mathbf{d^3x}\psi^\dagger_\sigma(\mathbf{x})(-\mathbf{i}\nabla)\psi_\sigma(\mathbf{x}) = \int \mathbf{d^3k} \sum_{\mathbf{r=1,2}} \mathbf{k}\Big(\alpha^{\mathbf{r}\dagger}_{\mathbf{k},\sigma}(\mathbf{t})\alpha^{\mathbf{r}}_{\mathbf{k},\sigma}(\mathbf{t}) - \alpha^{\mathbf{r}\dagger}_{-\mathbf{k},\sigma}(\mathbf{t})\alpha^{\mathbf{r}}_{-\mathbf{k},\sigma}(\mathbf{t})\Big), \tag{63}$$

for $\sigma = e, \mu$. We have $\mathbf{P}_\sigma(\mathbf{t}) = \mathbf{G}^{-1}_\theta(\mathbf{t})\mathbf{P_i}\mathbf{G}_\theta(\mathbf{t})$ and the conservation of total momentum:

$$\mathbf{P_e(t)} + \mathbf{P}_\mu(\mathbf{t}) = \mathbf{P_1} + \mathbf{P_2} \equiv \mathbf{P} \quad , \quad [\mathbf{P}, \mathbf{G}_\theta(\mathbf{t})] = \mathbf{0} \quad , \quad [\mathbf{P}, \mathbf{H}] = \mathbf{0}. \tag{64}$$

At time $t = 0$, the flavor state $|\alpha^r_{\mathbf{k},\mathbf{e}}\rangle \equiv |\alpha^r_{\mathbf{k},\mathbf{e}}(0)\rangle$ is an eigenstate of the momentum operator $\mathbf{P_e(0)}$:

$$\mathbf{P_e(0)}|\alpha^{\mathbf{r}}_{\mathbf{k},\mathbf{e}}\rangle \equiv \mathbf{k}|\alpha^{\mathbf{r}}_{\mathbf{k},\mathbf{e}}\rangle. \tag{65}$$

At time $t \neq 0$ the expectation value of the momentum in such a state (normalized to initial time value) gives [25]:

$$\mathcal{P}^e_{\mathbf{k},\sigma}(t) \equiv \frac{\langle\alpha^r_{\mathbf{k},\mathbf{e}}|\mathbf{P}_\sigma(\mathbf{t})|\alpha^{\mathbf{r}}_{\mathbf{k},\mathbf{e}}\rangle}{\langle\alpha^r_{\mathbf{k},\mathbf{e}}|\mathbf{P}_\sigma(\mathbf{0})|\alpha^r_{\mathbf{k},\mathbf{e}}\rangle} = |\{\alpha^r_{\mathbf{k},\mathbf{e}}(t), \alpha^{r\dagger}_{\mathbf{k},\mathbf{e}}(t')\}|^2 + |\{\alpha^{r\dagger}_{-\mathbf{k},\mathbf{e}}(t), \alpha^{r\dagger}_{\mathbf{k},\mathbf{e}}(t')\}|^2, \tag{66}$$

for $\sigma = e, \mu$, and the flavor vacuum expectation value of the momentum operator $\mathbf{P}_\sigma(\mathbf{t})$ vanishes at all time :

$$_{e,\mu}\langle 0|\mathbf{P}_\sigma(\mathbf{t})|\mathbf{0}_{\mathbf{e},\mu}\rangle \quad , \qquad \sigma = \mathbf{e}, \mu. \tag{67}$$

The explicit calculation of the oscillating quantities $\mathcal{P}^e_{\mathbf{k},\sigma}(t)$ is found to be [25]:

$$\mathcal{P}^e_{\mathbf{k},\mathbf{e}}(t) = 1 - \sin^2 2\theta\Big[|U_\mathbf{k}|\sin^2\frac{\omega_{k,2}-\omega_{k,1}}{2}t + |V_\mathbf{k}|\sin^2\frac{\omega_{k,2}+\omega_{k,1}}{2}t\Big], \tag{68}$$

$$\mathcal{P}^e_{\mathbf{k},\mu}(t) = \sin^2 2\theta\Big[|U_\mathbf{k}|\sin^2\frac{\omega_{k,2}-\omega_{k,1}}{2}t + |V_\mathbf{k}|\sin^2\frac{\omega_{k,2}+\omega_{k,1}}{2}t\Big]. \tag{69}$$

These formulas give the neutrino oscillation formulas in the QFT formalism and are different from the quantum mechanical approximation provided by the Pontecorvo formulas, to which they reduce in the large $|\mathbf{k}|$ limit, namely in $|U_\mathbf{k}| \to 1$ limit. In such a limit the Bogoliubov transformations and the algebra doubling are washed away. But then, as well, there is no NCSG construction anymore.

5. Conclusion

The study of the neutrino mixing carried on in the QFT formalism has shown that the mixing transformation is not just a simple rotation, but a transformation made of a rotation "nested" with a Bogoliubov transformation. This makes the mass vacuum and the flavor vacuum unitarily inequivalent. The Bogoliubov transformation is in turn the result of the algebra doubling $\mathcal{A} = \mathcal{A}_1 \otimes \mathcal{A}_2$ acting on the space $\mathcal{H} = \mathcal{H}_1 \otimes \mathcal{H}_2$, as we have discussed in Section 2. Then we have shown that Bogoliubov transformations are built in in the algebra doubling in the NCSG construction. In fact the doubling of the algebra is a characterizing feature at the basis of the NCSG construction. The neutrino mixing mechanism appears thus naturally incorporated in the NCSG construction. Since Majorana neutrinos appear in the NCSG construction, we consider their mixing and show how the neutrino oscillations emerge. We also show that in the limit of large $|\mathbf{k}|$ the quantum mechanical approximation of the Pontocorvo mixing is obtained and the Bogoliubov transformation reduces to the identity transformation. In such a limit the doubling of the algebra, characterizing feature of the NCSG construction, is also lost.

References

[1] Connes A 1994 *Noncommutative Geometry* (New York : Academic Press)
[2] Connes A 2006 *J.H.E.P.* **0611** 081
[3] Chamseddine A H, Connes A and Marcolli M 2007 *Adv. Theor. Math. Phys.* **11** 991
[4] Chamseddine A H and Connes A 2007 *Phys. Rev. Lett.* **99** 191601
[5] Connes A and Marcolli M 2008 *Noncommutative Geometry, Quantum Fields and Motives* (India: Hindustan Book Agency)
[6] Nelson W and Sakellariadou M 2009 *Phys. Lett.* B **680** 263
[7] Nelson W and Sakellariadou M 2010 *Phys. Rev.* D **81** 085038
[8] Nelson W, Ochoa J and Sakellariadou M 2010 *Phys. Rev.* D **82** 085021
[9] Nelson W, Ochoa J and Sakellariadou M 2010 *Phys. Rev. Lett.* **105** 101602
[10] Buck M, Fairbairn M and Sakellariadou M 2010 *Phys. Rev.* D **82** 043509
[11] Sakellariadou M 2011 *Int. J. Mod. Phys.* D **20** 785
[12] Sakellariadou M 2011 *PoS CNCFG* **2010** 028
[13] Sakellariadou M 2011 *PoS CORFU* **2011** 053
[14] van den Dungen K and van Suijlekom W D 2012 *Rev. Math. Phys.* **24** 1230004
[15] Chamseddine A H and Connes A 2012 *J.H.E.P.* **1209** 104
[16] Lambiase G, Sakellariadou M and Stabile A 2013 *JCAP* **12** 020
[17] Capozziello S, Lambiase G, Sakellariadou M, and Stabile A 2015 *Phys. Rev.* D **91** 044012
[18] Sakellariadou M, Stabile A and Vitiello G 2011 *Phys. Rev.* D **84** 045026
[19] Gargiulo M V, Sakellariadou M and Vitiello G 2014 *Eur. Phys. J.* C **74** 2695
[20] Celeghini E, De Martino S, De Siena S, Iorio A, Rasetti M and Vitiello G 1998 *Phys. Lett.* A **244** 455
[21] Umezawa H 1993 *Advanced field theory: Micro, macro, and thermal physics* (New York, USA: AIP)
[22] Blasone M and Vitiello G 1995 *Ann. Phys.* **244** 283
[23] Blasone M, Henning P A and Vitiello G 1999 *Phys. Lett.* B **451** 140
[24] Blasone M, Capolupo A and Vitiello G 2002 *Phys. Rev.* D **66** 025033

[25] Blasone M and Palmer J 2004 *Phys. Rev.* D **69** 057301
[26] Blasone M, Jizba P and Vitiello G 2011 *Quantum field theory and its macroscopic manifestations* (London: Imperial College)

Journal of Physics: Conference Series **626** (2015) 012015 doi:10.1088/1742-6596/626/1/012015

How to (path-) integrate by differentiating

Achim Kempf[1], **David M Jackson**[2] **and Alejandro H Morales**[3]

[1]Departments of Applied Mathematics and Physics, University of Waterloo, Ontario N2L 3G1, Canada

[2]Department of Combinatorics and Optimization, University of Waterloo, Ontario N2L 3G1, Canada

[3]Department of Mathematics, UCLA, Los Angeles, CA 90095, USA

E-mail: akempf@perimeterinstitute.ca

Abstract. Path integrals are at the heart of quantum field theory. In spite of their covariance and seeming simplicity, they are hard to define and evaluate. In contrast, functional differentiation, as it is used, for example, in variational problems, is relatively straightforward. This has motivated the development of new techniques that allow one to express functional integration in terms of functional differentiation. In fact, the new techniques allow one to express integrals in general through differentiation. These techniques therefore add to the general toolbox for integration and for integral transforms such as the Fourier and Laplace transforms. Here, we review some of these results, we give simpler proofs and we add new results, for example, on expressing the Laplace transform and its inverse in terms of derivatives, results that may be of use in quantum field theory, e.g., in the context of heat traces.

1. Introduction

Path integrals over fields, such as that for a scalar field,

$$Z[J] = \int e^{iS[\phi]+i\int d^4x\ J(x)\phi(x)} D[\phi], \tag{1}$$

are at the very heart of quantum field theory. Notice that Equ.1 can also be viewed as the Fourier transform of $e^{iS[\phi]}$. In spite of their seeming simplicity, such functional integrals or functional Fourier transforms are notoriously difficult to define and evaluate, see, e.g., [1]-[6].

In comparison to functional integration, functional differentiation is relatively straightforward and it is of course widely used, for example, in extremization problems. This has motivated recent work, [7], in which it was shown that in certain circumstances functional integration can be expressed through functional differentiation. In fact, the results show that integrals in general can be expressed in terms of differentiations. This includes proper and improper integrals and also integral transforms such as the Fourier and Laplace transforms. The new methods of [7] are not meant to replace the existing integration methods. They merely add to the toolbox of techniques for integrals and integral transforms. The new tools have the advantage that when they apply they are often quick and simple to use, work naturally with distributions and can even sidestep the need for contour integrations.

Here, we will review the new tools, add simpler proofs and we also add some new results.

2. Integration by differentiation

In [7], the following representations of integration through differentiation were shown:

$$\int_a^b f(x)\, dx = \lim_{\epsilon \to 0} f(\partial_\epsilon)\, \frac{e^{\epsilon b} - e^{\epsilon a}}{\epsilon} \tag{2}$$

$$\int_{-\infty}^{\infty} f(x)\, dx = \lim_{\epsilon \to 0} 2\pi f(-i\partial_\epsilon)\, \delta(\epsilon) \tag{3}$$

$$\int_{-\infty}^{\infty} f(x)\, dx = 2\pi\delta(i\partial_\epsilon)\, f(\epsilon) \tag{4}$$

(Notice that, as was shown in [7], in Equ.4 it is not necessary to take ϵ to a fixed value because the ϵ-dependence drops out.) Here, $f(\nu\partial_\epsilon)$ with $\nu \in \{\pm 1, \pm i\}$ means in the simplest cases that when f is expanded in a MacLaurin or Laurent series, its argument is to be replaced by $\nu\partial_\epsilon$. More generally, f can be interpreted as a function on the spectrum of the operator ∂_ϵ. This means, in particular, that when $f(\nu\partial_\epsilon)$ acts on one of its eigenfunctions, $e^{\epsilon a}$, we have:

$$f(\nu\partial_\epsilon)\, e^{\epsilon a} = f(\nu a) e^{\epsilon a} \tag{5}$$

While we will give proofs below that apply to fairly large classes of functions f, the above equations also hold in a distributional sense. The conditions and exact size of the space of generalized functions for which the above equations hold is not yet known.

Let us now add the following equations which follow straightforwardly from Equ.2 by taking the limits $a \to 0, b \to \infty$ and $a \to -\infty, b \to 0$ respectively:

$$\int_0^{\infty} f(x)\, dx = \lim_{\epsilon \to 0^+} f(-\partial_\epsilon)\, \frac{1}{\epsilon} \tag{6}$$

$$\int_{-\infty}^{0} f(x)\, dx = \lim_{\epsilon \to 0^+} f(\partial_\epsilon)\, \frac{1}{\epsilon} \tag{7}$$

$$\int_{-\infty}^{\infty} f(x)\, dx = \lim_{\epsilon \to 0} \left(f(\partial_\epsilon) + f(-\partial_\epsilon)\right)\, \frac{1}{\epsilon} \tag{8}$$

3. Examples of integration by differentiation

In [7], the aim was to derive new methods for integrals and integral transform for applications to quantum field theoretic path integrals. Here, we will look at these integration methods' general utility. To this end, in order to illustrate the above tools let us perform an integral that is usually considered nontrivial in the sense that it is usually done using contour integration. Namely, using the new Equ.6:

$$\int_0^{\infty} \frac{\sin(x)}{x}\, dx = \lim_{\epsilon \to 0^+} \frac{1}{2i}\left(e^{-i\partial_\epsilon} - e^{i\partial_\epsilon}\right) \frac{1}{-\partial_\epsilon}\, \frac{1}{\epsilon} \tag{9}$$

$$= \lim_{\epsilon \to 0^+} \frac{-1}{2i}\left(e^{-i\partial_\epsilon} - e^{i\partial_\epsilon}\right)(\ln(\epsilon) + c) \tag{10}$$

$$= \lim_{\epsilon \to 0^+} \frac{-1}{2i}\left(\ln(\epsilon - i) + c - \ln(\epsilon + i) - c\right) \tag{11}$$

$$= \frac{-1}{2i}\left(\frac{-i\pi}{2} - \frac{i\pi}{2}\right) = \frac{\pi}{2} \tag{12}$$

Here, we used that the antiderivative of $1/\epsilon$ is the logarithm with an integration constant c, and we used that the exponentiated derivate translates: $e^{a\partial_\epsilon} g(\epsilon) = g(\epsilon + a)$. For comparison, we

Journal of Physics: Conference Series **626** (2015) 012015 doi:10.1088/1742-6596/626/1/012015

now perform a similar integral using Equ.3:

$$\int_{-\infty}^{\infty} \frac{\sin(x)}{x}\,dx = 2\pi \lim_{\epsilon\to 0} \frac{1}{2i}\left(e^{\partial_\epsilon} - e^{-\partial_\epsilon}\right)\frac{1}{-i\partial_\epsilon}\,\delta(\epsilon) \tag{13}$$

$$= \pi \lim_{\epsilon\to 0}\left(e^{\partial_\epsilon} - e^{-\partial_\epsilon}\right)\left(\Theta(\epsilon) + c'\right) \tag{14}$$

$$= \pi \lim_{\epsilon\to 0}\left(\Theta(\epsilon+1) + c' - \Theta(\epsilon-1) - c'\right) \tag{15}$$

$$= \pi \tag{16}$$

Here, Θ is the Heaviside function and c' is an integration constant. Let us close this section with a quick derivation of Equ.2:

$$\int_a^b f(x)\,dx = \lim_{\epsilon\to 0}\int_a^b f(x)e^{-\epsilon x}\,dx \tag{17}$$

$$= \lim_{\epsilon\to 0} f(-\partial_\epsilon)\int_a^b e^{-\epsilon x}\,dx \tag{18}$$

$$= \lim_{\epsilon\to 0} f(-\partial_\epsilon)\,\frac{e^{\epsilon b} - e^{\epsilon a}}{\epsilon} \tag{19}$$

4. Forward and inverse Laplace transforms through differentiation

As was shown in [7], the representation of integration over the real line given in Equ.3 is a special case of a new representation of Fourier transformation by differentiation. Namely, with the following definition of the Fourier transform $F[f]$ of a function f,

$$F[f](x) = \frac{1}{\sqrt{2\pi}}\int_{-\infty}^{\infty} e^{ixy}\,f(y)\,dy, \tag{20}$$

we have the new representation:

$$F[f](x) = \sqrt{2\pi}\,f(-i\partial_x)\,\delta(x) \tag{21}$$

It is straightforward to prove, see [7], that Equ.21 holds for the basis of plane waves, and they span the function space. This means that the new representation for the integral over the real line given in Equ.3 follows from Equ.21 because integration is the zero frequency limit of the Fourier transform.

Our aim now is to show that our new Equ.6 is, similarly, the "zero-frequency" special case of a new and very simple representation of the Laplace transform, expressed entirely in terms of differentiation. Consider the usual definition of the Laplace transform, $L[f]$, of a function f, for $x > 0$:

$$L[f](x) = \int_0^{\infty} f(y)\,e^{-xy}\,dy \tag{22}$$

We claim that:

$$L[f](x) = f(-\partial_x)\,\frac{1}{x} \tag{23}$$

$$L^{-1}[f](x) = f(\partial_x)\,\delta(x) \tag{24}$$

For example, using Equ.23, we can immediately read off the Laplace transforms of monomials, $f(x) = x^n$:

$$L[f](x) = (-\partial_x)^n\,\frac{1}{x} = \frac{n!}{x^{n+1}} \tag{25}$$

Journal of Physics: Conference Series **626** (2015) 012015 doi:10.1088/1742-6596/626/1/012015

The new representation, Equ.24, of the inverse Laplace transform may be particularly useful in practice. This is because, with the usual methods, the inverse Laplace transform is somewhat tedious as it involves analytic continuation. Equ.24 can be comparatively straightforward to evaluate. For example, let us calculate the inverse Laplace transform of the function $f(x) = 1/(x - a)$. Using Equ.24, we obtain:

$$
\begin{aligned}
L^{-1}[f](x) &= \frac{1}{\partial_x - a}\ \delta(x) && (26)\\
&= \int_0^\infty e^{-w(\partial_x - a)}\ dw\ \delta(x) && (27)\\
&= \int_0^\infty e^{aw}\ \delta(x-w)\ dw = e^{ax} && (28)
\end{aligned}
$$

Here, we used:

$$\frac{1}{k} = \int_0^\infty e^{-wk}\ dw \tag{29}$$

5. Proofs for the forward and inverse Laplace transforms

Equ.23 for the forward Laplace transform is easy to derive:

$$L[f](x) = \int_0^\infty e^{-xy}\ f(y)\ dy = f(-\partial_x) \int_0^\infty e^{-xy}\ dy = f(-\partial_x)\ \frac{1}{x} \tag{30}$$

As a consequence, we have also proven Equ.6, namely as the special case $x \to 0^+$. Let us now show that Equ.24 indeed inverts the action of L, i.e., that $L^{-1} \circ L = id$:

$$
\begin{aligned}
L^{-1} \circ L[f](x) &= L^{-1}\left[f(-\partial_x)\frac{1}{x}\right] = \lim_{\epsilon\to 0} L^{-1}\left[f(-\partial_\epsilon)\ \frac{1}{\epsilon + x}\right] && (31)\\
&= \lim_{\epsilon\to 0} f(-\partial_\epsilon)\ L^{-1}\left[\frac{1}{\epsilon + x}\right] = \lim_{\epsilon\to 0} f(-\partial_\epsilon)\ \frac{1}{\epsilon + \partial_x}\ \delta(x) && (32)\\
&= \lim_{\epsilon\to 0} f(-\partial_\epsilon) \int_0^\infty e^{-(\epsilon + \partial_x)w}\ dw\ \delta(x) && (33)\\
&= \lim_{\epsilon\to 0} \int_0^\infty f(-\partial_\epsilon)\ e^{-\epsilon w}\ \delta(x - w)\ dw && (34)\\
&= f(x) && (35)
\end{aligned}
$$

In Equ.33, we used Equ.29 and the final step used Equ.5.

6. Example: trace of the heat kernel

A prominent occurrence of the Laplace transform in quantum field theory is in the context of the trace of the heat kernel, see, e.g., [1, 10, 11]:

$$h(t) = \sum_n e^{-\lambda_n t} \tag{36}$$

Here, the λ_n are the (assumed positive) eigenvalues of the heat operator. Given the function $h(t)$ it is possible to recover the spectrum $\{\lambda_n\}$ that generated it. With the new methods, this is particularly simple to see. To this end, let us introduce the spectral comb function:

$$s(\lambda) = \sum_n \delta(\lambda - \lambda_n) \tag{37}$$

Clearly, the trace of the heat kernel, $h(t)$, is its Laplace transform:

$$h(t) = \sum_n e^{-\lambda_n t} = \int_0^\infty e^{-\lambda t}\ s(\lambda) = L[s](t) \tag{38}$$

Using the new representation for the inverse Laplace transform, Equ.24, we can invert this relation straightforwardly:

$$L^{-1}[h](\lambda) \quad = \quad L^{-1}\left[\sum_n e^{-\lambda_n t}\right](\lambda) = \sum_n e^{-\lambda_n \partial_\lambda}\ \delta(\lambda) = \sum_n \delta(\lambda - \lambda_n) = s(\lambda) \tag{39}$$

7. Outlook

The methods we have presented here are meant as additions to the toolbox for integration and for integral transforms such as the Fourier and Laplace transforms. Being general purpose methods, they may possess applications beyond physics, e.g., to communication engineering and signal processing, [7, 8, 9]. Within mathematics, there is the possibility that these methods may be extensible covariantly to integration over higher-dimensional curved manifolds. There, a new relationship between integration and differentiation could yield new insights into Stokes' theorem. Here, in one dimension, Stokes' theorem is simply the first fundamental theorem of calculus. We have for the derivative, $f'(x) = df(x)/dx$, as is straightforward to verify:

$$f'(\partial_\epsilon) = f(\partial_\epsilon)\epsilon - \epsilon f(\partial_\epsilon) \tag{40}$$

With this and Equ.2, the first fundamental theorem of calculus takes the form:

$$\int_a^b f'(x)\ dx \quad = \quad \lim_{\epsilon\to 0}\left(f(\partial_\epsilon)\epsilon - \epsilon f(\partial_\epsilon)\right)\ \frac{e^{\epsilon b} - e^{\epsilon a}}{\epsilon} \tag{41}$$

$$= \quad \lim_{\epsilon\to 0}\left(f(b)\ e^{\epsilon b} - f(a)e^{\epsilon a} - \epsilon f(\partial_\epsilon)\frac{e^{\epsilon b} - e^{\epsilon a}}{\epsilon}\right) \tag{42}$$

$$= \quad f(b) - f(a) - \lim_{\epsilon\to 0}\epsilon \int_a^b f(x)\ dx = f(b) - f(a) \tag{43}$$

The challenge is to generalize the present methods to higher dimensions so that, for example, a new perspective into the algebraic workings of boundary and coboundary operators may be gained. For the algebraic background for formal power series, see, for example, [12].

We started this paper by considering the analytical difficulties of quantum field theoretical path integrals. In this context, the new methods could inspire, for example, new approaches to perturbative expansions. Indeed, as was shown in [7], integration by differentiation methods yield not only the small coupling expansion but also naturally the strong coupling expansion. In general, the Dirac Deltas in integration-by-differentiation equations such as Equs.3,4, can be regularized in multiple ways, such as the Gaussian and sinc regularisations. Each choice yields another kind of perturbative expansion.

Also, the new integration methods presented here may provide a new perspective on anomalies in quantum field theories. This is because anomalies in quantum field theory originate in nontrivial transformation properties of the measure in the path integral, see e.g., [13].

Acknowledgments
AK and DMJ acknowledge support from NSERC's Discovery program.

Journal of Physics: Conference Series **626** (2015) 012015 doi:10.1088/1742-6596/626/1/012015

References

[1] Weinberg S 1996 *The Quantum Theory of Fields II* (Cambridge, UK: Cambridge University Press)
[2] Klauder J R 2000 *Beyond Conventional Quantization* (Cambridge, UK: Cambridge University Press)
[3] Ticciati R 1999 *Quantum Field Theory for Mathematicians* (Cambridge, UK: Cambridge University Press)
[4] Bender C M, Cooper F, Guralnik G S and Sharp D H 1979 *Phys. Rev.* D **19** 1865
[5] Arfken G B and Weber H J 2000 *Mathematical Methods for Physicists* 5th ed. (Boston: Academic Press)
[6] Ito K and Nihon S 2000 *Encyclopedic Dictionary of Mathematics* Second Ed. (Cambridge MA, USA: MIT Press)
[7] Kempf A, Jackson D M and Morales A H 2014 *J. Phys. A: Math. Theor.* **47** 415204
[8] Puetter R C, Gosnell T R and Yahil A 2005 Digital image reconstruction: Deblurring and Denoising *Annual Review of Astronomy and Astrophysics* **43** 139-194
[9] Gunturk B K and Lin X (eds.) 2013 *Image restoration: fundamentals and advances* (Boca Raton: CRC Press)
[10] Davies E B 1990 *Heat Kernels and Spectral Theory* (Cambridge, UK: Cambridge University Press)
[11] Kempf A 2009 *Phys. Rev. Lett.* **103** 231301
[12] Goulden I P and Jackson D M 2004 *Combinatorial Enumeration* (Mineola NY: Dover)
[13] Fujikawa K 1979 *Phys. Rev. Lett.* **42** 1195

Journal of Physics: Conference Series **626** (2015) 012016 doi:10.1088/1742-6596/626/1/012016

The Möbius symmetry of quantum mechanics

Alon E Faraggi[1] **and Marco Matone**[2]

[1] Department of Mathematical Sciences, University of Liverpool, Liverpool L69 7ZL, UK

[2] Dipartimento di Fisica, Universit'a di Padova, Via Marzolo 8, I–35131 Padova, Italy

E-mail: `alon.faraggi@liv.ac.uk; marco.matone@pd.infn.it`

Abstract.
The equivalence postulate approach to quantum mechanics aims to formulate quantum mechanics from a fundamental geometrical principle. Underlying the formulation there exists a basic cocycle condition which is invariant under D–dimensional Möbius transformations with respect to the Euclidean or Minkowski metrics. The invariance under global Möbius transformations implies that spatial space is compact. Furthermore, it implies energy quantisation and undefinability of quantum trajectories without assuming any prior interpretation of the wave function. The approach may be viewed as conventional quantum mechanics with the caveat that spatial space is compact, as dictated by the Möbius symmetry, with the classical limit corresponding to the decompactification limit. Correspondingly, there exists a finite length scale in the formalism and consequently an intrinsic regularisation scheme. Evidence for the compactness of space may exist in the cosmic microwave background radiation.

1. Introduction

The mathematical modelling of observational data on the smallest and largest distance scales currently rests on two main theories, quantum mechanics and general relativity. On the smallest scales quantum mechanics, through its incarnation as the Standard Model of Particle Physics, accounts for all subatomic data with a high degree of precision. The vindication of the Standard Model received strong support with the observation of a scalar boson resonance at the Large Hadron Collider (LHC) with the expected Standard Model properties. Higgs studies will dominate the experimental particle physics program in the next few decades, *i.e.*: measuring its couplings to the Standard Model fermions and vector bosons; measuring its self–couplings; and constraining deviation from the Standard Model by constraining higher dimensional nonrenormalisable operators suppressed by a higher energy scale.

On the largest scales general relativity receives ample support from observations in celestial, galactical and cosmological data. In seeking extensions to the two theories, it is hard to compare the two sets of data as the particle physics experiments rely on billions of controlled events and a thorough understanding of the background, whereas the astrophysical and, in particular, the cosmological data, rely on a few, or a single, event, and poor understanding of the background. This distinction is particularly important when seeking to synthesise the two theories, and the weight placed on observations in the two domains. It is also vital in the consideration of future experimental facilities and the prospect that they will lead to improved understanding of fundamental physics.

While quantum mechanics and general relativity are successful in accounting for observational data in their respective domains, their synthesis is nothing but settled. Furthermore, the two formalisms follow conceptually distinct approaches. The principles of equivalence and covariance with respect to general diffeomorphism underly general relativity. That is, general relativity follows from a fundamental geometrical principle. On the other hand, quantum mechanics does not follow from a geometrical principle. The main tenant of the axiomatic formulation of quantum mechanics is the probability interpretation of the wave–function.

Thus, quantum mechanics and general relativity follow fundamentally distinct approaches. However, the two theories are also fundamentally incompatible. This incompatibility is most clearly seen in the treatment of the vacuum and the vacuum energy. In quantum mechanics we have to define a vacuum state on which the quantum operators operate and create the physical states of the Hilbert space. The vacuum is the state which is annihilated by all annihilation operators. The vacuum should be bounded from below. One issue arises in quantum field theories due to the existence of an infinite number of states and the normal ordering ambiguity[1]. The zero point energy of the vacuum state therefore leads to an infinite contribution to the vacuum energy. In particle physics experiment we only measure energy differences and absorb additional infinities in physical parameters that are measured experimentally. If the number of parameters needed to absorb the infinities is finite the theory is said to be renormalizable. Otherwise it is said to be nonrenomalisable. The triumph of the Standard Model is precisely because it contains a relatively small number of such parameters, and is able to account for a much larger number of experimental observations. This opens the possibility that the Standard Model is a valid description of the physical data not only at a scale which is within reach of contemporary experiments, but up to a much larger energy scale. It is clear then that the first priority of forthcoming experiments is to continue to test the validity of the Standard Model at increasing energy scales, by using effective field theory approach to parameterise possible deviations. Furthermore, the set of parameters associated with the Higgs sector are particularly ripe for experimental picking. Gravity on the other hand is known to be nonrenomalizable. Furthermore, gravitational measurements are sensitive to the absolute energy scale. Observations dictate that the vacuum energy is miniscule compared to what we would expect from particle physics. The dichotomy between the two theories motivates much of the contemporary research in theoretical particle physics. Different approaches are pursued to develop a consistent theory that synthesises quantum mechanics and gravity. These include: effective field theories [2]; Euclidean quantum gravity [3]; asymptotic safety [4]; causal dynamical triangulation [5]; twistor theory [6]; noncommutative geometry [7]; loop quantum gravity [8]; string theory [9]. What are we to learn from this very partial list? First, we note that all these approaches aim at quantising general relativity, *i.e.* quantising spacetime. Second, non of these theories produces an unequivocal signature that has been confirmed experimentally, and in that respect, all these attempts should be regarded on equal footing. Nevertheless, the most developed effort is undoubtedly that of string theory. The main successes of string theory is that: 1. it provides a viable perturbative approach to quantum gravity; 2. it produce the gauge and matter structures that underly the Standard Model. As such it provides a framework for the construction of phenomenologically realistic models, and is therefore relevant for experimental observations. In that respect, by unifying the gauge and matter sectors with gravity, string theory provides a framework to study how the parameters of the Standard Model may arise from a more fundamental theory, and goes beyond the field theoretic Grand Unified Theories. Moreover, string theory accommodates consistently one of the most intriguing properties of the observed particles, that of chirality and with it parity violation. It achieves that by its most important property, modular invariance, which underlies the anomaly cancellation. By achieving that string theory demonstrated its potential relevance for experimental data, though we may still be a long way off before we can describe in a detailed and rigorous way how this relevance

[1] *For lucid introduction to quantum field theories, see e.g.* [1].

Journal of Physics: Conference Series **626** (2015) 012016 doi:10.1088/1742-6596/626/1/012016

is realised in the material world. The state of the art in this endeavour is the derivation of the Minimal Supersymmetric Standard Model from heterotic–string theory [10]. These models demonstrate how the detailed structures of the Standard Model may arise from string theory. In particular, the models enabled the calculation of the Yukawa couplings of the top quark, bottom quark and tau lepton in terms of the gauge coupling at the unification scale. Assuming consistency with the low scale gauge data then enables extrapolation of the Yukawa coupling to the low scale and prediction of the top quark mass of order $O(175GeV)$ [11] in the vicinity of the observed value. These results unambiguously demonstrate the potential relevance of string theory for low scale experimental data. Furthermore, the string consistency conditions dictate that additional extra degrees of freedom beyond the Standard Model, are needed to obtain a consistent theory of perturbative quantum gravity, which is an unambiguous prediction of the theory. These may be interpreted as extra spacetime dimensions, and/or as additional gauge symmetries beyond those observed in the Standard Model. It is a secondary question whether we possess the technological tools to detect experimentally these additional degrees of freedom. The myriad of string solutions which may in principle be compatible with the low energy data places an additional hindrance on extracting unambiguous experimental signatures from string theory. Given the hierarchical separation between the strength of the gravitational and gauge couplings it may be a long while before such an unambiguous correlation may be extracted. In that respect progress in string theory is likely to be incremental and slow. Another limitation is that our current understanding of string theory is limited to asymptotic stable string solutions, and we lack a good understanding of it as a dynamical theory. This hinders the development of cosmological string scenarios and of a possible dynamical vacuum selection mechanism in string theory. One then has to resort to analysis in the effective low energy field theory limit of the string vacua, which is deficient because it misses the massive string spectrum and the possible roll that it may play in the string vacuum selection mechanism. However, as long as the low scale data does not indicate departure from perturbative Standard Model parameterisation of the experimental observations, string theory continues to provide the most detailed framework to calculate the Standard Model parameters from a more fundamental theory.

String theory therefore provides a viable perturbative framework to explore how the Standard Model parameters may arise from a fundamental theory and to develop a phenomenological approach to quantum gravity. However, string theory does not provide a fundamental physical principle, akin to the equivalence principle of general relativity, that we may use as the basic hypothesis, and formulate quantum gravity rigoursely staring from that hypothesis. For that we may need an entirely new approach. In that respect string theory may be viewed as an effective theory, perhaps in a similar spirit to the view of effective quantum field theories as effective theories. However, string theory may hint at a possible basic hypothesis from which string theory may be derived as an effective limit.

An important property of string theory is the relation that it exhibits between different vacua by perturbative and non–perturbative transformations, and the existence of self–dual states under the duality transformation. We may envision that the self–dual states, which represent enhanced symmetry points in the space of vacua, may play a role in the string vacuum selection mechanism. We may also imagine that the duality structures underlying string theory may provide a hint at the basic hypothesis from whence string theory may be derived. In that context, however, it is noted that the string dualities correspond to specific dualities between concrete vacua and detailed mathematical structures. In that respect they provide examples that possess too much structure and are not sufficiently rudimentary to provide a basic physical hypothesis. It therefore does not provide a sufficiently basic physical hypothesis. Rather, we may consider that the perturbative and non–perturbative dualities exhibited in string theory may all be regarded as dualities in an extended phase space.

Phase space represents the key transition from the classical to the modern description of

Journal of Physics: Conference Series **626** (2015) 012016 doi:10.1088/1742-6596/626/1/012016

physical experiments. In that respect we need to establish what is the usefulness of a physical theory. We should first prioritise physics as an experimental science. We may define physics as "mathematical modelling of experimental observations". The aim of a mathematical model is to predict the outcome of experiments. Starting with some initial conditions, which are fed into the mathematical model, the predicted outcomes are calculated, and are then confronted with experiments. Physics is a practical field and an accepted mathematical model is the one most successful in accounting for a wide range of experimental observations. From that perspective, as our technological tools develop, our capacity to make experimental observations advances with time. Consequently, our mathematical modelling evolves with time to accommodate the expanding body of experimental data. This process is traditionally termed as reductionism. Namely, as experimental tools evolve we can resolve more refined physical distances. From celestial in the Galilean era to the sub-nuclear in the LHC era. The process of adapting our mathematical models to the increased body of experimental data may be labelled as "unification", and is likewise a constant theme in the mathematical modelling of experimental data. Thus, for example, Newton unified terrestrial and celestial observations in Newtonian mechanics; Maxwell combined the electric and magnetic forces into electromagnetism; Einstein synthesised mechanics and electromagnetism in special relativity; Glashow, Salam and Weinberg meshed quantum electrodynamics with Marshak and Sudarshan's vector minus axial–vector theory of the weak interactions into the electroweak model. The Standard Model that describe all the subatomic interactions as gauge theories may also be regarded as unifying the subatomic interactions under a single physical principle.

The list above represent the steps in the mathematical modelling of experimental data in which diverse observations are described in a common framework. The next stops on this route are the ones for which we do not yet have experimental evidence. They may therefore not necessarily be realised in nature. These include Grand Unified Theories which unify the electroweak and strong interactions in a simple or semi–simple group. The main prediction of Grand Unified Theories is proton decay. Global supersymmetry which combines fermions and bosons in common multiplets. The generic prediction of the simplest realisations of supersymmetry is the existence of superpartners. However, non–linear realisations of supersymmetry may exist, in which this will not be the case. Local supersymmetry implies the existence of a spin 3/2 particle, and consequently the existence of a spin 2 particle. Local supersymmetry therefore unifies gravity with the gauge interactions, albeit in a nonrenormalisable theory.

In Newtonian mechanics the mathematical modelling of the physical systems uses the position coordinates of a particle in space and their derivative with respect to time, *i.e.* the velocities. Given the position and velocity at some initial time $t = t_i$, and given some force field $\vec{F}(\vec{x})$, the evolution of the position and velocity coordinates is determined by Newton's equation of motion. A fully equivalent classical description is provided by exchanging the velocities of the particles with their momenta. The transformation is from the configuration space to the phase space. An equivalent representation of Newton's equations of motion in classical mechanics is given given in terms of the Euler–Lagrange equations of motion, which are derived by defining a Lagrangian function in configuration space. By transforming from the configuration space to the phase space, we transform the Lagrangian to the Hamiltonian function in phase space by a Legendre transformation. In the process we transform the second order Euler–Lagrange equations to the set of first order Hamilton equations of motion. The price that we have to pay is the doubling of the number of equations. Namely for every second equation we have two first order equations. The Hamilton equations of motion are invariant, up to a sign, under exchange of coordinates and momenta. Similar to the Newton equations of motion, the Hamilton equations of motion account for the time evolution of the phase space variables.

The relevance of a physical theory is established by confronting a mathematical model with experimental data. Typically the experimental setup and the variables of the mathematical model may evolve with time. Therefore, an important set of variables in the mathematical model are those that do not change as the physical system evolves. These are the constants of the motions and are related to the symmetries of the physical systems.

2. The Hamilton–Jacobi Theory

A general method to solve a physical problem in classical mechanics is given by the Hamilton–Jacobi formalism. In the classical Hamilton–Jacobi formalism the solution of the physical problem is obtained by transforming the Hamiltonian from one set of phase–space variables to a new set of phase–space variables, such that Hamiltonian is mapped to a trivial Hamiltonian. We may refer to these transformations as trivialising transformations. Consequently, the new phase–space variables are constants of the motion, *i.e.*,

$$H(q,p) \longrightarrow K(Q,P)\equiv 0 \qquad \Longrightarrow \qquad \dot{Q}=\frac{\partial K}{\partial P}\equiv 0\ ,\ \dot{P}=-\frac{\partial K}{\partial Q}\equiv 0. \tag{1}$$

The solution to this problem is given by the Classical Hamilton–Jacobi Equation (CHJE), which in the stationary case is given by

$$\frac{1}{2m}\left(\frac{\partial S_0}{\partial q}\right)^2 + V(q) - E \equiv \frac{1}{2m}\left(\frac{\partial S_0}{\partial q}\right)^2 + W(q) = 0. \tag{2}$$

The phase space variables are taken to be independent and their functional dependence is only extracted from the solution of the Hamilton–Jacobi equation via the relation

$$p=\frac{\partial S(q)}{\partial q}, \tag{3}$$

where the generating function $S(q)$ is Hamilton's principal function. The key property of quantum mechanics is that the phase space variables are not independent. Namely, represented as quantum operator they do no commute. We may therefore envision posing a similar problem to the Hamilton–Jacobi procedure, but one in which the phase–space variables are not treated as independent variables, but are rather related by (3).

3. The cocycle condition

We first consider the stationary problem in the one dimensional case [12, 13, 15, 16, 17]. This reveals the general properties of the formalism and paves the way for the general cases. We therefore assume that there always exist coordinate transformations such that $H\to K\equiv 0$, *i.e.* such that in the new system both the potential energy and the kinetic energy vanish. More generally, impose that there exist coordinate transformations such that in the transformed system $W(Q)=V(Q)-E\equiv 0$. We consider the transformations on

$$(\ q\ ,\ S_0(q)\ ,\ p\ =\ \frac{\partial S_0}{\partial q})\ \longrightarrow\ (\ \tilde{q}\ ,\ \tilde{S}_0(\tilde{q})\ ,\ \tilde{p}\ =\ \frac{\partial \tilde{S}_0}{\partial \tilde{q}}\ ), \tag{4}$$

such that $W(q) \longrightarrow \tilde{W}(\tilde{q}) = 0$, exist for all $W(q)$. We refer to this proposition as the "equivalence postulate of quantum mechanics". It implies that all physical systems, labelled by a potential function $W(q)$, can be connected by coordinate transformations. More narrowly, we may regard it as adaptation of the classical Hamilton–Jacobi formalism to quantum mechanics. Irrespective of the concrete interpretation, it reveals the Möbius symmetry that underlies quantum mechanics. The equivalence postulate implies that the Hamilton–Jacobi (HJ) equation

Journal of Physics: Conference Series **626** (2015) 012016 doi:10.1088/1742-6596/626/1/012016

has to be covariant under coordinate transformations. However, from eq. (2) it is seen that the first term transforms as a quadratic differential, whereas the potential function $W(q)$, in general, does not. Furthermore, the state $W(q) \equiv 0$ is fixed under the transformations $q \to \tilde{q}(q)$. We therefore assume that the HJ equation is modified by adding a yet to be determined function $Q(q)$. The modified HJ equation then takes the form

$$\frac{1}{2m}\left(\frac{\partial S_0}{\partial q}\right)^2 \; + \; W(q) \; + \; Q(q) \; = \; 0, \tag{5}$$

where under the transformations $q \to \tilde{q}(q)$ the two functions $W(q)$ and $Q(q)$ transform as

$$\begin{aligned} \tilde{W}(\tilde{q}) &= \left(\frac{\partial \tilde{q}}{\partial q}\right)^{-2} W(q) + (\tilde{q}; q), \\ \tilde{Q}(\tilde{q}) &= \left(\frac{\partial \tilde{q}}{\partial q}\right)^{-2} Q(q) - (\tilde{q}; q), \end{aligned}$$

with $\tilde{S}_0(\tilde{q}) \; = \; S_0(q)$. It is seen that each of the functions $W(q)$ and $Q(q)$ transforms as a quadratic differential up to an additive term and that the combination $W(q) + Q(q)$ transforms as a quadratic differential. Starting with the trivial state $W^0(q^0) \equiv 0$, all other physical states arise from the additive inhomogeneous term as $W(q) \equiv (q; q^0)$. Furthermore, considering the transformations $q^a \to q^b \to q^c$ and $q^a \to q^c$ and the induced transformations $W^a(q^a) \to W^b(q^b) \to W^c(q^c)$ and $W^a(q^a) \to W^c(q^c)$ gives rise to a cocycle condition on the inhomogeneous term given by

$$(q^a; q^c) \; = \; \left(\frac{\partial q^b}{\partial q^c}\right)^2 \left[\; (q^a; q^b) \; - \; (q^c; q^b)\right]. \tag{6}$$

It can then be shown that the cocycle condition is invariant under Möbius transformations $\gamma(q^a)$. In the one dimensional case the Möbius symmetry uniquely fixes the functional form of the inhomogeneous term to be given by the Schwarzian derivative, *i.e.* $(q^a; q^c) \;\; \sim \;\; \{q^a; q^c\}$,where the Schwarzian derivative is defined by

$$\{f(q), q\} = \frac{f'''}{f'} - \frac{3}{2}\left(\frac{f''}{f'}\right)^2. \tag{7}$$

4. The Quantum Hamilton–Jacobi Equation

Considering the Schwarzian identity

$$\left(\frac{\partial S_0}{\partial q}\right)^2 = \frac{\beta^2}{2}\left(\{\mathrm{e}^{\frac{i2S_0}{\beta}}; q\} - \{S_0; q\}\right), \tag{8}$$

we note that the quadratic differential on the left–hand side of the equation is written as the difference of two Schwarzian derivatives. As we will see, the cocycle condition eq. (6) and the Schwarzian identity eq. (8) lay down the key ingredients for the generalisations to higher dimensions. Equally fundamental is the invariance of the cocycle condition and of the Schwarzian derivative under Möbius transformations. It is proven rigoursely [18] that the corresponding cocycle condition in D–dimensions is invariant under D–dimensional Möbius transformations, with respect to the Euclidean or Minkowski metrics. Similarly, the generalisation of the

Schwarzian identity eq. (8) entails writing the quadratic differential on the left–hand side as a generalised identity on the right–hand side. Making the identifications

$$\begin{aligned} W(q) &= -\frac{\beta^2}{4m}\{e^{\frac{i2S_0}{\beta}};q\} = V(q) - E, \\ Q(q) &= \frac{\beta^2}{4m}\{S_0;q\}, \end{aligned}$$

the modified Hamilton–Jacobi equation becomes,

$$\frac{1}{2m}\left(\frac{\partial S_0}{\partial q}\right)^2 + V(q) - E + \frac{\beta^2}{4m}\{S_0;q\} = 0. \tag{9}$$

The key property of quantum mechanics is gleaned from eq (9). In the CSHJE admits the solution $S_0 = constant$ for the state $W^0(q^0) \equiv 0$. The Quantum Stationary Hamilton–Jacobi Equation (QSHJE), eq. (9), admits the solutions,

$$S_0 = \pm\frac{\beta}{2}\log q \neq Aq + B,$$

where A and B are constants. Thus, in quantum mechanics S_0 is never a constant, and more generally is never a linear function of q. This is a key property of quantum mechanics in this formalism, and is intimately related to consistency of phase–space duality. Furthermore, from the properties of the Schwarzian derivative we know that the function $W(q) = V(q) - E$ is a potential of a second order differential equation given by

$$\left(-\frac{\beta^2}{2m}\frac{\partial^2}{\partial q^2} + V(q) - E\right)\Psi(q) = 0, \tag{10}$$

which we identify as the Schrödinger equation, and $\beta = \hbar$ as the covariantising parameter of the Hamilton–Jacobi equation. The general solutions of (10) and (9) are given by

$$\Psi(q) = \psi_1 + \psi_2 = \frac{1}{\sqrt{S_0'}}\left(Ae^{+\frac{i}{\hbar}S_0} + Be^{-\frac{i}{\hbar}S_0}\right) \tag{11}$$

and

$$e^{+\frac{i2S_0}{\hbar}} = e^{i\alpha}\frac{w + i\bar{\ell}}{w - i\ell} \tag{12}$$

where $w = \psi_1/\psi_2$, $\ell = \ell_1 + i\,\ell_2$, and $\ell_1 \neq 0$, $\alpha \in R$. The condition $\ell_1 \neq 0$ is synonymous to the condition that $S_0 \neq constant$.

It is noted that the Schrödinger equation serves as a tool to solve the Quantum Hamilton–Jacobi Equation (QHJE). Hence, the more fundamental equation is the QHJE, and captures the symmetry properties that underly quantum mechanics. In that respect the solutions of the Schrödinger equation, eq. (11), facilitate the solution of the QHJE via eq. (12).

It is further seen that consistency of the formalism necessitates that both solutions of the Schrödinger equation are used. This is evident in eq. (12) from which we see that the ratio of the two solutions of the Schrödinger equation is used to extract the solution for the the QHJE. It is a reflection of the Möbius symmetry that underlies the formalism, and has implication on the global geometry that underlies quantum mechanics and quantum gravity. The basic point is that the Möbius symmetry includes a symmetry under inversions. In the case of D–dimensional spacetime this can be implemented via inversions with respect to the unit sphere.

More generally, the global geometry, whatever it may be, has to be invariant under the global Möbius transformations. In the case of spatial space, *i.e.* with Euclidean signature of the metric, the implication is that spatial space must be compact. Otherwise the invariance under the Möbius symmetry cannot be applied consistently. Therefore, at the basic level what we face in comparison to conventional quantum mechanics, is a question of boundary conditions. Namely, in the case of bound physical systems in conventional quantum mechanics the solutions of the Schrödinger equation include a square integrable solution and a solution that diverges at infinity. Therefore, the physical solution is retained, whereas the solution that diverges at infinity is non–physical and hence discarded. In essence, however, it is a question of boundary conditions. Namely, if space is infinite then the diverging solution may be discarded. However, if space is compact, as implied by the Möbius symmetry, then the diverging solution cannot be discarded and must be retained. The wave function is necessarily a combination of the two solutions, albeit the coefficient of the diverging solution may be unobservationally small. In a sense therefore the equivalence postulate approach may be regarded as conventional quantum mechanics plus the condition that spatial space is compact, as dictated by the Möbius symmetry that underlies quantum mechanics. Consistency of the formalism therefore entails that the transformation $W(q) = V(q) - E \longrightarrow \tilde{W}(\tilde{q}) = 0$ always exists, and is given by $q \to \tilde{q} = \psi_1/\psi_2$. Applying the transformation then entails that

$$\left(-\frac{\beta^2}{2m}\frac{\partial^2}{\partial q^2} + V(q) - E\right)\psi(q) = 0 \rightarrow \frac{\partial^2}{\partial \tilde{q}^2}\tilde{\psi}(\tilde{q}) = 0 \tag{13}$$

where

$$\tilde{\psi}(\tilde{q}) = \left(\frac{dq}{d\tilde{q}}\right)^{-\frac{1}{2}}\psi(q).$$

5. Energy quantisation

In conventional quantum mechanics the probability interpretation implies that the wave function and its derivative are continuous and is square integrable. Consequently, for bound states the the energy level are quantised. The question is therefore, how is it replicated in the equivalence postulate approach? As the trivialising transformation is given by

$$q^0 = w = \frac{\psi_1}{\psi_2} = \frac{\psi^D}{\psi} \tag{14}$$

and is a solution of

$$\{w, q\} = -\frac{4m}{\hbar^2}(V(q) - E) \tag{15}$$

we have that $w \neq const;\ w \in C^2(R)$ *and* w'' differentiable on R. From the properties of the Schwarzian derivative under inversions it follows that $\{w, q^{-1}\} = q^4\{w, q\}$. Hence, the consistency conditions on the trivialising transformation must be imposed not merely on the real line, but on the extended real line, *i.e.* on the real line plus the point at infinity. That is,

$$w \neq const \ ; \ w \in C^2(\hat{R}) \ \textit{and} \ w'' \text{ differentiable on } \hat{R}, \quad \text{where } \hat{R} = R \cup \{\infty\}.$$

Consequently, the equivalence postulate implies the continuity of the two solutions of the Schrödinger equation and their derivatives. Furthermore, a general theorem that states that if the potential function $W(q)$ is bounding in some interval, then the ratio $w = \psi_1/\psi_2$ is continuous on the extended real line if and only if the Schrödinger equation admits a square integrable solution, is proven rigoursely in [15, 17, 19]. Rather than go through the theorem it is instructive to consider the simple problem of a particle in a potential well with

$$V(q) = \begin{cases} 0, & |q| \leq L, \\ V_0, & |q| > L. \end{cases} \tag{16}$$

Setting $k = \sqrt{2mE}/\hbar$ and $K = \sqrt{2m(V_0 - E)}/\hbar$, the general solutions in and outside the potential well are given by

$$\begin{array}{lll} |q| \leq L & \Psi_1^1 = \cos kq & \Psi_2^1 = \sin kq \\ q > L & \Psi_1^2 = e^{-Kq} & \Psi_2^2 = e^{Kq} \end{array}$$

The solutions at $q < -L$ are fixed by parity. For $q > L$ we can choose the solution to be Ψ_1^2 or Ψ_2^2 or a linear combination. Continuity across the boundary at $q = L$ implies that $\Psi_i^1(L) = \Psi_j^2(L)$ and $\partial_q\Psi_i^1(L) = \partial_q\Psi_j^2(L)$. Denoting such solutions as (i,j), in the $(1,1)$ case imposing continuity on the solution and its derivative yields the quantisation condition $k\ \tan kL = K$. In this case

$$\psi = \begin{cases} \cos(kL)\exp[K(q+L)], & q < -L, \\ \cos(kq), & |q| \leq L, \\ \cos(kL)\exp[-K(q-L)], & q > L, \end{cases} \tag{17}$$

and a linearly independent solution is given by

$$\psi^D = [2k\sin(kL)]^{-1} \cdot \begin{cases} \cos(2kL)\exp[K(q+L)] - \exp[-K(q+L)], & q < -L, \\ 2\sin(kL)\sin(kq), & |q| \leq L, \\ \exp[K(q-L)] - \cos(2kL)\exp[-K(q-L)], & q > L. \end{cases} \tag{18}$$

The trivialising map $w = \psi^D/\psi$ associated with the $(1,1)$ solution is therefore given by

$$\frac{\psi^D}{\psi} = [k\sin(2kL)]^{-1} \cdot \begin{cases} \cos(2kL) - \exp[-2K(q+L)], & q < -L, \\ \sin(2kL)\tan(kq), & |q| \leq L, \\ \exp[2K(q-L)] - \cos(2kL), & q > L. \end{cases} \tag{19}$$

In this case $\lim_{q\longrightarrow\pm\infty} \psi^D/\psi = \pm\infty$. Hence, in the $(1,1)$ case the trivialising map is continuous on $\hat{R}$ as required by consistency of the equivalence postulate. The solutions imposed by the continuity conditions $k\ \tan kL = K$ are therefore physical energy levels. Next, considering the case $(1,2)$ the constraint that Ψ, Ψ' are continuous across at $q = L$ imposes the condition $k\ \tan(kL) = -K$. Applying similar analysis to the $(1,1)$ case, the trivialising transformation w is give by

$$\frac{\psi^D}{\psi} = [k\sin(2kL)]^{-1} \cdot \begin{cases} \cos(2kL) - \exp[2K(q+L)], & q < -L, \\ \sin(2kL)\tan(kq), & |q| \leq L, \\ \exp[-2K(q-L)] - \cos(2kL), & q > L, \end{cases} \tag{20}$$

whose asymptotic behaviour is

$$\lim_{q\longrightarrow\pm\infty} \frac{\psi^D}{\psi} = \mp k^{-1}\cot(2kL). \tag{21}$$

The only possibility of having $w(-\infty) = w(+\infty)$ is that $k^{-1}\cot(2kL) = 0$. However, this is not compatible with the condition $k\ \tan(kL) = -K$. Hence, we have $w(-\infty) \neq w(+\infty)$. It follows that the energy eigenvalues associated with this solution are not consistent with the equivalence postulate and are therefore not physical. Therefore, the same physical eigenstates that are selected in convential quantum mechanics by the probability interpretation of the wave function, are selected in the equivalence postulate approach by mathematical consistency. Essentially respecting the Möbius symmetry that underlies quantum mechanics. We further note that the requirement that the trivialising transformation is continuous on the extended real line, amounts to the requirement that the real line is compact. Therefore, energy quantisation and square integrability arises from the consistency of the equivalence postulate and the compactness of space, which is mandated by the Möbius symmetry underlying quantum mechanics. However, the probabilistic nature of quantum mechanics, rather than a deterministic parameterisation of a particle propagation is indicated by many experiments. A viable question is whether such parameterisation is consistent with the quantum Hamilton–Jacobi formalism and the Möbius symmetry that underlies it.

6. Time parameterisation

There are two primary approaches to define parameterisation of trajectories in quantum mechanics. The first is Bohmian mechanics [20, 21] in which time parameterisation is defined by identifying the conjugate momentum with the mechanical momentum, *i.e.* $p = \partial_q S = m\dot{q}$, where S is the solution of the QHJE. The second is Floyd's definition of time parameterisation by using Jacobi's theorem, $t = \partial_E S_0^{QM}$, where S_0^{QM} is the solution of the QSHJE. In classical mechanics the two definitions are compatible. Namely, setting $p = \partial_q \mathcal{S}_0^{\rm cl} = m\dot{q}$ gives

$$t - t_0 = m\int_{q_0}^{q} \frac{dx}{\partial_x \mathcal{S}_0^{\rm cl}} = \int_{q_0}^{q} dx \frac{\partial}{\partial E}\partial_x \mathcal{S}_0^{\rm cl} = \frac{\partial \mathcal{S}_0^{\rm cl}}{\partial E}, \tag{22}$$

and by inverting we can solve the equation of motion for $q = q(t)$. However, in quantum mechanics the two definitions are not compatible as

$$t - t_0 = \frac{\partial S_0^{\rm qm}}{\partial E} = \frac{\partial}{\partial E}\int_{q_0}^{q} dx \partial_x S_0^{\rm qm} = \left(\frac{m}{2}\right)\int_{q_0}^{q} dx \frac{1 - \partial_E Q}{(E - V - Q)^{1/2}}. \tag{23}$$

and the mechanical momentum is given by

$$m\frac{dq}{dt} = m\left(\frac{dt}{dq}\right)^{-1} = \frac{\partial_q S_0^{\rm qm}}{(1 - \partial_E \mathcal{V})} \neq \partial_q S_0^{\rm qm}, \tag{24}$$

where $\mathcal{V}$ denotes the combined potential $\mathcal{V} = V(q) + Q(q)$. Thus in quantum mechanics the definition of time by using Jacobi's theorem does not coincide with the Bohmian definition of time by identifying the conjugate momentum with the mechanical momentum. Furthermore, the Bohmian time definition is not compatible with the Möbius symmetry that underlies quantum mechanics, and the compactness of space. In Bohmian mechanics the wave function is set as

$$\psi(\vec{q}, t) = R(q) e^{iS/\hbar}, \tag{25}$$

where $R(q)$ and $S(q)$ are the two real functions of the QHJE, and $\psi(q)$ is a solution of the Schrödinger equation. The conjugate momentum is given by

$$p = \hbar {\rm Im}\frac{\nabla\psi}{\psi},$$

which is used to define trajectories by setting $p = m\dot{q}$. However, the Möbius and compactness of space dictate that the wave function cannot be identified by (25) but must be a linear combination of the two solutions of the Schrödinger equation,

$$\psi = R(q)\left(A e^{\frac{i}{\hbar}S} + B e^{-\frac{i}{\hbar}S}\right). \tag{26}$$

Consequently, in this case

$$\nabla S \neq \hbar {\rm Im}\frac{\nabla\psi}{\psi}$$

and the Bohmian definition of trajectories is invalid [23].

Floyd proposed to define time by using Jacobi's theorem [22], *i.e.*

$$t - t_0 = \frac{\partial S_0^{\rm qm}}{\partial E}, \tag{27}$$

which provides a time parameterisation of the trajectories by inverting $t(q) \to q(t)$. However, if space is compact, as dictated by the Möbius symmetry that underlies the QHJE, then the energy levels are always quantised. Therefore, differentiation with respect to energy is ill defined, and the definition of time parameterisation of trajectories by using Jacobi's theorem is inconsistent in the Quantum Hamilton–Jacobi formalism. In quantum mechanics time can only be used as a classical background parameter. The trajectory representation can only be used in a semi–classical approximation and in that context provides a useful tool to study different physical systems [21].

7. Generalisations

The discussion so far focussed on the one dimensional stationary case. This led to the cocycle condition, eq. (6), and the Schwarzian identity, eq. (8), and their invariance under M"obius transformations. These are the key ingredients that pave the way for the generalisations to higher dimensions in Euclidean or Minkowski space, *i.e.* for the non-relativistic Schrödinger equation or the relativistic Klein–Gordon equation [18]. Furthermore, these generalisations are invariant under Möbius transformations in those spaces [18]. Thus, the Möbius symmetry captures the global symmetry structure that underlies quantum mechanics and dictates that spatial space is compact. Considering the transformations between two D–dimensional coordinate systems given by

$$q \to q^v = v(q) \tag{28}$$

and the induced transformations on the conjugate momenta

$$p_k = \frac{\partial S_0}{\partial q_k}, \tag{29}$$

with $S_0^v(q^v) = S_0(q)$. Consequently,

$$p_k \to p_k^v = \sum_{i=1}^{D} J_{ki} p_i \tag{30}$$

and the Jacobian matrix J is given by

$$J_{ki} = \frac{\partial q_i}{\partial q_j^v}. \tag{31}$$

Adopting the notation

$$(p^v|p) = \frac{\sum_k (p_k^v)^2}{\sum_k p_k^2} = \frac{p^t J^t J p}{p^t p}. \tag{32}$$

the cocycle condition takes the form

$$(q^a; q^c) = (p^c|p^b)\left[(q^a; q^b) - (q^c; q^b)\right]. \tag{33}$$

The cocycle condition, eq. (33) is invariant under D–dimensional Möbius transformations, which include translations, rotations, dilatations, and reflections with respect to the unit sphere [18].

In the case of Minkowski spacetime the invariance is with respect to the $D+1$–dimensional conformal group, where $q \equiv (ct, q_1, \cdots, q_D)$. The second key ingredient of the one dimensional formalism is the Schwarzian identity, equation (8). The generalisation of this identity provides the key for the extension of the formalism to higher dimensions in Euclidean and Minkowski spaces. In the non–relativistic case the identity is given by

$$\alpha^2(\nabla S_0)^2 = \frac{\Delta(Re^{\alpha S_0})}{Re^{\alpha S_0}} - \frac{\Delta R}{R} - \frac{\alpha}{R^2}\nabla\cdot(R^2\nabla S_0), \tag{34}$$

which holds for any two real functions R and S_0 and any constant α. Similarly, in the relativistic case the generalisation is given by [18]

$$\alpha^2(\partial S)^2 = \frac{\Box(Re^{\alpha S})}{Re^{\alpha S}} - \frac{\Box R}{R} - \frac{\alpha}{R^2}\partial\cdot(R^2\partial S), \tag{35}$$

whereas if we include minimal coupling in the form of a vector potential the identity takes the form

$$\alpha^2(\partial S - eA)\cdot(\partial S - eA) = \frac{D^2(Re^{\alpha S})}{Re^{\alpha S}} - \frac{\partial^2 R}{R} - \frac{\alpha}{R^2}\partial\cdot\left(R^2(\partial S - eA)\right), \tag{36}$$

where $D^\mu = \partial^\mu - \alpha e A^\mu$. Similarly to the one dimensional case, the higher dimensional cases are related to the conventional quantum mechanical equations. For example, in the case of eq. (35), setting $\alpha = i/\hbar$, we note that

$$\partial(R^2\cdot\partial S) = 0, \tag{37}$$

and

$$\frac{1}{2m}(\partial S)^2 = -\frac{\hbar^2}{2m}\frac{\Box(Re^{\frac{i}{\hbar}S})}{Re^{\frac{i}{\hbar}S}} + \frac{\hbar^2}{2m}\frac{\Box R}{R}. \tag{38}$$

In analogy to the one dimensional identity we set

$$\mathcal{W}_{rel} = \frac{\hbar^2}{2m}\frac{\Box(Re^{\frac{i}{\hbar}S})}{Re^{\frac{i}{\hbar}S}}, \tag{39}$$

that by (38) implies

$$Q_{rel} = -\frac{\hbar^2}{2m}\frac{\Box R}{R}. \tag{40}$$

Setting $\mathcal{W}_{rel} = 1/2mc^2$ reproduces the Klein–Gordon equation and Eq. (38) corresponds to the Relativistic Quantum Hamilton–Jacobi Equation (RQHJE) [18]. We further remark that the two particle case was considered in [24].

8. Phase space duality and Legendre transformations

We noted from eq. (5) that the modified Hamilton–Jacobi equation allows for non–trivial solutions for the physical system with $W(q) \equiv 0$. The QHJE therefore enables all physical states labelled by the potential function $W(q)$ to be connected to the trivial state via coordinate transformations, and facilitates the covariance of the QHJE. These properties are intimately related to phase space duality [12], which is implemented by the involutive nature of Legendre transformations. Manifest phase–space duality may therefore provide the fundamental physical principle that is sought as the axiomatic principle for formulating quantum gravity. In this respect we note that perturbative and nonperturbative dualities play an important role in attempts to develop a fundamental understanding of string theory, with T–duality being an important perturbative property of string theory [25]. T–duality in toroidal spaces exchanges momentum modes with winding modes. We may therefore view T–duality as phase–space duality in compact space. Furthermore, we may question whether T–duality reflects a property of string theory which is also valid in non–toroidal spaces. An additional important property of T duality in string theory is the existence of self–dual states under T–duality.

Manifest phase–space duality is implemented by the involutive nature of Legendre transformations. Recalling the relation between the momenta and coordinates via a generating function $p = \partial_q S$, we define a dual relation via a new generating function $T(p)$ as $p = \partial T$, where the generating functions are related by Legendre transformations as

$$S = p\frac{\partial T}{\partial p} - T \quad , \quad T = q\frac{\partial S}{\partial q} - S, \tag{41}$$

which in the stationary case reduces to

$$S_0 = p\frac{\partial T_0}{\partial p} - T_0 \quad , \quad T_0 = q\frac{\partial S_0}{\partial q} - S_0. \tag{42}$$

Remarkably, the left–hand side of eq. (42) is invariant under Möbius transformations

$$q \longrightarrow q^v = \frac{Aq + B}{Cq + D},$$

with the induced transformations on p and T_0

$$\begin{aligned} p &\longrightarrow p^v = \rho^{-1}(Cq + D)^2 p \ , \quad \rho = AD - BC \\ T_0 &\longrightarrow T_0^v(p^v) = T_0(p) + \rho^{-1}(ACq^2 + 2BCq + BD)p. \end{aligned}$$

We define a general coordinate transformation $q \to q^v = v(q)$ by the property that S_0 is a scalar function under v, *i.e.* $S_0^v(q^v) = S_0(q)$. With each Legendre transformation we associate a second order differential equation [12, 17], which in the stationary case is given by

$$\left(\frac{\partial^2}{\partial S_0^2} + U(S_0)\right)\begin{pmatrix} q\sqrt{p} \\ \sqrt{p} \end{pmatrix} = 0 \tag{43}$$

where $U(S_0)$ is given by the Schwarzian derivative of q with respect to S_0,

$$U(S_0) = \frac{1}{2}\{q, S_0\} = \frac{1}{2}\left(\frac{q'''}{q'} - \frac{3}{2}\left(\frac{q''}{q'}\right)^2\right).$$

We may associate a second order differential equation with the involutive Legendre transformation. We therefore obtain manifest $p \leftrightarrow q$ – $S_0 \leftrightarrow T_0$ duality with

$$\begin{aligned} p &= \frac{\partial S_0}{\partial q} & q &= \frac{\partial T_0}{\partial p} \\ S_0 &= p\frac{\partial T_0}{\partial p} - T_0 & T_0 &= q\frac{\partial S_0}{\partial q} - S_0 \end{aligned}$$

$$\left(\frac{\partial^2}{\partial S_0^2} + U(S_0)\right)\begin{pmatrix} q\sqrt{p} \\ \sqrt{p} \end{pmatrix} = 0 \qquad \left(\frac{\partial^2}{\partial T_0^2} + \mathcal{V}(T_0)\right)\begin{pmatrix} p\sqrt{q} \\ \sqrt{q} \end{pmatrix} = 0$$

However, the crucial point is the existence of self–dual states, with the property that $pq = \gamma = constant$, which are simultaneous solutions of the two pictures. In these cases $S_0 = -T_0 + constant$, and

$$S_0(q) = \gamma \ln \gamma_q q \qquad T_0(p) = \gamma \ln \gamma_p . p$$

Hence, $S_0 + T_0 = pq = \gamma$, where $\gamma_q\gamma_p\gamma = \mathrm{e}$ and γ_q, γ_p are constants. The remarkable point is that the self–dual states coincide with the $W^0(q^0) \equiv 0$ states of the Quantum Hamilton–Jacobi Equation (QHJE), which render the consistency of the equivalence postulate for all physical states, and the compatibility of quantum mechanics with the underlying Möbius symmetry of the QHJE. In the case of the Classical Hamilton–Jacobi Equation (CHJE), it was noted that the solution in the case with $W^0(q^0) \equiv 0$ correspond to $S_0 = constant$, or more generally $S_0 = Aq^0 + B$, with constants A and B. The Legendre transformation is not defined for linear functions and therefore classically in these case the phase–space duality would not be well defined. The existence of the self–dual quantum mechanical solutions, *i.e.* in the case with $W^{sd} = W^0 = 0$ and with $\gamma^{sd} = \pm\hbar/(2i)$, facilitate the consistency of phase–space duality for all physical states, as well as the consistency of the equivalence postulate, and compatibility with the underlying Möbius symmetry. It is crucial to appreciate that these properties merely accommodate the basic quantum mechanical properties, and in that sense they are not esoteric at all. Namely, in this approach the emergence of $\hbar$ as the basic quantum mechanical parameter, arises as the covariantising parameter in the QHJE, and enables the consistency of the formalism. Furthermore, it is noted that distinction between the classical and quantum mechanical cases in this approach is primarily reflected in the distinction of the $W^0 \equiv 0$ state. In this respect it will not be surprising if the formalism offers an intrinsic regularisation scheme. Furthermore, this intrinsic regularisation scheme is rooted in the global Möbius symmetry that underlies quantum mechanics. This is again not a surprise because the Möbius symmetry implies that spatial space is compact. In turn this implies the existence of a finite length scale in the formalism and therefore an intrinsic regularisation scale. Therefore, a deeper understanding of the implications of the Möbius symmetry that underlies quantum mechanics, brings forth the fertile soil on which the seeds of quantum gravity may grow.

9. Intrinsic length scale

The Schrödinger equation in the physical state with $W^0(q^0) = 0$ is given by

$$\frac{\partial^2\Psi}{\partial q^2} = 0 ,$$

and has two solutions $\psi_1 = q^0$ and $\psi_2 = constant$, which by the Möbius symmetry must both be included in the formalism. The duality, manifested by the invariance under the Möbius transformations, therefore imply the existence of a length scale in the formalism. The corresponding solution of the QHJE is given by [12, 17, 16]

$$\mathrm{e}^{\frac{2i}{\hbar}S_0^0} = \mathrm{e}^{i\alpha}\frac{q^0 + i\bar{\ell}_0}{q^0 - i\ell_0}, \tag{44}$$

where the constant ℓ_0 has the dimension of length [17, 16], and the conjugate momentum takes the form

$$p_0 = \partial_{q^0} S_0^0 = \pm\frac{\hbar(\ell_0 + \bar{\ell}_0)}{2|q^0 - i\ell_0|^2}. \tag{45}$$

It is seen that p_0 is null only when $q^0 \to \pm\infty$. As we noted in section 4 the condition that $\mathrm{Re}\ell_0 \neq 0$ is synonymous to the condition that $S_0 \neq constant$, which is the basic quantum mechanics property in the equivalence postulate formalism. We can represent this nonvanishing length parameter as a combination of some fundamental constants in nature. *e.g.* $\hbar$, c and G. The requirement that $\lim_{\hbar\to 0} p_0 = 0$ in the classical limit implies that we can identify $\mathrm{Re}\,\ell_0$ with the Planck length, [16, 17]

$$\mathrm{Re}\,\ell_0 = \lambda_p = \sqrt{\frac{\hbar G}{c^3}}, \tag{46}$$

The reason being that this identification has the correct scaling properties to reproduce the correct classical limit. Furthermore, we note from eq. (45) that the condition $\mathrm{Re}\ell_0 \neq 0$ serves as an ultraviolet cutoff, *i.e.* p_0 is maximal for $q^0 = -\mathrm{Im}\ell_0$, with

$$\mathrm{Max}|p_0| \;=\; \frac{\hbar}{\mathrm{Re}\ell_0}.$$

Therefore, the consistency of the equivalence postulate formalism with the underlying Möbius symmetry, implies the existence of an intrinsic regularisation scale in quantum mechanics. Additionally, we may identify the quantum potential as an intrinsic curvature term of elementary particles. This provides further evidence that in this approach quantum mechanics regularises itself and a possible connection with theories of extended objects. This is a mere reflection of the Möbius symmetry and the compactness of spatial space.

10. Quantum potential as a curvature term

Using the property of the Schwarzian derivative

$$\{S_0; q\} = -\left(\frac{\partial S_0}{\partial q}\right)^2 \{q; S_0\},$$

we can rewrite the Quantum Stationary Hamilton Jacobi Equation as,

$$\frac{1}{2m}\left(\frac{\partial S_0}{\partial \hat{q}}\right)^2 + V(\hat{q}) - E = 0,$$

where

$$\hat{q} = \int^q \frac{dx}{\sqrt{1 - \frac{\hbar^2}{2}\{q; S_0\}}}.$$

Flanders [26] have shown that the Schwarzian derivative can be interpreted as a curvature term of an equivalence problem for curves in P^1. Thus, the quantum potential, which is never vanishing, can be regarded as an intrinsic curvature term of elementary particle, and a deformation of the space geometry. Furthermore, we note from eq. (36) that the quantum potential as a universal character, which distinguishes it from the gauge interactions that are dependent on the gauge charges. In higher dimensions the quantum potential corresponds to the curvature of the function $R(q)$ as $Q(q) \sim \Delta R(q)/R$. We can estimate the scale of the quantum potential [27]. In the case $W^0(q^0) \equiv 0$ and using eq. (44) we obtain

$$Q^0 \;=\; \frac{\hbar^2}{4m}\{S_0^0, q^0\} \;=\; -\frac{\hbar^2(\mathrm{Re}\,\ell_0)^2}{2m}\frac{1}{|q^0 - i\ell_0|^4}. \tag{47}$$

Taking $m \sim 100GeV$; $\mathrm{Re}\ell_0 = \lambda_p \approx 10^{-35}m$ and q^0 as the size of the observable universe $q^0 \sim 93\mathrm{Ly}$, gives $|Q| \sim 10^{-202}eV$. The expected contribution of the quantum potential to the vacuum energy is very small. Nevertheless, we see from eq. (47) that the classical limit $Q(q) \to 0$ in fact correspond to the decompactification limit $q^0 \to \infty$.

11. Conclusions

The observation of a scalar resonance at the the LHC reinforces the picture that the Standard Model provides a viable effective description of all subatomic data up to the Planck scale. The synthesis of gravity with quantum mechanics remains an open problem. An alternative approach to quantising general relativity is to formulate a geometric approach to quantum mechanics. This is precisely the aim in the equivalence postulate approach to quantum mechanics. What is revealed is the key role of the Möbius symmetry that underlies quantum mechanics in this formalism. In turn the Möbius symmetry provides the key to a proper understanding of the geometry of the quantum spacetime. In this respect the formalism is intimately related to phase space duality manifested by the involutive nature of Legendre transformations. Recalling their role in thermodynamics we may envision that the Legendre transformations merely transfer from one set of variables to another, and neither set should be thought of as more fundamental. In this context the ubiquity of the variable themselves allows us to consider transformations between the space coordinates and the wave function itself [28], without considering one as being primary and the other secondary. Their only merit is their usefulness for a particular physical measurement. In this respect possible observational signatures of the Möbius symmetry underlying quantum mechanics may exist in the microwave background radiation [29]. Additionally, the quantum potential may lead to modified dispersion relations [30] with possible observational consequences [31].

Acknowledgments
This work is supported in part by the STFC under contract ST/G00062X/1.

References
[1] Majore M 2005 *A modern introduction to quantum field theories* (Oxford University Press)
[2] See *e.g.*: Donoghue J F 2005 *Preprint* arXiv:gr-qc/9512024
[3] See *e.g.*: Hawking S 1978 *NATO Sci. Ser.* **44** 145
[4] See *e.g.*: Weinberg S 1979 in *General Relativity: An Einstein centenary survey* ed. S W Hawking and W Israel p. 790
[5] See *e.g.*: Ambjorn J, Jurkiewicz J and Loll R 2013 *Int. J. Mod. Phys.* **A22** 1330019
[6] See *e.g.*: Penrose R 1972 *Phys. Rep.* **6** 241
[7] See *e.g.*: Chamseddine A H, Connes A and Mukhanov V 2014 *JHEP* **1412** 098
[8] See *e.g.*: Ashtekar A 2013 *Lect. Notes Phys.* **863** 31
[9] See *e.g.*: Becker K, Becker M and Schwarz J H 2007 *Strinbg theory and M-theory: A modern introduction* (Cambridge University Press)
[10] Faraggi A E, Nanopoulos D V and Yuan K 1990 *Nucl. Phys.* **B335** 347;
Faraggi A E1992 *Phys. Lett.* **B278** 131; 1992 *Nucl. Phys.* **B387** 239;
Cleaver G B, Faraggi A E and Nanopoulos D V 1999 *Phys. Lett.* **B455** 135;
Faraggi A E, Manno E and Timiraziu C 2007 *Eur. Phys. Jour.* **C50** 701;
Christodoulides K, Faraggi A E and Rizos J 2011 *Phys. Lett.* **B702** 81
[11] Faraggi A E 1992 *Phys. Lett.* **B274** 47; 1997 *Nucl. Phys.* **B487** 55
[12] Faraggi A E and Matone M 1999 *Phys. Lett.* **B450** 34
[13] Faraggi A E and Matone M 1998 *Phys. Lett.* **B437** 369
[14] Faraggi A E and Matone M 1998 *Phys. Lett.* **A249** 180
[15] Faraggi A E and Matone M 1999 *Phys. Lett.* **B445** 357
[16] Faraggi A E and Matone M 1998 *Phys. Lett.* **B445** 77
[17] Faraggi A E and Matone M 2000 *Int. J. Mod. Phys.* **A15** 1869
[18] Bertoldi G, Faraggi A E and Matone M 2000 Class. Quant. Grav. **17** 3965
[19] Faraggi A E and Matone M, arXiv:0912.1225.
[20] *For review and references see e.g.:* Holland P R 1993 *Phys. Rep.* **224** 95
[21] Wyatt R E 2005 *Quantum dynamics with trajectories. Introduction to quantum hydrodynamics* (Springer)
[22] Floyd E R 1982 *Phys. Rev.* **D26** 1339
[23] Faraggi A E and Matone M 2014 *Eur. Phys. Jour.* **C74** 2694
[24] Matone M 2002 *Found. Phys. Lett.* **15** 311
[25] Giveon A, Porrati M and Rabinovici E 1994 *Phys. Rep.* **244** 77
[26] Flanders H 1970 *J. Diff. Geom.* **4** 515
[27] Faraggi A E 2013 Adv. High Energy Phys. **2013**, 957394
[28] Faraggi A E and Matone M 1997 *Phys. Rev. Lett.* **78** 163
[29] Aslanyan G and Manohar A V 2012 *JCAP* **1206** 003;
Ben-David A, Rathaus B and Itzhaki N 2012 *JCAP* **1211** 020
[30] Faraggi A E 2012 *Eur. Phys. Jour.* **C72** 1944
[31] Amelino-Camelia G, Ellis J, Mavromatos N E, Nanopoulos D V and Sarkar S 1998 *Nature* **393** 763

The quantum sweeper effect

G Grössing, S Fussy, J Mesa Pascasio and H Schwabl
Austrian Institute for Nonlinear Studies, Akademiehof, Friedrichstr. 10, 1010 Vienna, Austria
E-mail: ains@chello.at

Abstract. We show that during stochastic beam attenuation in double slit experiments, there appear unexpected new effects for transmission factors below $a \lesssim 10^{-4}$, which can eventually be observed with the aid of weak measurement techniques. These are denoted as *quantum sweeper* effects, which are characterized by the bunching together of low counting rate particles within very narrow spatial domains. We employ a "superclassical" modeling procedure which we have previously shown to produce predictions identical with those of standard quantum theory. Thus it is demonstrated that in reaching down to ever weaker channel intensities, the nonlinear nature of the probability density currents becomes ever more important. We finally show that the resulting unexpected effects nevertheless implicitly also exist in standard quantum mechanics.

1. Introduction

In his criticism of David Bohm's causal interpretation of the quantum mechanical formalism, Isidor Rabi made the following statement in the 1950ies which is still shared by quite some researchers today: "I do not see how the causal interpretation gives us any line to work on other than the use of the concepts of quantum theory. Every time a concept of quantum theory comes along, you can say yes, it would do the same thing as this in the causal interpretation. But I would like to see a situation where the thing turns around, when you predict something and we say, yes, the quantum theory can do it too." [1]

Although doubtlessly the project of a causal interpretation á la de Broglie–Bohm has gained momentum in recent years, with many of its results exhibiting more detailed illustrations via particle trajectories, in fact no experimental prediction unknown to orthodox quantum theory has yet arisen from this approach.

However, in the present paper we discuss a finding based on a causal view of quantum mechanics that amounts to a new effect which has so far eluded orthodox quantum mechanics. Specifically, we recently reported theoretical results from extreme beam attenuation techniques in double-slit experiments whose phenomenology is described by what we term the *quantum sweeper effect.* [2] The discovery of this effect is due to our causal subquantum model for quantum systems, whose central features can be shown to exactly match with the de Broglie–Bohm theory [3].

In Chapter 2 of this paper, we firstly present the quantum sweeper effect in said subquantum scenario. Chapter 3, then, discusses the consequences of the mentioned beam attenuation techniques in purely quantum mechanical language, thus showing that the latter does provide statements in line with, but not as detailed as, the effect described in subquantum terms.

Finally, in Chapter 4 we shall discuss the significance of the sweeper effect. Essentially, we show that in employing ever weaker channel intensities, nonlinear effects become ever more

Journal of Physics: Conference Series **626** (2015) 012017 doi:10.1088/1742-6596/626/1/012017

important, which are generally not considered – although implicitly present – in ordinary quantum mechanics. The latter are a crucial characteristic of subquantum models as the one developed by our group, with experimental tests becoming feasible through the use of weak measurement techniques. Consequences are also discussed with respect to the meaning of the complementarity principle in the light of our new findings.

Before we discuss the sweeper effect, however, some preliminaries are necessary in order to provide the appropriate context. To begin with, we recall from [1, 2] that a beam chopper was employed as a deterministic absorber in one arm of a two-armed interferometer, whereas for stochastic absorption semitransparent foils of various materials were used. Despite the net effect of the same percentage of neutrons being attenuated, the quantum mechanical formalism predicts the following different behaviors for the two cases. Introducing the *transmission factor* a as the beam's transmission probability, in the case of a (deterministic) chopper wheel it is given by the temporal open-to-closed ratio, $a = \frac{t_{\text{open}}}{t_{\text{open}}+t_{\text{closed}}}$, whereas for a (stochastic) semitransparent material defined by its absorption cross section, it is simply the relation of the intensity I with absorption compared to the intensity I_0 without, i.e. $a = I/I_0$. Thus the beam modulation behind the interferometer is obtained in the following two forms. For the deterministic chopper system the intensity is, with φ denoting the phase difference, given by

$$I \propto (1-a)\,|\Psi_1|^2 + a\,|\Psi_1 + \Psi_2|^2 \propto 1 + a + 2a\cos\varphi, \tag{1}$$

whereas for stochastic beam attenuation with the semitransparent material it is

$$I \propto |\Psi_1 + \Psi_2|^2 \propto 1 + a + 2\sqrt{a}\cos\varphi. \tag{2}$$

In other words, although the same number of neutrons is observed in both cases, in the first one the contrast of the interference pattern is proportional to a, whereas in the second case it is proportional to $\sqrt{a}$.

In [2] we accounted for the just described attenuation effects, and we chose the usual double slit scenario, primarily because this turned out to be very useful when discussing more extreme intensity hybrids (i.e., combinations of high and very low transmission factors). Moreover, in this way we could make simple use of the essentials of our model which are employed in our explanation of the beam attenuation phenomena.

Throughout the last years we have developed an approach to quantum mechanics within the scope of theories on "Emergent Quantum Mechanics". (For the proceedings of the first two international conferences devoted to this subject, see [6, 7].) Essentially, we consider the quantum as a complex dynamical phenomenon, rather than as representing some ultimate-level phenomenon in terms of, e.g., pure formalism, wave mechanics, or strictly particle physics only. Our assumption is that a particle of energy $E = \hbar\omega$ is actually an oscillator of angular frequency ω phase-locked with the zero-point oscillations of the surrounding environment, the latter of which containing both regular undulatory and fluctuating components and being constrained by the boundary conditions of the experimental setup via the emergence of standing waves. In other words, the particle in this approach is an off-equilibrium steady-state maintained by the throughput of zero-point energy from its vacuum surroundings. This is in close analogy to the bouncing/walking droplets in the experiments of Couder and Fort's group [8–10], which in many respects can serve as a classical prototype guiding our intuition. However, we denote our whole ansatz as "superclassical" [7], because it connects the classical physics at vastly different scales, i.e. the ordinary classical one and an assumed subquantum one, with "new" effects emergent on intermediate scales, which we have come to know and describe as quantum ones.

In fact, we have succeeded in reproducing a number of quantum mechanical results with our superclassical model, i.e. without any use of the quantum mechanical formalism, like states, wave functions, *et cetera*. Note, moreover, that a Gaussian emerging from, say, a single slit with rounded edges (so as to avoid diffraction effects) is in our model the result of statistically collecting the effects of the aleatory bouncing of our particle oscillator. Rather, the Gaussian stands for the statistical mean of the "excitation" (or "heating up") of the medium within the confines of the slit, and later, as the bouncer/walker progresses, further away from it. We have described this in terms of a thermal environment that represents stored kinetic energy in the vacuum and that is responsible for where the particle is being guided to. For example, consider particle propagation coming out from a Gaussian slit. For a particle exactly at the center of the Gaussian, the diffusive momentum contributions from the heated up environment will on average cancel each other for symmetry reasons. However, the further off the particle is from that center, the stronger the symmetry will be broken, thus resulting in a position-dependent net acceleration or deceleration, respectively – in effect, resulting in the decay of the wave packet. (For a detailed analysis, see [11].) In other words, due to wave-like diffusive propagations originating from the particle's bounces to stir up the medium of the vacuum, particle paths can be influenced by the agitations of the vacuum even in places where no other particle is around.

As already mentioned, we have shown that the spreading of a wave packet can be exactly described by combining the forward (convective) with the orthogonal diffusive velocity fields. The latter fulfill the condition of being unbiased w.r.t. the convective velocities, i.e. the orthogonality relation for the *averaged* velocities derived in [11] is $\overline{\mathbf{vu}} = 0$, since any fluctuations $\mathbf{u} = \delta\left(\nabla S/m\right)$ are shifts along the surfaces of action $S = \text{const}$. Moreover, the fluctuations can be directed towards the left or towards the right from the mean (i.e. from the Ehrenfest trajectory), which leads us to introduce the notations $\mathbf{u}_{i\mathrm{L}}$ and $\mathbf{u}_{i\mathrm{R}}$, respectively.

Reducing the general case discussed in [12] to the double-slit case, one notes for the first and second channels the emergent velocity vectors $\mathbf{v}_{1(2)}$, $\mathbf{u}_{1(2)\mathrm{R}}$, and $\mathbf{u}_{1(2)\mathrm{L}}$, together with associated amplitudes $R_{1(2)}$, respectively. In order to completely accommodate the totality of the system of currents present, one obtains a local wave intensity for any velocity component, e.g. for $\mathbf{v}_1$, by the pairwise projection on the unit vector $\hat{\mathbf{v}}_1$ weighted by R_1 of the totality of all amplitude weighted unit velocity vectors being operative at $(\mathbf{x}, t)$:

$$P(\mathbf{v}_1) = R_1\hat{\mathbf{v}}_1 \cdot (\hat{\mathbf{v}}_1 R_1 + \hat{\mathbf{u}}_{1\mathrm{R}} R_1 + \hat{\mathbf{u}}_{1\mathrm{L}} R_1 + \hat{\mathbf{v}}_2 R_2 + \hat{\mathbf{u}}_{2\mathrm{R}} R_2 + \hat{\mathbf{u}}_{2\mathrm{L}} R_2). \tag{3}$$

The local current attributed to each velocity component is defined as the corresponding "local" intensity-weighted velocity, e.g. for $\mathbf{v}_1$ it is given as $\mathbf{J}(\mathbf{v}_1) = \mathbf{v}_1 P(\mathbf{v}_1) = \mathbf{v}_1\left(R_1^2 + R_1 R_2 \cos\varphi\right)$. The local intensity of a partial current is dependent on all other currents, and the total current itself is composed of all partial components, thus constituting a representation of what we call *relational causality*. After a short calculation, the total current turns out as $\mathbf{J}_{\mathrm{tot}} = \mathbf{v}_1 P(\mathbf{v}_1) + \mathbf{u}_{1R} P(\mathbf{u}_{1\mathrm{R}}) + \mathbf{u}_{1\mathrm{L}} P(\mathbf{u}_{1\mathrm{L}}) + \mathbf{v}_2 P(\mathbf{v}_2) + \mathbf{u}_{2R} P(\mathbf{u}_{2\mathrm{R}}) + \mathbf{u}_{2\mathrm{L}} P(\mathbf{u}_{2\mathrm{L}})$, which, by identifying the resulting diffusive velocities $\mathbf{u}_{i\mathrm{R}} - \mathbf{u}_{i\mathrm{L}}$ with the effective diffusive velocities $\mathbf{u}_i$ for each channel, finally leads to

$$\mathbf{J}_{\mathrm{tot}} = R_1^2\mathbf{v}_1 + R_2^2\mathbf{v}_2 + R_1 R_2\left(\mathbf{v}_1 + \mathbf{v}_2\right)\cos\varphi + R_1 R_2\left(\mathbf{u}_1 - \mathbf{u}_2\right)\sin\varphi. \tag{4}$$

The trajectories or streamlines, respectively, are given by

$$\dot{\mathbf{x}} = \mathbf{v}_{\mathrm{tot}} = \frac{\mathbf{J}_{\mathrm{tot}}}{P_{\mathrm{tot}}}. \tag{5}$$

As first shown in [13], by re-inserting the expressions for convective and diffusive velocities, respectively, i.e.

$$\mathbf{v}_i = \frac{\nabla S_i}{m}, \quad \text{and} \quad \mathbf{u}_i = -\frac{\hbar}{m}\frac{\nabla R_i}{R_i}, \tag{6}$$

one immediately identifies Eq. (5) with the Bohmian guidance equation and Eq. (4) with the quantum mechanical pendant for the probability density current [14]. As we have shown also

Journal of Physics: Conference Series **626** (2015) 012017 doi:10.1088/1742-6596/626/1/012017

the latter identity, we are assured that our results are the same as those of standard quantum mechanics – provided, of course, that generally (i.e. in the standard quantum as well as in our ansatz) the idealization of using Gaussians or similar regular distribution functions is applicable for the high degrees of attenuation studied here. However, as in our model we can also make use of the velocity field to plot the averaged particle trajectories, we can in principle provide a more detailed picture, in similar ways to the Bohmian one, but still not relying on any quantum mechanical tool like a wave function, for example.

In interpreting their results of the beam attenuation experiments with neutrons, Rauch *et al.* found evidence in support of the complementarity principle. That is, the more pronounced the visibility of the interference fringes, the less which-path knowledge one can have of the particle propagation, and *vice versa*: the higher the probability is for a particle to take a path through one certain slit, the less visible the interference pattern becomes. This was in fact confirmed in the above-mentioned neutron interferometry experiments, albeit to a lesser degree for very low counting rates. In particular, the authors of [4, 5] often use expressions such as the "particle-like" or the "wave-like" nature of the quantum system studied, depending on whether which-path information or interference effects are dominant, respectively. While this is all correct as far as the mentioned papers are concerned, an extrapolation of the use of "particle-like" or "wave-like" attributed to more extreme intensity hybrids is not guaranteed. In fact, we shall show below a particular effect which undermines said dichotomy of "particle-like" and "wave-like" features, thereby calling for an improved, more general analysis of possible relationships between particle and wave features.

2. The quantum sweeper effect during stochastic beam attenuation in the double slit and its superclassical modeling

Let us start with a discussion of stochastic attenuation using a coherent beam in a double-slit experiment. With the intensity distribution being recorded on a screen, we are going to discuss a particular effect of the attenuation of one of the two emerging Gaussians at very small transmission factors. With the appropriate filtering of the particles going through one of the two slits, the recorded probability density on a screen in the surroundings of the experiment will appear differently than what one would normally expect. That is, even if one had a low beam intensity coming from one slit, one would expect the following scenario according to the usual quantum mechanical heuristics: The interference pattern would more and more become asymmetric in the sense that the contributions from the fully open slit would become dominant until such a low counting rate from the attenuated slit is arrived at that essentially one would have a one-slit distribution of recorded particles on the screen.

Interestingly, this is not exactly what one obtains at least for very low values of a when going through the calculations and/or computer simulations with our superclassical bouncer model. The latter consists, among other features, in an explicit form of the velocity field emerging from the double slit, as well as of the probability density current associated with it.

Fig. 1 shows the *quantum sweeper effect*: a series of probability density distributions plus averaged trajectories for the case that the intensity in slit 2 is gradually diminished. We use the same model as in [13], or [15], respectively: particles (represented by plane waves in the forward y-direction) from a coherent source passing through "soft-edged" slits in a barrier (located along the x-axis) and recorded at a screen in the forward direction, i.e. parallel to the barrier. This situation is described by two Gaussians representing the totality of the effectively "heated-up" path excitation field, one for slit 1 and one for slit 2, whose centers have the same distances from the plane spanned by the source and the center of the barrier along the y-axis, respectively. Now, with ever lower values of the transmission factor a during beam attenuation, one can see a steadily growing tendency for the low counting rate particles of the attenuated beam to become swept aside. In our model, this is straightforward to understand, because we have the analytical

(a) $a = 10^{-1}$ (b) $a = 10^{-4}$ (c) $a = 10^{-10}$

Figure 1: Trajectory behavior during the *quantum sweeper effect* for different transmission factors a at the right slit of a double slit setup: With ever lower values of a, one can see a steadily growing tendency for the originally dominant interference fringes to disappear and for the low counting rate particles of the attenuated beam to become swept aside. In our model, this phenomenology is explained by processes of diffusion due to the presence of accumulated kinetic energy mainly in the "strong" beam. The sweeper effect is thus the result of the vacuum heat sweeping aside the very low intensity beam, with a "no crossing" line defined by the balancing out of the osmotic momenta coming from the two beams, respectively. Throughout this paper, to demonstrate the effect more clearly, the same number of trajectories for each slit is displayed.

tools to differentiate between the forward propagations $\mathbf{v}_i$ and the diffusive influences of velocities $\mathbf{u}_i$, as distinguishable contributions from the different slits i. Thus, it is processes of diffusion which are seen in operation here, due to the presence of accumulated heat (i.e. kinetic energy), primarily in the "strong" beam, as discussed in the previous Section. So, in effect, we understand Fig. 1 as the result of the vacuum heat sweeping aside the very low intensity beam, with a "no crossing" line defined by the balancing out of the diffusive momenta, $m\,(\mathbf{u}_1 + \mathbf{u}_2) = 0$.

Importantly, for certain slit configurations and sizes of the transmission factor, the sweeper effect leads to a bunching of trajectories which may become deflected into a direction almost orthogonal to the original forward direction. In other words, one would need much wider screens in the forward direction to register them, albeit then weakened due to a long traveling distance. On the other hand, if one installed a screen orthogonal to the "forward screen", i.e. one that is parallel to the original forward motion (and thus to the y-axis), one could significantly improve the contrast and thus register the effect more clearly. Further, we note that changing the distance between the two slits does not alter the effect, but demonstrates the bunching of the low counting rate arrivals in essentially the same narrow spatial area even more drastically. So, again, if one places a screen not in the forward direction parallel to the barrier containing the double slit, but orthogonally to the latter, one registers an increased local density of particle arrivals in a narrow spatial area under an angle that is independent of the slit distance.

Let us now turn to the case of decoherent beams. For, although we shall refrain from constructing a concrete model of decoherence and implementing it in our scheme, we already have the tools of an effective theory, i.e. to describe decoherence without the need of a specified mechanism for it. Namely, as full decoherence between two (Gaussian or other) beams is characterized by the complete absence of the interference term in the overall probability distribution of the system, this means that $P_{\text{tot}} = R_1^2 + R_2^2$, since the interference term

$$R_1 R_2\,(\mathbf{v}_1 + \mathbf{v}_2)\cos\varphi = 0. \tag{7}$$

Journal of Physics: Conference Series **626** (2015) 012017 doi:10.1088/1742-6596/626/1/012017

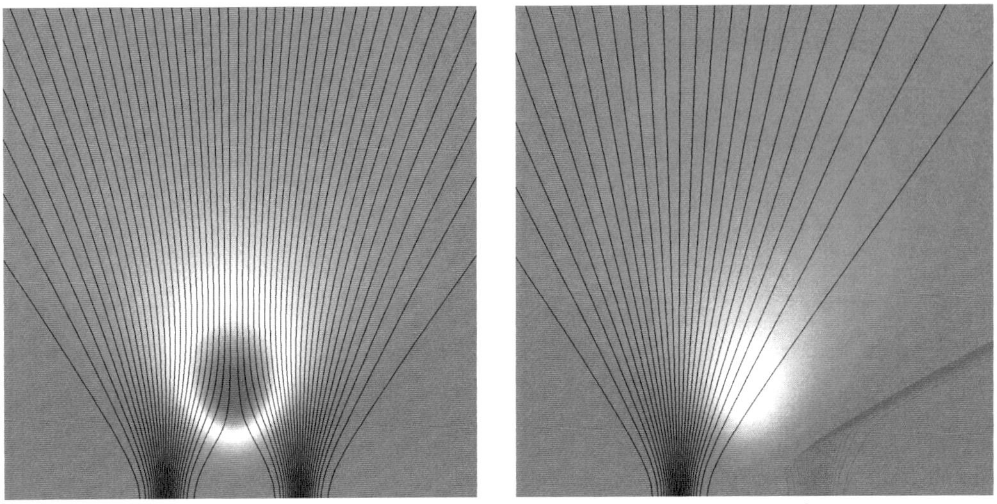

(a) $a = 1$ (b) $a = 10^{-8}$

Figure 2: Double-slit experiment with completely incoherent channels. (a) The average trajectories never cross the central symmetry line, a fact due to the diffusion related "hot spot" indicated in red-to-yellow-to-white (depicting both interference terms of the total current (4)), which represents a kinetic energy reservoir that effectively gives particles a push in the forward direction. (b) The current's interference term is now weakened by the factor $a = 10^{-8}$, which is why it does not affect the "strong" beam. However, it is sufficient for the attenuated beam to become deflected and thus to produce the sweeper effect.

If we therefore choose that on average one has $\cos\varphi = 0$, a situation with $\varphi = \frac{\pi}{2}$ effectively describes two incoherent beams in the double-slit system. What about the two interference terms in the probability density current (4), then? Well, the first term is identical with the vanishing (7), but the second term, with $\mathbf{u}_i = -\frac{\hbar}{m}\frac{\nabla R_i}{R_i}$ and $\varphi = \frac{\pi}{2}$ explicitly reads as

$$\frac{\hbar}{m} R_1 R_2 \left(\frac{\nabla R_2}{R_2} - \frac{\nabla R_1}{R_1} \right) = \frac{\hbar}{m} \left(R_1 \nabla R_2 - R_2 \nabla R_1 \right). \tag{8}$$

As the distributions R_i may have long wiggly tails – summing up, after many identical runs, to a Gaussian with no cutoff, but spreading throughout the whole domain of the experimental setup [16] –, the expression (8) is not at all guaranteed to vanish. In fact, a look at Fig. 2 shows that there is an effect even for incoherent beams: Although the product $R_1 R_2$ is negligible and therefore leads to no interference fringes on the screen, nevertheless expression (8) has the effect of "bending" average trajectories so as to obey the "no crossing" rule well known from our model as well as from Bohmian theory.

In sum, then, performing a double-slit experiment with decoherent beams leads to an emergent behavior of particle propagation which can be explained by the effectiveness of diffusion waves with velocities u_i interacting with each other, thereby creating a "hot spot" where the intensity of the diffusive currents is highest and leads to a deflection into the forward direction such that no crossing of the average velocities beyond the symmetry line is made possible (Fig. 2a). This is therefore in clear contradiction to the scenario where only one slit is open for the particle to go through. If the slits are not open simultaneously, the particles could propagate to locations beyond the symmetry line, i.e. to locations forbidden in the case of the second slit being open [17].

As our velocity fields $\mathbf{v}_i$ and $\mathbf{u}_i$ (6) are identical with the Bohmian and the "osmotic" momentum, respectively, one can relate them also to the technique of weak measurements. The latter have turned out [18–20] to provide said velocities as "weak values", which are just given by the real and complex parts of the quantum mechanical expression $\frac{\langle \mathbf{r}|\hat{p}|\Psi(t)\rangle}{\langle \mathbf{r}|\Psi(t)\rangle}$, i.e. the weak values associated with a weak measurement of the momentum operator $\hat{p}$ followed by the usual ("strong") measurement of the position operator $\hat{r}$ whose outcome is $\mathbf{r}$. In other words, in principle the trajectories for intensity hybrids generally, and for the quantum sweeper in particular, are therefore accessible to experimental confirmation.

3. The quantum mechanical description of the sweeper effect

Let us now consider the stochastic attenuation discussed above in purely quantum mechanical terms. As already mentioned, the probability density distribution is given by Equation (2). A graphic representation of this distribution in a distance of 5m from the double slit is shown in Fig. 3. Two cases of the attenuation factor at one of the two slits of a double slit system are shown, i.e. $a = 10^{-4}$ and $a = 10^{-8}$ affecting the right slit, respectively. As is to be expected, on a linear scale the distribution will appear as if practically the whole intensity goes through the left un-attenuated slit (Fig. 3a). Zooming in with a factor of 1,000 as shown in Fig. 3b, one can see the faint rest of interference phenomena for the case of $a = 10^{-4}$ (blue), whereas for $a = 10^{-8}$ (red) apparently smooth behavior is seen. Still, the full effect is best visible on the logarithmic scale shown in Fig. 3c. Compared to the dotted initial distributions for the cases of $a = 10^{-4}$ (blue) and $a = 10^{-8}$ (red), respectively, the whole distribution clearly shows interference phenomena which have been "swept aside" far to the right. Thus, the quantum sweeper effect is confirmed also via orthodox language.

The bunching together of low counting rate particles within a very narrow spatial domain, or channel, respectively, counters naive expectations that with ever higher beam attenuation nothing interesting may be seen any more. The reason why these expectations are not met is given by the explicit appearance of the nonlinear structure of the probability density current (4) in these domains for very low values of a.

4. Summary and implications

As for implications of our finding presented here, we shall now briefly remark on its relevance with respect to the issue of wave-particle duality. Considering the appearance of compressed interference fringes in the attenuated beam in Fig. 1, it is indisputable that one has to do with the result of a wave-like behavior. This is confirmed in Fig. 2b where the decoherent scenario is characterized by the complete absence of such wave-like behavior like interference fringes. This means, however, that an often used argument to describe the complementarity between wave- and particle-like behavior in the double slit experiment, or in interferometry, respectively, has only limited applicability, as it does not apply to intensity hybrids, since in our model the wave-like contributions due to diffusion are always present. Specifically, the relation for pure states [21]

$$D^2 + V^2 = 1, \tag{9}$$

with distinguishability

$$D = \left| \frac{R_1^2 - R_2^2}{R_1^2 + R_2^2} \right| \tag{10}$$

representing the particle-related which-path information and visibility

$$V = \frac{| R_1 + R_2 |^2 - | R_1 - R_2 |^2}{| R_1 + R_2 |^2 + | R_1 - R_2 |^2} = \frac{2R_1R_2}{R_1^2 + R_2^2} \tag{11}$$

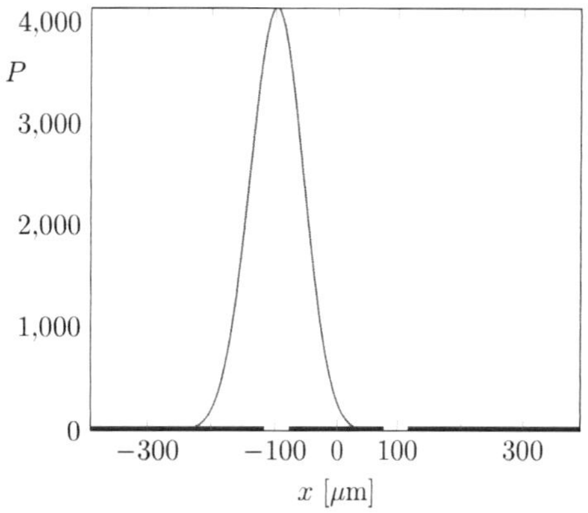

(a) Probability density distribution P in a distance of 5 m from the double slit. The two cases of the attenuation factor at the right slit of a double slit system, i.e. $a = 10^{-4}$ and $a = 10^{-8}$, respectively, essentially provide the same distribution at moderate resolution.

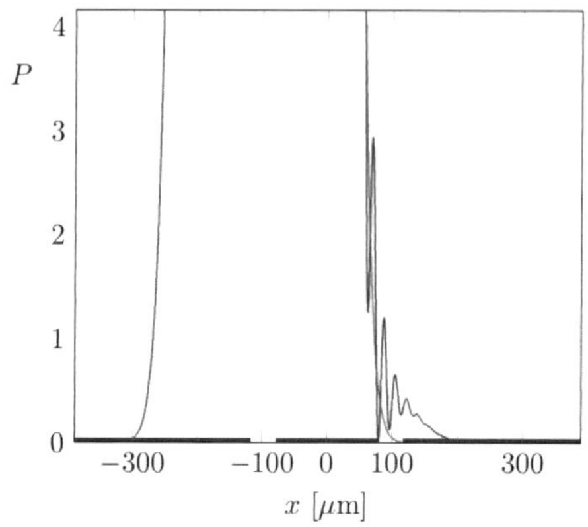

(b) Same as in (a), but zooming in with a factor of 1,000. Now the two cases are discernible: note the faint rest of interference phenomena for the case of $a = 10^{-4}$ (blue), whereas for $a = 10^{-8}$ (red) apparently smooth behavior is seen.

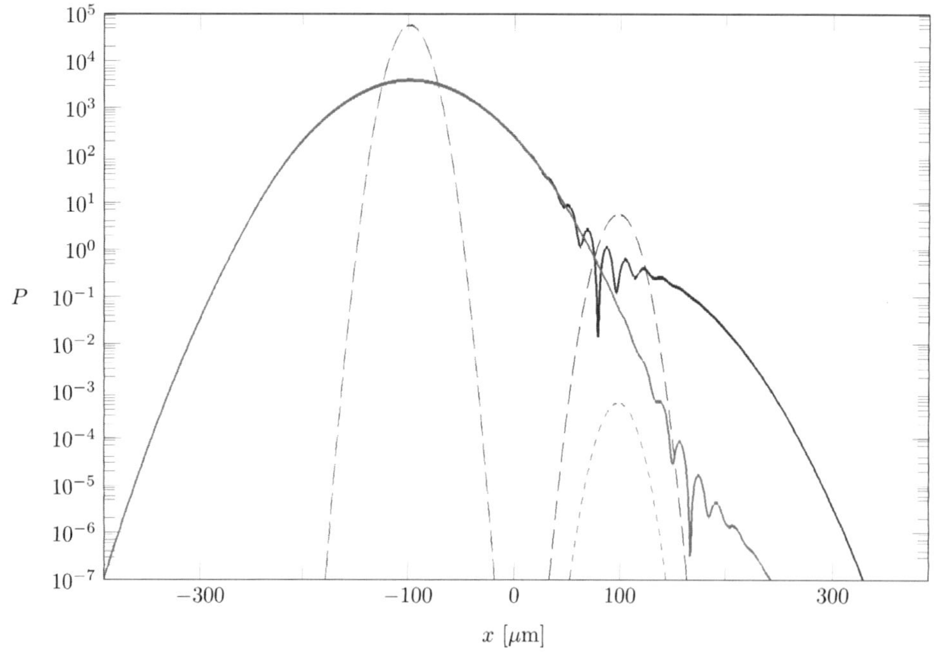

(c) Same as in (a), but now on a logarithmic scale. Dotted initial distributions for the cases of $a = 10^{-4}$ (blue) and $a = 10^{-8}$ (red), respectively, develop into distributions clearly showing interference phenomena which have been "swept aside" far to the right. This sweeper effect is due to the explicit appearance of the nonlinear structure of the probability density current in these domains for very low values of a.

Figure 3: The sweeper effect as described by quantum mechanics the contrast of the interference fringes in the standard quantum mechanical double slit scenario, suggests that with ever lower values of a ever lower values of V are implied in a constantly decreasing manner. In our case, by considering the superclassical nature of the sweeper effect, we find a deviating, characteristic signature at very low values of $a \lesssim 10^{-4}$. In this domain, the usual expectation would be that practically one has arrived at the "particle" side of the complementarity principle, i.e. essentially a one-slit distribution, with wave-like phenomena having almost disappeared. However, if one has a very strongly attenuated beam, the emerging behavior of its outgoing trajectories is different from a one-slit particle distribution scenario if the other slit is open and un-attenuated. The increased local relative contrast corresponding to a bunching of trajectories and due to said nonlinear effect of the probability density current is to be captured by a vertical screen (i.e. parallel to the y-axis) for optimal visibility.

We have thus shown that for transmission factors below $a \lesssim 10^{-4}$ in intensity hybrids, new effects appear which are not taken into account in a naive, i.e. linear, extrapolation of expectations based on higher-valued transmission factors. We have described the phenomenology of these *quantum sweeper* effects, including the bunching together of low counting rate particles within a very narrow spatial domain, or channel, respectively. However, we also stress that these results are in accordance with standard quantum mechanics, since we just used a re-labeling and re-drawing of the constituent parts of the usual quantum mechanical probability density currents. However, concerning the explicit phenomenological appearances due to the nonlinear structure of the probability density current in the respective domains for very low values of a, our subquantum model is better equipped to deal with these appearances explicitly.

With the discovery of the quantum sweeper effect on the basis of a superclassical causal approach to quantum mechanics, we claim to have presented a first example as it was demanded by Rabi. We are optimistic that through further developments, both in superclassical theory employing subquantum mechanics and in weak measurement techniques capable of probing the latter regime, more unexpected new effects can be predicted and eventually be confirmed in experiment.

Acknowledgments

We thank Jan Walleczek for many enlightening discussions and the Fetzer Franklin Fund for partial support of the current work.

References

[1] Freire Jr. O 2005 Science and exile: David Bohm, the cold war, and a new interpretation of quantum mechanics *HSPS* **36** 1–34, arXiv:physics/0508184 [physics.hist-ph]

[2] Grössing G, Fussy S, Mesa Pascasio J and Schwabl H 2015 Extreme beam attenuation in double-slit experiments: Quantum and subquantum scenarios *Ann. Phys.* **353** 271–281, arXiv:1406.1346 [quant-ph]

[3] Grössing G, Fussy S, Mesa Pascasio J and Schwabl H 2015 Implications of a deeper level explanation of the deBroglie–Bohm version of quantum mechanics *Quantum Stud.: Math. Found.* , arXiv:1412.8349 [quant-ph] in press

[4] Rauch H and Summhammer J 1984 Static versus time-dependent absorption in neutron interferometry *Phys. Lett.* A **104** 44–46

[5] Rauch H, Summhammer J, Zawisky M and Jericha E 1990 Low-contrast and low-counting-rate measurements in neutron interferometry *Phys. Rev. A* **42** 3726–3732

[6] Grössing G *ed* 2012 *Emergent Quantum Mechanics 2011* No. 361/1 (Bristol: IOP Publishing) Url: http://iopscience.iop.org/1742-6596/361/1

[7] Grössing G, Elze H-T, Mesa Pascasio J and Walleczek J *ed* 2014 *Emergent Quantum Mechanics 2013* No. 504/1 (Bristol: IOP Publishing) Url: http://iopscience.iop.org/1742-6596/504/1

[8] Fort E, Eddi A, Boudaoud A, Moukhtar J and Couder Y 2010 Path-memory induced quantization of classical orbits *PNAS* **107** 17515–17520

[9] Couder Y and Fort E 2006 Single-particle diffraction and interference at a macroscopic scale *Phys. Rev. Lett.* **97** 154101

[10] Couder Y and Fort E 2012 Probabilities and trajectories in a classical wave-particle duality *J. Phys.: Conf. Ser.* **361** 012001

[11] Grössing G, Fussy S, Mesa Pascasio J and Schwabl H 2010 Emergence and collapse of quantum mechanical superposition: Orthogonality of reversible dynamics and irreversible diffusion *Physica A* **389** 4473–4484, arXiv:1004.4596 [quant-ph]

[12] Fussy S, Mesa Pascasio J, Schwabl H and Grössing G 2014 Born's rule as signature of a superclassical current algebra *Ann. Phys.* **343** 200–214 arXiv:1308.5924 [quant-ph]

[13] Grössing G, Fussy S, Mesa Pascasio J and Schwabl H 2012 An explanation of interference effects in the double slit experiment: Classical trajectories plus ballistic diffusion caused by zero-point fluctuations *Ann. Phys.* **327** 421–437, arXiv:1106.5994 [quant-ph]

[14] Sanz A S and Miret-Artés S 2008 A trajectory-based understanding of quantum interference *J. Phys. A: Math. Gen.* **41** 435303, arXiv:0806.2105 [quant-ph]

[15] Holland P R 1993 *The Quantum Theory of Motion* (Cambridge, UK: Cambridge Univ. Press)

[16] Grössing G, Fussy S, Mesa Pascasio J and Schwabl H 2013 'Systemic nonlocality' from changing constraints on sub-quantum kinematics *J. Phys.: Conf. Ser.* **442** 012012, arXiv:1303.2867 [quant-ph]

[17] Sanz A S and Borondo F 2009 Contextuality, decoherence and quantum trajectories *Chem. Phys. Lett.* **478** 301–306, arXiv:0803.2581 [quant-ph]

[18] Leavens C R 2005 Weak measurements from the point of view of Bohmian mechanics *Found. Phys.* **35** 469

[19] Wiseman H M 2007 Grounding Bohmian mechanics in weak values and bayesianism *New J. Phys.* **9** 165–176, arXiv:0706.2522 [quant-ph]

[20] Hiley B J 2012 Weak values: Approach through the Clifford and Moyal algebras *J. Phys.: Conf. Ser.* **361** 012014, arXiv:1111.6536v1 [quant-ph]

[21] Greenberger D M and Yasin A 1988 Simultaneous wave and particle knowledge in a neutron interferometer *Phys. Lett. A* **128** 391–394

Classical probabilistic realization of "Random Numbers Certified by Bell's Theorem"

Andrei Khrennikov
International Center for Mathematical Modeling
in Physics, Engineering, Economics, and Cognitive Science
Linnaeus University, Växjö, S-35195, Sweden

E-mail: Andrei.Khrennikov@lnu.se

Abstract. We question the commonly accepted statement that random numbers certified by Bell's theorem carry some special sort of randomness, so to say, quantum randomness or intrinsic randomness. We show that such numbers can be easily generated by classical random generators.

1. Introduction

The idea that quantum randomness (QR) differs crucially from classical randomness (CR) is due to von Neumann who pointed that QR is intrinsic and CR is reducible to variability in an ensemble [1]. At the first stages of the development of quantum theory (QM) von Neumann's viewpoint to QR and its exceptional nature was merely of the purely foundational value. It was used to support various no-go statements, starting with von Neumann's no-go theorem [1] (in fact, this no-go statement was called by von Neumann "ansatz" and "theorem" as it is common in modern literature), see [2], [3] for discussions. However, with the development of quantum information theory and its technological implementation the inter-relation between classical and quantum randomness started to play the important role in real applications.

In this note we shall discuss one very novel and important application of quantum randomness, namely, in the form of random number generators certified by Bell's theory, see, e.g., [4] for details and review. As was claimed in [4]:

"It is thereby possible to design of a new type of cryptographically secure random number generator which does not require any assumption on the internal working of the devices. This strong form of randomness generation is impossible classically and possible in quantum systems only if certified by a Bell inequality violation."

It seems that this statement is in contradiction with the recent results [5], see also [6], on embedding of the data generated in experimental test demonstrating violations of Bell's inequality into the classical probability model (based on the Kolmogorov measure-theoretic model [7], 1933). As well as in [4], we consider the case of CHSH inequality which were tested and confirmed in numerous experiments (although it might happen that the final loophole free test will be performed not for Bell's inequality not in the CHSH form, but in the Eberhard form [8]).

2. Classical probability space for data violating the CHSH inequality

For readers convenience, we present shortly the model from [5].

In the CHSH-test we operate with probabilities $p_{ij}(\epsilon, \epsilon'), \epsilon, \epsilon' = \pm 1,$ – to get the results $a_{\theta_i} = \epsilon, b_{\theta'_j} = \epsilon'$ in the experiment with the fixed pair of orientations (θ_i, θ'_j). We borrow probabilities from the mathematical formalism of quantum mechanics:

$$p_{ij}(\epsilon, \epsilon) = \frac{1}{2}\cos^2\frac{\theta_i - \theta'_j}{2}, p_{ij}(\epsilon, -\epsilon) = \frac{1}{2}\sin^2\frac{\theta_i - \theta'_j}{2}. \qquad (1)$$

Let us now consider the set of points Ω (the space of "elementary events" in Kolmogorov's terminology):

$$\Omega = \{(\epsilon_1, 0, \epsilon'_1, 0), (\epsilon_1, 0, 0, \epsilon'_2), (0, \epsilon_2, \epsilon'_1, 0), (0, \epsilon_2, 0, \epsilon'_2)\},$$

where $\epsilon = \pm 1, \epsilon' = \pm 1.$

We define the following probability measure[1] on Ω

$$\mathbf{P}(\epsilon_1, 0, \epsilon'_1, 0) = \frac{1}{4}p_{11}(\epsilon_1, \epsilon'_1), \mathbf{P}(\epsilon_1, 0, 0, \epsilon'_2) = \frac{1}{4}p_{12}(\epsilon_1, \epsilon'_2)$$

$$\mathbf{P}(0, \epsilon_2, \epsilon'_1, 0) = \frac{1}{4}p_{21}(\epsilon_2, \epsilon'_1), \mathbf{P}(0, \epsilon_2, 0, \epsilon'_2) = \frac{1}{4}p_{22}(\epsilon_2, \epsilon'_2).$$

We now define random variables $A^{(i)}(\omega), B^{(j)}(\omega)$:

$$A^{(1)}(\epsilon_1, 0, \epsilon'_1, 0) = A^{(1)}(\epsilon_1, 0, 0, \epsilon'_2) = \epsilon_1, A^{(2)}(0, \epsilon_2, \epsilon'_1, 0) = A^{(2)}(0, \epsilon_2, 0, \epsilon'_2) = \epsilon_2; \qquad (2)$$

$$B^{(1)}(\epsilon_1, 0, \epsilon'_1, 0) = B^{(1)}(0, \epsilon_2, \epsilon'_1, 0) = \epsilon'_1, B^{(2)}(\epsilon_1, 0, 0, \epsilon'_2) = B^{(2)}(0, \epsilon_2, 0, \epsilon'_2) = \epsilon'_2. \qquad (3)$$

and we put these variables equal to zero in other points.

We find two dimensional probabilities

$$\mathbf{P}(\omega \in \Omega : A^{(1)}(\omega) = \epsilon_1, B^{(1)}(\omega) = \epsilon'_1) = \mathbf{P}(\epsilon_1, 0, \epsilon'_1, 0) = \frac{1}{4}p_{11}(\epsilon_1, \epsilon'_1), \ldots,$$

$$\mathbf{P}(\omega \in \Omega : A^{(2)}(\omega) = \epsilon_2, B^{(2)}(\omega) = \epsilon'_2) = \mathbf{P}(0, \epsilon_2, 0, \epsilon'_2) = \frac{1}{4}p_{22}(\epsilon_2, \epsilon'_2).$$

We also consider the random variables which are responsible for selections of pairs settings (θ_i, θ'_j). For "the settings at LHS":

$$\eta_L(\epsilon_1, 0, 0, \epsilon'_2) = \eta_L(\epsilon_1, 0, \epsilon'_1, 0) = 1, \eta_L(0, \epsilon_2, 0, \epsilon'_2) = \eta_L(0, \epsilon_2, \epsilon'_1, 0) = 2.$$

For "the settings at RHS":

$$\eta_R(\epsilon_1, 0, \epsilon'_1, 0) = \eta_R(0, \epsilon_2, \epsilon'_1, 0) = 1, \eta_R(0, \epsilon_2, 0, \epsilon'_2) = \eta_R(\epsilon_1, 0, 0, \epsilon'_2) = 2.$$

The points of the space of "elementary events" Ω have the following interpretation. Take,for example, $\omega = (\epsilon_1, 0, \epsilon'_1, 0)$. Here $\eta_L = 1$, i.e, the LHS setting is selected as θ_1, and $\eta_R = 1$, i.e, the RHS setting is selected as θ'_1. Then the random variable $A^{(1)} = \epsilon_1, A^{(1)} = 0$ and $B^{(1)} = \epsilon'_1, B^{(2)} = 0$. Consider now, e.g., $\omega = (\epsilon_1, 0, 0, \epsilon'_2)$. Here $\eta_L = 1$, i.e, the LHS setting is selected as θ_1, and $\eta_R = 2$, i.e, the RHS setting is selected as θ'_2. Then the random variable $A^{(1)} = \epsilon_1, A^{(1)} = 0$ and $B^{(1)} = 0, B^{(2)} = \epsilon'_2$.

[1] To match completely with Kolmogorov's terminology, we have to select a σ-algebra $\mathcal{F}$ of subsets of Ω representing events and define the probability measure on $\mathcal{F}$. However, in the case of a finite $\Omega = \{\omega_1, ..., \omega_k\}$ the system of events $\mathcal{F}$ is always chosen as consisting of all subsets of Ω. To define a probability measure on such $\mathcal{F}$, it is sufficient to define it for one-point sets, $(\omega_m) \to \mathbf{P}(\omega_m), \sum_m \mathbf{P}(\omega_m) = 1$, and to extend it by additivity: for any subset O of $\Omega, \mathbf{P}(O) = \sum_{\omega_m \in O} \mathbf{P}(\omega_m)$.

3. Classical random generation of "Random Numbers Certified by Bell's Theorem"

By using the previous classical probabilistic representation of CHSH-probabilities, we can now construct the classical random generator (at least theoretically) producing 6-dimensional vectors $\xi_m = (A_m^{(1)}, A_m^{(2)}, B_m^{(1)}, B_m^{(2)}, \eta_{Lm}, \eta_{Rm}), m = 1, 2, ...$ The last two coordinates take values 1,2 and the first four coordinates take the values $\{-1, 0, +1\}$. Now if we filter from these 6-dimensional vectors the zero coordinates, we shall get 4-dimensional vectors $(a_m, b_m, \eta_{Lm}, \eta_{Rm}$ which are indistinguishable from the viewpoint of Bell's test from those obtained, e.g., in the quantum optics experiments. The corresponding conditional probabilities, conditioning with respect to selection the fixed pairs of settings (θ_i, θ'_j), violate the CHSH inequality.

Finally we remark that our random generator is "local", see [6] for details.

4. Conclusion

In the light of recent author's results on embedding of quantum probabilistic data into the classical Kolmogrov model of probability von Neumann's statement about irreducible and intrinsic quantum randomness as opposed to classical reducible randomness which based only on the lack of knowledge has to be seriously reanalyzed (as well as its quantum information technological implementations). In fact, already careful reading of von Neumann's book [1] makes the impression that his conclusion about novel features of quantum probability and randomness comparing with their classical counter-parts was not completely logically justified. Roughly speaking he said. See, each electron is individually random! However, at the same time he pointed out that this individual randomness can be gained only from an ensemble of electrons. The latter ensemble approach to probability is purely classical. Therefore it would be natural to expect that one might be able to embed quantum probability into the classical probability model. Precisely this was done in [5], [6].

In our approach the CHSH-probabilities appear as *classical conditional probabilities.* In general I claim that all quantum probabilities can be modeled as classical conditional probabilities. In such a situation it is difficult to believe in some mystical and non-classical features of quantum random generators.

Finally, I remark that in this paper I do not question the quality of "Random Numbers Certified by Bell's Theorem". If they pass the standard tests for randomness, e.g., the NIST-test, they are good, if not, then they are bad, irrelative to non/violation of the CHSH-inequality and irrelative to the problem of closing of experimental loopholes [8]. My main concern is about claims that quantum random generators are in some way better than classical ones.

Acknowledgments

This paper was written with support by the grant Modeling of Complex Hierarchic Systems, the Faculty of Technology, Linnaeus University, and the grant of Quantum BioInformatic Center of Tokyo University of Science. The author would like to thank Luigi Accardi, Irina Basieva, Masanori Ohya, and Anton Zeilinger for critical discussions on comparison of classical and quantum random generators.

[1] von Neumann J 1955 *Mathematical Foundations of Quantum Mechanics* (Princeton: Princeton Univ. Press)
[2] Khrennikov A 2010 *Interpretations of Probability* (Berlin: De Gruyter)
[3] Khrennikov A 2009 *Contextual Approach to Quantum Formalism* (Berlin-Heidelberg-New York: Springer)
[4] Pironio S, Acin A, Massar S, Boyer de la Giroday A, Matsukevich D N, Maunz P, Olmschenk S, Hayes D, Luo L, Manning T A and Monroe C 2010 *Nature* **464** 1021
[5] Avis D, Fischer P, Hilbert A and Khrennikov A 2009 Single, Complete, Probability Spaces Consistent With EPR-Bohm-Bell Experimental Data, *Foundations of Probability and Physics-5* AIP Proc. Vol. **750** (Melville, NY: AIP) pp 294-301
[6] Khrennikov A 2014 Classical probability model for Bell inequality, *EmQM13: Emergent Quantum Mechanics, 3-6 October 2013, Vienna, Austria* J. Phys.: Conf. Ser. **504**

[7] Kolmogoroff A N 1933 *Grundbegriffe der Wahrscheinlichkeitsrechnung* (Berlin: Springer Verlag); English translation: Kolmogorov A N 1956 *Foundations of Theory of Probability* (New York: Chelsea Publishing Company)

[8] Giustina M, Mech Al, Ramelow S, Wittmann B, Kofler J, Beyer J, Lita A, Calkins B, Gerrits Th, Woo Nam S, Ursin R and Zeilinger A 2013 *Nature* **497** 227

Journal of Physics: Conference Series **626** (2015) 012019 doi:10.1088/1742-6596/626/1/012019

The time as an emergent property of quantum mechanics, a synthetic description of a first experimental approach

E Moreva[1,2], G Brida[1], M Gramegna[1], V Giovannetti[3], L Maccone[4] and M Genovese[1*]

[1]INRIM, strada delle Cacce 91, 10135 Torino, Italy
[2]International Laser Center of M.V.Lomonosov Moscow State University, 119991, Moscow, Russia
[3]NEST, Scuola Normale Superiore and Istituto Nanoscienze-CNR, piazza dei Cavalieri 7, I-56126 Pisa, Italy
[4]Dip. Fisica "A. Volta", INFN Sez. Pavia, Univ. of Pavia, via Bassi 6, I-27100 Pavia, Italy

E-mail: *m.genovese@inrim.it

Abstract. The "problem of time" in present physics substantially consists in the fact that a straightforward quantization of the general relativistic evolution equation and constraints generates for the Universe wave function the Wheeler-De Witt equation, which describes a static Universe. Page and Wootters considered the fact that there exist states of a system composed by entangled subsystems that are stationary, but one can interpret the component subsystems as evolving: this leads them to suppose that the global state of the universe can be envisaged as one of this static entangled state, whereas the state of the subsystems can evolve. Here we synthetically present an experiment, based on PDC polarization entangled photons, that shows a practical example where this idea works, i.e. a subsystem of an entangled state works as a "clock" of another subsystem.

1. Introduction

"...an infinite series of times, in a dizzily growing, ever spreading network of diverging, converging and parallel times. This web of time–the strands of which approach one another, bifurcate, intersect or ignore each other through the centuries–embraces every possibility." [Borges]

"Quid est ergo tempus? si nemo ex me quaerat, scio; si quaerenti explicare velim, nescio." [1] As Augustinus Hipponensis present physicists are in the situation where time is an essential physical parameter whose meaning is intuitively clear, but several problems arise when they try to provide a clear definition of time. First of all the definition of time is different in different branches of physics as classical and non-relativistic quantum mechanics (a fixed background parameter), special relativity (a proper time for each observer, time as fourth coordinate; set of privileged inertial frames) or general relativity (time is a general spacetime coordinate: not at all an absolute time, time non orientability, closed timelike curves)[2]. More specifically, the problem of time [2, 3, 4, 5, 6, 7, 8, 9, 10, 11] in physics consists in the fact that a straightforward quantization of the general relativistic evolution equation and constraints generates the Wheeler-

Journal of Physics: Conference Series **626** (2015) 012019 doi:10.1088/1742-6596/626/1/012019

De Witt equation $H_{tot} |\Psi\rangle = 0$, where H_{tot} is the global Hamiltonian of the universe and $|\Psi\rangle$ is its state. Obviously, this means that the state of the universe must be static, which clashes with our everyday experience of an evolving world.

Apart from the Wheeler-De Witt equation, one can also consider that a time shift of the state of the whole universe must be unobservable from a purely physical consideration: if shifting the state of the whole universe, there is nothing external that can keep track of this shift, so this shift must be unobservable (it is a trivial application of Mach's principle). Same considerations would apply, of course, also to the spatial degrees of freedom: a global shift in position of the whole state of the universe must similarly be unobservable, so the universe must be in an eigenstate of its momentum operator. In the following we will only consider temporal degrees of freedom.

Incidentally, this idea of a static universe surprisingly re-proposes, of course rephrased in modern language, ideas stemming from the work of Parmenides of Elea ["the phenomena of movement and change are simply appearances of a static, eternal reality"] and then diffused in roman-hellenistic word ["Tempus item per se non est, sed rebus ab ipsis consequitur sensus, transactum quid sit in aevo, tum quae res instet, quid porro deinde sequatur" [1] Titus Lucretius Carus, De Rerum Natura].

The Page and Wootters scheme [12, 13], which developed some previous ideas [14], is based on the fact that there exist states of a system composed by entangled subsystems that are stationary, but one can interpret the component subsystems as evolving. One can then suppose that the global state of the universe is one of this static entangled state, whereas the state of the subsystems (us, for example) can evolve. This solves in an extremely elegant way the problem of time. Incidentally, Page and Wootters' proposal naturally embodies the philosophy of relationalism [23, 24] and operationalism, since time is only defined in relation to clocks and to its measurement procedure [33, 34].

Up to now, all these considerations were of theoretical character. Here we epitomize an experimental approach [37] to this problem by providing an emblematic example of Page and Wootters idea at work, visualizing how time could emerge from a static (with respect to an abstract external time) entangled state. Even though the total state of a system is static, time is recovered as correlations between a subsystem that acts as a clock and the rest of the system that evolves according to such a clock. We use a system composed of two entangled photons (a paradigmatic system for several emblematic experiments [35]): the rotation of the polarization of the first acts as a clock for proving an evolution of the polarization of the second. Nonetheless, we demonstrate that the joint polarization state of both photons does not evolve. The Page-Wooters mechanism has been criticised [4, 25], and Gambini et al. have proposed some new ideas [21, 31] to overcome these criticisms. Our experiment demonstrates also some aspects of these ideas.

2. The experiment

The experimental setup is schematically depicted in Figure 1. It consists of two blocks, "preparation" and "measurement". The preparation block allows producing a family of ququarts biphoton polarization states [36] of the form:

$$|\Psi\rangle = \cos\theta\,|HH\rangle + e^{i\varphi}\sin\theta\,|VV\rangle \qquad (1)$$

It includes two orthogonally oriented BBO crystals ($1mm$) that, pumped by a 700 mW cw Ar laser operating at 351 nm, generate a pair of the basic ququart states via type-I spontaneous parametric down conversion in collinear, frequency degenerate regime around the central wavelength of 702 nm. The basic state amplitudes are controlled with the help of a

[1] Time also exists not by itself, but simply from the things which happen the sense apprehends what has been done in time past, as well as what is present and what is to follow after

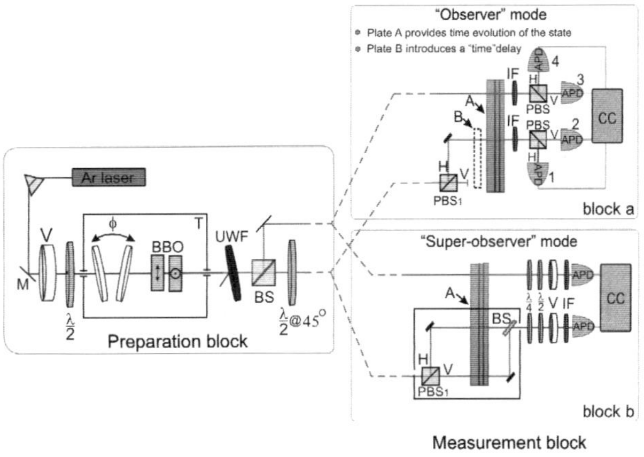

Figure 1. Scheme for the two experiments. The experimenter in observer mode (block a) can prove the time evolution of the first photon (upper path) using only correlation measurements between it and the clock photon (lower path) without access to an external clock. The experimenter in super-observer mode (block b) proves that the global state of the system is static through state tomography. The blue boxes (A) represent different thicknesses of birefringent plates which evolve the photons by rotating their polarization: different thicknesses represent different time evolutions. The dashed box (B) represents a (known) phase delay of the clock photon only; PBS stands for polarizing beam splitter in the H/V basis; BS for beam splitter.

Thompson prism (V), oriented verticaly and the half-wave plate $\lambda/2$ at an angle θ. Two 1 mm quartz plates (QP), that can be rotated along the optical axis, introduce a phase shift φ between horizontally and vertically polarized type-I entangled [40] biphotons. Non-linear crystals and quartz plates are placed into a temperature-stabilized closed box for achieving stable phase-matching conditions all the time. The preparation block also includes a non-polarizing beam splitter (BS), which is used to split the initial (collinear) biphoton field into two spatial modes 1, 2. For preparing the singlet Bell state

$$|\Psi^-\rangle = \frac{1}{\sqrt{2}}\left(|H_1\rangle|V_2\rangle - |V_1\rangle|H_2\rangle\right) \tag{2}$$

parameters $\theta = 45°, \varphi = 0°$, were selected and an additional half-wave plate $\lambda/2$ at 45° was introduced in the transmitted arm.

The measurement part can be mounted in two different modes "observer" and "super-observer". "Observer" mode (Figure 1, block a) serves as proving that one subsystem (polarization of the upper photon) evolves with respect to a clock constituted by the other subsystem (polarization of the lower photon). In the "super-observer" mode ((Figure 1, block b)) we show that the global state of the two photons is static with respect to any "external" clock. Let us analyze the observer mode first.

"Observer" mode: Each arm of a measurement block contains interference filters (IF) with central wavelength 702 nm (FWHM 1 nm) and polarizing beam splitter (PBS). Four avalanche photodiode detectors (APD) are placed at the ends and connected to a coincidence count scheme (CC) with 1.5 ns time window.

In this mode the polarization of the photon in the lower arm is used as a clock for proving an evolution of the polarization of the photon in the upper arm. In other words, the first polarizing beam splitter PBS acts as a non-demolition measurement that measures (in the H/V basis) the polarization of the second photon in the state $|\Psi^-\rangle$ (2). Once it is initialized in the H state, the polarization of both photons evolves in the birefringent quarts plates (A) at 45° as $|H\rangle \to \frac{1}{\sqrt{2}}\left(|H\rangle + e^{i\delta}|V\rangle\right), |V\rangle \to \frac{1}{\sqrt{2}}\left(|H\rangle - e^{i\delta}|V\rangle\right)$, where δ is the materials optical thickness. The clock readout is performed on the lower detectors, which measure whether the photon is $|H\rangle$ or $|V\rangle$. This clock is extremely primitive, since it has a dial with only two values: either it found with H polarization, corresponding to time t_1 or with V polarization, corresponding to time t_2.

In the "observer" mode we do not have access to any external clock, the only information we obtain are correlations (coincidences) between the detectors. Thus, this primitive clock allows extracting the information about the polarization of the upper photon in a defined state, for example $|V\rangle$, only from registering coincidence counts between the detectors $1-3$ (that corresponds to the measurement at the moment time t_1) or 2-3 (that corresponds to time t_2). Furthermore, one does not have any knowledge about the past time after the clock initialization (knowledge about this time is equivalent to the presence of an external clock), so we have to record all events taking place in selected points of time t_1 and t_2. From the experimental point of view it means measurements of coincidences between the detectors 1-3 and 2-3 for all possible thicknesses of the birefringent plates A [21, 26, 20]. Averaging the number of counts from detectors 1-3 and 2-3 over the total number of photon pairs from all detectors 1-2-3-4, gives the probability of finding the upper photon in polarization state $|V\rangle$ $p(t_1)$ and $p(t_2)$ at two points of time.

To obtain a more interesting clock, we perform the same conditional probability measurement introducing varying time delays to the clock photon, implemented through quartz plates of variable thickness (dashed box B in Figure 1, block b). [Even though he has no access to abstract coordinate time, he can have access to systems that implement known time delays, that he can calibrate separately.] Now, we obtain a sequence of time-dependent values for the conditional probability: $p(t_1+\tau_i) = P^{\tau_i}_{3|1}$ and $p(t_2+\tau_i) = P^{\tau_i}_{3|2}$, where $\tau_i = \delta_i/\omega$ is the time delay of the clock photon obtained by inserting the quartz plate B with thickness δ_i in the clock photon path. The experimental results are presented in Fig. 2, where each colour represents a different delay: the yellow points refer to τ_0; the red points to τ_1, etc.

Our results are in very good agreement with the theory (dashed line) [37]. The reduction in visibility of the sinusoidal time dependence of the probability is caused by the decoherence effect due to the use of a low-resolution clock (our clock outputs only two possible values), a well known effect [31, 26, 32].

"Super-observer" mode: The former parter of the experiment is addressed to show that one photon "evolves" when using the other photon as a clock. Now we have to show that the global state of two photons is static with respect to any "external" clock. To do so, we need to repeat the whole experiment in "super-observer" mode, and show that for arbitrary thicknesses of the birefringent plates which represent an "external" time the state is invariant. This setup of the experiment is shown in Figure 1(block b). Here the 50/50 beam splitter (BS) on the output of a balanced interferometer is performing a quantum erasure of the polarization measurement performed by the polarizing beam splitter PBS1. In the ideal case the beam splitter should coherently transmit a photon that passes in the upper arm and reflects a photon that passes in the lower arm of the interferometer without affecting its polarization. Without loss of generality, in our setup we use a conventional beam splitter and post-selection on one of the exits. For temporal stability the interferometer was placed into a closed box and narrow interference filters (702 nm, FWHM=1 nm) were inserted before the detectors. A set of quartz plates (A) with different thicknesses are used to introduce an "external" time.

For analyzing an output state of transformed ququarts we used the quantum state tomography procedure. Necessary projective measurements were realized by polarization filters placed in

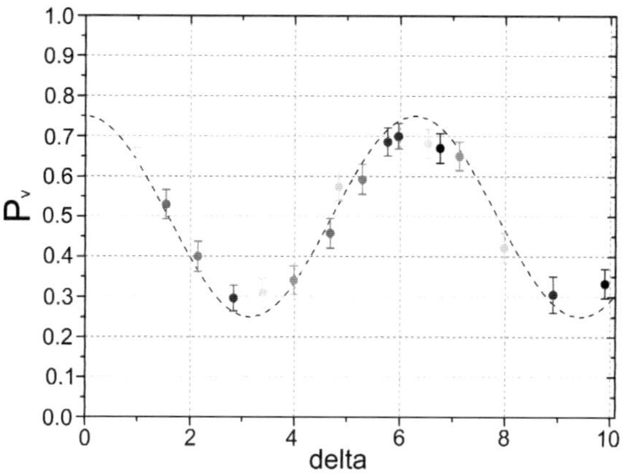

Figure 2. Experimental results. Probability $p(t)$ that the upper photon is V as a function of the time t recovered from the lower photon. The points with matching colors represent $p(t_1 + \tau_i)$ and $p(t_2 + \tau_i)$: yellow, red, blue, etc., for $i = 0, 1, 2, \cdots$, respectively. Here nine different values of τ_i are obtained from a thick quartz plate rotated by nine different angles. The dashed line is the theoretical value.

Table 1. Bell state reconstruction. The table contains values of "external" time and fidelities.

"external" time	Fidelity
0	0.953
$\pi/2\omega$	0.928
$\pi/4\omega$	0.940
$3\pi/2\omega$	0.932

front of detectors. Each filter consists of a sequence of quarter- and half-wave plates and a polarization prism which transmits vertical polarization. Two APDs linked to a coincidence scheme with 1.5 ns time window were used as single photon detectors. In our experiment we used the protocol of Ref. [15, 39]. Registering the coincidence rate for 16 different projections, that were realized by half and quarter plates and a fixed analyzer, it was possible to reconstruct the arbitrary polarization state of ququarts.

The accuracy of quantum tomography is estimated through the fidelity $F = \left(Tr\sqrt{\rho_{in}^{1/2}\rho_{out}\rho_{in}^{1/2}}\right)^2$ where ρ_{in} is the initial (theoretical) density matrix and ρ_{out} is the reconstructed density matrix after transformation. Our results are summarized in Table I, demonstrating the stationarity of the global state respect to the evolution.

3. Conclusions

In summary, by running our experiment in two different modes ("observer" and "super-observer" mode) we have experimentally shown how the same energy-entangled Hamiltonian eigenstate can be perceived as evolving by the internal observers that test the correlations between a clock subsystem and the rest (also when considering two-time measurements), whereas it is static for the super-observer that tests its global properties. Our experiment is a practical implementation of the Page and Wooters[19] and Gambini et al. [21] mechanisms but, obviously, it cannot discriminate between this and other proposed solutions for the problem of time [4, 5, 6, 2, 7], representing only an "illustration" of this phenomenon. Developments to higher dimensional systems could eventually be envisaged by using atoms [41].

In closing, we note that the time-dependent graphs of Fig. 2 have been obtained without any reference to an external time (or phase) reference, but only from measurements of correlations between the clock photon and the rest representing an implementation of a 'relational' measurement of a physical quantity (time) relative to an internal quantum reference frame

[33, 34], i.e. an example of relational metrology.

Acknowledgments

This research was supported by the John Templeton Foundation (the opinions expressed in this publication are those of the authors and do not necessarily reflect the views of the John Templeton Foundation), E.V.Moreva acknowledges the support from the Dynasty Foundation and Russian Foundation for Basic Research (project 13-02-01170-a).

References

[1] Aurelius Augustinus Hipponensis Quid est ergo tempus? si nemo ex me quaerat, scio; si quaerenti explicare velim, nescio Confessiones, XI, 14
[2] Anderson E 2012 *Ann. Phys.* **524** 757 (*Preprint* arXiv:gr-qc/1009.2157v3)
[3] Sorkin R D 1994 *Int. J. Theor. Phys.* **33** 523; Kiefer C 2009 *Preprint* arXiv:gr-qc/0909.3767; Vedral V 2014 *Preprint* arXiv:quant-ph/1408.6965
[4] Kuchař K V 1992 Time and interpretations of quantum gravity *Proceedings of the 4th Canadian Conference on General Relativity and Relativistic Astrophysics* (Singapore: World Scientific) 1-92
[5] Isham C J 1993 Canonical quantum gravity and the problem of time *Integrable Systems, Quantum Groups and Quantum Field Theories* (Dordrecht: Kluwer)
[6] Ashtekar A 2005 Gravity and the quantum *New J. Phys.* **7** 198
[7] Merali Z 2013 The origins of space and time *Nature* **500** 516-519
[8] Nikolic H 2014 *International Journal of Quantum Information* in press, DOI: 10.1142/S0219749915600011; *do.* 2012 *International Journal of Quantum Information* **10** 1250024
[9] Nicolis S 2014 *Preprint* arXiv:quant-ph/1411.7293
[10] Briggs J S and Rost J M 2001 *Found. Phys.* **31** 693; Pegg D T 1991 *J. Phys.* A **24** 3031
[11] Aharonov Y, Popescu S and Tollaksen J 2013 *Preprint* arXiv:quant-ph/1305.1615
[12] Wootters W K 1984 *Int. J. Theor. Phys.* **23** 701
[13] Page D N and Wootters W K 1983 *Phys. Rev.* D **27** 2885
[14] Vedral V and Mott N 1931 *Proc. Camb. Phil. Soc.* **27** 553; Banks T 1985 *Nucl. Phys.* B **249** 332; Brout R 1987 *Found. Phys.* **17** 603; Brout R, Horwitz G and Weil D 1987 *Phys. Lett.* B **192** 318; Brout R 1987 *Z. Phys.* B **68** 339
[15] James D F V, Kwiat P G, Munro W J and White A G 2001 *Phys. Rev.* A **64** 052312
[16] Kwiat P, Waks E, White A, Appelbaum I and Eberhard P 1991 *Phys. Rev.* A **60** 773(R)
[17] DeWitt B S 1967 Quantum theory of gravity. I. the canonical theory *Phys. Rev.* **160** 1113
[18] Hartle J B and Hawking S W 1983 Wave function of the Universe' *Phys. Rev.* D **28** 2960
[19] Page D N and Wootters W K 1983 Evolution without evolution: dynamics described by stationary observables *Phys. Rev.* D **27** 2885
[20] Page D N 1993 Clock time and entropy *Physical Origins of Time Asymmetry* (Cabridge: Cambridge Univ. Press)
[21] Gambini R, Porto R A, Pullin J and Torterolo S 2009 Conditional probabilities with Dirac observables and the problem of time in quantum gravity *Phys. Rev.* D **79** 041501(R)
[22] Peres A 1980 Measurement of time by quantum clocks *Am. J. Phys.* **48** 552
[23] Rovelli C 1996 Relational Quantum Mechanics *Int. J. Theor. Phys.* **35** 1637
[24] Rovelli C 1991 *Phys. Rev.* D **43** 442
[25] Unruh W G and Wald R M 1989 Time and the interpretation of canonical quantum gravity *Phys. Rev.* D **40** 2598
[26] Gambini R, Pintos L P G and Pullin J 2011 An axiomatic formulation of the Montevideo interpretation of quantum mechanics *Studies His. Phil. Mod. Phys.* **42** 256
[27] Einstein A 1905 Zur Elektrodynamik bewegter Körper *Ann. Phys.* **17** 891; translated in *do.*1923 *The Principle of Relativity* (London: Methuen & Co.)
[28] Scully M O, Englert B G and Walther H 1991 Quantum optical tests of complementarity *Nature* **351** 111
[29] Dürr S, Nonn T and Rempe G 1998 Origin of quantum-mechanical complementarity probed by a 'which-way' experiment in an atom interferometer *Nature* **395** 33
[30] Preskill J *Lecture Notes for Physics* 229 (http://www.theory.caltech.edu/people/preskill/ph229, p 68)
[31] Gambini R, Pintos L P G and Pullin J 2010 Undecidability and the problem of outcomes in quantum measurements it Found. Phys. **40** 93
[32] Egusquiza I L, Garay L J and Raya J M 1999 Quantum evolution according to real clocks *Phys. Rev.* A **59** 3236

[33] Bartlett S D, Rudolph T and Spekkens W 2007 Reference frames, superselection rules, and quantum information *Rev. Mod.Phys.* **79** 555

[34] Wiseman H M 2004 Defending continuous variable teleportation: why a laser is a clock, not a quantum channel *J.Opt. B: Quantum Semiclass. Opt.* **6** S849

[35] Tiefenbacher F *et al* 2007 Entanglement-based quantum communication over 144 km *Nature Physics* **3** 481; Boschi D *et al* 1998 Experimental Realization of Teleporting an Unknown Pure Quantum State via Dual Classical and Einstein-Podolsky-Rosen Channels *Phys. Rev. Lett.* **80** 1121; Kwiat P G *et al* 2001 Experimental entanglement distillation and hidden non-locality *Nature* **409** 1014; Yamamoto T *et al* 2003 Experimental extraction of an entangled photon pair from two identically decohered pairs *Nature* **421** 343; Pan J W *et al* 2001 Entanglement purification for quantum communication *Nature* **410** 1067; Brida G *et al* 2012 *Phys. Rev.* A **68**, 033803; Brida G *et al* 2006 *Phys. Rev. Lett.* **96** 143601

[36] Genovese M 2005 *Phys. Rep.* **413** 319

[37] Moreva E V, Brida G, Gramegna M, Giovannetti V, Maccone L and Genovese M 2014 *Phys. Rev.* A **89** 052122

[38] Bogdanov Yu I, Brida G, Genovese M, Shurupov A P, Moreva E V and Kulik S P 2010 *Phys. Rev. Lett.* **105** 010404

[39] Bogdanov Yu I, Moreva E V, Maslennikov G A, Galeev R F, Straupe S S and Kulik S P 2006 *Phys. Rev.* A **73** 063810

[40] Brida G, Degiovanni I P, Florio A, Genovese M, Giorda P, Meda A, Paris M and Shurupov A 2010 *Phys. Rev. Lett.* **104** 100501

[41] Massar S, Spindel P, Varn A F and Wunderlichet C 2014 *Preprint* arXiv:quant-ph 1410.6683

Causal fermion systems as a candidate for a unified physical theory

Felix Finster and Johannes Kleiner

Fakultät für Mathematik, Universität Regensburg, D-93040, Germany

E-mail: finster@ur.de; Johannes.Kleiner@ur.de

Abstract. The theory of causal fermion systems is an approach to describe fundamental physics. Giving quantum mechanics, general relativity and quantum field theory as limiting cases, it is a candidate for a unified physical theory. We here give a non-technical introduction.

This article is an introduction to causal fermion systems which is intended to explain the basic concepts and the general physical picture behind the approach. The article is organized as follows. In Section 1 we define the basic objects of the theory. In Section 2 we proceed by explaining how those objects appear naturally in the familiar physical situation of Dirac particles in Minkowski space. In Section 3 it is shown how the objects of quantum mechanics are encoded in a causal fermion system. Section 4 explains in the example of the Minkowski vacuum how the causal fermion system encodes the causal structure. In Section 5 we exemplify how to describe other physical situations or more general space-times. In Section 6 we outline a limiting case in which the causal fermion system can be described by a second-quantized Dirac field coupled to classical gauge fields and gravity. Section 7 gives the resulting perspective on the foundations of quantum mechanics, in particular on the measurement problem. In Section 8 we conclude with a few clarifying remarks.

1. The theory

The general structure of the theory of causal fermion systems can be understood in analogy to general relativity. In general relativity, our universe is described by a four-dimensional space-time (Lorentzian manifold) together with particles and fields. However, not every configuration of Lorentzian metric, particles and fields is considered to be "physical" in the sense that it could be realized in nature. Namely, for the configuration to be physically realizable, the Einstein equations must hold. Moreover, the particles must satisfy the equations of motion, and the additional fields must obey the field equations (like Maxwell's equations). This means that in general relativity, there are two conceptual parts: on the one hand one has mathematical objects describing possible configurations, and on the other hand there is a principle which singles out the physical configurations.

The theory of causal fermion systems has the same conceptual structure consisting of mathematical objects and a principle which singles out the physical configurations. We first introduce the mathematical objects:

Journal of Physics: Conference Series **626** (2015) 012020 doi:10.1088/1742-6596/626/1/012020

Definition 1.1 (Causal fermion system)

- Let $(\mathcal{H}, \langle .|. \rangle_{\mathcal{H}})$ be a separable complex Hilbert space.
- Given a parameter $n \in \mathbb{N}$ (the *spin dimension*), let $\mathcal{F} \subset \mathrm{L}(\mathcal{H})$ be the set of all self-adjoint operators on $\mathcal{H}$ of finite rank, which (counting multiplicities) have at most n positive and at most n negative eigenvalues.
- Let ρ be a positive measure on $\mathcal{F}$ (the *universal measure*).

Then $(\mathcal{H}, \mathcal{F}, \rho)$ is a *causal fermion system.*

Here separable means that the Hilbert space has an at most countable orthonormal basis. Mapping the basis vectors to each other, one sees that any two Hilbert spaces are isomorphic, provided that their dimensions coincide. Therefore, the structure $(\mathcal{H}, \mathcal{F})$ is completely determined by the parameters $n \in \mathbb{N}$ and $f := \dim \mathcal{H} \in \mathbb{N} \cup \{\infty\}$. Apart from these parameters, the only object specifying a causal fermion system is the universal measure ρ.

It will be outlined below that this definition indeed generalizes mathematical structures used in contemporary physics. The picture is that one causal fermion system describes a space-time together with all structures and objects therein (including the metric, particles and fields).

Next, we state the principle which singles out the physical configurations. Similar to the Lagrangian formulation of contemporary physics, we work with a variational principle, referred to as the *causal action principle.* It states that a causal fermion system which can be realized in nature should be a minimizer of the so-called causal action. In order to formulate the causal action principle, we assume that the Hilbert space $(\mathcal{H}, \langle .|. \rangle_{\mathcal{H}})$ and the spin dimension n have been chosen. Let $\mathcal{F}$ be as in Definition 1.1 above. Then for any $x, y \in \mathcal{F}$, the product xy is an operator of rank at most $2n$. We denote its non-trivial eigenvalues (counting algebraic multiplicities) by $\lambda_1^{xy}, \ldots, \lambda_{2n}^{xy} \in \mathbb{C}$. The *spectral weight* $|\,.\,|$ of the operator products xy and $(xy)^2$ is defined by

$$|xy| = \sum_{i=1}^{2n} |\lambda_i^{xy}| \qquad \text{and} \qquad \big|(xy)^2\big| = \sum_{i=1}^{2n} |\lambda_i^{xy}|^2 \,.$$

Next, the *Lagrangian* $\mathcal{L} : \mathcal{F} \times \mathcal{F} \to \mathbb{R}_0^+$ is defined by

$$\mathcal{L}(x,y) := \big|(xy)^2\big| - \frac{1}{2n} |xy|^2 = \frac{1}{4n} \sum_{i,j=1}^{2n} \Big(|\lambda_i^{xy}| - |\lambda_j^{xy}| \Big)^2 \,. \tag{1}$$

The particular form of this Lagrangian is the result of research carried out over several years (see Section 8.2).

Definition 1.2 (Causal action principle) The *causal action* $\mathcal{S}$ is obtained by integrating the Lagrangian with respect to the universal measure,

$$\mathcal{S}(\rho) = \iint_{\mathcal{F} \times \mathcal{F}} \mathcal{L}(x,y)\, d\rho(x)\, d\rho(y) \,.$$

The *causal action principle* is to minimize $\mathcal{S}$ under variations of the universal measure, taking into account the following constraints:

$$\textit{volume constraint:} \qquad \rho(\mathcal{F}) = \text{const} \tag{2}$$

$$\textit{trace constraint:} \qquad \int_{\mathcal{F}} \mathrm{tr}(x)\, d\rho(x) = \text{const} \tag{3}$$

$$\textit{boundedness constraint:} \qquad \mathcal{T} := \iint_{\mathcal{F} \times \mathcal{F}} |xy|^2\, d\rho(x)\, d\rho(y) \leq C \,, \tag{4}$$

where C is a given constant (and tr denotes the trace of a linear operator on $\mathcal{H}$).

In mathematical terms, the measure ρ is varied within the class of positive regular Borel measures on $\mathcal{F}$, where on $\mathcal{F}$ one takes the topology induced by the sup-norm on $\mathrm{L}(\mathcal{H})$ (for basic definitions see for example [1, Chapters 2 and 5] or [2, Chapter X]). The volume and trace constraints are needed in order to avoid trivial minimizers and are important for the analysis of the corresponding Euler-Lagrange equations because they give rise to Lagrange multiplier terms. The boundedness constraint is needed in order to ensure the existence of minimizers. In most applications, it does not give rise to a Lagrange multiplier term. Therefore, it does not seem to have any physical consequences.

This concludes the mathematical definition of the theory. In order to obtain a physical theory, we need to give the mathematical objects a physical interpretation. It is one of the main objectives of the next sections to do so by explaining how the above mathematical objects relate to the common notions in physics. The conclusion will be that causal fermion systems are indeed a candidate for a fundamental physical theory.

2. Example: Dirac wave functions in Minkowski space

As a first step towards explaining how causal fermion systems relate to contemporary physics, we now explain how the familiar physical situation of Dirac particles in Minkowski space can be described by a causal fermion system.

Let $\mathscr{M}$ be Minkowski space and μ the natural volume measure thereon, i.e. $d\mu = d^4x$ if $x = (x^0, x^1, x^2, x^3)$ is an inertial frame (we use the signature convention $(+,-,-,-)$). We consider a finite number of f Dirac particles described by one-particle wave functions $\psi_1, \ldots, \psi_f$ which are solutions of the Dirac equation,

$$\left(i\gamma^j \partial_j - m\right) \psi_k = 0, \qquad k = 1, \ldots, f\,, \tag{5}$$

where m is the rest mass, and γ^j are Dirac matrices in the Dirac representation. For simplicity, we assume that the wave functions $\psi_1, \ldots, \psi_f$ are continuous.

Before going on, we remark that this description of the f-particle system by f one-particle wave functions departs from the usual Fock space description. The connection to Fock spaces will be established later in this article (see Section 7). For the moment, it is preferable to work with the one-particle wave functions. We also remark that the assumption of considering a finite number of continuous wave functions merely is a technical simplification for our presentation. All constructions can be extended to an infinite number of possibly discontinuous wave functions (for details see [3, Section 4] or [4, Chapter 1]).

The wave functions ψ_k span a vector space which we denote by $\mathcal{H}$.

$$\mathcal{H} := \mathrm{span}(\psi_1, \ldots, \psi_f)\,. \tag{6}$$

On $\mathcal{H}$ we consider the usual scalar product on solutions of the Dirac equation

$$\langle \psi | \phi \rangle_{\mathcal{H}} := 2\pi \int_{t=\text{const}} (\overline{\psi} \gamma^0 \phi)(t, \vec{x})\, d^3x \tag{7}$$

(here $\overline{\psi} = \psi^\dagger \gamma^0$ is the adjoint spinor, where the dagger denotes complex conjugation and transposition). If one evaluates (7) for $\phi = \psi$, the integrand can be written as $(\overline{\psi}\gamma^0\psi)(t,\vec{x}) = (\psi^\dagger\psi)(t,\vec{x})$, having the interpretation as the probability density of the Dirac particle corresponding to ψ to be at the position $\vec{x}$. In view of the conservation of probability (being a consequence of current conservation), the integral in (7) is time independent. Since the probability density is positive, the inner product (7) is indeed positive definite. We thus obtain an f-dimensional Hilbert space $(\mathcal{H}, \langle .|. \rangle_{\mathcal{H}})$.

For any $x \in \mathcal{M}$, we now introduce the sesquilinear form

$$b_x : \mathcal{H} \times \mathcal{H} \to \mathbb{C} , \qquad b_x(\psi, \phi) = -(\overline{\psi} \phi)(x) ,$$

which maps two solutions of the Dirac equation to their inner product at x. The sesquilinear form b_x can be represented by a self-adjoint operator $F(x)$ on $\mathcal{H}$, which is uniquely defined by the relations

$$\langle \psi | F(x) \phi \rangle_{\mathcal{H}} = b_x(\psi, \phi) \qquad \text{for all } \psi, \phi \in \mathcal{H} .$$

More concretely, in the basis $(\psi_k)_{k=1,\ldots,f}$ of $\mathcal{H}$, the last relation can be written as

$$\langle \psi_i | F(x) \psi_j \rangle_{\mathcal{H}} = -\big(\overline{\psi_i} \psi_j \big)(x) . \tag{8}$$

If the basis is orthonormal, the calculation

$$F(x)\, \psi_j = \sum_{i=1}^{f} \langle \psi_i | F(x) \psi_j \rangle_{\mathcal{H}}\, \psi_i = -\sum_{i=1}^{f} \big(\overline{\psi_i} \psi_j \big)(x)\, \psi_i$$

(where we used the completeness relation $\phi = \sum_i \langle \psi_i | \phi \rangle\, \psi_i$), shows that the operator $F(x)$ has the matrix representation

$$\big(F(x) \big)^i_j = -\big(\overline{\psi_i} \psi_j \big)(x) .$$

In physical terms, the matrix element $-(\overline{\psi_i} \psi_j)(x)$ gives information on the correlation of the wave functions ψ_i and ψ_j at the space-time point x. Therefore, we refer to $F(x)$ as the *local correlation operator* at x.

Let us analyze the properties of $F(x)$. First of all, the calculation

$$\langle F(x)\, \psi \,|\, \phi \rangle_{\mathcal{H}} = \overline{\langle \phi \,|\, F(x)\, \psi \,\rangle_{\mathcal{H}}} = -\overline{(\overline{\phi} \psi)(x)} = -(\overline{\psi} \phi)(x) = \langle \psi \,|\, F(x)\, \phi \rangle_{\mathcal{H}}$$

shows that the operator $F(x)$ is self-adjoint (where we denoted complex conjugation by a bar). Furthermore, since the pointwise inner product $(\overline{\psi} \phi)(x)$ has signature $(2, 2)$, we know that b_x has signature (p, q) with $p, q \leq 2$. As a consequence, the operator $F(x)$ has at most two positive and at most two negative eigenvalues (counting multiplicities). It follows immediately, that $F(x) \in \mathcal{F}$ if the spin dimension in Definition 1.2 is chosen as $n = 2$.

Constructing the operator $F(x) \in \mathcal{F}$ for every space-time point $x \in M$, we obtain the mapping

$$F : \mathcal{M} \to \mathcal{F} , \qquad x \mapsto F(x) .$$

This allows us to introduce a measure ρ on $\mathcal{F}$ as follows. For any $\Omega \in \mathcal{F}$, one takes the pre-image $F^{-1}(\Omega) \subset \mathcal{M}$ and computes its space-time volume,

$$\rho(\Omega) := \mu\big(F^{-1}(\Omega) \big) .$$

This gives rise to the so-called *push-forward measure* which in mathematics is denoted by $\rho = F_* \mu$ (we remark for the mathematically oriented reader that the σ-algebra of ρ-measurable sets is defined as all sets $\Omega \subset \mathcal{F}$ whose pre-image $F^{-1}(\Omega)$ is μ-measurable).

Putting the above structures together, we obtain a causal fermion system $(\mathcal{H}, \mathcal{F}, \rho)$ of spin dimension two. Thus we have succeeded in constructing a causal fermion system starting from a system of Dirac wave functions in Minkowski space. But it is not obvious how much of the information on the physical system is encoded in the causal fermion system. In other words, taking the causal fermion system $(\mathcal{H}, \mathcal{F}, \rho)$ as the starting point, the question is which structures of the original system can be recovered. For example, is the Minkowski metric still determined?

Is it possible to reconstruct the Dirac wave functions? Precise answers to these questions will be given in Section 3 below. In preparation, we now give a few hints.

We first explain what the points of Minkowski space correspond to in our causal fermion system. Recall that to every space-time point $x \in \mathscr{M}$ we associated a linear operator $F(x) \in \mathcal{F}$. Hence the space-time points correspond to the subset $F(\mathscr{M}) \subset \mathcal{F}$. This subset can also be characterized as the set where the measure ρ is non-zero. In mathematical terms, this is captured in the notion of the *support* of the universal measure, defined as the set of all the points of $\mathcal{F}$ such that every open neighborhood of this point has a non-zero measure. Then (for details see [4, Chapter 1])

$$\operatorname{supp} \rho = \overline{F(\mathscr{M})}\,, \tag{9}$$

where the bar denotes the closure. In all situations of physical interest, the mapping F will be injective and its image closed (see again [4, Chapter 1]). Provided that this is the case, identifying $x \in \mathscr{M}$ with the corresponding operator $F(x) \in \mathcal{F}$ makes it possible to identify Minkowski space with the support of ρ as a topological space. Under suitable smoothness and non-degeneracy assumptions, one can identify $\mathscr{M}$ with $\operatorname{supp} \rho$ even as a differentiable manifold. We make this identification manifest by using the letter x for the operator $F(x)$. In order to avoid confusion, we use two different fonts, making it possible for the reader to distinguish a point $x \in \mathscr{M}$ of Minkowski space from the corresponding point $x \in M := \operatorname{supp} \rho$. Once the reader has become familiar with our concepts, the different fonts will be unnecessary.

This consideration shows that the topological and differentiable structures of our space-time are encoded in the causal fermion system. Clearly, Minkowski space also has metric and causal structures, which we have not yet addressed. The general idea for recovering these structures is to take operators $x, y \in \operatorname{supp} \rho$ and to analyze the eigenvalues of the operator product xy. The eigenvalues of such operator products contain plenty of information, inducing relations and structures between the space-time points. This will be explained more concretely in the next section.

3. Inherent structures

Let $(\mathcal{H}, \mathcal{F}, \rho)$ be a causal fermion system of spin dimension n (see Definition 1.1). We now introduce additional objects which will turn out to generalize familiar notions in physics. All of these structures are inherent in the sense that we only use information already encoded in the causal fermion system.

Motivated by the consideration above (see the paragraph before (9)), we define *space-time* M as the support of the universal measure,

$$M := \operatorname{supp} \rho \subset \mathcal{F}\,.$$

On M we introduce the following notion of causality. Recall that for $x, y \in M$, the product xy is an operator of rank at most $2n$. We again denote its non-trivial eigenvalues (counting algebraic multiplicities) by $\lambda_1^{xy}, \ldots, \lambda_{2n}^{xy} \in \mathbb{C}$.

Definition 3.1 (Causality) The space-time points x and y are defined to be

- *spacelike* separated if all the λ_j^{xy} have the same absolute value.
- *timelike* separated if the λ_i^{xy} do not all have the same absolute value and are all real.
- *lightlike* separated if the λ_i^{xy} do not all have the same absolute value and are not all real.

This definition is compatible with the causal action in the following sense. If the points x and y are spacelike separated, then all the λ_j^{xy} have the same absolute value, so that the

Lagrangian $\mathcal{L}(x,y)$ vanishes according to (1). In a more physical language, this means that no interaction takes place between regions with spacelike separation (this does not exclude nonlocal correlations and entanglement, as will be discussed in Section 7). In this way, our setting incorporates a general version of the principle of causality.

The next step is to introduce wave functions. The construction is guided by the usual structure of a Dirac wave function ψ, which to every space-time point x associates a spinor $\psi(x)$. The latter is a vector in the corresponding spinor space $S_x\mathcal{M} \simeq \mathbb{C}^4$, which is endowed with the inner product $\overline{\psi}\phi$ of signature $(2,2)$. In the setting of causal fermion systems, for a space-time point $x \in M$ we define the *spin space* $S_x \subset \mathcal{H}$ as the image of the operator x,

$$S_x := x(\mathcal{H})\,.$$

It is a subspace of $\mathcal{H}$ of dimension at most $2n$. On S_x we introduce the inner product

$$\prec .|.\succ_x : S_x \times S_x \to \mathbb{C}\,, \qquad \prec u|v\succ_x := -\langle u|xv\rangle_{\mathcal{H}}\,, \tag{10}$$

referred to as the *spin scalar product*. Since x has at most n positive and at most n negative eigenvalues, the spin scalar product is an indefinite inner product of signature (p,q) with $p,q \leq n$. A *wave function* ψ is defined as a function which to every $x \in M$ associates a vector of the corresponding spin space,

$$\psi \,:\, M \to \mathcal{H} \qquad \text{with} \qquad \psi(x) \in S_x \quad \text{for all } x \in M\,.$$

Clearly, it is not sufficient to define wave functions abstractly, but we need to specify those wave functions which are realized in the physical system. Using a familiar physical language, we need to declare which one-particle states are occupied (for the connection to multi-particle Fock states see Section 7). To this end, to every vector $u \in \mathcal{H}$ of the Hilbert space we associate a wave function ψ^u by projecting the vector u to the spin spaces, i.e.

$$\psi^u \,:\, M \to \mathcal{H}\,, \qquad \psi^u(x) := \pi_x u \in S_x\,, \tag{11}$$

where π_x is the orthogonal projection in $\mathcal{H}$ on the subspace $x(\mathcal{H}) \subset \mathcal{H}$. We refer to ψ^u as the *physical wave function* corresponding to the vector $u \in \mathcal{H}$.

Finally, we define the *kernel of the fermionic projector* $P(x,y)$ for any $x,y \in M$ by

$$P(x,y) = \pi_x\, y|_{S_y} \,:\, S_y \to S_x \tag{12}$$

(where $|_{S_y}$ denotes the restriction to the subspace $S_y \subset \mathcal{H}$). This object is useful for analyzing the relations and structures between space-time points. In particular, the kernel of the fermionic projector encodes the causal structure and makes it possible to compute the eigenvalues $\lambda_1^{xy}, \ldots, \lambda_{2n}^{xy}$ which appear in the Lagrangian (1). In order to see how this comes about, we first define the *closed chain* as the product

$$A_{xy} = P(x,y)\,P(y,x) \,:\, S_x \to S_x\,. \tag{13}$$

Computing powers of the closed chain and using that $y\pi_y = y$ (because the image and kernel of self-adjoint operators are orthogonal), we obtain

$$A_{xy} = (\pi_x y)(\pi_y x)|_{S_x} = \pi_x\, yx|_{S_x} \qquad \text{and thus} \qquad (A_{xy})^p = \pi_x\,(yx)^p|_{S_x}\,.$$

Taking the trace, we obtain for all $p \in \mathbb{N}$,

$$\begin{aligned} \mathrm{Tr}_{S_x}\big((A_{xy})^p\big) &= \mathrm{Tr}_{S_x}\big(\pi_x\,(yx)^p|_{S_x}\big) = \mathrm{tr}\big(\pi_x\,(yx)^p|_{S_x}\big) \\ &= \mathrm{tr}\big((yx)^p \pi_x\big) = \mathrm{tr}\big((yx)^p\big) = \mathrm{tr}\big((xy)^p\big) \end{aligned}$$

Journal of Physics: Conference Series **626** (2015) 012020 doi:10.1088/1742-6596/626/1/012020

(where tr again denotes the trace of a linear operator on $\mathcal{H}$). Since the coefficients of the characteristic polynomial of an operator can be expressed in terms of traces of powers of the corresponding matrix, we conclude that the eigenvalues of the closed chain coincide with the non-trivial eigenvalues $\lambda_1^{xy}, \ldots, \lambda_{2n}^{xy}$ of the operator xy in Definition 1.2. In this way, one can recover the $\lambda_1^{xy}, \ldots, \lambda_{2n}^{xy}$ as the eigenvalues of a $(2n \times 2n)$-matrix. In particular, the kernel of the fermionic operator encodes the causal structure of M.

The kernel of the fermionic projector is the starting point for constructions which unveil the geometric structures of a causal fermion system. More specifically, this kernel gives rise to a spin connection and corresponding curvature. Moreover, one can introduce tangent spaces endowed with a Lorentzian metric together with a corresponding metric connection and curvature. For brevity, we cannot enter these topics here. Instead we refer the interested reader to [5, 6], where also questions concerning the topology of causal fermion systems are treated. The important point to keep in mind is that all these constructions are tailored in order to understand the meaning of information contained in the causal fermion system. No additional input is required. The system is completely determined by the causal fermion system $(\mathcal{H}, \mathcal{F}, \rho)$. In particular, when varying the universal measure in the causal action principle, one also varies all the derived structures mentioned above.

4. The Minkowski vacuum

In order to illustrate the above inherent structures, we now return to the example of Dirac particles in Minkowski space introduced in Section 2. In this example, the Hilbert space $\mathcal{H}$ is spanned by solutions of the Dirac equation. Thus a vector $u \in \mathcal{H}$ is a Dirac wave function, which at a point $x \in \mathscr{M}$ of Minkowski space takes values in the corresponding spinor space, $u(x) \in S_x\mathscr{M}$. On the other hand, in the previous section we introduced the corresponding physical wave function ψ^u, which at a point $x = F(x) \in M \subset \mathcal{F}$ takes values in the corresponding spin space, $\psi^u(x) \in S_x$. We now show that these objects can be identified. Indeed, for any $u, v \in S_x \subset \mathcal{H}$,

$$\prec\!\psi^u(x) \,|\, \psi^v(x)\!\succ_x \overset{(10)}{=} -\langle \pi_x u \,|\, x\, \pi_x v\rangle_{\mathcal{H}} = -\langle u \,|\, x\, v\rangle_{\mathcal{H}} = -\langle u \,|\, F(x)\, v\rangle_{\mathcal{H}} \overset{(8)}{=} \overline{u(x)} v(x)\,.$$

This shows that the inner products on $S_x\mathscr{M}$ and S_x are compatible. It implies that, after choosing suitable bases, one can indeed identify $S_x\mathscr{M}$ with S_x (for details see [4, Section 1.2] or [5, Section 4]). This identification implies that $\psi^u(x) = u(x)$ for all $u \in \mathcal{H}$ and $x \in M$ respectively $x \in \mathscr{M}$.

Next, it is instructive to bring the kernel of the fermionic projector (12) into a more tractable form. To this end, we choose an orthonormal basis $u_1, \ldots, u_f$ of $\mathcal{H}$. Then for any $\phi \in S_y$,

$$\begin{aligned} P(x,y)\,\phi = \pi_x\, y\, \phi &\overset{(\star)}{=} \sum_{\ell=1}^{f} \big(\pi_x u_\ell\big)\, \langle u_\ell |\, y\, \phi\rangle_{\mathcal{H}} \\ &\overset{(10)}{=} -\sum_{\ell=1}^{f} \big(\pi_x u_\ell\big) \prec\!\pi_y u_\ell | \phi\!\succ_y \overset{(11)}{=} -\sum_{\ell=1}^{f} \psi^{u_\ell}(x) \prec\!\psi^{u_\ell}(y) \,|\, \phi\!\succ_y\,, \end{aligned}$$

where in $(\star)$ we used the completeness relation. Using the above identifications of spinors and their inner products, we can write this formula in the shorter form

$$P(x,y) = -\sum_{\ell} u_\ell(x)\, \overline{u_\ell(y)}\,. \tag{14}$$

This shows that the kernel of the fermionic projector is composed of all the physical wave functions of the system.

In order to work in a more concrete example, we next consider the *Minkowski vacuum.* To this end, we want to implement the concept of the Dirac sea which in non-technical terms states that in the vacuum all the negative-energy states of the Dirac equation should be occupied (see Section 8.4 for further explanations of this point). In order to implement this concept, one needs to consider an infinite number of physical wave functions. This can be achieved simply by letting $\mathcal{H}$ in Definition 1.1 be an infinite-dimensional Hilbert space. However, a difficulty arises in the construction of the local correlation operators, because the Dirac wave functions (being square-integrable functions) are in general not defined pointwise, so that the right side of (8) is ill-defined. In order to resolve this problem, one needs to introduce an ultraviolet regularization. For conceptual clarity, we postpone the explanation of the ultraviolet regularization to Section 6 and now merely mention that an ultraviolet regularization amounts to modifying the Dirac wave functions on a microscopic scale ε, which can be thought of as the Planck scale. In order to avoid the technical issues involved in the regularization, we here simply use the formula (14), but now sum over all negative-energy solutions of the Dirac equation. This sum can be rewritten as an integral over the lower mass shell (see again [4, Section 1.2]),

$$P(x,y) = \int \frac{d^4k}{(2\pi)^4}\, (k_j\gamma^j + m)\, \delta(k^2 - m^2)\, \Theta(-k_0)\, e^{-ik(x-y)} \,. \tag{15}$$

In this formula, the necessity for an ultraviolet regularization is apparent in the fact that the Fourier integral is not defined pointwise, but only in the distributional sense More precisely, the distribution $P(x,y)$ is singular if the vector $\xi := y - x$ is lightlike, but it is a smooth function otherwise (as can be verified for example by explicit computation). As a consequence, a typical ultraviolet regularization will affect the behavior of $P(x,y)$ only in a small neighborhood of the light cone of the form $\big||\xi^0| - |\vec{\xi}|\big| \lesssim \varepsilon$. With this in mind, for the following argument we may disregard the ultraviolet regularization simply by restricting attention to the region outside this neighborhood.

The representation (15) allows us to understand the relation between the Definition 3.1 and the usual notion of causality in Minkowski space: Since the expression (15) is Lorentz invariant and is composed of a vector and a scalar component, the function $P(x,y)$ can be written as

$$P(x,y) = \alpha\, \xi_j\gamma^j + \beta\, \mathbb{1}$$

with two complex-valued functions α and β (where again $\xi = y - x$). Taking the conjugate with respect to the spin scalar product, we see that

$$P(y,x) = P(x,y)^* = \overline{\alpha}\, \xi_j\gamma^j + \overline{\beta}\, \mathbb{1} \,.$$

As a consequence,

$$A_{xy} = P(x,y)\, P(y,x) = a\, \xi_j\gamma^j + b\, \mathbb{1}$$

with two real-valued functions a and b given by

$$a = \alpha\overline{\beta} + \beta\overline{\alpha} \,, \qquad b = |\alpha|^2\, \xi^2 + |\beta|^2 \,.$$

Applying the formula $(A_{xy} - b\mathbb{1})^2 = a^2\, \xi^2\, \mathbb{1}$, the roots of the characteristic polynomial of A_{xy} are computed by

$$b \pm \sqrt{a^2\, \xi^2} \,.$$

Thus if the vector ξ is timelike, the term ξ^2 is positive, so that the λ_j are all real. By explicit computation one sees that the coefficients a and b are non-zero (see [4, Section §1.2.5]), implying that the eigenvalues λ_j do not all have the same absolute value. Conversely, if the vector ξ is spacelike, then the term ξ^2 is negative. Thus the λ_j form a complex conjugate pair, implying that they all have the same absolute value. We conclude that the notions of spacelike and timelike as defined for causal fermion systems in Definition 3.1 indeed agree with the usual notions in Minkowski space. We remark that this simple argument cannot be used for lightlike directions because in this case the distribution $P(x,y)$ is singular, making it necessary to consider an ultraviolet regularization (the reader interested in the technical details is referred to [7]).

To summarize, we have seen that the inherent structures of a causal fermion system give back the usual causal structure if one considers the Dirac sea vacuum in Minkowski space. Indeed, a more detailed analysis reveals that the additional inherent structures mentioned at the end of Section 3 also give back the geometric structures of Minkowski space (like the metric and the connection).

5. Description of more general space-times

The constructions explained above also apply to more general physical situations. First, one can consider systems involving *particles* and *anti-particles* by occupying additional states and removing states from the Dirac sea, respectively. Moreover, our construction also apply in *curved space-time* (see [3, 8]) or in the presence of an *external potential* (see [9]). In all these situations, the resulting causal fermion systems again encode all the information on the physical system (see [5, 4]).

The framework of causal fermion systems also allows to describe generalized space-times (sometimes referred as quantum space-times). We now illustrate this concept in the simple example of a *space-time lattice*. Thus we replace Minkowski space by a four-dimensional lattice $\mathscr{M} := (\varepsilon\mathbb{Z})^4$ of lattice spacing ε. Likewise, the volume measure d^4x is replaced by a counting measure μ (thus $\mu(\Omega)$ is equal to the number of lattice points contained in Ω). Restricting the Dirac spinors of Minkowski space to the lattice, one gets a spinor space $S_x\mathscr{M}$ at every point $x \in \mathscr{M}$. Dirac wave functions $\psi_1, \ldots, \psi_f$ can again be introduced as mappings which to every $x \in \mathscr{M}$ associate a vector in the corresponding spinor space. These Dirac wave functions can be chosen for example as solutions of a discretized version of the Dirac equation. Again choosing $\mathcal{H}$ as the span of the wave functions (6) and choosing a suitable scalar product $\langle .|. \rangle_{\mathcal{H}}$, one defines the local correlation operators again by (8). Introducing the universal measure as the push-forward of the counting measure μ, we obtain a causal fermion system $(\mathcal{H}, \mathcal{F}, \rho)$ of spin dimension two. The only difference to the causal fermion system in Minkowski space as constructed in Section 2 is that now the universal measure is not a continuous but a discrete measure.

When describing the Dirac sea vacuum on the lattice, the lattice spacing gives rise to a natural ultraviolet regularization on the scale ε. For example, one may consider all plane-wave solutions $\psi(x) \sim e^{ikx}$ of the Dirac equation whose four-momenta lie in the first Brillouin zone, i.e. $-\pi < \varepsilon\, k_j \leq \pi$ for all $j = 0, \ldots, 3$. Then one introduces $\mathcal{H}$ as the Hilbert space generated by all these plane-wave solutions restricted to the lattice.

Other examples of discrete or singular space-times are described in [6].

6. The continuum limit

In the previous sections we saw that a causal fermion system has inherent structures which generalize corresponding notions in quantum theory and relativity. The next task is to analyze the dynamics of these objects as described by the causal action principle. To this end, one considers the Euler-Lagrange (EL) equations corresponding to the causal action. These equations have a mathematical structure which is quite different from conventional physical equations (see [10]). Therefore, the main difficulty is to reexpress the EL equations in terms of the inherent structures so as to make them comparable with the equations of contemporary physics. This can indeed be accomplished in the so-called continuum limit. Since the mathematical methods needed for the analysis of the continuum limit go beyond the scope of this paper (for details see [4, 11]), here we can only explain the general concept and discuss the obtained results.

We outlined in Sections 2 and 4 how to describe the Minkowski vacuum by a causal fermion system. Recall that the construction required an *ultraviolet regularization* on a microscopic scale ε. Such a regularization can be performed in many different ways. The simplest method is to smooth out the wave functions on the microscopic scale by convolution with a test function. Another method is to introduce a cutoff in momentum space on the scale ε^{-1}. Alternatively, one can regularize by putting the system on a four-dimensional lattice with lattice spacing ε (for example as explained in Section 5 above). It is important to note that each regularization gives rise to a different causal fermion system, describing a physical space-time with a different microstructure. Thus in the context of causal fermion systems, the regularization has a physical significance. The freedom in regularizing reflects our lack of knowledge on the microstructure of physical space-time. When analyzing the EL equations corresponding to the causal action, it is not obvious why the effective macroscopic equations should be independent of the regularization details. Therefore, it is necessary to consider a sufficiently large class of regularizations, and one needs to analyze carefully how the results depend on the regularization. This detailed analysis, referred to as the *method of variable regularization* (for more explanations see [11, §4.1]), reveals that for a large class of regularizations, the structure of the effective macroscopic equations is indeed independent of the regularization (for details see [4, Chapters 3-5]).

The *continuum limit* is a method for evaluating the EL equations corresponding to the causal action in the limit $\varepsilon \searrow 0$ when the ultraviolet regularization is removed. The effective equations obtained in this limit can be evaluated conveniently in a formalism in which the unknown microscopic structure of space-time (as described by the regularization) enters only in terms of a finite (typically small) number of so-called *regularization parameters.*

It turns out that the causal fermion system describing the Minkowski vacuum satisfies the EL equations in the continuum limit (for any choice of the regularization parameters). If one considers instead a system involving additional particles and anti-particles, it turns out the EL equations in the continuum limit no longer hold. In order to again satisfy these equations, we need to introduce an interaction. In mathematical terms, this means that the universal measure ρ must be modified. Expressed in terms of the inherent structures of a causal fermion system, all the physical wave functions $\psi^{u_k}(x)$ must be changed collectively. The analysis shows that this collective behavior of all physical wave functions (including the states of the Dirac sea) can be described by inserting a potential $\mathcal{B}$ into the Dirac equation (5),

$$\left(i\partial\!\!\!/ + \mathcal{B} - m\right) u_k(x) = 0, \qquad k = 1, \ldots, f \tag{16}$$

(where as usual $\partial\!\!\!/ = \gamma^j \partial_j$). Moreover, the EL equations in the continuum limit are satisfied if and only if the potential $\mathcal{B}$ satisfies field equations. Before specifying these field equations, we point out that in the above procedure, the potential $\mathcal{B}$ merely is a convenient device in order to describe the collective behavior of all physical wave functions. It should not be considered as a fundamental object of the theory. We also note that, in order to describe variations of the physical wave functions, the potential in (16) can be chosen arbitrarily. Each choice of $\mathcal{B}$ describes a different variation of the physical wave functions. The EL equations in the continuum limit single out the physically admissible potentials as being those which satisfy the field equations.

In [4] the continuum limit is worked out in several steps beginning from simple systems and ending with a system realizing the fermion configuration of the standard model. For each of these systems, the continuum limit gives rise to effective equations for second-quantized fermion fields coupled to classical bosonic gauge fields (for the connection to second-quantized bosonic fields see Section 8.5 below). To explain the structure of the obtained results, it is preferable to first describe the system modelling the leptons as analyzed in [4, Chapter 4]. The input to this model is the configuration of the leptons in the standard model without interaction. Thus the fermionic projector of the vacuum is assumed to be composed of three generations of Dirac particles of masses $m_1, m_2, m_3 > 0$ (describing e, μ, τ) as well as three generations of Dirac particles of masses $\tilde{m}_1, \tilde{m}_2, \tilde{m}_3 \geq 0$ (describing the corresponding neutrinos). Furthermore, we assume that the regularization of the neutrinos breaks the chiral symmetry (implying that we only see their left-handed components). We point out that the definition of the model does not involve any assumptions on the interaction.

The detailed analysis in [4, Chapter 4] reveals that the effective interaction in the continuum limit has the following structure. The fermions satisfy the Dirac equation coupled to a left-handed SU(2)-gauge potential $A_L = \left(A_L^{ij}\right)_{i,j=1,2}$,

$$\left[i\partial\!\!\!/ + \begin{pmatrix} A\!\!\!/_L^{11} & A\!\!\!/_L^{12}\, U_{\text{MNS}}^* \\ A\!\!\!/_L^{21}\, U_{\text{MNS}} & -A\!\!\!/_L^{11} \end{pmatrix} \chi_L - mY \right] \psi = 0 \,,$$

where we used a block matrix notation (in which the matrix entries are 3×3-matrices). Here mY is a diagonal matrix composed of the fermion masses,

$$mY = \text{diag}(\tilde{m}_1, \tilde{m}_2, \tilde{m}_3,\ m_1, m_2, m_3)\,, \tag{17}$$

and U_{MNS} is a unitary 3×3-matrix (taking the role of the Maki-Nakagawa-Sakata matrix in the standard model). The gauge potentials A_L satisfy a classical Yang-Mills-type equation, coupled to the fermions. More precisely, writing the isospin dependence of the gauge potentials according to $A_L = \sum_{\alpha=1}^3 A_L^\alpha \sigma^\alpha$ in terms of Pauli matrices, we obtain the field equations

$$\partial^k \partial_l (A_L^\alpha)^l - \Box (A_L^\alpha)^k - M_\alpha^2\, (A_L^\alpha)^k = c_\alpha\, \overline{\psi}\big(\chi_L \gamma^k\, \sigma^\alpha\big)\psi\,, \tag{18}$$

valid for $\alpha = 1, 2, 3$ (for notational simplicity, we wrote the Dirac current for one Dirac particle; for a second-quantized Dirac field, this current is to be replaced by the expectation value of the corresponding fermionic field operators). Here M_α are the bosonic masses and c_α the corresponding coupling constants. The masses and coupling constants of the two off-diagonal components are equal, i.e. $M_1 = M_2$ and $c_1 = c_2$, but they may be different from the mass and coupling constant of the diagonal component $\alpha = 3$. Generally speaking, the mass ratios M_1/m_1, M_3/m_1 as well as the coupling constants c_1, c_3 depend on the regularization. For a given regularization, they are computable.

Finally, our model involves a gravitational field described by the Einstein equations

$$R_{jk} - \frac{1}{2}\, R\, g_{jk} + \Lambda\, g_{jk} = \kappa\, T_{jk}\,, \tag{19}$$

where R_{jk} denotes the Ricci tensor, R is scalar curvature, and T_{jk} is the energy-momentum tensor of the Dirac field. Moreover, κ and Λ denote the gravitational and the cosmological constants, respectively. We find that the gravitational constant scales like $\kappa \sim \delta^{-2}$, where $\delta \geq \varepsilon$ is the length scale on which the chiral symmetry is broken.

In [4, Chapter 5] a system is analyzed which realizes the configuration of the leptons and quarks in the standard model. The result is that the field equation (18) is replaced by field equations for the electroweak and strong interactions after spontaneous symmetry breaking (the dynamics of the corresponding Higgs field has not yet been analyzed). Furthermore, the system again involves gravity (19).

A few clarifying remarks are in order. First, the above field equations come with corrections which for brevity we cannot discuss here (see [4, Sections 3.8, 4.4 and 4.6]). Next, it is worth noting that, although the states of the Dirac sea are explicitly taken into account in our analysis,

Journal of Physics: Conference Series **626** (2015) 012020 doi:10.1088/1742-6596/626/1/012020

they do not enter the field equations. More specifically, in a perturbative treatment, the divergences of the Feynman diagram describing the vacuum polarization drop out of the EL equations of the causal action. Similarly, the naive "infinite negative energy density" of the sea drops out of the Einstein equations, making it unnecessary to subtract any counter terms. We finally remark that the only free parameters of the theory are the masses in (17) as well as the parameter δ which determines the gravitational constant. The coupling constants, the bosonic masses and the mixing matrices are functions of the regularization parameters which are unknown due to our present lack of knowledge on the microscopic structure of space-time. The regularization parameters cannot be chosen arbitrarily because they must satisfy certain relations. But except for these constraints, the regularization parameters are currently treated as free empirical parameters.

To summarize, the dynamics in the continuum limit is described by Dirac spinors coupled to classical gauge fields and gravity. The effective continuum theory is manifestly covariant under general coordinate transformations. The only limitation of the continuum limit is that the bosonic fields are merely classical. However, as will be briefly mentioned in Section 8.5, a detailed analysis which goes beyond the continuum limit gives rise even to second-quantized bosonic fields. Based on these results, the theory of causal fermion systems seems to be a promising candidate for a unified physical theory.

7. Connection to foundations of quantum theory

As explained above, in the continuum limit the dynamics of quantum mechanical wave functions is described by the Dirac equation, which in the non-relativistic limit reduces to the Pauli equation or the Schrödinger equation (cf. [12, 13]).

It is well-known in the foundations of quantum theory that the linearity of the Schrödinger equation per se seems to be in conflict with the experimental observation of definite measurement outcomes (one observes a dead cat or an alive cat, but no superposition of the two). This conflict, referred to as the *measurement problem* (cf. [14]), can be remedied in several ways, leading to modifications of the original quantum mechanics as formulated by von Neumann [15]. Since the theory of causal fermion systems is a candidate for a unified physical theory, this raises the question of what its implications are on the measurement problem. This question is presently under investigation. We here outline how it can be addressed and formulate a conjecture.

We first point out that the EL equations corresponding to the causal action are *nonlinear*. This is obvious in the abstract setting simply because the positive Borel measures do not form a vector space (the linear combination of two such measures is in general no longer positive). More explicitly, this nonlinearity can be seen by reexpressing the EL equations in terms of inherent structures like the physical wave functions. Namely, according to (14), the fermionic projector is quadratic in the wave functions. Thus the closed chain (13) is of fourth order in the wave functions. Computing the eigenvalues of A_{xy} involves square roots, also we need to take absolute values of these eigenvalues. Therefore, the Lagrangian (1) depends on the wave functions in a highly nonlinear way. Nevertheless, *linear* evolution equations are obtained by considering linear perturbations of the Dirac sea vacuum. This approximation is justified in many situations because the contribution of, for example, a single electron wave function to the causal action is very small compared to the total contribution of all the sea states. This is the underlying reason why the dynamics of the wave functions as obtained in the continuum limit is linear. However, this linear dynamics is only an approximation.

Two corrections to the linear Schrödinger dynamics seem to be most relevant. First, a nonlinear correction to the Dirac equation arises by taking into account the perturbation of the Dirac sea vacuum in the EL equations to second order. Another effect is due to the microscopic structure of space-time. Taking into account fluctuations of this microstructure (which are disregarded in the continuum limit) seems to give rise to a stochastic correction term to the

Journal of Physics: Conference Series **626** (2015) 012020 doi:10.1088/1742-6596/626/1/012020

Dirac equation. This combination of a nonlinear and a stochastic correction term is reminiscent of the modifications of the Schrödinger equation used for example in the spontaneous localization model (see [16, 17, 18] or [19, Chapter 8]). This leads us to the following

> *Conjecture.* The theory of causal fermion systems gives rise to an effective dynamical collapse theory.

By "dynamical collapse theory" we mean a Schrödinger or Dirac equation involving nonlinear and stochastic terms which resolve the measurement problem. "Effective" means that the nonlinear equations are not taken as the starting point, but they arise in an effective description of the dynamics as determined by the EL equations of the causal action.

Other points of interest with respect to foundations of quantum theory are *nonlocality* and *entanglement*. Both are experimentally tested features of quantum theory which need to be explained by the theory of causal fermion systems. To understand the role of nonlocality, one should keep in mind that in a causal fermion system, a fermion is described by a physical wave function $\psi^u(x)$. As in standard quantum mechanics, these wave functions are nonlocal objects spread out in space-time, giving rise to the usual nonlocal correlations for one-particle measurements.

In order to describe entanglement, one needs to work with a multi-particle wave function. The simplest method to obtain the connection is to choose an orthonormal basis $u_1, \ldots, u_f$ of $\mathcal{H}$ and to form the f-particle Hartree-Fock state

$$\Psi := \psi^{u_1} \wedge \cdots \wedge \psi^{u_f} . \tag{20}$$

Clearly, the choice of the orthonormal basis is unique only up to the unitary transformations

$$u_i \to \tilde{u}_i = \sum_{j=1}^{f} U_{ij}\, u_j \quad \text{with} \quad U \in \mathrm{U}(f) .$$

Due to the anti-symmetrization, this transformation changes the corresponding Hartree-Fock state only by an irrelevant phase factor,

$$\psi^{\tilde{u}_1} \wedge \cdots \wedge \psi^{\tilde{u}_f} = \det U \; \psi^{u_1} \wedge \cdots \wedge \psi^{u_f} .$$

Thus the configuration of the physical wave functions can be described by a fermionic multi-particle wave function.

The shortcoming of the above construction is that the Hartree-Fock state (20) does not allow for the description of entanglement. But entanglement arises naturally if the effect of *microscopic mixing* is taken into account, as we now briefly outline. Microscopic mixing is based on the observation that the causal action of a Dirac sea configuration is smaller if the physical wave functions have fluctuations on the microscopic scale. To be more precise, one constructs a universal measure ρ which consists of L components, i.e. $\rho = \rho_1 + \cdots + \rho_L$. This also gives rise to a decomposition of the corresponding space-time, i.e. $M = M_1 \cup \cdots \cup M_L$ with $M_\ell := \operatorname{supp} \rho_\ell$. Now one considers variations of the measures ρ_ℓ obtained by modifying the phases of the physical wave functions in the sub-space-times M_ℓ. Minimizing the causal action under such variations, one sees that the kernel of the fermionic projector $P(x, y)$ becomes very small if x and y are in different sub-space-times. This effect can be understood similar to a dephasing of the physical wave functions in different sub-space-times.

The resulting space-time M has a structure which cannot be understood classically. One way of visualizing M is that it consists of different global space-times M_ℓ which are interconnected by relations between them. An alternative intuitive picture is to regard M as a single space-time which is "fine-grained" on the microscopic scale by the sub-space-times M_ℓ. For the physical

Journal of Physics: Conference Series **626** (2015) 012020 doi:10.1088/1742-6596/626/1/012020

wave functions, the above dephasing effect means that every physical wave function $\psi^u(x)$ has *fluctuations* on the microscopic scale. Moreover, comparing $\psi^u(x)$ and $\psi^u(y)$ for x and y in the same sub-space-time, one finds *nonlocal correlations* on the macroscopic scale. A detailed analysis shows that taking averages over the sub-space-times gives rise to an effective description of the interaction in terms of multi-particle wave functions and Fock spaces (see [20, Sections 5, 6 and 8]). In particular, this gives agreement with the usual description of entanglement.

To summarize, entanglement arises naturally in the framework of causal fermion systems when taking into account the effect of microscopic mixing. The reader who wants to understand the concept of microscopic mixing on a deeper quantitative level is referred to [20].

8. Clarifying remarks

This section aims to address some of the questions which might have come to the mind of the reader.

8.1. Where does the name "causal fermion system" come from?

The term *"causality"* in the name causal fermion system refers to the fact that there are causal relations among the space-time points (see Definition 3.1). The causal action is "causal" because it vanishes for space-time points with spacelike separation. In this way, the notion of causality is intimately connected with the framework of causal fermion systems. The term "fermion" refers to the fact that a causal fermion system encodes physical wave functions $\psi^u(x)$ (see (11)) which are interpreted as *fermionic* wave functions (like Dirac waves). This interpretation as fermionic wave functions is justified because, rewriting the configuration of the physical wave functions in the Fock space formalism, one obtains a totally anti-symmetric multi-particle state (see (20)). *Bosonic* fields appear in the causal fermion systems merely as a device to describe the collective behavior of the fermions (see (16)). But they should not be considered as fundamental objects of the theory.

8.2. Why this form of the causal action principle?

The first attempts to formulate a variational principle in space-time in terms of fermionic wave functions can be found in the unpublished preprint [?]. The variational principle proposed in [11, Section 3.5] coincides with the causal action principle, except that it is formulated in the setting of discrete space-times and that the constraints (2) and (3) are missing. The general structure of the Lagrangian (1) can be understood from the requirements that it should be non-negative and that it should vanish for spacelike separation. The detailed form of the Lagrangian (1) is determined uniquely by demanding that the Dirac sea vacuum should be a stable minimizer of the variational principle (as is made precise by the notion of "state stability"; see [11, Section 5.6]). The necessity and significance of the constraints (2) and (3) became clear when analyzing the existence theory [21, 22] and deriving the EL equations [10]. It should also be noted that the so-called identity constraint considered in [22] has turned out to be a too strong condition which is not compatible with the so-called spatial normalization of the fermionic projector as discussed in [23, Section 2.2] and established in [24].

8.3. Why the name "continuum limit"?

Causal fermion systems were first analyzed in the more restrictive formulation of discrete space-times (see [11, Section 3.3]). In this setting, the continuum limit as introduced in [11, Chapter 4] arises as the limit when the discretization scale ε tends to zero, meaning that the discrete space-time goes over to a space-time continuum. The more general notion of causal fermion systems given here allows for the description of both continuous and discrete space-times. Then the parameter ε should be regarded as a regularization length, but space-time could very well be continuous on this scale. In this more general context, the notion "continuum limit" merely means that we take the limit $\varepsilon \searrow 0$ in which space-time $M := \operatorname{supp} \rho$ goes over to the *usual* space-time continuum $\mathscr{M}$ (i.e. Minkowski space or a Lorentzian manifold, without microscopic mixing).

8.4. Connection to the notion of the Dirac sea

The concept of the Dirac sea was introduced by Dirac in order to remedy the problem of the negative-energy solutions of the Dirac equation. Dirac's original conception was that in the vacuum all negative-energy states are occupied. Due to the Pauli exclusion principle, additional particles must occupy states of positive energy. This concept led to the prediction of anti-particles, which are described as "holes" in the sea.

If taken literally, the concept of the Dirac sea leads to problems such as an infinite negative energy density or an infinite charge density. This is the main reason why in modern quantum field theory, the concept of the Dirac sea is no longer apparent. It has been replaced by Wick ordering and the reinterpretation of creation and annihilation operators corresponding to the negative-energy states. Therefore, it is a common view that the Dirac sea is merely a historical relic which is no longer needed.

In the theory of causal fermion system, Dirac's original concept is revived. Namely, when constructing a causal fermion system starting from a classical space-time the states of the Dirac sea need to be taken into account (cf. (14) and (15) in the Minkowski vacuum). This can be understood as follows. It is a general concept behind causal fermion systems that all structures in space-time should be encoded in the physical wave functions. This concept only works if there are "sufficiently many" physical wave functions. More specifically, this is the case if the causal fermion system is composed of a regularized Dirac sea configuration, possibly with additional particles and/or anti-particles.

In contrast to the problems in the naive Dirac sea picture, in the description with causal fermion systems the ensemble of the sea states does *not* give rise to an infinite negative energy density or an infinite charge density. Namely, due to the specific form of the causal action principle, the sea states drop out of the Euler-Lagrange equations in the continuum limit.

8.5. Connection to quantum field theory

The continuum limit gives an effective description of the interaction on the level of second-quantized fermionic fields coupled to classical bosonic fields. A full quantum field theory, in which also the bosonic fields are quantized, arises if the effect of microscopic mixing is taken into account. We refer the reader to Section 7 as well as the article [20]. The detailed analysis of the resulting Feynman diagrams, renormalization and a comparison with standard quantum field theory is work in progress.

8.6. Which physical principles are incorporated in a causal fermion system?

Causal fermion systems evolved from an attempt to combine several physical principles in a coherent mathematical framework. As a result, these principles appear in the framework in a specific way:

- The **principle of causality** is built into a causal fermion system in a specific way, as explained in Section 8.1 above.
- The **Pauli exclusion principle** is incorporated in a causal fermion system, as can be seen in various ways. One formulation of the Pauli exclusion principle states that every fermionic one-particle state can be occupied by at most one particle. In this formulation, the Pauli exclusion principle is respected because every wave function can either be represented in the form ψ^u (the state is occupied) with $u \in \mathcal{H}$ or it cannot be represented as a physical

wave function (the state is not occupied). But it is impossible to describe higher occupation numbers. When working with multi-particle wave functions, the Pauli exclusion principle becomes apparent in the total anti-symmetrization of the wave function (see (20)).

- A **local gauge principle** becomes apparent once we choose basis representations of the spin spaces and write the wave functions in components. Denoting the signature of $(S_x, \prec .|.\succ_x)$ by $(p(x), q(x))$, we choose a pseudo-orthonormal basis $(\mathfrak{e}_\alpha(x))_{\alpha=1,\ldots,p+q}$ of S_x. Then a wave function ψ can be represented as

$$\psi(x) = \sum_{\alpha=1}^{p+q} \psi^\alpha(x)\, \mathfrak{e}_\alpha(x)$$

with component functions $\psi^1, \ldots, \psi^{p+q}$. The freedom in choosing the basis $(\mathfrak{e}_\alpha)$ is described by the group $\mathrm{U}(p,q)$ of unitary transformations with respect to an inner product of signature (p,q). This gives rise to the transformations

$$\mathfrak{e}_\alpha(x) \to \sum_{\beta=1}^{p+q} U^{-1}(x)_\alpha^\beta\, \mathfrak{e}_\beta(x) \qquad \text{and} \qquad \psi^\alpha(x) \to \sum_{\beta=1}^{p+q} U(x)^\alpha_\beta\, \psi^\beta(x)$$

with $U \in \mathrm{U}(p,q)$. As the basis $(\mathfrak{e}_\alpha)$ can be chosen independently at each space-time point, one obtains *local gauge transformations* of the wave functions, where the gauge group is determined to be the isometry group of the spin scalar product. The causal action is *gauge invariant* in the sense that it does not depend on the choice of spinor bases.

- The **equivalence principle** is incorporated in the following general way. Space-time $M := \operatorname{supp}\rho$ together with the universal measure ρ form a topological measure space, being a more general structure than a Lorentzian manifold. Therefore, when describing M by local coordinates, the freedom in choosing such coordinates generalizes the freedom in choosing general reference frames in a space-time manifold. Therefore, the equivalence principle of general relativity is respected. The causal action is *generally covariant* in the sense that it does not depend on the choice of coordinates.

8.7. Philosophical remarks

Since causal fermion systems are a candidate for a unified physical theory, one may take a consistent realist point of view and assume that our universe is a causal fermion system. Here by "realist point of view" we mean that one assumes that there is a reality independent of human observation and that one can describe this reality in a mathematical language. "Consistent" means that this point of view does not lead to contradictions or inconsistencies. Finally, by "universe is a causal fermion system" we mean that the fundamental entities of our universe are the causal fermion system $(\mathcal{H}, \mathcal{F}, \rho)$ as well as its inherent structures.

This position could be investigated from a philosophical point of view. We find the following points interesting:

- Space-time is a set of operators. The relations between space-time points are all encoded in properties of products of these operators. No additional structures need to be specified.
- Similar to the picture in dynamical collapse theories, the basic object to describe a fermion is the physical wave function $\psi^u(x)$. The particle character, however, comes about merely as a consequence of the dynamics as described by the causal action principle.
- The structures of space-time and matter are described in terms of a single object: the universal measure. In particular, it is no longer possible to separate space-time from the matter content therein. This seems to go a step further than relativity: In relativity, space

and time do not exists separately, but are combined to space-time. In the approach of causal fermion systems, space-time does not exist without the matter content (including the Dirac sea). Space-time and the matter content are combined in one object.

A further investigation of these and related points might offer new insight on questions in the philosophy of science.

Acknowledgments

J.K. gratefully acknowledges support by the "Studienstiftung des deutschen Volkes." We would like to thank Meinard Kuhlmann, Olaf Müller, Christian Röken and Jan-Hendrik Treude for helpful comments on the manuscript. We are grateful to Thomas Elze for organizing the inspiring conference in Castiglioncello.

References

[1] Rudin W 1987 *Real and Complex Analysis* 3rd ed (New York: McGraw-Hill Book Co.)
[2] Halmos P 1974 *Measure Theory* (New York: Springer)
[3] Finster F and Reintjes M 2015 *Adv. Theor. Math. Phys.* to appear (*Preprint* arXiv:1301.5420 [math-ph])
[4] Finster F *The Continuum Limit of Causal Fermion Systems* book in preparation based on *Preprint* arXiv:0908.1542 [math-ph], *Preprint* arXiv:1211.3351 [math-ph], *Preprint* arXiv:1409.2568 [math-ph]
[5] Finster F and Grotz A 2012 *Adv. Theor. Math. Phys.* **16** 1197–1290 (*Preprint* arXiv:1107.2026 [math-ph])
[6] Finster F and Kamran N 2014 *Preprint* arXiv:1403.7885 [math-ph]
[7] Finster F 2008 *J. Math. Phys.* **49** 032304, 60 (*Preprint* arXiv:math-ph/0612003)
[8] Finster F and Reintjes M 2013 *Preprint* arXiv:1312.7209 [math-ph]
[9] Finster F, Murro S and Röken C *Preprint* arXiv:1501.05522 [math-ph]
[10] Bernard Y and Finster F 2014 *Adv. Calc. Var.* **7** 27–57 (*Preprint* arXiv:1205.0403 [math-ph])
[11] Finster F 2006 *The Principle of the Fermionic Projector* (*AMS/IP Studies in Advanced Mathematics* vol 35) (Providence, RI: American Mathematical Society) (*Preprint* hep-th/0001048, *Preprint* hep-th/0202059, *Preprint* hep-th/0210121)
[12] Bjorken J and Drell S 1964 *Relativistic Quantum Mechanics* (New York: McGraw-Hill Book Co.)
[13] Peskin M and Schroeder D 1995 *An Introduction to Quantum Field Theory* (Reading, MA: Addison-Wesley Publishing Company Advanced Book Program)
[14] Peres A 1993 *Quantum Theory: Concepts and Methods* (*Fundamental Theories of Physics* vol 57) (Kluwer Academic Publishers Group, Dordrecht)
[15] von Neumann J 1932 *Mathematische Grundlagen der Quantenmechanik* reprinted 1955 (Berlin: Springer Verlag); English translation by RT Beyer 1981 *Mathematical Foundations of Quantum Mechanics* (Princeton University Press)
[16] Pearle P 1989 *Phys. Rev. A* **39** 2277–2289
[17] Ghirardi G, Pearle P and Rimini A 1990 Stochastic processes in Hilbert space: a consistent formulation of quantum mechanics *Foundations of quantum mechanics in the light of new technology (Tokyo, 1989)* (Tokyo: Phys. Soc. Japan) pp 181–189
[18] Bassi A, Lochan K, Satin S, Singh T and Ulbricht H 2013 *Rev. Mod. Phys.* **85**(2) 471–527
[19] Joos E, Zeh H, Kiefer C, Giulini D, Kupsch J and Stamatescu I O 2003 *Decoherence and the Appearance of a Classical World in Quantum Theory* 2nd ed (Berlin: Springer-Verlag)
[20] Finster F 2014 *J. Math. Phys.* **55** 042301 (*Preprint* arXiv:1310.4121 [math-ph])
[21] Finster F 2007 *Calc. Var. Partial Differential Equations* **29** 431–453 (*Preprint* arXiv:math-ph/0503069)
[22] Finster F 2010 *J. Reine Angew. Math.* **646** 141–194 (*Preprint* arXiv:0811.2666 [math-ph])
[23] Finster F and Tolksdorf J 2014 *J. Math. Phys.* **55** 052301 (*Preprint* arXiv:1401.4353 [math-ph])
[24] Finster F and Kleiner J *Noether-like theorems for causal variational principles* in preparation

7th International Workshop DICE2014 Spacetime – Matter – Quantum Mechanics IOP Publishing
Journal of Physics: Conference Series **626** (2015) 012021 doi:10.1088/1742-6596/626/1/012021

Irreversibility and decoherence in an ideal gas

Janos Polonyi

Strasbourg University, CNRS-IPHC, 23 rue du Loess, BP28 67037 Strasbourg Cedex 2, France

E-mail: polonyi@iphc.cnrs.fr

Abstract. Different models are described where non-interacting particles generate dissipative effective forces by the mixing of infinitely many soft normal modes. The effective action is calculated for these models within the Closed Time Path formalism. This is a well known scheme for quantum systems but its application in classical mechanics presents a new, more unified derivation and treatment of dissipative forces within classical and quantum physics.

1. Introduction

The derivation of dissipative forces is a challenge for us on two counts. First, these forces are not covered by the fundamental equations of motion. Dissipation, a special kind of breakdown of the time reversal invariance, arises from the restriction of our attention to a subset of a closed system, in short it is an effective force. Second, the usual formalism of classical or quantum mechanics, based on Lagrangians and Hamiltonians, can not cope with such non-conservative forces.

Some attempts to derive friction dissipative forces and decoherence are surveyed here by calculating the effective action for the observed system within the Closed Time Path (CTP) formalism, well known in quantum field theory [1, 2, 3, 4]. Its extension for classical mechanics [5, 6, 7, 8], which provides a unified treatment of effective forces on the classical and the quantum level and offers a fresh look into the quantization of open systems, is surveyed in the first part of this paper. The second half contains a brief discussion of dissipative effective forces in two classical models, in a set of harmonic oscillators and in classical electrodynamics, followed by the quantum mechanical examples, a test particle in an ideal gas and the dynamics of the probability current in an ideal fermi gas.

2. Dissipative forces in effective theories

Let us consider a classical model, described by the system and the environment coordinates x and $y_1, \ldots, y_N$, respectively. The closed dynamics, defined by the action $S[x, y]$, is supposed to be symmetric under the inversion of and the translation in the time. The solution of the equations of motion,

$$\frac{\delta S[x,y]}{\delta x} = 0, \qquad \frac{\delta S[x,y]}{\delta y} = 0, \tag{1}$$

are made unique by some auxiliary condition, to be chosen as initial conditions. To find the effective system equation of motion one solves the second equation for a general system trajectory and inserts the solution, $y[x]$, into the first equation. One can gain more insight into the effective dynamics by calculating the effective action, $S_{eff}[x] = S[x, y[x]]$, whose variational equation,

$$\frac{\delta S_{eff}[x]}{\delta x} = 0, \tag{2}$$

is the effective equation of motion.

But this procedure is formal an incomplete. The problem comes from the way the auxiliary conditions are handled in the variational method: The Euler-Lagrange equation is derived by fixing the initial and final coordinates but we have initial conditions in this problem. This is an important issue for dissipative effective dynamics where the final environment coordinates are extremely complicated. An extension of the traditional variation method is needed at this point. The proposition, put forward in this Section, is to let ourself guided in establishing an extension of the classical variational method by the already well known solution of this problem in quantum theory, namely by the use of mixed states. This strategy not only lead us to an action principle which covers dissipative forces but also provides a fresh view of the standard quantization rules when applied for open systems.

2.1. Semiholonomic force

One of the main obstacle to describe effective dynamics is that it contains non-conservative forces. Analytic mechanics starts with d'Alembert principle, stating that the virtual work of the external and the inertial forces is always vanishing along the trajectory,

$$(F_{ext} - m\ddot{x})\delta x = 0. \tag{3}$$

This equation can be converted into a much more powerful form by assuming that the force is holonomic, namely it can be obtained from a scalar function, $U(x, \dot{x})$ by derivation,

$$F(x, \dot{x})\delta x = -\delta x \partial_x U(x, \dot{x}) - \delta \dot{x} \partial_{\dot{x}} U(x, \dot{x}). \tag{4}$$

Hamilton considered the time integral of d'Alambert principle for a holonomic force and noticed that it can be written as a variational equation,

$$0 = \delta \int_{t_i}^{t_f} dt \left[\frac{m}{2}\dot{x}^2 - U(x, \dot{x})\right] - \delta x(m\dot{x} + \partial_{\dot{x}} U)\Big|_{t_i}^{t_f}. \tag{5}$$

Note that the initial and final variations cancel in this equation. To underline the formal similarities of the classical and quantum dynamics we continue with the usual variational principle,

$$0 = \delta \int_{t_i}^{t_f} dt \left[\frac{m}{2}\dot{x}^2 - U(x, \dot{x})\right]. \tag{6}$$

The holonomic forces are conservative and we need a generalization of (4) for effective forces. The coordinate and velocity dependence dependence of $U(x, \dot{x})$ in (4) provides us an expression of the force and assures the energy conservation. We separate these two roles by introducing passive and active copies of the degrees of freedom,

$$x \to \hat{x} = (x^+, x^-), \tag{7}$$

the active coordinate being used to calculate the force and the passive coordinate controlling the potential in an independent manner. The result is expression,

$$F(x, \dot{x})\delta x = -[\delta x \partial_{x^+} U(\hat{x}, \dot{\hat{x}}) + \delta \dot{x} \partial_{\dot{x}^+} U(\hat{x}, \dot{\hat{x}})]_{|x^+ = x^- = x}, \tag{8}$$

for the semiholonomic forces. These forces are non-conservative and it will be argued below that they cover all effective forces, arising within a subsystem of a closed system.

2.2. Initial conditions

Another complication of the effective dynamics is related to the missing auxiliary conditions. Let us suppose that the environment is finite, $N < \infty$, when the effective equation of motion,

$$\left[c_0 + c_2\partial_t^2 + \cdots + c_{2(N+1)}\partial_t^{2(N+1)}\right] x = 0. \tag{9}$$

can easily be derived in harmonic models by the use of normal modes. The general solution of this equation requires the knowledge of $2(N+1)$ auxiliary conditions of which two are known, $x(t_i)$ and $\dot{x}(t_i)$, the environment being unobserved. How to solve this equation with the insufficient number of auxiliary conditions at hand? It is easy to see that this problem is not restricted to harmonic models.

The missing initial conditions may lead to yet another, more particular problem of the effective equation of motion (9). This equation is formally invariant under time reversal but this symmetry is bound to be broken in the effective theory. In fact, to check the the time reversal invariance of the effective theory one records the motion of the system and checks whether the motion, seen by playing the record backward in time, satisfies the same effective equation of motion. The point is that the environment initial conditions influence the effective equation of motion in an implicit manner and the initial environment conditions of the time reversed record are the final environment conditions of the original motion.

We propose the use of variational principle as a remedy of these problems. More precisely, the solution of the effective equation of motion will be constructed by the help of the retarded Green function, derived from the effective action.

2.3. Classical chrono-dynamics

To avoid the need of fixing the final coordinates and to allow the dynamical breakdown of time reversal invariance we follow the motion in an extended time interval. We start the motion at t_i with some initial conditions, perform a time reversal at t_f and follow our system back in time until it reaches its initial state, c.f. Fig. 1. The result is a closed time path, $\tilde{x}(t) = x(t)$, for $t_i < t < t_f$ and $\tilde{x}(t) = x(2t_f - t)$ when $t_f < t < 2t_f - t_i$. This notation is rather cumbersome and a simpler book-keeping is provided by the introduction of the CTP doublers, (7), with trajectory $\hat{x}(t) = (x^+(t), x^-(t)) = (\tilde{x}(t), \tilde{x}(2t_f - t))$. This step sheds more light on the role of the active and passive copies, introduced above: The non-conservative part of the effective force arises from the new segment of the extended motion and the active and the passive coordinates belong to the same degrees of freedom because the equation $x^+(t) = x^-(t)$ holds after having imposed the equation of motion. Note that the time arrow is oriented in the opposite way for the two trajectory. This is the main achievement of this formalism, the possibility of following the motion in both directions of the time, and is the source of the name chronon, introduced for the doublet $\hat{x}$.

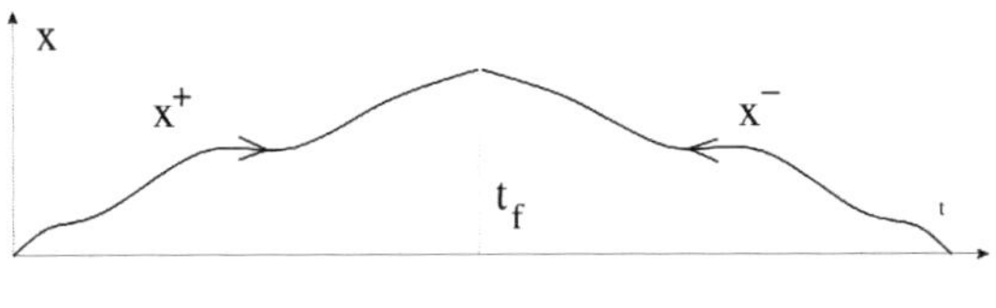

Figure 1. The system undergoes a time reversal transformation, the time arrow is flipped at $t = t_f$, and the motion is followed back to the initial state.

Both member of the chronon, $x^\pm$, are dynamical variables therefore one needs an action, $S[\hat{x}]$, to generate their dynamics. The obvious choice for a closed system is

$$S[\hat{x}] = \int_{t_i}^{t_f} dt[L(x^+(t), \dot{x}^+(t)) - L(x^-(t), \dot{x}^-(t))], \tag{10}$$

Journal of Physics: Conference Series **626** (2015) 012021 doi:10.1088/1742-6596/626/1/012021

where $L(x, \dot{x})$ denotes the traditional Lagrangian. The minus sign arises from the opposite time arrow of the trajectories $x^+(t)$ and $x^-(t)$ and introduces a symplectic structure for the CTP index. The same initial conditions are imposed for both members of the chronon. The boundary term in the variational equation of motion, (6), should not be canceled in the usual way, by requiring $\delta x^{\pm}(t_f) = 0$, rather by imposing the constraint

$$x^+(t_f) = x^-(t_f), \tag{11}$$

which leaves the final state of the motion free, an essential requirement for the environment of dissipative systems.

There is a degeneracy in the action at $x^+(t) = x^-(t)$ which has to be split in order to arrive at unique Green functions. For this end we add an infinitesimal imaginary part to the CTP Lagrangian,

$$L(\hat{x}, \dot{\hat{x}}) = L(x^+, \dot{x}^+) - L(x^-, \dot{x}^-) + L_{spl}(\hat{x}, \dot{\hat{x}}), \tag{12}$$

with $L_{spl}(\hat{x}, \dot{\hat{x}}) = i\epsilon(x^{+2} + x^{-2})/2$. The advantage of removing the degeneracy with an imaginary term is the CTP symmetry, the transformation

$$S[\tau\hat{x}] = -S^*[\hat{x}] \tag{13}$$

of the action under the exchange, $\tau(x^+, x^-) = (x^-, x^+)$, which flips the time arrow.

2.4. *Green functions*

Let us first consider the Green function of a harmonic system, given by the action $S[\hat{x}] = \hat{x}\hat{K}\hat{x}/2 + \hat{x}\hat{j}$ where $\hat{j} = (j^+, j^-)$ is an external source. We use a condensed notation where the trajectories are handled as vectors, the scalar product denotes the time integration and $\hat{x}\hat{j} = \sum_\sigma x^\sigma j^\sigma$. The CTP Green function is defined as $\hat{D} = \hat{K}^{-1}$ and allows us to write the trajectory, corresponding to a physically realizable source, $j^+ = -j^-$, as

$$x(t) = -\sum_{\sigma'} \int dt' D_0^{\sigma\sigma'}(t, t')\sigma' j(t'). \tag{14}$$

This expression is independent of σ and leads to the CTP exchange symmetry, $D^{++} + D^{--} = D^{+-} + D^{-+}$, of the Green function. Similar relation is satisfied by $\hat{K}$, as well. These together yield the block structure

$$\hat{D} = \begin{pmatrix} D^n & -D^f \\ D^f & -D^n \end{pmatrix} + iD^i \begin{pmatrix} 1 & 1 \\ 1 & 1 \end{pmatrix} \tag{15}$$

and

$$\hat{K} = \hat{\sigma} \left[\begin{pmatrix} K^n & -K^f \\ K^f & -K^n \end{pmatrix} + iK^i \begin{pmatrix} 1 & 1 \\ 1 & 1 \end{pmatrix} \right] \hat{\sigma}, \tag{16}$$

where

$$\hat{\sigma} = \begin{pmatrix} 1 & 0 \\ 0 & -1 \end{pmatrix} \tag{17}$$

is the "metric tensor" of CTP. The application of these rules for the electromagnetic field suggests the definition $D^{\overset{r}{a}} \equiv D^n \pm D^f$ for the retarded and advanced Green functions. One can define these components for the inverse, $K^{\overset{r}{a}} \equiv K^n \pm K^f$, as well, and the relations $K^{\overset{r}{a}} = (D^{\overset{r}{a}})^{-1}$ and $K^i = -(D^a)^{-1}D^i(D^r)^{-1}$ follow. The symmetry, $K^{\sigma\sigma'}(t, t') = K^{\sigma'\sigma}(t', t)$, introduces similar equation for $\hat{D}$ and the relations $D^n(t, t') = D^n(t', t)$, $D^i(t, t') = D^i(t', t)$ and $D^f(t, t') = -D^f(t', t)$.

Journal of Physics: Conference Series **626** (2015) 012021 doi:10.1088/1742-6596/626/1/012021

The translation invariance in time is recovered by performing the limit $t_i \to -\infty$, $t_f \to \infty$ and the Green function of the harmonic oscillator, $L = m\dot{x}^2/2 - m\Omega^2 x^2/2$, turns out to be [5]

$$\hat{D}(t,t') = \int \frac{d\omega}{2\pi} e^{-i\omega(t-t')} \hat{D}_\omega(\Omega), \tag{18}$$

with

$$\hat{D}_\omega(\Omega) = \frac{1}{m} \begin{pmatrix} \frac{1}{\omega^2-\Omega^2+i\epsilon} & -2\pi i\Theta(-\omega)\delta(\omega^2-\Omega^2) \\ -2\pi i\Theta(\omega)\delta(\omega^2-\Omega^2) & -\frac{1}{\omega^2-\Omega^2-i\epsilon} \end{pmatrix}. \tag{19}$$

This result can be used to define the action in this limit, $K^n = m(\omega^2 - \Omega^2)$, $K^f = i\text{sign}(\omega)\epsilon$, $K^i = \epsilon$, $S[\hat{x}] = S[x^+] - S[x^-] + S_{spl}[\hat{x}]$, where the infinitesimal imaginary part,

$$S_{spl}[\hat{x}] = \frac{i\epsilon}{2} \int_{-\infty}^{\infty} dt[x^+(t) - x^-(t)]^2 + \frac{\epsilon}{\pi} P \int_{-\infty}^{\infty} dt dt' \frac{x^+(t)x^-(t')}{t-t'}, \tag{20}$$

handles the boundary conditions in time, and P stands for the principal value prescription. Note that the coupling between the chronons at the final time, (11), is replaced by infinitesimal, time translation invariant mixing terms.

Let us consider now a function of the coordinate, $A(x)$, and perform the Legendre transformation, $W[\hat{j}] = S[\hat{x}] + \hat{j}\hat{A}(\hat{x})$, where $\delta\{S[\hat{x}] + \hat{j}\hat{A}(\hat{x})\}/\delta\hat{x} = 0$. The Green functions of $A(x)$ for an interacting system are defined as the coefficient functions of the expansion

$$W[\hat{j}] = \sum_{n=0}^{\infty} \frac{1}{n!} \sum_{\sigma_1,\ldots,\sigma_n} \int dt_1 \cdots dt_n D^{\sigma_1,\ldots,\sigma_n}(t_1,\ldots,t_n) j^{\sigma_1}(t_1) \cdots j^{\sigma_n}(t_n), \tag{21}$$

and can be calculated by iteration. The inverse transformation, $S_{eff}[\hat{A}] = W[\hat{j}] - \hat{j}\hat{A}(\hat{x})$ and $\delta W[\hat{j}]/\delta\hat{j} = \hat{A}(\hat{x})$, gives the effective action. The solution of its equation of motion,

$$\hat{A}(\hat{x}(t)) = \sum_{n=0}^{\infty} \frac{1}{n!} \sum_{\sigma,\sigma_1,\ldots,\sigma_n} \int dt_1 \cdots dt_n D^{\sigma,\sigma_1,\ldots,\sigma_n}(t, t_1,\ldots,t_n) j^{\sigma_1}(t_1) \cdots j^{\sigma_n}(t_n), \tag{22}$$

is expressed by the help of the Green functions. Such a method to construct the solution of the effective equation of motion does not produce runaways trajectories as long as we are allowed to use the residuum theorem to calculate frequency integrals of the Green functions. This is how this scheme realizes Dirac's idea about the elimination of runaway solution for a point charge in classical electrodynamics [9].

2.5. *Effective action*

After having included the initial conditions into the variational principle we return to the effective action in the generic model, defined by the action $S[x,y] = S_s[x] + S_e[x,y]$. The steps, leading to eq. (2) are now well defined and yield

$$\begin{aligned} S_{eff}[\hat{x}] &= S_s[x^+] + S_e[x^+, y^+[\hat{x}]] - S_s[x^-] - S_e[x^-, y^-[\hat{x}]] + S_{spl}[\hat{x}] + S_{spl}[\hat{y}[\hat{x}]] \\ &= S_s[x^+] - S_s[x^-] + S_{infl}[\hat{x}] + S_{spl}[\hat{x}], \end{aligned} \tag{23}$$

where the trajectory $\hat{y}[\hat{x}]$ satisfies the equation of motion, $\delta S_e[\hat{x},\hat{y}]/\delta\hat{y} = 0$. The second equation defines the influence functional [10],

$$S_{infl}[\hat{x}] = S_e[x^+, y^+[\hat{x}]] - S_e[x^-, y^-[\hat{x}]]. \tag{24}$$

A physically better motivated form of the effective action is

$$S_{eff}[\hat{x}] = S_1[x^+] - S_1[x^-] + S_2[\hat{x}] + S_{spl}[\hat{x}] \tag{25}$$

where the separation of the terms is rendered well defined by the condition $S_2[0, x^-] = S_2[x^+, 0] = 0$. The role of the different actions, introduced in this manner, becomes clear in the parametrization, $x^\pm = x \pm x^d/2$, whose advantage in classical mechanics is that it is sufficient to know the action in $S = \mathcal{O}(x^d)$ since $x^+(t) = x^-(t)$ holds for the solution of the equations of motion. The variational equation of motion for x^d at $x^d = 0$,

$$0 = \frac{\delta S_1[x]}{\delta x} + \frac{\delta S_2[x^+, x^-]}{\delta x^+}_{|x^+=x^-=x}, \tag{26}$$

represents the holonomic and the semiholonomic forces by the help of S_1 and S_2, respectively, and realizes the scheme of Section 2.1: The dynamics, generated by the single coordinate, S_1, is conservative and obeys Noether theorem and the coupling of the doubler, S_2, introduces the dissipative forces [11]. The chronon couplings are due to the influence of the final environment coordinate by the system trajectory and the CTP scheme reduces to two independent traditional action principles if the final environment coordinates are kept fixed during variation. The CTP symmetry, (13), is inherited by the effective action and provides the proof that the family of semiholonomic forces is closed with respect to generating effective interactions.

A distinguishing feature of the CTP formalism is that the system-environment interactions are mapped into the interactions within the chronon. This is an unexpected and a highly non-trivial result, holding for arbitrary large and complex environment. We shall see in the quantum case, discussed below, that the same chronon coupling encodes the system-environment entanglement, as well, making entanglement and semiholonomic forces identical.

2.6. *Quantum chrono-dynamics*

The CTP formalism was introduced by Schwinger to carry out the perturbation expansion for an expectation value $\bar{A}(t) = \langle\psi(0)|U^\dagger(t)AU(t)|\psi(0)\rangle$ [1], where the need of the reduplication of the degrees of freedom arises from the simultaneous use of bra and ket, two virtually independent but physically identical state vectors. One is tempted to follow the naive quantization procedure of Schrödinger for the chronon and to introduce the chronon wave function, $\psi(x^+, x^-)$. This function is actually the density matrix, $\psi(x^+, x^-) = \rho(x^+, x^-)$, shedding a new light on Gleason theorem.

The more systematical, perturbative treatment is facilitated by the generating functional,

$$e^{\frac{i}{\hbar}W[\hat{j}]} = \mathrm{Tr}T[e^{-\frac{i}{\hbar}\int dt(H(t)-j^+(t)x(t))}]\rho_i T^*[e^{\frac{i}{\hbar}\int dt(H(t)+j^-(t)x(t))}], \tag{27}$$

where ρ_i denotes the initial density matrix. Its path integral representation is

$$e^{\frac{i}{\hbar}W[\hat{j}]} = \int D[\hat{x}]e^{\frac{i}{\hbar}S[\hat{x}]+\frac{i}{\hbar}\int dt\hat{j}(t)\hat{x}(t)}, \tag{28}$$

where the integration extends over chronon trajectories which satisfy the condition (11) and the action is the same as in the classical case.

The effective dynamics can be found in a manner, similar to the classical case. The (Wilsonian) effective action is defined by

$$e^{\frac{i}{\hbar}S_{eff}[\hat{x}]} = te^{\frac{i}{\hbar}S_s[x^+]-\frac{i}{\hbar}S_s[x^-]}\int D[\hat{y}]e^{\frac{i}{\hbar}S_e[x^+,y^+]-\frac{i}{\hbar}S_e[x^-,y^-]+\frac{i}{\hbar}S_{slp}[\hat{x}]}, \tag{29}$$

and the form of eq. (25) applies again.

One may consider a generalization where one leaves out the trace in the generator functional (27). The result is the Open Time Path scheme, an extension of the CTP formalism, based on the density matrix,

$$\rho(x_f^+, x_f^-) = \langle x_f^+|U(t_f,t_i)\rho_i U^\dagger(t_f,t_i)|x_f^-\rangle \tag{30}$$

The path integral representation of the reduced density matrix,

$$\rho(x_f^+, x_f^-) = \int D[\hat{x}]D[\hat{y}]e^{\frac{i}{\hbar}S[x^+,y^+]-\frac{i}{\hbar}S[x^-,y^-]+\frac{i}{\hbar}S_{spl}[\hat{x}]+\frac{i}{\hbar}S_{spl}[\hat{y}]}, \tag{31}$$

where the trajectories satisfy the final conditions $\hat{x}(t_f) = \hat{x}_f$ and $y^+(t_f) = y^-(t_f)$. Noteworthy the effective action is the same as in the CTP scheme,

$$\rho(x_f^+, x_f^-) = \int D[\hat{x}]e^{\frac{i}{\hbar}S_1[x^+]-\frac{i}{\hbar}S_1^*[x^-]+\frac{i}{\hbar}S_2[\hat{x}]+\frac{i}{\hbar}S_{spl}[\hat{x}]}. \tag{32}$$

This form reveals that decoherence in the coordinate (field) diagonal representation consists of the suppression of the contributions to the path integral of well separated trajectories within a chronon and is generated by $\Im S_{infl}$. The couplings within a chronon, S_2, are due to the simultaneous contributions of several final environment states in the trace of (27), they render the system state mixed and encode the system-environment entanglement. An important role of the CTP formalism in quantum field theory is to visualize the elementary processes which contribute to the full reduced density matrix, rather than to the transition amplitudes between pure states, in terms of Feynman graphs.

3. Effective models

Now we look into model calculations where irreversibility, and decoherence in the quantum case, are generated by non-interacting modes. Irreversibility and decoherence will be detected by the dissipative terms in the effective equation of motion and the suppression of a chronon trajectory contribution to the path integral with well separated trajectories, respectively.

3.1. Classical toy model

The simplest, non-trivial classical effective model corresponds to a set of linearly coupled harmonic oscillators, given by the Lagrangian [12],

$$L = \frac{m}{2}\dot{x}^2 - \frac{m\omega_0^2}{2}x^2 + \sum_{n=1}^{N}\left(\frac{m}{2}\dot{y}_n^2 - \frac{m\omega_n^2}{2}y_n^2 - g_n y_n x\right). \tag{33}$$

It is sometime more advantageous to write it in a form where the system action alone is already a good approximation for the slow system dynamics, $\dot{x} \to 0$,

$$L = \frac{m}{2}\dot{x}^2 - \left(\frac{m\omega_0^2}{2} - \sum_n \frac{g_n^2}{2m\omega_n^2}\right)x^2 + \sum_n\left[\frac{m}{2}\left(\dot{z}_n - \frac{g_n}{m\omega_n^2}\dot{x}\right)^2 - \frac{m\omega_n^2}{2}z_n^2\right]. \tag{34}$$

The action will be written in the form

$$S[\hat{x},\hat{y}] = \frac{1}{2}\hat{x}\hat{D}_0^{-1}\hat{x} + \frac{1}{2}\hat{y}\hat{G}^{-1}\hat{y} - \hat{x}\sigma g\hat{y}, \tag{35}$$

in condensed notation. The elimination of the environment coordinates by means of their equation of motion leads to the effective action, $S_{eff}[\hat{x}] = \hat{x}\hat{D}^{-1}\hat{x}/2$, with $\hat{D}^{-1} = \hat{D}_0^{-1} - \sigma\hat{\Sigma}\sigma$, where the self energy, $\hat{\Sigma} = g\hat{G}g$, contains the influence functional,

$$S_{infl}[\hat{x}] = -\frac{1}{2}\sum_{\sigma\sigma'}\sigma\sigma'\int\frac{d\omega}{2\pi}dte^{-i\omega(t'-t)}\Sigma_\omega^{\sigma\sigma'}x^\sigma(t')x^{\sigma'}(t). \tag{36}$$

The spectral function,

$$\rho(\Omega) = \sum_n \frac{g_n^2}{2m\omega_n}\delta(\omega_n - \Omega), \tag{37}$$

can be used to parametrize the model. The self energy assumes the form

$$\hat{\Sigma}_\omega = \frac{1}{m}\int_0^\infty d\Omega 2\Omega\rho(\Omega)\hat{D}_\omega(\Omega), \tag{38}$$

where $\hat{D}_\omega(\Omega)$ is given by eq. (19), in particular

$$\Sigma_\omega^n = \frac{2}{m}P\int_0^\infty d\Omega\frac{\Omega\rho(\Omega)}{\omega^2 - \Omega^2}, \tag{39}$$

$\Sigma_\omega^f = -i\pi\text{sign}(\omega)\rho(|\omega|)$, and $\Sigma_\omega^i = -\pi\rho(|\omega|)$.

The influence functional is non-local in time but the expansion of the exponent in $t' - t$ and the self energy in ω renders it local [13],

$$L_{infl} = -\frac{1}{2}(x\vec{\Sigma}^n x^d + x^d\vec{\Sigma}^n x + x^d\vec{\Sigma}^f x - x\vec{\Sigma}^f x^d + x^d i\vec{\Sigma}^i x^d), \tag{40}$$

where

$$\vec{\Sigma}^{\sigma\sigma'} = \sum_{\ell=0}^\infty \frac{(-1)^\ell}{\ell!}\partial_{i\omega}^\ell\Sigma_0^{\sigma\sigma'}\partial_t^\ell. \tag{41}$$

The equation of motion, generated by the variation of x^d, at $x^d = 0$, is

$$m\ddot{x} = -(m\omega_0^2 + \vec{\Sigma}^r)x, \tag{42}$$

with $\vec{\Sigma}^r = \vec{\Sigma}^n + \vec{\Sigma}^f$, and contains the following terms up to $\mathcal{O}(\partial_t^2)$: The harmonic oscillator frequency is renormalized, $\omega_0^2 \to \omega_0^2 - \Delta\omega^2$, in $\mathcal{O}(\partial_t^0)$ with

$$\Delta\omega^2 = \frac{2}{m^2}\int_0^\infty d\Omega\frac{\rho(\Omega)}{\Omega}, \tag{43}$$

realizing the system potential, seen in eq. (34). A Newtonian friction force, $F = -k\dot{x}$, arises in $\mathcal{O}(\partial_t)$ with $k = \pi\rho'(0)$. The $\mathcal{O}(\partial_t^2)$ mass renormalization, $m \to m + \delta m$, takes place with

$$\delta m = \frac{4}{m}\int_0^\infty d\Omega\frac{\rho(\Omega)}{\Omega^3}, \tag{44}$$

reflecting that the dressing by the environment increases the inertia ($\rho \geq 0$).

The origin of dissipative forces in this harmonic model can be found by noting that the energy, received by x, spreads over the whole system [14]. This is similar to the spread of energy within an interactive system and leads to dissipation if there are sufficient strength at low frequencies. A more physical insight can be found by viewing irreversibility as a spontaneous symmetry breaking, an adiabatic approximation which becomes exact in the thermodynamical limit: To test reversibility one has to monitor all normal modes. The observation of a normal mode needs time at least in the order of magnitude of its period length. Hence the observations, carried out in an arbitrary long, but finite time leave infinitely many normal modes unresolved and are unable to establish reversibility if there is a condensation point in the spectrum at vanishing frequency. The infinitely many unresolved slow normal mode drive a gradual loss of system energy, diffusion.

3.2. Classical point charge

The spectral function of the toy model can freely be chosen and different dissipative features can be modeled. This freedom is due to the lack of local space-time structure. When the degrees of freedom are distributed in the space-time then their spectral weight and the emerging dissipative forces are severely restricted by the space-time symmetries. This motivates our next model, the effective theory of a classical point charge [15]. The electrodynamics of a single charge is defined by the action $S = S_{ch} + S_{EMF} + S_i$, where a point charge, the electromagnetic field (EMF) dynamics and their interactions are described by

$$\begin{aligned} S_{ch} &= -m_B c \sum_\sigma \sigma \int ds \sqrt{\dot{x}^{\sigma 2}(s)} \\ S_{EMF} &= -\frac{1}{8\pi c} \sum_\sigma \sigma \int \frac{d^4 p}{(2\pi)^4} Q(p^2) p^2 A^\sigma_\mu(-p) \left(g^{\mu\nu} - \frac{p^\mu p^\nu}{p^2} \right) A^\sigma_\nu(p), \\ S_i &= -\frac{e}{c} \sum_\sigma \sigma \int ds \dot{x}^{\sigma\mu}(s) A^\sigma_\mu(x^\sigma(s)), \end{aligned} \tag{45}$$

$Q(p^2)$ being the regulator for the singular UV regime of the EMF. The influence functional with the real part,

$$\Re S_{infl} = \frac{2\pi e^2}{c} \int ds ds' \dot{\hat{x}}(s) \Re \hat{D}(x(s) - x(s')) \dot{\hat{x}}(s'), \tag{46}$$

can easily be found by eliminating of the EMF. It goes beyond the usual action-at-a-distance action [16, 17, 18, 19] insofar as it contains the radiation field contribution, too. The EMF Green function,

$$\Re \hat{D}(x) = \frac{\delta_\ell(x^2)}{4\pi} \begin{pmatrix} -1 & -\mathrm{sign}(x^0) \\ \mathrm{sign}(x^0) & 1 \end{pmatrix}, \tag{47}$$

needs some regularization. The diagonal, near field singularities are removed by smearing the Dirac-delta, $\delta(z) \to \delta_\ell(z)$, over a distance ℓ. The far field singularity calls for special attention: First, because it arises from the coincidence of the singular points of the two distributions in the off-diagonal blocks. Second, any smearing of $\mathrm{sign}(x^0)$ breaks Lorentz invariance. We keep $\mathrm{sign}(x^0)$ unchanged and require $\delta_\ell(0) = 0$. Note that we do not need this latter condition in QED, set in the usual formalism, to calculate transition amplitudes between pure states. A simple regularization which satisfies all requirements is $\delta_\ell(z) = \delta(z - \ell^2)$.

The expansion of (46) in $(s - s')/\ell$ is straightforward and produces the non-local influence Lagrangian

$$L_{infl}(s) = x^d(s) \frac{4e^2}{c} \int_{-\infty}^0 du \delta'_\ell(u^2)[x(s+u) - u\dot{x}(s+u) - x], \tag{48}$$

up to terms $\mathcal{O}(x^2)$. The corresponding equation of motion for x^d, at $x^d = 0$, is

$$m_B c \ddot{x}^\mu = -\frac{e^2}{c\ell} \ddot{x}^\mu + (g^{\mu\nu} - \dot{x}^\mu \dot{x}^\nu) \left[K_\nu + \mathcal{O}(x^3) \right] \tag{49}$$

with

$$K(s) = \frac{2e^2}{3c}\left\{\dddot{x}(s) - 6\int_{-\infty}^{0} du\delta'_\ell(u^2)\left[x(s+u) - u\dot{x}(s+u) - x(s) + \frac{u^2}{2}\ddot{x}(s) + \frac{u^3}{3}\dddot{x}(s)\right]\right\}. \quad (50)$$

The convergence of the integral in the influence Lagrangian, (48), is not uniform when the cutoff is removed, $\ell \to 0$, and the first term of (50), the well known Abraham-Lorentz force, owes its existence to this irregularity. The emergence of the cutoff-independent Abraham-Lorentz force is reminiscent of quantum anomalies: Both arise as finite cutoff effects, left behind after the removal of the cutoff. In the case of the renormalized classical EMF one finds a light cone anomaly, as well, the modification of the retarded Green function, $D_{r\ell}(x,y) \to D_{r\ell}(x,y)[1 + (x-y)^2/s_0^2]$, amounts to an $\mathcal{O}(e^2/c^2 s_0)$ change of the mass of the charge. In other words, the EMF field which is placed slightly off shell by the regulator remains sensitive to the off-shell modifications of its dynamics even after the removal of the cutoff. The second term in (50), the integral is $\mathcal{O}(\ell)$ and generates a crossover, reminiscent of the Landau pole of QED, in the scaling laws at the classical electron radius, $r_0 = e^2/mc^2$, the only scale parameter of the model. The usual electromagnetic interaction is recovered in the IR side and an acausal theory emerges in the UV regime of this hypothetical model, without quantum physics.

3.3. Test particle in an ideal quantum gas

What changes in the previously discussed models when their quantum version is considered? The classical equation of motion remains an operator equation for harmonic systems, hence the effective equation of motion, derived for the harmonic toy model, remains valid for the expectation values of the coordinate. The toy model has already extensively been studied in the CTP formalism [20]. Similarly trivial extension of the classical results to the quantum case is possible for the effective theory of a point charge, too. One has naturally to restrict the motion of the charge to the non-relativistic domain to keep its path integral Gaussian.

Let us now move to a more realistic problem of a quantum test particle, interacting with an ideal gas and determine its effective Lagrangian in the leading order of the Landau-Ginzburg double expansion [13]. The effective dynamics of a particle which interacts with an ideal gas has already been considered by using the traditional effective action approach in imaginary time [21] and by means of the CTP formalism [22]. The quantum master equation [23] has been used extensively, too: The dephasing has been described in the pioneering work [24], followed by the inclusion of relaxation [25, 26, 27]. The calculation, reviewed here, goes beyond these works insofar as it is based on the CTP formalism, needed to keep track of dissipative forces, and retains all contributions of the CTP self energy and offers a simpler and more flexible alternative to arrive at the effective dynamics.

The action of this model, $S = S_p + S_g + S_i$, is given by

$$\begin{aligned} S_p[\boldsymbol{x}] &= \int dt\left[\frac{M}{2}\dot{\boldsymbol{x}}^2(t) - V(\boldsymbol{x}(t))\right], \\ S_g[\psi^\dagger,\psi] &= \int dt d^3y \psi^\dagger(t,\boldsymbol{y})\left[i\hbar\partial_t + \frac{\hbar^2}{2m}\Delta + \mu\right]\psi(t,\boldsymbol{y}), \\ S_i[\boldsymbol{x},\psi^\dagger,\psi] &= \int dt d^3y U(\boldsymbol{y} - \boldsymbol{x}(t))\psi^\dagger(t,\boldsymbol{y})\psi(t,\boldsymbol{y}), \end{aligned} \quad (51)$$

where M denotes the mass of the test particle which propagates under the influence of an external potential V. The gas is described by the field ψ and its density is coupled to the test particle by the potential U. The influence functional of the test particle,

$$e^{\frac{i}{\hbar}S_{infl}[\hat{\boldsymbol{x}}]} = \int D[\hat{\psi}]D[\hat{\psi}^\dagger]e^{\frac{i}{\hbar}\hat{\psi}^\dagger(\hat{F}^{-1}+\hat{\Gamma}[\hat{\boldsymbol{x}}])\hat{\psi}}, \quad (52)$$

can be found by integrating over the gas degrees of freedom. The propagator of the gas, the inverse of the quadratic form of S_g, is

$$\hat{F}(t,\boldsymbol{y},t',\boldsymbol{y}') = \int \frac{d\omega}{2\pi}\frac{d^3q}{(2\pi)^3} e^{-i\omega(t-t')+i\boldsymbol{q}(\boldsymbol{y}-\boldsymbol{y}')}\hat{F}_{\omega\boldsymbol{q}}, \tag{53}$$

with

$$\hat{F}_{\omega\boldsymbol{q}} = \begin{pmatrix} \frac{1}{\omega-\frac{\hbar\boldsymbol{q}^2}{2m}+i\epsilon} & 0 \\ -i2\pi\delta(\omega-\omega_{\boldsymbol{q}}) & \frac{1}{\frac{\hbar\boldsymbol{q}^2}{2m}-\omega+i\epsilon} \end{pmatrix} - \xi n_{\boldsymbol{q}} i2\pi\delta\left(\omega-\frac{\hbar\boldsymbol{q}^2}{2m}\right)\begin{pmatrix} 1 & 1 \\ 1 & 1 \end{pmatrix}, \tag{54}$$

where $\xi = 1$ for bosons and $\xi = -1$ for fermions and $n_{\boldsymbol{q}}$ stands for the occupation number. The temperature should be high enough for bosons to suppress the condensate. The test particle-gas interaction is represented by the term $\hat{\psi}^\dagger\hat{\Gamma}[\hat{\boldsymbol{x}}]\hat{\psi} = \sum_\sigma \sigma S_i[\boldsymbol{x}^\sigma,\psi^{\sigma\dagger},\psi^\sigma]$. We assume that the potential localizes the particle strong enough to justify the expansion in the coordinate and seek the leading, $\mathcal{O}(\boldsymbol{x}^2)$ contributions to the influence functional. Hence we retain the $\mathcal{O}(\hat{\Gamma}^2)$ part,

$$S_{infl}[\hat{\boldsymbol{x}}] = -\frac{1}{2}\hat{j}\hat{\sigma}\hat{G}\hat{\sigma}\hat{j}, \tag{55}$$

only. The tadpole contribution is canceled by introducing a homogeneous, neutralizing classical background charge and

$$G^{\sigma_1\sigma_2}(x_1,x_2) = i\xi\hbar n_s \hat{F}^{\sigma_1\sigma_2}(x_1,x_2)\hat{F}^{\sigma_2\sigma_1}(x_2,x_1) \tag{56}$$

denotes the density two-point function and $j^\sigma(t,\boldsymbol{y}) = U(\boldsymbol{y}-\boldsymbol{x}^\sigma(t))$. The two-point function is given by well known one-loop integrals, $G^n_{\omega\boldsymbol{q}} = G^+_{\omega\boldsymbol{q}} + G^+_{-\omega\boldsymbol{q}}$, $G^f_{\omega\boldsymbol{q}} = G^-_{\omega\boldsymbol{q}} - G^-_{-\omega\boldsymbol{q}}$ and $iG^i_{\omega\boldsymbol{q}} = G^-_{\omega\boldsymbol{q}} + G^-_{-\omega\boldsymbol{q}}$, with

$$\begin{aligned} G^+_{\omega\boldsymbol{q}} &= -\xi\frac{n_s}{\hbar^2}P\int\frac{d^3q}{(2\pi)^3}\frac{n_{\boldsymbol{q}}}{\omega-\omega_{\boldsymbol{q}+\boldsymbol{q}}+\omega_{\boldsymbol{q}}}, \\ G^-_{\omega\boldsymbol{q}} &= -i\xi\pi\frac{n_s}{\hbar^2}\int\frac{d^3q}{(2\pi)^3}n_{\boldsymbol{q}}(n_{\boldsymbol{q}+\boldsymbol{q}}+\xi)\delta(\omega-\omega_{\boldsymbol{q}+\boldsymbol{q}}+\omega_{\boldsymbol{q}}), \end{aligned} \tag{57}$$

where $n_s = 2s+1$ and s denotes the spin of the particles of the gas. These integrals are analytic functions of the dimensionless variables $x = \omega/|\mathbf{q}|v_F$, $y = |\boldsymbol{q}|/k_F$ where $v_F = \hbar k_F/m$, $k_F = \sqrt[3]{6\pi^2 n/n_s}$, n being the density of the gas, if the gas is in thermal equilibrium at finite temperature. The characteristic frequency of the particle-gas interaction is $|\boldsymbol{q}\dot{\boldsymbol{x}}|$ hence the second expansion of the Landau-Ginzburg scheme, the expansion in the time derivative, is possible for $|x| \ll 1$, if the test particle moves slower than the gas particles, $|\dot{\boldsymbol{x}}| \ll v_F$.

In calculating the influence Lagrangian we follow the strategy of Section 3.1 and start with the influence functional

$$S_{infl} = -\frac{1}{2}\sum_{\sigma\sigma'}\sigma\sigma'\int\frac{d\omega}{2\pi}\frac{d^3q}{(2\pi)^3}dtdt'U^2_{\boldsymbol{q}}e^{-i\omega(t'-t)+i\boldsymbol{q}(\boldsymbol{x}^\sigma(t')-\boldsymbol{x}^{\sigma'}(t))}G^{\sigma,\sigma'}_{\omega\boldsymbol{q}}, \tag{58}$$

whose $\mathcal{O}(\boldsymbol{x}^2)$ part gives the influence Lagrangian,

$$L_{infl} = \frac{1}{4}\sum_{\sigma\sigma'}\sigma\sigma'\int_{\boldsymbol{q}}U^2_{\boldsymbol{q}}[\boldsymbol{q}(\boldsymbol{x}^\sigma-\boldsymbol{x}^{\sigma'}+\Delta\boldsymbol{x}^\sigma)]^2G^{\sigma\sigma'}_{0\boldsymbol{q}}, \tag{59}$$

with

$$\Delta \boldsymbol{x} = \sum_{n=1}^{\infty} \frac{\boldsymbol{x}^{(n)}}{n!} \partial_{i\omega}^{n}, \tag{60}$$

and $\boldsymbol{x} = \boldsymbol{x}(t)$, $\boldsymbol{x}^{(n)} = d^n \boldsymbol{x}/dt^n$. The influence Lagrangian can be brought into the form

$$\begin{aligned} L_{infl} &= \frac{1}{24\pi^2} \int_0^\infty dq q^2 U_q^2 [2\Delta \boldsymbol{x} \Delta \boldsymbol{x}^d G_{0q}^n + (2\Delta \boldsymbol{x} \Delta \boldsymbol{x}^d + 4\boldsymbol{x}^d \Delta \boldsymbol{x}) G_{0q}^f \\ &\quad -2(\boldsymbol{x}^{d2} + \boldsymbol{x}^d \Delta \boldsymbol{x}^d) i G_{0q}^i], \end{aligned} \tag{61}$$

by the help of the block structure (15). The leading order terms in the expansion of the time derivative lead to

$$L_{infl} = -k\boldsymbol{x}^d \dot{\boldsymbol{x}} + \delta M \dot{\boldsymbol{x}}^d \dot{\boldsymbol{x}} + i d_0 \boldsymbol{x}^{d2} - i d_2 \dot{\boldsymbol{x}}^{d2} + \mathcal{O}(\partial_t^3) + \mathcal{O}(\boldsymbol{x}^4), \tag{62}$$

with the coefficients

$$\begin{aligned} k &= -\frac{1}{6\pi^2 v_F} \int_0^\infty dq q^3 U_q^2 \partial_{ix} G^f(x, y)_{|x=0}, \\ \delta M &= \frac{1}{12\pi^2 v_F^2} \int_0^\infty dq q^2 U_q^2 \partial_{ix}^2 G^n(x, y)_{|x=0}, \\ d_0 &= -\frac{1}{12\pi^2} \int_0^\infty dq q^4 U_q^2 G^i(x, y)_{|x=0}, \\ d_2 &= \frac{1}{24\pi^2 v_F^2} \int_0^\infty dq q^2 U_q^2 \partial_{ix}^2 G^i(x, y)_{|x=0}. \end{aligned} \tag{63}$$

The expectation value of the real part of the Euler-Lagrange equation for $\boldsymbol{x}^d$ at $\langle \boldsymbol{x}^d \rangle = 0$,

$$M_R \langle \ddot{\boldsymbol{x}} \rangle = -\langle \boldsymbol{\nabla} V(\boldsymbol{x}) \rangle - k \langle \dot{\boldsymbol{x}} \rangle + \mathcal{O}(\partial_t^3) + \mathcal{O}(\langle \boldsymbol{x}^3 \rangle), \tag{64}$$

comes from $\Re L_{infl}$ and includes a mass renormalization, $M_R = M + \delta M$, and a friction force. The imaginary part of L_{infl} generates decoherence, a suppression factor, $\exp -\Im S_{infl}$, in the path integral. The terms involving $\boldsymbol{x}^{d2}$ and $\dot{\boldsymbol{x}}^{d2}$ represent the decoherence strength in the coordinate and in the momentum basis, respectively.

Note that the Newtonian friction force allows us to determine the velocity with respect to the gas. This is possible due to the breakdown of the boost invariance for the test particle by its environment. In our calculation of the effective dynamics of a point charge the EMF started in a Lorentz invariant intial state, $A_\mu = 0$, and the dissipative effective interaction, the Abraham-Lorentz force, must contain higher, odd-order time derivative.

3.4. Quantum ideal gas

The emergence of dissipative forces and decoherence when a particle interacts with a gas raises the question whether dissipative forces could be found within the ideal gas itself, without any test particle [28].

The simplest, hand-waving way to argue about the possibility of finding a non-trivial effective dynamics for a non-linear combination of the fundamental coordinates in a harmonic system is to recall that a non-linear coordinate transformation generates inertial forces. For instance, if one observe the function $y = x^2/2$ of the coordinate of a harmonic oscillator, defined by the Lagrangian $L = m\dot{x}^2/2 - m\omega^2 x^2/2$ then the effective dynamics is given by the Lagrangian $L = m\dot{y}^2/4y - m\omega^2 y$ which seems to generate a non-trivial dynamics. The inertial forces superficially seem as complicated as the forces, arising from genuine interactions.

To find a more systematical approach to the effective forces in a quantum gas we exploit the stability of the particles within the gas. Owing to the conserved particle number the elementary excitations, driven by some coupling to the environment, are handled by a bi-local operator, $\Phi(x,y) = \psi(x)\psi^\dagger(y)$, where the elementary field, $\psi(x)$, belongs to the gas particles. By borrowing from the idea of operator product expansion we can replace the non-local operator by infinitely many local composite operators,

$$\Phi_n^{\mu_1,\dots,\mu_n}(x) = \frac{\partial^n}{\partial z^{\mu_1}\cdots\partial z^{\mu_n}}\psi(x+z)\psi^\dagger(x-z). \tag{65}$$

These operators serve as possible terms in describing the interactions with the environment and as possible observables to use in diagnosing the gas dynamics. When they appear in the gas-environment coupling then the energy-momentum, received from the environment, is spread over infinitely many normal modes. If they are used as observables then they receive contributions from equally infinitely many normal modes.

The simplicity of the ideal gas dynamics appears only if the elementary fields, ψ and $\psi^\dagger$, are used to diagnose the system. In fact, the Wick theorem, stating the factorization of the n-point functions of ψ and $\psi^\dagger$ into the product of the free propagators, implies the absence of 1PI vertex functions. But there are non-factorizable, 1PI composite operator n-point functions for arbitrary n! When can we use the elementary fields, ψ or $\psi^\dagger$, and uncover the non-interacting particle picture and when are we forced to deal with the bi-local or composite operators? The answer is easy to find if the symmetry, behind the conservation of the gas particle number, is gauged, as in the case for the electron gas, whatever weak the gauge coupling constant might be. The dynamics of a single quasi-particle of such a model is given in terms of non-local observables, such as the scattering amplitudes. The local observables are constructed by the help of the composite operator family, Φ_n. Thus the price of using local observables is the loss of the simplicity, offered by the quasi-particle picture. When is the locality a more important feature than the simplicity in describing the interactive system as a gas of quasi-particles with weak residual interactions? When we want to study the structure of a single quasi-particle, rather than the long distance correlations. The local fields which appear in the microscopic, local interaction Hamiltonian are the natural variables to understand the internal structure of the quasi-particles, c.f. the restricted hydrodynamical regime of the ideal gas collective modes in an interacting gas, given by eq. (75) below.

It is natural to inquire about the effective dynamics of a restricted subset of the composite operators. We follow up this question by considering the effective dynamics of the $n = 0$, ultra-local operators in a non-interacting electron gas. There are 16 possibilities and we retain 4, $j^\mu = \bar\psi\gamma^\mu\psi$, the current. Note that the current has an environment, provided by the remaining local composite operators, Φ_n, even if the Fock-space is not a system$\otimes$environment type direct product [29].

The generator functional of the current Green functions is

$$e^{\frac{i}{\hbar}W[\hat a]} = \int D[\hat\psi]D[\hat{\bar\psi}]e^{\frac{i}{\hbar}\hat{\bar\psi}(\hat F^{-1}+\hat{\not a})\hat\psi}, \tag{66}$$

where $\hat{\bar\psi}\hat{\not a}\hat\psi = \sum_\sigma a^\sigma_\mu\bar\psi^\sigma\gamma^\mu\psi^\sigma$. The Gaussian integral is easy to perform,

$$W[\hat a] = -i\hbar\mathrm{Tr}[\hat F^{-1} + \hat{\not a}], \tag{67}$$

and indeed we find 1PI Green functions at arbitrary orders. However there is an important difference between the ideal gas dynamics and a truly interactive system: The former has one-loop 1PI graphs only and the vertex functions receive contributions from arbitrarily high orders of the loop-expansion in the latter case. Nevertheless the functional $W[\hat{a}]$ of the ideal gas, together with its Legendre transform present a formidable problem, they contain arbitrary high order contributions, without adjustable coupling constant to organize an expansion. The correlations, generated by the composite operators, used in diagnosing the gas, generate strong long range correlations. Fortunately, one can consider the weak external perturbations limit where one finds

$$W[\hat{a}] = -\frac{1}{2}\hat{a}\hat{G}\hat{a}, \tag{68}$$

by ignoring the $\mathcal{O}(\hat{a}^3)$ contributions, with

$$G^{\sigma_1\sigma_2}_{\mu_1\mu_2}(x_1, x_2) = -i\hbar\,\mathrm{tr}[\gamma^{\mu_1}\hat{F}^{\sigma_1\sigma_2}(x_1, x_2)\gamma^{\mu_2}\hat{F}^{\sigma_2\sigma_1}(x_2, x_1)], \tag{69}$$

denoting the current two-point function. The corresponding effective action,

$$\Gamma[\hat{J}] = \frac{1}{2}\hat{J}\hat{G}^{-1}\hat{J}, \tag{70}$$

can be written as

$$\Gamma[J, J^d] = \frac{1}{2}J^d K^r J + \frac{1}{2}J K^a J^d - \frac{i}{2}J^d K^r K^a J^d, \tag{71}$$

where $J = (J^+ + J^-)/2$, $J^d = J^+ - J^-$, $\hat{K} = \hat{\sigma}\hat{G}^{-1}\hat{\sigma}$ and the linearized equation of motion for a physically realizable external source, $\hat{a} = (a, -a)$,

$$a = -K^r J, \tag{72}$$

sometime called as the resistivity formula [30], is the inverse of Kubo's linear response equation, $J = -G^r a$.

The Green function, (69), can be calculated in a straightforward manner for zero temperature fermion gas and leads to the equation of motion for the inhomogeneous part of the current, $j^\mu = (n_0 + n, \boldsymbol{j})$,

$$\begin{aligned} -\frac{v_F}{\pi^2\hbar}\phi &= \left[\sum_{jk=0}^{\infty} b_{\ell,j,k}\left(\frac{i\omega}{v_F|\boldsymbol{q}|}\right)^j\left(\frac{\boldsymbol{q}^2}{k_F^2}\right)^k\right] n \qquad (73)\\ \frac{T^2}{v_F\hbar}\boldsymbol{a} - \frac{b_{t,2,0}}{b_{\ell,2,0}}\frac{v_F}{\pi^2\hbar}\frac{\omega\boldsymbol{q}}{\boldsymbol{q}^2}\phi &= \left[\sum_{jk=0}^{\infty} b_{t,j,k}\left(\frac{i\omega}{v_F|\boldsymbol{q}|}\right)^j\left(\frac{\boldsymbol{q}^2}{k_F^2}\right)^k + \frac{\boldsymbol{q}\otimes\boldsymbol{q}}{\boldsymbol{q}^2}\sum_{jk=0}^{\infty} b_{j,k}\left(\frac{i\omega}{v_F|\boldsymbol{q}|}\right)^j\left(\frac{\boldsymbol{q}^2}{k_F^2}\right)^k\right]\boldsymbol{j}, \end{aligned}$$

where $a^\mu = (\phi, \boldsymbol{a})$, k_F denotes the Fermi wave number, $v_F = \hbar k_F/m$, and $b_{\ell,j,k}$, $b_{t,j,k}$ and $b_{j,k}$ are dimensionless constants. This equation holds for

$$q^\mu = (\omega, \boldsymbol{q}) \in \mathcal{D}^0_{hydr} = \left\{(\omega, \boldsymbol{q})| \left|\frac{|\omega|}{v_F|\boldsymbol{q}|} - \frac{|\boldsymbol{q}|}{k_F}\right| < 1\right\}. \tag{74}$$

It is instructive to compare eq. (73) with the Navier-Stokes equation of phenomenological hydrodynamics. Despite the obvious similarities, both mix the odd and the even powers of ω and describe dissipative dynamics, there are important differences:

(i) Hydrodynamics refers to the energy-momentum current, rather than the Noether current, appearing in the effective action (70).

(ii) The hydrodynamical equations are closed by relying on local equilibrium and thermodynamic considerations while the assumption of the stability of the ground state, the smallness of the inhomogeneity of the current, is enough to arrive at the closed equation (72).

(iii) The gradient expansion of the phenomenological hydrodynamics assumes that the equations are analytic at $q^\mu = 0$ and the IR dynamics in the hydrodynamical regime is independent of the way the point $q^\mu = 0$ is approached. The Fermi-level makes the ω-dependence of the the Green function to appear through the combination $\omega/|\boldsymbol{q}|$ thereby making the IR point, $q^\mu = 0$, singular and the IR dynamics dependent on the slope of the curve at this point on the $(\omega, |\boldsymbol{q}|)$ plane. The result is a Laurent series in $|\boldsymbol{q}|$. The factor $1/|\boldsymbol{q}|$, accompanying each $i\omega$ in the equation of motion, can not be foreseen in the phenomenological approach.

(iv) The $\mathcal{O}(\partial_t)$ phenomenological hydrodynamical equations describe collective modes, sound waves. The effective equation of motion, (73), describes sound waves, as well, but it has to be truncated at least at $\mathcal{O}(\partial_t^2)$ to yield a qualitatively stable, truncation independent form. This composite sound, neither the known zero nor first sound, is damped, displays long range correlations and dominates the current decoherence. The long range correlations arise because the effective equation of motion, written in the space-time, contains the powers of Δ and the convolutions of the Fourier transforms of $\omega/|\boldsymbol{q}|$, the latter being divergent at large distance beyond the order $\mathcal{O}(\omega^2)$.

The physics of the ideal gas hydrodynamics, described by eqs. (73) seems to be rather different from the usual Fermi-liquid picture. To resolve this conflict we return to a remark, made previously, namely that interactions lead to higher order 1PI graphs. One of their effects, the finite life-time of the quasi-particles, restores the analycity at $q^\mu = 0$. The collisions in a realistic, interacting gas form the usual collective modes and quasi-particles at the distance scale of the mean-free path, $r_{mfp} > 1/k_F$. The ideal gas description with its composite sound is thereby limited to the regime

$$\mathcal{D}^{int}_{hydr} = \left\{(\omega, \boldsymbol{q})\left|\left|\frac{|\omega|}{v_F|\boldsymbol{q}|} - \frac{|\boldsymbol{q}|}{k_F}\right| < 1, \frac{|\boldsymbol{q}|}{k_F} > \frac{1}{k_F r_{mfp}}\right.\right\}. \tag{75}$$

Therefore the composite sound is relevant at distances between $1/k_F$ and r_{mfp} and plays an important role in forming the structure of the quasi-particles, in agreement with the remark, made above. It is interesting to look at the QCD vacuum from this point of view, where the single gluon modes play the role of composite sound within the quasi-particles of the vacuum, glueballs, and display strong correlation beyond the quasi-particle size.

References

[1] Schwinger J 1961 *J. Math. Phys.* **2** 407
[2] Mahanthappa, K T 1962 *Phys. Rev.* **126** 329
[3] Bakshi P M and Mahanthappa K T 1963 *J. Math. Phys.* **4** 1; 1963 *J. Math. Phys.* **4** 12
[4] Keldysh L V 1964 *Zh. Eksp. Teor. Fiz.* **47** 1515
[5] Polonyi J 2012 Environment Induced Time Arrow *Preprint* arXiv:1206.5781
[6] Galley C R 2013 *Phys. Rev. Lett.* **110** 174301
[7] Polonyi J 2014 *Phys. Rev.D* **90** 065010
[8] Galley C R, Tsang D and Stein L C 2014 The principle of stationary nonconservative action ofr classical mechanics and field theory *Preprint* arXiv:14123082
[9] Dirac P A M 1938 *Proc. Roy. Soc. A* **167** 148
[10] Feynman R P and Vernon F L 1963 *Ann. Phys.* **24** 118
[11] Grozdanov S and Polonyi J 2013 Viscosity and dissipative hydrodynamics from effective field theory *Preprint* arXiv:1302.3670
[12] Caldeira A O and Legget A J 1981 *Phys. Rev. Lett.* **46** 211; 1983 *Ann. Phys. (N.Y.)* **149** 374
[13] Polonyi J 2015 Dissipation by an ideal gas *Preprint* arXiv:1501.06620
[14] Polonyi J 2011 *Phys. Rev. D* **84** 105021
[15] Polonyi J 2014 *Ann. Phys.* **342** 239
[16] Schwarzschild K 1903 *Göttinger Nachrichten* **128** 132
[17] Tetrode H 1922 *Zeits. f. Physik* **10** 317
[18] Fokker A D 1929 *Zeits. f. Physik* **58** 386; *do.* 1929 *Physica* **9** 33; *do.* 1932 *Physica* **12** 145

[19] Wheeler J A and Feynman R 1945 *Rev. Mod. Phys.* **17** 157
[20] Grabert H, Schramm P and Ingold G L 1988 *Phys. Rep.* **168** 115
[21] Guinea F 1984 *Phys. Rev. Lett.* **53** 1268
[22] Hedegård P and Caldeira A O 1987 *Phys. Scr.* **35** 609
[23] Vacchini B and Hornberger K 2009 *Phys. Rep.* **478** 71
[24] Joos E and Zeh H D 1985 *Z. Phys. B* **59** 223
[25] Diosi L 1995 *Europhys. Lett.* **30** 63
[26] Hornberger K 2006 *Phys. Rev. Lett.* **97** 060601
[27] Vacchini B 2001 *Phys. Rev. E* **63** 066115
[28] Grozdanov S and Polonyi J 2015 Dynamics of the electric current in an ideal electron gas *Preprint* arXiv:1501.06620
[29] Diosi L 2013 private communication.
[30] Toda M, Kubo R, Saito N and Hashitsume N 1991 *Statistical Physics II.: Nonequilibrium Statistical Mechanics*, Series C (Berlin: Springer)

Journal of Physics: Conference Series **626** (2015) 012022 doi:10.1088/1742-6596/626/1/012022

Elementary particles, the concept of mass, and emergent spacetime

Piotr Żenczykowski

Institute of Nuclear Physics, Radzikowskiego 152, 31-342 Kraków, Poland

E-mail: piotr.zenczykowski@ifj.edu.pl

Abstract. It is argued that the problem of space quantization should be considered in close connection with the problem of mass quantization. First, the nonlocality of quantum physics suggests that if spacetime emerges from the underlying quantum layer, this emergence should occur simultaneously at all distance and momentum scales, and not just at the Planck scale. Second, the spectrum of elementary particles provides us with a lot of important information, experimentally inaccessible at the Planck scale, that could be crucial in unravelling the mechanism of emergence. Accordingly, we start with a brief review of some fundamental issues appearing both in the spectroscopy of excited baryons and in connection with the concept of quark mass. It is pointed out that experiment suggests the inadequacy of the description of baryonic interior in terms of ordinary spacetime background. Thus, it is argued that one should be able to learn about the emergence of space from the studies of the quark/hadron transition. The problem of mass is then discussed from the point of view of nonrelativistic phase space and its Clifford algebra, which proved promising in the past. Connection with the Harari-Shupe explanation of the pattern of a single Standard Model generation is briefly reviewed and a proposal for the reintroduction of relativistic covariance into the phase-space scheme is presented.

1. Introduction

Since the time quantum theory was originally formulated there have been many attempts to unite it with general relativity. As gravitational interactions are a consequence of malleable spacetime, proposals were put forward that spacetime is a purely macroscopic concept that somehow emerges from the underlying microscopic quantum degrees of freedom [1, 2, 3]. For example, the spin network idea of Penrose [3] posits that the continuous array of directions in 3D space emerges out of the quantum concept of spin in the limit of large quantum numbers. Since elementary particles exhibit other quantized properties beside that of spin, a simple extension of Penrose's idea would be to take them into account as well. Thus, one might hope that the inclusion of other spatial quantum numbers, internal quantum numbers and quantized particle masses should somehow lead to the emergence of further features of classical spacetime [1]. In particular, with mass being the source of gravity-inducing space distortions, it seems quite natural that the quantized mass should play an important role in the emergence of classical spacetime out of the quantum layer. Since a successful theory of mass and space quantization should express all particle masses in terms of one mass scale (say, Planck mass), elementary particle masses give us important experimental clues on space quantization itself - clues that are completely inaccessible at the minuscule Planck length scale. Consequently, an understanding of

Journal of Physics: Conference Series **626** (2015) 012022 doi:10.1088/1742-6596/626/1/012022

quantized particle masses, even if very limited, could be of great help in the construction of the sought emergence scheme. In particular, the very concept of mass should be critically studied as well. For this reason, we start with a brief description of some relevant problems associated with the concept of mass, and quark mass in particular.

2. The problem of mass

After putting aside the propaganda, it should be clear that the Higgs mechanism does not solve the problem of mass. In the author's opinion this mechanism may be regarded only as a low-energy field-theoretical approximative description of the problem of mass generation. Indeed, there are many unanswered fundamental questions concerning the issue of mass in both the lepton and quark sectors. First, there is the intriguing observation of Koide that relates the e, μ, and τ masses in a formula involving two small integers only [4]. Putting in the central values of experimentally measured masses, this constraint on the masses of the three charged leptons works perfectly with the accuracy of five decimal digits (which is better than 1 standard deviation). It does not seem to be an 'accident' and thus it strongly suggests an algebraic origin of mass. While neutrino and quark masses are not known with sufficient precision to test for the analogues of Koide's charged lepton formula, the quark sector elicits important conceptual questions related to the standard 'unproblematic' notion of quark mass. Specifically, the macroscopic concept of mass, naturally applicable to leptons, cannot be extended to quarks in a simpleminded way. Indeed, the masses of leptons are directly measurable from leptons' free propagation over macroscopic distances. For quarks this route cannot be used since, due to confinement, individual free quarks do not appear in the asymptotic states.

Given this qualitative difference between leptons and quarks, a question emerges what meaning should be assigned to the concept of mass for a quark. In the Standard Model, quark mass is a parameter in the QCD Lagrangian. A calculational technique to deal with the issue of confinement is then offered by lattice QCD . Yet, one may ask questions that QCD (with or without lattice) answers in a somewhat carefree way. Specifically:

Can one asssume the existence of the ordinary background spacetime within hadrons? After all, no clocks and rods — needed in the operational Einsteinian definition of space distances and time intervals — may be inserted into hadrons. In fact, as argued by Salecker and Wigner [5], the concept of spacetime cannot be consistently attributed to individually observable elementary particles at the strictly microscopic level. The situation with quarks is obviously worse since they are not individually observable. Are we then right when we imagine quarks as moving freely at small 'distances within hadrons'? Can one meaningfully talk about the concept of quark mass if the ordinary spacetime background within hadrons — in which quarks supposedly propagate — may not exist?

If QCD worked perfectly everywhere, we would have to assume that the QCD assumption of the underlying background spacetime corresponds well to reality. Yet, in spite of what one hears around, we will argue on experimental grounds that quantum chromodynamics does not work everywhere. In fact, we believe that QCD needs important modifications, at least at large distances.

2.1. Hadron spectroscopy problems

Quark confinement leads to the appearance of quark conglomerates such as $q\bar{q}$ (meson) or qqq (baryon) states. Consequently, an acceptable theory of large-distance quark interactions should describe the salient features of the observed meson and baryon spectra. It appears, however, that there are serious difficulties with the description of the spectrum of baryons. The problem is with the excited baryons. Within the standard $SU(6)$ (flavor-spin) $\times$ $O(3)$ (orbital) classification of baryon states in the old constituent quark model, the first levels with the three smallest consecutive values of the principal quantum number N are decomposed as [6] [1]:

$$\begin{aligned} N&=0 \qquad && (56,0^+), \\ N&=1 && (70,1^-), \\ N&=2 && (56',0^+),\ (70,0^+),\ (56,2^+),\ (70,2^+),\ (20,1^+). \end{aligned} \tag{1}$$

Now, while almost all the $N = 0, 1$ states are experimentally seen, a large proportion of the $N = 2$ states is not observed. In particular there are no experimental candidates for the $(20, 1^+)$ multiplet.

There are two possible explanations of this shortage of states. The first possibility is that one of the two internal spatial degree of freedom is frozen (i.e. that, for unknown reasons, it does not get excited). This is the resolution suggested e.g. by the diquark model. In this model one accepts that from the two $SU(6)$ multiplets, a priori possible for the diquark (on account of $6\otimes6 = 21\oplus15$), only the symmetric 21-plet is for some reasons permitted. Since $21\otimes6 = 70\oplus56$, the absence of the 20-plet (and several other $N = 2$ states) is thus qualitatively explained. The spectrum of the diquark model is definitely closer to the one we observe in experiments. The second possible explanation is that many of the $N = 2$ baryons are not seen simply because they do not decay to the experimentally accessible decay channels.

Although the above problem was originally identified in the old constituent quark model, it reappears in essentially the same way in lattice QCD. In particular, lattice QCD does not exhibit diquark clustering or freezing of one internal spatial degree of freedom [7]. Consequently, there is no satisfactory agreement between the calculated and the observed spectra of the excited baryons. [2] A possible way out would be to prove the unlikely decoupling of *all* (or most of) the unseen states. Unfortunately, due to the enormous complexity of such calculations in lattice QCD, they are not feasible now.

Given the questionability of the existence of the ordinary background space within hadrons, our tentative conclusion is that the QCD description of excited baryons misses 'the problem of missing baryons' and, consequently, is not fully adequate. The situation is so serious that in their review on baryon spectroscopy Capstick and Roberts wrote [6]:

> *'These questions about baryon physics are fundamental. If no new baryons are found, both QCD and the quark model will have made incorrect predictions, and it would be necessary to correct the misconceptions that led to these predictions. Current understanding of QCD would have to be modified, and the dynamics within the quark model would have to be changed.'*

Accordingly, we view QCD as appropriate for the description of hadron scattering at large momentum transfers, but deviating from reality for low momentum transfers and composite states, i.e. beyond the region of original idealization. There is nothing wrong in accepting that QCD is an idealization that works in a restricted range only. After all, all our theories are idealizations and approximations with limited ranges of their applicability [8]. We should keep this in mind and commit not the error of identifying an abstract description of certain aspects of nature with nature itself.

With lattice QCD description of baryon spectroscopy judged as not fully adequate, the detailed values of quark masses, as extracted via lattice QCD technique, cannot be trusted. Furthermore, since the shortage of excited baryons constitutes a hint that the concept of

[1] The notation is (dimension of $SU(6)$ representation, L^P), where L, P are orbital angular momentum and parity.

[2] QCD (either via one-gluon exchange models or in lattice calculations) does describe the spectrum of ground-baryons successfully. Still, these baryons may be equally well described in approaches based on strictly hadron-level dynamics (i.e. with pion and other meson exchanges or the related idea of meson cloud used in place of quark-gluon interactions [9, 10, 11]. The real challenge lies in the description of excited baryons.

background spacetime need not be fully applicable within hadrons, the concept of quark propagation within hadrons is additionally questioned. Consequently, one has to inquire into the very concept of quark mass itself.

2.2. Current quark masses

What can be therefore said about the concept of quark mass if we leave QCD considerations aside? Can we conceive of quark masses and extract them from the hadron-level data with reasonable accuracy and without using the uncertain QCD baggage, i.e. without adopting a specific concept of quark motion 'within' hadrons? The answer to that question is positive. There is a simple way in which the concept of mass was originally attributed to quarks. This simple way does not utilize QCD at all. The idea is based on the 50 years old algebra of *hadronic* currents ('Current algebra') that satisfy quark-level *symmetries* but do not treat quarks as ordinary particles. In particular, current algebra uses global and spacetime concepts defined at the *hadron* (not quark) level.

The current algebra way of extracting quark masses from experiment uses the Gell-Mann–Oakes–Renner (GMOR) formula [12]:

$$m_\pi^2 \delta_{ab} = -\frac{1}{f_\pi^2}\langle 0|[Q_5^a,[Q_5^b,H(0)]]|0\rangle. \tag{2}$$

Here m_π, f_π are pion mass and decay constant, a, b – isospin indices, Q_5 is the axial charge, and $|0\rangle$ is the hadronic vacuum. The contribution of the quark mass term $m_q\bar{q}q$ in the Hamiltonian $H(0)$ is evaluated from:

$$\begin{aligned}[][Q_5^b,\bar{q}q] &= \bar{q}\tau^b\gamma_5 q,\\ [Q_5^a,\bar{q}\tau^b\gamma_5 q] &= \delta_{ab}\bar{q}q.\end{aligned} \tag{3}$$

Assuming that all other terms in $H(0)$ commute with axial charges, one derives:

$$m_\pi^2 = -\frac{1}{f_\pi^2}(m_u\langle 0|\bar{u}u|0\rangle + m_d\langle 0|\bar{d}d|0\rangle), \tag{4}$$

where $\langle 0|\bar{q}q|0\rangle$ are expectation values of quark bilinears in the hadronic vacuum.

After extending isospin $SU(2)$ to flavor $SU(3)$ (for decay constants and vacuum expectation values, but not for the masses [3]), one can use Eq. (4) together with an analogous formula for kaons to derive:

$$\begin{aligned}\frac{2m_s}{m_u+m_d} &= \frac{m^2(K^0)-m^2(K^\pm)+m^2(\pi^\pm)}{m^2(\pi^0)} = 25.9,\\ \frac{m_u}{m_d} &= \frac{2\,m^2(\pi^0)}{m^2(K^0)-m^2(K^\pm)+m^2(\pi^\pm)} - 1 = 0.56.\end{aligned} \tag{5}$$

The ratios of quark masses are here extracted from the hadron-level data without any assumption concerning the existence of spacetime background within hadrons. The absolute values of masses are then *approximately* determined by accepting that the strangeness-induced splitting of ground-state baryons requires $m_s - m_{u,d} \approx 140-150$ MeV. This leads to quark mass values $m_u \approx 4\ MeV$, $m_d \approx 8\ MeV$, $m_s \approx 150\ MeV$ which are in half-quantitative agreement with those given in Particle Data Group tables [13] [4].

[3] Such a way of applying and breaking $SU(3)$ has ample phenomenological support.

[4] The values given in Ref. [13] are all smaller than our estimates by a single overall factor of around 1.5. The reason is that they are defined in a different way. Specifically, the masses given in Ref. [13] are defined at the field-theoretical mass scale of $\mu = 2$ GeV^2. On the other hand, within the field-theoretical language the larger mass values extracted here are thought to correspond to $\mu \approx 1$ GeV^2.

Journal of Physics: Conference Series **626** (2015) 012022 doi:10.1088/1742-6596/626/1/012022

The above-described way of an approximate extraction of quark mass does not use the Dirac equation $(\not{p} - m)q = 0$. This is highly welcome, as the use of the on-mass-shell formulas for confined quarks is very questionable: these formulas and the standard propagator poles correspond to spatial infinity which cannot be reached by confined quarks. In fact, one may point out places, where the contribution from the on-mass-shell formulas and propagators leads to artefacts that are in direct contradiction with experiment [14, 15].

The GMOR-based extraction of quark masses is based solely on the global chiral properties of the fermion mass term. It does not use the classical relativistic connection between four-momentum and mass. It does not use the concept of quark momentum at all. It does not specify how quarks 'propagate' in spacetime. It does not use the concept of quark color [5]. This leaves a lot of freedom for the connection between the concept of quark mass and the concept of macroscopic spacetime.

3. Spacetime, phase space, and quantum

3.1. Emergent spacetime

With gravity described as a consequence of spacetime distortion induced by the presence of masses, i.e. with the reduction of gravitational interactions to geometry, the idea of emergent spacetime is usually linked with general relativity. Since from the gravitational constant G, Planck constant h, and the velocity of light c one can form an expression of the dimension of length, it is at the resulting dimunitive Planck length scale that the quantum nature of space is generally supposed to appear and spacetime is supposed to emerge.

Yet, the problems at the intersection of the quantum and the classical do not appear solely when general relativity is considered. It is well known that a kind of tension exists also between quantum physics and special relativity. This tension manifests itself as Bell's nonlocality. It indicates that the mismatch between quantum physics and the concept of classical space appears at *all* distance scales, not just at the Planck length scale. Indeed, Norsen writes [16]:

> *A much higher-level inconsistency between quantum theory and (general) relativity has been the impetus for enormous efforts (...) spent pursuing "presently fashionable string theories of everything". How might a resolution of the more basic inconsistency identified by Bell shed light on (or radically alter the motivation and context for) attempts to quantize gravity?*

In fact, the problem shows up even earlier than the intersection of quantum physics and special relativity. It appears also when nonrelativistic classical and quantum physics are juxtaposed. The problem exists because the two approaches are formulated on two completely different arenas: the ordinary 3D space and the Hilbert space. In addition, one is deterministic, the other one — probabilistic. Consequently, the quantum-classical mismatch appears already at the strictly nonrelativistic level and should be first attacked at this level. Indeed, this is the level at which Penrose's spin-network idea was originally formulated. There are therefore at least two reasons for working with the nonrelativistic approach. First, one should always start with the simplest things, and the interplay of quantum and classical physics is certainly simpler at the nonrelativistic level. Second, the existence of spatial quantum numbers (spin, parity, C-parity [6]) may be deduced already at the nonrelativistic level. Thus, connecting particle masses and internal quantum numbers with the emerging macroscopic classical arena of events should be first attempted in the nonrelativistic scheme. Our insistence on the need to establish such a

[5] If color is introduced, the GMOR extraction refers to color-blind expressions involving the sum over three colors.

[6] Charge conjugation parity. Although the existence of particles and antiparticles was originally predicted by the relativistic Dirac equation, it may also be deduced without relativity via the linearization of the nonrelativistic Schrödinger equation [17].

Journal of Physics: Conference Series **626** (2015) 012022 doi:10.1088/1742-6596/626/1/012022

connection need not be in disagreement with the expected space and time emergence at all the distance and interval scales: the scale of particle masses may be just most *useful* for unraveling the mechanism of this emergence.

3.2. From space and time to phase space

The discussion of the concept of emergence usually starts by challenging the statement that the ordinary 3D space is a background canvas on which things move and interact. In fact, this Democritean view was denied already by Aristotle who considered space 'as an accident of matter'. Many great physicists and philosophers, including Leibnitz, Mach and Einstein, subscribed later to this Aristotelian '*... old and great idea that space and time are, so to speak, stretched out by matter...*' [18]. In other words, space and time should be considered as defined by things. Yet, time is not defined by things in the same way space is. Speaking more precisely, time is defined by the *change* of things. This is how (astronomical) time was originally introduced by Ptolemy, and how it is understood by Mach and Barbour (see e.g. [19]).

Alternatively, one may say that space and time are defined by things and processes. One can claim then on philosophical grounds that things and processes should be treated on the same footing. This idea, when expressed in the language of physics, suggests a symmetric treatment of positions (defined by things) and momenta (defined by changes of things). Thus, one may argue that one should replace the Democritean description in terms of the background space and time by a more Heraclitean description in terms of the background position and momentum spaces, i.e. in terms of the background phase space. The original argument against the treatment of space and time as a background translates then into the argument against the treatment of phase space as a background. Therefore, if space and time are both emergent, their emergence may be regarded as a consequence of space and momentum space being emergent. The idea that it is the phase space that actually emerges first from the quantum layer seems to be particularly attractive as quantum mechanics 'lives and works in phase space' [20]. The known symmetries of space and time must then originate from the properties of phase space. Accordingly, all spatial quantum numbers must be interpretable in the phase-space language. Yet, as the number of phase-space dimensions is larger than that of 3D space and time, the language of nonrelativistic phase space may admit more symmetries than the language of space and time. While in the latter language these 'new' symmetries and the related quantum numbers (if any) cannot manifest as ordinary spatial ones, they might be still identifiable in experiments. Obviously, they would have to appear to us as the non-spatial ones.

3.3. Born's reciprocity

The idea that positions and momenta should be treated in a way more symmetric than the one actually used in current standard approaches, was originally proposed by Max Born [21, 22]. He noticed that various physical formulas, such as the Hamilton's equations of motion, the position-momentum commutation relations, or the expression for the orbital angular momentum are invariant under 'reciprocity transformation', i.e. under the interchange:

$$\mathbf{x} \rightarrow \mathbf{p}, \qquad \mathbf{p} \rightarrow -\mathbf{x}. \tag{6}$$

At the same time, Born observed that the symmetry, expected on account of (6), between the relativistic momentum space invariant $P = E^2 - \mathbf{p}^2$ and the relativistic position space invariant $R = t^2 - \mathbf{x}^2$, is completely broken by the concept of (quantized) mass. In his own words:

> *(...) the distance P in momentum space is capable of an infinite number of discrete values (...) while the distance R in coordinate space is not an observable quantity at all.*

In other words, quantized mass is always associated with momentum, never with position coordinates. Deeply dissatisfied with this asymmetric state of affairs, Born concluded:

'This lack of symmetry seems to me very strange and rather improbable'.

In fact, reciprocity transformation (6) suggests the existence of a new phase-space constant κ of dimension [momentum/distance], which, when the quantum constant h and the velocity of light c are added, defines the phase-space-related mass scale. With this scale being possibly very different from the Planck mass scale, its appearance indicates that the problem of quantized mass could be attacked in the language of phase-space, i.e. from an angle completely different from the seemingly more adequate language of curved spacetime. It might appear that the violation of the reciprocity symmetry by the concept of mass is an argument against the phase-space route. Yet, the failure of the reciprocity idea does not mean that symmetry between momentum and position could not be introduced in another way. Below we will argue that one can consider a different $\mathbf{x} \leftrightarrow \mathbf{p}$ symmetry as more appropriate for the generalization of the concept of mass.

3.4. Mass and phase-space heuristic

We start by noting that of the six phase-space coordinates the unknown mass-defining principle singles out the physical three-momentum as associated with the standard concept of mass. Indeed, the energy of any individually observable free particle is expressed (whether relativistically or nonrelativistically) in terms of a formula involving mass and three-momentum. The position variables are absent (violation of translational invariance could be avoided by admitting position differences only). We will call the 'canonical momentum' any three phase-space variables appearing (without their canonical partners) alongside mass in such a dispersion formula (i.e. in a Hamiltonian). The application of the reciprocity transformation (6) to such an expression for a free particle changes the three-momentum into a position three-vector, thus singling out the other three of the six phase-space coordinates as the possible coordinates of the canonical momentum. As Born noted, this leads to a formula that does not seem to describe any existing 'particle'. Still, six different choices for the canonical momentum exist in addition to the two just discussed. We list all eight of them below.

$$
\begin{array}{cccccc}
\multicolumn{2}{c}{\text{canonical}} & & \multicolumn{2}{c}{\text{canonical}} & \\
\text{position} & \text{momentum} & & \text{position} & \text{momentum} & \\
(x_1,x_2,x_3) & (p_1,p_2,p_3) & \leftarrow \text{Born} \rightarrow & (p_1,p_2,p_3) & (x_1,x_2,x_3) & \\
(x_1,p_2,p_3) & (p_1,x_2,x_3) & & (p_1,x_2,x_3) & (x_1,p_2,p_3) & \\
(p_1,x_2,p_3) & (x_1,p_2,x_3) & & (x_1,p_2,x_3) & (p_1,x_2,p_3) & \\
(p_1,p_2,x_3) & (x_1,x_2,p_3) & & (x_1,x_2,p_3) & (p_1,p_2,x_3) & (7) \\
\multicolumn{2}{c}{\text{even permutations}} & & \multicolumn{2}{c}{\text{odd permutations}} &
\end{array}
$$

The l.h.s (r.h.s.) pairs of entries in Eq. (7) constitute even (odd) permutations of the standard (top-left) choice. As indicated, the left- and right-hand sides of the first line above are connected (up to a sign) by Born's reciprocity argument. It was noted above that no quantized mass is associated with the top-right entry. We observe that this violation of reciprocity could actually originate from the distinction between even and odd permutations. If this is so, one would expect that quantized mass could also be associated with the second, third, and fourth canonical momenta on the left hand side, but not with the corresponding ones on the right hand side. Such a generalization of the concept of mass clearly violates its rotational invariance. Since our classical experience requires rotational invariance to hold, the corresponding 'particles' certainly cannot exist as individual objects in the macroscopic world. However, we do not see any reason why such 'particles' could not exist 'inside' conglomerates built of several such objects, provided these conglomerates exhibit all the macroscopically expected covariance and invariance properties.

The appearance of new 'particles' which exist in triplicate and cannot be observed individually bears such a striking resemblance to quarks of the Standard Model that we feel forced to put forward a conjecture that the four options on the l.h.s of Eq. (7) actually correspond to a lepton and three colored quarks. We have to stress that for the above conjecture to hold one does not have to question the approximate adequacy of the field-theoretical description of quarks by bispinor fields $q(x)$. However, these fields are clearly not expected to satisfy the classically motivated constraint $(\not{p} - m)q(x) = 0$. In other words, it is solely the adequacy of the standard concept of quark mass (and its connection to momentum) that is questioned here.

If the masses of the individual colored quarks are associated with the rotationally non-covariant canonical momenta, there is obviously no rationale for the description of quark behavior 'inside' hadrons with the help of standard fermion propagators. On the other hand, the replacement of physical momenta by the canonical ones does not affect the GMOR extraction of quark masses. As explained earlier at some length, the GMOR extraction is based on the properties of the fermion mass term only. The concept of quark momentum is not used.

More specifically, the conjecture of Eq. (7) suggests how the lepton Hamiltonian, which in the standard language is composed of two additive terms: the momentum term $H_l(p_1, p_2, p_3) = \boldsymbol{\alpha} \cdot \mathbf{p}$ and the mass term $H_l^0(m) = \beta m$, should be transformed into the Hamiltonians for three colored quarks. For example, for quark #1 it should be built as a sum of some canonical momentum term $H_{q1}(p_1, x_2, x_3)$ and some mass term $H_{q1}^0(m)$. Since the mass term used by GMOR refers to a sum over colors, the corresponding mass term of the phase-space language is expected to be built as a sum over colors as well, i.e. $H_{q1}^0(m) + H_{q2}^0(m) + H_{q3}^0(m)$. It is this sum that should exhibit all the necessary invariance properties. The actual construction and the explicit forms of all the relevant terms will be discussed in the last Section.

4. Clifford algebra of nonrelativistic phase space

The quantum concept of spin may be arrived at from the classical level by the linearization of the 3D space invariant $\mathbf{p}^2$. The most $\mathbf{x} \leftrightarrow \mathbf{p}$ symmetric extension of this procedure to the case of the 6D classical phase space leads us to consider the linearization of the expression $\mathbf{p}^2 + \mathbf{x}^2$ and, consequently, to Clifford algebra $Cl_{6,0}$ [23, 24, 25, 27]. Since position and momentum do not commute in the quantum case, the relevant Clifford algebra calculations have to be supplemented with the subsidiary condition $[x_j, p_k] = i\delta_{jk}$ (we choose units such that $\hbar = 1$). Then, with $\mathbf{A}$ and $\mathbf{B}$ being six mutually anticommuting elements of $Cl_{6,0}$, one derives

$$(\mathbf{A} \cdot \mathbf{p} + \mathbf{B} \cdot \mathbf{x})(\mathbf{A} \cdot \mathbf{p} + \mathbf{B} \cdot \mathbf{x}) = \mathbf{p}^2 + \mathbf{x}^2 + R, \tag{8}$$

where the R term (classically absent) appears because position and its conjugated momentum do not commute. The elements $\mathbf{A}$ and $\mathbf{B}$ may be represented by eight-dimensional matrices:

$$\begin{aligned} A_k &= \sigma_k \otimes \sigma_0 \otimes \sigma_1, \\ B_j &= \sigma_0 \otimes \sigma_j \otimes \sigma_2. \end{aligned} \tag{9}$$

One calculates that

$$R = -\frac{i}{2} \sum_k [A_k, B_k] = \sum_k \sigma_k \otimes \sigma_k \otimes \sigma_3 \equiv \sum_k R_k. \tag{10}$$

The 7-th anticommuting element of our Clifford algebra is:

$$B = iA_1A_2A_3B_1B_2B_3 = \sigma_0 \otimes \sigma_0 \otimes \sigma_3. \tag{11}$$

4.1. Gell-Mann–Nishijima formula

From B and R we construct now the elements I_3 and Y which, as it will turn out, have simple physical interpretation:

$$I_3 = \frac{1}{2}B, \qquad Y = \frac{1}{3}RB. \tag{12}$$

It is straightforward to check that I_3 and Y are invariant under 3D rotations and reflections (generated respectively by S_k and P, which are defined by their actions on both $\mathbf{A}$ and $\mathbf{B}$):

$$\begin{aligned} [S_k, I_3] = [S_k, Y] &= 0, \\ [P, I_3] = [P, Y] &= 0. \end{aligned} \tag{13}$$

Since in addition the commutator $[I_3, Y]$ vanishes, and therefore the eigenvalues of I_3 and Y can be simultaneously specified, these elements constitute candidates for two internal quantum numbers. In order to proceed, we rescale the expression in Eq. (8) and take the lowest (vacuum) eigenvalue of $\mathbf{p}^2 + \mathbf{x}^2$:

$$Q \equiv \frac{1}{6}\left[(\mathbf{p}^2 + \mathbf{x}^2)_{vac} + R\right] B = I_3 + \frac{Y}{2}. \tag{14}$$

With the eigenvalues of I_3 and Y being

$$\begin{aligned} I_3 &= \pm\frac{1}{2} \qquad \text{(four times)}, \\ Y &= \sum_k Y_k \equiv \frac{1}{3}\sum_k R_k B = -1, +\frac{1}{3}, +\frac{1}{3}, +\frac{1}{3} \quad \text{(twice)}, \end{aligned} \tag{15}$$

one naturally identifies Eq. (14) with the Gell-Mann–Nishijima formula for the electric charges of the eight fermions from a single generation in the Standard Model. Elements I_3 and Y denote, respectively, (weak) isospin and (weak) hypercharge. For the Y_k's we introduce the name 'partial hypercharges'.

4.2. The Harari-Shupe model

The three partial hypercharges commute among themselves, i.e. $[Y_k, Y_m] = 0$ $(k, m = 1, 2, 3)$. Consequently, their eigenvalues may be simultaneously specified. It is then instructive to analyse in detail the way in which the four eigenvalues of hypercharge Y are constructed out of the eignenvalues of Y_k's. This is shown in Table 1.

Table 1. The structure of weak hypercharge in terms of its component partial hypercharges

Y_1	Y_2	Y_3	Y
$-\frac{1}{3}$	$+\frac{1}{3}$	$+\frac{1}{3}$	$+\frac{1}{3}$
$+\frac{1}{3}$	$-\frac{1}{3}$	$+\frac{1}{3}$	$+\frac{1}{3}$
$+\frac{1}{3}$	$+\frac{1}{3}$	$-\frac{1}{3}$	$+\frac{1}{3}$
$-\frac{1}{3}$	$-\frac{1}{3}$	$-\frac{1}{3}$	-1

It appears that the pattern of Table 1 is in one-to-one correspondence with the charge structure of the celebrated Harari-Shupe (HS) rishon model [28, 29], a model in which leptons

Table 2. Rishon structure of the $I_3 = +1/2$ members of a single SM generation in the Harari-Shupe model

ν_e	u_R	u_G	u_B	e^+	$\bar{d}_R$	$\bar{d}_G$	$\bar{d}_B$
VVV	TTV	TVT	VTT	TTT	VVT	VTV	TVV

and quarks are built of fermionic subcomponents (i.e. 'rishons' V and T of charges $Q_V = 0$ and $Q_T = +1/3$) as specified in Table 2 [23, 24].

The correspondence in question is:

$$Y_k = -1/3 \leftrightarrow V, \qquad Y_k = +1/3 \leftrightarrow T, \tag{16}$$

where k labels the position in the ordering of rishons in Table 2 (e.g. $[(Y_1, Y_2, Y_3) = (-1/3, +1/3, +1/3)] \leftrightarrow VTT$). Thus, rishons may be roughly understood as partial hypercharges. It has to be stressed that the correspondence between the HS model and the phase-space scheme refers to the charge structure *only*. Consequently, *the shortcomings of the HS model* (and there are many of them [7]) *are all absent in the phase-space scheme*. The shortcomings of the rishon model do not appear in the phase-space scheme because the partial hypercharges are algebraic components of the hypercharge operator only and do not reside on any subparticles. In other words, in the phase-space scheme we have symmetry-related sectors without symmetry-'explaining' subparticles ('preons'). Therefore, contrary to naive understanding, the phase-space approach *is not a preon model*. It does not provide a theoretical justification for the introduction of preons. Its success in the explanation of the working part of the HS model constitutes an argument *against* preons.

4.3. Even subalgebra of Clifford alegbra

The even subalgebra of our Clifford algebra contains 32 elements, which may be divided into two sets of 16 elements each, acting in sectors with $I_3 = \pm 1/2$ [25, 26]. Within each such sector the relevant 16-element set consists of a unit element and 15 generators of $SO(6)$. The set of 15 generators decomposes further into $SU(3)$ multiplets as $15 = 1 \oplus 8 \oplus 3 \oplus 3^*$. When looked at from the point of view of the ordinary 3D rotations, the singlet is a scalar, while the octet contains three familiar $SO(3)$ generators. Multiplets 3 and 3^* generate 'genuine' 6D rotations in phase-space [8], and lead to lepton-quark transformations. For example, a rotation by $\pi/2$ induced by the generator $F_{+2} \equiv -\frac{i}{4}\epsilon_{2kl}[A_k, B_l]$ permutes quark #2 with a lepton while leaving quarks #1,3 unaffected [23, 24]. The corresponding transformation in phase space is

$$(p_1, p_2, p_3) \rightarrow (-x_1, p_2, x_3). \tag{17}$$

Thus, Clifford algebra provides a highly non-trivial connection between the internal quantum numbers (i.e. weak isospin and hypercharge, Eq. (15)) and the phase-space (and mass) heuristic discussed in connection with Eq. (7). From our point of view, therefore, the three colored quarks should be regarded as leptons with canonical momenta rotated in phase space in three possible ways (cyclic counterparts of the r.h.s. of Eq. (17)).

[7] The unwanted properties of the original HS model include: a problem with rishon statistics, the lack of explanation why TTT states are free but TVV are confined, various predictions (of unobserved fundamental spin-3/2 fermions, of unobserved $TT\bar{T}$ states, of the violation of baryon number,...) etc.

[8] By 'genuine' 6D rotations we mean in particular rotations that change some momenta coordinates into position coordinates and vice versa.

4.4. Odd part of Clifford algebra

The odd part of our Clifford algebra consists of 16+16=32 elements linking the sectors of opposite values of I_3. Each 16-element subset decomposes in $SU(3)$ as $16 = 1 \oplus 3 \oplus 3 \oplus 3^* \oplus 6$ (or its conjugate). The $SU(3)$ singlet (and an $SO(3)$ scalar) works in the lepton subspace $Y = -1$ only. Its explicit form is

$$G_0 \propto (1 - \sum_k \sigma_k \otimes \sigma_k) \otimes (\sigma_1 + i\sigma_2). \tag{18}$$

Element G_0 and its hermitian conjugate (which belongs to the other 16-element subset) constitute the only odd elements of Clifford algebra which (1) have $Y = -1$ and (2) are $SO(3)$-scalars. Consequently, it is only from them that a candidate for an algebraic counterpart of the lepton mass term may be formed.

Application of the previously discussed finite rotation generated by F_{+2} transforms G_0 into

$$G_{22} \propto (1 + \sum_k \sigma_k \otimes \sigma_k - 2\sigma_2 \otimes \sigma_2) \otimes (\sigma_1 + i\sigma_2). \tag{19}$$

It may be checked that G_{22} and its hermitian conjugate have $Y = +1/3$ and thus they both operate in the quark subspace. Furthermore, as expected, G_{22} and $G_{22}^\dagger$ are not rotationally invariant. It is only from them that a (rotationally still noninvariant) candidate for an algebraic counterpart of the mass term may be formed for quark of color #2. Elements G_{22} and its two analogues (G_{11} and G_{33}) belong to the $SU(3)$ sextet $G_{\{kl\}}$. This sextet decomposes in $SO(3)$ into a scalar (the trace $\sum_k G_{kk}$) and a quintet. Thus, as originally anticipated, the summation over the color degree of freedom would lead to the restoration of rotational invariance for the algebraic counterpart of the effective quark mass term.

Although the basic element of the mechanism of quark conspiracy seems to work, one cannot directly use G_0 and G_{kk} (together with their hermitean conjugates) as the algebraic counterparts of lepton and quark mass. Specifically, it turns out that the 64-element Clifford algebra of nonrelativistic phase space is too small to reproduce the standard algebra leading to Dirac equation for a lepton [9]. In the next Section we will enlarge the algebra and show how the problem could be solved.

5. Compositeness and additivity

If the phase-space approach is to constitute a successful idealization of certain aspects of nature, there should be a way to apply it to the description of the emergence and spatial behavior of colorless composite states (mesons and baryons). In the standard field-theoretical language this behavior is described via hadron propagation in the underlying background space of Democritus. On the other hand, within our Aristotelian philosophy of emergent phase space, the actual relation between hadrons (or leptons) and the classical arena of events is inverted: the macroscopic space is expected to be well defined only at the level of individually observable (colorless) particles. If space emerges in full starting only at the hadronic level (recall the issue of the frozen degree of freedom discussed in Section 2), the idea of space emergence should be regarded as constituting an unorthodox point of view on the problem of confinement. In other words, the emergence of the 'background' space (which space constitutes an input into QCD) and the problem of confinement (the emergence of composite states) might be considered as two faces of the same issue. Unfortunately, we have no detailed idea on how to construct the classical space from the quantum layer of spins, internal quantum numbers, etc.; we have no idea what are the rules that govern the emergence. Still, some expectations may be presented. These expectations are based on the concept of additivity which constitutes a basic element in the determination of the properties of composite systems of particles. Indeed, additivity is

[9] The hypercharge projector in G_0 conflicts with the algebraic counterparts of momentum.

Journal of Physics: Conference Series **626** (2015) 012022 doi:10.1088/1742-6596/626/1/012022

applied both to their quantum numbers (via the additivity of spins, flavors, etc.) and — in the case of individually observable particles — to their physical momenta. For example, the momentum of a system composed of leptons A and B is given as $\mathbf{p} = \mathbf{p}^A + \mathbf{p}^B$. The same formula is used for the determination of the momentum of a hadronic resonance formed when two hadrons collide. Such additivity prescriptions must constitute important ingredients in the way our classical world emerges out of its quantum components. Although they seem trivial and are generally not given much thought, in the case of quarks we must be more careful.

5.1. *Additivity of quark canonical momenta*

The problem is with the transition from the hadron to the quark level. It seems that for quarks a natural extension of the principle of the additivity of physical momenta consists in the additivity of the canonical momenta [10]. Now, the canonical momenta of red, green, and blue quarks are (in order to distingush the positions of green and blue quarks, we renamed x_1 in Eq. (17) into y_1, with similar cyclic changes in other places):

$$\begin{aligned} \mathbf{p}^R &= (p_1, x_2, -y_3), \\ \mathbf{p}^G &= (-y_1, p_2, x_3), \\ \mathbf{p}^B &= (x_1, -y_2, p_3). \end{aligned} \tag{20}$$

Addition of the canonical momenta of three quarks of different colors (the baryon case) leads to a translationally invariant expression:

$$\mathbf{p}^R + \mathbf{p}^G + \mathbf{p}^B = (p_1, p_2, p_3, x_1 - y_1, x_2 - y_2, x_3 - y_3). \tag{21}$$

If different quarks conspire, the expression on the r.h.s. above can be made rotationally covariant as well. The r.h.s. of Eq. (21) is then expected to contain the total physical momentum $\mathbf{p}$ of the resulting baryon. This weird conspiration constitutes a glimpse into the anticipated physics of space emergence which the phase-space approach suggests. The r.h.s. of Eq. (21) contains also the expression $\mathbf{\Delta} = \mathbf{x} - \mathbf{y}$ formed out of selected components of interquark displacements. Since $\mathbf{p}$ is a vector, $\mathbf{\Delta}$ is expected to be a vector as well. If quarks were located in the 3D background space, there would be two independent displacements (spatial degrees of freedom) in each spatial direction. In our case, however, the constraint that $\mathbf{\Delta}$ is built from three particular perpendicular vectorial components leads to an unexpected conclusion. To see what happens, one has to analyse the situation in the position representation. For example, one has to think of the blue quark as located at (x_1, y_2, z_3), with analogous expressions for the other two quarks. Consider the three components of $\mathbf{\Delta}$ as fixed perpendicular vectors (straight 'strings') connecting the relevant quarks. It turns then out that together they form an 'impossible triangle'. Therefore, the whole construction is noncontradictory only if this triangle converges to a point: each one of the three components of $\mathbf{\Delta}$ must vanish. Thus, one of the two usually expected internal spatial degrees of freedom (for each of the three spatial directions) is necessarily frozen. We conclude that there appears an interesting analogy between our additivity-based reasoning and the existence of a phenomenologically established spatial constraint 'inside' the excited baryons.

The above argument is based on an attempt to extend the applicability of the macroscopic classical concept of space into the 'interior' of baryons, an idea we already argued to be wanting. The bizzare character of the obtained picture is assigned to the inapplicability of this extension [11]. In our view, therefore, the background space can be used 'inside' hadrons

[10] Additivity of physical momenta of individual colored quarks cannot be experimentally tested, as all probes used for our studies of quark behavior (photons, weak bosons) are necessarily color-blind.

[11] We do not consider the seemingly crazy character of our construction to constitute sufficient grounds for its rejection. After all, our assumptions (extension to phase space, connection with internal quantum numbers via its Clifford algebra, extension of the additivity principle) look extremely solid.

solely as a rough appproximation. It cannot be used as a visualisation of what 'really' happens there. What is obviously badly needed is a more specific proposal for the mechanism of space emergence. Unfortunately, such a proposal is currently lacking (our arguments here, based on the additivity of canonical momenta, may be considered as hints only, indicative of the direction in which one should actually proceed). Yet, given the conceptual problems with the notion of quark mass and the phenomenological problems with the spectroscopy of excited baryons, we believe that important information on the details of this mechanism is hidden in the spectroscopy and properties of hadrons, i.e. in the emergence of hadrons out of quarks. In other words we think that one can learn about space emergence from a deeper understanding of the quark/hadron transition. After translating this idea into the standard language it means that the quark-confining interactions should be understood in geometro-algebraic (or pregeometric) terms. Embarking on such an ambitious agenda is clearly beyond our goal here.

In order to apply our additivity proposal to the discussion of mesons, we must interpret charge conjugation (i.e. the existence of antiparticles) in phase space terms. The more familiar operations of space inversion P and time reversal T are obviously represented as:

$$\begin{aligned} P: &\quad (\mathbf{p},\mathbf{x}) \;\rightarrow\; (-\mathbf{p},-\mathbf{x}), \\ T: &\quad (\mathbf{p},\mathbf{x}) \;\rightarrow\; (-\mathbf{p},+\mathbf{x}). \end{aligned} \tag{22}$$

If one looks at the quantum case, the invariance of $[x_m, p_n] = i\delta_{mn}$ requires that time reversal be accompanied by complex conjugation $i \rightarrow -i$. The product of P and T is:

$$PT: \quad (\mathbf{p},\mathbf{x}) \rightarrow (\mathbf{p},-\mathbf{x}), \qquad \text{and} \qquad i \rightarrow -i. \tag{23}$$

Assuming that $CPT = 1$, the operation (23) is then identified with charge conjugation C. Indeed, for an ordinary particle the transition to the corresponding antiparticle is achieved by keeping the particle's momentum unchanged but reversing its charge Q. [12]

Adding the canonical momenta of a red quark ($\mathbf{p}^R = (p_1^R, x_2^R, -y_3^R)$) and its antiquark gives then

$$\mathbf{p}^R + \mathbf{p}^{\bar{R}} = (p_1^R + p_1^{\bar{R}}, x_2^R - x_2^{\bar{R}}, -y_3^R + y_3^{\bar{R}}). \tag{24}$$

Thus, translationally invariant expressions are obtained also for quark-antiquark systems. [13] In short, the requirement of translational invariance suggests the acceptability of qqq, $q\bar{q}$, and $\bar{q}\bar{q}\bar{q}$ systems as individually observable particles, and the nonacceptability of q, qq, $qq\bar{q}$, etc. This conclusion, which actually corroborates our assumption on the additivity of canonical momenta, is very similar to the analogous one obtained via the color-singlet argument of the standard quark model. Note that the requirement of translational invariance explains why in the HS model the combination TVV is confined but TTT or $TVV \otimes VTV \otimes VVT$ are not.

5.2. Additivity of colored quark Hamiltonians

We proceed now to the issue of the additivity of quark Hamiltonians. We postpone the discussion of the mass terms for the moment and start from the momentum part of the lepton Hamiltonian, i.e. from the (nonrelativistic) expression

$$H^L = \mathbf{A} \cdot \mathbf{p}. \tag{25}$$

[12] Since all our theories provide idealized descriptions of certain aspects of reality only and must not be identified with nature, it should not be considered strange (or unacceptable) that antiparticles may be recognized and described in a language different from that of the relativistic field theory.

[13] Rotational covariance requires taking into account green and blue quarks as well.

Genuine 6D rotations of $\mathbf{A}$ and $\mathbf{p}$ (see e.g. Eq.(17)), when combined with appropriate sign changes in the matrix part, transform (25) into the following Hamiltonians for the three colored quarks:

$$\begin{aligned} H^R &= A_1p_1 + B_2x_2 - B_3y_3, \\ H^G &= -B_1y_1 + A_2p_2 + B_3x_3, \\ H^B &= B_1x_1 - B_2y_2 + A_3p_3. \end{aligned} \tag{26}$$

While ordinary 6D rotations necessarily yield plus signs in the above formulas, we have used here the possibility of redefining some of the elements A_k, B_k in a way that does not affect the (even in A_k, B_k) operators of weak isospin and hypercharge. [14] Summation over quark colors gives the total quark Hamiltonian:

$$H^{Q-total} = A_1p_1 + A_2p_2 + A_3p_3 + B_1(x_1 - y_1) + B_2(x_2 - y_2) + B_3(x_3 - y_3). \tag{27}$$

If quarks of different colors are taken to be located at the same point, one obtains

$$H^{Q-total} = \mathbf{A} \cdot \mathbf{p}. \tag{28}$$

This resembles the momentum part of the (nonrelativistic) lepton Hamiltonian (25) completely. The difference is that here a *sum over colors* is taken and contributions from quarks of different colors conspire to yield the momentum part of the total Hamiltonian *as if of a lepton*.

5.3. Special relativity

We are now in a position to discuss the issue of how to include mass terms and special relativity. For leptons this means that we want to extend the nonrelativistic algebraic equality

$$(\mathbf{A} \cdot \mathbf{p})(\mathbf{A} \cdot \mathbf{p}) = \mathbf{p}^2 \tag{29}$$

to a relativistic form, i.e. to an expression in which the r.h.s. above would be replaced with $\mathbf{p}^2 + m^2$. In its essence, we face the Dirac's problem. Indeed, as remarked earlier in Section 4.4, our tentative identification of a Hermitean combination of the $Y = -1$ elements G_0 and $G_0^\dagger$ with the algebraic counterpart of the lepton mass term suffers from an incompatibility similar to the one encountered by Dirac. Specifically, there is a problem with the product of elements A_k and G_0 ($G_0^\dagger$): the mixed mp_k terms do not cancel.

The solution of the Dirac's problem consisted in the extension of the nonrelativistic Pauli algebra via a tensor product construction. We propose here a similar construction that generalizes our nonrelativistic Clifford algebra to an algebra that admits a relativistic generalization of (29). Specifically, we move the $Y = -1$ projector $y_{-1} \equiv \frac{1}{4}(1_4 - \sum_k \sigma_k \otimes \sigma_k)$ in Eq. (18) to a new tensor factor so that

$$G_0 - G_0^\dagger \propto M_0 \equiv y_{-1} \otimes \sigma_2 \rightarrow G_0' - G_0'^\dagger \propto M_0' \equiv 1_4 \otimes \sigma_2 \otimes y_{-1}, \tag{30}$$

and accept the following generalization of $\mathbf{A} \cdot \mathbf{p}$ in the lepton sector:

$$p_kA_k \rightarrow p_kA_k' \equiv p_kA_k \otimes y_{-1}. \tag{31}$$

Since the $Y = -1$ projector in M_0' and $\sigma_k \otimes \sigma_0$ in A_k' work now in different tensor factors, therefore upon squaring

$$\mathbf{A}' \cdot \mathbf{p} + M_0' m, \tag{32}$$

[14] For example, for the red quark the redefinition consists in $A_3, B_3 \rightarrow -A_3, -B_3$ while keeping the remaining A_k and B_k unchanged.

one readily reproduces the relativistic form $\mathbf{p}^2 + m^2$.

In order to extend our discussion to quarks, we consider a fully symmetric counterpart of (32) with the $Y = -1$ projector y_{-1} in the rightmost factor replaced by 1_4. We separate then the color singlet and triplet ($Y = -1$ and $Y = +1/3$) subspaces via

$$\begin{aligned} & p_k A_k \otimes 1_4 + m 1_4 \otimes \sigma_2 \otimes 1_4 \\ = & \ (p_k A_k + m 1_4 \otimes \sigma_2) \otimes y_{-1} \\ + & \ (p_k A_k + m 1_4 \otimes \sigma_2) \otimes y_{+\frac{1}{3}} \end{aligned} \tag{33}$$

where

$$y_{+\frac{1}{3}} = \frac{1}{4}(3 \cdot 1_4 + \sum_m \sigma_m \otimes \sigma_m) \tag{34}$$

is the projector onto the color triplet subspace. Finally, we decompose the second term on the r.h.s. of Eq. (33) into contributions from quarks of definite color:

$$\begin{aligned} & p_k A_k \otimes y_{+\frac{1}{3}} + m 1_4 \otimes \sigma_2 \otimes y_{+\frac{1}{3}} \\ = & \ (p_1 A_1 + x_2 B_2 - x_3 B_3) \otimes y_{+\frac{1}{3}} + m 1_4 \otimes \sigma_2 \otimes y_{+\frac{1}{3},1} \\ + & \ (-x_1 B_1 + p_2 A_2 + x_3 B_3) \otimes y_{+\frac{1}{3}} + m 1_4 \otimes \sigma_2 \otimes y_{+\frac{1}{3},2} \\ + & \ (+x_1 B_1 - x_2 B_2 + p_3 A_3) \otimes y_{+\frac{1}{3}} + m 1_4 \otimes \sigma_2 \otimes y_{+\frac{1}{3},3}, \end{aligned} \tag{35}$$

where $y_{+\frac{1}{3},k}$ are projectors onto the subspaces of red, green and blue quarks (see also Eq. (19)):

$$y_{+\frac{1}{3},k} = \frac{1}{4}(1_4 + \sum_m \sigma_m \otimes \sigma_m - 2\sigma_k \otimes \sigma_k), \tag{36}$$

that satisfy $\sum_k y_{+\frac{1}{3},k} = y_{+\frac{1}{3}}$. Eq. (35) constitutes the proposed extension beyond the nonrelativistic case discussed in Eqs. (27, 28). For the canonical momentum terms the rotationally noninvariant distinction between quark colors resides only in the first factor of the tensor product, while for the mass term this distinction resides only in the last factor. The total quark Hamiltonian looks like a Hamiltonian for a free relativistic particle used in pre-QCD current quark mass extraction procedures. According to our proposal the old (pre-QCD) form should be viewed as involving an explicit summation over colors for both the canonical momentum terms and the mass terms. As a result, translational, rotational and relativistic invariances are restored and the standard relativistically invariant couplings to external colorless objects may be introduced at the level of observable combinations of quark contributions. It is only at this colorless level, the level at which special relativity has to be recovered, that the concept of spacetime point becomes meaningful and the standard gauge interactions may be introduced. Since the concept of spacetime point constitutes a prerequisite for any talk of local gauge structure, it is the issue of spacetime emergence from the quantum layer that has to be successfully addressed before any attempt to introduce interactions is made.

6. Concluding remarks

In various approaches to quantum gravity, macroscopic spacetime is often imagined as emerging from the quantum layer at the minuscule Planck scale. The nonlocality of quantum physics suggests, however, that this emergence occurs simultaneously at all distance scales, from the Planck scale to the cosmological distances. The quantum-mechanically-related momentum scales span a correspondingly enormous range. Since this range includes in particular the masses of all elementary particles, their spectrum could provide us with important hints on the actual mechanism of spacetime emergence. This relevance of the problem of mass to the idea of space quantisation was supported with a brief review of fundamental issues related to the concept of particle masses, both at the hadron and quark levels. The encountered problems suggest that important information on the idea of spacetime emergence could be accessed via the studies of quark/hadron transition.

An attempt to address such ideas, based on the Clifford algebra of nonrelativistic phase space, was briefly reviewed. It was recalled that the proposed scheme reproduces important features of the observed spectrum of elementary particles, including the appearance of internal quantum numbers of weak isospin and hypercharge that characterise leptons and quarks, and providing an explanation of the Harari-Shupe model. A definite asset of this explanation is that it does not assume the heavily criticized existence of fermionic 'preons' inside fundamental fermions of the Standard Model. It was also pointed out that the scheme offers an unorthodox, alternative view on the problems of quark and hadron masses and the issue of quark confinement. In particular, possible connections of phase-space ideas to quark and hadron phenomenology were indicated. A way to introduce relativistic covariance at the level of individually observable particles (leptons and hadrons) was also proposed and discussed.

References

[1] Zimmerman E J 1962 *Amer. J. Phys.* **30** 97
[2] Finkelstein D 1969 *Phys. Rev.* **184** 1261
[3] Penrose R 1971 Angular momentum: an approach to combinatorial space-time, in *Quantum Theory and beyond* ed T Bastin (Cambridge Univ. Press) p 151
[4] Koide Y 1983 *Phys. Lett.* B **120** 161
[5] Salecker H and Wigner E P 1958 *Phys. Rev.* **109** 571
[6] Capstick S and Roberts W 2000 *Progr. Theor. Part. Nucl. Phys.* **45** S241
[7] Edwards R G, Dudek J J, Richards D G and Wallace S J 2011 *Phys. Rev.* D **84** 076508
[8] Heisenberg W 1958 *Physics and Philosophy: The Revolution in Modern Science* (New York: Harper and Row) p 200
[9] Törnqvist N A and Żenczykowski P 1984 *Phys. Rev.* D **29** 2139
[10] Żenczykowski P 1986 *Ann. Phys. (NY)* **169** 453
[11] Glozman L Y and Riska D O 1996 *Phys. Rep.* **268** 263
[12] Gell-Mann M, Oakes R J and Renner B 1968 *Phys. Rev. Lett.* **175** 2195
[13] Beringer J *et al.* (Particle Data Group) 2012 *Phys. Rev.* D **86** 010001
[14] Żenczykowski P and Lach J 1995 *Int. J. Mod. Phys.* A **10** 3817
[15] Żenczykowski P 2006 *Phys. Rev.* D **73** 076005
[16] Norsen T 2009 *Found. Phys.* **39** 273
[17] Horzela A and Kapuścik E 2003 *Electromagnetic Phenomena* V **3** 63
[18] Heisenberg W 1979 in *Philosophical problems of quantum physics* (Woodbridge, CT: Ox Bow Press)
[19] Barbour J 2003 *Lect. Notes. Phys.* **633** 15
[20] Zachos C 2002 *Int. J. Mod. Phys.* A **17** 297
[21] Born M 1938 *Proc. Royal Soc. London* A**165** 291
[22] Born M 1949 *Rev. Mod. Phys.* **21** 463
[23] Żenczykowski P 2007 *Acta Phys. Polon.* B **38** 2053
[24] Żenczykowski P 2008 *Phys. Lett.* B **660** 567
[25] Żenczykowski P 2009 *J. Phys.* A **42** 045204
[26] Żenczykowski P 2010 *Int. J. Theor. Phys.* **49** 2246
[27] Żenczykowski P 2014 *Elementary Particles and Emergent Phase Space* (Singapore: World Scientific)
[28] Harari H 1979 *Phys. Lett.* B **86** 83
[29] Shupe M 1979 *Phys. Lett.* B **86** 87

Quantum criticality at the origin of life

Gábor Vattay[1], Dennis Salahub[2], István Csabai[1], Ali Nassimi[2,3] and Stuart A Kaufmann[2,4]

[1] Department of Physics of Complex Systems, Eötvös University, 1117 Budapest, Pázmány P. s. 1/A, Hungary

[2]Department of Chemistry, CMS - Centre for Molecular Simulation and IQST - Institute for Quantum Science and Technology, University of Calgary, 2500 University Drive NW, Calgary, Alberta, Canada T2N 1N4

[3] Chemical Physics Theory Group, Department of Chemistry, University of Toronto, 80 Saint George Street, Toronto, ON M5S 3H6, Canada

[4] Institute for Systems Biology, Seattle, WA 98109, USA

E-mail: vattay@elte.hu

Abstract. Why life persists at the edge of chaos is a question at the very heart of evolution. Here we show that molecules taking part in biochemical processes from small molecules to proteins are critical quantum mechanically. Electronic Hamiltonians of biomolecules are tuned exactly to the critical point of the metal-insulator transition separating the Anderson localized insulator phase from the conducting disordered metal phase. Using tools from Random Matrix Theory we confirm that the energy level statistics of these biomolecules show the universal transitional distribution of the metal-insulator critical point and the wave functions are multifractals in accordance with the theory of Anderson transitions. The findings point to the existence of a universal mechanism of charge transport in living matter. The revealed bio-conductor material is neither a metal nor an insulator but a new quantum critical material which can exist only in highly evolved systems and has unique material properties.

Advances in the theory of complex systems over the last quarter century reinforced that living systems exist at the edge of chaos[1, 2, 3, 4] and order, poised at criticality[5]. Finding the detailed mechanism behind this apparent self-organized criticality[6, 7] is still a tantalizing problem. One of the fascinating aspects of life is the highly organized molecular machinery taking care of myriads of complex processes such as DNA replication, protein synthesis, cell division and metabolism, to mention only a few. Electric forces animating the parts require a perpetual and precise motion of charges throughout the system for perfect execution of biochemical tasks. In this paper we show that practically all biomolecules, from small signalling molecules to proteins taking part in biochemical electronic processes, belong to a fundamentally new class of conducting material. This is a disordered conductor where the strength of the disorder is tuned exactly to the metal-insulator transition point and it is consequently in a permanent critical quantum state.

Our initial perspective is that of condensed-matter physics. The unique properties of the critical quantum state at the localization-delocalization transition point have been described first in the the well known Anderson[8] model, which represents the current paradigm for understanding conduction in condensed matter. In the Anderson Hamiltonian $H = \sum_j \epsilon_i a_i^+ a_i - \sum_{<ij>} a_j^+ a_i$ on a 3D lattice with random uniformly distributed on-site energies $\epsilon_i \in [-W, W]$, the critical level of disorder[9] is $W_c = 16 \pm 0.5$. For $W < W_c$ the system is a disordered metal with extended states and for $W > W_c$ the states are localized and the system is an insulator. It has been shown[10] that in the exact critical point of the Anderson transition ($W = W_c$) the electron eigenfunctions are extended, but strongly inhomogeneous multifractals[11]. Similar metal-insulator transitions (MIT) exist in a wide range of physical systems at various dimensionalities[12] including the 1D Harper-Hofstadter model[13] and quantum-Hall-type transitions.

For example, the 1D Harper model[14] describes the energy spectrum of an electron in a 2D lattice, whose graphic representation is widely known as the Hofstadter butterfly[13]. In the related Aubry-André Hamiltonian[15] $H = \sum_j \lambda \cos(2\pi\sigma i)a_i^+ a_i - \sum_{<ij>} a_j^+ a_i$, where $\sigma = (\sqrt{5}-1)/2)$ is the Golden Mean Ratio, the critical value separating localized and extended phases is $\lambda_c = 2$. The wave functions are also multifractals[16] there. Other examples include[12] quantum-Hall-type transitions in disordered conductors and superconductors in strong magnetic fields, the spin quantum Hall effect, the thermal quantum Hall effect, Dirac fermions in random vector potentials and Bethe lattices.

MIT-like transitions exist also in low-dimensional quantum chaos. Quantum counterparts of strongly chaotic systems[17] share the properties of delocalized systems, while integrable systems[18] have localized wave functions in quantum numbers corresponding to conserved quantities. Pseudointegrable systems[19] lie at the border of chaos and integrability, where classical trajectories diverge only slowly (with zero Lyapunov exponent) but their dynamics is complex and the periodic orbits proliferate in their phasespace exponentially. Quantized pseudointegrable systems also show all the key features of critical systems, including multifractal wavefunctions[20].

Criticality can also be observed in the energy spectrum of systems at the transition point. Random Matrix Theory[21, 22, 23, 24] (RMT) is the main tool for the characterization of the universal statistical properties of Hamiltonians of complex systems. The distance between consecutive energy levels fluctuates in the spectrum. The raw distance between levels $\sigma_n = E_{n+1} - E_n$ can be normalized using the average separation of levels $\Delta(E)$ at a given energy window around E. The ratio $s_n = \sigma_n/\Delta(E_n)$ is called the level spacing. Random matrix theory has certain predictions for the form of the distribution $P(s)$ of level spacings.

It has been established[25] that in the 3D Anderson model the localized, delocalized and the critical states each have a distinct level spacing distribution. These three distributions are believed to be universal, i.e. independent of the microscopic details of the disordered system. In the delocalized metallic phase ($W < W_c$) the distribution coincides with the level statistics of the Gaussian Orthogonal Ensemble (GOE), which is the ensemble of real symmetric random matrices with identically distributed Gaussian elements. The level spacing distribution is the Wigner surmise[21]

$$P_W(s) = \frac{\pi s}{2} \exp\left(-\frac{\pi s^2}{4}\right).$$

In the localized insulating phase ($W > W_c$) the energy levels form a random Poisson process and the level spacing distribution is exponential

$$P_P(s) = \exp(-s).$$

It has been shown[17, 12, 26] that not only in the 3D Anderson model, but in all other examples of Anderson-like transitions, the energy level statistics in the delocalized phase is universal and corresponds to the proper Gaussian Random Matrix Ensemble reflecting the symmetries of the system, while in the localized phase it is always random Poissonian[18, 12, 26].

At the transitional point ($W = W_c$) a third kind of intermediate spectral statistics $P_T(s)$ exists, which is the hallmark of the critical state[25]. In solid state models theoretical

Journal of Physics: Conference Series **626** (2015) 012023 doi:10.1088/1742-6596/626/1/012023

arguments[27, 28] and numerical studies[29, 30] suggest the general form

$$P_T(s) = c_1 s \exp\left(-c_2 s^{1+\gamma}\right),$$

where c_1 and c_2 are γ dependent normalization constants. In the case of the 3D Anderson model[30] $\gamma \approx 0.2$. In the Harper model[26] and in other pseudointegrable models[31] showing critical quantum chaos a numerical value of $\gamma \approx 0$ has been found[26], which supports the semi-Poissonian distribution

$$P_T(s) = P_{SP}(s) = 4s \exp(-2s).$$

The semi-Poissonian distribution has also been predicted from a short range plasma model[32] of energy levels introduced in RMT.

While in physical systems the critical state can be reached only upon a careful tuning of the strength of the disorder, in the following we show that certain biomolecules are precisely at the critical state without any external tuning.

The exact numerical solution of the Schrödinger equation for the electronic states of large molecules such as proteins is a prohibitive task. Various approximations have been developed, which reduce the problem to a one-electron problem in the effective field of the remaining electrons. Wave functions of molecules are usually written in the form of Linear Combinations of Atomic Orbitals (LCAO) $\phi_i = \sum_r C_{ir}\chi_r$ where ϕ_i is a Molecular Orbital (MO) represented as the sum of atomic orbital (AO) contributions χ_r. The one-electron eigenenergies and eigenvectors then can be determined from the generalized eigenvalue equation $HC = ESC$, where $S_{rs} = \langle \chi_r \mid \chi_s \rangle$ and $H_{rs} = \langle \chi_r \mid \hat{H}_{eff} \mid \chi_s \rangle$ are the overlap and effective Hamiltonian matrices respectively. The effective Hamiltonian depends on the coefficients which makes the problem nonlinear in C. This is the case in Hartree-Fock and Density Functional Theory (DFT) calculations, which then cannot be routinely carried out for proteins involving thousands of atoms. If we restrict our interest to the localization-delocalization problem in valence electrons and treat the two-electron part of the Hamiltonian as in the case of the electrons in metals, in an average sense only, we can apply semi-empirical methods. Once the positions of the atoms are known the Extended Hückel (EH) Molecular Orbital Method[33] is quite successful in calculating the MOs of organic molecules. The diagonal part $H_{rr}^{(EH)}$ of the EH Hamiltonian is given by the ionization energies of the AOs[34], while the off-diagonal elements are calculated from the diagonal elements and the overlap matrix

$$H_{rs}^{(EH)} = \frac{1}{2} K \left(H_{rr} + H_{ss}\right) S_{rs},$$

where the common choice for the empirical constant is $K = 1.75$. This is similar in spirit to other tight binding Hamiltonians in various models of the Anderson transition.

Myoglobin is the first[35] and one of the best studied protein structures. It plays a central role in the oxygen storage of muscles. It consists of 153 amino acids and weighs about 18000 Daltons. For the numerical studies we selected NMR data of a solution form[36] (PDB ID:1MYF) from the Protein Data Bank[37] of RCSB, as it is captured in the "living" state and contains the positions of hydrogen atoms essential for the calculations. The EH calculations have been carried out by the numerical package YAeHMOP. There are $N = 6329$ valence electron AOs and the EH Hamiltonian and overlap matrices are sufficiently large of dimension 6329×6329, which makes it possible to make a good numerical comparison with similar calculations in solid state physics[16, 30, 25, 24] and quantum chaos[26, 20, 31, 24]. Löwdin transformation of the coefficient vector $C' = S^{1/2}C$ has been applied to transform the EH Hamiltonian to a real symmetric self-adjoint operator $H^L = S^{-1/2}H^{(EH)}S^{-1/2}$, which satisfies a normal eigenequation $H^L C^L = EC^L$.

The Highest Occupied Molecular Orbital (HOMO) and the Lowest Unoccupied Molecular Orbital (LUMO) play a key role both in electron transport and reactions of organic molecules.For the visual demonstration of the fractal nature of the eigenfunctions we show the HOMO and the LUMO of Myoglobin in Fig. 1. Absolute values of coefficients C_r^L are shown such that the index r is ordered in the sequence of appearance of atoms and orbitals in the amino acid sequence of the protein. Statistical similarity of the magnified part of the wave functions to the entire function is a visual indication of a fractal.

The multifractal analysis of the protein wave functions is based on the standard box counting procedure[38], dividing the 1D index space of ordered AO indices along the protein sequence into $N_l \approx N/l$ boxes of size l and determining the box probability of the wave function in the kth box,

$$\mu_k(l) = \sum_{n=0}^{l-1} |C^L_{(k-1)l+n}(E_i)|^2, k = 1, \ldots, N_l,$$

as a suitable measaure. If the qth moments of this measure are counted in all boxes and is proportional to some power $\tau(q)$ of the box size,

$$\chi_q = \left\langle \sum_k \mu_k^q(l) \right\rangle_E \sim l^{\tau(q)},$$

multifractal behaviour might be derived. For a simple monofractal $\tau(q) = (q-1)D$, where D is the fractal dimension. For multifractals the $\tau(q)$ curve is nonlinear and the generalized fractal dimensions D_q can be recovered $\tau(q) = (q-1)D_q = \lim_{L\to 0} \ln \chi_q / \ln l$.

We expect that for extended wave functions in the conducting phase the coefficients are evenly distributed around their mean, which is $\left\langle |C_r^L|^2 \right\rangle_E = (1/N)\sum_{k=1}^N |C_r^L(E_k)|^2 = 1/N$, and are independent of the position due to the normalization of the wave functions. The measure then scales like $\mu_k(l) \sim l/N$ and the moments scale as $\chi_q \sim (l/N)^{(q-1)}$, yielding $D_q = D = 1$ independent of q. For localized states in the insulating phase the coefficients are nearly zero except in a short interval of the size of the localization length ξ. The localization length is much smaller than the system size $\xi \ll N$, therefore at intermediate length scales $\xi \ll l \ll N$ an interval of size l either contains the localization interval and almost the full probability $\mu_k(l) \approx 1$ or it is practically empty $\mu_k(l) \approx 0$. The moments do not scale with the length $\chi_q \sim 1$ and $\tau(q) = 0$ yielding $D_q = D = 0$.

The numerical values of the fractal dimension of the protein wave function can be determined more conveniently by calculating the box probability for all possible boxes of length l and performing an additional averaging to smooth out statistical fluctuations

$$\left\langle \sum_k \mu_k^q(l) \right\rangle_E = N_l \times \frac{1}{N_l} \sum_k \langle \mu_k^q(l) \rangle_E \approx \frac{N}{l} \frac{1}{N-l} \sum_{r=1}^{N-l} \left\langle \left(\sum_{n=1}^{l} |C_{r+n}^L|^2 \right)^q \right\rangle_E .$$

In Fig. 2. we show the generalized fractal dimensions for Myoglobin obtained numerically. The most significant value is the correlation dimension $D_2 \approx 0.5$ which is just midway between localization $D = 0$ and delocalization $D = 1$ confirming that the system is critical and the wave functions are multifractals. We note that the same numerical value $D_2 = 0.5$ has been obtained also for critical quantum chaos[26, 20].

Next we show that the level statistics of Myoglobin is also transitional. The energy levels computed with the EH method have been analyzed with a statistical method of RMT suitable for the analysis of a relatively low number of eigenvalues. The distance between two consecutive levels $\sigma_i = E_{i+1} - E_i$ is normalized with the average of k level spacings to the left and to the right $\Delta_i = \frac{1}{2k+1}\sum_{j=-k}^{j=+k} \sigma_{i+j}$ giving the unfolded level spacing $s_i = \sigma_i/\Delta_i = (2k+1)(E_{i+1} -$

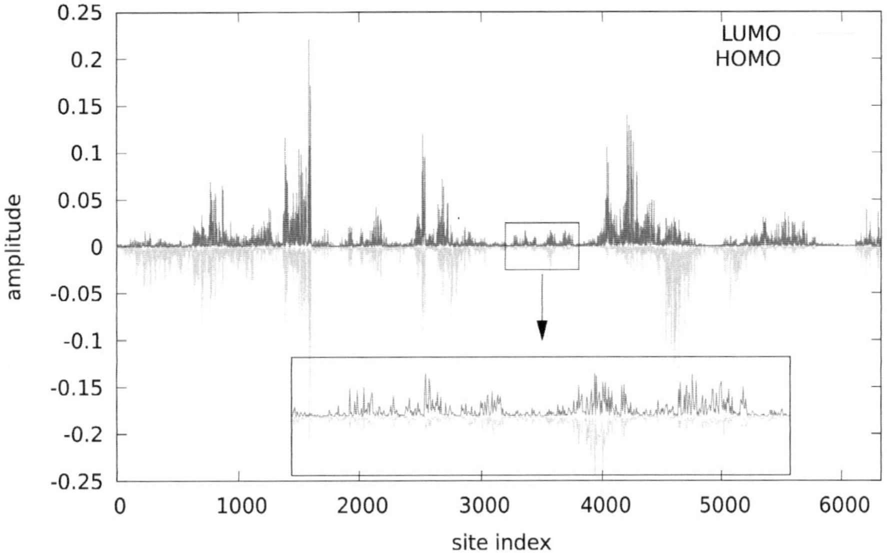

Figure 1. The HOMO/LUMO orbitals for Myoglobin (PDB ID:1MYF) calculated with the Extended Hückel method. Vertical axis: the absolute value of C_r^L in red for the HOMO and LUMO in green (flipped). Horizontal axis: the index sequence r of AOs ordered along the amino acid sequence. Inset: enlarged part of the box.

$E_i)/(E_{i+k+1} - E_{i-k})$. The choice of k depends on the variability of the density of energy levels. Statistical averaging would require large k values, while fast variation (especially singularities) in the density of energy levels restricts our choice to low values. We have found that a choice of $k = 2\ldots5$ ensures the stability of the distribution in our examples. Next, following standard procedures[31], the cummulative spacing $I(S) = \#\{s_i < S\}/N = \int_0^S P(s)ds$ is calculated and compared to the theoretical predictions. For the Poissonian statistics $I_P(S) = 1 - \exp(-S)$, for the Wigner surmise $I_W(S) = 1 - \exp(-\pi S^2/4)$ and for the semi-Poissonian transitional statistics $I_{SP}(S) = 1-(2S+1)\exp(-2S)$. In the main part of Fig. 3. we show the cummulative level spacing for the 6328 spacings in the spectrum of Myoglobin and in the inset we show the difference to $I_{SP}(S)$. We can see that without any parameter fitting the spacings for Myoglobin follow the critical theoretical curve with astonishing precision. Note, that no parameter fitting is involved in the procedure, the calculated spacing distribution has a less than 3% error like in the case of systems of critical quantum chaos[26, 20], which are purely theoretical models as opposed to our case, where the positions of atoms in the protein are obtained experimentally.

In addition to Myoglobin, we included in Fig. 3. two other proteins with known biochemical functions, selected randomly from the PDB just by size (close to 10000 valence AOs) and by the availability of the necessary structural data including the coordinates of Hydrogen atoms. Human profilin (PDB ID:1PFL)[39] in solution form has $N = 5232$ valence AOs. It is a ubiquitous eukaryotic protein that binds to both cytosolic actin and the phospholipid phosphatidylinositol-4,5-bisphosphate. Human apolipoprotein E (PDB ID:2L7B)[40] has $N =$ 11980 valence AOs. It is one of the major determinants in lipid transport, playing a critical role in atherosclerosis and other diseases. One can see that these other randomly picked proteins are also on the critical curve with the same precision as Myoglobin. Further analysis (not shown here) reveals essentially the same $D_2 \approx 0.5$ values and the same generalized dimension D_q

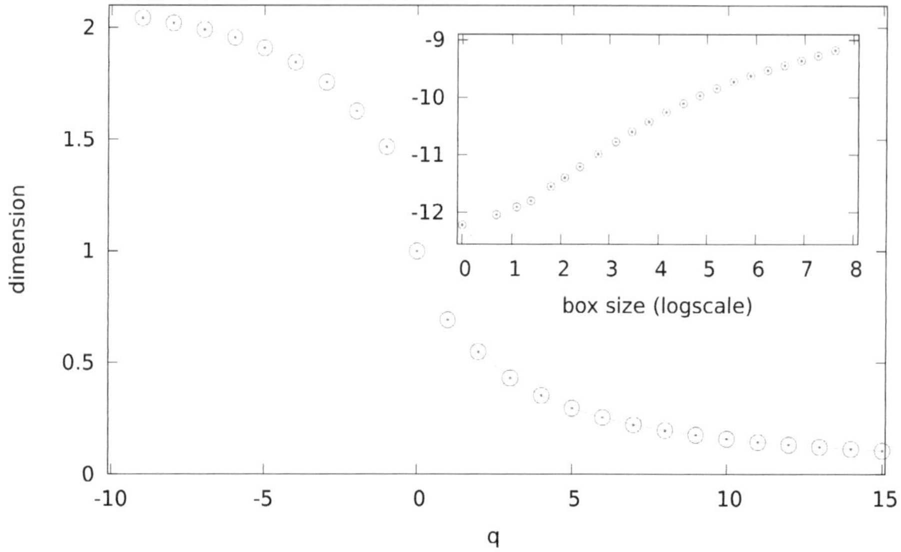

Figure 2. Generalized fractal dimensions D_q of wave functions of the protein Myoglobin (PDB ID:1MYF) averaged for all energies from the Extended Hückel calculation. Inset: Scaling of χ_q for $q = 2$ as a function of the box length l on a double logarithmic plot. The green linear function is fitted to the scaling region and its slope yields the correlation dimension $D_2 = 0.5 \pm 0.01$.

spectra for these proteins as well.

We should emphasize again, that finding a large tight binding Hamiltonian tuned exactly or almost exactly to the critical point by random chance can happen only with an astronomically low probability. So, finding just a single protein with more than 100 amino acids having this property at random is impossible.

Next, we investigate whether criticality is restricted to certain proteins only or is it a more wide spread phenomenon. The verification of fractality is not possible for smaller molecules as it requires a length scale of two decades (say $l \sim 10 - 1000$) to fit a reliable power law to the curve $\chi_q(l) \sim l^{\tau(q)}$. In the case of level statistics we need much less data to verify the shape of the cumulative level spacing $I(S)$. In the RMT analysis in solid state physics and quantum chaos normally on the order of a thousand levels is sufficient. Here we have found that using the technique developed for smaller data sets and the usage of cumulative level statistics $I(S)$ instead of the distribution function $P(s)$, which requires the binning of the data, jointly allow us to verify molecules with as low as $N = 80$ valence AOs reliably. In Fig. 4. we show a collection of level statistics coming from organic molecules of various size. The 3D structures of small molecules were taken from PubChem[41] and the energies are calculated with the EH method (YAeHMOP).

Our first and most striking observation is that each molecule investigated from the biological domain belongs to one of the "clean" categories I_P, I_W or I_{SP}. The reason is by no means obvious. In RMT[24] and especially in quantum chaos[42] intermediate distributions between the classes occur in finite systems. It seems that the Hamiltonians of these biomolecules are not random, they are tuned firmly to one of the classes.

Our first example is Silk (PDB ID:1SLK)[43], a protein which serves as a structural material

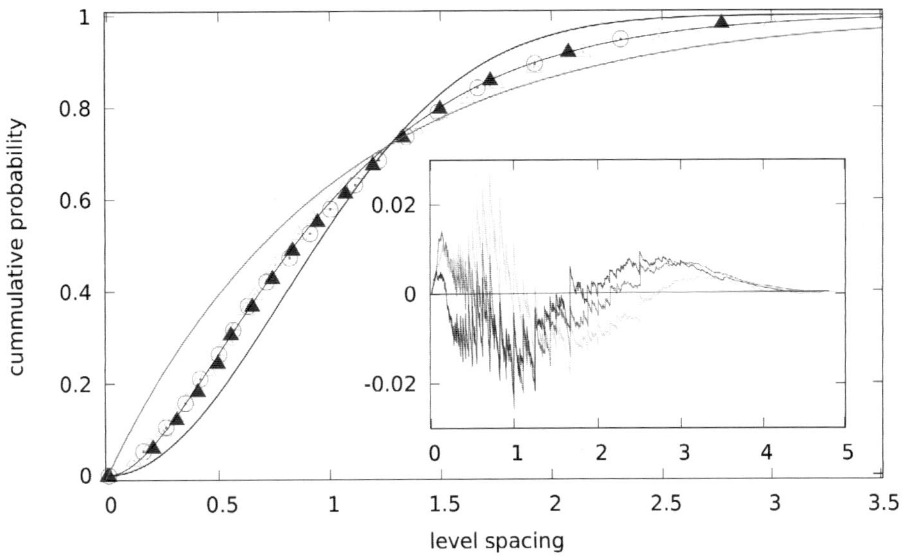

Figure 3. Cummulative level spacing distribution $I(S)$ for Myoglobin (blue triangles), Profilin (red circles) and Human Apolipoprotein E (green circles) are shown. Only every 300 value is shown for legibility. For comparison the theoretical curves $I_P(S)$ (red line), $I_W(S)$ (blue line) and $I_{SP}(S)$ (black line) are also shown. Inset: Difference between the data curves and $I_{SP}(S)$ are shown with the colour of the data set. Each value is plotted. The error is below 3% in probability.

and does not take an active part in biochemical processes. In Fig. 4. we can see that it has Poissonian level statistics and it belongs to the localized class. This is in line with the fact that silk is a very good insulator. This example confirms that criticality is not the property of individual amino acids. Amino acid sequences can produce not just critical materials but insulators as well.

Other examples for structural biomaterials in Fig. 4. are Dextrin (CID 62698), which is a gum like substance and Octadecane (CID 11635) which is an alkane hydrocarbon found in mineral oil and Gefarnate (CID 5282182) a water insoluble terpene fatty acid. They all show Poissonian level statistics and are good insulators.

It is less obvious why DNA belongs to this category. In Fig. 4. we show level statistics for a 21 basis pair DNA sequence (NDB ID:2JYK)[44] which is clearly Poissonian.

The conductivity properties of DNA or RNA sequences are highly debated, however there seems to be a consensus that native DNA is a wide band gap semiconductor, practically an insulator[45].

We picked a few representatives from essential classes of biomolecules. From each class we used the molecule with the largest number of AOs possible among all possible molecules having complete 3D crystallographic data in PubChem. They all show criticality and semi-Poissonian statistics. The list includes Linoleic acid (CID 5280450), Primary fluorescent chlorophyll catabolite (CID 54740347), Sucrose (CID 5988), Vitamin D3 (CID 25245915), Vitamin B12 (CID 16212801) and the largest amino acid Leucine (CID 6106). We investigated a dozen more biomolecules of lesser size. For all practical purposes they all showed semi-Poissonian statistics, but the statistics had larger errors due to the small number of levels. This list includes

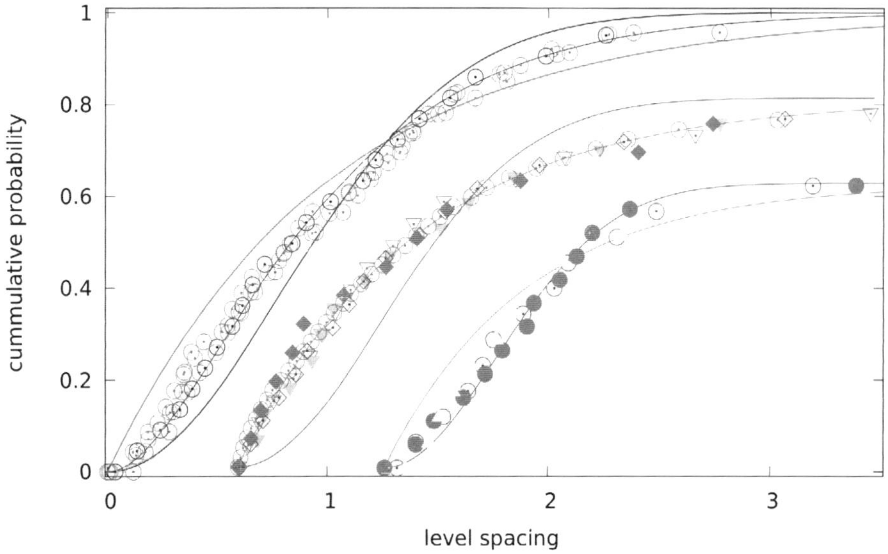

Figure 4. Cummulative level spacing distribution for various molecules. (Only 10-15 data points for each molecule is shown for legibility.) The main part: Molecules with critical cummulative level statistics $I_{SP}(S) = 1 - (2S+1)e^{-2S}$. The molecules shown are Vitamin B12, Vitamin D3, Linoleic Acid, Primary fluorescent chlorophyll catabolite, Sucrose and Leucine. Middle part: Molecules with Poissonian cummulative level statistics $I_P(S) = 1 - e^{-S}$ (Notice, that the inset is zoomed and the main axis scales don't apply). The molecules shown are Dextrin, Silk, Octadecane, Gefarnate and a 21 base pair DNA sequence (NDB ID: 2JYK). Small part: Molecules with Wigner cummulative level statistics $I_W(S) = 1 - e^{-\pi S^2/4}$ (Notice, that the inset is zoomed and the main axis scales don't apply). The molecules shown are Testosterone, Progesterone, Dibenzo(a,e)pyrene and Aristolochic Acid.

Nicotine, Adenosine, Caffeine, Amphetamine, Benzoanthracene, Chlorpozamine, Glucose, Fatty acids omega 3 and omega 6, Picrotin, Picrotoxin, Theophylline, Thiactin, Xanthine. We also carried out the analysis for the 20 amino acids coded by the universal genetic code. They are generally too small for the analysis of the level statistics to the level shown in Fig. 4. but some of them can already be classified by inspection. Based on this it is likely that Arginine, Cysteine, Sele Histidine, Isoleucine, Leucine, Methionine, Phenylalanine, Proline, Serine, Threonine and Tryptophan are critical and Alanine, Asparagine, Aspartic-acid, Glutamic-acid, Lysine, Tyrosine and Valine show Poissonian level statistics. For Glutamine and Glycine the results are inconclusive.

We can also find molecules which belong to the good conductor class with Wigner level spacing statistics. We could find this only in polycyclic molecules with delocalized wave functions spreading through the entire molecule. In Fig. 4. we show Testosterone (CID 6013), Progesterone (CID 5994), Dibenzo(a,e)pyrene (CID 9126) and Aristolochicacid (CID 2236). Among biomolecules we could find only Steroids in this class, the rest of such molecules were involved in combustion such as polycyclic aromatic hydrocarbons and were toxic or carcinogenic.

We can summarize these findings as follows: Most of the molecules taking part actively in biochemical processes are tuned exactly to the transition point and are critical conductors.

Journal of Physics: Conference Series **626** (2015) 012023 doi:10.1088/1742-6596/626/1/012023

There is only the special class of polycyclic molecules with closely packed aromatic rings which show metallic behaviour and delocalization and the class of "structural materials" which play a role in the mechanical stiffness of biological systems. Individual amino acids can be Poissonian or critical but they seem to form proteins and polypeptides which are also either Poissonian or critical.

These findings suggest an entirely new and universal mechanism of conductance in biology very different from the one used in electrical circuits. In metallic conductors charges float due to voltage differences. The electrical field accelerates electrons while scattering on impurities dissipates their energy fixing a constant average propagation velocity. In biological systems we seldom see examples for this. A more likely scenario is that a charge entering a critical conductor biomolecule will be under the joint influence of the quantum Hamiltonian and the excessive decoherence caused by the environment[46]. Such conductance mechanism has been found for the excitons in light harvesting systems[47] and it is currently in the focus of research in Quantum Biology[48]. In these systems Environment-assisted Quantum Transport[49] (ENAQT) is dominant and facilitates the fast quantum spreading of excitations over the system. We think that this mechanism is more universal in biological systems and charges in biological conductors are also subjects of this transport mechanism. Recently we have shown[50] that ENAQT is the most effective at the critical point of the localization-delocalization transition and the excitonic Hamiltonians of light harvesting systems are also at or near the critical point[51]. In the localized regime transport is hindered by strong quantum effects, while in delocalized systems decoherence destroys quantum propagation and the anti-Zeno effect[49] slows down diffusion. At the mobility edge the existence of extended multifractal wave functions throughout the system ensure end-to-end transport while coherence decays only algebraically[50] ensuring a longer coherence time and supressing the anti-Zeno effect.

Our results also suggest that quantum transport played a distinguished role in evolution and selection. Both the number of known small molecules and proteins is about 10^8 and the number of chemically feasible small ($< 500Da$) organic compounds is astronomical, estimated[52] to be 10^{60}. The number of proteins grows exponentially with the number n of amino acids as $\sim 20^n$, and the largest known has about $n \approx 26000$. This shows that chemical and biological evolution selected only a tiny fraction $p \sim 10^{-50}$ of possible small biomolecules and even less for proteins. As the probability of finding a critical molecule or protein by random chance is also astronomically low, the large number of critical molecules and proteins found by quasi random browsing of major databases just by size and availability of the 3D crystallographic data suggests that criticality of the quantum Hamiltonian is prevailing in the evolutionary selection of biomolecules. Besides, the fact that some proteins are natural critical conductors may open up new avenues in materials science as well.

References

[1] Kauffman S A 1993 *The origins of order: Self-organization and selection in evolution* (Oxford Univ. Press)
[2] Crutchfield J P and Young K 1988 *The Santa Fe Institute, Westview* (Citeseer)
[3] Langton C G 1990 *Physica D: Nonlinear Phenomena* **42** 12–37
[4] Maynard Smith J 1995 *The New York review of books* **42** 28–30
[5] Mora T and Bialek W 2011 *Journ. Statist. Phys.* **144** 268–302
[6] Lewin R and Bak P 1993 *Am. Journ. Phys.* **61** 764–765
[7] Bak P, Tang C and Wiesenfeld K 1988 *Phys. Rev. A* **38** 364
[8] Anderson P W 1958 *Phys. Rev.* **109** 1492
[9] MacKinnon A and Kramer B 1981 *Phys. Rev. Lett.* **47** 1546
[10] Wegner F 1981 *Z. Phys. B: Condensed Matter* **44** 9–15
[11] Castellani C and Peliti L 1986 *Journ. Phys. A: Mathematical and General* **19** L429
[12] Evers F and Mirlin A D 2008 *Rev. Mod. Phys.* **80** 1355
[13] Hofstadter D R 1976 *Phys. Rev. B* **14** 2239
[14] Harper P 1955 *Proc. Phys. Soc.: Section A* **68** 879

[15] Aubry S and André G 1980 *Ann. Israel Phys. Soc.* **3** 37
[16] Evangelou S 1990 *Journ. Phys. A: Mathematical and General* **23** L317
[17] Bohigas O, Giannoni M J and Schmit C 1984 *Phys. Rev. Lett.* **52** 1
[18] Berry M V and Tabor M 1977 *Proc. Roy. Soc. London. A. Mathematical and Physical Sciences* **356** 375–394
[19] Richens P J and Berry M V 1981 *Physica D: Nonlinear Phenomena* **2** 495–512
[20] Bogomolny E and Schmit C 2004 *Phys. Rev. Lett.* **92** 244102
[21] Wigner E P 1955 *Ann. Mathematics* **62** 548–564
[22] Dyson F J 1962 *Journ. Math. Phys.* **3** 140–156
[23] Brody T A, Flores J, French J B, Mello P A, Pandey A and Wong S S M 1981 *Rev. Mod. Phys.* **53**(3) 385–479
[24] Guhr T, MuellerGroeling A and Weidenmueller H A 1998 *Phys. Rep.* **299** 189–425
[25] Shklovskii B I, Shapiro B, Sears B R, Lambrianides P and Shore H B 1993 *Phys. Rev. B* **47**(17) 11487–11490
[26] Evangelou S and Pichard J L 2000 *Phys. Rev. Lett.* **84** 1643
[27] Aronov A G, Kravtsov V E and Lerner I V 1994 *JETP Lett.* **59** 40
[28] Aronov A G, Kravtsov V E and Lerner I V 1995 *Phys. Rev. Lett.* **74** 1174
[29] Evangelou S 1994 *Phys. Rev. B* **49** 16805
[30] Varga I, Hofstetter E, Schreiber M and Pipek J 1995 *Phys. Rev. B* **52** 7783
[31] Bogomolny E B, Gerland U and Schmit C 1999 *Phys. Rev. E* **59** R1315
[32] Bogomolny E, Gerland U and Schmit C 2001 *Europ. Phys. Journ. B-Condensed Matter and Complex Systems* **19** 121–132
[33] Hoffmann R 1963 *Journ. Chem. Phys.* **39** 1397–1412
[34] Pople J A and Beveridge D L 1970 *Approximate molecular orbital theory* vol 30 (New York: McGraw-Hill)
[35] Watson H C 1969 *Prog. Stereochem.* **4** 299–333
[36] Ösapay K, Theriault Y, Wright P E and Case D A 1994 *Journ. Molec. Biol.* **244** 183–197
[37] Berman H M, Westbrook J, Feng Z, Gilliland G, Bhat T N, Weissig H, Shindyalov I N and Bourne P E 2000 *Nucleic acids research* **28** 235–242
[38] Schreiber M and Grussbach H 1991 *Phys. Rev. Lett.* **67** 607
[39] Metzler W J, Farmer B T, Constantine K L, Friedrichs M S, Mueller L and Lavoie T 1995 *Protein Science* **4** 450–459
[40] Chen J, Li Q and Wang J 2011 *Proc. Nat. Acad. Sci.* **108** 14813–14818
[41] Bolton E E, Wang Y, Thiessen P A and Bryant S H 2008 *Annual reports in computational chemistry* **4** 217–241
[42] Berry M V and Robnik M 1984 *Journ. Phys. A: Mathematical and General* **17** 2413
[43] Fossey S A, Némethy G, Gibson K D and Scheraga H A 1991 *Biopolymers* **31** 1529–1541
[44] Masliah G, René B, Zargarian L, Fermandjian S and Mauffret O 2008 *Journ. Molec. Biol.* **381** 692–706
[45] Mallajosyula S S and Pati S K 2010 *Journ. Phys. Chem. Lett.* **1** 1881–1894
[46] de la Lande A, Babcock N S, Řezáč J, Lévy B, Sanders B C and Salahub D R 2012 *Phys. Chem. Chem. Phys.* **14** 5902–5918
[47] Engel G S, Calhoun T R, Read E L, Ahn T K, Mančal T, Cheng Y C, Blankenship R E and Fleming G R 2007 *Nature* **446** 782–786
[48] Huelga S F and Plenio M B *Nature Physics* **10** 621–622
[49] Rebentrost P, Mohseni M, Kassal I, Lloyd S and Aspuru-Guzik A 2009 *New Journ. Phys.* **11** 033003
[50] Vattay G, Kauffman S and Niiranen S 2014 *PloS one* **9** e89017
[51] Vattay G and Kauffman S A 2013 *Preprint arXiv:1311.4688*
[52] Reymond J L and Awale M 2012 *ACS chemical neuroscience* **3** 649–657

Journal of Physics: Conference Series **626** (2015) 012024 doi:10.1088/1742-6596/626/1/012024

The hidden geometry of electromagnetism

Y Hadad[1], E Cohen[2], I Kaminer[3] and A C Elitzur[4]

[1] Department of Mathematics, University of Arizona, Tucson, AZ, 85721, USA
[2] School of Physics and Astronomy, Tel Aviv University, Ramat-Aviv, Tel Aviv 6997801, Israel
[3] Department of Physics, Massachusetts Institute of Technology, 77 Massachusetts Avenue, Cambridge, Massachusetts 02139, USA
[4] Iyar, The Israeli Institute for Advanced Research, Rehovot, Israel

E-mail: yaron.hadad@gmail.com

Abstract. Nearly all field theories suffer from singularities when particles are introduced. This is true in both classical and quantum physics. Classical field singularities result in the notorious self-force problem, where it is unknown how the dynamics of a particle change when the particle interacts with its own (self) field. Self-force is a pressing issue and an active research topic in gravitational phenomena, as well as a source of controversies in classical electromagnetism. In this work, we study a hidden geometrical structure manifested by the electromagnetic field-lines that has the potential of eliminating all singularities from classical electrodynamics. We explore preliminary results towards a consistent way of treating both self- and external fields.

1. Introduction

The notion of a field-line was introduced by Michael Faraday in his work on magnetism [1]. Faraday argued that the forces of electricity, magnetism and gravity are better described by fields, populated with field-lines. From Faraday's point of view, a field is not a mere mathematical abstract but a real physical object.

Today, despite Faraday's groundbreaking work, field-lines are commonly used as a mere pedagogical tool for visualizing classical fields rather than providing a genuine clue about the field's physical nature.

We suggest to revive the field's ontological status in light of several open problems in physics. Classical field theories suffer from a fundamental flaw when particles are introduced. If we naively treat the particle as point-like, the field will inevitably have singularities. For example, Maxwell's equations predict the electromagnetic field to be singular at the location of point-like electrons. Consequently, the force exerted on the electron must be infinite due to its own self-field. This is the notorious *self-force problem.*

The self-force problem is ubiquitous in classical field theories. In general relativity, (gravitational) self-force is actively studied as a correction to the geodesic motion for sufficiently small bodies [2]. In classical electromagnetism, self-force introduces analogous corrections to the Lorentz force equation, one of which is the radiation-reaction force. Radiation-reaction is one of the oldest debates in physics for almost a century, with more than 15 different models suggested so far [3]-[18]. To date there is no compelling *experimental* reason to prefer any of them. New experiments to test radiation-reaction were suggested by one of us [19] following Labun et al., as well as other investigators [20], and will hopefully shed light on this issue in the near future.

There are many ways to try and address the self-force problem in the classical regime. One is to endow the electron with an internal structure. However, such an internal structure cannot maintain its stability unless one assumes additional unknown forces [21, 22]. A second approach is to modify Maxwell's equations to prevent the field from becoming singular in the first place. The most famous attempt of this kind is the Born-Infeld model [23], in which nonlinear effects produce an upper bound on the electromagnetic field in the vicinity of the point particle. Nevertheless, here too there is no compelling empirical reason to support the belief that the field equations are indeed nonlinear.

One may argue that the self-force problem is beyond the realm of our classical theories, hence quantum physics must be employed. In this work we present preliminary results which hint that self-forces can be treated consistently within classical physics.

Since self-force is such an elementary difficulty, it should not come as a surprise to find it strongly linked to the very concept of field-lines. By definition, a *test* charge will accelerate in the direction of the tangent to the field-line that passes through it. However, pure test particles are only an idealization. In reality, every charge changes the field-lines configuration irrespectively of how small it is (see figure 3). This means that *any (test) charge will interfere with the formation of the same field-lines it is expected to reveal.* In particular, the very concept of field-lines is ill-defined at the location of the charge, and field-lines are also inflicted by the self-force problem.

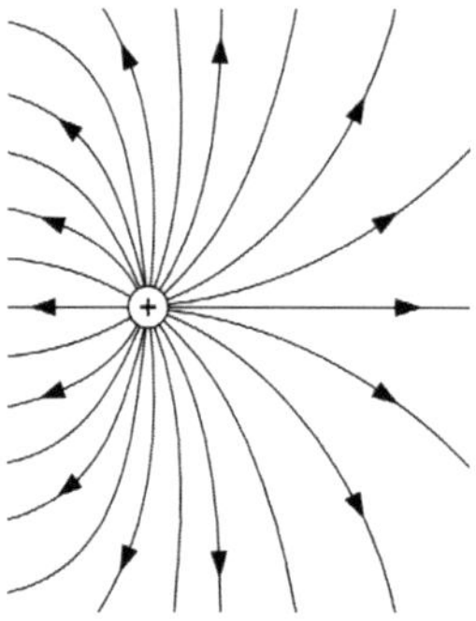

Figure 1. (Image credits: BillC, Wikimedia Commons) The field-lines in the vicinity of a positive charge. By definition, a *test* charge will accelerate in the direction of the tangent to the field-line at the position of the charge. The arrows point in the direction of acceleration of a positive test charge.

In the present work we argue that, paradoxically, it is the neglected notion of field-lines that offers a classical solution to the self-force problem. In earlier works, two of us [24, 25] adopted Faraday's viewpoint and elevated the field-lines from a mathematical abstract to a real physical entity. Just like material objects, the dynamics of the field-lines depends on their *mechanical* properties. The field's dynamics is determined by the curvature of its field-lines and the stress this curvature induces. For the sake of simplicity, we proceed by studying electric field-lines in non-relativistic systems, although the ideas presented here can be generalized to other fields.

2. Curvature and acceleration

A free charge in (flat) empty space may be in one of two states, either at rest or traveling with a constant velocity. In the former state, the field-lines around the resting charge are isotropic.

In the latter, they are denser in the directions perpendicular to the particle's velocity direction due to Lorentz contraction.

Consider next a charge changing its state from the first (resting) state to the second (rectilinearly moving) state. Its velocity changes abruptly from zero to $\vec{v} = v_0\hat{x}$, thereby emitting an electromagnetic wave. For a sufficiently far observer the charge still appears to be at rest, its field-line still being isotropic outside of the forward light-cone of the instant of acceleration. Inside the light-cone, the charge is already moving, hence the field-lines appear denser in the direction perpendicular to the direction of motion. To ensure continuity of the field-lines, they appear to be partially 'broken-and-redrawn' on the light-cone's surface, as shown in figure 2a.

This simple account can be extended to reveal a connection between the curvature of the field-lines and the charge's dynamics. In practice, the charge's velocity cannot change from zero to another value instantly but only continuously. Consider for example a velocity changing in constant rate from zero to $\vec{v} = v_0\hat{x}$. By the same logic used above, nearly all field-lines must curve as in figure 2b. The only field-lines that do not curve are the two that are directed along the charge's acceleration vector. The lines remain straight because they are on the only axis on which symmetric constraints act from both side, 'pealing off' the curved lines.

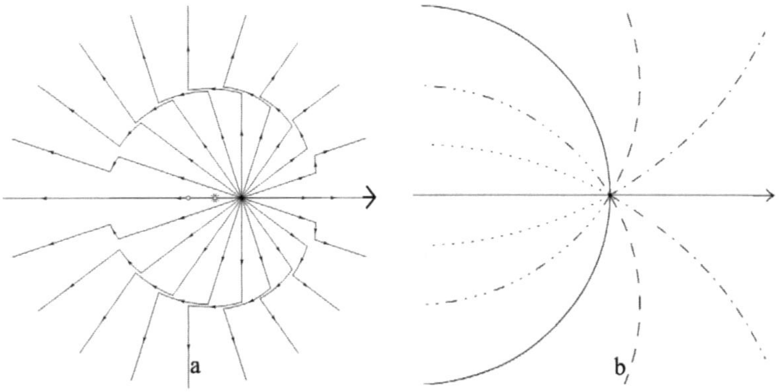

Figure 2. (a) Field-lines of a charge after an instantaneous velocity change: (b) Field-lines of a charge under constant acceleration. The field-lines curve, with only those pointing in the direction of acceleration having curvature zero.

This leads to the following observations:

- The field-lines around an accelerating charge curve.
- If the charge is accelerating, then the acceleration is along field-lines with zero curvature.

In the next section we will show that this observation is true not only in this specific case, but holds in general. Thus, it should be elevated to a postulate, relating the geometry of the electric field-lines to the dynamics of the charges. Under the influence of an electrostatic field, *charges must travel along field-lines with zero curvature.*

3. Minimal (zero) curvature

We start by considering a physical system which includes charges traveling under the influence of an electric field. For the sake of simplicity, we assume no magnetic fields or other forces. Our goal is to study the geometrical properties of the electric field-lines and their relation to the dynamics of the charges.

An electric field-line is defined to be a curve in $\mathbb{R}^3$ for which the direction of the curve is equal to the direction of the electric field. In other words, if $\vec{\gamma}(t,s)$ is an electric field-line at time t with parametrization s, then $\vec{\gamma}$ must satisfy the *electric field-line equation*:

$$\frac{d}{ds}\vec{\gamma}(t,s) = \vec{E}(t,\vec{\gamma}(s)). \tag{1}$$

The curvature κ of the curve $\vec{\gamma}$ in $\mathbb{R}^3$ is given by a well-known formula from differential geometry:

$$\kappa = \frac{|\vec{\gamma}' \times \vec{\gamma}''|}{|\vec{\gamma}'|^3}, \tag{2}$$

where the prime denotes differentiation with respect to the parametrization s. It should be clear that both functions κ and $\vec{\gamma}$ depend on the time t and the parametrization s, but to keep the notation concise we will omit these arguments.

By replacing the electric field-line Eq. (1) in the curvature Eq. (2), we receive the curvature $\kappa(t,\vec{x})$ of the unique field-line that passes through $\vec{x}$ at time t,

$$\kappa(t,\vec{x}) = \frac{|\vec{E} \times (\vec{E}\cdot\nabla)\vec{E}|}{|\vec{E}|^3} \tag{3}$$

where the electric field $\vec{E}$ is sampled at the point $\vec{x}$ and time t.

In order to assess the validity of our observation of the previous section we need to evaluate the curvature near the position $\vec{x_0}(t)$ of the charge. In classical electromagnetism it is common to omit the self-fields of the charges, to avoid the self-force problem that leads to the notorious singularities as well as the radiation-reaction problem [21]. But here the situation is fundamentally different. *When dealing with the geometry of the field-lines, one must include the self-fields of the charge to obtain meaningful results.* In other words, we must treat the electric field as a whole.

Let us separate the electric field into two parts. The first is the external electric field $\vec{E}_{\text{ext}}$. Here 'external' refers to the contribution of the electric field from all sources other than the charge located at the point $\vec{x_0}(t)$. The self-field of this charge will be denoted by $\vec{E}_{\text{self}}$. The total electric field is the superposition,

$$\vec{E} = \vec{E}_{\text{self}} + \vec{E}_{\text{ext}}. \tag{4}$$

Substituting Eq. (4) in the curvature Eq. (2) yields

$$\kappa(t,\vec{x}) = \frac{|(\vec{E}_{\text{self}} + \vec{E}_{\text{ext}}) \times ((\vec{E}_{\text{self}} + \vec{E}_{\text{ext}})\cdot\nabla)(\vec{E}_{\text{self}} + \vec{E}_{\text{ext}})|}{|\vec{E}_{\text{self}} + \vec{E}_{\text{ext}}|^3}. \tag{5}$$

Coulomb law shows that in the immediate vicinity of the charge the self-field is greater in magnitude than the external field, while farther away the external field gives a greater contribution. This means that in the nearby neighborhood of the charge, the dominating factor in the last equation is $\vec{E}_{\text{self}}$. We may therefore expand the curvature κ about the charge's position $\vec{x}_0(t)$ by treating the self-field as a dominating term.

In the zeroth order, the curvature is determined by the self-field alone, and we have

$$\kappa(t,\vec{x}) \approx \frac{|\vec{E}_{\text{self}} \times (\vec{E}_{\text{self}}\cdot\nabla)\vec{E}_{\text{self}}|}{|\vec{E}_{\text{self}}|^3}. \tag{6}$$

By Coulomb's law the self field-lines are straight. Hence this term vanishes for all $\vec{x} \neq \vec{x}_0(t)$, and one must consider the next order in the perturbation expansion.

The next order in the perturbation gives the first non-zero contribution to the curvature and is given by

$$\kappa(t, \vec{x}) = \frac{|\vec{k}_1 + \vec{k}_2 + \vec{k}_3|}{|\vec{E}_{\text{self}}|^3}, \tag{7}$$

where

$$\vec{k}_1 = \vec{E}_{\text{ext}} \times (\vec{E}_{\text{self}} \cdot \nabla)\vec{E}_{\text{self}}, \tag{8}$$

$$\vec{k}_2 = \vec{E}_{\text{self}} \times (\vec{E}_{\text{ext}} \cdot \nabla)\vec{E}_{\text{self}}, \tag{9}$$

$$\vec{k}_3 = \vec{E}_{\text{self}} \times (\vec{E}_{\text{self}} \cdot \nabla)\vec{E}_{\text{ext}}. \tag{10}$$

Note that each of the terms $\vec{k}_1$, $\vec{k}_2$ and $\vec{k}_3$ has two appearances of the self-field and a single appearance of the external field.

Let us assume that we are in the instantaneous reference frame of the charge. In this reference frame Coulomb's law holds and the self-field is

$$\vec{E}_{\text{self}}(t, \vec{x}) = q\frac{\vec{x}_0(t) - \vec{x}}{|\vec{x}_0(t) - \vec{x}|^3}. \tag{11}$$

Plugging this result into Eqs. (8, 9) gives

$$\vec{k}_1(t, \vec{x}) = \frac{2q^2}{|\vec{x}_0(t) - \vec{x}|}\vec{E}_{\text{ext}}(t, \vec{x}) \times (\vec{x}_0(t) - \vec{x}), \tag{12}$$

$$\vec{k}_2(t, \vec{x}) = \frac{q^2}{|\vec{x}_0(t) - \vec{x}|}\vec{E}_{\text{ext}}(t, \vec{x}) \times (\vec{x}_0(t) - \vec{x}). \tag{13}$$

Similarly, Eq. (10) shows that $\vec{k}_3$ is of a smaller magnitude $(1/r^4)$ than $\vec{k}_1$ and $\vec{k}_2$ $(1/r^5)$ due to the appearance of the derivative of external field in Eq. (10). $\vec{k}_3$ may therefore be omitted.

The above calculation shows that

$$\kappa(t, \vec{x}) \approx \frac{3}{q}|\vec{E}_{\text{ext}}(t, \vec{x}) \times (\vec{x}_0(t) - \vec{x})| \tag{14}$$

about the charge and in its rest frame. Remarkably, we see that any direct dependence on the self-field of the charge actually vanished. In the neighborhood of the charge, the curvature of the electric field-lines depends on the external field (as well as the vector $\Delta\vec{x} = \vec{x}_0(t) - \vec{x}$). This has profound consequences, as Eq. (14) shows that the electric field-lines curvature is regular nearby the charge. This result holds for all electric charges, and therefore shows that the electric field-lines curvature is always regular. In other words, even *for point-like charged particles the electric curvature is a well-defined property.*

Eq. (14) provides several immediate results. First, the limiting case $q \to \infty$ gives $\kappa \to 0$. This is not surprising, as in this case the self-field dominates the motion and the field-lines are straight.

The second limiting case $q \to 0$ is singular for this formula. In this case Eq. (14) is no longer valid as the assumptions we used in its derivation no longer hold. This is because we assumed $q \neq 0$ when we started this perturbation expansion, and for this case the equation is no longer valid since the external field dominates over the self-field, and the relative orders of terms k_1,k_2,k_3 changes. Furthermore, the relative magnitudes of E_{ext} and q provides a natural length dimension (including the curvature radius) which separates between the 'near' and the 'far' zones.

In the last section we observed that a charge accelerates in the direction of field-lines with zero curvature. We are now ready to prove that this is a general phenomenon. Let Δt be an

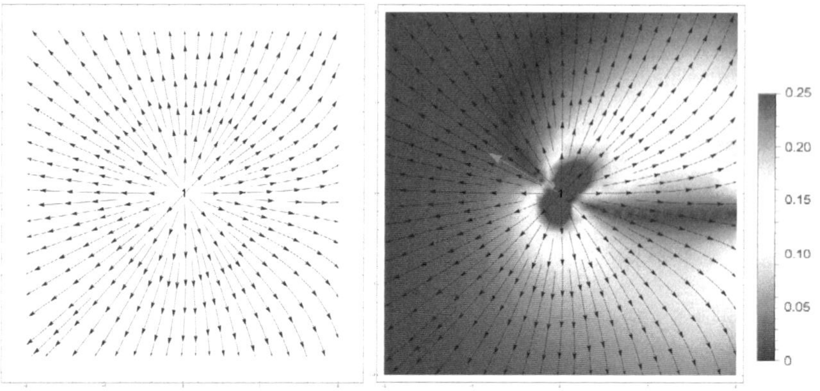

Figure 3. (a) The field-lines around a positive charge. In the immediate vicinity of the charge the field-lines are mostly determined by its self-field, where farther away it is the external field that predominantly determines the field-lines. (b) Each field-line is a curve in $\mathbb{R}^3$, and in this figure the color of each point represents the value of the field-lines' curvature. The blue color indicates regions where the field-line curvature vanishes, and it is clear that the acceleration vector of the charge (the orange arrow) points in the direction of zero curvature.

infinitesimal time duration. Consider evaluating the curvature of the field at the next time step of the charge,

$$\vec{x} = \vec{x}_0(t + \Delta t). \tag{15}$$

For small Δt we have,

$$\vec{x}_0(t + \Delta t) - \vec{x}_0(t) \approx \vec{v}_0(t)\Delta t + \frac{1}{2}\vec{a}_0(t)(\Delta t)^2. \tag{16}$$

The curvature formula (2) was derived in the instantaneous rest frame of the charge, in which $\vec{v}_0 = \vec{0}$. The curvature (2) at the position of the charge at time $t + \Delta t$ can be estimated using Eq. (16) (where $\vec{v}_0(t) = \vec{0}$) to give

$$\begin{aligned} \kappa(t, \vec{x}_0(t + \Delta t)) &\approx \frac{3}{q}|\vec{E}_{\text{ext}}(t, \vec{x}) \times (\vec{x}_0(t) - \vec{x}_0(t + \Delta t))| \\ &\approx \frac{3(\Delta t)^2}{2q}|\vec{E}_{\text{ext}}(t, \vec{x}_0(t + \Delta t)) \times \vec{a}_0(t)|. \end{aligned} \tag{17}$$

This means that $\kappa(t, \vec{x}_0(t + \Delta t)) = 0$ if and only if the acceleration is parallel to the external field. *The charge will always accelerate in the direction of a field-line that has zero curvature.*

The most striking result from the curvature Eq. (2) is that the electric curvature is regular nearby point charges. This is a surprising result, as most electromagnetic quantities are singular near point particles [26]. This preliminary result hints on a consistent way of dealing with point particles without the renormalization techniques that are typically employed in the literature [27].

4. Maximal curvature

The acceleration of the charge is a vector. By its definition, the acceleration is uniquely determined by its direction and magnitude. In the last section we saw that the geometrical properties (i.e. curvature) of the field-lines suffice to determine the direction of the acceleration. The goal of this section is to demonstrate that the field-lines also determine the acceleration's magnitude.

Consider figure 2a again. The greater the difference between the initial and final velocities, the more broken the field-lines will appear on the light cone's surface. This implies that there is a relationship between the magnitude of the charge's acceleration and the curvature of the field-lines in the vicinity of the accelerating charge. In other words, as the acceleration becomes higher the field-lines nearby the charge are more curved (see 2b).

There are several approaches to measure the total curvature of the field-lines about the charge. Here we present only one, not necessarily the most natural nor the most general. It is considered only as a case study that shows that *the field-lines contain all the information needed to know the dynamics of the charge.*

In the last section, the field-lines curvature κ in the neighborhood of the charge was shown to determine the direction of the charge's acceleration vector. This was done by requiring that the charge accelerates along a vector of *zero electromagnetic curvature.* It is clear that lines of zero curvature cannot determine the magnitude of the acceleration vector. Since the electromagnetic curvature is always non-negative, the zero curvature condition isolates the *minimal* curvature in the neighborhood of the charge. A natural contra-distinctive way of determining the magnitude of the acceleration is to use the *maximum* curvature in the neighborhood of the charge.

The curvature of the electromagnetic field at the point $\vec{x}$ was computed in Eq. (14), and can be rewritten as

$$\kappa(t,\vec{x}) \approx \frac{3}{q}|\vec{E}_{\text{ext}}(t,\vec{x})||(\vec{x}_0(t)-\vec{x})||\sin\theta|, \tag{18}$$

where θ is the angle between the electric field-line at the point $\vec{x}$ and the vector $\vec{x}_0(t)-\vec{x}$. Assume that the radius of the ball is small enough so that the electric field barely varies inside the ball,

$$\vec{E}_{\text{ext}}(t,\vec{x}) \approx \vec{E}_{\text{ext}}(t,\vec{x}_0). \tag{19}$$

Under this assumption, Eq. (18) shows that the maximal curvature in a ball of radius r about $x_0(t)$ is obtained along the angle $\theta = \pi/2$, corresponding to a perpendicular direction to the acceleration vector and is given by

$$\max_{|\vec{x}-\vec{x}_0(t)|\leq r} \kappa = \frac{3}{q}|\vec{E}_{\text{ext}}(t,\vec{x}_0)|r. \tag{20}$$

Eq. (20) shows that the maximal curvature is proportional to the magnitude of the total electric field. It is well-known that in the absence of other forces, the total electric field is proportional to the magnitude of the acceleration of the charge.

We therefore postulate that *the charge accelerates in the direction of zero curvature field-lines with acceleration magnitude proportional to the maximal curvature of the field (line).* The proportionality constant can be determined using dimensional analysis. The dimensions of curvature are one over length. To obtain dimensions of acceleration we need to multiply the curvature κ by a constant with dimensions of velocity squared, and it is only natural to multiply it by c^2 (here c denotes the speed of light propagation in vacuum).

The only variable left to determine is the radius r used in Eq. (20). Any result produced by the method suggested here must be consistent with the Lorentz force in the absence of magnetic fields,

$$m\vec{a} = q\vec{E}. \tag{21}$$

It is now easy to see that setting

$$r = \frac{1}{3}\frac{q^2}{mc^2} \tag{22}$$

Journal of Physics: Conference Series **626** (2015) 012024 doi:10.1088/1742-6596/626/1/012024

in Eq. (20) produces the expected results (21) in the leading order. Interestingly, for electrons the radius we obtained in Eq. (22) is

$$r = \frac{1}{3} r_e, \tag{23}$$

where

$$r_e = \frac{e^2}{m_e c^2} = 2.8179403267(27) \times 10^{-13}\,\text{cm} \tag{24}$$

is the classical electron radius. Here e and m_e are the electric charge and mass of the electron respectively.

5. Conclusions

The notion of field-lines is widely discussed in textbooks as a tool for visualizing fields. In this work, we argued to the contrast. Rather than a mere visualization tool, field-lines may causally determine the dynamics of moving particles. This postulate was proven for electric fields, as it was shown that the electric field-lines completely determine the dynamics of electric charges. The curvature of the field-lines as curves in $\mathbb{R}^3$ contains all the information needed to determine the acceleration of electric charges under an electrostatic force.

We studied an electric charge q moving under the influence of both its self-field $\vec{E}_{\text{self}}$ and an external electric field $\vec{E}_{\text{ext}}$. The curvature of the field-lines was always computed by considering the total electric field $\vec{E} = \vec{E}_{\text{self}} + \vec{E}_{\text{ext}}$. The acceleration of the charge was shown to be directed along the minimum (zero) field-line curvature, and its magnitude is proportional to the maximum field-line curvature. Written explicitly,

$$\hat{a} = \text{direction of } \min(\kappa) = \hat{E}, \tag{25}$$

and

$$|\vec{a}| = c^2 \max(\kappa) = \frac{q}{m} |\vec{E}_{\text{ext}}(t, \vec{x}_0)| \tag{26}$$

where c is the speed of light, m is the mass of the charge, $\vec{x}_0$ is its position, and both the minimum and maximum curvature are computed within a ball of a radius given in Eq. (22). For electrons, this radius is equal to one third of the classical electron radius.

The most surprising result from this analysis is that although we studied the curvature of the total electric field $\vec{E}$, the self-field produced no singularities in the total curvature. This is apparent from Eq. (14) which only depends on the external field $\vec{E}_{\text{ext}}$. This result holds independently of the electric charge's structure, and continues holding for point charges. This hints that the formalism studied here is capable of treating point particles consistently without the need for renormalization. Since field singularities constitute an inherent issue when studying field theories that contain particles, this analysis appears to be a promising research direction for eliminating singularities from field theories in general.

The results provided here are still preliminary, and many interesting questions arise:

- How can the above mechanism be extended to treat fields other than the electric field? The most natural next step would be to study the magnetic field. Should that prove successful, venturing straight ahead to gravity would be the next natural step.
- How can these results be derived in Minkowski spacetime using the 4-vector definition of curvature on general electromagnetic fields? The results produced here are indeed consistent with the special theory of relativity, and a robust relativistic formulation of them can be very insightful.
- Can the field-lines be used to obtain an action principle for classical electrodynamics that treats the self-fields on the same footing as the external fields without apparent singularities?

- What about quantum fields?

It is important to remember that in this work we approximated the maximal curvature in Eq. (20) in order to derive the Lorentz force Eq. (21). It is only expected that the full non-approximated expression will include correction terms. We hope that these correction terms will shed light on the problem of radiation-reaction, which remains an open problem in classical electrodynamics.

Acknowledgments
We thank Ofek Birnholtz, Paz Beniamini, Michael Bialy, Doron Grossman and Larry Horwitz for very helpful comments and discussions. E.C. was supported in part by Israel Science Foundation Grant No. 1311/14. I.K. was supported by a Marie Curie Grant No. 328853-MC-BSiCS.

[1] Faraday M 1846 *Philos. Mag.* **28** 345
[2] Barack L 2014 *Fund. Theor.* **177** 147
[3] Dirac P A M 1938 *Proc. R. Soc. Lon. Ser. A* **167** 148
[4] Eliezer C 1948, *Proc. R. Soc. Lon. Ser. A* **194** 543
[5] Landau L D and Lifshitz E M 1975 *The Classical Theory of Fields* (Oxford: Elsevier)
[6] Prigogine I and Henin F 1962 *Physica* **28** 667
[7] Prigogine I and Henin F 1963 *Physica* **29** 286
[8] Nodvik J S 1964 *Ann. Phys.* **28** 225
[9] Teitelboim C 1971 *Phys. Rev. D* **4** 45
[10] Mo T C and Papas C H 1971 *Phys. Rev. D* **4** 3566
[11] Gonzalez-Gascon F 1976 *Nuovo Ciemnt. B* **11** 333
[12] Petzold J and Sorg M 1977 *Z. Phys. A* **283** 207
[13] Caldirola A 1979 *Riv. Nuovo Cimento* **2** 1
[14] Ford G W and R F O'Connell 1991 *Phys. Lett. A* **157** 217
[15] Yaghjian A D 1992 *Relativistic Dynamics of a Charged Sphere* (Berlin: Springer-Verlag)
[16] Sokolov I V, Naumova N M, Nees J A, Mourou G A and Yanovsky V P 2009 *Phys. Plasmas* **16** 093115
[17] Hammond R T 2010 *Elect. J. Theor. Phys.* **6** 221
[18] de Oca A C M and Castineiras J 2013 On radiation reaction and the Abraham-Lorentz-Dirac equation *Preprint* arXiv:1304.2203
[19] Hadad Y, Labun L Rafelski J, Elkina N, Klier C and Ruhl H 2010 *Phys. Rev. D* **82** 096012
[20] Di Piazza A, Müller C, Hatsagortsyan K Z and Keitel C H 2012 *Rev. Mod. Phys.* **84** 1177
[21] Rohrlich F 1997 *Am. J. Phys.* **65** 1051
[22] Isoyama S and Poisson E 2012 *Classical Quant. Grav.* **29** 155012
[23] Born M and Infeld L 1934 *Proc. R. Soc. Lon. Ser. A* **144** 425
[24] Elitzur A C, Cohen E and Beniamini P 2012 Charge Acceleration and Field-Lines Curvature: A Fundamental Symmetry and Consequent Asymmetries *AIP Conf. Proc.* **1411** 211 (*Preprint* arXiv:1208.5164)
[25] Cohen E, Beniamini P, Grossman D, Horwitz L and Elitzur A C 2013 Mechanical Properties of the Electric Field: A Novel Prediction derived from the Field's Mass and Stress *Preprint* arXiv:1304.5598
[26] Dyson F J 1952 *Phys. Rev.* **85** 631
[27] Peskin M E and Schroeder D V 1995 *An introduction to quantum field theory* (Boulder: Westview)

Journal of Physics: Conference Series **626** (2015) 012025 doi:10.1088/1742-6596/626/1/012025

Entropy is a consequence of a discrete time

Roland Riek

Laboratory of Physical Chemistry, ETH Zurich, Switzerland

E-mail: roland.riek@phys.chem.ethz.ch

Abstract. While the basic microscopic physical laws are time reversible, the arrow of time and time irreversibility appears only at the macroscopic physical laws by the second law of thermodynamics with its entropy term S. It is the attempt of the present work to bridge the microscopic physical world with its macroscopic one with an alternative approach than the statistical mechanics theory of Gibbs and Boltzmann. For simplicity a "classical", single particle in a one dimensional space is selected. In addition, it is assumed that time is discrete with constant step size. As a consequence time irreversibility at the microscopic level is obtained if the present force is of complex nature ($F(r) \neq const$). In order to compare this discrete time irreversible mechanics with its classical Newton analog, time reversibility is reintroduced by scaling the time steps for any given time step n by the variable s_n leading to the Nosé-Hoover Lagrangian comprising a term $N_{df}\, k_B\, T\, ln\, s_n$ (k_B the Boltzmann constant, T the temperature, and N_{df} the number of degrees of freedom) which is defined as the microscopic entropy S_n at time point n multiplied by T. Upon ensemble averaging of the microscopic entropy in a many particles system in thermodynamic equilibrium it approximates its macroscopic counterpart known from statistical mechanics. The presented derivation with the resulting analogy between the ensemble averaged microscopic entropy and its statistical mechanics analog suggests that the entropy term itself has its root not in statistical mechanics but rather in the discreteness of time.

1. Introduction

The second law of thermodynamics claims that the entropy S of an isolated system increases over time until the system is in its thermodynamic equilibrium. Following the systematic formulation of statistical mechanics by Gibbs and Boltzmann, entropy S^G is a measure of the number of accessible micro-states of the system of interest in its thermodynamic equilibrium. For a micro canonical ensemble the entropy is given by $S^G = k_B\, ln\, \Omega$ with k_B the Boltzmann constant, and Ω the number of accessible micro-states [1][2]. Interestingly and somewhat strangely, this probability-based entity exerts its presence as being part of the total free energy with $G = U + pV - TS$ with U is the inner energy, p the pressure, V the volume, and T the temperature [1][2]. While it is not obvious that a pure statistical number times $k_B T$ is part of the total free energy of the system, the statistical mechanics argument is sound, that a system has the tendency to evolve towards its most probable state, which is the equilibrium state. However, also for the latter argument there are powerful protests such as the Loschmidt and Zermelo's reversal and reoccurrence objections [3][4]. At the heart of these objections is that a single event does not follow statistics and thus the ensemble-based entropy is not relevant and thus can not be calculated. Another oddity is that microscopic physics is time reversible, but that in accordance to statistical mechanics an arrow of time comes apparently only into play at the macroscopic level with ensemble averaging over many repeated experiments. Consequently, it is the notion of the author and many other scientists (for example reference [5] and [6]) that the arrow of time is such an important physical measure that it has likely its root deep in physics at the microscopic level and not just at the macroscopic statistical level.

But from where could the arrow of time, time irreversibility, and entropy be originated from? As we shall see, it is our attempt to derive the arrow of time and the entropy part of the total free energy from the assumption that time is discrete. The discreetness of time is sound because experimentally measured time is composed of an array of events, which can be exemplified in physics only by an energy-consuming clock measurement, and thus can not be measured continuously because of the uncertainty principle between energy and time ($\Delta E \Delta t > h/2$, with h being the Planck constant). Furthermore, upon energy quantization for quantum mechanics the introduction of a discrete time appears to be straightforward (see for example refernces [7-13]). It is however not a very popular concept, because the continuity of time and space enables powerful mathematical tools.

To describe the evolution of a non-relativisitc classical particles system in presence of a discrete time, we follow here in part the approach by Lee [10], who introduced time as a dynamical discrete variable yielding a scaling of time that depends on the potential present. The consequences of scaling time onto the law of energy conservation requests a reformulation of the Lagrangian as established by Nosé for isothermal molecular dynamics simulations of macroscopic systems [14-18]. There is a logarithmic term in the Nosé-Hoover Lagrangian which is dependent on the scaling of time. As we shall see, this term is defined as the microscopic entropy and if ensemble-averaged is equivalent to the macroscopic entropy in thermodynamic equilibrium. After the introduction of the discrete time (3.1), and the Nosé-Hoover Lagrangian (3.2), the Newton equation is derived (3.3), and time irreversibility in a microscopic system is discussed (3.4), followed by the definition of the microscopic entropy both for a single particle (3.5), as well as a many non-interacting particles system (3.6). It is then further shown that the averaged microscopic entropy corresponds to the Boltzmann entropy (3.7). Thus, the presented work is following in part our original article on this subject [19].

2. Theory

In classical physics time t is a continuos parameter. A non-relativistic particle with mass m in a one dimensional space at the position $r(t)$ at a given time t in a potential $V(r)$ has the following Lagrangian $L = \frac{1}{2} m \dot{r}^2 - V(r)$ with $\dot{r} = \frac{dr}{dt}$ being the velocity. The equation of motion, i.e. the Newtonian law, can then be derived from the Lagrangian equation $\frac{\partial}{\partial t} \frac{\partial L}{\partial \dot{r}} = -\frac{\partial L}{\partial r}$.

If time is a discrete variable t_n the continuous function r(t) is replaced by a sequence of discrete values [10]:

$$(r_0, t_0), (r_1, t_1), ..., (r_n, t_n), ..., (r_{N+1}, t_{N+1}) \tag{1}$$

with (r_0, t_0) the initial and (r_{N+1}, t_{N+1}), the final position. In this description r_n is still continuous, while t_n is discrete and as requested a dynamic variable. The dynamic part of the time can be described by a time scaling variable s_n defined by

$$\boxed{s_n \Delta t_n = t_n - t_{n-1}} \tag{2}$$

with $\Delta t_n = const.$ (Please note, that the s_n is defined as the corresponding variable s of Nosé with $dt = \frac{d\tilde{t}}{s}$ in the Nosé-Hoover thermostat, see below; [2, 14-16,17,18,20]) yielding a description of the evolution of the particle with $t_0 = 0$ by the following sequence of discrete values:

$$(r_0, t_0 = 0), (r_1, s_1 \Delta t), ..., (r_n, \sum_{i=1}^{n} s_i \Delta t), ..., (r_{N+1}, \sum_{i=1}^{N+1} s_i \Delta t) \tag{3}$$

Journal of Physics: Conference Series **626** (2015) 012025 doi:10.1088/1742-6596/626/1/012025

In discrete mechanics there are many possible definitions of the velocity $\dot{r_n}$. Since it is the attempt of the present work to bridge the gap between microscopic and macroscopic physics, also the microscopic physics should have the property to be time irreversible under certain conditions. Thus, the velocity at time point n $\dot{r_n}$ is defined time asymmetric. Furthermore, it is defined backward in time permitting a forward progressing description of the system from past and present information. This approach is in line with our daily experience that the presence is determined by the past and presence (i.e. to describe a present state of a system information can experimentally-derived only from the past and presence):

$$\dot{r_n} = \dot{r_n}(r_n, r_{n-1}, t_n, t_{n-1}) = \frac{r_n - r_{n-1}}{t_n - t_{n-1}} \tag{4}$$

2.1. The Nosé-Hoover Lagrangian

In presence of a discrete dynamic time variable described by the time scaling variable s_n a Lagrangian must be derived that is able to describe adequately the system. Such a Lagrangian under an artificial continuous scaled time s has been introduced by Nosé and Hoover in the so called Nosé-Hoover thermostat [2, 14-16,17,18,20]. It allows the molecular dynamics simulation at constant temperature of a system and samples the configurations of a canonical ensemble. Hence, the Nosé-Hoover thermostat introduces a Lagrangian that describes a system on the microscopic level, which yields upon ensemble averaging a macroscopic thermodynamic description of the system bridging the microscopic with the macroscopic world, which is the scope of the present work [14,21].

The discrete analog of the Nosé-Hoover Lagrangian (in real space) is given by the following expression:

$$\begin{aligned} L_n^N &= L_n^N(r_n, r_{n-1}; t_n, t_{n-1}) = L_n^N(r_n, r_{n-1}; \dot{s}_n, s_n)L_n \\ &= s_n(\frac{1}{2} m \dot{r}_n^2 - V(r_n) + \frac{1}{2} Q \frac{\dot{s}_n^2}{s_n^2} - N_{df}\, k_B T \, ln\, s_n) \end{aligned} \tag{5}$$

where k_B is the Boltzmann constant, Q is a constant, which has been described as a "mass"-like term for the motion of s_n with $Q > 0$ and with actual units of energy time squared, and N_{df} is the degree of freedom of the single particle system. The first two terms of the Lagrangian represent the kinetic energy minus the potential energy of the system, while the third and fourth terms represent the kinetic energy minus the potential energy associated with the s_n variable. The third term can be interpreted as a kinetic energy of a heat bath coupled to the system of interest, and the fourth term describes the heat transfer between the heat bath and the system of interest. The heat bath mimics a bath composed of an infinite collection of harmonic oscillators and thus shows an infinite heat capacity [22]. The heat or temperature bath is however of another nature than usually defined in thermodynamics because it is a heat bath of the unitless time scaling variable s_n with a mass-like Q with units Js^2 (energy*seconds*seconds) and a velocity $\dot{s}_n = \frac{s_{n+1} - s_n}{t_n - t_{n-1}}$ with the unit s^{-1}. Nonetheless, it enables the use of a temperature as highlighted by the explicit presence of the temperature T in the fourth term of the Lagrangian although the system described is of microscopic nature.

2.2. The Newton equation for a discrete time

If the discrete analog of the Lagrangian equation is defined as

$$\frac{1}{\Delta t_n}(\frac{L_{n+1}^N(r_{n+1}, \dot{r}_{n+1},)}{\partial\, \dot{r}_{n+1}} - \frac{\partial L_n^N(r_n, \dot{r}_n)}{\partial \dot{r}_n}) = -\frac{\partial L_n^N}{\partial r_n} \tag{6}$$

Journal of Physics: Conference Series **626** (2015) 012025 doi:10.1088/1742-6596/626/1/012025

the discrete Newton' law can be obtained using the discrete Lagrangian from eq. 5

$$\frac{1}{\Delta t_n}(m\,\dot{r}_{n+1}s_{n+1} - m\,\dot{r}_n s_n) = s_n\,F(r_n) \tag{7}$$

$$\frac{1}{\Delta t_n}(\,\dot{r}_{n+1} - \,\dot{r}_n + \dot{r}_n - \dot{r}_n\frac{s_n}{s_{n+1}}) = \frac{s_n}{m\,s_{n+1}}F(r_n) \tag{8}$$

$$\frac{\dot{r}_{n+1} - \,\dot{r}_n}{\Delta t_n} = \frac{s_n}{m\,s_{n+1}}F(r_n) - \dot{r}_n(1 - \frac{s_n}{s_{n+1}})\frac{1}{\Delta t_n} \tag{9}$$

$$\frac{\dot{r}_{n+1} - \,\dot{r}_n}{\Delta t_n} = \ddot{r}_n = \frac{s_n}{m\,s_{n+1}}F(r_n) - \dot{r}_n(1 - \frac{s_n}{s_{n+1}})\frac{1}{\Delta t_n} \tag{10}$$

$$\boxed{\ddot{r}_n{=}\frac{s_n}{m\,s_{n+1}}F(r_n) - \gamma_n\dot{r}_n} \tag{11}$$

with $\gamma_n = \frac{\dot{s_n}}{s_{n+1}}$and $\dot{s_n} = \frac{s_{n+1}-s_n}{\Delta t_n}$.
An alternative derivation used below is

$$\frac{1}{\Delta t_n}(\,\dot{r}_{n+1}\frac{s_{n+1}}{s_n} - \,\dot{r}_{n+1} + \dot{r}_{n+1} - \dot{r}_n) = \frac{1}{m}F(r_n) \tag{12}$$

$$\ddot{r}_n = \frac{1}{\Delta t_n}(\dot{r}_{n+1} - \dot{r}_n) = \frac{1}{m}F(r_n) - \dot{r}_{n+1}(\frac{s_{n+1}}{s_n} - 1)\frac{1}{\Delta t_n} \tag{13}$$

Eqs. 11 and 13, respectively, describe the Newton's law under a discrete variable time.
If there is no scaling of time (i.e. $s_i = 1$ for all $i = 1...N+1$) the Newton's law under a discrete time with constant time steps is of the form

$$m\ddot{r}_n = F(r_n) \tag{14}$$

which resembles its continuous analog. In presence of a scaling of time unequal to 1 however, the Newton's law has an additional term, which can be regarded a friction term (see also below).

2.3. Time reversibility/irreversibility in a one dimensional space

Having established the Newton's law (eqs. 11 and 13) in presence of a discrete dynamic time, we would like to pursuit the issue of time reversibility or irreversibility. Time reversibility can be described by a two step process having one step forward followed by a step backward. Let us consider the evolution of the discrete Newton's law of a single particle with two step forwards in a one dimensional space. Following eq. 11

(i)

$$\dot{r}_{n+1} = \dot{r}_n + \frac{s_n}{ms_{n+1}}\Delta t_n F(r_n) - (1 - \frac{s_n}{s_{n+1}})\dot{r}_n \tag{15}$$

For the second time step we take the second expression of the discrete Newton's law (eq. 13)

(ii)

$$\dot{r}_{n+2} = \dot{r}_{n+1} + \frac{1}{m}\Delta t_{n+1}F(r_{n+1}) - \dot{r}_{n+2}(\frac{s_{n+2}}{s_{n+1}} - 1) \tag{16}$$

If the second step is now backward in time

$$\dot{r}_{n+2} = \dot{r}_{n+1} - \frac{1}{m}\Delta t_{n+1} F(r_{n+1}) - \dot{r}_{n+2}(\frac{s_{n+2}}{s_{n+1}} - 1) \tag{17}$$

and if time reversibility is requested (i.e. $\dot{r}_n = \dot{r}_{n+2}$ and $s_{n+2} = s_n$)

$$\dot{r}_n = \dot{r}_{n+2} = \dot{r}_{n+1} - \frac{1}{m}\Delta t_{n+1} F(r_{n+1}) - \dot{r}_n(\frac{s_n}{s_{n+1}} - 1) \tag{18}$$

$$\dot{r}_n = \dot{r}_n + \frac{s_n}{m s_{n+1}}\Delta t_n F(r_n) - (1 - \frac{s_n}{s_{n+1}})\dot{r}_n - \frac{1}{m}\Delta t_{n+1} F(r_{n+1}) - \dot{r}_n(\frac{s_n}{s_{n+1}} - 1) \tag{19}$$

$$\frac{s_n}{m s_{n+1}}\Delta t_n F(r_n) = \frac{1}{m}\Delta t_{n+1} F(r_{n+1}) \tag{20}$$

and since $\Delta t_n = \Delta t_{n+1} = const$

$$\boxed{\frac{s_n}{s_{n+1}} = \frac{F(r_{n+1})}{F(r_n)}} \tag{21}$$

This equation is defined as the reversibility axiom. Following the request from the reversibility axiom the discrete time mechanics is time reversible if time is a dynamic variable. Time reversibility is gained by the use of the scaling of time (i.e. s_i), which introduces a friction term into the discrete Newtonian' law such, that if in the time forward step the force increases ($F(r_{n+1}) > F(r_n)$)) this force is scaled down by the friction term in the step backward in order to fulfill reversibility.

If however the time steps are constant in size (i.e. $s_i = 1$ for all $i = 1...N+1$) the microscopic system evolves time irreversible (unless $F(r_n) = const = F(r_{n+1})$) because the reversibility axiom (eq. 21) is not fulfilled. That complex processes with many particles under a complex force (i.e. $F(r_n) \neq const$) are time irreversible is in line with our daily experiences (such as a glass vessel that falls from a table and breaks into pieces), and thus sound. It is therefore the view of the author that the presented discrete time mechanics with $\Delta t_n = const = \Delta t$ (with Δt probably equal to the Planck time $5.4\,10^{-44}$ s and $s_i = 1$) may well describe nature, while the continuous, time reversible Newton mechanics is only an approximation. Most importantly, under this assumption time reversibility of a system under a complex force is lost and hence an arrow of time at the microscopic physics level is introduced automatically without the need to introduce a macromolecular ensemble description that follows thermodynamics.

If the presented discrete time mechanics is a more profound theory than the Newton's mechanics, on the one hand it should under certain boundary conditions asymptotically approximate the Newton's mechanics and thus time reversibility (which easily can be shown by $lim\, \Delta t \to 0$ in eq. 14) and on the other hand it should reflect irreversible properties (i.e. the entropy) of a thermodynamic state in its thermodynamic equilibrium which is described next.

2.4. The microscopic entropy of a single particle

Following Nosé and Hoover [2, 14, 17,18 20] from the discrete Lagrangian (eq. 5) a constant of motion evaluating to the total energy of the entire system is given by

$$H_n^N = \frac{1}{2} m\dot{r}_n^2 + V(r_n) + \frac{1}{2} Q \frac{\dot{s}_n^2}{s_n^2} + N_{df} k_B T\, ln\, s_n \tag{22}$$

This constant of motion is composed of the inner energy of the system U described by the first two terms followed by the energy of the bath and the exchange energy (i.e. the latter two terms having the variable s_n). Because of the resemblance of this term with the free energy $A = U - TS$ a microscopic entropy of a single particle at time point t_n is defined to be

$$S_n = -N_{df}\, k_B\, ln\, s_n \tag{23}$$

yielding

$$H_n^N = \frac{1}{2} m\dot{r}_n^2 + V(r_n) + \frac{1}{2} Q \frac{\dot{s}_n^2}{s_n^2} - T\, S_n = U + \frac{1}{2} Q \frac{\dot{s}_n^2}{s_n^2} - T\, S_n \tag{24}$$

Please note that this microscopic entropy S_n is a non ensemble-averaged term and concomitantly is in general of reversible character and thus not a monotonously increasing quantity. Furthermore, it can be calculated for a single particle system for a single event without the boundary of the system to be in thermodynamic equilibrium.

2.5. The microscopic entropy of a system with many particles

To extend the system from one particle to a system with Z non interacting particles the potential $V_i(r_{i,n}))$ for each particle i and a particle individual scaling of time denoted $s_{i,n}$, following the reversibility axiom eq. 21: $\frac{s_{i,n+1}}{s_{i,\,n}} = \frac{F_i(r_{i,n})}{F_i(r_{i,n+1})}$, are introduced yielding a constant of motion evaluating to the total energy of the Z particles system of type

$$H_n = \sum_{i=1}^{Z} \frac{1}{2} m_i \dot{r}_{i,n}^2 + V_i(r_{i,n}) + \frac{1}{2} Q_i \frac{\dot{s}_{i,n}^2}{s_{i,n}^2} + N_{i,df}\, k_B T\, ln\, s_{i,n} \tag{25}$$

In accordance with eq. 23, the microscopic entropy of the Z particles system is defined as

$$S_n = -\sum_{i=1}^{Z} N_{i,df}\, k_B\, ln\, s_{i,n} \tag{26}$$

In an attempt to simplify eqs. 25 and 26 by reducing amongst others the number of scaling factors $s_{i,n}$ it is assumed that there are J groups of particles and within each group the particles have undistinguishable properties but are still distinct since the system is microscopically described (i.e. within a group each particle has the same degrees of freedom, the same response to the potential, the same scaling factor, and the same velocity and mass). We define $p_j = \frac{\#\, of\, particles\, in\, group\, j}{Z} = \frac{Z_j}{Z}$ with $\sum_{j=1}^{J} p_j = 1$. For very large systems (i.e. at the thermodynamic limit) p_j is also the probability of a particle to be in the j group. Thus, the derived microscopic entropy of the Z particles system at time point n is given by

$$S_n = -Z \sum_{j=1}^{J} p_j\, N_{j,df}\, k_B\, ln\, s_{j,n} =< -k_B\, g\, ln\, s_n >= -k_B\, N_{df} < ln\, s_n > \tag{27}$$

with $N_{df} = Z\, N_{i,df}$ requesting for simplicity reasons that all particles have the same degree of freedom. N_{df} describes thus the degree of freedom of the entire system.

In the next step, it is assumed that the scaling of time is very small (i.e. $s_{j,n}$ is very close to 1). This assumption is called the "slow changing force limit". It is valid if the change of the force from one time step to the next is very small (i.e. $F(r_{n+1}) \approx F(r_n) + \Delta$ with Δ small). Under this assumption the ln of eq. 27 can be described by a Taylor expansion of first order and the averaging can be put inside the ln, which results in

$$S_n = -k_B\, N_{df}\; < ln\, s_n > \approx -k_B\, N_{df} \sum_{j=1}^{J} p_j\; (s_{j,n} - 1) = -k_B\, N_{df}\, [(\sum_{j=1}^{J} p_j\; s_{j,n}) - 1] \tag{28}$$

$$S_n \approx -k_B\, N_{df}\, ln \sum_{j=1}^{J} p_j\; s_{j,n} = -k_B\, N_{df}\; ln < s_n > \tag{29}$$

Correspondingly, the microscopic entropy difference between two time points n and m for a Z particles system is given by

$$\Delta S \;=\; S_m - S_n = -k_B\, N_{df}(< ln\, s_m > - < ln\, s_n >) \approx -k_B\, N_{df}\, ln \frac{< s_m >}{< s_n >} \tag{30}$$

It is interesting to note that this description can be used to calculate for a Z particles system at the thermodynamic limit the microscopic entropy difference between two time points by using averaged time scaling factors. Furthermore it is reiterated, that in contrast to the macroscopic thermodynamic entropy, the microscopic entropy can also be calculated for a system with a single particle or a few particles at any given time point n. In addition, the microscopic entropy difference can be calculated without the request that the system in both states is in thermodynamic equilibrium. Furthermore, the averaging is not over all possible states as in statistical mechanics, but over the observed state.

2.6. Boltzmann entropy versus the microscopic entropy of a many particles system

In the following it is the aim to show for a large system of Z independent, identical, indistinguishable particles (such as the ideal gas or diluted gas) a profound relationship between the microscopic entropy and the corresponding macroscopic one, i.e. the Boltzmann entropy. For the system of interest each micro-state of the macroscopic system in thermodynamic equilibrium has the same probability yielding the Boltzmann entropy to be

$$S^B = -k_B\; < ln\, p_j > = -k_B N_{df} \sum_{j=1}^{J} p_j\, ln\, p_j \tag{31}$$

with $p_j = \frac{Z_j}{Z}$ being the probability of a single particle to be in the state j and the average is a single particle average taken over all the possible states J of the particle.

In comparison, the microscopic entropy of this Z particles system is given by $S_n = -k_B\, N_{df} \sum_{j=1}^{J} p_j\, ln\, s_{j,n}$. Because the description is microscopic, each particle is distinct and thus requests its own scaling of time (i.e. $s_{j,n}$). In the "slow exchanging force limit" the averaging can be put inside the ln

$$S_n \approx -\, k_B\, Z\, ln < s_n > = -k_B\, N_{df}\, ln \sum_{j=1}^{J} p_j\, s_{j,n} \tag{32}$$

Since the Boltzmann entropy is only defined for an ensemble, for a comparison between the microscopic and the Boltzmann entropy, single particle averaging of the microscopic entropy is necessary. This can be obtained as follows

$$< S_n > \approx -k_B\, N_{df} < ln \sum_{j=1}^{J} p_j\, s_{j,n} > \approx -k_B\, N_{df} \sum_{j=1}^{J} p_j\, ln\, (p_j\; s_{j,n}) \tag{33}$$

A classical thermodynamic system in equilibrium for which the entropy is calculated can be assumed to be in the "slow changing force" limit with $s_{l,n}$ close to 1, yielding

$$< S_n >\approx -k_B\, N_{df} \sum_{l=1}^{J} p_l\, ln(p_l\, s_{l,n}) \approx -k_B\, N_{df} \sum_{l=1}^{J} p_l\, ln\, p_l \;=\; S^B \tag{34}$$

Thus, for a system in the "slow changing force limit" with independent non interacting particles, the average microscopic entropy in equilibrium approximates the Boltzmann entropy of the system. By setting the scaling factor to 1, also the time reversibility property of the microscopic entropy is lost (i.e. the reversibility axiom of eq. 21 is not fulfilled anymore, please also note, that time irreversibility of the microscopic entropy can also be obtained by setting Q to infinity, for more details see reference [19]).

3. Conclusion

An alternative microscopic derivation of entropy is given, which originates from the quantization of time, which is a concept that has only occasionally been investigated in the past. This theory introduces an arrow of time already at the microscopic physics level. Furthermore, it connects the microscopic physics with the macroscopic one without a statistical argument and can thus be calculated for any system (and thus also for a system with a small number of particles), which is not possible for the Boltzmann entropy. In contrast to the thermodynamic entropy, the Boltzmann entropy gets extensive and thus a state function only for a system at the thermodynamic limit (with number of particles -> infinity) [23]. It also resolves the many objections that have been raised for the Boltzmann entropy, since it can be calculated for a single system without ensemble averaging and without the request that the system under study must be in thermodynamic equilibrium.

In addition, in the presented derivation the presence of the entropy term being part of the total energy of the system is a consequence of a friction term in the discrete Newton's equation that is a consequence of the request for time reversibility, while in the Boltzmann approach oddly a statistical argument is part of the total energy of the system.

In summary, it is suggested that the original description of the entropy by Boltzmann is just an ensemble averaging of the time scaling variable s_n which is in equilibrium close to 1, but that the entropy term itself has its root not in statistical mechanics but rather in the discreteness of time.

Acknowledgments

The present manuscript is adapted in part from paper [19] published by the author.

References

[1] Landau L D and Lifshitz E M 1980 *Statistical Physics* 5 (3 ed) (Oxford: Pergamon Press) ISBN 0-7506-3372-7, translated by J B Sykes and M J Kearsley
[2] Hoover W G 1999 *Adv. Series in nonlinear dynamics* **13** 1 (ISBN 981-02-4073-2)
[3] Loschmidt J 1876 *Sitzungsber. Kais. Akad. Wiss. Wien, Math. Naturwiss. Classe* 73 128
[4] Poincaŕe H 1890 *Acta Math.* **13** 1
[5] Prigogine I 1997 *The End of Certainty: Time, Chaos and the New Laws of Nature* (New York: The Free Press)
[6] Prize H 1996 *Time's arrow and Archimedes' Point: New directions for the physics of time* ISBN 0-19-510095-6
[7] Farias R A H and Recami E 2007 *Preprint* arXiv:quant-ph/9706059
[8] Jaroszkiewicz G and Norton K 1997 *J. Phys. A.: Math. Gen* **30** 3115
[9] Jaroszkiewicz G and Norton K 1998 *J. Phys. A.: Math. Gen* **31** 977
[10] Lee T D 1983 *Phys. Lett.* B **122** 217
[11] Levi R 1927 *Journal de Physique et le Radium* **8** 182
[12] Thomson J J 1925 *Proc. Roy. Soc. of Edinburgh* **46** 90

[13] Yang C N 1947 *Phys. Rev.* **72** 874
[14] Nosè S 1984 *J. Chem. Phys.* **81** 511
[15] Nosè S 1984 *Mol. Phys.* **52** 255
[16] Nosè S 1986 *Mol. Phys.* **57** 187
[17] Hoover W G 1985 *Phys. Rev.* A **31** 1695
[18] Hoover W G 2007 *Mol. Simulation* **33** 13.
[19] Riek R 2014 *Entropy* **16** 3149
[20] Martyna G J, Klein M L and Tuckerman M 1992 *J. Chem. Phys.* **97** 2635
[21] Hünenberger P H 2005 *Adv. Polym. Sci.***173** 105
[22] Campisi M and Hänggi P 2013 *J. Phys Chem.* B **117** 12829
[23] Styer DF 2004 *Am. J. Phys.* **72** 25

Journal of Physics: Conference Series **626** (2015) 012026 doi:10.1088/1742-6596/626/1/012026

Disentangling mass and angle dependence in neutrino mixing

Massimo Blasone[1,2], **Maria Vittoria Gargiulo**[1,2,3] **and Giuseppe Vitiello**[1,2]

[1] Dipartimento di Fisica "E.R.Caianiello", Università degli Studi di Salerno, Via Ponte don Melillo, I-84084 Fisciano (SA), Italy
[2] INFN Sezione di Napoli, Gruppo collegato di Salerno, Italy
[3] Department of Physics, King's College, University of London, Strand WC2R 2LS, London, U.K.

E-mail: blasone@sa.infn.it

Abstract. We show that it is possible to decompose the generator of mixing transformations for Dirac fields, in terms of a rotation and two ordinary Bogoliubov transformations, accounting for mass shift in the fermion fields. This result clarifies the nature of the mechanism which leads to the unitary inequivalence of the flavor and mass vacua.

1. Introduction

The phenomenon of neutrino mixing and oscillations [1, 2] is certainly one of the hottest topics in the physics beyond the Standard Model of elementary particle physics [3]. The study of particle mixing, in particular of neutrino mixing, has therefore attracted much attention in the past decades, especially in connection with the experimental observation of the mixing phenomenon. After Pontecorvo's pioneer work, the theoretical basis of neutrino mixing has been studied in many details and the quantum field theory (QFT) formalism for mixed fields has been much developed [4]-[13]. The mixing phenomenon appears today to be a truly quantum field phenomenon, whose quantum mechanical approximation reproduces the Pontecorvo formalism and its results. Far from being a simple minded rotation among massive neutrinos, the field mixing appears to be the result of all the complex mathematical structure of QFT. The discovery of the unitary inequivalence between the massive neutrino vacuum and the flavored neutrino vacuum has displayed the condensate nature of the flavored vacuum. While the origin of the non vanishing neutrino masses remain a still open problem, the nature of the mixed, flavored vacuum seems to be clarified in many details. In the present report, by resorting to recent results [14], we analyze the apparently puzzling dependence of the mixing mechanism both, on the neutrino mass values and on the mixing angle. As well known, since Pontecorvo first formulation, in order to occur the neutrino mixing, neutrino masses need to be non zero and different among themselves, from one side; on the other side, the mixing angle must be, of course, non zero. The question then arises if any relation exists between the masses and the mixing angle, and which one is such a relation, if any. The result we obtain, to be presented in this report, is that a possibility exists to disentangle in the mixing transformation the dependence of the mixing generator on the angle form the one on the masses. The way this occurs shows the crucial role played by the condensate of the flavored vacuum, thus reinforcing the QFT nature of the mixing phenomenon. Once the vacuum has acquired its condensate structure, it is no more invariant under rotation by the mixing angle θ.

Our analysis is limited to the case of two Dirac neutrinos. Extension to three neutrinos is in our plans. However, we have good reasons to believe that the present results are general, since our arguments are of algebraic nature. The results also extend to the mixing phenomenon of any particle, and are not limited to the case of Dirac neutrinos.

Since we will use them in our discussion, we need to introduce very few aspects and notations of the QFT mixing formalism, which is done in Section 2. In Section 3 the decomposition of the mixing generator is obtained in terms of elementary transformations and in Section 4 the energies of the various vacuum states associated to these transformations are considered. Section 5 is devoted to conclusions.

2. Fermion mixing

The mixing relations of two (Dirac) neutrino fields with definite flavors ν_e, ν_μ are

$$\nu_e(x) = \cos\theta\;\nu_1(x) + \sin\theta\;\nu_2(x)\,; \qquad \nu_\mu(x) = -\sin\theta\;\nu_1(x) + \cos\theta\;\nu_2(x)\,. \tag{1}$$

ν_1, ν_2 are the (free) neutrino fields with definite masses m_1, m_2, respectively. θ is the mixing angle. The fields ν_1 and ν_2 are expanded as

$$\nu_i(x) = \frac{1}{\sqrt{V}}\sum_{\mathbf{k},r}\left[u^r_{\mathbf{k},i}(t)\alpha^r_{\mathbf{k},i} + v^r_{-\mathbf{k},i}(t)\beta^{r\dagger}_{-\mathbf{k},i}\right]e^{i\mathbf{k}\cdot\mathbf{x}}, \qquad i=1,2\,. \tag{2}$$

where $u^r_{\mathbf{k},i}(t) = e^{-i\omega_{k,i}t}u^r_{\mathbf{k},i}$ and $v^r_{\mathbf{k},i}(t) = e^{i\omega_{k,i}t}v^r_{\mathbf{k},i}$, with $\omega_{k,i} = \sqrt{\mathbf{k}^2+m_i^2}$. The $\alpha^r_{\mathbf{k},i}$ and the $\beta^r_{\mathbf{k},i}$ $(r=1,2)$, are the annihilation operators for the vacuum state $|0\rangle_{1,2} \equiv |0\rangle_1 \otimes |0\rangle_2$: $\alpha^r_{\mathbf{k},i}|0\rangle_{1,2} = \beta^r_{\mathbf{k},i}|0\rangle_{1,2} = 0$. The standard anticommutation relations are:

$$\{\nu_i^\alpha(x), \nu_j^{\beta\dagger}(y)\}_{t=t'} = \delta^3(\mathbf{x}-\mathbf{y})\delta_{\alpha\beta}\delta_{ij}\;, \qquad \alpha,\beta = 1,..,4\,, \tag{3}$$

$$\{\alpha^r_{\mathbf{k},i}, \alpha^{s\dagger}_{\mathbf{q},j}\} = \delta_{\mathbf{kq}}\delta_{rs}\delta_{ij}; \qquad \{\beta^r_{\mathbf{k},i}, \beta^{s\dagger}_{\mathbf{q},j}\} = \delta_{\mathbf{kq}}\delta_{rs}\delta_{ij}, \quad i,j=1,2\,. \tag{4}$$

Eqs.(1) can be recast as [4]:

$$\nu_e^\alpha(x) = G_\theta^{-1}(t)\;\nu_1^\alpha(x)\;G_\theta(t)\,; \qquad \nu_\mu^\alpha(x) = G_\theta^{-1}(t)\;\nu_2^\alpha(x)\;G_\theta(t)\,, \tag{5}$$

where the generator $G_\theta(t)$ is given by

$$G_\theta(t) = \exp\left[\theta\int d^3\mathbf{x}\left(\nu_1^\dagger(x)\nu_2(x) - \nu_2^\dagger(x)\nu_1(x)\right)\right]\,, \tag{6}$$

G_θ is an element of $SU(2)$ [4]. It can be written as

$$G_\theta(t) = \exp[\theta(S_+(t) - S_-(t))] \quad \text{with} \quad S_+(t) = S_-^\dagger(t) \equiv \int d^3\mathbf{x}\;\nu_1^\dagger(x)\nu_2(x)\,. \tag{7}$$

By introducing

$$S_3 \equiv \frac{1}{2}\int d^3\mathbf{x}\left(\nu_1^\dagger(x)\nu_1(x) - \nu_2^\dagger(x)\nu_2(x)\right)\,, \tag{8}$$

the $su(2)$ algebra is closed (at fixed t):

$$[S_+(t), S_-(t)] = 2S_3\quad, \quad [S_3, S_\pm(t)] = \pm S_\pm(t)\,, \tag{9}$$

By use of $G_\theta(t)$, the flavor fields can be expanded as:

$$\nu_\sigma(x) = \sum_{r=1,2} \int \frac{d^3\mathbf{k}}{(2\pi)^{\frac{3}{2}}} \left[u^r_{\mathbf{k},i}(t)\alpha^r_{\mathbf{k},\sigma}(t) + v^r_{-\mathbf{k},i}(t)\beta^{r\dagger}_{-\mathbf{k},\sigma}(t)\right] e^{i\mathbf{k}\cdot\mathbf{x}}, \tag{10}$$

with $(\sigma, i) = (e,1),(\mu,2)$. The flavor annihilation operators are defined as $\alpha^r_{\mathbf{k},\sigma}(t) \equiv G^{-1}_\theta(t)\alpha^r_{\mathbf{k},i}G_\theta(t)$ and $\beta^{r\dagger}_{-\mathbf{k},\sigma}(t) \equiv G^{-1}_\theta(t)\beta^{r\dagger}_{-\mathbf{k},i}G_\theta(t)$. For $\mathbf{k} = (0,0,|\mathbf{k}|)$, we have :

$$\alpha^r_{\mathbf{k},e}(t) = \cos\theta\;\alpha^r_{\mathbf{k},1} \;+\; \sin\theta\;\left(U^*_{\mathbf{k}}(t)\;\alpha^r_{\mathbf{k},2} \;+\; \epsilon^r\;V_{\mathbf{k}}(t)\;\beta^{r\dagger}_{-\mathbf{k},2}\right) \tag{11}$$

$$\alpha^r_{\mathbf{k},\mu}(t) = \cos\theta\;\alpha^r_{\mathbf{k},2} \;-\; \sin\theta\;\left(U_{\mathbf{k}}(t)\;\alpha^r_{\mathbf{k},1} \;-\; \epsilon^r\;V_{\mathbf{k}}(t)\;\beta^{r\dagger}_{-\mathbf{k},1}\right) \tag{12}$$

$$\beta^r_{-\mathbf{k},e}(t) = \cos\theta\;\beta^r_{-\mathbf{k},1} \;+\; \sin\theta\;\left(U^*_{\mathbf{k}}(t)\;\beta^r_{-\mathbf{k},2} \;-\; \epsilon^r\;V_{\mathbf{k}}(t)\;\alpha^{r\dagger}_{\mathbf{k},2}\right) \tag{13}$$

$$\beta^r_{-\mathbf{k},\mu}(t) = \cos\theta\;\beta^r_{-\mathbf{k},2} \;-\; \sin\theta\;\left(U_{\mathbf{k}}(t)\;\beta^r_{-\mathbf{k},1} \;+\; \epsilon^r\;V_{\mathbf{k}}(t)\;\alpha^{r\dagger}_{\mathbf{k},1}\right) \tag{14}$$

where $\epsilon^r = (-1)^r$ and

$$U_{\mathbf{k}}(t) \equiv u^{r\dagger}_{\mathbf{k},2}(t)u^r_{\mathbf{k},1}(t) = v^{r\dagger}_{-\mathbf{k},1}(t)v^r_{-\mathbf{k},2}(t) = |U_{\mathbf{k}}|\,e^{i(\omega_{k,2}-\omega_{k,1})t} \tag{15}$$

$$V_{\mathbf{k}}(t) \equiv \epsilon^r\;u^{r\dagger}_{\mathbf{k},1}(t)v^r_{-\mathbf{k},2}(t) = -\epsilon^r\;u^{r\dagger}_{\mathbf{k},2}(t)v^r_{-\mathbf{k},1}(t) = |V_{\mathbf{k}}|\;e^{i(\omega_{k,2}+\omega_{k,1})t} \tag{16}$$

$$|U_{\mathbf{k}}| = \frac{|\mathbf{k}|^2 + (\omega_{k,1}+m_1)(\omega_{k,2}+m_2)}{2\sqrt{\omega_{k,1}\omega_{k,2}(\omega_{k,1}+m_1)(\omega_{k,2}+m_2)}}, \qquad |U_{\mathbf{k}}|^2 + |V_{\mathbf{k}}|^2 = 1. \tag{17}$$

The action of the mixing generator on the vacuum $|0\rangle_{1,2}$ is non-trivial; at finite volume V we have:

$$|0(t)\rangle_{e,\mu} \equiv G^{-1}_\theta(t)\;|0\rangle_{1,2}\,. \tag{18}$$

In the infinite volume limit one obtains [4]

$$\lim_{V\to\infty}\;{}_{1,2}\langle 0|0(t)\rangle_{e,\mu} = \lim_{V\to\infty}\;e^{\frac{V}{(2\pi)^3}\int d^3\mathbf{k}\;ln\;(1-\sin^2\theta\;|V_{\mathbf{k}}|^2)} = 0 \tag{19}$$

Eq.(19) expresses the unitary inequivalence between the flavor and the mass representations and shows the non-trivial nature of the mixing transformations (1) resulting in the condensate structure of the flavor vacuum [4, 13].

The form of the flavor vacuum (at $t = 0$) is the following one:

$$\begin{aligned}|0\rangle_{e,\mu} \;=\; \prod_{\mathbf{k},r}\Big[&(1-\sin^2\theta|V_{\mathbf{k}}|^2) - \epsilon^r\sin\theta\cos\theta|V_{\mathbf{k}}|(\alpha^{r\dagger}_{\mathbf{k},1}\beta^{r\dagger}_{-\mathbf{k},2} + \alpha^{r\dagger}_{\mathbf{k},2}\beta^{r\dagger}_{-\mathbf{k},1}) + \\ &+\epsilon^r\sin^2\theta|V_{\mathbf{k}}||U_{\mathbf{k}}|(\alpha^{r\dagger}_{\mathbf{k},1}\beta^{r\dagger}_{-\mathbf{k},1} - \alpha^{r\dagger}_{\mathbf{k},2}\beta^{r\dagger}_{-\mathbf{k},2}) + \sin^2\theta|V_{\mathbf{k}}|^2\alpha^{r\dagger}_{\mathbf{k},1}\beta^{r\dagger}_{-\mathbf{k},2}\alpha^{r\dagger}_{\mathbf{k},2}\beta^{r\dagger}_{-\mathbf{k},1}\Big]\;|0\rangle_{1,2}\end{aligned} \tag{20}$$

from which it is evident the condensate nature made of particle-antiparticle pairs with same or different masses.

The condensation density of the flavor vacuum is given by

$${}_{e,\mu}\langle 0(t)|\alpha^{r\dagger}_{\mathbf{k},i}\alpha^r_{\mathbf{k},i}|0(t)\rangle_{e,\mu} = {}_{e,\mu}\langle 0(t)|\beta^{r\dagger}_{\mathbf{k},i}\beta^r_{\mathbf{k},i}|0(t)\rangle_{e,\mu} = \sin^2\theta\;|V_{\mathbf{k}}|^2, \qquad i = 1,2\,, \tag{21}$$

with the same result for antiparticles. Note that the $|V_{\mathbf{k}}|^2$ has a maximum at $\sqrt{m_1 m_2}$ and $|V_{\mathbf{k}}|^2 \simeq \frac{(m_2-m_1)^2}{4|\mathbf{k}|^2}$ for $|\mathbf{k}| \gg \sqrt{m_1 m_2}$.

Finally, let us consider the explicit expression of the mixing generator, obtained by use of the expressions for the free fields ν_i:

$$S^{\mathbf{k},r}_{+} = \left(U^*_{\mathbf{k}}\;\alpha^{r\dagger}_{\mathbf{k},1}\alpha^r_{\mathbf{k},2} - \epsilon^r\;V^*_{\mathbf{k}}\;\beta^r_{-\mathbf{k},1}\alpha^r_{\mathbf{k},2} + \epsilon^r\;V_{\mathbf{k}}\;\alpha^{r\dagger}_{\mathbf{k},1}\beta^{r\dagger}_{-\mathbf{k},2} + U_{\mathbf{k}}\;\beta^r_{-\mathbf{k},1}\beta^{r\dagger}_{-\mathbf{k},2}\right) \tag{22}$$

where we defined $S_\pm = \sum_{\mathbf{k}} S^{\mathbf{k}}_\pm$, with $S^{\mathbf{k}}_+ = S^{\mathbf{k}\dagger}_- \equiv \sum_r S^{\mathbf{k},r}_+$

3. Decomposition of the mixing generator

The mixing generator $G_\theta(t)$, Eq(6), is a function of m_1, m_2, and θ. Our purpose is now to disentangle the mass dependence of $G_\theta(t)$ from its dependence on the rotation angle[1]. In order to do that, let us define [14]:

$$R(\theta) = \exp\Big\{\theta\sum_{\mathbf{k},r}\Big[\Big(\alpha^{r\dagger}_{\mathbf{k},1}\alpha^{r}_{\mathbf{k},2}+\beta^{r\dagger}_{-\mathbf{k},1}\beta^{r}_{-\mathbf{k},2}\Big)e^{i\psi_k}-\Big(\alpha^{r\dagger}_{\mathbf{k},2}\alpha^{r}_{\mathbf{k},1}+\beta^{r\dagger}_{-\mathbf{k},2}\beta^{r}_{-\mathbf{k},1}\Big)e^{-i\psi_k}\Big]\Big\}, \quad (23)$$

$$B_i(\Theta_i) = \exp\Big\{\sum_{\mathbf{k},r}\Theta_{\mathbf{k},i}\,\epsilon^r\Big[\alpha^{r}_{\mathbf{k},i}\beta^{r}_{-\mathbf{k},i}e^{-i\phi_{k,i}}-\beta^{r\dagger}_{-\mathbf{k},i}\alpha^{r\dagger}_{\mathbf{k},i}e^{i\phi_{k,i}}\Big]\Big\}, \qquad i=1,2. \quad (24)$$

Here the $B_i(\Theta_i)$, $i = 1, 2$, are the generators of Bogoliubov transformations and $R(\theta)$ is the rotation generator. Since $[B_1(\Theta_1), B_2(\Theta_2)] = 0$, we may also define $B(\Theta_1,\Theta_2) \equiv B_2(\Theta_2)B_1(\Theta_1)$ and then calculate how ladder operators transform under $B(\Theta_1,\Theta_2)$:

$$\tilde{\alpha}^{r}_{\mathbf{k},i} \equiv B^{-1}(\Theta_1,\Theta_2)\,\alpha^{r}_{\mathbf{k},i}\,B(\Theta_1,\Theta_2) = \cos\Theta_{\mathbf{k},i}\,\alpha^{r}_{\mathbf{k},i} - \epsilon^r\,e^{-i\phi_{k,i}}\,\sin\Theta_{\mathbf{k},i}\,\beta^{r\dagger}_{-\mathbf{k},i} \quad (25)$$

$$\tilde{\beta}^{r}_{-\mathbf{k},i} \equiv B^{-1}(\Theta_1,\Theta_2)\,\beta^{r}_{-\mathbf{k},i}\,B(\Theta_1,\Theta_2) = \cos\Theta_{\mathbf{k},i}\,\beta^{r}_{-\mathbf{k},i} + \epsilon^r\,e^{i\phi_{k,i}}\,\sin\Theta_{\mathbf{k},i}\,\alpha^{r\dagger}_{\mathbf{k},i} \quad (26)$$

On the other hand, the action of the rotation generator on the annihilation operators gives:

$$R(\theta)^{-1}\alpha^{r}_{\mathbf{k},1}R(\theta) = \cos\theta\,\alpha^{r}_{\mathbf{k},1} + e^{i\psi_k}\sin\theta\,\alpha^{r}_{\mathbf{k},2} \quad (27)$$

$$R(\theta)^{-1}\alpha^{r}_{\mathbf{k},2}R(\theta) = \cos\theta\,\alpha^{r}_{\mathbf{k},2} - e^{-i\psi_k}\sin\theta\,\alpha^{r}_{\mathbf{k},1} \quad (28)$$

and similar ones for the $\beta^r_{\mathbf{k},i}$.

The action of the above transformations on the vacuum for the free fields ν_1, ν_2 is given by

$$|\tilde{0}\rangle_{1,2} \equiv B^{-1}(\Theta_1,\Theta_2)|0\rangle_{1,2} = \prod_{i=1,2}\prod_{\mathbf{k},r}\Big[\cos\Theta_{\mathbf{k},i} + \epsilon^r e^{i\phi_{k,i}}\sin\Theta_{\mathbf{k},i}\,\alpha^{r\dagger}_{\mathbf{k},i}\beta^{r\dagger}_{-\mathbf{k},i}\Big]|0\rangle_{1,2} \quad (29)$$

$$R^{-1}(\theta)|0\rangle_{1,2} = |0\rangle_{1,2}\,. \quad (30)$$

Define now $\tilde{R} \equiv \tilde{R}(\theta,\Theta_1,\Theta_2) = B^{-1}(\Theta_1,\Theta_2)R^{-1}(\theta)B(\Theta_1,\Theta_2)$, which can be written as

$$\tilde{R} = \exp\Big\{\theta\sum_{\mathbf{k},r}\Big[\Big(\tilde{\alpha}^{r\dagger}_{\mathbf{k},1}\tilde{\alpha}^{r}_{\mathbf{k},2}+\tilde{\beta}^{r\dagger}_{-\mathbf{k},1}\tilde{\beta}^{r}_{-\mathbf{k},2}\Big)e^{i\psi_k}-\Big(\tilde{\alpha}^{r\dagger}_{\mathbf{k},2}\tilde{\alpha}^{r}_{\mathbf{k},1}+\tilde{\beta}^{r\dagger}_{-\mathbf{k},2}\tilde{\beta}^{r}_{-\mathbf{k},1}\Big)e^{-i\psi_k}\Big]\Big\}, \quad (31)$$

Thereby, using Eqs.(25), (26) and imposing the following constraint

$$U_{\mathbf{k}}(t) = e^{-i\psi_k}\cos(\Theta_{\mathbf{k},1}-\Theta_{\mathbf{k},2}) \quad (32)$$

$$V_{\mathbf{k}}(t) = e^{\frac{(\phi_{k,1}+\phi_{k,2})}{2}}\sin(\Theta_{\mathbf{k},1}-\Theta_{\mathbf{k},2}), \quad (33)$$

we obtain

$$\tilde{R} = \exp\Big\{\ \sum_r\Big(U^*_{\mathbf{k}}\,\alpha^{r\dagger}_{\mathbf{k},1}\alpha^{r}_{\mathbf{k},2} - \epsilon^r\,V^*_{\mathbf{k}}\,\beta^{r}_{-\mathbf{k},1}\alpha^{r}_{\mathbf{k},2} + \epsilon^r\,V_{\mathbf{k}}\,\alpha^{r\dagger}_{\mathbf{k},1}\beta^{r\dagger}_{-\mathbf{k},2} + U_{\mathbf{k}}\,\beta^{r}_{-\mathbf{k},1}\beta^{r\dagger}_{-\mathbf{k},2}\Big)$$
$$-\ \sum_r\Big(U_{\mathbf{k}}\,\alpha^{r\dagger}_{\mathbf{k},2}\alpha^{r}_{\mathbf{k},1} + \epsilon^r\,V^*_{\mathbf{k}}\,\beta^{r}_{-\mathbf{k},2}\alpha^{r}_{\mathbf{k},1} - \epsilon^r\,V_{\mathbf{k}}\,\alpha^{r\dagger}_{\mathbf{k},2}\beta^{r\dagger}_{-\mathbf{k},1} + U^*_{\mathbf{k}}\,\beta^{r}_{-\mathbf{k},2}\beta^{r\dagger}_{-\mathbf{k},1}\Big)\Big\} \quad (34)$$

[1] A first (partial) attempt in doing this is contained in Ref.[4].

Journal of Physics: Conference Series **626** (2015) 012026 doi:10.1088/1742-6596/626/1/012026

which indeed coincides with the expression of the mixing generator $G_\theta(t)$, cf. Eq.(22), provided we make the following identifications:

$$\psi_k = (\omega_{k,1} - \omega_{k,2})\, t \tag{35}$$

$$\phi_{k,i} = 2\,\omega_{k,i}\, t \tag{36}$$

$$\Theta_{\mathbf{k},i} = \frac{1}{2}\cot^{-1}\left(\frac{|\mathbf{k}|}{m_i}\right) \tag{37}$$

In definitive, we have shown that it is possible to decompose the mixing generator G_θ in the following way (for $t = 0$)

$$G(\theta, m_1, m_2) = B^{-1}(\Theta_1, \Theta_2)\, R(\theta)\, B(\Theta_1, \Theta_2) \tag{38}$$

i.e., we have been able to write G as a product of operators depending only on the masses or on the mixing angle. From the above decomposition, it is quite evident that the non-trivial nature of the mixing generator arises as a consequence of the non-commutativity of the rotation generator with the generator of Bogoliubov transformation(s).

From Eq.(35), we see that $\Theta_{\mathbf{k},i}$ are functions of the masses and the momentum only. Thus we can regard the generator $B(\Theta_1, \Theta_2)$, where the momentum has been integrated out, as dependent on the mass parameters, i.e. as $B(m_1, m_2)$. Then we rewrite Eq.(39) as

$$G_t(\theta, m_1, m_2) = B_t^{-1}(m_1, m_2)\, R_t(\theta)\, B_t(m_1, m_2) \tag{39}$$

where we have included now the dependence on time as an index t.

Finally, let us consider the flavor vacuum (for $t = 0$), which can be written as

$$|0\rangle_{e,\mu} \equiv G^{-1}|0\rangle_{1,2} = |0\rangle_{1,2} + \left[B(m_1, m_2)\, ,\, R^{-1}(\theta)\right] |\widetilde{0}\rangle_{1,2} \tag{40}$$

where we see as the condensate structure arises as a consequence of the non vanishing commutator $[B, R^{-1}]$.

4. Vacuum energies

Having shown that the mixing generator can be written as $G = B^{-1}\, R\, B$, we use now such a decomposition to investigate the nature of the flavor vacuum, starting from $|0\rangle_{1,2}$ and acting on it with the above Bogoliubov transformations and rotation generators. In the following table we report the vacuum expectation values of the (unordered) Hamiltonian on the various *vacua* obtained by acting step-by-step with the above generators.

$\langle H_{\mathbf{k},1} + H_{\mathbf{k},2}\rangle$	State
$-(\omega_{k,1} + \omega_{k,2})$	$\|0\rangle_1 \otimes \|0\rangle_2 \equiv \|0\rangle_{1,2}$
$-(k + k)$	$B^{-1}(m_1, m_2)\|0\rangle_{1,2} \equiv \|\widetilde{0}\rangle_{1,2}$
$-k(2\cos^2\theta + (\frac{\omega_{k,1}}{\omega_{k,2}} + \frac{\omega_{k,2}}{\omega_{k,1}})\sin^2\theta)$	$R^{-1}(\theta)B^{-1}(m_1, m_2)\|0\rangle_{1,2} = R^{-1}(\theta)\|\widetilde{0}\rangle_{1,2}$
$-(\omega_{k,1} + \omega_{k,2})(1 - 2\sin^2\theta\sin^2(\Theta_{\mathbf{k},1} - \Theta_{\mathbf{k},2}))$	$B(m_1, m_2)R^{-1}(\theta)B^{-1}(m_1, m_2)\|0\rangle_{1,2} \equiv \|0\rangle_{e,\mu}$

From the results shown in the table we can understand the steps that lead to the non-equivalence between the two vacua. Starting from the vacuum for the free fields $|0\rangle_{1,2}$, the first Bogoliubov

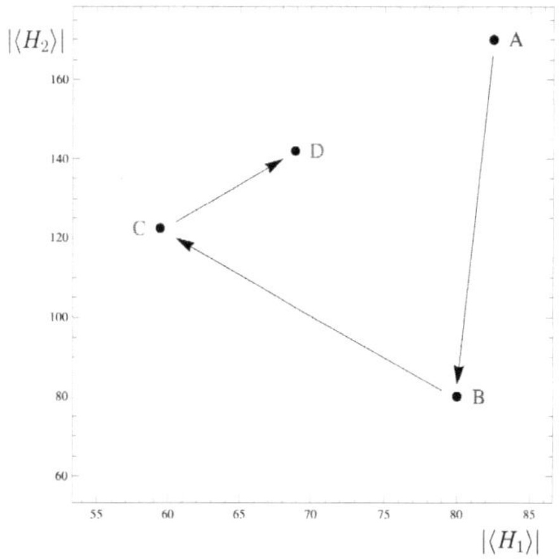

Figure 1. Plot of vacuum expectation values of H_1 and H_2 for the states of Table 1. Sample values of parameters are chosen for $\theta = \pi/4$, $k = 80$, $m_1 = 20$, $m_2 = 150$. We put $A = |0\rangle_{1,2}$, $B = |\widetilde{0}\rangle_{1,2}$, $C = R^{-1}(\theta)|\widetilde{0}\rangle_{1,2}$, and $D = |0\rangle_{e,\mu}$.

transformation(s) acts as a mass shift that brings the two species of neutrinos to have a massless vacuum. The second transformation that acts is a rotation, then it follows the inverse Bogoliubov transformation which however cannot put back masses at their original position, since in the meanwhile they have been "rotated" by R. This mechanism is essentially due to the non vanishing commutator $[B, R^{-1}]$.

The situation is clarified by Fig.1, where we plot, for sample value of the parameters, the (absolute values of) expectation values of $H_{\mathbf{k},1}$ and $H_{\mathbf{k},1}$ for the above vacua. The arrows indicate the "way" from $|0\rangle_{1,2}$ to $|0\rangle_{e,\mu}$, making clear the origin of the energy gap between these two vacua.

In the following figures, we consider other values of the parameters, showing various limits in which the effects of the condensate tends to be less important and eventually to disappear.

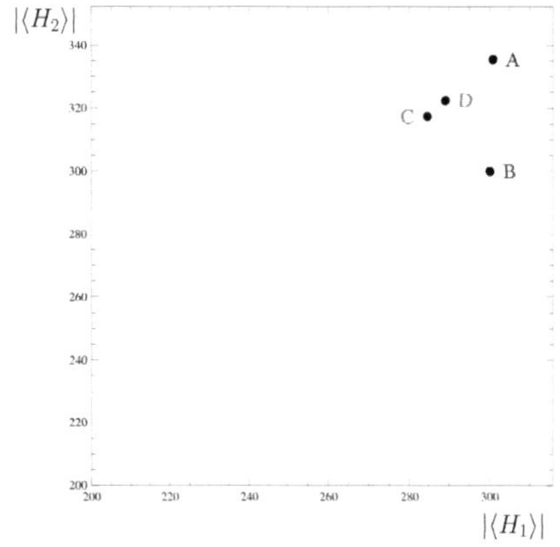

Figure 2. Plot of vacuum expectation values of H_1 and H_2 for the states of Table 1. Sample values of parameters are chosen for $\theta = \pi/4$, $k = 300$, $m_1 = 20$, $m_2 = 150$.

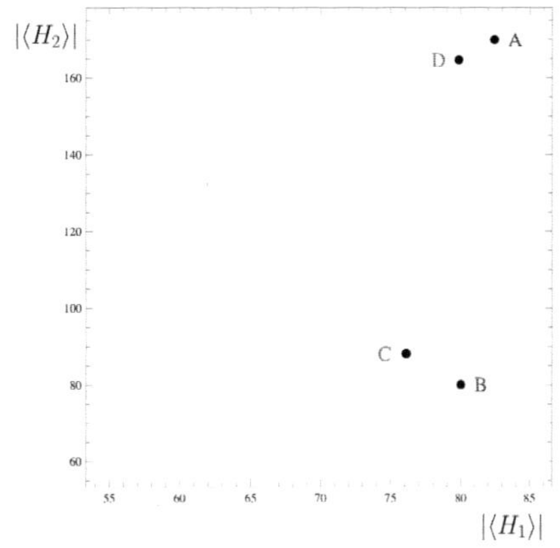

Figure 3. Plot of vacuum expectation values of H_1 and H_2 for the states of Table 1. Sample values of parameters are chosen for $\theta = \pi/10$, $k = 80$, $m_1 = 20$, $m_2 = 150$.

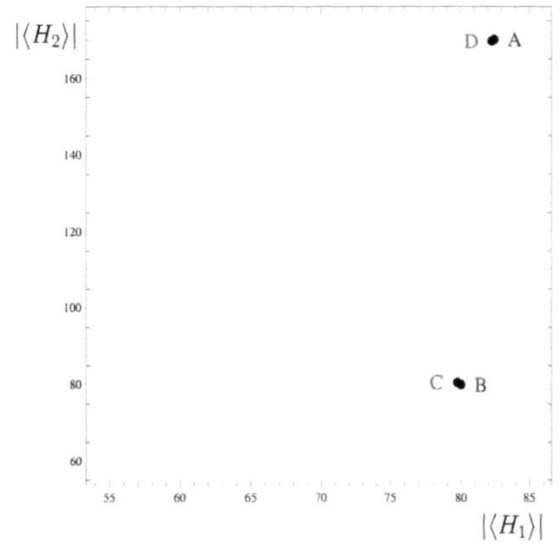

Figure 4. Plot of vacuum expectation values of H_1 and H_2 for the states of Table 1. Sample values of parameters are chosen for $\theta = \pi/40$, $k = 80$, $m_1 = 20$, $m_2 = 150$.

5. Conclusions

We have shown that it is possible to decompose the generator of flavor mixing transformations for two Dirac fields in terms of a rotation, parameterized only by the mixing angle θ, and two Bogoliubov transformations, depending only on the masses m_1 and m_2. This result allows for a better understanding of the mixing phenomenon and the associated inequivalence among the mass and flavor representations (Hilbert spaces), in terms of the non-commutativity of the two transformations (Bogoliubov and rotation).

It is interesting to note that the Bogoliubov transformations appearing in the above decomposition are responsible for mass shift in the fermion fields, and are of the same type of those used in Ref.[15], where the generation of masses and mixing was studied in the context of dynamical symmetry breaking.

Finally, the energy gap between the flavor and mass vacua can be associated to the entropy of the flavor vacuum. Such an issue will be discussed in detail in a separate publication [14].

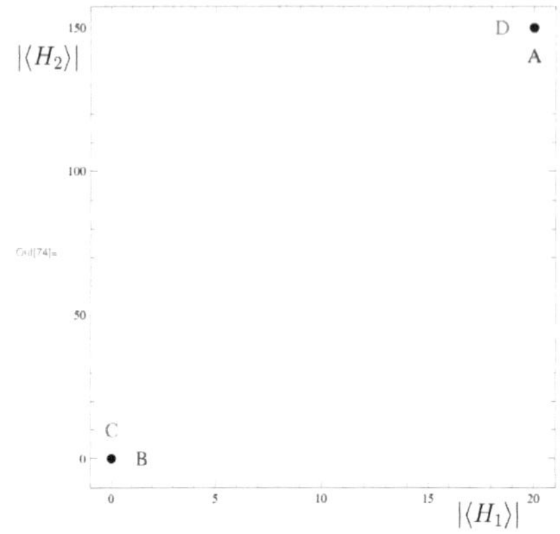

Figure 5. Plot of vacuum expectation values of H_1 and H_2 for the states of Table 1. Sample values of parameters are chosen for $\theta = \pi/4$, $k = 0$, $m_1 = 20$, $m_2 = 150$.

References

[1] Bilenky S M and Pontecorvo B 1978 *Phys. Rept.* **41** 225
[2] Giunti C and Kim C W 2007 *Fundamentals of Neutrino Physics and Astrophysics*, (Oxford University Press)
[3] Cheng T and Li L 1989 *Gauge Theory of Elementary Particle Physics* (Clarendon Press)
[4] Blasone M and Vitiello G 1995 *Ann. Phys. (N.Y.)* **244** 283
[5] Blasone M, Henning P A and Vitiello G 1996 In *La Thuile 1996, Results and perspectives in particle physics* 139 [hep-ph/9605335].
[6] Binger M and Ji C R 1999 *Phys. Rev.* D **60** 056005
[7] Blasone M, Henning P A and Vitiello G 1999 *Phys. Lett.* B **451** 140
[8] Blasone M, Jizba P and Vitiello G 2001 *Phys. Lett.* B **517** 471
[9] Blasone M, Capolupo A and Vitiello G 2002 *Phys. Rev.* D **66**, 025033
[10] Beuthe M 2003 *Phys. Rept.* **375** 105
[11] Blasone M, Di Mauro M and Vitiello G 2011 *Phys. Lett.* B **697** 238
[12] Blasone M 2011 *J. Phys. Conf. Ser.* **306** 012037
[13] Blasone M, Jizba P and Vitiello G 2011 *Quantum Field Theory and its Macroscopic Manifestations* - (World Scientific & ICP)
[14] Blasone M, Gargiulo M V and Vitiello G 2015, in preparation
[15] Blasone M, Jizba P, Lambiase G and Mavromatos N E 2014 *J.Phys.Conf.Ser.* **538** 012003.

7th International Workshop DICE2014 Spacetime – Matter – Quantum Mechanics IOP Publishing
Journal of Physics: Conference Series **626** (2015) 012027 doi:10.1088/1742-6596/626/1/012027

Experimental search for the "impossible atoms" Pauli Exclusion Principle violation and spontaneous collapse of the wave function at test

C Curceanu[abi], S Bartalucci[a], A Bassi[c], S Bertolucci[d], C Berucci[ae], A M Bragadireanu[ab], M Cargnelli[e], A Clozza[a], L De Paolis[a], S Di Matteo[f], S Donadi[c], A d'Uffizi[a], J-P Egger[g], C Guaraldo[a], M Iliescu[a], T Ishiwatari[e], M Laubenstein[h], J Marton[e], E Milotti[c], D Pietreanu[ab], K Piscicchia[ai], T Ponta[b], E Sbardella[a], A Scordo[a], H Shi[a], D L Sirghi[ab], F Sirghi[ab], L Sperandio[a], O Vazquez Doce[j], J Zmeskal[e]

[a]INFN, Laboratori Nazionali di Frascati, CP 13, Via E. Fermi 40, I-00044, Frascati (Roma), Italy
[b] "Horia Hulubei"National Institute of Physics and Nuclear Engineering, Str. Atomistilor no. 407, P.O. Box MG-6, Bucharest - Magurele, Romania
[c]Dipartimento di Fisica, Università di Trieste and INFN– Sezione di Trieste, Via Valerio, 2, I-34127 Trieste, Italy
[d]CERN, CH-1211, Geneva 23, Switzerland
[e]The Stefan Meyer Institute for Subatomic Physics, Boltzmanngasse 3, A-1090 Vienna, Austria
[f]Institut de Physique UMR CNRS-UR1 6251, Université de Rennes1, F-35042 Rennes, France
[g]Institut de Physique, Université de Neuchâtel, 1 rue A.-L. Breguet, CH-2000 Neuchâtel, Switzerland
[h]INFN, Laboratori Nazionali del Gran Sasso, S.S. 17/bis, I-67010 Assergi (AQ), Italy
[i]Museo Storico della Fisica e Centro Studi e Ricerche "Enrico Fermi", Roma, Italy
[j]Excellence Cluster Universe, Technische Universität München, Garching, Germany

E-mail: catalina.curceanu@lnf.infn.it

Abstract. Many experiments investigated the possible violation of the Pauli Exclusion Principle (PEP) since its discovery in 1925. The VIP(Violation of the Pauli Principle) experiment tested the PEP by measuring the probability for an external electron to be captured and undergo a 2p to 1s transition during its cascading process, with the 1s state already occupied by two electrons. This transition is forbidden by the PEP. The VIP experiment resulted in an upper limit for the probability of PEP violation of 4.7×10^{-29}. Currently a setup for the follow-up experiment VIP2 is under preparation. The goal of this experiment is to improve the upper limit for the violation of the PEP by two orders of magnitude, by using new X-ray detectors and by implementing an active shielding. We then present the idea of using an analogous experimental technique to search for X rays as a signature of the spontaneous collapse of the wave function, predicted by the continuous spontaneous localization theories, and discuss some very encouraging preliminary results.

1. Introduction

Formulated by Wolfgang Pauli in 1925 [1], the Pauli Exclusion Principle (PEP) is one of the building blocks of quantum mechanics. It is the foundation for our understanding of nuclear physics, particle physics, condensed-matter physics, and astrophysics where many-fermion systems are concerned. By far the principle does not have an intuitive explanation for its physical cause, and there is possibility that high precision experiment may discover small violation that could reveal more fundamental principles. However the experimental test is difficult because there has been no well-established parameter that can account for PEP violation quantitatively in a theory.

In the reviews given by Greenberg and Mohapatra [2, 3], they surveyed over the searches for a phenomenology framework for possible small violation of the PEP, and then pointed out that no satisfactory solution could be found to be consistent within a local field theory. However, they argued that following the parameterization proposed by Ignatiev and Kuzmin (IK) [4], in an extended model of a single fermion-like oscillator which allows double occupancy with a small amplitude of β, one can discuss about the phenomenology of a small violation of the PEP with a parameter quantitatively derivable from experiments.

The first precision measurement done by Ramberg and Snow [5] follows a method Greenberg and Mohapatra [6] proposed after they extended the IK model. The method first used by Goldhaber and Scharff-Goldhaber [7] back in 1948, was initially intended to check if the beta rays from beta decay are identical to ordinary electrons. Their idea was that, if not identical to electrons, the beta rays absorbed by a block of metal (in this case lead) will neglect all the electrons occupying the atomic states and deexcite via the cascade process. The $2p$-$1s$ transition will have a different energy with respect to the normal $2p$ - $1s$ transition, due to the shielding effect of an additional electron in the ground state [11]. Based on the non-existence of the anomalous X-rays, they first concluded the equivalence of beta ray to electron, and, more interestingly, they later pointed out the experiment can be interpreted as a test for PEP. A quantitative evaluation based on the result of the experiment was done by Greenberg [3], who deduced explicitly that the possibility that the PEP can be violated is less than 0.03.

The idea of introducing external "fresh" electrons to the target system as applied by this pioneering experiment and the Ramberg-Snow experiment is crucial in the method of testing the PEP for electrons. Without the "fresh" electrons, two experiments in the 1970s [8, 9] looked for prohibited X-rays or γ rays from stable atomic or nucleus systems, and argued the null results served as tests for the PEP violation. However this type of measurement does not validate to be a test, because it has assumed that the transitions between different permutation group could occur. Such assumption violates in the first place the superselection rule separating states in different presentations of the symmetric group [10]. On the other hand, external electrons that had no interaction with the target system not only make source of electrons in large population possible, they are also the pre-requisite that small violation of the PEP can be discussed in the framework of quantum mechanics as Greenberg proposed [3]. Because the newly captured electron and the copper atom have the possiblility of forming a "mixed" symmetry state that is highly excited, anomalous X-rays can be observed. To represent the probability of a small violation of the PEP in the absence of a field theory, Ramberg and Snow used the β parameter introduced first in the IK model. For a random pair of electrons, $1-\frac{1}{2}\beta^2$ is the possibility of the pair in the normal antisymmetric state, and $\frac{1}{2}\beta^2$ the probability in the anomalous symmetric state. In the IK model, β is explicitly defined with the zero, one, and two particle states of $|0\rangle$, $|1\rangle$, and $|2\rangle$, together with the creation operator $a^\dagger$ and the annihilation operator a as :

$$a^\dagger|0\rangle = |1\rangle, a^\dagger|1\rangle = \beta\ |2\rangle, a^\dagger|2\rangle = 0. \quad a\ |0\rangle = 0, a\ |1\rangle = |0\rangle, a\ |2\rangle = \beta\ |1\rangle; \qquad (1)$$

Following the IK model, Greenberg and collaborators constructed the "quon" algebra [16] with q parameter:

$$a_k a_l^\dagger - q\, a_l^\dagger a_k = \delta_{kl}, \tag{2}$$

which can be understood as the average of the Bose and Fermi commutation relations:

$$\frac{1+q}{2}[a_k, a_l^\dagger]_- + \frac{1-q}{2}[a_k, a_l^\dagger]_+ = \delta_{kl}, \tag{3}$$

and the β parameter can be written in terms of the q parameter as:

$$\frac{1}{2}\,\beta^2 = \frac{1}{2}\,(1+q). \tag{4}$$

Although still having open questions to solve, the "quon" theory is by far the best attempt to violate by a small amount the Fermi and the Bose statistics. However, for a direct comparison of the experimental results, the VIP experiment used and will use the same notation of β^2 by Ramberg and Snow.

In next section, we describe the experimental method used by VIP experiment, and show the improvements in the sensitivity achieved with respect to the past experiments. Afterwards we introduce our follow-up experiment VIP2 and its progress in the ongoing preparation.

2. VIP experimental method

With the same idea of searching for anomalous transition X-rays, Ramberg and Snow improved drastically the sensitivity by changing the source of electrons from beta decay to constant electric current. Performed at the ground floor of the Muon building at Fermilab, they used a proportional tube counter as the X-ray detector with a resolution of 1 keV at 8 keV, and a large array of plastic scintillators to veto possible signals from charged cosmic rays. A thin strip copper as target was connected to a 50 A power supply. By comparing the X-ray spectra from measurements with and without power supply, the excess of events in the forbiden transition energy region when current is supplied, will be due to the violation of the PEP.

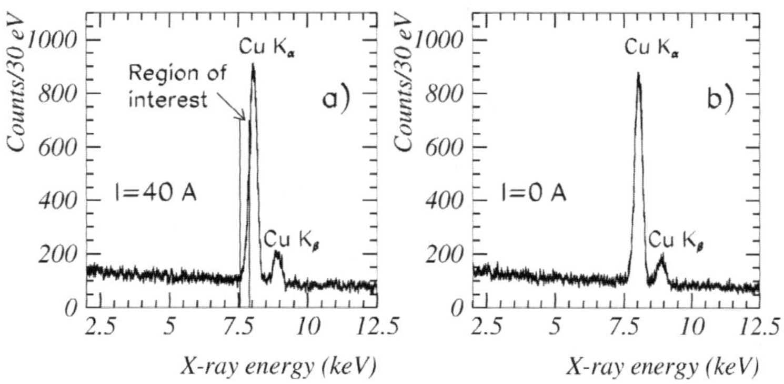

Figure 1. Energy spectra for the VIP experiment [12] : (a) with 40 A current, (b) without current, from part of the data set. Normal K transitions of Cu present are background, and they are due to excitation of target by cosmic rays and environmental radiations.

Table 1. Limits of the Pauli violation probability for electrons from recent high precision experiments:

Experiment	Target	Upper limit of $\beta^2/2$	reference
Ramberg-Snow	Copper	1.7×10^{-26}	[5]
S.R. Elliott et al.	Lead	1.5×10^{-27}	[14]
VIP(2006)	Copper	4.5×10^{-28}	[12]
VIP(2012)	Copper	4.7×10^{-29}	[13]
VIP2(goal)	Copper	$\times 10^{-31}$	[15]

2.1. VIP experiment and results

The VIP experiment followed the method of the Ramberg-Snow experiment, and used the same definition of the parameter $\frac{1}{2}\beta^2$ to represent the violation to the PEP for a direct comparison of the experimental results. The improvement in sensitivity was achieved firstly due to the site of the experiment at the underground laboratory in Laboratori Nazionali del Gran Sasso (LNGS), which has the advantage of the excellent shielding against cosmic rays [12]. The other reason is the use of Charge Coupled Device (CCD) as the X-ray detector which had a typical resolution of 320 eV at 8 keV, that increased the precision in the definition of the region of interest to search for anomalous X-rays.

In Table 1, all the results from experiments using "fresh" electrons are listed, together with the goal of the planned VIP2 experiment at LNGS.

2.2. VIP2 experiment

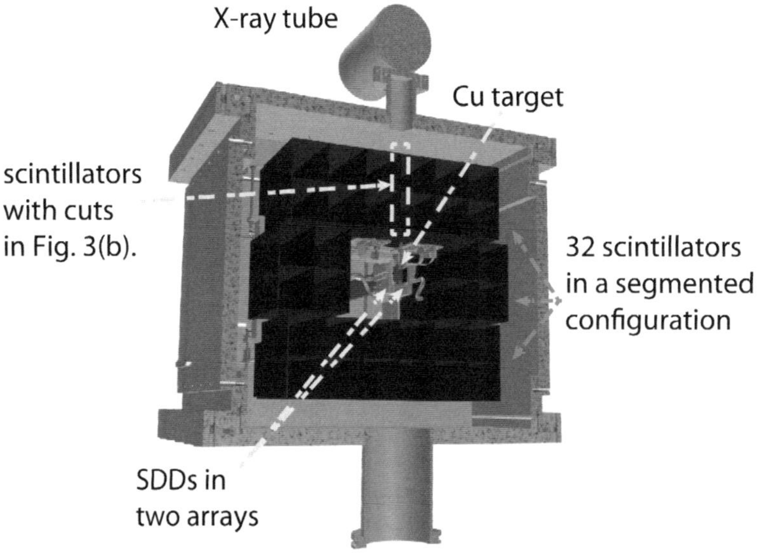

Figure 2. An artist presentation for the cutaway view of the setup. Over 90% of the solid angle for the SDDs acceptance is covered by 32 plastic scintillators as active shielding. The timing capability of the SDDs will allow us to reduce most of the background of the Cu K-series X-rays induced by cosmic rays impinging on the target.

2.2.1. Design In the follow-up VIP2 experiment, we aim to improve the sensitivity of VIP experiment by two orders of magnitude [17]. A detailed list for the features that will contribute to the overall improvement is summarized in Table 2 [15]. The dominant factor of background reduction will come from the use of the Silicon Drift Detectors (SDDs) as the X-ray detector and to an active shielding with arrays of plastic scintillator as veto counters. Compared to the readout time of the order of seconds for CCD, the SDD has a charge collection time of less than one microsecond. This allows to use the time correlation between the X-ray events and the events at the veto counters, to exclude all the X-rays, including the K-series X-rays of Cu from the target excited by cosmic rays or by the environmental radiation, as the energy spectra in Fig. 1 show.

We plan to use six SDD detectors with a total active area of 6 cm^2 mounted close to the pure Cu target in the shape of a strip 3 cm in length. Surrounding the SDDs and readout electronics, as shown in Fig. 2, 32 pieces of plastic scintillators each with a dimension of 40 mm $\times$ 32 mm $\times$ 250 mm will be mounted in a segmented configuration, covering about 90% of the solid angle for the acceptance of SDDs. To readout the light output of each scintillator, we attach with optic cement solid-state Silicon Photo-Multipliers (SiPMs) directly to the scintillators. More information for the plastic scintillator and SiPM and its electronic board can be found in Ref. [15].

Table 2. The improvement factors for VIP2 in comparison to the features of VIP [15] :

Changes in VIP2	value VIP2 (VIP)	expected gain
acceptance	12 %	12
increase current	100 A (40 A)	> 2
reduced length	3 cm (8.8 cm)	1/3
total linear factor		8
energy resolution	170 eV (320 eV) @ 8 keV	4
reducecd active area	6 cm ^{2} (114 cm ^{2}	20
better shielding and veto		5-10
higher SDD efficiency		1/2
background reduction		200 - 400
overall improvement		> 120

3. Future perspectives: tests of collapse models

We are presently considering the possibility to perform in the future measurements of X rays (having such excellent X-ray detectors, as the CCDs and SDDs) generated as spontaneous radiation predicted by (some) collapse models. The collapse models deal with the "measurement problem" in quantum mechanics by introducing a new physical dynamics that naturally collapses the state vector. Collpase models make predictions which differ from those of standard quantum mechanics [18]. One of the most exciting task is to perform cutting-edge experiments, in order to asses whether quantum mechanics is exact, or an approximation of a deeper level theory.

In the nonrelativistic collapse model developed by Ghirardi, Rimini, Weber [19] and Pearle [20] (see also ref. [21] for a review), namely the continuous spontaneous localization (CSL) model, the state vector undergoes a nonunitary evolution in which particles interact with a fluctuating scalar field. This interaction has not only the effect of collapsing the state vector towards the particle number density eigenstates in position space, but it increases the expectation value of particle's energy as well. This means, for a free charged particle (as the electron) electromagnetic radiation. This type of phenomenon is predicted by the CSL and is totally absent in the standard quantum mechanics.

In the paper [22] a pioneering work on this spontaneous emission of radiation was performed - the authors analysed X-ray data measured in an underground experiment and interpreted them as a limit for the combination of the CSL parameter λ/a^2. It was shown that the highest sensitivity is at few keV X-rays, exactly in the range where our detectors are ideal. We have done a similar analysis, with a very preliminary results [23]. We are presently performing a feasibility study to define a dedicated experiment to measure X rays coming from the spontaneous collapse models and improve by few orders of magnitude the actual limit.

Acknowledgments

We thank H. Schneider, L. Stohwasser, and D. Stückler from Stefan-Meyer-Institut for their fundamental contribution in designing and building the VIP2 setup and to the INFN-LNGS laboratory staff for the support during all phases of preparation, installation and data taking. We acknowledge the support from the: HadronPhysics FP6(506078), HadronPhysics2 FP7 (227431), HadronPhysics3 (283286) projects, EU COST Action MP1006, Fundamental Problems in Quantum Physics, Austrian Science Foundation (FWF) which supports the VIP2 project with the grant P25529-N20 and Centro Fermi ("Problemi aperti nella meccania quantistica" project).

References

[1] Pauli W 1925 *Z. Phys.* **31** 765
[2] Greenberg O W and Mohapatra R N 1989 *Phys. Rev.* D **39** 2032
[3] Greenberg O W 1989 *Nucl. Phys.* B (Proc. Suppl.) **6** 83
[4] Ignatiev A Yu and Kuzmin V A 1987 *Yad. Fiz.* **46** 786
[5] Ramberg E and Snow G A 1990 *Phys. Lett.* B **238** 438
[6] Greenberg O W and Mohapatra R N 1987 *Phys. Rev. Lett.* **59** 2507
[7] Goldhaber M and Scharff-Goldhaber G 1948 *Phys. Rev.* **73** 1472
[8] Reines F and Sobel H W 1974 *Phys. Rev. Lett.* **32** 954
[9] Logan B A and Ljubicic A 1979 *Phys. Rev.* C **20** 1957
[10] Messiah A M I and Greenberg O W 1964 *Phys. Rev.* **136** B248
[11] Curceanu C *et al.* 2013 *INFN report, INFN-13-21/LNF*
[12] Bartalucci S *et al.* 2006 *Phys. Lett.* B **641** 18
[13] Curceanu C *et al.* 2012 *AIP Conf. Proc.* **1508** 136
[14] Elliott S R *et al.* 2012 *Found. Phys.* **42** 1015
[15] Marton J *et al.* 2013 *J. Phys.: Conf. Ser.* **447** 012070
[16] Greenberg O W 1991 *Phys. Rev.* D **43** 4111
[17] Marton J *et al.* 2011 *J. Phys.: Conf. Ser.* **335** 012060
[18] Adler S L and Bassi A 2009 *Science* **325** 275
[19] Ghirardi G C, Rimini A and Weber T 1986 *Phys. Rev.* D **34** 470
[20] Pearle P 1989 *Phys. Rev* A **39** 2277; Ghirardi G C, Pearle P and Rimini A 1990 *Phys. Rev.* A **42** 78
[21] Bassi A 2007 *Journal of Physics* **67** 012013
[22] Fu Q 1997 *Phys. Rev* A **56** 1806
[23] Curceanu C *et al.* 2015 *Phys. Scr.* **90** 028003

7th International Workshop DICE2014 Spacetime – Matter – Quantum Mechanics IOP Publishing
Journal of Physics: Conference Series **626** (2015) 012028 doi:10.1088/1742-6596/626/1/012028

The geometry of singularities and the black hole information paradox

O C Stoica

Department of Theoretical Physics, National Institute of Physics and Nuclear Engineering – Horia Hulubei, Bucharest, Romania.

E-mail: `cristi.stoica@theory.nipne.ro`

Abstract. The information loss occurs in an evaporating black hole only if the time evolution ends at the singularity. But as we shall see, the black hole solutions admit analytical extensions beyond the singularities, to globally hyperbolic solutions. The method used is similar to that for the apparent singularity at the event horizon, but at the singularity, the resulting metric is degenerate. When the metric is degenerate, the covariant derivative, the curvature, and the Einstein equation become singular. However, recent advances in the geometry of spacetimes with singular metric show that there are ways to extend analytically the Einstein equation and other field equations beyond such singularities. This means that the information can get out of the singularity. In the case of charged black holes, the obtained solutions have non-singular electromagnetic field. As a bonus, if particles are such black holes, spacetime undergoes dimensional reduction effects like those required by some approaches to perturbative Quantum Gravity.

1. Introduction

There are two outstanding problems in General Relativity (GR): the problem of singularities, and the quantization of gravity problem. The theory predicts singularities both inside the black holes, and at the big-bang. It was hoped that they are an artifact of the symmetry of the solutions, but the singularity theorems of Penrose and Hawking show that they are unavoidable [1–6]. The occurrence of singularities was regarded from the beginning by many as undesirable for GR, because the field equations don't make sense there, but they are not necessarily that bad. The big-bang singularity is in the past, and is no longer a threat for the fields. The singularities inside the uncharged and non-rotating black holes are spacelike, and they are in the future of any observer that may cross the event horizon, so they are not a problem either. To make harmless the singularities of rotating and charged black holes too, Penrose proposed the *cosmic censorship hypothesis* [7, 8], according to which all singularities are either in the past (like the big-bang one) or are always hidden beyond event horizons.

Singularities became more dangerous when Hawking realized that black holes evaporate and the singularity seems to become naked [9, 10]. In this case, the information contained in the initial data is partially lost at the singularity, leading to a violation of the unitary evolution. This is the *black hole information paradox*.

Hawking evaporation takes place near the event horizon. In fact, particle creation is not limited to the event horizon, but near the horizon the particle-antiparticle pairs may become separated, one of them crossing the horizon and the other one escaping.

Because Hawking evaporation takes place near the horizon, and also because the area of the event horizon is proportional to the entropy of the black hole, it was hypothesized that the event horizon plays a special role, unlike any other light cone. The most investigated approaches to the information loss paradox try to solve the problem at the horizon: this is the case of the *black hole complementarity* proposal [11–13], and the *firewall* proposal [14–16]. However, assigning to the event horizon a special place among other lightlike surfaces in GR is at odds with the *principle of equivalence*, especially since we know that black holes evaporate, and the event horizons are only apparent. In addition, if the information is lost at the singularity, why would the event horizon come to rescue it? It is as if GR has a problem in one place, and we hope to have another problem or perhaps a miracle in another place, so that they cancel one another. If the information is thought to be lost at the singularity, shouldn't we see if we can recover it at the singularity?

We will explore here the possibility that the evolution equations can be extended beyond the singularities, leading to a new possibility, that the information goes out of the singularities. The main mathematical tool is *singular semi-Riemannian geometry*, first introduced in [17–22]. It was developed for the case of degenerate metric with constant signature in [23, 24], but the method was not invariant, requiring the choice of special splits of the tangent bundle. In addition, it could not be extended to metrics with variable signature. The approach I introduced in [25–27] overcame these problems. This allowed the extension of the Einstein equation at a large class of singularities [25, 28]. It could be successfully applied to singularities in GR where the metric is smooth but degenerate, such as the FLRW big-bang model [29, 30], leading to a general model of the big-bang which is not necessarily isotropic or homogeneous, but satisfies the Weyl curvature hypothesis [31]. The more resisting situation was that of black hole singularities, but as I explain below, this case could be worked out too [32–35]. In the following, I will show how these results suggest that the information can survive the singularities.

2. What is the real problem with singularities in General Relativity?

Let's first identify two different situations when the metric tensor becomes singular. If the metric tensor is smooth, but degenerate (*i.e.* it has vanishing determinant), the singularity is called *benign*. If the metric tensor has components that become infinite, then the singularity is called *malign*.

There are some methods to identify the metric singularities, based on symptoms like geodesic incompleteness, or scalar invariants which become infinite only if the metric is not regular (*i.e.* smooth and non-degenerate). These symptoms are used only to identify singularities, but they don't mean the singularities can't be resolved. The reason why singularities are a problem is not the existence of singular scalar invariants, because there are such invariants which become singular even at points where the metric is regular. Also it is not merely the geodesic incompleteness. These are just symptoms.

The reason why there are problems at singularities is that the field equations as usually written don't work there. These equations involve the Ricci curvature, and the Levi-Civita connection. The Levi-Civita connection is given in terms of the Christoffel symbols of the second kind, which involve both the components of the metric g_{ab}, and of its inverse, g^{ab}. For the malign singularities, some of the coefficients g_{ab} are singular, while for the benign ones, they are not, but the inverse g^{ab} is not defined. Hence, for PDE on curved spacetimes the covariant derivatives blow up, $\Gamma^c{}_{ab} = \frac{1}{2}g^{cs}(\partial_a g_{bs} + \partial_b g_{sa} - \partial_s g_{ab})$, even in the benign case. Einstein's equation blows up in addition because the Einstein tensor

$$G_{ab} = R_{ab} - \frac{1}{2}Rg_{ab} \quad (1)$$

blows up, being expressed in terms of the curvature, which is defined in terms of the covariant derivative: $R_{ab} = R^{s}{}_{asb}$, $R = g^{pq}R_{pq}$, $R^{d}{}_{abc} = \Gamma^{d}{}_{ac,b} - \Gamma^{d}{}_{ab,c} + \Gamma^{d}{}_{bs}\Gamma^{s}{}_{ac} - \Gamma^{d}{}_{cs}\Gamma^{s}{}_{ab}$.

Even if g_{ab} are all finite, these equations are also in terms of g^{ab}, and $g^{ab} \to \infty$ when $\det g \to 0$.

So, if the field equations we normally use are not defined at the singularities, it seems natural to conclude that GR itself predicts its own breakdown, as it was often said [10, 36–40].

The good news is that the field equations can be written in an apparently equivalent way, but which works also for a large class of singularities. This means that the conclusion that GR predicts its own breakdown was too rushed, and there are still unexplored ways. The secret is to rewrite the equations by replacing the singular fields with non-singular ones, which are equivalent to them as long as the metric is non-degenerate. Table 1 contains a dictionary which allows the replacement of singular fields with non-singular ones for some important classes of singularities.

Singular	**Non-Singular**	**When g is...**
$\Gamma^{c}{}_{ab}$ (2-nd)	Γ_{abc} (1-st)	smooth
$R^{d}{}_{abc}$	R_{abcd}	semi-regular [25]
$R_{ab} = R^{s}{}_{asb}$	$R_{ab}\sqrt{\lvert\det g\rvert}^{W}$, $W \leq 2$	semi-regular
$R = g^{st}R_{st}$	$R\sqrt{\lvert\det g\rvert}^{W}$, $W \leq 2$	semi-regular
Ric	$\mathrm{Ric} \circ g$	quasi-regular [28]
R	$Rg \circ g$	quasi-regular

Table 1. A dictionary used to replace singular with non-singular fields.

In Table 1, the symbol $\circ$ denotes the *Kulkarni-Nomizu product*, defined as

$$(h \circ k)_{abcd} := h_{ac}k_{bd} - h_{ad}k_{bc} + h_{bd}k_{ac} - h_{bc}k_{ad}. \tag{2}$$

The idea to rewrite the field equations in ways which are equivalent outside the singularity, but are free of infinities at singularities too, is not completely new. In [37,41], Ashtekar proposed an equivalent Hamiltonian formulation of the Einstein equation, to deal with quantization of gravity, which had some advantages over the ADM formalism [42]. As a side effect, it was believed that rewriting the equations in terms of the "new variables" will also resolve the singularities, but this turned out not to be the case for at least one of the variables (see for example [43]). The idea is even older, being proposed in the 1935 article of Einstein and Rosen, who credited W. Mayer for it [44]. The idea was to resolve a singularity which was benign, by multiplying Einstein's equations with the determinant of the metric as many times as necessary, and it was only mentioned, and not used explicitly or justified in an invariant way.

3. Singular semi-Riemannian geometry

In [25–27] I extended the mathematical apparatus behind GR, semi-Riemannian geometry, to a large class of benign singularities. We recall here some notions about singular semi-Riemannian manifolds, and some of the main results from [25], which will be used in the remainder of the article.

A *singular semi-Riemannian manifold* (M, g) is a differentiable manifold M endowed with a symmetric bilinear form $g \in \mathcal{T}^{0}_{2}M$ named *metric*. Note that the metric is not required to be non-degenerate. In particular, if the metric is non-degenerate, (M, g) is a *semi-Riemannian manifold*. If in addition g is positive definite, (M, g) is a *Riemannian manifold*.

Let $(T_pM)_\circ = T_pM^\perp$ denote the *radical* of T_pM. The metric g_p on T_pM is non-degenerate if and only if $(T_pM)_\circ = \{0\}$. The radical of TM is defined as $T_\circ M = \cup_{p\in M}(T_pM)_\circ$. We define

$T^{\bullet}M := \bigcup_{p\in M}(T_pM)^{\bullet}$, where $(T_pM)^{\bullet} := (T_pM)^{\flat}$, and $X_p^{\flat}(Y_p) := \langle X_p, Y_p\rangle$, $X_p, Y_p \in T_pM$. The module of sections of $T^{\bullet}M$ is defined as

$$\mathcal{A}^{\bullet}(M) := \{\omega \in \mathcal{A}^1(M) | \omega_p \in (T_pM)^{\bullet} \text{ for any } p \in M\}. \tag{3}$$

The metric g induces on $T^{\bullet}M$ a unique non-degenerate inner product $g_{\bullet}$, defined by $g_{\bullet}(\omega, \tau) := \langle X, Y\rangle$, where $X^{\flat} = \omega$, $Y^{\flat} = \tau$, for two vector fields $X, Y \in \mathfrak{X}(M)$.

A tensor T of type (r, s) is called *radical-annihilator* in the l-th covariant slot if $T \in \mathcal{T}^r_{l-1}M \otimes_M T^{\bullet}M \otimes_M \mathcal{T}^0_{s-l}M$. The metric $g_{\bullet}$ can be used to define uniquely the *covariant contraction* or *covariant trace* between two covariant indices in which a tensor is radical-annihilator. If $T \in \mathcal{T}^r_s$ is a tensor field, then we denote the contraction $C_{kl}T$ by $T(\omega_1, \ldots, \omega_r, v_1, \ldots, \bullet, \ldots, \bullet, \ldots, v_s)$.

The Koszul form is defined as $\mathcal{K} : \mathfrak{X}(M)^3 \to \mathbb{R}$,

$$\mathcal{K}(X, Y, Z) := \frac{1}{2}\{X\langle Y, Z\rangle + Y\langle Z, X\rangle - Z\langle X, Y\rangle - \langle X, [Y, Z]\rangle + \langle Y, [Z, X]\rangle + \langle Z, [X, Y]\rangle\}. \tag{4}$$

For non-degenerate metric, the Koszul form defines the Levi-Civita connection, by raising the 1-form $\mathcal{K}(X, Y, \_)$, $\langle\nabla_X Y, Z\rangle = \mathcal{K}(X, Y, Z)$. Properties of the Koszul form of a singular semi-Riemannian manifold are very similar to those of the Levi-Civita connection, and can be found in [25, 27]. The important difference is that, unlike the Levi-Civita connection, the Koszul form is well defined and finite even if the metric is degenerate.

Let $X, Y \in \mathfrak{X}(M)$. The differential 1-form $\nabla^{\flat}_X Y \in \mathcal{A}^1(M)$, $(\nabla^{\flat}_X Y)(Z) := \mathcal{K}(X, Y, Z)$ for any $Z \in \mathfrak{X}(M)$, is called the *lower covariant derivative* of Y in the direction of X. The *lower covariant derivative operator* is defined as $\nabla^{\flat} : \mathfrak{X}(M) \times \mathfrak{X}(M) \to \mathcal{A}^1(M)$, and it associates to each $X, Y \in \mathfrak{X}(M)$ the differential 1-form $\nabla^{\flat}_X Y$.

A singular manifold (M, g) is *radical-stationary* if, for any $X, Y \in \mathfrak{X}(M)$, $\mathcal{K}(X, Y, \_) \in \mathcal{A}^{\bullet}(M)$. Let $X \in \mathfrak{X}(M)$, $\omega \in \mathcal{A}^{\bullet}(M)$, where (M, g) is radical-stationary. The covariant derivative of ω in the direction of X is defined as $\nabla : \mathfrak{X}(M) \times \mathcal{A}^{\bullet}(M) \to A_d{}^1(M)$,

$$(\nabla_X \omega)(Y) := X(\omega(Y)) - g_{\bullet}(\nabla^{\flat}_X Y, \omega), \tag{5}$$

where $A_d{}^1(M)$ denotes the set of 1-forms which are smooth on the regions of constant signature. For a radical-stationary singular semi-Riemannian manifold (M, g), we define:

$$\mathscr{A}^{\bullet 1}(M) = \{\omega \in \mathcal{A}^{\bullet}(M) | (\forall X \in \mathfrak{X}(M))\ \nabla_X \omega \in \mathcal{A}^{\bullet}(M)\}. \tag{6}$$

The *Riemann curvature tensor* of a singular semi-Riemannian manifold (M, g) is defined as $R : \mathfrak{X}(M) \times \mathfrak{X}(M) \times \mathfrak{X}(M) \times \mathfrak{X}(M) \to \mathbb{R}$,

$$R(X, Y, Z, T) := (\nabla_X \nabla^{\flat}_Y Z - \nabla_Y \nabla^{\flat}_X Z - \nabla^{\flat}_{[X,Y]} Z)(T) \tag{7}$$

for any vector fields $X, Y, Z, T \in \mathfrak{X}(M)$.

A *semi-regular semi-Riemannian manifold* is defined by the condition $\nabla^{\flat}_X Y \in \mathscr{A}^{\bullet 1}(M)$, for any vector fields $X, Y \in \mathfrak{X}(M)$. A radical-stationary semi-Riemannian manifold (M, g) is semi-regular if and only if for any $X, Y, Z, T \in \mathfrak{X}(M)$, $\mathcal{K}(X, Y, \bullet)\mathcal{K}(Z, T, \bullet) \in \mathscr{F}(M)$.

The Riemann curvature of a semi-regular semi-Riemannian manifold (M, g) is a smooth tensor field $R \in \mathcal{T}^0_4 M$. It satisfies for any vector fields $X, Y, Z, T \in \mathfrak{X}(M)$

$$\begin{aligned} R(X, Y, Z, T) &= X\mathcal{K}(Y, Z, T) - Y\mathcal{K}(X, Z, T) - \mathcal{K}([X, Y], Z, T) \\ &\quad + \mathcal{K}(X, Z, \bullet)\mathcal{K}(Y, T, \bullet) - \mathcal{K}(Y, Z, \bullet)\mathcal{K}(X, T, \bullet). \end{aligned} \tag{8}$$

The *densitized Einstein tensor* $G_{ab}\det g$ is smooth on semi-regular spacetimes of dimension 4 even if G_{ab} is singular [25]. Hence, we can write a *densitized Einstein equation*

$$G_{ab}\sqrt{|\det g|}^{W} + \Lambda g_{ab}\sqrt{|\det g|}^{W} = \kappa T_{ab}\sqrt{|\det g|}^{W}, \tag{9}$$

where T_{ab} is the stress-energy tensor, $\kappa := \dfrac{8\pi\mathcal{G}}{c^4}$, $\mathcal{G}$ is Newton's constant and c the speed of light, and $W \in \{0, 1, 2\}$. In many cases it is enough to take $W = 1$ [29], or even $W = 0$ [32].

4. Resolving the Schwarzschild singularity

The problem of singularities dates back to the very beginnings of GR, when Schwarzschild wrote his famous solution to Einstein's equation, representing the spacetime metric outside of a spherically symmetric, non-rotating and uncharged body of mass m [45, 46]. In the coordinates originally found by Schwarzschild, the metric is

$$\mathrm{d}s^2 = -\frac{r-2m}{r}\mathrm{d}t^2 + \frac{r}{r-2m}\mathrm{d}r^2 + r^2\mathrm{d}\sigma^2, \tag{10}$$

where $\mathrm{d}\sigma^2 = \mathrm{d}\theta^2 + \sin^2\theta\mathrm{d}\phi^2$.

Schwarzschild was dissatisfied with this solution, because the metric appears to be singular at $r = 0$ and $r = 2m$. He therefore replaced the coordinate r with $R = r - 2m$, and imposed the condition $R > 0$. However, the singularity $r = 2m$ turned out to be not a singularity of the metric tensor, but of the coordinates. In suitable coordinates, the metric tensor becomes regular at $r = 2m$, as shown by Eddington [47] and Finkelstein [48]. So the event horizon $r = 2m$ should also be included in the solution, together with the interior region of the black hole, $r < 2m$.

One major remark about the solution proposed by Eddington and Finkelstein is the understanding that, by moving to another atlas, a metric that looks singular may be made regular. This is not at odds with the principle of general relativity, which states that the physical laws are independent on the coordinates, nor it is at odds with known simple differential geometry observations that the metric remains regular after a non-singular coordinate change, because the coordinate transformations involved here are singular. In the case of the Schwarzschild metric, because of the constrained imposed by Schwarzschild to the solution, the coordinates are singular at $r = 0$ and $r = 2m$, and we have to move to non-singular ones, which are the correct ones. This is what Eddington and Finkelstein did for the event horizon.

This raises the question whether we can apply the same idea of using a singular coordinate transformation to remove the $r = 0$ singularity. This is not the case, because the Kretschmann scalar $R_{abcd}R^{abcd}$ is singular at $r = 0$, and no coordinate transformation, not even a singular one can change it, hence either the Riemann tensor R_{abcd} or R^{abcd} is singular in any coordinates, and so is the metric.

However, it is enough to find coordinates in which the singularity is made benign, and not necessarily completely removed. In [32] I showed that changing the coordinates to

$$\begin{cases} r &= \tau^2 \\ t &= \xi\tau^4 \end{cases} \tag{11}$$

leads to the following expression for the four-metric:

$$\mathrm{d}s^2 = -\frac{4\tau^4}{2m-\tau^2}\mathrm{d}\tau^2 + (2m-\tau^2)\tau^4\left(4\xi\mathrm{d}\tau + \tau\mathrm{d}\xi\right)^2 + \tau^4\mathrm{d}\sigma^2, \tag{12}$$

which is *analytic* and *semi-regular* at $r = 0$. The metric extends analytically beyond the singularity $r = 0$. In the case of evaporating black holes, this leads to the Penrose diagram in Fig. 1 B. In the new coordinates, the metric becomes not only analytic, but also semi-regular. This means that the Einstein equation can be rewritten in a form which holds at the singularity $r = 0$ too. This suggests that information is not necessarily destroyed by singularities.

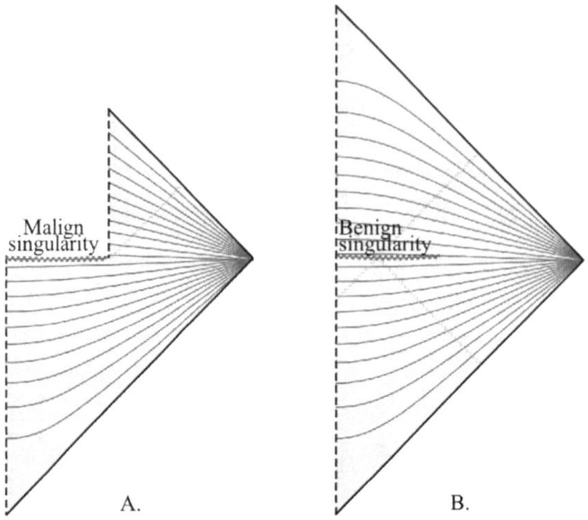

Figure 1. A. Standard evaporating black hole, whose singularity destroys the information. **B.** Evaporating black hole extended through the singularity preserves information.

5. Discussion

We have seen that a large class of benign singularities (which are due only to the degeneracy of the metric) have reasonable geometric properties, and the equations can be rewritten without infinities. This includes some very general big-bang singularities [29–31]. Also, we have seen that the Schwarzschild black hole singularity, although malign, can be made benign by changing the atlas [32]. Similarly, in [33, 34] I found coordinates that make the metric analytic for Reissner-Nordström and Kerr-Newman black holes, and in [35] that these solutions can be used to obtain globally hyperbolic solutions of stationary and evaporating, charged and neutral, rotating or non-rotating black holes.

Moreover, in the coordinates found for the charged black holes in [33,34], the electromagnetic field and potential are smooth even at the singularities. But these solutions have spherical or cylindrical symmetry. What happens if we assume such a solution as background, and consider other fields that cross the singularity? For example, the Maxwell and the Yang-Mills equations can be written in many equivalent ways, by using the covariant derivative, or using the Hodge dual, or using the codifferential operator δ. All these need the metric to be non-degenerate, but in [49] I found conditions which allow us to rewrite these equations without infinities even if the metric becomes degenerate. But the question of what generic initial conditions lead to solutions satisfying such conditions remains open.

Another question concerns the physical interpretation of the formulations mentioned in this article. More precisely, what motivates the preference for the quantities in the second column in Table 1, over those in the first column? As long as the metric is non-degenerate, these quantities are equivalent. For example, we move from Γ_{abc} to $\Gamma^c{}_{ab}$, or from R_{abcd} to $R^d{}_{abc}$, by contracting with g^{ab}. So in the non-degenerate case, the two have the same physical content. But when the metric is degenerate, why would we consider Γ_{abc} and R_{abcd} more fundamental than $\Gamma^c{}_{ab}$ and $R^d{}_{abc}$? Is the fact that they remain finite and smooth at singularities an enough reason? Also, why would we use a version of Einstein's equation in terms of $G_{ab}\sqrt{|\det g|}^W$ and $T_{ab}\sqrt{|\det g|}^W$, $W \in \{0,1,2\}$? If the fact that $\sqrt{|\det g|}^W$ vanishes enough to cancel the divergence of R_{ab} and R is not enough, here are some additional possible reasons, supporting the case $W = 1$. First, the Lagrangian density for the Hilbert-Einstein action is $R\sqrt{|\det g|}\mathrm{d}x^0 \wedge \mathrm{d}x^1 \wedge \mathrm{d}x^2 \wedge \mathrm{d}x^3$, and not just R. It is a differential form, and not a scalar, and if we insist to consider only the scalar part may lead to infinities at singularities, but the differential form may still remain finite. Also, from the action we obtain the stress-energy tensor density, and not the stress-energy tensor T_{ab}. A concrete example can be found in [29], where it is explained that at the big bang in the FLRW model, although the scalars ρ and p, usually named *energy density* and *pressure density* become infinite at the singularity, the actual densities are $\rho\sqrt{|\det g|}\mathrm{d}x^0 \wedge \mathrm{d}x^1 \wedge \mathrm{d}x^2 \wedge \mathrm{d}x^3$ and $p\sqrt{|\det g|}\mathrm{d}x^0 \wedge \mathrm{d}x^1 \wedge \mathrm{d}x^2 \wedge \mathrm{d}x^3$, and they remain finite and smooth even at the singularity, and so remain $T_{ab}\sqrt{|\det g|}$, $R_{ab}\sqrt{|\det g|}$, and $R\sqrt{|\det g|}$. Quantities like mass, energy, pressure are defined on extended regions, by integrating such densities, which always have to include in their definitions the quantity $\sqrt{|\det g|}$. On differentiable manifolds we don't integrate scalars, we integrate differential forms, which include together with the scalar, also $\sqrt{|\det g|}$, in the volume form, and remain non-singular. Nevertheless, further investigations of the physical interpretation of the quantities proposed here as more fundamental should be made, to deepen our understanding of why some quantities remain finite and others don't, and which of them are more fundamental than the others, if any.

To conclude, the results mentioned in this article and in the ones cited here suggest that singularities not only are less harmful than it was initially thought, but may have even healthy effects to other problems in GR, like Quantum Gravity [50].

References

[1] Penrose R 1965 *Phys. Rev. Lett.* **14** 57–59
[2] Penrose R 1969 *Revista del Nuovo Cimento; Numero speciale 1* 252–276
[3] Hawking S W and Penrose R 1970 *P. Roy. Soc. A-Math. Phy.* **314** 529–548
[4] Hawking S W 1966 *P. Roy. Soc. A-Math. Phy.* **294** 511–521
[5] Hawking S W 1966 *P. Roy. Soc. A-Math. Phy.* **295** 490–493
[6] Hawking S W 1967 *P. Roy. Soc. A-Math. Phy.* **300** 187–201
[7] Penrose R 1979 in *General relativity: an Einstein centenary survey* vol 1 ed Hawking S W and Israel W (Cambridge: CUP Archive) pp 581–638
[8] Penrose R 1998 in *Black holes and relativistic stars* ed Wald R M (Chicago, Ill.: University of Chicago Press) pp 233–248
[9] Hawking S W 1975 *Comm. Math. Phys.* **43** 199–220
[10] Hawking S W 1976 *Phys. Rev. D* **14** 2460
[11] Susskind L, Thorlacius L and Uglum J 1993 *Phys. Rev. D* **48** 3743
[12] Susskind L and Lindesay J 2005 *An Introduction to Black Holes, Information and the String Theory Revolution: The Holographic Universe* (Singapore: World Scientific)
[13] Susskind L 2008 *The black hole war: My battle with Stephen Hawking to make the world safe for quantum mechanics* (Hachette Digital, Inc.)
[14] Almheiri A, Marolf D, Polchinski J and Sully J 2013 *Journal of High Energy Physics* **2013**(2) 1–20
[15] Almheiri A, Marolf D, Polchinski J, Stanford D and Sully J 2013 *Journal of High Energy Physics* **2013**(9) 1–32
[16] Marolf D and Polchinski J 2013 *Phys. Rev. Lett.* **111** 171301
[17] Moisil G C 1940 *Bull. Math. Soc. Roumaine Sci.* **42** 33–52
[18] Strubecker K 1941 *Sitzungsber. Akad. Wiss. Wien, Math.-Naturw. Kl., Abt. IIa* **150** 1–53
[19] Strubecker K 1942 *Math. Z.* **47** 743–777
[20] Strubecker K 1942 *Math. Z.* **48** 369–427
[21] Strubecker K 1944 *Math. Z.* **50** 1–92
[22] Vrănceanu G 1942 *Disqu. Math. Phys. Bucureşti* **2** 253–281
[23] Kupeli D 1987 *Geom. Dedicata* **23** 259–290
[24] Kupeli D 1996 *Singular Semi-Riemannian Geometry* (Amsterdam: Kluwer Academic Publishers Group)
[25] Stoica O C 2014 *Int. J. Geom. Methods Mod. Phys.* **11**(5) 1450041
[26] Stoica O C 2011 *Preprint gr-qc/1105.3404*
[27] Stoica O C 2014 *Balkan J. Geom. Appl.* **19** 118–126
[28] Stoica O C 2014 *Cent. Eur. J. Phys* **12**(2) 123–131
[29] Stoica O C 2015 to appear in *Int. J. Theor. Phys.*
[30] Stoica O C 2012 *Commun. Theor. Phys.* **58** 613–616

[31] Stoica O C 2013 *Ann. of Phys.* **338** 186–194
[32] Stoica O C 2012 *Eur. Phys. J. Plus* **127**(7) 1–8
[33] Stoica O C 2012 *Phys. Scr.* **85** 055004
[34] Stoica O C 2015 *U.P.B. Sci Bull. Series A* **77**(1)
[35] Stoica O C 2012 *An. Şt. Univ. Ovidius Constanţa* **20** 213–238
[36] Hawking S W and Penrose R W 1970 *Proc. Roy. Soc. London Ser. A* **314** 529–548
[37] Ashtekar A 1991 *Lectures on non-perturbative canonical gravity, Notes prepared in collaboration with Ranjeet S. Tate* vol 17 (Singapore: World Scientific)
[38] Hawking S W and Penrose R W 1996 *The Nature of Space and Time* (Princeton and Oxford: Princeton University Press)
[39] Ashtekar A 2009 *J. Phys. Conf. Ser.* **189** 012003
[40] Ashtekar A and Wilson-Ewing E 2009 *Phys. Rev. D* **79** 083535
[41] Ashtekar A 1987 *Phys. Rev. D* **36** 1587–1602
[42] Arnowitt R, Deser S and Misner C W 1962 *Gravitation: An Introduction to Current Research* (New York: Wiley) pp 227–264
[43] Yoneda G, Shinkai H and Nakamichi A 1997 *Phys. Rev. D* **56** 2086–2093
[44] Einstein A and Rosen N 1935 *Phys. Rev.* **48** 73
[45] Schwarzschild K 1916 *Sitzungsber. Preuss. Akad. d. Wiss.* 189–196
[46] Schwarzschild K 1916 *Sitzungsber. Preuss. Akad. d. Wiss.* 424–434
[47] Eddington A S 1924 *Nature* **113** 192
[48] Finkelstein D 1958 *Phys. Rev.* **110** 965
[49] Stoica O C 2014 *X-th International Conference on Finsler Extensions of Relativity Theory, Braşov, Romania, FERT 2014* (*Preprint* gr-qc/1408.3812)
[50] Stoica O C 2014 *Ann. of Phys.* **347** 74–91

Journal of Physics: Conference Series **626** (2015) 012029 doi:10.1088/1742-6596/626/1/012029

How many futures on Finsler spacetime?

E. Minguzzi

Dipartimento di Matematica e Informatica "U. Dini", Università degli Studi di Firenze, Via S. Marta 3, I-50139 Firenze, Italy

E-mail: ettore.minguzzi@unifi.it

Abstract. Some recent results by the author on the geometry and dynamics of Finsler spacetimes are reviewed. It is shown that in Finslerian generalizations of general relativity the number of predicted lightlike cones is two, one past and one future, as in general relativity. This result is non-trivial as it can fail, for instance, in spacetime dimension two. It is also shown that suitable versions of the reverse Cauchy-Schwarz and reverse triangle inequalities hold on Finsler spacetimes. Finally, a long standing problem of Finslerian gravity concerns the development of dynamical equations which imply a conservation law. We make some progress following a recent proposal by the author according to which physical Finsler spacetimes have affine sphere indicatrices of hyperbolic type.

1. Introduction

Although Finsler geometry is a quite venerable subject it is usually a good idea to remind the reader what it is all about. Shiing-Shen Chern tried to popularize this theory publishing an article entitled "Finsler geometry is just Riemannian geometry without the quadratic restriction". In fact Riemann suggested to consider spaces for which the arc-length of curves is given by $\int_\gamma \sqrt{2|\mathscr{L}(x,\dot{x})|}\mathrm{d}t$ where

$$\mathscr{L}(x,v)=\frac{1}{2}\,g_{\mu\nu}(x,v)v^\mu v^\nu, \tag{1}$$

namely where g although positive definite depends on v and hence does not define a quadratic form. Observe that the arc-length calculated with the above formula must be independent of the parametrization, and so it is natural to require $\mathscr{L}$ to be positive homogeneous of degree two

$$\forall s>0, \qquad \mathscr{L}(x,sv)=s^2\mathscr{L}(x,v). \tag{2}$$

In this way $g_{\mu\nu}$ can be defined as the Hessian with respect to the velocities of $\mathscr{L}$. In index free notation we shall also write $g_v=g_{\mu\nu}\mathrm{d}x^\mu\otimes \mathrm{d}x^\nu$, $g_v\colon E\mapsto T^*M\otimes_M T^*M$, $E=TM\backslash 0$, with a v index to stress the dependence on the velocity. It can be easily shown, differentiating twice Eq. (2) with respect to s, that g_v cannot be extended to the whole TM while remaining continuous otherwise it would be independent of v and we would be back to the Riemannian case. The Finsler Lagrangian $\mathscr{L}$ is said to be reversible if $\mathscr{L}(x,-v)=\mathscr{L}(x,v)$.

While it is true that Finsler geometry is obtained dropping the quadratic restriction on $\mathscr{L}$, this fact does not really explain why this theory is interesting and why we should study it. After all any theory could be made more general dropping some axiom. Instead, we want the theory to be simple and rich while being based on few axioms. The importance and naturalness of Finsler geometry is better understood with the following observation:

Riemannian geometry gives each tangent space of a manifold the structure of a finite dimensional Hilbert space; Finsler geometry gives each tangent space the structure of a Banach space whose unit balls are strongly convex.

In this work we shall be interested in Lorentz-Finsler geometry which is obtained from the above axioms replacing the positive definite signature of g with the Lorentzian signature $(-,+,\cdots,+)$. John Beem tried this option [1, 2] but could not establish the existence of two cones, namely the fact that the set of timelike vectors at $p \in M$

$$I_p := \{v \in T_pM\backslash 0 : \mathscr{L}(p,v) < 0\},$$

consists of two convex cones to be interpreted as the future and the past causal cones. Other authors [3] that considered the causal aspects of the theory simply stipulated the existence of a future timelike convex cone I_p^+ where $\mathscr{L} < 0$.

It seems that most authors working in Finsler geometry did not care very much of the signature of the metric, as it was simpler to leave aside, at least temporarily, the difficulties connected with the causal and geometrical interpretation of the theory. Also, they were essentially looking for tensorial generalizations of the Einstein's equations so they were not very much concerned with the underlying geometry [4–10]. Actually, understanding the geometry of the theory turned out to be very important since, as in Finsler geometry there are many connections, there were several tensorial candidates for a generalization of Einstein's equations and their physical interpretation was not entirely clear.

The investigations by Beem confirmed that causality in Finsler space is non-trivial. Although each connected component of the the timelike region I_p had to be a convex cone [1, 2, 11, 12], he found that I_p could consist of more than two convex cones. For instance, he gave the example of a 2-dimensional spacetime $\mathbb{R}^2$ endowed with a Finsler Lagrangian

$$\mathscr{L}(x,v) = \frac{(v^1)^3 v^2 - v^1 (v^2)^3}{(v^1)^2 + (v^2)^2},$$

where I_p is independent of $x \in M$ and has four convex cone components. Actually, Beem believed to have found similar examples for higher-dimensional spacetimes [2], but his examples can be shown to be incorrect since the metric has vanishing determinant at some point of the slit tangent bundle.

In this work I shall outline further progress by the author which has shown that, at least for reversible Lagrangians, Beem's pathological examples can be found only in two dimensional spacetimes. In the physical 4-dimensional spacetime, the Lorentzianity condition on g_v implies that I_p consists precisely of two convex cones. Thus, Finsler causality predicts that there is just one future as desired. This is really a good news since the mathematical theory of Finsler connections, sprays, exponential maps and convex neighborhoods has been developed using the slit tangent bundle $E = TM\backslash 0$. We do not have to restrict the domain of $\mathscr{L}$ to a convex cone thus we can use the standard theory of Finsler connections in the study of Lorentz-Finsler geometry.

In our recent works we have also solved other problems. In fact we proved for the first time the reverse Cauchy-Schwarz and reverse triangle inequalities for Lorentz-Finsler geometry. These results are fundamental in many geometrical arguments. They have been used to establish that lightlike geodesics locally maximize the Lorentzian length [13], a fact which implies by the usual interpolation argument [14, Prop. 2.8] that any causal geodesic which is not an achronal lightlike geodesic can be deformed to a timelike curve keeping endpoints fixed. Let us denote with J_p^α

Journal of Physics: Conference Series **626** (2015) 012029 doi:10.1088/1742-6596/626/1/012029

a connected component of the set of causal vectors, namely the closed set of the slit tangent bundle given by

$$J_p = \{v \in T_pM\backslash 0 : \mathscr{L}(p, v) \le 0\}.$$

Then the inequalities read as follows ($V = T_pM$).

Theorem 1.1 (Finslerian reverse Cauchy-Schwarz inequality). *Let $v_1, v_2 \in J^\alpha_p$ then*

$$-g_{v_1}(v_1, v_2) \ge \sqrt{-g_{v_1}(v_1, v_1)}\,\sqrt{-g_{v_2}(v_2, v_2)},$$

with equality if and only if v_1 and v_2 are proportional. In particular, if $v_1, v_2 \in J^\alpha_p$ then $g_{v_1}(v_1, v_2) \le 0$ with equality if and only if v_1 and v_2 are proportional and lightlike.

Theorem 1.2 (Finslerian reverse triangle inequality). *Let $v_1, v_2 \in J^\alpha_p$ then defined $v = v_1 + v_2$, we have $v \in J^\alpha_p$ and*

$$\sqrt{-g_v(v, v)} \ge \sqrt{-g_{v_1}(v_1, v_1)} + \sqrt{-g_{v_2}(v_2, v_2)}.$$

with equality if and only if v_1 and v_2 are proportional. In particular, if v_1 is timelike and v_2 is causal then v is timelike.

We also showed that the Legendre map is a diffeomorphism.

Theorem 1.3 (The Legendre map is a diffeomorphism for spacetime dimension $n + 1 \ge 3$). *Suppose that $\dim V \ge 3$, then the map $\ell \colon V\backslash 0 \to V^*\backslash 0$ defined by*

$$v \mapsto g_v(v, \cdot) = \partial\mathscr{L}/\partial v$$

is a bijection and hence a diffeomorphism. Its extension to the whole V with $\ell(0) := 0$ is a Lipeomorphism (locally Lipschitz homeomorphism with locally Lipschitz inverse). Moreover, without any condition on dimension, it is always true that ℓ establishes a bijection between J^α_p and its polar cone

$$J^{\alpha *}_p = \{q \in T^*_M\backslash 0 : \ q(v) \le 0, \ \textit{for every } v \in J^\alpha_p\}.$$

Since this results plays a role in the proof that I_p has two components we sketch the proof.

Idea of proof. Let $V = T_pM$ and let us introduce the equivalence relation on $V\backslash 0$ "$v_1 \sim v_2$ if there is $s > 0$ such that $v_1 = sv_2$", namely let us regard $V\backslash 0$ as a radial bundle over a base Q diffeomorphic to S^n. Analogously, let us introduce the quotient Q^* of $V^*\backslash 0$ with respect to the radial directions. The map ℓ satisfies $\ell(sv) = s\ell(v)$ for every $s > 0$, thus it passes to the quotient to a map $\tilde{\ell} \colon Q \to Q^*$. Since Q and Q^* are closed manifolds with the same dimension, $\tilde{\ell}$ is actually a covering. Since $Q^* \sim S^n$ is simply connected (here we use $n \ge 2$) and $Q \sim S^n$ is connected this covering is actually a homeomorphism. From here the proof is straightforward. □

Lemma 1.4. *Let $\dim V \ge 3$ and let $\mathscr{L}$ be reversible. Given any two distinct components J^α and J^β we have $J^\alpha \cap (-J^\beta) \ne \emptyset$.*

Idea of proof. The proof is by contradiction. Suppose that $J^\alpha \cap (-J^\beta) = \emptyset$ then there is a hyperplane $W \subset V$ passing through the origin which separates J^α and $J^{-\beta} = -J^\beta$ and denoting with u^α and u^β the points of tangency of the translations of W with the indicatrix $\{v : 2\mathscr{L}(p, v) = -1\}$ which stays inside the cones

$$W = \ker g_{u^\alpha}(u^\alpha, \cdot) = \ker g_{u^\beta}(u^\beta, \cdot)$$

and from here it is easy to find $s > 0$ such that

$$g_{u^\alpha}(u^\alpha, \cdot) = g_{su^\beta}(su^\beta, \cdot)$$

contradicting the injectivity of the Legendre transform. □

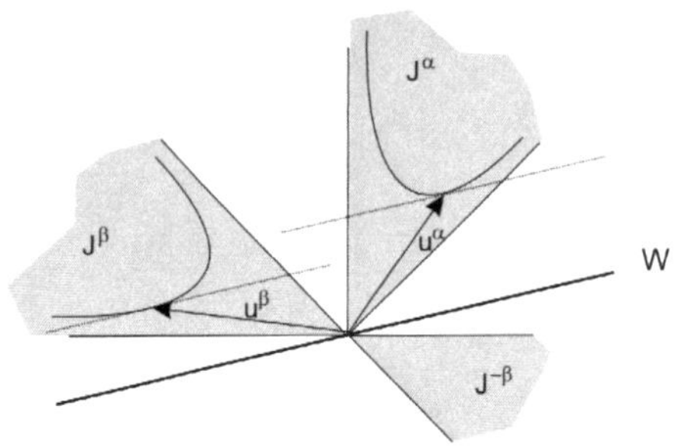

Figure 1. The idea of Lemma 1.4.

We arrive at the final step of the proof

Theorem 1.5. *Let* $dim V \geq 3$ *and suppose that* $\mathscr{L}$ *is reversible, then* I_p *has two components.*

Proof. Let I_p^α be a component of I_p, and let $I_p^{-\alpha} = -I_p^\alpha$ be its opposite. Suppose that there is another component I_p^β with $\beta \neq \alpha, -\alpha$, then I^β cannot intersect neither I_p^α nor $I_p^{-\alpha}$ since the components of I_p are disjoint, however it must intersect both of them. The contradiction proves that there are just two components. □

2. A proposal for a Finslerian gravitational dynamics

The variational principle

$$\delta \int \mathscr{L}(x, \dot{x}) \mathrm{d}t = 0,$$

gives the geodesic equation

$$\ddot{x}^\mu + 2G^\mu(x, \dot{x}) = 0,$$

where the spray is determined by

$$2G^\alpha(x, v) = g^{\alpha\eta}\left(\frac{\partial^2 \mathscr{L}}{\partial x^\nu \partial v^\eta} v^\nu - \frac{\partial \mathscr{L}}{\partial x^\eta}\right). \tag{3}$$

The geodesic equation determines the motion of free falling particles in the theory. As we mentioned a lot of work was devoted to the study of the Finslerian generalization of Einstein's equations. The main difficulty is that of finding a divergence free stress-energy tensor, so as to establish a suitable conservation law. This result has been obtained under additional conditions, e.g. for spaces of constant curvature [6] or for spaces of scalar curvature [9].

Let us write G^α_β for $\partial G^\alpha / \partial v^\beta$ and $G^\alpha_{\beta\gamma}$ for $\partial G^\alpha_\beta / \partial v^\gamma$. Let us define

$$\frac{\delta}{\delta x^\mu} := \frac{\partial}{\partial x^\mu} - G^\alpha_\mu(x, v)\frac{\partial}{\partial v^\alpha},$$

and

$$\Gamma^\alpha_{\mu\nu}(x, v) = \frac{1}{2} g^{\alpha\eta}\left(\frac{\delta}{\delta x^\nu} g_{\eta\mu} + \frac{\delta}{\delta x^\mu} g_{\eta\nu} - \frac{\delta}{\delta x^\eta} g_{\nu\mu}\right),$$

so that the spray (3) can be written $2G^\alpha = G^\alpha_{\beta\gamma} v^\beta v^\gamma = \Gamma^\alpha_{\beta\gamma} v^\beta v^\gamma$. The coefficients $G^\alpha_{\beta\gamma}$ and $\Gamma^\alpha_{\beta\gamma}$ are examples of horizontal connection coefficients of a Finsler connection: the Berwald connection in the former case and the Chern-Rund connection in the latter case.

We assume that the mean Cartan torsion vanishes

$$I_\gamma := \frac{1}{2} g^{\alpha\beta} \frac{\partial}{\partial v^\gamma} g_{\alpha\beta} = \frac{\partial}{\partial v^\gamma} \left(\ln \sqrt{|\det g|} \right), \tag{4}$$

namely that for every $x \in M$, the indicatrix $\{v : 2\mathscr{L}(x,v) = -1\}$ is an affine sphere of hyperbolic type.

Remark 2.1. These spaces are really important, indeed using some standard results in affine differential geometry [15] it is not difficult to show that they are equivalent to a pair given by a distribution of convex sharp cones (the future light cones) and a volume form (the spacetime measure).

Moreover, we assume that the space is Landberg, namely the Berwald horizontal derivative is compatible with the metric ($\nabla^{HB} g = 0$)

$$\frac{\delta}{\delta x^\mu} g_{\alpha\beta} - G^\nu_{\alpha\mu} g_{\nu\beta} - G^\nu_{\beta\mu} g_{\alpha\nu} = 0. \tag{5}$$

This condition is equivalent to $G^\alpha_{\beta\gamma} = \Gamma^\alpha_{\beta\gamma}$, that is, the Berwald and Chern-Rund connections coincide. In particular we can simply write ∇^H for the horizontal derivative. The HH-curvature of the Berwald-Chern-Rund connection is

$$R^\alpha_{\ \beta\gamma\delta} = \frac{\delta}{\delta x^\gamma} G^\alpha_{\beta\delta} - \frac{\delta}{\delta x^\delta} G^\alpha_{\beta\gamma} + G^\alpha_{\mu\gamma} G^\mu_{\beta\delta} - G^\alpha_{\mu\delta} G^\mu_{\beta\gamma}.$$

Whenever needed we shall lower the first index to the left. Our suggestion [16] for the Einstein tensor is

$$E_{\alpha\beta} := \frac{1}{2} (R_\alpha{}^\mu{}_{\beta\mu} + R_\beta{}^\mu{}_{\alpha\mu} - g_{\alpha\beta} R^{\mu\nu}{}_{\mu\nu}), \tag{6}$$

This tensor is symmetric and does indeed satisfy the conservation law (for a proof see [16])

$$\nabla^H_\beta (E^\beta{}_\alpha v^\alpha) = 0, \tag{7}$$

namely, the energy-momentum is conserved for all observers. Indeed, using a divergence theorem due to Rund [17] for any section $s\colon M \to TM\backslash 0$, and vector field $Z^\gamma(x,v)$

$$\nabla^{s^*g} \cdot s^* Z = s^*(\nabla^{HC} \cdot Z) + s^*(I_\beta Z^\gamma + \frac{\partial Z^\gamma}{\partial v^\beta}) D_\gamma s^\beta, \tag{8}$$

where ∇^{s^*g} is the Levi-Civita connection of the pullback metric $s^*g(x) := g(x, s(x))$, ∇^{HC} is the horizontal Cartan (or Chern-Rund) Finslerian covariant derivative and D is the non-linear covariant derivative $D_\gamma s^\beta = \frac{\partial s^\beta}{\partial x^\gamma} + G^\beta_\gamma(x, s(x))$. Thus

$$\nabla^{s^*g}_\gamma (E^\gamma_\alpha(x, s(x)) s^\alpha) = s^*(\frac{\partial}{\partial v^\beta}(E^\gamma_\alpha(x,v) v^\alpha)) D_\gamma s^\beta, \tag{9}$$

which is the analog of the general relativistic almost conservation equation $(T^{\alpha\beta} u_\beta)_{;\alpha} = T^{\alpha\beta} u_{\beta;\alpha}$.

Remark 2.2. Concerning the problem of constructing dynamical equations for Finsler gravity which imply a conservation law, Ishikawa wrote [8]

> There have been several attempts to construct the Einstein tensor in a Finsler space itself. [...] However, it seems difficult or seems to be impossible to construct the Einstein tensor according to this line of approach (by using the Finslerian Bianchi's identity).

We have solved the problem though we had to impose some additional conditions, namely Eqs. (4)-(5). We wish to remark a key idea which played an important role in our arguments. Contrary to previous approaches we did not try to find a divergence free stress-energy *tensor* but rather a divergence free energy-momentum *current* dependent on the observer. In fact, in our opinion what really matters is the map $v \mapsto \xi$ where $\xi^\alpha(v) = -E^\alpha_\beta(v)v^\beta$. One can even try to define the notion of *dominant energy condition* using this map (assume $\Lambda = 0$ for simplicity). There are two inequivalent possibilities: (a) at every $p \in M$ the map $v^\alpha \mapsto -E_{\alpha\beta}(v)v^\beta$ sends the future causal cone J^+_p to its polar cone $(J^+_p)^*$, that is: for every v, w f.d.-causal, $-E_{\alpha\beta}(v)v^\beta w^\alpha \le 0$; or (b) the map $v^\alpha \mapsto -E^\alpha_\beta(v)v^\beta$ sends J^+_p into J^+_p.

Let us end this work comparing our equations with previous proposals. Let $Ric(v) = R^\alpha_{\ \beta\alpha\gamma}v^\beta v^\gamma$. It can be observed that upon contraction with $v^\alpha v^\beta$ our vacuum equation $E_{\alpha\beta} = 0$ gives

$$Ric(v) = 0. \tag{10}$$

Since $Ric(v)$ is positive homogeneous of degree two, Eq. (10) can also be rewritten

$$\frac{1}{2}\frac{\partial^2 Ric(v)}{\partial v^\alpha \partial v^\beta} = 0 \tag{11}$$

where the expression on the left-hand side is the Akbar-Zadeh Ricci tensor [18]. All authors seem to agree on the validity of this equation in the vacuum case since it is implied by almost every choice of dynamical equations that has been proposed so far, starting from the first proposal by Horvath [4]. Not all authors obtained this equation from a tensorial generalization of Einstein's. Rutz [19], for instance, argued for its validity using an analogy based on the the Jacobi deviation equation. Its consequences have also been explored by Li and Chang [20].

Acknowledgments

This work has been partially supported by GNFM of INDAM.

References

[1] Beem J K 1970 *Can. J. Math.* **22** 1035–1039
[2] Beem J K 1973 *Atti Accad. Naz. Lincei Rend. Cl. Sci. Fis. Mat. Natur. (8)* **54** 385–392 (1974)
[3] Asanov G S 1985 *Finsler geometry, relativity and gauge theories* (Dordrecht: D. Reidel Publishing Co.)
[4] Horváth J I 1950 *Phys. Rev.* **80** 901
[5] Horváth J I and Moór A 1952 *Z. Physik* **131** 544–570
[6] Rund H 1962 *Monatsh. Math.* **66** 241–251
[7] Takano Y 1974 *Lettere al Nuovo Cimento* **10** 747–750
[8] Ishikawa H 1980 *Il Nuovo Cimento* **56** 252–262
[9] Ishikawa H 1980 *Ann. Physik (7)* **37** 151–154
[10] Ishikawa H 1981 *J. Math. Phys.* **22** 995–1004
[11] Perlick V 2006 *Gen. Relativ. Gravit.* **38** 365–380
[12] Minguzzi E 2015 *Commun. Math. Phys.* **334** 1529–1551 arXiv:1403.7060
[13] Minguzzi E 2013 *Monatshefte für Mathematik* DOI:10.1007/s00605-014-06. arXiv:1308.6675
[14] Lerner D 1972 *Springer Lecture Notes in Physics* vol 14 pp 1–44
[15] Nomizu K and Sasaki T 1994 *Affine differential geometry* (Cambridge: Cambridge University Press)
[16] Minguzzi E 2014 *Int. J. Geom. Meth. Mod. Phys.* **11** 1460025 arXiv:1405.0645
[17] Rund H 1975 *Monatsh. Math.* **79** 233–252
[18] Bao D and Robles C 2004 *A sampler of Riemann-Finsler geometry* (*Math. Sci. Res. Inst. Publ.* vol 50) (Cambridge Univ. Press, Cambridge) pp 197–259
[19] Rutz S F 1993 *Gen. Relativ. Gravit.* **25** 1139–1158
[20] Li X and Chang Z 2014 *Phys. Rev. D* **90** 064049 arXiv:1401.6363v1

On the Hamiltonian and energy operators in a curved spacetime, especially for a Dirac particle

Mayeul Arminjon

Laboratory "Soils, Solids, Structures, Risks" (CNRS & Université Grenoble-Alpes), Grenoble, France

E-mail: `Mayeul.Arminjon@3sr-grenoble.fr`

Abstract. The definition of the Hamiltonian operator H for a general wave equation in a general spacetime is discussed. We recall that H depends on the coordinate system merely through the corresponding reference frame. When the wave equation involves a gauge choice and the gauge change is time-dependent, H as an operator depends on the gauge choice. This dependence extends to the energy operator E, which is the Hermitian part of H. We distinguish between this ambiguity issue of E and the one that occurs due to a mere change of the "representation" (e.g. transforming the Dirac wave function from the "Dirac representation" to a "Foldy-Wouthuysen representation"). We also assert that the energy operator ought to be well defined in a given reference frame at a given time, e.g. by comparing the situation for this operator with the main features of the energy for a classical Hamiltonian particle.

1. Introduction

The quantum effects in the classical gravitational field which have been observed on Earth for neutrons and for atoms (e.g. [1, 2, 3, 4, 5]) are quantum-mechanical effects. That is, they were predicted (before their observation) by using "first-quantized" theory. They are, to this author's knowledge, the only observed effects of the gravity-quantum coupling. Thus, QM in a curved spacetime covers all currently available experiments about the interaction between gravity and the quantum. Neutrons are spin half particles, and spin half particles are normally described by the Dirac equation. Moreover, besides observing directly the quantum behaviour of particles in a gravitational field, one also would like to account for the influence of the gravitational field on, for example, the quantum behaviour of particles in an electromagnetic field, e.g. the behaviour of electrons in an atom. (This influence is very small in the very weak gravitational field that we have in the solar system, of course. Yet it is expected to be important in strong gravitational fields.) Therefore, QM of the (generally-)covariant Dirac equation is quite an important chapter of curved-spacetime QM. The covariant Dirac equation involves the choice, at any point X in spacetime (and depending smoothly on X), of an orthonormal basis of the tangent space at that point X. I.e., it involves the choice of a *tetrad field* [6, 7]. This plays the role of a gauge field. Indeed, the two realizations of the covariant Dirac equation that are got with two different tetrad fields are equivalent, at least locally [8]. However, it has been proved in recent years that the Hamiltonian operator associated, in a given coordinate system, with the covariant Dirac equation, does depend on the choice of the tetrad field. In particular, the energy spectrum and the energy mean values depend on that choice [9, 10], and so does the presence or absence of Mashhoon's spin-rotation coupling term [11, 12].

In the present conference paper, we summarize the main part of the recent results [13] related with the non-uniqueness problem of the Hamiltonian and energy operators. We add some new remarks, especially we add a new discussion in Sect. 4. In Sect. 2, the definition of the Hamiltonian operator H for a general wave equation in a general spacetime is discussed. We recall that H depends on the coordinate system merely through the corresponding reference frame. Then in Sect. 3 we consider the situation in which the wave equation involves a gauge choice (as is the case for the covariant Dirac equation). We show that, when the gauge transformation depends on the time coordinate, H as an operator depends on the gauge choice. This extends the previous results [9, 10] from the covariant Dirac equation to any quantum wave equation admitting a gauge choice. Section 4 introduces the energy operator and discusses its ambiguity in the presence of a gauge choice. We distinguish between this ambiguity issue and the one that occurs due to a mere change of the "representation" — e.g. from the "Dirac representation" to a "Foldy-Wouthuysen representation". In Sect. 5, we assert that the energy operator ought to be well defined. In particular, recalling that in QM the meaning of the energy is inherited from classical Hamiltonian mechanics, we note that the classical Hamiltonian energy is in general not conserved and depends exactly on the reference frame as we define it. The same is true for the quantum-mechanical energy as it arises from the energy operator E, as long as there is no gauge choice.

2. Hamiltonian operator in a general spacetime V

Consider a wave equation for a quantum particle in the spacetime manifold V. In order to define a Hamiltonian operator, one needs first to choose a local chart (coordinate system). This is in general defined only in some subdomain U of V. To any point X in U, the chart χ associates a quadruplet of real numbers: $\chi(X) = \mathbf{X} \equiv (x^\mu) \in \mathbb{R}^4$. In any such chart χ, the wave operator gets a local expression, i.e., an expression as an usual linear differential operator acting on local wave functions $\Psi = \Psi(\mathbf{X})$ defined on the open subset $\chi(\mathrm{U})$ of $\mathbb{R}^4$ and taking values in some $\mathbb{C}^n$ space. [1] Assume this local expression contains only first-order derivatives with respect to $t \equiv x^0/c$. In that case, we may rewrite the wave equation as the Schrödinger equation (setting $\hbar = 1$ in this paper):

$$\mathrm{i}\frac{\partial \Psi}{\partial t} = \mathrm{H}\Psi, \tag{1}$$

in which the Hamiltonian operator H is a linear differential operator that contains no derivative with respect to t. This definition makes it clear that H depends on the coordinate system as do the local expressions of the wave function and of the wave operator. Also, H is defined only over the domain of the chart, $\mathrm{U} \subset \mathrm{V}$. But in practice, U can be taken large enough that we may forget about the values of Ψ outside U (or rather, outside $\chi(\mathrm{U})$). More exactly, $\Psi(t, \mathbf{x})$ may be assumed to vanish when the triplet $\mathbf{x}$ made with the spatial coordinates x^j $(j = 1, 2, 3)$ is on the spatial boundary of $\chi(\mathrm{U})$. (See Ref. [13] for more detail.) If one changes the chart: $\mathbf{X} = (x^\mu) \hookrightarrow (x'^\mu) = \mathbf{X}'$, one gets another operator $\mathrm{H}' \neq \mathrm{H}$, whose relation with H is not simple in general [14]. However, H is essentially unchanged under a *purely spatial* coordinate change:

$$x'^j = f^j((x^k)) \ (j, k = 1, 2, 3), \qquad \text{and} \qquad x'^0 = x^0. \tag{2}$$

E.g., when Ψ behaves as a scalar, $\Psi'(\mathbf{X}') = \Psi(\mathbf{X})$, we have:

$$(\mathrm{H}'\Psi')(\mathbf{X}') = (\mathrm{H}\Psi)(\mathbf{X}). \tag{3}$$

We call "reference frame" an *equivalence class* F *of charts* defined on the same domain U and exchanging by (2) [15]. A reference frame F in that sense amounts to the data of:

[1] We may assume that the "intrinsic" wave function, say φ, takes value directly in this $\mathbb{C}^n$ space: Otherwise, if φ is a section of some vector bundle with base V, the components Ψ^a $(a = 1, ..., n)$ of the local expression of φ are got from the expansion of φ on some local frame field (e_a).

- the domain U of the spacetime,
- The set of the reference world lines "$\mathbf{x} \equiv (x^j) =$ Constant, x^0 variable",
- A time coordinate map $X \mapsto x^0(X)$ defined for events $X \in \mathrm{U}$.

So this extends the notion of reference frame of classical mechanics [13].
The operator H *depends precisely on the reference frame in this sense* [14].

3. Hamiltonians with a gauge choice

In this section and in the next one, we fix the reference frame, but we suppose from now that the wave operator depends on the choice of some "gauge" field G. This happens e.g. for the covariant Dirac equation, for which G is the tetrad field. We assume [13] that:

- (i) For any choice of the gauge field G, there is a Hilbert space H of "states" $\Psi = \Psi(\mathbf{x})$, with scalar product: $(\Psi \mid \Phi)$. Thus for another choice $\widetilde{G}$ of the gauge field, we get another Hilbert space $\widetilde{\mathsf{H}}$, and another scalar product $(\Xi \,\widetilde{\mid}\, \Omega)$. [2]
- (ii) Given any two gauge fields, there is (for any value of t) a *unitary transformation* $\mathcal{U}$ from H onto $\widetilde{\mathsf{H}}$:
$$\forall t,\ \forall \Psi, \Phi \in \mathsf{H}, \quad (\mathcal{U}\Psi \,\widetilde{\mid}\, \mathcal{U}\Phi) = (\Psi \mid \Phi). \tag{4}$$
- (iii) The wave equation is covariant under any unitary transformation $\mathcal{U}$ arising due to an admissible change of the gauge field: $G \hookrightarrow \widetilde{G}$.

These assumptions are indeed valid in the case of the covariant Dirac equation [10]. When changing the gauge field: $G \hookrightarrow \widetilde{G}$, we get a new form of the wave equation. Thus, when rewriting it as the Schrödinger equation, we get a new Hamiltonian operator: $\mathrm{H} \hookrightarrow \widetilde{\mathrm{H}}$. The covariance of the wave equation under the corresponding unitary transformation $\mathcal{U}$ means this:

$$\text{If} \quad \Xi \equiv \mathcal{U}\Psi, \quad \text{then} \quad \mathrm{i}\frac{\partial \Psi}{\partial t} = \mathrm{H}\Psi \Leftrightarrow \mathrm{i}\frac{\partial \Xi}{\partial t} = \widetilde{\mathrm{H}}\Xi\,. \tag{5}$$

In turn, it is easy to check that (5) is true if and only if:

$$\underline{\widetilde{\mathrm{H}} = \mathcal{U}\mathrm{H}\mathcal{U}^{-1} - \mathrm{i}\mathcal{U}\left[\partial_t \left(\mathcal{U}^{-1}\right)\right]}. \tag{6}$$

Equation (6) is well known (in a somewhat different context, see Section 4) [16, 17]. It means that H, seen as the *generator of the time evolution,* transforms consistently.

But, seen as an *operator acting on the states* $\Psi = \Psi(\mathbf{x})$, H would transform consistently if and only if, at any time t, all mean values were invariant under the unitary transformation. Thus, $\mathcal{D}$ being the domain of H, if and only if:

$$\forall \Psi \in \mathcal{D}, \quad \langle \widetilde{\mathrm{H}} \rangle \equiv (\mathcal{U}\Psi \,\widetilde{\mid}\, \widetilde{\mathrm{H}}\,(\mathcal{U}\Psi)) = (\Psi \mid \mathrm{H}\,\Psi) \equiv \langle \mathrm{H} \rangle. \tag{7}$$

Whether H is the Hamiltonian or not, one may prove [10] that (7) is true if and only if

$$\widetilde{\mathrm{H}} = \mathcal{U}\mathrm{H}\mathcal{U}^{-1}. \tag{8}$$

Note that, using the unitarity of $\mathcal{U}$ (4), it is trivial to check that the condition (8) is *sufficient* to ensure the validity of Eq. (7), i.e., to ensure that all mean values of H and $\widetilde{\mathrm{H}}$ are equal.

[2] In general, the definition of the scalar product involves some coefficient *fields*, e.g. the field of the curved-spacetime Dirac matrix γ^0 in the case of the covariant Dirac equation: for instance, see Eq. $(26)_1$ in Ref. [13]. It follows that in general, strictly speaking, the scalar product and even the Hilbert space H depend on the value of the time coordinate t. We will keep this dependence implicit.

Journal of Physics: Conference Series **626** (2015) 012030 doi:10.1088/1742-6596/626/1/012030

Comparing (6) and (8), we find immediately that the *Hamiltonian* operators H and $\widetilde{\mathrm{H}}$ are equivalent as operators if and only if

$$\partial_t \mathcal{U} = 0. \tag{9}$$

Thus: *for any wave equation which transforms covariantly under a unitary gauge transformation* $\mathcal{U}$ *depending on the time coordinate, the Hamiltonian operator* H *and its mean values depend on the gauge choice.* Again, note that it is not surprising that by adding a non-zero term $-\mathrm{i}\mathcal{U}\left[\partial_t\left(\mathcal{U}^{-1}\right)\right]$ to $\mathcal{U}\mathrm{H}\mathcal{U}^{-1}$, one gets different mean values. Indeed, Goldman [17] stated without a general proof (but giving an example) that Eq. (6) leads to different mean values for H and $\widetilde{\mathrm{H}}$ in the time-dependent case. However, it was not obvious that the condition (8) is indeed *necessary* to the equality of all mean values, i.e., to the validity of Eq. (7).

4. Ambiguity of the energy operator

In usual quantum mechanics, the energy operator is simply the Hamiltonian operator H. However, in a time-dependent gravitational field, it happens that H is in general not Hermitian (at least for the covariant Dirac equation [14, 18]). To account for this fact, one defines then the energy operator E as the Hermitian part of H. It is not hard to guess that the gauge dependence of H implies a gauge dependence of E, and this is confirmed by a careful examination in the case of the covariant Dirac equation [9, 10]. Actually, for that equation, one may even prove [10] that the equality of all mean values of the energy operators E and $\widetilde{\mathrm{E}}$ *up to a constant that is independent of the state* Ψ implies $\widetilde{\mathrm{E}} = \mathcal{U}\mathrm{E}\mathcal{U}^{-1}$ (which is in general not verified when $\mathcal{U}$ is the unitary transformation associated with a change of the tetrad field). Besides this fact, it was also proved [9] that the *eigenvalue spectrum* of the energy operator depends on the choice of the tetrad field for the covariant Dirac equation. Both of these results are surprising facts.

This is not a small effect: for the covariant Dirac equation, the difference in the mean values of E for different choices of the tetrad field can be made arbitrarily large [10]. This is also true with an electromagnetic field [12]. All of this is already true in a Cartesian chart in a Minkowski spacetime [10, 12]. It means in particular this [12]: Unlike the genuine Dirac equation (Dirac's), *the covariant Dirac equation cannot predict the energy levels of the hydrogen atom* — as long as the freedom in the choice of the tetrad field is left too large. In particular, the ambiguity in the energy operator and its mean values is still present, as proved in detail in Ref. [10], if one restricts the choice of the tetrad field merely by imposing the "Schwinger gauge". This is contrary to what had been claimed before by other authors [19, 20]. [3]

In a very recent paper [22], Silenko also investigates the problem arising from a difference in the mean values of two energy operators (Hermitian Hamiltonians), in a given coordinate system. In his work, the second energy operator is got from the first one by changing the "representation", i.e., by applying a unitary transformation to the wave function. (That wave function corresponds to some quantum wave equation which *a priori* is not necessarily the covariant Dirac equation.) The change of representation that Silenko especially investigates is the one that occurs when going to specifically a Foldy-Wouthuysen representation [16]. Thus, the unitary transformation does not arise due to the change of a gauge field. This same problem has been investigated a long time ago by Goldman [17] and by Nieto [23]. On the other hand, recall that for the covariant Dirac equation, the tetrad field plays the role of a gauge field

[3] In brief: in the "Schwinger gauge", the tetrad field can still be subjected to an arbitrary time-dependent rotation. Around Eq. (35) in Ref. [21] (whose content is otherwise very close to that of the preprint [20]), these authors object on my explicit calculation [10] of the difference in the mean values of E for two Schwinger tetrads that, in their opinion, "one should also perform averaging over the spin states". But the states relevant to the Dirac equation are four-component states that include the "spin attribute". The mean value for such a state ψ includes the appropriate "averaging over the spin states": namely, over those that (in general) are "mixed" in the state ψ [13].

and its data determines point-dependent Dirac matrices, which do change when one changes that gauge field. Hence, the problem investigated in Refs. [17, 22, 23] is not the same as the problem investigated in Refs. [9, 10]. There, we looked to what happens when one changes the point-dependent Dirac matrices (e.g. through the change of the tetrad field) in specifically the covariant Dirac equation. However, it turns out that changing the coefficient fields in the covariant Dirac equation implies that a new representation is got for the Dirac wave function, related to the starting one by a unitary transformation $\mathcal{U}$: see Eqs. (1)–(5) in Ref. [10]. Thus, Eqs. (4) to (9) above do apply to the situations investigated in both groups of works: Refs. [17, 22, 23] as well as Refs. [9, 10].

Equation (31) in Ref. [22] gives "the Dirac Hamiltonian" for a particle in a (possibly nonstationarily) rotating frame [in a spacetime that is flat at least in the domain of the rotating coordinate system, see Eq. (30) there]. To avoid misinterpretations, one should note that this equation gives actually the Dirac Hamiltonian in this rotating frame with a *particular choice* of the tetrad field. Namely (in the case of a stationary rotation, $\omega(t) = \text{Constant}$), this is exactly the Hamiltonian got in the rotating frame with "Ryder's rotating tetrad", Eq. (72) in Ref. [12]. That Hamiltonian, denoted H_3 there, has Mashhoon's "spin-rotation coupling" term. Another choice of tetrad is the "Cartesian tetrad" (the natural basis of a Cartesian inertial coordinate system, with respect to which the rotating frame is indeed only rotating). Then in the rotating frame one gets the Hamiltonian H_1 given by Eq. (33) in Ref. [12]. That one does not have Mashhoon's term and is not physically equivalent to H_3, i.e. to the Hamiltonian in Eq. (31) of Ref. [22]. (The physical inequivalence of H_1 and H_3 has been proved quickly in Ref. [10], the paragraph after Eq. (32) there: the proof of inequivalence is indeed the same as the detailed proof given for the case of the Hamiltonians in the inertial frame.)

To solve the problem of the dependence of the energy mean values on the chosen representation in the non-stationary case, Silenko [22] proposes to select the Foldy-Wouthuysen representation as the correct one, essentially because it has a nice semi-classical limit (a point which he demonstrates in detail). He states that the exact definition of a Foldy-Wouthuysen(-type) transformation proposed by Eriksen [24] and further elaborated by Eriksen & Kolsrud [25] and de Vries & Jonker [26] can be extended to the nonstationary case. [4] His proposal [22] is at odds (as also noted by him) with the analysis that was made by Goldman [17] in the archetypical case of the (genuine) Dirac equation in an inertial frame in a Minkowski spacetime but in a time-dependent electromagnetic field. In the latter case, it seems natural to admit, as did Goldman (albeit with caution), that the correct energy mean values are the ones got from the starting Hamiltonian, which is the (genuine) Dirac Hamiltonian in the electromagnetic field, Eq. (8) in Ref. [17]. These mean values differ from those got after the Foldy-Wouthuysen transformation. In our opinion, the fact that in the Foldy-Wouthuysen representation the semi-classical limit is closer to the "classical" equations of motion (those for a particle with spin yet) than it is in the initial ("Dirac") representation, is not a decisive argument for the choice. Obtaining the classical equations of motion in some limit needs precisely to restrict oneself to some limiting behaviour, which by definition is only at the boundary of the possible behaviours in the quantum domain. Anyway, let us repeat that the energy ambiguity arising from changing the representation (from the Dirac one to the Foldy-Wouthuysen one) is a different problem than the dependence of the energy mean values on the gauge choice.

[4] Actually, the work of Eriksen & Kolsrud [25] also contains an extension to the case with a time-dependent external field, and that extension seems to differ from the one considered in Ref. [22]: in the work [25], the unitary transformation U leading to the new representation is expressed in terms of an operator Λ, such that $U\Lambda U^\dagger = \beta$, and in general Λ is not equal to the sign operator $\lambda \equiv H.H^{-1/2}$ of the Hamiltonian operator H. In the work [22], the transformation U is defined in terms of that sign operator λ. On the other hand, in the presence of a time-dependent metric and tetrad field (as considered in the latter work), the Hamiltonian does not have directly the form considered in the work [25], Eqs. (54)–(55) there. A further discussion could be worth.

However, we note that in a still recent paper [27], the energy ambiguity issues due to the presence of a gauge choice (specifically the choice of a tetrad field in the Schwinger gauge) and to a change of the representation (specifically the change to a Foldy-Wouthuysen representation) are simultaneously present. In Ref. [13], we have discussed in detail how these inequivalence issues do affect the work [27]. If one would adopt the view [22] that the Foldy-Wouthuysen representation is the correct one, then of course he would see no problem in the second inequivalence issue, but the first one would remain.

5. Energy operator vs. classical energy

In this section we summarize a few reasons why the energy operator *ought to be* well defined in a given reference frame [13], so that it is a real problem that it is not in the presence of a gauge choice. Note first that, even in the time-dependent case, the eigenvalues of E at some (coordinate) time t are the observable values of the energy of the quantum-mechanical system at hand, at this time t [9]. Moreover, in the stationary case, the eigenvalues of E are associated with stationary solutions of the Schrödinger equation (when $\mathrm{H} = \mathrm{E}$ is Hermitian). Also, recall that QM has a strong relation to *classical Hamiltonian mechanics.* E.g. the standard non-relativistic Schrödinger equation is got from the non-relativistic classical Hamiltonian of a test particle:

$$e = H(\mathbf{p}, \mathbf{x}, t) = \frac{\mathbf{p}^2}{2m} + V(\mathbf{x}, t) \equiv T + V, \qquad \mathbf{p} \equiv m\mathbf{v}, \tag{10}$$

by applying the classical-quantum correspondence

$$e \to +\mathrm{i}\frac{\partial}{\partial t}, \qquad p_j \to -\mathrm{i}\frac{\partial}{\partial x^j}. \tag{11}$$

The correspondence $(11)_1$ leads also directly to interpret the Hamiltonian operator H (or its Hermitian part E) as the *energy operator.* The meaning of the energy operator of QM is hence inherited from classical Hamiltonian mechanics.

In Newtonian physics and in special relativity, energy is conserved: there is a local energy conservation equation involving the energy density w and its flux $\mathbf{\Phi}$,

$$\frac{\partial w}{\partial t} + \mathrm{div}\, \mathbf{\Phi} = 0. \tag{12}$$

Within that continuum with conserved energy, a small piece of matter (its smallness being relative to the problem that is considered) may be modeled as a *test particle.* In Newtonian mechanics more specifically, the energy of a test particle is the sum (10) of its kinetic energy and its potential energy in the force field (the latter being assumed to derive from the potential V). Clearly, e is *not* conserved unless V is time-independent. Also, e depends exactly on the reference frame: e.g., changing the inertial frame changes T, but does not change V.

Similarly, in a general spacetime in the presence of an electromagnetic potential V_μ, the energy of a test particle is defined as

$$e \equiv c\breve{p}_0, \tag{13}$$

where

$$\breve{p}^\mu \equiv mc\frac{\mathrm{d}x^\mu}{\mathrm{d}s} + \frac{q}{c}V^\mu \tag{14}$$

and $\breve{p}_\mu \equiv g_{\mu\nu}\breve{p}^\nu$ [signature $(+---)$], with q the charge of the particle [28]. (Also, $s \equiv c\tau$ where τ is the proper time along the trajectory of the particle.) Thus, $\breve{p}_\mu$ being a covector, it follows that $e = c\breve{p}_0$ again depends exactly on the reference frame. Moreover, using the dispersion relation

$$g^{\mu\nu}\left(\breve{p}_\mu - \frac{q}{c}V_\mu\right)\left(\breve{p}_\nu - \frac{q}{c}V_\nu\right) - m^2c^2 = 0, \tag{15}$$

the energy (13) can be expressed as a function

$$H(\mathbf{p}, \mathbf{x}, t) \equiv e \equiv c\breve{p}_0 \qquad (16)$$

with $\mathbf{p} = (p_j) \equiv (-\breve{p}_j)$. This is a Hamiltonian function for the motion of the particle [28] — as was already known [29] in the case without electromagnetic field, i.e., for a free particle in a curved spacetime. Note that in that case there is no potential ($V_\mu = 0$), hence the definition (13)–(14) of the energy is completely unique for a free test particle in a given reference frame. The Hamiltonian (16) was also already known in the case without gravitational field, i.e., for a charged particle in a Minkowski spacetime [30]. Clearly, also in the present general case the energy (13) of the test particle is in general not conserved. In conclusion, the fact that the energy operator depends on time in a general spacetime is no justification for it not being unique in a given reference frame at a given time, for the energy of a free test particle also depends on time and it *is* unique in a given reference frame at a given time.

6. Conclusion

The energy operator E of a quantum particle is the quantum equivalent of the Hamiltonian energy e of a classical test particle. Just as does e, it depends precisely on the reference frame, and it is in general not constant. However, in the presence of a gauge choice associated with time-dependent unitary transformations, E is not well defined in a given reference frame. This indicates that the gauge freedom must be reduced to avoid this case. Two proposals along this line have been made for the covariant Dirac equation, each of which provides us with a solution of this non-uniqueness problem [31, 32]. However, these two solutions are not equivalent. They might be discriminated by the observation or non-observation of the Mashhoon effect [12].

Acknowledgement. I am grateful to Alexander Silenko for a discussion.

[1] Colella R Overhauser A W and Werner S A 1975 Observation of gravitationally induced quantum interference *Phys. Rev. Lett.* **34** 1472–1474
[2] Werner S A Staudenmann J A and Colella R 1979 Effect of Earth's rotation on the quantum mechanical phase of the neutron *Phys. Rev. Lett.* **42** 1103–1107
[3] Riehle F Kisters Th Witte A Helmcke J and Bordé Ch J 1991 Optical Ramsey spectroscopy in a rotating frame, Sagnac effect in a matter-wave interferometer *Phys. Rev. Lett.* **67** 177–180
[4] Kasevich M and Chu S 1991 Atomic interferometry using stimulated Raman transitions *Phys. Rev. Lett.* **67** 181–184.
[5] Nesvizhevsky V V *et al.* 2002 Quantum states of neutrons in the Earth's gravitational field *Nature* **415** 297–299
[6] Brill D R and Wheeler J A 1957 Interaction of neutrinos and gravitational fields *Rev. Modern Phys.* **29** 465–479. Brill D R and Wheeler J A 1961 Erratum *Rev. Modern Phys.* **33** 623–624
[7] Chapman T C and Leiter D J 1976 On the generally covariant Dirac equation *Am. J. Phys.* **44** 858–862
[8] Fock V 1929 L'équation d'onde de Dirac et la géométrie de Riemann *J. Phys. Rad.* **10**, No. 11, 392–405
[9] Arminjon M and Reifler F 2011 A non-uniqueness problem of the Dirac theory in a curved spacetime *Ann. Phys. (Berlin)* **523** 531–551
[10] Arminjon M 2013 On the non-uniqueness problem of the covariant Dirac theory and the spin-rotation coupling *Int. J. Theor. Phys.* **52** 4032–4044
[11] Ryder L 2008 Spin-rotation coupling and Fermi-Walker transport *Gen. Relativ. Gravit.* **40** 1111–1115
[12] Arminjon M 2014 Should there be a spin-rotation coupling for a Dirac particle? *Int. J. Theor. Phys.* **53** 1993–2013
[13] Arminjon M 2015 Some remarks on quantum mechanics in a curved spacetime, especially for a Dirac particle *Int. J. Theor. Phys.* **54** (print version to appear) DOI: 10.1007/s10773-014-2439-4
[14] Arminjon M and Reifler F 2010 Basic quantum mechanics for three Dirac equations in a curved spacetime *Braz. J. Phys.* **40** 242–255

[15] Arminjon M and Reifler F 2011 General reference frames and their associated space manifolds *Int. J. Geom. Meth. Mod. Phys.* **8** 155–165
[16] Foldy L L and Wouthuysen S A 1950 On the Dirac theory of spin 1/2 particles and its non-relativistic limit *Phys. Rev.* **78** 29–36
[17] Goldman T 1977 Gauge invariance, time-dependent Foldy-Wouthuysen transformations, and the Pauli Hamiltonian *Phys. Rev. D* **15** 1063–1067
[18] Leclerc M. 2006 Hermitian Dirac Hamiltonian in the time-dependent gravitational field, *Class. Quant. Grav.* **23** 4013–4020.
[19] Gorbatenko M V and Neznamov V P 2011 Uniqueness and self-conjugacy of Dirac Hamiltonians in arbitrary gravitational fields *Phys. Rev. D* **83**, 105002
[20] Gorbatenko M V and Neznamov V P 2013 Absence of the non-uniqueness problem of the Dirac theory in a curved spacetime. Spin-rotation coupling is not physically relevant *Preprint* arXiv:1301.7599v2 (gr-qc)
[21] Gorbatenko M V and Neznamov V P 2014 On the uniqueness of the Dirac theory in curved and flat spacetime *Ann. Phys. (Berlin)* **526** 195–200
[22] Silenko A J 2015 Energy expectation values of a particle in nonstationary fields *Phys. Rev. A* **91** 012111
[23] Nieto M M 1977 Hamiltonian expectation values for time-dependent Foldy-Wouthuysen transformations: implications for electrodynamics and resolution of the external-field πN ambiguity *Phys. Rev. Lett.* **38** 1042–1045
[24] Eriksen E 1958 Foldy-Wouthuysen transformation. Exact solution with generalization to the two-particle problem *Phys. Rev.* **111** 1011–1016
[25] Eriksen E and Kolsrud M 1960 Canonical transformations of Dirac's equation to even forms. Expansion in terms of the external fields *Suppl. Nuovo Cim.* **18** 987–1025
[26] de Vries E and Jonker J E 1968 Non-relativistic approximations of the Dirac Hamiltonian *Nuclear Physics B* **6** 213-225
[27] Obukhov Yu N, Silenko A J and Teryaev O V 2013 Spin in an arbitrary gravitational field *Phys. Rev. D* **88** 084014
[28] Arminjon M and Reifler F 2013 Equivalent forms of Dirac equations in curved spacetimes and generalized de Broglie relations *Braz. J. Phys.* **43** 64–77
[29] Bertschinger E 1999 Hamiltonian dynamics of particle motion *Online course*
[30] Johns O D 2005 *Analytical Mechanics for Relativity and Quantum Mechanics* (Oxford: Oxford University Press)
[31] Arminjon M 2011 A solution of the non-uniqueness problem of the Dirac Hamiltonian and energy operators *Ann. Phys. (Berlin)* **523** 1008–1028
[32] Arminjon M 2013 A simpler solution of the non-uniqueness problem of the covariant Dirac theory *Int. J. Geom. Meth. Mod. Phys.* **10**, No. 7, 1350027

An introduction to the general boundary formulation of quantum field theory

Daniele Colosi

Escuela Nacional de Estudios Superiores, Unidad Morelia, Universidad Nacional Autónoma de México, Campus Morelia, C.P. 58190, Morelia, Michoacán, México

E-mail: dcolosi@enesmorelia.unam.mx

Abstract. We give a brief introduction to the so-called general boundary formulation (GBF) of quantum theory. This new axiomatic formulation provides a description of the quantum dynamics which is manifestly local and does not rely on a metric background structure for its definition. We present the basic ingredients of the GBF, in particular we review the core axioms that assign algebraic structures to geometric ones, the two quantisation schemes so far developed for the GBF and the probability interpretation which generalizes the standard Born rule. Finally we briefly discuss some of the results obtained studying specific quantum field theories within the GBF.

1. Introduction

One of the most relevant obstacle to the realisation of the synthesis of general relativity and quantum mechanics, in the yet-to-be-found theory of quantum gravity, stems from the different ways in which the two theories are formulated. In particular, general relativity describes the geometry of space-time as a dynamical field that evolves according to equations of motion, the Einstein's field equations. Such dynamics is not described with respect to a non-dynamical reference system. This property is sometimes called background independence. In contrast, a fixed, non-dynamical space-time metric (flat or curved) is an essential ingredient in the standard formulation of quantum theory from the point of view of the mathematical and conceptual foundations of the theory. In particular, in the framework of quantum field theory, quantum fields evolve in agreement with the causal structure determined by the space-time on which they are defined. For example, they are required to satisfy the so-called 'microcausility relation' expressed as the commutation of fields evaluated at space like separated points.

One possibility in order to construct a quantum theory of gravity, is to learn how to incorporate in quantum field theory the background independence of general relativity, namely how to describe the dynamics of quantum fields without reference to a fixed background space-time metric. A first step in this direction is represented by the development since the late 1980s of the so-called 'Topological Quantum Field Theory' (TQFT) [1] which has provided the first examples of background independent quantum theories in the sense that the quantum evolution (involving also topological change) is defined on manifolds not carrying a metric structure. In particular, TQFT has been successful in constructing quantum gravity models in three space-time dimensions where the classical theory has only global but not local degrees of freedom, namely it reduces to a topological theory[1].

In this note we present an introduction to a new axiomatic formulation of quantum theory called the general boundary formulation (GBF) [3, 4] which is based on two main ingredients: The mathematical framework of TQFT and a generalisation of the Born rule that guaranties a consistent physical interpretation. In particular the axioms of the GBF provide a consistent assignment of algebraic structures to geometric ones. State spaces are associated with hypersurfaces, namely oriented $(n-1)$-dimensional manifolds, and amplitude maps are associated with regions, namely oriented n-dimensional manifolds. Hypersurfaces arise as boundaries of regions. The physical interpretation of these structures is obtained by generalising the ordinary Born rule of quantum mechanics [5]: Probabilities and expectation values of observables are extracted from maps associated to regions. A remarkable aspects of these structures is that no background space-time metric is required for their implementation.

When considering quantum fields defined on manifold carrying a metric structure, the GBF does not impose any special choice for the boundaries of regions where dynamics takes place. For example, in the standard formulation of quantum field theory in Minkowski space-time, states are defined on equal time hyperplanes. Their evolution from an initial hyperplane at time t_i to a final one at time t_f is described by a temporal evolution operator, as in the formula $\langle\psi_f|U(t_i,t_f)|\psi_i\rangle$ representing the transition amplitude from the initial state ψ_i to the final state ψ_f. So, the space-time region under consideration is the time-interval region that we denote as $[t_i, t_f]$. In the case of quantum fields on curved spaces, hyperplanes are replaced by Cauchy surfaces, but the structure of transition amplitudes (in terms of initial and final states and an evolution operator) is unchanged. The standard S-matrix is then constructed by taking the temporal asymptotic limit, namely $t_i \to -\infty$, $t_f \to \infty$. As has been mentioned, the GBF is not bounded to a specific region even in Minkowski space-time. Its versatility allows to consider more general regions, e.g. regions whose boundary is timelike, or connected or even compact are allowed. The implementation of the standard formulation of the quantum field theory fails in these situations. Concrete examples of such kind of boundaries have been considered in the GBF and will be discussed later.

The new way to describe the dynamics of quantum fields provided by the GBF appears to be interesting in many respects: First of all, the GBF can be viewed as an extension and a generalisation of quantum theory, offering new perspectives on aspects of the theory, as will be shown when discussing the quantization of scalar field in Minkowski space-time in non-time-interval regions. Moreover, the GBF offers useful tools for implementing theoretical treatments more close to the experimental setups. Indeed, experiments often involve an apparatus which occupies a certain volume in space and is only active during a temporal interval. Therefore, the boundary of space-time region concerned by real experiments is limited in space and in time, namely is a compact region in four dimensions; precisely such region is of the kind the GBF can deal with no conceptual difficulties (although the technical ones could be non-trivial).

Secondly, since general space-time regions are considered in the GBF, it may solve the difficulties that arise from restricting the notion of dynamics to the evolution of initial state to a final one, i.e. from the restriction to time-interval regions. In Anti-de Sitter space, no asymptotic temporal regions exist, and consequently the construction of the standard S-matrix appears to be unfeasible due to the lack of temporal asymptotic free states. The peculiar geometry of Anti-de Sitter space does not represent an obstacle for a general boundary quantum field theory: Indeed, an amplitude for *spatially*, instead of temporally, asymptotic states can be implemented (and has been implemented, as discussed later).

Thirdly, some of the conceptual problems of background independent theory can be solve in the GBF. We briefly mention the problem of locality, understood as the problem of isolating the

[1] One of the most studied TQFT model of quantum gravity in three dimensions is the Turaev-Viro model [2].

physical system of interest from the rest of the universe. In standard Minkowski-based quantum field theory, the S-matrix of two spacelike separated systems factorizes in the product of two S-matrices, one for each system. This is the so called cluster decomposition of the S-matrix and it relies on the existence of a fixed causal structure, provided by the Minkowski metric. But in a background independent theory no fixed metric or causal structure is available. Then, how can we separate the system from the rest of the universe? Given a region and its boundary, the GBF treatment rests only on the dynamics unfolding in the interior of that region and on the configurations of the fields on the boundary hypersurface(s). No additional data are needed, in particular data referring to the exterior of the region. In other words, locality is naturally implemented within the GBF.

Finally, the GBF may contribute in the resolution of the problem of quantum gravity, which has been originally the main motivation for its development. In that respect it is interesting to notice that the GBF is certainly compatible with quantum gravity in three dimensions, where the theory is topological and the GBF is precisely based on the mathematical framework of TQFT. Moreover, current approaches to quantum gravity like Spin Foam[2] [6] models and Group Field Theory [7] fits into the GBF.

2. Basic axioms, state spaces and amplitude maps

We present in this section a reduced and simplified version of the core axioms of the GBF and refer the reader to [8] for the complete list.

Let denote with M an oriented topological manifold of dimension n that we call 'region'. We indicate with Σ an oriented manifold without boundary of dimension $(n-1)$, namely a 'hypersurface'. We can think of hypersurfaces as boundaries of regions (and the orientations of the regions induce those of the boundaries). Moreover, regions can be glued along (components of) their common boundaries to produce other regions.

To each hypersurface Σ is associated a Hilbert space $\mathcal{H}_\Sigma$ with inner product $\langle\cdot,\cdot\rangle_\Sigma$. The dual Hilbert space $\mathcal{H}^*_\Sigma$ is then associated with the same manifold but with opposite orientation, which is denoted by $\overline{\Sigma}$, namely $\mathcal{H}_{\overline{\Sigma}} = \mathcal{H}^*_\Sigma$. The following composition rule is satisfied by the Hilbert spaces: If the hypersurface Σ decomposes into a disjoint union of hypersurfaces as $\Sigma = \Sigma_1 \cup \cdots \cup \Sigma_m$ then the Hilbert space $\mathcal{H}_\Sigma$ results to be given by the tensor product of Hilbert spaces: $\mathcal{H}_{\Sigma_1} \otimes \cdots \otimes \mathcal{H}_{\Sigma_m}$.

To each region M is associated an amplitude map $\rho_M : \mathcal{H}_{\partial M} \to \mathbb{C}$, from the Hilbert space $\mathcal{H}_{\partial M}$ defined on boundary ∂M of the region M. According to the gluing rule, if two regions M_1 and M_2 share a common boundary Σ and are glued along it, the amplitudes maps ρ_{M_1} and ρ_{M_2} can be composited into the map $\rho_{M_1 \cup M_2}$ for the union region, $\rho_{M_1\cup M_2} = \rho_{M_1} \circ \rho_{M_2}$. The symbol $\circ$ represents composition of maps, which involves a complete sum over an orthonormal basis of the Hilbert space $\mathcal{H}_\Sigma$ associated with the common hypersurface Σ,

$$\rho_{M_1\cup M_2}(\psi_1 \otimes \psi_2) = \sum_{i\in I} \rho_{M_1}(\psi_1 \otimes \xi_i)\rho_{M_2}(\overline{\xi_i} \otimes \psi_2), \tag{1}$$

where $\psi_1 \in \mathcal{H}_{\partial M_1 \setminus \Sigma}$, $\psi_2 \in \mathcal{H}_{\partial M_2 \setminus \Sigma}$ and $\{\xi_i\}_{i\in I}$ is an orthonormal basis of $\mathcal{H}_\Sigma$. In contrast, if the two regions are disjoint, the amplitude map of the union region is equal to the product of the amplitude maps of the two regions, namely $\rho_{M_1\cup M_2}(\psi_1 \otimes \psi_2) = \rho_{M_1}(\psi_1)\rho_{M_2}(\psi_2)$. Finally, if the region M reduces to the empty region $\hat{\Sigma}$, the boundary of which is given by the union of a hypersurface and a copy of this same hypersurface but with opposite orientation, i.e. $\partial\hat{\Sigma} = \Sigma \cup \overline{\Sigma}'$, ($\overline{\Sigma}'$ being a copy of $\overline{\Sigma}$) then the amplitude map reduces to the inner product of the Hilbert space associated to the hypersurface: $\rho_{\hat{\Sigma}}(\psi_1 \otimes \psi_2) = \langle\psi_1, \psi_2\rangle_\Sigma$.

[2] See Oeckl's talk at the Mexilazos Conference, held in Morelia, 10 November 2012, available at http://matmor.unam.mx/eventos/mexilazos2012/files/Oeckl-ML2012.pdf.

The amplitude map ρ_M can be seen as a generalisation of the standard transition amplitude, which is recovered in the standard setting, namely when the space-time region M reduces to the standard time-interval region $[t_i, t_f]$. In this case the boundary hypersurface ∂M is given by the disjoint union of flat equal time hyperplanes Σ_{t_i} and $\overline{\Sigma}_{t_f}$ (with opposite orientation) labeled by the time instants t_i and t_f. According to the composition rule of Hilbert spaces, we have $\mathcal{H}_{\partial M} = \mathcal{H}_{t_i} \otimes \mathcal{H}^*_{t_f}$. Then, the amplitude map for the boundary state $\psi_i \otimes \psi_f \in \mathcal{H}_{t_i} \otimes \mathcal{H}^*_{t_f}$ results to be: $\rho_M(\psi_i \otimes \psi_f) = \langle \psi_f | U(t_i, t_f) | \psi_i \rangle$, which is the expression of the standard transition amplitude for the evolution from the state ψ_i to the state ψ_f determined be the temporal evolution operator $U(t_i, t_f)$.

In the GBF quantum observables are associated with regions M and implemented as maps, called observable maps, from the Hilbert space $\mathcal{H}_{\partial M}$ to the complex number, [9]. A set of axioms prescribes how different observables (associated with different regions) compose in analogy with the composition of amplitude maps. Indeed, amplitude maps can be considered as special cases of observable maps.

The implementation of all these algebraic structures for a specific theory can be realized through the application of a quantization procedure to a classical field theory. So far, two quantization schemes have been developed within the GBF and we shall present them in the next section.

3. Quantization schemes

We consider a scalar field in an N-dimensional Lorentzian manifold with metric g described by the linear action $S_M(\phi) = \int_M \mathrm{d}^N x \mathcal{L}(\phi, \partial\phi, x)$ in the space-time region M. We do not require the boundary hypersurface ∂M to be a Cauchy or a spacelike hypersurface. The classical theory equips the boundary ∂M with a symplectic vector space $(L_{\partial M}, \omega_{\partial M})$, $L_{\partial M}$ being the space of (germs of) solutions of the Euler-Lagrange equations in a neighbourhood of ∂M, and $\omega_{\partial M}$ is the symplectic structure given by the anti-symmetric bilinear form $\omega_{\partial M}(\xi, \eta) = \frac{1}{2}[\xi, \eta]_{\partial M} - \frac{1}{2}[\eta, \xi]_{\partial M}$ with

$$[\xi, \eta]_{\partial M} = \int_{\partial M} \mathrm{d}^{N-1}\sigma\, \eta(x(\sigma)) \left(n^\mu \frac{\delta \mathcal{L}}{\delta \partial_\mu \xi} \right)(x(\sigma)), \tag{2}$$

where n^μ is the unit normal vector to the hypersurface ∂M. An additional ingredient is needed to quantize the theory: A compatible complex structure $J_{\partial M}$ defined on $L_{\partial M}$ such that $J^2_{\partial M} = -\mathrm{id}_{\partial M}$ and is compatible with the symplectic structure, i.e. $\omega_{\partial M}(J_{\partial M}\xi, J_{\partial M}\eta) = \omega_{\partial M}(\xi, \eta)$.

Based on the above ingredients two quantization schemes have been developed so far for the GBF: The holomorphic quantization, inspired by geometric quantisation, and the Schrödinger-Feynman one where the path integral quantization is implemented.

The symplectic and complex structures define a real and complex inner product on the space $L_{\partial M}$ respectively as

$$g_{\partial M}(\xi, \eta) = 2\omega_{\partial M}(\xi, J_{\partial M}\eta), \qquad \{\xi, \eta\}_{\partial M} = g_{\partial M}(\xi, \eta) + 2\mathrm{i}\omega_{\partial M}(\xi, \eta). \tag{3}$$

The complex inner product $\{\cdot, \cdot\}_{\partial M}$, upon completion, makes $L_{\partial M}$ a complex Hilbert space.

In the holomorphic representation, the Hilbert space $\mathcal{H}^h_{\partial M}$ is given by the set of square integrable holomorphic functions on $L_{\partial M}$.[3] The amplitude map for a coherent state $\psi_\xi \in \mathcal{H}^h_{\partial M}$, defined by $\xi \in L_{\partial M}$, is

$$\rho_M(\psi) = \exp\left(\frac{1}{4} g_{\partial M}(\hat{\xi}, \hat{\xi}) \right), \tag{4}$$

where $\hat{\xi}$ is a complex solution of the equations of motion of the form $\hat{\xi} = \xi^R - \mathrm{i}\xi^I$. The elements ξ^R and ξ^I are global solutions in the region M mapped to the corresponding (germs of) solutions in $L_{\partial M}$.

[3] We are neglecting some mathematical technicalities; the rigorous construction can be found in Ref. [8].

Journal of Physics: Conference Series **626** (2015) 012031 doi:10.1088/1742-6596/626/1/012031

Within the Schrödinger-Feynman quantization scheme, quantum states are in the Schrödinger representation, namely the Hilbert space $\mathcal{H}^S_{\partial M}$ is the set of wave functionals defined on the space of field configurations. The amplitude map for a state $\psi \in \mathcal{H}^S_{\partial M}$ is formally given by

$$\rho_M(\psi) = \int \mathcal{D}\varphi\, \psi(\varphi) \int_{\phi|_{\partial M}=\varphi} \mathcal{D}\phi\; e^{\mathrm{i}S_M(\phi)}, \tag{5}$$

where the inner integral is extended over the space-time field configurations ϕ that reduce to φ on the boundary hypersurface ∂M and the outer integral is over all the configurations φ.

The two representations turn out to be equivalent, in the sense that the Hilbert spaces $\mathcal{H}^h_{\partial M}$ and $\mathcal{H}^S_{\partial M}$ are related by an isomorphism [10]. Moreover, under such isomorphism the amplitude maps (4) and (5) coincide.

The two quantization prescriptions have been designed in order to obtain a general boundary quantum field theory. Indeed, it can be shown that the Hilbert spaces and the amplitude maps, in both representations, satisfy the axioms of the GBF.

4. Probability interpretation

Probabilities are extracted from the amplitude map by selecting appropiate subspaces of the Hilbert space associated to the boundary hypersurface, $\mathcal{H}_{\partial M}$, in agreement with the following idea. Imagine we perform a measurement on a quantum system. First we prepare the system in an initial state, i.e. the initial state encodes the information of the system prior to the measurement process. Then, after the measurement, we ask the question about the probability to observe the system on a final state, i.e. the final state encodes the information relative to observation. The result is a standard quantum mechanical probability which is a conditional one: The probability of observing the final state given that the system was prepared in the initial state.

In the GBF, initial and final states ere elements of $\mathcal{H}_{\partial M}$, and information about preparation and observation is consequently encoded on subspaces of $\mathcal{H}_{\partial M}$. We call $\mathcal{S} \subset \mathcal{H}_{\partial M}$ the closed subspace of the boundary Hilbert space that represents 'preparation of the system', in the sense that $\mathcal{S}$ encodes the information concerning what can be considered as resulting form the preparation of the system. The closed subspace $\mathcal{A} \subset \mathcal{H}_{\partial M}$ represents 'observation of the system', i.e. this subspace encodes the information on what is interpreted as the result of the observation of the system. The conditional probability $P(\mathcal{A}|\mathcal{S})$ for observing A given that S has been prepared is defined as

$$P(\mathcal{A}|\mathcal{S}) = \frac{\langle \rho_M \circ P_\mathcal{S}, \rho_M \circ P_\mathcal{A}\rangle_{\partial M}}{\langle \rho_M \circ P_\mathcal{S}, \rho_M \circ P_\mathcal{S}\rangle_{\partial M}}, \tag{6}$$

where $\langle\cdot,\cdot\rangle_{\partial M}$ is the inner product in $\mathcal{H}_{\partial M}$, $P_\mathcal{S}$ and $P_\mathcal{A}$ are the orthogonal projectors onto the subspaces $\mathcal{S}$ and $\mathcal{A}$ respectively and $\circ$ denotes composition of maps. This expression makes sense if the maps $\rho_M \circ P_\mathcal{S}$ and $\rho_M \circ P_\mathcal{A}$ are linear continuous maps from $\mathcal{H}_{\partial M}$ to $\mathbb{C}$ and also $\rho_M \circ P_\mathcal{S}$ does not vanish.

The conditional probability $P(\mathcal{A}|\mathcal{S})$ is a generalisation of the standard Born rule of quantum theory and reduces to it in the standard setting, namely when the space-time M is the time-interval region $[t_i, t_f]$. In this case, as in Sec. 2, we have $\mathcal{H}_{\partial M} = \mathcal{H}_{t_i} \otimes \mathcal{H}^*_{t_f}$. We now choose the subspaces $\mathcal{S}$ and $\mathcal{A}$ as follows, $\mathcal{S} = \psi_i \otimes \mathcal{H}^*_{t_f}$ and $\mathcal{A} = \mathcal{H}_{t_i} \otimes \psi_f$. The inner products of the composition of maps in (6) result to be: $\langle \rho_M \circ P_\mathcal{S}, \rho_M \circ P_\mathcal{A}\rangle_{\partial M} = |\rho_M(\psi_i \otimes \psi_f)|^2$ and $\langle \rho_M \circ P_\mathcal{S}, \rho_M \circ P_\mathcal{S}\rangle_{\partial M} = \sum_{k\in J} |\rho_M(\psi_i \otimes \xi_k)|^2 = 1$ where $\{\xi_k\}_{k\in J}$ is an orthonormal basis of the space $\mathcal{H}^*_{t_f}$. Finally, the conditional probability (6) reads $P(\mathcal{A}|\mathcal{S}) = |\rho_M(\psi_i \otimes \psi_f)|^2 = |\langle\psi_f|U(t_i,t_f)|\psi_f\rangle|^2$, which is the standard Born rule.

Journal of Physics: Conference Series **626** (2015) 012031 doi:10.1088/1742-6596/626/1/012031

5. Results from the GBF

This section presents a partial list of results obtained by implementing the GBF for specific quantum field theories in the space-time and for different regions.

(i) A minimal requirement any new formulation of quantum theory is the ability to reproduce known and well established results. In [11, 12] we apply the GBF for a general interacting real massive scalar field in Minkowski space-time in two different settings: the standard time-interval region $[t_i, t_f]$ and the hypercylinder region, namely a ball of radius R centred at the origin of space and extended over all of time. In the region $[t_i, t_f]$ standard results as the Feynman propagator, the perturbative treatment, the S-matrix were recovered. In particular the S-matrix arises as the temporal asymptotic limit of the amplitude map $\rho_{[t_i,t_f]}$. The boundary of the hypercylinder region is timelike and connected, two aspects that render the standard formulation of quantum field theory of difficult applicability. In contrast, a mathematically well defined quantization of the field was realised within the GBF. Notice that the connectedness of the boundary prevents to talk about transition. Nevertheless the GBF provides a consistent description of dynamics. The standard S-matrix is obtained as the spatial asymptotic limit $R \to \infty$ of the amplitude map ρ_R for the hypercylinder region. An isomorphism exists between the Hilbert spaces $\mathcal{H}_{[t_i,t_f]}$ and $\mathcal{H}_R$ associated with the boundary of the two regions. Consequently the tensor product of an n-incoming particles state and an m-outgoing particles state in $\mathcal{H}_{[t_i,t_f]}$ is mapped into an $(m+n)$-particles state in $\mathcal{H}_R$. It is worth noticing that an n-particles state on the hypercylinder is neither 'incoming' nor 'outgoing'. Consequently the crossing symmetry of the S-matrix is automatically implemented by the amplitude map ρ_R. In particular, crossing symmetry could be viewed as a prediction (which is verified) of the dynamics described in the spatially asymptotic hypercylinder region. An additional result from the quantum theory in the hypercylinder is the manifest quantization of the so-called evanescent modes, which are modes of the field exponentially decaying in space, and which are completely hidden in the time-interval region.

(ii) The results obtained in Minkowski space-time have been extended in de Sitter space in [13], where the quantum theory of a real massive scalar field in the presence of a general interaction was studied perturbatively in the time-interval region and the hypercylinder region. In particular the amplitude map for the hypercylinder region leads, in the spatially asymptotic limit, to a new representation of the S-matrix and the Feynman propagator. Moreover, a new type of propagator proposed by Polyakov in [14] can be derived in a simple way from the time-interval region by sending one of the boundary to infinity[4].

(iii) As mentioned in Sec. 1, the standard formulation of the S-matrix in Anti-de Sitter space is obstructed by the lack of temporally asymptotic regions, and consequently temporally asymptotic free states. However, spatially asymptotic free states of a quantum scalar field theory can be defined within the GBF. In this respect, the hypercylinder region turns out to be a useful tool, as realised in [15] where the asymptotic amplitude map defined for the hypercylinder region was proposed as definition of S-matrix in Anti-de Sitter space.

(iv) The spacetime regions of main interest in the GBF are those with a compact boundary since in such regions the GBF will provide a manifestly local description of quantum dynamics. The applications mentioned in the previous points involve infinitely extended regions. Instead, in [16] a real scalar field was considered in a two dimensional Riemannian space-time, both in the time-interval region and in a disk region. The disk region has a circle of radius R centred at origin as its boundary hypersurface. Analogous results to those obtained in Minkowski were recovered. In particular, Hilbert spaces were defined on the boundary of the disk region and amplitude maps for the regions.

[4] Colosi D, in preparation.

(v) The dynamics of a real scalar field in two dimensional Rindler space was studied in [17]. The GBF was successfully implemented in the usual time-interval and then in a region bounded by a timelike connected hypersurface determined by a constant value of the Rindler spatial coordinate, namely an hyperbola. The quantum theories in the two regions were compared and shown to be equivalent in the sense discussed in the first point of this list.

(vi) Considerable work has been devoted to the understanding of the so-called Unruh effect (see the review [18]). In [19] this effect has been interpreted as the coincidence of expectation values of a class of observables computed on two different states: On the one hand on the vacuum state of a scalar field in Minkowski space-time and on the other hand on a mixed state in Rindler space. In particular these expectation values are obtained from the observable maps associated with two time-interval regions (in Minkowski and in Rindler spaces). The observables considered are the Weyl observables, which are linear functionals of the field ϕ of the form $F(\phi) = \int \mathrm{d}x\, \mu(x)\phi(x)$ where $\mu(x)$ has compact support in the right wedge of Minkowski.

The formal development as well as the examples of its application show that the GBF provides a powerful and versatile formulation of quantum theory. The GBF is still a work in progress and many lines of further research may be pursued in the future, among which we mention: The relation between the quantization schemes presented in Sec. 3 and the canonical quantization prescription; the consideration of more general regions, in particular of regions with compact boundaries in Lorentzian space-time; the definition of KMS states and the treatment of the Hawking effect; a proposal for a new candidate theory of quantum gravity.

Acknowledgments

This work was supported in part by UNAM-DGAPA-PAPIIT through project grant IA-102314.

References

[1] Atiyah M 1988 Topological quantum field theories *Inst. Hautes Études Sci. Publ. Math.* **68** 175-86
[2] Turaev V 1994 *Quantum invariants of knots and 3-manifolds* (New York: De Gruyter)
[3] Oeckl R 2003 A 'general boundary' formulation for quantum mechanics and quantum gravity *Phys. Lett.* B **575** 318-24 (*Preprint* hep-th/0306025)
[4] Oeckl R 2008 General boundary quantum field theory: Foundations and probability interpretation *Adv. Theor. Math. Phys.* **12** 319-52 (*Preprint* hep-th/0509122)
[5] Oeckl R 2007 Probabilities in the general boundary formulation *J. Phys. Conf. Ser.* **67** 012049 (*Preprint* hep-th/0612076)
[6] Perez A 2013 The Spin Foam Approach to Quantum Gravity *Living Rev. Rel.* **16** 3 (*Preprint* arXiv:1205.2019 [gr-qc])
[7] Freidel L 2005 Group field theory: An overview *Int. J. Theor. Phys.* **44** 1769-83 (*Preprint* arXiv:hep-th/0505016)
[8] Oeckl R 2012 Holomorphic quantization of linear field theory in the general boundary formulation *SIGMA* **8** 050 (*Preprint* arXiv:1009.5615 [hep-th])
[9] Oeckl R 2012 Observables in the General Boundary Formulation *Quantum Field Theory and Gravity (Regensburg, 2010)* (Basel: Birkhauser) pp. 137-156 (*Preprint* arXiv:1101.0367 [hep-th])
[10] Oeckl R 2012 The Schrödinger representation and its relation to the holomorphic representation in linear and affine field theory *J. Math. Phys.* **53** 072301 (*Preprint* arXiv:1109.5215 [math-ph])
[11] Colosi D and Oeckl R 2008 S-matrix at spatial infinity *Phys. Lett.* B **665** 310-13 (*Preprint* arXiv:0710.5203 [hep-th])
[12] Colosi D and Oeckl R 2008 Spatially asymptotic S-matrix from general boundary formulation *Phys. Rev.* D **78** 025020 (*Preprint* arXiv:0802.2274 [hep-th])
[13] Colosi D 2009 General boundary quantum field theory in de Sitter spacetime (*Preprint* arXiv:1010.1209 [hep-th])
[14] Polyakov A M 2008 De Sitter space and eternity *Nucl. Phys.* B **797** 199-217 (*Preprint* arXiv:0709.2899 [hep-th])
[15] Colosi D, Dohse M and Oeckl R 2012 S-Matrix for AdS from General Boundary QFT *J. Phys. Conf. Ser.* **360** 012012 (*Preprint* arXiv:1112.2225 [hep-th])

[16] Colosi D and Oeckl R 2009 States and amplitudes for finite regions in a two-dimensional Euclidean quantum field theory *J. Geom. Phys.* **59** 764-80 (*Preprint* arXiv:0811.4166 [hep-th])
[17] Colosi D and Rätzel D 2013 Quantum field theory on timelike hypersurfaces in Rindler space *Phys. Rev.* D **87** 12, 125001 (*Preprint* arXiv:1303.5873 [hep-th])
[18] Crispino L C B, Higuchi A and Matsas G E A 2008 The Unruh effect and its applications *Rev. Modern Phys.* **80** 787-838 (*Preprint* arXiv:0710.5373)
[19] Colosi D and Rätzel 2013 The Unruh effect in general boundary quantum field theory *SIGMA* **9** 019 (*Preprint* arXiv:1204.6268 [hep-th])

Journal of Physics: Conference Series **626** (2015) 012032 doi:10.1088/1742-6596/626/1/012032

Emission of cosmic radio–waves, X– or γ–rays by moving unstable particles at late times

Krzysztof Urbanowski

University of Zielona Góra, Institute of Physics, ul. Prof. Z. Szafrana 4a, 65–516 Zielona Góra, Poland.

E-mail: K.Urbanowski@if.uz.zgora.pl

Abstract. A new quantum effect connected with the late time behavior of decaying states is described and its possible observational consequences are analyzed: It is shown that charged unstable particles as well as neutral unstable particles with non–zero magnetic moment which live sufficiently long may emit electromagnetic radiation. This mechanism is due to the nonclassical behavior of unstable particles at late times (at the post exponential time region). Analyzing the transition times region between exponential and non-exponential form of the survival amplitude it is found that the instantaneous energy of the unstable particle can take very large values, much larger than the energy of this state at times from the exponential time region. Based on the results obtained for the model considered, it is shown that this new purely quantum mechanical effect may be responsible for causing unstable particles produced by astrophysical sources and moving with relativistic velocities to emit electromagnetic–, X– or γ–rays at some time intervals from the transition time regions.

1. Introduction

Late time properties of unstable states arouse interest since the theoretical predictions of deviations from the exponential form of the decay law at suitably long times. The problem was that the predicted effect is almost negligible smal and therefore very difficult to its confirmations in laboratory experiments. On the other hand, numbers of created unstable particles during some astrophysical processes are so large that some of them can survive up to times t at which the survival probability depending on t transforms from the exponential form into the inverse power–like form. It appears that at this time region a new quantum effect is observed: A very rapid fluctuations of the instantaneous energy of unstable particles take place. These fluctuations of the instantaneous energy should manifest itself as fluctuations of the velocity of the particle. Based on results presented in [1] we show in this report that this effect may cause unstable particles to emit electromagnetic radiation of a very wide spectrum: from radio frequencies through X–rays to γ–rays of very high energies and that the astrophysical processes are the place where this effect should be observed.

2. Preliminaries

Analyzing properties of unstable states $|\phi\rangle \in \mathcal{H}$ (where $\mathcal{H}$ is the Hilbert space of states of the considered system) one usually starts from studies of their decay law. The decay law, $\mathcal{P}_{\phi}(t)$ of

Journal of Physics: Conference Series **626** (2015) 012032 doi:10.1088/1742-6596/626/1/012032

an unstable state $|\phi\rangle$ decaying in vacuum is defined as follows

$$\mathcal{P}_\phi(t) = |a(t)|^2, \tag{1}$$

where $a(t)$ is the probability amplitude of finding the system at the time t in the initial state $|\phi\rangle$ prepared at time $t_0 = 0$,

$$a(t) = \langle\phi|\phi(t)\rangle. \tag{2}$$

and $|\phi(t)\rangle$ is the solution of the Schrödinger equation for the initial condition $|\phi(0)\rangle = |\phi\rangle$:

$$i\hbar\frac{\partial}{\partial t}|\phi(t)\rangle = H|\phi(t)\rangle, \tag{3}$$

where H denotes the total self–adjoint Hamiltonian for the system considered.

From basic principles of quantum theory it is known that the amplitude $a(t)$, and thus the decay law $\mathcal{P}_\phi(t)$ of the unstable state $|\phi\rangle$, are completely determined by the density of the energy distribution $\omega(\mathcal{E})$ for the system in this state [2]

$$a(t) = \int_{Spec.(H)} \omega(\mathcal{E})\, e^{-\frac{i}{\hbar}\,\mathcal{E}\,t}\, d\mathcal{E}. \tag{4}$$

where $\omega(\mathcal{E}) \geq 0$ and $a(0) = 1$.

In [3] assuming that the spectrum of H must be bounded from below, ($Spec.(H) = [E_{min}, \infty)$ and $E_{min} > -\infty$), and using the Paley–Wiener Theorem [4] it was proved that in the case of unstable states there must be $|a(t)| \;\geq\; A\,\exp\,[-b\,t^q]$, for $|t| \to \infty$. Here $A > 0$, $b > 0$ and $0 < q < 1$. This means that the decay law $\mathcal{P}_\phi(t)$ of unstable states decaying in the vacuum, (1), can not be described by an exponential function of time t if time t is suitably long, $t \to \infty$, and that for these lengths of time $\mathcal{P}_\phi(t)$ tends to zero as $t \to \infty$ more slowly than any exponential function of t. This effect was confirmed in experiment described in the Rothe paper [5], where the experimental evidence of deviations from the exponential decay law at long times was reported. In the light of this result the following problem seems to be worth considering: If (and how) deviations from the exponential decay law at long times affect the energy of the unstable state and its decay rate at this time region.

Studying properties of unstable states at late times it is useful to express $a(t)$ in the following form $a(t) = a_{exp}(t) + a_{lt}(t)$, where $a_{exp}(t)$ is the exponential part of $a(t)$, that is $a_{exp}(t) = N\,\exp\,[-i\frac{t}{\hbar}(E_\phi^0 - \frac{i}{2}\,\Gamma_\phi^0)]$, ($E_\phi^0$ is the energy of the system in the state $|\phi\rangle$ measured at the canonical decay times, i.e. when $\mathcal{P}_\phi(t)$ has the exponential form, Γ_ϕ^0 is the decay width, N is the normalization constant), and $a_{lt}(t)$ is the late time non–exponential part of $a(t)$. From the literature it is known that the characteristic feature of survival probabilities $\mathcal{P}_\phi(t)$ is the presence of sharp and frequent fluctuations at the transition times region, when contributions from $|a_{exp}(t)|^2$ and $|a_{lt}(t)|^2$ into $\mathcal{P}_\phi(t)$ are comparable and that the amplitude $a_{lt}(t)$ and thus the probability $\mathcal{P}_\phi(t)$ exhibits inverse power–law behavior at the late time region for times t much later than the crossover time T. The crossover time T can be found by solving the following equation,

$$|a_{exp}(t)|^2 = |a_{lt}(t)|^2. \tag{5}$$

In general $T \gg \tau_\phi$, where $\tau_\phi = \hbar/\Gamma_\phi^0$ is the live–time of ϕ. Formulae for T depend on the model considered (i.e. on $\omega(E)$) in general). A typical form of the decay curve, that is the form of the non–decay probability $\mathcal{P}_\phi(t)$ as a function of time t is presented in Fig. (1). The results of calculations presented in this Figure were obtained for the Breit–Wigner energy distribution function, $\omega(E) \equiv \omega_{BW}(E)$, where

$$\omega_{BW}(E) \stackrel{\text{def}}{=} \frac{N}{2\pi}\,\Theta(E - E_{min})\,\frac{\Gamma_\phi^0}{(E - E_\phi^0)^2 + (\frac{\Gamma_\phi^0}{2})^2}, \tag{6}$$

and $\Theta(E)$ is the unit step function. The calculations leading to graphics presented in Fig. (1) were performed for $s_0 \stackrel{\text{def}}{=} (E_\phi^0 - E_{min})/\Gamma_\phi^0 = 20$.

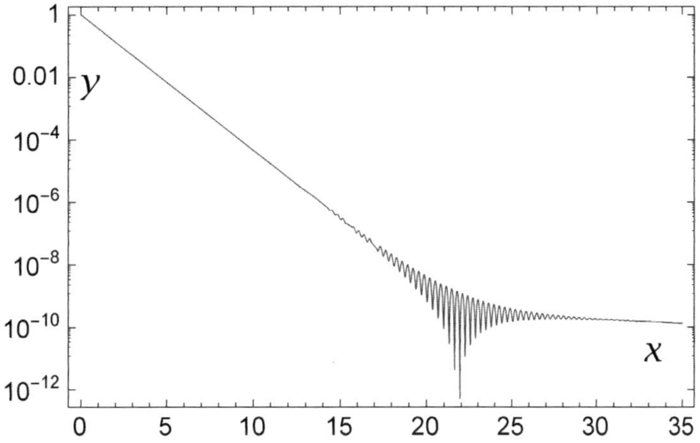

Figure 1. The general, typical form of the decay curve $\mathcal{P}_\phi(t)$. Axes: $x = t/\tau_\phi, y = \mathcal{P}_\phi(t)$ — the logarithmic scale.

3. Late times and energy of unstable states

The energy and the decay rate of the system in the state $|\phi\rangle$ under considerations, (to be more precise the instantaneous energy and the instantaneous decay rate), can be found using the following relations (for details see [6, 7, 8]).

$$\mathcal{E}_\phi \equiv \mathcal{E}_\phi(t) = \Re\,(h_\phi(t)), \quad \gamma_\phi \equiv \gamma_\phi(t) = -2\,\Im\,(h_\phi(t)), \tag{7}$$

where $\Re\,(z)$ and $\Im\,(z)$ denote the real and imaginary parts of z respectively and $h_\phi(t)$ denotes the "effective Hamiltonian" for the one–dimensional subspace of states $\mathcal{H}_{||}$ spanned by the normalized vector $|\phi\rangle$,

$$h_\phi(t) \stackrel{\text{def}}{=} i\hbar\,\frac{\partial a(t)}{\partial t}\,\frac{1}{a(t)}. \tag{8}$$

It is easy to show that equivalently

$$h_\phi(t) \equiv \frac{\langle\phi|H|\phi(t)\rangle}{\langle\phi|\phi(t)\rangle}. \tag{9}$$

There is $\mathcal{E}_\phi(t) = E_\phi^0$ and $\Gamma_\phi(t) = \Gamma_\phi^0$ at the canonical decay times and at asymptotically late times [1, 6, 8, 9]

$$\mathcal{E}_\phi(t) \simeq E_{min} + \frac{c_2}{t^2} + \frac{c_4}{t^4}\ldots, \quad (\text{for } t \gg T), \tag{10}$$

$$\Gamma_\phi(t) \simeq \frac{c_1}{t} + \frac{c_3}{t^3} + \ldots \quad (\text{for } t \gg T), \tag{11}$$

where $c_i = c_i^*$, $i = 1, 2, \ldots$, ($c_1 > 0$ and the sign of c_i for $i \geq 2$ depends on the model considered), so $\lim_{t\to\infty} \mathcal{E}_\phi(t) = E_{min}$ and $\lim_{t\to\infty} \varGamma_\phi(t) = 0$. Results (10) and (11) are rigorous.

The basic physical factor forcing the amplitude $a(t)$ to exhibit inverse power law behavior at $t \gg T$ is the boundedness from below of $\sigma(H)$. This means that if this condition is satisfied and $\int_{-\infty}^{+\infty} \omega(E)\,dE \equiv \int_{E_{min}}^{+\infty} \omega(E)\,dE < \infty$, then all properties of $a(t)$, including the form of the time–dependence at times $t \gg T$, are the mathematical consequence of them both. The same applies by (8) to the properties of $h_\phi(t)$ and concerns the asymptotic form of $h_\phi(t)$ and thus of $\mathcal{E}_\phi(t)$ and $\varGamma_\phi(t)$ at $t \gg T$.

Using relations (2), (4), (8) and assuming the form of $\omega(E)$ and performing all necessary calculations numerically one can find $\mathcal{E}_\phi(t)$ for all times t including times $t \gg T$. A typical behavior of the instantaneous energy $\mathcal{E}_\phi(t)$ at the transition time region is presented in Fig (2). In this Figure calculations were performed for the Breit–Wigner energy distribution function (6) and function $\kappa(t)$ is defined as follows

$$\kappa(t) = \frac{\mathcal{E}_\phi(t) - E_{min}}{E_\phi^0 - E_{min}}. \tag{12}$$

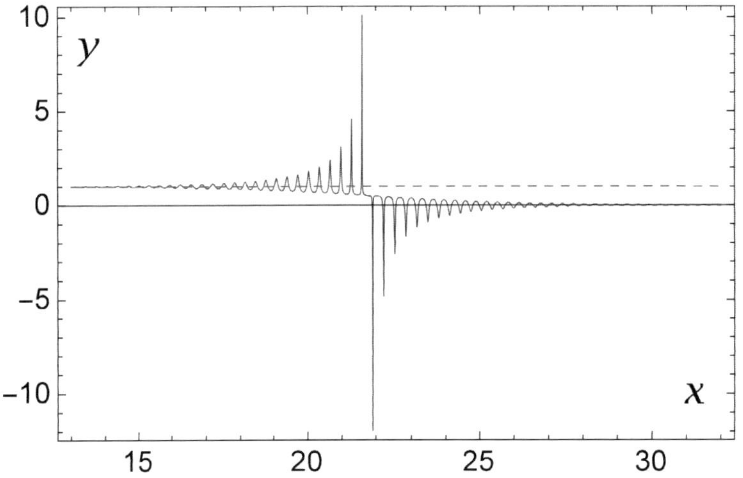

Figure 2. The instantaneous energy $\mathcal{E}_\phi(t)$ in the transitions time region: The case $s_0 = 20$. Axes: $y = \kappa(t)$, $x = t/\tau_\phi$. The dashed line denotes the straight line $y = 1$.

As it can be seen in Fig. (1) the characteristic feature decay curves is the presence of sharp and frequent oscillations at the transition times region. This means that derivatives of the amplitude $a(t)$ may reach extremely large values for some times from the transition time region and the modulus of these derivatives is much larger than the modulus of $a(t)$, which is very small for these times. This explains why in this time region the real and imaginary parts of $h_\phi(t) \equiv \mathcal{E}_\phi(t) - \frac{i}{2}\gamma_\phi(t)$, which can be expressed by the relation (8), ie. by a large derivative of $a(t)$ divided by a very small $a(t)$, reach values much larger than the energy $\mathcal{E}_\phi^0$ of the the unstable state measured at times for which the decay curve has the exponential form.

Journal of Physics: Conference Series **626** (2015) 012032 doi:10.1088/1742-6596/626/1/012032

4. Possible observable effects

From the point of view of a frame of reference in which the time evolution of the unstable system was calculated the Rothe experiment as well as the picture presented in Figs (1), (2) refer to the rest coordinate system of the unstable system considered.

Properties of the ratio $\kappa(t)$ taking place for some time intervals mean that the instantaneous energy $\mathcal{E}_{\phi}(t)$ at these time intervals differs from the energy E_{ϕ}^{0} measured at times from the exponential decay time region (i.e. at the canonical decay times). The relation (9) explains why such an effect can occur. Indeed, if to rewrite the numerator of the righthand side of (9) as follows,

$$\langle\phi|H|\phi(t)\rangle \equiv \langle\phi|H|\phi\rangle\, a(t) \,+\, \langle\phi|H|\phi(t)\rangle_{\perp}, \tag{13}$$

where $|\phi(t)\rangle_{\perp} = Q|\phi(t)\rangle$, $Q = \mathbb{I} - P$ is the projector onto the subspace od decay products, $P = |\phi\rangle\langle\phi|$ and $\langle\phi|\phi(t)\rangle_{\perp} = 0$, then one can see that decay products described by $|\phi(t)\rangle_{\perp}$ contribute permanently to the energy of the unstable state considered with the intensity depending on time t. This contribution into the instantaneous energy is practically constant in time at canonical decay times whereas at the transition times, when $t \sim T$, it is fluctuating function of time and the amplitude of these fluctuations may be significant.

From astronomical observations it is known that there are sources of unstable particles in space that emit them with relativistic or ultra–relativistic velocities in relation to an external observer so many of these particles move in space with ultra high energies. In such cases one has to consider the following problem: What effects can be observed by an external observer when the unstable particle, say ϕ, which survived up to the transition times region, $t \sim T$, or longer is moving with a relativistic velocity in relation to this observer. The distance d from the source reached by this particle is of order $d \sim d_T$, where $d_T = v^{\phi} \cdot T'$ and $T' = \gamma T$, $\gamma \equiv \gamma(v^{\phi}) = (\sqrt{1-\beta^2})^{-1}$, $\beta = v^{\phi}/c$, v^{ϕ} is the velocity of the particle ϕ. (For simplicity we assume that there is a frame of reference common for the source and observer both and that they do not move with respect to this frame of reference). So, in the case of moving particles created in astrophysical processes one should consider the effect shown in Fig (2) together with the fact that the particle gains extremely huge energy, W^{ϕ}, which have to be conserved.

Let us assume that the unstable particle under considerations has a constant momentum $\vec{p}$, $|\vec{p}| = p > 0$. Next, let us assume for simplicity that $\vec{v} = (v_1, 0, 0) \equiv (v^{\phi}, 0, 0)$ then there is $\vec{p} = (p, 0, 0)$. Now, let Λ_p be the Lorentz transformation from the reference frame $\mathcal{O}$, where the momentum of unstable particle considered is zero, $\vec{p} = 0$, into the frame $\mathcal{O}'$ where the momentum of this particle is $\vec{p} \equiv (p, 0, 0) \neq 0$ or, equivalently, where its velocity $\vec{v}^{\phi}$ equals $\vec{v}^{\phi} = \vec{v}_p^{\phi} \equiv \vec{p}/m\gamma$, and let E be the energy of the particle in the reference frame $\mathcal{O}$ and E' be the energy of the moving particle and having the momentum $\vec{p} \neq 0$. In this case the corresponding 4–vectors are: $\wp = (E/c, 0, 0, 0) \in \mathcal{O}$ and $\wp' = (E'/c, p, 0, 0) = \Lambda_p\, \wp \in \mathcal{O}'$. There is $\wp' \cdot \wp' \equiv (\Lambda_p \wp) \cdot (\Lambda_p \wp) = \wp \cdot \wp$ in Minkowski space, which is an effect of the Lorentz invariance. (Here the dot "·" denotes the scalar product in Minkowski space). Hence, in our case: $\wp' \cdot \wp' \equiv (E'/c)^2 - p^2 = (E/c)^2$ because $\wp \cdot \wp \equiv (E/c)^2$ and thus $(E')^2 = c^2 p^2 + E^2$. One can show that in the case of a free particle $p \equiv \frac{E}{c}\gamma\beta$. Therefore $c^2 p^2 + E^2 \equiv E^2 \gamma^2 \beta^2 \,+\, E^2 = E^2(1 \,+\, \beta^2 \gamma^2) \equiv E^2 \gamma^2$. The last relation means that the energy, E', of moving particles equals

$$E' \,=\, \sqrt{E^2 + c^2 p^2} \equiv E\,\gamma \stackrel{\text{def}}{=} W^{\phi}. \tag{14}$$

So in the case of the instantaneous energy, $\mathcal{E}_{\phi}(t)$ considered earlier we can write that

$$W^{\phi} \equiv \mathcal{E}_{\phi}(t_k)\, \gamma_k = W^{\phi}(t_k), \tag{15}$$

at the instant t_k (where $\gamma_k = \gamma(t_k)$.

We have $\mathcal{E}_\phi(t) \equiv E^0_\phi$ at canonical decay times. Thus, $W^\phi \equiv E^0_\phi \gamma$, at these times (where $\gamma = \gamma(t \sim \tau_\phi)$) and at times $t \gg \tau_\phi$, $t \sim T$ we have $\mathcal{E}_\phi(t) \neq E^0_\phi$. Invoking now the principle of conservation of energy, we can conclude that in the case illustrated in Fig (3), we have

$$W^\phi_1 = W^\phi_2 = W^\phi_3, \tag{16}$$

where $W^\phi_i = W^\phi(t_i)$.

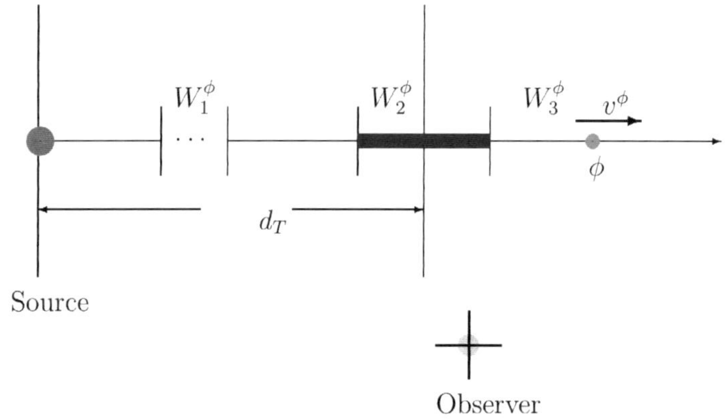

Figure 3. Here $d_T = v^\phi \cdot T'$, $T' = \gamma T$, $W^\phi_i = W^\phi(t_i)$, $(i = 1, 2, 3)$ and $W^\phi_i(t_i)$ is the energy of moving relativistic particle ϕ, $t_1 \ll t_2 \ll t_3$, $W^\phi(t_i) = \mathcal{E}_\phi(t_i)\,\gamma(t_i)$, $\gamma = \gamma(t) = (\sqrt{1-\beta^2})^{-1}$, $\beta = v^\phi(t)/c$.

The most interesting relation can be obtained from (16) for $t_1 \ll T'$ and $t_2 \sim T'$, (where $T' = \gamma T$): $W^\phi_1 \equiv E^0_\phi \gamma = W^\phi_2$, or the equivalent one,

$$E^0_\phi \gamma_1 \equiv \mathcal{E}_\phi(t_2)\,\gamma(t_2). \tag{17}$$

It is convenient to rewrite the last relation as follows,

$$\gamma_1 = \frac{\mathcal{E}_\phi(t_2)}{E^0_\phi}\gamma(t_2). \tag{18}$$

The ratio $\frac{\mathcal{E}_\phi(t_2)}{E^0_\phi}$ can be extracted from Fig (2):

$$\frac{\mathcal{E}_\phi(t)}{E^0_\phi} = 1 + \big(\kappa(t) - 1\big)\,\frac{E^0_\phi - E_{min}}{E^0_\phi}. \tag{19}$$

Thus if $\kappa(t) \neq 1$ then $\gamma_2 \neq \gamma_1$, which means that $v^\phi_2 \neq v^\phi_1$ or that $a_{average} = \frac{v^\phi_2 - v^\phi_1}{t_2 - t_1} \neq 0$. So, the moving charged unstable particle ϕ has to emit electromagnetic radiation at the transition time region. The same concerns neutral unstable particles with non–zero magnetic moment. The energy of the electromagnetic radiation emitted by such a charged particle in a unit of time can be found using the Larmor formula:

$$P = \frac{1}{6\pi\epsilon_0}\,\frac{q^2\,\dot{v^\phi}^2}{c^2}\,\gamma^6, \tag{20}$$

Journal of Physics: Conference Series **626** (2015) 012032 doi:10.1088/1742-6596/626/1/012032

where P is total radiation power, q is the electric charge, ϵ_0 – permittivity for free space. The known analogy from classical physics is the following one: A conservation of the angular momentum and a pirouette like effect.

5. Some estimations

From the results presented in Fig (2) one can find coordinates of, e.g., the highest maximum: $(x_{mx}, y_{mx}) = (21.60, 10.27)$. They allow one to find that $\kappa(t_{mx}) = y_{mx} = 10.27$. Coordinates of points of the intersection of this maximum with the straight line $y = 1$ are equal: $(x_1, y_1) = (21.58, 1.0)$ and $(x_2, y_2) = (21.62, 1.0)$. Thus $\Delta x = x_2 - x_1 = 0.04$. (Here $x = t/\tau_\phi$), and this Δx allows one to find $\Delta t = t_{mx} - t_1$. Let us consider now μ meson as an example. In such a case: $E_\mu^0 - E_{min} = m_{\mu^\pm} - (m_e + m_{\bar{\nu}_e} + m_{\nu_\mu}) \simeq 105$ [MeV], ($\tau_\mu = 2,198 \times 10^{-6}$[s] and the crossover time equals: $T^\mu \simeq 165\tau_\mu = 0,37 \times 10^{-3}$[s], [1]). Using these data one finds that $\gamma(t_1) \simeq 10.21\gamma(t_{mx})$. From this relation one can extract Δv^μ. Next using such obtained Δv^μ and $\Delta t = t_{mx} - t_1$ extracted from Fig (2) one can use the Larmor formula (20) and to calculate the total radiation power P. The result is $P \sim 4.6$ [eV/s].

Astrophysical and cosmological processes in which extremely huge numbers of unstable particles are created seem to be a possibility for the above discussed effect to become manifest. The fact is that the probability $\mathcal{P}_\phi(t) = |a(t)|^2$ that an unstable particle ϕ survives up to time $t \sim T$ is extremely small. According to estimations of the luminosity of some γ–rays sources the energy emitted by these sources can even reach a value of order 10^{52} [erg/s], and it is only a part of the total energy produced there. If to consider a source emitting energy 10^{50} [erg/s] then, eg., an emission of $\mathcal{N}_0 \simeq 6.25 \times 10^{47}$ [1/s] particles of energy 10^{18} [eV] is energetically allowed. The same source can emit $\mathcal{N}_0 \simeq 6.25 \times 10^{56}$ [1/s] particles of energy 10^9 [eV] and so on.

For the model defined by $\omega_{BW}(E)$ the cross–over time T is given by the following approximate relation (valid for $E_\phi^0/\varGamma_\phi^0 \gg 1$),

$$\varGamma_\phi^0 T \equiv \frac{T}{\tau_\phi} \sim 2 \ln \Big[2\pi\Big(\frac{E_\phi^0 - E_{min}}{\varGamma_\phi^0}\Big)^2\Big], \tag{21}$$

(see, eg. [6, 8, 10]). As it was mentioned, e.g. in [11], a typical value of the ratio $(E_\phi^0 - E_{min})/\varGamma_\phi^0$ is $(E_\phi^0 - E_{min})/\varGamma_\phi^0 \geq O(10^3 - 10^6)$. Taking $(E_\phi^0 - E_{min})/\varGamma_\phi^0 = 10^6$ one obtains from (21) that $\mathcal{N}_\phi(T) \equiv \mathcal{P}_\phi(T)\mathcal{N}_0 \sim 2.53 \times 10^{-26}\mathcal{N}_0$. So there are $\mathcal{N}_\phi(T) \sim 14 \times 10^{21}$ particles per second of energy $W^\phi = 10^{18}$ [eV] or $\mathcal{N}_\phi(T) \sim 14 \times 10^{30}$ particles of energy $W^\phi = 10^9$ [eV] in the case of the considered example and T calculated using (21).

6. Conclusions

- Fluctuations of the instantaneous energy, $\mathcal{E}_\phi(t)$, of the moving unstable particles at transition times region, $t \sim T$, (where T is the crossover time — see (5) and (21)) cause fluctuations of the velocity of these particles at this time region, which forces the charged particles to emit electromagnetic radiation.
- Similarly to the charged particle, a moving neutral unstable particle with non–zero magnetic moment have to emit electromagnetic radiation in the transition time region.
- Ultra relativistic unstable charged particles or unstable particles with non–zero magnetic moment can emit X– or γ–rays in the transition time region.
- Astrophysical sources are able to create such numbers $\mathcal{N}_0$ of unstable particles that sufficiently large number $\mathcal{N}_\phi(T) \gg 1$ of them has to survive up to times T and therefore to emit electromagnetic radiation at transition times. The expected spectrum of this radiation can be very wide: From radio frequencies up to γ–rays.

Acknowledgments
The work was supported by the NCN grant No DEC-2013/09/B/ST2/03455.

References
[1] Urbanowski K and Raczyńska K 2014 *Phys. Lett.* B **731** 236–241 (*Preprint* 1303.6975 [astro–ph.HE])
[2] Krylov S and Fock V A 1947 *Zh. Eksp. Teor. Fiz.* **17** 93
[3] Khalfin L A 1957, *Zh. Eksp. Teor. Fiz.* **33** 1371; [1958 *Sov. Phys. – JETP* **6** 1053]
[4] Paley R E A C and Wiener N 1934 *Fourier transforms in the complex domain* (New York: American Mathematical Society)
[5] Rothe C, Hintschich S I and Monkman A P 2006 *Phys. Rev. Lett.* **96** 163601
[6] Urbanowski K 2009 *Cent. Eur. J. Phys.* **7** 696–703
[7] Urbanowski K 1991 *Int. J. Mod. Phys.* A **6** 1051–1061
[8] Urbanowski K 2009 *Eur. Phys. J.* D **54** 25–29
[9] Urbanowski K 2011 *Phys. Rev. Lett.* **107** 209001
[10] Urbanowski K 2008 *Eur. Phys. J.* C **58** 151–157
[11] Krauss L M and Dent J 2008 *Phys. Rev. Lett.* **100** 171301

Cosmology with a decaying vacuum energy parametrization derived from quantum mechanics

M Szydłowski[1], **A Stachowski**[1] **and K Urbanowski**[2]

[1] Astronomical Observatory, Jagiellonian University, Orla 171, 30-244 Kraków, Poland
[2] Institute of Physics, University of Zielona Góra

E-mail: marek.szydlowski@uj.edu.pl

Abstract. Within the quantum mechanical treatment of the decay problem one finds that at late times t the survival probability of an unstable state cannot have the form of an exponentially decreasing function of time t but it has an inverse power-like form. This is a general property of unstable states following from basic principles of quantum theory. The consequence of this property is that in the case of false vacuum states the cosmological constant becomes dependent on time: $\Lambda - \Lambda_{\text{bare}} \equiv \Lambda(t) - \Lambda_{\text{bare}} \sim 1/t^2$. We construct the cosmological model with decaying vacuum energy density and matter for solving the cosmological constant problem and the coincidence problem. We show the equivalence of the proposed decaying false vacuum cosmology with the $\Lambda(t)$ cosmologies (the $\Lambda(t)$CDM models). The cosmological implications of the model of decaying vacuum energy (dark energy) are discussed. We constrain the parameters of the model with decaying vacuum using astronomical data. For this aim we use the observation of distant supernovae of type Ia, measurements of $H(z)$, BAO, CMB and others. The model analyzed is in good agreement with observation data and explain a small value of the cosmological constant today.

1. Introduction

The nature and origin of an accelerating expansion of the Universe is the basic problem of modern cosmology. The most natural explanation of the acceleration of the Universe seems to be the cosmological constant parameter interpreted as vacuum energy. Although such an explanation is simple it brings a conundrum that the values of the cosmological constant required by quantum theory ($\sim 10^{71}$ GeV^4) and obtained from type Ia supernovae observations ($\rho_\Lambda = \frac{\Lambda c^2}{8\pi G} \sim 10^{-47}$ GeV^4) differ about 100 orders of magnitude [1]. This problem is called the cosmological problem.

One approach toward the solving this problem is to consider the cosmological model with the time varying cosmological constant $\Lambda(t)$, where t is the cosmological time. The parametrization for decaying vacuum is usually taken by hand to the cosmological model [2, 3, 4, 5]. Our idea is to derive the parametrization of decaying vacuum energy directly from the first principle, namely from the quantum mechanics, then to construct the cosmological model and to test it statistically using astronomical data. Therefore our approach to decaying vacuum cosmology is not purely phenomenological and is motivated by the fundamental theory of quantum mechanics.

Decaying false vacuum states from the point of view of the quantum theory of unstable states evolve in time. From basic principles of quantum theory it is known that the amplitude $A(t)$, and thus the decay law $\mathcal{P}_M(t) = |A(t)|^2$ of the unstable state $|M\rangle$, are completely determined by the density of the energy distribution function $\omega(E)$ for the system in this state: $A(t) = \int_{E_{min}}^{\infty} \omega(E) \exp[-\frac{i}{\hbar} E\, t]\, dE$, where $\omega(E) \geq 0$ and $\omega(E) = 0$ for $E < E_{\text{min}}$. From this last condition and from the Paley–Wiener theorem it follows that there must be $|A(t)| \geq A_1 \exp[-A_2\, t^q]$, for $|t| \to \infty$. Here $A_1 > 0$, $A_2 > 0$ and $0 < q < 1$. This means that the decay law $\mathcal{P}_M(t)$ of unstable states decaying in the vacuum can not be described by an exponential function of time t if time t is suitably long, $t \to \infty$, and that for these lengths of time $\mathcal{P}_M(t)$ tends to zero as $t \to \infty$ more slowly than any exponential function of t. It appears that these deviations from the exponential decay law at long times affect the energy of the unstable state and its decay rate at this time region and thus they affect the energy of the unstable false vacuum states at these times. It is shown in [6, 7] that at transition times $t \sim T$, where T denotes time when contributions of the exponential part of the survival probability and of its late time non–exponential part are equal, the instantaneous energy of the false vacuum states fluctuates and at late times, much latter then transition times, $t \gg T$, it tends to the energy of the true vacuum state as $1/t^2$ for $t \to \infty$. The asymptotically late time behavior of the energy of the system in the false vacuum state is given by the following relation

$$E_0^{\text{false}}(t) \simeq E_0^{\text{true}} \pm \frac{\alpha^2}{t^2} \cdots, \quad \text{for } t \gg T. \tag{1}$$

This means that in the case of such false vacuum states the cosmological constant becomes time depending. The standard relation is $\rho_0^{\text{true}} \equiv \rho_{\text{bare}} = \frac{\Lambda_{\text{bare}}}{8\pi G}$. So the fluctuations of $\rho_0^{\text{false}}(t) = E_0^{\text{false}}(t)/\mathbf{V}$ at the transition times region and the asymptotically late behaviour of $\rho_0^{\text{false}}(t)$ at $t \gg T$ mean that identical behaviour of Λ have to be observed at these times.

There are two possible scenarios. In the cosmological terminology if the universe is in a false vacuum state then the cosmological constant is time-dependent in the following way

$$\Lambda(t) = \Lambda_{\text{bare}} + \frac{3\beta}{t^2}, \quad \text{for } t \gg T. \tag{2}$$

In the model with decaying dark energy and matter, as in most cosmological models, we act on assumptions that the Universe is homogeneous and isotropic at the large scale (the cosmological principle), the structure and evolution of the Universe is governed by the Einstein theory of relativity, and the source of gravity is described by energy-momentum tensor for matter and decaying vacuum. It is also assumed for simplicity that the Universe is flat.

Then cosmological evolution is determined by the scale factor a as a function of the universal cosmological time t, which satisfies the Friedmann equation

$$3\frac{\dot{a}^2}{a^2} = \rho_{\text{m}} + \rho_{\text{vac}} = \rho_{\text{m}} + \Lambda_{\text{bare}} + \frac{3\beta}{t^2} \tag{3}$$

where matter density fulfils the conservation condition

$$\bar{T}^{\alpha\beta}{}_{;\beta} = T_{\text{m}}^{\alpha\beta} + \Lambda(t) g^{\alpha\beta} = 0. \tag{4}$$

This condition for the flat homogeneous and isotropic Universe has the form

$$\dot{\rho}_{\text{m}} + 3H\rho_{\text{m}} = -\dot{\Lambda}(t) \tag{5}$$

In the case of the ΛCDM model ($\beta = 0$) it would be useful to rewrite (3) to the new form after substitution $a^3 = x^2$

$$\dot{x}^2 = \frac{3}{4}\rho_{\text{m},0} + \frac{3}{4}\Lambda_{\text{bare}} x^2. \tag{6}$$

In cosmology we use instead of density of fluids dimensionless parameters called density parameters

$$\Omega_{i,0} = \frac{\rho_{i,0}}{3H_0^2}$$

Journal of Physics: Conference Series **626** (2015) 012033 doi:10.1088/1742-6596/626/1/012033

where ρ_i is energy density of the fluid i, H_0 is the Hubble function $H = \dot{a}/a$ and the index 0 denotes the present epoch.

Therefore eq. (6) assumes the form

$$\left(\frac{dx}{d\tau}\right)^2 = \frac{9}{4}\Omega_{\text{m},0} + \frac{9}{4}\Omega_{\Lambda_{\text{bare}},0}x^2 \tag{7}$$

where $\Omega_{\Lambda_{\text{bare}}} = \frac{\Lambda}{3H_0^2}$, and $\tau \equiv |H_0|t$.

Then we obtain the solution in the form

$$x(t) = \left(\frac{\Omega_{\text{m},0}}{\Omega_{\Lambda_{\text{bare}},0}}\right)^{1/2} \sinh\left(\frac{3}{2}\sqrt{\Omega_{\Lambda_{\text{bare}},0}}\ H_0 t\right). \tag{8}$$

After dividing both sides of eq. (3) by $3H_0^2$, we obtain the relation

$$\left(\frac{H}{H_0}\right)^2 = \Omega_{\text{m},0}a^{-3} + \Omega_{\Lambda_{\text{bare}},0}. \tag{9}$$

This relation can be rewritten in the terms of redshift z as $1 + z = a^{-1}$

$$\left(\frac{H}{H_0}\right)^2 = \Omega_{\text{m},0}(1+z)^3 + \Omega_{\Lambda_{\text{bare}},0}. \tag{10}$$

Today ($z = 0$ and $H = H_0$) density parameters satisfy the constraint relation $\Omega_{\Lambda,0} = 1 - \Omega_{\text{m},0}$. This relation is fundamental for cosmography which analyzes the observational effects of photons propagation along the zero geodesics.

For the model with vacuum decaying

$$\left(\frac{H}{H_0}\right)^2 = \Omega_{\text{m}}(z) + \Omega_{\Lambda_{\text{bare}}} + \frac{\beta}{H_0^2}T(z)^{-2} \tag{11}$$

where

$$T(z) = -\int_{\infty}^{z} \frac{dz}{(1+z)H(z)}$$

is the age of the Universe up to the redshift z.

For estimation of the model parameters we assume that

$$H(z) = H_0(1+z)^{\gamma}, \qquad \gamma = \text{const} > 0$$

which gives

$$T(z) = \frac{1}{\gamma H_0}(1+z)^{-\gamma} \propto H^{-1} = t_H,$$

where t_H is a Hubble scale. Then

$$\left(\frac{H}{H_0}\right)^2 = \Omega_{\text{m}} + \Omega_{\Lambda_{\text{bare}}} + \frac{\beta\gamma^2}{H_0^2}H^2. \tag{12}$$

The acceleration equation has the form

$$\dot{H} = \frac{\Lambda_{\text{bare}}}{2} - \delta H^2, \qquad \delta = \frac{3}{2}(1-\beta). \tag{13}$$

After changing the variable $t \to a$ we obtain the acceleration equation in the form

$$\frac{dH}{da}\frac{da}{dt} = \frac{\Lambda_{\text{bare}}}{2} - \delta H^2 \tag{14}$$

or

$$\frac{dH}{da} = \frac{1}{Ha}\left(\frac{\Lambda_{\text{bare}}}{2} - \delta H^2\right). \tag{15}$$

The first integral of eq. (15) is

$$H^2(a) = \left(H_0^2 - \frac{\Lambda_{\text{bare}}}{2\delta}\right)\left(\frac{a}{a_0}\right)^{-2\delta} + \frac{\Lambda_{\text{bare}}}{2\delta}. \tag{16}$$

From this relation we obtain $H^2(z)$ formula which will be used in model parameter estimation

$$\left(\frac{H}{H_0}\right)^2 = \frac{3}{2\delta}\Omega_{\text{m}}(1+z)^{2\delta} + \frac{3}{2\delta}\Omega_{\Lambda_{\text{bare}}} \tag{17}$$

where (further let write down $\Lambda_{\text{bare}} = \Lambda$)

$$\Omega_{\text{m},0} + \Omega_{\Lambda,0} = \frac{2}{3}\delta.$$

When $\delta = 3/2$ the model reduces to the standard ΛCDM model.

Finally, we obtain the following model

$$\left(\frac{H}{H_0}\right)^2 = \frac{\Omega_{\text{m},0}}{\Omega_{\text{m},0}+\Omega_{\Lambda,0}}(1+z)^{3(\Omega_{\text{m},0}+\Omega_{\Lambda,0})} + \frac{\Omega_{\Lambda,0}}{\Omega_{\text{m},0}+\Omega_{\Lambda,0}}. \tag{18}$$

After the generalization of the corresponding substitution $a^{3(\Omega_{\text{m},0}+\Omega_{\Lambda,0})} = x^2$, we obtain the expression for the scale factor in following form

$$a(t) = \left(\frac{\Omega_{\text{m},0}}{\Omega_{\Lambda,0}}\right)^{\frac{1}{3(\Omega_{\text{m},0}+\Omega_{\Lambda,0})}}\left[\sinh\left(\frac{3}{2}\sqrt{\Omega_{\Lambda,0}}H_0 t\right)\right]^{\frac{2}{3(\Omega_{\text{m},0}+\Omega_{\Lambda,0})}}.$$

2. Data

For the estimation of parameters of the model (18) we used the observations of SNIa data, BAO, CMB, measurements of $H(z)$, Alcock-Paczyński test.

First, we consider the SNIa data. The likelihood function is

$$\ln L_{\text{SNIa}} = -\frac{1}{2}\sum_{i=1}^{N}\left(\frac{\mu_i^{\text{obs}} - \mu_i^{\text{th}}}{\sigma_i}\right)^2, \tag{19}$$

where the summing is over the SNIa sample; the distance modulus $\mu^{\text{obs}} = m - M$ (where m is the apparent magnitude and M is the absolute magnitude of SNIa stars) and $\mu^{\text{th}} = 5\log_{10} D_L + 25$ (where the luminosity distance is $D_L = c(1+z)\int_0^z \frac{dz'}{H(z')}$ and σ is the uncertainties. The Union 2.1 sample of 580 supernovae was used as the data here [8].

The BAO (baryon acoustic oscillation) data were taken from the Sloan Digital Sky Survey Release 7 (SDSS R7) dataset which consists of 893 319 galaxies [9]. The likelihood function is given by

$$\ln L_{\text{BAO}} = \frac{\left(\frac{r_s(z_d)}{D_V(z)} - d(z)\right)^2}{\sigma^2}$$

where $r_s(z_d)$ is the sound horizon at the drag epoch and $z = 0.275$, $d(z) = 0.1390$, $\sigma = 0.0037$ [10].

Planck observations of cosmic microwave background (CMB) radiation were also used [11]. With addition information on lensing from the Planck and low-ℓ polarization from the WMAP (WP), we obtain the combined likelihood function in the form

$$\ln L_{\text{CMB+lensing+WP}} = -\frac{1}{2}\sum_{ij}(x_i^{\text{th}} - x_i^{\text{obs}})\mathbb{C}^{-1}(x^{\text{th}} - x^{\text{obs}}), \tag{20}$$

where $\mathbb{C}$ is the covariance matrix with the errors, x is a vector of the acoustic scale l_A, the shift parameter R and $\Omega_b h^2$ where

$$l_A = \frac{\pi}{r_s(z^*)} c \int_0^{z^*} \frac{dz'}{H(z')} \tag{21}$$

$$R = \sqrt{\Omega_{\text{m}} H_0^2} \int_0^{z^*} \frac{dz'}{H(z')} \tag{22}$$

where z^* is the recombination redshift.

The Alcock-Paczynski test is the comparison of the radial and tangential size of an object, which is isotropic in the correct choice of model [12, 13]. The likelihood function is independent on the H_0 parameter and has the following form

$$\ln L_{AP} = -\frac{1}{2}\sum_i \frac{\left(AP^{\text{th}}(z_i) - AP^{\text{obs}}(z_i)\right)^2}{\sigma^2}. \tag{23}$$

where $AP(z)^{\text{th}} \equiv \frac{H(z)}{z}\int_0^z \frac{dz'}{H(z')}$ and $AP(z_i)^{\text{obs}}$ are observational data [14, 15, 16, 17, 18, 19, 20, 21, 22].

At the end it is also valuable to add the constraints on the Hubble parameter, i.e. $H(z = 0) \equiv H_0$. Data of $H(z)$ for samples of different galaxies were also used [23, 24, 25]

$$\ln L_{H(z)} = -\frac{1}{2}\sum_{i=1}^{N}\left(\frac{H(z_i)^{\text{obs}} - H(z_i)^{\text{th}}}{\sigma_i}\right)^2. \tag{24}$$

The final likelihood function for the observational Hubble function is

$$L_{\text{tot}} = L_{\text{SNIa}} L_{\text{BAO}} L_{\text{CMB+lensing+WP}} L_{\text{AP}} L_{H(z)}. \tag{25}$$

3. Estimation of the model

To estimate the model parameters we use our own code CosmoDarkBox implementing the Metropolis-Hastings algorithm [26, 27] and using the Pade approximants [28] for the calculation of the likelihood function [29, 10].

We use observation data of 580 supernovae of type Ia, selected subsets of the data points of Hubble function, the measurements of BAO from SDSS+2dSGRS. We also use data for the application of the Alcock-Paczynski test – 18 observational points. At last, we estimated model parameters with CMB data from Planck, low-ℓ polarization from WMAP and lensing from Planck.

The results of statistical analysis are represented in Figures 1–4. The values of estimated parameter are completed in Table 1. In Figures 1 and 2 it is shown the likelihood function with 68% and 95% confidence levels projection on the $(1 - \Omega_{\text{m}} - \Omega_\Lambda, \Omega_{\text{m}})$ plane and the

Table 1. The values of estimated parameters – best fitted values of parameters with errors. We consider two cases for errors estimation, first H_0 is assumed as 68.22 km/(s Mpc), then Ω_m is assumed as the best fit value.

	parameter	best fit	68% CL	95% CL
Case 1 – assumed $H_0 = 68.22$	$1 - \Omega_{\rm m} - \Omega_\Lambda$	0.0068	$^{+0.0261}_{-0.0276}$	$^{+0.0419}_{-0.0461}$
	$\Omega_{\rm m}$	0.2926	$^{+0.0251}_{-0.0239}$	$^{+0.0416}_{-0.0388}$
Case 2 – assumed $\Omega_{\rm m} = 0.2926$	$1 - \Omega_{\rm m} - \Omega_\Lambda$	0.0068	$^{+0.0200}_{-0.0203}$	$^{+0.0324}_{-0.0331}$
	H_0	68.22	$^{+0.92}_{-0.91}$	$^{+1.50}_{-1.45}$

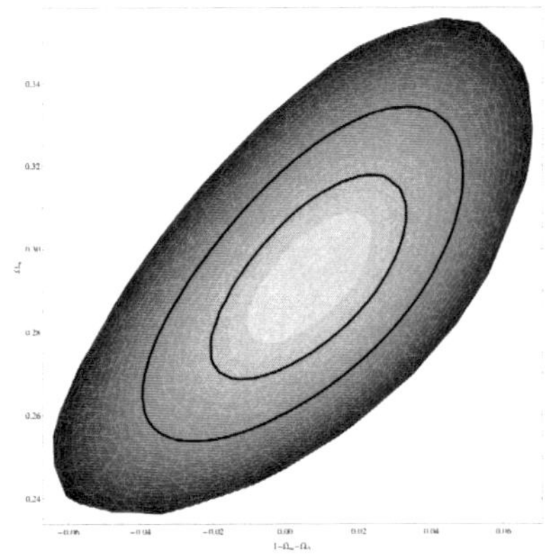

Figure 1. The likelihood function of two model parameters $(1 - \Omega_{\rm m} - \Omega_\Lambda, \Omega_{\rm m})$ with the marked 68% and 95% confidence levels. The value of Hubble constant is estimated from the data as best fit value $H_0 = 68.22$ km/(s Mpc) and then the figure is obtained for this value.

$(1 - \Omega_{\rm m} - \Omega_\Lambda, H_0)$ plane, respectively. In figures 3 and 4 the intersections with respect to fixed H_0 and Ω_Λ for Ω_m; and fixed H_0 and Ω_m for Ω_Λ have been presented, respectively.

We obtain, that the effect of decaying vacuum is rather small and errors one order higher. Therefore the effect cannot be detected from available observations data. Second, the errors for parameter $1 - \Omega_m - \Omega_\Lambda$ are the same order like Ω_m, Ω_Λ and we are looking for small parameter close to zero.

4. Conclusion

(i) The cosmological model with decaying vacuum explains while the value of cosmological constant is so small — possible solution of the cosmological constant problem.

(ii) The model with decaying vacuum (cosmological constant) was tested using astronomical data (SNIa, BAO, CMB, $H(z)$).

(iii) The evidence of decaying dynamical vacuum effect for the current Universe is equivalent to the sum of $\Omega_{m,0} + \Omega_{\Lambda,0} \neq 1$. For the standard cosmological model this sum is equal one.

(iv) The value of sum of Ω_m and Ω_Λ is close to 1 (the obtained value is 0.993) — the effect of decaying vacuum is very weak.

(v) The cosmological models with decaying vacuum be can treated as an extension of the

Journal of Physics: Conference Series **626** (2015) 012033 doi:10.1088/1742-6596/626/1/012033

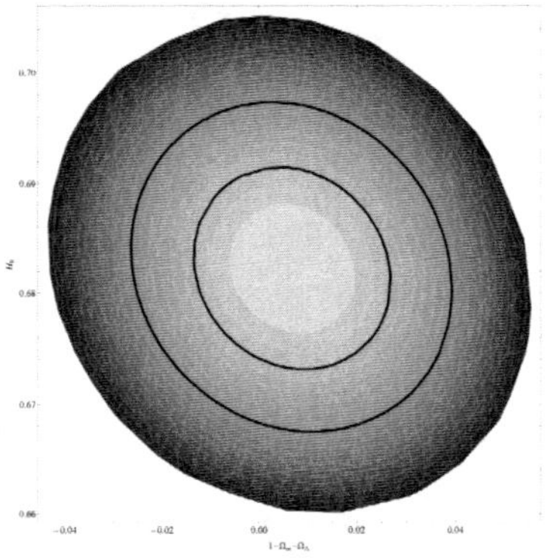

Figure 2. The likelihood function of two model parameters $(1 - \Omega_{\rm m} - \Omega_\Lambda, H_0)$ with the marked 68% and 95% confidence levels. The value of $\Omega_{\rm m}$ is estimated from the data as best fit value 0.2926 and then figure is obtained for this value. Note that the effect of decaying vacuum appears as the value of the parameter $1 - \Omega_{\rm m} - \Omega_\Lambda$ is different from zero.

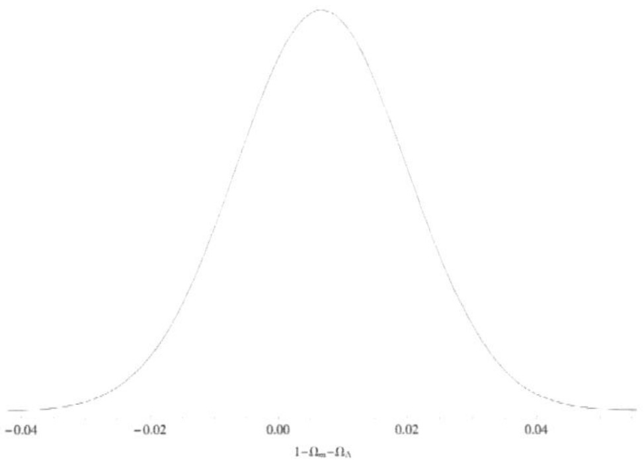

Figure 3. Diagram of PDF for parameter $1 - \Omega_{\rm m} - \Omega_\Lambda$ obtained as an intersection of the likelihood function. Two planes of intersection likelihood function are $H_0 = 68.22$ km/(s Mpc) and $\Omega_{\rm m} = 0.2926$. The planes of intersection are constructed from the best fitting value of the model parameters. The maximum of PDF is reached for $1 - \Omega_{\rm m} - \Omega_\Lambda = 0.0068$.

standard cosmological model. This model is in a good agreement with astronomical data, and it offers the solution of the cosmological constant conundrum.

Acknowledgments

The work was supported by the grant NCN DEC-2013/09/B/ST2/03455.

References

[1] Weinberg S 1989 *Rev.Mod.Phys.* **61** 1–23
[2] Alcaniz J S and Lima J 2005 *Phys.Rev.* **D72** 063516 (*Preprint* astro-ph/0507372)
[3] Wang P and Meng X H 2005 *Class.Quant.Grav.* **22** 283–294 (*Preprint* astro-ph/0408495)
[4] Bessada D and Miranda O D 2013 *Phys.Rev.* **D88** 083530 (*Preprint* 1310.8571)
[5] Graef L, Costa F and Lima J 2014 *Phys.Lett.* **B728** 400–406 (*Preprint* 1303.2075)

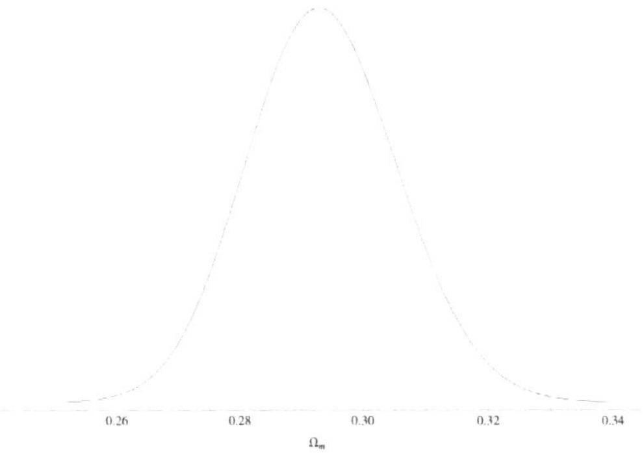

Figure 4. Diagram of PDF for parameter $\Omega_{\rm m}$ obtained as an intersection of the likelihood function. Two planes of intersection likelihood function are $H_0 = 68.22$ km/(s Mpc) and $1 - \Omega_{\rm m} - \Omega_\Lambda = 0.0068$. The planes of intersection are constructed from the best fitting value of the model parameters. The maximum of PDF is reached for $\Omega_{\rm m} = 0.2926$.

[6] Urbanowski K and Szydlowski M 2012 *AIP Conf.Proc.* **1514** 143–146 (*Preprint* 1304.2796)
[7] Urbanowski K and Raczyska K 2014 *Phys.Lett.* **B731** 236–241 (*Preprint* 1303.6975)
[8] Suzuki N, Rubin D, Lidman C, Aldering G, Amanullah R *et al.* 2012 *Astrophys.J.* **746** 85 (*Preprint* 1105.3470)
[9] Percival W J *et al.* (SDSS Collaboration) 2010 *Mon.Not.Roy.Astron.Soc.* **401** 2148–2168 (*Preprint* 0907.1660)
[10] Eisenstein D J and Hu W 1998 *Astrophys.J.* **496** 605 (*Preprint* astro-ph/9709112)
[11] Ade P *et al.* (Planck Collaboration) 2014 *Astron.Astrophys.* **571** A16 (*Preprint* 1303.5076)
[12] Alcock C and Paczynski B 1979 *Nature* **281** 358–359
[13] Lopez-Corredoira M 2014 *Astrophys.J.* **781** 96 (*Preprint* 1312.0003)
[14] Sutter P, Lavaux G, Wandelt B D and Weinberg D H 2012 *Astrophys.J.* **761** 187 (*Preprint* 1208.1058)
[15] Blake C, Glazebrook K, Davis T, Brough S, Colless M *et al.* 2011 *Mon.Not.Roy.Astron.Soc.* **418** 1725–1735 (*Preprint* 1108.2637)
[16] Ross N P, da Angela J, Shanks T, Wake D A, Cannon R D *et al.* 2007 *Mon.Not.Roy.Astron.Soc.* **381** 573–588 (*Preprint* astro-ph/0612400)
[17] Marinoni C and Buzzi A 2010 *Nature* **468** 539–541
[18] da Angela J, Outram P and Shanks T 2005 *Mon.Not.Roy.Astron.Soc.* **361** 879–886 (*Preprint* astro-ph/0505469)
[19] Outram P, Shanks T, Boyle B, Croom S, Hoyle F *et al.* 2004 *Mon.Not.Roy.Astron.Soc.* **348** 745 (*Preprint* astro-ph/0310873)
[20] Anderson L, Aubourg E, Bailey S, Bizyaev D, Blanton M *et al.* 2013 *Mon.Not.Roy.Astron.Soc.* **427** 3435–3467 (*Preprint* 1203.6594)
[21] Paris I, Petitjean P, Aubourg E, Bailey S, Ross N P *et al.* 2012 *Astron.Astrophys.* **548** A66 (*Preprint* 1210.5166)
[22] Schneider D P *et al.* (SDSS Collaboration) 2010 *Astron.J.* **139** 2360–2373 (*Preprint* 1004.1167)
[23] Simon J, Verde L and Jimenez R 2005 *Phys.Rev.* **D71** 123001 (*Preprint* astro-ph/0412269)
[24] Stern D, Jimenez R, Verde L, Kamionkowski M and Stanford S A 2010 *JCAP* **1002** 008 (*Preprint* 0907.3149)
[25] Moresco M, Cimatti A, Jimenez R, Pozzetti L, Zamorani G *et al.* 2012 *JCAP* **1208** 006 (*Preprint* 1201.3609)
[26] Metropolis, N and Rosenbluth, A W and Rosenbluth, M N and Teller, A H and Teller, E 1953 *J. Chem. Phys.* **21** 1087–1092
[27] Hastings W K 1970 *Biometrika* **57** 97–109
[28] Aviles A, Bravetti A, Capozziello S and Luongo O 2014 *Phys.Rev.* **D90** 043531 (*Preprint* 1405.6935)
[29] Hu W and Sugiyama N 1996 *Astrophys.J.* **471** 542–570 (*Preprint* astro-ph/9510117)

Journal of Physics: Conference Series **626** (2015) 012034 doi:10.1088/1742-6596/626/1/012034

Quantum coherent states in cosmology

Houri Ziaeepour

Institut UTINAM-CNRS, Université de Franche Comté, Besançon, France

E-mail: houriziaeepour@gmail.com

Abstract. Coherent states consist of superposition of infinite number of particles and do not have a classical analogue. We study their evolution in a FLRW cosmology and show that only when full quantum corrections are considered, they may survive the expansion of the Universe and form a global condensate. This state of matter can be the origin of accelerating expansion of the Universe, generally called dark energy, and inflation in the early universe. Additionally, such a quantum pool may be the ultimate environment for decoherence at shorter distances. If dark energy is a quantum coherent state, its dominant contribution to the total energy of the Universe at present provides a low entropy state which may be necessary as an initial condition for a new Big Bang in the framework of bouncing cosmology models.

1. Introduction

Cosmological observations have demonstrated that at least during two epochs the Universe has had an accelerating expansion. The first era is called *inflation* [1] and occurred after the Big Bang. The second era is the present accelerating expansion which its effect has become significant since redshift 0.5. Its origin is unknown and is given the generic name of *dark energy* [2]. It is generally believed that inflation is caused by quantum phenomena such as symmetry breaking and phase transition, most probably related to the breaking of supersymmetry or supergravity. The smallness of limits put on primordial non-gaussianity by observations [3] is probably an evidence for a high energy scale of inflation $E_{inf} \sim 10^{-3} - 10^{-6} M_P$ where M_P is the Planck mass [4].

Dark energy may be explained in the same manner, that is induced by condensation of a quantum scalar field - or a vector field in some models. Although in the last few years or so gravity models have become a favorite candidate for the origin of dark energy, the fact that inflation and recent acceleration of the Universe seem to be very similar encourage the hypothesis of a quantum origin rather than geometry for dark energy. Obviously the underlying quantum model may be related to quantum gravity - if there is at all such a model. But it is not a necessarily so. In fact it has been shown that condensation of quantum fields at cosmological scales can occur after inflation and reheating, that is during radiation and matter domination epochs [5, 6, 7]. Moreover, a non-gravitational interaction in the dark sector is a physically motivated possibility. For instance the decay of a heavy dark matter which is produced during preheating and has a small branching ratio to a very light scalar field, provides a *quintessence field* to play the role of dark energy. Moreover, it explains in a natural way and without violation of null energy principle the effective equation of state of dark energy w_{ed} which seems to be $\lesssim -1$ [8].

Journal of Physics: Conference Series **626** (2015) 012034 doi:10.1088/1742-6596/626/1/012034

Presently, we first review condensates and their relation with coherent states. The latter can be used to introduce a frame independent definition for vacuum of quantum fields in curved spacetimes. We show that according to this definition the energy of the real vacuum is always zero. This solves the apparent problem of too small observed vacuum if dark energy is interpreted as the energy of the vacuum. Assuming that dark energy is the condensate of one or more quantum fields, we briefly review its evolution in an expending Universe and discuss the necessary conditions for its survival. Finally we present preliminary results from an on going study which its aim is to investigate the evolution of cosmological condensates in the frame work of non-equilibrium quantum field theory.

2. Quantum condensates and coherent states

2.1. Definition of quantum condensates

Condensates of quantum scalar fields are very special states of matter. They are defined as states $|\psi\rangle$ for which:[1]

$$\langle\psi|\Phi|\psi\rangle \neq 0 \tag{1}$$

By decomposing the field Φ, which evolves according to free Hamiltonian of the theory, to annihilation and creation operators, it is straightforward to see that the state $|\psi\rangle$ cannot contain a finite number of particles. Condensate states are quantum field theoretical extension of Bose-Einstein Condensate (BEC) in condensed matter when the number of particles sharing the same energy state in very large. The list of the best observed examples of such states include: the condensate of Cooper pair of electrons in superconductors, condensate of electron-hole called exciton in semiconductors, quark-antiquark condensate in hadronic matter leading to the breaking of chiral symmetry, and condensation of Higgs field leading to the breaking of Electroweak symmetry and mass acquisition Standard Model (SM) particles. Finally, the most recent observations of Cosmic Microwave Background (CMB) is consistent with an spinorial condensate, that is a condensate with effective negative mass and two minima in its effective potential [10]. There is no general expression for a condensate state. Nonetheless, coherent states fulfill the definition (1).

2.2. Coherent States

R.J. Glauber [9] has proposed 3 equivalent definitions for condensate states [11]:

(i) Eigen state of annihilation operator a_α:

$$a_\alpha|\psi\rangle = C_\alpha|\psi\rangle, \quad [a_\alpha, a'^{\dagger}_\alpha] = \delta_{\alpha\alpha'} \tag{2}$$

(ii) State generated by application of displacement operator $D_\alpha(C_\alpha)$ on the vacuum (a reference state):

$$|C_\alpha\rangle = D_\alpha(C_\alpha)|0\rangle, \quad D_\alpha(C_\alpha) \equiv \exp(C_\alpha a_\alpha - C^*_\alpha a^\dagger_\alpha) \tag{3}$$

(iii) State with minimum uncertainty:

$$(\Delta q)^2(\Delta p)^2 = (\hbar/2)^2, \quad \hat{q} \equiv (a + a^\dagger)/\sqrt{2}, \quad \hat{p} \equiv (a - a^\dagger)/i\sqrt{2},$$
$$(\Delta f)^2 \equiv \langle\psi|(\hat{f} - \langle\hat{f}\rangle)^2|\psi\rangle, \quad \langle\hat{f}\rangle \equiv \langle\psi|\hat{f}|\psi\rangle \quad \text{for any operator } \hat{f} \tag{4}$$

where α indicates indices that classify different modes, including symmetry indices in multi-field models. Coherent states that satisfy (2) fulfill also the condition for a condensate state. In particular, Glauber coherent state [9] is defined as:

$$|C_\alpha\rangle \equiv e^{-|C_\alpha|^2} e^{C_\alpha a^\dagger_\alpha} = e^{-|C_\alpha|^2} \sum_{i=0} C^i_\alpha / i! (a^\dagger_\alpha)^i \tag{5}$$

[1] Here we present fields and states in interaction picture unless explicitly specified.

and satisfies all the definitions (2-4) as well as the condensate condition (1). On the other hand, various extensions of this state may satisfy only some of the definitions [11]. The generalization that we will review in the next subsection is based on (2) because it explicitly satisfies the definition of a condensate state.

2.3. A generalized coherent state

In [7] we described a generalized coherent state for a scalar field based on Glauber coherent state:

$$|\Psi_{GC}\rangle \equiv \sum_k A_k e^{C_k a_k^\dagger}|0\rangle = \sum_k A_k \sum_{i=0}^{N\to\infty} \frac{C_k^i}{i!}(a_k^\dagger)^i|0\rangle \tag{6}$$

$$a_k|\Psi_{GC}\rangle = C_k|\Psi_{GC}\rangle \qquad \langle\Psi_{GC}|N_k|\Psi_{GC}\rangle = |A_k C_k|^2 \tag{7}$$

In (6) we have restricted mode indices to momentum k. In this case the condensate in (6) can be considered as superposition of Glauber condensates moving with respect to mode $k = 0$ with a momentum equal to k.

One can further extend the definition (6) by considering products of states of type $|\Psi_{GC}\rangle$ which may in addition belong to different field species indicated by index i in the following expression:

$$|\Psi_G\rangle \equiv \sum_{k_1,k_2,\cdots} \left(\prod_{k_i} A_{k_i}\right) e^{\sum_i C_{k_i} a_{k_i}^\dagger}|0\rangle \tag{8}$$

Such a state includes products of states in which particles do not have the same momentum and consists of all combinations of states with any number of particles and momenta. Because (8) includes linear superposition and direct product of Glauber states, it satisfies at least the first two definitions of a coherent state. It is less straightforward to prove the third property.

3. Definition of vacuum using coherent states

The aim for definition of an extended coherent state [7, 12] is finding a frame independent vacuum state for applications in curved spacetimes. In quantum field theory vacuum is defined as particle-less state of a system:

$$\hat{N}_\alpha|0\rangle = 0 \;\; \forall \; \{\alpha\} \tag{9}$$

In Minkowski space this state is both physically and mathematically well defined. However, as it is well known in curved spacetimes $|0\rangle$ is not preserved under a general coordinate transformation and it is projected to a state with infinite number of particles moving with a variable speed with respect to each others, see e.g. [13] and references therein. Another definition of vacuum used in the literature is the lowest available energy state. This definition has many practical applications. For instance, in condensed matter the lowest energy level of conducting Cooper pairs may be called the *vacuum*. But is only a relative vacuum and does not mean the absence of matter. On the other hand, in cosmology and in the frame work of Einstein gravity, every matter component is observable through its gravity and contributes to the evolution of spacetime. Therefore, a true vacuum state must be *a state with no contribution in semi-classical Einstein equation*. Moreover, such a state must be at least asymptotically measurable by all observers. Such a state can be used as an absolute *reference* for verification of the presence of *matter*. Evidently, the standard definition of vacuum state does not fulfill these conditions because it is frame dependent.

For defining a frame independent vacuum we use the general condensate state (8). When $C_{k_i} \to 0 \; \forall \; k_i$, this state is neutralized by all annihilation operators and the expectation value of particle number approaches zero for all modes. Therefore, this state satisfies the condition (9) for a vacuum state. Coefficients A_k are relative amplitude of modes k and are determined by underlying model. Therefore, in general they are not zero. The coherent state (8) has the interesting property that under a Bogolubov transformation it is projected to itself:

$$a_{k_i} = \sum_j \sum_{k_j} \mathcal{A}_{k_j k_i} a'_{k_j} + \sum_j \sum_{k_j} \mathcal{B}_{k_j k_i} a'^{\dagger}_{k_j} \qquad a^{\dagger}_{k_i} = \sum_j \sum_{k_j} \mathcal{A}^*_{k_j k_i} a'^{\dagger}_{k_j} + \sum_j \sum_{k_j} \mathcal{B}^*_{k_j k_i} a'_{k_j} \tag{10}$$

Replacing $a^{\dagger}_{k_i}$ in (8) with the corresponding expression in (10) leads to an expression for $|\Psi_G\rangle$ similar to (8) but with respect to the new operator $a'^{\dagger}_k$ and $C'_{k_j} = \sum_i \sum_{k_i} \mathcal{A}^*_{k_j k_i} C_{k_i}$. For $C_{k_i} \to 0 \ \forall \ k_i$ and finite $\mathcal{A}^*_{k_j k_i}$, $C'_{k_i} \to 0 \ \forall \ k_i$. Note that here we assume that the Bogolubov transformation changes $|0\rangle$ to a similar state which is neutralized by $a'_k \ \forall \ k$. Therefore, in contrast to the null state of the Fock space, $|\Psi_G\rangle$ is frame-independent. However, it is easy to verify that this new definition of vacuum does not solve the singularity of naive determination of the expectation value of energy-momentum tensor $\hat{T}^{\mu\nu}$ without operator ordering [14, 12]. The singularity of $\hat{T}^{\mu\nu}$ originates from extension of classical $T^{\mu\nu}$ to a quantum field theory operator functional. Due to crossing ambiguity, this operator is not well defined and needs a regularization or operator ordering, but this operation is considered to be not appropriate in the context of Einstein model of gravity in which there is an absolute energy reference and all non-zero energies affect the spacetime metric.

Rather than using $\hat{T}^{\mu\nu}$, it is proposed to use number operator $\sum_k \hat{N}_k$ to determine the energy density of quantum states and vacuum [13]. Notably, frame-independent of vacuum defined above using general coherent state (8) is neutralized by the number operator $\hat{N}_k|\Psi_G\rangle = 0 \ \forall \ k$. Previously, this alternative method for measuring energy density of vacuum has been considered to be a poor replacement because the standard definition of vacuum in (9) is frame dependent. In addition to an ambiguity-free determination of vacuum energy, the new definition is more consistent with many-particle nature of quantum field theory and can be experimentally realized as a condensate with an amplitude approaching zero.

4. Dark energy and condensates

The absolute and frame independent vacuum defined above has a null energy according to the prescription explained in the previous section. Therefore, the observed accelerating expansion of the Universe cannot be due to such a vacuum and we must find alternative explanations. Many of suggested alternatives are based on classical scalar fields. Considering the fact that we live in a quantum universe, ultimately the origin of the classical field is quantum. Even in modified gravity models in which the classical scalar field is associated to geometry, presumable dilaton in a scalar-tensor gravity models, the origin of deviation from Einstein gravity is assumed to be quantum gravity, e.g. supergravity, models inspired by string theory, etc.

The only way to obtain a classical field $\varphi(x)$ from a quantum one is its condensation, that is a state for which $\langle\psi|\Phi(x)|\psi\rangle \equiv \varphi(x) \neq 0$. As we discussed in the previous sections $|\psi\rangle$ is a condensate and we must study the evolution of a quantum scalar field in a FLRW geometry to see whether a condensate with properties of dark energy can form. Many authors have studied the evolution of quantum perturbations of scalar field(s) and particle production in the framework of quantum field theory in de Sitter spacetime as a good approximation for geometry of the Universe during inflation [15]. The evolution of condensate(s) is also studied for some inflation model [16], but its formation and evolution pre-inflation era and backreaction of its evolution and particle production on metric is usually ignored.

The study of the evolution of dark energy candidate scalar fields is more sophisticated because it must cover more recent epochs of the Universe, namely radiation and matter domination eras, and present epoch in which dark energy is dominant but matter also has yet a non-negligible contribution. In these eras the geometry is not as symmetric as de Sitter and thereby it is

Journal of Physics: Conference Series **626** (2015) 012034 doi:10.1088/1742-6596/626/1/012034

more difficult to solve dynamical equations. Moreover, to have a consistent solution which can be compared with observation one needs to consistently determine the backreaction of various components on each others.

We have begun the programme of studying the condensation of a quintessence field[2] with [7] in which we studied the process of condensation in radiation and matter domination eras using formulation and techniques of non-equilibrium quantum field theory. In this work we limited ourselves to lowest order of quantum corrections, i.e. tree diagrams. A full formulation using 2-Particle-Irreducible (2PI) technique and its numerical simulation is at present an on going work. Here we briefly review conclusions of analytical investigations and preliminary results from numerical simulations.

4.1. Formation of a condensate

We study, in the framework of 2PI formalism, the evolution of a model with 3 scalars each in one of 3 physically important scales of particle physics: **A decaying massive particle X:** Depending on the initial condition it may be considered as inflaton or a decaying dark matter, for initial condition fixed at pre-inflation epoch or post inflation and preheating, respectively. In both cases the mass scale of the particle is assumed to be (sub)GUT (Grand Unification Theory) scale; **A scalar field with electroweak scale mass A:** This is assumed to be one of the two remnants of the decay of X particles, presumably a Standard Model particle or one which ultimately decay to SM particles. It can be also a collective notation for all remnants except a light axion-like scalar field; **A light axion-like (pseudo-)scalar Φ:** The condensate of this field should play the role of a classical quintessence field.

We assume that heavy fields X and A do not have self-interaction and produce negligible condensate. Indeed in [7] it is shown that the amplitude of condensate decays very quickly with increasing mass. Because a condensate includes states with infinite number of particles, the amplitude of condensate must decreases very quickly with mass, otherwise even a very small probability for highly occupied states may make the Universe over-dense.

Evolution equations of the condensate and propagators can be obtained from integration of effective action $\Gamma[\varphi, G_{\alpha\beta}]$, $\alpha,\ \beta \in X,\ A,\ \Phi$ and are the followings:

$$\frac{1}{\sqrt{-g}}\partial_\mu(\sqrt{-g}g^{\mu\nu}\partial_\nu\varphi) + m_\Phi^2\varphi + \frac{\lambda}{n!}\sum_{i=0}^{n-1}(i+1)\binom{n}{i+1}\varphi^i\langle\phi^{n-i-1}\rangle - \mathrm{g}\langle XA\rangle = 0 \quad (11)$$

$$\frac{1}{\sqrt{-g}}\partial_\mu(\sqrt{-g}g^{\mu\nu}\partial_\nu + M_i^2(x))\, G_i^F(x,y) = -\int_\infty^{x^0} d^4z\sqrt{-g(z)}\,\Pi_{ij}^\rho(x,z)\, G_{ij}^F(z,y) + \int_\infty^{y^0} d^4z\sqrt{-g(z)}\,\Pi_{ij}^F(x,z)\, G_{ij}^\rho(z,y) \quad (12)$$

$$\frac{1}{\sqrt{-g}}\partial_\mu(\sqrt{-g}g^{\mu\nu}\partial_\nu + M_i^2(x))\, G_i^\rho(x,y) = -\int_{y^0}^{x^0} d^4z\sqrt{-g(z)}\,\Pi_{ij}^\rho(x,z)\, G_{ij}^\rho(z,y) \quad (13)$$

$$M_\Phi^2(x) = m_\Phi^2 + \frac{-i\lambda}{(n-2)!}\sum_{j=0}^{[n/2]-1} C_{2j}^{n-2}C_2^{2j}\varphi^{n-2j}(x)(G_\Phi^F(x,x))^j, \quad M_{X,A}^2 = m_{X,A}^2 \quad (14)$$

$$\Pi(\varphi, G) \equiv 2i\frac{\partial\Gamma_2[\varphi, G]}{\partial G} \quad (15)$$

[2] For the sake of simplicity, here we call all dark energy candidate scale fields *quintessence*, even when the field belong to a modified gravity model, as long as it is considered to be a quantum field.

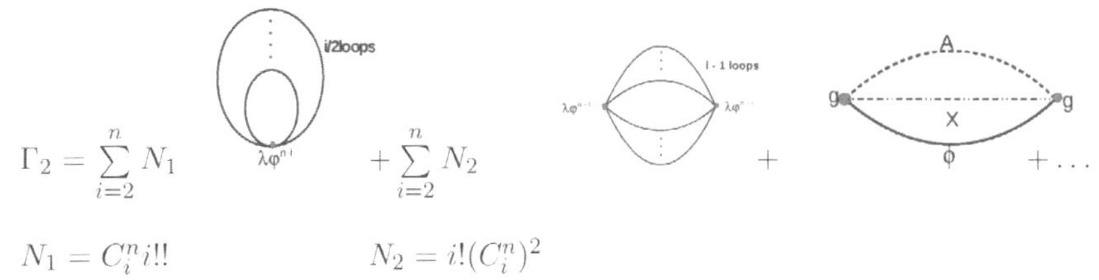

Figure 1. Diagrams contributing to $\Gamma_2(\varphi, G)$.

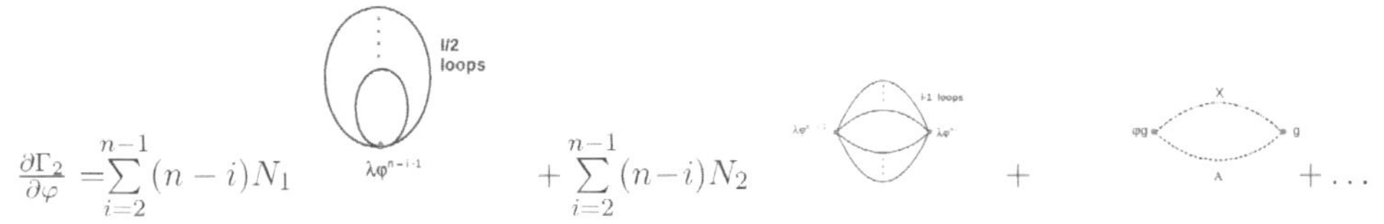

Figure 2. Diagrams contributing to $\partial\Gamma_2(\varphi, G)/\partial\varphi$. Coefficients N_1 and N_2 are defined in Fig. 1.

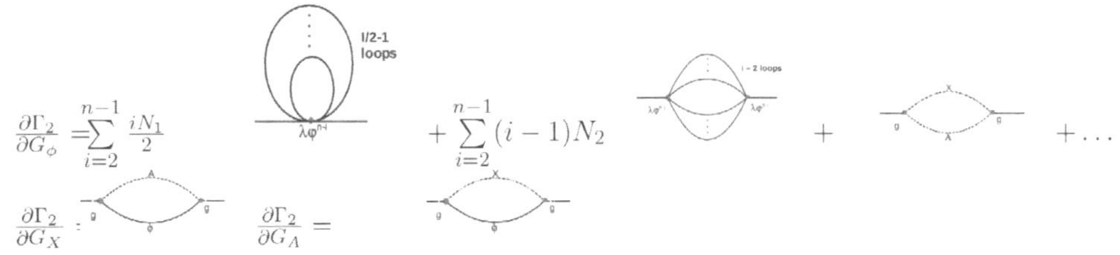

Figure 3. Diagrams contributing to $\partial\Gamma_2(\varphi, G)/\partial G_\phi$, $\partial\Gamma_2(\varphi, G)/\partial G_X$, and $\partial\Gamma_2(\varphi, G)/\partial G_A$. Coefficients N_1 and N_2 are defined in Fig. 1. The tadpole diagrams only contribute to the mass term (14) and do not appear in the r.h.s. of equations (12) and (13).

where $\Pi(\varphi, G)$ is self-energy and Γ_2 is the contribution of 2PI diagrams in the effective action. Indices i, j define the field species $i = X,\ A,\ \phi$ and F and ρ indicate symmetric and antisymmetric propagators, respectively. We should remind that above equations are exact at all perturbative order and truncation to a limited order arises when one calculates self-energies and effective masses up to a limited number of loops and vertices. In (14) $[n/2]$ means the integer part of $n/2$ and C_j^i is the combinatory coefficient. These equations are written for an arbitrary metric $g_{\mu\nu}$ which is treated as a classical field and we use semi-classical Einstein equation [17] for its evolution. Figure 1 shows lowest order 2PI diagrams contributing to the effective action. Each diagram should be understood as a closed double path and propagators and vertices has indices indicating on which path they are determined. Here, for the sake of simplicity we omit path indices. Figures 1 and 2 shows 2PI diagrams contributing in evolution of condensate and propagators, respectively.

4.2. Approximate analytical solution in radiation and matter domination eras

Equations (11-13) cannot be solved analytically. The lowest order of correction is when free propagators are used, which corresponds to neglecting the right hand side of equations (12) and (13), and considering only tree diagrams (which are not shown here) for determination of expectation values in the condensate evolution equation (11). Moreover, rather than evolving the metric consistently, we assume that it is controlled by a component external to the model.

The reason is the coupling of the expansion factor to fields, even for a homogeneous FLRW geometry. This makes analytical solution of evolution equations impossible.

With these simplifications we solve evolution equations (see [7] for details) with an initial condition at radiation domination era and zero amplitude for the condensate. We observe that the amplitude of the condensate grows exponentially in a similar manner to preheating resonance. Obviously, this exponential gross cannot continue for long time otherwise it over closes the Universe. However, under simplifications explained above it is impossible to take the backreaction into account. Notably, interaction between condensate and non-condensed component, that is quantum fluctuations behaving approximately as free particles, is crucial for backreaction process and control of unlimited gross of condensate.

Next we assume a matter domination era and evolve the expansion rate accordingly. In this epoch even the simplest evolution equations in which all interactions and quantum corrections are neglected can be solved analytically only for special cases of zero mass or zero momentum Fourier modes. Using these special solutions and WKB approximation, it is shown [7] that when interactions are neglected or linearized the amplitude of spatial Fourier transform of the condensate decreases with time as $\varphi_k \propto t_0/t$ meaning that condensate does not survive acceleration. Indeed this evolutionary behaviour leads to an equation of state $w \lesssim 0.7$ for the condensate - presumably dark energy - which is already excluded by observations. Therefore, to have a better idea about the exact evolution of condensate one has to solve the full nonlinear evolution equation of condensate. Evidently this is impossible, but some simplification plus power counting show that the equation includes terms similar to an effective potential with negative power which is a necessary condition for existence of tracking solutions [18]. We find that for a self-interacting polynomial potential of order n, quantum corrections of order i have negative exponents and close to constant coefficients if:

$$17 - 6n + 2i \geqslant 0, \quad i < n - 1 \tag{16}$$

It is easy to verify that this condition is satisfied for $n \lesssim 4$, i.e. the only renormalizable scalar field models in 4D spacetimes. This conclusion which arises without renormalization of the model is by itself very interesting. However, giving large amount of simplifications performed for obtaining it, reduces the confidence, and therefore it needs confirmation with more robust analytical or numerical calculations.

4.3. Numerical simulations

Numerical simulation of the model described in the previous subsections is a work in progress. Figure 4 shows an example of these simulations in which the initial condition is fixed for pre-inflationary Universe, with $\langle\phi(t_0)\rangle = 0$ and no free particle, i.e. a vacuum state. It shows that in contrast to the case of radiation domination universe, after an initial increase the effective potential of condensate arrives to a saturated state and produces an inflation. Ultimately simulations should be continued for enough long time to achieve an $N \sim 60$ e-folding before it decreases and transfer its energy to a free Φ particles and other species. For the time being, due to numerical errors our simulations cannot achieve this goal and we are working on its improvement. Notably, it is plausible that limited order of quantum corrections considered in the theoretical formulation is not enough for dealing with ultra soft modes. Indeed theoretical investigations show that infrared (IR) modes of massless and light fields in de Sitter space may have an IR singularity [19] when treated with perturbative techniques and need non-perturbative treatment. An example of an analogue situation is *small-x* regime in Deep Inelastic Scattering of hadrons and pair-production in strong electric or magnetic fields [20]. A method analogue to Color Glass Condensate (CGC) [21] used in QCD is developed for curved spacetimes [22] and may be applicable to cosmological IR problem.

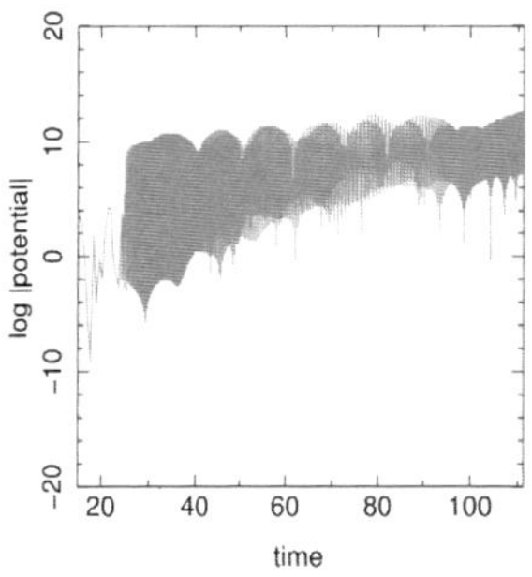

Figure 4. Effective potential of a light scalar field. Initial conditions of the simulation are defined for pre-inflationary epoch with a null initial value for the amplitude of the condensate.

5. Outline

Coherent states are fully quantum states that behave collectively like a classical field. They may had a crucial role during inflation and reheating, and present expansion of the Universe may be also due to coherent superposition at cosmological distances. In this case, in one hand cosmological coherent states may be considered as ultimate environment for decoherence of quantum states. Moreover, if dark energy is a quantum condensate, its present dominant contribution to total energy of the Universe means that its constituents are mainly in a quantum superposition state rather than being decohered and classical. The verification of such a claim is not easy. Nonetheless, better understanding of early Universe physics through observations of the CMB anisotropies and properties of dark energy, as well as theoretical works, may allow to find a mean for verification of large scale quantum properties of the Universe.

References

[1] Linde A 2006 *Prog. Theor. Phys. Suppl.* **163** 295 (*Preprint* hep-th/0503195)
[2] Silvestri A and Trodden M 2009 *Rept. Prog. Phys.* **72** 096901 (*Preprint* arXiv:0904.0024)
[3] Planck Collaboration 2015 *Preprint* arXiv:1502.01592
[4] Álvarez-Gaumé L, Gómez C and Jimenez R 2013 *Preprint* arXiv:1307.0696
[5] Ziaeepour H 2004 *Phys. Rev.* D **69** 063512 (*Preprint* astro-ph/0308515)
[6] Ziaeepour H 2005 in *Progress in Dark Matter Research* (New York: Nova Science) (*Preprint* astro-ph/0406079)
[7] Ziaeepour H 2010 PRD **81** 103526 (*Preprint* arXiv:1003.2996)
[8] Ziaeepour H 2000 *Preprint* astro-ph/0002400
[9] Glauber R J 1963 *Phys. Rev.* **131** 2766
[10] Planck Collaboration (2015 *Preprint* arXiv:1502.02114
[11] Zhang W M, Feng D H and Gilmore R 1990 *Rev. Mod. Phys.* **62** 897
[12] Ziaeepour H 2012 *Mod. Phys. Lett.* A **27** 1250154 (*Preprint* arXiv:1205.3304)
[13] Birrell N D and Davies P C W 1982 *Quantum fields in curved space* (Cambridge University Press)
[14] Weinberg S 1989 *Rev. Mod. Phys.* **91** 1
[15] Garbrecht B and Rigopoulos G 2011 *Phys. Rev.* D **84** 063516 (*Preprint* arXiv:1105.0418)
[16] Boyanovsky D and De Vega H 2004 *Phys. Rev.* D **70** 063508 (*Preprint* astro-ph/0406287)
[17] Ramsey S A and Hu B L 1997 *Phys. Rev.* D **56** 661 (*Preprint* gr-qc/9706001)
[18] Steinhardt P J, Wang L and Zlatev I 1999 *Phys. Rev.* D **59** 123504 (*Preprint* astro-ph/9812313)
[19] Boyanovsky D and Holman R 2011 *J. High Ener. Phys.* **5** 47 (*Preprint* arXiv:1103.4648)
[20] Anderson P R and Mottola E 2014 *Phys. Rev.* D **89** 104038 (*Preprint* arXiv:1310.0030)
[21] Iancu E, Leonidov A and McLerran L 2001 *Nucl. Phys.* A **692** 583 (*Preprint* hep-ph/0011241)
[22] Ziaeepour H 2005 *Mod. Phys.* A **20** 419 (*Preprint* hep-ph/0407046)

Journal of Physics: Conference Series **626** (2015) 012035 doi:10.1088/1742-6596/626/1/012035

What after CERN?
Opportunities from co-responding systems

Alfredo Iorio

Faculty of Mathematics and Physics
Charles University
V Holešovičkách 2, 18000 Prague 8
Czech Republic

E-mail: alfredo.iorio@mff.cuni.cz

Abstract. We offer some arguments in favor of the construction of an experimental facility, where to test fundamental theories of Nature by using co-responding systems. Co-responding systems are physical systems such that certain behaviors of one system are clearly related to certain behaviors of the other system. Physical systems available at our energy scales, co-responding to the unreachable high-energy systems, are what we need, to attack from an experimental perspective the open questions beyond the reach of CERN. The focus here is on two scenarios with which we have some familiarity: hadron production in high energy scattering processes as a Unruh phenomenon, and graphene as a quantum field theory in a curved spacetime, the latter being our prime bet.

1. Introduction

There are many indications that this might be the right time to gather into one place theoretical, numerical and experimental efforts to study fundamental aspects of Nature reproduced in systems within our reach.

Sometimes this research is referred to as "analogue" or "artificial" physics [1]. Instead, we take the view that when, for instance, Hawking radiation is produced in a Bose-Einstein condensate [2], that is... Hawking radiation, with no further attribute.

We do not claim that testing, for instance, black hole physics at the black hole location it is not worthwhile. Surely, many astrophysical issues are better understood by studying the astrophysical black hole than studying some realization in condensed matter systems [3]. Nonetheless: first, this is hard; second, there are too many fundamental questions awaiting an answer since decades [4]; third, certain fundamental issues (like the mentioned Hawking radiation, the related Bekenstein-Hawking entropy, and all the way till the quantum gravity scenarios), can well be addressed by identifying their essential features, and by reproducing these features here and now.

We shall not venture in a proper scientific defense of this approach as a whole. We take, instead, a practical view-point, by presenting two cases where certain important fundamental questions can be addressed in co-responding systems. The two examples are singled-out purely on the basis of our own familiarity with them.

Journal of Physics: Conference Series **626** (2015) 012035 doi:10.1088/1742-6596/626/1/012035

2. Entanglement vs. Bekenstein-Hawking entropies. Co-responding system: high energy collisions

Let us here illustrate how, by considering hadron production in high energy collisions as the co-responding system of the Unruh radiation [5], can serve the scope of investigating fundamental questions, such as the nature of the Bekenstein-Hawking entropy.

What is customarily done, in the attempt to explain the mysteries of the gravitational entropy, that appears to give us information on the inaccessible interior region of a black hole (see, e.g., [6] and [7]), is to consider the black hole and the surrounding quantum relativistic matter as two separate systems, so that the entropy of the composite system is viewed as the sum of the two

$$S_{tot} = S_{BH} + S_{matt} \tag{1}$$

S_{matt} is customarily seen as an entanglement entropy, between modes of the quantum field that are near the horizon, on both sides, in and out, see, e.g., [8]. Less clear, if not fully obscure, is the origin of the S_{BH} part. There are proposals on the latter being also an entanglement entropy, [8], [9], but there are several issues that need to be solved before considering this proposal a tenable solution [7]. Although the holographic description holds for the entanglement entropy as well as for S_{BH} (see [9] and [10], for two different but related approaches), the most important obstacle to this interpretation is that standard quantum field computations lead to infinite entanglement entropy, unless one introduces an ad hoc cutoff to regulate the ultraviolet (UV) divergencies.

Let us, then, remove altogether the black hole from the picture, but let us retain an event horizon and quantum relativistic matter surrounding it. A way to do so is to accelerate the quantum matter relativistically, so that the spacetime becomes a Rindler spacetime [11]. According to the generally accepted wisdom of (1), and using a well known application to the Rindler spacetime of the Gibbons-Hawking Euclidean action formula, see [12], that gives $S_{Rindler} = \frac{1}{4}A/r_P^2$, this situation should lead to

$$S_{tot} = S_{Rindler} + S_{matt} = \frac{1}{4}\frac{A}{r_P^2} + \alpha\frac{A}{r^2}\,, \tag{2}$$

where r_P is the Planck length, r is the length scale characterizing the quantum dynamics, α is a numerical constant, and $A = \int dydz$ is a surface of constant acceleration, i.e., of constant Rindler coordinate ξ, and of constant Rindler time η, [11] (we suppose that, the topology of the specific Rindler spacetime under consideration, is spherically symmetric, with radius R, hence, $A = 4\pi R^2$. This topology is precisely the one of interest for the hadronization mechanism [5]).

Formula (2) is mysterious in many respects. In the first term, the appearance of r_P, hence of G, makes no sense in a context where there is no gravitational field behind the acceleration of the Rindler spacetime. In other words, while, of course, the formula makes perfect sense for the near-horizon approximation of a black hole (on which black-hole to use for the hadron production case, see [13]), the converse is no longer true. I.e., if the dynamics (force) behind the Rindler acceleration is not of gravitational origin, it makes no sense to have r_P, on the contrary, one expects r instead. In fact, r enters only in the second contribution to (2), i.e. to the entanglement entropy of the matter fields around the horizon, see [8]. The second term a) has a numerical constant α that cannot be fixed from the theory, and, most importantly, b) if one cannot fix the length scale r to a nonzero value, it presents the UV divergence we have recalled earlier. The other mystery of (2) is that, even if one can find a natural scale r, it would be natural to have $r >> r_P$, and this means that the "gravitational" entropy of the Rindler spacetime would dominate over any form of matter entropy.

A solution to these puzzles comes from the hypothesis that entanglement entropy coincides with the "gravitational" entropy (that would be, then, better called "geometric" entropy), and that there is one entropy all the time, at least in the Rindler case

$$S = \frac{1}{4}\frac{A}{r^2} \tag{3}$$

The factor 1/4 is the signature of the geometric entropy. The scale fixed by r, and not by r_P, is the signature that this is the matter/entanglement entropy. Furthermore, since no gravity is involved, the Bekenstein upper bound [14] can be invoked all the time. This serves the important scope of fixing a nonzero r, all the time. Finally, at least in this spherically symmetric Rindler spacetime, one can also invoke the saturation of the Bekenstein bound, and this would fix all the parameters involved.

These theoretical arguments we now offer to the co-responding system of hadron production in high energy collisions, at zero chemical potential $\mu = 0$. The sequential quark-antiquark pair-productions, which is the basic ingredient for hadron production, generate a series of decelerations of constant magnitude that are a realization of the Unruh phenomenon in a Rindler spacetime [5]. By using $r = 1/k_T$, with k_T the transverse momentum (the characteristic small length scale of the quantum dynamic, here the quark-antiquark dynamics), in [15] it is shown that, the experimental result [16]

$$s/T \simeq 7 \, , \tag{4}$$

with s entropy density and T freeze-out temperature, is well matched by the formula (3), that gives

$$s/T^3 = 3\pi^2/4 \simeq 7.4 \, . \tag{5}$$

3. Curved spacetimes singularities and nonunitarity. Co-responding system: graphene

In this second example, we point to the relation among curvature of the spacetime, singularities, and inequivalent quantizations/Hilbert spaces. The co-responding system is graphene physics [17]. The fundamental physics arena is the long-ongoing debate on unitarity, equivalence principle, quantum field theory (QFT) in curved spacetimes, and black-hole physics, see, e.g., [18].

Consider a metric that can encode intrinsic curvature, and that is the next-to-trivial with respect to the (1+1)-dimensional ((1+1)d) case,

$$g_{\mu\nu}(q) = \begin{pmatrix} 1 & 0 & 0 \\ 0 & & \\ 0 & & g_{ij} \end{pmatrix} , \tag{6}$$

with $\mu, \nu, \lambda, ... = 0, 1, 2$ (time and space), $i, j, k, l, ... = 1, 2$ (space only), $\partial_t g_{ij} = 0$, and we are in a coordinate frame $q^\mu = (t, x, y)$.

There are various reasons to choose the set-up (6). First, this way we can take advantage of the well-known differential geometry of surfaces. Then, since we focus on how quantization is affected in the process of going from a flat to a curved spacetime, and since we want to be in a simple but not trivial case, we choose the conformally flat (CF) sub-set of (6). This way, on the one hand, we shall be able to use the geometry of surfaces and the richness of the structure there, see [19]. On the other hand, being in (2+1)d, we know that very interesting spacetimes exist, e.g., the black-hole of [20], or the gravitational kink of [21].

The conformal structure is used in an "anomalous" sense, i.e., it is a classical symmetry, but quantum mechanically we are able to distinguish among Weyl-related configurations [19]. To single-out this phenomenon, we want to avoid the well-known scale/trace anomaly [22]. Being in (2+1)d, we can ignore the latter altogether. The issue becomes more involved when some sort of horizon is present, like the conformal Killing horizon of [3], for more details see [17].

In that case, an effective (1+1)d theory emerges, and anomalies of the standard kind (general relativistic or trace), will indeed play an important role. If one requires that no horizon is formed, the "quantum anomalies" we shall isolate persist in all cases where intrinsic curvature is present (hence, *also* when an horizon is present).

In two dimensions all metrics are CF [23], i.e. there exists $\tilde{q}^\mu \equiv (t, \tilde{x}, \tilde{y})$, such that

$$g_{\mu\nu}(\tilde{q}) = \begin{pmatrix} 1 & 0 & 0 \\ 0 & -e^{2\sigma(\tilde{x},\tilde{y})} & 0 \\ 0 & 0 & -e^{2\sigma(\tilde{x},\tilde{y})} \end{pmatrix} . \quad (7)$$

Notice that we acted only on the 2d spatial part of the coordinates q^μ: $(x, y) \to (\tilde{x}, \tilde{y})$. The latter coordinates are usually called "isothermal". What we need is a CF metric in (2+1)d, which is obtained iff the surface has constant Gaussian (intrinsic) curvature $\mathcal{K}$, i.e.

$$\sigma = \sigma_H \quad \text{or} \quad \sigma = \sigma_L , \quad (8)$$

where $(\partial^2_{\tilde{x}} + \partial^2_{\tilde{y}})\sigma_H = 0$, i.e. σ_H is an harmonic function, corresponding to an intrinsically flat (7), and with $(\partial^2_{\tilde{x}} + \partial^2_{\tilde{y}})\sigma_L = -\mathcal{K}\ e^{2\sigma_L}$, i.e. σ_L is a Liouville function, corresponding to an intrinsically constant curved (7). On this see [19]. Further attention is necessary when dealing with the global properties of the surface, like the ones we are interested in here.

When σ is given by (8), then exists a third coordinate frame $Q^\mu = (T, X, Y)$, such that the metric (6) is

$$g_{\mu\nu}(Q) = e^{2\Sigma(Q)}\ \eta_{\mu\nu} , \quad (9)$$

where, of course, $\Sigma(\sigma)$. This is as far the classical background for the quantum field is concerned.

For the quantum field, the best setting to single-out the effect we are looking for, is to have a field that classically has the same dynamics in the flat and in the curved case, whereas, quantum mechanically the difference will show-up only for the quantum inequivalence we are investigating. As explained earlier, this is not trace anomaly. All of this points clearly to the use of the Dirac field that, for $m = 0$, enjoys the full local Weyl invariance, at the classical level [24] (here $\hbar = c = 1$)

$$\mathcal{A}_\Sigma = i \int d^3Q \sqrt{g} \bar{\psi}_\Sigma \not{\nabla} \psi_\Sigma = i \int d^3Q \bar{\psi}\ \not{\partial} \psi = A , \quad (10)$$

when

$$g_{\mu\nu}(Q) = e^{2\Sigma(Q)}\eta_{\mu\nu} \quad \text{and} \quad \psi_\Sigma(Q) = e^{-\Sigma(Q)}\psi(Q) . \quad (11)$$

Clearly, with the use of the quantum implementation of Weyl symmetry, through the operator

$$U[\Sigma] : \psi \to \psi_\Sigma , \quad (12)$$

and by focusing on the conformally flat cases, we can have a clear understanding of the relation between the Hilbert/Fock spaces in the flat and in the curved cases.

What one needs to investigate is how intrinsic (Gaussian) curvature $\mathcal{K}$ affects the operator $U[\Sigma]$, making it ill-defined/essentially singular, etc.. This would mean that quantum vacua, in the curved and in the flat case, are not mapped one to the other with a unitary operator, just like predicted by the failure of the Stone-von Neumann theorem in the presence of a non trivial topology [25]. The line of thoughts here is to link intrinsic curvature with a topological defect, hence with a singularity. Since, on the one hand, we are in a setting mimicking three-dimensional gravity [26], and since, on the other hand, the hexagonal lattice of graphene requires disclination defects for an intrinsically curved membrane, such a link should be possible along the lines of [27], where three dimensional gravity and defects in elastic media are merged (notice that inequivalent quantizations, in the presence of a singularity, were already found in [28]).

Work is in progress in this direction [29].

4. Conclusions

It might be truly worthwhile that the scientific community makes an effort towards the construction of an experimental facility, with the same motivations of CERN, but entirely dedicated to test fundamental phenomena of Nature by reproducing their key aspects on co-responding systems.

To name things is a crucial part of making them happens [30], thus, let us do it here. We name this facility HELIOS, for High Energy Laboratory for Indirect ObservationS. Of course, we would see no harm if the facility is called in any other way...

Seeds of HELIOS are probably in many places, right now. For what matters our own work in that direction, we are committed to set under control all possible aspects of graphene as a quantum field theory in curved spacetimes, and as a quantum gravity playground [17]. The enterprize is a tall-order, and we are proceeding by incremental steps. The first is to set under control the geometries of the graphene membranes that are of interest for the studies of fundamental physics [31] (see also [32], and [33] for an introductory overview).

Acknowledgments

The financial support of the Czech Science Foundation (GAČR), under contract no. 14-07983S, is acknowledged.

References

[1] Novello M, Visser M and Volovik G E 2002 *Artifical Black Holes* (Singapore: World Scientific); Barceló C, Liberati C and Visser M 2011 Analogue Gravity *Liv. Rev. Rel.* **14** 3
[2] Steinhauer J 2014 *Nature Phys.* **10** 864
[3] Iorio A and Lambiase G 2014 *Phys. Rev.* D **90** 025006; 2012 *Phys. Lett.* B **716** 334
[4] de Haro S, Dieks D., 't Hooft G and Verlinde E (eds) 2013 *Found. Phys.* **43** (1)
[5] Castorina P, Kharzeev D and Satz H 2007 *Eur. Phys. J.* C **52** 187
[6] Bousso R 2002 *Rev. Mod. Phys.* **74** 825
[7] Solodukhin S N 2011 *Liv. Rev. Rel.* **14** 8
[8] Iorio A, Lambiase G and Vitiello G 2004 *Ann. Phys.* **309** 151
[9] Terashima H 2000 *Phys. Rev.* D **61** 104016
[10] Srenidcky M 1993 *Phys. Rev. Lett.* **71** 666
[11] Wald R M 1984 *General relativity* (Chicago: Univ. Chicago Press)
[12] Laflamme R 1987 *Phys. Lett.* B **196** 449
[13] Castorina P, Grumiller D and Iorio A 2008 *Phys. Rev.* D **77** 124034
[14] Bekenstein J D 1981 *Phys. Rev.* D **23** 287
[15] Castorina P, Iorio A and Satz H 2014 Hadron Freeze-Out and Unruh Radiation *Preprint* arXiv1409.3104
[16] Bazazov A et al. (HotQCD Collaboration) 2014 The equation of state in (2+1)-flavor QCD *Preprint* arXiv:1407.6387
[17] Iorio A 2015 *Int. J. Mod. Phys.* D **24** 1530013; *do.* 2014 *Weyl Symmetry in Graphene* Lecture Notes for the Workshop "Modeling Graphene-like Systems" (Univ. College Dublin, Ireland, May 2014), available at http://mathsci.ucd.ie/~miguel/GRAPHENE_WORKSHOP_2014/Alfredo_Iorio/
[18] Hawking S W 2014 Information preservation and weather forecasting for black holes *Preprint* arXiv:1401.5761 and references therein. See the debate in http://astrobites.org/2014/01/31/information-preservation-and-weather-forecasting-for-black-holes/
[19] Iorio A 2011 *Ann. Phys.* **326** 1334; *do.* 2012 *Eur. Phys. J. Plus* **127** 156
[20] Bañados M, Teitelboim C and Zanelli J 1992 *Phys. Rev. Lett.* **69** 1849
[21] Guralnik G, Iorio A, Jackiw R and Pi S Y 2003 *Ann. Phys.* **308** 222
[22] Christensen S M 1978 *Phys. Rev.* D **17** 946
[23] Chern S 1955 *Proc. Amer. Math. Society* **6** 771
[24] Iorio A, O'Raifeartaigh L, Sachs S and Wiesendanger C 1997 *Nucl. Phys.* B **495** 433
[25] Kastrup H A 2006 *Phys. Rev.* A **73** 052104
[26] Deser S, Jackiw R and 't Hooft G 1984 *Ann. Phys.* **152** 220
[27] Katanaev M O and Volovich I V 1992 *Ann. Phys.* **216** 1
[28] Gupta K S and Sen S 2008 *Phys. Rev.* B **78** 205429
[29] Capolupo A, Iorio A, Lambiase G and Vitiello G 2015 in progress

[30] Toraldo di Francia G 1986 *Le cose e i loro nomi* (Bari: Laterza)
[31] Gabbrielli R, Iorio A, Pugno N, Simonucci S and Taioli S 2015 in progress
[32] Iorio A 2015 *Front. Mater.* **1** 36; *do.* 2013 *J. Phys.: Conf. Series* **442** 012056
[33] Moldoveanu F 2014 The amazing Graphene *Preprint* http://fmoldove.blogspot.cz/2014/10/the-amazing-graphene-continuing.html

Journal of Physics: Conference Series **626** (2015) 012036 doi:10.1088/1742-6596/626/1/012036

Quantum models of classical systems

P Hájíček

Institute for Theoretical Physics, University of Bern, Sidlerstrasse 5, CH-3012 Bern, Switzerland

E-mail: `hajicek@itp.unibe.ch`

Abstract. Quantum statistical methods that are commonly used for the derivation of classical thermodynamic properties are extended to classical mechanical properties. The usual assumption that every real motion of a classical mechanical system is represented by a sharp trajectory is not testable and is replaced by a class of fuzzy models, the so-called maximum entropy (ME) packets. The fuzzier are the compared classical and quantum ME packets, the better seems to be the match between their dynamical trajectories. Classical and quantum models of a stiff rod will be constructed to illustrate the resulting unified quantum theory of thermodynamic and mechanical properties.

1. Introduction

There are some features of the classical world that seem to be incompatible with quantum mechanics:

Realism Properties such as position and momentum can be ascribed to a chair, say, independently of whether they are observed or not.

Sharp Trajectories By a common interpretation of classical mechanics, the real chair is even at a sharp point of its phase space at each time. Attempts to model this property by a quantum state with minimum uncertainty leads to coherent states that are pure.

No Superpositions The chair is never observed in a linear superposition of being, e.g., simultaneously in the kitchen as well as in the bedroom. However, pure states in quantum mechanics can be superposed in this manner.

Robustness Measurement of every classical observable can be done in such a way that the state of the observed system is arbitrarily weakly disturbed. However, pure quantum states are not disturbed only by measurements of very few very special observables.

Thus, attempts to solve the problem of Sharp Trajectories aggravate problems of Robustness and of No Superpositions.

There is a vast literature about the problems. Let us list examples of the most popular ideas: macroscopic systems do not obey quantum mechanics [1]; quantum decoherence theory [2]; only coarse-grained operators represent classical measurements [3]; Coleman-Hepp theory [4]; dynamical collapse theory [5, 6]. The list is incomplete.

Our theory is different. It rejects sharp trajectories and seeks quantum mechanical derivation of classical properties possessed by fuzzy mechanical states. The present paper is a short review of [7, 8] as well as of some new results.

2. Hypothesis of high entropy states

To motivate our approach, let us briefly recapitulate some ideas of statistical thermodynamics. Consider rarefied equilibrium gas in a vessel. There is a classical model $\mathcal{S}_c$ of this gas offered by phenomenological thermodynamics, called "ideal gas", and the properties of $\mathcal{S}_c$ are examples of classical properties. They are described by thermodynamic quantities such as internal energy E, volume Ω, pressure, entropy, temperature, specific heats, etc.

To obtain the values of such quantities from quantum mechanics, we need a quantum model $\mathcal{S}_q$ of the gas. As $\mathcal{S}_q$, we can choose a system of N spin-zero point particles, each with mass μ, in a deep potential well of volume Ω with Hamiltonian

$$\mathsf{H} = \sum_{k=1}^{N} \frac{|\vec{\mathsf{p}}_k|^2}{2\mu} ,$$

where $\vec{\mathsf{p}}_k$ is the momentum of k-th particle in the rest system of Ω. H is then the operator of the internal energy and the classical internal E energy is an average of H.

The most important assumption of the quantum model is the choice of state. It is the state that maximises the (von Neumann) entropy for fixed value E of the average of the internal energy. It is called "Gibbs state". All properties of the $\mathcal{S}_c$ can then be calculated from $\mathcal{S}_q$ as properties of the Gibbs states.

The main (heuristic) principle of our theory is a generalisation of this idea to all classical properties, including the mechanical ones. Thus, we state the following hypothesis:

Assumtion 1 *Let a real system $\mathcal{S}$ have a classical model $\mathcal{S}_c$. Then, there is a quantum model $\mathcal{S}_q$ of $\mathcal{S}$ such that all properties of $\mathcal{S}_c$ are selected properties of some high-entropy states of $\mathcal{S}_q$.*

An important reason for accepting this hypothesis is that it suggests ways in which all four problems mentioned in the introduction can be solved. Indeed, the Realism Problem could be approached as follows. Our theory of objective properties of quantum systems [9, 10] justifies the assumption that quantum states are objective. If classical properties are properties of some states of the quantum model, they will also be objective. The No-Superposition Problem is based of some properties of pure quantum states. But high-entropy states are not pure: they cannot be superposed. As for the robustness problem, we can use the fact that very many quantum states correspond to a single classical state. Even if quantum states may be disturbed by observation, the corresponding classical states need not be. Finally, there is no Sharp-Trajectories Problem for thermodynamics. All these points are just suggestions and must be more carefully studied on some mathematically well-defined models.

Assumption 1 might work for thermodynamics, but what could be the high-entropy states for Newtonian mechanics?

3. Classical ME packets

Let us consider the classical mechanical model $\mathcal{S}_c$ defined as a system with a single degree of freedom and Hamiltonian

$$H = \frac{p^2}{2\mu} + V(q) . \tag{1}$$

The classical equations of motion are

$$\dot{q} = \frac{p}{\mu} , \quad \dot{p} = -\frac{dV}{dq} \tag{2}$$

Journal of Physics: Conference Series **626** (2015) 012036 doi:10.1088/1742-6596/626/1/012036

and their solution is a sharp trajectory

$$q = q(t) \ , \quad p = p(t)$$

for every initial values $q(0)$ and $p(0)$.

Let us choose the corresponding quantum model $\mathcal{S}_q$ to be a system of one degree of freedom with position operator q, momentum operator p and spin 0. Let the Hamiltonian be

$$\mathsf{H} = \frac{\mathsf{p}^2}{2\mu} + V(\mathsf{q}) \ . \tag{3}$$

The Heisenberg equations of motion are

$$\dot{\mathsf{q}} = \frac{\mathsf{p}}{\mu} \ , \quad \dot{\mathsf{p}} = -\frac{dV}{d\mathsf{q}} \ . \tag{4}$$

Then the time dependence of position and momentum averages $Q = \langle \mathsf{q} \rangle$ and $P = \langle \mathsf{p} \rangle$ in a state $|\psi\rangle$ is

$$\dot{Q} = \frac{P}{\mu} \ , \quad \dot{P} = -\left\langle \frac{dV}{d\mathsf{q}} \right\rangle \ .$$

To evaluate the right-hand side of the second equation, let us expand the potential function in powers of $\mathsf{q} - Q$:

$$V(\mathsf{q}) = V(Q) + (\mathsf{q} - Q)\frac{dV}{dQ} + \frac{1}{2}(\mathsf{q} - Q)^2\frac{d^2V}{dQ^2} + \ldots$$

so that

$$\frac{dV}{d\mathsf{q}} = \frac{dV}{dQ} + (\mathsf{q} - Q)\frac{d^2V}{dQ^2} + \frac{1}{2}(\mathsf{q} - Q)^2\frac{d^3V}{dQ^3} + \ldots \ .$$

If we take the average of the last equation and use relations $\langle(\mathsf{q} - Q)\rangle = 0$ and $\langle(\mathsf{q} - Q)^2\rangle = \Delta Q^2$, where ΔQ is the variance of q in state $|\psi\rangle$, we obtain

$$\left\langle \frac{dV}{d\mathsf{q}} \right\rangle = \frac{dV}{dQ} + \frac{1}{2}\Delta Q^2\frac{d^3V}{dQ^3} + \ldots \ .$$

Let us assume that coordinate q and momentum p of $\mathcal{S}_c$ are obtained from the quantum model by formulas

$$q = Q \ , \quad p = P \ .$$

Then, already for potentials of the third order, the quantum equations of motion for averages deviate from classical equation of motion for sharp trajectories. This deviation would be negligible for small ΔQ, that is, the spread of the wave packet $|\psi\rangle$ over the space must be as small as possible. However, if the variance ΔP is large, ΔQ will quickly increase with time. This implies that the minimum-uncertainty wave packets may give the best approximation to classical sharp trajectories.

Let us stop here and ask: what is the reason for trying to get sharp trajectories from quantum mechanics? Clearly, it is the popularity of the specific form of classical realism mentioned in the Introduction: a *real* mechanical system possesses a sharp position and momentum at any instant of time. Let us call this assumption **Sharp Trajectory Hypothesis** (STH). There are many tacitly assumed consequences of STH, for example that a probability distributions on phase space is only an expression of insufficient knowledge of the real state.

However, there is *no evidence* supporting STH: indeed, as yet, any real observation of macroscopic bodies has been compatible with

$$2\Delta Q \Delta P \gg \hbar \ , \tag{5}$$

where "$\gg$" represents many orders of magnitude. *This* is well known but there can be *two* attitudes to Eq. (5):

(i) With improving techniques, the left-hand side of Eq. (5) will approach zero. This must be false if quantum mechanics holds true.
(ii) Sharp trajectory is just a handy model of a real, fuzzy, one. That is, it lies within a tube associated with the fuzzy trajectory. But then, a more realistic model of any Newtonian motion would be a probability distribution.

References [9, 8] assume the second attitude. For us, the most important consequence is that it is sufficient to approximate *fuzzy* Newtonian trajectories by quantum mechanics, where fuzzy trajectories are some probability distributions on the phase space of the system. Such a theory of classical properties can do without pure states. Of course, *this probability distribution is not completely knowable and measurable: in any case, the sharp points do not exist* [11, 12]. The fact that the points of the phase space do exist mathematically and must be used for mathematical description of a real state is only an unrealistic feature of Newtonian mechanics.

Thus, instead of a sharp point of the phase space a distribution on the phase space can be considered as the real state of a mechanical system. It is determined by preparation similarly as in quantum theory. In this way, we preserve the realism (for more details on realism, see [10]) as such but change the form of it as expressed by STH.

This opens the problem to application of Bayesian methods, see, e.g., [13]. These methods recommend maximising entropy in the cases of missing knowledge. Let us define a fuzzy state called *maximum-entropy packet* (ME packet) as a phase-space distribution maximising entropy for given averages and variances of mechanical state coordinates.

More precisely, for the classical model $\mathcal{S}_c$, we consider the states described by distribution function $\rho(q,p)$ on the phase space spanned by q and p. The function $\rho(q,p)$ is dimensionless and normalised by

$$\int \frac{dq\,dp}{v}\rho = 1 \ ,$$

where v is an auxiliary phase-space volume to make ρ dimensionless. The entropy of $\rho(q,p)$ can be defined by

$$S := -\int \frac{dq\,dp}{v}\rho \ln \rho \ .$$

The value of entropy will depend on v but most other results will not. Classical mechanics does not offer any idea of how to fix v. We shall get its value from quantum mechanics.

Definition 1 *ME packet is the distribution function ρ that maximizes the entropy subject to the conditions:*

$$\langle q\rangle = Q \ , \quad \langle q^2\rangle = \Delta Q^2 + Q^2 \ , \tag{6}$$

and

$$\langle p\rangle = P \ , \quad \langle p^2\rangle = \Delta P^2 + P^2 \ , \tag{7}$$

where Q, P, ΔQ and ΔP are given values.

We have used the abbreviation

$$\langle x\rangle = \int \frac{dq\,dp}{v} x\rho \ .$$

The explicit form of ρ can be found using the Lagrange-multiplier and partition-function method [7]:

Theorem 1 *The distribution function of the classical ME packet for a one-degree-of-freedom system with given averages and variances Q, ΔQ of coordinate and P, ΔP of momentum, is*

$$\rho[Q,P,\Delta Q,\Delta P](q,p) = \left(\frac{v}{2\pi}\right)\frac{1}{\Delta Q\Delta P}\exp\left[-\frac{(q-Q)^2}{2\Delta Q^2} - \frac{(p-P)^2}{2\Delta P^2}\right] \ . \tag{8}$$

In this way, to describe the mechanical degrees of freedom, we need twice as many variables as the standard mechanics. The doubling of state coordinates is due to the necessity to define a fuzzy distribution rather than a sharp trajectory.

The model can be generalised to any number of degrees of freedom. Also, the ME packet could be defined by different pairs of conjugate variables. It seems plausible that our main results would then remain valid.

4. Quantum ME packets

Definition 2 *State* T *of quantum model* $\mathcal{S}_q$ *with one degree of freedom that maximizes von Neumann entropy*

$$S = -tr(\mathsf{T}\ln\mathsf{T})$$

under the conditions

$$tr[\mathsf{Tq}] = Q\ , \quad tr[\mathsf{Tq}^2] = Q^2 + \Delta Q^2\ ,$$
$$tr[\mathsf{Tp}] = P\ , \quad tr[\mathsf{Tp}^2] = P^2 + \Delta P^2\ ,$$

where Q, P, ΔQ *and* ΔP *are given numbers, is called* quantum ME packet.

The following theorem can be proved by the method of Lagrange multipliers and partition function, but the proof is non-trivial because of non-commuting factors [7]:

Theorem 2 *The state operator of the ME packet of a one-degree-of-freedom system with given averages and variances* Q, P, ΔQ *and* ΔP *is*

$$\mathsf{T}[Q,P,\Delta Q,\Delta P] = \frac{2}{\sqrt{\nu^2-1}}\exp\left(-\frac{\nu}{2}\ln\frac{\nu+1}{\nu-1}\mathsf{K}\right)\ , \tag{9}$$

where

$$\mathsf{K} = \frac{(\mathsf{q}-Q)^2}{2\Delta Q^2} + \frac{(\mathsf{p}-P)^2}{2\Delta P^2}$$

and

$$\nu = \frac{2\Delta P\Delta Q}{\hbar}\ . \tag{10}$$

Generalisation to any number of degrees of freedom is easy. It is amusing to observe how the forms of Eqs. (8) and (9) approach each other in the limit $\Delta P\Delta Q \to \infty$. Indeed,

$$\lim_{\nu\to\infty}\frac{\nu}{2}\ln\frac{\nu+1}{\nu-1} = 1\ .$$

The entropy of state (9) can be shown [7] to be an increasing function of $\nu \in (1,\infty)$ diverging for $\nu \to \infty$. For $\nu = 1$ (minimum quantum uncertainty), T is a pure state with wave function

$$\psi(q) = \left(\frac{1}{\pi}\frac{1}{2\Delta Q^2}\right)^{1/4}\exp\left[-\frac{1}{4\Delta Q^2}(q-Q)^2 + \frac{iPq}{\hbar}\right]\ .$$

This is just a Gaussian wave packet and the entropy is zero. Thus, quantum ME packets are generalization of Gaussian wave packets.

5. Comparing classical and quantum evolutions

Let us consider the time evolution of the averages and variances for ME packet (8) with initial data $Q, P, \Delta Q, \Delta P$ at $t = 0$ and let us define the *classical trajectory* of the classical model $\mathcal{S}_c$ by the quadruple $Q_c(t), P_c(t), \Delta Q_c(t), \Delta P_c(t)$. Let $Q_q(t), P_q(t), \Delta Q_q(t), \Delta P_q(t)$ be an analogous trajectory for the quantum model $\mathcal{S}_q$ starting in state (9). Each of the two trajectories is described by four real functions so that they can be compared.

Let us first study the special case of at most quadratic potential:

$$V(q) = V_0 + V_1 q + \frac{1}{2} V_2 q^2 ,$$

where V_k are constants with suitable dimensions. If $V_1 = V_2 = 0$, we have a free particle, if $V_2 = 0$, it is a particle in a homogeneous force field and if $V_2 \neq 0$, it is an harmonic or anti-harmonic oscillator. In these cases, exact solutions can be found:

$$\begin{aligned} Q_c(t) = Q_q(t) &= f_0(t) + Qf_1(t) + Pf_2(t) , \\ \Delta Q_c(t) = \Delta Q_q(t) &= \sqrt{f_1^2(t)\Delta Q^2 + f_2^2(t)\Delta P^2} , \\ P_c(t) = P_q(t) &= g_0(t) + Qg_1(t) + Pg_2(t) , \\ \Delta P_c(t) = \Delta P_q(t) &= \sqrt{g_1^2(t)\Delta Q^2 + g_2^2(t)\Delta P^2} . \end{aligned}$$

If $V_2 \neq 0$, the functions are

$$f_0(t) = -\frac{V_1}{V_2}(1 - \cos\omega t) , \quad f_1(t) = \cos\omega t , \quad f_2(t) = \frac{1}{\xi}\sin\omega t ,$$

$$g_0(t) = -\xi\frac{V_1}{V_2}\sin\omega t , \quad g_1(t) = -\xi\sin\omega t , \quad g_2(t) = \cos\omega t ,$$

where

$$\xi = \sqrt{\mu V_2} , \quad \omega = \sqrt{\frac{V_2}{\mu}} .$$

If $V_2 = 0$, we obtain

$$f_0(t) = -\frac{V_1}{2\mu}t^2 , \quad f_1(t) = 1 , \quad f_2(t) = \frac{t}{\mu} ,$$
$$g_0(t) = -V_1 t , \quad g_1(t) = 0 , \quad g_2(t) = 1 .$$

Hence, for at-most-quadratic potentials, classical and quantum trajectories coincide. However, already for a third order potentials, there are non-trivial quantum corrections. For example, if $V = V_3\mathsf{q}^3/6$, one can use Heisenberg equations of motion (4) to calculate the 9-th time derivative of p where there is finally a term,

$$\frac{d^9\mathsf{p}}{dt^9} = \ldots - \frac{125}{4}\frac{V_3^5}{\mu^4}\mathsf{q}^6 ,$$

that has a quantum correction, namely

$$\langle \mathsf{q}^6 \rangle_q = \langle \mathsf{q}^6 \rangle_c + 9\Delta Q^6 \nu^{-1} - 3\Delta Q^6 \nu^{-3} ,$$

where ν is defined by Eq. (10) and

$$\langle \mathsf{q}^6 \rangle_c = Q^6 + 15Q^4\Delta Q^2 + 45Q^2\Delta Q^4 + 15\Delta Q^6 .$$

One can see from these equations that the quantum correction becomes negligible in the limit $\Delta Q \to \infty$. In general, one can show:

Theorem 3 *For general polynomial potential, the classical and quantum trajectories satisfy:*

$$\lim_{\Delta Q\to\infty,\Delta P\to\infty}\frac{Q_q(t)-Q_c(t)}{Q_c(t)}=0\ ,\quad \lim_{\Delta Q\to\infty,\Delta P\to\infty}\frac{P_q(t)-P_c(t)}{P_c(t)}=0\ ,$$

and

$$\lim_{\Delta Q\to\infty,\Delta P\to\infty}\frac{\Delta Q_q(t)-\Delta Q_c(t)}{\Delta Q_c(t)}=0\ ,\quad \lim_{\Delta Q\to\infty,\Delta P\to\infty}\frac{\Delta P_q(t)-\Delta P_c(t)}{\Delta P_c(t)}=0\ ,$$

for all t for which the formulas make sense.

The proof is based on the calculation of averages of polynomials in q and p described in [7] and will be published elsewhere.

Hence, the fuzzier the compared ME packets are, the better their dynamical evolutions match each other. In other words, the classical limit for mechanical degrees of freedom is

$$\Delta Q\to\infty\ ,\quad \Delta P\to\infty\ . \tag{11}$$

This seems to contradict the usual belief that the classical physics is best approximated by minimum-uncertainty ($\nu = 1$) quantum states. But the explanation of this paradox is simple. The two answers to the question which states best approximate classical physics are different because the questions asked are, in fact, different: the first one compares two fuzzy states, the second one compares a quantum state with a sharp classical trajectory.

The conjecture that the classical limit is given by equation (11) seems to be plausible, but it has yet to be shown for non-polynomial potential function, such as the Coulomb potential (Kepler orbits).

6. Mechanical and thermostatic properties unified

We have tried to prove that classical mechanical properties of an object can be obtained from its quantum model as properties of high-entropy quantum states. However, this also holds for classical thermostatic properties, such as internal energy, temperature, entropy, specific heats etc., which suggests that the quantum theory of classical properties can be based on a single principle. In the present section, we try to show in more detail how such "unified" theory could look like.

We use a very simple model so that its quantum equations are exactly solvable and we can concentrate on conceptual questions. As a real object $\mathcal{S}$, consider a thin stiff rod of mass M and length L extended and moving freely in one space dimension. Let its classical model $\mathcal{S}_c$ be a one-dimensional continuum. Its (classical) state is determined by the values of 5 quantities: internal energy E_{int}, average X and variance ΔX of its centre-of-mass coordinate as well as average P and variance ΔP of its total momentum.

Let the quantum model $\mathcal{S}_q$ be a chain of $N+1$ particles, each of mass μ. We denote the position operator of the n-th particle by x_n and that of its momentum by p_n, $n = 1,\dots N+1$. Let the Hamiltonian of the quantum model be

$$\mathsf{H}=\frac{1}{2\mu}\sum_{n=1}^{N+1}\mathsf{p}_n^2+\frac{\kappa^2}{2}\sum_{n=2}^{N+1}(\mathsf{x}_n-\mathsf{x}_{n-1}-\xi)^2\ .$$

The potential represents nearest-neighbour elastic forces, κ being the oscillator strength and ξ the equilibrium inter-particle distance. The algebra of observables of $\mathcal{S}_q$ is generated by x_n and p_n, a set of $2N+2$ operators. We assume that $N \approx 10^{23}$. This implies that the quantum state contains much more information than the classical one.

Journal of Physics: Conference Series **626** (2015) 012036 doi:10.1088/1742-6596/626/1/012036

A linear (in fact, Fourier) transformation of variables x_n and p_n to normal modes u_n and q_n diagonalizes the Hamiltonian [9, 8]. Moreover, it becomes the sum of the total momentum part and the internal energy part E_{int} (see [9, 8]):

$$\mathsf{H} = \frac{1}{2M}\mathsf{P}^2 + \mathsf{E}_{\text{int}} ,$$

where $M = (N+1)\mu$ is the total mass of the chain, P its total momentum,

$$\mathsf{E}_{\text{int}} = \frac{1}{2\mu}\sum_{m=1}^{N}\mathsf{q}_m^2 + \frac{\mu}{2}\sum_{m=1}^{N}\omega_m^2\mathsf{u}_m^2$$

and ω_m, $m = 1,\ldots,N$ are "phonon" frequencies:

$$\omega_m = \frac{2\kappa}{\sqrt{\mu}}\sin\left(\frac{\pi}{2(N+1)}m\right) .$$

The mechanical evolution thus decouples from the thermodynamics.

As the state of $\mathcal{S}_q$, the tensor product $\mathsf{T}_{\text{therm}} \otimes \mathsf{T}_{\text{mech}}$ can, therefore, be chosen. Next, we apply the unifying principle: $\mathsf{T}_{\text{therm}}$ is the maximum entropy quantum state for a given value E of the averages of E_{int} and T_{mech} is the maximum entropy state for given averages of X and P and variances $\Delta\mathsf{X}$ and $\Delta\mathsf{P}$. It then follows that $\mathsf{T}_{\text{therm}}$ is the Gibbs state and T_{mech} is an ME packet.

The phonons of species m form statistically independent subsystems. Hence, the Gibbs state factorizes and the factors are Gibbs states T_m of the species:

$$\mathsf{T}_m = \sum_{r=0}^{\infty}|r\rangle \mathrm{P}_r^m \langle r| ,$$

where r is the number of phonons of species m,

$$\mathrm{P}_r^m = \left(1 - e^{-\lambda\hbar\omega_m}\right)e^{-\lambda\hbar\omega_m r}$$

and λ is the Lagrange multiplier of the variational problem for the conditional maximum of entropy. The variational principle couples the value of λ with the value of the energy average E. As λ is interpreted as $1/kT$, k being the Boltzmann constant and T the temperature, the relation between internal energy and temperature results. The average number $\langle r\rangle$ of photons in state T_m is the Bose distribution.

All properties of the classical model (such as the temperature and length of the rod, its dynamical trajectory etc.) have been obtained, in a good approximation, from the quantum one ([9, 8]). For example the length L of the rod is the average of a natural rod-length operator $\mathsf{x}_{N+1} - \mathsf{x}_1$. The calculation in [9, 8] yields

$$L = N\xi .$$

This is independent of the parameter E of the Gibbs state. Hence, the model describes a rigid rod. The relative variances of the internal energy and length are indirectly proportional to N.

7. Conclusion and outlook

Our results suggest that there is a unified theory for both thermostatic and mechanical properties. It is based on the assumption that the states of quantum system that exhibit classical properties are some states with high entropy.

The fuzzier are the compared mechanical states, the better is the match between classical and quantum mechanical trajectories, if the ν-Conjecture holds true. This would confirm the feeling that quantum mechanics is more accurate and finer than Newtonian mechanics. We hope to be able to prove this statement for non-polynomial potentials later.

The paper suggests promising ideas of how all four conceptual problems can be solved. More detailed models that would describe such solutions for some simple cases ought to be constructed.

The project is in its beginnings. Only extremely simple models have been studied. Also, a generalisation of the idea to classical electro- and magnetostatic properties, as well as a generalisation to the relativistic classical electrodynamics is missing as yet.

Acknowledgments

The author is indebted to Petr Jizba, Jiří Tolar and Uwe-Jens Wiese for useful discussions.

References

[1] Leggett A J 2002 *J. Phys.: Condens. Matter* **14** R415
[2] Schlosshauer M 2004 *Rev. Mod. Phys.* **76** 1267
[3] Kampen van N G *Physica* 1995 **194** 542
[4] Hepp K 1972 *Helvetica Phys. Acta* **45** 237
[5] Pearle P 1989 *Phys. Rev. A* **39** 2277
[6] Ghirardi G C 2002 *The Stanford Encyclopedia of Philosophy* http://plato.stanford.edu/entries/qm-collapse
[7] Hájíček P 2009 *Found. Phys.* **39** 1072
[8] Hájíček P 2013 *Entropy* **15** 789
[9] Hájíček P and Tolar J 2009 *Found. Phys.* **39** 411
[10] Hájíček P 2013 *J. Phys.: Conf. Ser.* **442** 012043
[11] Exner F 1922 *Vorlesungen über die physikalischen Grundlagen der Naturwissenschaften* (Deuticke: Leipzig)
[12] Born M 1955 *Phys. Blätter* **11** 49
[13] Jaynes E T 2003 *Probability Theory. The Logic of Science* (Cambridge University Press: Cambridge, UK)

Journal of Physics: Conference Series **626** (2015) 012037 doi:10.1088/1742-6596/626/1/012037

Classical and Quantum Stability of Branched Flow

Bo Liu[1,2,3]

[1] Department of Physics, Harvard University, Cambridge, MA 02138, USA
[2] Institute for Applied Computational Science, Harvard University, Cambridge, MA 02138, USA
[3] Center for Integrated Quantum Materials, Cambridge, MA 02138, USA

E-mail: `boliu.physics@post.harvard.edu`

Abstract. Branched flow is a universal phenomenon of wave propagation in random media. From a classical point of view, branched flow is the overall pattern of classical electron trajectories moving in a potential with randomly placed weak scatterers. Individually, each electron trajectory is exponentially unstable to small perturbations due to the chaotic nature of the classical dynamics of electrons moving in a random potential. However, the overall pattern, branched flow, displays strong stability against large perturbations. In this paper, we discuss both the classical and quantum stability of branched flow against perturbations.

1. Discovery of Branched Flow

Coherent electron transport in disordered media is a widely studied topic in condensed matter physics. Different physics phenomena arise in different regimes. When the disorder is strong enough or the sample is long enough such that the sample size is larger than the localization length(the mean free path of the electrons adjusted by the number of open conduction modes), it leads to the Anderson Localization[1]. When the size of the sample is smaller than the localization length, but much larger than the mean free path, it leads to the Universal Conductance Fluctuations[2]. Branched flow arises in a yet different regime with much weaker disorders. In this case, the mean free path of the electrons is even larger than the sample size. Therefore, it falls under the category of coherent ballistic transport.

Branched flow was first observed in an experiment by Prof. Westervelt[3, 4] where he measured the two dimensional electron transport pattern through a random potential in a GaAs heterostructure. The system is cooled down to liquid Helium temperature(4.2K) to ensure coherent transport. In the experiment, the electrons are confined to move only in two dimensions due to a lateral confinement in the third dimension. The randomly placed donor atoms and impurities generate a weakly correlated random potential that can scatter the electrons. The potential due to the donor atoms is usually weak compared to the energy of the electrons, leading only to deflection of the electron trajectories. The impurities, on the other hand, can contribute to a strong potential that results in backscattering. Overall, the random potential is still weak and has a standard deviation of only 8% the energy of the electrons. Thus, the electron can in theory fly through the whole sample without being deflected much.

In the experiment, these two dimensional electrons are launched into the random potential through a narrow injector called the Quantum Point Contact(QPC). The QPC can be thought as an electron waveguide with only one or two open modes. The electron wavelength in the

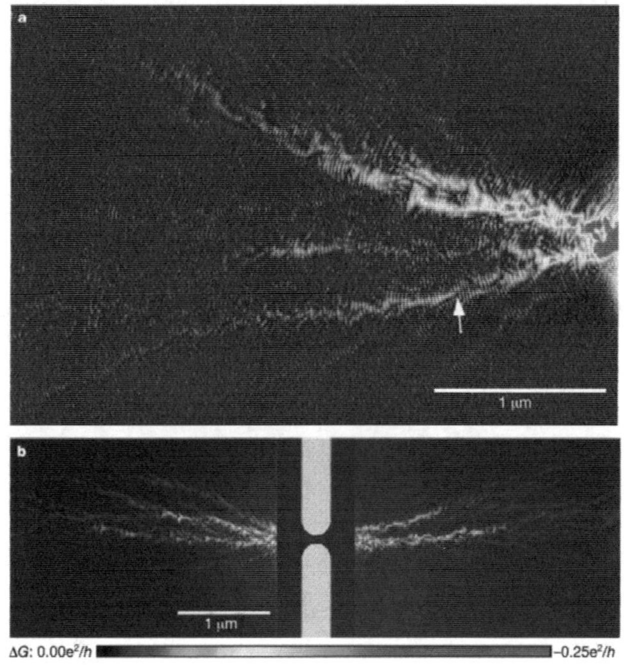

Figure 1. Experimentally observed branched flow. This figure is taken from Ref. 2 with copyright permissions from Nature Publishing Group. In this figure, we see the experimentally observed branched flow pattern. The two patterns in (b) correspond to two different random potential. This figure is obtained with the copyright permission from Nature Publishing Group.

experiment is 37nm, and the narrowest opening in the QPC is on the same order of magnitude. Semiclassically, one can imagine the effect of this QPC as aligning the initial momenta of the electrons in one direction while confining their positions down to nanoscales in the other direction[5]. This effectively creates an initial condition resembling that from a "point" injector. These highly aligned electrons are then launched into the random potential where they are being scattered. Magnified by the chaotic classical dynamics, the tiny differences in initial conditions lead to different classical electron trajectories, the overall pattern of which is called the branched flow.

As the name suggests, the electrons move along a small number of narrow branches in the random potential. Fig. 1 shows the experimentally measured electron flow pattern that displays the branching behavior and this figure is taken from Ref.[4] with copyright permissions from Nature Publishing Group. This branched flow pattern is measured by a technique called the Scanning Gate Microscopy(SGM). In the SGM, a negatively charged metallic tip is held on top of the sample, generating a potential barrier that can backscatter the electrons impinging on it. These backscattered electron flux leads to a reduction in the measured conductance, which is measured as a function of the position of the metallic tip. If the tip is placed in a region with high electron flux, a larger reduction in conductance will be recorded, while a smaller change will be recorded if the tip is placed in a region with small electron flux. By moving the tip across the surface of the sample and recording the changes in conductance, it produces a conductance map that reveals information of the electron flux pattern in the random potential.

What makes branched flow particularly interesting is the contrast between the relatively weak random potential and the significant effect on the overall electron flow pattern. As mentioned before, the random potential has a mean energy of zero and standard deviation of 8% the energy of the electrons. Moreover, the sample size of the system is smaller than the classical mean free path of the electrons. If one thinks in terms of perturbation theory, it would not be expected that the random potential will be able to cause branching. Moreover, the classical dynamics of electrons moving a random potential is chaotic. Naively, chaos should separate electron trajectories with small differences in initial conditions exponentially fast, leading to a more uniform flow pattern rather than a branched flow pattern.

After its initial discovery in two dimensional electron gases(2DEGs), branched flow has later been observed and studied in other systems with much different length scales. For 2DEGs, the wavelength of the electrons is on the scale of nanometers[3, 4]. On the micrometer scales, branched flow has been experimentally observed in quasi-two-dimensional microwave resonators[6]. Branched flow is also used to study sound propagation in oceans with megameter length scales[7] and is found to have implications for the formation of freak waves in oceans[8].

2. Classical Stability of Branched Flow: "Unexpected Features"

The formation of branched flow has long been explained using classical theory. In Ref.[4], the authors reported a numerical simulation based on classical electron trajectories that reproduces many of the key features of branched flow. Moreover, with a clever choice of the initial conditions, the classical simulation closely resembles the result of a quantum simulation on the same random potential.

According to the classical theory, branched flow can form due to two mechanisms. One mechanism is by forming caustics in phase space. Caustics forms after electrons pass through a potential dip, which focuses parallel electron trajectories. Immediately after the focal point, electrons travel along caustics with high flux density. One theory based on this caustics idea is called the "kick and drift" model[9]. In this model, the random potential is modeled as a series of potential dips that focus electron trajectories. As the electrons move across the random potential, a series of random caustics form and they are where branches appear. Detailed statistical analysis of the distribution of these caustics has also been reported[10, 11].

Another mechanism that could lead to branched flow is called stable regions in the phase space[9, 12]. Similarly to caustics, stable regions in phase space form by chance. Qualitatively, stable regions in phase space refer to regions in phase space that is less sensitive to changes in initial conditions. One way to understand those stable regions is like this: one fundamental property of any phase space manifold is that the total electron count can not change over time. In the case of branched flow, the total area being occupied by the electrons in the phase space can not expand exponentially over time since the sample size is smaller than the mean free path. As a result, when certain region's electron density is diminished, this decrease in electron density has to be compensated with a higher electron density somewhere else and these more densely populated regions are stable regions. Quantitatively, a rarefaction exponent[9, 13], which is a variant of the Lyapunov exponent, is developed to characterize those stable regions. More detailed account of these two mechanisms can be found in Ref.[12, 13, 14].

One consequence of this classical theory is that the dynamics becomes chaotic, meaning that each individual classical trajectory is exponentially unstable to perturbations. This chaotic nature of the classical dynamics motivated a new experiment to test the stability of branched flow[15]. The setup of this new experiment is almost identical to the setup used to observe branched flow in the first set of experiments[3, 4]. The only difference is that the branched flow patterns are observed for two sets of initial conditions. This difference in initial conditions is achieved by shifting the position of the QPC, the electron injector. In the experiment, the authors first measure the branched flow for a fixed QPC position. After the branched flow pattern has been mapped out, the QPC is shifted away from the previous position by about 60 nanometers in the direction of confinement and the same procedure of measurement of the branched flow pattern is repeated with the new shifted QPC position over the same random potential. This shifting of the QPC position corresponds to a change in the initial conditions of the electrons. Moreover, this 60nm change in initial conditions is actually very significant in terms of both the classical dynamics and the quantum dynamics. Classically, this 60nm change in initial conditions is comparable to the correlation length of the random potential, which measures how fast the self-correlation of the random potential decays over distance. Quantum mechanically, this 60nm change is also comparable to the narrowest opening in the QPC. Thus, if one imagines channeling two wavepackets through two different QPCs separated by 60nm, the coherent overlap between the two wavepackets is estimated to be less than five percent[15].

According to the classical theory, branched flow is the overall pattern of individually chaotic classical trajectories moving through a random potential. Since each classical trajectory is chaotic and exponentially unstable to perturbation, it seems reasonable to assume that this large change in initial conditions due to the shifting of the QPC will result in a significant change in the measured branched flow pattern. However, the experimental results display an unexpected long-range stability of the branched flow. It was observed that about seventy correlation lengths away from the injection points, the electrons move along almost identical branches for the two sets of initial conditions. This recovery of branches was named "unexpected features of branched flow" and was suspected to be of pure quantum origin without any classical analogy. However, as we explained in Ref.[5], this long range stability of branched flow is actually of classical origin and we reported in that paper classical simulations based on chaotic classical trajectories that reproduce the long range stability. Our theory shows that this long range stability of branched flow derives its origin from the QPC's ability to correlate classical trajectories. We showed in Ref.[5] that even if one starts with two initial conditions that have zero overlap to begin with in the phase space, the QPCs can boost their overlap to almost eighty percent for the experimental conditions. The key point to notice is that shifting the QPC corresponds to a change in the Hamiltonian of the physical system. Even though the overlap between any two manifolds in the phase space is guaranteed to be preserved over time if they are propagated according to the same Hamiltonian, there is no mathematical constraint on how the overlap will evolve if the Hamiltonians are different. In this particular example, we have shown that the overlap in phase space is monotonically increased as the electrons move through the QPCs. Thus, this unexpected feature of branched flow has nothing to do with the chaotic nature of the classical dynamics of electrons moving in a random potential and is completely generated by the ability of the QPC to produce overlap in the phase space.

3. Quantum Stability of Branched Flow

In Ref.[5], we reported additional stability of branched flow against the change in the open modes of QPC. We showed that the branched flow pattern for the first open mode in the QPC bears close similarity to the pattern for the second mode. In terms of quantum mechanics, the two modes are orthogonal to each other and thus the overlap is zero to start with. Moreover, the two modes are propagated under the same Hamiltonian, so even the QPC could not boost their coherent overlap. We gave a qualitative explanation of this additional stability in terms of the Wigner quasiprobability distribution[16] and attributed its formation to overlap in the classical phase space.

We have also studied the stability of branched flow against energy changes in Ref.[12] and have shown how this additional stability is also compatible with the classical interpretation of branched flow. In Ref.[12], we presented numerical simulations showing that the classical branched flow pattern is stable against a 22% change in the electron's energy and argued how this stability is not due to quantum interference, which is destroyed by thermal averaging. In

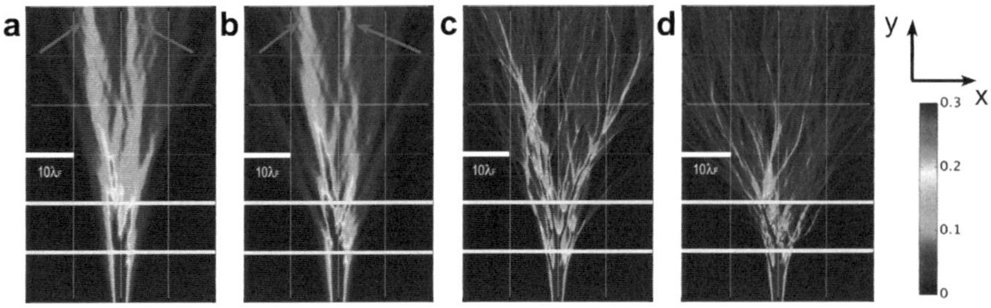

Figure 2. Quantum Flux v.s. Classical Flux Patterns Along the y Direction The quantum and classical flux patterns over the unperturbed random potential are shown in (**a**) and (**c**) respectively. (**b**) and (**d**) show the respective quantum and classical flux patterns after the same perturbations are added to the regions between the yellow lines. The white reference grid denotes the same location in all images and all images starts at $15\lambda_F$ from the injection point. The color axis shows the flux density per wavelength. As we can see, the same branches pointed to by the red arrows remain in (**a**) and (**b**), while all strong branches in (**c**) are destroyed by the perturbations in (**d**).

this section, we report one more stability of branched flow that we believe is of quantum origin. More detailed quantitative analysis of its origin will be left for a later publication[17] and we will present only some preliminary results here.

In this case, we are interested in the stability of branched flow against strong perturbations without moving the QPC. We numerically simulate both the quantum flow patterns and the classical branched flow patterns after an area of strong perturbations is introduced into the systems. Our numerical methods are primarily based on the thermal wavepacket approach developed in Ref.[18] and more detailed information about its implementation can be found in Ref.[5].

The results are shown in Fig.2. All parameters(Fermi energy E_F, wavelength λ_F, donor to two-dimensional electron gases(2DEGs) distance, sample mobility etc.) are chosen to match that in a previous experiment[15]. The random potential has a standard deviation of $8\%E_F$ and correlation length $0.9\lambda_F$, as estimated in that experiment. The perturbation in this case is introduced by imposing a region of random potential on top of the original random potential. This region of strong perturbations is introduced at $25\lambda_F$ from the injection point and lasts for $10\lambda_F$ as shown in Fig.2. In this region, a random potential of twice the original standard deviation($16\%E_F$) and the same correlation length is superimposed on the original random potential. This perturbation is classically strong enough such that it destroys all the classical branches in the long range(long in terms of the correlation length of the random potential), as shown in Fig.2c and Fig.2d. However, for the quantum simulations in Fig.2a and Fig.2b, the branches are only distorted inside and near the region of perturbations and recover themselves in the long range. It is our belief that this recovery of branches in the long range for the quantum dynamics is closely tied to the piecewise nature of interference in chaotic systems, which is a key difference between the quantum and classical dynamics in chaotic systems[17].

Acknowledgments
Bo Liu was supported by the IACS fellowship and the Van Vleck Travel Fellowship by Harvard Physics Department.

References

[1] Anderson P W 1958 *Phys. Rev.* **109** 1492
[2] Lee P A and Stone A D 1985 *Phys. Rev. Lett.* **55** 1622

Journal of Physics: Conference Series **626** (2015) 012037 doi:10.1088/1742-6596/626/1/012037

[3] Topinka M A, LeRoy B J, Shaw S E J, Heller E J, Westervelt R M, Maranowski K D and Gossard A C 2000 *Science* **289** 2323
[4] Topinka M A, LeRoy B J, Westervelt R M, Shaw S E J, Fleischmann R, Heller E J, Maranowski K D and Gossard A C 2001 *Nature* **410** 183
[5] Liu B and Heller E J 2013 *Phys. Rev. Lett.* **111** 236804
[6] Hohmann R, Kuhl U, Stockmann H J, Kaplan L and Heller E J 2010 *Phys. Rev. Lett* **104** 093901
[7] Wolfson M A and Tomsovic S 2011 *J. Acoust. Soc. Am.* **109** 2693
[8] Heller E J, Kaplan L, and Dahlen A 2008 *J. Geophys.Res.* **113** C09023
[9] Heller E J and Shaw S E J 2003 *Int. J. Mod. Phys.* B **17** 3977
[10] Kaplan L 2002 *Phys. Rev. Lett.* **89** 184103
[11] Metzger J J, Fleischmann R and Geisel T 2010 *Phys. Rev. Lett.* **105** 020601
[12] Liu B 2015 *Selected Topics in Scattering Theory: From Chaos to Resonance* (Ph.D. thesis, Harvard University)
[13] Shaw S E J 2002 *Propagation in Smooth Random Potentials* (Ph.D. thesis, Harvard University)
[14] Metzger J J 2011 *Branched Flow and Caustics in Two-Dimensional Random Potentials and Magnetic Fields* (Ph.D. thesis, University of Göttingen)
[15] Jura M P, Topinka M A, Urban L, Yazdani A, Shtrikman H, Pffer L N, West K W and Goldhaber-Gordon D 2007 *Nat. Phys.* **3** 841
[16] Heller E J 1976 *J. Chem. Phys.* **65** 1289
[17] Liu B 2014 *Preprint* cond-mat/1409.0574
[18] Heller E J, Aidala K A, LeRoy B J, Bleszynski A C, Kalben A, Westervelt R M, Maranowski K D and Gossard A C 2005 *Nano Lett.* **5** 1285

Journal of Physics: Conference Series **626** (2015) 012038 doi:10.1088/1742-6596/626/1/012038

Violation of Leggett–Garg inequalities for quantum–classical hybrids

Marcin Łobejko[1] **and Jerzy Dajka**[1,2]

[1]Institute of Physics, University of Silesia, Katowice, Poland
[2]Silesian Center for Education and Interdisciplinary Research, University of Silesia, Chorzów, Poland

Abstract. Violation of Leggett–Garg inequalities can serve as a signature of a failure of (macroscopic) realism. We investigate violation of the simplest Leggett–Garg inequality for a qubit coupled to an integer j spin (angular momentum). Such a system effectively reveals quantum–classical hybrid behavior in the limit of large j values. We show that a maximal violation of the Leggett– Garg inequality is larger for quantum–classical hybrids than for fully quantum systems.

1. Introduction

Discoveries of entanglement [1] and other types of quantum correlations like quantum discord [2] play essential role in a way of seeing and understanding of the microscopic and macroscopic reality. Well known violation of Bell inequalities [3] is the best evidence of the existence of such spatial correlations which cannot be understand in terms of some "classical" local hidden variables.

Moreover, recently also temporal quantum correlations are studied. In analogy to the spatial correlations quantified by the Bell inequalities [3] one predicts and observes violations of the so called Leggett-Garg inequalities (LGI) [4]. LGI are derived under two assumptions: *macroscopic realism* and *non–invasive measurability* of the system. The second requirement is subtle and often leads to conceptual difficulties related to the clumsiness loophole [4]. The first requirement, the macroscopicity, also is often abandoned as the LGI are interesting by its own also in microscopic systems [5].

It was shown that violation of LGI by free spin particle with dichotomous measurement increases with respect to absolute value of spin [5]. It is interesting fact since systems with higher spin are naturally considered as more macroscopic (like atomic nucleus) and hence probably 'more realistic'. In this paper we consider a simple composite system consisting of a qubit coupled to an integer spin (angular momentum) exhibiting well defined classical limit. In other words, we consider quantum–classical hybrid system with a qubit as a quantum part. It is known that the spatial correlations (entanglement) in quantum–classical hybrid systems can exhibit unexpected properties [6, 7]. Here we show that it is also the case for LGIs which are more violated by hybrid systems than by those which operate in fully quantum regime.

2. Model

Leggett–Garg inequalities are constructed from time correlation functions of a series of measurements of some dichotomous variable carried on a considered system. We study the simplest among LGIs [4] when the system undergoes the measurement only thrice:

$$K_3 \ := \ C_{21} + C_{32} - C_{31} \leq 1 \tag{1}$$

where

$$C_{\beta\alpha} \ = \ \sum_{m,l} q_m q_l \mathrm{Tr}\left[\Pi_m U_{\beta\alpha} \Pi_l U_{\alpha 0} \rho_0 U^\dagger_{\alpha 0} \Pi_l U^\dagger_{\beta\alpha}\right] \tag{2}$$

is the time correlation function for some dichotomous variable Q and and $q_l, q_m = \pm 1$ represent (projective) measurement outputs given by eigenvalues of the observable corresponding to Q and associated with a set of projectors Π_l. Here ρ_0 is the initial state of the system and $U_{\beta\alpha} = U(t_\beta, t_\alpha) = e^{-i(t_\beta - t_\alpha)H}$ is the unitary time-evolution operator with the Hamiltonian H. For non–invasive measurement of Q, violation of the LGI in Eq.(1) indicates possible failure of the (macroscopic) realism.

For our purposes we assume a simple spin–orbit type of interaction:

$$H \ = \ \alpha\, \vec{\sigma} \cdot \vec{J} \tag{3}$$

typical for Hydrogen–like atomic objects. Here $\vec{\sigma} = (\sigma_x, \sigma_y, \sigma_z)$ is a vector of Pauli matrices acting on a qubit subspace, $\vec{J} = \frac{1}{j}(J_x, J_y, J_z)$ is a vector of an integer spin operator ($j = 1, 2, \ldots$) and α is a coupling constant. In the limit of $j \to \infty$ there is a well defined classical limit of the integer spin part of the system [8]. As the qubit part, given by $\vec{\sigma}$, remains quantum, the total system effectively mimics in large j limit quantum–classical hybrid. Let us notice that there are alternative approaches for studying quantum–classical hybrid systems [9, 10].

For the quantity which is measured to obtain the correlation function Eq.(2) we set an x–component of the qubit spin (i.e. $m, l = \pm 1$ in Eq.(2)). Such an observable is represented by a pair of projectors

$$\Pi_\pm \ = \ |\pm x\rangle \langle \pm x| \otimes \mathcal{I} \tag{4}$$

where $|\pm x\rangle$ is an eigenstate of the x–component of σ_x and the identity $\mathcal{I}$ acts on the integer spin part of the system.

3. Results

In the absence of qubit–spin coupling ($\alpha = 0$) there is no time evolution, always $K_3 = 1$ and the LGI Eq.(1) is never violated. For $\alpha > 0$ let us consider the following initial preparation of the system:

$$|\psi\rangle = \frac{1}{\sqrt{2}} (|0\rangle + |1\rangle) \otimes |m\rangle \tag{5}$$

where the states $|0\rangle$ and $|1\rangle$ are the eigenstates of σ_z operator (with eigenvalues $+1$ and -1) and $|m\rangle$ is one of eigenstates of J_z operator, i.e. $J_z |m\rangle = m |m\rangle$.

In the following we assume that the measurements required to calculate K_3 in Eq.(1) appearing in Eq.(2) are equally separated in time i.e. $t_1 = \tau, t_2 = 2\tau$ and $t_3 = 3\tau$.

Journal of Physics: Conference Series **626** (2015) 012038 doi:10.1088/1742-6596/626/1/012038

3.1. One-dimension coupling
Let us first consider a simplified analytically solvable model with interaction restricted only to one spin direction:

$$H = \frac{\alpha}{j}\,\sigma_z J_z \tag{6}$$

For the initial state (5) the K_3 function reads as follows:

$$K_3 = 2\cos(\omega\tau) - \cos(2\omega\tau) \tag{7}$$

where $\omega = 2\alpha m/j$ and the period of K_3 function is $T_1 = j\pi/\alpha m$ where the subscript '1' indicates one-dimensional interaction Eq.(6).

Any changes of coupling strength α and integer spin value j affects just the period of the K_3 function. In the (semi)classical limit of large j, the smaller are the values of m of the integer spin projection on quantization axis, the longer is the period of K_3 oscillation. Moreover, with increasing m the period of K_3 function becomes minimal (given by $T_1 = \pi/\alpha$) and independent on the dimension of the Hilbert space. It holds true for all initial preparations of the form Eq.(5) with $|m = j\rangle$. One can conclude that the interaction between a qubit and a (semi)classical object results [for small m in Eq.(5)] in stretching of time periods of the LGI violation. In particular, an amplitude of the K_3 function remain the same with an exception of the initial state Eq.(5) with $|m = 0\rangle$ when $K_3 \equiv 1$.

3.2. Three-dimension coupling
In order to examine violation of the LGI of the system evolving according to the full Hamiltonian Eq. (3) we applied numeric simulations using Python toolbox QuTiP [11]. As in simplified one-dimensional model Eq.(6), the multiplication constant in a front of Hamiltonian Eq.(3) representing a strength of an interaction affects only the period of the K_3 function (and not its amplitude). The period of K_3 reads as follows:

$$T_3 = \frac{2\pi j}{\alpha N} \tag{8}$$

where $N = 2j + 1$ is the dimension of the spin j Hilbert subspace. In the classical limit we obtain:

$$T_3^c \;:=\; \lim_{j\to\infty} T_3 = \frac{\pi}{\alpha} \tag{9}$$

i.e. the same value as the minimal period in the case one-dimension coupling T_1. However, in the three dimensional case, the time period of the quantum–classical hybrid system Eq.(9) is maximal i.e. $T_3^c \geq T_3$. Moreover, in this case, the T_3–period does not depend on m in the initial state Eq.(5) and hardly increases with respect to the j value.

Let us notice the *qualitative* difference between one– and three–dimensional case. In the later one not only the period of K_3 function but also its amplitude changes with respect to the quantum numbers m and j in the initial state Eq.(5) as in Fig.1(a) and Fig.1(b).

Particularly interesting is of *maximal violation* of the LGI when the j–component of the system approaches classical limit i.e. when the qubit–spin system starts to be a quantum–classical hybrid. Maximum of the LGI violation for different values of j and initial preparations is presented in Fig.2. One can infer that the LGI violation is stronger in the system interacting with higher spin values i.e. for quantum–classical hybrids when the LGI violation approaches its maximal possible value $K_3 = 3/2$ [4].

Journal of Physics: Conference Series **626** (2015) 012038 doi:10.1088/1742-6596/626/1/012038

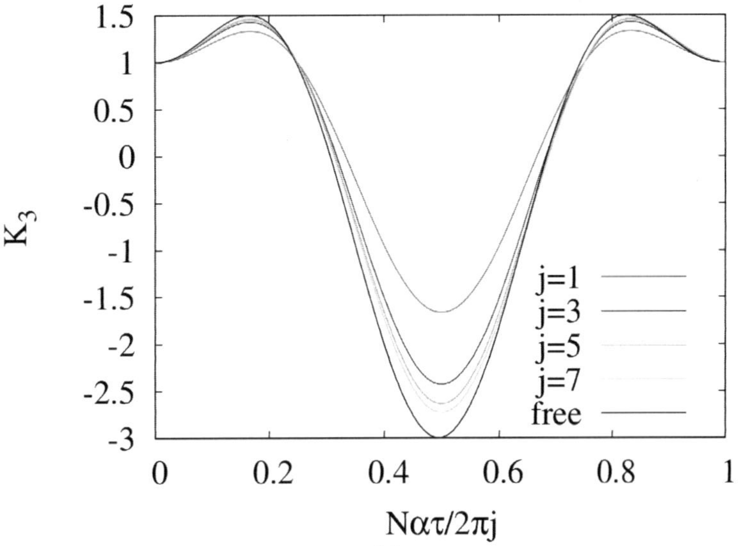

(a)

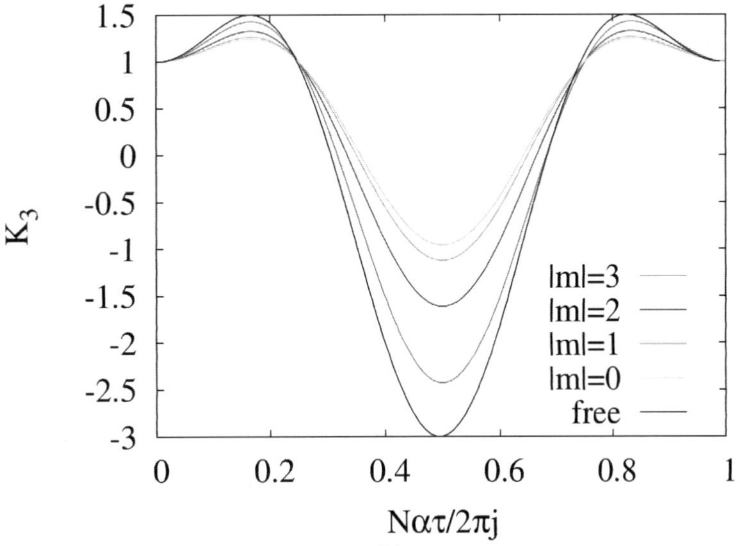

(b)

Figure 1. (Color online) K_3 function for a qubit interacting with integer j spin in the initial state $|m\rangle$ in Eq.(5): *(a)* for different values of j and $m = j$; *(b)* for different values of m for fixed $j = 3$. Time interval τ between measurements is rescaled with respect to the period T_3 in Eq.(8).

4. Conclusion

Violation of the Leggett–Garg inequalities, as in the case of Bell inequalities [3], provide an insight into very counter–intuitive properties of Nature. In our work we investigate a composite

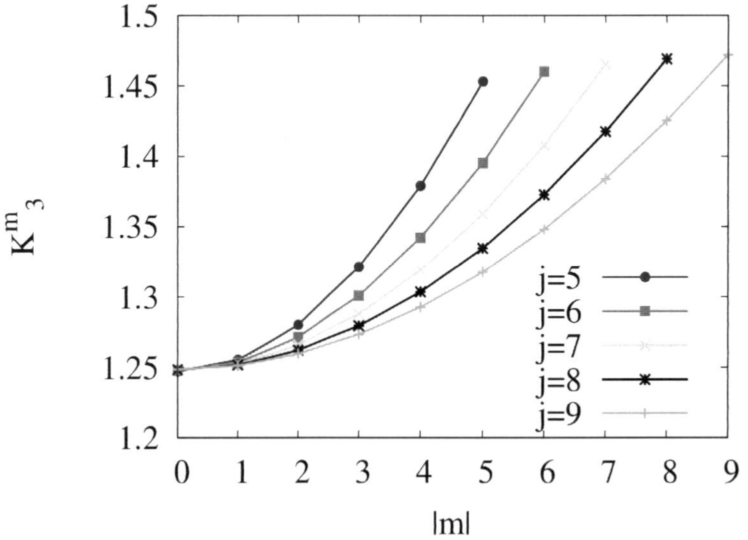

Figure 2. (Color online) Maximal value K_3^m of K_3 function of the qubit coupled to integer spin j according to Eq.(3) prepared in the state Eq.(5) versus m and for different values of j and initial state Eq.(5).

system consisting of a qubit, a basic building block for quantum information, and the integer spin, angular momentum, applicable in various models of atomic physics. We considered dichotomous measurement of the x–component of the qubit corresponding to projecting the system on eigenstates of σ_x operator tensorized with identity acting on integer spin system. Our aim was to study violation of the LGIs in the limit of large j when the composite qubit–spin system can mimic quantum–classical hybrid. For the simplest one–dimensional qubit–spin coupling we obtain analytic results for the K_3 function with a period (but not an amplitude) changing with increasing j. In general three–dimensional case, requiring numerical treatment, the amplitude of the K_3 function became j–dependent in a rather counter–intuitive fashion: maximal violation of the LGI occurring for the quantum–classical hybrid system is *larger* than in the fully quantum systems with small values of j. In particular, for $j \to \infty$ $K_3 \to 3/2$ i.e. its value approaches largest possible value.

Since systems with high spin values are naturally considered as more macroscopic we show that, counter–intuitively, microscopic qubit violates LGI more when it interacts with more macroscopic object.

Acknowledgments

The work has been supported by the NCN project DEC-2013/09/B/ST3/01659 (JD) and the Forszt project co-financed by EU from the European Social Fund (MŁ).

References

[1] Horodecki R et al. 2009 *Rev. Mod. Phys.* **81** 865
[2] Modi K et al. 2012 *Rev. Mod. Phys.* **84** 1655
[3] Brunner N et al. 2014 *Rev. Mod. Phys.* **86** 839
[4] Emary C, Lambert N, and Nori F 2014 *Rep. Prog. Phys.* **77** 016001

[5] Budroni C and Emary C 2014 *Phys. Rev. Lett.* **113** 050401
[6] Fratino L, Lampo A, and Elze H-T 2014 *Phys. Scr.* **2014** 014005
[7] Dajka J 2014 *Int. J. Theor. Phys.* **53** 870
[8] Haake F 1991 Quantum Signatures of Chaos (Wiley)
[9] Elze H-T 2012 *Phys. Rev. A* **85** 052109
[10] Lampo A, Fratino L, and Elze H-T 2014 *Phys. Rev. A* **90** 042120
[11] Johansson J R, Nation P D, and Nori F 2013 *Comp. Phys. Comm.* **184** 1234

Journal of Physics: Conference Series **626** (2015) 012039 doi:10.1088/1742-6596/626/1/012039

Environmental gravitational decoherence and a tensor noise model

Fumika Suzuki and Friedemann Queisser

Department of Physics, The University of British Columbia, Vancouver, BC V6T 1Z1, Canada

E-mail: fumika@physics.ubc.ca, queisser@phas.ubc.ca

Abstract. We discuss decrease of coherence in a massive system due to the emission of gravitational waves. In particular we investigate environmental gravitational decoherence in the context of an interference experiment. The time-evolution of the reduced density matrix is solved analytically using the path-integral formalism. Furthermore, we study the impact of a tensor noise onto the coherence properties of massive systems. We discuss that a particular choice of tensor noise shows similarities to a mechanism proposed by Diósi and Penrose.

1. Introduction

In recent years, there has been a growing interest in testing gravitational decoherence or possible gravitational effects on quantum mechanics (QM) in condensed matter and quantum-optical systems [1, 2, 3]. Decoherence can be studied in the framework of quantum mechanics and it does not require any additional assumptions. The dynamics of a system which is coupled with the environment follows from the Schrödinger equation. An observer who has only access to system degrees of freedom observes a nonunitary dynamics which can be obtained by tracing out the environmental degrees of freedom from the total density matrix. This averaging generically reduces the coherence of the reduced density matrix describing the system. Thus, a part of the phase information is distributed over the environmental variables. Whether the phase information can be restored or not depends crucially on the form of the coupling and the properties of the environment. We will address decoherence via emission of gravitational waves in section 2.

Interestingly, it has also been argued that gravity might lead to a loss of coherence which cannot be discussed in terms of standard quantum mechanics [4, 5, 6, 7, 8, 9, 10, 11]. A prominent example of this "intrinsic decoherence" was discussed in [4, 5, 6]. There, it was argued that superpositions of static configurations have finite life-times and decay on a time-scale $T \sim 1/E_g$, where E_g denotes the gravitational self-energy of the difference of the mass-distributions which are involved in the superposition. The main motivation for such an effect is the lack of a canonical time-like killing vector when superpositions of space-times are considered. In section 3, we want to address the question whether it is possible to derive such a decoherence rate from a tensor noise model. We point out several problems which come along when one introduces a tensor noise "by hand".

2. Environmental gravitational decoherence

Every physical system is coupled to gravity via its energy-momentum tensor. In the weak-field limit, the metric $g_{\mu\nu}$ can be expanded around the Minkowski background $\eta_{\mu\nu}$ according to $g_{\mu\nu} = \eta_{\mu\nu} + h_{\mu\nu}$. The resulting action is quadratic in the metric perturbations $h_{\mu\nu}$. Using the path integral formalism, the influence functional can be evaluated exactly. When the metric perturbations have a time-dependent quadrupole moment, the environmental modes can carry away phase-information about the system which will lead to a real loss decoherence [12]. In the following, we discuss the loss of coherence of a particle which does not move on a geodesic and radiates gravitational waves. The analogue effect in electrodynamics due to emission of photons (bremsstrahlung) has been studied in [13, 14].

2.1. Influence functional

In this section, we follow notations used in [15]. The total action of a mass distribution which is centered around $\mathbf{x}_0$ and coupled to gravity can be written as

$$S = S_{\rm sys}(\mathbf{x}_0) + S_{\rm grav}(h_{\mu\nu}) + S_{\rm int}(h_{\mu\nu}, \mathbf{x}_0). \tag{1}$$

The action $S_{\rm grav}(h_{\mu\nu})$ contains in general also gravitational self-interaction terms which are of higher order and will not be considered here. We choose the transverse-traceless gauge, i.e., $h_{0\mu} = h^{ij}{}_{,i} = h^i{}_i = 0$. Then the graviton action adopts the simple quadratic form

$$S_{\rm grav}(h_{\mu\nu}) = \frac{1}{4\pi G}\int d^4x h_{\mu\nu,\alpha}h^{\mu\nu,\alpha} = \frac{1}{4\pi G}\int d^4x(\partial_t h_{ij}\partial^t h^{ij} - h_{ij,k}h^{ij,k}). \tag{2}$$

The interaction between the energy-momentum tensor $T^{ij}_{\mathbf{x}_0}$ of the system and the external graviton field is bilinear, i.e.,

$$S_{\rm int}(h_{\mu\nu}, \mathbf{x}_0) = \int d^4x h_{ij} T^{ij}_{\mathbf{x}_0}. \tag{3}$$

Here only the spatial components of the energy-momentum tensor contribute due to the gauge choice above. The graviton field can be expanded into plane waves according to

$$h_{ij}(\mathbf{x}, t) = \int \frac{d^3k}{(2\pi)^3}\sum_\lambda \epsilon^\lambda_{ij}(\mathbf{k})\left(q^\lambda_s(t)\cos(\mathbf{k}\cdot\mathbf{x}) + q^\lambda_a(t)\sin(\mathbf{k}\cdot\mathbf{x})\right) \tag{4}$$

where we introduced the transverse-traceless polarization tensors $\epsilon^\lambda_{ij}(\mathbf{k})$. The time-evolution of the system's density matrix is determined by the action $S_{\rm sys}(\mathbf{x}_0)$ and the influence functional $\mathcal{F}(\mathbf{x}_0, \mathbf{x}'_0)$. Then the propagator of the reduced density matrix is given by

$$W(\mathbf{x}_{0,i}, \mathbf{x}_{0,f}; \mathbf{x}'_{0,i}, \mathbf{x}'_{0,f}) = \int_{\mathbf{x}_{0,i}}^{\mathbf{x}_{0,f}} D\mathbf{x}_0 \int_{\mathbf{x}'_{0,i}}^{\mathbf{x}'_{0,f}} D\mathbf{x}'_0 \exp(iS_{\rm sys}(\mathbf{x}_0) - iS_{\rm sys}(\mathbf{x}'_0))\mathcal{F}(\mathbf{x}_0, \mathbf{x}'_0) \tag{5}$$

where the subscripts i, f indicate the initial and final values respectively. The influence functional adopts the explicit form

$$\begin{aligned}\mathcal{F}(\mathbf{x}_0, \mathbf{x}'_0) = \exp(i\phi) &= \exp\Bigg[i\int_0^T d^4x \int_0^t d^4x'(T^{ij}_{\mathbf{x}_0}(x) + T^{ij}_{\mathbf{x}'_0}(x))\gamma_{ij,kl}(x - x')(T^{kl}_{\mathbf{x}_0}(x') - T^{kl}_{\mathbf{x}'_0}(x')) \\ &\quad - \int_0^T d^4x \int_0^t d^4x'(T^{ij}_{\mathbf{x}_0}(x) - T^{ij}_{\mathbf{x}'_0}(x))\eta_{ij,kl}(x - x')(T^{kl}_{\mathbf{x}_0}(x') - T^{kl}_{\mathbf{x}'_0}(x'))\Bigg]\end{aligned} \tag{6}$$

with the correlation functions

$$\gamma_{ij,kl}(x-x') = \frac{G}{8\pi^2}\int\frac{d^3k}{|\mathbf{k}|}\sin(|\mathbf{k}|(t-t'))e^{i\mathbf{k}\cdot(\mathbf{x}-\mathbf{x}')}\Pi_{ij,kl}(\mathbf{k}),$$
$$\eta_{ij,kl}(x-x') = \frac{G}{8\pi^2}\int\frac{d^3k}{|\mathbf{k}|}\cos(|\mathbf{k}|(t-t'))\coth\left(\frac{\beta|\mathbf{k}|}{2}\right)e^{i\mathbf{k}\cdot(\mathbf{x}-\mathbf{x}')}\Pi_{ij,kl}(\mathbf{k}). \tag{7}$$

We assumed the environmental modes to be in a thermodynamical state with the inverse temperature β. The sum over polarizations gives the polarization tensor $\sum_\lambda \epsilon^\lambda_{ij}(\mathbf{k})\epsilon^{\lambda *}_{kl}(\mathbf{k}) = \Pi_{ij,kl}(\mathbf{k}) = P_{ik}P_{jl} + P_{il}P_{jk} - P_{ij}P_{kl}$ where the projection operators are $P_{ij} = \delta_{ij} - \frac{k_ik_j}{|\mathbf{k}|^2}$.

2.2. Time-dependent behaviour of decoherence functional

The real and imaginary parts of the phase ϕ in the influence functional (6) lead to dissipation and decoherence, respectively. In this section, we calculate

$$\mathrm{Im}(\phi) = -\int_0^T d^4x\int_0^t d^4x'(T^{ij}_{\mathbf{x}_0}(x) - T^{ij}_{\mathbf{x}'_0}(x))\eta_{ij,kl}(x-x')(T^{kl}_{\mathbf{x}_0}(x') - T^{kl}_{\mathbf{x}'_0}(x')) \tag{8}$$

from which the dependence of the decoherence rate on the various parameters of the model follows. As particular example, we discuss the evaluation of (8) using an interference device setup (Figure 1). A matter distribution is in superposition of either following a right trajectory $\mathbf{x}_0$ (blue) or a left one $\mathbf{x}'_0$ (red). For a small angle θ, Figure 1 can be interpreted as the double-slit experiment, whereas $\theta = \frac{\pi}{2}$ corresponds to an interferometer. On both trajectories, the system changes its direction at time $t = t_{\text{kick}}$ due to a slit or mirror. The energy-momentum tensor of a particle in flat space-time takes the form

$$T^{\mu\nu}_{\text{part},\mathbf{x}_0}(x) = \frac{mv^\mu(t)v^\nu(t)}{\sqrt{1-|\mathbf{v}_0(t)|^2}}\delta_\sigma(\mathbf{x}-\mathbf{x}_0(t)) \tag{9}$$

with the velocity vector $v_\mu(t) = (1, \mathbf{v}_0(t))$. We assume that a particle is smeared out over a space-region determined by σ, i.e., $\delta_\sigma(\mathbf{x}) = \frac{1}{(2\pi\sigma^2)^{3/2}}\exp\left[-\frac{\mathbf{x}^2}{2\sigma^2}\right]$.

It is well-known that the divergence equation of the energy-momentum tensor gives the geodesic equation with respect to the background space-time. However, the two possible trajectories of a particle in Figure 1 do not correspond to geodesics in Minkowski space-time, hence $\partial_\nu T^{\mu\nu}_{\text{part},\mathbf{x}_0} \neq 0$. Generally, one would need to model a mirror, the interaction between a particle and a mirror and the interaction of gravity with the particle-mirror system. Here, we follow a simpler approach by adding momentum densities which kick a particle at $t = t_{\text{kick}}$. Thus, our system can be described now as "particle + momentum density such that the energy-momentum tensor is divergence-free along the trajectories". In particular, we choose

$$T^{01}_{\text{mom},\mathbf{x}_0}(x) = -m\int_{-\infty}^t ds\delta_\sigma(\mathbf{x}-\mathbf{x}_0(s))\frac{d}{ds}\left(\frac{v^1(s)}{\sqrt{1-|\mathbf{v}_0(s)|^2}}\right), \tag{10}$$

$$T^{02}_{\text{mom},\mathbf{x}_0}(x) = -m\int_{-\infty}^t ds\delta_\sigma(\mathbf{x}-\mathbf{x}_0(s))\frac{d}{ds}\left(\frac{v^2(s)}{\sqrt{1-|\mathbf{v}_0(s)|^2}}\right). \tag{11}$$

The time-time component $T^{00}_{\text{mom},\mathbf{x}_0}(x)$ can be chosen such that

$$\partial_\alpha(T^{\alpha 0}_{\text{mom},\mathbf{x}_0}(x) + T^{\alpha 0}_{\text{part},\mathbf{x}_0}(x)) = 0. \tag{12}$$

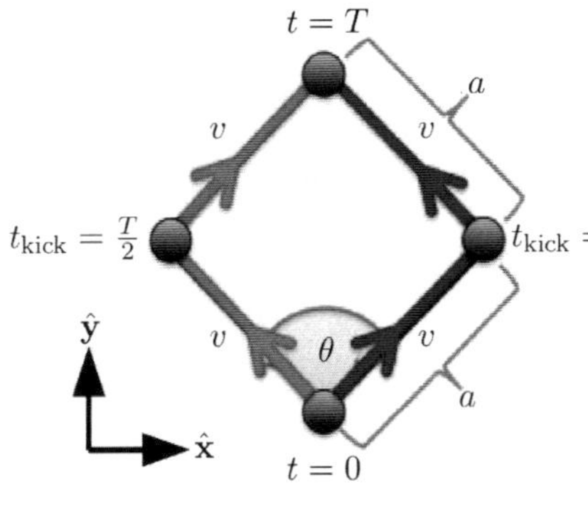

Figure 1. Typical trajectories followed by a system in a quantum superposition experiment [13, 14]. In this diagram, $t = t_{\text{kick}} = \frac{T}{2}$.

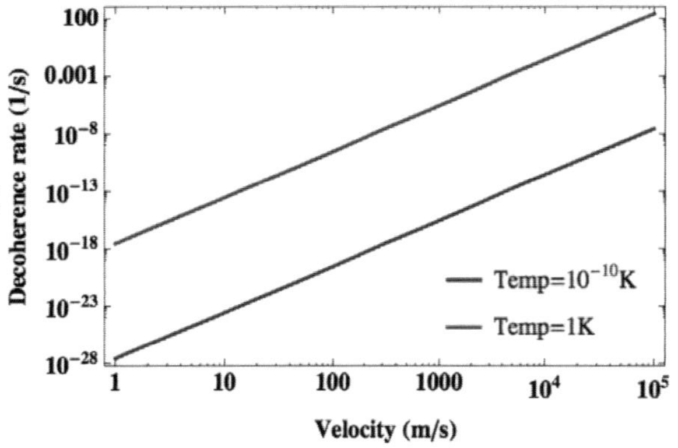

Figure 2. The dependence of decoherence rate on velocity v of an object with temperature of gravitational waves $T = 1\text{K}$ and $T = 10^{-10}\text{K}$. We chose the mass of an object $m = 10^{-5}\text{kg}$.

The additional momentum densities ensure that the divergence of the energy-momentum tensor $T^{\mu\nu}_{\mathbf{x}_0} = T^{\mu\nu}_{\text{part},\mathbf{x}_0} + T^{\mu\nu}_{\text{mom},\mathbf{x}_0}$ vanishes along the trajectory.

In order to obtain an interpretation of the momentum densities introduced, let us consider $\mathbf{v}_0(s) = v_1(s)\hat{\mathbf{x}} + v_2\hat{\mathbf{y}}$ with $v_1(s) = v\sin(\theta/2)\theta(t_{\text{kick}} - s) - v\sin(\theta/2)\theta(s - t_{\text{kick}})$ and $v_2 = v\cos(\theta/2)$. Then we have

$$-m\int_{-\infty}^{t} ds\delta_\sigma(\mathbf{x} - \mathbf{x}_0(s))\frac{d}{ds}\left(\frac{v^1}{\sqrt{1 - |\mathbf{v}_0(s)|^2}}\right) = -\frac{2mv\sin(\theta/2)}{\sqrt{1 - v^2}}\theta(t - t_0)\delta_\sigma(\mathbf{x} - \mathbf{x}_0(t_0)). \quad (13)$$

We interpret this expression as follows: When $t < t_{\text{kick}}$, a particle moves straight in positive $\hat{\mathbf{x}}$- and $\hat{\mathbf{y}}$-directions. At time t_{kick}, it gets a momentum transfer which changes its velocity in $\hat{\mathbf{x}}$-direction from $v\sin(\theta/2)$ to $-v\sin(\theta/2)$. This corresponds to the right trajectory $\mathbf{x}_0$ in Figure 1. A left trajectory $\mathbf{x}'_0$ which has a kick at t_{kick} towards the right can be constructed in the same way by defining $\mathbf{v}'_0(s) = -v_1(s)\hat{\mathbf{x}} + v_2\hat{\mathbf{y}}$.

By substituting $T^{\mu\nu}_{\mathbf{x}_0} = T^{\mu\nu}_{\text{part},\mathbf{x}_0} + T^{\mu\nu}_{\text{mom},\mathbf{x}_0}$ into (8) and evaluating the thermal part of the expression (8) along the trajectories, we find for velocities $v \ll 1$

$$\begin{aligned}\text{Im}(\phi)_{\text{therm}} = \;& \tfrac{4Gm^2v^4}{105\pi}\left[42(1 + \cos(\theta)) + (63 + 64\cos(\theta) + 5\cos(2\theta))v^2\right]\sin^2\left(\tfrac{\theta}{2}\right) \\ & \times\left(4\ln\left[\tfrac{\sinh(\pi T/(2\beta))}{\pi T/(2\beta)}\right] - \ln\left[\tfrac{\sinh(\pi T/\beta)}{\pi T\beta}\right]\right) + \mathcal{O}(v^8) \end{aligned} \quad (14)$$

where T is a time when two trajectories meet again. With an angle $0 < \theta < \pi$, the effect is approximately proportional to v^4. The authors of [16] found the same velocity-dependence in the non-relativistic single-mode limit from a master equation approach. When $\pi T/(2\beta) \gg 1$ and $\theta = \pi/2$, we have

$$\text{Im}(\phi)_{\text{therm}} = \frac{4Gm^2Tv^4}{5\beta}. \quad (15)$$

The decoherence factor $\text{Im}(\phi)$ vanishes when two trajectories are parallel $\theta = 0$, that is, the trajectories coincide with each other. With $\theta = \pi$, $\text{Im}(\phi)$ is proportional to v^6,

$$\text{Im}(\phi)_{\text{therm}} = \frac{16Gm^2Tv^6}{105\beta}. \quad (16)$$

Apart from thermal gravitational waves, vacuum fluctuations also contribute to the decoherence rate. For realistic parameters, this contribution is very small that it can be neglected. Its expression diverges logarithmically with the inverse of the spread of the mass distribution, $1/\sigma$.

In Figure 2, we plotted the decoherence rate with the temperature of the gravitational waves being 1K and 10^{-10}K. Since gravitational waves decoupled earlier from photons and have therefore a lower temperature than the Cosmic Microwave background. However, the actual temperature of the thermal gravitational waves background can be much lower and it depends on the details of reheating after inflation [17]. Here, we assume a thermal contribution with 1K as the upper bound and vacuum contribution would be the lower bound on the decoherence rate.

2.3. Gravitational radiation

We claim that the decoherence effect comes along with the emission of gravitational waves. The emitted energy of a particle along one of the trajectories depicted in Figure 1 can be calculated classically. The emission occurs when the direction of a particle is changed by a mirror. As long as the emitted radiation can be distinguished for trajectories, the environment measures which path a particle is moving to some extent, and this leads to environmental gravitational decoherence.

For a trace-reversed perturbation, $\bar{h}^{\rm pert}_{\mu\nu} = \bar{h}^{\rm pert}_{\mu\nu} - \bar{h}^{\rm pert}\eta_{\mu\nu}/2$, the solution to the linearized Einstein equations gives the expression

$$\bar{h}^{\rm pert}_{\mu\nu}(t, \mathbf{x}) = 4G \int d^3y \frac{T_{\mu\nu}(t - |\mathbf{x} - \mathbf{y}|, \mathbf{y})}{|\mathbf{x} - \mathbf{y}|} . \quad (17)$$

When the particle is slowly moving, the far field is dominated by quadrupole moment tensor $I_{ij} = \int d^3x\, x^i x^j T_{00,\mathbf{x}_0}(x)$. In terms of the trace-free quadrupole moment, $J_{ij} = I_{ij} - \delta_{ij}\delta^{kl}I_{kl}/3$, the emitted power is

$$P(t, r) = -\frac{G}{5}\frac{d^3J_{ij}(t - r)}{dt^3}\frac{d^3J^{ij}(t - r)}{dt^3} . \quad (18)$$

Here we will assume a smooth trajectory of a particle which moves along one path

$$\mathbf{v}_0(t) = v\sin(\theta/2)[f(t, 0) - 2f(t, T/2) + f(t, T)]\hat{\mathbf{x}} + v\cos(\theta/2)[f(t, 0) - f(t, T)]\hat{\mathbf{y}} \quad (19)$$

with $f(t, s) = (\tanh((t-s)/t_0)-1)/2$. The time t_0 determines how fast the direction of a particle is changed at a mirror, how fast a particle is accelerated initially from a source and decelerated at a detector finally. In the limit $T \gg t_0$, the radiated energy is crucially determined by t_0,

$$E \sim -\frac{v^4Gm^2}{t_0^5} . \quad (20)$$

For vacuum fluctuations, the minimal possible time t_0 is determined by the extension of the particle, that is $t_{0,\rm min} \sim \sigma$. For finite temperatures, the effective cutoff of the gravitational modes is determined by $1/\beta$, so one should expect $t_{0,\rm min} \sim \beta$.

3. Tensor noise model

It has been argued that the incompatibility between general relativity and quantum mechanics may lead to a form of intrinsic decoherence [4, 5, 6]. Such an effect, if it exists, is not related to any form of environmental degrees of freedom which carry away phase information from the system. The conflict between the principle of general covariance and the superposition principle of quantum mechanics is fundamental: When gravity is dynamical, the geometry of space-time must depend on the quantum state of the matter which it contains. Thus, different matter states live in different space-times which cannot be identified pointwise with each other. One might argue that the difference of three-forces in the superposed space-times give an estimate for the typical life-time of a superposition. This life-time $t = 1/\Gamma$ is given by $\sim 1/E_g$ with

$$E_g = \int d^3\mathbf{x}(\mathbf{f}(x) - \mathbf{f}'(x)) \cdot (\mathbf{f}(x) - \mathbf{f}'(x)) \\ = -4\pi G \int d\mathbf{x} d\mathbf{x}' \frac{[\rho(\mathbf{x}) - \rho'(\mathbf{x})][\rho(\mathbf{x}') - \rho'(\mathbf{x}')]}{|\mathbf{x} - \mathbf{x}'|} . \tag{21}$$

Although the relation (21) has not been derived from fundamental physics and although it is not clear yet, whether an intrinsic decoherence exists in nature at all [18], we want to address the question, whether this decoherence rate can be derived in some limit from a noise model. Scalar noise models have been studied in the literature, see, for example [19]. However, we want to couple the energy-momentum tensor to the noise which requires a tensor random current. Furthermore, we assume that the life-time (21) is the Newtonian limit of a more general expression. A possible generalization would be

$$\Gamma \sim \int dt' \int d\mathbf{x} \int d\mathbf{x}' T_{\mu\nu}(\mathbf{x}, t) D^{\mu\nu,\lambda\rho}(\mathbf{x} - \mathbf{x}', t - t') T_{\lambda\rho}(\mathbf{x}', t') \tag{22}$$

where $D_{\mu\nu,\lambda\rho}$ is the Green's function for linearized gravitational waves. Since $D_{\mu\nu,\lambda\rho}$ is not positive definite, we are immediately confronted with a problem: The expression (22) can also adopt negative values, which invalidates the interpretation of a decoherence rate. In the context of quantum mechanics, this would violate the positivity of the density matrix since the off-diagonal elements would be unbounded,

$$\langle i|\hat{\rho}(T)|j\rangle \approx \exp\left[-\int_0^T dt\Gamma(t)\right] \langle i|\hat{\rho}(0)|j\rangle, \qquad i \neq j\,. \tag{23}$$

However, it might be possible that terms such as (21) or (22) are only a part of a more general *positive* definite quantity. The simplest covariant terms which can be added to the Fierz-Pauli action take the form

$$S_j = \int d^4x j^{\mu\nu} \left(a\, h_{\mu\nu} + b\, T_{\mu\nu}\right) \tag{24}$$

with arbitrary constants a and b. Here we introduced the tensor noise $j_{\mu\nu}$ which couples to the linearized gravitational field and to the energy-momentum tensor. The first term would correspond to a stochastic contribution of the energy-momentum tensor whereas the second part resembles stochastic gravitational fluctuations. The gauge-invariance of the graviton field requires $\partial_\mu j^{\mu\nu} = 0$. Furthermore we assume the Gaussian distribution $\mathcal{P}$ of the tensor noise to be

$$\mathcal{P}[j_{\mu\nu}] = \frac{1}{N} \exp\left(-\frac{1}{2}\int d^4x j_{\mu\nu} j^{\mu\nu}\right) . \tag{25}$$

Contrary to section 2, we will use the de-Donder gauge. This includes also the instantaneous interaction between stationary matter distributions in addition to the radiation part of the gravitational perturbations. After performing the integrals over $j_{\mu\nu}$ and $h_{\mu\nu}$, additional contributions to the influence functional arise due to the presence of the noise. In the non-relativistic limit, we find for large times the expression

$$\mathcal{F} = \mathcal{F}_0 \times \exp\left[-(\Gamma_1 + \Gamma_2 + \Gamma_3)T\right] , \tag{26}$$

where $\mathcal{F}_0$ denotes the influence functional due to gravitational waves. Assuming an energy-momentum tensor of the form $T_{\mu\nu} = \rho u_\mu u_\nu$, the decoherence rates Γ_i have a simple interpretation. Γ_1 measures the difference of the Newtonian gravitational potentials in the respective spacetimes,

$$\Gamma_1 = \frac{\kappa^2 a^2}{8\pi^2} \int d^3\mathbf{x} (\Phi(\mathbf{x}) - \Phi'(\mathbf{x}))^2 . \tag{27}$$

Γ_2 is equal to the expression (21) up to a numerical factor and accounts for the difference of the Newtonian forces,

$$\Gamma_2 = -\frac{2ab}{\kappa} \int d^3\mathbf{x} (\mathbf{f}(\mathbf{x}) - \mathbf{f}'(\mathbf{x})) \cdot (\mathbf{f}(\mathbf{x}) - \mathbf{f}'(\mathbf{x})) . \tag{28}$$

Dimensional arguments might suggest that $ab \sim \kappa$. Finally, Γ_3 incorporates the difference of the mass densities,

$$\Gamma_3 = \frac{b^2}{2} \int d^3\mathbf{x} (\rho(\mathbf{x}) - \rho'(\mathbf{x}))^2 . \tag{29}$$

It is not yet clear whether this result is a gauge-independent statement. The time-evolution of the density matrix requires a definite choice of the time-parameter t. (We are not interested in scattering matrix elements where the interaction decreases adiabatically to zero for $t \to \pm\infty$ and Lorentz-invariant quantities can be defined.) A change in the time-parameter, corresponding to an transformation of the metric components, might alter the expression for the decoherence rate.

4. Conclusion

We studied the decrease of coherence due to the emission of gravitational waves and gave an estimate for the emitted gravitational energy. The dominant contribution of the decoherence rate is given by thermal gravitational waves. Due to the smallness of the gravitational interaction, this effect is negligible for elementary particles. If it would be possible to send an object with mass $m = 10^{-6}$kg and velocity $v = 10^5$m/s through an interferometer, the decoherence rate would be roughly $\Gamma = 3\text{s}^{-1}$ with temperature of gravitational waves being 1K.

Furthermore, we discussed a tensor noise model which partly resembles, in the Newtonian limit, a form of intrinsic decoherence which was discussed in the literature before.

The coherence decrease due to this noise model depends on the difference of the mass density, the gravitational potential and the Newtonian 3-force of the superposition.

Acknowledgments

The authors would like to thank G. W. Semenoff, D. Carney, D. Scott, B. L. Hu, C. Anastopoulos and W. G. Unruh for helpful discussions, and express our gratitude to P. C. E. Stamp for support. F. S. is partially supported by a UBC international tuition award, UBC faculty of science graduate award, the research foundation for opto-science and technology, NSERC Discovery grant and CFI funds for CRUCS. The part of the work was supported by the Templeton foundation (grant number JTF 36838).

References

[1] Marshall W, Simon C, Penrose R and Bouwmeester D 2003 *Phys. Rev. Lett.* **91** 130401
[2] Kleckner D et al. 2008 *New Journal of Physics.* **10** 095020
[3] Kaltenbeck R et al. 2012 *Exp Astron.* **34** 123
[4] Penrose R 1996 *General Relativity and Gravitation.* **28** 581-599
[5] Penrose R 1986, In *Quantum Gravity 2: A Second Oxford Symposium*, Oxford University Press
[6] Penrose R 1986, In *Quantum Concepts in Space and Time*, Oxford University Press
[7] Diósi L 1987 *Phys. Lett.* **120** A 377-381
[8] Károlyházy F 1966 *Nuovo Cim.* **52** 390
[9] Károlyházy F 1974, Magyar Fizikai Polyoirat **12** 24
[10] Komar A B 1969, Int. J. Theor. Phys. **2** 157
[11] Omnès R 1992, Rev. Mod. Phys. **64** 339
[12] Unruh W G 2000 *In Relativistic quantum measurement and decoherence* eds Breuer H P and Petruccione F (Springer) pp 125-140
[13] Breuer H and Petruccione F 2001 *Phys. Rev.* A **63** 032102-1
[14] Breuer H and Petruccione F 2002 *The Theory of Open Quantum Systems.* (Oxford)
[15] Anastopoulos C 1996 *Phys. Rev.* D **54** 1600
[16] Anastopoulos C and Hu B L 2013 *Class. Quantum Grav.* **30** 165007
[17] Hu B L 2014 *J. Phys.: Conf. Ser.* **504** 012021
[18] Anastopoulos C and Hu B L 2008 *Class. Quantum Grav.* **25** 154003
[19] Cloutier J and Semenoff G W 1991 *Phys. Rev.* D **44** 3218-3229

A comparison between models of gravity induced decoherence of the wavefunction

S Bera[1], S Donadi[2,3], K Lochan[4] and TP Singh[1]

[1] Tata Institute of Fundamental Research, Homi Bhabha Road, Mumbai 400 005, India
[2] Department of Physics, University of Trieste, Strada Costiera 11, 34151 Trieste, Italy
[3] Istituto Nazionale di Fisica Nucleare, Trieste Section, Via Valerio 2, 34127 Trieste, Italy
[4] IUCAA, Post Bag 4, Pune University Campus, Ganeshkhind, Pune 411 007, India

E-mail: sayantani.bera@tifr.res.in

Abstract. It has already been suggested that quantum theory needs to be reformulated or modified in order to explain the measurement process and the successive collapse of the wave-function. However, there are also models of another type which keep quantum theory intact and instead modify the classical gravity by introducing stochasticity to it. These models suggest that there is a fluctuation in the background gravitational field which eventually results in the decoherence of the wavefunction. These fluctuations limit the precision with which one can measure the properties of a spacetime geometry with a quantum probe. Two similar models along this line have been suggested by Karolyhazy (K-model) and Diósi (D-model). They are based upon apparently different spacetime bounds. The results obtained for the coherence length are also somewhat different. In this article, we show that, given certain conditions apply, the minimal spacetime bounds in these two models are equivalent. We also derive the two-point correlation for the fluctuation potential in K-model which turns out to be non-white, unlike in D-model, where the corresponding correlation is white noise in time. In our opinion, this is the origin of discrepancy in the predictions of the two models. We argue that the noise correlation cannot be determined uniquely from a given spacetime bound.

**Paper presented by Sayantani Bera at the DICE meeting 2014, Italy*

1. Introduction

Quantum theory is a linear theory governed by the Schrödinger equation that allows a superposition of two possible quantum states of an object. This feature has been observed in Young's double slit experiment and various other interference experiments. But one problem that still needs an explanation is the *"quantum measurement problem"* as we know it. What happens when we attempt to measure the properties of a "quantum" particle using a classical apparatus? Why do we get one of the possible states with a definite probability instead of observing a superposed state of the possible outcomes as predicted by quantum theory? Clearly, there is an inherent randomness in this theory, the origin of which is not quite obvious. The probability of getting a particular outcome is given by the Born probability rule. Why does the collapse occur during a measurement process of a quantum object and what is the reason of emergence of the Born probability: these two questions together are referred to as the *"quantum measurement problem"* [1, 2].

Many works on this have been going on to explain properly how the randomness came into a theory which seems to be deterministic and the superposition is "lost" when a classical object is involved. One of the first resolutions came in the form of Copenhagen interpretation [3] which says that quantum mechanics and classical theory - they have a mutually exclusive domain of validity i.e. laws of quantum mechanics do not hold for macro-objects, without giving proper explanation of what a macro-object is and where the divide lies. Later on, von Neumann [4] gave a more refined form of this interpretation. Next explanation was the decoherence of the wave function [5–7] which is the role played by the environment to destroy the coherence of the states through interaction with the system. This was supplemented by the many-worlds interpretation [8, 9] saying that observers belonging to different branches of the universe see different outcomes. These two together was an attempt to explain decoherence along with Born's rule. Though this theory is far from perfection, it is still an accepted view because it does not modify quantum theory. Bohmian mechanics is also an alternative explanation [10, 11] according to which the quantum evolution is deterministic, the randomness being attached to the choice of initial states itself. Another view is that quantum theory is not complete, it is only approximation to a more general nonlinear dynamical theory whose dynamics is stochastic. The most convincing nonlinear model is known as Continuous Spontaneous Localization (CSL) which has been able to explain Born's rule successfully by invoking a non-linear and stochastic modification to the Schrödinger equation [12]. It seems very promising apart from the fact that it does not explain the origin of such modifications.

A more fundamental reason for collapse is still being searched and various proposed models are being investigated. One possible solution seems to be the connection of gravity with the collapse process. As gravity is a universal force, it has effects on both micro and macro objects. It is proposed that macro-objects, being heavier, are affected by gravity more and that is somehow connected to the spontaneous collapse whereas in micro-objects, the effect is minimum so that they are still governed by quantum mechanics. Various gravity induced collapse models have been proposed, the central idea of which is that the intrinsic quantum fluctuations in the motion of an object induce a haziness in the structure of spacetime. This fluctuation manifests itself in the form of a universal stochastic noise that results in the decoherence of the wave function. While Karolyhazy and others introduce a family of metrics to represent such noise [13–16], Diósi brings in the concept of a spacetime averaged classical Newtonian field [17–19] and the uncertainty in the measurement of this field by a quantum probe as we shall discuss in the next section. The first one will be called as the K-model and the second one will be referred to as the D-model. Both the models start with a similar spirit but end up getting different results. In our work, we find what could be responsible for such discrepancies. It is worthwhile to note that these models can only predict localization, but they cannot explain the Born probability rule. Beyond the scope of our present work is the ramification [19] of the D-model [18], similar to CSL, which can reproduce the Born rule.

2. Wavefunction collapse and gravity

Even though the idea may not seem intuitive, there is a possibility that gravity has an important role to play in the reduction of a wave vector and thus resulting collapse. Many theories have been suggested on this as possible mechanism of the collapse none of which has yet been confirmed through experiments. In this section we shall give a brief overview of two such models: K-model (proposed by F. Karolyhazy) and D-model (proposed by L. Diósi).

2.1. K-model (Karolyhazy et. al)

Karolyhazy shows that to measure a certain length $s = cT$ in flat spacetime using a clock, there will be an inherent uncertainty involved with this measurement arising from the quantum uncertainty of the clock and is given by

$$\Delta s^3 \approx l_p^2 s \tag{1}$$

where l_p is Planck's length. To formulate a mathematical model for this, he describes the spacetime metric as fluctuating around the flat Minkowski metric. In his model, the fluctuation has been described by a family of matter-free metrics $\{g_{\mu\nu}^{\beta}\}$, each containing a perturbation $\gamma_{\mu\nu}^{\beta}$ around the Minkowskian. The uncertainty in the line segment $s = cT$ is defined in this scheme as

$$\Delta s = [\langle (s - s_\beta)^2 \rangle]^{1/2} \tag{2}$$

where s is the average length over the family of metrics and $\langle\rangle$ denotes the stochastic average. The fluctuation term $\gamma_{\mu\nu}^{\beta}$ is assumed to satisfy the wave equation and hence has a solution of the form,

$$\gamma_\beta(\mathbf{x}, t) = \frac{1}{\sqrt{l^3}} \sum_{\mathbf{k}} \left[c_\beta(\mathbf{k}) e^{i(\mathbf{k}\cdot\mathbf{x} - \omega t)} + c.c \right] \tag{3}$$

where l the length of an arbitrarily chosen large box and $\omega = ck$. The model assumes that the coefficients $c_\beta(\mathbf{k})$ vary around an average value zero while moving through the family and take values independently from each other i.e.

$$\langle c_\beta(\mathbf{k}) \rangle = 0, \qquad \langle c_\beta^2(\mathbf{k}) \rangle = 0, \qquad \langle c_\beta(\mathbf{k}) c_\beta^*(\mathbf{k}') \rangle = \delta_{\mathbf{k},\mathbf{k}'} (f(k))^2 \tag{4}$$

where $f(k)$ needs to be determined. It was shown that in order to get the K-model uncertainty (1) correctly we must have

$$f(k) = l_p^{2/3} k^{-5/6}. \tag{5}$$

Due to this metric fluctuation, when a wavefunction propagates through this spacetime, a phase will be introduced to the wavefunction with respect to the free wavefunction in the absence of any fluctuation. One can calculate the variance of relative phase between two spacetime points over the family of metrics and this gives a measure of decoherence. When this becomes of the order of π, we say that coherence is lost between the parts of the wavefunction at the two spacetime points. The corresponding spatial separation between these points is called the coherence length a_c and the time required for it is called the decoherence time τ_c. Karolyhazy and Frenkel have calculated the coherence length and decoherence time for point masses and extended objects [16].

For a proton, it turns out

$$a_c \approx 10^{25} cm, \quad \tau_c \approx 10^{53} s. \tag{6}$$

On the other hand, for a ball of radius, say, 1 cm and having density 1g/cc, the quantities calculated are

$$a_c \approx 10^{-16} cm, \quad \tau_c \approx 10^{-4} s. \tag{7}$$

2.2. D-model (L. Diósi)

Diósi model works with the weak field Newtonian limit of GR where the metric becomes

$$g_{ij} = \begin{cases} 1 + \frac{2\phi}{c^2} & i = j = 0 \\ -\delta_{ij} & i, j = 1, 2, 3 \end{cases} \tag{8}$$

ϕ representing the background potential field.

This model questions the accuracy with which a Newtonian gravitational field can be measured using a quantum probe which obeys uncertainty principle. The gravitational field which is to be measured by a quantum probe is given by

$$\mathbf{g}(\mathbf{r}, t) = -\nabla\phi(\mathbf{r}, t). \tag{9}$$

The quantum probe obeys uncertainty principle and this uncertainty in its motio back-reacts onto the gravitational field which introduces an uncertainty in the measured field. The uncertainty in the measured field $\tilde{\mathbf{g}}$ is given by $\delta\tilde{\mathbf{g}}$ and its variance follows the Diósi bound

$$(\delta\tilde{\mathbf{g}})^2 \geq G\hbar/VT \tag{10}$$

where the measured field $\tilde{\mathbf{g}}$ is the average of the gravitational field $\mathbf{g}$ over the entire volume V of the measuring probe and over the measuring time T.

The two point correlation function is then determined as

$$\langle\phi(\mathbf{r}, t)\phi(\mathbf{r}', t')\rangle = \frac{\hbar G}{|\mathbf{r} - \mathbf{r}'|}\delta(t - t'). \tag{11}$$

Given this noise correlation, one can write down the Schrödinger equation and derive the master equation for the density matrix between two configurations $|X\rangle$ and $|X'\rangle$ as

$$\langle X|\,\dot{\hat{\rho}}\,|X'\rangle = -\frac{i}{\hbar}\langle X|\,[\hat{H}, \rho(t)]\,|X'\rangle - [\tau_d(X, X')]^{-1}\langle X|\,\hat{\rho}(t)\,|X'\rangle \tag{12}$$

where $\hat{H}$ is the Hamiltonian, the decay time $\tau_d(X, X')$ between the two configurations is given by the following equation

$$[\tau_d(X, X')]^{-1} = \frac{G}{2\hbar}\int\int d^3r d^3r' \frac{[f(\mathbf{r}|X) - f(\mathbf{r}|X')][f(\mathbf{r}'|X) - f(\mathbf{r}'|X')]}{|\mathbf{r} - \mathbf{r}'|}. \tag{13}$$

The same decay time was independently proposed by Penrose [20].

Corresponding to this decay time, there is a critical length a_{crit} beyond which the damping term becomes important. Assuming a spherical mass distribution of radius R, Diósi has calculated both the decoherence time and the critical length [18]. Surprisingly, the results obtained in this model are quantitatively very different from the K-model predictions. For a proton, the critical length and decay time come out to be

$$a_{crit} \approx 10^6 cm, \quad \tau_d \approx 10^{15} s. \tag{14}$$

In the rest of this paper, we shall try to understand the origin of this discrepancy and shall investigate whether any of these models gives the estimates correctly or both are lacking in some way.

3. Comparison of the two models

In the previous section we have seen that the predictions of the K-model and D-model differ by orders of magnitude despite relying on the same underlying physics. At this point, it is worthwhile to note that even though the background physics is same for these two models, they are apparently different in the methodology. Also the spacetime bounds look completely different and seem unrelated. In this section, we will systematically investigate both the models to find at what stage the actual difference sets in. We will first start by comparing the two apparently different spacetime bounds to see if they are related. Then we will compare the two different approaches taken in the two models and lastly we will compare the nature of the noise correlations assumed in these models.

3.1. Equivalence of K-model and D-model bounds

Here we show explicitly that the spacetime averaged potential in the D-model also implies the K-model spacetime bound, provided the white noise correlation is assumed [21]. We calculate the stochastic world line length similar to s_β in K-model but now with a spacetime averaged potential:

$$s = \int_0^T \sqrt{1 + \frac{2\tilde{\phi}}{c^2}}\, c\, dt. \tag{15}$$

Hence,

$$s - s_{avrg} \simeq c \int_0^T \frac{\tilde{\phi}}{c^2}\, dt \tag{16}$$

where $s_{avrg} = cT$. Note that here $\tilde{\phi}$ still represents a c-number stochastic variable even though it has been averaged over space and time. Next we calculate $\Delta s^2 = \langle (s - s_{avrg})^2 \rangle$ as before:

$$\Delta s^2 = \frac{1}{c^2} \left\langle \int_0^T \tilde{\phi}\, dt' \int_0^T \tilde{\phi}\, dt'' \right\rangle$$

. Assuming that $\tilde{\phi}$ does not change much within the measuring time T, we can directly integrate over time

$$\Delta s^2 = \frac{T^2}{c^2} \langle \tilde{\phi}^2 \rangle$$

. Using the white noise correlation, $\langle \tilde{\phi}^2 \rangle$ is calculated which has been described in [21]. Using the result, we get the uncertainty in measuring the length $s = cT$ as

$$\Delta s^2 \approx \frac{l_p^2}{R} s. \tag{17}$$

At this stage, we make the claim that R has to be of the order of Δs to get the minimal uncertainty. Increasing R will reduce the uncertainty but R cannot go beyond Δs as that would make the imprecision to be R itself, not Δs. So, $R \sim \Delta s$ would be the optimal choice which reduces the above bound essentially to the K-bound i.e,

$$\Delta s^3 \sim l_p^2 s. \tag{18}$$

We thus see that the minimal spacetime bounds in the D-model and in the K-model, even though they appear quite different, are essentially equivalent.

Journal of Physics: Conference Series **626** (2015) 012040 doi:10.1088/1742-6596/626/1/012040

3.2. Phase variance and master equation

In Karolyhazy model, we have seen that the phase variance method was adopted to get the coherence length and time [16]. Diósi, in his model, on the other hand followed the conventional method of deriving the master equation for the density matrix. The two approaches are completely different. In this section we try to see if they produce the same result. We check whether phase variance method is a valid approach which is able to produce similar results obtained from master equation.

We start with the Diósi's choice of potential and, instead of writing the master equation, we now calculate the phase accumulated by the wave function. In Diósi's model, the potential for a given configuration X is given as follows:

$$U(X,t) = \int_{vol} \phi(\mathbf{x},t) f(\mathbf{x}|X)\, d^3x. \tag{19}$$

The phases accumulated at time t for two different configurations X and X' are,

$$\delta(X,t) = -\frac{1}{\hbar}\int_0^t \int_{vol} \phi(\mathbf{x}',t') f(\mathbf{x}'|X)\, d^3x'\, dt' \tag{20}$$

$$\delta(X',t) = -\frac{1}{\hbar}\int_0^t \int_{vol} \phi(\mathbf{x}'',t'') f(\mathbf{x}''|X')\, d^3x''\, dt'' \tag{21}$$

We now follow the phase variance approach followed by Karolyhazy and collaborators and evaluate the variance $\langle[\delta(X,t) - \delta(X',t)]^2\rangle$. When this becomes $\sim \pi^2$, we get an estimate for the decoherence time. Calculations show [21] that the decoherence time becomes,

$$\tau_d = \left[\frac{G}{\pi^2\hbar}\int\int d^3x\, d^3x' \frac{[f(\mathbf{x}|X) - f(\mathbf{x}|X')][f(\mathbf{x}'|X) - f(\mathbf{x}'|X')]}{|\mathbf{x}-\mathbf{x}'|}\right]^{-1} \tag{22}$$

which is same as the decoherence time obtained by Diósi using the master equation apart from some constant factors [22]. Thus, use of master equation or phase variance method gives similar results.

3.3. Noise two-point correlation in K-model

Having tested all possible sources of discrepancy, we now compare the noise two-point correlations of the two models. K-model confines its calculations to the Fourier space without referring to any noise correlation. In this section, we calculate the noise correlation for this model and figure out differences with the white noise prediction of D-model. Now the family γ_β is to be thought of as a stochastic potential with zero mean whose two point correlation is such that when a length $s = cT$ is measured in the presence of such a potential, it exhibits an uncertainty given by (1).

Using the Fourier expansion (3) we compute $\langle\gamma_\beta(\mathbf{x},t)\gamma_\beta(\mathbf{x}',t')\rangle$ (see [21] for the detailed calculations) and show that in the limit $l \to \infty$, this becomes,

$$\langle\gamma_\beta(\mathbf{x},t)\gamma_\beta(\mathbf{x}',t')\rangle = \frac{l_p^{4/3}}{4\pi^2 r}\Gamma(1/3)\left[\frac{1}{(r+c|\tau|)^{1/3}} + \frac{sign(r-c|\tau|)}{|r-c|\tau||^{1/3}}\right] \tag{23}$$

where we have used $r = |\mathbf{x}-\mathbf{x}'|$ and $\tau = t - t'$. This form was also reported earlier in [23]. This is evidently not white noise, and is predicted to be the feature responsible for the difference in

Journal of Physics: Conference Series **626** (2015) 012040 doi:10.1088/1742-6596/626/1/012040

the results obtained for localization length in the K-model and the D-model.

The Schrödinger equation now can be written for a mass distribution without projecting it to the position basis,

$$i\hbar\frac{\partial}{\partial t}\left|\psi\left(t\right)\right\rangle = \left[H + \frac{c^2}{2}\int d\mathbf{x}'\hat{f}\left(\mathbf{x}'\right)\gamma_\beta\left(\mathbf{x}',t\right)\right]\left|\psi\left(t\right)\right\rangle \tag{24}$$

where $\hat{f}$ represents the local mass density operator.

The corresponding non-Markovian master equation can be written using the perturbative results of [24] as follows,

$$\frac{d\rho\left(t\right)}{dt} = -\frac{i}{\hbar}\left[\hat{H},\rho\left(t\right)\right] - \left(\frac{c^2}{2\hbar}\right)^2\int d\mathbf{x}d\mathbf{x}'\int_0^t ds\left\langle\gamma_\beta(\mathbf{x},t)\gamma_\beta(\mathbf{x}',s)\right\rangle\left[\hat{f}\left(\mathbf{x}\right),\left[\hat{f}\left(\mathbf{x}';s-t\right),\rho\left(t\right)\right]\right] \tag{25}$$

where $\hat{f}(\mathbf{x};s-t)$ is the local mass density operator in the interaction picture evolved upto time $s-t$. Calculation of decoherence time from this master equation is not very straightforward.

4. Conclusion

In this paper we tried to compare the K-model and D-model by treating them in a similar approach. We argued that the spacetime uncertainty bounds in the two models are essentially equivalent to each other given that the uncertainty is minimal. We also treated the Diósi potential with the phase variance method approach adopted by Karolyhazy and showed that it generates the same results as the master equation scheme does. Up to this point, both the models do not exhibit any discrepancy. The major discrepancy shows up when we calculate the two point correlation function in the K-model. We showed that this noise correlation is non-white noise unlike in D-model where the correlation is white noise in time. Consequently, the master equation for the density matrix in K-model becomes non-Markovian. We argued that this basic difference in the choice of the noise is possibly the reason which leads to different results predicted by these models.

We argue that the noise correlation cannot be predicted uniquely from the spacetime bound and many choices are possible, each of which is likely to give different results for the decoherence time scale. White noise maybe the simplest choice, but there seems no physical reason why gravitational effects must conform to white noise. Thus it appears that additional criteria, apart from the minimal bound, are essential to precisely define a model of gravity induced decoherence with a unique choice of the noise potential. Nonetheless, it can be said that the role of gravity in decoherence is fundamentally suggested, and further investigation of this problem is highly desirable.

References

[1] Bell J S 1987 *Speakable and unspeakable in quantum mechanics* (Cambridge: Cambridge University Press) ISBN 0-521-33495-0

[2] Ghirardi G 2005 *Sneaking a Look at God's Cards: Unraveling the Mysteries of Quantum Mechanics* (Princeton University Press)

[3] Bohr N 1928 *Nature* **121** 580

[4] von Neumann J 1955 *Mathematical Foundations of Quantum Mechanics* (Princeton, NJ: Princeton University Press)

[5] Harris R A and Stodolsky L 1981 *J. Chem. Phys.* **74** 2145

[6] Zurek W H 2003 *Rev. Mod. Phys.* **75** 715

[7] Zeh H D 1970 *Found. Phys.* **1** 69

[8] Everett III H 1957 *Rev. Mod. Phys.* **29** 454–462

[9] DeWitt D S and Graham N 1973 *The many-worlds interpretation of Quantum Mechanics* (Princeton, NJ: Princeton University Press)

[10] Bohm D 1952 *Phys. Rev.* **85** 166

[11] Bub J 1997 *Interpreting the quantum world* (Cambridge University Press)

[12] Ghirardi G C, Pearle P and Rimini A 1990 *Phys. Rev. A* **42** 78–89 ISSN 1050-2947

[13] Karolyhazy F 1966 *Nuovo Cimento* **42A** 390

[14] Karolyhazy F, Frenkel A and Lukács B 1986 On the possible role of gravity in the reduction of the wave function *Quantum concepts in space and time* ed Penrose R and Isham C J (Oxford: Clarendon)

[15] Karolyhazy F 1990 *Sixty-two years of uncertainty* ed Miller A (Plenum, New York)

[16] Frenkel A 2002 *Found. Phys.* **32** 751

[17] Diósi L and Lukacs B 1987 *Annalen der Physik* **44** 488

[18] Diósi L 1987 *Physics Letters A* **120** 377 – 381 ISSN 0375-9601

[19] Diósi L 1989 *Phys. Rev. A* **40** 1165–1174

[20] Penrose R 1996 *Gen. Rel. Grav.* **28** 581–600 ISSN 0001-7701

[21] Bera S, Donadi S, Lochan K and Singh T P 2014 A comparison between models of gravity induced decoherence *Preprint* arXiv:1408.1194

[22] Bassi A, Lochan K, Satin S, Singh T P and Ulbricht H 2013 *Rev. Mod. Phys.* **85** 471

[23] Diósi L and Lukács B 1989 *Phys. Lett. A* **142** 331

[24] Adler S L and Bassi A 2007 *J. Phys. A* **40** 15083

Journal of Physics: Conference Series **626** (2015) 012041 doi:10.1088/1742-6596/626/1/012041

Chameleon fields, wave function collapse and quantum gravity

A Zanzi

Dipartimento di Fisica e Scienze della Terra, Universitá di Ferrara - Italy

E-mail: andrea.zanzi@unife.it

Abstract. Chameleon fields are quantum (usually scalar) fields, with a density-dependent mass. In a high-density environment, the mass of the chameleon is large. On the contrary, in a small-density environment (e.g. on cosmological distances), the chameleon is very light. A model where the collapse of the wave function is induced by chameleon fields is presented. During this analysis, a Chameleonic Equivalence Principle (CEP) will be formulated: in this model, quantum gravitation is equivalent to a conformal anomaly. Further research efforts are necessary to verify whether this proposal is compatible with phenomenological constraints.

1. Introduction

In this work we discuss a connection between the collapse of the wave function, quantum gravity and chameleon fields. A more detailed discussion can be found in the original proposal [1]. We start with the question: what are chameleon fields? They are quantum fields, typically scalar, with a density-dependent mass and a coupling to baryons [2,3]: in a large-density environment they are massive, while in a small-density environment (for example on cosmological distances), chameleon fields are very light. Here is one of the possible actions that can give us this result:

$$S = \int d^4x\sqrt{-g}\left\{\frac{M_{Pl}^2 R}{2} - \frac{(\partial\phi)^2}{2} - V(\phi)\right\} + \int d^4x\mathcal{L}_m(\psi_m^{(i)}, g_{\mu\nu}^{(i)}) \tag{1}$$

where the g-metric is the Einstein frame (E-frame) metric, ϕ is the chameleon field, V is a potential for the chameleon and the last term is a matter lagrangian where several different species of particles know gravity through several different metrics. So this is a multi-metric theory by construction. The crucial point about chameleon fields is to understand that their dynamical behaviour in the Einstein frame is not determined only by a potential V but it is determined by an *effective* potential that is the sum of two different contributions. On the one hand, a bare potential V, on the other hand, a matter branch written in the form $\rho B(\phi)$ where ρ is the matter density and B is a function encoding the coupling between matter and chameleon. If there is a competition between V and the matter branch, then we can construct a minimum where the scalar field can sit. Consequently, if we change the matter density, we modify the position of the minimum and also the mass of the chameleon. The ground state and its energy are related to the matter density. The harmonic approximation is valid around one single ground state but the theory is *non-linear*. These comments will be useful in the remaining part of this work.

2. The model

Let us discuss the model for the collapse of the wave function. This model can be obtained, at least partially, from heterotic-M-theory [4]. The non-perturbative string frame (S-frame) lagrangian is

$$\mathcal{L} = \mathcal{L}_{\rm SI} + \mathcal{L}_{\rm SB} \tag{2}$$

where

$$\mathcal{L}_{\rm SI} = \sqrt{-g}\left(\frac{1}{2}\xi\phi^2 R - \frac{1}{2}\epsilon g^{\mu\nu}\partial_\mu\phi\partial_\nu\phi - \frac{1}{2}g^{\mu\nu}\partial_\mu\Phi\partial_\nu\Phi - \frac{1}{4}f\phi^2\Phi^2 - \frac{\lambda_\Phi}{4!}\Phi^4\right) \tag{3}$$

is a Scale Invariant lagrangian, ϕ is the string dilaton, Φ is another scalar field representative of matter fields and

$$\mathcal{L}_{\rm SB} = \sqrt{-g}\left(a\phi^2 + b + c\frac{1}{\phi^2}\right) \tag{4}$$

is a Symmetry Breaking (SB) part, namely a stabilizing S-frame potential which breaks scale invariance. All the parameters of the lagrangian are of the order of one (no fine-tuning is present). The set-up we obtain in the S-frame is given by a stabilized dilaton with a very large vacuum energy (much larger than the Dark Energy scale).

Now we move to the E-frame and we ask: what happens to the mass of the dilaton and to the vacuum energy? It has been shown [5] that in the E-frame the dilaton is a chameleon in this model and the E-frame vacuum energy is small in the IR region including all quantum corrections and without fine-tuning. To illustrate these points, let us further analyze the model in the E-frame. Since our intention is to quantize the lagrangian, we want to regularize the theory and we will use dimensional regularization. Therefore, let us write the E-frame lagrangian in D=2d dimensions:

$$\mathcal{L}_* = \sqrt{-g_*}\left(\frac{1}{2}R_* - \frac{1}{2}g_*^{\mu\nu}\partial_\mu\sigma\partial_\nu\sigma + \mathcal{L}_{*matter}\right), \tag{5}$$

where $\mathcal{L}_{*matter}$ turns out to be

$$\mathcal{L}_{*matter} = -\frac{1}{2}g_*^{\mu\nu}D_\mu\Phi_*D_\nu\Phi_* - e^{2\frac{d-2}{d-1}\zeta\sigma}(\xi^{-1}\frac{f}{4}M_p^2\Phi_*^2 + \frac{\lambda_\Phi}{4!}\Phi_*^4) \tag{6}$$

and $D_\mu = \partial_\mu + \zeta\partial_\mu\sigma$. σ is the E-frame dilaton. As we can see from formula (6), when the number of spacetime dimensions is different from 4, a direct interaction between dilaton and matter is present and, for this reason, we can write many one-loop Feynman diagrams exploiting this interaction. An explicit calculation shows that, taking into account the various diagrams, a finite result in 4 dimensions is obtained. In other words, when we perform the limit to $D = 4$ in the loop calculation, a factor $D - 4$ cancels a pole in a gamma function and, consequently, a finite interaction vertex in $D = 4$ between dilaton and matter is obtained. This is one of the possible ways an anomaly may present itself. This conformal anomaly induced vertex can be written as

$$\mathcal{L}_{\Phi\Phi\sigma} = -\frac{1}{2}\frac{I}{M_p}\tilde{\Phi}^2\sigma, \tag{7}$$

where, I is a factor coming from the evaluation of the integrals related to the loops and $\tilde{\Phi}$ comes from the splitting of the matter field in a expectation value and a fluctuating component, $\Phi_* = v + \tilde{\Phi}$. The anomaly induced interaction vertex has the form of a matter branch $\rho\sigma$, where ρ is the matter density. This vertex can be exploited to obtain an infinite number of Feynman diagrams that are an expansion where all loops are taken into account and the final result is

Journal of Physics: Conference Series **626** (2015) 012041 doi:10.1088/1742-6596/626/1/012041

that the E-frame dilaton is a chameleon (see [5] for a detailed discussion of these points). In this model, the E-frame dilaton parametrizes the amount of scale invariance: in this room scale invariance is broken and the dilaton is heavy because the matter density is large, but, on the contrary, in a small density background scale invariance is almost restored and the dilaton is very light. In this way the cosmological constant is under control in this model including all quantum corrections and without fine-tuning of the parameters. For a detailed discussion of these results the reader is referred to [5].

Remarkably, the cosmological constant is under control only in the E-frame and, for this reason, the non-equivalence of different conformal frames at the quantum level has been pointed out in [5]. The E-frame is the physical one. One more remark is necessary. The careful reader may be worried by the absence of a ϕ^4 term in the S-frame lagrangian, because this term is expected from loop calculations. However, as already mentioned in [6], this is not a problem, because the Planck mass is running in this model. In other words, we can include a ϕ^4 term because, after the conformal transformation to the E-frame, this term will correspond to M_p^4 and the (renormalized) Planck mass is exponentially suppressed in the IR [6]. Therefore, ϕ^4 does not clash with the cosmological constant.

3. Chameleonic wave function collapse

Let us discuss the connection between chameleon fields and wave function collapse in this model. How can we give a mass to σ? The mass of the E-frame dilaton is related to an explicit breaking of scale invariance and this result can be obtained in various ways. The first possibility (let us call it PATH1) is to exploit the anomaly-induced interaction vertex, however, the matter density is an exponential function of σ. Is this run-away behaviour problematic? It seems, at first sight, that it might be difficult to obtain the chameleonic minima. Happily, this is not the case, because we have a second possibility (let us call it PATH2) to stabilize σ: we can exploit gauge fields. Indeed, σ is coupled to the F^2-term of gauge fields and performing the variation of the action we find

$$\delta S = \int d^4x\sqrt{-g_*}\{-V_{,\sigma}(\sigma)\delta\sigma - \frac{1}{16\pi}\frac{\partial\alpha_g^{-1}}{\partial\sigma}2(B^2 - E^2)\delta\sigma\} \tag{8}$$

where we introduced explicitly the electric and magnetic fields through the formula $F_{\mu\nu}F^{\mu\nu} = 2(B^2 - E^2)$. We define $\rho_\gamma = \frac{B^2-E^2}{8\pi}$. Do not confuse $\rho_\gamma \propto B^2 - E^2$ with the electromagnetic energy density (proportional to the *sum* $B^2 + E^2$). If we require $\delta S = 0$ we thus have the following term in the equation of motion for σ:

$$-V_{,\sigma}(\sigma) + \frac{1}{\alpha^2(\sigma)}\frac{\partial\alpha}{\partial\sigma}\rho_\gamma. \tag{9}$$

From this formula we infer that the chameleonic competition with the F^2-term of gauge fields is present granted that we switch off the couplings in the large σ region (this a rather natural requirement in this model because the large σ region is the weak coupling region of the string). In this way we can construct chameleonic minima, for example, inside standard matter the chameleonic minimum is guaranteed by the F^2-term of gluons.

These comments are related to the collapse of the wave function. When we perform a quantum measurement, we induce a chameleonic jump from one ground state to a different ground state and we break the superposition principle for a short time because the harmonic approximation is not valid during the jump. PATH1 is useful when a shift of matter density is part of a measurement. For example, in a electronic diffraction experiment, when electrons enter into the screen, the matter density is shifted to a larger value and, consequently, the dilaton jumps from a low-density ground state to a high-density one and the superposition principle is broken during

Journal of Physics: Conference Series **626** (2015) 012041 doi:10.1088/1742-6596/626/1/012041

the jump. PATH2 is useful to rediscuss the Stern-Gerlach (SG) experiment. Indeed, when particles enter into the magnetic field of a SG, ρ_γ is switched on and the chameleon mechanism will induce a dilatonic jump. In both cases (PATH1 and PATH2), the chameleonic jump can be summarized by a conformal anomaly, because when we modify the value of σ, we change also the matter density ($\rho \simeq e^{-4k\sigma}$, where k is a constant) and, therefore, we modify the anomaly induced interaction vertex $\rho\sigma$.

We know that during a measurement the superposition principle is broken. However, this is not enough to claim the existence of a (non-unitary) projection operator. Let us further elaborate this point. Consider a no-particles vacuum. To create particles we exploit creation operators. Once particles have been created, the number density (hence the local vacuum energy in the chameleonic effective potential) is non-vanishing and the state of minimum energy (in this room) is *not* annihilated by all the annihilation operators. Our system is similar to the Free Electron Gas Model (FEGM) granted that we compare (1) the electron in FEGM with the matter particle in the chameleon model and (2) the Fermi energy in FEGM with the non-vanishing local vacuum energy. Now the matter energy density is non-vanishing. We can create more particles, but we have to use creation operators which are different with respect to the creation operators we started with. The two sets of operators (hence the two sets of quantum fields, before and after measurement) are related to each other by a NON-UNITARY (see e.g. [7]) Bogoliubov transformation: this is the projection operator for the collapse. Happily, an alternative route gives the same result: projective representations of the conformal group [1]. In this way a superselection rule is obtained [1] when the conformal anomaly is non-negligible (i.e. when the dilaton is stabilized).

4. Chameleonic Equivalence Principle

The lagrangian of this chameleonic model satisfies a Chameleonic Equivalence Principle (CEP) pointed out in [1]. Here is the CEP:

For each pair of vacua V1 and V2 allowed by the theory there is a conformal transformation that connects them and such that the mass of matter fields m0, V1 is mapped to m0, V2. When a conformal transformation connects two vacua with a different amount of conformal symmetry, an additional term (in the form of a conformal anomaly) must be included in the field equations and this additional term is equivalent to the quantum gravitational field.

Some comments are in order. 1) This is not a postulate, it is a principle. In other words, the CEP is a property satisfied by our lagrangian. 2) The CEP is reminiscent of the Faraggi-Matone equivalence principle for quantum mechanics [8]. 3) There are a number of consequences of the CEP, for example, we can discuss (a) chameleonic matter [1], (b) quantum helioseismology [9] and (c) quantum gravity. Let us further analyze quantum gravitation.

How can we describe a complete absence of gravity? Our general-relativity-based intuition tells us that we must remove all the masses and all the energy sources (including vacuum energy). We can switch on a small *gravitational field* by adding a massive (or even massless) particle, the chameleon mechanism will slightly reduce the amount of conformal symmetry and the jump can be summarized by a *conformal anomaly*. In this way a connection between gravity and conformal anomaly is present in this model. In general relativity (GR) when we move from an inertial to a non-inertial frame, additional terms must be included and these terms describe gravity in harmony with Einstein's equivalence principle. To proceed further, let us consider the following strategy: let us replace the inertial frame of GR with a conformal ground state of the chameleonic model and let us summarize this replacement writing symbolically

Inertial frame $\Longrightarrow$ Conformal ground state.

Moreover, let us consider various replacements together (once again GR is before the arrow, while the chameleon model is after the arrow):

Non inertial frame $\Longrightarrow$ Non conformal ground state

General coordinate transformation $\Longrightarrow$ Conformal transformation
Metric and connection (i.e. the additional terms mentioned above) $\Longrightarrow$ Conformal anomaly

In this way a dictionary between GR and chameleonic quantum gravity is obtained. In the chameleonic model, conformal anomaly describes quantum gravitation in harmony with the CEP.

5. Conclusions

In this model, quantum gravitation is equivalent to a conformal anomaly and the collapse of the wave function is, in this model, a quantum gravity effect.

Acknowledgments

This article is the text of a talk given at the 7th international workshop DICE2014 in Castiglioncello (September 2014). I am grateful to the organizers for the invitation.

References

[1] Zanzi A 2014 *Preprint* quant-ph/1404.1942
[2] Khoury J and Weltman A 2004 *Phys. Rev. Lett.* **93** 171104
[3] Khoury J and Weltman A 2004 *Phys. Rev.* D **69** 044026
[4] Zanzi A 2012 *Preprint* hep-th/1210.4615
[5] Zanzi A 2010 *Phys. Rev.* D **82** 044006
[6] Zanzi A 2012 *Preprint* hep-th/1206.4463
[7] Strocchi F 1985 *Elements of quantum mechanics of infinite systems* (World Scientific Publishing)
[8] Faraggi A and Matone M 1999 *Phys. Lett.* B **450** 34-40
[9] Zanzi A and Ricci B 2014 *Preprint* hep-ph/1405.1581

Journal of Physics: Conference Series **626** (2015) 012042 doi:10.1088/1742-6596/626/1/012042

Interpreting quantum discord in quantum metrology

Davide Girolami

Department of Atomic and Laser Physics, University of Oxford, Parks Road, Oxford OX1 3PU, UK

E-mail: davegirolami@gmail.com

Abstract. Multipartite quantum systems show properties which do not admit a classical explanation. In particular, even nonentangled states can enjoy a kind of quantum correlations called quantum discord. I discuss some recent results on the role of quantum discord in metrology. Given an interferometric phase estimation protocol where the Hamiltonian is initially unknown to the experimentalist, the quantum discord of the probe state quantifies the minimum precision of the estimation. This provides a physical interpretation to a widely investigated information-theoretic quantity.

1. Introduction

Developing new technology requires to encode an increasing amount of information in a decreasing amount of space, and then to process it as fast as possible. This demand for miniaturisation and for speed makes quantum effects relevant for information manipulation, as the physical carriers of the information reach the quantum regime, e.g. atomic scales. In spite of a century of successes of the quantum mechanical predictions, characterizing the boundary between the classical and quantum worlds is still and intriguing challenge for quantum physicists. Apart from improving our ability to store, manipulate and transmit data, a better understanding of the transition between classical and quantum scenarios may convey a deeper grasp of some biological processes and more generally of complex systems. In order to achieve these objectives, we need to identify the key particulars of a quantum phenomenon. Coherence seems to be the premier evidence of quantumness, emerging from wave-like probability distributions of measurement outcomes. Yet, subtler quantum features emerge when multipartite systems are considered. Quantum entanglement is a statistical dependence between parts of composite systems which does not admit an explanation in classical Physics. It has been intensively studied, and it is usually considered the most promising resource to provide speed-up to information protocols [1]. Taken for granted the ability to prepare systems in pure states, which is a somehow idealised situation, entanglement is the only kind of quantum correlations (in fact, all but a null measure set of pure states is entangled).

However, large scale technology needs to deal with high temperature, high disorder systems, whose description is provided by mixed states. Quantum correlations in mixed states are generally of difficult characterization, as an hierarchical structure emerges, see Fig. 1. Also, it is unclear if entanglement keeps the status of premier quantum resource for mixed state quantum computing. Reaching a final answer to this question is an extremely challenging task, as well as to identify alternative quantum resources for mixed state quantum technology.

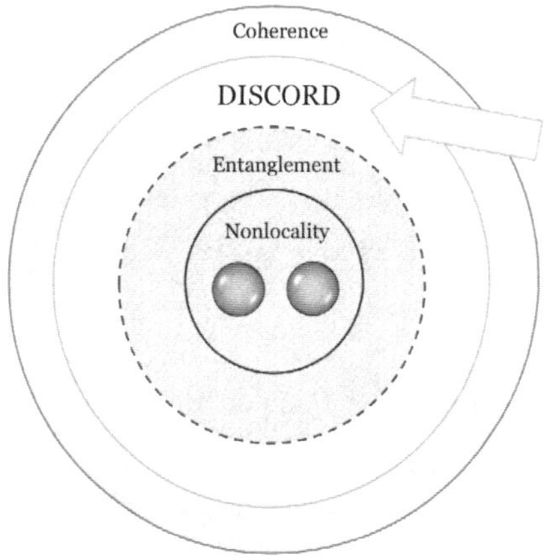

Figure 1. Hierarchy of quantum features for bipartite mixed states, here depicted as two spheres, whilst the argument can be generalised to the multipartite case. Coherence is the necessary ingredient for any quantum behaviour. While pure states are divided in entangled and classically correlated states, the correlation structure of mixed states is richer. Non-locality, i.e. violation of Bell inequalities, is a property enjoyed by a subset of entangled states, and (almost all) separable states show a nonclassical statistical dependence called quantum discord (from now on, I drop the quantum label). Here I will discuss an operational interpretation of discord in quantum metrology (steering and contextuality, other spooky features of compound quantum systems, are not here depicted for the sake of simplicity).

A more modest goal is to determine if nonentangled states still enjoy some form of quantumness. Technically, almost all separable states of multipartite systems are inherently quantum: they show a kind of quantum correlations called discord [2, 3], which is not observable when the state of the system is described by a classical probability distribution.
For pure states, discord and entanglement coincide, while for mixed states the interplay between these two quantum features has been extensively investigated (see [4] and references therein). More generally, discord has been studied on its own because of some interesting properties: it can be created by local operations and classical communications (LOCC) and it is intrinsically robust under noisy dynamics. Several operational interpretations of discord have been proposed in the recent years [4]. I here report the original definition [2, 3]. Given, for the sake of simplicity, a bipartite system AB, one makes a local measurement on one of the subsystems, say A. The state of the system is classically correlated if and only if the total correlations between A and B, measured by the mutual information, are invariant. In the quantum case, the mutual information is nonincreasing under local operations, i.e. the measurement destroys correlations. The discord of the state is the amount of correlations destroyed when the most classical, i.e. the least disturbing local measurement, is performed.
Other interesting arguments to explain the emergence of discord as an intrinsically quantum phenomenon have been proposed. In this work, I discuss a recently developed interpretation of discord in the context of quantum metrology, i.e. the discipline studying how Quantum Mechanics limits or improves the precision of measurements. In Section 2, I report two recent contributions on this line [5, 6]. First, bipartite discord is redefined as the minimum amount of genuine quantum uncertainty on a local observable. Then, I discuss a standard metrology task,

i.e. an interferometric phase estimation protocol, where building up discord in the input state guarantees nonvanishing precision for any Hamiltonian generating the phase shift, a condition not satisfied by classically correlated states. This theoretical result has been experimentally tested in a Nuclear Magnetic Resonance system [6]. Moreover, I provide a simple numerical example to clarify the role of discord in phase estimation. I draw the final remarks in Section 3.

2. Discord as quantum uncertainty on local observables

2.1. Local Quantum Uncertainty

The uncertainty principle is one of the pillars of Quantum Mechanics. It states that independently of the efficiency of our measurement apparatus, the product of the uncertainties on measurement outcomes of noncommuting observables cannot be arbitrarily small. In a recent work [5], it has been reported that whenever a system A shares discord with a system B, then even the measurement of any *single* quantity on A is intrinsically uncertain. Let us consider a bipartite system described a density matrix ρ_{AB}. One wants to measure the local observable $K^A = K_A \otimes \mathbb{I}_B$. For technical reasons explained in [5], it is important here to require the observable to be bounded and nondegenerate. The information content of ρ_{AB} about K^A is here quantified by how much the measurement of K^A on the state is uncertain. If the state is an eigenvector of the observable, the measurement outcome is certain. If it is a mixture of eigenvectors of K^A, the uncertainty is only due to imperfect knowledge of the state. Any departure from these situations implies a genuinely quantum contribution to the measurement uncertainty. Such quantum effect is provably measured by a nonnegative, convex function of the commutator $[\rho, K^A]$. The simplest function satisfying these two properties, to the best of my knowledge, is the Wigner-Yanase skew information $\mathcal{I}_{wy}(\rho, K^A) = -1/2\text{Tr}\{[\sqrt{\rho}, K^A]^2\}$ [7, 8]. While it is always possible to find a global observable K^{AB} commuting with the global state, for local observables with a fixed spectrum Λ one can show that the Local Quantum Uncertainty defined as

$$\mathcal{D}^A(\rho_{AB}) = \min_{K_\Lambda^A} \mathcal{I}_{wy}(\rho_{AB}, K_\Lambda^A) \tag{1}$$

is vanishing only for the null measure set of states taking the form $\rho_{AB}^C = \sum_i c_i |i\rangle\langle i|_A \otimes \rho_B^i, \sum_i c_i = 1$, where $\{|i\rangle\}$ is an orthonormal basis [5]. This is exactly the set of states of zero discord, also called classical-quantum states [4]. Moreover, $\mathcal{D}^A(\rho_{AB})$ satisfies all the other properties identifying reliable discord-like measures: it is invariant under local unitary transformations; it is non increasing under quantum operations on the unmeasured party B; for pure states, it is an entanglement monotone. This is somehow surprising, as discord was originally introduced in an information-theoretic context as the minimum decrease of the mutual information $\mathcal{I}(\rho_{AB})$ after a local measurement identified by the Kraus operators $\{M_i^A\}$: $\mathcal{I}(\rho_{AB}) - \max_{M_i^A} \mathcal{I}(\rho_{M_i^A(A)B})$. Conversely to entanglement, discord is in general asymmetric. The quantity $\mathcal{D}^A$ has been defined by considering a local measurement on A. For pure states, discord reduces to entanglement, thus it is symmetric, yet in general for mixed states one has $\mathcal{D}^A \neq \mathcal{D}^B$. To sum up, Quantum Mechanics predicts ineludible uncertainty even on a single observable whenever the system of interest shares discord with another system. Classical uncertainty on a measurement outcome is brought about by the state mixedness, while discord entails a different kind of uncertainty due to the disturbance induced by a local measurement. The classical-quantum states are the only states which are not affected by at least one (local) measurement.

2.2. Quantum Interferometric Power

In the previous section, I showed a novel link between two apparently unrelated quantum features as (global) correlations of states and (local) uncertainty on observables, shedding a new light on the ultimate meaning of discord: it triggers local quantum uncertainty. It is legitimate to

ask if this result is of any practical usefulness for quantum metrology. This discipline consists of exploring the limits of measurement precision dictated by Quantum Mechanics. There are in fact methods to take advantage of quantumness for improving the sensitivity of measurement devices in a number of case studies. I refer the reader to a recent review [9], mentioning here the progresses in noisy interferometry for optical and atomic setups, and a rather fascinating programme to improve gravitational wave detectors by quantum squeezing.
Many of the most important metrology tasks require to estimate the value of an unknown parameter in an interferometric configuration. This is the quantum measurement apparatus *par excellence*: the effect of a phase shift applied to one of the arms, while it is not directly measurable as no Hermitian operator is associated to the phase parameter, is estimated by (a function of) the visibility at the output. Here quantumness manifests in measurement statistics entailing quantum uncertainty and an interference fringe pattern at the output. Since discord guarantees quantum uncertainty on local observables (for example, the Hamiltonian H which generates the phase transformation), a discordant probe state always implies nonvanising precision for the measurement of the phase parameter. I provide here a formalisation of the above argument, showing that discord is a resource for phase estimation.
The estimation protocol splits into three steps: preparation of an input probe ρ, which has to be sensitive to the parameter variations; encoding of the information about the parameter in the probe, which it is assumed to be a unitary evolution, i.e. a phase shift $\rho_\theta = U_\theta \rho U_\theta^\dagger, U_\theta = e^{iH\theta}$; measurement of an appropriate observable in the output state ρ_θ, which allows to build up an (unbiased) estimator $\hat{\theta}$. The aim is to minimise the variance of the estimator $\hat{\theta}$. The fundamental limit imposed by Quantum Mechanics for ν runs of the protocol is given by the quantum Cramér-Rao bound:

$$\mathrm{Var}_\rho(\hat{\theta}) \geq 1/[\nu \mathcal{F}(\rho, H)], \tag{2}$$

where $\nu \gg 1$ is the number of repetitions of the experiment, and $\mathcal{F}$ is the quantum Fisher Information (QFI) [10, 11], which evaluates the sensitivity of the probe state to the phase shift assuming that the best measurement strategy is performed. For single parameter estimation, the best estimator statistical spread saturates (in the asymptotic limit) the quantum Cramér-Rao bound: $\mathrm{Var}_\rho(\hat{\theta}_{\mathrm{best}}) = 1/[\nu \mathcal{F}(\rho, H)]$.

Let me now suppose that the estimation is blind, in the sense that the experimenter, call it Alice, must prepare the input state ρ_A without any prior information about the Hamiltonian. I assume that the phase direction is unveiled to her at the output, thus Alice is still allowed to carry out the most informative measurement and build the best possible estimator. As the sensitivity of the probe is given by the amount of coherence with respect to the eigenbasis of the Hamiltonian, there is no input which guarantees any degree of precision in the estimation. However, the situation is different if Alice collaborates with a second player Bob and implements a two-arm interferometer to perform the estimation. While a classically correlated two-party probe ρ_{AB} is still insufficient to ensure the success of the estimation regardless of the phase direction, it is provable that implementing an input state with discord always provides a fixed amount of precision. Indeed, the minimum of the QFI over all the possible same spectrum local Hamiltonians, i.e.

$$\mathcal{P}^A(\rho_{AB}) = 1/4 \min_{H_\Lambda^A} \mathcal{F}(\rho_{AB}, H_\Lambda^A), \tag{3}$$

called Quantum Interferometric Power (QIP), is a *bona fide* measure of discord-like correlations (1/4 is just a normalisation factor). This is not surprising, as the QFI and the Wigner-Yanase information are parent quantities, being two quantum versions of the classical Fisher Information and satisfying the relation $\mathcal{I}_{wy} \leq \mathcal{F} \leq 2\mathcal{I}_{wy}$ [8]. The Wigner-Yanase information is a brute force quantisation of the classical Fisher Information definition, where a density matrix replaces a probability distribution [8]. The QFI is obtained by a further maximisation of the information retrieved by the measurement at the output of the estimation protocol [10, 11].

2.3. A numerical example
Here I provide a simple numerical example, which is a slightly modified version of the one reported in [6]. I compare the performance in a single run phase estimation protocol of two-qubit probe states which show a different kind of correlations between their parts.

The first one is a one-parameter Bell-diagonal state

$$\rho_Q = \begin{pmatrix} \frac{p+1}{4} & 0 & 0 & \frac{1}{4}p(p+1) \\ 0 & \frac{1-p}{4} & -\frac{1}{4}(p-1)p & 0 \\ 0 & -\frac{1}{4}(p-1)p & \frac{1-p}{4} & 0 \\ \frac{1}{4}p(p+1) & 0 & 0 & \frac{p+1}{4} \end{pmatrix},$$

which is prepared by initialising two single qubit states in $\rho_i = 1/2(\mathbb{I}_2 + p\sigma_z), i = A, B, p \in [0, 1]$, then applying a Hadamard gate on the first quibt followed by a two-qubit CNOT gate [12].

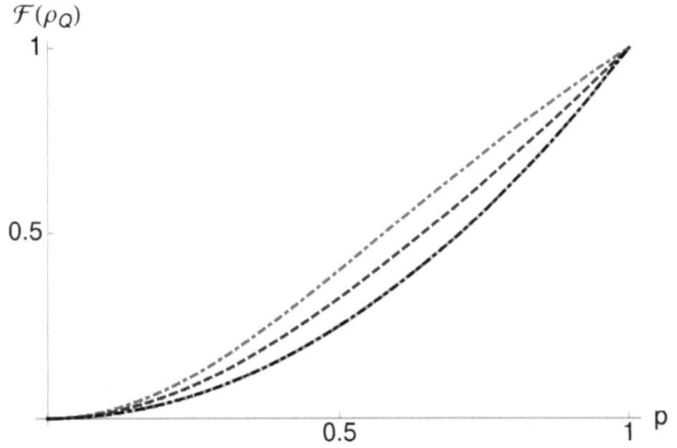

Figure 2. (Colors online) Quantum Fisher Information of the quantum correlated state ρ_Q with respect to four local Hamiltonians $H^A = H_A \otimes \mathbb{I}_B, H^A = \sigma_x^A, y, z, (\sigma_x^A + \sigma_y^A)/\sqrt{2}$. The parameter p determines the purity of the state. I depict the functions $\mathcal{F}(\rho_Q, \sigma_x^A) = p^2$ (brown, continuous line), $\mathcal{F}(\rho_Q, \sigma_y^A) = 2p^2/(1+p^2)$ (red dot-dashed line), $\mathcal{F}(\rho_Q, \sigma_z^A) = p^2$ (green dotted line), $\mathcal{F}(\rho_Q, (\sigma_x^A + \sigma_y^A)/\sqrt{2}) = p^2(p^2+3)/(2(p^2+1))$ (blue dashed line), and the QIP given by the formula in [6] (black dot-dashed line). Note that the QFI is lower bounded by the discord of the state, which therefore ensures a minimum precision to the estimation.

The second probe is a classically correlated state:

$$\rho_C = \begin{pmatrix} \frac{1}{4} & \frac{p^2}{4} & \frac{p}{4} & \frac{p}{4} \\ \frac{p^2}{4} & \frac{1}{4} & \frac{p}{4} & \frac{p}{4} \\ \frac{p}{4} & \frac{p}{4} & \frac{1}{4} & \frac{p^2}{4} \\ \frac{p}{4} & \frac{p}{4} & \frac{p^2}{4} & \frac{1}{4} \end{pmatrix}. \quad (4)$$

Note that the comparison is fair as the probes have the same purity: $\mathrm{Tr}\{\rho_{C,Q}^2\} = 1/4(1+p^2)^2$. I calculate the QFI for the two states with respect to four local observables $H_A \otimes \mathbb{I}_B, H_A =$

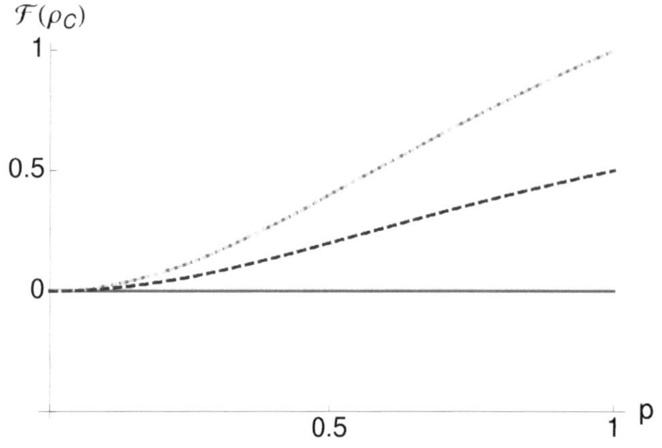

Figure 3. (Colors online) Quantum Fisher Information of the classically correlated state ρ_C with respect to four local Hamiltonians $H^A = H_A \otimes \mathbb{I}_B, H^A = \sigma_x^A, y, z, (\sigma_x^A + \sigma_y^A)/\sqrt{2}$. The parameter p determines the purity of the state. I depict the functions $\mathcal{F}(\rho_C, \sigma_x^A) = 0$ (brown, continuous line), $\mathcal{F}(\rho_C, \sigma_y^A) = 2p^2/(1+p^2)$ (red dot-dashed line), $\mathcal{F}(\rho_C, \sigma_z^A) = 2p^2/(1+p^2)$ (green dotted line), $\mathcal{F}(\rho_C, (\sigma_x^A + \sigma_y^A)/\sqrt{2}) = p^2/(1+p^2)$ (blue dashed line). Note that the state is unable to store information about the phase shift along the x-direction.

$\sigma_{x,y,z}, (\sigma_x + \sigma_y)/\sqrt{2}$. This is given by $\mathcal{F}(\rho_{Q,C}, H^A) = 4\sum_{p_i+p_j\neq 0} \frac{(p_i-p_j)^2}{p_i+p_j} |\langle\psi_i|H^A|\psi_j\rangle|^2$, where $\{p_i\}, \{|\psi_i\rangle\}$ are the eigenvalues and eigenvectors of $\rho_{Q,C}$. The QIP for the quantum correlated state is obtained by applying the closed formula derived in the appendix of [6], while of course one has $\mathcal{P}^A(\rho_C) = 0$. A study of how the QFI and the QIP change with variations of the noise parameter p is presented in Figs. 2,3. The results highlight the peculiarity of the discordant state: it guarantees a nonvanishing precision in the estimation for any phase direction.

3. Conclusions

I here discussed recent advances on the interpretation of discord in metrology. The peculiar asymmetry of this kind of quantum correlations between parts of a compound system is explained by means of an interferometric phase estimation scheme. Whenever a system shares discord, Quantum Mechanics predicts that any local measurement has a degree of uncertainty which translates into an improved sensitivity in parameter estimation. The Local Quantum Uncertainty and the Quantum Interferometric Power are parent discord-like measures which quantify the minimum amount of precision in interferometric phase estimation. An interesting question is to establish if the metrologic measures of discord, which have been introduced to catch bipartite statistical dependence, can be extended to quantify multipartite correlations. We want to find out when and how physical systems undergo genuinely quantum processes, and when such effects may boost the performance of information processing protocols. The discussed result may lead to new ways of exploiting quantum physics for delivering entanglement-free quantum technology and provide conceptual advances for the understanding of complex systems.

Acknowledgments

The work was supported by the UK Engineering and Physical Sciences Research Council (EPSRC) under the Grant No. EP/L01405X/1, the Foundational Questions Institute (FQXi) and the Wolfson College, University of Oxford. I thank the organisers of the Conference *DICE*

2014 – Spacetime - Matter - Quantum Mechanics held in Castiglioncello for the kind hospitality.

References

[1] Horodecki R, Horodecki P, Horodecki M and Horodecki K 2009 *Rev. Mod. Phys.* **81** 865
[2] Ollivier H and Zurek W H 2001 *Phys. Rev. Lett.* **88** 017901
[3] Henderson L and Vedral V 2001 *J. Phys.* A **34** 6899
[4] Modi K, Brodutch A, Cable H, Paterek T and Vedral V 2012 *Rev. Mod. Phys.* **84** 1655
[5] Girolami D, Tufarelli T and Adesso G 2013 *Phys. Rev. Lett.* **110** 240402
[6] Girolami D, Souza A M, Giovannetti V, Tufarelli T, Filgueiras J G, Sarthour R S, Soares-Pinto D O, Oliveira I S and Adesso G 2013 *Phys. Rev. Lett.* **112** 210401
[7] Wigner E P and Yanase M M 1963 *Proc. Natl. Acad. Sci. U.S.A.* **49** 910
[8] Luo S. 2003 *Phys. Rev. Lett.* **91** 180403
[9] Giovannetti V, Lloyd S and Maccone L 2011 *Nature Photon.* **5** 222
[10] Helstrom C W 1976 *Quantum Detection and Estimation Theory* (New York: Academic Press)
[11] Holevo A S 1982 *Probabilistic and Statistical Aspect of Quantum Theory* (Amsterdam: North-Holland)
[12] Nielsen M A and Chuang I L 2000 *Quantum Computation and Quantum Information* (Cambridge: University Press)

Journal of Physics: Conference Series **626** (2015) 012043 doi:10.1088/1742-6596/626/1/012043

Geometric phase and quantum correlations for a bipartite two-level system

Fernando C Lombardo and Paula I Villar

Departamento de Física *Juan José Giambiagi*, FCEyN UBA and IFIBA CONICET-UBA, Facultad de Ciencias Exactas y Naturales, Ciudad Universitaria, Pabellón I, 1428 Buenos Aires, Argentina

Abstract. We calculate the geometric phase of a bipartite two-level system coupled to an external environment. We compute the correction to the unitary geometric phase through a kinematic approach. To this end, we analyse the reduced density matrix of the bipartite system after tracing over the environmental degrees of freedom, for arbitrary initial states of the composite system. In all cases considered, the correction to the unitary phase has a similar structure as a function of the degree of the entanglement of the initial state. In the case of a maximally entangled state (MES), the survival phase is only the topological phase, and there is no correction induced by the environments. Further, we compute the quantum discord and concurrence of the bipartite state and analyse possible relations among these quantities and the geometric phase acquired during the non-unitary system's evolution.

1. Introduction

For a bipartite quantum system, it is important to know whether the system is entangled, separable, classically correlated or quantum correlated. It is well known that entanglement makes possible tasks in quantum information that could not be possible for its classical counterpart. As a valuable resource in quantum information processing, entanglement attracts much attention from researchers either in theory or in experiment and much progress concerning entanglement has been achieved [1]. Due to its unique property in the sense that it has no classical counterpart, entanglement has been applied to the implementation of quantum teleportation [2] and quantum cryptography [3].

A bipartite quantum state contains both classical $\mathcal{C}(\rho^{\rm AB})$ and quantum correlations $\mathcal{Q}(\rho^{\rm AB})$. These correlations are justified jointly by their "quantum mutual information" $\mathcal{I}(\rho^{\rm AB})$, since it is written as the sum of both $\mathcal{I}(\rho^{\rm AB}) = \mathcal{C}(\rho^{\rm AB}) + \mathcal{Q}(\rho^{\rm AB})$. This quantum part is known as quantum discord [4]. Even for the simplest case of two entangled qubits, the relation between quantum discord, entanglement and classical correlation is not yet clear. For pure states quantum correlation is exactly equal to entanglement, whereas classical correlation attains its maximal value 1. However, for a general two-qubit mixed state, the situation is more complicated. Qubit-qubit entanglement has been characterised completely, while quantum discord has only been quantified for particular cases [5].

From another point of view, a system can storage the information of its motion when it undergoes a cyclic evolution, in the form of a geometric phase (GP), which was first put forward by Pancharatnam in optics [6] and later studied explicitly by Berry in a general quantal system [7]. Since then, great progress has been achieved in this field. The geometric phase has been extended to the case of non-adiabatic evolutions [8]. As an important evolvement, the application of the geometric phase has been proposed in many fields, such as the geometric quantum computation. Due to its global properties, the geometric phase is propitious to construct fault tolerant quantum gates. In this line of work, many physical systems have been investigated to realise geometric quantum computation, such as NMR (Nuclear Magnetic Resonance) [9], Josephson junction [10], Ion trap [11] and semiconductor quantum dots [12]. The quantum computation scheme for the geometric phase has been proposed based on the Abelian or non-Abelian geometric phase, in which geometric phase has been shown to be intrinsic against faults in the presence of some kind of external noise due to the geometric nature of Berry phase. It was therefore seen that the interactions play an important role for the realisation of some specific operations. Consequently, study of the geometric phase was soon extended to open quantum systems. Following this idea, many authors have analysed the correction to the geometric phase under the influence of an external environment using different approaches [13, 14, 15, 16, 17, 18, 19, 20].

In this context, we shall study the geometric phase acquired by a bipartite system in the presence of an external environment. We shall consider both the presence of a bosonic and spin environment. We shall also study the relation between classical and quantum correlations when the geometric phase is affected by the environment.

2. Model

We shall consider a bipartite system, that is to say, two interacting two-level systems, both coupled to an external reservoir. In terms of the Hamiltonians, the model can be mathematically described by the Hamiltonian of the free bipartite system $H_{\rm S}$ is

$$H_{\rm S} = \frac{\hbar\Omega_1}{2}\sigma_z^1 + \frac{\hbar\Omega_2}{2}\sigma_z^2 + \gamma\ \sigma_z^1 \otimes \sigma_z^2, \tag{1}$$

and the Hamiltonian of interaction between the bipartite and the external bath $H_{\rm I}$

$$H_{\rm I} = \sigma_z^1 \otimes \sum_{n=1}^{N} \lambda_n q_n + \sigma_z^2 \otimes \sum_{n=1}^{N} g_n q_n \tag{2}$$

where the constants λ_n and g_n couple the system to each oscillator in the environment, and γ is the coupling strength between both spin-1/2 particles. Here we have assumed that the coupling constant of the two-level systems with the environment is different being λ_n for the spin 1 and g_n for spin 2. The external bath $H_{\rm B}$ can be either considered as set of delocalised bosonic field modes ($H_{\rm B} = \sum_{n=1}^{N} \hbar\omega_n a_n^\dagger a_n$) or spin environments which are typically the appropriate model in the low temperature regime. In the case of having a spin environment, the interaction Hamiltonian is given by

$$H_{\rm I} = \sigma_z^1 \otimes \sum_{i=1}^{N} \epsilon_i \sigma_{zi} + \sigma_z^2 \otimes \sum_{i=1}^{N} \lambda_i \sigma_{zi}. \tag{3}$$

In the most general case, for an initial state of the system $|\Phi(0)\rangle = \alpha|00\rangle + \beta|01\rangle + \zeta|10\rangle + \delta|11\rangle$, the reduced density matrix for this model can be written as,

$$\rho_{\rm r}(t) = \begin{pmatrix} |\alpha|^2 & \alpha\beta^* e^{-i(2\gamma+\Omega_2)t}\Gamma_{12} & \alpha\zeta^* e^{-i(2\gamma+\Omega_1)t}\Gamma_{13} & \alpha\delta^* e^{-i(\Omega_1+\Omega_2)t}\Gamma_{14} \\ \beta\alpha^* e^{i(2\gamma+\Omega_2)t}\Gamma_{21}^* & |\beta|^2 & \beta\zeta^* e^{-i(\Omega_1-\Omega_2)t}\Gamma_{23} & \beta\delta^* e^{-i(\Omega_1-2\gamma)t}\Gamma_{24} \\ \zeta\alpha^* e^{i(2\gamma+\Omega_1)t}\Gamma_{31}^* & \zeta\beta^* e^{i(\Omega_1-\Omega_2)t}\Gamma_{32}^* & |\zeta|^2 & \zeta\delta^* e^{-i(\Omega_2-2\gamma)t}\Gamma_{34} \\ \delta\alpha^* e^{i(\Omega_1+\Omega_2)t}\Gamma_{41}^* & \delta\beta^* e^{i(\Omega_1-2\gamma)t}\Gamma_{42}^* & \delta\zeta^* e^{i(\Omega_2-2\gamma)t}\Gamma_{43}^* & |\delta|^2 \end{pmatrix},$$

The effect of the environment is encoded in the $\Gamma_{ij}(t)$ functions. They are deduced for bosonic and fermionic environments in [21]. In the case of having an environment composed by an infinite set of harmonic oscillators at temperature $T = 0$, the coefficients Γ_{ij} are given by

$$\Gamma(t) = e^{-2\gamma_0 \log(1+\Lambda^2 t^2)}, \tag{4}$$

where it is supposed that all the coupling strengths are the same, therefore we have omitted subindex in the expression for the decoherence factors (detailed calculations can be found in Ref. [21]). In this case, γ_0 is a dissipation constant (related to the coupling constant between the system and the environment) and Λ is the frequency cutoff of the bath. In the case of a spin-environment, the decoherence factors can be written as

$$\Gamma(t) = \prod_{i=1}^{N} \left\{ 1 - \left(\frac{2(\varepsilon_i \pm \lambda_i)^2}{h_i^2 + (\varepsilon_i \pm \lambda_i)^2} \right) \sin^2(t\sqrt{h_i^2 + (\varepsilon_i \pm \lambda_i)^2}) \right\}, \tag{5}$$

where signs ($\pm$) in the parenthesis, are related with the type of initial state selected. These cases will be defined in the next Section.

3. Geometric phase

In order to compute the geometric phase for the open system, we shall use a kinematic approach proposed by [13], defined as

$$\phi_G = \arg\left\{ \sum_k \sqrt{\varepsilon_k(0)\varepsilon_k(\tau)} \langle \Psi_k(0)|\Psi_k(\tau)\rangle \times e^{-\int_0^\tau dt \langle \Psi_k | \frac{\partial}{\partial t} | \Psi_k \rangle} \right\}, \tag{6}$$

where $\varepsilon_k(t)$ are the eigenvalues and $|\Psi_k\rangle$ the eigenstates of the reduced density matrix ρ_r (obtained after tracing over the reservoir degrees of freedom). In the last definition, τ denotes a time after the total system completes a cyclic evolution when it is isolated from the environment. Taking into account the effect of the environment, the system no longer undergoes a cyclic evolution. However, we shall consider a quasi cyclic path $\mathcal{P} : t \in [0, \tau]$, with $\tau = 2\pi/\Omega$ (Ω is the system's characteristic frequency). When the system is open, the original GP, i.e. the one that would have been obtained if the system had been closed ϕ_G^U, is modified. This means, in a general case, the phase is $\phi_G = \phi_G^U + \delta\phi_G$, where $\delta\phi_G$ depends on the kind of environment coupled to the main system [14, 15, 16].

In this manuscript we shall consider different initial states to compute the correction to the unitary geometric phase when these states evolve in a no-unitary way.

- *Case 1.* We consider the initial state $\rho = a|\phi^+ >< \phi^+| + (1-a)|1,1 >< 1,1|$ ($0 < a \leq 1$), where $|\phi^+ >= (|0,0> + |1,1>)/\sqrt{2}$ is a maximally entangled state. In this case, we can compute the geometric phase by extracting the eigenvalues from the reduced density matrix. In order to know how the geometric phase of this family of states is affected by the presence of an environment, we shall present the rate between the open geometric phase and the unitary geometric phase in a density plot in Fig.1, where the vertical axis corresponds to the coupling constant γ_0 for the oscillators bath, and ϵ (or λ) in the case of spin environment. The horizontal is for the parameter a which set the degree of entanglement. Therein we consider both cases: (a) a bosonic environment and (b) a fermonic environment.

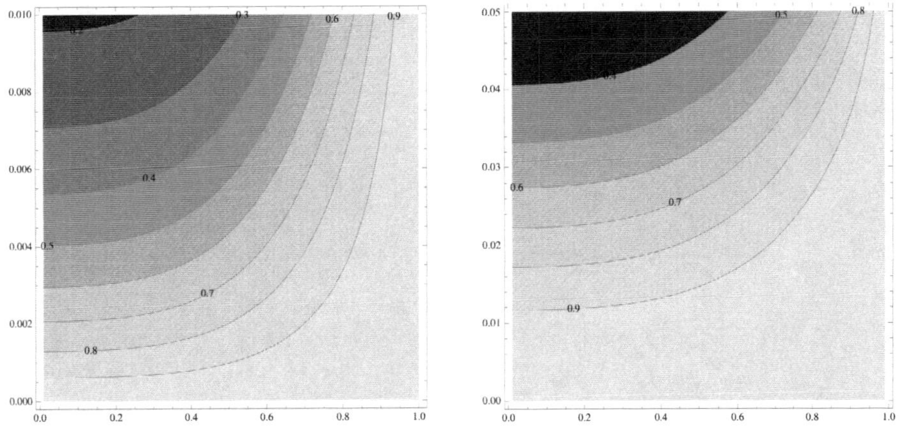

Fig1.(a) ϕ/ϕ_u for the bosonic environment. There is no correction to the GP for the MES.

Fig1.(b) ϕ/ϕ_u in the case of a spins environment.There is no correction to the GP for the MES.

In this case, the parameters used are: $\Lambda = 100$ for a set of harmonic oscillators at zero temperature while we used and $h = 1$ for an environment composed of $N = 100$ spins. In all cases, we considered that the interaction between each spin and the environment was equivalent. In both plots of Fig.1 we can see that the correction to the geometric phase $\phi/\phi_u \approx 1$ (which means $\delta\phi_G \approx 0$) when the parameter a $\rightarrow 1$, which corresponds to a state approaching a maximally entangled state (MES). In this case, the survival phase is only the topological phase, and there is no correction induced by the environments. These results has also been showed in [21] with another set of states, which reveals that the robustness of the topological phase under the influence of decoherence is a more general result, not restricted to some states or kind of environmental influence. The topological phase, which is indeed a consequence of the geometry of the entangled two-level system, has been studied for MES and it is at the origin of singularities appearing in the phase of MES during a cyclic evolution. In Ref. [22], it is studied the phase dynamics of entangled qubits under unitary cyclic evolutions. Therein, it is shown that, after a cyclic evolution, the combination of the different phases always leads to a global phase of an entire multiple of π. This result, already known and verified experimentally for a single qubit is recovered here for an entangled qubit with maximal degree of entanglement in the presence of an environment. The total phase gained by a state in a closed evolution is a combination of not only the dynamical and geometrical phase but also the topological phase. Similarly to one qubit states, MES also gain a total phase of π (or $n\pi$) under a cyclic evolution. However, this phase is of topological origin. In the case of a MES there is no correction to the unitary phase. For other states, the correction to the phase increases with the smaller values of parameter a.

- *Case 2.* We take the class of states defined as $\rho = a|\psi^+ >< \psi^+| + (1-a)|1,1 >< 1,1|$ (0<a ≤ 1), where $|\psi^+ >= (|0,1> + |1,0>)/\sqrt{2}$ is a maximally entangled state.
  In Fig.2. we present the results for this case, where decoherence does not affect the system. Therefore, the geometric phase is the unitary geometric phase for all set of states given for any value of a. These states are fully robust under the influence of external conditions such the ones presented in these examples. In all cases, we considered that the interaction between each spin and the environment was equivalent. We have considered the same

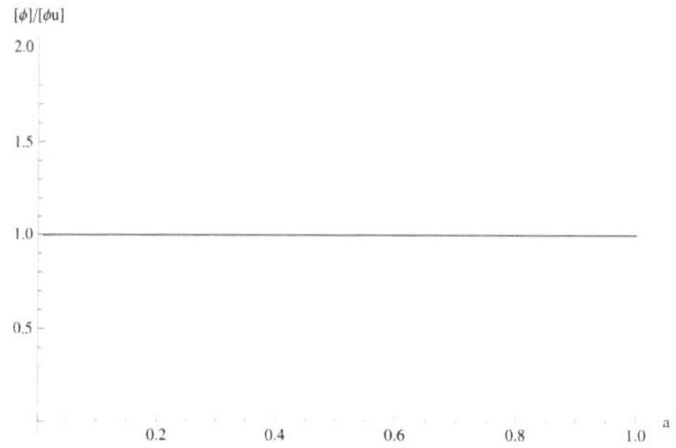

Fig2. ϕ/ϕ_u for this type of state under either a bosonic or fermonic environment. There is no correcction to the GP for these states.

parameters as in Case 1.

4. Correlations

In this section we can analyse the quantum and classical correlations that exist for both family of states considered in Case 1 and Case 2 above, for an isolated system (the effect of the environment on these quantities will be presented elsewhere). A measure of the classical relations can be written as $\mathcal{C}(\rho) := \sup_{B_k}\mathcal{I}(\rho|B_k)$ where B_k means all the possible von Neumann measurements of the qubit B. As the quantum discord is defined as $\mathcal{Q}(\rho) := \mathcal{I}(\rho)-\mathcal{C}(\rho)$, the obstacle to computing this quantity lies in this complicated maximisation procedure. For a general two-qubit X state, the quantification of quantum discord is still missing with some particular results available. Herein, we shall follow the method developed in [5] to evaluate the classical correlation and quantum discord of the two-qubit X states considered above. Classical and quantum correlations of the bipartite quantum system are quantified jointly by the mutual information. If ρ^{AB} is the density matrix (operator) of the bipartite system AB, ρ^A (ρ^B) is the density matrix associated to the part A (B). Thus, one can define the quantum mutual information as

$$\mathcal{I}(\rho^{AB}) = S(\rho^A) + S(\rho^B) - S(\rho^{AB}), \tag{7}$$

where $S(\rho^A) = -\mathrm{tr}(\rho \log_2 \rho)$ is the von Neumann entropy.

In Fig3. we present the classical correlation ($\mathcal{C}$ dotted black line), the quantum discord ($\mathcal{Q}$ dashed red line), and the concurrence (C in blue solid line), for each family of states considered above. In these examples, the quantum correlation is measured by the concurrence (see for instance Ref. [21]).

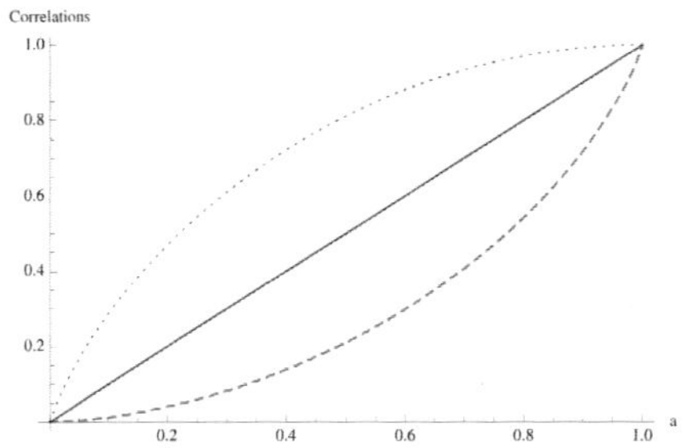

Fig.3-a) Classical correlations are bigger when we consider initial state is $\rho = a|\phi^+ >< \phi^+| + (1-a)|1,1 >< 1,1|$.

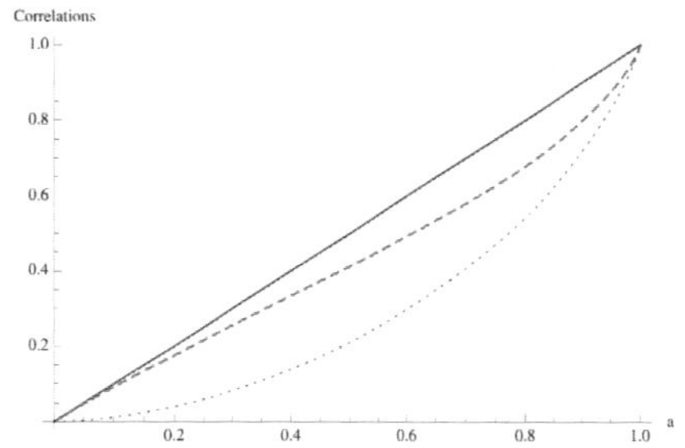

Fig.3-b) Quantum correlations are an upper bound when the initial state is $\rho = a|\psi^+ >< \psi^+| + (1-a)|1,1 >< 1,1|$.

From the plots, we can see that in *Case 1* there is a given hierarchy in the correlations. Classical correlations are always much bigger than the quantum discord, which is also smaller than the concurrence of these states. In the case of a MES, all measure of correlations coincide in the same value. On the other side, it is possible to note that the quantum correlation (concurrence in our example) results in an upper bound for all type of correlations, classical or even the quantum discord, in the *Case 2* example. In this case we have shown that there is not decoherence effects coincidentally with the fact that the quantum correlation remains bigger than the others.

5. Conclusion

As can be noticed, the environment has a stronger influence on the geometric phase when the state is less correlated, say small values of a. In those cases, the family of states proposed in *Case 1*, have a considerable correction to the unitary geometric phase. However, this correction decreases as the states become more correlated. For the family states of *Case 2*, we see that decoherence does not affect the geometric phase at all. In that case, we can see that the quantum correlations are very important being always bigger than classical correlations. Maybe a further insight into the quantum correlations of the quantum states can set light on the way the geometric phase is corrected for a bipartite quantum state under no-unitary evolution.

Acknowledgments

This work was supported by UBA, CONICET and ANPCyT - Argentina. We would like to thank H.-Thomas Elze for the organization of DICE2014. FCL acknowledges B. Lok Hu, Nick Mavromatos, and M. Blasone for the very interesting conversations about topology, decoherence and geometric phases.

References

[1] Nielsen M A and Chuang I L 2000 *Quantum Computation and Quantum Information* (Cambridge: Cambridge Univ. Press)

[2] Bennett C H *et. al* 1993 Teleporting an unknown quantum state via dual classical and Einstein-Podolsky-Rosen channels *Phys. Rev. Lett.* **70** 1895

[3] Gisin N *et. al* 2002 Quantum cryptography *Rev. Mod. Phys.* **74** 145

[4] Ollivier H and Zurek W H 2001 Quantum Discord: A Measure of the Quantumness of Correlations *Phys. Rev. Lett.* **88** 017901

[5] Ali M *et. al.* 2010 Quantum discord for two-qubit X states *Phys. Rev.* A **81** 042105

[6] Pancharatnam S 1956 Generalized Theory of Interference, and its Applications. Part I. Coherent Pencils *Proc. Indian Acad. Sci.* A **44** 247

[7] Berry M Y 1984 Quantal Phase Factors Accompanying Adiabatic Changes *Proc. R. Soc. Lond.* A **392** 45

[8] Aharonov Y and Anandan J 1988 Phase change during a cyclic quantum evolution *Phys. Rev. Lett.* **58** 1593; Anandan J and Aharonov Y 1988 Geometric quantum phase and angles *Phys. Rev.* D **38** 1863

[9] Jones J A, Vedral V, Ekert A and Castagnoli G 2000 Geometric quantum computation using nuclear magnetic resonance *Nature* **403** 869

[10] Faoro L, Siewert J and Fazio R 2003 Non-Abelian Holonomies, Charge Pumping, and Quantum Computation with Josephson Junctions *Phys. Rev. Lett.* **90** 028301

[11] Duan L M *et. al.* 2001 Geometric Manipulation of Trapped Ions for Quantum Computation *Science* **292** 1695

[12] Solinas P, Zanardi P, Zangh N and Rossi F 2003 Semiconductor-based geometrical quantum gates *Phys. Rev.* B **67** 121307

[13] Tong D M, Sjoqvist E, Kwek L C and Oh C H 2004 Kinematic Approach to the Mixed State Geometric Phase in Nonunitary Evolution *Phys. Rev. Lett.* **93** 080405; *do.* 2005 Erratum: Kinematic Approach to the Mixed State Geometric Phase in Nonunitary Evolution [*Phys. Rev. Lett.* 93, 080405 (2004)] *Phys. Rev. Lett.* **95** 249902

[14] Lombardo F C and Villar P I 2006 Geometric phases in open systems: A model to study how they are corrected by decoherence *Phys. Rev.* A **74** 042311

[15] Lombardo F C and Villar P I 2008 Environmentally induced corrections to the geometric phase in a two-level system *Int. J. Qu. Info. (IJQI)* **6** 707713

[16] Villar P I 2009 Spin bath interaction effects on the geometric phase *Phys. Lett.* A **373** 206

[17] Whitney R S and Gefen Y 2003 Berry Phase in a Nonisolated System *Phys. Rev. Lett.* **90** 190402; Whitney R S, Makhlin Y, Shnirman A and Gefen Y 2005 Geometric Nature of the Environment-Induced Berry Phase and Geometric Dephasing *Phys. Rev. Lett.* **94** 070407

[18] Carollo A, Fuentes-Guridi I, Franca Santos M and Vedral V 2003 Geometric Phase in Open Systems *Phys. Rev. Lett.* **90** 160402; *do.* 2004 Spin-1/2 Geometric Phase Driven by Decohering Quantum Fields *Phys. Rev. Lett.* **92** 020402

[19] De Chiara G, Lozinski A and Palma G M 2007 Berry phase in open quantum systems: a quantum Langevin equation approach *Eur. Phys. J.* **D 41** 179-183; De Chiara and G Palma G M 2003 Berry Phase for a Spin 1/2 Particle in a Classical Fluctuating Field *Phys. Rev. Lett.* **91** 090404

[20] Cucchietti F M, Zhang J F, Lombardo F C, Villar P I and Laflamme R 2010 Geometric Phase with Nonunitary Evolution in the Presence of a Quantum Critical Bath *Phys. Rev. Lett.* **105** 240406

[21] Lombardo F C and Villar P I 2010 Environmentally induced effects on a bipartite two-level system: Geometric phase and entanglement properties *Phys. Rev.* A **81** 022115

[22] Milman P and Mosseri R 2003 Topological Phase for Entangled Two-Qubit States *Phys. Rev. Lett.* **90** 230403; Milman P 2006 Phase dynamics of entangled qubits *Phys. Rev.* A **73** 062118

Twisted spectral geometry for the standard model

Pierre Martinetti
Università di Trieste, via Valerio 12/1, I-34127
E-mail: pmartinetti@units.it

Abstract. In noncommutative geometry, the spectral triple of a manifold does not generate bosonic fields, for fluctuations of the Dirac operator vanish. A Connes-Moscovici twist forces the commutative algebra to be multiplied by matrices. Keeping the space of spinors untouched, twisted-fluctuations then yield perturbations of the spin connection. Applied to the spectral triple of the Standard Model, a similar twist yields the scalar field needed to stabilize the vacuum and to make the computation of the Higgs mass compatible with its experimental value.

1. Introduction

Noncommutative geometry [1] provides a description of the standard model of elementary particles [SM] as a purely gravitational theory [2]. By this, one means that assuming space (time) is described by a slightly non-commutative generalization of a manifold, then Einstein-Hilbert action (in Euclidean signature) together with the bosonic action of the SM, including the Higgs sector, are obtained from one single action formula, the *spectral action* [3]. The bosonic Lagrangian is the noncommutative counterpart of the Einstein-Hilbert action, and the Higgs field comes out on the same footing as the other bosons as a connection 1-form, but associated to the noncommutative part of the geometry.

More precisely, the Einstein-Hilbert action together with the various pieces of the SM are obtained as the asymptotic expansion $\Lambda \to \infty$ of the spectral action, where Λ is a cut-off parameter. The spectral action thus provides some boundary conditions between the parameters of the SM at a putative energy of unification, and physical predictions are obtained by running these parameters under the renormalization group flow down to the electroweak breaking scale [4]. The mass of the Higgs boson is then function of the inputs of the theory (mainly the Yukawa coupling of the fermions and the mixing angles for quarks and neutrinos). Assuming there is no new physics between the electroweak and the unification scales, it is computed around 170 Gev, a value ruled out by Tevatron in 2008.

Since then the Higgs boson has been discovered around 125 GeV, which is below the threshold of stability of the electroweak vacuum: the later is a metastable state. It is not clear whether this is a problem or not, since the life-time of this metastable state is far larger than the age of the universe. However the estimation of the probability that somewhere in our past light-cone the Higgs field has tunneled down to its true vacuum - liberating a quantity of energy that should have destroyed the whole universe - depends on the model of Inflation at the tip of the light-cone. So even if not problematic, it is at least intriguing that the electroweak vacuum is metastable, but on the edge of stability (see [5] for a recent update on these issues).

To cure this instability, a long known solution proposed by particle physicists is to postulate another heavy scalar field, say σ, suitably coupled to the Higgs. In noncommutative geometry,

Journal of Physics: Conference Series **626** (2015) 012044 doi:10.1088/1742-6596/626/1/012044

bosonic fields are obtained by so called *fluctuations of the metric*, roughly speaking a way to turn the constant parameters of the theory (the Yukawa couplings) into fields (there is in reality a intricate re-parametrization and the correspondance Yukawa couplings/bosonic fields is more subtle). Chamseddine and Connes noticed in [6] that by turning into a field one of the constant entry of the generalized Dirac operator describing the SM, namely the Majorana mass k_R of the neutrino, then one gets exactly the field σ suitably coupled to the Higgs. As a bonus, σ modifies the flow of the renormalisation group and makes the computation of the Higgs mass compatible with its experimental value. The question is then to understand how to turn the constant parameter k_R into a field respecting the framework of noncommutative geometry. The problem is that unlike the other bosonic fields, σ cannot be obtained by a fluctuation of the metric because the constant parameter k_R only fluctuates to a constant field. This impossibility has its origin in of one mathematical requirements of noncommutative geometry, namely the condition asking that the generalized Dirac operator is a first-order differential operator.

Various models have been proposed to justify the turning of k_R into a field. In [7, 8] the first order condition is relaxed, yielding to a Pati-Salam generalization of the standard model. The later is retrieved dynamically, as a minimum of the spectral action. Earlier models pre-2012 had already shown how to lower the Higgs mass thanks to extra-scalar fields, but they required also new fermions [9, 10]. Recently, in [11] a variation on the notion of symmetry in NCG yields a model with extra bosonic and scalar fields carrying a $B - L$ charge; in [12] new fields are obtained as a consequence of a non-standard grading.

In [13] we proposed to generate the field σ in a way satisfying, at least partially, the first order condition. The key idea is to allow the commutative algebra $C^\infty(\mathcal{M})$ of smooth functions on a compact spin manifold $\mathcal{M}$ to act non-trivially on the space of spinors $L^2(\mathcal{M}, S)$. The drawback is that the commutator

$$[\not{\partial}, f] \quad f \in C^\infty(\mathcal{M}) \tag{1}$$

with the free Dirac operator $\not{\partial}$ is no longer bounded [14], in contradiction with one of the primary requirements of noncommutative geometry. There exists however a variation of these requirements, introduced in [15] to deal precisely with this kind of problem. Given an algebra $\mathcal{A}$ and a generalized Dirac operator, rather than the boundedness of $[D, a]$ one asks that there exists an automorphism ρ of $\mathcal{A}$ such that the *twisted commutator*

$$[D, a]_\rho := Da - \rho(a)D \tag{2}$$

is bounded. Such twists have mathematical motivations that have nothing to do with physics, but we showed in [16] how a very simple twist of the model proposed in [13] - in fact a chiral transformation - permits to obtain a coherent picture of the SM in which the field σ is generated by a fluctuation of k_R that satisfies a twisted version of the first-order condition.

In this note we give a non-technical account of these results. Rather than the up-bottom approach developed in [16] (i.e. twists as solution to the unboundeness of the commutator coming from the non-trivial action of $C^\infty(\mathcal{M})$ on spinors), we propose a bottom-up approach: after some generalities in section 2, we show in section 3 how requiring a non-trivial twist forces the manifold to be multiplied by a matrix geometry . We discuss the physical consequences, in particular the generation of new fields. Finally in section 4 we state the results of [16] on the standard model of elementary particles.

2. Twisted almost-commutative geometry

2.1. Almost commutative geometry in a nutshell

By "slightly noncommutative generalization of a manifold", one intends a "space" such that the set of functions defined on it is no longer commutative, but is of the kind

$$C^\infty(\mathcal{M}) \otimes \mathcal{A}_F \tag{3}$$

Journal of Physics: Conference Series **626** (2015) 012044 doi:10.1088/1742-6596/626/1/012044

where $\mathcal{A}_F$ is a finite dimensional algebra. This is called an *almost-commutative geometry* because the center $C^\infty(\mathcal{M})$ of the algebra (3) is infinite dimensional (as an algebra). It is represented on

$$L^2(\mathcal{M}, S) \otimes \mathcal{H}_F \tag{4}$$

where $\mathcal{H}_F$ is a finite dimensional space (whose basis are the fermions of the model) carrying a representation of $\mathcal{A}_F$, while $L^2(\mathcal{M}, S)$ is the space of spinors on which functions act by multiplication: $(\pi(f)\psi)(x) = f(x)\psi(x)$ for any $x \in \mathcal{M}$. That is f is represented as the operator

$$\pi(f) := f\mathbb{I} \tag{5}$$

where $\mathbb{I}$ is the identity matrix of dimension n the numbers of components of a spinor.

For the standard model, $\mathcal{H}_F = \mathcal{H}_{sm}$ has dimension the number of fermions (2 colored quarks $+1$ electron $+1$ neutrino $= 8$ multiplied by 2 (chirality), 2 (antiparticles) and 3 (generations) $= 96$). The space (4) is then the space of fermionic fields of the SM. Notice the overcounting of degrees of freedom: the distinction between chirality and anti/particles is taken into account both in the finite dimensional space $\mathcal{H}_F$ and by the number of components of the spinors in $L^2(\mathcal{M}, S)$. This *fermion doubling* is projected out on the fermionic action thanks to a Pfaffian [4], thus it is not a real nuisance, except maybe from an aesthetic point of view. From our perspective, it provides in fact a solution for generating the field σ.

The key idea of noncommutative geometry is that all the geometrical information of the manifold $\mathcal{M}$ is encoded within the Dirac operator $\partial\!\!\!/ = -i\gamma^\mu\nabla_\mu$, where γ^μ's are the Dirac matrices and $\nabla_\mu := \partial_\mu + \omega_\mu$ is the covariant derivative associated to the spin connection ω_μ. Connes worked out a purely algebraic characterization of the Dirac operator, that he then exported to the noncommutative setting [2]. Hence the notion of *spectral triple*, that is an algebra $\mathcal{A}$ (non necessarily commutative), acting through a representation π on a Hilbert space $\mathcal{H}$, together with an operator D with compact resolvent (or a generalization of this condition in case the algebra is not unital) called (generalized) Dirac operator such that

$$||[D, \pi(a)]|| \text{ is bounded for any } a \in \mathcal{A}. \tag{6}$$

One also asks that $\mathcal{H}$ is a graded Hilbert space, that is there exists an operator Γ such that $\Gamma^2 = 1$ which, furthermore, anticommutes with D and commutes with $\mathcal{A}$. For the standard model one has

$$\Gamma = \gamma^5 \otimes \Gamma_{sm} \tag{7}$$

where, writing

$$\mathcal{H}_{sm} = \mathcal{H}_R \oplus \mathcal{H}_L \oplus H^c_R \oplus \mathcal{H}^c_L, \tag{8}$$

as the sum of four copies of $\mathbb{C}^{24}$ labelled by left/right and anti/particles indices, one has $\Gamma_{sm} = \text{diag}(\mathbb{I}_{24}, -\mathbb{I}_{24}, -\mathbb{I}_{24}, \mathbb{I}_{24})$ while γ^5 is the product of the Dirac matrices.

Adding other conditions, one shows that given any spectral triple with unital commutative algebra $\mathcal{A}$, then there exists a compact spin manifold $\mathcal{M}$ such that $\mathcal{A} = C^\infty(\mathcal{M})$ [17]. The conditions on the analytic properties of the operator D are automatically satisfied in the finite dimensional case. As well, we shall not take into account here the orientability condition and Poincaré duality, but will focus on the *real structure* and the already mentioned *first order condition*. The real structure J is an antilinear operator ($J(\lambda\psi) = \bar{\lambda}J\psi$ for $\lambda \in \mathbb{C}$, $\psi \in \mathcal{H}$) whose square is $\pm\mathbb{I}$, and which commutes or anticommutes with the graduation Γ and the operator D. The three signs

$$J^2 = \epsilon\mathbb{I},\ JD = \epsilon' DJ,\ J\Gamma = \epsilon''\Gamma J \tag{9}$$

determines the so called KO-dimension of the spectral triple.

All these conditions are satisfied by the triple $(C^\infty(\mathcal{M}), L^2(\mathcal{M}, S), \partial\!\!\!/)$, with real structure the charge conjugation $\mathcal{J}$, grading the chirality γ^5. The KO-dimension is then the dimension of the manifold $\mathcal{M}$ (modulo 8). A classification of finite dimensional spectral triples satisfying these conditions as been made in [4, 18] and yields an (almost) unique choice $(\mathcal{A}_{sm}, \mathcal{H}_{sm}, D_{sm})$ relevant for the standard model. The choice of the algebra $\mathcal{A}_{sm}$ is discussed in §4, the Hilbert space is $\mathcal{H}_{sm} = \mathbb{C}^{96}$ described below (5), the constant entries of 96×96 matrix D_{sm} are the Yukawa coupling of fermions and the mixing matrices for quarks and neutrinos. A general formula of products of spectral triples yields the almost-commutative spectral triple

$$C^\infty(\mathcal{M}) \otimes \mathcal{A}_{sm}, \quad L^2(\mathcal{M}, S) \otimes \mathcal{H}_{sm}, \quad D := \partial\!\!\!/ \otimes \mathbb{I} + \gamma^5 \otimes D_{sm}. \tag{10}$$

The graduation is given in (7) and the real structure is $\mathcal{J} \otimes J_{sm}$ where J_{sm} is the operators that exchange particles with antiparticles in $\mathcal{H}_{sm}$.

2.2. Bosonic fields and spectral action

Bosonic fields are generated by the so-called *fluctuations of the metric*. Given an arbitrary spectral triple $(\mathcal{A}, \mathcal{H}, D)$, those are defined as the substitution of the Dirac operator with a *covariant operator*

$$D_A := D + A + JAJ^{-1} \tag{11}$$

where A is a selfadjoint element of the set of (generalized) 1-forms

$$\Omega^1_D(\mathcal{A}) := \left\{\pi(a^i)[D, \pi(b_i)], \quad a^i, b_i \in \mathcal{A}\right\}. \tag{12}$$

The name is justified because for $\mathcal{A}_F = M_N(\mathbb{C})$ and $D_F = 0$, then D_A is nothing but the covariant Dirac operator of a $U(n)$ gauge theory on $\mathcal{M}$. For the almost-commutative geometry (10) of the standard model, these fluctuations generate the bosonic fields and the Higgs.

The spectral actions consists in counting the eigenvalue of D_A smaller than an energy scale Λ,

$$S = \operatorname{Tr} f\left(\frac{D_A^2}{\Lambda^2}\right) \tag{13}$$

where f is a smooth approximation of the characteristic function of the interval $[0, 1]$. As explained in introduction, for the spectral triple (10) fluctuated as in (11), the asymptotic expansion of S yields Einstein-Hilbert action and the SM bosonic action, including the Higgs.

2.3. Twisted spectral triple

In the definition (11) of the covariant Dirac operator, it is important that the commutators $[D, \pi(a)]$ are bounded, otherwise bosons would be described by unbounded operators. Whatever the finite dimensional spectral triple $(\mathcal{A}_F, \mathcal{H}_F, D_F)$, the commutator $[D_F, \pi(a)]$ is automatically bounded. The same is true for the commutative part $[\partial\!\!\!/, \pi(f)] = (\partial\!\!\!/ f)\mathbb{I}$. Hence as long as the almost-commutative algebra (3) acts on (4) with the trivial action (5) on spinors, the commutator $[D, \pi(a)]$ is always bounded.

Nevertheless, as explained in [15], there are situations where the requirement (6) is too strong, like the lift $\partial\!\!\!/'$ to $\partial\!\!\!/$ of a conformal map. Then $[\partial\!\!\!/', \pi(f)]$ is no longer bounded, but there exists an automorphism ρ of $C^\infty(\mathcal{M})$ such that the twisted commutator $[\partial\!\!\!/', \pi(f)]_\rho$ is bounded. More generally, requiring the boundedness of the twisted commutator makes sense mathematically and allows to treat cases (type III spectral triple) where the usual commutator is never bounded [15]. This yields the definition of a *twisted spectral triple* $(\mathcal{A}, \mathcal{H}, \mathcal{D}, \rho)$, similar to a spectral triple except that $[D, \pi(a)]$ bounded is replaced by $[D, \pi(a)]_\rho$ bounded for some automorphism ρ.

To the best of our knowledge, the conditions for the reconstruction theorem [17] (in particular the real structure) have not been adapted to the twisted case yet. This does not matter because in the commutative case $(C^\infty(\mathcal{M}), L^2(\mathcal{M}, S), \partial\!\!\!/)$ with representation (5) (which is the one adressed by the reconstruction theorem), twists are not relevant. Indeed, whatever the automorphism ρ of $C^\infty(\mathcal{M})$,

$$[\partial\!\!\!/, \pi(f)]_\rho = (\partial\!\!\!/\,\pi(f)) - i[\gamma^\mu \pi(f) - \pi(\rho(f))\gamma^\mu]\nabla_\mu \tag{14}$$

is bounded if and only if the differential part vanishes, that is

$$[\gamma^\mu, \pi(f)]_\rho = 0 \quad \forall f \in C^\infty(\mathcal{M}). \tag{15}$$

This is equivalent to $f = \rho(f)$ for any f, that is ρ is the trivial automorphism.

3. Need for twist

The precedent section shows that for the usual spectral triple of a manifold, not only there is no need for a twist because the usual commutator is bounded, but there is no space for it. One may wonder what the minimal modifications are, so that to authorize a non-trivial twist. We already know one answer [15]: the lift to the Dirac operator of a conformal map. But sticking to the idea that physics is contained within the Dirac operator, we want to keep the usual Dirac operator $\partial\!\!\!/$ and rather play with the other parameters: the algebra and its representation.

3.1. Degenerate representation and almost-commutative algebra

We first try a degenerate representation $\pi(f) = fp$ for some projection $p \neq \mathbb{I}$. Condition (15) becomes

$$\gamma^\mu fp - \rho(f)p\gamma^\mu = 0 \quad \forall f \in C^\infty(\mathcal{M}). \tag{16}$$

For $f = \rho(f) = 1$, this implies $[\gamma^\mu, p] = 0\,\forall \mu$. Only the multiple of the identity commute with all Dirac matrices, hence $p = \lambda\mathbb{I}$ for some $\lambda \in \mathbb{C}$. $\pi(1) = \pi(1)^2$ fixes $\lambda = 1$, and one is back to (5).

A possibility to have a non-trivial twist would be to act with the automorphism on ρ on p rather than on f in (16). To understand this better, let us assume $\mathcal{M}$ has dimension 4 to fix notations. In the chiral basis, the Euclidean Dirac matrices are

$$\gamma^\mu = \begin{pmatrix} 0_2 & \sigma^\mu \\ \tilde{\sigma}^\mu & 0_2 \end{pmatrix} \tag{17}$$

where $\sigma^\mu = \{\mathbb{I}, -i\sigma_i\}$, $\tilde{\sigma}^\mu = \{\mathbb{I}, i\sigma_i\}$ with $\sigma_{i=1,2,3}$, the Pauli matrices. Consider the representation of the algebra $\mathbb{C}$ on the Hilbert space $\mathbb{C}^2$ as $\mathbb{C} \ni \lambda \to \lambda\mathbb{I}_2$. Let ρ be the automorphism of $\mathbb{C}^2$ that permutes the two terms,

$$\rho(\lambda_1, \lambda_2) = (\lambda_2, \lambda_1) \qquad \forall(\lambda_1, \lambda_2) \in \mathbb{C}^2. \tag{18}$$

For π the representation of $\mathbb{C}^2$ on $\mathbb{C}^4$ given by $\pi(\lambda_1, \lambda_2) = \lambda_1\mathbb{I}_2 \oplus \lambda_2\mathbb{I}_2$, one has

$$[\gamma^\mu, \pi(\lambda_1, \lambda_2)]_\rho = \begin{pmatrix} 0_2 & [\sigma^\mu, \lambda_2\mathbb{I}_2] \\ [\tilde{\sigma}^\mu, \lambda_1\mathbb{I}_2] & 0_2 \end{pmatrix} = 0. \tag{19}$$

If one could work with two representations $\pi(f) = fp$, $\pi'(f) := fp'$ of $C^\infty(\mathcal{M})$ where p, p' are two orthogonal projections in $L^2(\mathcal{M}, S)$, then the algebra isomorphism $\tau : \pi(f) \to \pi'(f)$ would define a modified-commutator

$$\partial\!\!\!/\pi(f) - \tau(\pi(f))\partial\!\!\!/ = -i\gamma^\mu\nabla_\mu(pf) - if(\gamma^\mu p - p'\gamma^\mu)\nabla_\mu \tag{20}$$

where $\nabla_\mu(pf) := \partial_\mu(pf) + [\omega_\mu, pf]$. This is bounded iff the second term vanishes, for instance when $p = \text{diag}\,(\mathbb{I}_2, 0_2)$, $p' = \text{diag}\,(0_2, \mathbb{I}_2)$. The point is that τ is not an automorphism of $C^\infty(\mathcal{M})$, since the algebra generated by $\pi(f)$ and $\pi'(f)$ for $f \in C^\infty(\mathcal{M})$ is not $C^\infty(\mathcal{M})$, but two copies of it, that is $C^\infty(\mathcal{M}) \otimes \mathbb{C}^2$.

In other terms, to have a non-trivial twist one needs to multiply the manifold by a matrix geometry. In this sense, a "raison d'être" of almost-commutative algebra is to allow non-trivial twists, which are forbidden in the case $(C^\infty(\mathcal{M}), L^2(\mathcal{M}, S), \partial\!\!\!/)$.

3.2. Twisted fluctuation of the metric

Consider $\mathcal{A} = C^\infty(\mathcal{M}) \otimes \mathcal{A}_F$ acting on $L^2(\mathcal{M}, S)$ in agreement with §3.1, that is such that there exists an automorphism ρ of $\mathcal{A}$ guaranteeing that (omiting the symbol of representation)

$$[\gamma^\mu, a]_\rho := \gamma^\mu a - \rho(a)\gamma^\mu = 0 \quad \forall \mu. \tag{21}$$

We define the *twisted-covariant free Dirac operator*

$$\eth_{A_\rho} := \eth + A_\rho + \mathcal{J} A_\rho \mathcal{J}^{-1} \tag{22}$$

where A_ρ is a element of the set of twisted 1-forms

$$\Omega^1_{\eth,\rho}(\mathcal{A}) := \left\{ a^i [\eth, b_i]_\rho, \; a^i, b_i \in \mathcal{A} \right\}. \tag{23}$$

By (21) one has $[\eth, a]_\rho = -i\gamma^\mu \nabla_\mu a$ where $\nabla_\mu a := (\partial_\mu a) + [\omega_\mu, a]$. So a twisted 1-form is

$$A_\rho = -i a^i \gamma^\mu \nabla_\mu b_i = -i\gamma^\mu X_\mu \quad \text{where} \quad X_\mu := \rho^{-1}(a^i)\nabla_\mu b_i. \tag{24}$$

Fluctuating $\eth$ by $C^\infty(\mathcal{M})$ acting as in (5) has no interest, because [19] $\eth_A = \eth$. This is no longer true for a twisted-fluctuation (22). To see it, consider $a^i = (f^i, g^i)$, $b_i = (f'_i, g'_i)$ in $\mathcal{A} = C^\infty(\mathcal{M}) \otimes \mathbb{C}^2$ acting as in (19) with $\rho = \rho^{-1}$ as in (18), that is

$$\rho(a^i) = \begin{pmatrix} g^i \mathbb{I}_2 & 0_2 \\ 0_2 & f^i \mathbb{I}_2 \end{pmatrix}, \quad b_i = \begin{pmatrix} f'_i \mathbb{I}_2 & 0_2 \\ 0_2 & g'_i \mathbb{I}_2 \end{pmatrix}, \quad X_\mu = \begin{pmatrix} g^i \partial_\mu f'_i \mathbb{I}_2 & 0_2 \\ 0_2 & f^i \partial_\mu g'_i \mathbb{I}_2 \end{pmatrix} \tag{25}$$

where we use that $\omega_\mu = -\frac{1}{4}\Gamma^\nu_{\mu\rho}\gamma^\rho\gamma_\nu$ (Γ the Christoffel symbol in the orthonormal basis) commutes with b_i. Since $[\mathcal{J}, i\gamma^\mu] = 0$, one has $\mathcal{J} A_\rho \mathcal{J}^{-1} = -i\gamma^\mu \mathcal{J} X_\mu \mathcal{J}^{-1}$. Furthermore,

$$\mathcal{J} X_\mu \mathcal{J}^{-1} = \mathcal{J}\rho(a^i)\mathcal{J}^{-1}\mathcal{J}\partial_\mu b_i \mathcal{J}^{-1} = \rho(a^i)^* \partial_\mu b_i^* = X_\mu^* \tag{26}$$

because $[\mathcal{J}, \partial_\mu] = 0$ and $\mathcal{J} a \mathcal{J}^{-1} = a^*$ commutes with $\mathcal{J}\rho(b)\mathcal{J}^{-1} = \rho(b)^*$. Therefore

$$A_\rho + \mathcal{J} A_\rho \mathcal{J}^{-1} = -i\gamma^\mu (X_\mu + X_\mu^*). \tag{27}$$

In the non-twisted case, X_μ, X_μ^* commute with γ^μ so (27) is selfadjoint iff $X_\mu = -X_\mu^*$, that is $A_\rho + \mathcal{J} A_\rho \mathcal{J}^{-1} = 0$ as announced. In the twisted case, X_μ, X_μ^* twisted-commute with γ^μ: the adjoint of (27) is $i\gamma^\mu \rho(X_\mu + X_\mu^*)$ and $\eth_{A_\rho}$ is selfadjoint as soon as $X_\mu + X_\mu^* = -\rho(X_\mu^* + X_\mu)$.

This condition holds for instance when $g^i \partial_\mu f'_i = -f^i \partial_\mu g'_i := f_\mu$ is a real function. Then

$$\eth_{A_\rho} = \eth - 2i\gamma^\mu \begin{pmatrix} f_\mu \mathbb{I}_2 & 0_2 \\ 0_2 & -f_\mu \mathbb{I}_2 \end{pmatrix}. \tag{28}$$

This simple example shows that a non-trivial twist may have interesting physical consequences: while fluctuations of the of the free Dirac operator by $C^\infty(\mathcal{M})$ are trivial, twisted fluctuations by $C^\infty(\mathcal{M}) \otimes \mathbb{C}^2$ generate a vector field X_μ. Its physical interpretation is delicate: by making functions acting non-trivially on spinors, one breaks the invariance of the representation of $C^\infty(\mathcal{M})$ under the spin group. In this sense, these models are "pre-geometric": the spin structure is not explicit in the representation, but is somehow "hidden" in the Dirac operator.

One may be puzzled by our lack of care in viewing $L^2(\mathcal{M}, S)$ as $L^2(\mathcal{M}) \otimes \mathbb{C}^4$. We argue in [13] that this makes sense in a local trivialization. Eventual non-local effect should be studied.

Also, a peculiarity of the simple model presented here is that the almost-commutative algebra acts on $L^2(\mathcal{M}, S)$, and not on $L^2(\mathcal{M}, S) \otimes \mathcal{H}_F$. The field X_μ is a perturbation of the spin connection, whereas in case there is an action of $\mathcal{A}_F$ on a finite dimensional Hilbert space $\mathcal{H}_F$ (like for the SM, in next section), then a twisted-fluctuation of $\eth \otimes \mathbb{I}$ yields a perturbation of the $U(\mathcal{A}_F)$ connection 1-form induced by a non-twisted fluctuation. One may wonder if the spin connection ω_μ itself could be generated by a twisted fluctuation of the flat Dirac operator.

Journal of Physics: Conference Series **626** (2015) 012044 doi:10.1088/1742-6596/626/1/012044

4. Twist for the standard model

Let us conclude by the applications to the standard model. We work with one generation only, so that the finite dimensional algebra $\mathcal{A}_{sm}$ (discussed below) acts on the finite dimensional space $\mathcal{H}_{sm} = \mathbb{C}^{96/3=32}$.

4.1. Grand symmetry

Independent considerations on the signature of the metric [20], the mass of neutrinos and the fermion doubling [21] indicates that the KO-dimension of the finite dimensional part of the spectral triple of the SM should be 6. Under natural hypothesis (irreducible actions of the algebra and of the real structure, existence of a separating vector) and an explicit ad-hoc "symplectic hypothesis", it has been shown in [18] that in order to accomodate a real structure J and a non-trivial grading Γ, the finite dimensional algebra of the almost-commutative geometry of the SM has to be

$$M_a(\mathbb{H}) \oplus M_{2a}(\mathbb{C}) \tag{29}$$

for a an integer greater than 1, acting on a space with dimension $d = 2(2a)^2$. For $a = 2$ the dimension $d = 32$ is precisely the number of particles per generation of the SM. By further imposing the grading condition ($[\Gamma, a] = 0$) and the first order condition ($[[D, a], Jb^*J^{-1}] = 0$), one arrives to the algebra of the standard model

$$\mathcal{A}_{sm} := \mathbb{C} \oplus \mathbb{H} \oplus M_3(\mathbb{C}). \tag{30}$$

In [13] we noticed that for $a = 4$, the dimension $d = 128$ was precisely 4 times the number of particles per generation. Viewing 4 as the dimension of spinors on a four-dimensional space (time) $\mathcal{M}$, one identifies locally $L^2(\mathcal{M}, S) \otimes \mathcal{H}_{sm}$ as $L^2(\mathcal{M}) \otimes (\mathcal{H}_{sm} \otimes \mathbb{C}^4)$, which provides precisely the space needed to represent the *grand algebra*

$$\mathcal{A}_G := M_4(\mathbb{H}) \oplus M_8(\mathbb{C}). \tag{31}$$

Any element of $M_4(\mathbb{H})$ and $M_8(\mathbb{C})$ are viewed as 2×2 matrices with entry in $M_2(\mathbb{H})$ and $M_4(\mathbb{C})$. These entries act on $\mathcal{H}_{sm}$ as does (29) for $a = 2$, so that at the end of the game one retrieves the action of $\mathcal{A}_{sm}$. The novelty is that the 2×2 block have a non trivial action on the remaining $\mathbb{C}^4$. In this way, one obtains a representation of $C^\infty(\mathcal{M}) \otimes \mathcal{A}_G$ with a non-trivial action on spinors.

The grading condition breaks $\mathcal{A}_G$ to

$$\mathbb{H}_L^l \oplus \mathbb{H}_R^l \oplus \mathbb{H}_L^r \oplus \mathbb{H}_L^r \oplus M_4(\mathbb{C}) \tag{32}$$

where L, R are the left-right indices in $\mathcal{H}_F$ and l, r the left-right indices of spinors. Because of the non-trivial action on spinors, the commutator $[\partial\!\!\!/ \otimes \mathbb{I}, a]$ is never bounded. But there exists a twist ρ such that $[\partial\!\!\!/ \otimes \mathbb{I}, a]_\rho$ is bounded, this is simply the exchange of the spinorial left-right indices:

$$\mathbb{H}_L^l \Longleftrightarrow \mathbb{H}_L^r, \quad \mathbb{H}_R^l \Longleftrightarrow \mathbb{H}_R^r. \tag{33}$$

By further considering the the twisted version of the first order condition

$$[[D, a]_\rho, Jb^*J^{-1}]_\rho = 0 \tag{34}$$

for the Dirac operator of the SM

$$D = \partial\!\!\!/ \otimes \mathbb{I} + \gamma_5 \otimes D_{sm}, \tag{35}$$

one works out a sub-algebra of (32) acting on $L^2(\mathcal{M}, S) \otimes (\mathbb{C}^4 \otimes \mathcal{H}_{sm})$, namely

$$\mathcal{B} := \mathbb{H}_L^l \oplus \mathbb{C}_R^l \oplus \mathbb{H}_L^r \oplus \mathbb{C}_R^r \oplus M_3(\mathbb{C}) \tag{36}$$

which, together with D, defines a twisted spectral triple [16].

4.2. Extra scalar field and additional vector fields

Twisted fluctuations of $\partial\!\!\!/ \otimes \mathbb{I}$ by $\mathcal{B}$ yields a vector field X_μ as in (25). Twisted fluctuations of $\gamma^5 \otimes D_R$, where D_R is the part of D_{sm} containing the Majorana mass k_R of the neutrino, generates a scalar field $\boldsymbol{\sigma}$ which coincides with the field σ studied in [18] up to a global γ^5 factor.

The spectral action yields a potential for these two fields, including an interaction term. This potential is minimum precisely when D is fluctuated by the sub-algebra of $C^\infty(\mathcal{M}) \otimes \mathcal{B}$ invariant under the twist, that is by $C^\infty(\mathcal{M}) \otimes \mathcal{A}_{sm}$ [16].

5. Conclusion

The twist (33) allows to build a twisted spectral triple for the standard model with a non-trivial action on spinors. The extra scalar field $\boldsymbol{\sigma}$ is generated by a twisted-fluctuation of the Majorana part of the Dirac operator, while twisted-fluctuations of the free Dirac operator generate an additional-vector field. All these fluctuations satisfy a twisted version of the first order condition. Furthermore, as in [8] the breaking to the standard model (in our cae: the un-twisting) is obtained dynamically by minimizing the spectral action.

Acknowledgments

Thanks to G. Landi for discussion. Supported by the italian "Prin 2010-11 - Operator Algebras, Noncommutative. Geometry and Applications".

References

[1] Connes A 1994 *Noncommutative Geometry* (Academic Press)
[2] Connes A 1996 *Commun. Math. Phys.* **182** 155–176
[3] Chamseddine A H and Connes A 1996 *Commun. Math. Phys.* **186** 737–750
[4] Chamseddine A H, Connes A and Marcolli M 2007 *Adv. Theor. Math. Phys.* **11** 991–1089
[5] Buttazzo D, Degrassi G, Giardino P P, Giudice G F, Sala F and Salvio A *arXiv:1307.3536 [hep-ph]*
[6] Chamseddine A H and Connes A 2012 *JHEP* **09** 104
[7] Chamseddine A H, Connes A and van Suijlekom W 2013 *J. Geom. Phy.* **73** 222–234
[8] Chamseddine A H, Connes A and van Suijlekom W 2013 *JHEP* **11** 132
[9] Stephan C A 2009 *Phys. Rev. D* **79** 065013
[10] Stephan C A 2013
[11] Farnsworth S and Boyle L 2015 *New J. Phys.* **17** 023021
[12] D'Andrea F and Dabrowski L 2015 *arXiv:1501.00156 [math-phys]*
[13] Devastato A, Lizzi F and Martinetti P 2014 *JHEP* **01** 042
[14] Devastato A, Lizzi F and Martinetti P 2014 *Fortschritte der Physik* **62** 863–868
[15] Connes A and Moscovici H 2008 *Traces in number theory, geometry and quantum fields, Aspects Math. Friedt. Vieweg, Wiesbaden* **E38** 57–71
[16] Devastato A and Martinetti P *arXiv 1411.1320 [hep-th]*
[17] Connes A 2013 *J. Noncom. Geom.* **7** 1–82
[18] Chamseddine A H and Connes A 2008 *J. Geom. Phys* **58** 38–47
[19] Landi G 1997 *Lecture Notes in Physics m51*
[20] Barrett J W 2007 *J. Math. Phys.* **48** 012303
[21] Connes A 2006 *JHEP* **081**

Gauge theories on noncommutative Kähler manifolds

[1] **Yoshiaki Maeda,** [2] **Akifumi Sako,** [3] **Toshiya Suzuki and** [3] **Hiroshi Umetsu**

[1] Tohoku Forum for Creativity, Tohoku University,
2-1-1, Katahira, Aoba-Ku, Sendai, Miyagi, 980-8577, Japan
[2] Department of Mathematics, Faculty of Science, Tokyo University of Science,
1-3 Kagurazaka, Shinjuku-ku, Tokyo 162-8601, Japan
[3] Kushiro National College of Technology,
2-32-1 Otanoshike-nishi, Kushiro, Hokkaido 084-0916, Japan

E-mail: sako@rs.tus.ac.jp

Abstract. We study gauge theories on noncommutative homogeneous Kähler manifolds. To make the noncommutative manifolds, we use the deformation quantization with separation of variables for Kähler manifolds. We construct models of noncommutative gauge theories that are connected with usual Yang-Mills theories in the commutative limits. It is expected that the models connecting to commutative gauge theories are uniquely determined. As examples, we give noncommutative $\mathbb{C}P^N$ and noncommutative $\mathbb{C}H^N$ and gauge theories on them. Some kinds of gauge symmetry breaking and topological symmetry breaking by noncommutative deformations are observed by concrete geometrical calculations.

1. Motivation

One of the most important problems in physics is to formulate a quantum gravity theory. String theories are the strongest candidates of the quantum gravity theory. However the string theories themselves have not been completed until now. It means that many string theories are defined as perturbative formulations, and we have to formulate them as non-perturbative formulations. There are some approaches to make non-perturbative string theories; string field theories, M theory, Matrix models, and so on, but they have never get the total win yet. IKKT(Type IIB) matrix theory is one of the candidates of non-perturbative string theory[1]. This matrix theory contains noncommutative gauge theories as classical solutions, for example, a gauge theory on the the Moyal plane, the one on the fuzzy $\mathbb{C}P^1$, and so on. The gauge theories on the Moyal spaces have been studied in detail, but research of gauge theories on the noncommutative $\mathbb{C}P^N$ or more generally gauge theories on noncommutative homogeneous Kähler manifolds are not so much as the ones on Moyal spaces. One reason is difficulty in constructing concretely calculable models of noncommutative gauge theories on such spaces.

In this article, we introduce concretely calculable gauge theories on noncomutative $\mathbb{C}P^N$ and $\mathbb{C}H^N$. Noncommutative deformations of the manifolds are performed by the deformation quantization. This method has an advantage that we can calculate geometrical quantities in the similar way of classical geometry, because all objects appearing in deformation quantizations are interpreted as formal power series of quantities on classical manifolds. Using this advantage, we can discuss various phenomena in the noncommutative field theories. As an example, we observe a gauge symmetry and topological symmetry breaking by a noncommutative deformation.

2. Deformation quantization of Kähler manifolds

Noncommutative spaces in this article are constructed by using the deformation quantization defined as follows. Let M be a smooth manifold.

Definition 2.1 *Let $\mathcal{F}$ be a set of formal power series of $\hbar \in \mathbb{R}$:*

$$\mathcal{F} := \left\{ f \;\middle|\; f = \sum_k f_k \hbar^k, \text{ where } f_k \text{ is a smooth function on } M \right\}.$$

A star product of $f, g \in \mathcal{F}$ is defined as

$$f * g = \sum C_k(f,g)\hbar^k,$$

such that the product satisfies the following conditions.

(i) $*$ is associative product.

(ii) C_k is a bidifferential operator.

(iii) C_0 and C_1 is defined as

$$C_0(f,g) = fg, \qquad C_1(f,g) - C_1(g,f) = i\{f,g\}.$$

where $\{f,g\}$ is the Poisson bracket.

(iv) $f * 1 = 1 * f = f$.

There is a star product called a star product with separation of variables, which is defined on Kähler manifolds. A Kähler manifold has a Kähler potential Φ which introduces a Kähler 2-form ω and a metric g as

$$\omega = i g_{i\bar{j}} dz^i \wedge d\bar{z}^j, \qquad g_{i\bar{j}} = \frac{\partial^2 \Phi}{\partial z^i \partial \bar{z}^j}.$$

$*$ is called a star product with separation of variables when

$$a * f = af$$

for locally holomorphic function a and

$$f * b = fb$$

for locally anti-holomorphic function b. Karabegov found a way to construct this star product with separation variables $*$ for arbitrary Kähler manifolds [2] (see also [3, 4]). Let us review how to construct the star product $*$.

In this method of deformation quantization, a star product is constructed as a formal power series of differential operators. Let L_f be a differential operator corresponding to a left $*$ multiplication by f:

$$L_f g := f * g.$$

Then L_f has the following form:

$$L_f = \sum_{n=0}^{\infty} \hbar^n A_n,$$

where A_n is given by

$$A_n = a_{n,\alpha}(f) \prod_i \left(D^{\bar{i}}\right)^{\alpha_i}, \qquad (D^{\bar{i}} := g^{\bar{i}j}\partial_j).$$

Journal of Physics: Conference Series **626** (2015) 012045 doi:10.1088/1742-6596/626/1/012045

We use $\partial_i = \dfrac{\partial}{\partial z^i}$ for simplicity. We also use $D^i := g^{i\bar{j}}\partial_{\bar{j}}$, where $\partial_{\bar{i}} = \dfrac{\partial}{\partial \bar{z}^i}$, in the following. It is required that L_f satisfies

$$L_f 1 = f * 1 = f,$$
$$L_f\left(L_g h\right) = f * (g * h) = (f * g) * h = L_{L_f g} h.$$

L_f which has the properties described above is determined by the following condition:

$$[L_f, \partial_{\bar{i}}\Phi + \hbar\partial_{\bar{i}}] = 0, \tag{1}$$

and $A_0 = f$. This condition is equivalent to the recursion relations:

$$[A_n,\ \partial_{\bar{i}}\Phi] = [\partial_{\bar{i}},\ A_{n-1}].$$

If one obtains the operator $L_{\bar{z}^i}$ $(L_{\bar{z}^l} f = \bar{z}^l * f)$, L_f is given by

$$L_f = \sum_\alpha \frac{1}{\alpha!}\left(\frac{\partial}{\partial \bar{z}}\right)^\alpha f\left(L_{\bar{z}} - \bar{z}\right)^\alpha.$$

Here, α is a multi-index, $\alpha = (\alpha_1, \cdots, \alpha_m)$. It is not easy to derive explicit expressions of star products in all order of $\hbar$ by solving the recursion relation.

3. Gauge theory on noncommutative homogeneous Kähler manifolds

3.1. Derivations

Field theories on noncommutative spaces can be constructed based on the derivations of the algebra $C^\infty(M)[[\hbar]]$ with its star product, where derivation $\mathbf{d}$ are linear operators satisfying the Leibniz rule, i.e. $\mathbf{d}(f * g) = \mathbf{d}f * g + f * \mathbf{d}g$. In commutative space, vector fields are derivations. However first order differential operators in noncommutative space do not satisfy the Leibniz rule in general, because star products depend on the local position. $\mathcal{L}$ is an inner derivations when $P \in C^\infty(M)[[\hbar]]$ exists such that $\mathcal{L}(f) = [P, f]_* := P * f - f * P$, for any $f \in C^\infty(M)[[\hbar]]$. Note that $[P, f]_*$ includes higher derivative terms of f for a generic P. So, it turns out that inner derivations corresponding to vector fields play an important role, when we construct physical field theories. What is the vector field having such the property. Here is an answer.

Proposition 3.1 [Müller-Bahns and N. Neumaier] [6, 7] *In deformation quantization defined above, the inner derivations given as vector fields are the Killing vector fields* $\mathcal{L}_a$.

An important point to read the later half of this article is that the first order differential operators satisfy the Leibniz rule are the Killing vector fields.

3.2. Noncommutative gauge theory

In the following, we consider $U(n)$ gauge theories for simplicity. All following results can be applied for any matrix groups. Proofs of theorems in this section are given in [6]

We introduce a noncommutative $U(n)$ transformations as a deformation of the unitary transformations. If $g \in U(n)$, then $g^\dagger g = I$, where $g^\dagger$ is the hermitian conjugate of g and I is the identity matrix.

As a natural extension, we define $G := M_n(C^\infty[[\hbar]])$ such that for $U = \sum_{k=0}^{\infty} \hbar^k U^{(k)}$ and

Journal of Physics: Conference Series **626** (2015) 012045 doi:10.1088/1742-6596/626/1/012045

$$U^\dagger = \sum_{k=0}^{\infty} \hbar^k U^{(k)\dagger} \in G,$$

$$U^\dagger * U = \sum_{n=0}^{\infty} \hbar^n \sum_{m=0}^{n} U^{(m)\dagger} * U^{(n-m)} = I.$$

For arbitrary $U^{(0)} : M \to U(n)$, This condition has solutions which are determined recursively at each order.

Let's construct a gauge theory on deformed homogeneous Kähler manifold $\mathcal{G}/\mathcal{H}$ whose covariant derivatives are determined by inner derivations corresponding to the Killing vector fields.

In a $\mathcal{G}/\mathcal{H}$, there are the holomorphic Killing vector fields $\mathcal{L}_a = \zeta_a^i(z)\partial_i + \zeta_a^{\bar{i}}(\bar{z})\partial_{\bar{i}}$ corresponding to the isometry group $\mathcal{G}$,

$$[\mathcal{L}_a, \mathcal{L}_b] = i f_{abc} \mathcal{L}_c,$$

where f_{abc} is a structure constant of the Lie algebra of $\mathcal{G}$. As we saw in the previous subsection, $\mathcal{L}_a$ satisfies the Leibniz rule.

$$\mathcal{L}_a(f * g) = (\mathcal{L}_a f) * g + f * (\mathcal{L}_a g).$$

The Killing vectors are normalized here as

$$\eta^{ab}\zeta_a^i\zeta_b^{\bar{j}} = g^{i\bar{j}}, \quad \eta^{ab}\zeta_a^i\zeta_b^j = 0, \quad \eta^{ab}\zeta_a^{\bar{i}}\zeta_b^{\bar{j}} = 0,$$

where η^{ab} is the inverse of the Killing form of the Lie algebra of $\mathcal{G}$.

In the commutative homogeneous Kähler manifolds, $M = \mathcal{G}/\mathcal{H}$, let us introduce a gauge connection $\mathcal{A}_a^{(0)}$ given as

$$\mathcal{A}_a^{(0)} = \zeta_a^\mu A_\mu = \zeta_a^i A_i + \zeta_a^{\bar{i}} A_{\bar{i}},$$

where A_μ or its complex expressions A_i and $A_{\bar{i}}$ are usual gauge fields on M. The curvature is also defined as

$$\mathcal{F}_{ab}^{(0)} := \mathcal{L}_a\mathcal{A}_b^{(0)} - \mathcal{L}_b\mathcal{A}_a^{(0)} - i[\mathcal{A}_a^{(0)}, \mathcal{A}_b^{(0)}] - i f_{abc}\mathcal{A}_c^{(0)},$$

where $[A, B] = AB - BA$. $\mathcal{F}_{ab}^{(0)}$ is related to the curvature of A_μ, $F_{\mu\nu} = \partial_\mu A_\nu - \partial_\mu A_\nu - i[A_\mu, A_\nu]$, as

$$\mathcal{F}_{ab}^{(0)} = \zeta_a^\mu \zeta_b^\nu F_{\mu\nu}.$$

It is shown easily that

$$\eta^{ac}\eta^{bd}\mathcal{F}_{ab}^{(0)}\mathcal{F}_{cd}^{(0)} = g^{\mu\rho}g^{\nu\sigma}F_{\mu\nu}F_{\rho\sigma}.$$

These fields $\mathcal{A}_a^{(0)}$ and $\mathcal{F}_{ab}^{(0)}$ are introduced to be convenient when we use the Killing vector fields as derivations to construct gauge theories.

Now, we consider a noncommutative deformation of the gauge theory. The formal power series of the gauge connections are written as

$$\mathcal{A}_a := \sum_{k=0}^{\infty} \hbar^k \mathcal{A}_a^{(k)}.$$

Journal of Physics: Conference Series **626** (2015) 012045 doi:10.1088/1742-6596/626/1/012045

The gauge transformation is defined by

$$\mathcal{A}_a \to \mathcal{A}'_a = iU^{-1} * \mathcal{L}_a U + U^{-1} * \mathcal{A}_a * U.$$

Let us define the curvature of $\mathcal{A}_a$ by

$$\mathcal{F}_{ab} := \mathcal{L}_a\mathcal{A}_b - \mathcal{L}_b\mathcal{A}_a - i[\mathcal{A}_a, \mathcal{A}_b]_* - if_{abc}\mathcal{A}_c.$$

We can show that the following lemma.

Lemma 3.2 *$\mathcal{F}_{ab}$ transforms covariantly:*

$$\mathcal{F}_{ab} \to \mathcal{F}'_{ab} = U^{-1} * \mathcal{F}_{ab} * U.$$

Using this lemma, we obtain the gauge invariant action.

Theorem 3.3 *A gauge invariant action for the gauge field is given by*

$$S_g := \int_{\mathcal{G}/\mathcal{H}} \mu_g \operatorname{tr}\left(\eta^{ac}\eta^{bd}\mathcal{F}_{ab} * \mathcal{F}_{cd}\right),$$

where μ_g is a trace density.

The gauge invariance of the action is obtained by the above lemma and the cyclic symmetry of the trace density. The existence of trace density, $\int_M f * g\mu_g = \int_M g * f\mu_g$, is guaranteed [5]. We expect that this is the unique action of noncommutative Yang-Mills theory connecting with commutative Yang-Mills theory in the commutative limit in our noncommutative homogeneous Kähler manifolds.

Other fields can be formulated with the gauge covariant way. For example, let us introduce a complex scalar field $\phi = \sum_k \phi^{(k)}\hbar^k$ and its hermitian conjugate scalar field $\phi^\dagger$ whose gauge transformation as the fundamental representation of the gauge group is given by

$$\phi \to \phi' = U^{-1} * \phi, \qquad \phi^\dagger \to \phi^{\dagger\prime} = \phi^\dagger * U.$$

The covariant derivative for this scalar field is defined by

$$\nabla_a\phi := \mathcal{L}_a\phi - i\mathcal{A}_a * \phi,$$

Using these, a gauge invariant action is obtained.

Theorem 3.4 *Gauge invariant action is given by*

$$S_\phi = \int_{\mathcal{G}/\mathcal{H}} \mu_g \left\{\eta^{ab}\nabla_a\phi^\dagger * \nabla_b\phi + V(\phi^\dagger * \phi)\right\},$$

where V is a potential as a function of one variable.

4. Noncommutative deformation of $\mathbb{C}P^N$

As examples of above models, we study noncommutative $\mathbb{C}P^N$ in this section and $\mathbb{C}H^N$ in the next section. The detail derivations of the following results are found in [8].

Journal of Physics: Conference Series **626** (2015) 012045 doi:10.1088/1742-6596/626/1/012045

4.1. Noncommutative $\mathbb{C}P^N$

Let z^i $(i = 1, 2, \cdots, N)$ be inhomogeneous coordinates of $\mathbb{C}P^N$. Then, the Kähler potential of $\mathbb{C}P^N$ is given by

$$\Phi = \ln\left(1 + |z|^2\right), \qquad (|z|^2 = \sum_i z^i \bar{z}^i).$$

The complex metric $(g_{i\bar{j}})$ is derived from the Kähler potential as

$$\begin{aligned} ds^2 &= 2g_{i\bar{j}} dz^i d\bar{z}^j, \\ g_{i\bar{j}} &= \partial_i \partial_{\bar{j}} \Phi = \frac{(1 + |z|^2)\delta_{ij} - z^j \bar{z}^i}{(1 + |z|^2)^2}. \end{aligned}$$

Its inverse metric $(g^{\bar{i}j})$ is given as

$$g^{\bar{i}j} = (1 + |z|^2)\left(\delta_{ij} + z^j \bar{z}^i\right).$$

Solving the recursion relation (1), we obtain $L_{\bar{z}^l}$;

$$\begin{aligned} L_{\bar{z}^l} &= \bar{z}^l + \hbar D^{\bar{l}} + \sum_{n=2}^{\infty} \hbar^n \sum_{m=2}^{n} a_m^{(n)} \partial_{\bar{j}_1} \Phi \cdots \partial_{\bar{j}_{m-1}} \Phi D^{\bar{j}_1} \cdots D^{\bar{j}_{m-1}} D^{\bar{l}} \\ &= \bar{z}^l + \sum_{m=1}^{\infty} \alpha_m(\hbar) \partial_{\bar{j}_1} \Phi \cdots \partial_{\bar{j}_{m-1}} \Phi D^{\bar{j}_1} \cdots D^{\bar{j}_{m-1}} D^{\bar{l}}. \end{aligned}$$

Here $\alpha_m(t)$ is defined as

$$\begin{aligned} \alpha_2(t) &= \sum_{n=2}^{\infty} t^n a_2^{(n)} = \sum_{n=2}^{\infty} t^n = \frac{t^2}{1-t}, \\ \alpha_m(t) &= t^m \prod_{n=1}^{m-1} \frac{1}{1 - nt} = \frac{\Gamma(1 - m + \frac{1}{t})}{\Gamma(1 + \frac{1}{t})}, \qquad (m \geq 2). \end{aligned}$$

This $\alpha_m(t)$ is a generating function of the Stirling number of the second kind $S(n, k)$.

Using these results, star products among z^i and $\bar{z}^i$ are obtained as

$$z^i * z^j = z^i z^j, \qquad z^i * \bar{z}^j = z^i \bar{z}^j, \qquad \bar{z}^i * \bar{z}^j = \bar{z}^i \bar{z}^j, \tag{2}$$

$$\begin{aligned} \bar{z}^i * z^j &= \bar{z}^i z^j + \hbar \delta_{ij} (1 + |z|^2) {}_2F_1\left(1, 1; 1 - 1/\hbar; -|z|^2\right) \\ &\quad + \frac{\hbar}{1-\hbar} \bar{z}^i z^j (1 + |z|^2) {}_2F_1\left(1, 2; 2 - 1/\hbar; -|z|^2\right). \end{aligned} \tag{3}$$

Furthermore, we can derive an explicit star product for arbitrary functions f and g:

$$f * g = \sum_{n=0}^{\infty} \frac{\alpha_n(\hbar)}{n!} g_{j_1 \bar{k}_1} \cdots g_{j_n \bar{k}_n} \left(D^{j_1} \cdots D^{j_n} f\right) \left(D^{\bar{k}_1} \cdots D^{\bar{k}_n} g\right).$$

This star product on $\mathbb{C}P^N$ is characterized by a function of $\hbar$, $\alpha_n(\hbar)$.

Journal of Physics: Conference Series **626** (2015) 012045 doi:10.1088/1742-6596/626/1/012045

4.2. Inner derivations and gauge theory on noncommutative $\mathbb{C}P^N$

Let us study differentials in a noncommutative $\mathbb{C}P^N$. The isometry group of this $\mathbb{C}P^N$ is $SU(N+1)$. In the following, we give concrete expressions of the Killing vectors corresponding to the generators of $su(N+1)$, the Lie algebra of $SU(N+1)$. We use homogeneous coordinates of $\mathbb{C}P^N$;

$$\{\xi^A | A = 0, 1, \cdots, N\} = \{\xi^0, \xi^i | i = 1, 2, \ldots, N\},$$

and inhomogeneous coordinates on the chart of $\xi^0 \neq 0$;

$$z^i = \frac{\xi^i}{\xi^0}, \qquad \bar{z}^i = \frac{\bar{\xi}^i}{\bar{\xi}^0}, \qquad (i = 1, 2, \ldots, N).$$

Generators $(T_a)_{AB}$ of $su(N+1)$ in the fundamental representation satisfy the following relations,

$$[T_a, T_b] = i f_{abc} T_c, \qquad \mathrm{Tr}\ T_a = 0,$$
$$\mathrm{Tr}\ T_a T_b = \delta_{ab},$$
$$(T_a)_{AB}(T_a)_{CD} = \delta_{AD}\delta_{BC} - \frac{1}{N+1}\delta_{AB}\delta_{CD},$$

where f_{abc} is the structure constant of $su(N+1)$, $a = 1, 2, \ldots, N^2 + 2N$, and $A, B = 0, 1, \ldots, N$. As you can find it from the fact that Kähler potential is given by $\Phi = \ln(1 + |z|^2) = \ln |\xi|^2$, the isometry $SU(N+1)$ transformation with the homogeneous coordinates is given by

$$\delta\xi^A = i\theta^a (T_a)_{AB}\xi^B, \qquad \delta\bar{\xi}^A = -i\theta^a \bar{\xi}^B (T_a)_{BA},$$

and its Lie derivative is given by

$$\mathcal{L}_a = -(T_a)_{AB}\left(\xi^B \frac{\partial}{\partial \xi^A} - \bar{\xi}^A \frac{\partial}{\partial \bar{\xi}^B}\right),$$
$$[\mathcal{L}_a, \mathcal{L}_b] = i f_{abc}\mathcal{L}_c.$$

The generators of the isometry $SU(N+1)$ is given as

$$\mathcal{L}_a = \zeta_a^i \partial_i + \zeta_a^{\bar{i}} \partial_{\bar{i}} = (T_a)_{00}\left(z^i\partial_i - \bar{z}^i\partial_{\bar{i}}\right) + (T_a)_{0i}\left(z^i z^j \partial_j + \partial_{\bar{i}}\right)$$
$$+ (T_a)_{i0}\left(-\partial_i - \bar{z}^i\bar{z}^j\partial_{\bar{j}}\right) + (T_a)_{ij}\left(-z^j\partial_i + \bar{z}^i\partial_{\bar{j}}\right),$$

and

$$\zeta_a^i := (T_a)_{00} z^i + (T_a)_{0j} z^j z^i - (T_a)_{i0} - (T_a)_{ij} z^j,$$
$$\zeta_a^{\bar{i}} := -(T_a)_{00}\bar{z}^i + (T_a)_{0i} - (T_a)_{j0}\bar{z}^j\bar{z}^i + (T_a)_{ji}\bar{z}^j.$$

The quadratic forms of ζ_a^i and $\zeta_a^{\bar{i}}$ become the metric,

$$\zeta_a^i \zeta_a^{\bar{j}} = -(1 + |z|^2)(\delta_{ij} + z^i\bar{z}^j) = -g^{i\bar{j}},$$
$$\zeta_a^i \zeta_a^j = 0, \qquad \zeta_a^{\bar{i}}\zeta_a^{\bar{j}} = 0.$$

For the noncommutative $\mathbb{C}P^N$ trace density μ_g is given by the Riemannian volume form. Thus, the Yang-Mills type action is constructed as

$$S_g := \int_{\mathbb{C}P^N} \sqrt{g} dz^1 \cdots dz^N d\bar{z}^1 \cdots d\bar{z}^N \ \mathrm{tr}\left(\mathcal{F}_{ab} * \mathcal{F}_{cd}\eta^{ac}\eta^{bd}\right), \tag{4}$$

where tr is trace for gauge group G. All classical calculations of this Yang-Mills theory can be done by using above Killing vectors and concrete expressions of the star products.

Journal of Physics: Conference Series **626** (2015) 012045 doi:10.1088/1742-6596/626/1/012045

5. Noncommutative deformation of $\mathbb{C}H^N$

Similar to the $\mathbb{C}P^N$, a noncommutative deformation of $\mathbb{C}H^N$ is constructed.

As similar to (2)-(3), star products between inhomogeneous coordinates are given as

$$\begin{aligned}
z^i * z^j =& z^i z^j, \\
z^i * \bar{z}^j =& z^i \bar{z}^j, \\
\bar{z}^i * \bar{z}^j =& \bar{z}^i \bar{z}^j, \\
\bar{z}^i * z^j =& \bar{z}^i z^j + \hbar\delta_{ij}(1-|z|^2)_2F_1\left(1,1;1+1/\hbar;|z|^2\right) \\
& - \frac{\hbar}{1+\hbar}\bar{z}^i z^j(1-|z|^2)_2F_1\left(1,2;2+1/\hbar;|z|^2\right).
\end{aligned}$$

The explicit representation of the star product with separation of variables on $\mathbb{C}H^N$ is given by

$$L_{\bar{z}^l} = \bar{z}^l + \sum_{m=1}^{\infty}(-1)^{m-1}\beta_m(\hbar)\partial_{\bar{j}_1}\Phi\cdots\partial_{\bar{j}_{m-1}}\Phi D^{\bar{j}_1}\cdots D^{\bar{j}_{m-1}}D^{\bar{l}},$$

with

$$\beta_n(t) = (-1)^n\alpha_n(-t) = \frac{\Gamma(1/t)}{\Gamma(n+1/t)}.$$

The generators of the isometry $SU(1,N)$ in the inhomogeneous coordinates are

$$\begin{aligned}
\mathcal{L}_a = \zeta_a^i\partial_i + \zeta_a^{\bar{i}}\partial_{\bar{i}} = & (T_a)_{00}\left(z^i\partial_i - \bar{z}^i\partial_{\bar{i}}\right) + (T_a)_{0i}\left(z^i z^j\partial_j - \partial_{\bar{i}}\right) \\
& + (T_a)_{i0}\left(-\partial_i + \bar{z}^i\bar{z}^j\partial_{\bar{j}}\right) + (T_a)_{ij}\left(-z^j\partial_i + \bar{z}^i\partial_{\bar{j}}\right),
\end{aligned}$$

and

$$\begin{aligned}
\zeta_a^i &:= (T_a)_{00}z^i + (T_a)_{0j}z^j z^i - (T_a)_{i0} - (T_a)_{ij}z^j, \\
\zeta_a^{\bar{i}} &:= -(T_a)_{00}\bar{z}^i - (T_a)_{0i} + (T_a)_{j0}\bar{z}^j\bar{z}^i + (T_a)_{ji}\bar{z}^j.
\end{aligned}$$

Using these data, we can construct a noncommutative gauge theory on $\mathbb{C}H^N$ in the similar way in the previous section.

6. Spontaneous gauge symmetry breaking

As an application of the above gauge theories, let us consider gauge symmetry breaking on noncommutative homogeneous Kähler manifolds $M(=\mathcal{G}/\mathcal{H})$.

The action functional (4) on an open subset in $M(=\mathcal{G}/\mathcal{H})$ does not have gauge symmetry, i.e. for $D\subset M, D\neq M$,

$$\begin{aligned}
S'_g|_D - S_g|_D =& \int_D \mu_g \,\mathrm{tr}U^{-1} * \left(\eta^{ac}\eta^{bd}\mathcal{F}_{ab} * \mathcal{F}_{cd}\right) * U - \int_D \mu_g \,\mathrm{tr}\left(\eta^{ac}\eta^{bd}\mathcal{F}_{ab} * \mathcal{F}_{cd}\right) \\
\neq& 0.
\end{aligned}$$

Since trace density μ_g does not have the cyclic symmetry on the subset D, i.e. $\int_D \mu_g\, f * g \neq \int_D \mu_g\, g * f$. This phenomenon is not special one appearing in our gauge theory. However owing to the concrete expressions of $*$-products we can observe what happen by the gauge symmetry breaking.

For example, let us observe that a topological invariant receives quantum corrections in the $\mathbb{C}P^1 \cong S^2$. Consider $\int_{S^2} \mu_g \mathrm{tr} \mathcal{L}_a \epsilon_{abc} \mathcal{F}_{bc}$. This is rewritten as $\int_{S^2} \mathrm{tr} \partial_\mu (\zeta_a^\mu \epsilon_{abc} \mathcal{F}_{bc})$, so we find this is a topological term. This is zero on commutative $S^2 \cong \mathbb{C}P^1$ since the Stokes' theorem, but we can show that this is non zero on the noncommutative $\mathbb{C}P^1$ in general and estimate the value of it concretely by using explicit expression of the star products (3) [9].

7. Summary and discussions

In this article, we introduced noncommutative homogeneous Kähler manifolds by using the Karabegov's star products with separation of variables, and we constructed gauge theories on them. The gauge theories are connected with the ordinary Yang-Mills theory on commutative homogeneous Kähler manifolds in the commutative limit. In our noncommutative manifolds, it is expected that such gauge theories are uniquely determined, because the inner derivations given as vector fields are only the Killing vector fields. Owing to the explicit expressions of the star products of $\mathbb{C}P^N$ and $\mathbb{C}H^N$, we can calculate physical quantities in the gauge theories, concretely. As an example, we observed that noncommutative deformation breaks gauge symmetry and topological symmetry. Someone might think we should omit this type of symmetry breaking. However, if we remind that the Seiberg-Witten transformation between a noncommutative gauge theory and a commutative gauge theory with background B field [10], we might realize such situations as real physics. Anyway further studying these phenomena are left for future as an important subjects.

Acknowledgments

Y.M. was supported in part by JSPS KAKENHI No.23340018 and No.22654011, and A.S. was supported in part by JSPS KAKENHI No.23540117.

[1] Ishibashi N, Kawai H, Kitazawa Y and Tsuchiya A 1997 A Large N reduced model as superstring *Nucl. Phys.* B **498** 467 (*Preprint* arXiv:hep-th/9612115)

[2] Karabegov A V 1996 Deformation quantizations with separation of variables on a Kahler manifold *Commun. Math. Phys.* **180** 745 (*Preprint* arXiv:arXiv:hep-th/9508013)

[3] Karabegov A V 1996 On deformation quantization, on a Kahler manifold, associated to Berezin's quantization *Funct. Anal. Appl.* **30** 142

[4] Karabegov A V 2011 An explicit formula for a star product with separation of variables *Preprint* arXiv:1106.4112 [math.QA]

[5] Karabegov A V 1998 On the canonical normalization of a trace density of deformation quantization *Lett. Math. Phys.* **45** 217

[6] Maeda Y, Sako A, Suzuki T and Umetsu H 2014 Gauge theories in noncommutative homogeneous Kähler manifolds *J. Math. Phys.* **55** 092301 (*Preprint* arXiv:arXiv:1403.5727 [hep-th])

[7] Müller-Bahns M and Neumaier N 2004 Invariant Star Products of Wick Type: Classification and Quantum Momentum Mappings *Lett. Math. Phys.* **70** 1

[8] Sako A, Suzuki T and Umetsu H 2012 Explicit Formulas for Noncommutative Deformations of $\mathbb{C}P^N$ and $\mathbb{C}H^N$ *J. Math. Phys.* **53** 073502 (*Preprint* arXiv:arXiv:1204.4030 [math-ph])

[9] Sako A, Suzuki T and Umetsu H in preparation

[10] Seiberg N and Witten E 1999 String theory and noncommutative geometry *JHEP* **9909** 032 (*Preprint* arXiv:hep-th/9908142)

Entanglement and separability in the noncommutative phase-space scenario

Alex E Bernardini[1], Catarina Bastos[2], Orfeu Bertolami[3], Nuno Costa Dias[4] and João Nuno Prata[4,5]

[1]Departamento de Física, Universidade Federal de São Carlos, PO Box 676, 13565-905, São Carlos, SP, Brasil
[2]Instituto de Plasmas e Fusão Nuclear, Instituto Superior Técnico Avenida Rovisco Pais 1, 1049-001 Lisboa, Portugal
[3]Departamento de Física e Astronomia, Faculdade de Ciências da Universidade do Porto, Rua do Campo Alegre, 687,4169-007 Porto, Portugal
[4]Departamento de Matemática, Universidade Lusófona de Humanidades e Tecnologias Avenida Campo Grande, 376, 1749-024 Lisboa, Portugal
[5]Grupo de Física Matemática, UL, Avenida Prof. Gama Pinto 2, 1649-003, Lisboa, Portugal

E-mail:
alexeb@ufscar.br,catarina.bastos@ist.utl.pt,orfeu.bertolami@fc.up.pt,ncdias@meo.pt, joao.prata@mail.telepac.pt

Abstract. Quantumness and separability criteria for continuous variable systems are discussed for the case of a noncommutative (NC) phase-space. In particular, the quantum nature and the entanglement configuration of NC two-mode Gaussian states are examined. Two families of covariance matrices describing standard quantum mechanics (QM) separable states are deformed into a NC QM configuration and then investigated through the positive partial transpose criterium for identifying quantum entanglement. It is shown that the entanglement of Gaussian states may be exclusively induced by switching on the NC deformation. Extensions of some preliminary results are presented.

1. Introduction

The most salient quantum features in the context of a phase-space noncommutative (NC) extension of quantum mechanics (QM) have already been quantified [1] and issues such as quantum entanglement [2], quantum beating, wave function collapse and loss of quantum coherence [3] have also been examined. These findings allow for a broader understanding of the correspondence between quantum and classical systems, as well as for probing the role of the NC deformation when tuning to standard QM predictions.

The NC extensions of QM show an impressive range of characteristic features when concerned on studies of the quantum Hall effect [4], on the the gravitational quantum well for ultra-cold neutrons [5], on the Landau level and the $2D$ harmonic oscillator problems in the phase-space [6, 7], and as a probe for quantum beating and missing information effects [3] as well as for quantum entanglement issues [2]. This NC QM scenario also admits putative violations of the Robertson-Schrödinger uncertainty relation [8]. On a broader context, noncommutativity is believed to be relevant in the formalism of quantum gravity and string theory [9, 10, 11]. Furthermore, in the framework of quantum cosmology, phase-space noncommutativity has shown to bring up novel features in what concerns the black hole singularity [12, 13, 14, 15], as well as to the equivalence principle [16].

From the theoretical perspective, the NC QM is supported by extensions of the Heisenberg-Weyl algebra [1, 5, 17, 18, 19]. The theory lives in a $2d$-dimensional phase-space where time is required to be a commutative parameter, and coordinate and momentum variables obey a NC algebra. In particular, the NC extension of QM is more suitably formulated in the Weyl-Wigner-Groenewold-Moyal (WWGM) formalism for QM [20, 21, 22]. From the subjacent algebra perspective, the phase-space NC generalizations of QM are based on extensions of the Heisenberg-Weyl algebra [1, 17, 23]. NC QM shows also to have a bearing on issues such as the Osawa's uncertainty principle (OUP)[24, 25, 26] as discussed in Ref. [27]. Relevantly, the OUP is claimed to be the one supported experimentally [28, 29].

The proposal of this contribution is to extend the formalism for describing quantumness and separability of two-mode Gaussian states, setting up a NC criterion for entanglement. We report on some previous results on the phase-space NC QM [2], more specifically on the role of phase-space noncommutativity to yield Gaussian entangled states. Concerning the framework for identifying the separability and the entanglement of quantum Gaussian states, one of the main results reported in this manuscript is related to the positive partial transposed (PPT) separability criterion [30, 31]. Generically, the PPT criterion provides a necessary and, in some cases, sufficient condition for distinguishing between separable and entangled states in discrete quantum systems. The framework has been extended to continuous variable systems [32] by implementing the partial transpose operation as a mirror reflection in the Wigner phase-space. The continuous PPT criterion presented in the theory of quantum information of Gaussian states [33, 34, 35] has been examined in the context of identifying the entanglement properties exclusively induced by a NC deformation of phase-space [2]. Besides recovering the results of Ref. [2], hereon novel results for typical Gaussian states are presented.

2. Noncommutative criterion for entanglement and separability

Consider a bipartite quantum system described in terms of a $2n_A$-dimensional subsystem A and a $2n_B$-dimensional subsystem B with $n_A+n_B = n$, one can write the collective degrees of freedom of the composite quantum system as $\hat{z} = (\hat{z}^A, \hat{z}^B)$, where $\hat{z}^A = (\hat{x}^A_1, \cdots, \hat{x}^A_{n_A}, \hat{p}^A_1, \cdots, \hat{p}^A_{n_A})$ and $\hat{z}^B = (\hat{x}^B_1, \cdots, \hat{x}^B_{n_B}, \hat{p}^B_1, \cdots, \hat{p}^B_{n_B})$, which correspond to the generalized coordinates of the two subsystems [2]. The assumed commutation relations given by

$$[\hat{z}_i, \hat{z}_j] = i\Omega_{ij}, \qquad i, j = 1, \cdots, 2n, \tag{1}$$

are described in terms of the associated matrix given by $\mathbf{\Omega} = [\Omega_{ij}] \equiv Diag\left[\mathbf{\Omega^A}, \mathbf{\Omega^B}\right]$, where $\mathbf{\Omega^K}$, with $K = A$ and B, is a real skew-symmetric non-singular $2n_K \times 2n_K$ matrix of the form

$$\mathbf{\Omega^K} = \begin{pmatrix} \mathbf{\Theta^K} & \mathbf{I^K} \\ -\mathbf{I^K} & \mathbf{\Upsilon^K} \end{pmatrix}, \tag{2}$$

where $\mathbf{\Theta^K} = \left[\theta^K_{ij}\right]$ and $\mathbf{\Upsilon^K} = \left[\eta^K_{ij}\right]$, respectively, measure the noncommutativity of the position-position and momentum-momentum sectors of the subsystems A and B, with $\mathbf{I^K}$ as the $n_K \times n_K$ identity matrix (and it has been set $\hbar = 1$). The NC structure can be formulated in terms of commuting variables by considering a linear Darboux transformation (DT), $\hat{z} = \mathbf{S}\hat{\zeta}$, where $\hat{\zeta} = (\hat{\zeta}^A, \hat{\zeta}^B)$, with $\hat{\zeta}^A = (\hat{q}^A_1, \cdots, \hat{q}^A_{n_A}, \hat{k}^A_1, \cdots, \hat{k}^A_{n_A})$ and $\hat{\zeta}^B = (\hat{q}^B_1, \cdots, \hat{q}^B_{n_B}, \hat{k}^B_1, \cdots, \hat{k}^B_{n_B})$ satisfying the usual QM commutation relations:

$$\left[\hat{\zeta}_i, \hat{\zeta}_j\right] = iJ_{ij}, \qquad i, j = 1, \cdots, 2n, \tag{3}$$

where $\mathbf{J} = [J_{ij}] = Diag\left[\mathbf{J^A}, \mathbf{J^B}\right]$ with

$$\mathbf{J^K} = -(\mathbf{J^K})^T = -(\mathbf{J^K})^{-1} = \begin{pmatrix} \mathbf{0} & \mathbf{I^K} \\ -\mathbf{I^K} & \mathbf{0} \end{pmatrix}, \tag{4}$$

which is a $2n_K \times 2n_K$ standard symplectic matrix for $K = A$ and B. The linear transformation $\mathbf{S} \in Gl(2n)$ is such that $\mathbf{S} = Diag\left[\mathbf{S^A}, \mathbf{S^B}\right]$, and $\mathbf{\Omega} = \mathbf{SJS}^T$, or in terms of submatrices labelled by K, $\mathbf{\Omega^K} = \mathbf{S^K J^K (S^K)}^T$. Giving that the map $\mathbf{S}$ is not uniquely defined, if one composes $\mathbf{S}$ with block-diagonal canonical transformations, an equally valid DT is obtained.

For a composite system described by the density matrix ρ, function of the NC variables $\widehat{z}$, the DT yields the density matrix $\widetilde{\rho}(\widehat{\zeta}) = \rho\left(\mathbf{S}\widehat{\zeta}\right)$, which is associated with the Wigner function:

$$W\widetilde{\rho}(\zeta) = \frac{1}{(2\pi)^n} \int_{R^{n_A}} dy^A \int_{R^{n_B}} dy^B e^{-i\left(y^A \cdot k^A + y^B \cdot k^B\right)} \left\langle q^A + \frac{y^A}{2}, q^B + \frac{y^B}{2} \,|\widetilde{\rho}|\, q^A - \frac{y^A}{2}, q^B - \frac{y^B}{2} \right\rangle. \tag{5}$$

Upon inversion of the DT, one obtains the NC Wigner function [1]:

$$W\rho(z) = (\det \mathbf{\Omega})^{-1/2} W\widetilde{\rho}(\mathbf{S}^{-1}z). \tag{6}$$

If $\mathbf{\Sigma}$ denotes the covariance matrix of $W\rho$ and $\widetilde{\mathbf{\Sigma}}$ that of $W\widetilde{\rho}$, then the two are related by:

$$\mathbf{\Sigma} = \mathbf{S}\widetilde{\mathbf{\Sigma}}\mathbf{S}^T. \tag{7}$$

A necessary condition for the phase-space function $W\widetilde{\rho}$ with covariance matrix $\widetilde{\mathbf{\Sigma}}$ to be an admissible (commutative) Wigner function is that it satisfies the Robertson-Schrödinger uncertainty principle (RSUP),

$$\widetilde{\mathbf{\Sigma}} + (i/2)\mathbf{J} \geq 0, \tag{8}$$

that is, a $2n \times 2n$ positive matrix in C^{2n}. From this condition and Eq. (7) one concludes that for $W\rho$ to be an equally admissible NC Wigner function, it has to satisfy the NC RSUP [8]:

$$\mathbf{\Sigma} + (i/2)\mathbf{\Omega} \geq 0. \tag{9}$$

For Gaussian states this condition is also sufficient [8].

A composite quantum system is separable if its density matrix takes the form $\rho = \sum_{i=1}^{\infty} \lambda_i \rho_i^A \otimes \rho_i^B$, where $0 \leq \lambda_i \leq 1$, for all $i \in N$, $\sum_{i=1}^{\infty} \lambda_i = 1$ and ρ_i^A (resp. ρ_i^B) is a density matrix which involves only A (resp. B) coordinates $\widehat{z}^A$ (resp. $\widehat{z}^B$). The associated Wigner function is:

$$W\rho(z) = \sum_{i=1}^{\infty} \lambda_i W\rho_i^A(z^A) W\rho_i^B(z^B). \tag{10}$$

and, through the DT, the commutative counterpart in terms of the commutative variables ζ is:

$$W\widetilde{\rho}(\zeta) = \sum_{i=1}^{\infty} \lambda_i W\widetilde{\rho}_i^A(\zeta^A) W\widetilde{\rho}_i^B(\zeta^B). \tag{11}$$

To simplify the manipulation of the above results, let one defines $\mathbf{\Lambda}$ to be the $2n \times 2n$ matrix $\mathbf{\Lambda} = Diag\left[\mathbf{I^A}, \mathbf{\Lambda^B}\right]$, with $\mathbf{\Lambda^B} = Diag\left[\mathbf{I}, -\mathbf{I}\right]$. Thus, the transformation $\zeta \mapsto \mathbf{\Lambda}\zeta$ amounts to a mirror reflection of Bob's momenta.

Thus, through the PPT criterion, if a Wigner function $W\widetilde{\rho}(\zeta)$ is that of a separable state, then the transformation

$$W\widetilde{\rho}(\zeta) \mapsto W\widetilde{\rho}'(\zeta) = W\widetilde{\rho}(\mathbf{\Lambda}\zeta), \tag{12}$$

leads to an equivalent Wigner function. Hence, if the state $\widetilde{\rho}$ is separable then

$$\widetilde{\mathbf{\Sigma}}' + (i/2)\mathbf{J} \geq 0 \tag{13}$$

where $\widetilde{\mathbf{\Sigma}}'$ is the covariance matrix of $W\widetilde{\rho}'(\zeta)$. Eq. (12) can then be rewritten in terms of the NC variables as follows:

$$W\rho(z) \mapsto W\rho'(z) = W\rho(\mathbf{D}z), \tag{14}$$

where $W\rho(z) = \frac{1}{\sqrt{\det \mathbf{\Omega}}} W\widetilde{\rho}(\mathbf{S}^{-1}z)$ and $W\rho'(z) = \frac{1}{\sqrt{\det \mathbf{\Omega}}} W\widetilde{\rho}'(\mathbf{S}^{-1}z)$ are defined accordingly to Eq. (6) and $\mathbf{D} = \mathbf{D}^{-1} = \mathbf{S}\mathbf{\Lambda}\mathbf{S}^{-1} = Diag[\mathbf{I}^{\mathbf{A}}, \mathbf{S}^{\mathbf{B}}\mathbf{\Lambda}^{\mathbf{B}}(\mathbf{S}^{\mathbf{B}})^{-1}]$. It follows from Eq. (7) that the covariance matrices $\mathbf{\Sigma}'$ and $\widetilde{\mathbf{\Sigma}}'$ of $W\rho'$ and $W\widetilde{\rho}'$ are related by $\mathbf{\Sigma}' = \mathbf{S}\widetilde{\mathbf{\Sigma}}'\mathbf{S}^T$. Hence, the separability condition Eq. (13) can then be written exclusively in terms of the NC objects

$$\mathbf{\Sigma}' + (i/2)\mathbf{\Omega} \geq 0 \ . \tag{15}$$

In addition, notice that $\mathbf{\Sigma}' = \mathbf{D}\mathbf{\Sigma}\mathbf{D}^T$ where $\mathbf{\Sigma}$ is the covariance matrix of $W\rho$. Defining $\mathbf{\Omega}' = \mathbf{D}^{-1}\mathbf{\Omega}(\mathbf{D}^T)^{-1}$ one obtains

$$\mathbf{\Sigma} + (i/2)\mathbf{\Omega}' \geq 0 \ . \tag{16}$$

We point out that the matrix $\mathbf{\Omega}'$ is simply given by

$$\mathbf{\Omega}' = Diag\left[\mathbf{\Omega}^{\mathbf{A}}, -\mathbf{\Omega}^{\mathbf{B}}\right], \tag{17}$$

where one has used the definition of $\mathbf{\Omega}$ and the fact that $\mathbf{\Lambda}^{\mathbf{B}}$ is an anti-symplectic transformation, i. e. $\mathbf{\Lambda}^{\mathbf{B}}\mathbf{J}^{\mathbf{B}}\mathbf{\Lambda}^{\mathbf{B}} = -\mathbf{J}^{\mathbf{B}}$. By itself, this result is an elegant generalization of the result from Ref. [32] which states that $\mathbf{J}$ is replaced by $\mathbf{J}' = Diag\left[\mathbf{J}^A, -\mathbf{J}^B\right]$ under PPT. We stress the fact that Eq. (17) is valid assuming that the DTs take the block-diagonal form $\mathbf{S} = Diag\left[\mathbf{S}^{\mathbf{A}}, \mathbf{S}^{\mathbf{B}}\right]$.

3. Results for 2-dim Gaussian states

A real normalizable phase-space Gaussian function $F(z)$ with covariance matrix $\mathbf{\Sigma}$ is the NC Wigner function of a quantum separable state if and only if $\mathbf{\Sigma}$ satisfies the Eqs. (9) and (16) with respect to quantum nature and separability, respectively. As suggested in [2], let one consider a Gaussian state written as

$$F(z) = \frac{1}{\pi^4\sqrt{\det \mathbf{\Sigma}}} \exp(-z^T\mathbf{\Sigma}^{-1}z), \tag{18}$$

which lives on a 8-dimensional NC phase-space with $n_A = n_B = 2$, $\mathbf{\Omega}^{\mathbf{A}} = \mathbf{\Omega}^{\mathbf{B}}$, $\theta_{ij} = \theta\epsilon_{ij}$ and $\eta_{ij} = \eta\epsilon_{ij}$, with i, $j = 1$, 2, and where the covariance matrix is given by,

$$\mathbf{\Sigma} = \frac{b}{2}\begin{pmatrix} \mathbf{I}_4 & \gamma^{\mathbf{T}} \\ \gamma & \mathbf{I}_4 \end{pmatrix}, \tag{19}$$

for which the DT corresponds to the map $\mathbf{S} = Diag[\mathbf{S}^{\mathbf{A}}, \mathbf{S}^{\mathbf{B}}]$, with

$$\mathbf{S}^{\mathbf{A}} = \mathbf{S}^{\mathbf{B}} = \begin{pmatrix} \lambda & 0 & 0 & -\theta/2\lambda \\ 0 & \lambda & \theta/2\lambda & 0 \\ 0 & \eta/2\mu & \mu & 0 \\ \eta/2\mu & 0 & 0 & \mu \end{pmatrix}, \tag{20}$$

subject to $\lambda\mu = (1+\sqrt{1-\eta\theta})/2$, which ensures the map invertibility. The NC parameters, θ and η, are real positive constants satisfying the condition $\theta\eta < 1$.

The NC quantumness and separability criteria are expressed in terms of the so-called NC symplectic spectrum. For a $2n \times 2n$ real symmetric positive-definite matrix $\tilde{\mathbf{\Sigma}}$ it is the set of eigenvalues of the matrix $2i\mathbf{J}^{-1}\tilde{\mathbf{\Sigma}}$ and these eigenvalues are called the Williamson invariants of $\tilde{\mathbf{\Sigma}}$. They are all positive and $\tilde{\nu}_-$ denotes the smallest Williamson invariant. By Williamson's Theorem [32], one can show that $\tilde{\mathbf{\Sigma}}$ complies with the RSUP ($\tilde{\mathbf{\Sigma}} + (i/2)\mathbf{J} \geq 0$) if and only if $\tilde{\nu}_- \geq 1$. By the same token, one calls the set of eigenvalues of $2i\mathbf{\Omega}^{-1}\mathbf{\Sigma}$ the NC symplectic spectrum of $\mathbf{\Sigma}$ and the eigenvalues are called the NC Williamson invariants. The smallest NC Williamson invariant of $\mathbf{\Sigma}$ is denoted by ν_-.

First example - One considers the non-block-diagonal elements of γ as

$$\gamma = \begin{pmatrix} n & 0 & m & 0 \\ 0 & n & 0 & -m \\ m & 0 & -n & 0 \\ 0 & -m & 0 & -n \end{pmatrix}, \tag{21}$$

where b, m and n are real parameters which are assumed, for simplicity, as to be constrained by the relations $R = \sqrt{m^2+n^2}$, $b = (1+R)/(1-R)$. In this case, the smallest NC Williamson invariants of $\mathbf{\Sigma}$ and $\mathbf{\Sigma}'$ are given by [2]

$$\nu_- = \frac{1}{1-\eta\theta}\frac{1+R}{1-R}\sqrt{\frac{\omega_-}{2} - \sqrt{\frac{\omega_-^2}{4} - (1-R^2)^2(1-\eta\theta)^2}}, \tag{22}$$

$$\nu'_- = \frac{1}{1-\eta\theta}\frac{1+R}{1-R}\sqrt{\frac{\omega_+}{2} - \sqrt{\frac{\omega_+^2}{4} - (1-R^2)^2(1-\eta\theta)^2}}, \tag{23}$$

respectively, where $\omega_\pm = 2(1\pm n^2) + (1\mp n^2)(\eta^2+\theta^2) \pm 2m^2(1+\eta\theta) + n(1\mp 1)|\eta^2-\theta^2| + 2m(1\pm 1)(\eta+\theta)$, which allows for examining the role of the NC parameters on the Gaussian entanglement. The NC quantum nature and the separability of Gaussian states are ensured for $\nu_- \geq 1$ and $\nu'_- \geq 1$, respectively. Entangled quantum states are found in the range $\nu_- \geq 1 > \nu'_-$. Setting θ(either η) equals to constant values, the obtained NC Williamson invariants, ν_- and ν'_-, depend exclusively on η(or θ), as depicted in Fig. 1. The standard (commutative) QM limit is obtained by setting $\theta = \eta = 0$, which implies that $\nu_- = (1+R)^{3/2}/(1-R)^{1/2}$ and $\nu'_- = 1+R$. This means that for $0 \leq R < 1$, all states are quantum and separable, given that $\nu_- \geq 1$ and $\nu'_- \geq 1$, respectively. Fig. 2 shows how the NC phase-space ($\theta \neq 0 \neq \eta$) induces the entanglement of the corresponding Gaussian states. The influence of the NC parameters, θ and η, on the quantum nature, separability and entanglement of states can be depicted for several values of R (with a degeneracy associated to $m \leftrightarrow n$).

Second example - One considers the non-block-diagonal elements of γ as

$$\gamma = \begin{pmatrix} n & 0 & 0 & -m \\ 0 & n & m & 0 \\ 0 & m & -n & 0 \\ -m & 0 & 0 & -n \end{pmatrix}, \tag{24}$$

where again the parameters are constrained by the relations $R = \sqrt{m^2+n^2}$, $b = (1+R)/(1-R)$. In this case, the expression for the smallest NC Williamson invariants are too extensive to be written explicitly. We just depict the results from Figs. 3 and 4 from which one can notice that the symmetry involving the dependence on $\eta \leftrightarrow \theta$ disappears.

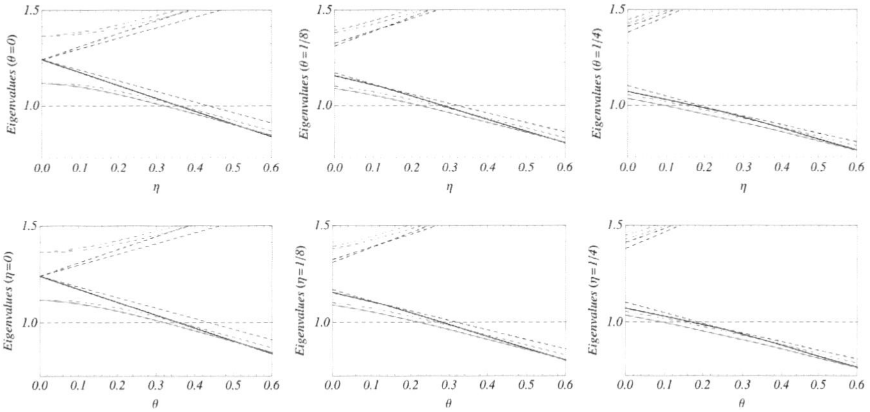

Figure 1. Eigenvalues ν (black lines) and ν' (red lines) for $\theta = 0$, 1/8, and 1/4 as function of η in the range $[0, 0.6]$ (first set) and for $\eta = 0$, 1/8, and 1/4 as function of θ in the range $[0, 0.6]$. Solid lines correspond to the respective smallest eigenvalues, ν_- and ν'_-. Notice that entanglement ($\nu'_- < 1$) coexists with quantum behavior ($\nu_- \geq 1$) for $\eta \neq 0$. Plots are for $m = R/10$ and $n = 3\sqrt{11}R/10$ and the results are in correspondence with the first example.

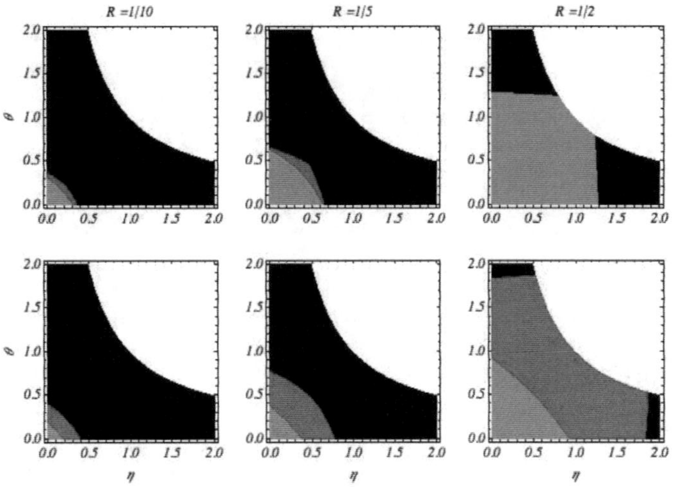

Figure 2. The entanglement properties of the NC Gaussian states with the covariance matrix constrained by $R = 1/10$, 1/5, and 1/2. Plots are for $m = R/10$ and $n = 3\sqrt{11}R/10$ (first line) and for $m = 3\sqrt{11}R/10$ and $n = R/10$ (second line). Gray and red regions correspond respectively to separable ($\tilde{\nu}_- \geq 1$) and entangled ($\nu'_- < 1$) quantum states ($\nu'_- \geq 1$). Black region denotes the violation of the RSUP ($\nu_- < 1$) and the white region is out of the region bound by $\theta\eta \leq 1$. The results are in correspondence with Fig. 1.

Journal of Physics: Conference Series **626** (2015) 012046 doi:10.1088/1742-6596/626/1/012046

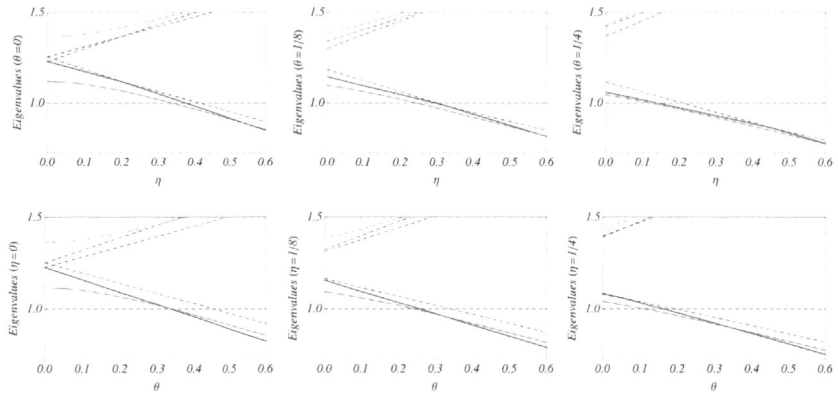

Figure 3. Eigenvalues ν (black lines) and ν' (red lines) for $\theta = 0$, $1/8$, and $1/4$ as function of η in the range $[0, 0.6]$ (first set) and for $\eta = 0$, $1/8$, and $1/4$ as function of θ in the range $[0, 0.6]$. Plots are for $m = R/10$ and $n = 3\sqrt{11}R/10$ and the results are in correspondence with the second example.

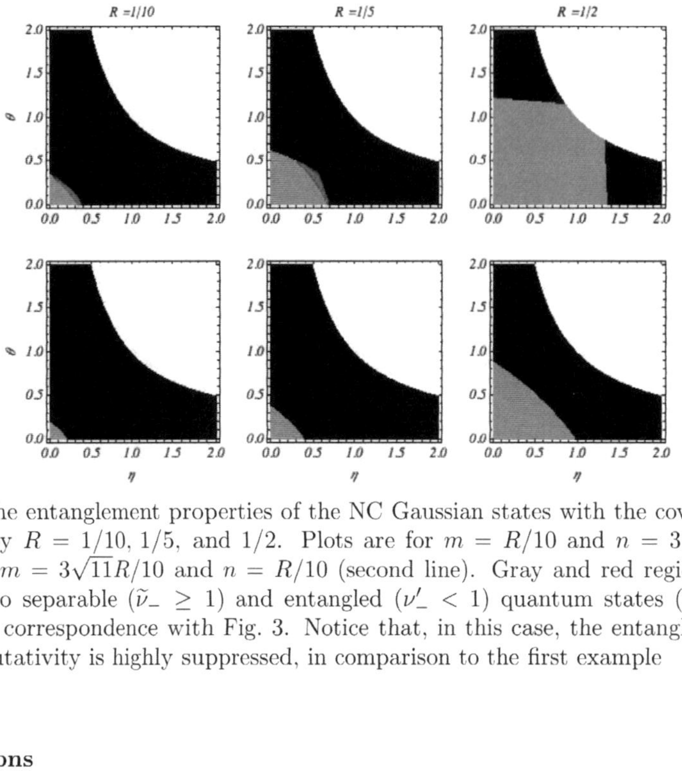

Figure 4. The entanglement properties of the NC Gaussian states with the covariance matrix constrained by $R = 1/10$, $1/5$, and $1/2$. Plots are for $m = R/10$ and $n = 3\sqrt{11}R/10$ (first line) and for $m = 3\sqrt{11}R/10$ and $n = R/10$ (second line). Gray and red regions correspond respectively to separable ($\tilde{\nu}_- \geq 1$) and entangled ($\nu'_- < 1$) quantum states ($\nu'_- \geq 1$). The results are in correspondence with Fig. 3. Notice that, in this case, the entanglement induced by noncommutativity is highly suppressed, in comparison to the first example

4. Conclusions

Results derived from an extension of the PPT criterion for investigating the entanglement and the separability in the phase-space NC QM are discussed and the framework has been considered in the analysis of a novel Gaussian state configuration. Typical constraints on the smallest symplectic eigenvalue of the partially transposed covariance matrix of the state were defined in order to quantify the effect of noncommutativity on the quantumness and separability properties of Gaussian states. Once again, separable standard QM two-mode Gaussian states have been shown to exhibit quantum entanglement, which is exclusively driven by the NC deformation of the phase-space. This means that noncommutativity can by itself induce the entanglement of Gaussian states. This also reinforces our previous results which imply that QM effects is a timed version of a more encompassing phase-space NC structure [1, 8, 26].

Acknowledgments

AEB thanks for the financial support from the Brazilian Agency CAPES (grant nr. AEX 4979/14-1).

References

[1] Bastos C, Bertolami O, Dias N C and Prata J N 2008 *J. Math. Phys.* **49** 072101
[2] Bastos C, Bernardini A E, Bertolami O, Dias N C and Prata J N 2013 *Phys. Rev.* **D 88** 085013
[3] Bernardini A E and Bertolami O 2013 *Phys. Rev.* **A 88** 012101
[4] Prange R and Girvin S 1987 *The Quantum Hall Effect* (New York: Springer)
[5] Bertolami O, Rosa J G, de Aragão C M L, Castorina P and Zappalà D 2005 *Phys. Rev.* **D72** 025010
[6] Douglas M R and Nekrasov N A 2001 *Rev. Mod. Phys.* **73** 977
[7] Rosenbaum M and David Vergara J 2006 *Gen. Rel. Grav.* **38** 607
[8] Bastos C, Bertolami O, Dias N C and Prata J N 2012 *Phys. Rev.* **D86** 105030
[9] Connes A, Douglas M R and Schwarz A 1998 *JHEP* **02** 003
[10] Douglas M R and Hull C 1998 *JHEP* **02** 008
[11] Seiberg N and Witten E 1999 *JHEP* **9909** 032
[12] Bastos C, Bertolami O, Dias N C and Prata J N 2008 *Phys. Rev.* **D78** 023516
[13] Bastos C, Bertolami O, Dias N C and Prata J N 2009 *Phys. Rev.* **D80** 124038
[14] Bastos C, Bertolami O, Dias N C and Prata J N 2010 *Phys. Rev.* **D82** 041502
[15] Bastos C, Bertolami O, Dias N C and Prata J N 2011 *Phys. Rev.* **D84** 024005
[16] Bastos C, Bertolami O, Dias N C and Prata J N 2011 *Class. Quant. Grav.* **28** 125007
[17] Dias N C and Prata J N 2009 *Annals Phys.* **324** 73
[18] Bastos C and Bertolami O 2008 *Phys. Lett.* **A372** 5556
[19] Gamboa J, Loewe M and Rojas J C 2001 *Phys. Rev.* **D64** 067901
[20] Groenewold H 1946 *Physica* **12** 405
[21] Moyal J 1949 *Proc. Camb. Phil. Soc.* **45** 99
[22] Wigner E 1932 *Phys. Rev.* **40** 749
[23] Dias N C, de Gosson M, Luef F and Prata J N 2010 *J. Math. Phys.* **51** 072101
[24] Ozawa M 1988 *Phys. Rev. Lett.* **60** 385
[25] Ozawa M 2002 *Phys. Lett.* **A299** 1
[26] Bastos C, Bernardini A E, Bertolami O, Dias N C and Prata J N 2014 *Phys. Rev.* **A 89** 042112
[27] Bastos C, Bernardini A E, Bertolami O, Dias N C and Prata J N 2014 *Phys. Rev.* **D 90** 045023
[28] Rozema L A, Darabi A, Mahler D H, Hayat A, Soudagar Y, Steinberg A M 2012 *Phys. Rev. Lett.* **109** 100404
[29] Sulyok G, Sponar S, Erhart J, Badurek G, Ozawa M and Hasegawa Y 2013 *Phys. Rev.* **A88** 022110
[30] Peres A 1996 *Phys. Rev. Lett.* **77** 1413
[31] Horodecki P 1997 *Phys. Lett.* **A232** 333
[32] R. Simon, E. C. G. Sudarshan and N. Mukunda 1987 *Phys. Rev.* **A 36** 3868
[33] Adesso G and Illuminati F 2007 *J. Phys. A: Math. Theor.* **40** 7821
[34] Adesso G, Serafini A and Illuminati F 2004 *Phys. Rev. Lett.* **92** 087901
[35] Weedbrook C, Pirandola S, Garcia-Patron R, Cerf N J, Ralph T C, Shapiro J H, and Lloyd S 2012 *Rev. Mod. Phys.* **84** 671

Journal of Physics: Conference Series **626** (2015) 012047 doi:10.1088/1742-6596/626/1/012047

A constructive presentation of rigged Hilbert spaces

Enrico Celeghini

Dipartimento di Fisica, Università di Firenze, I50019 Sesto Fiorentino, Firenze, Italy
Depatamento de Fisica Téorica, Universidad de Valladolid, E-47005, Valladolid, Spain

E-mail: `celeghini@fi.infn.it`

Abstract. We construct a rigged Hilbert space for the square integrable functions on the line $L^2(\mathbb{R})$ adding to the generators of the Weyl-Heisenberg algebra a new discrete operator, related to the degree of the Hermite polynomials. All together, continuous and discrete operators, constitute the generators of the projective algebra $io(2)$. $L^2(\mathbb{R})$ and the vector space of the line $\mathbb{R}$ are shown to be isomorphic representations of such an algebra and, as both these representations are irreducible, all operators defined on the rigged Hilbert spaces $L^2(\mathbb{R})$ or $\mathbb{R}$ are shown to belong to the universal enveloping algebra of $io(2)$. The procedure can be extended to orthogonal and pseudo-orthogonal spaces of arbitrary dimension by tensorialization.

Circumventing all formal problems the paper proposes a kind of toy model, well defined from a mathematical point of view, of rigged Hilbert spaces where, in contrast with the Hilbert spaces, operators with different cardinality are allowed.

1. introduction

Physical quantities we take into account in quantum mechanics are both discrete and continuous.

In the free particle case, position and energy have both the cardinality $\aleph_1$ but, when we consider localized states, the energy has a discrete spectrum i.e. the cardinality $\aleph_0$, while position remains continuous.

In a formal description, this causes problems as, in a Hilbert space (HS), dimensions are well defined so that we cannot in the same HS have operators with a spectrum of different cardinality.

For this reason the standard approach depicts localized systems in a countable HS where energy is diagonalizable while position, that is continuous, is described by means of elaborated limits on functions with compact support. We are indeed compelled to a complicated and formally unsatisfactory description of position that cannot be described by an operator.

In reality at the beginning of quantum mechanics this problem was not there, as a physical state was not described by a vector in a HS but as a ray in such a Hilbert space i.e. it was associated to a whole family of vectors defined up to an arbitrary complex number. The one-to-one correspondence was thus between a physical state and an element of a rigged Hilbert space (RHS), an intricate concept involving a Gelfand triple $\phi \subseteq \mathcal{H} \subseteq \phi'$, where $\mathcal{H}$ is a Hilbert space, ϕ (dense subset of $\mathcal{H}$) is the "*ket*" space and ϕ' (dual of ϕ) is the "*bra*" space [1]. However quickly RHS has been considered an unnecessary complication as the results can be found, at least at the formal level usually accepted by physicists, in the Hilbert space obtained representing, by means of the axiom of choice, each entire ray with a vector of norm one and phase zero. In concrete the expectation value of the operator Ω in the state $|\psi\rangle$ was defined as [2]:

$$\langle\,\Omega\,\rangle \;=\; \frac{\langle\psi|\Omega|\psi\rangle}{\langle\psi|\psi\;\rangle}\;\;;$$

where the norm of the vector $|\psi\rangle$ is arbitrary and the operator Ω can relate vectors not necessarily of the same norm. In the evolution of the theory this freedom has been disregarded and a one-to-one relation was established between physical states and normalized vectors in a Hilbert space, preserving only the phase freedom related to the gauge theories.

Furthermore also group theory suggests that the "right" space could be the RHS and not the simpler HS. For instance in [3] continuous bases are described jointed to the discrete one into the same representation of $SU(1,1)$ and the HS defined by the discrete basis is implemented introducing the space of differentiable vectors and its dual ϕ' that contains the "generalized eigenvectors" of the non compact generators [4].

Moreover by means of rigged Hilbert spaces the Dirac formalism can be reproduced, for the harmonic oscillator, as shown by Böhm in his book [1].

We do not attempt to discuss here the formal properties of RHS , as did Böhm [1], but we exhibit a one dimensional example where all formal problems are circumvented using special functions. Anyway this example is in reality quite more then an example as it can be easily extended, by tensorial construction, to orthogonal and pseudo-orthogonal spaces of any dimension.

In conclusion, let us stress that the difference between RHS formalism and the usual HS one appears to be minor from the physicists point of view but is essential from the mathematical point of view and leads to a tremendous mathematical simplification: in fact it justifies the mathematically undefined operations that the physicists have been accustomed to in their calculations.

Our fundamental statement is thus that the restriction from RHS to HS is unjustified in the sense that we lose more of what we gain in the reduction, because in RHS all is mathematically well defined and all observables can be correctly described. In addition, as shown in the following, RHS allow to include inside Lie algebras and Lie universal enveloping algebras operators with spectrum of different cardinality.

The fundamental ingredients of the paper are well known:

1) Hermite functions with their discrete label and continuous variable that are described by observables of different cardinality.

2) Lie algebras, groups and representations of special functions.

A more detailed discussion of the algebraic properties of special functions and of their role as transition matrices between discrete and continuous bases can be found in [5], while in [6] the technical aspects of the introduction of RHS by means of special functions are discussed, showing the solidity of the approach.

Here, on the contrary, our attention is addressed to stress that the RHS is an enough simple and mathematically satisfactory formalism.

2. Hermite functions $\{\psi_n(x)\}$ as localized wave functions on the line

As a first step, let us introduce the Hermite functions [7]

$$\psi_n(x) := \frac{e^{-x^2/2}}{\sqrt{2^n n! \sqrt{\pi}}}\, H_n(x)\,.$$

As [8]

$$\int_{-\infty}^{\infty} \psi_n(x)\;\psi_{n'}(x)\;dx = \delta_{n,n'}\;, \qquad\qquad \sum_{n=0}^{\infty} \psi_n(x)\;\psi_n(x') \;=\; \delta(x-x')\,,$$

Journal of Physics: Conference Series **626** (2015) 012047 doi:10.1088/1742-6596/626/1/012047

$\{\psi_n(x)\}$ is a basis of the space of real/complex square integrable functions on the line $L^2((-\infty,\infty)) \equiv L^2(\mathbb{R})$.

Using the freedom of fixing the origin as well as the scale of position and momentum, Hermite functions can be generalized to $\psi[n, x_0, s, x]$ so that

$$\psi[n, x_0, s, x] := \frac{e^{-\frac{(x-x_0)^2}{2s^2}}}{\sqrt{2^n n! s\sqrt{\pi}}} H_n\left[(x-x_0)/s\right]$$

$$\psi[n, p_0, s, p] := \frac{e^{-\frac{(p-p_0)^2 s^2}{2}}\sqrt{s}}{\sqrt{2^n n!\sqrt{\pi}}} H_n\left[(p-p_0)s\right]$$

$$\int_{-\infty}^{\infty} \psi[n, x_0, s, x]\ \psi[n', x_0, s, x]\ dx \ = \ \delta_{nn'}$$

$$\sum_{n=0}^{\infty} \psi[n, x_0, s, x]\ \psi[n, x_0',, s, x'] \ = \ s\ \delta[(x-x_0)-(x'-x_0')] \,.$$

Among the ∞-many bases of square integrable functions on the line $L^2(\mathbb{R})$, Hermite functions are particularly suitable to connect classical and quantum physics. Indeed they describe wave packets in the position and, as $\psi_n(x)$ is an eigenvector of the Fourier transform, also in the momentum; so that they correspond to our intuitive vision of quantum mechanics and, in the appropriate limit, allow to reconstruct the classical pattern.

Note also that they allow to describe systems with the appropriate behavior in function of energy. The minimal indetermination corresponds to the state with $n=0$, while as n increases, and with n the energy of the state, also the indetermination increases.

Indeed as

$$\int_{-\infty}^{\infty} (x-x_0)^2\ \psi[n, x_0, s, x]^2\ dx \ = \ (n+1/2)\ s^2\ ,$$

$$\int_{-\infty}^{\infty} \psi[n, x_0, s, x]\ \partial_x^2\, \psi[n, x_0, s, x]\ dx \ = \ -(n+1/2)/s^2\ ,$$

we have

$$\Delta X = \sqrt{n+1/2}\ s\,, \qquad \Delta P = \sqrt{n+1/2}/s$$

and thus

$$\Delta X\ \ \Delta P \ = n+1/2\ .$$

3. Algebra of Hermite functions

The basic idea is to introduce the operator N that read the label n of the $\{\psi_n(x)\}$ [5]. In addition to X and $P \equiv \mathbf{i}D_x$ we put thus in the space $L^2(\mathbb{R})$ the operators N and $\mathbb{I}$ such that

$$X\psi_n(x) := x\,\psi_n(x)\,, \quad D_x\psi_n(x) := \psi_n'(x)\,, \quad N\psi_n(x) := n\,\psi_n(x)\,, \quad \mathbb{I}\,\psi_n(x) := \psi_n(x). \qquad (1)$$

This allows to rewrite the recurrence relations of Hermite polynomials

$$H_n'(x) - 2\,x\,H_n(x) \ = \ H_{n+1}(x), \qquad H_n'(x) \ = \ 2\,n\,H_{n-1}(x)$$

as

$$a^\dagger\ \psi_n(x) \;=\; \sqrt{n+1}\ \psi_{n+1}(x)\,, \qquad a\ \psi_n(x) \;=\; \sqrt{n}\ \psi_{n-1}(x), \tag{2}$$

where $a^\dagger$ and a are defined in terms of the hermitian operators X and P

$$a^\dagger := \frac{1}{\sqrt{2}}\,(X - \mathbf{i}P) \quad , \qquad a := \frac{1}{\sqrt{2}}\,(X + \mathbf{i}P) \ .$$

The algebra contains the rising and lowering operators on the Hemite functions [5, 9]

$$[N, a^\dagger] = a^\dagger, \qquad [N, a] = -a\ , \qquad [a, a^\dagger] = \mathbb{I}\ , \qquad [\mathbb{I}, \bullet] = 0$$

and is isomorphic to the projective algebra $io(2)$ [10, 11]:

$$[N, X] = -\mathbf{i}P\ , \qquad [N, P] = \mathbf{i}X\ , \qquad [X, P] = \mathbf{i}\,\mathbb{I}\ , \qquad [\mathbb{I}, \bullet] = 0\,. \tag{3}$$

As discussed in [12], eqs.(1,2) do not define the algebra (3) but only one of its representations. The Casimir operator

$$C \;\equiv\; (X^2 - D_x^2)/2 - (N + 1/2)\,\mathbb{I} \;=\; \{a^\dagger, a\}/2 - (N + 1/2)\,\mathbb{I} \tag{4}$$

has indeed zero value on the square integrable functions on the line $L^2(\mathbb{R})$, where we can assume $\mathbb{I} = 1$ and write

$$C\,\psi_n(x) \;=\; \Big[\{a^\dagger, a\}/2 - N - 1/2\Big]\,\psi_n(x) \;=\; 0$$

or, alternatively,

$$C\,\psi_n(x) \;=\; \left[(X^2 - D_x^2)/2 - N - 1/2\right]\ \psi_n(x) \;=\; 0$$

that, by inspection, are equivalent to the Hermite differential equation:

$$H_n''(x) - 2\,x\,H_n'(x) + 2\,n\,H_n(x) = 0\,.$$

The eq. $C = 0$ can be thus also considered as the operatorial identity that defines $L^2(\mathbb{R})$:

$$N \;\equiv\; \left(X^2 - D_x^2 - 1\right)/2 \;\equiv\; \{a, a^\dagger\}/2 - 1/2\,. \tag{5}$$

Let us stress that, usually, the operator N is included inside the UEA of the Weyl-Heisenberg algebra as $N := a^\dagger a$ [13], while here it has been introduced starting from the label of Hermite polynomials and it has the role of an independent generator of the algebra $io(2)$. Only when the representation $C = 0$ is considered (the one of the Hermite functions) the results of the two approachs coincide.

4. The line $\mathbb{R}$ and its bases

To construct the bases of $\mathbb{R}$, we move now to group theory [14].

We start from the unitary irreducible representations of the translation group T^1

$$P\,|p\rangle \;=\; p\,|p\rangle\,, \qquad\qquad U^p(x)\,|p\rangle \;=\; e^{-\mathbf{i}px}\,|p\rangle\,.$$

The regular representation $\{|p\rangle\}$ $(-\infty < p < \infty)$ is such that

$$\langle\, p\,|\,p'\,\rangle \;=\sqrt{2\pi}\;\delta(p - p')\,, \qquad\qquad \frac{1}{\sqrt{2\pi}}\int_{-\infty}^{+\infty} |p\rangle\,dp\,\langle p| \;=\; \mathbb{I}$$

Now we can move, by means of the Fourier transform, to the continuous group element label x from the irreducible representation label p (also continuous) resorting, in this way, the strict connection of group theory with harmonic analysis.

The conjugate basis $\{|x\rangle\}$, defined by the operator X, is obtained, indeed, as

$$|x\rangle := \left[\frac{1}{\sqrt{2\pi}}\int_{-\infty}^{+\infty} dp\; e^{-\mathbf{i}px}\right]|p\rangle\,, \qquad \langle\, x\,|\,x'\,\rangle = \sqrt{2\pi}\;\delta(x-x')\,, \qquad \frac{1}{\sqrt{2\pi}}\int_{-\infty}^{+\infty}|x\rangle\, dx\,\langle x| = \mathbb{I}$$

and the operators X and P close, together with $\mathbb{I}$, the Weyl-Heisenberg algebra.

Consistently with the previous section, we introduce now the set of vectors $\{|n\rangle\}$

$$|n\rangle := (2\pi)^{-1/4}\int_{-\infty}^{\infty} dx\;\psi_n(x)\,|x\rangle \qquad\qquad n\in\mathbb{N}\,, \tag{6}$$

that, by inspection, is an orthonormal and complete set in $\mathbb{R}$

$$\langle\, n\,|\,n'\,\rangle = \delta_{n\,n'}\,, \qquad\qquad \sum_{n=0}^{\infty}|n\rangle\,\langle n| = \mathbb{I}\,.$$

$\{|n\rangle\}$ is thus a discrete basis in the real line $\mathbb{R}$, i.e. $\mathbb{R}\equiv\{|p\rangle\}\equiv\{|x\rangle\}\equiv\{|n\rangle\}$ and $\{\psi_n(x)\}$ are the transition matrices between $\{|n\rangle\}$ and $\{|x\rangle\}$:

$$\langle n|x\rangle = (2\pi)^{1/4}\;\psi_n(x)\,.$$

Relations among the three bases are easily established, as $\{\psi_n(x)\}$ are eigenvectors of Fourier transform,

$$\left[\frac{1}{\sqrt{2\pi}}\int_{-\infty}^{\infty} dx\; e^{\,\mathbf{i}px}\right]\psi_n(x) = \mathbf{i}^n\;\psi_n(p)\,,$$

$$|\,x\,\rangle = (2\pi)^{1/4}\sum_{n=0}^{\infty}\psi_n(x)\,|\,n\,\rangle\,, \qquad\qquad |p\,\rangle = \left[\frac{1}{\sqrt{2\pi}}\int_{-\infty}^{\infty} dx\, e^{\,\mathbf{i}px}\right]|x\,\rangle\,,$$

$$|n\rangle = \mathbf{i}^n\,(2\pi)^{-1/4}\int_{-\infty}^{\infty} dp\;\psi_n(p)\;|p\rangle\,, \qquad\qquad |\,p\,\rangle = (2\pi)^{1/4}\sum_{n=0}^{\infty}\mathbf{i}^n\;\psi_n(p)\,|\,n\,\rangle\,.$$

For an arbitrary vector $|f\rangle\in\mathbb{R}$ we thus have

$$|f\rangle = \frac{1}{\sqrt{2\pi}}\int_{-\infty}^{+\infty} dx\; f(x)\,|x\rangle = \frac{1}{\sqrt{2\pi}}\int_{-\infty}^{+\infty} dp\; f(p)\;|p\rangle = \sum_{n=0}^{\infty} f_n\,|n\rangle\,,$$

$$f(x) := \langle x|f\rangle = (2\pi)^{1/4}\sum_{n=0}^{\infty}\psi_n(x)\,f_n\,, \qquad f(p) := \langle p|f\rangle = (2\pi)^{1/4}\sum_{n=0}^{\infty}(-\mathbf{i})^n\,\psi_n(p)\,f_n\,,$$

$$f_n := \langle n|f\rangle = (2\pi)^{-1/4}\int_{-\infty}^{+\infty} dx\;\psi_n(x)\,f(x) = \mathbf{i}^n(2\pi)^{-1/4}\int_{-\infty}^{+\infty} dp\;\psi_n(p)\,f(p)$$

and the wave functions $f(x)\,, f(p)$ and the sequence $\{f_n\}$ describe $|f\rangle$ in the three bases.

All seems trivial, but $\{|n\rangle\}$ has the cardinality of the natural numbers $\aleph_0$ and, as all bases in a Hilbert space have the same cardinality, the structure we have constructed (the quantum space on the line $\mathbb{R}$) is not a Hilbert space but a rigged Hilbert space, where

$$|x\rangle\in\phi'\,, \qquad |p\rangle\in\phi'\,, \qquad |n\rangle\in\phi\,.$$

Journal of Physics: Conference Series **626** (2015) 012047 doi:10.1088/1742-6596/626/1/012047

Because of Eq.(6), $\mathbb{R}$ and $L^2(\mathbb{R})$) are isomorphic and we can write the algebra $io(2)$ on $\{|n\rangle\}$ that is, as on $\{\psi_n(x)\}$, a representation of $io(2)$ with zero value of the Casimir operator (4).

Again on the representations we can assume $\mathbb{I} = 1$ and write

$$X|n\rangle = (a^\dagger + a)|n\rangle = \sqrt{n+1}|n+1\rangle + \sqrt{n}|n-1\rangle \, ,$$

$$P|n\rangle = \mathbf{i}(a^\dagger - a)|n\rangle = \mathbf{i}\sqrt{n+1}|n+1\rangle - \mathbf{i}\sqrt{n}|n-1\rangle,$$

$$N|n\rangle = n|n\rangle \, , \qquad \mathbb{I}|n\rangle = |n\rangle \qquad , a^\dagger|n\rangle = \sqrt{n+1}|n+1\rangle \, , \; a|n\rangle = \sqrt{n}|n-1\rangle \, ,$$

$$C\,|n\rangle \;=\; \left[(X^2 - D_x^2)/2 - (N+1/2)\right]|n\rangle \;=\; \left[\{a, a^\dagger\}/2 - N - 1/2\right]|n\rangle \;=\; 0\,,$$

and the identity (5) defines not only the vector space $L^2(\mathbb{R})$ but also the vector space $\mathbb{R}$.

Note that, if we restrict ourselves to a Hilbert space, half of the above presented relations are meaningless.

5. Universal enveloping algebra and operators in a RHS

Let us discuss now the implications of the algebraic discussion of previous section on the operators defined on $L^2(\mathbb{R})$ and $\mathbb{R}$.

In a RHS we have no problems to consider X and P as generators of a Lie algebra together with the number operator N [1, 5]. This allows us to include differential operators inside the algebraic structure and to extend the set of operators defined in the universal enveloping algebra.

Both on $\{\psi_n(x)\}$ and on $\{|n\rangle\}$, the representations are irreducible, so that -on both spaces $L^2(\mathbb{R})$ and $\mathbb{R}$- all operators of the UEA[$io(2)$] are defined and an isomorphism exists between the UEA[$io(2)$] and the space of the operators $\{\mathcal{O}[L^2(\mathbb{R})]\}$ and $\{\mathcal{O}[\mathbb{R}]\}$:

$$\{\mathcal{O}[L^2(\mathbb{R})]\} \;\equiv\; UEA[io(2)] \;\equiv\; \{\mathcal{O}[\mathbb{R}]\} \, ,$$

i.e. each operator $\mathcal{O}$ can be written

$$\mathcal{O} \;= \sum c_{\alpha\,\beta\,\gamma}\, X^\alpha\, D_x{}^\beta\, N^\gamma \;\; = \;\; \sum d_{\alpha\,\beta\,\gamma}\; a^{\dagger\alpha}\, N^\beta\, a^\gamma.$$

¿From the analytical point of view, an ordered monomial $X^\alpha\, D_x{}^\beta\, N^\gamma \in$ UEA[$io(2)$] is an order β differential operator but, because of the operatorial identity (5), we have

$$D_x^2 \;\equiv\; X^2 - 2N - 1,$$

and any operator in $\{\mathcal{O}[L^2(\mathbb{R})]\}$ and in $\{\mathcal{O}(\mathbb{R})\}$ can be written in the form

$$\mathcal{O} \;=\; f_0(X)\, g_0(N) \;+\; f_1(X)\, D_x\, g_1(N)$$

as all higher power of D_x can be removed and substituted by functions of X and N. In particular, on the vector $\psi_n(x)$ we have thus

$$\mathcal{O}\,\psi_n(x) \;=\; f_0(x)\, g_0(n)\, \psi_n(x) \;+\; f_1(x)\, g_1(n)\, \psi'{}_n(x)\,.$$

6. Conclusions

Rigged Hilbert spaces are shown to be more effective than Hilbert spaces in quantum physics as operators of different cardinality can be considered together.

In rigged Hilbert spaces discrete and continuous bases exist togheter. In particular discrete and continuous bases coexist such that special functions are transformation matrices between them.

In RHS -consistently with group theory- operators of different cardinality can be together generators of a Lie algebra or elements of an universal enveloping Lie algebra.

The fundamental point is that, while in a HS hermitian operators with spectrum of different cardinality lead to undefined operations (that, anyway, physicists are used to perform without too much hesitation), in a rigged Hilbert space the theory is mathematically consistent.

We have discussed here the complex rigged Hilbert space of quantum mechanics. The discussion of the real Hilbert space used in signal processing follows exactly the same lines.

References

[1] Böhm A 1978 *The Rigged Hilbert Space and Quantum Mechanics* (Berlin: Springer-Verlag)
[2] Roman P 1965 *Advanced Quantum Theory* (Reading Mass.: Addison-Wesley)
[3] Lindblad G and Nagel B 1970 *Ann. Inst. Henry Poincaré* **13** 27-56
[4] Maurin K 1968 *General eigenfunction expansions and unitary representations of topological groups* (Warzawa: PWN)
[5] Celeghini E and del Olmo M A 2013 *Annals of Physics* **335** 78-85
[6] Celeghini E and del Olmo M A 2014 *Preprint* arXiv:1411.3263 [math-ph], in press in J. Phys.: Conf. Series
[7] Olver F et al. 2010 *NIST Handbook of Mathematical Functions* (Cambridge Mass.: Cambridge Univ. Press)
[8] Folland G 1992 *Fourier Analysis and its Applications* (Belmont, California: Wadsworth)
[9] Lorente M 2001 *J. Phys. A: Math Gen.* **34** 569-88
[10] Hamermesh M 1962 *Group Theory* (Reading Mass.: Addison-Wesley)
[11] Bargmann V 1954 *Ann. of Math.* **59** 1-46
[12] Celeghini E and Tarlini M 1981 *Nuovo Cimento* **65B** 172-180
[13] Coutinho S C 1995 *A Primer of Algebraic D-modules* (Cambridge Mass.: Cambridge Univ. Press)
[14] Wu-Ki Tung 1985 *Group Theory in Physics* (Philadelphia: World Scientific)

Journal of Physics: Conference Series **626** (2015) 012048 doi:10.1088/1742-6596/626/1/012048

Feynman checkerboard picture and neutrino oscillations

Petr Jizba

FNSPE, Czech Technical University in Prague, Břehová 7, 115 19 Praha 1, Czech Republic
ITP, Freie Universität Berlin, Arnimallee 14 D-14195 Berlin, Germany

E-mail: p.jizba@fjfi.cvut.cz

Abstract. In this paper we show how the Feynman checkerboard picture for the 1+1-dimensional Dirac equation can be extended to accommodate the neutrino mixing and ensuing neutrino oscillations.

1. Introduction

The relativistic checkerboard (also chessboard) picture [1, 2, 3] was an attempt by R.P. Feynman to generalize his space-time view of quantum mechanics to include special relativity. In particular, Feynman hoped that he could explain particle's spin as a result of the space-time structure alone. He abandoned this program in 60's partially because he felt dissatisfaction with not being able to extend the picture to higher dimensions and partially because Grassmann integral (invented in 60's be Berezin) became just the right tool to describe fermions in line with his path-integral philosophy. In spite of (or perhaps because of) Feynman's abandonment of the checkerboard paradigm, there has been over the years a growing number of works indicating that the checkerboard picture is not merely an interesting mathematical curiosity but it may have more substance than originally thought [4, 5, 6, 7].

In its essence the FCP represents a relativistic quantum random walk in $1 + 1$ dimensions. The underlying (Euclidean-space) stochastic process for the FCP represents a specific form of a Poisson process in which the direction of motion and helicity are simultaneously changed at random Poisson-distributed times. Here we adopt the FCP paradigm as a conceptual playground that will allow us to treat two-fermion oscillations from a rather unusual, and very instructive, point of view.

The structure of the paper is as follows. To set the stage we recall in the next section some fundamentals of the Feynman checkerboard picture (FCP) in the Euclidean regime. This will be needed in the main body of the text. In Section 3 the "checkerboard prescription" for the sum over histories will be applied to two-fermion mixing. From this, the phenomenologically observed neutrino oscillations will emerge as an important byproduct. The Euclidean formulation of the FCP will prove instrumental in guiding our intuition about the underlying inner workings of neutrino mixing. In particular, we will see that the mixing can be envisaged as a Poisson process with two types of events with mutually complementary probabilities. Various remarks and generalizations are proposed in the concluding section.

Journal of Physics: Conference Series **626** (2015) 012048 doi:10.1088/1742-6596/626/1/012048

2. Feynman checkerboard picture

The original formulation of the Feynman checkerboard picture was done by Feynman in $1+1$ dimensions [1]. Though the higher-dimension generalizations are also available [6, 7], they loose the intuitive appeal of the original formulation. In order to be as close as possible to the original Feynman's intuitive view, we will confine ourselves to 1+1 dimensions.

2.1. *Dirac equation in 1+1 dimensions*

In ordinary 3+1 dimensions the Dirac equation can be written in a Schrödinger-like form:

$$i\hbar\frac{\partial\psi}{\partial t} = mc^2\gamma^0\psi - i\hbar c\gamma^0\gamma^i\frac{\partial\psi}{\partial x^i} = mc^2\beta\psi - i\hbar c\alpha^i\frac{\partial\psi}{\partial x^i}, \tag{1}$$

with the γ-matrices fulfilling the Clifford algebra $\{\gamma^\mu,\gamma^\nu\} = 2\eta^{\mu\nu}$. In passing to 1+1 dimensions the Clifford algebra keeps its structure, but μ and ν can be only 0 or 1. The representation of the two γ-matrices is well known. In particular, one can chose $\gamma^0 = \gamma_0 = \sigma_1$ and $\gamma_1 = -\gamma^1 = i\sigma_2$. With this the Dirac equation in 1+1 dimensions can be cast in the form

$$i\hbar\frac{\partial\psi}{\partial t} = mc^2\sigma_1\psi - i\hbar c\sigma_1(-i\sigma_2)\frac{\partial\psi}{\partial x} = mc^2\sigma_1\psi - i\hbar c\sigma_3\frac{\partial\psi}{\partial x} \tag{2}$$

we have used the identity $\sigma_i\sigma_j = \delta_{ij} + i\varepsilon_{ijk}\sigma_k$. The propagator, say $G(x,t_x;y,t_y)$, for ψ is a 2×2 matrix function of space and time and it fulfills the defining equation

$$\psi(x,t_x) = \int_{\mathbf{R}} dy\, G(x,t_x;y,t_y)\,\psi_0(y,t_y)\,. \tag{3}$$

Euclidean version of (2) is easily obtained when we analytically continue the time t to imaginary times and assume the Clifford algebra in the form $\{\gamma_E^\mu,\gamma_E^\nu\} = 2\delta^{\mu\nu}$. By setting $t = -i\tau$ ($\tau\in\mathbf{R}$) and using $\gamma_E^0 = \sigma_1$ and $\gamma_E^1 = -\sigma_2$ we obtain

$$\frac{\partial\psi_E}{\partial\tau} = \frac{mc^2}{\hbar}\sigma_1\psi_E - c\sigma_3\frac{\partial\psi_E}{\partial x}\,. \tag{4}$$

Here $\psi_E(x,\tau) = \psi(x,t=-i\tau)$. Among others, the equation (4) implies the correct Euclidean Klein–Gordon equation in 1+1 dimensions. The Euclidean Green function $G(x,-i\tau_x,y,-i\tau_y)\equiv \mathcal{P}(x,\tau_x|y,\tau_y)$ satisfies the Fokker–Planck-like equation:

$$(\partial_\tau + c\sigma_3\partial_x - \sigma_1 mc^2/\hbar)\mathcal{P}(x,\tau|x',\tau') = \delta(x-x')\delta(\tau-\tau')\,. \tag{5}$$

Note also that $mc^2/\hbar = 1/\tau_C$ where τ_C is the Compton time, i.e., time in which light crosses a distance equal to particle's Compton wave length λ_C.

2.2. *Euclidean checkerboard picture*

Here we start by introducing the so-called *Euclidean checkerboard picture*. To this end we consider a particle with a fixed speed v moving on a line. We suppose, further, that from time to time the particle suffers a complete reversal of the direction (and hence also $v\leftrightarrow -v$). Let these reversals be random with a fixed rate, say a, of the reversal and with the probability for the reversal in a time interval $d\tau$ being $ad\tau$. Such a process represents the Poisson stochastic.

Let $p_+(x,\tau)$ and $p_-(x,\tau)$ be probability density functions (PDF's) for the particle being at the positions x at time τ and moving to the *right* and *left*, respectively. By looking at Fig. 1 we

Journal of Physics: Conference Series **626** (2015) 012048 doi:10.1088/1742-6596/626/1/012048

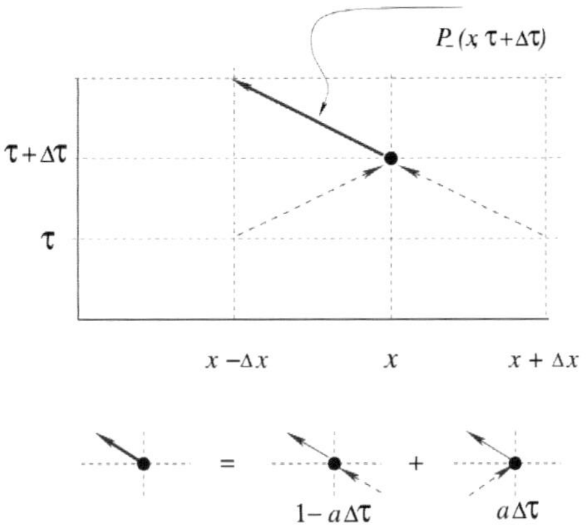

Figure 1. Graphical representation of the master equation (6).

can easily persuade ourselves that the following difference master equations hold

$$p_{\pm}(x, \tau + \Delta\tau) \;=\; p_{\pm}(x \mp \Delta x, \tau)(1 - a\Delta\tau) \;+\; p_{\mp}(x \pm \Delta x, \tau)a\Delta\tau\,. \tag{6}$$

In the continuum limit, we receive two interlocked differential equations

$$\frac{\partial p_{\pm}(x,\tau)}{\partial \tau} \;=\; -a\left[p_{\pm}(x,\tau) - p_{\mp}(x,\tau)\right] \;\mp\; v\frac{\partial p_{\pm}(x,\tau)}{\partial x}\,, \tag{7}$$

where the identification $v = dx/d\tau$ was made. At this point we introduce the probability doublet

$$P(x,\tau) \;=\; \left(\begin{array}{c} p_{+}(x,\tau) \\ p_{-}(x,\tau) \end{array}\right). \tag{8}$$

In terms of P the equation (7) reads as

$$\frac{\partial P(x,\tau)}{\partial \tau} \;=\; -aP(x,\tau) + a\sigma_1 P(x,\tau) - v\sigma_3\frac{\partial P(x,\tau)}{\partial x}\,. \tag{9}$$

The latter implies that the underlying stochastic process is the Poisson process. Equation (9) can be brought into a simpler form by performing the substitution

$$P(x,\tau) \;=\; e^{-a\tau}\mathcal{P}(x,\tau)\,. \tag{10}$$

For $\mathcal{P}$ we then have the master equation

$$\frac{\partial \mathcal{P}(x,\tau)}{\partial \tau} \;=\; a\sigma_1 \mathcal{P}(x,\tau) - v\sigma_3\frac{\partial \mathcal{P}(x,\tau)}{\partial x}\,. \tag{11}$$

Before we proceed further, two comments are in order. First, the substitution (10) represents more than just a simple mathematical trick. In fcat, back in the Minkowski picture the substitution (10) represents just a multiplication of a wave function by phase factor $\exp\left[(imc^2/\hbar)t\right]$. The latter is the usual relativistic phase factor which is responsible for a shift of the ground-state energy by mc^2 [6]. Second, by comparing Eq. (4) with Eq. (11) we see that the equations are identical provided we set $\mathcal{P}(x,\tau) = \psi_E(x,\tau)$, $a = mc^2/\hbar = 1/\tau_C$ and $v = c$.

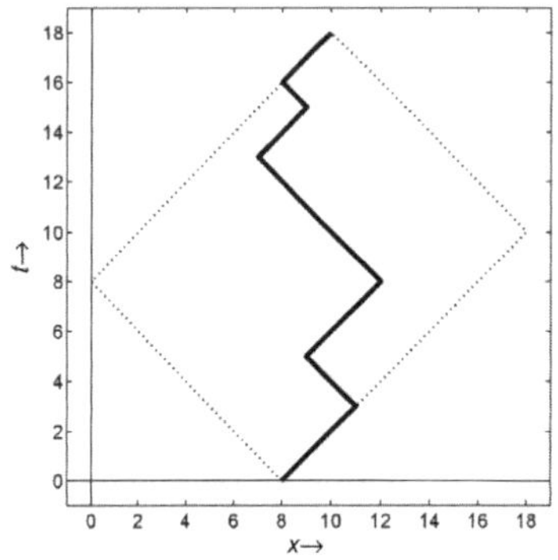

Figure 2. An example of a sample "++" trajectory that enters Feynman checkerboard picture for a spin 1/2 particle in (1 + 1)-dimensional spacetime.

The Euclidean propagator (or conditional probability) for Eq. (11) can be calculated by various means. In particular, the Feynman–Kac formula allows to phrase the propagator via path integral. There one should perform a weighted sum over *all* continuous trajectories running from the initial position (x', τ') to the finite position (x, τ) (see, e.g., Ref. [14] for a detailed exposition of this approach). It is thus interesting, that in fact not all trajectories are needed. Indeed, we have seen that (11) arises only by considering a special sub-class of zig-zag paths. From (10) follows that the propagator can be then written as

$$\mathcal{P}_{ij}(x,\tau|x',\tau') \;=\; e^{aT} P_{ij}(x,\tau|x',\tau')\,, \tag{12}$$

where $T = \tau - \tau'$ and $P(x,\tau|x',\tau')$ is the matrix-valued conditional probability density tying together the initial probabilities $P(x', \tau')$ and the final (marginal) probability $P(x, \tau)$, i.e.

$$P_i(x,\tau) \;=\; \int_{\mathbf{R}} dx' P_{ij}(x,\tau|x',\tau')\, P_j(x',\tau')\,. \tag{13}$$

Both in (12) and (13) the indices i, j run through the set $\{+, -\}$. Let us define X_{ij} as a set of all Poisson processes (with the rate a) that start at the state described by values $\{x', \tau', j\}$ and end in the state $\{x, \tau, i\}$, and let us denote $n(T)$ as the number of events that happen in our Poisson process in time T. With these we can write the transitional probability for our Poisson process by summing (marginalizing) in the the joint probability distribution $P[(n(T) = k) \cap X_{ij}]$ over all possible events k. In particular

$$P_{ij}(x,\tau|x',\tau') \;=\; \sum_k P[(n(T)=k)\cap X_{ij}] \;=\; \sum_k \mathfrak{P}[n(T)=k]P(X_{ij}|n(T)=k)\,. \tag{14}$$

The second equality in (14) is just the Bayes rule with $\mathfrak{P}[n(T) = k]$ describing the probability that in time T one observes k events. The analysis can be carried further when we discretize the time interval into N equidistant pieces. Such a regularization allows to evaluate the above conditional probability explicitly for any N. As with any regulating scheme, the limit $N \to \infty$ should be performed *after* the calculations with the fixed N are done. Consequently we can

Journal of Physics: Conference Series **626** (2015) 012048 doi:10.1088/1742-6596/626/1/012048

write

$$
\begin{aligned}
P_{ij}(x,\tau|x',\tau') &= \lim_{N\to\infty}\sum_k \mathfrak{P}[n(T)=k]\,\frac{\phi^{(N)}[(n(T)=k)\cap X_{ij}]}{\phi^{(N)}[n(T)=k]} \\
&= \lim_{N\to\infty} e^{-aT}\sum_k \left(\frac{aT}{N}\right)^k \phi^{(N)}[(n(T)=k)\cap X_{ij}]\,. \qquad (15)
\end{aligned}
$$

In the first equality we have utilized the definition of the conditional probability. The function $\phi^{(N)}[\ldots]$ denotes the *number* of distinct realizations of the random process in its argument. In the second equality we have used the fact that in the large N limit Stirling's formula gives

$$
\phi^{(N)}(n(T)=k) \;=\; \frac{N!}{k!(N-k)!} \;\cong\; \frac{N^k}{k!}\,. \qquad (16)
$$

This in turn allows to write the Euclidean propagator (12) in the form

$$
\mathcal{P}_{ij}(x,\tau|x',\tau') \;=\; \lim_{N\to\infty}\sum_k \left(\frac{mc^2T}{N\hbar}\right)^k \phi_{ij}^{(N)}(k)\,. \qquad (17)
$$

Here we have employed the simplifying notation $\phi_{ij}^{(N)}(k) \equiv \phi^{(N)}[(n(T)=k)\cap X_{ij}]$. Eq. (17) is (in spirit of Feynman's PI) a sum over alternatives, each summand being a (weighted) number of the Poisson processes (with the rate $a = mc^2/\hbar$) with exactly k reversals (events) running from the initial state $\{x',\tau',j\}$ to the final state $\{x,\tau,i\}$.

Since $\phi_{ij}^{(N)}(k)$ represents the number of paths it does not change when we pass from Euclidean to Minkowski picture. The only change is in the path weighting factor $(mc^2T/N\hbar)^k$ where the Euclidean T should change to the Minkowski iT. Consequently we can write the associated Minkowski Green function in the form

$$
G_{ij}(x,t|x',t') \;=\; \lim_{N\to\infty}\sum_k \left(i\frac{mc^2T}{N\hbar}\right)^k \phi_{ij}^{(N)}(k)\,, \qquad (18)
$$

where now $T = t - t'$. This is precisely Feynman's original result [1]. By its very construction should (18) coincide with the usual Bessel-functions based formula for the Dirac propagator. Taking into account the explicit form of $\phi_{ij}^{(N)}(k)$ this can be, indeed, explicitly checked. Following Refs. [2, 3], we can calculate, for instance, $\phi_{++}^{(N)}(k)$ as follows: first we realize that for any "++" (and also "−−") configuration k is an even number and there are precisely $k/2$ turns to the left and $k/2$ turns to the right [1]. By splitting the time axis into N equidistant pieces, we define the elementary time step $\Delta\tau \equiv T/N$. We further denote the total time (in the units of $\Delta\tau$) that corresponds to the left turns as L and the time for right turns as R. With this we can write

$$
x - x' \;=\; c(R-L)\Delta\tau \;\equiv\; c\Delta\tau M \quad \text{and} \quad R+L \;=\; N\,, \qquad (19)
$$

$$
\Rightarrow\; R \;=\; \frac{N+M}{2}\,, \quad L \;=\; \frac{N-M}{2}\,. \qquad (20)
$$

On Fig. 3 we see that the total number of "++" paths starting at (x',τ') and ending at (x,τ)

[1] For "−+" and "+−" configurations k is odd. For "−+" configuration there $(k-1)/2+1$ turns to the left and $(k-1)/2$ turns to the right. For "+−" configuration is the role of *left* and *right* reversed.

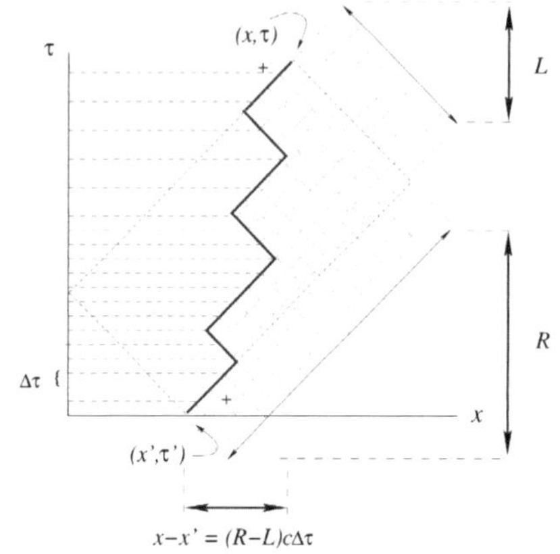

Figure 3. Conventions used in the $\phi_{++}^{(N)}(k)$ calculation.

with k turns can be written as the product of the number of different ways in which one can distribute $k/2$ right turns among $R-2$ potential time occasions for the right turn (-2 enters because the initial and final time steps are fixed by the "++" boundary conditions), times the number in which one can distribute $k/2-1$ left turns (-1 enters because one turn to the left is compulsory for $m \neq 0$) among $L-1$ potential time occasions for the left turn (-1 enters because ne turn to the left is compulsory for $m \neq 0$) , i.e.

$$\phi_{++}^{(N)}(k) \ = \ \begin{pmatrix} R-2 \\ k/2 \end{pmatrix} \begin{pmatrix} L-1 \\ k/2-1 \end{pmatrix} \ \cong \ \frac{L^{k/2-1}}{(k/2-1)!} \frac{R^{k/2}}{(k/2)!} \ = \ \frac{(LR)^{(k-1)/2}}{(k/2-1)!(k/2)!}\sqrt{\frac{R}{L}} \, . \tag{21}$$

Using further the fact that

$$LR \ = \ \frac{1}{4} \ = \ \left(\frac{N}{2\gamma}\right)^2, \quad R/L \ = \ \frac{\gamma^2}{(\Delta\tau^2 N)^2}\left[T + (x-x')/c\right]^2, \tag{22}$$

(here $v = (x-x')/T$ and γ is the Lorentz factor) and plugging (22) to (21) we receive

$$\mathcal{P}_{++}(x,t|x',t') \ \cong \ \frac{\gamma}{T}\left[T + (x-x')/c\right] \sum_{k=1}^{\infty} \left(\frac{mc^2T}{N\hbar}\right)^{2k} \left(\frac{N}{2\gamma}\right)^{2k-1} \frac{1}{(k-1)!k!} \tag{23}$$

$$= \ \Delta\tau \, \frac{mc^2(T + (x-x')/c)}{s\hbar} \, I_1(mc^2 s/\hbar) \, , \tag{24}$$

where we have used abbreviation $s = T/\gamma = [T^2 - (x-x')^2/c^2]^{1/2}$. Similarly we would arrive at the result

$$\mathcal{P}_{-+}(x,t|x',t') \ \cong \ \Delta\tau \, \frac{mc^2}{\hbar} \, I_0(mc^2 s/\hbar) \, . \tag{25}$$

The correct normalization in the continuum limit can be obtained from the probability normalization. Going back to the Poisson process described by $P_{ij}(x,\tau|x',\tau')$ (cf. Eq. (12)) we should require the normalization

$$e^{-at}\int_{-ct}^{ct} dx \, \left[\mathcal{P}_{-+}(x,t|0,0) + \mathcal{P}_{++}(x,t|0,0)\right] \ = \ \wp(T_1 \leq t) \ = \ 1 - e^{-at} \, , \tag{26}$$

(similar normalization should hold if we have started from the "−" initial configuration). Here we have used the normalization condition for conditional distributions, i.e., $\sum_i P(A_t^i|B_0) = 1$. The summation is taken over all possible states A_t^i that the state $B_0 \equiv B(t=0)$ can reach at time t. Note that (26) does not equal "1" because for *massive* particle the future position of the particle should stay *within* the future light cone of the initial particle position. In the integral in (26) we have, however, accounted also for the boundary of the light cone. In particular, when the particle starts at $x = 0, t = 0$ the future position in the time t should be somewhere in the interval $(-ct, ct)$ (at least for times that are substantially larger than the Compton time τ_C) but not at the boundary points $x = \pm tc$. So in our case we should subtract from the *unit* probability the probability that the particle will end up at the light-cone boundary, i.e., "1" minus probability that no reversal happens (which is nothing but the probability that during t will happen at least one reversal). Explicit calculation gives, however

$$e^{-at}\int_{-ct}^{ct} dx\ [\mathcal{P}_{-+}(x,t|0,0) + \mathcal{P}_{++}(x,t|0,0)] \ = \ 2\Delta\tau(1-e^{-at})\,. \tag{27}$$

This implies that both (24) and (25) should be divided by $2\Delta\tau$ in order to fulfill the normalization condition (26).

The passage from the Euclidean transition probability (24) to the Minkowski Green function is established by changing T to iT in the path-weighting factor. Consequently we can write

$$G_{++}(x,t|x',t') \ \cong \ -\frac{mc^2(T+(x-x')/c)}{2s\hbar}\,J_1(mc^2s/\hbar)\,. \tag{28}$$

Analogous reasonings can be done also for other components. Finally we can write for the full matrix propagator

$$G_{ij}(x,t|x',t') \ = \ \frac{mc^2}{2s\hbar}\begin{pmatrix} -(T+(x-x')/c)J_1(mc^2s/\hbar) & s\,J_0(mc^2s/\hbar) \\ s\,J_0(mc^2s/\hbar) & (-T+(x-x')/c)J_1(mc^2s/\hbar)\end{pmatrix}. \tag{29}$$

This indeed coincides with the known result for the 1+1 dimensional Dirac's fermion propagator (see, e.g. Ref. [2]).

Eq. (17) offers a conceptually interesting interpretation of a representative trajectory for an Euclideanized Dirac particle in two dimensions. In particular, a massless(!) particle propagates over an average distances λ_C with velocity c before it reverts its direction. It is only on much larger spatial scales (after many directional reversals take place) where the Brownian motion with a sub-luminal average velocity emerges. So in quantum mechanics one must sum over all such zig-zag trajectories subject given Dirichlet boundary condition in order to obtain correct Feynman causal propagator for Dirac fermion.

It is quite feasible that the Minkowski metric might be just a way of interpreting our experience, while the actual underlying structure of space-time is Euclidean. This line of thoughts has been advocated, for instance, by Polyakov [12] and Hawking [13]. The original Feynman's hope was that a similar type of reasoning that we have used in this section could provide a novel explanation of the geometrical origin of spin, and potentially it could also shape a new intuition about quantum dynamics. It will be this last point that will guide us in the following section to discuss the phenomenon of two-fermion oscillation.

3. Two-fermion oscillations

Let us briefly summarize basic aspects of the traditional approach to fermion oscillations. For explicitness we will use in our exposition the phenomenologically most relevant fermion

Journal of Physics: Conference Series **626** (2015) 012048 doi:10.1088/1742-6596/626/1/012048

oscillations, namely neutrino oscillations. First, states in which neutrinos propagate are called "mass eigenstates", while the states that take part in interaction (and detection) are called "flavor eigenstates". Second, mass eigenstates propagate at different speeds and so the components get out of phase with each other. In particular, neutrino born as one flavor will soon start looking like another one. Third, for neutrinos there are 3 flavors $(\nu_e, \nu_\mu, \nu_\tau)$ but mixing with ν_τ is typically very weak and in the first approximation is neglected. In consequence, mass eigenstates (ν_1, ν_2) associated with (ν_e, ν_μ) have different speeds and hence different on-shell masses. Finally, because the flavor states are superpositions of mass states they must oscillate in time because the quantum mechanical phases of the involved mass states advance at slightly different rates. This is an analogue of the well known phenomenon from wave physics; when two waves with similar frequency add, they interfere and produce *beats*.

Above quantum-mechanical description is not satisfactory for various reasons and it is believed that only full quantum field theoretical approach is free from conceptual difficulties [9, 10]. We shall now demonstrate that the two-fermion oscillations can be accommodated in the FCP rather simply. In fact, it is a fortunate feature of the 1 + 1-dimensional nature of the FCP that we do not need to invoke the field theory concepts (see also comment in Conclusions).

Denote a flavor doublet $\nu_f = (\nu_e, \nu_\mu)^T$ and mass doublet $\nu_m = (\nu_1, \nu_2)^T$. Then the corresponding Dirac equations read

$$\begin{aligned} i\partial_0\nu_e &= (-i\alpha\cdot\nabla + \beta m_e)\nu_e + \beta m_{e\mu}\nu_\mu , \\ i\partial_0\nu_\mu &= (-i\alpha\cdot\nabla + \beta m_\mu)\nu_\mu + \beta m_{e\mu}\nu_e , \end{aligned} \tag{30}$$

which can be succinctly written as

$$i\partial_0\nu_f = (-i\alpha\cdot\nabla + \beta M_f)\nu_f . \tag{31}$$

Similarly for the mass doublet we have $i\partial_0\nu_m = (-i\alpha\cdot\nabla + \beta M_d)\nu_m$. Here $M_d = \text{diag}(m_1, m_2)$ is a diagonal mass matrix. Mixing relations for two Dirac fermions are parametrized via the mixing angle θ which is conventionally defined via the implicit ralation $m_{e\mu} = m_{\mu e} = (m_\mu - m_e)\tan(2\theta)$. With this

$$\nu_e = \cos\theta\nu_1 + \sin\theta\nu_2, \quad \nu_\mu = -\sin\theta\nu_1 + \cos\theta\nu_2 , \tag{32}$$

which directly gives

$$\begin{aligned} m_e &= m_1\cos^2\theta + m_2\sin^2\theta, \quad m_\mu = m_1\sin^2\theta + m_2\cos^2\theta, \\ m_{e\mu} &= (m_2 - m_1)\sin\theta\cos\theta . \end{aligned} \tag{33}$$

To accommodate Dirac fermion oscillations in the FC picture we notice that by analogy with (6) the difference master equations for flavor states (probabilities) read (see Fig. 4)

$$\begin{aligned} p_\pm^{(e)}(x, \tau + \Delta\tau) &= p_\pm^{(e)}(x \mp \Delta x, \tau)(1 - a\Delta\tau) + p_\mp^{(e)}(x \pm \Delta x, \tau)pa\Delta\tau \\ &\quad + p_\mp^{(\mu)}(x \pm \Delta x, \tau)(1 - p)a\Delta\tau , \\ p_\pm^{(\mu)}(x, \tau + \Delta\tau) &= p_\pm^{(\mu)}(x \mp \Delta x, \tau)(1 - a'\Delta\tau) + p_\mp^{(\mu)}(x \pm \Delta x, \tau)p'a'\Delta\tau \\ &\quad + p_\mp^{(e)}(x \pm \Delta x, \tau)(1 - p')a'\Delta\tau . \end{aligned} \tag{34}$$

What this all means. Let a Poisson process representing Dirac neutrino (say ν_e) has rate a, and suppose then each time an even occurs, it can be classified into two types. First type

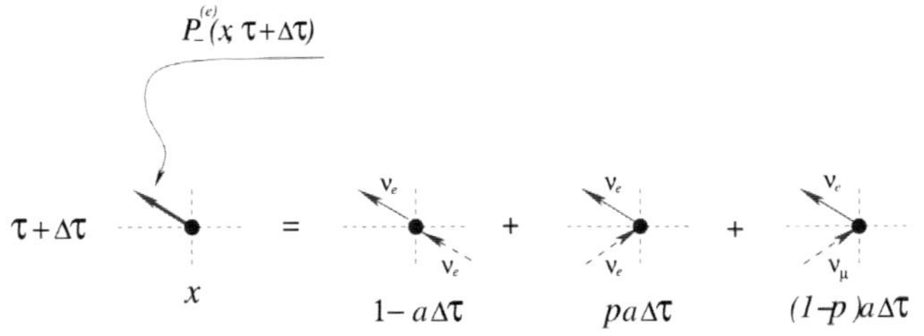

Figure 4. Graphical representation of the master equation for $p_-^{(e)}(x,\tau+\Delta\tau)$. Similar representations can be drawn also for $p_+^{(e)}$ and $p_\mp^{(\mu)}$.

corresponding to change of direction without change of particle's identity has probability p and a second type corresponding to change of direction with the change of identity has probability $(1-p)$. In close analogy with (6) we now define the probability doublet

$$P^{(e),(\mu)}(x,\tau) \ = \ \left(\begin{array}{c} p_+^{(e),(\mu)}(x,\tau) \\ p_-^{(e),(\mu)}(x,\tau) \end{array}\right), \tag{35}$$

and set $pa = m_e c^2/\hbar$, $p'a' = m_\mu c^2/\hbar$ and $(1-p)a = (1-p')a' = m_{\mu e}c^2/\hbar$. In addition, we define $P^{(e)}(x,\tau) = e^{-a\tau}\mathcal{P}^{(e)}(x,\tau)$ and $P^{(\mu)}(x,\tau) = e^{-a'\tau}\mathcal{P}^{(\mu)}(x,\tau)$, the master equations (34) will then boil down to

$$\begin{aligned} \frac{\partial\mathcal{P}^{(e)}(x,\tau)}{\partial\tau} &= -c\sigma_3\frac{\partial\mathcal{P}^{(e)}(x,\tau)}{\partial x} + m_e c^2/\hbar\sigma_1\mathcal{P}^{(e)}(x,\tau) + m_{\mu e}c^2/\hbar\sigma_1\mathcal{P}^{(\mu)}(x,\tau)\,, \\ \frac{\partial\mathcal{P}^{(\mu)}(x,\tau)}{\partial\tau} &= -c\sigma_3\frac{\partial\mathcal{P}^{(\mu)}(x,\tau)}{\partial x} + m_\mu c^2/\hbar\sigma_1\mathcal{P}^{(\mu)}(x,\tau) + m_{e\mu}c^2/\hbar\sigma_1\mathcal{P}^{(e)}(x,\tau)\,. \end{aligned} \tag{36}$$

Our requirement that $m_{\mu e} = m_{e\mu}$ is a consequence of a micro-reversibility, namely that the probability for event $\nu_\mu \to \nu_e$ should be in the interval $\Delta\tau$ the same as $\nu_e \to \nu_\mu$. We have also neglected the term $e^{\pm\tau(m_e-m_\mu)c^2/\hbar}$ in front of the $m_{e\mu}$ term. This in particular shows, that the outlined Poisson process is a good description of the neutrino oscillations if $\tau(m_e - m_\mu)c^2/\hbar \ll 1$ for typical observational times τ — which indeed is the case. Equations (36) is the Euclidean version of the Cabibbo two-flavor mixing where for the Cabibbo angle one has

$$\theta \ = \ \arctan[(1-p)(1-p')/pp']\,. \tag{37}$$

In the FCP there is no need for the mass-states, since freely evolving ν_μ and ν_e propagate with speed of light, i.e., they represent particles with a sharp zero-mass value. As before, the actual inertial mass is due to a zig-zag nature of the representative trajectories in the FCP. This is akin to massless neutrinos experiencing a friction (via random noise or collisions) within the vacuum medium. The effect of this mass-giving environment is phenomenologically described via random Poisson-distributed interactions with massless neutrinos. In a sense, this is a quantum-mechanical analogue of the Higgs mechanism. In this picture oscillations are not due to different propagation speeds of mass eigenstates but they result from two different types of events that a massless neutrino in a Poisson process can experience. These two distinct events can be attributed to two different interaction channels through which massless neutrinos can interact with the mass-giving environment.

4. Conclusions

In this paper we have presented a novel heuristic picture for neutrino oscillations. A *modus operandi* we have employed is the standpoint that a stochastic picture of two-fermion oscillations in Euclidean space-time is ontologically more fundamental than the usual quantum-mechanical description. Actual inner workings of our approach were based on Feynman's checkerboard picture, i.e., a Poisson-process based random walk allowing to address quantum dynamics of Dirac particles in 1 + 1 dimensions. We have shown here that two-fermion oscillations (e.g., $\nu_\mu \leftrightarrow \nu_e$) can be accommodated in the FCP easily by assuming that each time an event occurs, it is classified either as a type one (directional change without identity change) or a type two (directional change with identity change).

Let us add few concluding comments. First, the logical consistency of the presented picture might make one wonder how we could dispense with quantum field theory. Well, this is a byproduct of the fact that we deal with 1 + 1 dimensional theory. Actually, because of the Mermin–Wagner–Coleman theorem the flavor-state vacuum and mass-state vacuum are unitary equivalent in $1+1$ dimensions [9] and hence the field-theoretical treatment does not produce any extra corrections as is the case in higher dimensions [9]. Extension of our approach to higher dimensions would be worth pursuing particularly in view of a naturalness with which the FCP handles fermionic particle systems. Second, checkerboard pictures are as a rule formulated only for fermions. It is, however, possible to formulate the checkerboard-like picture also for a Klein–Gordon (i.e., spinless relativistic) particle in 1 + 1 dimensions. To this end one should use the first-order version of the Klein–Gordon equation known as the Feshbach–Villars representation of a scalar particle [6, 8]. Work along these lines will be presented elsewhere.

Acknowledgments
I would like acknowledge stimulating discussions with L. Schulman and I am grateful to H.-T. Elze and organizers of the Seventh International Workshop DICE2014 for their cordial hospitality. This work was supported by GAČR Grant No. GA14-07983S

References

[1] Feynman R P and Hibbs A R 1965 *Quantum Mechanics and Path Integrals* (New York: McGraw-Hill)
[2] Gaveau B, Jacobson T, Kac M and Schulman L S 1984 *Phys. Rev. Lett.* **53** 419; Jacobson T and Schulman L S 1984 *J. Phys.* A **17** 375
[3] Karmanov V A 1993 *Phys. Lett.* A **174** 371
[4] Ord G N 1992 *Int., J. Theor. Physics* **31** 1177
[5] McKeon D G C and Ord G N 1992 *Phys. Rev. Lett.* **69** 3
[6] Jizba P and Kleinert H 2010 *Phys. Rev.* D **82** 085016
[7] Jizba P and Scardigli F 2012 *Phys. Rev.* D **86** 025029
[8] Feshbach H and Villars F 1958 *Rev. Mod. Phys.* **30** 24
[9] Blasone M, Jizba P and Vitiello G 2011 *Quantum Field Theory and its Macroscopic Manifestations* (London: World Scientific & ICP)
[10] Cheng T and Li L 2000 *Gauge theory of elementary particle physics* (Oxford: Clarendon Press)
[11] Kröger H 2000 *Phys. Rep.* **323** 81
[12] Polyakov A 1989 Fields, Strings and Critical Phenomena, in *Les Houches, Session XLIX* ed E Brézin and J Zinn-Justin (Berlin: Elsevier, 1989)
[13] Hawking S W 1978 Euclidean quantum gravity, in: *Recent developments in gravitation, Cargese Lectures* ed M Levy and S Deser (New York: Plenum, 1978)
[14] Kleinert H 2009 *Path Integrals in Quantum Mechanics, Statistics, Polymer Physics and Financial Markets* (Singapore: World Scientific)
[15] Hsu H P 2011 *Probability, Random Variables, and Random Processes* (London: McGraw Hill)

Measurements of observables replaced by "evaluations" in Quantum Theory

Giuseppe Nisticò and Angela Sestito[1]

Dipartimento di Matematica, Università della Calabria, Italy and INFN – gruppo collegato di Cosenza, Italy

E-mail: gnistico@unical.it, sestito@mat.unical.it

Abstract. In quantum physics there are circumstances where the direct measurement of particular observables encounters difficulties; in some of these cases, however, its value can be *evaluated*, i.e. it can be inferred by measuring *another* observable characterized by perfect correlation with the observable of interest. Though an *evaluation* is often interpreted as a *measurement* of the evaluated observable, we prove that the two concepts cannot be identified in quantum physics, because the identification yields contradictions. Then, we establish the conceptual status of evaluations in Quantum Theory and the role can be ascribed to them.

1. Introduction

In quantum physics there are circumstances where some difficulties encountered in measuring observables are outflanked by exploiting the correlations existing among observables.

As an example, we can consider a typical Stern–Gerlach experiment for a spin-1/2 particle, where the gradient of the magnetic field is oriented along z. The z component S_z of the spin is an observable which pertains to an *internal* degree of freedom of the particle, so that it is difficult to concretely design a direct measurement of S_z. However, the observable T_{up} which localizes the particle in the upper exit of the magnet is perfectly correlated, according to the laws of Quantum Mechanics, with the values of S_z; hence, the value $+1/2$ of S_z is inferred for the out-coming particles localized in the upper exit by a measurement of T_{up}.

Such a method is used also to outflank the obstacles raised by the fundamental principles of Quantum Mechanics. For instance, in a typical double-slit experiment the observable W that indicates the slit the particle passes through is represented by a localization operator $\hat{W}$ which *does not commute* with the operator $\hat{Q}_F$ representing the impact position on the final screen, because W and Q_F are positions at different times. Therefore, the measurement of W is *in principle* forbidden for a particle whose final position Q_F is measured. However, under suitable conditions [1], an observable T_W exists such that $[\hat{T}_W, \hat{Q}_F] = \mathbf{0}$, and whose outcomes are perfectly correlated with the outcomes of W in every simultaneous measurement of T_W and W; so, by measuring T_W and Q_F together, which slit information is inferred from the outcome of T_W, *via* the perfect correlation between T_W and W without measuring W, while Q_F is directly measured.

[1] Part of Angela Sestito's work is supported by the European Commission, European Social Fund and by the Calabria Region, Regional Operative Program (ROP) Calabria ESF 2007/2013 - IV Axis Human Capital - Operative Objective M2- Action D.5

A priori, to assign an observable E the value obtained as actual outcome of a correlated observable T should be distinguished from a *direct measurement of E*; so, we call it *evaluation* of E by T. The different statuses of *measurement* and *evaluation* in Quantum Theory are established in section 2.1 and 2.2.

In this work we address the problem of establishing to what extent evaluations can be interpreted as real measurements. In fact, in the experiment of Stern&Gerlach, for instance, the localization of the particle by T_{up} is interpreted as a valid measurement of the spin S_z. But, from a theoretical point of view the problem exists. Indeed in section 2.3 we show that to identify the evaluation of an observable with its measurement *leads to contradictions* in Quantum Mechanics.

Now, since evaluations are diffusely practiced in quantum physics, the task of establishing how they are related to measurements cannot be overlooked in Quantum Theory.

Then, in section 3 we shows that evaluations behave as *perfect simulations* of the measurement of the evaluated observable; more precisely, we show that the physical consequences of the occurrence of every measurement's outcome of an observable are physically indistinguishable from the consequences of the occurrence of the same outcome for its evaluation.

However, in section 4 we point out that the interpretation of evaluations as simulations does not apply if the evaluation of an observable E by T is performed together with the measurement of another observable F which does not commute with the evaluated observable E. In this more general case we prove that a unique joint probability $p_\rho(E\&F)$ exists which rules over a value assignment for E consistent with the simultaneous occurrence of outcomes of actually performed measurements of F; furthermore, we prove that just to assign E the values evaluated by T realizes such a unique probability. As a consequence, though E does not commute with F, the evaluated observable E can be assigned the value evaluated by T without violating logical consistency and such an assignment consistently extends the actual measurements' outcomes.

The impossibility of identifying evaluations with measurement proved in section 2 is then explained on the basis of our results.

2. "Evaluations" in Quantum Theory

Here, in section 2.2, we formally establish the concept of *evaluation* within Quantum Theory. To do this, in section 2.1 we have to make explicit some implications of the standard interpretation of Quantum Theory. Section 2.3 shows that evaluations cannot be identified with measurements, because the identifications provokes contradictions in Quantum Mechanics.

2.1. Basic formalism

Let $\mathcal{H}$ be the Hilbert space of the quantum theory of the investigated physical system. Given any observable A, let $\hat{A}$ be the corresponding self-adjoint operator, the expected value of A is $Tr(\rho\hat{A})$, where the density operator ρ is the *quantum state* of the system [2].

Given a quantum state ρ, by *support* of ρ we mean any *concrete* subset $\mathcal{S}(\rho)$ of specimens of the physical system [3], whose quantum state is ρ. Given a support $\mathcal{S}(\rho)$, by $\mathbf{A}(\mathcal{S}(\rho))$ we denote the concrete subset of all specimens in $\mathcal{S}(\rho)$ which *actually* undergo a measurement of A. In the following, we shall write simply $\mathbf{A}$ instead of $\mathbf{A}(\mathcal{S}(\rho))$ to avoid a cumbersome notation, whenever no confusion is likely.

By *elementary* observable we mean any observable E having only 0 or 1 as possible outcomes, and hence represented by a projection operator $\hat{E}$; the expected value $Tr(\rho\hat{E})$ of an elementary observable E coincides with the *probability* that outcome 1 occurs in a measurement of E. By $\mathcal{E}$ we denote the set of all elementary observables, and by $\hat{\mathcal{E}}(\mathcal{H})$ the set of all projection operators of $\mathcal{H}$.

Fixed any support $\mathcal{S}(\rho)$, in correspondence with every elementary observable E we define the following extensions of E in $\mathcal{S}(\rho)$.

- the set $\mathbf{E}$ of the specimens in $\mathcal{S}(\rho)$ which *actually* undergo a measurement of E;

Journal of Physics: Conference Series **626** (2015) 012049 doi:10.1088/1742-6596/626/1/012049

- the set $\mathbf{E}_1 \subseteq \mathbf{E}$ (resp., $\mathbf{E}_0 \subseteq \mathbf{E}$) for which the outcome 1 (resp., 0) of E has been obtained.

In agreement with Quantum Mechanics, we assume that the following statements hold [3].

(2.1.i) If E is an elementary observable, then for every ρ a support $\mathcal{S}(\rho)$ exists such that $\mathbf{E} \neq \emptyset$.

(2.1.ii) For every support $\mathcal{S}(\rho)$, $\mathbf{E}_1 \cap \mathbf{E}_0 = \emptyset$ and $\mathbf{E}_1 \cup \mathbf{E}_0 = \mathbf{E}$, for every ρ.

(2.1.iii) If $Tr(\rho\hat{E}) \neq 0$ then a support $\mathcal{S}(\rho)$ exists such that $\mathbf{E}_1 \neq \emptyset$, and
if $Tr(\rho\hat{E}) \neq 1$, then a support $\mathcal{S}(\rho)$ exists such that $\mathbf{E}_0 \neq \emptyset$.

In Quantum Theory [2], if $\hat{B} = f(\hat{A})$ holds for two self-adjoint operators $\hat{A}$ and $\hat{B}$, then a measurement of the observable B, henceforth denoted by $f(A)$, can be performed by measuring A and then transforming the outcome a by the function f into the outcome $b = f(a)$ of B. As a consequence, the following statements hold in Quantum Theory.

$$\text{If} \quad \hat{B} = f(\hat{A}) \quad \text{then} \quad x \in \mathbf{A} \text{ implies } x \in \mathbf{B}. \tag{2.2}$$

If $[\hat{A}, \hat{B}] = \mathbf{0}$, then a third self-adjoint operator $\hat{C}$ and two functions f and g exist so that $\hat{A} = f(\hat{C})$ and $\hat{B} = g(\hat{C})$ [2]. Therefore, A and B can be measured together if the corresponding operators commute with each other. Thus, the following implications hold.

(2.1.iv) $\{E_j\} \subseteq \mathcal{E}$ and $[\hat{E}_j, \hat{E}_k] = \mathbf{0}\ \forall j, k$ imply $\forall\rho\ \exists\mathcal{S}(\rho)$ such that $\cap_j \mathbf{E_j} \neq \emptyset$.

(2.1.v) If $[\hat{A}, \hat{B}] = \mathbf{0}$ and $\hat{D} = f(\hat{A}, \hat{B})$ then $x \in \mathbf{A} \cap \mathbf{B}$ implies $x \in \mathbf{D}$, $\forall\mathcal{S}(\rho)$.

(2.1.vi) If $F, G \in \mathcal{E}$ and $\hat{F}\hat{G} = \mathbf{0}$, i.e. if $\hat{F} \perp \hat{G}$, then $\mathbf{F}_1 \cap \mathbf{G}_1 = \emptyset$, $\forall\mathcal{S}(\rho)$, $\forall\rho$,
i.e. in every simultaneous measurement of F and G the outcome 1 for F and 1 for G are mutually exclusive; in this case the projection $\hat{E} = \hat{F} + \hat{G}$ belongs to $\hat{\mathcal{E}}(\mathcal{H})$;
conversely, if $\hat{F} \perp G$ then $\hat{E} = \hat{F} + \hat{G}$ represents an elementary observable E whose measurement's outcome can be the sum of the simultaneous outcomes of F and G.

2.2. Evaluations of elementary observables in Quantum Theory

In general, given two elementary observables E and T we say that E can be *evaluated* by T if, according to Quantum Mechanics, the following perfect correlation holds:

"the outcome of T is 1 if and only if the outcome of E is 1 in every simultaneous measurement".

By *evaluation* of E by T we mean to assign E the value obtained as outcome of an actual measurement of T. Then we can give the following formal definition.

Definition 2.1. *Given $E, T \in \mathcal{E}$, the elementary observable E can be evaluated by T when the system is assigned the state ρ, written $E \prec \rho \succ T$, if*

(i) *a support $\mathcal{S}(\rho)$ exists such that $\mathbf{T} \cap \mathbf{E} \neq \emptyset$ (simultaneous measurability),*

(ii) *if $x \in \mathbf{E} \cap \mathbf{T}$ then $[x \in \mathbf{T}_1$ if and only if $x \in \mathbf{E}_1]$ and $[x \in \mathbf{T}_0$ if and only if $x \in \mathbf{E}_0]$, $\forall\mathcal{S}(\rho)$.*

If an evaluation of E by T were identifiable with a measurement of E in the state ρ, then $\mathbf{T}_1 \subseteq \mathbf{E}_1$ and $\mathbf{T}_0 \subseteq \mathbf{E}_0$ should hold for any $\mathcal{S}(\rho)$. But the relation $E \prec \rho \succ T$ is symmetric; thus the identification would be fully expressed by the following statement.

$$E \prec \rho \succ T \qquad \text{if and only if} \qquad \mathbf{T}_1 = \mathbf{E}_1 \text{ and } \mathbf{T}_0 = \mathbf{E}_0\,,\ \forall\mathcal{S}(\rho). \tag{2.3}$$

Another relation $E \overset{\rho}{\leftrightarrow} T$ can be defined as follows

Definition 2.2. *The relation $T \overset{\rho}{\leftrightarrow} E$ holds if $[\hat{T}, \hat{E}] = \mathbf{0}$ and $\hat{T}\rho = \hat{E}\rho$ hold.*

Relation $\overset{\rho}{\longleftrightarrow}$ is stronger than $\prec \rho \succ$; indeed the following proposition holds.

Proposition 2.1. *Given $E, T \in \mathcal{E}$, if $E \overset{\rho}{\longleftrightarrow} T$ holds then $E \prec \rho \succ T$ holds too.*

Proof. Condition (i) in def.2.1 follows from (2.1.iv). Then condition (ii) holds if the probability of the pairs $(1, 0)$ and $(0, 1)$ in a simultaneous measurement of T and F is zero, i.e., if $Tr(\rho\hat{T}[\mathbf{1} - \hat{E}]) = \mathbf{0}$ and $Tr(\rho[\mathbf{1} - \hat{T}]\hat{E}) = \mathbf{0}$, i.e. if $\hat{T}\rho = \hat{T}\hat{E}\rho$ and $\hat{E}\rho = \hat{E}\hat{T}\rho$.

2.3. Evaluations are not measurements

Now we shall single out seven elementary observables $E^\alpha, E^\beta, F, G^\alpha, G^\beta, L^\alpha, L^\beta$ of a particular quantum system, chosen so that they are all measurable together *if the identification (2.3) of evaluations with measurements holds.* Then we show that their simultaneous outcomes $\eta^\alpha, \eta^\beta, \phi, \gamma^\alpha, \gamma^\beta, \lambda^\alpha, \lambda^\beta$ must satisfy the following constraints, where f is the function $f(\xi) = 2\xi - 1$.

$$\left\{ \begin{array}{lll} \text{i)} & f(\eta^\alpha)f(\phi) & = -f(\gamma^\alpha)f(\lambda^\alpha), \\ \text{ii)} & f(\eta^\beta)f(\phi) & = -f(\gamma^\beta)f(\lambda^\alpha), \\ \text{iii)} & f(\eta^\beta)f(\phi) & = -f(\gamma^\alpha)f(\lambda^\beta), \\ \text{iv)} & f(\eta^\alpha)f(\phi) & = f(\gamma^\beta)f(\lambda^\beta), \end{array} \right. \tag{2.4}$$

Since each factor $f(\xi)$ in (2.4) must be -1 or $+1$, these constraints are contradictory, because, by elementary algebra, they imply $f(\gamma^\alpha)f(\gamma^\beta) = -f(\gamma^\alpha)f(\gamma^\beta)$. Thus, in Quantum Mechanics evaluations cannot be identified with measurements.

To realize such a program, we consider a quantum system described in the Hilbert space $\mathcal{H} = \mathcal{H}_1 \otimes \mathcal{H}_2 \otimes \mathcal{H}_3 \otimes \mathcal{H}_4$, where each $\mathcal{H}_k$ is $\mathbf{C}^2$. The following projection operators represent seven elementary observables[2].

$$\hat{E}^\alpha = \tfrac{1}{2}\begin{bmatrix} 1 & 1 \\ 1 & 1 \end{bmatrix}_1 \otimes \mathbf{1}_2 \otimes \mathbf{1}_3 \otimes \mathbf{1}_4; \qquad \hat{E}^\beta = \tfrac{1}{2}\begin{bmatrix} 1 & -i \\ i & 1 \end{bmatrix}_1 \otimes \mathbf{1}_2 \otimes \mathbf{1}_3 \otimes \mathbf{1}_4;$$

$$\hat{F} = \mathbf{1}_1 \otimes \tfrac{1}{2}\begin{bmatrix} 1 & 1 \\ 1 & 1 \end{bmatrix}_2 \otimes \mathbf{1}_3 \otimes \mathbf{1}_4;$$

$$\hat{G}^\alpha = \mathbf{1}_1 \otimes \mathbf{1}_2 \otimes \tfrac{1}{2}\begin{bmatrix} 1 & 1 \\ 1 & 1 \end{bmatrix}_3 \otimes \mathbf{1}_4; \qquad \hat{G}^\beta = \mathbf{1}_1 \otimes \mathbf{1}_2 \otimes \tfrac{1}{2}\begin{bmatrix} 1 & -i \\ i & 1 \end{bmatrix}_3 \otimes \mathbf{1}_4;$$

$$\hat{L}^\alpha = \mathbf{1}_1 \otimes \mathbf{1}_2 \otimes \mathbf{1}_3 \otimes \tfrac{1}{2}\begin{bmatrix} 1 & 1 \\ 1 & 1 \end{bmatrix}_4; \qquad \hat{L}^\beta = \mathbf{1}_1 \otimes \mathbf{1}_2 \otimes \mathbf{1}_3 \otimes \tfrac{1}{2}\begin{bmatrix} 1 & -i \\ i & 1 \end{bmatrix}_4;$$

Let the physical system be assigned the pure state $\rho_0 = |\psi_0\rangle\langle\psi_0|$, where

$$\psi_0 = \frac{1}{\sqrt{2}}\left(\begin{bmatrix} 1 \\ 0 \end{bmatrix}_1 \otimes \begin{bmatrix} 1 \\ 0 \end{bmatrix}_2 \otimes \begin{bmatrix} 0 \\ 1 \end{bmatrix}_3 \otimes \begin{bmatrix} 0 \\ 1 \end{bmatrix}_4 - \begin{bmatrix} 0 \\ 1 \end{bmatrix}_1 \otimes \begin{bmatrix} 0 \\ 1 \end{bmatrix}_2 \otimes \begin{bmatrix} 1 \\ 0 \end{bmatrix}_3 \otimes \begin{bmatrix} 1 \\ 0 \end{bmatrix}_4 \right).$$

The four projection operators $\hat{E}^\alpha, \hat{F}, \hat{G}^\beta, \hat{L}^\alpha$ commute with each other; hence, by (2.1.iv), all the corresponding elementary observables can be measured together, i.e. a support $\mathcal{S}(\rho_0)$ and a specimen $x_0 \in \mathcal{S}(\rho_0)$ exist such that

$$x_0 \in \mathbf{E}^\alpha \cap \mathbf{F} \cap \mathbf{G}^\beta \cap \mathbf{L}^\alpha.$$

Let $\eta^\alpha, \phi, \gamma^\beta, \lambda^\alpha$ be their respective outcomes relative to x_0. Now, the projection operator

$$\hat{M} = \frac{\mathbf{1} - f(\hat{E}^\alpha)f(\hat{F})f(\hat{L}^\alpha)}{2}$$

[2] Our argument makes use of the mathematical setting adopted by Greenberger, Horne, Shimony and Zeilinger (GHSZ) [4] to prove that a given set of premises are in contradiction with Quantum Theory. In [3] we proved that if the premises of GHSZ theorem are modified, then the contradiction does not necessarily arise. The hypotheses of the present argument, i.e. identification (2.3), are different from the premises of GHSZ. Thus the occurrence of a contradiction needs an explicit proof.

is a function of $\hat{E}^\alpha, \hat{F}, \hat{L}^\alpha$; therefore $x_0 \in \mathbf{M}$ because of (2.1.v) and $\mu = \frac{1}{2}[1 - f(\eta^\alpha)f(\phi)f(\lambda^\alpha)]$ must be the outcome of the elementary observable M so measured on x_0.

But $[\hat{M}, \hat{G}^\alpha] = \mathbf{0}$ trivially holds; moreover, a direct calculation shows that the equation $\hat{M}\rho_0 = \hat{G}^\alpha\rho_0$ is satisfied; then $M \overset{\rho_0}{\longleftrightarrow} G^\alpha$ holds and Prop.2.1 implies $M \prec \rho_0 \succ G^\alpha$. If the identification (2.3) holds, then x_0 belongs to $\mathbf{G}^\alpha$ and $\gamma^\alpha = \mu = \frac{1}{2}[1 - f(\eta^\alpha)f(\phi)f(\lambda^\alpha)]$ is to be identified as the outcome of the measurement of G^α on x_0. Then, from $f(\gamma^\alpha) \equiv 2\gamma^\alpha - 1$ we find (2.4.i). Then, $x_0 \in \mathbf{E}^\alpha \cap \mathbf{F} \cap \mathbf{G}^\alpha \cap \mathbf{G}^\beta \cap \mathbf{L}^\alpha$ and the constraint (2.4.i) must hold.

Now we derive (2.4.ii). By defining $\hat{N} = \frac{\mathbf{1} - f(\hat{F})f(\hat{G}^\beta)f(\hat{L}^\alpha)}{2}$, we can verify that $N \overset{\rho_0}{\longleftrightarrow} E^\beta$. Then, following the argument which led us to (2.4.i), we obtain that $x_0 \in \mathbf{E}^\alpha \cap \mathbf{E}^\beta \cap \mathbf{F} \cap \mathbf{G}^\alpha \cap \mathbf{G}^\beta \cap \mathbf{L}^\alpha$ and (2.4.ii) holds.

Similarly, by defining $\hat{R} = \frac{\mathbf{1} - f(\hat{E}^\beta)f(\hat{F})f(\hat{G}^\alpha)}{2}$, it turns out that $R \overset{\rho_0}{\longleftrightarrow} L^\beta$ holds. Then, we can imply that $x_0 \in \mathbf{E}^\alpha \cap \mathbf{E}^\beta \cap \mathbf{F} \cap \mathbf{G}^\alpha \cap \mathbf{G}^\beta \cap \mathbf{L}^\alpha \cap \mathbf{L}^\beta$ and (2.4.iii) hold.

But we can also define $\hat{S} = \frac{\mathbf{1} + f(\hat{E}^\alpha)f(\hat{F})f(\hat{G}^\beta)}{2}$. The elementary observable S turns out to satisfy the relation $S \overset{\rho_0}{\longleftrightarrow} L^\beta$, which implies (2.4.iv).

Then all the constraints (2.4) must hold for the simultaneous measurement of $E^\alpha, E^\beta, F, G^\alpha, G^\beta, L^\alpha, L^\beta$ on the specimen x_0. Thus, identification (2.3) cannot hold in Quantum Mechanics.

3. Evaluations as simulations of measurements

The impossibility of identifying evaluations with measurements, proved in section 2, opens the question of establishing the conceptual status of the evaluations in Quantum Theory; since evaluations are diffusely practiced in quantum physics, it is important give an answer to such a question. In this section we show that an evaluation of E by T works as a *perfect simulation* of a measurement of E.

A way to understand which is the conceptual status of an evaluation of E by T is to compare the *physical implications* of the occurrences of the outcomes of E with the physical implications of the occurrences of the corresponding outcomes of T. Such a comparison amounts to compare the conditional probabilities $P(F \mid E) = Tr(\rho\hat{F}\hat{E})/Tr(\rho\hat{E})$, $P(F \mid E') = Tr(\rho\hat{F}\hat{E}')/Tr(\rho\hat{E}')$ established by Quantum Theory with the conditional probabilities $P(F \mid T) = Tr(\rho\hat{F}\hat{T})/Tr(\rho\hat{T})$, $P(F \mid T') = Tr(\rho\hat{F}\hat{T}')/Tr(\rho\hat{T}')$, where $E' \equiv 1 - E$ and $T' \equiv 1 - T$.

These conditional probabilities are defined whenever $[\hat{F}, \hat{T}] = [\hat{F}, \hat{E}] = \mathbf{0}$; therefore the domain of the comparison is $\mathcal{F}(E,T) = \{F \in \mathcal{E} \mid [\hat{F}, \hat{T}] = [\hat{F}, \hat{E}] = \mathbf{0}\}$. The following statement follows from prop. 2.1.

$$\text{If } T \overset{\rho_0}{\longleftrightarrow} E \text{ then } P(F \mid T) = \frac{Tr(\rho\hat{F}\hat{T})}{Tr(\rho\hat{T})} = \frac{Tr(\rho\hat{F}\hat{E})}{Tr(\rho\hat{E})} = P(F \mid E),\ \forall F \in \mathcal{F}(E,T). \tag{3.1}$$

Now, from Def.2.2 it follows that $T \overset{\rho}{\longleftrightarrow} E$, holds iff $T' \overset{\rho}{\longleftrightarrow} E'$; so (3.1) extends to

$$\text{If } T \overset{\rho_0}{\longleftrightarrow} E \text{ then } P(F \mid T') = \frac{Tr(\rho\hat{F}\hat{T}')}{Tr(\rho\hat{T}')} = \frac{Tr(\rho\hat{F}\hat{E}')}{Tr(\rho\hat{E}')} = P(F \mid E'),\ \forall F \in \mathcal{F}(E,T). \tag{3.2}$$

Therefore, in the case that $T \overset{\rho_0}{\longleftrightarrow} E$ holds, the effects of an evaluation of E by T are indistinguishable from the effects of the occurrence of the same outcome in a direct measurement of E. In other words, the measurement of E is perfectly *simulated* by a measurement of a evaluating observable T.

Remark 3.1. In fact, once fixed the evaluated observables E, our results are obtained for evaluating observables T such that $T \overset{\rho}{\longleftrightarrow} E$; these observables form a subset of all evaluating observables evaluating E which have to satisfy the weaker condition $T \prec \rho \succ E$. However, the converse of Prop.2.1, i.e. the implication $T \prec \rho \succ E$ implies $T \overset{\rho}{\longleftrightarrow} E$, immediately follows from the principle of Quantum Mechanics which establishes that T and E are measurable together if and only if $[\hat{T}, \hat{E}] = \mathbf{0}$. Therefore, if such a principle is assumed, our results holds for all evaluating observables.

4. To evaluate E while incompatible observables are measured

Let T, E and F be elementary observables such that $T \overset{\rho}{\longleftrightarrow} E$ and $[\hat{T}, \hat{F}] = \mathbf{0}$, so that T can be measured together with F; in the case in which $F \notin \mathcal{F}(E,T)$, i.e. if $[\hat{F}, \hat{E}] \neq \mathbf{0}$, a measurement of T simultaneous to a measurement of F cannot be interpreted as a simulation of a measurement of E, according to section 3.1, because there is no conditional probability $P(F \mid E)$ to be compared with $P(F \mid T)$.

The double slit experiment described in the introduction is an emblematic example of this circumstance. In that case it is possible to evaluate the which slit observable W by T_W; but if also the elementary observable $F(\Delta)$, which indicates the localization in the region Δ of the final screen, is measured, then the evaluation cannot be interpreted as simulation of a measurement of W because $P(F(\Delta) \mid W)$ does not exist since $[\hat{W}, \hat{F}(\Delta)] \neq \mathbf{0}$.

In the present section we address the interpretative lack which occurs in a general situation where

$$T \overset{\rho}{\longleftrightarrow} E, \quad F \in \mathcal{F}(T) \equiv \{F \in \mathcal{E} \mid \quad [\hat{T}, \hat{F}] = \mathbf{0}\}, \quad \text{but} \quad [\hat{E}, \hat{F}] \neq \mathbf{0}, \tag{4.1}$$

and E is evaluated by T simultaneously to a measurement of F.

In section 4.1 we establish results which allow us to provide the problem with an answer we formulate in section 4.2. Remark 3.1 applies also in this more general case.

4.1. Consistency of assignment by evaluations

Let us start with the following result of Cassinelli and Zanghì [5].

Lemma 4.1. *Let $\hat{\mathcal{A}}$ be a Von Neumann algebra*[3], *and let $\Pi(\hat{\mathcal{A}})$ be the set of all projection operators in $\hat{\mathcal{A}}$. Given a projection operator $\hat{E} \in \Pi(\hat{\mathcal{A}})$, for every density operator ρ the function*

$$p_\rho(E\& : \Pi(\hat{\mathcal{A}}) \to [0,1], \quad p_\rho(E\&F) = Tr(\rho\hat{E}\hat{F}\hat{E})$$

is the unique functional which satisfies the following conditions.

(i) *If $\hat{F} \in \Pi(\hat{\mathcal{A}})$ and $[\hat{F}, \hat{E}] = \mathbf{0}$ then $p_\rho(E\&F) = Tr(\rho\hat{E}\hat{F})$;*

(ii) *if $\{\hat{F}_j\}_{j\in J} \subseteq \Pi(\hat{\mathcal{A}})$ is any countable family of projection operators such that $\sum_{j\in J}\hat{F}_j \equiv \hat{F} \in \Pi(\hat{\mathcal{A}})$, then $\quad p_\rho(E\&F) = \sum_{j\in J} p_\rho(E\&F_j)$.*

Since the set $\hat{\mathcal{F}}(\hat{T}) = \{\hat{F} \in \mathcal{E}(\mathcal{H}) \mid [\hat{F}, \hat{T}] = \mathbf{0}\}$ is just the set of all projection operators of the Von Neumann algebra $\hat{\mathcal{A}}(\hat{T}) = \{\hat{T}\}'$, and hence $\hat{\mathcal{F}}(\hat{T})$ generates $\hat{\mathcal{A}}(\hat{T})$, the following theorem can be proved by means of Lemma 4.1.

Theorem 4.1. *Let T be an elementary observable. Given $E \in \mathcal{F}(T)$, for every quantum state ρ the mappings*

$p_\rho(E\& : \mathcal{F}(T) \to [0,1]$, $p_\rho(E\&F) = Tr(\rho\hat{E}\hat{F}\hat{E})$, *and*

$p_\rho(E'\& : \mathcal{F}(T) \to [0,1]$, $p_\rho(E'\&F) = Tr(\rho\hat{E}'\hat{F}\hat{E}')$

are the unique functionals which satisfy the following conditions.

[3] A Von Neumann algebra [6] is a subset $\hat{\mathcal{A}}$ of bounded linear operators of the Hilbert space $\mathcal{H}$ such that $\hat{\mathcal{A}} = (\hat{\mathcal{A}}')' \equiv \hat{\mathcal{A}}''$, where $\hat{\mathcal{A}}'$ denotes the *commutant* of $\hat{\mathcal{A}}$, i.e. the set of all bounded linear operators $\hat{B}$ of $\mathcal{H}$ such that $[\hat{B}, \hat{A}] = \mathbf{0}$ for all $\hat{A} \in \hat{\mathcal{A}}$. The theory of Von Neumann algebras [6] shows that if $\hat{\Pi}(\hat{\mathcal{A}})$ is the set of all projection operators in the Von Neumann algebra $\hat{\mathcal{A}}$, then $\hat{\mathcal{A}} = \hat{\Pi}(\hat{\mathcal{A}})''$.

(C.1) *If* $F \in \mathcal{F}(T)$ *and* $[\hat{F},\hat{E}] = \mathbf{0}$ *then* $p_\rho(E\&F) = Tr(\rho\hat{E}\hat{F})$ *and* $p_\rho(E'\&F) = Tr(\rho\hat{E}'\hat{F})$;

(C.2) *if* $\{F_j\}_{j\in J} \subseteq \mathcal{F}(T)$ *is any countable family such that* $\sum_{j\in J}\hat{F}_j \equiv \hat{F} \in \hat{\mathcal{F}}(\hat{T})$, *then* $p_\rho(E\&F) = \sum_{j\in J} p_\rho(E\&F_j)$ *and* $p_\rho(E'\&F) = \sum_{j\in J} p_\rho(E'\&F_j)$.

Theorem 4.1 implies that if E can be assigned values which extend its measured values to the case that E is not measured, in a way which is logically consistent with the outcomes of whatever actually measured observable $F \in \mathcal{F}(T)$, then it is necessary that such an assignment be ruled over by the probability

$$p_\rho(E\&F) = Tr(\rho\hat{E}\hat{F}\hat{E}) \quad (\text{resp., } p_\rho(E'\&F) = Tr(\rho\hat{E}'\hat{F}\hat{E}'))$$

of assigning E value 1 (resp., 0) and obtaining value 1 for F.

However, the agreement with probability P_ρ is not sufficient. Logical consistency requires that the condition must hold too.

(C.3) $$p_\rho(E'\&F) + p_\rho(E\&F) = Tr(\rho\hat{F}) \quad \text{for all } F \in \mathcal{F}(T)$$

In general, (C.3) does not hold. Indeed, we have
$Tr(\rho\hat{F}) = Tr(\rho[\hat{E}+\hat{E}']\hat{F}[\hat{E}+\hat{E}']) = p_\rho(E\&f) + p_\rho E'\&F) + 2\mathbf{Re}Tr((\hat{E}\hat{F}\hat{E}')$.
Therefore, if $[\hat{F},\hat{E}] \neq \mathbf{0}$ the presence of a non-vanishing *interference* term $2\mathbf{Re}Tr((\hat{E}\hat{F}\hat{E}')$ cannot be excluded, and (C.3) is violated.

The following theorem proves that to assign E just the outcome of the evaluation of E by T guarantees all conditions (C.1), (C.2), (C.3).

Theorem 4.2. *Given two elementary observables* $E, T \in \mathcal{E}$, *if* $E \overset{\rho}{\longleftrightarrow} T$, *so that* T *can evaluate* E, *then for all* $F \in \mathcal{F}(T)$ *the equalities* $p_\rho(E\&F) = Tr(\rho\hat{T}\hat{F})$ *and* $p_\rho(E'\&F) = Tr(\rho\hat{T}'\hat{F})$ *hold. Furthermore, also (C.3) holds.*

Proof. If $E \overset{\rho}{\longleftrightarrow} T$, then $\hat{E}\rho = \hat{T}\rho$ and $\rho\hat{E} = \rho\hat{T}$ hold; therefore $p_\rho(E\&F) = Tr(\rho\hat{E}\hat{F}\hat{E}) = Tr(\hat{E}\rho\hat{E}\hat{F}) = Tr(\hat{T}\rho\hat{E}\hat{F}) = Tr(\hat{T}\rho\hat{T}\hat{F}) = Tr(\rho\hat{T}\hat{F})$ holds because $[\hat{F},\hat{T}] = \mathbf{0}$. Analogously, since $E \overset{\rho}{\longleftrightarrow} T$ implies $E' \overset{\rho}{\longleftrightarrow} T'$, $p_\rho(E'\&F) = Tr(\rho\hat{T}'\hat{F})$ is proved. Then $p_\rho(E\&F)+p_\rho(E'\&F) = Tr(\rho\hat{T}\hat{F}) + Tr(\rho\hat{T}'\hat{F}) = Tr(\rho\hat{F})$; thus also (C.3) is proved.

4.2. Conclusions

In virtue of theorem 4.2 we can state the following conclusions, which are valid whenever $E \overset{\rho}{\longleftrightarrow} T$.

(Ev.1) It is logically consistent to assign E the value obtained by an evaluation by T together with simultaneous measurement of whatsever family $\{F_j\}$ of observables from $\mathcal{F}(T)$, also if $[\hat{E},\hat{F}] \neq \mathbf{0}$.

(Ev.2) The probability which rules over such an assignment and the occurrences of the outcomes extends the probability established by Quantum Theory for actually measured outcomes.

(Ev.3) The consistency is guaranteed within the domain $\mathcal{F}(T) \subseteq \mathcal{E}$, not elsewhere.

According to these conclusions, to assign E the value evaluated by T cannot provoke contradiction if the evaluation is performed together with whatsever measurement of an observable F, also if $[\hat{F},\hat{E}] \neq \mathbf{0}$. Thus, our conclusions (Ev) contribute to fulfill the interpretative lack about evaluations which arises in the circumstances where conditions (4.1) hold.

As further implications of our results, we can deduce that it is possible to exploit evaluations for consistently assigning non-commuting observables simultaneous values. Indeed, let us consider two elementary observables E_1, E_2 such that $E_1 \overset{\rho}{\longleftrightarrow} T_1$ and $E_2 \overset{\rho}{\longleftrightarrow} T_2$, so that they can be evaluated by T_1 and T_2, respectively, in the state ρ. If $[\hat{T}_1, \hat{T}_2] = \mathbf{0}$ and $E_1, E_2 \in \mathcal{F}(T_1) \cap \mathcal{F}(T_2)$, the evaluations of both E_1 and E_2 can be obtained simultaneously by measuring together T_1 and T_2. Therefore E_1 and E_2 can be consistently assigned simultaneous values, also if $[\hat{E}_1, \hat{E}_2] \neq \mathbf{0}$. Meaningful examples are developed in [1],[7],[8].

These results do no conflict with the impossibility of identifying evaluations with measurement proved in section 2.3; on the contrary, they explain it. Therein the observables actually measured are $E^\alpha, F, G^\beta, L^\alpha$. Hence, also the elementary observables M and N represented by the projection operators $\hat{M} = \frac{\mathbf{1} - f(\hat{E}^\alpha) f(\hat{F}) f(\hat{L}^\alpha)}{2}$ and $\hat{N} = \frac{\mathbf{1} - f(\hat{F}) f(\hat{G}^\beta) f(\hat{L}^\alpha)}{2}$ are actually measured according to (2.1.v). Since $\hat{N} = \frac{\mathbf{1} - f(\hat{F}) f(\hat{G}^\beta) f(\hat{L}^\alpha)}{2}$, $N \xleftrightarrow{\rho_0} E^\beta$ and $G^\alpha, E^\beta \in \mathcal{F}(M) \cap \mathcal{F}(N)$ hold, G^α and E^β can be consistently assigned values. However, L^β cannot be consistently assigned the value evaluated by R, though $R \xleftrightarrow{\rho_0} L^\beta$, because R is not actually measured. Indeed, since $\hat{R} = \frac{\mathbf{1} - f(\hat{E}^\beta) f(\hat{F}) f(\hat{G}^\alpha)}{2}$, the measurement of R entails the measurement of E^β and G^α; but $[\hat{E}^\beta, \hat{E}^\alpha] \neq \mathbf{0}$, $[\hat{G}^\alpha, \hat{G}^\beta] \neq \mathbf{0}$ and E^α, G^β are actually measured. Thus, according to our results, a consistent value assignment also to E^β is not guaranteed, and indeed an inconsistency arises in attempting it.

References

[1] Nisticò G 2008 *J. Phys. A: Math. Theor.* **41** 125302
[2] Von Neumann J 1955 *Mathematical Foundations of Quantum Mechanics* (Princeton: Princeton Univ. Press)
[3] Nisticò G and Sestito A 2011 *Found. Phys.* **41** 1263
[4] Greenberger D M, Horne M A, Shimony A and Zeilinger A 1990 *Am. J. Phys.* **58** 1131
[5] Cassinelli G and Zanghì N 1983 *Il Nuovo Cim.* B **73** 237
[6] Dixmier J 1957 *Les algebres d'opérateurs dans l'espace Hilbertien* (Paris: Gauthier-Villars)
[7] Sestito A 2008 *Found. Phys.* **38** 935
[8] Nisticò G and Sestito A 2012 *Int. J. Qu. Info* **10** 1250055

Journal of Physics: Conference Series **626** (2015) 012050 doi:10.1088/1742-6596/626/1/012050

Robertson-Schrödinger formulation of Ozawa's uncertainty principle

Catarina Bastos[1], Alex E. Bernardini[2], Orfeu Bertolami[3], Nuno Costa Dias[4] and João Nuno Prata[4,5]

[1]Instituto de Plasmas e Fusão Nuclear, Instituto Superior Técnico Avenida Rovisco Pais 1, 1049-001 Lisboa, Portugal

[2]Departamento de Física, Universidade Federal de São Carlos, PO Box 676, 13565-905, São Carlos, SP, Brasil.

[3]Departamento de Física e Astronomia, Faculdade de Ciências da Universidade do Porto, Rua do Campo Alegre, 687,4169-007 Porto, Portugal

[4]Departamento de Matemática, Universidade Lusófona de Humanidades e Tecnologias Avenida Campo Grande, 376, 1749-024 Lisboa, Portugal

[5]Grupo de Física Matemática, UL, Avenida Prof. Gama Pinto 2, 1649-003, Lisboa, Portugal

E-mail: `catarina.bastos@ist.utl.pt`, `alexeb@ufscar.br`, `orfeu.bertolami@fc.up.pt`, `ncdias@meo.pt`, `joao.prata@mail.telepac.pt`

Abstract. A more general measurement disturbance uncertainty principle is presented in a Robertson-Schrödinger formulation. It is shown that it is stronger and having nicer properties than Ozawa's uncertainty relations. In particular it is invariant under symplectic transformations. One shows also that there are states of the probe (measuring device) that saturate the matrix formulation of measurement disturbance uncertainty principle.

1. Introduction

In this work we present a matrix formulation of Ozawa's measurement disturbance uncertainty principle (OUP), which takes into account position-momentum correlations [1]. This type of formulation is to the OUP [2, 3] as Robertson-Schroedinger formulation of uncertainty principle is to the Robertson-like inequality. The latter is usually associated to a set of kinematical restrictions that can be imposed to a quantum system,

$$\sigma(A,\psi)\sigma(B,\psi) \geq \frac{|\langle\psi|\ [A,B]\ |\psi\rangle|}{2}\ , \tag{1}$$

where $\sigma(C,\psi)$ denotes the standard mean deviation for an observable C: $\sigma^2(C,\psi) = \langle\psi|(C - \langle\psi|C|\psi\rangle)^2|\psi\rangle$. Here, $[A,B] = AB - BA$ is the commutator and $\{A,B\} = \frac{AB+BA}{2}$ is the anti-commutator. These uncertainty relations can be written in terms of the positivity of a matrix, the covariance matrix of the quantum state $\mathbf{\Sigma}$, the so-called Robertson-Schrödinger uncertainty principle (RSUP)

$$\mathbf{\Sigma} + \frac{i\hbar}{2}\mathbf{J} \geq 0\ . \tag{2}$$

Let $A = X$ be the position of the particle and $B = P$, its momentum, then $\mathbf{\Sigma}$ is given by

$$\mathbf{\Sigma} = \begin{pmatrix} \sigma(X,\psi)^2 & \sigma(X,P,\psi) \\ \sigma(P,X,\psi) & \sigma(P,\psi)^2 \end{pmatrix} , \tag{3}$$

where $\sigma(X,P,\psi) = \sigma(P,X,\psi) = \langle\psi|\ \{\Delta X, \Delta P\}\ |\psi\rangle$ are the covariance elements for position-momentum correlations, and

$$\mathbf{J} = \begin{pmatrix} 0 & 1 \\ -1 & 0 \end{pmatrix} \tag{4}$$

is the standard symplectic matrix.

Why this matrix formulation should be better than the inequality (1)? In fact, the RSUP formulation, Eq. (2), has several advantages. It implies inequality (1), thus it is stronger. However, the converse is not true and the RSUP accounts for the correlations between position and momentum. For Gaussian states, the RSUP is a necessary and sufficient condition of quantumness, and it also establishes the separability or entanglement of 2-mode Gaussian states, when a reflection transformation is done [4]. Finally, in contrast with inequalities (1), the RSUP is invariant under symplectic transformations. An experimental violation of the uncertainty principles (1) and (2) it would imply two possibilities. The violation should be due to the failure of the Hilbert space in correctly describing quantum systems, which clearly would have important implications for the theoretical structure of quantum mechanics. Alternatively, it could be a consequence of a deformation of the algebra defined by the pair of observables, position and momenta. In a recent work [5], this kind of deformation was studied, even though there is no evidence that such an experimental violation has ever been observed.

When considering measurement-disturbance uncertainty principles, one can relate the accuracy and disturbance in the measurement of a pair of observables. These type of relations are known as error-disturbance trade-off relations and are dynamical uncertainty principles. An example of this relation was related by Heisenberg in his well-known γ-ray experiment [6], where he obtained a relation between the accuracy of an appropriate position measurement, $\epsilon(X,\psi)$ and the disturbance of the particle+s momentum, $\chi(P,\psi)$,

$$\epsilon(X,\psi)\chi(P,\psi) \geq \frac{|\langle\psi|\ [X,P]\ |\psi\rangle|}{2} . \tag{5}$$

Recently, there has been a heated discussion about the status and the interpretation of this Heisenberg's formulation of uncertainty principle [2, 3, 7, 8, 9]. In fact, experimental violations of this inequality have been found using or weak measurements and polarized entangled photons [10], or through neutron spin measurements [11], or even through approximate joint measurements of incompatible polarizations observables on single photons [12]. Theoretically, the different between the two main approaches is based on the definition of the error and disturbance quantities. On one hand, Busch et al. [7] considered the measures of error and disturbance as state independent. On the other hand, others considered state dependent quantities [2, 9, 13]. The most discussed approach is the OUP, which defines $\epsilon(A,\psi)$ and $\chi(B,\psi)$ as the noise in the measurement of observable A and the disturbance of B due to that measurement, respectively. It considers a composite system of an object and a probe, and initially prepared in a product state $\Psi = \psi \otimes \xi$, where ψ and ξ describe the object and the probe, respectively. In the Heisenberg picture, a noise operator $N(A)$ associated with observable A and a disturbance operator $D(B)$ have been introduced. They are self-adjoint operators, defined by

$$N(A) = M^{out} - A^{in} \ , \ \ D(B) = B^{out} - B^{in} \ , \tag{6}$$

where $A^{in} = A \otimes I, B^{in} = B \otimes I$ are observables A, B prior to the measurement interaction, $B^{out} = U^{\dagger}(B \otimes I)U$ is the observable B immediately after the measurement and M is the probe observable. U is the unitary time evolution operator during the measuring interaction. Clearly, $M^{in} = I \otimes M$ and $M^{out} = U^{\dagger}(I \otimes M)U$ [3]. Ozawa argued that the noise $\epsilon(A, \psi)$ and disturbance $\eta(B; \psi)$ are defined by:

$$\epsilon(A, \psi)^2 = \langle \Psi | N(A)^2 | \Psi \rangle \ , \ \eta(B, \psi)^2 = \langle \Psi | D(B)^2 | \Psi \rangle \ . \tag{7}$$

Then, as M and B are observables in different systems, they commute $\left[M^{out}, B^{out}\right] = 0$. Thus, using these last definitions, the triangle and Cauchy-Schwartz inequalities one has finally the OUP,

$$\epsilon(A, \psi)\eta(B, \psi) + \epsilon(A, \psi)\sigma(B, \psi) + \sigma(A, \psi)\eta(B, \psi) \geq \frac{|\langle \psi | \ [A, B] \ | \psi \rangle}{2} \ . \tag{8}$$

Clearly, this relation does not take into account noise-disturbance correlations. Thus, in what follows one shows how to obtain a more general and stronger formulation of this OUP and where one has information about these correlations.

2. Matrix formulation of Ozawa's uncertainty principle

One considers a multidimensional system, where A_i^{in} and B_j^{in}, $i, j = 1, \cdots, n$ such that

$$\left[A_i^{in}, A_j^{in}\right] = \left[B_i^{in}, B_j^{in}\right] = 0 \ , \ \left[A_i^{in}, B_j^{in}\right] = iC_{ij} \ , \tag{9}$$

for $i, j = 1, \cdots, n$, and where $\{C_{ij}\}$ are some self-adjoint operators. If A and B are the particle's position and momentum, one simply has $C_{ij} = \hbar\delta_{ij}$. These self adjoint operators can be collectively written as $Z^{in} = (A_1^{in}, \cdots, A_n^{in}, B_1^{in}, \cdots, B_n^{in})$ satisfying the commutation relations

$$\left[Z_\alpha^{in}, Z_\beta^{in}\right] = iG_{\alpha\beta} \ , \ \alpha, \beta = 1, \cdots, 2n \ . \tag{10}$$

$\mathbf{G} = \{G_{\alpha\beta}\}$ is the self-adjoint operator-valued skew-symmetric matrix

$$\mathbf{G} = \begin{pmatrix} \mathbf{0} & \mathbf{C} \\ -\mathbf{C} & \mathbf{0} \end{pmatrix} , \tag{11}$$

with $\mathbf{C} = \{C_{ij}\}$. Again, if A and B are the position and momentum operators, then one simply has $2n \times 2n$ standard symplectic matrix $\hbar\mathbf{J}$. The same strategy can de used to write collectively the noise and disturbance operators,

$$N = N(A) = (N_1, \cdots, N_n) \ , \ D = D(B) = (D_1, \cdots, D_n) \ , \tag{12}$$

into $K = (N_1, \cdots, N_n, D_1, \cdots, D_n)$. Finally, let

$$M^{out} = \left(M_1^{out}, \cdots, M_n^{out}\right) \ , \ B^{out} = \left(B_1^{out}, \cdots, B_n^{out}\right) \ , \tag{13}$$

denote the output of the (commuting) probe observables and the output of B, respectively. They can be written collectively into $Z^{out} = (M_1^{out}, \cdots, M_n^{out}, B_1^{out}, \cdots, B_n^{out})$. Thus, as before one has $\left[Z_\alpha^{out}, Z_\beta^{out}\right] = 0 \ , \ \alpha, \beta = 1, \cdots, 2n$ and $Z^{out} = Z^{in} + K$.

Considering an arbitrary set of complex numbers denoted by $\{\lambda_\alpha\}_{1 \leq \alpha \leq 2n}$, one has [1]

$$\begin{aligned} 0 &= \sum_{\alpha,\beta=1}^{2n} \overline{\lambda}_\alpha \lambda_\beta \langle \left[\Delta Z_\alpha^{out}, \Delta Z_\beta^{out}\right] \rangle \\ &= \sum_{\alpha,\beta=1}^{2n} \overline{\lambda}_\alpha \lambda_\beta \left(i\langle G_{\alpha\beta} \rangle + \langle \left[Z_\alpha^{in}, K_\beta\right] + \left[K_\alpha, Z_\beta^{in}\right] \rangle + \langle [K_\alpha, K_\beta] \rangle \right) \ . \end{aligned} \tag{14}$$

Journal of Physics: Conference Series **626** (2015) 012050 doi:10.1088/1742-6596/626/1/012050

Writing $\mathcal{K} = \sum_\alpha \lambda_\alpha K_\alpha$, and assuming that $\mathcal{K}\mathcal{K}^\dagger$ is a positive operator, one has

$$\sum_{\alpha,\beta=1}^{2n} \overline{\lambda}_\alpha \lambda_\beta \langle [K_\alpha, K_\beta] \rangle = \langle \mathcal{K}^\dagger \mathcal{K} \rangle - \langle \mathcal{K}\mathcal{K}^\dagger \rangle \ \leq \ \langle \mathcal{K}^\dagger \mathcal{K} \rangle \ . \tag{15}$$

Thus,

$$\sum_{\alpha,\beta=1}^{2n} \overline{\lambda}_\alpha \lambda_\beta \langle [K_\alpha, K_\beta] \rangle \ \leq \ \sum_{\alpha,\beta=1}^{2n} \overline{\lambda}_\alpha \lambda_\beta \langle K_\alpha K_\beta \rangle = \sum_{\alpha,\beta=1}^{2n} \overline{\lambda}_\alpha \lambda_\beta \left(\langle \{K_\alpha, K_\beta\} \rangle + \frac{1}{2} \langle [K_\alpha, K_\beta] \rangle \right) . \tag{16}$$

And so,

$$\sum_{\alpha,\beta=1}^{2n} \overline{\lambda}_\alpha \lambda_\beta \langle [K_\alpha, K_\beta] \rangle \ \leq \ 2 \sum_{\alpha,\beta=1}^{2n} \overline{\lambda}_\alpha \lambda_\beta \langle \{K_\alpha, K_\beta\} \rangle \ . \tag{17}$$

Upon substitution of Eq. (17) into Eq. (14) one finally obtains

$$0 \leq \sum_{\alpha,\beta=1}^{2n} \overline{\lambda}_\alpha \lambda_\beta \left(i\langle G_{\alpha\beta} \rangle + \langle \left[Z_\alpha^{in}, K_\beta \right] + \left[K_\alpha, Z_\beta^{in} \right] \rangle + 2 \langle \{K_\alpha, K_\beta\} \rangle \right) . \tag{18}$$

Then, defining the $2n \times 2n$ real symmetric positive-definite matrix

$$\mathbf{K}_{\alpha\beta} = \langle \{K_\alpha, K_\beta\} \rangle \ , \tag{19}$$

the $2n \times 2n$ real skew-symmetric matrices $\mathcal{G} = \langle \mathbf{G} \rangle$ and

$$\mathbf{\Gamma}_{\alpha\beta} = \frac{1}{i} \langle \left[Z_\alpha^{in}, K_\beta \right] + \left[K_\alpha, Z_\beta^{in} \right] \rangle \ , \tag{20}$$

it is possible to rewrite (18) in the matrix form:

$$\mathbf{K} + \frac{i}{2} \left(\mathbf{\Gamma} + \mathcal{G} \right) \geq 0 \ , \tag{21}$$

which is a matrix version of OUP. If a measuring interaction is of independent intervention, then $\mathbf{\Gamma} = 0$, and one recovers the matrix generalization of Heisenberg's noise disturbance relation Eq. (5),

$$\mathbf{K} + \frac{i}{2} \mathcal{G} \geq 0 \ . \tag{22}$$

Now one shows that the uncertainty principle, Eq. (21), is in fact stronger than the OUP. In the case of $n = 1$, one finds

$$\mathbf{K} = \begin{pmatrix} \langle N^2 \rangle & \langle \{N, D\} \rangle \\ \langle \{D, N\} \rangle & \langle D^2 \rangle \end{pmatrix} \ , \ \mathbf{\Gamma} = \frac{1}{i} \begin{pmatrix} 0 & \langle [A^{in}, D] + [N, B^{in}] \rangle \\ \langle [D, A^{in}] + [B^{in}, N] \rangle & 0 \end{pmatrix} . \tag{23}$$

Clearly, if inequality (21) holds, then the matrix $\mathbf{K} + \frac{i}{2} \left(\mathbf{\Gamma} + \mathcal{G} \right)$ must have non-negative determinant, and then [1]

$$\begin{aligned} \langle N^2 \rangle \langle D^2 \rangle \ & \geq \ \langle \{N, D\} \rangle^2 + \frac{1}{4} \left| \langle \left[A^{in}, D \right] + \left[N, B^{in} \right] \rangle + \langle \left[A^{in}, B^{in} \right] \rangle \right|^2 \\ & \geq \ \frac{1}{4} \left| \langle \left[A^{in}, D \right] + \left[N, B^{in} \right] \rangle + \langle \left[A^{in}, B^{in} \right] \rangle \right|^2 \ . \end{aligned} \tag{24}$$

Taking the square root of this last inequality,

$$\langle N^2\rangle^{1/2}\langle D^2\rangle^{1/2} \geq \frac{1}{2}\left|\langle\left[A^{in},D\right]+\left[N,B^{in}\right]\rangle - \langle\left[B^{in},A^{in}\right]\rangle\right| \,, \tag{25}$$

considering the definitions $\epsilon(A) = \langle N^2\rangle^{1/2}$ and $\eta(B) = \langle D^2\rangle^{1/2}$, and using the inequality $|a-b| \geq \big|\,|a|-|b|\,\big|$, one obtains

$$\epsilon(A)\eta(B) \geq \frac{1}{2}\Big|\ |\langle\left[A^{in},D\right]+\left[N,B^{in}\right]\rangle| - |\langle\left[A^{in},B^{in}\right]\rangle|\ \Big| \,. \tag{26}$$

And in particular

$$\epsilon(A)\eta(B) \geq \frac{1}{2}|\langle\left[A^{in},B^{in}\right]\rangle| - \frac{1}{2}\left|\langle\left[A^{in},D\right]+\left[N,B^{in}\right]\rangle\right| \,, \tag{27}$$

which is OUP, Eq. (8).

3. Properties of the matrix formulation of the measurement disturbance uncertainty principle

In order to demonstrate that the matrix formulation of OUP, inequality (21), is in fact stronger than OUP, one uses the example of a backaction evading quadrature amplifier (BAE) [1, 3]. The system is described by a set of quadrature operators X_a, Y_a and the probe by the operators X_b, Y_b with

$$[X_a, Y_a] = [X_b, Y_b] = \frac{i}{2} \,. \tag{28}$$

Then, the measuring interaction is given by:

$$\begin{cases} X_a^{out} = X_a^{in} \\ X_b^{out} = X_b^{in} + GX_a^{in} \\ Y_a^{out} = Y_a^{in} - GY_b^{in} \\ Y_b^{out} = Y_b^{in} \,, \end{cases} \tag{29}$$

where G is the gain. The probe observable is then set to $M = \frac{1}{G}X_b$, and thus

$$M^{out} = X_a^{in} + \frac{1}{G}X_b^{in} \,. \tag{30}$$

Moreover

$$N(X_a) = \frac{1}{G}X_b^{in} \,,\ D(X_a) = 0 \,,\ D(Y_a) = -GY_b^{in} \tag{31}$$

or in the notation introduced above

$$K = \left(\frac{1}{G}X_b^{in}, -GY_b^{in}\right) \,. \tag{32}$$

Clearly, $\mathbf{\Gamma} = 0$ and the measuring interaction is of independent intervention for the pair (X_a, Y_a). Also from Eq. (28), one finds $\mathcal{G} = \frac{1}{2}\mathbf{J}$.

One considers the state of the probe ξ to have vanishing expectation values $\langle\xi|X_b^{in}|\xi\rangle = \langle\xi|Y_b^{in}|\xi\rangle = 0$, and a covariance matrix

$$\mathbf{\Sigma^b} = \begin{pmatrix} \mathbf{\Sigma^b}_{11} & \mathbf{\Sigma^b}_{12} \\ \mathbf{\Sigma^b}_{12} & \mathbf{\Sigma^b}_{22} \end{pmatrix} = \begin{pmatrix} 1/4 & 1/2 \\ 1/2 & 1/4 \end{pmatrix} \,. \tag{33}$$

Journal of Physics: Conference Series **626** (2015) 012050 doi:10.1088/1742-6596/626/1/012050

One obtains the elements of $\mathbf{K}$ from Eq. (32) [1]. For instance,

$$\langle N^2\rangle = \frac{1}{G^2}\langle (X_b^{in})^2\rangle = \frac{1}{G^2}\mathbf{\Sigma^b}_{11} = \frac{1}{4G^2} \ . \tag{34}$$

The other elements are obtained in an analogous way. Thus,

$$\mathbf{K} = \begin{pmatrix} \frac{1}{4G^2} & -\frac{1}{2} \\ -\frac{1}{2} & \frac{G^2}{4} \end{pmatrix} . \tag{35}$$

We then have from $\mathcal{G} = \frac{1}{2}\mathbf{J}$ and Eq. (35):

$$\mathbf{K} + \frac{i}{2}(\mathbf{\Gamma} + \mathcal{G}) = \mathbf{K} + \frac{i}{4}\mathbf{J} = \begin{pmatrix} \frac{1}{4G^2} & -\frac{1}{2} + \frac{i}{4} \\ -\frac{1}{2} - \frac{i}{4} & \frac{G^2}{4} \end{pmatrix} . \tag{36}$$

Since $\det(\mathbf{K} + \frac{i}{4}\mathbf{J}) = -\frac{1}{4}$, then it is not a positive matrix and the uncertainty principle (21) is violated. However, since $\epsilon(A) = \langle N^2\rangle^{1/2} = \frac{1}{2G}$ and $\eta(B) = \langle D^2\rangle^{1/2} = \frac{G}{2}$, one concludes that $\epsilon(A)\eta(B) = \frac{1}{4}$, which exactly saturates the OUP. This shows that in fact the inequality (21) is indeed stronger that the OUP.

Next, one shows that the uncertainty principle, Eq. (21), can be saturated if there is a set of complex numbers $\{\lambda_\alpha\}_{1\le\alpha\le 2n}$ and a state Ψ such that

$$\sum_{1\le\alpha\le 2n} \lambda_\alpha K_\alpha \Psi = 0 \ . \tag{37}$$

This may or may not be possible, depending on the commutation relations of the operators $\{K_\alpha\}_{1\le\alpha\le 2n}$. In the case of BAE described above this is not possible. Indeed, to have a saturating state Ψ, Eq. (37), would have to satisfy:

$$\left(\frac{\lambda_1}{G}X_b^{in} + \frac{i}{2}G\lambda_2\frac{\partial}{\partial X_b^{in}}\right)\Psi = 0 \ . \tag{38}$$

for some constants $(\lambda_1, \lambda_2) \in C^2\backslash\{(0,0)\}$. In the representation of quadrature operators and using their commutation relations, one concludes that there are no square-integrable solutions, and thus there is no saturating state [1].

Alternatively, if one considers a noiseless quadrature transducer [3] with the following input-output relations for the quadrature operators (28)

$$\begin{cases} X_a^{out} = X_a^{in} - X_b^{in} \ , \\ X_b^{out} = X_a^{in} \ , \\ Y_a^{out} = -Y_b^{in} \ , \\ Y_b^{out} = Y_b^{in} + Y_a^{in} \ . \end{cases} \tag{39}$$

and a probe observable $M = X_b$, then

$$K = (0, -Y_a^{in} - Y_b^{in}) \ . \tag{40}$$

And thus, a saturating state satisfies

$$-\lambda_2(Y_a^{in} + Y_b^{in})\Psi = 0 \ . \tag{41}$$

Journal of Physics: Conference Series **626** (2015) 012050 doi:10.1088/1742-6596/626/1/012050

This can be achieved by any state Ψ for $\lambda_2 = 0$.

Finally, one should notice that the matrix formulation of the noise-disturbance uncertainty principle, Eq. (21) remains unchanged under symplectic transformations. To show that one considers that $[K_\alpha, K_\beta] = i\gamma J_{\alpha\beta}$, where $\gamma \neq 0$ is some real constant and $\mathbf{J}$ is the standard symplectic matrix. Suppose like in the BAE casa that $(\mathbf{\Gamma} = 0)$. Then Eq. (21) becomes $\mathbf{K} + \frac{i\gamma}{2}\mathbf{J} \geq 0$. Notice that this looks formally like the RSUP Eq. (2). If the system undergoes a linear symplectic transformation as $K_\alpha \mapsto K'_\alpha = \sum_{1\leq\beta\leq 2n} S_{\alpha\beta}K_\beta$, where $\mathbf{S} \in Sp(2n; R)$, then the noise-disturbance covariance matrix transforms by similarity $\mathbf{K} \mapsto \mathbf{K}' = \mathbf{SKS}^T$. Since $\mathbf{S}^{-1}\mathbf{J}(\mathbf{S}^{-1})^T = \mathbf{J}$, one concludes that the matrix uncertainty principle remains unchanged: $\mathbf{K}' + \frac{i\gamma}{2}\mathbf{J} \geq 0$.

In contrast, the OUP, when considering for instance the previous example of the BAE, becomes simply the Heisenberg inequality

$$\epsilon\eta \geq \frac{1}{4} \, . \tag{42}$$

Rewriting this now as a noise-disturbance covariance matrix $\mathbf{K} = diag(\epsilon^2, \eta^2)$ and considering that the probe (and possibly the object) is subjected to a symplectic transformation such that:

$$K \mapsto K' = \frac{\sqrt{2}}{2}\begin{pmatrix} 1 & 1 \\ -1 & 1 \end{pmatrix} K \, . \tag{43}$$

One clearly sees that the noise-disturbance vector K is rotated through an angle of $\frac{\pi}{4}$. Such a transformation can be implemented by a certain unitary transformation $U(S)$ generated by an appropriate hermitian operator, quadratic in the variables X_b, Y_b of the probe. Thus, the OUP is modified to

$$(\epsilon')^2 + (\eta')^2 \geq \sqrt{1 + 4\langle\{N', D'\}\rangle^2}, \tag{44}$$

which is manifestly different from Eq. (42).

4. Conclusions

We have presented here an universal matrix formulation of the measurement disturbance uncertainty principle. It is more stringent than the Ozawa's measurement disturbance relation as it in fact implies the OUP. Then we have shown that there are states that saturate the OUP but violates the matrix formulation. Furthermore, it is possible to obtain states that saturate also the matrix formulation of OUP. Finally, it is worth to mention that it could be interesting to investigate the experimental validity of this more general formulation of measurement disturbance uncertainty relation. In fact, the OUP proved to be valid through experimental work performed using weak measurements by Rozema et al. [10] and considering 3-state mode systems by Sulyok et al. [11] and by Ringbauer et al. [12].

Acknowledgments

The work of CB is supported by the FCT (Portugal) grant SFRH/BPD/62861/2009.

References

[1] Bastos C, Bernardini A E, Bertolami O, Dias NC and Prata JN 2014 *Phys. Rev.* A **89** 042112
[2] Ozawa M 1988 *Phys. Rev. Lett.* **60** 385; 2002 *Phys. Lett.* A **299** 1; *Squeezed and Nonclassical Light* ed P Tombesi and E R Pike (New York: Plenum, 1989) 263-268
[3] Ozawa M 2003 *Phys. Rev.* A **67** 042105
[4] Simon R 2000 *Phys. Rev. Lett.* **84** 2726
[5] Bastos C, Bertolami O, Dias NC and Prata JN 2012 *Phys. Rev.* D **86** 105030
[6] Heisenberg W 1927 *Z. Phys.* **43** 172

[7] Busch P, Lahti P and Werner RF 2013 *Phys. Rev. Lett.* **111** 160405; *do.* 2013 Noise Operators and Measures of RMS Error and Disturbance in Quantum Mechanics *Preprint* arxiv: quant-ph/1312.4393
[8] Fujikawa K 2012 *Phys. Rev.* A **85** 062117
[9] Branciard C 2013 *Proc. Nat. Acad. Sci.* (USA) **110** 6742
[10] Rozema LA, Darabi A, Mahler DH, Hayat A, Soudagar Y and Steinberg AM 2012 *Phys. Rev. Lett.* **109** 100404
[11] Sulyok G, Sponar S, Erhart J, Badurek G, Ozawa M and Hasegawa Y 2013 *Phys. Rev.* A **88** 022110
[12] Ringbauer M, Biggerstaff D, Broome M, Fedrizzi A, Branciard C and White A 2014 *Phys. Rev. Lett.* **112** 020401
[13] Korzekwa K, Jennings D and Rudolph T 2014 *Phys. Rev.* A **89** 052108

Journal of Physics: Conference Series **626** (2015) 012051 doi:10.1088/1742-6596/626/1/012051

Initial entanglement between detectors allows violating Heisenberg's uncertainty relations

Antonio Di Lorenzo

E-mail: dilorenzo@infis.ufu.br
Universidade Federal de Uberlândia, Uberlândia, MG, Brazil

Abstract. Physics textbooks introduce an inequality derived by Kennard that concerns the impossibility of preparing a quantum state with well defined momentum and position. However, Heisenberg had formulated two different inequalities: one stating the impossibility of a simultaneous detection of position and momentum with arbitrary precision, and another one imposing a tradeoff between the precision on the measurement of a variable and the backaction on a subsequent measurement of a conjugated variable. Here, we explore the connections among these three inequalities, and we show that the latter can be violated, if the detectors are initially entangled. The results, besides being of fundamental interest, can be useful for building up an ideal momentum, or position, detector (i.e. one that introduces no noise in the measurement, besides the intrinsic statistical noise of the input state).

1. Introduction

Recently, there have been several experiments claiming to have violated Heisenberg's uncertainty relation [1–4]. At first sight, this may seem as revolutionary, since the uncertainty principle is one of the pillars of quantum mechanics. However, there are two important considerations to make: (1) Heisenberg [5] formulated an inequality about the uncertainty of a position measurement and the disturbance in a subsequent, or simultaneous, momentum measurement. The inequality demonstrated in quantum mechanics textbooks is due to Kennard [6], who proved it shortly after Heisenberg's semi–qualitative formulation, and it concerns the impossibility of preparing an ensemble of systems such that a measurement of momentum in a subensemble yields a variance σ_P^2 and a measurement of position in another subensemble, at the same time, yields a variance $\sigma_Q^2 < \hbar^2/4\sigma_P^2$. (2) The inequality violated in the experiments concerns finite–dimensional systems, and was not formulated by Heisenberg. Instead, this inequality was retrofitted, following the reasoning: Heisenberg formulated an uncertainty relation for uncertainty in position and disturbance in a later measurement of momentum and another relation for the uncertainties in a joint measurement of momentum and position

$$\varepsilon_Q \eta_{P|Q} \gtrsim \hbar, \quad \text{for sequential measurements,} \tag{1a}$$

$$\varepsilon_{Q|P} \varepsilon_{P|Q} \gtrsim \hbar, \quad \text{for joint measurements;} \tag{1b}$$

Kennard proved a rigorous inequality for the preparation variances

$$\sigma_Q \sigma_P \geq \frac{\hbar}{2}; \tag{1c}$$

Robertson and Schrödinger generalized Kennard's result to finite–dimensional systems [7, 8]

$$\sigma_A \sigma_B \geq \frac{1}{2} |\langle [\hat{A}, \hat{B}] \rangle|; \tag{1d}$$

therefore, Heisenberg *would have* endorsed the corresponding inequality for uncertainty in a variable and disturbance in the other variable, at least according to Ozawa [9]

$$\varepsilon_A \eta_{B|A} \overset{?}{\geq} \frac{1}{2} |\langle [\hat{A}, \hat{B}] \rangle|, \tag{1e}$$

$$\varepsilon_{A|B} \varepsilon_{B|A} \overset{?}{\geq} \frac{1}{2} |\langle [\hat{A}, \hat{B}] \rangle|. \tag{1f}$$

However, a simple inspection reveals that equations (1e) and (1f) cannot hold. Indeed, for a finite–dimensional system, both the uncertainty ε and the disturbance η, whatever its definition is, must be finite. On the other hand, the precision of a measurement can be made, in principle, arbitrary small. Hence the products $\varepsilon_A \eta_{B|A}$ and $\varepsilon_{A|B} \varepsilon_{B|A}$ cannot have a lower bound. This consideration is to be contrasted with the original inequalities formulated in (1a) and (1b), for an infinite–dimensional Hilbert space. In this case, indeed, one can imagine that the disturbance $\eta_{P|Q}$ or the uncertainty $\varepsilon_{P|Q}$ diverge if the uncertainty $\varepsilon_{Q|P}$ tends to zero, so that the inequalities make sense, and cannot be discarded *a priori*.

The goal of the present manuscript, therefore, is to verify whether the inequalities (1a) and (1b) can be violated. In order to do this, it is necessary to define the uncertainty of a measurement, and the disturbance of a subsequent or simultaneous measurement. It turns out that the latter definition is a very controversial subject [10–13]. Therefore, the violation of the Heisenberg inequality in the uncertainty–disturbance formulation depends fatally on one's definition of disturbance. Here, we shall argue that our definition translates the intuitive notion of disturbance of a measurement. Furthermore, independently of the definition, we shall prove a somewhat surprising corollary: it is possible to make a very precise measurement of momentum by preceding it with an imprecise measurement of position. This results is uncontroversial, and it opens up a new possibility for realizing a precise measurement apparatus. So far, a precise apparatus should be prepared in a very sharp state of the pointer variable J, i.e. a state very close to the ideal state $|J = 0\rangle$. The result of the present manuscript, however, allows to make a very precise measurement, say, of momentum, even though the detector is not in such an initial state, provided that its output variable is (almost) perfectly anticorrelated to the input variable of the first measurement. In quantum mechanics, perfect anticorrelations can be achieved only by means of entanglement. Hence, our main result can be summarized as in the title of this contribution: *Initial entanglement between detectors allows violating Heisenberg's uncertainty relations.*

A short account of the results presented here was given in Ref. [12]. Bullock and Busch have expanded the subject further [14]. Here, we shall present a more detailed derivation, in particular we shall justify the additional term due to the finite resolution of the readout that contributes to the uncertainty of the measurement. By making only the hypothesis of a uniform resolution of the detectors, we prove that the finite resolution contributes with additive terms to the cumulants of the observable outputs. We also include an appendix on the connection with the POVM approach that so far was published only in the pre–print version of [12].

2. Definitions

In the following, we define what we mean by 'uncertainty' and 'disturbance' in operational terms. Our definitions do not rely on any specific theory, so that they can be translated into different mathematical formulas depending on the underlying theory. In the next section, we shall provide the translation of these definitions in quantum mechanical terms.

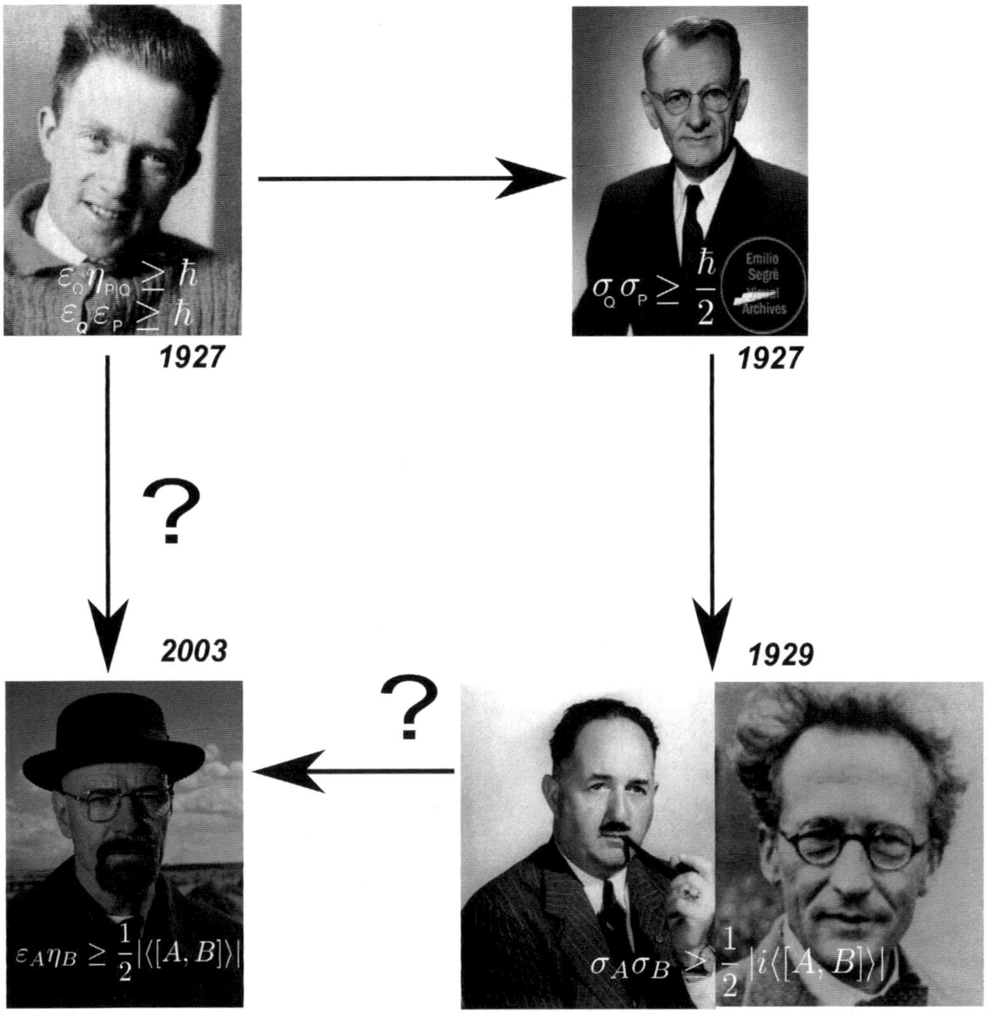

Figure 1. Schematic view of the history of uncertainty relations.

2.1. Uncertainty

The uncertainty in a measurement is not in any way connected to the sample variance. A simple illustration is the following: suppose you are measuring the height of a large group of people. Say that the histogram is well approximated by a Gaussian with average $\bar{h} = 175.$cm and a variance $\sigma_h = 10.$cm. Let us consider a single person, whose height is reported by our apparatus as $h = 169.3$cm. What is the uncertainty in the height of this person? Certainly not 10.cm! The uncertainty on an individual measurement depends on the resolution of the meter, and on how the meter interacted with the measured system. The uncertainty cannot be lower than the resolution, which for a typical tape is 1mm (if we exclude interpolation due to the human eye being a meter *per se*). For instance, was the 0 of the tape placed at the feet of the person, or by mistake a non-zero value was used? Was the tape held straight? And so on. So, how do we estimate the uncertainty in an individual measurement? Usually, we do by calibration. We measure a large set of identically prepared systems, all of them having a value $[A_0-\delta/2, A_0+\delta/2]$, where δ is in principle the best precision that we can achieve in the preparation, and we observe the outputs A of the apparatus. Unless our measuring apparatus is better than our preparation apparatus, we shall observe a distribution centered in $\langle A\rangle$ with a spread $\Delta > \delta$. The quadratic difference $\varepsilon^2 = \Delta^2 - \delta^2$ is the statistical uncertainty of our apparatus. Usually, δ is negligible compared to Δ. We note that in principle the uncertainty varies with the output h.

We also introduce the systematic uncertainty (error, in the common terminology): it is the difference between the average value $\langle A\rangle$ in the limit of a large number of observations $N \to \infty$, and the 'true' value A_0, $U_{A_0} = |\langle A\rangle - A_0|$. This values is also known a priori, in principle.

2.2. Disturbance

The disturbance is the most controversial quantity to be defined. Since in quantum mechanics it is problematic to assume that a system possesses a property, unless it is prepared in an eigenstate of the corresponding observable, we avoid interpreting the disturbance as being a perturbation in an observable, but we interpret this concept as a disturbance of a measurement of an observable, i.e. we maintain an operational stance, in line with the Bohr's approach to quantum mechanics, and we choose to speak only of results of observations.

We say that a measurement of a quantity B at a time t_B is disturbed by a measurement of a quantity A at time t_A if the variance increases or the average changes with respect to the case when the measurement of A is not performed. Let $\langle B\rangle$ and Δ_B^2 the average value and the variance observed when the measurement of A is not performed, and $\langle B\rangle_A$ and $\Delta^2_{B|A}$ the corresponding values when A is measured.[1] We shall also define the same values for the case when the measurement A is performed and the corresponding output is A_0, $\langle B\rangle_{A_0}$ and $\Delta^2_{B|A_0}$.

We define the average statistical disturbance and the average systematic disturbance as $\eta^2_{B|A} = \Delta^2_{B|A} - \Delta^2_B$ and $D_{B|A} = |\langle B\rangle_A - \langle B\rangle|$. We define also the statistical disturbance and the systematic disturbance as $\eta^2_{B|A_0} = \Delta^2_{B|A_0} - \Delta^2_B$ and $D_{B|A_0} = |\langle B\rangle_{A_0} - \langle B\rangle|$.

It is worth noting that Busch, Lahti and Werner [11] propose a different definition of disturbance. The statistics of the B measurement when the A measurement is performed is compared to the statistics of an *ideal* B measurement when A is not performed, $\tilde{\eta}^2_{B|A} = \Delta^2_{B|A} - \Delta^2_{B(ideal)}$. This formulation is in line with the interpretation that what is disturbed is the observable, not the measurement of the observable. Besides this difference in the definition, our results agree with those of [11].

Our definition, on the other hand, should be taken with a grain of salt. Let us imagine a broken detector, which always yields the same output B_0. This measurement would not be disturbed by any measurement performed before it, trivially, and the Heisenberg relation (1*a*) would be violated. Our definition should be applied only to a 'faithful' measurement. By this term, we mean a measurement that yields different average outputs for different inputs, provided that the inputs are sufficiently far from each other (if their difference is below the resolving power of the measuring apparatus, of course the output will be the same).

3. Methods

In the theory of sequential or joint measurements it is usually assumed that the initial state of the system and of the two probes factorizes as

$$\rho_0 = \rho_A \otimes \rho_B \otimes \rho_{sys}. \tag{2}$$

[1] Clearly, causality implies that $\langle B\rangle = \langle B\rangle_A$ and $\Delta^2_B = \Delta^2_{B|A}$ if the measurement of A is performed after B (or better, if the measurement B is either in the past light cone or outside the light cone of the measurement of A). Our definition, however, holds also if causality were violated.

The main idea of the present work is to question the preceding assumption. While it is uncontroversial that the initial state of the system must be uncorrelated to that of the detectors (otherwise, we would have a measurement before the interaction started), there is no compelling reason to assume that the two detectors are initially uncorrelated. Therefore, we shall write, in full generality,

$$\rho_0 = \rho_{A,B} \otimes \rho_{sys}. \tag{3}$$

We assume a von Neumann interaction, leading to the following time–evolution of the system and the meters

$$U = \exp[i\hat{P}\hat{\Phi}_B]\exp[i\hat{Q}\hat{\Phi}_A], \text{ for sequential measurements,} \tag{4}$$

$$U = \exp[i\hat{P}\hat{\Phi}_B + i\hat{Q}\hat{\Phi}_A], \text{ for joint measurements,} \tag{5}$$

with $\hat{\Phi}_\alpha$ representing the input variable of the detector α. We consider the readout states of the detectors $\hat{F}_A(\mu)$, described in the appendix, as Gaussian. It is convenient to work with the characteristic function Z of the readouts, i.e. the Fourier transform of the joint probability $\mathcal{P}(\mu_A, \mu_B)$.

For later convenience, we introduce the column vectors (indicated by horizontal lists within square brackets) $s = [q, p]$, $\phi = [\phi_P, \phi_Q]$, $j = [j_P, j_Q]$. The components of the vectors are labeled by an index α taking values in the set $\{P, Q\}$, so that $s_P = q$ and $s_Q = p$. We also introduce the vector $S = [P, Q]$. With these definitions, e.g. the term $pQ + qP$ can be written as $S \cdot s$.

For a 1–dimensional system, the Moyal quasi-characteristic function $M(q, p)$ [15] is defined as the Fourier transform of the Wigner quasiprobability $W(P, Q)$ [16].

$$M(q,p) = \int dPdQ\, e^{i(Pq+Qp)}\, W(P,Q). \tag{6}$$

We recall that $M(q, 0) = Z(q)$ is the characteristic function for the probability distribution in K, i.e. $M(q, 0) = \int dP \exp(iPq)\langle P|\rho|P\rangle$, and $M(0, q)$ is the characteristic function for the probability in Q. Let $M_{sys}(s)$ the initial Moyal function of the system, and $M_{det}(\phi, j)$ that of the probes. If the probes are initially uncorrelated, then $M_{det}(\phi, j) = M_Q(\phi_Q, j_Q)M_P(\phi_P, j_P)$. However, we shall consider the general case when the probes are initially correlated, so that $M_{det}(\phi, j) \neq M_Q(\phi_Q, j_Q)M_P(\phi_P, j_P)$. The variances and covariances of the probes can be expressed as logarithmic derivatives of their Moyal function:

$$\delta_\alpha^2 = \langle \hat{J}_\alpha^2\rangle - \langle \hat{J}_\alpha\rangle^2 = -\left.\frac{\partial^2 \ln M_{det}(\phi,j)}{\partial \phi_\alpha^2}\right|_{\phi=j=[0,0]}, \tag{7a}$$

$$\tilde{\delta}_\alpha^2 = \langle \hat{\Phi}_\alpha^2\rangle - \langle \hat{\Phi}_\alpha\rangle^2 = -\left.\frac{\partial^2 \ln M_{det}(\phi,j)}{\partial j_\alpha^2}\right|_{\phi=j=[0,0]}, \tag{7b}$$

$$\xi = \langle \hat{\Phi}_P \hat{J}_Q\rangle - \langle \hat{\Phi}_P\rangle\langle \hat{J}_Q\rangle = -\left.\frac{\partial^2 \ln M_{det}(\phi,j)}{\partial \phi_Q \partial j_P}\right|_{\phi=j=[0,0]}, \tag{7c}$$

$$\kappa = \langle \hat{\Phi}_Q \hat{J}_P\rangle - \langle \hat{\Phi}_Q\rangle\langle \hat{J}_P\rangle = -\left.\frac{\partial^2 \ln M_{det}(\phi,j)}{\partial \phi_P \partial j_Q}\right|_{\phi=j=[0,0]}. \tag{7d}$$

We defined as $\tilde{\delta}_\alpha^2$ the initial variance in the input variable Φ_α and as δ_α^2 the initial variance in the output variable J_α. By Kennard relation, applied to the probes,

$$\tilde{\delta}_\alpha \delta_\alpha \geq \frac{1}{2}. \tag{8}$$

We shall use the relation between the final Moyal function for system and probes in terms of the initial ones [17, 18]

$$M(s;\phi,j) = M_{sys}(s+\phi)M_{det}(\phi, j + 2\alpha_0 s + \alpha_\tau \phi), \tag{9}$$

where

$$\alpha_+ = \begin{pmatrix} 0 & 0 \\ 1 & 0 \end{pmatrix}, \ \alpha_- = \begin{pmatrix} 0 & -1 \\ 0 & 0 \end{pmatrix}, \ \alpha_0 = \frac{\alpha_+ + \alpha_-}{2}. \tag{10}$$

The derivation of this equation is provided in the Supplemental Material of Ref. [18]. The symbol $\tau \in \{+, -, 0\}$ specifies the order of the measurements: $\tau = 0$ means joint measurements, $\tau = +$ means a position measurement followed by a momentum measurement, and $\tau = -$ means the vice versa. We recall that the Moyal function in Eq. (9) provides a complete description of the system+probes, and it is equivalent to knowing the density matrix. The characteristic function of the readout probability $Z(\phi) = \int d\mu \exp[iJ \cdot \phi]P(J)$ is obtained by tracing out the system ($s = 0$) and by putting $j = 0$ in Eq. (9). We note that Z can be inferred directly from the experimental data.

References [17, 18], in deriving (9), assumed as readout states the projectors $\hat{F}_{det}(\mu) = |J = \mu\rangle\langle J = \mu|$. In a realistic measurement with finite resolution, however, $\hat{F}_{det}(\mu) = \int dJ\, p(\mu|J)\hat{\Pi}(J)$, where $p(\mu|J)$ is the probability of observing an output μ given that the value of the pointer is J. We recall that for a uniform resolution this latter probability depends only on the difference of its arguments, $p(\mu|J) = f(\mu - J)$. Here, we make only this hypothesis on the resolution probability, we do not assume e.g., that it is a Gaussian. The results of Refs. [17, 18] are then easily generalized thanks to the convolution theorem: the characteristic function, i.e. the Fourier transform of the joint probability of observing $\mu = (\mu_P, \mu_Q)$ as the output,

$$Z(\phi) = \int d\mu\, e^{i\mu\cdot\phi} pr(\mu) = \int d\mu\, dJ\, e^{i\mu\cdot\phi}\ f(\mu - J)\Pi(J), \tag{11}$$

is obtained by introducing a factor $z(\phi) = \int d\mu \exp[i\mu \cdot \phi] f(\mu)$,

$$Z(\phi) = z(\phi) M_{sys}(\phi) M_{det}(\phi, \alpha_\tau \phi). \tag{12}$$

For definiteness, we consider $\tau = +$, i.e. a measurement of position is made first, and a measurement of momentum follows. Then, $Z(\phi) = z(\phi)M_{sys}(\phi)M_{det}(\phi, [0, \phi_P])$. Notice the appearance of the term $[0, \phi_P]$ as the second argument of M_{det}. By taking the second derivatives of $\ln(Z)$, we obtain the statistical variances of the readout

$$\Delta_Q^2 = -\left.\frac{\partial^2 \ln Z(\phi)}{\partial \phi_Q^2}\right|_{\phi=0} = \sigma_Q^2 + \delta_Q'^2 + \delta_Q^2, \tag{13}$$

$$\Delta_{P|Q}^2 = -\left.\frac{\partial^2 \ln Z(\phi)}{\partial \phi_P^2}\right|_{\phi=0} = \sigma_P^2 + \delta_P'^2 + \delta_P^2 + \tilde{\delta}_Q^2 + 2\kappa. \tag{14}$$

Since the finite resolution enters with a multiplicative term, we find that its contribution to the cumulants is simply of adding a term. The last two terms in Eq. (14) arise because of the derivation rule for a function $g(u) = f(u, u)$

$$\frac{d^2 g(u)}{du^2} = \left[\frac{\partial^2 f(u,v)}{\partial u^2} + \frac{\partial^2 f(u,v)}{\partial v^2} + 2\frac{\partial^2 f(u,v)}{\partial u \partial v}\right]_{v=u}, \tag{15}$$

with $f(u, v) \to \ln\{M_{det}([\phi_P, 0], [0, j_Q])\}$.

If the Q measurement were not performed, the statistical variance in the P measurement would be

$$\Delta_P^2 = \sigma_P^2 + \delta_P^2 + \delta_P'^2. \tag{16}$$

Therefore the statistical disturbance is

$$\eta_{P|Q}^2 \equiv \Delta_{P|Q}^2 - \Delta_P^2 = \tilde{\delta}_Q^2 + 2\kappa. \tag{17}$$

Here, κ is the initial covariance between the input variable Φ_Q of the first probe and the output variable J_P of the second probe.

The statistical noise for the position measurement is

$$\epsilon_Q^2 \equiv \Delta_Q^2 - \sigma_Q^2 = \delta_Q^2 + \delta_Q'^2, \tag{18}$$

while for the momentum measurement

$$\epsilon_P^2 \equiv \Delta_P^2 - \sigma_P^2 = \delta_P^2 + \delta_P'^2 + \tilde{\delta}_Q^2 + 2\kappa \tag{19}$$

We can make the systematic error $D_Q = \langle J_Q \rangle$ and disturbance $D_{P|Q} = \langle J_P \rangle + \langle \Phi_Q \rangle$ vanish, by applying two appropriate translations to the initial state of the probe (since we are looking for minimal disturbance and error, it is sensible to assume that the probes are not biased).

4. Results and applications

Inspection of (17) reveals that the square of the disturbance can not only be arbitrary small, but even negative, This means that the first measurement actually did not disturb the second one, but rather it helped make it sharper. At the same time, while the disturbance reaches zero, the error (or uncertainty) in the first measurement does not diverge. Therefore, the error–disturbance relation can be violated. In order for $\eta_{P|Q}^2$ to become negative, it is necessary that the correlator κ defined in (7*d*) becomes negative. The lowest disturbance, and therefore the lowest error in the second measurement, is achieved when the input variable Φ_Q of the first probe is perfectly anti–correlated to the output variable J_P of the second probe.

On the other hand, multiplying the errors in (18) and (19) yields

$$\epsilon_Q^2 \epsilon_P^2 = (\delta_Q^2 + \delta_Q'^2)(\delta_P^2 + \delta_P'^2 + \tilde{\delta}_Q^2 + 2\kappa) \geq \delta_Q^2 \delta_{J_P+\Phi_Q}^2 \geq \frac{1}{4}, \tag{20}$$

because of Kennard relation applied to the variable $\hat{J}_Q$ and $\hat{J}_P + \hat{\Phi}_Q$, which are canonically conjugated.

In summary, we found that the error–disturbance relation can be violated, with our definition of disturbance, if the detectors have initial correlations, while the error–error relation cannot be violated. We note that what we define as the error is equivalent to the disturbance of [11, 14], and therefore, using this alternative definition of disturbance, one would conclude that the error–disturbance relation cannot be violated. Irrespectively of the definition, there is a useful consequence of our result: it is possible to make a measurement of momentum infinitely sharp by preceding it with a position measurement. This is because the former measurement perturbs the momentum by a quantity Φ_Q, the write–in variable of the position probe. If this variable is perfectly anti–correlated to the readout variable J_P of the second probe, the outcome of the latter will be automatically corrected. Now, in quantum mechanics, two variables can be perfectly anti–correlated only if the two probes are in a maximally entangled state. Specifically, they need to be in an Einstein–Podolsky–Rosen state [19].[2]

Acknowledgments

This work was performed as part of the Brazilian Instituto Nacional de Ciência e Tecnologia para a Informação Quântica (INCT–IQ) and it was supported by Fundação de Amparo à Pesquisa de Minas Gerais (FAPEMIG).

[2] We are referring to the original formulation with continuous variables of the so–called EPR paradox, not to the more commonly known discrete version of Bohm and Aharonov [20].

Journal of Physics: Conference Series **626** (2015) 012051 doi:10.1088/1742-6596/626/1/012051

Appendix A. Description of the measurement

In a measurement, a meter interacts with a quantum system. What sets apart the meter from the system is its ability to cause a sensation in a human being, if not directly, through a chain of amplifications and further interactions with the visible electromagnetic field. Here, the effect of this von Neumann chain [21] is accounted for by considering the readout state of the probe to be a mixed state.

We can justify this assumption in the following way: There are many microscopic states that correspond to the same perceived readout J of a detector, let them $|J,m\rangle$, with m a collective quantum number. The probability of observing an output J is therefore

$$\mathcal{P}(J) = \sum_m \langle J,m| \otimes \langle s|\, U(\hat{\rho}_{sys} \otimes \hat{\rho}_{det})U^\dagger |J,m\rangle \otimes |s\rangle = \mathrm{Tr}\{(\mathbb{1} \otimes \hat{\Pi}_{det}(J))U(\hat{\rho}_{sys} \otimes \hat{\rho}_{det})U^\dagger\}, \quad \text{(A.1)}$$

where U is the time-evolution operator, and $\hat{\Pi}_{det}(J)$ is the projection operator on the subspace spanned by $|J,m\rangle$ when m varies for fixed J. Next, we consider that there is always some finite resolution in the observation and that the observation may fail, both these processes being characterized by a conditional probability $p(\mu|J)$ of observing a value μ when the actual value is J. Strictly speaking, the observed μ should take discrete values, however, since we are more familiar with integrals than with sums, it is often treated as a continuous variable.

The probability of observing μ is therefore

$$\mathcal{P}(\mu) = \int dJ p(\mu|J)\mathrm{Tr}\{(\mathbb{1} \otimes \hat{\Pi}_{det}(J))U(\hat{\rho}_{sys} \otimes \hat{\rho}_{det})U^\dagger\}, \quad \text{(A.2)}$$

$$\mathcal{P}(\mu) = \mathrm{Tr}\{(\mathbb{1} \otimes \hat{F}_{det}(\mu))U(\hat{\rho}_{sys} \otimes \hat{\rho}_{det}?)U^\dagger\}, \quad \text{(A.3)}$$

where $\hat{F}_{det}(\mu) = \int dJ p(\mu|J)\hat{\Pi}_{det}(J)$ is a family of positive operators describing the readout states. These must satisfy the normalization $\int d\mu\, \hat{F}_{det}(\mu) = \mathbb{1}$, with $d\mu$ a Lebesgues-Stieltjes measure, corresponding to a discrete or continuous distribution of outputs μ.

The readout states $\hat{F}_{det}(\mu)$ are not normalized, but in general $\mathrm{Tr}[\hat{F}_{det}(\mu)] = pr(\mu) \neq 1$. The quantities $pr(\mu)$ are the priors [22], and they are not necessarily one, unless $\hat{F}_{det}(\mu)$ is a rank–1 projector. Instead, they represent the unnormalized probability of obtaining an outcome μ in absence of any other information. They are not a conventional probability, since

$$\int d\mu\, pr(\mu) = \int d\mu \mathrm{Tr}[\hat{F}_{det}(\mu)] = \mathrm{Tr}\,\mathbb{1} = d, \quad \text{(A.4)}$$

with d the dimension of the Hilbert space of the meter (which could be infinite). However, the only relevant quantity are the ratios $pr(\mu_1)/pr(\mu_2)$ representing relative probabilities.

Furthermore, it follows from their definition that the readout states are classical, i.e. $[\hat{F}_{det}(\mu_1), \hat{F}_{det}(\mu_2)] = 0,\ \forall \mu_1, \mu_2$. This hypothesis allows to individuate one, or more, privileged basis, $|J\rangle$, the one that diagonalizes at once all $\hat{F}_{det}(\mu)$. The existence of such a basis can also be justified by the decoherence approach [23, 24]. The $\hat{F}_{det}(\mu)$ are labeled by the average of the operator $\hat{J} = \int dJ J\,|J\rangle\langle J|$ and they have a spread $\delta'(\mu)$ in J, i.e.,

$$\mu = \mathrm{Tr}_{det}\{\hat{J}\hat{\rho}_{det}(\mu)\}, \quad \text{(A.5)}$$

$$\delta'^2(\mu) = \mathrm{Tr}_{det}\{\hat{J}^2\hat{\rho}_{det}(\mu)\} - \Big(\mathrm{Tr}_{det}\{\hat{J}\hat{\rho}_{det}(\mu)\}\Big)^2, \quad \text{(A.6)}$$

with $\hat{\rho}_{det}(\mu) = \hat{F}_{det}(\mu)/pr(\mu)$ the normalized readout states. The spread $\delta'(\mu)$ is but the resolution of the probe, and in principle it can be different for different outputs.

Journal of Physics: Conference Series **626** (2015) 012051 doi:10.1088/1742-6596/626/1/012051

For instance, we could choose

$$\hat{F}_{det}(\mu_n) = \int_{\mu_n-\delta'_n/2}^{\mu_n+\delta'_n/2} dJ\,|J\rangle\langle J| \tag{A.7}$$

and let the readout μ take discrete values μ_n spaced by $(\delta'_n + \delta'_{n+1})/2$ from one another (then $d\mu$ is a distribution formed by a sum of Dirac deltas). In this case the priors are $pr(\mu_n) = \delta'_n$.

In general, we can write

$$\hat{F}_{det}(\mu) = \int dJ\, p(\mu|J)\,|J\rangle\langle J|, \tag{A.8}$$

with $p(\mu|J)$ a probability distribution for μ. The priors are thus $pr(\mu) = \int dJp(\mu|J)$. Assuming a uniform resolution $\delta'(\mu) = \delta'$, Eq. (A.5) is satisfied for $p(\mu|J) = f(\mu - J)$, where $f(\mu)$ is a probability distribution having zero average and spread δ'. An important example is given by the Gauss-Laplace distribution,

$$\hat{F}_{det}(\mu) = \int_{-\infty}^{+\infty} dJ\, \frac{\exp\left[-(J-\mu)^2/2\delta'^2\right]}{\sqrt{2\pi}\delta'}|J\rangle\langle J|, \tag{A.9}$$

with $\mu \in \mathbb{R}$ and uniform priors $pr(\mu) = 1$.

The conditional state of the system for given μ is

$$\hat{\rho}_{sys|\mu} = \mathcal{P}(\mu)^{-1}\mathrm{Tr}_{det}\{(\mathbb{1} \otimes \hat{F}_{det}(\mu))U(\hat{\rho}_{sys} \otimes \hat{\rho}_{det})U^\dagger\}. \tag{A.10}$$

Equivalent formulas arise when treating postselected weak measurement [25, 26], but with the difference that U is expanded in a perturbation series and the rôles of the system and the probe are reversed. Let us introduce the preparation basis $|I\rangle$, the one in which $\hat{\rho}_{det} = \int dI\, w(I)|I\rangle\langle I|$. We can rewrite the conditional state of the system

$$\hat{\rho}_{sys|\mu} = \frac{\int dJdI M_{J,I}(\mu)\hat{\rho}_{sys}M^\dagger_{J,I}(\mu)}{\int dJdI\mathrm{Tr}_{sys}\{E_{J,I}(\mu)\hat{\rho}_{sys}\}}. \tag{A.11}$$

We defined the generalized operations

$$M_{J,I}(\mu) = \sqrt{p(\mu|J)w(I)}\langle J|\hat{U}|I\rangle, \tag{A.12}$$

and the generalized effects $E_{J,I}(\mu) = M^\dagger_{J,I}(\mu)M_{J,I}(\mu)$. Both $M_{J,I}(\mu)$ and $E_{J,I}(\mu)$ are operators on the Hilbert space of the system alone. If we do not make the hypothesis of classical readout, Eq. (A.12) becomes $M_{J,I}(\mu) = \sqrt{p(\mu|J)w(I)}\langle \mu : J|\hat{U}|I\rangle$, where $|\mu : J\rangle$ is the basis that diagonalizes $\hat{F}_{det}(\mu)$.

The measurement can also be described in the language of positive-operator valued measures (POVM) [27–32]. The linear operations

$$\mathcal{I}_D(\rho_{sys}) \equiv \int_D d\mu \int dJdI M_{J,I}(\mu)\hat{\rho}_{sys}M^\dagger_{J,I}(\mu), \tag{A.13}$$

where D is a measurable domain in the space of possible outcomes (i.e. an element of a σ–algebra in the Kolmogorov formalism), are called *instruments* if they preserve the non-negativity of ρ. In the POVM formalism, the instruments are introduced as the basic objects describing the measurement. Besides preserving the nonnegativity of the density operator, they must also satisfy [27, 33]

(i) $\mathcal{I}_\emptyset = 0$, i.e., the measurement always gives an outcome (notice, however, that the outcome is not necessarily a number, but it could consist, e.g., in the observation that the apparatus has failed, etc.).

(ii) $\mathcal{I}_{\cup D_j} = \sum_j \mathcal{I}_{D_j}$ for any family of disjoint elements of the σ–algebra.

(iii) $\mathrm{Tr}_{sys}[\mathcal{I}_\Omega(\rho_{sys})] = 1$, where Ω is the whole set of possible outcomes.

The instruments defined in Eq. (A.13) automatically satisfy the three above properties. An interesting question is whether any instrument can be written as in Eq. (A.13).

The instruments describe the change of the density matrix one assigns to the system, conditioned on the information that the outcome of the measurement was in D. In other words, the new density matrix is

$$\rho_{sys|D} = \frac{\mathcal{I}_D(\rho_{sys})}{P(D|\rho_{sys})}, \tag{A.14}$$

with the normalization $P(D|\rho_{sys})$ being the probability of obtaining an outcome in D,

$$P(D|\rho_{sys}) = \mathrm{Tr}_{sys}[E(D)\rho_{sys}], \tag{A.15}$$

$$E(D) = \mathcal{I}^*_D(\mathbb{1}). \tag{A.16}$$

The family of conjugate operators $\mathcal{I}^*_D$is defined univocally by

$$\mathrm{Tr}_{sys}[\rho\mathcal{I}^*_D(\sigma)] = \mathrm{Tr}_{sys}[\mathcal{I}_D(\rho)\sigma]. \tag{A.17}$$

When both $\hat{F}_{det}(\mu)$ and $\hat{\rho}_{det}$ are pure states, then only one term survives in the double sum of Eq. (A.13) and the instruments are called Lüders instruments. The operators $E(\mu)$ can be inferred from the observed probabilities $\mathcal{P}(\mu)$ by changing the preparation $\hat{\rho}_{sys}$. Recovering $E_{J,I}(\mu)$, however, is a nontrivial task, and requires the knowledge of the measurement apparatus, as different $M_{J,I}(\mu)$ correspond to the same instrument. If $E_{J,I}(\mu)$ are known, the operations are determined modulo a unitary operator: $M_{J,I}(\mu) = V_{J,I}(\mu)E^{1/2}_{J,I}(\mu)$, where $E^{1/2}_{J,I}$ is univocally defined as a positive operator. In other words, we consider the eigenvalues of the positive operator $E_{J,I}(\mu)$, take their arithmetic square root, and define $E^{1/2}_{J,I}(\mu)$ as the operators having the same eigenstates as $E_{J,I}(\mu)$, and as eigenvalues the said arithmetic root. The additional operations $V_{J,I}(\mu)$ represent an unwanted and avoidable feedback on the system. For instance, in a Stern-Gerlach apparatus, the beam exiting the magnetic field gradient could pass through a uniform magnetic field, which rotates the spin, so that a second measurement would not confirm the result of the first one. Ideally, $V_{J,I}(\mu) = \mathbb{1}$. Only by making subsequent measurements on the system, is it possible to check whether the measurement process is introducing this unwanted feedback.

We are interested in particular to the case of two sequential measurements. A probe measuring $\hat{A}$ couples first to the system, then, after the first probe decouples, a second probe measuring $\hat{B}$ couples. The time-evolution operator factorizes then as $U = U_B U_A$, where U_A (respectively, U_B) acts on the Hilbert space of the system and the probe A (respectively, B). As we are making two measurements, it is natural to assume that the readout states factor, $\hat{F}_{det}(\mu) = \hat{F}_A(\mu_A)\hat{F}_B(\mu_B)$, each factor acting on a different Hilbert space. A fundamental consideration is that, if we assume as well that the initial state of the two probes factors as $\rho_{det} = \rho_A \otimes \rho_B$, i.e. if we assume no initial correlations between the probes, then the probability of observing the outcomes μ_A, μ_B can be written

$$\mathcal{P}(\mu_A, \mu_B) = \mathrm{Tr}_{sys}\left\{\mathcal{I}_{\mu_B}\left[\mathcal{I}_{\mu_A}(\rho_{sys})\right]\right\} \tag{A.18}$$

and the conditional state of the system is

$$\rho_{sys|\mu} \propto \mathcal{I}^B_{\mu_B}\left[\mathcal{I}^A_{\mu_A}(\rho_{sys})\right], \tag{A.19}$$

Journal of Physics: Conference Series **626** (2015) 012051 doi:10.1088/1742-6596/626/1/012051

where the composing instruments are defined by

$$\mathcal{I}^{k}_{\mu_k}(\rho) = \int dI_k dJ_k M^{k}_{J_k,I_k}(\mu_k)\rho M^{k\dagger}_{J_k,I_k}(\mu_k), \tag{A.20}$$

$$M^{k}_{J,I}(\mu) = \sqrt{p(\mu|J)w(I)}\langle J|\hat{U}_k|I\rangle. \tag{A.21}$$

However, if we do assume initial correlations, then the total instrument cannot be written as a composition of two instruments, contrary to what is commonly assumed (see e.g. [31]), and we can only associate one set of instrument, defined on the probability space $M_A \times M_B$.

Finally, we note that the marginal probabilities are

$$\mathcal{P}(\mu_A) = \int d\mu_B \mathrm{Tr}\left\{\left[\mathbb{1}\otimes\hat{F}_A(\mu_A)\otimes\hat{F}_B(\mu_B)\right]U_B U_A(\rho_{sys}\otimes\rho_{\mathrm{det}})U_A^\dagger U_B^\dagger\right\}$$

$$= \mathrm{Tr}_{sys,A}\{[\mathbb{1}\otimes F_A(\mu_A)]U_A(\rho_{sys}\otimes\rho_A)U_A^\dagger\}\,, \qquad \text{with } \rho_A = \mathrm{Tr}_B(\rho_{det}) \tag{A.22}$$

$$\mathcal{P}(\mu_B) = \int d\mu_A \mathrm{Tr}\left\{\left[\mathbb{1}\otimes\hat{F}_A(\mu_A)\otimes\hat{F}_B(\mu_B)\right]U_B U_A(\rho_{sys}\otimes\rho_{\mathrm{det}})U_A^\dagger U_B^\dagger\right\}$$

$$= \mathrm{Tr}_{sys,B}\{[\mathbb{1}\otimes\hat{F}_B(\mu_B)]U_B\rho_{sys,B}U_B^\dagger\}\,, \qquad \text{with } \rho_{sys,B} = \mathrm{Tr}_A[U_A(\rho_{sys}\otimes\rho_{det})U_A^\dagger]. \tag{A.23}$$

Notice that the correlations are swapped: from correlations between A and B, they become correlations between B and the system, after tracing out the first probe. Applying Bayes' rule, we see that neither conditional probability

$$\mathcal{P}(\mu_B|\mu_A) = \frac{\mathcal{P}(\mu_A,\mu_B)}{\mathcal{P}(\mu_A)} = \frac{\mathrm{Tr}[E(\mu_A,\mu_B)\rho_{sys}]}{\mathrm{Tr}_{sys,A}\{[\mathbb{1}\otimes F_A(\mu_A)]U_A(\rho_{sys}\otimes\rho_A)U_A^\dagger\}}\,, \tag{A.24}$$

$$\mathcal{P}(\mu_A|\mu_B) = \frac{\mathcal{P}(\mu_A,\mu_B)}{\mathcal{P}(\mu_B)} = \frac{\mathrm{Tr}[E(\mu_A,\mu_B)\rho_{sys}]}{\mathrm{Tr}_{sys,B}\{[\mathbb{1}\otimes\hat{F}_B(\mu_B)]U_B\rho_{sys+B}U_B^\dagger\}}\,, \tag{A.25}$$

admits a conditional state $\rho_{sys|\mu_A}$ or $\rho_{sys|\mu_B}$ such that

$$\mathcal{P}(\mu_B|\mu_A) = \mathrm{Tr}_{sys,B}\{[\mathbb{1}\otimes\hat{F}_B(\mu_B)]U_B(\rho_{sys|\mu_A}\otimes\rho_B)U_B^\dagger\}\,, \tag{A.26}$$

$$\mathcal{P}(\mu_A|\mu_B) = \mathrm{Tr}_{sys,A}\{[\mathbb{1}\otimes\hat{F}_A(\mu_A)]U_A(\rho_{sys|\mu_B}\otimes\rho_A)U_A^\dagger\}\,, \tag{A.27}$$

unless $\rho_{det} = \rho_A\otimes\rho_B$. This is because, for a given output μ_A (respectively, μ_B), the system becomes correlated with the probe B (respectively, A).

References

[1] Erhart J, Sponar S, Sulyok G, Badurek G, Ozawa M and Hasegawa Y 2012 Experimental demonstration of a universally valid error–disturbance uncertainty relation in spin measurements *Nat. Phys.* **8** 185–189 URL `http://dx.doi.org/10.1038/nphys2194`

[2] Rozema L A, Darabi A, Mahler D H, Hayat A, Soudagar Y and Steinberg A M 2012 Violation of Heisenberg's measurement-disturbance relationship by weak measurements *Phys. Rev. Lett.* **109**(10) 100404 URL `http://link.aps.org/doi/10.1103/PhysRevLett.109.100404`

[3] Baek S Y, Kaneda F, Ozawa M and Edamatsu K 2013 Experimental violation and reformulation of the Heisenberg's error-disturbance uncertainty relation *Scientific Reports* **3** 2221

[4] Weston M M, Hall M J W, Palsson M S, Wiseman H M and Pryde G J 2013 Experimental test of universal complementarity relations *Phys. Rev. Lett.* **110**(22) 220402 URL `http://link.aps.org/doi/10.1103/PhysRevLett.110.220402`

[5] Heisenberg W 1927 Über den anschaulichen inhalt der quantentheoretischen kinematik und mechanik *Zeitschr. f. Phys.* **43**(3-4) 172–198 URL http://dx.doi.org/10.1007/BF01397280

[6] Kennard E H 1927 Zur Quantenmechanik einfacher Bewegungstypen *Zeitschr. f. Phys.* **44**(4) 326–352 URL http://dx.doi.org/10.1007/BF01391200

[7] Robertson H P 1929 The uncertainty principle *Phys. Rev.* **34**(1) 163–164 URL http://link.aps.org/doi/10.1103/PhysRev.34.163

[8] Schrödinger E 1930 Zum Heisenbergschen Unschärfeprinzip *Sitzungsber. Preuss. Akad. Wiss., Phys.-Math. Kl.* **19** 296–303 [About Heisenberg uncertainty relation, (A. Angelow and M. C. Batoni transl.), Bulg. J. Phys. 26, 193–203 (1999)]

[9] Ozawa M 2004 Quantum measurements, instruments, and uncertainty principle *Interdiscip. Inform. Sci.* **10** 69–81 URL http://dx.doi.org/10.4036/iis.2004.69

[10] Ozawa M 2003 Universally valid reformulation of the Heisenberg uncertainty principle on noise and disturbance in measurement *Phys. Rev.* A **67**(4) 042105 URL http://link.aps.org/doi/10.1103/PhysRevA.67.042105

[11] Busch P, Lahti P and Werner R F 2013 Proof of Heisenberg's error-disturbance relation *Phys. Rev. Lett.* **111**(16) 160405 URL http://link.aps.org/doi/10.1103/PhysRevLett.111.160405

[12] Di Lorenzo A 2013 Correlations between detectors allow violation of the Heisenberg noise-disturbance principle for position and momentum measurements *Phys. Rev. Lett.* **110**(12) 120403 URL http://link.aps.org/doi/10.1103/PhysRevLett.110.120403

[13] Distler J and Paban S 2013 Uncertainties in successive measurements *Phys. Rev. A* **87**(6) 062112 URL http://link.aps.org/doi/10.1103/PhysRevA.87.062112

[14] Bullock T J and Busch P 2014 Focusing in Arthurs-Kelly-type joint measurements with correlated probes *Phys. Rev. Lett.* **113**(12) 120401 URL http://link.aps.org/doi/10.1103/PhysRevLett.113.120401

[15] Moyal J E 1949 Quantum mechanics as a statistical theory *Math. Proc. Cambr. Phil. Soc.* **45** 99–124 URL http://dx.doi.org/10.1017/S0305004100000487

[16] Wigner E 1932 On the quantum correction for thermodynamic equilibrium *Phys. Rev.* **40**(5) 749–759 URL http://link.aps.org/doi/10.1103/PhysRev.40.749

[17] Di Lorenzo A 2011 Strong correspondence principle for joint measurement of conjugate observables *Phys. Rev.* A **83**(4) 042104 URL http://link.aps.org/doi/10.1103/PhysRevA.83.042104

[18] Di Lorenzo A 2013 Sequential measurement of conjugate variables as an alternative quantum state tomography *Phys. Rev. Lett.* **110**(1) 010404 URL http://link.aps.org/doi/10.1103/PhysRevLett.110.010404

[19] Einstein A, Podolsky B and Rosen N 1935 Can quantum-mechanical description of physical reality be considered complete? *Phys. Rev.* **47**(10) 777–780 URL http://link.aps.org/doi/10.1103/PhysRev.47.777

[20] Bohm D and Aharonov Y 1957 Discussion of experimental proof for the paradox of Einstein, Rosen, and Podolsky *Phys. Rev.* **108** 1070–1076 URL http://link.aps.org/doi/10.1103/PhysRev.108.1070

[21] von Neumann J 1932 *Mathematische Grundlagen der Quantenmechanik* (Berlin: Springer) [*Mathematical Foundations of Quantum Mechanics* (Princeton University Press, Princeton, 1996)]

[22] Jaynes E 2003 *Probability Theory: The Logic of Science* (Cambridge University Press) ISBN 9780521592710 URL http://books.google.com.br/books?id=tTN4HuUNXjgC

[23] Zurek W H 1981 Pointer basis of quantum apparatus: Into what mixture does the wave packet collapse? *Phys. Rev.* D **24**(6) 1516–1525 URL `http://link.aps.org/doi/10.1103/PhysRevD.24.1516`

[24] Zurek W H 1982 Environment-induced superselection rules *Phys. Rev.* D **26**(8) 1862–1880 URL `http://link.aps.org/doi/10.1103/PhysRevD.26.1862`

[25] Aharonov Y, Albert D Z and Vaidman L 1988 How the result of a measurement of a component of the spin of a spin-1/2 particle can turn out to be 100 *Phys. Rev. Lett.* **60**(14) 1351–1354 URL `http://link.aps.org/doi/10.1103/PhysRevLett.60.1351`

[26] Di Lorenzo A 2012 Full counting statistics of weak-value measurement *Phys. Rev.* A **85**(3) 032106 URL `http://link.aps.org/doi/10.1103/PhysRevA.85.032106`

[27] Davies E B 1976 *Quantum Theory of Open Systems* (London: Academic Press)

[28] Helstrom C W 1976 *Quantum Detection and Estimation Theory* (London: Academic Press)

[29] Holevo A S 1982 *Probabilistic and statistical aspects of quantum theory* (Amsterdam: North Holland)

[30] Kraus K 1983 *States, effects, and operations: fundamental notions of quantum theory* (Springer-Verlag) ISBN 9780387127323

[31] Ludwig G 1983 *Foundations of Quantum Mechanics I* (Springer London, Limited) ISBN 9783540168843

[32] Busch P, Grabowski M and Lahti P J 1995 *Operational Quantum Physics* (Berlin: Springer) second corrected printing

[33] Davies E and Lewis J 1970 An operational approach to quantum probability *Comm. Math. Phys.* **17**(3) 239–260 URL `http://dx.doi.org/10.1007/BF01647093`

[34] Carmeli C, Heinosaari T and Toigo A 2011 Sequential measurements of conjugate observables *J. Phys. A: Math. Gen.* **44** 285304 URL `http://stacks.iop.org/1751-8121/44/i=28/a=285304`

Journal of Physics: Conference Series **626** (2015) 012052 doi:10.1088/1742-6596/626/1/012052

Constraining extra space dimensions using precision molecular spectroscopy

Beatriz Gato-Rivera[1,2,3]

[1] Instituto de Física Fundamental, IFF-CSIC, Serrano 123, 28006 Madrid, Spain
[2] Nikhef, Science Park 105, 1098 XG Amsterdam, The Netherlands

E-mail: t38@nikhef.nl

Abstract. Highly accurate measurements of quantum level energies in molecular systems provide a test ground for new physics, as such effects could manifest themselves as minute shifts in the quantum level structures of atoms and molecules. For the lightest molecular systems, neutral molecular hydrogen (H_2, HD and D_2) and the molecular hydrogen ions (H_2^+, HD^+ and D_2^+), weak force effects are several orders weaker than current experimental and theoretical results, while contributions of Newtonian gravity and the strong force at the characteristic molecular distance scale of 1 Å can be safely neglected. Comparisons between experiment and QED calculations for these molecular systems can be interpreted in terms of probing large extra space dimensions, under which gravity could become much stronger than in ordinary 3-D space. Under this assumption, using the spectra of H_2 we have derived constraints on the compactification scales for extra dimensions within the Arkani-Hamed-Dimopoulos-Dvali (ADD) framework, and constraints on the brane separation and bulk curvature within the Randall-Sundrum (RS-I and RS-II) frameworks.

1. Introduction

This presentation is based on work done in collaboration with Bert Schellekens, from the Institute of Subatomic Physics Nikhef (Amsterdam), together with Edcel Salumbides and Wim Ubachs, from the Vrije Universiteit of Amsterdam. In the present study, accurate results from precision molecular spectroscopy are exploited to put constraints on some higher-dimensional theories. The idea is that effects of new physics could manifest themselves as minute shifts in the quantum level structures of atoms and molecules. To detect such phenomena, spectroscopic experiments must be performed at extreme precision, where key ingredients are ultra-stable lasers and laser metrology techniques, and comparisons must be made on systems for which very accurate first-principles calculations can be performed.

In recent years great progress has been made on the calculation of level energies in the neutral hydrogenic molecules (H_2, HD and D_2) and their ions (H_2^+, HD^+ and D_2^+). The results from the variety of experimental precision measurements on both the ionic and neutral hydrogenic molecules are in perfect agreement with the QED-calculations, within combined uncertainty limits from theory and experiment. This agreement between experiment and first-principles calculations on the level energies of hydrogenic molecules can be used to constrain the parameters associated with Physics Beyond the Standard Model, such as possible fifth

[3] Also known as B. Gato

forces between nucleons or the existence of extra space dimensions. Since weak, strong and (Newtonian) gravitational forces have negligible contributions to the molecular quantum level structure, electromagnetism is the sole force acting between the charged particles within light molecules. This makes it possible to derive bounds on possible fifth forces and extra dimensions from molecular precision experiments compared with QED-calculations [2, 3].

Although in the framework of String Theory extra space dimensions were found to be necessary from the very begining in the middle seventies of last century, it took much longer to appreciate the beauty and usefulness of extra dimensions in ordinary Quantum Field Theory. Only in the late nineties, after the seminal works of Antoniadis, Arkani-Hamed, Dimopoulos, and Dvali (ADD model) [4, 5, 6] on one side, and Randall and Sundrum (RS-I and RS-II models) [7, 8] on the other, the particle physics community started considering extra dimensions as a real scientific possibility deserving further investigation. The mathematical formalisms of these models can be applied to molecular physics test bodies, from which constraints on the compactification scales can be derived for the ADD scenario, whereas constraints on the brane separation and bulk curvature can be derived for the RS-I and RS-II scenarios.

2. The ADD Scenario

In the model put forward in 1998 by Arkani-Hamed, Dimopoulos, and Dvali [4, 5] (plus Antoniadis [6] in its implementation in string theory) our observable universe consists of a 3-brane embedded in a higher-dimensional cosmos. There are n extra space dimensions where only gravity can escape as the particles and interactions of the Standard Model remain confined in the ordinary 3-D space (up to a small, negligible penetration inside the bulk). The extra dimensions are flat and compactified with the same radius R, just for simplicity. Therefore the extradimensional space is an n-dimensional torus where gravitation propagates rendering the observed gravitational strength much weaker as the flux lines dilute into the higher-dimensional volume. Moreover, the propagation of gravity across the extradimensional space results in a modification of Newton's law. The exact ADD gravitational potential reads [5]:

$$V_{\mathrm{ADD}}(r) = -G\frac{m_1 m_2}{r} \sum_{(k_1,k_2,\ldots k_n)} e^{(2\pi|k|/R)r} \tag{1}$$

It has the contribution from the usual massless gravitons plus a tower of Yukawa potentials mediated by all the massive modes: the Kaluza-Klein higher-dimensional gravitons, with momenta in the extra dimensions quantized in units of $2\pi/R$, and therefore with mass splittings of $2\pi/R$ from our 3-D space perspective (from the higher-dimensional perspective they are massless). For distances much larger than the compactification length range, $r \gg R$, the Newtonian potential is recovered:

$$V_{\mathrm{ADD}}(r) = V_{\mathrm{N}}(r) = -G\frac{m_1 m_2}{r} = -\frac{\hbar c}{M_{\mathrm{Pl}}^2}\frac{m_1 m_2}{r}, \quad r \gg R \tag{2}$$

where M_{Pl} is the Planck mass (10^{16} TeV). For distances of the same size or shorter than the compactification length range, $r \leq R$, the ADD potential takes the simple form:

$$V_{\mathrm{ADD}}(r) = -G_{(4+n)}\frac{m_1 m_2}{r^{n+1}} = -\frac{\hbar c}{M_{(4+n)}^{(n+2)}}\frac{m_1 m_2}{r^{n+1}}, \quad r \leq R \tag{3}$$

where $M_{(4+n)}$ is the fundamental mass in the $(4+n)$-dimensional spacetime. Observe that this expression for the gravitational potential can be interpreted in the context of the Gauss law, without the need to invoke quantization of gravity. The Planck mass M_{Pl} is then related to the fundamental higher-dimensional mass $M_{(4+n)}$ via:

$$M_{\mathrm{Pl}}^2 = M_{(4+n)}^{n+2}(R)^n. \tag{4}$$

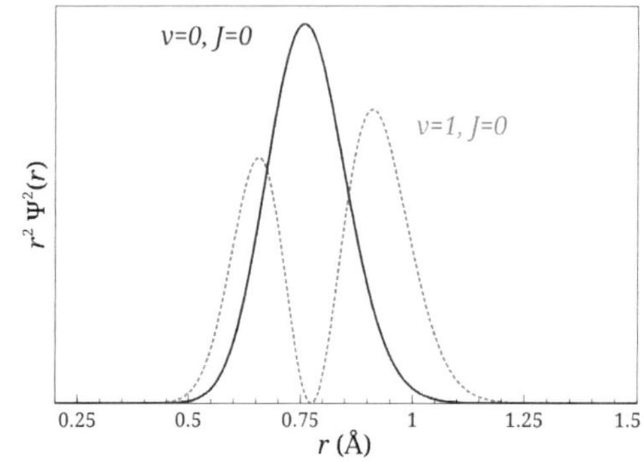

Figure 1. (Color online) Wave functions for H_2 in the electronic ground state with $J = 0$ and vibrational levels $v = 0, 1$.

Thus the fundamental mass $M_{(4+n)}$ may still be small and $M_{\rm Pl}$ becomes large due to the compactified volume of extra dimensions. Arkani-Hamed *et al.* proposed that $M_{(4+n)}$ equals the electroweak scale M_{EW} in order to solve the hierarchy problem, explaining in this manner the desert between the electroweak scale of 1 TeV and the Planck scale of 10^{16} TeV. In the present study, since it is not our intention to solve the hierarchy problem, we leave $M_{(4+n)}$ as a free parameter, together with R and n, and we attempt to constrain these parameters from molecular physics experiments.

The gravitational potential has an effect on the level energy of a molecular quantum state with wave function $\Psi(r)$, written as an expectation value [1]:

$$\langle V_{\rm ADD}\rangle = Gm_1m_2\left[\ \int_R^\infty \Psi^*(r)\frac{1}{r}\Psi(r)r^2dr + R^n\int_0^R \Psi^*(r)\frac{1}{r^{n+1}}\Psi(r)r^2dr\ \right] \tag{5}$$

$\Psi(r)$ represents the probability that the internuclear separation is r, the vibrational coordinate that probes the forces between the nucleons. The first integral represents the ordinary gravitational attraction, which for protons is 8×10^{-37} times weaker than the electrostatic repulsion, and can therefore be neglected. The second integral represents the effect of modified gravity and is evaluated using accurate wave functions for the molecules. The wave functions of the H_2 ground electronic state for $J = 0$ and vibrational levels $v = 0, 1$ are plotted in Fig. 1 as obtained from *ab initio* calculations [14, 15]. In practice, the integration is performed up to $r = 10$ Å since the wave function amplitude is negligible beyond that. In addition, at distances $r < 0.1$ Å the wave function amplitude becomes negligible so that the second integral in Eq. (5) converges without additional assumptions.

For spectroscopic transitions between quantum states Ψ_1 and Ψ_2 in molecules, the expectation value for a higher-dimensional gravity contribution is given by a differential effect:

$$\langle\Delta V_{\rm ADD}(n,R)\rangle = Gm_1m_2R^n\times\left[\left\langle\Psi_1\left|\frac{1}{r^{n+1}}\right|\Psi_1\right\rangle - \left\langle\Psi_2\left|\frac{1}{r^{n+1}}\right|\Psi_2\right\rangle\right] \tag{6}$$

Here the ADD-expectation value is written explicitly as a function of the two relevant parameters: the number n of extra space dimensions and the compactification scale R. Obviously, measurements on the dissociation limit of molecules, where Ψ_1 is the lowest energy bound state and Ψ_2 is the non-interacting two-atom limit at $r = \infty$, are the most sensitive probes as the diference between Ψ_1 and Ψ_2 is larger.

3. The Randall-Sundrum Scenarios

Let us now consider the Randall-Sundrum scenarios, RS-I [7] and RS-II [8], presented by the authors in 1999. They consist of two parallel 3-branes separated by a distance y_c along one extra dimension y. The branes have opposite tension T (vacuum energy) and the bulk has a negative cosmological constant. According to General Relativity, where all types of energy are gravity sources, the branes and the bulk produce gravitational fields, giving rise to the AdS metric:

$$ds^2 = e^{-2k|y|}\eta_{\mu\nu}dx^\mu dx^\nu + dy^2, \tag{7}$$

where $\exp(-k|y|)$ is the warp factor and k is the bulk curvature. The strength of gravity is highest on the positive $T > 0$ tension brane and falls off along the extra dimension.

In analogy with the ADD model, in the RS models the particles and interactions of the Standard Model are confined in one 3-brane, that we denote the SM-brane, whereas gravity propagates everywhere. Observe that the warped metric crucially differentiates the RS models from the ADD model with a flat metric where $k = 0$.

The modified gravitational potential between two masses separated a distance r in the SM-brane can be expressed as:

$$V_{\mathrm{RS}}(r) = -G\frac{m_1 m_2}{r}\left(1 + \Delta_{\mathrm{RS}}\right), \tag{8}$$

where Δ_{RS} is the RS-I or RS-II correction to the Newtonian potential.

In the RS-I scenario the SM-brane is the negative $T < 0$ tension brane. The authors did not present any gravitational potential for this model, however, and it was Callin [9], in 2004, who presented its computation. For short distances, $kr \ll 1$, he obtained:

$$\Delta_{\mathrm{RS-I}}(r) \simeq \frac{4}{3\pi kr}\frac{1 - e^{-2ky_c}}{1 + \frac{1}{3}e^{-2ky_c}}, \quad kr \ll 1, \tag{9}$$

where one can distinguish two regimes, $ky_c \ll 1$ and $ky_c \gg 1$, with the result at leading orders:

$$\Delta_{\mathrm{RS-I}}(r) \simeq \begin{cases} \frac{2y_c}{\pi r} + ..., & ky_c \ll 1, \\ \\ \frac{4}{3\pi kr} + ..., & ky_c \gg 1. \end{cases} \tag{10}$$

The potential for long distances, $kr \gg 1$, does not apply for the scales probed by molecules (it gives inconsistent results) and we do not consider it further.

In the RS-II scenario the SM-brane is the positive $T > 0$ tension brane. The $T < 0$ tension brane can be set to infinity, $y_c \to \infty$, as the massless gravitons are localized very close to the $T > 0$ brane, their wave function decreasing drastically along the extra dimension y. Therefore, in the RS-II scenario the extra dimension needs not be compactified. The authors derived a potential with the long distance limit that has been applied to astrophysics (see for example [11]). For short distances, $kr \ll 1$, Callin and Ravndal [10] obtained:

$$\Delta_{\mathrm{RS-II}}(r) \simeq \frac{4}{3\pi kr} + ..., \quad kr \ll 1. \tag{11}$$

Note the correspondence of this result with that of Eq. (10) in the limit $ky_c \gg 1$. This limit together with $kr \ll 1$ seems to indicate a smooth transition from RS-I to RS-II for small bulk curvature $k \ll 1/r$ and $y_c \gg 1/k$, as the brane separation y_c is increased to infinity [10].

From the RS corrections to the Newtonian potential the expectation values of the leading-order shifts of transitions in molecules, for short distances $kr \ll 1$, are therefore:

$$\langle \Delta V_{\mathrm{RS}}(k) \rangle = Gm_1 m_2 \mathcal{F}\left(\frac{4}{3\pi k}\right) \times \left[\left\langle \Psi_1 \middle| \frac{1}{r^2} \middle| \Psi_1 \right\rangle - \left\langle \Psi_2 \middle| \frac{1}{r^2} \middle| \Psi_2 \right\rangle\right], \tag{12}$$

where $\mathcal{F} = (1 - e^{-2ky_c})/(1 + \frac{1}{3}e^{-2ky_c})$ for the RS-I scenario and $\mathcal{F} = 1$ for the RS-II scenario. Using these expressions, we derive bounds on the brane separation y_c and the bulk curvature k from molecular spectroscopy data.

The RS-II scenario with a single brane can be generalized to n extra dimensions provided that the mixing effects of the warp factors for each extra dimension are small compared with the leading order terms. In this case the resulting correction to the Newtonian potential is, to a good approximation, the product of the leading corrections of each extra dimension. For n extra dimensions, if the bulk curvatures have the same value k, the RS-II correction is then $\Delta_{\mathrm{RS-II},n} = (\Delta_{\mathrm{RS-II}})^n$, where we use $\Delta_{\mathrm{RS-II}}$ in Eq. (11).

4. Constraints on extra dimensions from molecular spectroscopy data

In the previous sections, the expectation value for a higher-dimensional gravity contribution to a transition frequency in a molecule was presented for both ADD and RS approaches to higher dimensions. This expectation value is interpreted as a contribution to the binding energy of molecules in certain quantum states. However, as discussed in Ref. [1], analysis of recent experiments with hydrogen neutral molecules and their ions, together with their stable isotopomers containing deuterons, showed that QED computations are in very good agreement with observations for these molecular systems. Consequently, one can derive constraints on the parameters underlying the higher-dimensional theories: the compactification scale R depending on the number of extra dimensions n, for the ADD scenario, and the bulk curvature k or brane separation y_c for the RS scenarios. For that purpose one can use the combined uncertainty δE for molecular systems, where $\delta E_{\mathrm{theory}}$ and δE_{exp} are uncertainties of theory and experiment:

$$\delta E = \sqrt{\delta E_{\mathrm{exp}}^2 + \delta E_{\mathrm{theory}}^2}. \tag{13}$$

In the ADD scenario the constraining relation

$$\langle \Delta V_{\mathrm{ADD}}(n, R) \rangle < \delta E \tag{14}$$

translates on constraints on the compactification scale R, depending on n, via:

$$(R)^n < \frac{\delta E}{G m_1 m_2 \Delta}, \tag{15}$$

where Δ is the difference in expectation values over the wave function densities in the molecule:

$$\Delta = \left[\left\langle \Psi_1 \left| \frac{1}{r^{n+1}} \right| \Psi_1 \right\rangle - \left\langle \Psi_2 \left| \frac{1}{r^{n+1}} \right| \Psi_2 \right\rangle \right] \tag{16}$$

As shown in Fig. 1, the wave functions for the lowest vibrational states, in the case of H_2 and for $J = 0$, are located in the same region of space. Consequently, the fundamental vibrational transition in the hydrogen molecule ($v = 0 \to v = 1$) probes only a differential effect. Constraints on R, obtained from a comparison with the fundamental vibrational transition of H_2 are presented in Table 1.

The experimental as well as the theoretical results for the fundamental vibration in the hydrogen molecule are known to the 10^{-4} cm^{-1} level, an order of magnitude more accurate than the values for the binding energies [18, 19]. However for a comparison of dissociation limits, it is no longer a small difference along the internuclear coordinate axis that is probed, but the difference between the molecular scale of 1 Å and infinite atomic separation. The expectation value for the ADD-contribution to the binding energy of the lowest bound state in the H_2 molecule, the D_0 binding energy, is:

$$\langle \Delta V_{\mathrm{ADD}}(n, R) \rangle = G m_1 m_2 R^n \left\langle \frac{1}{r^{n+1}} \right\rangle_{\Psi_0} \tag{17}$$

Table 1. Constraints on the compactification scale R of extra dimensions (in units of m) as derived from both the measurement of the fundamental ($v = 0 \to 1$) vibration in H_2 and of the dissociation limit D_0 of H_2. The constraints are derived within the ADD-scenario assuming that the n extra dimensions are of equal size. The higher-dimensional Planck length $R_{\mathrm{Pl},(4+n)}$ (in units of m) and Planck mass $M_{(4+n)}$ (in units of GeV) for D_0 are also tabulated.

n	$R(v = 0 \to 1)$	$R(D_0)$	$R_{\mathrm{Pl},(4+n)}(D_0)$	$M_{(4+n)}(D_0)$
2	2.2×10^{4}	1.0×10^{4}	4.0×10^{-16}	4.9×10^{-1}
3	7.7×10^{-1}	1.9×10^{-1}	4.5×10^{-15}	4.4×10^{-2}
4	1.1×10^{-3}	8.5×10^{-4}	2.3×10^{-14}	8.7×10^{-3}
5	3.3×10^{-5}	3.2×10^{-5}	7.1×10^{-14}	2.8×10^{-3}
6	3.4×10^{-6}	3.7×10^{-6}	1.7×10^{-13}	1.2×10^{-3}
7	6.9×10^{-7}	7.8×10^{-7}	3.3×10^{-13}	6.0×10^{-4}

By comparing to the experimental findings on $D_0(H_2)$ [16] this leads to another set of constraints on R for n extra dimensions, which are also listed in Table 1. In Ref. [1] more constraints on R are presented corresponding to the dissociation limits of the deuterium molecule D_2 and the ($v = 0 \to 4$) R(2) ro-vibrational transition in HD^+.

Similarly, one can derive constraints pertaining to corrections in the RS models and the combined uncertainty δE for a specific molecular transition

$$\langle \Delta V_{\mathrm{RS}}(k) \rangle < \delta E. \tag{18}$$

Using δE for the dissociation limit $D_0(H_2)$ we obtain the following constraints. For the RS-I scenario in the short distance ($kr \ll 1$) regime, using Eq. (10) one finds a constraint on the brane separation of $y_c < 2 \times 10^{18}$ m in the limit $ky_c \ll 1$, and a constraint for the inverse of the curvature of $1/k < 3 \times 10^{18}$ m in the limit $ky_c \gg 1$.

For the RS-II scenario with one extra dimension in the short distance ($kr \ll 1$) regime, using Eq. (12) one finds a constraint for the inverse of the curvature of $1/k < 3 \times 10^{18}$ m. For the case of $n > 1$ extra dimensions with small curvatures with the same value k, using results from the $D_0(H_2)$ analysis one obtains constraints on the inverse of the curvature $1/k$ depending on n. For $n = 2$, $1/k < 1.5 \times 10^{+4}$ m; for $n = 3$, $1/k < 2.6 \times 10^{-1}$ m; for $n = 4$, $1/k < 1.0 \times 10^{-3}$ m; for $n = 5$, $1/k < 3.9 \times 10^{-5}$ m; for $n = 6$, $1/k < 4.3 \times 10^{-6}$ m and for $n = 7$, $1/k < 8.8 \times 10^{-7}$ m.

5. Comparison with other constraints

The constraints obtained from molecular systems probe length scales in the order of Angstroms. This complements bounds probing subatomic to astronomical length scales obtained from other studies using distinct methodologies. The micrometer to millimeter range are probed in torsion-balance type experiments, with the tightest constraint obtained by Kapner *et al.* [20] for a single extra dimension of $R < 4.4 \times 10^{-5}$ m.

Precision spectroscopies of hydrogen and muonic atoms have been interpreted in terms of the ADD-model [21] resulting in typical constraints of $R < 10^{-5}$ m. The interpretation is not straightforward because of the proton size puzzle [22, 23]; in fact, the argument has been turned around, and the existence of extra dimensions are instead invoked as a possible solution to the puzzle [24]. In addition, unlike for molecules, in the treatment of atoms some assumptions had to be made on the wave function density at $r = 0$, typical for the s-states involved, causing problems in calculating the second integral of Eq. (5) over the *electronic* wave function that has a significant amplitude at $r = 0$ in atoms.

Journal of Physics: Conference Series **626** (2015) 012052 doi:10.1088/1742-6596/626/1/012052

To probe length scales in the subatomic range, one is ultimately limited by the increasing contributions from nuclear structure and the strong interaction. In contrast to QED calculations, the most accurate lattice-QCD calculation of light hadron masses only achieves relative accuracies in the order of a few percent.

In high-energy particle collisions, higher-dimensional massive gravitons may be produced that could escape into the bulk, leading to events with missing energy in (3+1)-dimensional spacetime [4, 25]. Based on this premise the phenomenology of the SN 1987A supernova was investigated, obtaining limits on extra dimensions of $R < 3 \times 10^{-6}$ m for $n = 2$, $R < 4 \times 10^{-7}$ m for $n = 3$ and $R < 2 \times 10^{-8}$ m for $n = 4$ [26]. Similarly, from a missing energy analysis of proton colliding events at the LHC, the constraint $R < 3.2 \times 10^{-4}$ m can be extracted for $n = 2$ imposing the $M_{4+n} = 1.93$ TeV bound [27]. Also for $n > 2$, the bounds derived from the LHC are more stringent than those from molecules. However, additional assumptions beyond the ADD potential in Eq. (3) are necessary in order to interpret the LHC missing energy signals. Such assumptions are not needed for the molecular physics bounds, which are not sensitive to physics at very short distances.

The higher-dimensional Planck mass $M_{(4+n)}$ and corresponding Planck length $R_{\mathrm{Pl},(4+n)}$ derived from the present constraints on R are also tabulated in Table 1. Note that in all cases the Planck energy scale in $(4+n)$ dimensions turns out to be in the range 1-500 MeV. Assuming a straightforward extrapolation of Eq. (3) from molecular distance scales to nuclear distance scales, for an effect to be observable at molecular physics scales, the gravitational interaction must grow so rapidly at smaller distances that gravity becomes strong and non-perturbative already in the MeV scale. One would generally expect that strong gravity effects would already have been observed in scattering experiments, either through black hole creation, missing energy due to particles moving into the extra dimensions or other strong gravity effects, or large contributions to binding energies of nucleons and quarks. However, these expectations depend on a theory of gravity in the strong regime, and furthermore hadronic and nuclear binding is not a high-precision science, unlike atomic and molecular physics. Here we take the attitude that we are constraining the possibility that the gravitational force has a higher-dimensional dependence on r in the weak regime, where the computations are reliable, and we do not commit ourselves to what happens in the strong domain, where we cannot compute reliably. Our constraints are relevant only if gravity has a very soft behavior at shorter distances, so that its direct effects have escaped detection so far.

The idea that gravity at short distances may be soft has been explored in four dimensions in several contexts. This includes asymptotically safe gravity [28], *fat gravitons* as a possible solution to the cosmological constant problem [29], a way of avoiding black holes and singularities in string theory [30], and bouncing or self-inflating universes [31]. Therefore, given the limitations of our current understanding it is not unthinkable that this could happen in higher dimensions as well.

6. Conclusion and outlook

The alternative approaches for constraining compactification scales of extra dimensions, partially surveyed here, are all complementary as they probe different length and energy scales. Some approaches serve to produce tighter limits, however, often at the expense of additional assumptions. In the present study, the constraints are derived from precision measurements on molecules, leading to straightforward interpretations. Current state-of-the-art experiments on neutral hydrogen determine vibrational splittings of the order of 10^{-4} cm^{-1}, or 3 MHz [18]. Since the lifetimes of ro-vibrational quantum states in H_2 are of the order of 10^6 s [32], measurements of vibrational splittings of the order of 10^{14} Hz could in principle be possible at more than 20-digit precision, which leaves room for improvement "at the bottom" of over 10 orders of magnitude, if experimental techniques can be developed accordingly.

Acknowledgments
This presentation is based on work done in collaboration with Bert Schellekens, from the Institute of Subatomic Physics, Nikhef (Amsterdam), and Edcel Salumbides and Wim Ubachs, from the Vrije Universiteit of Amsterdam. I am very grateful to all of them. The author is partially supported by funding from the Spanish Ministerio de Economía y Competitividad, Research Project FIS2012-38816, and by the Project CONSOLIDER-INGENIO 2010, Programme CPAN (CSD2007-00042).

References

[1] Salumbides E J, Schellekens A N, Gato-Rivera B and Ubachs W 2015 Constraints on extra dimensions from precision molecular spectroscopy *Preprint* arXiv:1502.02838
[2] Salumbides E J, Koelemeij J C J, Komasa J, Pachucki K, Eikema K S E, and Ubachs W 2013 Bounds on fifth forces from precision measurements on molecules *Phys. Rev.* D **87** 112008
[3] Salumbides E J, Ubachs W and Korobov V I 2014 Bounds on fifth forces at the sub-Angstrom length scale *J. Mol. Spectrosc.* **300** 65
[4] Arkani-Hamed N, Dimopoulos S, and Dvali G 1998 The hierarchy problem and new dimensions at a millimeter *Phys. Lett.* B **429** 263
[5] Arkani-Hamed N, Dimopoulos S, and Dvali G 1998 Phenomenology, astrophysics and cosmology of theories with submillimeter dimensions and TeV scale quantum gravity *Phys. Rev.* D **59** 086004
[6] Antoniadis I, Arkani-Hamed N, Dimopoulos S, and Dvali G 1998 New dimensions at a millimeter to a Fermi and superstrings at a TeV *Phys. Lett.* B **436** 257
[7] Randall L and Sundrum R 1999 Large mass hierarchy from a small extra dimension *Phys. Rev. Lett.* **83** 3370
[8] Randall L and Sundrum R 1999 An alternative to compactification *Phys. Rev. Lett.* **83** 4690
[9] Callin P 2004 Corrections to the Newtonian potential in the two-brane Randall-Sundrum model *Preprint* arXiv:hep-ph/0407054
[10] Callin P and Ravndal F 2004 Higher order corrections to the Newtonian potential in the Randall-Sundrum model *Phys. Rev.* D **70** 104009
[11] Garriga J and Tanaka T 2000 Gravity in the Randall-Sundrum brane world *Phys. Rev. Lett.* **84** 2778
[12] Pachucki K and Komasa J 2009 Nonadiabatic corrections to rovibrational levels of H_2 *J. Chem. Phys.* **130** 164113
[13] Pachucki K 2005 Higher-order effective Hamiltonian for light atomic systems *Phys. Rev.* A **71** 012503
[14] Piszczatowski K, L ach G, Przybytek M, Komasa J, Pachucki K and Jeziorski B 2009 Theoretical determination of the dissociation energy of molecular hydrogen *J. Chem. Theory Comp.* **5** 3039
[15] Komasa J, Piszczatowski K, L ach G, Przybytek M, Jeziorski B and Pachucki K 2011 Quantum electrodynamics effects in rovibrational spectra of molecular hydrogen *J. Chem. Theory Comp.* **7** 3105
[16] Liu J, Salumbides E J, Hollenstein U, Koelemeij J C J, Eikema K S E, Ubachs W and Merkt F 2009 Determination of the ionization and dissociation energies of the hydrogen molecule *J. Chem. Phys.* **130** 174306
[17] Liu J, Sprecher D, Jungen C, Ubachs W and Merkt F 2010 Determination of the ionization and dissociation energies of the deuterium molecule (D_2) *J. Chem. Phys.* **132** 154301
[18] Dickenson G D, Niu M L, Salumbides E J, Komasa J, Eikema K S E, Pachucki K and Ubachs W 2013 Fundamental vibration of molecular hydrogen *Phys. Rev. Lett.* **110** 193601
[19] Niu M L, Salumbides E J, Dickenson G D, Eikema K S E and Ubachs W 2014 Precision spectroscopy of the $X\,^1\Sigma_g^+$, $v = 0 \to 1\,(J = 0-2)$ rovibrational splittings in H_2, HD and D_2 *J. Mol. Spectrosc.* **300** 44
[20] Kapner D J, Cook T S, Adelberger E G, Gundlach J H, Heckel B R, Hoyle C D and Swanson H E 2007 Tests of the gravitational inverse-square law below the dark-energy length scale *Phys. Rev. Lett.* **98** 021101
[21] Li Z-G, Ni W-T and Pulido Patón A 2007 Extra dimensions and atomic transition frequencies *Chinese Phys.* B **17** 70
[22] Pohl R *et al.* 2010 The size of the proton *Nature* **466** 213
[23] Antognini A *et al.* 2013 Proton structure from the measurement of 2S-2P transition frequencies of muonic hydrogen *Science* **339** 417
[24] Wang L-B and Ni W-T 2013 Proton radius puzzle and large extra dimensions *Mod. Phys. Lett.* A **28** 1350094
[25] Mirabelli E A, Perelstein M and Peskin M E 1999 Collider signatures of new large space dimensions *Phys. Rev. Lett.* **82** 2236
[26] Cullen S and Perelstein M 1999 SN1987A constraints on large compact dimensions *Phys. Rev. Lett.* **83** 268
[27] Aad G *et al.* (Atlas Collaboration) 2013 Search for dark matter candidates and large extra dimensions in events with a photon and missing transverse momentum in *pp* collision data at $\sqrt{s} = 7$ TeV with the ATLAS detector *Phys. Rev. Lett.* **110** 011802

[28] Weinberg S 1979 *Ultraviolet Divergences in Quantum Theories of Gravitation* (General Relativity: an Einstein centenary survey) ed S H Hawking and W Israel (Cambridge: Cambridge University Press) p 790
[29] Sundrum R 2004 Fat gravitons, the cosmological constant and submillimeter tests *Phys. Rev.* D **69** 044014
[30] Siegel W 2003 Stringy gravity at short distances *Preprint* arXiv:hep-th/0309093
[31] Khoury J 2007 Fading gravity and self-inflation *Phys. Rev.* D **76** 123513
[32] Black J H and Dalgarno A 1976 Interstellar H_2 - population of excited rotational states and infrared response to ultraviolet-radiation *Astrophys. J.* **203** 132
[33] Sprecher D, Jungen C, Ubachs W and Merkt F 2011 Towards measuring the ionisation and dissociation energies of molecular hydrogen with sub-MHz accuracy *Faraday Disc.* **150** 51
[34] Karr J-P 2014 H_2^+ and HD^+: Candidates for a molecular clock *J. Mol. Spectrosc.* **300** 37
[35] Tran V Q, Karr J-P, Douillet A, Koelemeij J C J and Hilico L 2013 Two-photon spectroscopy of trapped HD^+ ions in the Lamb-Dicke regime *Phys. Rev.* A **88** 033421

Journal of Physics: Conference Series **626** (2015) 012053 doi:10.1088/1742-6596/626/1/012053

EPR & Klein Paradoxes in Complex Hamiltonian Dynamics and Krein Space Quantization

Farrin Payandeh

Department of Physics, Payame Noor University (PNU), P.O. BOX, 19395-3697 Tehran, Iran

E-mail: payandehfarrin92@gmail.com , f_payandeh@pnu.ac.ir

Abstract. Negative energy states are applied in Krein space quantization approach to achieve a naturally renormalized theory. For example, this theory by taking the full set of Dirac solutions, could be able to remove the propagator Green function's divergences and automatically without any normal ordering, to vanish the expected value for vacuum state energy. However, since it is a purely mathematical theory, the results are under debate and some efforts are devoted to include more physics in the concept. Whereas Krein quantization is a pure mathematical approach, complex quantum Hamiltonian dynamics is based on strong foundations of Hamilton-Jacobi (H-J) equations and therefore on classical dynamics. Based on complex quantum Hamilton-Jacobi theory, complex spacetime is a natural consequence of including quantum effects in the relativistic mechanics, and is a bridge connecting the causality in special relativity and the non-locality in quantum mechanics, i.e. extending special relativity to the complex domain leads to relativistic quantum mechanics. So that, considering both relativistic and quantum effects, the Klein-Gordon equation could be derived as a special form of the Hamilton-Jacobi equation. Characterizing the complex time involved in an entangled energy state and writing the general form of energy considering quantum potential, two sets of positive and negative energies will be realized. The new states enable us to study the spacetime in a relativistic entangled "space-time" state leading to 12 extra wave functions than the four solutions of Dirac equation for a free particle. Arguing the entanglement of particle and antiparticle leads to a contradiction with experiments. So, in order to correct the results, along with a previous investigation [1], we realize particles and antiparticles as physical entities with positive energy instead of considering antiparticles with negative energy. As an application of modified descriptions for entangled (space-time) states, the original version of EPR paradox can be discussed and the correct answer can be verified based on the strong rooted complex quantum Hamilton-Jacobi theory [2-27] and as another example we can use the negative energy states, to remove the Klein's paradox without the need of any further explanations or justifications like backwardly moving electrons. Finally, comparing the two approaches, we can point out to the existence of a connection between quantum Hamiltonian dynamics, standard quantum field theory, and Krein space quantization [28-43].

1. Introduction

Complex spacetime actually stems from complex time, as first proposed by El Naschie [2], according to a special case of E^{∞} theory [3-5] and then applied by C. D. Yang in a series of papers [9-18]. El Naschie's theory provides the probability of constructing a valid relativistic quantum mechanics for particles which are moving faster than light [7] and the two dimensional time introduced in El Naschie's theory can cause the equipotential effects of gravity [8]. However, the complex spacetime was proposed by C. D. Yang in a series of papers in the form

Journal of Physics: Conference Series **626** (2015) 012053 doi:10.1088/1742-6596/626/1/012053

$x^\mu = x^{\mu_R} + ix^{\mu_I}$, $x^{\mu_R}, x^{\mu_I} \in R$; $x^\mu = (ct, x, y, z)$, mentioning that the boundary between classical and quantum mechanics could be broken, since the quantum operators are derived from Hamilton equation of motion [9,10]. These complex Hamilton equations were used to investigate wave-particle duality [11], tunneling [12], atomic shell structure of Hydrogen atom [13], multiple trajectories of particle beams [14], quantum state transition [15] and spin [16]. Therefore, it was shown the relativistic quantum mechanics is described by the complex quantum Hamilton-Jacobi (H-J) equations, i.e. the complex spacetime is a natural consequence of including quantum effects in the relativistic mechanics, and is a bridge connecting the causality in special relativity and the non-locality in quantum mechanics, so that extending special relativity to the complex domain leads to relativistic quantum mechanics and considering both relativistic and quantum effects, the Klein-Gordon equation could be derived as a special form of the Hamilton-Jacobi equation. It was also pointed out that the entangled state, causing the faster-than-light links, is a consequence of an entangled energy, plus a quantum potential, i.e. $E^2+2m_0c^2Q$, resulting in a constant quantity. This quantity if used in the relativistic H-J equations, may describe quantum multiple trajectories. In this paper, the basis of our work is on a paper of Yang [17] in which after characterizing the complex time involved in an entangled energy state and writing the general form of energy considering quantum potential, two sets of positive and negative energies could be realized. However, we will realize new states for both positive and negative values of energy and momentum, and we will discuss the complex spacetime in a relativistic entangled "space-time" state leading to 12 extra wave functions than the four solutions of Dirac equation for a free particle. We will see that the entanglement of two particles or two antiparticles could be done only with opposite momenta and the entanglement of particle and antiparticle could be done only with the same momenta, where the latter is in contradiction with experiments. Therefore, in the means of modifying the results, we will realize particles and antiparticles as physical entities with positive energy instead of considering antiparticles with negative energy. Moreover, we will introduce unphysical particle and antiparticle with negative energy, as the complement of the sets of solutions for Dirac equation, in accordance to the concept of Krein quantization. Then, we will rewrite and modify the first introduced 16 wave functions in terms of physical and unphysical particles and antiparticles for different values of positive and negative values of momentum. Finally, we will discuss about the original version of Einstein-Podolsky-Rosen (EPR) paradox and we will verify the correct answer which is a unique solution due to the new results based on quantum Hamiltonian dynamics approach and then, using the unphysical negative energy states, we will remove the Klein's paradox without the need of any further explanations or justifications like backwardly moving electrons.

2. Complex spacetime in a relativistic entangled "space-time" state

Our starting point is the relation (3.6) in ref. [17] by C. D. Yang:

$$\psi(t,x,y,z) = (C_0^+ e^{i(k_0/\hbar)t} + C_0^- e^{-i(k_0/\hbar)t})(C_1^+ e^{ik_1x} + C_1^- e^{-ik_1x})$$
$$(C_2^+ e^{ik_2y} + C_2^- e^{-ik_2y})(C_3^+ e^{ik_3z} + C_3^- e^{-ik_3z}) \tag{1}$$

the relation (4.1) of ref. [17]:

$$\psi(t,x,y,z) = (C_0^+ e^{i(k_0/\hbar)t} + C_0^- e^{-i(k_0/\hbar)t})e^{-i(k_1x+k_2y+k_3z)} \tag{2}$$

which can be rewritten in terms of energy and momentum, respectively as:

$$\psi(t,\vec{r}) = (C_0^+ e^{i(E/\hbar)t} + C_0^- e^{-i(E/\hbar)t})(C^+ e^{i\vec{p}.\vec{r}/\hbar} + C^- e^{-i\vec{p}.\vec{r}/\hbar}) \tag{3}$$

$$\psi(t,\vec{r}) = (C_0^+ e^{i(E/\hbar)t} + C_0^- e^{-i(E/\hbar)t})e^{-i\vec{p}.\vec{r}/\hbar} \tag{4}$$

and finally, the identity (4.20) of ref. [17]: $E^2(t) + 2m_0c^2Q(t) = k_0^2 = Const.$ in which, both and are time dependent, but their summation appears to be a constant. Now, we can write the Eq. (5) as:

$$E(t) = \pm\sqrt{k_0^2 - 2m_0c^2Q(t)} = \pm\sqrt{(m_0c^2)^2 + c^2p^2 - 2m_0c^2Q(t)} \equiv \pm E_\pm \qquad (5)$$

It is clear that for any time t, there are two momenta $(p > 0, p < 0)$ and two energies $(E_+ > 0, E_- < 0)$, and we can see that in this general form of energy, the quantum potential $Q(t)$ is nonzero. Also, we should notice that in Eqs. (3.13)-(3.16) of ref. [17]: $\psi_1(t,x,y,z) = Ce^{i(Et-\vec{p}.\vec{r})/\hbar}$, $\psi_2(t,x,y,z) = Ce^{i(-Et-\vec{p}.\vec{r})/\hbar}$, $\psi_3(t,x,y,z) = Ce^{i(Et+\vec{p}.\vec{r})/\hbar}$, $\psi_4(t,x,y,z) = Ce^{i(-Et+\vec{p}.\vec{r})/\hbar}$, the derived four wave functions were the eigenfunctions solved from the Dirac equation for a free particle and there was considered no entanglement and $Q(t) = 0$; we call it space-time state. In Eq.(2) or (4), the entanglement of time was considered and the wave function was written only for positive values of energy and momentum with $Q(t) \neq 0$; we call it space-entangled time state. In the following, we will realize another states for both values of positive and negative energy and momentum due to Eq.(5). Then, by using Eq.(5) we can write them down in terms of both positive and negative values of energy and momentum (i.e. E_+,E_-,$+p$,$-p$): $\psi(t,\vec{r}) = (C_0^+e^{i(E_\pm/\hbar)t} + C_0^-e^{-i(E_\pm/\hbar)t})(C^+e^{i(\pm\vec{p}).\vec{r}/\hbar} + C^-e^{-i(\pm\vec{p}).\vec{r}/\hbar})$ and $\psi(t,\vec{r}) = (C_0^+e^{i(E_\pm/\hbar)t} + C_0^-e^{-i(E_\pm/\hbar)t})e^{-i(\pm\vec{p}).\vec{r}/\hbar}$.

Now, we can realize four states: (a) Space-time state $(\psi_1,\psi_2,\psi_3,\psi_4)$; $Q(t) = 0$ (b) Space-entangled time state $(\psi_5,\psi_6,\psi_7,\psi_8)$; $Q(t) \neq 0$ (c) Entangled space-time state $(\psi_9,\psi_{10},\psi_{11},\psi_{12})$; $Q(t) \neq 0$ (d) Entangled space-entangled time state $(\psi_{13},\psi_{14},\psi_{15},\psi_{16})$; $Q(t) \neq 0$, leading to the following 16 wave functions:

$$\psi_1 = Ce^{\frac{i}{\hbar}(Et-\vec{p}.\vec{r})} \quad (E>0\ ,\ p>0) \quad ; \quad \psi_2 = Ce^{\frac{i}{\hbar}(Et-(-\vec{p}).\vec{r})} \quad (E>0\ ,\ p<0)$$
$$\psi_3 = Ce^{\frac{i}{\hbar}((-E)t-\vec{p}.\vec{r})} \quad (E<0\ ,\ p>0) \quad ; \quad \psi_4 = Ce^{\frac{i}{\hbar}((-E)t-(-\vec{p}).\vec{r})} \quad (E<0\ ,\ p<0)$$
$$\psi_5 = (C_0^+e^{i(E_+/\hbar)t} + C_0^-e^{-i(E_+/\hbar)t})C^-e^{-\frac{i}{\hbar}\vec{p}.\vec{r}/\hbar} \quad (E_+>0\ ,\ p>0)$$
$$\psi_6 = (C_0^+e^{i(E_+/\hbar)t} + C_0^-e^{-i(E_+/\hbar)t})C^-e^{-\frac{i}{\hbar}(-\vec{p}).\vec{r}/\hbar} \quad (E_+>0\ ,\ p<0)$$
$$\psi_7 = (C_0^+e^{i(E_-/\hbar)t} + C_0^-e^{-i(E_-/\hbar)t})C^-e^{-\frac{i}{\hbar}\vec{p}.\vec{r}/\hbar} \quad (E_-<0\ ,\ p>0)$$
$$\psi_8 = (C_0^+e^{i(E_-/\hbar)t} + C_0^-e^{-i(E_-/\hbar)t})C^-e^{-\frac{i}{\hbar}(-\vec{p}).\vec{r}/\hbar} \quad (E_-<0\ ,\ p<0)$$
$$\psi_9 = C_0^+e^{i(E_+/\hbar)t}(C^+e^{\frac{i}{\hbar}\vec{p}.\vec{r}/\hbar} + C^-e^{-\frac{i}{\hbar}\vec{p}.\vec{r}/\hbar}) \quad (E_+>0\ ,\ p>0)$$
$$\psi_{10} = C_0^+e^{i(E_+/\hbar)t}(C^+e^{\frac{i}{\hbar}(-\vec{p}).\vec{r}/\hbar} + C^-e^{-\frac{i}{\hbar}(-\vec{p}).\vec{r}/\hbar}) \quad (E_+>0\ ,\ p<0)$$
$$\psi_{11} = C_0^+e^{i(E_-/\hbar)t}(C^+e^{\frac{i}{\hbar}\vec{p}.\vec{r}/\hbar} + C^-e^{-\frac{i}{\hbar}\vec{p}.\vec{r}/\hbar}) \quad (E_-<0\ ,\ p>0)$$
$$\psi_{12} = C_0^+e^{i(E_-/\hbar)t}(C^+e^{\frac{i}{\hbar}(-\vec{p}).\vec{r}/\hbar} + C^-e^{-\frac{i}{\hbar}(-\vec{p}).\vec{r}/\hbar}) \quad (E_-<0\ ,\ p<0)$$
$$\psi_{13} = (C_0^+e^{i(E_+/\hbar)t} + C_0^-e^{-i(E_+/\hbar)t})(C^+e^{\frac{i}{\hbar}\vec{p}.\vec{r}/\hbar} + C^-e^{-\frac{i}{\hbar}\vec{p}.\vec{r}/\hbar}) \quad (E_+>0\ ,\ p>0)$$
$$\psi_{14} = (C_0^+e^{i(E_+/\hbar)t} + C_0^-e^{-i(E_+/\hbar)t})(C^+e^{\frac{i}{\hbar}(-\vec{p}).\vec{r}/\hbar} + C^-e^{-\frac{i}{\hbar}(-\vec{p}).\vec{r}/\hbar}) \quad (E_+>0\ ,\ p<0)$$
$$\psi_{15} = (C_0^+e^{i(E_-/\hbar)t} + C_0^-e^{-i(E_-/\hbar)t})(C^+e^{\frac{i}{\hbar}\vec{p}.\vec{r}/\hbar} + C^-e^{-\frac{i}{\hbar}\vec{p}.\vec{r}/\hbar}) \quad (E_-<0\ ,\ p>0)$$
$$\psi_{16} = (C_0^+e^{i(E_-/\hbar)t} + C_0^-e^{-i(E_-/\hbar)t})(C^+e^{\frac{i}{\hbar}(-\vec{p}).\vec{r}/\hbar} + C^-e^{-\frac{i}{\hbar}(-\vec{p}).\vec{r}/\hbar}) \quad (E_-<0\ ,\ p<0)$$

The wave functions $\psi_1, \psi_2, ..., \psi_{16}$ after doing the multiplications could be described as: (a) ψ_1 : particle with positive momentum, ψ_2 : particle with negative momentum, ψ_3 : antiparticle with positive momentum, ψ_4 : antiparticle with negative momentum. (b) ψ_5, ψ_7 : entangled particle-antiparticle with positive momenta, ψ_6, ψ_8 : entangled particle-antiparticle with negative momenta. (c) ψ_9, ψ_{10} : entangled particle-particle with opposite momenta. (d) ψ_{11}, ψ_{12} : entangled antiparticle-antiparticle with opposite momenta. (e) $\psi_{13}, \psi_{14}, \psi_{15}, \psi_{16}$: entanglement of (entangled particle-antiparticle with positive momenta) to (entangled particle-antiparticle with negative momenta) or entanglement of (entangled particle-particle with opposite momenta) to (entangled antiparticle-antiparticle with opposite momenta). It can be seen that the entanglement of two particles or two antiparticles could be done only with opposite momenta, but the entanglement of particle and antiparticle could be done only with the same momenta. Furthermore, we can deduce from (b) and (c) that the entanglement of time causes the entanglement of particle with antiparticle, and the entanglement of space causes the particle-particle or antiparticle-antiparticle entanglement. However, we encounter to a confliction, here. Empirical experiments demonstrate the quantum correlation at a distance of a particle-antiparticle system like kaon and antikaon system which are entwined. Therefore, at this stage, we want to introduce a parallel approach to correct all the being results. It seems as if something is missed here. According to the theory of Dirac, antiparticles are believed to be particles of negative energy. But, due to the fact that antiparticles are detectable, so the physical antiparticles must be of positive energies. Moreover, according to Eq. (5) there are both positive and negative energy states. However, it seems that taking the negative energies as antiparticles, is not covering all the underlying physics [1]. So, we propose that it is rational to accept that positive energy belongs to physical particles and negative energy belongs to unphysical particles. Then, we can say that the solutions of Dirac equation describe both physical particles and antiparticles with positive energy and both unphysical particles and antiparticles with negative energy. Consequently, we should modify the descriptions of wave functions $\psi_1, \psi_2, ..., \psi_{16}$ according to the following:

(aa) ψ_1 : physical particle or antiparticle with $E_+ > 0$, $p > 0$. ψ_2 : physical particle or antiparticle with $E_+ > 0$, $p < 0$. ψ_3 : unphysical particle or antiparticle with $E_- < 0$, $p > 0$. ψ_4: unphysical particle or antiparticle with $E_- < 0$, $p < 0$. (bb) ψ_5, ψ_6 : entangled physical (particle-particle or antiparticle-antiparticle or particle-antiparticle or antiparticle-particle with $E_+ > 0$ and $p > 0$, or $p < 0$. ψ_7, ψ_8 : entangled unphysical (particle-particle or antiparticle-antiparticle or particle-antiparticle or antiparticle-particle) with $E_- < 0$ and $p > 0$, or $p < 0$. (cc) ψ_9, ψ_{10} : entangled physical (particle-particle or antiparticle-antiparticle or particle-antiparticle or antiparticle-particle with $E_+ > 0$ and opposite momenta. ψ_{11}, ψ_{12} : entangled unphysical (particle-particle or antiparticle-antiparticle or particle-antiparticle or antiparticle-particle) with $E_- < 0$ and opposite momenta. ψ_{13}, ψ_{14} : entanglement of two entangled physical (particle-particle or antiparticle-antiparticle or particle-antiparticle or antiparticle-particle) with $E_+ > 0$ and the same or opposite momenta. ψ_{15}, ψ_{16} : entanglement of two entangled unphysical (particle-particle or antiparticle-antiparticle or particle-antiparticle or antiparticle-particle) with $E_- < 0$ and the same or opposite momenta.

3. Discussion about EPR and Klein Paradoxes in Complex Hamiltonian Dynamics

In this section, first we discuss about the original version of Einstein-Podolsky-Rosen (EPR) paradox and find out the unique solution, due to the wave functions of ψ_9, ψ_{10} in part (cc). In quantum mechanics, the position of a particle and its momentum have the commutation relation $[\widehat{x}, \widehat{p}] = i\hbar$. According to EPR, if we consider a two-particle system, the operators $(\widehat{x_1} - \widehat{x_2})$ and $(\widehat{p_1} + \widehat{p_2})$ will commute:

$$[\widehat{x_1} - \widehat{x_2} \, , (\widehat{p_1} + \widehat{p_2})] = 0 \tag{6}$$

meaning that these operators have a common eigenstate with eigenvalues $x_1 - x_2 = a = const.$ and $p_1 + p_2 = 0$, $p_2 = -p_1$. However, the above mentioned two-system implies a strange state, because the two particles are moving in opposite momentum directions while keeping their distance unchanged, and this is an unimaginable state. In a paper by G-j Ni et al. [19], it was mentioned that by the EPR definition of a two-particle system with the commutation relation (6), it seems that the original version of EPR paradox still remains to be answered. Then, for solving the problem, they propose the commutation relation $[\widehat{x_1} + \widehat{x_2} \, , (\widehat{p_1} - \widehat{p_2})] = 0$ instead of the relation (6), and conclude that the correct answer turns out to be a particle and its antiparticle moving in opposite direction with momentum p_1 and $p_2 = -p_1$ and position x_1 and $x_2 = -x_1$, which is a unique solution to the original version of EPR paradox. As we can see, the wave functions in part (cc) can describe an entangled particle-antiparticle with $E_+ > 0$

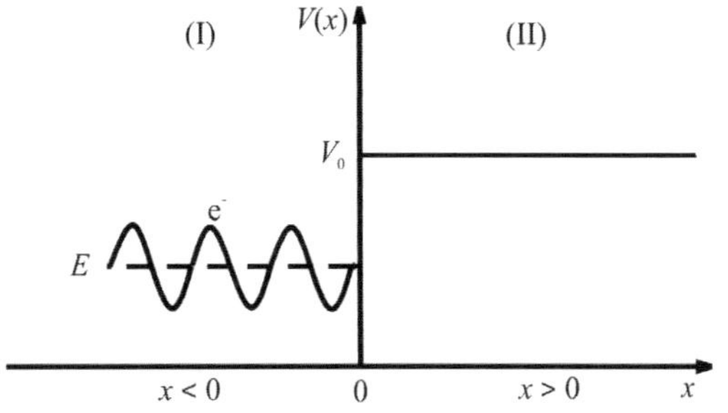

Figure 1. The incident beam of electrons of energy E, in region I, coming down on the potential barrier of strength V_0, located in region II.

and opposite momenta that can verify the correct answer which is a unique solution due to the new results based on quantum Hamiltonian dynamics approach.

As another example, we can treat Klein's paradox with full set of Dirac solutions. In 1929, Oskar Klein [20] calculated the reflection and transmission coefficients for an incident beam of electrons of energy E , falling on a potential barrier of strength V_0 . He found out that the unexpected amount of reflected electrons or transmitted electrons with a steady rate causes paradoxical results. Briefly speaking, according to Fig. 1, The momentum of free particles, i.e. the incident current in region (I) would be $p = \frac{1}{c}\sqrt{E^2 - m_0{}^2c^4}$, whereas in region (II), it becomes $q = \frac{1}{c}\sqrt{(eV_0 - E)^2 - m_0{}^2c^4}$. Using Dirac's relativistic theory [21], Klein found that the reflection and transmission coefficients for large potentials are respectively $R = \left(\frac{1-\kappa}{1+\kappa}\right)^2$, $T = \frac{4\kappa}{(1+\kappa)^2}$ and one can see $R + T = 1$. Also, κ is a kinematical factor, defined as $\kappa = \frac{q}{p}\frac{E+m_0c^2}{E-eV_0+m_0c^2}$ and choosing $e = 1$, the common form $\kappa = \frac{q}{p}\frac{E+m_0c^2}{E-V_0+m_0c^2}$ is obtained which becomes negative, when $E < V_0 - m_0c^2$. This implies a "paradoxical issue" since in this case, we get $R > 1$ while $T < 0$; the number of reflected particles from the barrier is greater than that of the incident ones on it. This is known as *Klein's Paradox*. Another important point here, corresponds to the group velocity. For relation $q^2c^2 = (V_0 - E)^2 - m_0{}^2c^4$, the group velocity becomes $v_{gr} = \frac{\mathrm{d}E}{\mathrm{d}q} = \frac{q}{E-V_0}$. So for left-to-right moving particles in region (II) of Figure 1, $v_{gr} > 0$, and since $E < V_0$, therefore q has to be negative, i.e. $q = -\frac{1}{c}\sqrt{(V_0 - E)^2 - m_0{}^2c^4}$. This expression for q, gives [22,23]

$$\kappa = \left[\frac{(V_0 - E + m_0c^2)(E + m_0c^2)}{(V_0 - E - m_0c^2)(E - m_0c^2)}\right]^{\frac{1}{2}} . \tag{7}$$

Now, this gives positive values for R and T since from (7) it is always $\kappa \geq 1$ and also for intermediate potentials ($m_0c^2 < V_0 < E + m_0c^2$), R and T satisfy $R + T = 1$. Nevertheless the incident current is fulfilled in this latest case, the paradox still exists, since for $V_0 \to \infty$ and a fixed E, Eq.(7) results in a finite value for κ and as a consequence, we get a non-zero finite limit for T. Therefore, it seems that the electrons could have passed the infinitely strong barrier, without any exponential damping. This is, once again, a paradoxical result which is in contrast to the damping, expected from the quantum tunnelling processes. This unexpected tunnelling was named "Klein Tunnelling" [24]. Now, we choose the one-dimensional Dirac equation $i\hbar c\gamma_x\frac{\mathrm{d}\psi}{\mathrm{d}x} + m_0c^2\psi = \gamma_0(E - V_0)\psi$, in which γ_x is the x-component of Dirac matrices and $\gamma_0 = \begin{pmatrix} \mathbf{1}_2 & \mathbf{0} \\ \mathbf{0} & -\mathbf{1}_2 \end{pmatrix}$. The four components of Dirac matrices obey the Clifford algebra $\gamma_\mu\gamma_\nu + \gamma_\nu\gamma_\mu = 2\eta_{\mu\nu}$, where $\eta_{\mu\nu} = \mathrm{diag}(+1, -1, -1, -1)$ is the Minkowskian flat metric. For

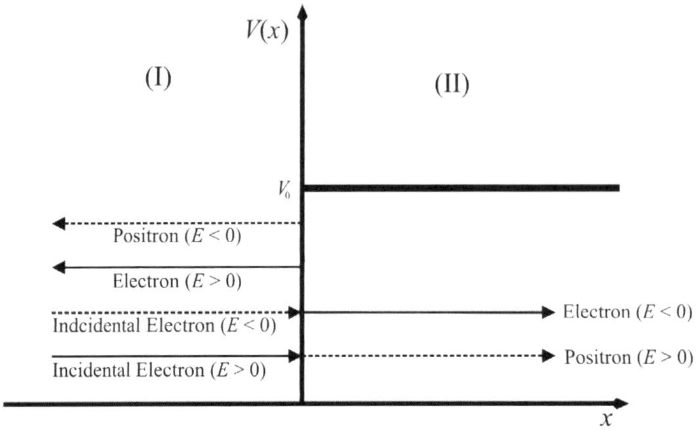

Figure 2. The incident, reflected and transmitted electrons or positrons, of positive and negative energies.

energies in the range $m_0c^2 < E < V_0 - m_0c^2$, the incident wave function [22] in region (I), would be $\psi_{\rm I}(x) = \begin{pmatrix} {\rm i}p \\ E - m_0c^2 \end{pmatrix} {\rm e}^{{\rm i}px} + A \begin{pmatrix} -{\rm i}p \\ E - m_0c^2 \end{pmatrix} {\rm e}^{-{\rm i}px}$, while this wave function for transmitted electrons in region (II) becomes $\psi_{\rm II}(x) = B \begin{pmatrix} -{\rm i}|q| \\ V_0 - E - m_0c^2 \end{pmatrix} {\rm e}^{-{\rm i}|q|x}$ Corresponding to the particle current $\boldsymbol{j}(x) = \psi^\dagger(x)\ \hat{\alpha}_x\ \psi(x)$, $\hat{\alpha}(x) = \gamma_x\gamma_0$. The reflected and transmitted currents satisfy

$$R = \frac{|\boldsymbol{j}_R|}{|\boldsymbol{j}_{inc}|} = \left(\frac{1+\kappa}{1-\kappa}\right)^2 \quad , \quad T = \frac{|\boldsymbol{j}_T|}{|\boldsymbol{j}_{inc}|} = \frac{4\kappa}{(1-\kappa)^2}, \tag{8}$$

where $\boldsymbol{j}_{inc}$ is the incident current. Once again, the paradoxical result appears $|\boldsymbol{j}_R| > |\boldsymbol{j}_{inc}|$.

What is important here is that both spin up or down electrons in the incident current, were supposed to be of positive energies. So the negative energies have been ignored in all available approaches. However, if we keep both positive and negative energies as viable values (although the negative energies are related to un-physical states) or in other words, if we take the full set of Dirac solutions or strictly speaking, the negative energies, then we can show that the reflected and transmitted currents are the same for both electrons and positrons and the paradox disappears and this result states that the Klein's paradox arises because of some lack in initial data, which leads us to consider some reflected electrons and transmitted positrons (Hole theory) [1]. To prove this mathematically, we get back to the kinematical parameter from Eq.(7) for positive energies ($E > 0$). Maintaining the negative mode states, we define another kinematical parameter κ' as $\kappa' = \left[\frac{(V_0+E+m_0c^2)(-E+m_0c^2)}{(V_0+E-m_0c^2)(-E-m_0c^2)}\right]^{\frac{1}{2}}$ for $E < 0$. Note that $\mathcal{K} = \kappa\kappa' = \left[\frac{(V_0+m_0c^2)-E^2}{(V_0-m_0c^2)-E^2)}\right]^{\frac{1}{2}} > 1$. This equation implies that for $V_0 \to \infty$, $\mathcal{K} \to 1$. According to equation (8), having κ and κ', one can obtain relations between incident, reflected and transmitted currents for electrons (and positrons) of negative energies. We have

$$R' = \frac{|\boldsymbol{j}'_R|}{|\boldsymbol{j}'_{inc}|} = \left(\frac{1+\kappa'}{1-\kappa'}\right)^2 = \left(\frac{1+\frac{\mathcal{K}}{\kappa}}{1-\frac{\mathcal{K}}{\kappa}}\right)^2 = \left(\frac{\mathcal{K}+\kappa}{\mathcal{K}-\kappa}\right)^2,$$

$$T' = \frac{|\boldsymbol{j}'_T|}{|\boldsymbol{j}'_{inc}|} = \frac{4\kappa'}{(1-\kappa')^2} = \frac{4\left(\frac{\mathcal{K}}{\kappa}\right)}{(1-\kappa)^2} = \frac{4\mathcal{K}\kappa}{(\mathcal{K}-\kappa)^2}. \tag{9}$$

Now, we can easily see that maintaining all modes could provides equal values for reflected and transmitted electrons and positrons [42]. According to processes in regions (I) and (II) in Figure 2, and using equation (8) together with (9), one can define $M \doteq R|_{E>0} + T'|_{E<0} = \left(\frac{1+\kappa}{1-\kappa}\right)^2 + \frac{4\mathcal{K}\kappa}{(\mathcal{K}-\kappa)^2}$ which is the summation of reflected electrons of positive energies and transmitted ones of negative energies, and $N \doteq R'|_{E<0} + T|_{E>0} = \left(\frac{\mathcal{K}+\kappa}{\mathcal{K}-\kappa}\right)^2 + \frac{4\kappa}{(1-\kappa)^2}$ which is the summation of the reflected positrons of negative energies and transmitted ones of positive energies, as it is in Figure 2. Dealing with numerators, it is simple to see that quantities M and N coincide, showing that the reflected and transmitted currents are the same for both electrons and positrons.

As a consequence, what we argue here is that negative energies could be as important as positive ones, since they contribute in variety of processes and some outstanding physicists have pointed out to their importance. According to Dirac himself: "*Negative energies and probabilities should not be considered as nonsense. They are well-defined concepts mathematically, like a negative sum of money, since the equations which express the important properties of energies and probabilities, can still be used when they are negative. Thus, negative energies and probabilities should be considered simply as things which do not appear in experimental results. The physical interpretation of relativistic theory involves these things and is thus in contradiction with experiment. We therefore, have to consider ways of modifying or supplementing this interpretation*" [25,26]. Also, Feynman has discussed the negative probabilities as viable concepts in quantum physics [27,28]. Negative energy states play the main role in the concept of Krein quantization, too, which is a naturally re-normalized theory. This theory by taking the full set of Dirac solutions, is able to remove the propagator Green's function's divergences and automatically without any normal ordering, cancel the expected value for vacuum state energy. Krein quantization was also used to re-interpret physical phenomena like Casimir effect. However, since it is a purely mathematical theory, the results are under debate and some efforts are devoted to include more physics in the concept [25-38]. Whereas Krein quantization is a pure mathematical approach, complex quantum Hamiltonian dynamics is based on the strong foundations of Hamilton-Jacobi equations and therefore on classical dynamics. Hence, it seems as if quantum Hamiltonian dynamics may construct a connecting bridge between standard quantum field theory and Krein space quantization in order to explain the reason for the practical ugly mathematics of renormalization and provide an answer when asked for what he had won the Nobel prize, Feynman replied, "For sweeping the infinities under the rug"?

4. Conclusion

Due to complex quantum Hamilton-Jacobi theory, complex spacetime is a natural consequence of including quantum effects in the relativistic mechanics. Characterizing the complex time involved in an entangled energy state and writing the general form of energy considering quantum potential, two sets of positive and negative energies could be realized that are in accordance with Krein space Quantization. Here, we realized new states for both positive and negative values of energy and momentum and then, we discussed the complex space-time in a relativistic entangled "space-time" state leading to 12 extra wave functions than the four solutions of Dirac equation for a free particle. We found that the entanglement of two particles or two antiparticles could be done only with opposite momenta and the entanglement of particle and antiparticle could be done only with the same momenta, where the latter was in contradiction with experiments, since experiments show the quantum correlation at a distance of a particle-antiparticle system like kaon and antikaon system which are entwined. Therefore, we came about to correct the obtained results. It seemed us as if something was missed there. Due to the theory of Dirac, antiparticles are believed to be particles of negative energy. But, according to the fact that antiparticles are detectable, so the physical antiparticles must be of positive energies i.e. taking the negative energies as antiparticles, is not covering all the underlying physics [1]. So, in order to correct the results, along with a previous investigation [1], we realized particles and antiparticles as physical entities with positive energy instead of considering antiparticles with negative energy. Then, as an application of modified descriptions for entangled (space-time) states, we discussed the original version of EPR paradox and verified the correct answer based on the strong rooted complex quantum Hamilton-Jacobi theory [2-27] and as another example we mentioned that applying negative energy states removes the Klein's paradox without the need of any further explanations or justifications like backwardly moving electrons [1,42]. Finally, comparing the two approaches, we pointed out the existence of a connection between quantum Hamiltonian dynamics, standard quantum field theory, and Krein space quantization [28-43].

Acknowledgement
I would like to thank the organizers of DICE 2014, especially Prof. Hans Thomas Elze.

References

[1] Payandeh F 2014 *Mod. Phys. Lett. A* **29** 18
[2] El Naschie M S 2005 *Int. J. Nonlinear. Sci. Simulat.* **6** 95
[3] El Naschie M S 1995 *Chaos Solitons Fract.* **5** 1031
[4] El Naschie M S 1995 *Chaos Solitons Fract.* **5** 1551
[5] El Naschie M S 2000 *Chaos Solitons Fract.* **11** 1149
[6] Mejias A, Sigalotti L, Di G, Sira E and de Felice F 2004 *Chaos Solitons Fract.* **19** 773
[7] Sigalotti L, Di G and Mejias A 2006 *Int. J. Nonlinear. Sci. Numer. Simulat.* **7** 467
[8] Czajko J 2000 *Chaos Solitons Fract.* **11** 1983
[9] Yang C D 2007 *Chaos Solitons Fract.* **33** 1073
[10] Yang C D 2007 *Chaos Solitons Fract.* **32** 274
[11] Yang C D 2005 *Ann. Phys.* **319** 444
[12] Yang C D 2007 *Chaos Solitons Fract.* **32** 312
[13] Yang C D 2005 *Ann. Phys.* **319** 399
[14] Yang C D and Wei CH 2007 *Chaos Solitons Fract.* **33** 118
[15] Yang C D 2007 *Int. J. Nonlinear. Sci. Numer. Simulat.* **8** 397
[16] Yang C D 2006 *Chaos Solitons Fract.* **30** 41
[17] Yang C D 2008 *Chaos Solitons Fract.* **38** 316
[18] Yang C D 2006 Ann. Phys. **321** 2876
[19] Ni G-j, Guan H 1999 arXiv:quant-ph/9901046v1
[20] Klein O 1929 *Z. Phys.* **53** 157
[21] Dirac P A M 1928 *Proc. Roy. Soc.* **117** 612
[22] Galogeracos A and Dombey N 1999 *Contemp. Phys.* **40** 313
[23] Dombey N and Galogeracos A 1999 *Phys. Rep.* **315** 41
[24] Galogeracos A and Dombey N 1999 *Int. J. Mod. Phys. A* **14**631
[25] Dirac P A M 1942 *Bakerian lecture Proc. Roy. Soc. Lond. A* **180** 1
[26] Dirac P A M 1933 *Nobel lecture*, December 12
[27] Scully M O and Walther H 1994 *Phys. Rev. A* **49** 3
[28] Calarco T, Cini M and Onofrio R 1999 *EPL* **47** 407
[29] Rouhani S and Takook M V 2009 *Int. J. Theor. Phys.* **48** 2740
[30] Takook M V 2001 *Mod. Phys. Lett A* **16** 1691
[31] Takook M V 2002 *Int. J. Mod. Phys. E* **11** 509
[32] Payandeh F 2009 *Rev. Cub. Fis.* **26** 232
[33] Payandeh F, Mehrafarin M and Takook M V 2009 *Sci. Chin. G: Phys. Mech. Astron.* **52** 212
[34] Payandeh F, Mehrafarin M, Rouhani S and Takook M V 2008 *Ukr. J. Phys.* **53** 12
[35] Payandeh F 2011 *J. Phys. Conf. Ser.* **306** 012054
[36] Payandeh F, Moghaddam Z Gh and Fathi M 2012 *Fortschr. Phys.* **60** 1086
[37] Payandeh F 2012 *ISRN High Energy Phys.* **2012** 714
[38] Pejhan H, Tanhayi M R and Takook M V 2014 *Ann. Phys.* **341** 195
[39] Refaei A and Takook M V 2011 *Phys. Lett. B* **704** 326
[40] Refaei A and Takook M V 2011 *Mod. Phys. Lett. A* **26** 31
[41] Forghan B, Takook M V and Zarei A 2012 *Ann. Phys.* **327** 2388
[42] Payandeh F, Mohammad Pur T, Fathi M and Moghaddam Z Gh 2013 *Chin. Phys. C* **37** 113103
[43] Payandeh F, Submitted to *Ann. Phys.*

Journal of Physics: Conference Series **626** (2015) 012054 doi:10.1088/1742-6596/626/1/012054

The principle of relativity, superluminality and EPR experiments.

"Riserratevi sotto coverta ..."

B Cocciaro

Department of Physics Enrico Fermi, Largo Pontecorvo 3, I-56127 Pisa, Italy

E-mail: cocciaro@df.unipi.it

Abstract. The principle of relativity claims the invariance of the results for experiments carried out in inertial reference frames if the system under examination is not in interaction with the outside world. In this paper it is analysed a model suggested by J. S. Bell, and later developed by P. H. Eberhard, D. Bohm and B. Hiley on the basis of which the *EPR* correlations would be due to superluminal exchanges between the various parts of the entangled system under examination. In the model the existence of a privileged reference frame (*PF*) for the propagation of superluminal signals is hypothesized so that these superluminal signals may not give rise to causal paradoxes. According to this model, in an *EPR* experiment, the entangled system interacts with the outer world since the result of the experiment depends on an entity (the reference frame *PF*) that is not prepared by the experimenter. The existence of this privileged reference frame makes the model non invariant for Lorentz transformations. In this paper, in opposition to what claimed by the authors mentioned above, the perfect compatibility of the model with the theory of relativity is strongly maintained since, as already said, the principle of relativity does not require that the results of experiments carried out on systems interacting with the outside world should be invariant.

1. Introduction

In the scientific community the impossibility to send superluminal signals has been accepted nearly unanimously for over a century. Generally this statement is considered a consequence of the theory of relativity. For example A. Einstein supports the impossibility to send superluminal signals in a paper of 1907 [1] and he bases his thesis exactly on claims that he considers consequences of the theory of relativity. The assumption of the non existence of superluminal signals plays a fundamental role in interpreting the results of the experiments about the violation of Bell's inequality. In those experiments measurements on two (or more) particles of an entangled system are performed. The measurements appear correlated in a way so as to violate Bell's inequality and the violation shows that the correlations cannot only be due to "local variables", that is "features" present in the particles since their creation. Something else must exist, and the simplest hypothesis to imagine is that there is a sort of communication among the various parts of the entangled system. In 1982, with Aspect's experiment, the violation of Bell's inequality is shown in such a condition to make the exchange of subluminal signals among the various parts of the system impossible: if the "something else" is the exchange of signals between particles of the entangled system or between parts of the experimental apparatus, then those signals must necessarily travel at superluminal speed. Once the impossibility of superluminal signals is accepted, the result of the Aspect's experiment has been seen by nearly all scientists as the final defeat of the supporters of local realism: the non local aspects of the quantum mechanics, that is the existence of distant correlations without local variables or communication, have been considered incontrovertible facts, experimentally proven.

In this paper I will analyse some statements usually given in support of the impossibility to send superluminal signals. I will start from the analysis of the arguments quoted by Einstein in the above mentioned paper of 1907, arguments that, in my opinion, should be considered wrong. I will then move on to examine the theorems about the alleged paradoxicality of superluminal signals which have been reported in literature since 1950s. As we will see, those theorems show the existence of causal paradoxes in the hypothesis that the superluminal signals are produced "below decks", that is their production and propagation depend exclusively on the operation the experimenter should perform in order to produce those signals. For instance, the superluminal signal would not appear paradoxical if there was a "given" privileged reference frame for them (as the air reference is for sound signals), independent of any operation the experimenter carries out to prepare an experiment. I will then analyse a realist and local model suggested by J. S. Bell [2], later developed by P. H. Eberhard [3], D. Bohm and B. Hiley [4], according to which the correlations observed in the experiments on the violation of Bell's inequality would be due to exchanges of superluminal signals between the various parts of the entangled system; superluminal signals for which the authors hypothesize the existence of a privileged reference frame just in order to avoid causal paradoxes. The existence of such a privileged reference frame makes the model non invariant for the Lorentz transformations and this, according to the above mentioned authors, would put the model in contrast with the theory of relativity. The main purpose of this paper is specifically to disprove this latter statement: I will try to explain why that model should be considered *in perfect accord* with the theory of relativity. I will start by reconsidering the principle of relativity presented by Galileo [5] in 1630, a principle which I believe should be put as basis to the theory of relativity as given by Galileo. The principle states the invariance of the results of the experiments performed in inertial reference frames if the experimental apparatus is, in Galileo's own words, "below decks". The need to be "below decks", that is with the experimental apparatus unaffected by entities "outside"[1] the reference frame in which the experiment is being carried out, might be considered trivial (obviously possible different external influences will produce different results), nevertheless it is a necessary hypothesis in order to state the invariance of the results of the experiments. It follows that possible models non invariant for the Lorentz transformations will be in accord with the principle of relativity in the event that the non invariant effects were, at least partly, caused by entities which were not "below decks". And this is exactly the case for the model under study whose non invariant aspects are due to the presence of a privileged reference frame which is not below decks, i.e. which cannot be prepared by the experimenter.

I have already presented a study [6] in which I claim the thesis contained in this paper. Although some parts of this paper are repetitions of others already presented in [6], I will try here not to dwell on those parts already developed in details in [6] but to focus on the ideas only hinted in my previous study.

[1] I believe that the concept of "below decks" introduced by Galileo should be further developed. In my opinion everything the experimenter can access should be considered "below decks". Entities the experimenter cannot affect are not below decks even if they are "present" inside the reference frame in which the experiments are performed. It is not to be considered below decks a possible privileged reference frame (like the one which, in the model examined in this paper, is assumed to support the propagation of superluminal signals) which cannot be either shielded or dragged, i.e. that can't be "prepared" by the experimenter at his will.

2. Conventionality of simultaneity

The issue of the conventionality of the simultaneity emerged with the appearance of the theory of relativity. Among the main champions of the conventionalist thesis there was definitely Hans Reichenbach who presented his ideas in several papers, among which *Philosophie der Raum-Zeit-Lehre* [7] of 1928 is probably the most famous. The debate became passionate only some years later, as M. Jammer pointed out [8] (page 192):

> It was only in the late 1950s that Reichenbach's philosophy of space and time, and his thesis of the conventionality of simultaneity, found the attention they deserve.

In my opinion, to this day, a considerable part of the scientific community ignores the principal points of the debate about the conventionality of simultaneity and we may often read scientific papers in which the anticonventionalist thesis is implicitly taken for granted. Such a tacit assumption could only be accepted if the debate on the issue had ended with the confutation of the conventionalist thesis, which is surely not true:

> The debate about conventionality of simultaneity seems far from settled, although some proponents on both sides of the argument might disagree with that statement.[9]

I believe that the studies in which the anticonventionalist thesis is implicitly assumed must be considered incorrect. We will later see remarkable examples of studies to be considered wrong in my opinion. Let's first sum up the terms of the matter.

2.1. The terms of the matter

In his very famous article of 1905 [10], A. Einstein initially assigns a triplet of coordinates $\vec{x} = (x_1, x_2, x_3)$ to every single point in an inertial reference frame by using rigid rulers. Then, in order to describe the motion of a body in the given reference frame, he sets a clock in every point and the motion of the body is described by the instant t marked by the clock set in $\vec{x}$ in the moment the body is in $\vec{x}$. To make this description physically relevant (so that, for example, the motion could be reproduced through the description) it is necessary *to synchronize* the various clocks set in the various points of the reference frame. Einstein clearly states his opinion on this matter pointing out that the synchronization is set *by definition*[2] establishing that the velocity of light from a certain point A towards a point B is the same as the velocity of light from B towards A, i.e. calling t_A the instant marked by the clock set in A when a light beam starts from A and t'_A the instant marked by the same clock when the beam returns in A after being reflected in B, we say that the clock in B is synchronized with the one in A if it marks the instant

$$t_B = t_A + \frac{t'_A - t_A}{2} \tag{1}$$

at the moment the beam is reflected in B. The "time" thus defined, that is the set of instants marked by the clocks fixed in the reference frame and synchronized according to (1), is called by Einstein "the time of the stationary system". Hans Reichenbach will later stress the conventional aspect of the "time of the stationary system" being the synchronization of the clocks conventional: the physical content of the relativity would not change if we put

$$t_B = t_A + \varepsilon \left(t'_A - t_A \right), \tag{2}$$

with ε free parameter, instead of (1). By choosing the (2) a non standard synchronization is said to be adopted whereas the (1), i.e. the particular case $\varepsilon = 1/2$, results in what it is called standard synchronization (that is the one Einstein adopted in his 1905 study). The time interval $t'_A - t_A$ is not conventional, since it is the result of a measurement (the one taken by the clock

[2] The words "durch Definition" are highlighted in the original text.

fixed in A from the moment the light beam starts from A to the moment it returns in A after being reflected in B); however, the interval $t_B - t_A$ which depends on the value chosen for ε is conventional. Reichenbach also points out that the second postulate of the special relativity does not claim that the speed of light from A to B is the same as the one we find from B to A (the speed from one point to the other is called one-way) as this statement does not have physical content, that is it is not experimentally verifiable, since the one-way speed is a conventional entity. The second postulate states that the back and forth speed of light is always the same, that is if two lightbeams start simultaneously from A, to return there after reflecting one in B and the other in C, then the two lightbeams will get back simultaneously if $\overline{AB} = \overline{AC}$. The back and forth speed of light is indicated by c. For instance in (2), we can set $t'_A - t_A = \frac{2\overline{AB}}{c}$. Actually Reichenbach does not leave the parameter ε completely free, he sets $0 < \varepsilon < 1$ so that, from his point of view, the time order of casually linked events always agrees with the causal order, i.e. it is always $t_E > t_C$ if the cause occurs in the point C when the clock fixed in C is marking the instant t_C and the effect occurs in the point E when the clock fixed in E is marking the instant t_E. Reichenbach obtains this result assuming the non existence of superluminal signals, assumption he considers "not an arbitrary assumption but a physical law based on experience. [...] In order to accelerate a body to the velocity of light, an infinite supply of energy would be required, and it is therefore physically impossible for any object to attain this speed" ([7] pag. 203). However, the fact that it is impossible to accelerate a mass particle to the velocity of light does not imply that superluminal signal could not exist. A superluminal signal could start as such and always remain superluminal until it is absorbed. It does not necessarily need to be generated as subluminal to become superluminal after acceleration. Mass particles must necessarily have subluminal velocity for the reason given by Reichenbach quoted above (they would need infinite energy to reach the light velocity); however, nothing imposes to associate mass to a superluminal signal. The condition $0 < \varepsilon < 1$ is no longer imposed in modern studies, at least not by all the supporters of the conventionalist thesis. For example in [11], page 130, we can read

> it should be pointed out that one should distinguish "spatially coincident causality" from "distant causality"; there is no contradiction if an occurrence at P at time t causes another occurrence at Q≠P at time $t' < t$ because the two different times are measured at spatially different locations. Indeed, as mentioned in Section 1.5.1, such apparent inconsistencies are familiar consequences of the International Date Line for airline travellers.
>
> The rejection of a temporal ordering in distant causality espouses a point of view in which time at any spatial point flows independently of time at other points, with there being no canonical prescription for the way one links the times at spatially separated points.

In substance, the (2) could be generalized in

$$t_B = t_A + \frac{\overline{AB}}{c} + \phi(\vec{x}_B) - \phi(\vec{x}_A) \quad (3)$$

and there is no "canonical prescription" for the function $\phi(\vec{x})$. In particular, the time sequence between two distant events can be set as wanted subject to the appropriate choice of the function $\phi(\vec{x})$. But the time order of distant events has nothing to do with the causal order. Causally linked events obviously maintain their causal link regardless of our descriptive choices. The example of the planes travelling westwards quoted in [11] in the extract above is quite clear: departure and arrival are causally linked being the former the cause and the latter the effect. This causal link remains obviously even if the departure takes place in New Zealand at three o'clock (time given by the clocks fixed in New Zealand) and the arrival in Rome at two o'clock (time marked by the clocks set in Rome). By choosing the (2) or the (3), instead of (1), there would only be descriptive complications although the physical content of the theory would be unaltered. In other words, it is not nature which "imposes" (1), it is us who choose it to deal with easier descriptions; we could as well choose the (2), or the (3), without facing any substantial problem since descriptive simplicity does not have any physical value. As an example, in the paragraph 2.3.3 of [11] we can find the equations of electromagnetism if we choose to synchronize employing (3) with $\phi(\vec{x}) = -\vec{k} \cdot \vec{x}$ being $\vec{k}$ a generic constant vector.

As presented above, the conventionalist thesis is not universally accepted at present. Surely it is not universally rejected, either. It is not my intention to join the debate and I refer the readers interested in the topic to the quoted references [8, 9, 11] as well as to the substantial bibliographical notes given in those papers. I just confine myself to offer some quotations and some final remarks.

In 1928 Albert Einstein published a review [12] of Reichenbach's book *Philosophie der Raum-Zeit-Lehre* [7] where he also said:

> special care has been taken to ferret out clearly what in the relativistic definition of simultaneity is a logically arbitrary decree and what in it is a hypothesis, i.e., an assumption about the constitution of nature. (quoted in [8] pag.181).

In the famous book *Relativity; the special and the general theory, a popular exposition* [13] Einstein wrote also:

> That light requires the same time to traverse the path A→M as for the path B→M [M being the midpoint of the distance AB] is in reality neither a supposition nor a hypothesis about the physical nature of light, but a stipulation which I can make of my own free will in order to arrive at a definition of simultaneity.

Marco Mamone Capria writes his ideas critical of the conventionalist thesis in a study of 2001 [14] where he also states (page 777):

> The debate on this issue has been going on for decades without quite reaching a satisfactory conclusion, though it appears that the CS [Conventionality of Simultaneity] thesis has somewhat conquered the status of a majority view - not always for good reasons, as we shall see.

So, according to a critic of the conventionalist thesis, this thesis "appears the majority view."

My personal opinions at the end of this paragraph represent the introduction to the following one. The time order between distant events induced by any synchronization can't be associated to the causal order without supporting such an association with a demonstration. For example we could follow Reichenbach's demonstration mentioned above: we could believe that the non existence of superluminal signals is experimentally verified and on the basis of this assumption we would demonstrate that the causal order between distant events is always in accord with the time order brought about through standard synchronization, as well as through all synchronizations induced by the (3) under the condition $\left|\vec{\nabla}\phi\right| < 1$. Certainly, because of a clear problem of circularity, the non existence of superluminal signals cannot be demonstrated simply because those signals might travel "backwards in time", that is they could give rise to effects in a certain point E at the instant t_E, since the cause occurred in a point $C \neq E$ at the instant $t_C > t_E$. The causal order is definitely not conventional. The time order between distant events brought about through standard synchronization, as well as through any other synchronization, arises like a conventional order to which it is a mistake to give the value of causal order without supporting it with any justification. We will see that outstanding examples of such mistakes can be often found in literature. The conventional origin of time order is ignored and an anticonventionalist approach is implicitly and often unconsciously taken: standard synchronization is given the status of "correct" synchronization and the time order induced through the standard synchronization is given the value of causal order.

2.2. Some mistakes

2.2.1. Einstein. In 1907 Einstein publishes a study in which he introduces a demonstration about the impossibility of the existence of superluminal signals later recalled by numerous authors such as: Tolman [15] (pagg. 54-55), Pauli [16] (pag. 16), Von Laue [17] (pag. 70). Adopting the standard synchronization, Einstein analyses two events, *caus* and *eff*, which are described through the coordinates $caus = (ct_c, x_c)$ and $eff = (ct_e, x_e)$ in a given inertial reference frame Σ. Einstein hypothesizes that it is

$$ct_e > ct_c \tag{4}$$

and that the two events are "space like", that is

$$c^2 (t_e - t_c)^2 - (x_e - x_c)^2 < 0. \tag{5}$$

From the (4) and (5), supposing it is $x_e > x_c$, it follows

$$x_e - x_c > ct_e - ct_c. \tag{6}$$

It is hypothesized that the two events are linked causally, i.e. *eff* is the effect and *caus* the cause. At this point Einstein describes the two events in a new inertial reference frame, Σ', in motion at the speed βc in respect to Σ, with $|\beta| < 1$. The coordinates of the two events in the reference frame Σ', $caus = (ct'_c, x'_c)$ and $eff = (ct'_e, x'_e)$, are obtained by using standard Lorentz transformations, i.e. Lorentz transformations in the form they take if standard synchronization is assumed[3]. In particular it is obtained

$$ct'_e - ct'_c = \gamma \left[(ct_e - ct_c) - \beta (x_e - x_c)\right] \tag{7}$$

with $\gamma = 1/\sqrt{1-\beta^2}$. From the (7) it is obtained:

$$\beta > \frac{ct_e - ct_c}{x_e - x_c} \Longleftrightarrow ct'_e < ct'_c \tag{8}$$

that is, considering the (6), if the two events are "space like", then it is always possible to obtain appropriate values of β, with $|\beta| < 1$, for which the time order between the events *eff* and *caus* will be discordant between Σ and Σ'. In the reference frame Σ' the cause *caus* is associated to a greater instant than the one associated to the effect *eff*. This is deemed impossible by Einstein (and by many authors who have resumed his demonstration) and hence it can be inferred that space like events can't be linked causally. Einstein's words are the following (in the translation quoted in [18]):

> If, as we have assumed $W > c$ [i.e. (6) in our notations], one can always choose v [β] such that $T < 0$ [$ct'_e < ct'_c$]. This result means that we would have to consider as possible a transfer mechanism whereby the achieved effect would precede the cause. Even though this result, in my opinion, does not contain any contradiction from a purely logical point of view, it conflicts with the character of all our experience to such an extent that this seems sufficient to prove the impossibility of the assumption $W > c$

[3] Assuming a certain non standard synchronization, Lorentz transformations change their form keeping their physical content. In [11] (pag. 129) there is the form taken by Lorentz transformations between two reference frames Σ and Σ' in which it is synchronized through the (3), with $\phi(\vec{x}) = -\vec{k} \cdot \vec{x}$ in Σ and $\phi(\vec{x}) = -\vec{k}' \cdot \vec{x}$ in Σ', being $\vec{k}$ e $\vec{k}'$ generic constant vectors.

Journal of Physics: Conference Series **626** (2015) 012054 doi:10.1088/1742-6596/626/1/012054

In 1905 Einstein defines conventionally the "time of the stationary system" then, in 1907, it is precisely on it that he bases his statement "the achieved effect would precede the cause". As if the causality might be derived from a "stipulation which I can make of my own free will", in the words Einstein himself will use, as we have seen, in 1924. It is like defining conventionally a certain synchronization for the clocks fixed on the Earth and then establishing that it is impossible that a plane departed from New Zealand at three o'clock arrives in Rome at two because, if that happened, "the achieved effect would precede the cause". Naturally an anticonventionalist approach might always be assumed and then claim that the clocks fixed on the Earth are not synchronized in the "correct" way and that if all the clocks were set at GMT, it would then be true that neither planes nor other signals of any kind departed from a certain point at the instant t_c in a given inertial reference frame Σ could reach another point at an instant $t_e < t_c$; nevertheless, such a statement would have to be demonstrated. Anyway Einstein never assumes an anticonventionalist approach; I would therefore say that the demonstration given above must necessarily be considered incorrect.

2.2.2. The principle of reinterpretation. From the 1960s studies [19–23] have been introduced in scientific literature in which the existence of superluminal particles (called "tachyons") is hypothesized and they claim that it is possible to overcome the "paradox" we have just seen on the basis of a principle they call principle of reinterpretation. The alleged paradox consists in the fact that it is always possible to find an appropriate reference frame Σ' so that, if a tachyon travels "forward in time" in the reference frame Σ, it will then travel "backward in time" in Σ'. It is totally overlooked the fact that the expressions "forward in time" and "backward in time" are based on the "time of the stationary system", in its turn defined through a synchronization (the standard one) assumed conventionally and those expressions are given as if they had an evident physical meaning. In particular, as Einstein had already done in his study in 1907, it is assumed, as something not to be justified at all, that the travels "backward in time" are not possible. This time, however, Einstein's statement is not deduced from this assumption. This time a mistake is made which, in my opinion, is far more serious. A principle is stated, on the basis of which cause and effect would be "written" in the time order of the "time of the stationary system" and if this order in Σ' is discordant from the one we have in Σ, then in Σ' the role of cause and effect must be "reinterpreted". Namely, the event which is cause in Σ is "reinterpreted" as effect in Σ'. The authors believe they can justify this principle on the basis that the tachyons travelling "backward in time" are always associated to negative energy and once they have been "reinterpreted", they would travel "forward in time" with positive energy. Personally I believe that the "principle of reinterpretation" is a true violence to the concept of causality. The causal order has physical value and it is the time order induced through any synchronization which must be "reinterpreted", if the existence of tachyons is to be hypothesized.

2.2.3. The multisimultaneity. In 1997 A.Suarez and V. Scarani [24] proposed an experiment aimed at evaluating a description of the phenomenon of the entanglement alternative to the one given by the standard quantum mechanics. The name multisimultaneity was given to such alternative description. The proposed experiment was then carried out by the Gisin group [25] and the result was in perfect accord with the standard quantum mechanics.
The issue is summarised perfectly on the webpage [26] from which some excerpts are quoted below:

> Entangled particles behave as if there was a *faster than light connection* between them. [...]
>
> I was strongly convinced that it should be possible to give a time-ordered causal explanation of nonlocal correlations, in terms of "before" and "after". Indeed, such

> a description is very well possible in conventional Bell experiments, in which all apparatuses are standing still in a laboratory frame. Since the emission time of the photons is not exactly the same, and the fibers guiding the photons from the source to the measuring devices don't have exactly the same length, according to the clock defined by the laboratory's inertial frame, one of the measurements always takes place before the other, and the particle arriving later can be considered to take account of the outcome of the one arriving before.[...]
>
> But what about experiments with moving apparatuses in which several relevant frames are involved? In this case, different clocks watch the arrival times, and what is "after" according to the laboratory clock may become "before" according to one moving clock. Then, it is possible to define other time orderings: If each measuring device in its own reference frame is the first to select the output of photons, we have before-before timing. If each measuring device in its own reference frame selects the photon output after the other, we have after-after timing. Is it also possible to give a time-ordered causal explanation for relativistic experiments using apparatuses in motion? I assumed it was, and developed an alternative nonlocal description termed "multisimultaneity", in which the time-ordered description of the nonlocal correlations extends to experiments with before-before and after-after timings. Consider, for instance, experiments in which the measuring devices are in motion in such a way that each of them, in its own reference frame, is the first to select the output of photons (before-before timing). Then each particle's choice will become independent of the other's and, according to multisimultaneity, the nonlocal correlations should disappear. In contrast, quantum mechanics requires that the particles stay non local, correlated independent of any timing, even in such a before-before situation.

Here too, the identity between causal order and time order is taken for granted without giving any supporting justification. It is hypothesized that the nonlocal correlations can have a "causal explanation", that the two measurements can be ordered causally with one occurring "before" the other, but the "before" is established by the "time of the stationary system". Here too, the causal order is "forced" to respect the time order induced through standard synchronization. So much to assume that the causal order must break up (bringing about the disappearance of nonlocal correlations) when the time orders of the two measurements in the corresponding reference frame are discordant. Here too, constraints are put to an order with physical content, the causal order, to safeguard an order which starts as conventional (and which continues, even after its appearance, to be considered necessarily conventional according to what "appears to be the majority view" among the people who publish papers on this topic).

3. Superluminal signals and causal paradoxes

The modern texts present the theorem about the link between superluminal signals and causal paradoxes in a different way from the one mentioned in the paragraph 2.2.1. What is considered paradoxical is the possibility to communicate with one's past [27] (pp. 52-53), [28] (pp. 186-189), [29] (p. 21), [30] (pp. 274-275), [31] (p. 55). This possibility would give rise to undeniable paradoxes which clearly have nothing to do with the procedure we decide to follow to synchronize the clocks (even if I decided to turn my clock back 48 hours, yesterday would still remain yesterday, it would not become tomorrow). The theorem presented by Møller [27] has a more general form; however, since I have already examined it in detail in [6], I will here analyse only the simplified form commonly quoted [28–31]. Recalling the example seen in the paragraph 2.2.1, we put

$$\beta_t \equiv \frac{x_e - x_c}{ct_e - ct_c}. \tag{9}$$

We can hypothesize that the events *caus* and *eff*, which in Σ have coordinates $caus = (ct_c, x_c)$ and $eff = (ct_e, x_e)$, are respectively emission and detection of a superluminal signal which has speed equal to $\beta_t c$ in Σ (from the (4)-(6) follows $\beta_t > 1$). We can call S_{go} this signal. From Lorentz transformations we obtain that, in the reference frame Σ', in motion at speed βc respect to Σ, the event *eff* has coordinates $eff = (ct'_e, x'_e)$ with

$$\begin{cases} ct'_e = \gamma\left(ct_e - \beta x_e\right) \\ x'_e = \gamma\left(x_e - \beta ct_e\right) \end{cases} . \tag{10}$$

At this point we *assume* that, in coincidence with the event *eff*, it is possible to send a new superluminal signal, S_{back}, in the negative direction of x', which has in Σ' speed $-\beta_t c$. More generally we could assume that,

(i) if it is possible to send a superluminal signal which has in Σ speed $\beta_t c$, then, in any other inertial reference frame, it must be possible to emit superluminal signals in any directions which have the same speed $\beta_t c$.

It is commonly believed that this assumption is just the direct consequence of the relativity principle. In Σ' the time law of the superluminal signal S_{back} is

$$x' = x'_e - \beta_t\left(ct' - ct'_e\right), \tag{11}$$

and from it, through Lorentz transformations, the time law of S_{back} in Σ can be easily obtained:

$$x = x_e - \frac{\beta_t - \beta}{1 - \beta\beta_t}\left(ct - ct_e\right). \tag{12}$$

Considering the (9), from the (12), the coordinates of a new event, eff_1, associated with the arrival of S_{back} in x_c, can be obtained:

$$eff_1 = \left(ct_{e_1} = ct_c + \frac{2\beta_t - \beta\left(1 + \beta_t^2\right)}{\beta_t\left(\beta_t - \beta\right)}\left(x_e - x_c\right), x_c\right). \tag{13}$$

So we have the chain of the following three events which are in causal order:
caus: departure of the signal S_{go} from the point which has abscissa x_c in Σ;
eff: arrival of S_{go} in the point which has abscissa $x_e > x_c$ in Σ, and simultaneous departure from the same point of the signal S_{back};
eff_1: arrival of S_{back} in the point which has abscissa x_c in Σ, i.e. the point from which S_{go} had departed.
If the instant ct_{e_1}, defined in the (13), could be less than ct_c, then it would be an evident causal paradox: in the point which has abscissa x_c in Σ, the effect eff_1 would occur "before" the cause *caus*. An observer set in Σ in the point with abscissa x_c could send a question through the signal S_{go} when his clock marks 3 and receive the answer through S_{back} when his clock marks 2. Here the situation is clearly different from the one described in the paragraph 2.2.1. Here the conclusion is totally independent of the procedure chosen to synchronize the clocks since now we do not compare instants marked by different clocks. Rather, instants marked by a single clock are compared, the clock fixed in Σ in the point with abscissa x_c. From the (13) it is easily obtained

$$ct_{e_1} < ct_c \Longleftrightarrow \beta > \frac{2\beta_t}{1 + \beta_t^2} \tag{14}$$

and, being $\frac{2\beta_t}{1+\beta_t^2} < 1 \forall \beta_t$, it follows that it is always possible to determine appropriate reference frames Σ', in motion at speed βc respect to Σ (with $|\beta| < 1$), so as to make the situation described above paradoxical. Hence we can say that it would always be possible to give rise to causal paradoxes if superluminal signals existed.

This demonstration is based on the assumption (i) reported above. We could say that it is based on the principle of relativity if we accepted that the (i) is the necessary consequence of the principle of relativity. If we accepted this approach we would have to say that, in order to make superluminal signals which don't give rise to causal paradoxes possible, we would have to deny the principle of relativity, thus questioning all the theories based on that principle like, for example, the theory of relativity which has exactly the principle of relativity among its postulates. In the following paragraph I will try to illustrate the reasons why, in my opinion, this approach is untenable: the assumption (i) can be violated without undermining even marginally the principle of relativity in any way.

4. The Principle of Relativity

The principle of relativity is stated by Galileo in *Dialogo Sopra i due Massimi Sistemi del Mondo* [5]:

> Shut yourself up with some friend in the main cabin below decks on some large ship, and have with you there some flies, butterflies, and other small flying animals. Have a large bowl of water with some fish on it; hang up a bottle that empties drop by drop into a narrow-mouthed vessel beneath it. With the ship standing still, observe carefully how the little animals fly with equal speed to all sides of the cabin. The fish swim indifferently in all direction; the drops fall into the vessel beneath; [...] When you have observed all these things carefully (though there is no doubt that when the ship is standing still everything must happen in this way), have the ship proceed with any speed you like, so long as the motion is uniform and not fluctuating this way and that. You will discover not the least change in all the effects named, nor could you tell from any of them whether the ship was moving or standing still.

In modern words, Galileo claims the invariance of the results for the experiments carried out in inertial reference frames[4]. Galileo points out that, in order to obtain the thesis, the experiments must be carried out "below decks". Galileo *does not* maintain the equivalence of all inertial reference frames! If we repeat a certain experiment on two different ships whose "motion is uniform and not fluctuating this way and that", we might as well obtain *different outcomes* if the experiment were not carried out "below decks" on both ships. The external environment, the one not directly connected with the experimental apparatus, might, as a matter of principle, interact somehow with some parts of the system causing asymmetries (the interaction with the outside will not necessarily be the same in both cases). For example, if a ship was aligned with the vertical and the other was leaning, even if both moved with uniform motion, experiments connected with gravity would have different results in the two ships: in one case a ball left stationary on a table would stay stationary, in the other it would start to roll. Although Galileo was not aware at his time, the gravitational effects result exactly from interactions with the outside which are not shielded[5]. We are not "below decks" in relation to gravitational interactions.

I believe that the concept of "below decks" should have the following meaning: every experiment on a certain physical system always consists in the preparation of some entities (causes) and the observation of some results from measurement operations (effects). It might be that the evolution

[4] After E. Fabri [32], with the words "reference frame" I mean a laboratory, a room, inside which all the instruments necessary to carry out any experiment are available (see also [6] par. 2).

[5] Here we are obviously using the metaphor of Galileo's ship from a point of view preceding the advent of the general relativity. According to the general relativity Galileo's ship cannot be considered an inertial reference frame since it is not in free fall.

of the physical system depends *only* on the causes appropriately prepared by the experimenter and, if that is true, then the outcomes of the measurement operations will always be the same. Being "below decks" means having shielded the system from every possible interaction with the external environment. In such a condition the evolution of the system will only depend on what is inside the reference frame, that is on what the experimenter can affect and has consequently been able to prepare appropriately. By giving this meaning to the concept of "below decks", claiming the validity of the principle of relativity would amount to claiming the validity of the causal determinism: if the same causes are repeated, the same effects will be obtained, regardless of "where" and also "when" the repetition of the causes takes place.
If a certain model should hypothesize the existence of a certain reference frame, privileged relatively to a certain set of phenomena *Ph*, then we obviously could not say that the model violates the principle of relativity. We should only not require the invariance for those experiments in which the phenomena *Ph* are somehow involved.

We might wonder if the principle of relativity, in the form above ascribed to Galileo, can be considered one of the postulates of the theory of relativity. In *The Meaning of Relativity* Einstein claims [33]:

> If *K* is an inertial system, then every other inertial system *K'* which moves uniformly and without rotation relatively to *K*, is also an inertial system; the laws of nature are in concordance for all inertial systems. This statement we shall call "principle of special relativity".

With "the laws of nature" Einstein clearly means all the natural laws, therefore also the law about the propagation of possible superluminal signals. Definitely this principle would be violated if we could demonstrate the existence of superluminal signals ruled by an "aether" which would determine a privileged reference frame for those signals. To hypothesize the existence of a privileged reference frame for the superluminal signals is the easiest way to imagine that the condition (i) does not count without incurring in the causal paradoxes examined above. The question we are posing is the following: in order to develop the theory of relativity do we need a "strong" postulate like the one stated by Einstein or a more general one like Galileo's is sufficient? Is the invariance to be required for all physical laws or only for those related to phenomena which can be reproduced below decks? In my opinion the answers are compulsory: in the form used by Einstein, the principle can be considered correct only if we imply that the reference frames *K* and *K'* are immersed in an environment without any role in the "laws of nature" we are speaking about. In other words, the principle stated by Einstein can be considered correct only if we imply that *K* an *K'* are "below decks". Following Einstein in the development of the theory of the special relativity we can see that he actually never uses the "strong" form of the principle of relativity. He assumes the non existence of a privileged reference frame for electromagnetic phenomena; he then deals with rulers and clocks as if there was not a privileged reference frame for them either, and on that basis he obtains a series of results. In substance we could say that the physics we do with rulers, clocks and electromagnetic phenomena is physics performed "below decks", therefore, on the basis of the principle of relativity, it will be physics for which the invariance of the results will be required. There will naturally be also other physics which can be performed below decks, for which the invariance will be required, but in no way does the principle of relativity impose that *all* physics must have the possibility to be performed below decks. The possible discovery of phenomena which cannot be reproduced below decks would leave unaltered the results of the theory of relativity, e.g. the fact that the descriptions of a phenomenon on two different inertial reference frames are linked through Lorentz transformations and not Galileo's ones.

The conclusion to be drawn is, in my opinion, inevitable: hypothesizing the existence of a privileged reference frame for the propagation of superluminal signals not only makes it possible to overcome cleverly the problems linked to the causal paradoxes mentioned above, it also makes those signals perfectly compatible with the theory of relativity. We will see in the next chapter that, on these basis, it is possible to build a model which enables us to represent *EPR* phenomena through a realist and local description.

5. A realist and local model for *EPR* phenomena

In an interview [2] John S. Bell gives the following answers to questions about Aspect's experiment:

> You can imagine that there is a preferred frame of reference, and in this preferred frame of reference things do go faster than light. But then in other frames of reference when they seem to go not only faster than light but backwards in time, that is an optical illusion [...] But that is certainly the cheapest solution. Behind the apparent Lorentz invariance of the phenomena, there is a deeper level which is not invariant. [...]
>
> The reason I want to go back to the idea of an aether here is because in these EPR experiments there is the suggestion that behind the scenes something is going faster than light [...]
>
> It introduces great problems, paradoxes of causality and so on. And so it' s precisely to avoid these that I want to say there is a real causal sequence which is defined in the aether (pagg 48-50).

Some years later P. H. Eberhard [3] and D. Bohm and B. Hiley [4] will present a model taking in Bell's suggestions. In [6] I have already quoted extracts from all these authors examining in details the various aspects of the model of which I will here only present the essential points. I will also introduce the main features the experiments designed to test the model must have. After N. Gisin [34], we will call these experiments "Salart-type" experiments.

The fundamental idea is that the collapse of the wave function is a process which originates in the point where a certain measurement is taken. If we imagine an *EPR* experiment on an entangled system with two particles, normally one will be measured before the other. From the point where the first measurement occurs a wave immediately propagates, a superluminal signal, which can reach the other particle before the corresponding measurement is performed on it. If that happens, the two measurements appear correlated according to what the quantum mechanics predicts, that is a violation of Bell's inequality occurs. The two measurements, instead, will not be correlated according to the predictions of the quantum mechanics (i.e. there will be no violation of Bell's inequality) if the "second" measurement is performed when the second particle still has not been hit by the superluminal signal which originated in the point in which the "first" measurement was taken[6]. Recent theoretical results [35, 36] show that, assuming the correctness of the model, even if we imagined these superluminal signals as "hidden", inaccessible to the macroscopic world, it would be possible to communicate at superluminal velocity: "the models based on hidden influences propagating at a finite speed $v > c$" allow "superluminal communication [which] does not require access to any hidden physical quantities, but only the manipulation of measurement devices"[35].

In order to overcome the problems connected to the causal paradoxes examined above the existence of a certain reference frame Σ' is hypothesized, which proves privileged as to the propagation of superluminal signals. The "privilege" consists in the fact that only in Σ' the propagation of the superluminal signals proves to be isotropic, that is, given two generic point A and B set in Σ', if from the point A a light beam Fl and a superluminal signal Ft are sent simultaneously, then Ft will reach B with a certain advance respect to Fl and the advance will be directly proportional to the length of the segment $\overline{AB}$. In particular the advance will prove independent of the direction of the segment $\overline{AB}$. In substance, it is hypothesized that an aether

[6] Actually in such a case it does not make much sense to order the two measurements since there is no causal relation between them. Neither measurement affects the other.

analogous to the one we have for sound signals exists for superluminal signals. If we hypothesize the standard synchronization in Σ' we can say that a superluminal signal has in Σ' a certain speed $\beta_t c$, with $\beta_t > 1$, regardless of its direction of propagation. The description of an *EPR* experiment on an entangled system with two particles proves particularly simple in Σ'. We can call e_A and e_B the events associated to the two measurements. The coordinates associated to the two events in the reference frame Σ' (in standard synchronization) are

$$\begin{cases} e_A : (ct'_A, \overrightarrow{x}'_A) \\ e_B : (ct'_B, \overrightarrow{x}'_B) \end{cases} . \tag{15}$$

In order to avoid that the two measurements may give rise to a violation of Bell's inequality the superluminal signal starting from $\overrightarrow{x}'_A$ at the instant ct'_A is to arrive in $\overrightarrow{x}'_B$ when the event e_B has already taken place, that is it has to be:

$$ct'_B < ct'_A + \frac{|\overrightarrow{x}'_B - \overrightarrow{x}'_A|}{\beta_t c}. \tag{16}$$

In the (16) the hypothesis of isotropy has been used, that is the fact that in Σ' the superluminal signals have speed $\beta_t c$ in any direction. In order not to have a violation of Bell's inequality also the superluminal signal starting from $\overrightarrow{x}'_B$ at the instant ct'_B is to arrive in $\overrightarrow{x}'_A$ when the event e_A has already occurred:

$$ct'_A < ct'_B + \frac{|\overrightarrow{x}'_A - \overrightarrow{x}'_B|}{\beta_t c}. \tag{17}$$

The (16) and (17) are summed up in

$$\left|ct'_B - ct'_A\right| < \frac{|\overrightarrow{x}'_B - \overrightarrow{x}'_A|}{\beta_t c}. \tag{18}$$

To verify the validity of the model under examination an *EPR* experiment should be performed in a condition so as to satisfy the (18) and the prediction of the model is that if the (18) is satisfied then the measurements will satisfy Bell's inequality in contrast with the predictions of quantum mechanics. The (18) will be surely satisfied if the two measurements are simultaneous (in standard synchronization) in Σ':

$$\Delta ct' \equiv ct'_B - ct'_A = 0. \tag{19}$$

However, experimental uncertainties will never allow that the (19) is verified exactly; it will always be

$$\left|\Delta ct'\right| < \delta' \tag{20}$$

where δ' is the experimental uncertainty with which the instants ct'_A and ct'_B can be equalized. From the (18) we gather that the experiment would allow the verification of the effect hypothesized by the model (the verification of Bell's inequality in contrast with the hypothesis of quantum mechanics) if

$$\beta_t c < \beta_{t,max} c \equiv \frac{|\overrightarrow{x}'_B - \overrightarrow{x}'_A|}{\delta'}. \tag{21}$$

Any experiment would give a null result if the superluminal signals were "too fast" i.e. if $\beta_t > \beta_{t,max}$. To improve an experiment, that is to enable to investigate increasingly greater values of β_t, the value of $\beta_{t,max}$ defined in the (21) must be made as great as possible, that is the farthest possible simultaneous measurements ($|\overrightarrow{x}'_B - \overrightarrow{x}'_A|$ great) must be performed with the smallest possible uncertainty about the simultaneity (δ' small).

Obviously the privileged reference frame Σ' is unknown and any experiment will be performed in the reference frame of the laboratory Σ which will move at a certain unknown speed $\vec{\beta} c$ in relation to Σ'. In Σ the coordinates associated to the events e_A and e_B will be:

$$\begin{cases} e_A : (ct_A, \vec{x}_A) \\ e_B : (ct_B, \vec{x}_B) \end{cases} . \tag{22}$$

From the Lorentz transformations we gather important information on the conditions to be followed in Σ so that the measurements are simultaneous in Σ'. Particularly we obtain

$$\Delta ct' = \gamma \left(\Delta ct + \vec{\beta} \cdot \Delta \vec{x} \right) \tag{23}$$

with $\gamma = 1/\sqrt{1-\beta^2}$, $\Delta ct \equiv ct_B - ct_A$ e $\Delta \vec{x} \equiv \vec{x}_B - \vec{x}_A$. To annul $\Delta ct'$ both addends to the right of the (23) can be annulled, however, being $\vec{\beta}$ unknown, it is not known the direction the vector $\Delta \vec{x}$ should have so as to annul the product $\vec{\beta} \cdot \Delta \vec{x}$. This problem can be solved taking advantage of the Earth's daily rotation: if we place the vector $\Delta \vec{x}$ in direction East-West, regardless of the direction of $\vec{\beta}$, every day there will always be two instants which will give $\vec{\beta} \cdot \Delta \vec{x} = 0$. The experimental procedure to comply with will then be the following: the points in which the measurements will be carried out, $\vec{x}_A$ and $\vec{x}_B$, are placed in direction East-West and simultaneous measurements (in standard synchronization) in Σ are performed continuously at least for 12 hours. We wait for the "right" instant, the one which should give rise to correlations which do not violate Bell's inequality. Every pair of measurements will have a time duration δt. In Σ too we will obviously have a certain experimental uncertainty δ about the simultaneity of the two measurements, i.e. it will be $\Delta ct = 0 \pm \delta$. Given $\bar{\rho} \equiv \frac{\delta}{|\Delta \vec{x}|}$ it is obtained [37, 38]:

$$\beta_{t,max} = \sqrt{1 + \frac{(1-\beta^2)(1-\bar{\rho}^2)}{\left(\bar{\rho} + \beta \sin\chi \sin\frac{\pi \delta t}{T}\right)^2}} \tag{24}$$

being $T = 24h$, $\beta = \left|\vec{\beta}\right|$ and χ the angle between $\vec{\beta}$ and the polar axis. From the (24) we obtain $\lim_{\beta \to 0^+} \beta_{t,max}(\beta) = 1/\bar{\rho}$ and $\lim_{\beta \to 1^-} \beta_{t,max}(\beta) = 1$. The minimum value of $\beta_{t,max}$ is obtained for $\chi = \pi/2$, besides, two regimes can be identified for the function $\beta_{t,max}(\beta, \chi = \pi/2)$:
regime 1, for $\beta \in \left(0, \bar{\rho}\frac{T}{\pi \delta t}\right)$, it is $\beta_{t,max}(\beta, \chi = \pi/2) \lesssim 1/\bar{\rho}$;
regime 2, for $\beta \in \left(\bar{\rho}\frac{T}{\pi \delta t}, 1\right)$, it is $\beta_{t,max}(\beta, \chi = \pi/2) \approx \frac{T}{\pi \delta t}\frac{1}{\beta}$.
Consequently it results that, in order to maximise $\beta_{t,max}$, we have to minimise $\bar{\rho}$ (to maximise the value obtained in the regime 1) and δt (to maximise the interval of β values which give rise to the regime 1). If $\delta t < \bar{\rho}\frac{T}{\pi}$ the regime 2 disappears and $\beta_{t,max}(\beta)$ keeps on the whole constant ($\beta_{t,max}(\beta) \approx 1/\bar{\rho}$) for nearly all the values of β then to approach 1 when β approaches 1. In figure 1 some graphs of the function $\beta_{t,max}(\beta, \chi = \pi/2)$ are shown.

At present three "Salart-type" experiments [37, 39, 40] have been carried out to which a fourth one could be added [41] which had, however, the problem of $\Delta \vec{x}$ being not parallel to the direction East-West. All these experiments have not highlighted any discrepancies with the predictions of the quantum mechanics. From the experimental tests performed so far, we can therefore infer that either the model is incorrect, or, if it is correct, the value β_t must be greater that the values $\beta_{t,max}$ investigated in the various experiments (see figure 1). We can note that only in [40] Bell's inequality is really evaluated in every measurement, in [37, 39] correlations that make the violation of Bell's inequality plausible are observed and no variation of those correlations occurs during the 24 hours' measurement.

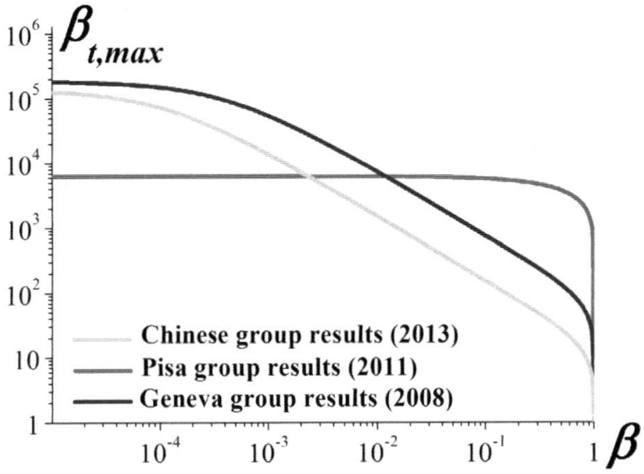

Figure 1: Graph of the function $\beta_{t,max}(\beta)$ defined in the (24) with $T = 24h$, $\chi = \frac{\pi}{2}$, $\bar{\rho}$ and δt obtained from the experimental conditions of the 3 Salart-Type experiments carried out so far. More specifically we have $\bar{\rho} = 5.4 \cdot 10^{-6}$ and $\delta t = 360s$ ($\bar{\rho}\frac{T}{\pi \delta t} \cong 4 \cdot 10^{-4}$) for the experiment of the Geneva group [37] (blue graph), $\bar{\rho} = 1.6 \cdot 10^{-4}$ and $\delta t = 4s$ ($\bar{\rho}\frac{T}{\pi \delta t} \cong 1$) for the experiment of the Pisa group [39] (red graph), $\bar{\rho} = 7.28 \cdot 10^{-6}$ and $\delta t = 1800s$ ($\bar{\rho}\frac{T}{\pi \delta t} \cong 1 \cdot 10^{-4}$) for the experiment of the Chinese group [40] (green graph).

6. Conclusions

J. S. Bell [2], P. H. Eberhard [3], D. Bohm and B. Hiley [4] agree on considering the model described above incompatible with the theory of relativity because of Lorentz non invariance of the model itself. I believe this conclusion is vitiated by an interpretation of the principle of relativity which cannot be considered correct in spite of its large diffusion. As pointed out by Galileo, the principle of relativity claims the invariance *only* for phenomena which can be reproduced below decks. Whether phenomena which cannot be reproduced below decks exist or not is a matter of competence for nature, certainly it is not a matter we can establish as matter of principle. Indeed the principle of relativity, in Galileo's form, does not claim the non existence of phenomena which cannot be reproduced below decks, it simply states that the invariance must be required only for phenomena which can be reproduced below decks. Models which hypothesize the existence of phenomena ruled by an "aether" which cannot be dragged in the lab reference frame will necessarily be non invariant; this non invariance, however, is in no way in contrast with the principle of relativity. Should possible future experiments prove the correctness of the model discussed above, important aspects of the current quantum mechanics would surely have to be modified; nevertheless nothing of the theory of relativity would have to be changed.

Acknowledgments

I would like to thank N. Gisin for his suggestions and C. Marmorale for the translation of the paper from the original Italian version.

References

[1] Einstein A 1907 *Annalen der Physik* **328** 371–384

[2] Davies P C W and Brown J R 1993 *The ghost in the atom: a discussion of the mysteries of quantum physics* canto ed. ed (Cambridge; New York: Cambridge University Press)

[3] Eberhard P H 1989 A realistic model for quantum theory with a locality property *Quantum theory and pictures of reality: foundations, interpretations, and new aspects* ed Schommers W (Berlin; New York: Springer-Verlag)

[4] Bohm D and Hiley B J 1991 *The undivided universe: an ontological interpretation of quantum mechanics* (Routledge)

[5] Galilei G 2001 *Dialogue concerning the two chief world systems, Ptolemaic and Copernican* (New York: Modern Library)

[6] Cocciaro B 2013 *Physics Essays* **26** 531–547
[7] Reichenbach H 1958 *The philosophy of space & time* (New York: Dover Publications)
[8] Jammer M 2006 *Concepts of simultaneity from antiquity to Einstein and beyond.* (Baltimore, Md.: Johns Hopkins University Press)
[9] Janis A 1998 URL `http://plato.stanford.edu/archives/fall2014/entries/spacetime-convensimul/`
[10] Einstein A 1905 *Annalen der Physik* **322** 891–921
[11] Anderson R, Vetharaniam I and Stedman G 1998 *Physics Reports* **295** 93 – 180 ISSN 0370-1573
[12] Einstein A 1928 *Deutsche Literaturzeitung* 19–20
[13] Einstein A 1924 *Relativity; the special & the general theory, a popular exposition,* (London: Methuen & Co.) translated by Robert W Lawson
[14] Capria M 2001 *Foundations of Physics* **31** 775–818
[15] Tolman R C 1917 *The theory of the relativity of motion,* (Berkeley: University of California press)
[16] Pauli W 1921 Relativitätstheorie *Encyklopädie der Mathematischen Wissenschaften mit Einschluss ihrer Anwendungen* vol 5 Physik ed Sommerfeld A (Vieweg+Teubner Verlag) translated as *Theory of Relativity* (London. Pergamon, 1958)
[17] Von Laue M 1922 *La Theorie de la Relativité* (Paris: Gauthier-Villars)
[18] Weinstein G 2012 arXiv: 1203.4954 URL `http://arxiv.org/abs/1203.4954`
[19] Bilaniuk O M P, Deshpande V K and Sudarshan E C G 1962 *American Journal of Physics* **30** 718–723
[20] Feinberg G 1967 *Physical Review* **159** 1089–1105
[21] Bilaniuk O M and Sudarshan E C G 1969 *Physics Today* **22** 43–51
[22] Recami E and Mignani R 1974 *La Rivista Del Nuovo Cimento Series 2* **4** 209–290
[23] Recami E 1978 *Foundations of Physics* **8** 329–340
[24] Suarez A and Scarani V 1997 *Physics Letters A* **232** 9 – 14
[25] Zbinden H, Brendel J, Gisin N and Tittel W 2001 *Phys. Rev. A* **63**(2) 022111
[26] Suarez A 2003 URL `http://www.quantumphil.org/history.htm`
[27] Møller C 1955 *The theory of relativity.* (Oxford: Clarendon)
[28] Bohm D 1996 *The special theory of relativity* (Routledge)
[29] Regge T 1981 *Cronache dell'universo* (Torino: P. Boringhieri)
[30] Penrose R 1989 *The emperor's new mind* (Oxford University Press)
[31] Rindler W 2006 *Relativity - Special, General, and Cosmological* 2nd ed (Oxford Univ. Press)
[32] Fabri E 2005 *Insegnare relatività nel XXI secolo - Dal 'navilio' di Galileo all'espansione dell' Universo* La Fisica nella Scuola, Quaderno 16, anno XXXVIII n.1 supplemento
[33] Albert Einstein 1922 *The Meaning of Relativity* (Gutenberg)
[34] Gisin N 2014 Quantum correlations in newtonian space and time: Faster than light communication or nonlocality *Quantum Theory: A Two-Time Success Story* ed Struppa D C and Tollaksen J M (Springer Milan) pp 185–203
[35] Bancal J D, Pironio S, Acin A, Liang Y C, Scarani V and Gisin N 2012 *Nat Phys* **8** 867–870
[36] Barnea T J, Bancal J D, Liang Y C and Gisin N 2013 *Phys. Rev. A* **88**(2) 022123
[37] Salart D, Baas A, Branciard C, Gisin N and Zbinden H 2008 *Nature* **454** 861–864
[38] Cocciaro B, Faetti S and Fronzoni L 2013 *Journal of Physics: Conference Series* **442** 012005
[39] Cocciaro B, Faetti S and Fronzoni L 2011 *Phys. Lett.* A **375** 379–384
[40] Yin J, Cao Y, Yong H L, Ren J G, Liang H, Liao S K, Zhou F, Liu C, Wu Y P, Pan G S, Li L, Liu N L, Zhang Q, Peng C Z and Pan J W 2013 *Phys. Rev. Lett.* **110**(26) 260407
[41] Scarani V, Tittel W, Zbinden H and Gisin N 2000 *Physics Letters* A **276** 1–7

Journal of Physics: Conference Series **626** (2015) 012055 doi:10.1088/1742-6596/626/1/012055

Unified description of classical and quantum behaviours in a variational principle

Tomoi Koide[1], Takeshi Kodama[1] and Kazuo Tsushima[2]

[1]Instituto de Física, Universidade Federal do Rio de Janeiro, C.P. 68528, 21941-972, Rio de Janeiro, Brazil
[2]International Institute of Physics, Federal University of Rio Grande do Norte, Natal 59078-400, RN, Brazil

E-mail: tomoikoide@gmail.com, kodama.takeshi@gmail.com, kazuo.tsushima@gmail.com

Abstract. We give a pedagogical introduction of the stochastic variational method and show that this generalized variational principle describes classical and quantum mechanics in a unified way.

1. Introduction

Variational approach conceptually plays a fundamental role in elucidating the structure of classical mechanics, clarifying the origin of dynamics and the relation between symmetries and conservation laws. In classical mechanics, the optimized function is characterized by Lagrangian, defined as $T - V$ with T and V being a kinetic and a potential terms, respectively.

We can still argue the variational principle in quantum mechanics, but the Lagrangian does not have any more the form of $T - V$, instead it is given by $\psi^*(i\hbar\partial_t - \hat{H})\psi$, where $\hat{H}$ is a Hamiltonian operator and ψ is a wave function. Therefore, at first glance, any clear or direct correspondence between classical and quantum mechanics does not seem to exist in the variational point of view, but it does exist. If we extend the idea of the variation to stochastic variable, the variational principle describes classical and quantum behaviors in a unified way.

This method is called stochastic variational method (SVM) and firstly proposed by Yasue [1, 2, 3, 4, 5] so as to reformulate Nelson's stochastic quantization [6, 7]. This framework is, however, based on special techniques attributed to stochastic calculus which is not familiar to physicists. In this paper, we give a pedagogical introduction of SVM in a self-contained manner, showing the unified description of classical and quantum mechanics. As another review, see, for example, Ref. [8].

Journal of Physics: Conference Series **626** (2015) 012055 doi:10.1088/1742-6596/626/1/012055

2. Variational method for stochastic variables

Because of the limitation of pages, we cannot explain all aspects of stochastic calculus in detail. See for example, Ref. [9] for standard techniques which are not explained here.

2.1. Forward and backward SDEs

In the variational principle for stochastic variables, a particle trajectory is not any more smooth and given by a zig-zag path in general. As the consequence, the evolution of a particle trajectory is defined by the following stochastic differential equation (SDE),

$$d\mathbf{r}(t) = \mathbf{u}(\mathbf{r}(t), t)dt + \sqrt{2\nu}d\mathbf{W}_t \quad (dt > 0). \tag{1}$$

In this paper, a difference $dA(t)$ is always defined by $A(t + dt) - A(t)$ independently of the sign of dt. The last term in Eq. (1) is the origin of the zig-zag motion and called noise term. The parameter ν characterizes the strength of this noise term. One can easily see that $\mathbf{u}(\mathbf{r}(t), t)$ is reduced to the usual classical definition of the particle velocity in the limit of vanishing ν. The property of $\mathbf{W}_t$ depends on the stochastic property of the noise term. In the present paper, we assume that $\mathbf{W}_t$ is the Wiener process, which is characterized by the following correlation properties,

$$\begin{aligned} E[d\mathbf{W}_t] &= 0, & (2)\\ E[(dW_t^i)(dW_t^j)] &= |dt|\delta^{ij}, \quad (i, j = x, y, z), & (3)\\ E[W_t^i dW_{t'}^j] &= 0 \text{ for } (t \le t'), & (4) \end{aligned}$$

where $E[\ ]$ indicates the average of stochastic events. It is clear from the above properties that $d\mathbf{W}_t$ behaves as the so-called Gaussian white noise.

Such a SDE (Langevin equation) has been used in statistical physics to discuss, for example, thermalization. This is essentially an irreversible process and we exclusively discuss the time evolution for a given initial condition. However, in the formulation of a variational method, we should fix not only an initial condition but also a final condition. If we consider a backward process in time, $dt < 0$, it should describe a stochastic process from the final condition to the initial condition.

Then, what is the time-reversed process corresponding to Eq. (1)? To discuss this, let us define the probability distribution as

$$\rho(\mathbf{x}, t) = \int d^3\mathbf{r}_i \ \rho_I(\mathbf{r}_i) E[\delta^{(3)}(\mathbf{x} - \mathbf{r}(t))], \tag{5}$$

where $\mathbf{r}(t)$ is the solution of Eq. (1) and $\rho_I(\mathbf{r}_i)$ is the initial particle distribution with $\mathbf{r}(t_i) = \mathbf{r}_i$ at an initial time t_i. As is well-known, the evolution equation is given by the Fokker-Planck equation,

$$\partial_t \rho(\mathbf{x}, t) = \nabla(-\mathbf{u}(\mathbf{x}, t) + \nu\nabla)\rho(\mathbf{x}, t). \tag{6}$$

If the probability distribution evolves from $\rho_I(\mathbf{r})$ to $\rho_F(\mathbf{r}) \equiv \rho(\mathbf{r}(t_f), t_f)$ at a final time t_f following Eq. (6), the corresponding time-reversed process should describe the evolution from ρ_F to ρ_I. Suppose that this process is described by

$$d\mathbf{r}(t) = \tilde{\mathbf{u}}(\mathbf{r}(t), t)dt + \sqrt{2\nu}d\mathbf{W}_t, \tag{7}$$

where, it should be noted, $dt < 0$. Then, differently from the classical dynamics, it generally holds,

$$\tilde{\mathbf{u}}(\mathbf{r}, t)dt \neq -\mathbf{u}(\mathbf{r}, t)|dt|. \tag{8}$$

To understand this reason, let us consider the case $\mathbf{u} = 0$, where Eq. (6) becomes a simple diffusion equation, and hence the corresponding time-reversed process should describe an accumulation (opposite of diffusion) process. However, if the condition (8) were satisfied, Eq. (7) coincides with the diffusion equation and cannot describe the accumulation process.

To obtain the precise relation instead of Eq. (8), we calculate the Fokker-Planck equation assuming Eq. (7) as

$$\partial_t \rho(\mathbf{x}, t) = \nabla(-\tilde{\mathbf{u}}(\mathbf{x}, t) - \nu\nabla)\rho(\mathbf{x}, t). \tag{9}$$

For the two Fokker-Planck equations (6) and (9) to be equivalent, we find the following condition

$$\mathbf{u}(\mathbf{x}, t) = \tilde{\mathbf{u}}(\mathbf{x}, t) + 2\nu\nabla \ln \rho. \tag{10}$$

This is the consistency condition for Eq. (7) to be the time-reversed process of Eq. (1). [1] In fact, for the diffusion case where $\mathbf{u} = 0$, we obtain by the consistency condition,

$$\tilde{\mathbf{u}}(\mathbf{x}, t) = -2\nu\nabla \ln \rho. \tag{11}$$

Substituting this, one can see that Eq. (7) indeed describes the accumulation process. Interestingly, this consistency condition can be derived even from the property of Bayes' theorem. See Ref. [16].

In the following, we call Eq. (1) forward stochastic differential equation (SDE), and Eq. (7) backward SDE, respectively.

For the later convenience, let us introduce the mean velocity,

$$\mathbf{v}(\mathbf{x}, t) = \frac{\mathbf{u}(\mathbf{x}, t) + \tilde{\mathbf{u}}(\mathbf{x}, t)}{2}. \tag{12}$$

Once this quantity is determined, one can easily find $\mathbf{u}$ and $\tilde{\mathbf{u}}$ by using the consistency condition. This mean velocity is parallel to the flow of the particle probability distribution. In fact, the above two Fokker-Planck equations are reduced to the following simple equation,

$$\partial_t \rho(\mathbf{x}, t) = -\nabla \cdot (\rho(\mathbf{x}, t)\mathbf{v}(\mathbf{x}, t)). \tag{13}$$

2.2. *Velocities in SVM and partial integration formula*

The action which we will optimize is the time integral of the Lagrangian, which depends on the particle velocity. When the trajectory is described by stochastic variables, however, the definition corresponding to the velocity is not trivial.

As is well-known, the time derivative of the trajectory described by SDE, for example Eq. (1), is not well-defined in the limit of $|dt| \to 0$. This can be seen from the fact that $d\mathbf{W}_t$ has a size proportional to $\sqrt{|dt|}$ from Eq. (3), and thus $d\mathbf{r}/dt \sim d\mathbf{W}_t/dt \sim 1/\sqrt{|dt|}$.

However, it is known that there are two possible definitions which have a well-defined limit of dt proposed by Nelson [6, 7]: One is the the mean forward derivative

$$D\mathbf{r}(t) = \lim_{dt\to 0+} E\left[\frac{\mathbf{r}(t+dt) - \mathbf{r}(t)}{dt}\Big|\mathcal{P}_t\right], \tag{14}$$

and the other the mean backward derivative,

$$\tilde{D}\mathbf{r}(t) = \lim_{dt\to 0-} E\left[\frac{\mathbf{r}(t+dt) - \mathbf{r}(t)}{dt}\Big|\mathcal{F}_t\right]. \tag{15}$$

[1] Strictly speaking, there is an ambiguity to obtain this relation. See Ref. [15].

These expectations are conditional averages, where $\mathcal{P}_t$ ($\mathcal{F}_t$) indicates to fix values of $\mathbf{r}(t')$ for $t' \le t$ ($t' \ge t$). For the Wiener process, one can easily find $D\mathbf{W}_t = 0$, but, in general, $\tilde{D}\mathbf{W}_t \neq 0$.[2] When $\mathbf{r}(t)$ is described by the forward and backward SDEs defined above, we find $D\mathbf{r}(t) = \mathbf{u}(\mathbf{r}(t), t)$ and $\tilde{D}\mathbf{r}(t) = \tilde{\mathbf{u}}(\mathbf{r}(t), t)$.

As a matter of fact, it is impossible to control the behavior of each trajectory completely because of the random noise. What we can adjust is, at best, only the trend of stochastic motions. Then the mean forward derivative defined above represents the most probable velocity forward in time when a particle is located at $\mathbf{r}(t)$, and the mean backward derivative is that of the backward in time. Therefore, what we should obtain by the variational procedure is the form of $\mathbf{u}$ (or equivalently $\tilde{\mathbf{u}}$), and it is natural to express the velocities appearing in an optimized function by these quantities.

Because of the two different time derivatives, the partial integration formula for the stochastic variable is modified as

$$\int_0^t dsE\left[(DX(s))Y(s)\right] = E\left[X(t)Y(t) - X(0)Y(0)\right] - \int_0^t dsE\left[X(s)\tilde{D}Y(s)\right]. \tag{16}$$

The derivation is given in Appendix A. One should notice that when the time derivative D moves from the left variable ($X(t)$) to the right ($Y(t)$), it is replaced by $\tilde{D}$.

3. Variation of stochastic action

As an example of SVM, let us consider the optimization of the one particle Lagrangian,

$$L = \frac{m}{2}\dot{\mathbf{r}}^2(t) - V(\mathbf{r}(t)), \tag{17}$$

where m is the mass of the particle and V is a potential. As is well-known, Newton's equations of motion is obtained when the usual variational method is applied to this. To implement the stochastic variation to this Lagrangian, we need to express each term by the corresponding stochastic quantities.

Due to the existence of the two possible definitions of the time derivatives, the most general quadratic form of the kinetic energy of the Lagrangian is given by [10]

$$\frac{m}{2}\dot{\mathbf{r}}^2(t) \longrightarrow \frac{m}{2}\left[B_+\{A_+(D\mathbf{r}(t))^2 + A_-(\tilde{D}\mathbf{r}(t))^2\} + B_-(D\mathbf{r}(t))\cdot(\tilde{D}\mathbf{r}(t))\right], \tag{18}$$

where $A_\pm = 1/2 \pm \alpha_1$ and $B_\pm = 1/2 \pm \alpha_2$ with α_1 and α_2 being arbitrary real constants. Note that the right hand side reduces to the left hand side in the limit $\nu \to 0$, independently of the values of α_i ($i = 1, 2$). If $\alpha_1 \neq 0$, the optimized dynamics violates the time reversal symmetry and we obtain, for example, the Navier-Stokes-Fourier equation [15]. In the present discussion, however, we focus on dynamics with the time-reversal symmetry, and choose $(\alpha_1, \alpha_2) = (0, 1/2)$. Then, the stochastic action corresponding to Eq. (17) is given by

$$I[\mathbf{r}(t)] = \int_{t_i}^{t_f} dtE\left[L(\mathbf{r}, D\mathbf{r}, \tilde{D}\mathbf{r})\right] = \int_{t_i}^{t_f} dtE\left[\frac{m}{4}((D\mathbf{r}(t))^2 + (\tilde{D}\mathbf{r}(t))^2) - V(\mathbf{r}(t))\right]. \tag{19}$$

It is known that there are several definitions for products of stochastic variables, for example, the Ito definition, Stratonovich definition and so on. However, there is no this ambiguity for, for example, $(D\mathbf{r}(t))^2$ in this formulation. See Appendix B for details.

The variation of the stochastic variable is introduced as

$$\mathbf{r}(t) \longrightarrow \mathbf{r}(t) + \mathbf{f}(\mathbf{r}(t), t). \tag{20}$$

[2] To understand this, the argument around Eq. (25) will be useful.

Here $f(\mathbf{x}, t)$ is an arbitrary infinitesimal function satisfying $\mathbf{f}(\mathbf{x}, t_i) = \mathbf{f}(\mathbf{x}, t_f) = 0$. Then, for example, the variation of the kinetic term is calculated as

$$\begin{aligned} &\int_{t_i}^{t_f} dtE\left[(D\mathbf{r}(t) + D\mathbf{f}(\mathbf{r}(t),t))^2\right] - \int_{t_i}^{t_f} dtE\left[(D\mathbf{r}(t))^2\right] \\ &= 2\int_{t_i}^{t_f} dtE\left[\mathbf{u}(\mathbf{r}(t),t)\cdot(D\mathbf{f}(\mathbf{r}(t),t))\right] + O(\mathbf{f}^2) \\ &= -2\int_{t_i}^{t_f} dtE\left[\{\tilde{D}\mathbf{u}(\mathbf{r}(t),t)\}\cdot\mathbf{f}(\mathbf{r}(t),t)\right]. \end{aligned} \tag{21}$$

Here we have first used the definition of the mean forward derivative, and then the stochastic partial integration formula. The potential part does not contain any time derivative terms, and its variation is the same as that in the classical variational method. Then the result of the variation is obtained as

$$\delta I = \int_{t_i}^{t_f} dtE\left[\left\{-\frac{m}{2}(\tilde{D}\mathbf{u}(\mathbf{r}(t),t) + D\tilde{\mathbf{u}}(\mathbf{r}(t),t)) - \nabla V(\mathbf{r}(t))\right\}\cdot\mathbf{f}(\mathbf{r}(t),t)\right]. \tag{22}$$

It is clear from the definition of the mean derivatives that $\mathbf{r}(t)$ in $\tilde{D}\mathbf{u}$ is described by the *backward* SDE. Then, substituting the definition of $\tilde{D}$ and applying Ito's lemma (Appendix C), we obtain

$$\tilde{D}\mathbf{u}(\mathbf{r}(t),t) = \left(\partial_t + \tilde{\mathbf{u}}(\mathbf{r}(t),t)\cdot\nabla - \nu\nabla^2\right)\mathbf{u}(\mathbf{r}(t),t). \tag{23}$$

Similarly, $\mathbf{r}(t)$ in $D\tilde{\mathbf{u}}$ is given by the *forward* SDE leading to

$$D\tilde{\mathbf{u}}(\mathbf{r}(t),t) = \left(\partial_t + \mathbf{u}(\mathbf{r}(t),t)\cdot\nabla + \nu\nabla^2\right)\tilde{\mathbf{u}}(\mathbf{r}(t),t). \tag{24}$$

Note that the last noise term in Ito's lemma disappears in the above expressions, because of the conditional average included in the definition of the mean derivatives. In fact,

$$E\left[\sqrt{2\nu}d\mathbf{W}_t\cdot\nabla\mathbf{u}(\mathbf{r}(t),t)\middle|\mathcal{F}_t\right] = \sqrt{2\nu}\nabla\mathbf{u}(\mathbf{r}(t),t)\cdot E\left[d\mathbf{W}_t\right] = 0 \quad (dt < 0). \tag{25}$$

In the variational principle of stochastic variables, we require that δI vanishes for 1) any choice of $\mathbf{f}(\mathbf{x}, t)$, and also 2) any distribution of the stochastic variable $\mathbf{r}(t)$. To satisfy these, $\mathbf{u}$ (or equivalently $\tilde{\mathbf{u}}$) should be the solution of

$$\left[-\frac{m}{2}(\tilde{D}\mathbf{u}(\mathbf{r}(t),t) + D\tilde{\mathbf{u}}(\mathbf{r}(t),t)) - \nabla V(\mathbf{r}(t))\right]_{\mathbf{r}(t)=\mathbf{x}} = 0. \tag{26}$$

Substituting Eqs. (23) and (24), we obtain

$$(\partial_t + \mathbf{v}(\mathbf{x},t)\cdot\nabla)\mathbf{v}(\mathbf{x},t) = -\frac{1}{m}\nabla V(\mathbf{x}) + 2\nu^2\nabla\rho^{-1/2}(\mathbf{x},t)\nabla^2\sqrt{\rho(\mathbf{x},t)}, \tag{27}$$

where the mean velocity $\mathbf{v}$ is defined by Eq. (12).

It is worth mentioning that Eq. (26) is formally expressed as

$$\left[\tilde{D}\frac{\partial L}{\partial D\mathbf{r}(t)} + D\frac{\partial L}{\partial\tilde{D}\mathbf{r}(t)} - \frac{\partial L}{\partial\mathbf{r}(t)}\right]_{\mathbf{r}(t)=\mathbf{x}} = 0. \tag{28}$$

Note that the stochastic variable $\mathbf{r}(t)$ is replaced by the position parameter $\mathbf{x}$ in the above, only after operating all mean derivatives. This is nothing but the stochastic generalization of the Euler-Lagrange equation.

4. Schrödinger equation

These two equations (13) and (27) determine the optimized dynamics of the action given by Eq. (19). However, these coupled equations can be cast into a more familiar form. Let us introduce the following complex function,

$$\psi(\mathbf{x},t) = \sqrt{\rho(\mathbf{x},t)}e^{i\theta(\mathbf{x},t)}, \tag{29}$$

where the phase is defined by

$$\mathbf{v}(\mathbf{x},t) = 2\nu\nabla\theta(\mathbf{x},t). \tag{30}$$

Then, from Eqs. (13) and (27), the evolution equation of this quantity is given by

$$i\partial_t\psi(\mathbf{x},t) = \left[-\nu\nabla^2 + \frac{1}{2\nu m}V(\mathbf{x})\right]\psi(\mathbf{x},t). \tag{31}$$

When we choose $\nu = \hbar/(2m)$, this is reduced to the Schrödinger equation, and $\psi(\mathbf{x},t)$ is identified with the wave function. Furthermore, one can easily find that $|\psi(\mathbf{x},t)|^2$ gives the probability density distribution, without introducing any quantum mechanical interpretations. In short, the procedure described above, can be regarded to give an alternative quantization scheme.

5. Stochastic Noether theorem

In the SVM quantization scheme, the physical operator is defined through Noether's theorem for the stochastic action [11]. Let us consider the spatial translation by an arbitrary constant spatial vector $\mathbf{A}$,

$$\mathbf{r}(t) \longrightarrow \mathbf{r}(t) + \mathbf{A}, \tag{32}$$

and require the invariance of the stochastic action for this transform. Then we find

$$\begin{aligned} 0 = \delta I &= \int_{t_i}^{t_f} dt E\left[\frac{\partial L}{\partial D\mathbf{r}}\cdot D\mathbf{A} + \frac{\partial L}{\partial \tilde{D}\mathbf{r}}\cdot \tilde{D}\mathbf{A} + \frac{\partial L}{\partial \mathbf{r}}\cdot\mathbf{A}\right] + O(\mathbf{A}^2) \\ &= \int_{t_i}^{t_f} dt\frac{d}{dt}E\left[\frac{\partial L}{\partial D\mathbf{r}}\cdot\mathbf{A} + \frac{\partial L}{\partial \tilde{D}\mathbf{r}}\cdot\mathbf{A}\right] \\ &= \frac{m}{2}\int_{t_i}^{t_f} dt\frac{d}{dt}E\left[D\mathbf{r} + \tilde{D}\mathbf{r}\right]\cdot\mathbf{A}. \end{aligned} \tag{33}$$

Here we have used the stochastic Euler-Lagrange equation (28) and

$$\frac{d}{dt}E\left[XY\right] = E\left[YDX + X\tilde{D}Y\right], \tag{34}$$

which is obtained from the stochastic partial integration formula. Because $\mathbf{A}$ is an arbitrary constant, one can deduce that the quantity, $mE\left[(D\mathbf{r}+\tilde{D}\mathbf{r})/2\right]$, is conserved. Using the solution obtained by the stochastic variation, this quantity is expressed as

$$\int d^3\mathbf{r}_i\ \rho_I\ \frac{m}{2}E\left[D\mathbf{r} + \tilde{D}\mathbf{r}\right] = \int d^3\mathbf{x}\rho(\mathbf{x},t)m\mathbf{v}(\mathbf{x},t) = \int d^3\mathbf{x}\psi(\mathbf{x},t)(-i\hbar\nabla)\psi(\mathbf{x},t). \tag{35}$$

Here the conserved quantity is integrated for the initial particle distribution. This is the well-known expression of momentum expectation value in quantum mechanics and $-i\hbar\nabla$ is identified with the momentum operator. Similarly, we can obtain the conservation laws of energy, angular momentum and charge by the stochastic Noether theorem.

6. Canonical equation

Although the canonical formulation of SVM has not yet been established, we can still formally write down the stochastic canonical equation [12]. Let us introduce quantities corresponding to the momenta as

$$\frac{1}{2}\mathbf{p} = \frac{\partial L}{\partial D\mathbf{r}}, \qquad \frac{1}{2}\bar{\mathbf{p}} = \frac{\partial L}{\partial \tilde{D}\mathbf{r}}. \tag{36}$$

Then the stochastic Hamiltonian can be introduced as the Legendre transform as

$$H(\mathbf{r},\mathbf{p},\bar{\mathbf{p}}) = \frac{1}{2}(\mathbf{p}\cdot D\mathbf{r} + \bar{\mathbf{p}}\cdot \tilde{D}\mathbf{r}) - L(\mathbf{r}, D\mathbf{r}, \tilde{D}\mathbf{r}). \tag{37}$$

The canonical equation is obtained by requiring that the invariance of the stochastic Hamiltonian by the following infinitesimal transforms,

$$\mathbf{r} \longrightarrow \mathbf{r} + \boldsymbol{\eta}, \qquad \mathbf{p} \longrightarrow \mathbf{p} + \boldsymbol{\zeta}, \qquad \bar{\mathbf{p}} \longrightarrow \bar{\mathbf{p}} + \bar{\boldsymbol{\zeta}}, \tag{38}$$

where $\boldsymbol{\eta}$, $\boldsymbol{\zeta}$ and $\bar{\boldsymbol{\zeta}}$ are infinitesimal constants. Substituting these into the both sides of Eq. (37) and keeping up to the first order of these infinitesimal variables, we find the following relations:

$$\frac{\partial H(\mathbf{r},\mathbf{p},\bar{\mathbf{p}})}{\partial \mathbf{r}} = -\frac{\partial L}{\partial \mathbf{r}}, \quad \frac{\partial H(\mathbf{r},\mathbf{p},\bar{\mathbf{p}})}{\partial \mathbf{p}} = D\mathbf{r}, \quad \frac{\partial H(\mathbf{r},\mathbf{p},\bar{\mathbf{p}})}{\partial \bar{\mathbf{p}}} = \tilde{D}\mathbf{r}. \tag{39}$$

Finally, using the relations above and the definitions of the momenta, the stochastic Euler-Lagrange equation is re-expressed as

$$\tilde{D}\frac{\partial L}{\partial D\mathbf{r}} + D\frac{\partial L}{\partial \tilde{D}\mathbf{r}} - \frac{\partial L}{\partial \mathbf{r}} = 0 \longrightarrow \frac{1}{2}\tilde{D}\mathbf{p} + \frac{1}{2}D\bar{\mathbf{p}} + \frac{\partial H(\mathbf{r},\mathbf{p},\bar{\mathbf{p}})}{\partial \mathbf{r}} = 0. \tag{40}$$

This corresponds to the canonical equation. By substituting this into Eq. (28) with the help of the second and the last equations in Eq. (39), one can confirm that Eq. (27) is indeed reproduced.

7. Concluding remarks

We have discussed the application of SVM to quantize classical particle systems. When we apply the stochastic variation of the action (19) assuming Eqs. (1) and (7), we obtain the Schrödinger equation. The result of the variation depends on the assumed form of Eqs. (1) and (7). If we use them with the limit of $\nu_i \to 0$, the stochastic variation of the same action leads to Newton's equation of motion. That is, the usual variational method is a special case of SVM, and both classical and quantum mechanics are described in the framework of this more generalized variational method. It is also possible to apply SVM to quantize Klein-Gordon field [13] and abelian gauge field [14]. As a related work associated with quantized fields and random fields, see Ref. [17].

The framework of SVM itself can be regarded more general than the method of quantization. In fact, it is possible to derive the Navier-Stokes-Fourier equation by applying SVM to the action, which leads to the Euler equation when the usual classical mechanical method of variation is applied [15]. It is interesting to note that the Gross-Pitaevskii equation also can be obtained in the framework of SVM [15].

There are various proposals for the non-conventional quantization scheme. One of them is the so-called stochastic quantization proposed by Parisi and Wu [18, 19]. In a similar way to SVM, the effect of quantum fluctuation is taken into account through SDE even in this method, but the philosophy for quantization seems to be completely different. For example, a fictitious time variable is introduced in the stochastic quantization. That is, to quantize a $3+1$ dimensional system, we need to consider $3+1+1$ dimension. Then SDE describes the evolution in this fictitious time. Moreover, what is calculated in this approach is propagators while the Schrödinger equation and physical operators are obtained in SVM. For other quantizations, See, for example, Refs. [16, 20, 21, 22, 23, 24].

Finally, we would like to list future problems to be studied by SVM:

- quantization of fermions;
- criticism by Takabayashi [25, 26] and Wallstrom [27, 28];
- canonical transform [29];
- extension to general curved coordinate systems;
- classicalization and quantum-classical hybrids [30];
- variational formulation of relativistic dissipative fluids;
- topology;
- anomaly.

Acknowledgments
This work is financially supported by CNPq. KT is supported by the Brazilian Ministry of Science, Technology and Innovation (MCTI-Brazil), and Conselho Nacional de Desenvolvimento Científico e Tecnológico (CNPq), project 550026/2011-8.

Appendix A. Stochastic partial integration formula

The time variable is discretized as

$$t_j = a + j\frac{b-a}{n}, \quad j = 0, 1, 2, \cdots, n. \tag{A.1}$$

Then, we can show the following with the notations such as $X_j \equiv X(t_j)$ etc.:

$$\begin{aligned}
\int_a^b dtE[\{DX(t)\}Y(t) + X(t)\tilde{D}Y(t)] &= \lim_{n\to\infty}\sum_{j=0}^{n-1} E\left[(X_{j+1}-X_j)\frac{Y_{j+1}+Y_j}{2}\right]\frac{b-a}{n} \\
&\quad + \lim_{n\to\infty}\sum_{j=1}^{n} E\left[\frac{X_j+X_{j-1}}{2}(Y_j - Y_{j-1})\right]\frac{b-a}{n} \\
&= \lim_{n\to\infty}\sum_{j=0}^{n-1} E[X_{j+1}Y_{j+1} - X_jY_j]\frac{b-a}{n} \\
&= \int_a^b dtE[\frac{d}{dt}\{X(t)Y(t)\}] \\
&= E[X(b)Y(b) - X(a)Y(a)].
\end{aligned} \tag{A.2}$$

This is called the stochastic generalization of the partial integration formula.

Appendix B. Ito definition or Stratonovich definition?

For usual (non-stochastic) numbers, an integral of a function $f(x)$ is defined by

$$\int_{x_a}^{x_b} dxf(x) = \sum_{i=0}^{N-1} f(x_i)dx_i, \tag{B.1}$$

where $dx_i = x_{i+1} - x_i$ and $x_i = x_a + \sum_{j=0}^{i-1} dx_j$. Here we have used $x_N = x_b$. However, the right hand side can be re-expressed as

$$\int_{x_a}^{x_b} dx f(x) = \sum_{i=0}^{N-1} f((x_i + x_{i+1})/2) dx_i, \tag{B.2}$$

because $f((x_i + x_{i+1})/2) = f(x_i + dx_i/2) = f(x_i) + O(dx_i)$ for a general smooth function $f(x)$. However, these two different definitions give different results when x is a stochastic variable.

Let us denote the Stieltjes integral for the Wiener process as $\int_0^t W_s dW_s$. Then corresponding to the argument above, we can define this integral in two different ways: one is the Ito definition,

$$(I) \int_0^t W_s dW_s = \sum_{i=0}^{N-1} W_i (W_{i+1} - W_i), \tag{B.3}$$

and the other is the Stratonovich definition

$$(S) \int_0^t W_s dW_s = \sum_{i=0}^{N-1} \frac{W_i + W_{i+1}}{2} (W_{i+1} - W_i). \tag{B.4}$$

These two different definitions are known to yield the difference in the results by $t/2$. Therefore, we must specify the definition of the product for the quantity like $f(W_t)dW_t$.

In the stochastic Lagrangian, we have the term such as $D\mathbf{r}(t) \cdot D\mathbf{r}(t)$. It is very similar to the quantity discussed above, and thus one might insist that we need to specify one of the definitions for this product. However, D is a conditional expectation value, and it does not contain any dW_t dependence! Therefore, we do not need to introduce special definitions for these type of products. This is also one of the advantages to introduce the mean derivatives.

Appendix C. Ito's lemma (Ito formula)

Let us assume an arbitrary function $g(\mathbf{x}, t)$ which is differentiable for $\mathbf{x}$ and t. Now we substitute the stochastic variable $\mathbf{r}(t)$, the solution of Eq. (1), to $\mathbf{x}$ in $g(\mathbf{x}, t)$. Ito's lemma tells us that this time derivative is given by

$$dg(\mathbf{r}(t), t) = \left(\partial_t + \mathbf{u}(\mathbf{r}(t), t) \cdot \nabla + \nu \frac{dt}{|dt|} \nabla^2 \right) g(\mathbf{r}(t), t) dt + \sqrt{2\nu} \nabla g(\mathbf{r}(t), t) d\mathbf{W}_t. \tag{C.1}$$

Note that this is reduced to the usual Taylor expansion when $\nu = 0$. For stochastic processes, $d\mathbf{W}_t$ has an order of $\sqrt{|dt|}$, and thus to keep $O(dt)$, we need to take into account a part of the second order contribution of the Taylor expansion. When $\mathbf{r}(t)$ follows Eq. (7), $\mathbf{u}$ in the above equation is replaced by $\tilde{\mathbf{u}}$.

References

[1] Yasue K 1981 Stochastic calculus of variation *J. Funct. Anal.* **41** 327
[2] Guerra F and Morato L M 1983 Quantization of Dynamical Systems and Stochastic Control Theory *Phys. Rev.* D **27** 1774
[3] Pavon M 1995 Hamilton's principle in stochastic mechanics *J. Math. Phys.* **36** 6774
[4] Nagasawa M 2000 *Stochastic Process in Quantum Physics* (Basel: Birkhäuser)
[5] Cresson J and Darses S 2007 Stochastic embedding of dynamical systems *J. Math. Phys.* **48** 072703
[6] Nelson E 1966 Derivation of the Schrödinger Equation from Newtonian Mechanics *Phys. Rev.* **150** 1079
[7] Nelson E 1985 *Quantum Fluctuations*, (Prinston, NJ: Princeton Univ. Press).
[8] Zambrini J C 1985 Stochastic Dynamics: A Review of Stochastic Calculus of Variations *Int. J. Theor. Phys.* **24** 277

[9] Gardiner C W 2004 *Handbook of Stochastic Methods for physics, Chemistry and the Natural Science*, (Berlin, Heidelberg: Springer-Verlag).
[10] Koide T 2013 How is an optimized path of classical mechanics affected by random noise? *J. Phys.: Conf. Ser.* **410** 012025
[11] Misawa T 1988 Noether's theorem in symmetric stochastic calculus of variations *J. Math. Phys.* **29** 2178
[12] As another derivation, see Koide T and Kodama T 2012 Uncertainty Relations in Viscous Dynamics and Quantum Dissipative Process *Preprint* arXiv:1208.0258
[13] Koide T and Kodama T, 2013 Stochastic Variational Method as Quantization Scheme I: Field Quantization of Complex Klein-Gordon Equation *Preprint* arXiv:1306.6922
[14] Koide T, Kodama T and Tsishima K 2014 Stochastic Variational Method as a Quantization Scheme II: Quantization of Electromagnetic Fields *Preprint* arXiv:1406.6295
[15] Koide T and Kodama T 2012 Navier-Stokes, Gross-Pitaevskii and generalized diffusion equations using the stochastic variational method *J. Phys. A: Math. Theor.* **45** 255204
[16] Caticha A 2011 Entropic dynamics, time and quantum theory *J. Phys. A: Math. Theor.* **44** 225303
[17] Morgan P 2009 Equivalence of the Klein-Gordon random field and the complex Klein-Gordon quantum field *Europhys. Lett.* **87** 31002
[18] Parisi G and Wu Yong-Shi 1981 Perturbation theory without gauge fixing *Sci. Cin.* **24** 483
[19] Namiki M 1992 *Stochastic Quantization*, (Heidelberg: Springer-Verlag)
[20] Beck C 1995 Chaotic quantization of field theories *Nonlinearity* **8** 423
[21] Callaway D and Bahman A 1982 Microcanonical ensemble formulation of lattice gauge theory *Phys. Bev. Lett.* **49** 613
[22] Iwazaki A Microcanonical formulation of quantum field theories *Phys. Lett.* B **141** 342
[23] Gogberashvili M 2011 Thermodynamic Gravity and the Schrödinger Equation *Int. J. Theor. Phys.* **50** 2391
[24] Kodama T and Koide T 2015 Variational Principle of Hydrodynamics and Quantization by Stochastic Process *Preprint* arXiv:1412.6472
[25] Takabayashi T 1952 On the Formulation of Quantum Mechanics associated with Classical Pictures *Prog. Theor. Phys.* **8** 143
[26] Takabayashi T 1953 Remarks on the Formulation of Quantum Mechanics with Classical Pictures and on Relations between Linear Scalar Fields and Hydrodynamical Fields *Prog. Theor. Phys.* **9** 187
[27] Wallstrom T C 1989 On the derivation of the Schrödinger equation from stochastic mechanics *Found. Phys. Lett.* **2** 113
[28] Wallstrom T C 1994 Inequivalence between the Schrödinger equation and the Madelung hydrodynamic equations *Phys. Rev.* A **49** 1613
[29] Misawa T and Yasue K 1987 Canonical stochastic dynamical systems *J. Math. Phys.* **28** 2569
[30] Koide T 2015 Classicalization of Quantum Variables *Preprint* arXiv:1412.6321

Journal of Physics: Conference Series **626** (2015) 012056 doi:10.1088/1742-6596/626/1/012056

Measurement theory for closed quantum systems

Michiel Wouters

TQC, Universiteit Antwerpen, Universiteitsplein 1, B-2610 Antwerpen, Belgium

E-mail: michiel.wouters@uantwerpen.be

Abstract. We introduce the concept of a "classical observable" as an operator with vanishingly small quantum fluctuations on a set of density matrices. Their study provides a natural starting point to analyse the quantum measurement problem. In particular, it allows to identify Schrödinger cats and the associated projection operators intrinsically, without the need to invoke an environment. We discuss how our new approach relates to the open system analysis of quantum measurements and to thermalization studies in closed quantum systems.

1. Introduction

'How does Hilbert space relate to our classical reality?' The consensus answer to this fundamental question is based on decoherence [1, 2, 3, 4]. Its starting point is a separation of the universe into system and environment. The decoherence induced by the environment kills quantum effects and provides the sought-for transition from the quantum to the classical world. While the separation of large systems into different parts clearly has its merits, the ambiguity in the division between system and environment remains awkward for a fundamental theory [6, 5]. Currently, there is a thriving research activity on closed quantum systems [7]. In particular, much progress has been made in understanding their thermalisation [8, 9, 10, 11], a phenomenon where environments traditionally play a central role. In this Letter, we will address the theory of quantum measurement for closed systems.

In order to define the relation between the quantum and classical worlds, we have to specify how the observable information relates to a quantum state. A crucial experimental fact is that all we know about the world are quantities with negligible quantum fluctuations. One may think here of an image on your computer screen, even though one does not have to go that far in the macroscopic world in practice. To make the connection between a quantum mechanical system and our knowledge about it, we therefore define the concept of a *classical observable* as an operator with vanishingly small quantum fluctuations. An experiment then corresponds to 'reading' its expectation value. This procedure corresponds in all cases to experimental practice, where there is always a link between the quantum system and our 'knowing it', that is described in terms of an expectation value and not in terms of a projection operator (think of the light emitted by the computer screen).

2. Classicality and signal to noise ratio

We quantify the *classicality* of an operator with respect to a density matrix ρ as

$$\mathcal{C}(A) = \frac{[\mathrm{tr}(A\rho)]^2}{\mathrm{tr}(A^2\rho)}. \tag{1}$$

It satisfies $\mathcal{C} \leq 1$ because of the Cauchy-Schwarz inequality. For a classical observable, the upper limit is closely approached ($\mathcal{C} \to 1$), where for an observable that is subject to large quantum fluctuations, $\mathcal{C}$ is substantially smaller. The unit operator clearly satisfies $\mathcal{C}(\mathbb{1}) = 1$, expressing that the norm of the wave function does not fluctuate. For the density matrix of a pure state $\rho = |\psi\rangle\langle\psi|$, also the projection operator $P = |\psi\rangle\langle\psi|$ is classical. However, when the wave function is evolved in time, its classicality will in general quickly decrease.

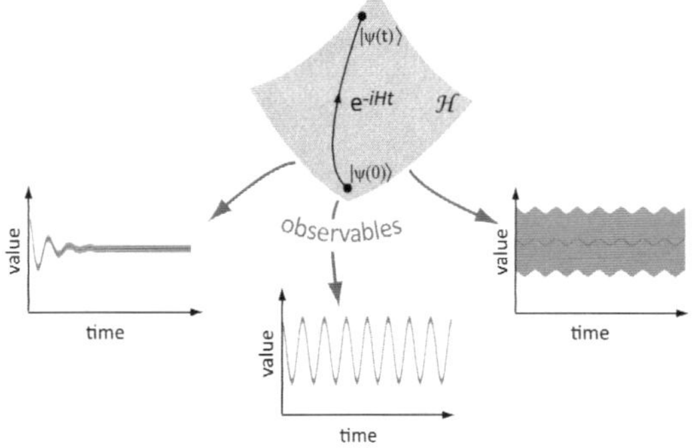

Figure 1. a) Illustration of the dynamics of observables for a wave function that evolves in Hilbert space $\mathcal{H}$ under Hamiltonian evolution. Shaded areas indicate the magnitude of quantum fluctuations. Most observables have most of the time a small signal to noise ratio and do not qualify as classical observables.

In order to find operators that do remain classical under time evolution, we maximize

$$\overline{\mathcal{C}}(A) = \frac{\int_0^T dt\{\mathrm{tr}[A\rho(t)]\}^2}{\int_0^T dt\, \mathrm{tr}[A^2\rho(t)]}. \tag{2}$$

The denominator can here be seen as a natural measure for the magnitude of the operator. In order to maximize $\overline{\mathcal{C}}$, we expand the operator A in a basis B_i of the linear Hermitian operators acting on the Hilbert space: $A = \sum_i a_i B_i$. Substituting this expansion in Eq. (2), one sees that for the eigenvectors of the generalized eigenvalue problem

$$\sum_j R_{ij} a_j^{(n)} = \overline{\mathcal{C}}_n \sum_j M_{ij} a_j^{(n)}, \tag{3}$$

the classicality corresponds to the generalized eigenvalue $\overline{\mathcal{C}}_n$. In Eq. (3) The matrices R_{ij} and M_{ij} are symmetric and defined as

$$R_{ij} = \frac{1}{T}\int_0^T dt\ \mathrm{tr}[B_i\rho(t)]\,\mathrm{tr}[B_j\rho(t)], \tag{4}$$

$$M_{ij} = \frac{1}{2}\mathrm{tr}[(B_iB_j + B_jB_i)\,\overline{\rho}]. \tag{5}$$

Two operator eigenvectors $A_n = \sum_i a_i^{(n)} B_i$ of Eq. (3) that belong to different eigenvalues are orthogonal in the following sense:

$$\frac{1}{2}\mathrm{tr}[(A_mA_n + A_nA_m)\,\overline{\rho}] = 0. \tag{6}$$

Because the unit operator is a generalized eigenvector, the other operators have all zero time averaged expectation value: $\mathrm{tr}(A_m\overline{\rho}) = 0$.

For an operator with zero expectation value, the square of its expectation value is a good measure of the signal. This motivates us to define the signal to noise ratio as

$$\mathrm{SNR}(A) = \frac{\int_0^T dt \{\mathrm{tr}[A\rho(t)]\}^2}{\int_0^T dt\, \mathrm{tr}[A^2\rho(t)] - \{\mathrm{tr}[A\rho(t)]\}^2}. \tag{7}$$

With the definition of the classicality (1), we can write it as

$$\mathrm{SNR}(A) = \frac{1}{1-\overline{\mathcal{C}}(A)}. \tag{8}$$

This relation tells us that classical observables, with $\overline{\mathcal{C}} \to 1$, show temporal variations that are much larger than their fluctuations, while the time dependence of the other operators is drowned in noise (see Fig. 1).

To complete the specification of the measurement problem, we have to choose an initial condition. If we only use the total Hamiltonian as an input for our analysis, we are restricted to formulate it in the energy eigenbasis $|n\rangle$. As an incoherent mixture of energy eigenstates precludes any dynamics, it is most natural to consider a pure state. We will assume that a finite number of N energy eigenstates, in an energy window $[E, E+\Delta E]$ is populated. Because there is no dynamics in the populations of the eigenstates, we should not lose physics by making the simplifying assumption that the initial state has equal overlap with all the energy eigenstates within the energy window [8, 9]. We do not loose generality by taking the overlap $\langle n|\psi(t=0)\rangle$ real for our specific initial condition, because the absolute phase of the energy eigenstates is arbitrary. We then have for the density matrix

$$\rho(t) = \frac{1}{N}\sum_{k,l=1}^{N} e^{-it(\omega_k-\omega_l)}|k\rangle\langle l|. \tag{9}$$

As a basis for the Hermitian operators on the Hilbert space, we choose

$$x_{m,n} = |m\rangle\langle n| + |n\rangle\langle m| \tag{10}$$
$$p_{m,n} = -i(|m\rangle\langle n| - |n\rangle\langle m|), \tag{11}$$

with the $m \neq n$ restricted to the energy levels that are populated according to the initial condition. We then have the following expectation values

$$\mathrm{tr}[x_{m,n}\rho(t)] = 2\cos[(\omega_m - \omega_n)t] \tag{12}$$
$$\mathrm{tr}[p_{m,n}\rho(t)] = 2\sin[(\omega_m - \omega_n)t] \tag{13}$$

For sufficiently long times $T \gg 2\pi \max(|\omega_m - \omega_n|^{-1})$, the matrix elements R_{ij}, where i is of the x-type and j is of the p type operator vanish. We can therefore restrict our search to operators of the x-type, and will automatically find a corresponding p-type operator. For long times and in the absence of degeneracies, we find that the matrix M in (5) is proportional to the unit matrix: $M = (2/N)\mathbb{1}$.

The elements of the matrix R, restricted to the x-operator space can be then written more explicitly as $R_{mn,m'n'}$. Equations (4) and (12) then show that off-diagonal elements of R vanish when the transition frequencies differ much more than the Heisenberg energy uncertainty, i.e. if $(|\omega_n - \omega_m| - |\omega_{n'} - \omega_{m'}|) \gg 2\pi/T$. The matrix R is thus approximately block-diagonal, with each block corresponding to some transition frequency. Within a block of size N_b, all the matrix elements are equal to $2/N^2$. The eigenvalue problem (3) has a single nonzero eigenvalue per block in R, with corresponding eigenvector in the block $V = (1, 1, \dots, 1)^T$ (all transitions act in phase). Because $RV = 2N_b/N^2 V$ and $MV = (2/N)V$, we obtain $\overline{\mathcal{C}} = N_b/N$.

3. Detection of Schrödinger cats

For a harmonic ladder of N equally spaced energy levels, we have $N_b = N - 1$, so that

$$\overline{\mathcal{C}}(X) = \overline{\mathcal{C}}(P) = 1 - \frac{1}{N}. \tag{14}$$

As expected, the operators X and P are (approximately) the harmonic oscillator position and momentum operators:

$$X = \sum_{n=1}^{N-1} \big(|n\rangle\langle n+1| + |n+1\rangle\langle n|\big), \tag{15}$$

$$P = -i\sum_{n=1}^{N-1} \big(|n\rangle\langle n+1| - |n+1\rangle\langle n|\big). \tag{16}$$

They can be made arbitrarily classical by increasing the number of levels. For a two-level system, $N = 2$, we simply recover the Pauli matrices and find that $\overline{\mathcal{C}}(\sigma_x) = \overline{\mathcal{C}}(\sigma_y) = 1/2$. The only nonzero elements of the commutator between position and momentum operators are $([X,P])_{1,1} = -([X,P])_{N,N} = 2i$. In the limit of a large number of levels $N \gg 1$, the commutator is thus negligible with respect to X and P. For large N, also powers of X and P have classicality close to one. Ordinary calculus can then be used for sufficiently smooth functions of X and P.

The fact that we do not precisely recover the usual harmonic oscillator position and momentum operators [a factor $\sqrt{n+1}$ is missing in Eq. (15)] can be attributed to the fact that the energy distribution is not Poissonian, but uniform in the interval $[E, E + \Delta E]$. Note that in the limit of a small energy window $\Delta E \lll E$ at high energy $E \gg \omega$, the variation of $\sqrt{n+1}$ becomes negligible and X approaches the usual harmonic oscillator positon operator.

The analysis becomes more interesting when we consider a direct sum of two incommensurate harmonic oscillator ladders. The total Hilbert space is then $\mathcal{H} = \mathcal{H}_1 \bigoplus \mathcal{H}_2$, each with $N/2$ levels, but with different energies: $\omega_n = n\omega_1$ for the first ladder and $\omega_n = n\omega_2$ for the second one (see Fig 2 c). It represents a qubit coupled to a harmonic oscillator, whose cavity-QED implementation has become a fruitful testing ground for quantum physics [12, 13, 14]. For short evolution times, $T \ll 2\pi/|\omega_1 - \omega_2|$, the two harmonic oscillator ladders are indistinguishable within the Heisenberg limited energy resolution (see Fig. 2 d) and we find that $X = X_1 \bigoplus X_2$ is still a classical observable, with $\overline{\mathcal{C}}(X) = 1 - 2/N$. The operators X_1 and X_2 are analogous to (15), but with the levels restricted to $\mathcal{H}_1$ and $\mathcal{H}_2$ respectively. For evolution times much longer than the Heisenberg uncertainty time $T \gg 2\pi/|\omega_1 - \omega_2|$ on the other hand (panel d), a classical observable can no longer be found. The most classical ones are X_1 and X_2, with $\overline{\mathcal{C}}(X_1) = \overline{\mathcal{C}}(X_2) = 1/2 - 1/N$.

In general, the highest classicality is obtained for operators that consist of all resonant transitions $|n\rangle\langle m|$. High classicality up to time T is obtained when all occupied states form part of a harmonic oscillator ladder with common transition ω_c, within the Heisenberg energy uncertainty $2\pi/T$ (see Fig. 2 c). The corresponding classical observable is then fully collective: all states participate in it as in the X and P operators in Eqns. (15) and(16). The stringent requirements of harmonicity and collectivity provide an elementary explanation of the fact that classical phase space is so much smaller than Hilbert space.

The lack of a classical observable up to late times on the full Hilbert space $\mathcal{H}$, reflects the fact that the system turns into a Schrödinger cat. Note that we have identified the cat state without separating the universe in system and environment. Our analysis thus provides a solution to the 'preferred-basis problem' [2, 15, 16] in Everett's relative-state approach [17]. Observing the expectation value of the X at times $T > 2\pi/|\omega_1 - \omega_2|$ will result in the collapse of the wave function. The density matrix is then projected to a single subspace $\mathcal{H}_1$ or $\mathcal{H}_2$, with a probability

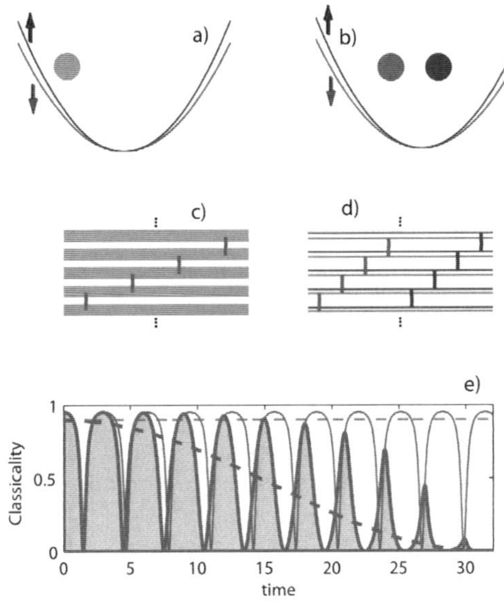

Figure 2. a) Illustration of the qubit coupled to a harmonic oscillator potential. At early times (a), the positions in the two potentials (disks) are the same. It corresponds to the poor energy energy resolution, indicated thick energy levels in panel c). At later times (panels b,d), when the two groups of transitions are dephased with respect to each other, Hilbert space 'breaks up'. Only on the red and blue harmonic subspaces, classical observables exist. Panel (e) shows the instantaneous classicality of the X-operator, $\mathcal{C}(X)$ (thick filled curve), and $(\mathcal{C}(X) + \mathcal{C}(P))/2$ (dashed thick). The thin lines show the same for the density matrix restricted to the subspace $\mathcal{H}_1$. Because of the limited number of frequencies in our toy model, the classicalities show a revival at later times and are periodically repeated.

proportional to their respective dimensions. This projection is precisely what is needed to restore the classicality of the observable X, as illustrated in Fig. 2. Where the classicality of the X-operator on the total Hilbert space (filled curve) decays as a function of time, its amplitude remains constant on the projected Hilbert space (thin line).

For more complex systems, the breakup of Hilbert space will occur for different subspaces on different time scales, as schematically illustrated in Fig. 3. At every branching point, a projection takes place. A possible scenario for surviving components is shown by the dark (red) regions. Because the different branches are eigenstates of the full Hamiltonian, elimination is definitive. A 'consistent history' [18, 19, 20] then appears naturally.

4. Discussion

The connection with the open system decoherence approach, where the environment consists of harmonic oscillators [2, 21], is straightforward. In case the system itself has a classical observable, this remains a classical observable under linear coupling to the environment. The most familiar example is a single harmonic oscillator coupled harmonic to a harmonic bath [21, 22, 23]. On the other hand, if the system does not have a classical observable, the coupling with the environment is essential to have classicality at all. Classical observables can then be constructed for the system combined with the environment after projection. The required projection operators correspond to the usual systems pointer states [24]. Our Schrödinger cat example in Fig. 2 is the simplest illustration of this mechanism. For the qubit alone $\overline{\mathcal{C}}(\sigma_{x,y}) = 1/2$, there is no classical observable. When it is coupled to a harmonic oscillator, the combined system does have classical observables after projection of the system on a pointer state. This mechanism can be generalized straightforwardly to multiple harmonic oscillators, as is the case in practice [25, 26], where the revival time (see caption of Fig. 2) tends to infinity.

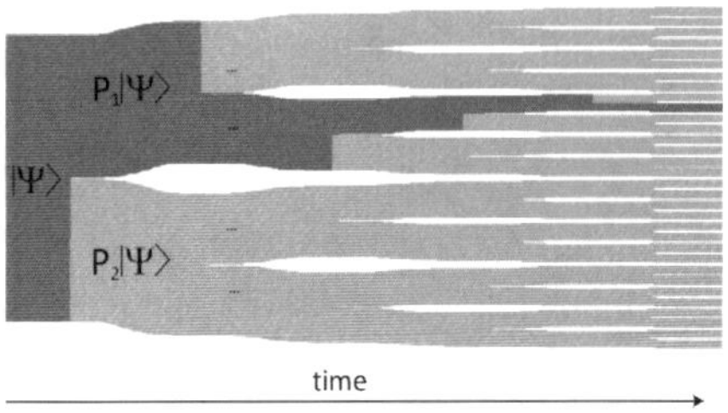

Figure 3. As time goes on, observables are classical only for a wave function projected in progressively smaller subspaces of Hilbert space. The dark (red) region shows a consistent choice for consecutively selected subspaces. The light (grey) regions are eliminated.

It is also worth discussing the connection to studies on thermalization in closed many-body quantum systems [8, 7, 9, 10]. For generic many-body systems, it is expected that no harmonic oscillator ladders exist, i.e. that from $\omega_n - \omega_m = \omega_{n'} - \omega_{m'}$ it follows that either $n = m$ and $n' = m'$ or $n = n'$ and $m = m'$ [9, 27, 28]. We then find that at late times, all classical dynamics with large SNR decay. Our investigation is thus complementary to the thermalization studies: where they are concerned with proving that most of the time for all observables the SNR is vanishingly small [9, 10], we focus on finding the observables that do show interesting dynamics in a given time window. Our analysis also sheds light on the thermalization time. Even though generic observables thermalize very quickly on the time scale of the Boltzmann time $\tau = \hbar/(k_B T)$ [29, 30], our construction allows to find atypical (i.e. classical) observables that do show much slower thermalization, set by the anharmonicity of the spectrum. Remember that the phase of the energy eigenstates was chosen such that the overlap $\langle n|\psi(t=0)\rangle$ was real. This can be done for any initial state, so that in every time window $[t, t+\tau_R]$, a similar set of classical observables can be constructed: if A is a classical observable in the interval $[0, \tau_R]$, then the Heisenberg backward evolved operator $A' = e^{-iHt}Ae^{iHt}$ is indeed trivially classical on the interval $[t, t+\tau_R]$. The relation between a specific set of classical observables and a certain time window could mean that our classical experiences are in this way related to the history of the universe. A more detailed analysis of this relation is presented in Ref. [31].

Acknowledgments
I gratefully acknowledge stimulating discussions with Dries Sels and Jacques Tempere.

5. References

[1] von Neumann J 1955, *Mathematical Foundations of Quantum Mechanics* (Princeton: Princeton University Press).
[2] Schlosshauer M 2007 *Decoherence and the quantum-to-classical transition* (Heidelberg: Springer-Verlag).
[3] Joos E, Zeh H D, Kiefer C, Giulini D, Kupsch J and Stamatescu I-O 2003 *Decoherence and the appearance of a classical world in quantum theory* (Berlin: Springer)
[4] Zurek W H 2003, Rev. Mod. Phys. **75** 715
[5] Bell J S 1989 *Speakable and unspeakable in quantum mechanics* (Cambridge: Cambridge University Press)
[6] Parcival I 2003 *Quantum State Diffusion* (Cambridge: Cambridge University Press)
[7] Polkovnikov A, Sengupta K, Silva A and Vengalattore M 2011 Rev. Mod. Phys. **83** 863
[8] Rigol M, Dunjko V, and Olshanii M 2008 Nature **452** 854
[9] Reimann P 2008 Phys. Rev. Lett. **101** 190403
[10] Short A J 2011 New J. Phys. **13** 053009

[11] Polkovnikov A 2011 Ann. Phys. **326** 486
[12] Haroche S and Raimond J-M 2006 *Exploring the quantum* (Oxford: Oxford University Press)
[13] Deléglise A *et al.* 2008 Nature **455** 510
[14] Vlastakis B *et al.* 2013 Science **342** 568
[15] Stapp H P 2002, Can. J. Phys. **80** 1043
[16] Wallace D 2002 Stud. Hist. Philos. Mod. Phys. **33** 637
[17] Everett H 1957 Rev. Mod. Phys. **29**, 454
[18] Griffiths R B 2002 *Consistent histories and the intrepretation of quantum mechanics* (Cambridge: Cambridge University Press)
[19] Gell-Mann M and Hartle J B 1993 Phys. Rev. D **47** 3345
[20] Omnès R 1992 Rev. Mod. Phys. **64**, 339
[21] Breuer H-P and Petruccione F 2003 *The theory of open quantum systems* (Oxford: Oxford university press)
[22] Feynman R P and Vernon F L 1963 Ann. Phys. **24** 118
[23] Caldeira A O and Leggett A J 1983 Physica A **121** 587
[24] Zurek W H 1981 Phys. Rev. D **24**, 1516
[25] Zurek W H 2009 Nat. Phys. **5** 181
[26] Riedel C J and Zurek W H 2010 Phys. Rev. Lett. **105**, 020404
[27] Peres A 1984 Phys. Rev. A **30**, 504
[28] Goldstein S, Hara T and Tasaki H 2013 Phys. Rev. Lett. **111**, 140401.
[29] Goldstein S, Hara T and Tasaki G 2014 Extremely quick thermalization in a macroscopic quantum system for a typical nonequilibrium subspace *Preprint* arXiv:1402.0324
[30] Malabarba A S L, García-Pintos L P, Linden N, Farrelly T C, Short A J 2014 Phys. Rev. E **90** 012121
[31] Sels D and Wouters M 2014 Quantum equivalence, the second law and emergent gravity *Preprint* arXiv:1411.3901

Journal of Physics: Conference Series **626** (2015) 012057 doi:10.1088/1742-6596/626/1/012057

How can the D-Wave machine exhibit long-time quantum behaviour

D Drakova[1] **and G Doyen**[2]

[1] Faculty of Chemistry, University of Sofia, Bulgaria
[2] Ludwig-Maximilians Universität, München, Germany

E-mail: nhdd@chem.uni-sofia.bg

Abstract. Extensive experiments have demonstrated quantum behaviour in the long-time operation of the D-Wave quantum computer. The decoherence time of a single flux qubit is reported to be on the order of nanoseconds [1], which is much shorter than the time required to carry out a computation on the timescale of seconds [2, 3]. Previous judgements of whether the D-Wave device should be thought of as a quantum computer have been based on correlations of the input-output behaviour of the D-Wave machine with a quantum model, called simulated quantum annealing, or classical models, called simulated annealing and classical spin dynamics [4]. Explanations for a factor of 10^8 discrepancy between the single flux qubit decoherence time and the long-time coherent quantum behaviour of many integrated flux qubits of the D-Wave device have not been offered so far. In our contribution we investigate a model of four qubits with one qubit coupled to a phonon and (optionally) to environmental particles of high density of states, called gravonons. The calculations indicate that when no gravonons are present, the current in the qubit is flipped at some time and adiabatic evolution is discontinued. The time dependent wave functional becomes a non-correctable superposition of many excited states. The results demonstrate the possibility of effectively suppressing the current flip and allowing for continued adiabatic evolution when the entanglement to gravonons is included. This adiabatic evolution is, however, a coherent evolution in high dimensional spacetime and cannot be understood as a solution of Schrödinger's time dependent equation in 4 dimensional spacetime. Compared to Schrödinger's time development in 4D, the evolution is considerably slowed down, though still adiabatic. The properties of our model reflect correctly the experimentally found behaviour of the D-Wave machine and explain the factor of 10^8 discrepancy between decoherence time and quantum computation time. The observation and our explanation are in anology to the 10^8 discrepancy factor found, when comparing experimental results on adsorbate quantum diffusion rate with predictions of Schrödinger's time dependent equation, which can also be resolved in a model with the coupling to gravonons included.

1. Introduction

The D-Wave machine is a quantum computer of the quantum annealing type [5]. It is designed to solve classical optimization problems, namely to optimize a cost function of many free parameters and problems, which can be mapped on the classical Ising model [6]. The way to achieve the solution is by the algorithm of quantum annealing: slow variation of external experimental parameters, so that, as the adiabatic theorem says, the system develops with time adiabatically in its energetically ground state. At the end of the quantum annealing time the D-wave presents the solution of the optimization problem. In this sense the D-wave machine is different from all other attempts to construct a quantum computer.

Journal of Physics: Conference Series **626** (2015) 012057 doi:10.1088/1742-6596/626/1/012057

The expectation that quantum computation will be faster than classical is not confirmed, but nevertheless there is sound proof that the D-Wave machine operates as a quantum computer. Numerous evidences, ranging from spectoscopic data [7] to theoretical simulations [4, 7, 8, 9], give sound basis to accept its quantum nature. This macroscopic construction, consisting of hundred to more than 500 flux qubits on a chip, interacting with each other and subjected to transverse magnetic fields, behaves coherently according to Schrödinger's time dependent equation and performs coherently quantum operations over long time of seconds and even minutes. A single flux qubit, the building block of the D-Wave device, a macroscopic object with $10^6 - 10^9$ Cooper pairs and dimensions of the order of 10^4 Å, "loses coherence" within nanoseconds in the photon field in Ramsey interference experiments [1]. The experimental observation is: the probability for switching between the two current states decays with time, interpreted as "loss of coherence", and reaches a finite value for the excitation probability larger than zero. The suggested causes for the damping of the amplitude of the oscillations range from experimental imperfections and poor control of the electromagnetic environment of the flux qubit, to external and internal noises of different nature and the involvement of various dissipative channels (charge noise, photon noise, quasiparticle excitations above the superconducting gap, radiative and dielectric loss, etc.).

The problem we focus on is the discrepancy of 8 orders of magnitude difference in the coherence time of a single flux qubit and the coherent operation of the D-Wave machine consisting of many flux qubits. We solve the time dependent Schrödinger equation with a Hamiltonian including the entanglement of the current states to gravonons, existing in high dimensional spacetime. The quantum theory lets Copenhagen quantum mechaniscs emerge, without postulating collapse, and provides a deterministic definite outcome of a "measurement", hence the name Emerging Quantum Mechanics [10]. It has been successfully used to solve the problem of localization of quantum particles and their appearance as classical particles [11, 12], the problem of quantum diffusion of atoms on solid surfaces [13] in a telegraph-signal like dynamics [14, 15] and wave-particle duality in matter wave diffraction [16]. Within the theory of Emerging Quantum Mechanics we explain the "short coherence time" of 20 ns of a single flux qubit (section 2) and the quantum behaviour of the D-Wave machine (section 4) over a period of time 8 orders of magnitude longer than that of a single flux qubit.

2. Constructing the Hamiltonian for a single flux qubit

The environmental excitations are the gravonons, massive bosons which emerge in the limit of weak and local gravitational interaction in high dimensional spacetime (11D) [10]. The coupling to gravonons is effective only within spacetime deformations called warp resonances. The model for the non-perturbed gravonons $\{| \, grav_i\rangle\}$ is a harmonic oscillator, whereas the gravonons perturbed by matter and force fields $\{| \, \kappa_i\rangle\}$ are the solution of the $\gamma - \eta$ model described in [13].

The Hamiltonian for the single flux qubit plus photon is:

$$\begin{aligned}
\mathrm{H}_{qubit} &= \mathrm{H}_o + \mathrm{H}_{phot} + \mathrm{H}_{qubit-grav} \\
&= \sum_{j=1}^{2} \left[E_{qubit_j} a^+_{qubit_j} a_{qubit_j} + E_{w_j} a^+_{w_j} a_{w_j} + V^{qubit_j}_{loc} (a^+_{qubit_j} a_{w_j} + a^+_{w_j} a_{qubit_j}) \right] \\
&+ \omega_{phot} b^+_{phot} b_{phot} + \sum_{j=1}^{2} \left[\varepsilon_{grav_j} c^+_{grav_j} c_{grav_j} + \sum_{\kappa_j} \varepsilon_{\kappa_j} c^+_{\kappa_j} c_{\kappa_j} \right] \\
&+ V_{phot,qubit} (a^+_{qubit_1} a_{qubit_2} + a^+_{qubit_2} a_{qubit_1})(b^+_{phot} + b_{phot}) \\
&+ \sum_{j=1}^{2} \sum_{\kappa_j} \left[W_{grav_j,w_j} n_{w_j} c^+_{grav_j} c_{\kappa_j} + W_{w_j,grav_j} n_{w_j} c^+_{\kappa_j} c_{grav_j} \right]. \qquad (1)
\end{aligned}$$

The meaning of the symbols is w_j: regions in the Josephson junction (warp resonances), where

Cooper pairs couple to gravonons. $a^+_{w_j}$, a_{w_j}, n_{w_j}: creation, annihilation and number operator for persistent current in the state $| w_j\rangle$ in the Josephson junction; $a^+_{qubit_j}$, a_{qubit_j}: creation and annihilation operator for persistent current in the state $| qubit_j\rangle$ in the flux qubit; b^+_{phot}, b_{phot}: creation and annihilation operator for the photon; $c^+_{grav_j}$, c_{grav_j}: creation and annihilation operator for the lowest energy gravonon in the potential, which is not perturbed by interaction between the matter field and the photon field; $c^+_{\kappa_j}$, c_{κ_j}: creation and annihilation operator for gravonons in the deformed manifold.

The terms in the second line of the Hamiltonian eq. (1) describe the two current states in the qubit (first term), in regions of the Josephson junction where the Cooper pairs couple to the gravonons w_j (second term) and their interaction $V^{qubit_j}_{loc}$ (third term). The first term in the third line of eq. (1) describes the photon field with frequency ω_{phot}, the second term the gravonons in the initial wave packet and the third term the excited gravonons due to interactions with the supercurrents in the Josephson junction. The interaction of the supercurrent with the photon is the standard coupling between a photon and matter, i.e. the common dipole coupling term (fourth line of eq. 1). The last term in eq. (1) is the coupling between gravonons and the supercurrents of the quadrupole coupling kind.

The time dependent Schrödinger equation is solved for the time development of the total wave functional: $i\hbar\frac{\partial}{\partial t}\Psi(t) = \mathrm{H}_{qubit}\Psi(t)$, which is represented in the basis of field configurations. These are tensor products of the field states of the supercurrent, the photon field and the gravonon field. In our theory the interaction with the environment is included in the Hamiltonian, the environment is accounted for, when we solve Schrödinger's equation. In contrast, stochastic approaches, starting from the Liouville equation for the dynamics of the density matrix with the total Hamiltonian of the system plus environment, revert to an equation where the system alone is included in the Hamiltonian, whereas the effects of the environment are in the Lindblad term [17].

The fields in the system are the persistent current of the qubit ϕ_{qubit}, the photon field ϕ_{phot}, the gravonon field comprising $\{| \ grav\rangle\}$, $\{| \ \kappa\rangle\}$ and $\{| \ \lambda\rangle\}$. The notation for the field configurations is $| \ \phi_{qubit}, \phi_{phot}, \kappa, \lambda\rangle$, for instance $| \ 1, 0, 0, 0\rangle$ for the initial wave packet, $| -1, 1, 0, 0\rangle$ for the switched current configuration plus photon, etc. According to the assumed principle that only those warped field configurations which arise from the flat field configurations (non-warped configurations) via first order transitions are taken into account, only configurations of the kind $| \ \phi_{qubit}, \phi_{phot}, \kappa, 0\rangle$ and $| \ \phi_{qubit}, \phi_{phot}, 0, \lambda\rangle$ are involved.

The physical nature of the gravonons as massive bosons arising from gravitons, as well as the derivation of an effective Schrödinger equation in high dimensional spacetime *yielding gravonons similar to the common quantum particles as its solution*, are described in detail in ref. [10]. The gravonons are the only quanta which reside not only in 4 dimensional spacetime, but in the additional compactified hidden spacial dimensions. The decisive features of the gravonons are the high density of the gravonon quanta and weak coupling to the matter fields.

3. Explaining the coherence time of a single flux qubit in the photon field

The solution of the time dependent Schrödinger equation gives the time dependent total wave functional as a superposition of the basis configurations, varying with time. The initial wave packet in interaction with the rest configurations, but not with the gravonons, splits into two components below and above its energy, corresponding mostly to the two current configurations, clockwise and anticlockwise. With the entanglement with the gravonons they acquire finite broadening and shifts as it is shown in the left panel of fig. 1. Each group of eigenstates has entangled to the nearly degenerate part of the gravonon continuum. Oscillations between the eigenstates below and above the energy of the initial wave packet occur in the absence of entanglement with the gravonons. This is the situation for short times. Since the total wave

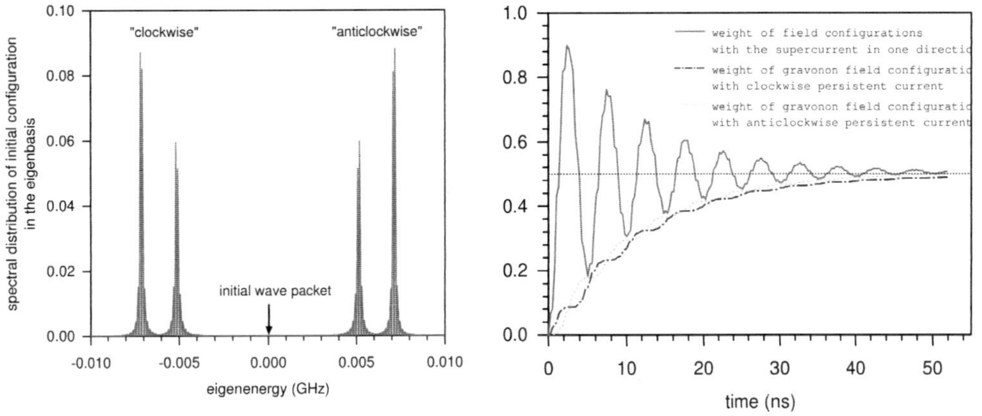

Figure 1. Left panel: spectral weight of the initial wave packet $| \ 1,0,0,0\rangle$ in the gravonon band resulting from the interactions in the local 4 dimensional subspace and the broadening due to entanglement of all local configurations with the gravonons. The configurations at energy lower than that of the initial wave packet are mostly with clockwise persistent current $| \ 1,0,0,0\rangle$ and $\{| \ 2,0,\kappa,0\rangle\}$. Those with higher energy are dominated by configurations with anticlockwise persistent current plus photon $| \ -1,1,0,0\rangle$ in the flux qubit and $\{| \ -2,1,0,\lambda\rangle\}$ in the warp resonance.
Right panel: The development with time of the weight of all field configurations with the supercurrent in one direction in a single flux qubit (full red curve) shows oscillations at short time which are exponentially damped, leading to a finite final value at long time. The reason for the damping of the oscillations is the development of the entanglement of the current states with gravonons which hinders and finally precludes the current switching. The weight of field configurations with gravonon components for the two current states are plotted as functions of time with green (short dashed) and blue (long dashed) curves.

functional has not yet acquired the components with the excited gravonons, it can be written as:

$$\Psi_1(t) = a(t) \mid 1,0,0,0\rangle + b(t) \mid 2,0,0,0\rangle + c(t) \mid -1,1,0,0\rangle + d(t) \mid -2,1,0,0\rangle \quad (2)$$

However, the entanglement with the gravonons develops with time. The weights of warped field configurations with the persistent current clockwise, as well as those with the persistent current in the anticlockwise direction, increase with time:

$$\Psi(t) = A(t)\Psi_1(t) + \sum_{\kappa} k_\kappa(t) \mid 2,0,\kappa,0\rangle + \sum_{\lambda} l_\lambda(t) \mid -2,1,0,\lambda\rangle \quad (3)$$

At $t = 50$ ns, when the entanglement to the gravonons has fully developed, only the warped field configurations in the total wave functional eq. (3) contribute, i.e. $A(t)$ tends to zero. They saturate at some final and finite values, i.e. the coefficients in the expansion of $\Psi(t)$ in the basis field configurations do not change with time any more. The single flux qubit is in a superposition of the two persistent current configurations, which does not change with time. In the time before the freezing of the current distribution in the flux qubit occurs, of course, the oscillations between the two current configurations are still discernible, as it shown by the full red curve on the right panel in fig. 1, though with damped amplitude. At time $t = 50$ ns a single

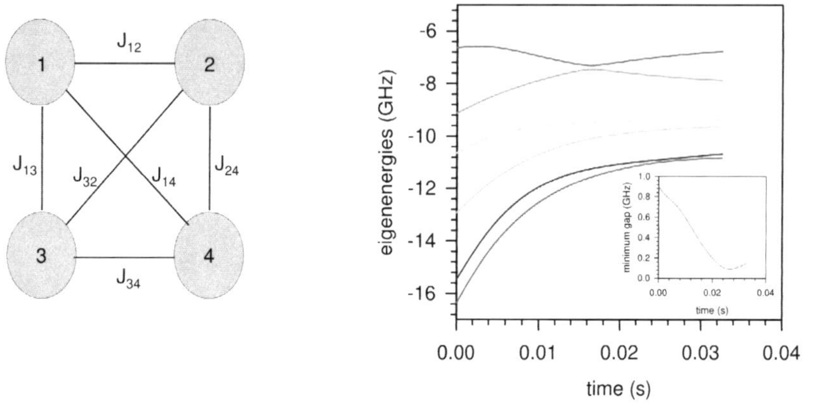

Figure 2. Left panel: a model of a four flux qubit processor implementing the time dependent Hamiltonian eqs. (4-6). The theoretical model of the D-Wave machine consists of 4 flux qubits interacting with each other with interaction strengths $J_{\alpha\beta}$ and subjected to external transverse magnetic fields with strength $ht_{\alpha}(t)$ varying with time.
Right panel: variation with time of the lowest eigenenergies of the 4 flux qubit system. The gap between the global energy minimum and the first excited field configuration is plotted in the inset, achieving a minimum value of less than 0.1 GHz at the time of avoided crossing.

flux qubit is frozen in a configuration in which the relative contributions of current clockwise and current anticlockwise in the superposition does not change. At that time, according to our theory, the flux qubit is maximally entangled with the gravonon environment in a coherent state. The reason not to change with time is that, due to entanglement with the gravonons, the components and their contributions to the total superimposed quantum configuration cannot change while the gravonons are in the hidden spacial dimensions. Recurrence of the gravonons back in three dimensional space, disentanglement from the degenerate gravonon band the local current configuration is originally entangled with, and entanglement with the gravonon band, degenerate with the current configuration of opposite direction, are needed to change the current configuration in the flux qubit. As these local field configurations are not energetically degenerate, the change of current requires energy.

The theoretical result in the right panel of fig. 1 is in satisfactory agreement with the experimental Ramsey fringes in ref. [1] (fig. 4A).

4. Explaining the coherence time of the D-Wave machine: four flux qubits in transverse magnetic fields

The theoretical model of the D-Wave machine consists of four qubits, as it is shown in the left panel of fig. 2, interacting with one another and subjected to external transverse magnetic fields which vary with time. The Hamiltonian for quantum annealing H_{QA} is now time dependent since the transverse external magnetic fields on the flux qubits vary with time:

$$\mathrm{H}_{QA}(t) = \sum_{\alpha=1}^{4} \mathrm{H}_{o,\alpha} + \mathrm{H}_{P} + V(t) + \mathrm{H}_{qubit-grav,3}. \tag{4}$$

Four terms describing the non-interacting flux qubits are taken from the Hamiltonian in eq. (1), however, exempting the interaction with the photon field in the fourth line of this equation.

Journal of Physics: Conference Series **626** (2015) 012057 doi:10.1088/1742-6596/626/1/012057

Gravonon coupling is introduced only within one qubit. H_{QA} includes terms describing the interaction between the flux qubits:

$$\mathrm{H}_P = \sum_{\alpha \neq \beta = 1}^{4} J_{\alpha\beta} \sigma_\alpha^z \sigma_\beta^z \tag{5}$$

and the interaction with the external magnetic fields:

$$V(t) = \sum_{\alpha=1}^{4} ht_\alpha(t) \sigma_\alpha^x \tag{6}$$

σ_α^z and σ_α^x are the Pauli matrices. The manipulation with time of the magnetic fields represents the quantum annealing procedure. Reducing the external transverse magnetic fields $\{ht_\alpha\}$ slowly ensures the adiabatic time development of the system towards the ground state of the Ising Hamiltonian, which is the solution of the optimization problem.

The time dependent Schrödinger equation is solved with a time dependent Hamiltonian:

$$\Psi(t) = \exp\left[-\frac{\mathrm{i} \int_0^t \mathrm{d}\tau \mathrm{H}_{QA}(\tau)\tau}{\hbar}\right] \Psi(0). \tag{7}$$

The computational basis consists of 16 flat field configurations denoted by 4 labels, which indicate the state of each flux qubit by 1 for current in the clockwise direction and -1 for current in the anticlockwise direction:

$$\begin{aligned} &|\,1111\rangle \;\; |\,111-1\rangle \;\; |\,11-11\rangle \;\; |\,1-111\rangle \;\; |\,{-1}111\rangle \;\; |\,11-1-1\rangle \;\; |\,1-11-1\rangle \;\; |\,{-1}11-1\rangle \\ &|\,1-1-11\rangle \;\; |\,{-1}1-11\rangle \;\; |\,{-1}-111\rangle \;\; |\,{-1}-1-11\rangle \;\; |\,{-1}-11-1\rangle \;\; |\,{-1}1-1-1\rangle \\ &|\,1-1-1-1\rangle \;\; |\,{-1}-1-1-1\rangle. \end{aligned} \tag{8}$$

They represent eigenstates of the Ising Hamiltonian. The notation for the gravonons is omitted, being in all cases the ground configurations $|\,00\rangle$ with gravonons in one qubit only. Thus without gravonons we have a 16×16 matrix of configurations to diagonalize to get the eigenenergies of the 4 flux qubit system. The lowest eigenenergies and their variation with time can be seen in the right panel of fig. 2.

At $t = 0$ the effect of $ht_\alpha(t = 0) \neq 0,\ \alpha = 1, ..., 4$ is to mix in all computational basis configurations, leading to the ground state configuration of the 4 flux qubit system in the presence of the transverse magnetic fields. At the annealing time $t = t_f$ the external local transverse magnetic fields are switched off, i.e. $ht_\alpha(t_f) = 0,\ \alpha = 1, ..., 4$ and the 4 flux qubit system provides the solution of the Ising Hamiltonian eq. (4) with $V(t_f) = 0$. As the coupling to the gravonons is very weak and local, the broadening into resonances of the eigenenergies of the 4 flux qubit system is orders of magnitude smaller than the energy differences between the eigenstates even at the time of avoided crossing (cf. the inset in the right panel of fig. 2). Hence, the energy broadening, due to entanglement to gravonons, cannot lead to superpositions of the eigenstates of the 4 flux qubit system, which would destroy the adiabaticity.

In analogy with the single flux qubit case, while entangled with the gravonons each flux qubit is frozen after 50 ns in a configuration, in which the relative contributions of the two current states in this qubit does not change with time any more. Each flux qubit is stabilized in this superposition, which is typical for the global ground state configuration. Hence, the global ground state of the D-Wave machine cannot change either. Once in the ground configuration, the system of many flux qubits will stay in the ground configuration for ever. Entanglement to gravonons stabilizes the global ground state configuration and helps the coherent adiabatic time development of the D-Wave machine towards the final solution.

Journal of Physics: Conference Series **626** (2015) 012057 doi:10.1088/1742-6596/626/1/012057

4.1. Destroyed adiabaticity of the D-Wave machine: four flux qubits in a phonon field

Thermal excitations cannot be avoided even at the millikelvin temperatures at which the D-Wave machine operates. It was experimentally demonstrated in ref. [3] for a 16 qubit D-Wave machine that the success probability at temperature in the range 20-100 mK is higher than at lower temperature in contrast to other attempts for quantum computing.

Let a phonon be excited in just one of the flux qubits, e.g. the third qubit in fig. 2. Terms are added to the quantum annealing Hamiltonian H_{QA} eq. (4) for the phonon and its interaction with the supercurrent of the kind:

$$\mathrm{H}_{phon} = V_{qubit,phon}\sigma_3^x(d_{phon}^+ + d_{phon}) + \frac{1}{2}\omega_{phon}d_{phon}^+ d_{phon}. \qquad (9)$$

The term H_{phon} is analogous to the Jaynes-Cummings Hamiltonian of a two-level atomic system, coupling to the phonon field with coupling strength $V_{qubit,phon}$. d_{phon}^+ and d_{phon} are creation and annihilation operators for the phonon mode and ω_{phon} is the phonon frequency. With the definition of H_{phon} in eq. (9) the effect of the phonons is to account just for switching the supercurrent in the third qubit.

The result of a sudden perturbation of the 4 flux qubit system by a single phonon mode is destroyed adiabaticity. Figure 3 (left panel) shows that a fast entanglement to the phonon field around $t = 0.01$ s perturbs the distribution of the supercurrents in the third flux qubit, which is typical for the global ground state configuration of the 4 flux qubit system. Dramatic and nonrepairable arbitrary switches between the currents in the flux qubit occur which means that the global ground configuration of the 4 flux qubit system is changed, hence its adiabatic time development is destroyed and cannot be recovered. The coupling to the phonon leads to destroyed adiabatic time development in the global ground configuration of the 4 flux qubit system, as it can be seen in the plot on the left panel in fig. 3. The quantity on the vertical axis is the number of the eigenstate of the non-perturbed 4 qubit system and not its energy. The areas of the points in the plot scale with the weight of the eigenstates of the 4 flux qubit system, which get involved in $\Psi(t)$. When the perturbation by the phonon is switched on excited field configurations of the 4 flux qubit system gain weight and a significant redistribution from the global ground configuration over several excited configurations occurs. Thus, coupling to a phonon mode destroys the adiabatic time development of the system in an irrepairable way.

4.2. Gravonons suppress the effect of the phonon: four flux qubits in phonon and gravonon fields

In the Josephson junction the tunnelling currents entangle with gravonons in high spacial dimensions. The result is suppression of the effect of the phonons, and retained coherence of the system in the adiabatic ground state configuration, as it can be seen in fig. 4. The length of the symbols in the plot scales with the weight of the initial wave packet in the gravonon band, i.e. it is the spectral distribution of the initial wave packet in the gravonon continuum. The entanglement with the gravonons quenches the transitions between the low lying eigenstates of the 4 flux qubit system due to the perturbation by the phonon. The spectral distribution of the initial wave packet in the gravonon continuum is not affected by the excitation of the phonon in one of the flux qubits. As a consequence the many-qubit ground state configuration is not modified by the interaction with the phonons. The 4 flux qubit system develops adiabatically in the global ground configuration for long time significantly exceding the life time of the phonon. The gravonons suppress the effect of the phonon. While being entangled with the gravonons in the hidden dimensions, the current cannot switch direction. Entanglement to gravonons stabilizes the adiabatic ground state configuration of the D-Wave machine.

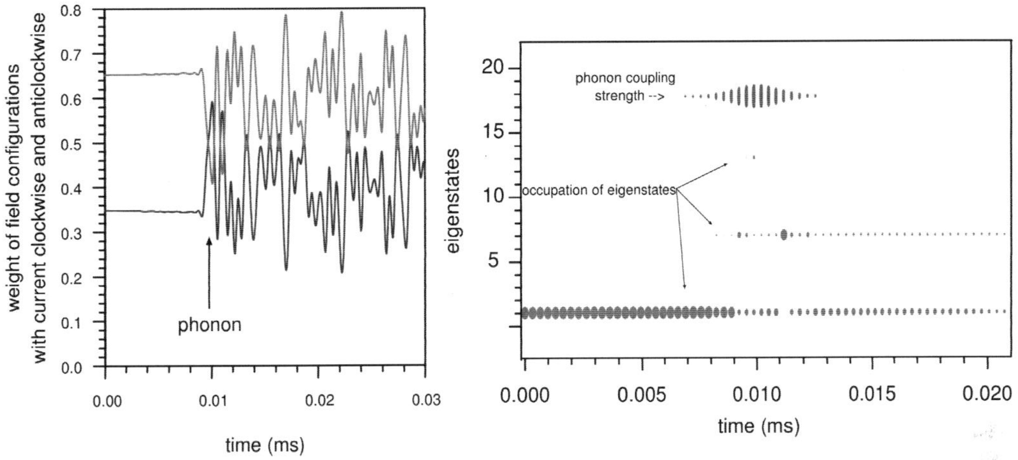

Figure 3. Destroyed current distribution and chaotic current flips in the third flux qubit as a result of the fast entanglement to the phonon field at $t \approx 0.01$ ms (left panel). On the right panel the destroyed adiabaticity in the global ground state configuration of the 4 flux qubit system as a result of the fast entanglement to the phonon field is illustrated with the variation of the weight of the low energy eigenstates at the time when the coupling in the third flux qubit to the phonon is effective. The weight of the non-perturbed eigenstates which are involved in $\Psi(t)$ is displayed with varying size of the points. When the perturbation by the phonon is switched on the variation of the weight of the low energy eigenstates of the 4 flux qubit system shows a significant redistribution from the global ground state configuration over several excited configurations. The quantity on the vertical axis is the number of the eigenstate of the non-perturbed 4 flux qubit system.

5. Conclusion

The present study of a quantum computer consisting of numerous flux qubits uses Emerging Quantum Mechanics, a method based on Schrödinger's time dependent quantum mechanics which accounts for the entanglement of fields in 4 dimensional spacetime with gravonons living in high dimensional spacetime [10]. The astonishing result is that a single flux qubit and the D-Wave computer, both macroscopic objects, retain coherence and behave as quantum objects according to the laws of quantum mechanics with long coherence time of the order of minutes. The clue to this result is the entanglement of the persistent currents in the flux qubits with the gravonons, the massive quanta of the gravitational field, which live in high spacial dimensions. The necessary condition for this result is weak coupling of the local quantum fields to an environmental continuum of high density of states, which is satisfied only by the gravonons in high spacial dimensions.

The major results of the present study of the D-Wave quantum computer can be summarized:

- The "coherence" time observed experimentally of a single flux qubit is explained, obtaining acceptable agreement with experiment. The explanation is based on entanglement with the gravonon continuum.
- The entanglement to gravonons also explains the coherence time of the D-Wave quantum computer.
- The entanglement to gravonons also explains why phonons do not perturb the quantum

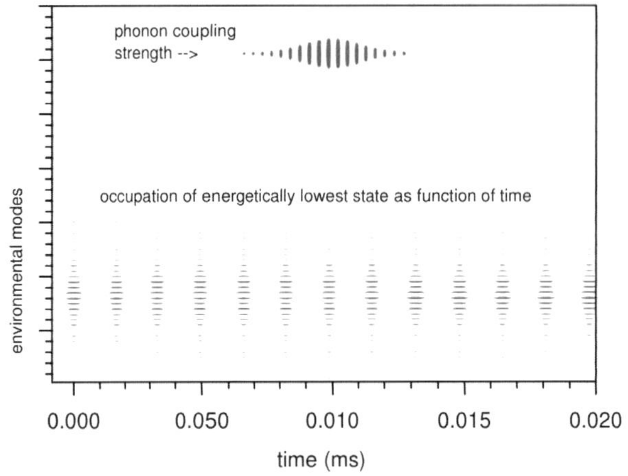

Figure 4. Entanglement of the currents with gravonons within the Josephson junction suppresses the effect of the phonon and recovers the adiabaticity of the global ground state configuration of the 4 flux qubit system. The quantity on the vertical axis is the number of gravonon state in the gravonon continuum. The length of the symbols in the plot scales with the weight of the initial wave packet in the gravonon band.

annealing process of the D-Wave machine.

These results as well as the experimental observations on the D-Wave quantum computer cannot be reproduced and explained either by stochastic quantum approaches or within conventional decoherence theory in four dimensions. This has, however, not been explicitely demonstrated in the paper so far.

Emerging Quantum Mechanics reproduces experimental observations of very different nature, where orders of magnitude discrepancies have been found by computations based on conventional quantum mechanics in three dimensional space. They include quantum diffusion of adsorbates on solid surfaces [13], and double slit diffraction experiments with massive molecules [16]. In these cases localization of quantum particles in three dimensional space is the result of the entanglement with the gravonons which live in high spacetime dimensions.

References

[1] Chiorescu I, Nakamura Y, Harmans C J P M and Mooij J E 2003 *Science* **299** 1869
[2] Johnson M W et al. 2011 *Nature* **473** 194
[3] Dickson N G et al. 2013 Nature Communications 4:1903, DOI: 10.1038/ncomms2920
[4] Boixo S, Rønnow T F, Isakov V, Wang Z, Wecker D, Lidar D A, Martinis J M and Troyer M 2014 *Nature Physics* **10** 218; *do.* 2013 *Preprint* arXiv:1304.4595v2
[5] cf. www.dwavesys.com
[6] Ising I 1924 *Contribution to the Theory of Ferromagnetism* PhD Thesis (Univ. Hamburg)
[7] Lanting T et al. 2014 *Phys. Rev.* X **4** 021041; *do.* 2014 *Preprint* arXiv:1401.3500
[8] Wang L, Rønnow T F, Boixo S, Isakov S V, Wang Z, Wecker D, Lidar D A, Martinis J M and Troyer M 2013 *Preprint* arXiv:1305.5837v1
[9] Albash T, Vinci W, Mishra A, Warburton P A, Lidar D A 2014 *Preprint* arXiv:1403.4228v3
[10] Doyen G and Drakova D 2014 *Preprint* arXiv:1408.2716
[11] Doyen G and Drakova D 2011 *J. Physics: Conf. Series* **306** 012033
[12] Drakova D and Doyen G 2011 *J. Physics: Conf. Series* **306** 012068
[13] Drakova D and Doyen G 2013 *J. Physics: Conf. Series* **442** 012049
[14] Drakova D and Doyen G 2012 *Preprint* arXiv:1204.5606 7

Journal of Physics: Conference Series **626** (2015) 012057 doi:10.1088/1742-6596/626/1/012057

[15] Doyen G and Drakova D 2013 *J. Physics: Conf. Series* **442** 012032
[16] Doyen G and Drakova D *present volume*
[17] Lindblad G 1976 *Commun. Math. Phys.* **48** 119

Journal of Physics: Conference Series **626** (2015) 012058 doi:10.1088/1742-6596/626/1/012058

Nonlinear electrodynamics and modification of initial singularities, and dark matter and dark energy affecting structure formation in the early and later universe

A W Beckwith

Physics Department, Chongqing University, Huxi Campus, College of Physics, No. 55 Daxuechen Nanlu, Shapingba District, Chongqing 401331, People's Republic of China

E-mail: Rwill9955b@gmail.com

Abstract. We find that having the scale factor close to zero due to a given magnetic field value in an early universe magnetic field affects how we would interpret Mukhanov's chapter on "self reproduction" of the universe in in his reference. The stronger an early-universe magnetic field is, the greater the likelihood of production of about 20 new domains of size $1/H$, with H the early-universe Hubble constant, per Planck time interval in evolution. We form DM from considerations as to a minimum time step, and then generate DM via axions. Through Ng's quantum infinite statistics, we compare a DM count, giving entropy. The remainder of the document is in terms of DE as well as comparing entropy in galaxies versus entropy in the universe, through a lens of Mistra's quantum theory of the big bang.

1. Introduction

This paper takes several routes to identifying nonlinear electrodynamic (NLED) phenomena pertinent to cosmological structure formation. First, we look at what Mukhanov writes as far as structure formation: Mainly, there is a so-called self-reproduction of inhomogeneity in terms of early universe conditions [1]. In this, the starting point is if one used the meme of chaotic inflation—i.e., inflation generated by a potential given by [1]:

$$V(potential) \sim \phi^2. \tag{1}$$

In this, Mukhanov writes that one can look at a scalar field at the end of (chaotic) inflation, with an amplitude given by ϕ, and the initial value of the inflaton such that

$$\delta_\phi^{\max} \sim m \cdot \phi_i^2, \tag{2}$$

where m will be determined by NLED inputs. In terms of the initial inflaton, inhomogenities do not form if the initial inflaton is bounded [1] as given by

$$m^{-1} > \phi_i > m^{-\frac{1}{2}}. \tag{3}$$

This leads to (low?) inhomogeneity in spacetime generated by inflation. Inflation is eternal [1] if there is only the inequality

$$\phi_i > m^{-\frac{1}{2}}. \tag{4}$$

Journal of Physics: Conference Series **626** (2015) 012058 doi:10.1088/1742-6596/626/1/012058

2. NLED applied to Eq. (4) plus details of structure formation added

What we will do is to look at the following treatment of mass, and this will be our starting point. That is, we will be looking at if l_p is a Planck length, and $\alpha > 0$, then

$$m \sim 10^{\alpha} \cdot l_p^{\alpha} \cdot \rho(density). \tag{5}$$

Then we can consider the following formulation of density given below. If we do not wish to consider a rotating universe, then Camara et al. [2] has an expression as to density, with a B field contribution to density, and we also can use the Weinberg result [3] of scaling density with one over the fourth power of a scale factor, which we remark upon in the general section, as well the Corda and Cuesta result of [4] for density (note [4] is for a star, [2] is for a universe). In addition, Corda and Cuesta [4] use quintessential density to falsify the null energy condition of a Penrose theorem cited in [4, 5]. Further details of what Penrose was trying to do as to this issue of GR can be seen in [4, 5]. To answer how to violate the null energy condition, one should go to [5] for quintessential density defined, with the constant in Eq. (4) greater than zero. Then in both the massive star and the early universe, the density result below is applicable.

$$\rho_\gamma = \frac{16}{3} \cdot c_1 \cdot B^4. \tag{6}$$

Keeping in mind what was said as to choices of what to do about density, and its relationship to Eq. (5) above, we then can reference what Mukhanov [1] says about structure formation as follows, namely look at how a Hubble parameter changes with respect to cosmic evolution. It changes with respect to H_{today} being the Hubble parameter in the recent era, and the scale factor a, with this scale factor being directly responsive to changes in density according to [3]—i.e.

$$\rho \sim a^{-4}. \tag{7}$$

In the next section, we will examine how [2] suggests how to vary the scale factor cited in Eq. (7), and we will in this section take note of what the scale factor cited in [2] does to the Hubble parameter given in Eq. (8) below, and then in the section afterwards review a possible reconciliation of what Eqs. (6) and (7) say about defining early-universe parameters. But to know why we are doing it, we should take into consideration what happens to the Hubble parameter, as given below.

$$H \sim \frac{H_{\text{today}}}{a^{\frac{3}{2}}}. \tag{8}$$

According to [1], if Eq. (4) holds, then inhomogeneous patches of spacetime appear in a causal region of spacetime for which

$$Causal - domain \sim H^{-1} \sim \frac{1}{\frac{H_{\text{today}}}{a^{\frac{3}{2}}}}. \tag{9}$$

Furthermore, [1] states that about 20 such domains are created in a Hubble time interval $\Delta t_H \propto H^{-1}$. As a function of say 10^a times Planck time, for a domain size given by Eq. (9) above and that this requires then a clear statement as to how the scale factor changes, due to considerations given by [3] and reconciling the density expression given in Eq. (6) and Eq. (7) above.

3. Showing a nonzero initial radius of the universe due to nonlinear spacetime E&M

What we are asserting is in [1] there exists a scaled parameter λ, and a parameter a_0, which is paired with α_0. For the sake of argument, we will set the $a_0 \propto \sqrt{t_{\text{Planck}}}$, with $t_{\text{Planck}} \sim 10^{-44}$ seconds. Also, Λ is a cosmological "constant" parameter which is described later, as in quintessence, via [6], and as in [2] via:

$$a_0 = \sqrt{\frac{4\pi G}{3\mu_0 c}} B_0 \text{ and} \tag{10}$$

$$\lambda = \frac{\Lambda c^2}{3}. \tag{11}$$

Then if, initially, Eq. (11) is large, due to a very large Λ the time, given in Eq. (53) of [2] is such that we can write, most likely, that, even though there is an expanding and contracting universe, the key time parameter may be set, due to very large Λ as

$$t_{\min} \approx t_0 \equiv t_{\text{Planck}} \sim 10^{-44}\text{s}. \tag{12}$$

Whenever one sees the coefficient like the magnetic field, with the small 0 coefficient, for large values of Λ, this is the initial coefficient at the beginning of spacetime which helps us make sense of the nonzero but tiny minimum scale factor [2]

$$a_{\min} = a_0 \cdot \left[\frac{\alpha_0}{2\lambda} \left(\sqrt{\alpha_0^2 + 32\lambda\mu_0\omega B_0^2} - \alpha_0 \right) \right]^{\frac{1}{4}}. \tag{13}$$

The minimum time, as referenced in Eq. (12), most likely means, due to large Λ that Eq. (13) is of the order of about 10^{-55}—i.e., 33 orders of magnitude smaller than the square root of the Planck time. We next will be justifying the relative size of Λ.

4. Showing how to obtain a varying Λ with a large initial value and its relationship to obtaining a scale factor value for the early universe via NLED methods

Notwithstanding the temperature variation in reference [2] for the cosmological Hubble parameter, we can also write

$$\Lambda(t) \sim (H_{\text{inflation}})^2. \tag{14}$$

In short, what we obtain, via looking at due to [6] that Eq. (14) is also equivalent to

$$\Lambda_{\max} \sim c_2 \cdot T^{\beta} \tag{15}$$

where T is temperature. Comparing Eq. (6) and Eq. (7) above, leads to the following constraints—i.e., if we extend [1, 2] we would get

$$\left(\rho \sim a^{-4}\right)^{-1} \sim a^4 \sim \frac{16}{3} \cdot c_1^{-1} \cdot B^{-4} \sim \frac{a_0^{-4}}{\left[\frac{\alpha_0}{2\lambda} \left(\sqrt{\alpha_0^2 + 32\lambda\mu_0\omega B_0^2} - \alpha_0 \right) \right]} \tag{16}$$

The above relationship will argue in favor of a large value for Eq. (15) and Eq. (16) B field and also the cosmological "constant" parameterized in Eq. (14) and Eq. (15). That is, once fully worked out, the allowed values of B, for initial conditions will be large but tightly constrained, and this in turn will allow for Eq. (9) having initially extremely small inhomogeneity behavior, in line with being proportional to the inverse of an allowed Hubble parameter based upon Eq.

(8). Note that Eqs. (11) to (13) are arguing in favor of a very small scale factor, implying a large initial density while Eq. (16) appears to give credence to a large Hubble parameter. Further work will come up with a set of constraints as to admissible early universe quintessence. Further refinements of NLED, and the relationship in Eq. (16), as to structure formation may give credence, or help falsify the conclusions of reference [1], with great refinements needed in defining the suggested relationships implied in Eq. (16)

5. Density in the case of certain strengths for the magnetic field: the weak energy condition versus a more specialized expression which violates the null energy condition of the Penrose theorem

In [2], there is a generalized density

$$\rho - \frac{1}{2\mu_0} \cdot B^2 \cdot \left(1 - 8 \cdot \mu_0 \cdot \omega \cdot B^2\right). \tag{17}$$

This density has a positive value only if [2]

$$B < \frac{1}{2 \cdot \sqrt{2\mu_0 \cdot \omega}}. \tag{18}$$

This density expression should be compared with the Corda and Cuesta [4] value of "quintessential density" given in Eq. (4). This Eq. (4) is how Corda and Cuesta violated the null energy condition of the Penrose theorem [4, 5]. Eq. (4) is part of the Corda and Cuesta pressure equation [4]:

$$p = \frac{\rho}{3} - \rho_\gamma. \tag{19}$$

We do believe that Eq. (16) is more general, although the magnetic-field dependence is far more complicated. The importance of the pressure in Eq. (19) goes back to temperature: The higher the pressure is, the more likely extreme temperatures exist that will let us talk about the role of entropy and DM/DE next. We assert that large pressure will influence a large temperature, which in turn has implications as to the following relationship. From [2], we derived the following quark–gluon result

$$\begin{aligned} \varepsilon(energy - density) \sim \varepsilon_{\text{QGP}} \simeq 47.5 \times \frac{\pi}{30} \times (k_B T)^4 \\ p = \frac{\varepsilon(energy - density)}{3} \sim 47.5 \times \frac{\pi}{30} \times (k_B T)^4 \\ p(high) \Leftrightarrow T(high) \end{aligned} \tag{20}$$

Our large scale value of temperature applies, next to the entropy of both DM and DE.

6. We discuss DM which we identify with axions, this after the following identification

In [7, 8], one million or more BHs in the center of an equal number of galaxies leads to an entropy

$$S_{\text{Total}} \sim 10^6 \times 10^{96} \sim Galazies(SMBH - center) \times [entropy, SMBH] \tag{21}$$

To understand what this means, we will review our version of Mishra's [9] quantum theory of the big bang to find linkages to axions, for DM and other issues. Begin now with Mach's principle, with M being the mass (of the universe), and R_0 being a radii of a presumed spherical space, then if R_0 is the presumed radii of the universe, and also the total following value of entropy:

$$\frac{GM}{R_0 c^2} \approx 1, \tag{22}$$

$$M_{\text{Total}} \sim M_{\text{DM}} + M_{\text{Baryon}} + M_{\text{DE}} = N_{\text{DM}} m_{\text{DM}} + N_{\text{Baryon}} m_{\text{Baryon}} + N_{\text{DE}} m_{\text{DE}} \approx S_{\text{DM}} m_{\text{DM}} + S_{\text{Baryon}} m_{\text{Baryon}} + S_{\text{DE}} m_{\text{DE}}. \quad (23)$$

Here, total entropy of the universe is assumed to be in the present era

$$S_{\text{Total}} \sim S_{\text{DM}} + S_{\text{Baryon}} + S_{\text{DE}}. \quad (24)$$

The use of entropy as akin to particle count comes from two sources. First source is due to "infinite quantum statistics" as given by Ng [10]:

$$Z_N \sim \left(\frac{1}{N!}\right) \cdot \left(\frac{V}{\lambda^3}\right)^N. \quad (25)$$

This, according to Ng, leads to entropy of the limiting value of

$$S \approx N \cdot \left(\log\left[\frac{V}{N\lambda^3}\right] + \frac{5}{2}\right). \quad (26)$$

But $V \approx R_H^3 \approx \lambda^3$, so unless N in Eq. (26) above is about 1, S (entropy) would be less than zero, which is a contradiction. Now this is where Jack Ng introduces removing the $N!$ term in Eq. (25) above. That is, inside the Log expression we remove the expression of N in Eq. (26) above. This is a way to obtain what Ng refers to as quantum infinite statistics, so then we obtain for sufficiently large N.

$$S \approx N. \quad (27)$$

Alternately, but in the limit of late time cosmological constant behavior, Cai writes [11]

$$S \leq N. \quad (28)$$

We will refer to a bound value as referenced by Cai [11], and also Bousso [12, 13], given by

$$N \equiv \left[\frac{3\pi}{G\Lambda}\right]. \quad (29)$$

Note, that the N of Eq. (29) refers to degrees of FREEDOM, which is interesting. The total degrees of freedom will be shown to become enormous for sufficiently small Λ [12, 13]. Pending a review of the situation, the following could be entertained.

$$N \equiv \left[\frac{3\pi}{G\Lambda}\right] \Leftrightarrow \Lambda(\text{large}) \Rightarrow N(\text{small}) \;\&\; \Lambda(\text{small}) \Rightarrow N(\text{large}) \quad (30)$$

6.1. Do Eq. (28) and Eq. (30) pertain to DM, to DE, or both entropy (numerical) counts?
What is being referred to is, the applicability of [12, 13]. Specifically, [12, 13] refers to DM, and we should in answering our question ascertain if DM is the preferred venue for explaining the behavior of Eq. (30). In fact, if there is quintessence in terms of the cosmological constant parameter, as by Eq. (9), then this may explain why there is

$$\Lambda(\text{field-theory}) \sim 10^{122} \times \Lambda(\text{actual-today}). \quad (31)$$

As given by [14], there IS a linkage of black hole entropy with Λ as vacuum energy, and L as a spatial length associated with black holes. Then

$$\begin{aligned} L^3\Lambda^3 &\leq S_{BH} = \pi L^2 M_p^2 \\ L^3\Lambda^4 &\leq L M_p^2 \\ \Leftrightarrow S_{\max} &\sim S_{BH}^{\frac{3}{4}}. \end{aligned} \quad (32)$$

Furthermore, Figure 1 of page 10 of [7], as well as [15], gives one an indication that DE may not be created in the beginning of spacetime evolution, but as an artifact of later cosmological evolution. If DE is not due to massive gravitons, then either quintessence (i.e., a varying vacuum energy over spacetime due to a background average temperature) should be considered with DM and DE as different facets of the same evolutionary cosmological dynamic. The multiverse is a way of making sense of the light value of a graviton mass and Mistra's quantum-cosmology conjecture. The quantum-cosmology conjecture is also dependent upon if entropy is determined by a counting algorithm. We argue that if there is a temperature dependence, in vacuum energy as given by Eq. (9) that high temperature means a different vacuum energy than today's "cosmological constant value." Also, that if N, as a bound to entropy, is inversely proportional to vacuum energy, that according to Figure 1 of [7], that DM will not be affected, but that DE is an artifact of vacuum energy. The author deduces from the above that the bound to entropy, which is called N, as given by Eq. (29) with Eqs. (9) and (11) as backup, is an artifact of DE, not DM. and that, as a result, Figure 1 of reference [7] is saying that a bound to entropy which changes over time is due to quintessence, at least in the beginning, as given by the dynamics of the aforementioned Figure 1 above. As of 13.7 billion years ago, the background temperature given by first light about 380,000 years after the big bang was 10^5 K, according to [15, 16]. As opposed to 3 K today [16]. So, if Eq. (9) for vacuum energy is used, then if we associate Λ with DE

$$\text{Eq. (9)} \Rightarrow \Lambda_{\text{DE}}(10^5\ \text{K}) \ll \Lambda_{\text{DE}}(3\ \text{K}) \Leftrightarrow N_{\text{DE}}(10^5\ \text{K}) \gg N_{\text{DE}}(10^5\ \text{K}) \tag{33}$$

The temperature scaling given in Eq. (9) plus Figure 1 of reference [7], argues strongly against DM being created by Λ.

6.2. *Refining Eq. (23) in lieu of Eq. (33)*

To summarize so far, based upon our modification of [9]: As of about 380,000 years after the big bang

$$\begin{aligned} M_{\text{Total}} &\sim M_{\text{DM}} + M_{\text{Baryon}} + M_{\text{DE}}(\text{no contribution 380,000 years after the big bang}) \\ &\sim M_{\text{DM}} + M_{\text{Baryon}} \sim N_{\text{DM}} m_{\text{DM}} + N_{\text{Baryon}} m_{\text{Baryon}} \approx S_{\text{DM}} m_{\text{DM}} + S_{\text{Baryon}} m_{\text{Baryon}}. \end{aligned} \tag{34}$$

The summed mass is our adaptation of [9], which is in the case that 380,000 years after the big bang, there was essentially a DM dominated universe, before DE became significant.

6.3. *Filling in DM contribution to entropy, 380,000 years after big bang*

From Subodha Mistra [9] his quantum model of the big bang has the following R is the presumed present radius of the universe, m is the mass of an "average" constituent particle, and N is the number of particles in a (model) universe, with τ being the time after the big bang, to the present era.

Table 1. Big bang values.

$m \times 10^{-35}$ g	$R \times 10^{28}$ cm	$N \times 10^{91}$	$M_{\text{total}} \times 10^{56}$ gm	$\tau_0 \times 10^9$ yr
1.07299	1.896	2.38429	2.5582	20
1.23891	1.422	1.54865	1.91875	15
1.51744	0.948	0.84297	1.27916	10
2.14598	0.474	0.29804	0.639588	5

Source: Subhodha Mistra [9], p. 212

Assume there is a situation analogous to the Figure 1 [7] circumstance 380,000 years after the big bang, assume then that M, as a total mass of the universe does not change. Then, according to Mistra [9], the average particle of the quantum universe is of the order of magnitude of an axion DM particle: 1.23×10^{-35} g $\sim 5.609 \times 10^{-2}$ eV, whereas we are assuming that the entropy is similar to a numerical count of "average" particles. Then axions have a range of $10^{-6} - 20$ eV in value and by Figure 1 [7] and Table 1,

$$S_{\text{DM}} \approx 10^{90} - 10^{91}. \tag{35}$$

Should we use axions for DM, we have to look at the role of DE. That is, DE is roughly 3–4 times more massive than DM (Figure 1 [7]).

6.4. The DE we identify with gravitons. How feasible is this choice?
Then there are several alternatives. If, say, massive gravitons are an active source for DE, as has been postulated by [17, 18]: If a graviton has a mass of 2×10^{-62} g $\sim 2.8 \times 10^{-30}$ eV, then

$$S_{\text{DE}} \approx 10^{117} - 10^{118}. \tag{36}$$

This value for Eq. (35) is for the present era. That is, if one is looking at say N in Eq. (29)–(31), with an initial temperature of, say 3 K, then Eq. (35) would hold. If 10^{32} K is used in Eqs. (29)–(31), then

$$S_{\text{DE}}(10^{32}\text{ K}) \approx \varepsilon^{+} \ll S_{\text{DE}} \approx 10^{117} - 10^{118}(\text{present}). \tag{37}$$

If there is no mass connected with gravitons, then they cannot be conflated with DE.

7. Conclusion

A graviton mass of 2×10^{-62} g $\sim 2.8 \times 10^{-30}$ eV, will lead to gravitons as a candidate for DE. If the gravitons are massless, then Mistra's procedure [9], with summation of the mass and information of Table 1 does not apply to DE. If so, then one can look at another representation of DE. The DM which is tabulated is consistent. We identify DM with axions, using the Mistra's formulation, and afterwards investigate what is done with the total entropy. Using Eq. (34), the DE term would increase entropy, if a graviton is of 2×10^{-62} g $\sim 2.8 \times 10^{-30}$ eV then using Mistra's figure of mass given in Table 1, there conceivably could be the following entropy tally, that there are possibly many more than 1 million galaxies—i.e., 10^6 to 10^{20} super-massive black holes in the center of galaxies—and that up to a point, the entropy of a super-massive black hole in the center of a galaxy is at most 10^{112}. To get an idea of what is going on, look at [19].

$$S_{\text{BH}}^{\text{Total}} \sim 3.2 \times 10^{101} \cdot \left(\frac{N}{10^{11}}\right) \times \left(\frac{M}{10^{7} \cdot M_{\text{Sun}}}\right) \sim S_{\text{universe}} \tag{38}$$

If so, then does the following make cosmological sense?

$$M \sim 10^{10} \cdot M_{\text{Sun}} \underset{\Leftrightarrow}{?} N \approx 10^{20} - 10^{23} \tag{39}$$

That is, the numerical factor is then so high, if there are indeed many more than 1 million galaxies contributing to entropy, we may be indeed looking at the confluence of multiple universe contributions to our present entropy, if gravitons have a small mass. The figure to consider is, if there are gravitons with mass, and this entropy from DE is so large that there may be a need to investigate multiverse contributions to our universe's evolution.

$$S_{\text{DE}}(10^{32}\text{ K}) \approx \varepsilon^{+} \ll S_{\text{DE}} \approx 10^{117} - 10^{118}(\text{present}). \tag{40}$$

The final take away from our investigation, as in [9] that, if gravitons have mass, there could be by [9] reasons to investigate if there are multiple universes for entropy. If not, then Table 1 argues for an axion type of DM contribution to entropy of the universe, with a value as given by Eq. (35) of $S_{\mathrm{DM}} \approx 10^{90} - 10^{91}$. More than that, Eq. (35) almost certainly precludes a universe, which would put a premium upon really understanding how Eq. (9) and Eq. (13) affect, or could affect, a choice between Eq. (40) which may lead to multiverses, as given by the end of [9], or a single repeating universe.

Acknowledgments
This work is supported in part by National Nature Science Foundation of China grant No. 11375279

References
[1] Mukhanov V 2005 *Physical Foundations of Cosmology* (Cambridge, UK: Cambridge University Press)
[2] Camara C S, de Garcia Maia M R, Carvalho J C and Lima J A S 2004 Nonsingular FRW cosmology and nonlinear dynamics *Preprint* arXiv:astro-ph/0402311
[3] Weinberg S 1972 *Gravitation and Cosmology, Principles and Applications of the General Theory of Relativity* (New York: John Wiley and Sons)
[4] Corda C and Cuesta H 2010 *Mod. Phys. A* **25** 2423–2429
[5] Corda C 2010 *Eur. Phys. J.* **C 65** 257–267 (*Preprint* arXiv:1007.40777)
[6] Park D K, Kim H and Tamarayan S 2002 *Phys. Lett.* **B535** 5–10
[7] Beckwith A 2014 Non linear electrodynamics and the question of minimum time length, modification of initial singularities in early universe cosmology, and DM/DE *Preprint*vixra:1410.0159
[8] Frampton P, Hsu S D H, Kephart T W and Reeb D 2009 *Class. Quant. Grav.* **26** Article 5005 (*Preprint*arXiv:0801.1847v3)
[9] Mistra S 2012 *The Big Bang: Theory, Assumptions and Problems* ed O'Connell J R and Hale A L (Beijing, China: Hauppauge, NY) pp 205–228
[10] Ng Y J 2008 *Entropy* **10**(4) 441–461 doi:10.3390/e10040441
[11] Cai R G 2008 *Superstring Theory* (*Advanced Lectures in Mathematics* vol 1) ed Liue K, Yau S T and Zhu C (Beijing: Higher Education Press of China) pp 234–250
[12] Bousso R, DeWolfe O and Meyers R C 2002 Unbounded entropy in spacetimes with positive cosmological constant *Preprint* arXiv:hep-th/0205080
[13] Bousso R 2000 Positive vacuum energy and the n-bound *Preprint* arXiv:hep-th/0010252
[14] Cohen A G, Kaplan D B and Nelson A E 1999 Effective field theory, black holes, and the cosmological constant *Preprint* arxiv:hep-th/9803132
[15] Das S 2012 Gravitational lensing of the CMBR, a new cosmological frontier *Preprint* chalonge.obspm.fr/Paris12_Das.pdf
[16] Ellis G, Maartens R and MacCallum M A H 2012 *Relativistic Cosmology* (Cambridge, UK: Cambridge University Press)
[17] Merali Z 2013 Fat gravity particle gives clues to dark energy September 10, 2013, www.nature.com
[18] Beckwith A W 2011 *J. Cosmol.* **13** Article 23 (journalofcosmology.com/BeckwithGraviton.pdf)
[19] Kephardt T and Ng Y J 2006 Black holes, mergers, and the entropy budget of the universe cds.cern.ch/record/549082/files/0204081.pdf

Journal of Physics: Conference Series **626** (2015) 012059 doi:10.1088/1742-6596/626/1/012059

Axion-photon mixing and geometric phase

A Capolupo, G Lambiase and G Vitiello

Dipartimento di Fisica "E.R. Caianiello" Universitá di Salerno, and INFN - Gruppo Collegato di Salerno, Italy.

Abstract. We report on recent results revealing the presence of a geometric phase in the axion-photon mixing. In laboratory observations its detection can reveal such mixing phenomenon and therefore prove the existence of axion like particles.

1. Introduction

The studies of ultra-light and weakly coupled particles is one of the main research fields of physics beyond the Standard Model. Particular attention has been devoted to the axion [1, 2] which is a hypothetical pseudo-Nambu-Goldstone boson with no electric charge and a mass expected to range between 10^{-6} and 10^{-2} eV. Cosmological analysis have shown that axions and axion like particles can contribute to the cold dark matter if their mass is $m_a \sim 80\mu$eV [3]–[5]. Another important property of such particles is their interaction with the electromagnetic field in the presence of strong magnetic fields. Such a coupling is responsible of the mixing and oscillation with photons [6].

Many experiments have attempted, in different ways, to reveal the presence of the axion [7]–[20], however without positive results.

On the other hand, great attention have attracted the studies of the geometric phases [21]–[35]. These have been detected in the evolution of many physical systems [36]–[40] and their possible applications in particle physics [41]–[44] and cosmology [45]–[49] have also been analyzed in detail.

In this paper we report the recent result according to which the axion-photon mixing phenomenon generate a geometric phase [50]. Such a phase appears during the time evolution of photons and axions only if the mixing between these two particles is present. Indeed, it is absent when the external magnetic field vanishes and the mixing vanishes. In such cases, the total phase coincides with the dynamical one. Then the geometric phase could be used to reveal the axions produced by photons passing through a magnetic field. By means of an interferometer, one in fact could measure the difference between the geometric phases associated to two rays of light propagating in the two arms of the interferometer, one crossing a strong magnetic field B, the other one passing through a region where $B = 0$. The $B = 0$ ray has geometric phase equal to zero, then the presence of a difference between the geometric phases of the two rays represents the manifestation of the axion-photon mixing phenomenon in the $B \neq 0$ ray.

In a recent paper, it was proposed an experimental method based on the study of the difference of phase in the interferometry to detect ALPs [51]. However no consideration concerning the geometric phase has been done in [51]. Here we consider a different physical interference phenomenon occurring only if the axion-photon mixing occurs, thus revealing the core of the origin of the interference phenomenon, since it survives even in the absence of the dynamical

Journal of Physics: Conference Series **626** (2015) 012059 doi:10.1088/1742-6596/626/1/012059

phase difference. The geometric phase difference represents a completely new tool for the analysis of the axion-photon mixing.

In our study we neglect the contribution to the geometric phase due to the birefringence of the vacuum produced by the presence of magnetic field [52]-[54]. Such a contribution is much smaller than the one due to axion-photon mixing. In the following we recall the formalism describing the axion-photon mixing and we compute the Mukunda-Simon phase [33] associated to the time evolution of photons in the presence of the mixing with the axion.

2. Axion-photon mixing

The lagrangian density of the axion-photon system [55] is $L = L_\gamma + L_a + L_{a\gamma\gamma}$, where L_γ and L_a are the lagrangian of free photon and axion, respectively, and $L_{a\gamma\gamma} = \frac{1}{4}\, g_{a\gamma\gamma}\, \epsilon_{\mu\nu\rho\sigma}\, a\, F^{\mu\nu} F^{\rho\sigma}$, is the interaction of two photons with the axion, which is responsible of the axion-photon mixing. $g_{a\gamma\gamma} \equiv g$ is the axion-photon coupling constant, $F^{\mu\nu}$ is the electromagnetic field strength and a is the axion field.

By considering a monochromatic light beam propagating along the z-direction in the presence of an arbitrary magnetic field and choosing the y-axis along the projection of $\mathbf{B}$ perpendicular to the z-axes, the propagation equations become

$$(\omega - i\partial_z + M)\begin{pmatrix} \gamma_{\|} \\ a \end{pmatrix} = 0\,. \tag{1}$$

Here M is the mixing matrix,

$$M = -\frac{1}{2\omega}\begin{pmatrix} \omega_P^2 & -g\omega B_T \\ -g\omega B_T & m_a^2 \end{pmatrix}, \tag{2}$$

with $\mathbf{B} = \mathbf{B}_T$ transverse field, ω_P plasma frequency and m_a axion mass. M is diagonalized by a rotation,

$$\begin{pmatrix} \gamma'_{\|}(z) \\ a'(z) \end{pmatrix} = \begin{pmatrix} \cos\theta & \sin\theta \\ -\sin\theta & \cos\theta \end{pmatrix}\begin{pmatrix} \gamma_{\|}(z) \\ a(z) \end{pmatrix}, \tag{3}$$

where $\theta = \frac{1}{2}\arctan\left(\dfrac{2g\omega B_T}{m_a^2 - \omega_P^2}\right)$ is the mixing angle and $\gamma_{\|}(z) = \gamma_{\|}(0)e^{-i\omega_\gamma z}$, $a(z) = a(0)e^{-i\omega_a z}$, with

$$\omega_a = \omega - \frac{\mu_-}{\omega}\,, \qquad \omega_\gamma = \omega - \frac{\mu_+}{\omega}\,,$$

and

$$\mu_\pm = \frac{\omega_P^2 + m_a^2}{2} \pm \frac{1}{2}\sqrt{(\omega_P^2 - m_a^2)^2 + (2g\omega B_T)^2}\,.$$

The conversion rate of photon in axion can be increased by introducing in the conversion region a buffer gas that induces a plasma frequency [56].

3. Mukunda-Simon phase and mixing

The Mukunda-Simon phase [33] for the mixed photons is

$$\Phi_\gamma(t) = \arg\langle\gamma'_{\|}(0)|\gamma'_{\|}(t)\rangle - \Im\int_0^t \langle\gamma'_{\|}(t')|\dot{\gamma}'_{\|}(t')\rangle dt'\,, \tag{4}$$

which explicitly is given by

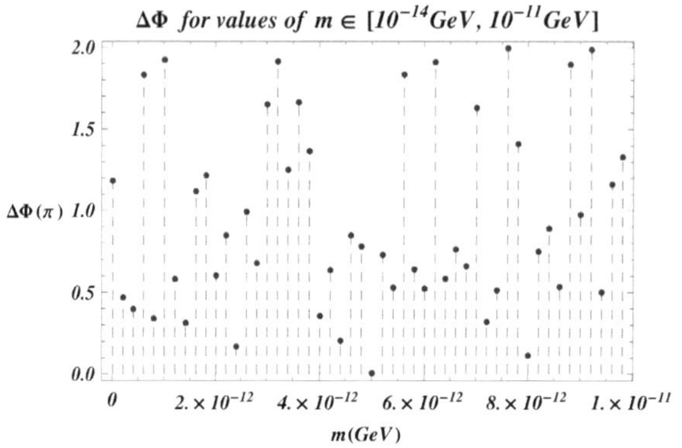

Figure 1. Plot of $\Delta\Phi = \Phi_\gamma$ as function of the axion masses m_a, for a time interval $t = 240h$ and for values of the coupling constant $g \in [10^{-12}, 10^{-10}]$ GeV^{-1} and $\omega_P = 0.9m_a$.

$$\Phi_\gamma(t) \;=\; \arg\left(\cos^2\theta e^{-i\omega_\gamma t} + \sin^2\theta e^{-i\omega_a t}\right) + \left(\omega_\gamma \cos^2\theta + \omega_a \sin^2\theta\right) t\,.$$

Non trivial geometric phase Φ_γ is obtained only in the presence the axion-photon mixing. Indeed, $\Phi_\gamma(t) = 0$, when $\theta = 0$.

Notice that for the estimated values of the axion masses, $m_a \in [10^{-6}, 10^{-2}]$ eV, the dynamical phases in the two arms of the interferometer, (the $B \neq 0$ branch and the $B = 0$ one), are almost equal, this implies that the difference of phase obtained in the crossing point of the two rays $\Delta\Phi$ is almost equal to the geometric phase, $\Delta\Phi \sim \Phi_\gamma$ [50].

In order to add many times the geometric phase and to make it visible, one can use suitable mirrors in the interferometer which make possible the traversing of the rays many times back and forth in the branches. In this way, one can obtain effective length of the arms arbitrarily long. Moreover, in order to enhance the sensibility of the experiment, Fabry-Perot cavity and squeezed light can be used in the interferometer [57]–[60].

We consider the following parameter for our estimations: the experiment last for a time $t = 240$ hours, the magnetic field $B = 10\,\mathrm{T} \simeq 1.95 \times 10^{-15}\,\mathrm{GeV}^2$, energy $\omega = 10$ eV, coupling constant and axion mass belonging to the intervals $g \in [10^{-12}, 10^{-10}]\mathrm{GeV}^{-1}$, $m_a \in [10^{-6}, 10^{-2}]$ eV, respectively, and plasma frequency $\omega_P = 0.9m_a$.

The results obtained are: a) the geometric phase is practically independent on the experimentally accepted value of the coupling constant g; b) on the contrary, Φ_γ is strongly dependent on the value of the axion mass and it changes suddenly for different mass values, as shown in Fig. 1. This fact implies not only that the presence of a non zero geometric phase of the photon demonstrates the presence of axion-photon mixing, but also that by means of the geometric phase one can achieve a precise estimation of the value of the axion mass.

4. Conclusion

We have revealed the existence of the geometric phase in the axion-photon mixing and we have proposed an interferometric measurement to reveal the occurrence of such a phenomenon and thus the existence of axion like particles. For estimated values of the axion mass and of the coupling constant g, the difference between the phases of the two rays, the one passing in the arm with $B = 0$ and that passing through the magnetic field region, is almost completely coincident with the geometric phase produced by the mixing in the $B \neq 0$ branch.

Therefore, the detection of the geometric phase associated with the time evolution of photons propagating in a magnetic field would implies the occurrence of the axion-photon mixing. We propose then the Mukunda-Simon phase as a novel tool in the detection of such a phenomena. In the system we propose and in the range of the parameter we considered, the geometric phase is practically independent on the estimated value of the coupling constant. However it is highly sensible to the axion masses. Therefore, the geometric phase could provide the demonstration of the existence of the axions and also a value of their masses, thus allowing a more deep understanding of their contributions to the cold dark matter and to the dark energy component given by the particle mixing phenomenon [61]–[66].

Acknowledgments
Partial financial support from MIUR and INFN is acknowledged. G.L. thanks the ASI (Agenzia Spaziale Italiana) for partial support through the contract ASI number I/034/12/0.

References

[1] Peccei R D and Quinn H 1977 *Phys. Rev. Lett.* **38** 1440
[2] Peccei R D and Quinn H 1977 *Phys. Rev.* D **16** 1791
[3] Sikivie P and Yang Q 2009 *Phys. Rev. Lett.* **103** 111301
[4] Sikivie P 2011 *Phys. Lett.* B **695** 22
[5] Di Valentino E, Giusarma E, Lattanzi M, Melchiorri A and Mena O 2014 *Phys. Rev.* D **90** 043534
[6] Raffelt G 1996 *Stars as laboratories for fundamental Physics* (University of Chicago Press); *do.* 2008 *Lecture Notes in Physics* **741** 51
[7] Zavattini E *et al.* (PVLAS Collaboration), 2006 *Phys. Rev. Lett.* **96** 110406
[8] Zavattini E *et al.* (PVLAS Collaboration), 2008 *Phys. Rev.* D **77** 032006
[9] Aune S *et al.* (CAST Collaboration), 2011 *Phys. Rev. Lett.* **107** 261302
[10] van Bibber K, McIntyre P M, Morris D E and Raffelt G G 1989 *Phys. Rev.* D **39** 2089
[11] Arik M, et al., 2011 *Phys. Rev. Lett.* **107** 261302
[12] Asztalos S *et al.* (ADMX Collaboration) 2004 *Phys. Rev.* D **69** 011101
[13] Asztalos S *et al.* (ADMX Collaboration)2010 *Phys. Rev. Lett.* **104** 041301
[14] Cameron R *et al.* 1993 *Phys. Rev.* D **47** 3707
[15] Fouche M *et al.* (BMV Collab.) 2008 *Phys. Rev.* D **78** 032013
[16] Pugnat P *et al.* (OSQAR Collab.) 2008 *Phys. Rev.* D **78** 092003
[17] Chou A *et al.* (GammeV T-969 Collab) 2008 *Phys. Rev. Lett.* **100** 080402
[18] Afanasev A *et al.* (LIPSS Collab.) 2008 *Phys. Rev. Lett.* **101** 120401
[19] Pugnat P *et al.* (OSQAR Collab.), 2014 *Eur. Phys. J.* C **74** 3027
[20] Ehret K *et al.* (ALPS Collab.), 2010 *Phys. Lett.* B **689** 149
[21] Berry M V 1984 *Proc. Roy. Soc. Lond.* A **392**
[22] Aharonov Y and Anandan J 1987 *Phys. Rev. Lett.* **58** 1593
[23] Samuel J and Bhandari R 1988 *Phys. Rev. Lett.* **60** 2339
[24] Anandan J and Aharonov Y 1990 *Phys. Rev. Lett.* **65** 1697
[25] Garcia de Polavieja G 1998 *Phys. Rev. Lett.* **81** 1
[26] Pancharatnam S 1956 *Proc. Indian Acad. Sci.* A **44** 1225
[27] Shapere A and Wilczek F 1989 *Geometric Phases in Physics* (Singapore: World Scientific)
[28] Simon B 1983 *Phys. Rev. Lett.* **51** 2167
[29] Garrison J C and Wright E M 1988 *Phys. Lett.* A **128** 177
[30] Garrison J C and Chiao R Y 1988 *Phys. Rev. Lett.* **60** 165
[31] Pati A K 1995 *J. Phys.* A **28** 2087
[32] Pati A K 1995 *Phys. Rev.* A **52** 2576
[33] Mukunda N and Simon R 1993 *Ann. Phys. (N.Y)* **228** 205
[34] Mostafazadeh A 1999 *J. Phys.* A **32** 8157
[35] Anandan J 1988 *Phys. Lett.* A **133** 171
[36] Tomita A and Chiao R Y 1986 *Phys. Rev. Lett.* **57** 937
[37] Jones J A, Vedral V, Ekert A and Castagnoli G 2000 *Nature* **403** 869
[38] Leek P J *et al.* 2007 *Science* **318** 1889
[39] Neeley M *et al.* 2009 *Science* **325** 722
[40] Pechal M *et al.* 2012 *Phys. Rev. Lett.* **108** 170401

[41] Capolupo A 2011 *Phys. Rev.* D **84** 116002
[42] Capolupo A and Vitiello G 2013 *Adv. High Energy Phys.* **2013** 850395
[43] Blasone M, Capolupo A, Celeghini E and Vitiello G 2009 *Phys. Lett.* B **674** 73
[44] Capozziello S and Lambiase G 2000 *Europhys. Lett.* **52** 15
[45] Capolupo A and Vitiello G 2013 *Phys. Rev.* D **88** 024027
[46] Capolupo A and Vitiello G 2015 *Vacuum condensate, geometric phase, Unruh effect and quantum thermometer*, submitted
[47] Bruno A, Capolupo A, Kak S, Raimondo G and Vitiello G 2011 *Mod. Phys. Lett.* B **25** 1661
[48] Hu J and Yu H 2012 *Phys. Rev.* A **85** 032105
[49] Martin-Martinez E, Fuentes I and Mann R B 2011 *Phys. Rev. Lett.* **107** 131301
[50] Capolupo A, Lambiase G and Vitiello G 2015 *Probing mixing of photons and axions by geometric phase*, submitted
[51] Tam H, Yang Q 2012 *Phys. Lett.* B **716** 435
[52] Heisenberg W and Euler H 1936 *Z. Phys.* **98** 714
[53] Schwinger J S 1951 *Phys. Rev.* **82** 664
[54] Dobrich B and Gies H 2009 *Europhys. Lett.* **87** 21002
[55] Mirizzi A, Raffelt G G and Serpico P D 2008 *Lect. Notes Phys.* **741** 115
[56] van Bibber K. *et al.* 1989 *Phys. Rev.* D **39** 2089
[57] Caves C M 1981 *Phys. Rev.* D **23** 1693
[58] The LIGO Scientific Collaboration 2011 *Nature Phys.* **7** 962
[59] Lang M D and Caves C M 2013 *Phys. Rev. Lett.* **111** 173601
[60] Eberle T *et al.* 2010 *Phys. Rev. Lett.* **104** 251102
[61] Capolupo A, Capozziello S and Vitiello G 2007 *Phys. Lett.* A **363** 53
[62] Capolupo A, Capozziello S and Vitiello G 2009 *Phys. Lett.* A **373** 601
[63] Capolupo A, Capozziello S and Vitiello G 2008 *Int. J. Mod. Phys.* A **23** 4979
[64] Blasone M, Capolupo A, Capozziello S and Vitiello G 2008 *Nucl. Instrum. Meth.* A **588** 272
[65] Blasone M, Capolupo A and Vitiello G 2010 *Prog. Part. Nucl. Phys.* **64** 451
[66] Blasone M, Capolupo A, Capozziello S, Carloni S and Vitiello G 2004 *Phys. Lett.* A **323** 182

Journal of Physics: Conference Series **626** (2015) 012060 doi:10.1088/1742-6596/626/1/012060

Geometric phase in vacuum condensates, application to Unruh effect and to quantum thermometer

Antonio Capolupo

Dipartimento di Fisica E.R.Caianiello Universitá di Salerno, and INFN Gruppo collegato di Salerno, Fisciano (SA) - 84084, Italy

Abstract.

We report on the recent results revealing the presence of the geometric phase in all the systems characterized by particle creation from vacuum and vacuum condensates. This fact makes the geometric phase a useful tool in the study and the understanding of disparate phenomena. Its possible application ranges from the dynamical Casimir effect to the Hawking effect, from quantum field theory in curved space to the study of CP and CPT symmetries, from the graphen physics to superconductivity and to the Bose Einstein condensate. Here, we consider the possibility of the detection of the Unruh effect and of the fabrication of a very precise quantum thermometer. We analyze the Mukunda-Simon phase for a two level atom system and consider two case: 1) atoms accelerated in electromagnetic field, and 2) atoms interacting with thermal states. The Mukunda-Simon phase generalizes the Berry phase to the case of non-cyclic and non-adiabatic evolutions; therefore it represents a more useful instrument in experimental implementations with respect to the Berry phase.

1. Introduction

Phenomena like Unruh [1], Hawking [2] and Parker effects [3, 4] and many aspects of the quantum field theory (QFT) in curved background are topics of particular interest in many field of the research. However, at the present, their detection is very difficult. In the next experiments, a key role in the attempt to reveal such phenomena could be played by the analysis of geometric phases. It has been shown that geometric phases and invariants can be used to test CPT symmetry in meson systems [5], to prove the existence of postulated particles like the axions [6], to test SUSY violation in thermal states [7], to reveal the Unruh effect [8, 9, 10] and to build a quantum thermometer [8, 11, 12]. The presence of geometric invariants has been shown also in other context [13, 14]. Here we present a more deep and general result. We show that all the systems in which vacuum condensates or particle creation from vacuum appear [15]–[20], are also characterized by the presence of the geometric phase [21]–[34] in their evolution [35]. In a previous paper [8], we have shown that in the above systems, and in general in all the phenomena represented by Bogoliubov transformations, the Aharonov–Anandan invariant (AAI) [36] is produced in their evolution. Such an invariant is however very hard to be experimentally observed. On the contrary, the geometric phase has been detected in many systems [37], [38], [39] and [40]. Therefore, the discovery of the existence of a relation between vacuum condensates and geometric phase can open a new way, based on the interferometry, to the investigation of disparate phenomena.

Journal of Physics: Conference Series **626** (2015) 012060 doi:10.1088/1742-6596/626/1/012060

We reveal the link between vacuum condensates and geometric phase [35], by using our results presented in Ref.[8], together with the results presented in [41, 42].

As example of possible application of the geometric phase produced by the systems listed above, here we study the realizability of an interferometer in which the Mukunda-Simon (MS) phase [24] can be detected in a two level system. In particular, we study the role which can be played by the MS phase in the detection of the Unruh effect and in the building of a quantum thermometer. The ideas of using geometric phases and invariants to probe the Unruh effect [8, 9, 10], and to build a thermometer [8, 11, 12] have been already presented in previous works. However our work represents a progress in several directions also in the two examples here considered. Indeed, besides the exhibition of the physically intriguing relation between boson condensation, AAI and the geometric phase, here we consider a realistic scheme for experimental implementations and a phase, the MS one which generalize the Berry phase, (used in [9, 11]) to the case of non-cyclic, non-adiabatic and non unitary evolution.

The results which we obtain are consistent with the ones presented in [9, 10] for the Unruh effect. In [10] a similar scheme, but a different definition of the geometric phase have been used [34]. The numerical results relative to the thermometer are consistent with to the ones presented in [11]. However, our approach can be used also in more wide contests.

Recently a further proposal has been presented for the realization of a quantum thermometer [12]. Such studies regard the analysis of the Berry phase and of the dynamical phase in a Bose-Einstein condensate .

In this paper, we propose to use a Mach-Zehnder interferometer in which a two level atom system with a nonunitary evolution is analyzed in two different cases. In the first case, we consider atoms interacting with an electromagnetic field, which are accelerated in one branch of the interferometer, and are inertial in the other arm. The difference between the MS phases produced in the two arms is due to the Unruh effect. In the second case, we consider atoms interacting with thermal states and we show that a very precise measurement of the temperature can be obtained by means of the MS phase.

In the following, we show the relation existing between the AAI, the geometric phase and the vacuum condensates. Then we study the possibility to use the geometric phases to detect the Unruh effect and to realize a quantum thermometer.

2. Geometric phase and AAI

Let us consider the state $|\psi\rangle$ of a quantum system and define the infinitesimal "reference distance" in the projective Hilbert space P [42],

$$dD^2 = \Big[\Big\langle \frac{d}{dt}\Big(\frac{\psi}{\|\psi\|}\Big)\Big|\frac{d}{dt}\Big(\frac{\psi}{\|\psi\|}\Big)\Big\rangle + \dot{\phi}^2 - 2i\dot{\phi}\Big\langle\frac{\psi}{\|\psi\|}\Big|\frac{d}{dt}\Big(\frac{\psi}{\|\psi\|}\Big)\Big\rangle\Big]dt^2 ,$$

where

$$\phi(t) = \frac{i}{2}\ln\left[\frac{\langle\psi(0)|\psi(t)\rangle}{\langle\psi(t)|\psi(0)\rangle}\right] .$$

Eq.(1) is gauge invariant and reparametrization invariant [42]. In a similar way, the Fubini-Study metric can be expressed as

$$dS^2 = \Big(\Big\langle \frac{d}{dt}\Big(\frac{\psi}{\|\psi\|}\Big)\Big|\frac{d}{dt}\Big(\frac{\psi}{\|\psi\|}\Big)\Big\rangle - \Big[i\Big\langle\frac{\psi}{\|\psi\|}\Big|\frac{d}{dt}\Big(\frac{\psi}{\|\psi\|}\Big)\Big\rangle\Big]^2\Big)dt^2 .$$

The total length of the S-path measured using the Fubini-Study metric differs from the AAI only by a numerical factor of two. By defining the quantity

$$d\Phi^2 = dD^2 - dS^2 \tag{1}$$

Journal of Physics: Conference Series **626** (2015) 012060 doi:10.1088/1742-6596/626/1/012060

where

$$d\Phi = \left[\left\langle \frac{\psi}{\|\psi\|} \middle| i\frac{d}{dt} - \dot{\phi}(t) \middle| \frac{\psi}{\|\psi\|} \right\rangle \right], \tag{2}$$

the geometric phase for an arbitrary evolution of a quantum system is then given by the difference between the D-path and the S-path in the projective Hilbert space P [42],

$$\Phi(\Gamma) = \int_\Gamma \sqrt{dD^2 - dS^2} = \int_\Gamma \left[\left\langle \frac{\psi}{\|\psi\|} \middle| i\frac{d}{dt} - \dot{\phi}(t) \middle| \frac{\psi}{\|\psi\|} \right\rangle \right] dt \, .$$

Such a formula matches the Aitchinson-Wanelik formula for the geometric phase and relates the Aharonov-Anandan geometric invariant (which is two times the Fubini-Study metric) to the geometric phase. Moreover, we note that, since the length of the S-path is the minimum length of the path measured by the "reference distance" function D [41, 42], the presence of the AAI in the evolution of a system implies the presence of the D invariant and consequentially the existence of the geometric phase (3). Therefore, the systems where vacuum condensates appear [16, 17, 18], for which the existence of the AAI has been revealed [8], are also characterized by the presence of the geometric phase, which can be detected much easier than the AAI.

3. Geometric phase and two level atoms

By following the discussion presented in [8, 10] and treating the atom as an open system in the reservoir of the electromagnetic field, one has a nonunitary evolution of the atom. In particular, considering the interaction of the atom with vacuum modes of the electromagnetic field in the multipolar scheme [43] and analyzing the evolution of the total density matrix in the frame of the atom (see [8] for details), one obtain the state describing the atomic system at time τ (with τ, atom proper time) [8]

$$|\phi_+(\tau)\rangle = \sin\left(\frac{\theta(\tau)}{2}\right)|+\rangle + \cos\left(\frac{\theta(\tau)}{2}\right)e^{i\Omega\tau}|-\rangle \, . \tag{3}$$

Here $\theta(\tau)$ is the angles reported in [8, 10].

We then study the Mukunda-Simon phase at time t for the state $|\phi_+(t)\rangle$. Such a geometric phase is defined as

$$\Phi(t) = \arg\langle\phi_+(0)|\phi_+(t)\rangle - \Im\int_0^t \langle\phi_+(\tau)|\dot{\phi}_+(\tau)\rangle d\tau \, ; \tag{4}$$

explicitly we have

$$\Phi(t) = \arg\left[\sin\frac{\theta}{2}\sin\frac{\theta(t)}{2} + \cos\frac{\theta}{2}\cos\frac{\theta(t)}{2}e^{i\Omega t}\right] - \Omega\int_0^t \cos^2\frac{\theta(\tau)}{2}d\tau \, , \tag{5}$$

with $\theta \equiv \theta(0)$. If one consider a time interval equal to $t = 2\pi\Omega$, for which $\arg\left[\sin\frac{\theta}{2}\sin\frac{\theta(t)}{2} + \cos\frac{\theta}{2}\cos\frac{\theta(t)}{2}e^{i\Omega t}\right] = 0$, the result of Eq.(5) coincide with Eq.(15) of ref.[10], in which another definition of the geometric phase has been used [34]. Notice that the phase introduced in [34] represents the geometric phase defined for nonunitary evolution. On the other hand, the MS phase can be used for systems undergoing both a unitary evolution and a nonunitary evolution [24]. Moreover it generalizes the Berry phase to the non-cyclic and non-adiabatic evolution case, therefore, the analysis of the MS phase is advantageous also in systems such as the one presented in [9], in which the behavior of the Berry phase has been studied.

Let us now use Eq.(5) for atoms accelerated and interacting with thermal states.

Journal of Physics: Conference Series **626** (2015) 012060 doi:10.1088/1742-6596/626/1/012060

4. Unruh effect

The Unruh effect consists in the fact that for an accelerated observer the ground state of an inertial system appears at a non-zero temperature depending on the acceleration of the observer. Such a phenomenon has not yet been detected. However, the new fact is that the use of geometric phases and invariants could allow the detection of the Unruh effect in table top experiments, a strategy that has been proposed already in other physically different settings [8, 9, 10].

Here, in particular, we show the possible realizability of a Mach-Zender interferometer in which paths of slightly different lengths can be chosen in order to let the geometric phase be dominating over the relative dynamical phase. We compute the Mukunda-Simon phase for the two level system in the presence of an acceleration and in the inertial case. The atom interaction with the electromagnetic field itself produces a geometric phase, however, the difference between the two phases is due only to the atom acceleration.

For a two-level atom uniformly accelerated in the x direction with acceleration a, the function $\sin\frac{\theta(t)}{2}$ in Eq.(5) becomes

$$\sin\frac{\theta_a(t)}{2} = \pm\sqrt{\frac{1}{2} + \frac{R_a - R_a e^{4A_a t} + \cos\theta}{2\sqrt{e^{4A_a t}\sin^2\theta + (R_a - R_a e^{4A_a t} + \cos\theta)^2}}}\,, \tag{6}$$

and similar for $\cos\frac{\theta(t)}{2}$. Here $R_a = B_a/A_a$, with A_a and B_a reported in [10, 8], moreover we consider the same values of the parameter as in [8].

For an inertial atom, $a = 0$, the geometric phase $\Phi_{a=0}$ assumes the identical expression of Eq. (5), with $\sin\frac{\theta_a(t)}{2}$ and $\cos\frac{\theta_a(t)}{2}$, replaced by $\sin\frac{\theta_0(t)}{2}$ and $\cos\frac{\theta_0(t)}{2}$, in which the coefficients A_a, B_a, R_a are replaced by $A_0 = B_0 = \gamma_0/4$, with γ_0 spontaneous emission rate, and $R_0 = 1$, respectively.

The phase difference between the accelerated and inertial atoms, $\Delta\Phi_U(t) = \Phi_a(t) - \Phi_{a=0}(t)$, gives the geometric phase in terms of the acceleration of the atom. We consider an angle $\theta(0) = \pi/5$ and transition frequencies of the atom in the microwave regime, $\omega_0 \sim 10^9 s^{-1}$. We have $\Delta\Phi_U \sim 10^{-4}\pi$ for accelerations of order of $10^{17} - 10^{18} m/s^2$ and time scale of order of $10^{-10}s$. Such values are accessible with the current technology. The dynamic phases can be made negligible compared to the geometric ones, if the branches of the interferometer are built in order that the two dynamical phases are almost equal. This result can be obtained for example, for atoms with accelerations of order of $10^{17}m/s^2$ and $\omega_0 = 4\times 10^9 s^{-1}$, with a Mach-Zehnder interferometer of length of 2 cm and a difference in the branch lengths of 10^{-1}mm.

5. Quantum thermometer

We show now that a very precise thermometer could be built by means an interferometer in which the Mukunda-Simon phase is analyzed. Indeed a phase similar to that in Eq.(5) appears also in the interactions of the atom with a thermal state. In this case the coefficients A_a and B_a are replaced by the coefficients A_T and B_T depending on the temperature and reported in [8]. Thus an interferometer in which a single atom follows two different paths and interacts with two thermal states at different temperatures can represent a very precise quantum thermometer. Indeed, if it is known the temperature of one sample, the temperature of the other one can be defined by measuring the difference between the geometric phases in the two paths.

For example, if one knows the temperature T_h of the hotter source, which can be used as reference temperature, by measuring $\Delta\Phi_T$, one can derive precise estimations of the temperature T_c of the colder cavity (in [8] the AAI invariant has been analyzed, which is very hard to be detected in an interferometer).

If one considers time intervals of the order of $t = 20\times 2\pi/\omega_0$, values of ω_0 of the order of the GHz and $T_h \sim 1K$, one can measure temperatures of the cold source of $\sim$ 2 orders of magnitude below the reference temperature of the hot source. If one could use atoms with energy gaps smaller than GHz, for example considering hyperfine structure of particular atoms and lower values of the reference temperature, one can have estimation T_c very low. Also in the case of the thermometer, paths of slightly different lengths can be chosen in order to let the geometric phase be dominating over the relative dynamical phase. A similar result has been obtained in Ref.[11], where the behavior of the Berry phase has been analyzed for an atom, described as an harmonic oscillator, interacting with a scalar field in a cavity, representing the thermal source. In ref.[12] a different kind of measurement has been proposed, consisting in revealing the dynamical phase in a Bose-Einstein condensate in a Ramsey interferometer, .

6. Conclusions

We have shown that since Aharonov-Anandan invariants are present in all the phenomena where vacuum condensates appear, then these phenomena are characterized by the presence also of geometric phases. Thus, the geometric phases appear to be a useful tool in the study and the understanding of phenomena in different fields of physics (see also [5]-[6]). Here we have studied the Mukunda-Simon phase for a two level atom system. In particular, we have shown that atoms with transition frequencies in the microwave regime accelerated in an Mach-Zender interferometer of length of $2\,cm$ could represent an efficient tool in the laboratory detection of the Unruh effect. On the other hand, we have shown that a two level atom system, interacting with two different thermal states can be utilized in an interferometer to build a very precise quantum thermometer.

Acknowledgement

Partial financial support from MIUR and INFN is acknowledged.

References

[1] Unruh W G 1976 *Phys. Rev. D* **14**, 870.
[2] Hawking S W 1975 *Commun. Math. Phys.* **43**, 199 [1976 Erratum-ibid. **46**, 206].
[3] Parker L 1968 *Phys. Rev. Lett.* **21**, 562.
[4] Schrödinger E 1939 *Physica* **6**, 899.
[5] Capolupo A 2011 *Phys. Rev. D* **84**, 116002.
[6] Capolupo A, Lambiase G and Vitiello G 2015 *Probing mixing of photons and axions by geometric phase*, submitted.
[7] Capolupo A, and Vitiello G 2013 *Adv. High Energy Phys.* **2013**, 850395.
[8] Capolupo A and Vitiello G 2013 *Phys. Rev. D* **88**, no. 2, 024027
[9] E. Martin-Martnez, I. Fuentes, R. B. Mann 2011 *Phys. Rev. Lett.* **107**, 131301.
[10] Hu J and Yu H 2012 *Phys. Rev. A* **85**, 032105.
[11] Martin-Martinez E, Dragan A, Mann R B, Fuentes I 2013 *New J. Phys.* **69**, 053036.
[12] Sabin S, White A, Hackermuller L, Fuentes L 2014 *Sci.Rep.* **4**, 6436.
[13] Bruno A,Capolupo A, Kak S, Raimondo G, Vitiello G 2011 *Mod. Phys. Lett. B* **25**, 1661.
[14] Blasone M, Capolupo A, Celeghini E, Vitiello G 2009 *Phys. Lett. B* **674**, 73.
[15] Umezawa H 1993 *Advanced field theory: Micro, macro, and thermal physics, New York, USA: AIP.*
[16] Blasone M, Henning P A and Vitiello G 1999 *Phys. Lett. B* **451**, 140; Blasone M, Capolupo A, Vitiello G 2002 *Phys. Rev. D* **66**, 025033; Blasone M, Capolupo A, Romei O, Vitiello G 2001 *Phys. Rev. D* **63**, 125015; Capolupo A, Ji C-R, Mishchenko Y, Vitiello G 2004 *Phys. Lett. B* **594**, 135.
[17] Capolupo A, Capozziello S, Vitiello G 2007 *Phys. Lett. A* **363**, 53; 2009 *Phys. Lett. A* **373**, 601; 2008 *Int. J. Mod. Phys. A* **23**, 4979; Blasone M, Capolupo A, Capozziello S, Vitiello G 2008 *Nucl. Instrum. Meth. A* **588**, 272; Blasone M, Capolupo A, Vitiello G 2010 *Prog. Part. Nucl. Phys.* **64**, 451. Blasone M, Capolupo A, Capozziello S, Carloni S, Vitiello G 2004 *Phys. Lett. A* **323**, 182.
[18] Capolupo A, Di Mauro M, Iorio A 2011 *Phys. Lett. A* **375**, 3415; Capolupo A, Di Mauro M 2012 *Phys. Lett. A* **376**, 2830; 2013 *Acta Phys. Polon. B* **44**, 81.
[19] Celeghini E, Rasetti M, Vitiello G 1992 *Annals Phys.* **215**, 156.
[20] Birrell N D, Davies P C W 1984 *Quantum Fields in Curved Space (Cambridge Monographs on Mathematical Physics*

[21] Berry M V 1984 *Proc. Roy. Soc. Lond. A* **392**, 45.
[22] Aharonov Y and Anandan J 1987 *Phys. Rev. Lett.* **58**, 1593.
[23] Samuel J and Bhandari R 1988 *Phys. Rev. Lett.* **60**, 2339.
[24] Mukunda N and Simon R 1993 *Ann.Phys.(N.Y)* **228**, 205.
[25] Garcia de Polavieja G 1998 *Phys. Rev. Lett.* **81**, 1.
[26] Pancharatnam S 1956 *Proc. Indian Acad. Sci. A* **44**, 1225.
[27] Shapere A and Wilczek F 1989 *Geometric Phases in Physics, World Scientific*, Singapore.
[28] Simon B 1983 *Phys. Rev. Lett.* **51**, 2167.
[29] Garrison J C and Wright E M 1988 *Phys. Lett.* A **128**, 177, Garrison J C and Chiao R Y 1988 *Phys. Rev. Lett.* **60**, 165.
[30] Pati A K 1995 *J. Phys. A* **28**, 2087.
[31] Pati A K 1995 *Phys. Rev. A* **52**, 2576.
[32] Anandan J 1988 *Phys. Lett. A* **133**, 171.
[33] Mostafazadeh A 1999 *J. Phys. A* **32**, 8157.
[34] Tong D M, Sjoqvist E, Kwek L C, and Oh C H 2004 *Phys. Rev. Lett.* **93**, 080405.
[35] Capolupo A, Vitiello G 2015 *Unruh effect, quantum thermometer and geometric phase* preprint.
[36] Anandan J and Aharonov Y 1990 *Phys. Rev. Lett.* **65**, 1697.
[37] Tomita A, Chiao R Y 1986 *Phys. Rev. Lett.* **57**, 937.
[38] Jones J A, Vedral V, Ekert A, and Castagnoli G 2000 *Nature* **403**, 869.
[39] Leek P J *et al.* 2007 *Science* **318**, 1889; Neeley M *et al.* 2009 *Science* **325**, 722.
[40] Pechal M *et al.* 2012 *Phys. Rev. Lett.* **108**, 170401.
[41] Pati A K 1991 *Phys. Lett. A* **159**, 105.
[42] Pati A K 1995 *Phys. Lett. A* **202**, 40.
[43] Compagno G, Passante R, Persico F 1995 *Atom-field interactions and dressed atoms* (Cambridge University Press, Cabridge).

Journal of Physics: Conference Series **626** (2015) 012061 doi:10.1088/1742-6596/626/1/012061

A model of the measurement process in quantum theory

H H Diel

Diel Software Entwicklung und Beratung, Seestr.102, 71067 Sindelfingen, Germany

E-mail: `diel@netic.de`

Abstract. The so-called measurement problem of quantum theory (QT) is still lacking a satisfactory, or at least widely agreed upon, solution. A number of theories, known as interpretations of quantum theory, have been proposed and found differing acceptance among physicists. Most of the proposed theories try to explain what happens during a QT measurement using a modification of the declarative equations that define the possible results of a measurement of QT observables or by making assumptions outside the scope of falsifiable physics. This paper proposes a solution to the QT measurement problem in terms of a model of the process for the evolution of two QT systems that interact in a way that represents a measurement. The model assumes that the interactions between the measured QT object and the measurement apparatus are "normal" interactions which adhere to the laws of quantum field theory.

1. Introduction

QT is a theory on (possible) processes in subatomic physics. It supports the computation of probabilities for a process to evolve from state s_i toward state s_f. To a limited degree, QT also allows the computation of intermediate states and paths taken to reach the final state s_f. Even less is known about how the multitude of alternative measurement results is converted into a definite result during a measurement. The details on how a measurement performs the transition from possible measurement results to definite measurement results are still in dispute among physicists. Alternative theories, called interpretations of QT (e.g., Copenhagen interpretation, many worlds interpretation (see [4])) have been proposed. Thus far, there does not seem to be general agreement among physicists of the correct, or at least most plausible, interpretation of QT.

The author claims that satisfying answers to the open questions related to QT measurements can only be obtained by considering the *process* that represents a measurement. The author furthermore assumes that the primary process elements involved in a measurement are *interactions* between the measured QT object and the measurement apparatus, and interaction processes are therefore the subject to be studied.

The QT area that addresses the interactions between particles and between particles and fields is quantum field theory (QFT). QFT addresses interactions between particles to a great extent (see [6], [8], [9]). QFT, with its Feynman diagrams and description of scattering processes, may also be viewed as containing some elements of a process-based description. The paper shows that the assumption that measurements consist of "normal" interactions that adhere to the laws of QFT partly explains some of the peculiarities of QT measurements if this is combined with a description of the interaction process.

Within physics the preferred method for the description of temporal relationships and of a physical process is the provision of differential equations. In [3], the author explains that more complicated temporal relationships and processes (which may also occur within physics) cannot be described by purely differential equations. Unfortunately, there does not exist a method (i.e., language) that is equally elegant as differential equations for the specification of more complex processes . Available description methods mainly originated from computer science (e.g., programming languages) and are often more complicated (because they are more powerful) than seems to be necessary for the description of physical processes.

In section 2. the basic aspects of the measurement process are discussed in relation to the measurement problem. Sections 3 and 4 give a summary of the functional model of QT within which the model of the measurement process is embedded.

2. The measurement problem

The measurement problem of QT can be expressed by a set of questions related to the overall question of what exactly happens during a measurement. The set of questions varies depending on selected basic assumptions to start with. A concise description of the measurement problem is given in [7] in the form of a trilemma. In [7], Maudlin shows that the following three claims are mutually inconsistent:

(i) The wave function of a system is complete.

(ii) The wave function always evolves in accord with a linear dynamical equation (e.g., Schrödinger equation).

(iii) Measurements always have a definite outcome.

Maudlin shows variations of these contradicting claims that are contradicting as well.
The proposed model of the QT measurement process has to be built on assumptions that avoid the above-described contradictions. This is the case with the following set of basic assumptions:

(i) The wave function is complete. This means the proposed model of the measurement process acts on the state of the system, and the description of this state of the system is derived purely from the wave function(s) and related parameters of the involved QT entities.

(ii) The wave function (and the state of the system of the process model) does not always evolve in accordance with a linear equation. The linear equations (e.g., Schrödinger equation) describe the evolution of probability amplitudes. The transition from probabilities to facts and the reduction to definite values are not (necessarily) described by linear equations.

(iii) Measurements of, for example, the spin of an electron always have determinate outcomes. A part of the measurement process is the reduction of the set of possible alternative outcomes to a single definite value.

Starting with these basic assumptions, further questions may be asked which have to be answered by a satisfactory model of the measurement process:

- Does measurement determine the state of a QT object?
- Why is it impossible to measure the non-definite state of the system?
- Why can certain observables not be measured simultaneously?
- When (in the sequence of process steps or under which circumstances) does the transition from probabilities to facts occur?
- Is a measurement always coupled with the collapse of a wave function?
- Is the transition from probabilities to facts related exclusively to measurements?
- Is the measurement process random or deterministic?

Journal of Physics: Conference Series **626** (2015) 012061 doi:10.1088/1742-6596/626/1/012061

3. The functional model of Quantum Theory

The model of the measurement process described in this paper is embedded in an overall functional model of QT (see [1]) and is based on the functional model of interactions in QT (see [2]). A functional description of a system describes the dynamical evolution of this system in terms of state changes.

The main characteristics of the functional model of QT described in [1] and [2] are:

- Discreteness of QT attributes
  The functional description of QT assumes discrete and coarse grain attributes not only for three-dimensional space, but for most other entities where standard QFT assumes differentiable attributes. This applies to the spatial extension of particles/waves and to their momentum. Also, the wave function is structured into a discrete set of alternative paths (see section 4.1 System State). This splitting into multiple paths is essential for the functional model of interactions as well as for the model of the measurement process.
- Transition from possibilities to facts
  The functional model of QT demonstrates the evolution of the wave function to generate probability amplitudes in accordance with the predictions of QT/QFT. However, it does not end with the determination of probability amplitudes but includes a model for the realization of the predictions represented by the probability amplitude. This process step is called "the transition from possibilities to facts". With standard QT, the transition from possibilities to facts is a non-deterministic process step that occurs exclusively with measurements.
  One of the key features of the functional description of QT is that the transition from possibilities to facts is not exclusively tied to measurements. The functional description assumes that measurements always imply interactions, more specifically, interactions that lead to a collapse of the wave function. The statement that the transition from possibilities to facts is not exclusively tied to measurements, first of all, means that the collapse of the wave function is not exclusively observed with measurement interactions. The collapse of the wave function also occurs with other ("normal") interactions.
- Particle fluctuations instead of virtual particles
  In the perturbation (Feynman) approach, virtual particles are an essential concept for describing interactions among particles. The functional description reinterprets the role of virtual particles; instead of the original QFT virtual particles, the functional description assumes "particle fluctuations" and "interaction channels". Particle fluctuations initiate the interaction, if the fluctuation location is shared by multiple particles. These particle fluctuations are assumed to actually occur (with a certain probability), whereas virtual particles are constructs which affect only the probability amplitudes of interactions.
- Splitting of a wave function *collection* into multiple paths
  The splitting of a wave function into multiple paths is a constituent part of the perturbation approach to QFT (see [5]). The overall effect of the wave function progression is then determined by the superposition (via path integrals) of the multiple paths. With the functional model of QT, the splitting into multiple paths is applied to **collections** of particles (see pw-collection below) which exit an interaction. This allows for the modeling of entanglement.

4. The interaction process

Two types of interactions between particles/waves are distinguished:

(i) "QFT interactions" - interactions that result in a change of the components and/or paths of one or both of the interacting quantum objects. This type of interaction is addressed by QFT in terms of, for example, the scattering matrix and Feynman diagrams.

Journal of Physics: Conference Series **626** (2015) 012061 doi:10.1088/1742-6596/626/1/012061

Table 1. Structure of a q-object consisting of two particle/waves pw1 and pw2.

paths	pw1-state	pw2-state	amplitude
path-1	pw1-state$_1$	pw2-state$_1$	ampl-1
path-2	pw1-state$_2$	pw2-state$_2$	ampl-2
...	...	...	...
path-N	pw1-state$_N$	pw2-state$_N$	ampl-N

(ii) "volatile interactions" - interactions which change only the values of attributes (e.g., momenta) of the interacting quantum objects.

For this description of the model of the measurement process, QFT interactions are of primary interest, because (1) QT measurements typically require QFT interactions to achieve observable effects [1] and (2) QFT interactions are the ones which result in the complications associated with the QT measurement problem. In the remaining paper only the QFT interactions are considered.

4.1. System state

Within the state of the system of the functional model, the wave function is represented by the q-object (quantum object). The q-object may consist of a single particle/wave or a collection of (entangled) particles/waves (pw-collection). In a second dimension, the q-object consists of multiple paths with associated probability amplitudes (see Table 1).

```
q-object :=
  path[1],
  ...
  path[NPATH];
```

4.2. Steps in an interaction process

The process steps occurring in the interaction process have to produce results that are compatible with the results predicted by standard QT/QFT, including the generation of result alternatives in terms of the probability amplitudes assigned to the S-matrix. With standard QFT, the treatment of an interaction is based on Feynman diagrams. Although Feynman diagrams already seem to contain certain process-oriented aspects, they are not a suitable basis for a functional (i.e., process-based) description. Typically, many Feynman diagrams are required to compute the result of an interaction. The results of the overall interaction are determined from the superposition of the individual diagrams. There are no reasonable intermediate states or any process structure derivable from the collection of diagrams. As described in section 3, the functional model of QT assumes particle fluctuations and interaction channels instead of Feynman diagrams. The proposed model subdivides the overall interaction process into four process steps:

```
perform-interaction ::= {
  Step1: Appearance and start of an interaction - particle/wave fluctuation;
  Step2: Generation of the interaction-object;
  Step3: Formation and processing of interaction channels;
  Step4: Generation of the "out"-pw-collection -  Information exchange;
}
```

Step1 (appearance and start of an interaction) and step4 (generation of the "out"-pw-collection) are key for the process model of QT measurement.

[1] Nevertheless, typical measurements contain both types of interactions.

Journal of Physics: Conference Series **626** (2015) 012061 doi:10.1088/1742-6596/626/1/012061

4.2.1. Appearance and start of an interaction - particle/wave fluctuation Concerning the appearance and start of an interaction, this model of the QT measurement process makes two assumptions that are essential for an explanation of the QT measurement process:

(i) The appearance of an interaction is an event that actually happens (with a certain probability). Insofar, the occurrence of an interaction is the first step in the transition from probabilities to facts, which is associated with QT measurements.

(ii) An interaction occurs at a definite position. If a q-object consist of multiple paths, only the path(s) that occupy the interaction position affect the interaction process further. The remaining paths are eliminated. This may be viewed as the collapse of the wave function. The elimination of the unaffected paths represents a reduction of the wave function and of the measurement result to a definite value (with respect to position).

Both assumptions cannot be derived from existing QT/QFT, but they are only in conflict with existing QT/QFT if one believes that the Schrödinger equation and similar equations of motion completely determine the dynamical evolution of a QT system, including the transition from probabilities to facts.

Thus, the appearance of an interaction, including its location, is determined randomly. Various mechanisms are imaginable for the random determination of the interaction position. The functional model assumes that an interaction is a possible consequence of a particle/wave fluctuation, which supports that the interaction occurs at a definite position.

4.2.2. Generation of the interaction-object At the beginning of the interaction process the information from the interacting particles is merged into the interaction-object. Further details on this process step are not important for the model of QT measurement, but can be obtained from [2].

4.2.3. Formation and processing of interaction channels The main processing of the interaction consists in the formation and processing of ia-channels (interaction channels). Ia-channels (like virtual particles) guide the possible flow of particle transitions during an interaction. Further details on this process step are not important for the model of QT measurement, but can be obtained from [2].

4.2.4. Generation of the "out"-pw-collection - Information exchange The processing of the ia-channels ends with a certain "out" particle/wave combination. With some types of interactions, different "out" particle/wave combinations may occur. For example, with Bhabha scattering (i.e., electron-positron scattering) the "out" combinations (electron, positron), (muon, antimuon), and (tauon, antitauon) are possible. The process model of QT interactions assumes that, from the possibly multiple alternative "out" particle/wave combinations, only one leads actually to an interaction. This is a further instance of transition from probability to facts. [2] The detailed mechanism for the selection of the "out" particle/wave combination is beyond the scope of the present paper. Several alternative mechanisms are imaginable.

At the end of the interaction, the ia-channels perform superpositions to produce the "out" pw-collection containing the "out" particle/waves. The "out" pw-collection consists of (a discrete set of) paths to reflect all possible interaction results for which the scattering matrix provides a non-zero probability amplitude.

Thus, the interaction may be said to transform the information from the "in" particles to the information of the "out" pw-collection.

[2] Because standard QT does not assume multiple points for the "transition from probability to facts", this possibly represents a deviation from standard QT, which however will be difficult to test.

Journal of Physics: Conference Series **626** (2015) 012061 doi:10.1088/1742-6596/626/1/012061

5. The model of the measurement process

Measurement may be viewed as the mapping of the state of the measured q-object to the state of some measurement device. For this purpose, at least one QFT interaction has to be performed between the measured particle and the measurement apparatus. Depending on the kind of measured state component, further interactions may be required to prepare for the main interaction, and still further interactions to transform the measurement result to usable information. Measurements are realized through normal interactions. Normal interactions are those types of interactions that are used to implement a measurement and may also occur in situations not related to measurements.

To understand the limitations of QT measurement (which result in the QT measurement problem), the understanding is that QFT interactions are not information preserving transformations for two reasons:

(i) Reduction
At the start of the interaction, the multitude of alternative paths is reduced to those paths which occupy the interaction position (see section 4.2.1).

(ii) QFT interactions do not represent a bijective mapping
Although it has been shown that the scattering matrix S is a unitary matrix (i.e. there exists an inverse matrix S^+ such that $SS^+ = 1$, see [9]), this does not mean that it is possible to deduce the complete "in" state of the interacting particles from the "out" state of the interaction.

5.1. Answers given by the process model to questions related to the measurement problem

Based on the above described functional model of QFT interactions and the limited information transformation with QFT interactions, the following answers to the questions related to the measurement problem are offered.

- Does measurement always have a definite outcome?
  The functional model assumes that measurements are always realized using QFT interactions. QFT interactions reduce the set of alternative paths to a single path as input (parameter) to the interaction process. If the QFT interaction is capable to map this single path to observable facts (e.g., the interaction position) in the "out"-state of the interaction, this may be called a definite outcome. A non-definite outcome is possible (it is even the default case), but is useless if it does not provide information on the complete set of alternative paths.
- Does measurement determine the state of a QT object?
  The purpose of a measurement is the determination of specific components of the system state. Unfortunately, a measurement, in general, also changes the state of the measured QT object?
- Why is it impossible to measure the non-definite state of the system?
  Measurement of the non-definite state of the system would mean a bijective mapping of the complete set of alternative paths of the q-object (rather than a single path) to an observable "out"-state of a QFT interaction. QFT does not support such interactions.
- Why can certain observables not be measured simultaneously?
  Generally, QT observables which do not commute (i.e., $[O_1 O_2] \neq 0$) are said to be non-simultaneously measurable with arbitrary precision. The standard example for QT observables which are not considered to be simultaneously measurable are the position X and the momentum P. According to the functional model of QFT interactions, a measurement (i.e., a QFT interaction) which maps the position as well as the momentum of the selected

Table 2. Structure of a q-object ψ consisting of two entangled particle/waves pw1 and pw2.

paths	pw1-state	pw2-state	amplitude
path-1	pw1-up	pw2-down	ampl-1
path-2	pw1-down	pw2-up	ampl-2
...	...	...	...
path-N	pw1-state$_N$	pw2-state$_N$	ampl-N

path of the measured q-object, is possible, but useless because the measurement results do not sufficiently reflect the state of the measured QT object.

- Is a measurement always coupled with the collapse of a wave function?
  To deliver observable facts, reasonable measurements have to be realized using QFT interactions. [3] QFT interactions imply a reduction of the set of alternative paths to a single path. This may be called a "collapse of a wave function".
- Is the transition from probabilities to facts related exclusively to measurements?
  In view of the q-objects of the state of the system described in section 3, the phrasing transition from probabilities to facts is misleading. More correctly, we should rather say transition from a set of alternative paths with associated probability amplitudes to a state of the system where one of the alternatives is selected and its properties mapped to observable facts in the "out" state of an interaction.
  As described in sections 3 and 4, this process is not exclusively related to measurements but occurs with each QFT interaction.
- When (in the sequence of process steps or under which circumstances) does the transition from probabilities to facts occur?
  The functional model assumes multiple points within the interaction process where the changes in the state of the system are a function of the probability amplitudes. The most important point is the occurrence of an interaction and the determination of the interaction position.
- Is the measurement process random or deterministic?
  The functional model assumes that the transition from probabilities to facts performed with QFT interactions is random based on probability amplitudes.

6. Discussions

6.1. Handling of entanglement

Entanglement is related to measurement. Two particles/waves pw1 and pw2 are entangled, if the result of a measurement of pw1 influences the possible results of a measurement of pw2. The typical example used in QT literature is the wave function

(1) $\psi = 1/\sqrt{2}\ (\mid pw1.up, pw2.down > + \mid pw1.down, pw2.up >)$

referring to a wave function ψ with entangled elements pw1 and pw2. With the functional model of QT ψ would be represented by a particle/wave-collection consisting of the two elements pw1 and pw2 and a discrete number of paths which reflect the correlation (see Table 2). [4] As described in sections 3 and 4, a measurement (using a QFT interaction) reduces the set of paths to a single path. This selects not only the definite value for the first measured object (e.g., pw1), but also the possible measurement results for the other element, e.g., pw2. [5]

[3] There may be experiments which do not include QFT interactions, but nevertheless deliver useful information. However, such experiments are not considered typical measurements by the author.

[4] Differing from eq. (1), the q-object has to state explicitly the whole set of alternative paths.

[5] This does not mean that the measurement value for pw2 is already uniquely determined.

6.2. Relation to standard QT/QFT

Although the process model aims for maximum compatibility with standard QT/QFT, there are exceptions in the process model. Examples include some small deviations to standard QFT that have been intentionally included, items where it is not clear what the QT/QFT conformal behavior exactly is, and items where there is no equivalent QT/QFT position because the level of detail is below the scope of QT/QFT. For all of these areas, verification by experiments is appropriate. There is one area in which the functional model is designed to deviate from standard QT/QFT: the concept of the transition from probabilities to facts. The process model assumes the transition from probabilities to facts to occur in multiple steps. As indicated in section 4, it will be very difficult to test possible deviations from standard QT in this area.

7. Conclusions

Measurements are realized by interactions between the measured object and parts of the measurement apparatus. With QT measurements, the possibilities for interactions to map the state of the measured QT object to the state of the measurement apparatus are limited. The greatest potential for a useful mapping is given with what in this paper is called QFT interactions. The transformation of the "in"-state of a QFT interaction to the "out"-state is supported by QFT in terms of scattering matrices and Feynman diagrams. However, QFT interactions, too, are severely limited in their capabilities to map the state of the measured QT object to the "out"-state of the interaction, for two reasons:

(i) QFT interactions change the state of the measured object. This reduces the possibilities to determine the state of the measured object through multiple (accumulative) interactions.

(ii) QFT interactions are not information preserving, i.e., it is not possible to deduce completely and exactly the "in"-state of the interaction from the "out"-state (see section 5).

These limitations are the cause of the peculiarities associated with the QT measurement. A more detailed explanation, including answers to questions associated with the QT measurement problem) can be given (and has been given in this this paper) by a model of the QT measurement *process*.

A key ingredient of the proposed process model of QT measurement is the assumption of a discrete set of alternative wave function paths. The evolution of this set of alternative wave function paths during the process of a QFT-based interaction explains many of the peculiarities of QT measurements.

As described in section 6, there exist a number of points where the functional model of QT interactions (within which the proposed model of QT measurement is embedded) possibly deviates from standard QT/QFT, either intentionally or where the QT/QFT position is unclear, or where the level of detail is below standard QT/QFT. From the list of (possible) deviations and refinements from standard QT/QFT, only a small part is essential for the described process model of QT measurement. The major part of the proposed process model of QT measurement is based on an extended application of existing QT, in particular QFT.

References

[1] Diel H 2013 A Functional Interpretation of quantum theory, Proceedings of ICCQMNP, Barcelona
[2] Diel H 2015 A Functional Model of interactions in quantum theory, arxiv:1405.6099 [quant-ph]
[3] Diel H 2013 The formulation of temporal relationships with physics theories, *Pensee Journal* **75**
[4] Everett III H 1957 Relative state' formulation of quantum mechanics. *Rev Mod Phys* **29** 454
[5] Feynman R P 1985 QED: *The Strange Theory of Light and Matter* (Princeton: Princeton University Press)
[6] Mandl F, Shaw G 1993 *Quantenfeldtheorie* (Wiesbaden: AULA Verlag) 351
[7] Maudlin T 1995 Three measurement problems. *Topoi-Int Rev Philos* **14**
[8] Ryder L H 1985 *Quantum Field Theory* chapter 3, (Cambridge: Cambridge University Press)
[9] Weinberg S 2005 *The Quantum Theory of Fields, Volume 1 Foundations* (Cambridge: Cambridge University Press)

Journal of Physics: Conference Series **626** (2015) 012062 doi:10.1088/1742-6596/626/1/012062

Testing cellular automata interpretation of quantum mechanics in carbon nanotubes and superconductivity

Donatello Dolce and Andrea Perali

University of Camerino, Piazza Cavour 19F, 62032 Camerino, Italy.

Abstract. Cellular Automata (CA) are represented at an effective level as intrinsic periodic phenomena, classical in the essence, reproducing the complete coherence (perfect recurrences) associated to pure quantum behaviours in condensed matter systems. By means of this approach it is possible to obtain a consistent, novel derivation of SuperConductivity (SC) essential phenomenology and of the peculiar quantum behaviour of electrons in graphene physics and Carbon Nanotubes (CNs), in which electrons cyclic dynamics simulate CA. In this way we will derive, from classical arguments, the essential electronic properties of these — or similar — graphene systems, such as energy bands and density of states. Similarly, in the second part of the paper, we will derive the fundamental phenomenology of SC by means of fundamental quantum dynamics and geometrical considerations, directly derived from the CA evolution law, rather than on empirical microscopical characteristics of the materials as in the standard approaches. This allows for a novel heuristic interpretation of the related gauge symmetry breaking and of the occurrence of high temperature superconductivity by means of simple considerations on the competition of quantum recurrence and thermal noise.

1. Introduction

As proven by G. 't Hoof in recent works, particular CA models have much in common with Quantum Mechanics (QM). This correspondence has originated a deterministic interpretation of fundamental aspects of QM [1, 2]. The idea considered in this paper stems from the model describing CA of generic number N of "ontic" sites on a circle and permutations among the neighbour sites and period T. The evolution law is $|t_l\rangle \to |t_{l+1} + \text{mod } T\rangle)$. Such a CA can be equivalently regarded as a "particle moving on a circle" of N lattice sites. For the continuous limit, 't Hooft has proven that "*there is a close relationship between a particle moving on a circle with period T and the Quantum Harmonic Oscillator (QHO) with the same period*" [1]. This is a fundamental result as the QHO is the essential ingredient of a second quantised field, and in turn of the whole Quantum Field Theory (QFT). This represents the central concept used in this paper to derive applications to condensed matter. We will name the continuum of ontic sites of the continuous CA model as "ontic" time t of period T. In this continuum limit the CA evolution law from a generic initial time t_i to a final time t_f is $|t_i\rangle \to |t_f + \text{mod } T\rangle$. To such a CA of period T is associated a fundamental energy $\hbar\omega = 2\pi\hbar/T$, according to the de Broglie phase harmony condition in the ontic time $\omega T = 2\pi$. According to 't Hooft, the CA periodic dynamics turn out to be described, at a statistical level, by a wave-function $\Phi(t)$, also named "CA physical state", whose generic solution is $\phi(t) = e^{-i\omega t}$, satisfying the Periodic Boundary

Conditions (PBCs) $\Phi(t) = \Phi(t+T)$. These PBCs encode the CA evolution law in time and determine the CA eigenstates $\phi_n(t) = e^{-i\omega_n t}/\sqrt{2\pi}$. They form a complete, orthogonal set with harmonic energy spectrum $\hbar\omega_n = n\hbar\omega = n\frac{2\pi\hbar}{T}$ ($n \in \mathbb{Z}$). Hence the CA defines an Hilbert space of basis $|n\rangle$ such that $\langle t|n\rangle = \phi_n(t)$ with induced inner product $\langle n|n'\rangle = \delta_{n,n'}$. Indeed the CA evolution law in time is $|t\rangle \to |t + \delta t + \bmod T\rangle$. The CA turns out to be represented by the superposition of these eigenstates $\Phi(t) = \sum_{n\in\mathbb{Z}} a_n \phi_n(t)$. That is, in the related Hilbert space representation, it is described by a point in the Hilbert space $|\Phi\rangle = \sum_n \alpha_n |n\rangle$. Since $i\partial_t \phi_n(t) = \omega_n \phi_n(t)$ the CA ontic time evolution is given, in this Hilbert space formalism, by the Schrödinger equation $i\partial_t|\Phi(t)\rangle = \mathcal{H}|\Phi(t)\rangle$ where the Hermitian operator $\mathcal{H}$ is defined as $\mathcal{H}|n\rangle = \omega_n|n\rangle$. Hence, its evolution is given by the Hilbert operator $\mathcal{U}(dt) = e^{-i\mathcal{H}dt}$. We must bear in mind that the fundamental topology of our CA is the topology of the circle $\mathbb{S}^1$. This determines, by means of the PBCs, the quantum number n.

The formalism and results presented in this paper are a direct generalisation to CA of the Elementary Cycles theory (EC) [3, 4, 5, 6, 7, 8]. EC theory is interpreted in terms of 't Hooft CA in [9] constituting an introduction to this paper. Indeed EC is an effective description of continuous periodic CA in which the CA evolution law is encoded in PBCs of the relativistic coordinates and the CA is thus represented as vibrations associated to this relativistic cyclic dynamics. The powerful formalism of EC theory allows in this way to test 't Hooft proposal in advanced aspects of theoretical physics and condensed matter physics [3, 4]. These results will be here explicitly interpreted in terms of CA theory.

Actually, the dynamics described by this CA model, similarly to the EC model, correspond to those of the time evolution of a Quantum Harmonic Oscillator (QHO) of period T, except for the fact that the quantum number can assume negative values, $n \in \mathbb{Z}$ (positive and negative frequencies). This seems to imply a non positively defined Hamiltonian operator for CA. Nevertheless, as pointed out by 't Hooft, the negative modes can be regarded as describing anti-particles. In confirmation of this we will in fact find that in graphene physics these negative modes correspond to the holes in the Fermi sea, i.e. to the negative modes of the Carbon Nanotubes (CN) energy bands. Indeed, the Elementary Charge Carriers (ECCs) in a CN behaves as "particle on a circle". In particular, a CN with N carbon atoms along the circumference corresponds to a CA of N sites on a circle. They both have the fundamental topology of the circle $\mathbb{S}^1$. We will use the covariant generalisation of this correspondence to derive in a simple way, and in agreement with present literature [10, 11], the essential electronic properties of CNs. CNs can be therefore used to test CA theory and foundational aspects of QM.

Similarly we will show that the CA evolution law, as well as the PBCs in EC theory, yields the most foudantional aspects of SC such as: quantisation of the magnetic flux in unit of $\varphi_0/2 = hc/2e$, the Meissner effect, the Josephson effect, the Little-Park effect, etc. Indeed, the CA evolution law directly implies discretised variations of the Goldstone field which in turn describe the explicit breaking of the ElectroMagnetic (EM) gauge invariance typical of SC [3, 12]. In short, the SC phenomenology will be derived by applying the simple requirement that the wave function of electrons in the superconductor must have closed space-time orbits, in analogy with the closed space-time orbits of electrons in the atomic orbitals (see Bohr atom or Bohr-Sommerfeld quantisation). According to this analogy, the electron in the atomic orbitals can be therefore regarded as being locally in a superconducting regime.

2. Testing CA theory by means of Carbon Nanotubes: Compton clock vibrational modes and controvariant modulations.

As well-known, the electrons in a graphene monolayer behave as massless charged particles (with pseudo-spin), i.e. as relativistic particles traveling at the Fermi velocity v_F (the analogous of the speed of light for graphene physics). Let us assume that the graphene layer is curled-up in the same direction of the electron motion. The result will be a "particles moving on a circle" at "light" speed v_F. Since the CN will have a finite number N of carbon atoms along the circumference C_h, such a "circle" will be on a lattice of N sites. In short, the electron moving at velocity v_F along a CN circumference characterised by N carbon atoms can be regarded as a CA with N sites and time period $T_C = C_h/v_F$. Indeed CNs are real physical systems simulating CA. For the sake of simplicity we start by investigating the continuous lattice limit. The effective CN electronic properties will be then obtained by passing to the lattice case. For instance the CA energy spectrum $\hbar\omega_n = n\hbar\omega = \frac{2\pi\hbar n}{T}$ in the lattice case is $\hbar\omega_n = \frac{N\hbar}{2T}\sin(\frac{\pi n}{N})$.

As a consequence of the compactification of one of the two spatial dimensions of the graphene layer, the resulting CN effective dynamics are characterised by a single spatial dimension, i.e. the axial direction $x_\parallel$. Therefore, in such an effective CN one-dimensional space, the electron moving along the circumference is effectively at rest with respect to the spatial dimension. We denote the momentum of the electron along the axial direct with $k_\parallel$ and the corresponding period of the electron moving along the CN circumference, at rest with respect to the axial direction, as $T(k_\parallel = 0) = T_C$. According to the CA model, this effective rest period corresponds to a rest energy $\hbar\omega(k_\parallel = 0)$, such that $T(k_\parallel = 0) = 2\pi/\omega(k_\parallel = 0)$. In relativity it is natural to associate a rest energy to a mass: $\hbar\omega(k_\parallel = 0) = m_* v_F^2$. Indeed the rest period T_C corresponds to the effective Compton time of the electron in the CN, which in turn fixes the effective mass scale of the ECCs in CN according to the Compton relation $m_* = \frac{2\pi\hbar}{T_C v_F^2}$ (that is, the Compton time of a relativistic particle of mass m is $T_C = \frac{2\pi\hbar}{mc^2}$). Thus, we find that the effective mass scale of the electron in the CNs is determined by the CN circumference according to

$$m_* = \frac{2\pi\hbar}{C_h v_F} \tag{1}$$

This effective mass scale is the fundamental rest energy of our CA of rest period $T(k_\parallel = 0) = T_C$. The ECCs at rest can be regarded as rest CA in order to simulate the effective de Broglie internal clock of the electron in the CNs. Notice that the Compton time of the electron $\sim 10^{-21}$s is effectively rescaled to about 10^{-15}s in CNs. Since the latter time scale is accessible to modern timekeepers, CN allows to indirectly test foundational quantum aspects of elementary particles, as pointed out in [4].

The CA is characterised not only by the fundamental energy considered so far, but by a whole spectrum of energy levels. In this rest case the CA rest energy spectrum is $\hbar\omega_n(k_\parallel = 0) = n\hbar\omega(k_\parallel = 0) = nm_*$. Since this is a rest energy spectrum, it is natural to associate it to a mass spectrum $m_n = nm_* = \frac{2\pi n\hbar}{C_h v_F}$, with $n \in \mathbb{Z}$. This yields interesting analogies to the Kaluza-Klein theory as described in [8]. Similarly to the general case, this harmonic spectrum denotes the eigenmodes associated to the PBCs in the proper time τ for the CA describing the ECC in the CN: $\Phi(\tau) = \Phi(\tau + T_C)$, where τ is the proper time of the ECC (i.e. the time coordinate for the electron at rest with respect to the axial direction). In short these PBCs encodes the evolution law of the CA evaluated at rest. In the derivation of the lattice limit we have to consider that, as a consequence of the hexagonal geometry of the graphene layer, two kind of PBCs are possible for the electron moving along the CN circumference. Besides the trivial PBCs described above, it is possible to have configurations symmetric under lattice rotations of $\pm\frac{2}{3}\pi$. In this case the PBCs are thus twisted by a corresponding factor. That is, in general, we can write the PBCs of the ECCs in CNs as

$$\Phi(\tau) = e^{i2\pi\alpha}\Phi(\tau + T_C) \tag{2}$$

where $\alpha = 0, \pm\frac{1}{3}$. In other words the evolution law in the proper time of the CA simulated by CNs is $|\tau_i\rangle \to e^{i2\pi\alpha}|\tau_f + \text{mod } T_C\rangle$. By considering this twist of the PBCs the resulting mass spectrum of the rest CA turns out to have a shift αm_*:

$$m_n = (n+\alpha)\frac{2\pi\hbar}{T_C} = (n+\alpha)m_* = (n+\alpha)\frac{2\pi\hbar}{C_h v_F}\ , \quad (n \in \mathbb{Z})\,. \tag{3}$$

This implies that for $\alpha = 0$ the CN is characterised by a massless mode $m_0 = \bar{m} = 0$, whereas all the other modes are massive: $m_n = nm_*$ with $n \neq 0$. Indeed this case describes metallic CNs. In the case $\alpha = \pm\frac{1}{3}$ all the modes are massive. This in fact corresponds to semiconducting CNs whose characteristic fundamental mode has effective mass

$$\bar{m} = \frac{2\pi\hbar}{3C_h v_F} . \tag{4}$$

in agreement with present literature [10]. The lattice limit correctly leads to the effective CN mass spectrum [11]. We find that, for ZigZag (ZZ) and ArmChair (AC) CNs of N carbon atoms in the perimeter, the resulting energy spectra turns out to be, respectively,

$$\text{ZZ)} \quad m_n = m_* \frac{N}{\pi} \sin(\frac{\pi n}{N}) \; ; \qquad \text{AC)} \quad m_n = m_* \frac{N}{3\pi} \left[1 + 2\cos(\frac{\pi n}{N})\right] . \tag{5}$$

In order to derive the CN energy bands we must consider the relativistic transformation of our CA of rest period T_C. This can be easily done in analogy with undulatory mechanics. Indeed a periodic continuous CA can be regarded as a "de Broglie periodic phenomenon" and the CA at rest describes the so called "de Broglie internal clock". As originally pointed out by de Broglie, the period and wavelength of a moving particle, and thus its energy and momentum, are determined, through Lorentz transformations, by the rest periodicity, i.e. by the Compton period of the particle. This de Broglie description, at the base of modern undulatory mechanics and QM, directly applies to CNs and CA (or EC).

The case of ECC with a non zero velocity along the CN axial direction corresponds to an effective Lorentz boost in the effective CN space-time. That is, the effective period $T(k_\parallel)$ and wave-length $\lambda_\parallel$ of an electron with non-vanishing momentum $\hbar k_\parallel$ in the axial direction of a CN is obtained from the Compton time T_C by means of the Lorentz transformation $T_C = \gamma_\parallel T(k_\parallel) - \beta_\parallel \gamma_\parallel \lambda_\parallel$ in the effective CN space-time, where the Lorentz factor is $\gamma_\parallel = 1/\sqrt{1-\beta_\parallel}$ with $\beta_\parallel = v_\parallel / v_F$ and $v_\parallel$ is the velocity along the axial direction. In short, the motion along the CN is described by boosting the rest CA described above. The effective Compton period $T_C = \frac{C_h}{v_F}$ implies, by means of Lorentz transformations, a time period $T(k_\parallel)$ and a spatial recurrence $\lambda_\parallel$ when when the electron moves along the axial direction — as a consequence, for instance, of an electric potential. These recurrences in time and space determine the effective energy $\hbar\omega(k_\parallel) = \frac{2\pi\hbar}{T(k_\parallel)}$ and the momentum $\hbar k_\parallel = \frac{2\pi\hbar}{\lambda_\parallel}$ of the electron in the CN, according to undulatory mechanics. Notice that the former relation is the relation between time period and fundamental energy for a CA in the rest frame $k_\parallel$, i.e. the CA phase harmony relation in the rest frame of the electron. In analogy to relativistic notations, by introducing the two tangent two-vector $k_\mu = \{\omega/v_F, -k_\parallel\}$ (covariant) and $\lambda^\mu = \{T v_F, \lambda_\parallel\}$ (controvariant), the CA phase harmony relation can be generalised to the invariant form $m_* v_F^2 T_C/\hbar = k_\mu \lambda^\mu = 2\pi$. Thus, the relativistic constraints for these two vectors are respectively $m_* v_F^2 = \hbar^2 k_\mu k^\mu$ and $\frac{1}{T_C^2} = v_F^2 \frac{1}{\lambda^\mu} \frac{1}{\lambda_\mu}$, denoting the dual relativistic dispersion relations for the fundamental energy and time period, respectively. The covariant CA evolution law for an electron moving in the effective CN space-time $x^\mu = \{t, x_\parallel\}$ is thus given by $|x_i^\mu\rangle \to e^{i2\pi\alpha}|x_f^\mu + \text{mod}\ \lambda^\mu\rangle$ (the CA evolution law in time and space are given by the component $\mu = 0$ and $\mu = 1$, respectively). In analogy to the relativistic Doppler effect, this implies that the time period $T(k_\parallel) = 2\pi/\omega(k_\parallel)$ varies with $k_\parallel$ in such a way that the electron fundamental energy turns out to satisfies the effective relativistic dispersion relation

$$\hbar^2\omega^2(k_\parallel) = \frac{(2\pi\hbar)^2}{T^2(k_\parallel)} = m_*^2 v_F^4 + \hbar^2 k_\parallel^2 v_F^2 . \tag{6}$$

In short we have obtained that the motion of the electrons along CNs is the analogous of the relativistic motion, provided that the mass of the electron is replaced by the effective mass m_* and the speed of light by the Fermi velocity v_F.

If we now apply this effective relativistic dispersion relation to every mass eigenmode m_n allowed for the electron effective Compton periodicity T_C we immediately obtain the energy bands structure of CNs in the continuous limit

$$\hbar\omega_n(k_\|) = (n+\alpha)\frac{2\pi\hbar}{T(k_\|)} = (n+\alpha)\sqrt{m_*^2 v_F^4 + \hbar^2 k_\|^2 v_F^2}\,, \quad (n \in \mathbb{Z}) \tag{7}$$

The CNs energy bands follow from this result by performing the lattice limit (and considering that the CN lattice has also a periodicity along the axil direction yielding Brillouin zones). For instance, in agreement we present literature, in the case ZZ CNs we find the energy bands structure — or similarly the density of states [4]:

$$\text{ZZ)} \quad \hbar^2\omega_n(k_\|) = m_*^2 v_F^4 \frac{N^2}{3\pi^2}\left(1 + 2\sin\frac{\pi n}{N}\right)^2 - \frac{2}{3}\hbar^2 k_\|^2 v_F^2 \cos\frac{\pi n}{N}\,. \tag{8}$$

This description is very interesting to test the validity of CA theory. Indeed one of the main problematics of the CA interpretation of QM is that the negative modes ($n = -1, -2, \dots$) associated to the CA evolution law implies a non positively defined Hamiltonian operator. 't Hooft as noticed however that these negative modes can be consistently interpreted as antiparticles, i.e. holes in the Dirac sea. This is fully confirmed by CN in which, actually, the negative energy bands describes holes in the Fermi sea.

3. Testing CA theory by means of superconductivity: the role of the temperature

Here we will introduce the role of the temperature in the EC theory, which can be directly generalised to CA. We must bear in mind that in CA, as in undulatory mechanics and EC theory, the time period and the energy are "two faces of the same coin". Therefore interactions, i.e. variations of energy $\hbar\omega = 2\pi\hbar/T$, correspond to local modulations of the period T. Since the temperature implies random collisions among particles (thermal noise), a quantum system at finite temperature is characterised by chaotic (Poissonian) decay of the "complete coherence" of the particles with a characteristic thermal time $\beta = \hbar/k_B\mathcal{T}$, i.e. a dumping $e^{-\hbar\omega/k_B\mathcal{T}} = e^{-\beta/T}$ of the cyclic behaviour of periodicity T of the system, being k_B the Boltzmann constant and $\mathcal{T}$ the temperature. While free CA at zero temperature, as well as the perfect coherence of pure quantum systems, are characterised by perfect periodicity T in the Minkowskian time, thermal quantum systems are characterised by Euclidean time periodicity of period $\beta = \hbar/k_B\mathcal{T}$. These two kinds of temporal periodicities (which can be related by Wick's rotation, as for instance in the Hawking theory of black holes) have fundamental different physical meanings: they are in competition. In the pure quantum systems at zero temperature we have a persistent exact recurrence in time, e.g. described by the phasor $e^{-i\omega_n t}$, whereas in the latter case, corresponding to a non zero temperature, we have a dissipation of the quantum recurrence by Boltzmann factors $e^{-\omega_n/k_B\mathcal{T}} = e^{-n\beta/T}$, i.e. a dumping factor of the Minkowskian periodicity. In other words, while the Minkowskian periodicity T of quantum mechanics tends to form perfect coherent states ("periodic phenomena"), the Euclidean periodicity β describes a dumping associated to the thermal noise which tends to destroy the perfect recurrences of the pure quantum systems. Thus, if the $T \ll \beta$ the system can autocorrelated and give rise to pure quantum phenomena such as superconductivity whereas in the opposite limit the thermal noise breaks the the quantum recurrence before it can give rise to autocorrelation, in this case we have the classical behaviour, e.g. ordinary electric resistance. Another aspect of the temperature which will be used in the following description is the Boltzmann distribution. In a few words, the dumping factor $e^{-\omega_n/k_B\mathcal{T}} = e^{-n\beta/T}$ can be also regarded as the probability to populate the nth vibrational mode of the system. Thus, at very low temperature $T \ll \beta$ only the fundamental mode $n = 1$ will be populated whereas at high temperature many vibrational modes must be considered (e.g.

as in the classical limit in Bohr atom). We will use this aspect to describe the EM symmetry breaking occurring in superconductivity (in analogy with the effective gauge symmetry breaking in extra dimensional extensions of the Standard Model).

These simple arguments, further motivated below, allow us to predict that high temperature SC happens in quantum systems characterised by very small quantum recurrences T, orders of magnitude smaller than in ordinary SC. This is confirmed by CNs in which, as we have seen, the quantum recurrence is explicitly determined by the CN circumference C_h, so that, in agreement with [10, 13], the critical temperature is inversely proportional to the circumference C_h.

Our description of CNs shows that covariant CA provide a model describing the "periodic phenomena" conjectured by de Broglie to describe the quantum behaviour of elementary particles. Indeed, in the continuous limit, a CA (or an EC) of rest period T_C describes the quantum behaviour of a relativistic particle of mass $m = 2\pi\hbar/T_C c^2$, where we have replaced the Fermi velocity with the speed of light c. In general a CA, similarly to the ideal case of a free quantum particle at zero temperature, is characterised by a "complete coherence" in space-time.

Generalising the CN description to the four-dimensional space-time of ordinary relativity, in a given reference frame the free CA as time period $T(\vec{k})$ and wave-length $\vec{\lambda}$ which determines its energy $\hbar\omega(\vec{k}) = 2\pi\hbar/T(\vec{k})$ and momentum $\hbar k_i = 2\pi\hbar/\lambda^i$ $(i = 1, 2, 3)$. In covariant notation we can introduce the four vectors $k_\mu = \{\omega/c, -\vec{k}\}$ and $\lambda^\mu = \{Tc, \vec{\lambda}\}$ related by the phase harmony relation $k_\mu \lambda^\mu = 2\pi\hbar$. This means that the evolution law is $|x_i^\mu\rangle \to |x_f^\mu + \text{mod } \lambda^\mu\rangle$ (notice that in this case a twist fact of $\alpha = 1/2$ reproduces the vacuum energy of the particles $\hbar\omega(\vec{k})/2$). The generic CA solution, in the free case, is $\Phi(x) = \sum_n e^{-ik_{(n)\mu}x^\mu}$ and the PBCs, encoding the evolution law, are $\Phi(x^\mu) = \Phi(x^\mu + \lambda^\mu)$. These imply the (normally ordered) energy-momentum spectrum $\vec{k}_{(n)\mu} = n\vec{k}_\mu$. As can be easily seen from time CA evolution, $\Phi(t) = \sum_n e^{i\omega_n t}$ and $\Phi(t) = \Phi(t+T)$ imply the ordinary (normally ordered) energy spectrum of a relativistic quantum particle $\omega(\vec{k}) = n\hbar\omega(\vec{k})$ (in analogy with a string vibrating with period $T(\vec{k})$). In short, the CA evolution law implies that the physical state Φ describing the wave function of a free relativistic particle at zero temperature must have closed orbits along its space-time evolution. That is, we have a correspondence with the Bohr-Sommerfeld quantisation. Indeed, an interacting CA is characterised by a locally modulated space-time period so that, in the interacting case, the generic solution has the form of a locally modulated wave $\phi(x) = e^{-i\int^x dy^\mu k^\mu(y)}$. The evolution law, i.e. the PBCs for this generic solution, in this case yields the relativistic Bohr-Sommerfeld quantisation condition $\oint_\lambda dy^\mu k_\mu(y) = 2\pi n\hbar$. It is well-known that in the case of a Coulombian potential this yields the Bohr energy levels of the atomic orbitals $E_n = -13.6$ eV$/n^2$.

Let us assume unitary gauge invariance for a CA (representing electrons in equilibrium with the EM field in the superconducting material) $\Phi(\vec{x}, t) = U(\vec{x}, t)\Phi(\vec{x}, t)$, where $U(\vec{x}, t) = e^{-i\frac{e}{\hbar c}\theta(\vec{x},t)}$. The Goldstone $\theta(\vec{x}, t)$ denotes the local invariance of the EM interaction and e is the electric charge. In analogy with the CA edescription, we now impose PBCs $\Phi(\vec{x}, t) = \Phi(\vec{x}, t + T(\vec{k}))$ (for the sake of simplicity we only consider the time component, furthermore we assume here that the topology is the orbifold $\mathbb{S}^1/\mathcal{Z}_2$ associated to the fundamental topology $\mathbb{S}^1$ of the CA, so that the phase invariance is $n\pi$ rather than $2\pi n$). Thus, in this case the PBCs in time of period $T(\vec{k})$ imply the following evolution law for the Goldstone mode

$$\frac{e}{\hbar c}\theta(\vec{x}, t) = \frac{e}{\hbar c}\theta(\vec{x}, t + T(\vec{k})) + \text{mod } n\pi. \tag{9}$$

As a consequence of the CA evolution law the Goldstone field can only vary by finite amounts

$$\Delta\theta(\vec{x}, t) = \frac{\varphi_0}{2}\,, \quad \text{with } \varphi_0 = \frac{hc}{e}\,. \tag{10}$$

This simple condition yields the characteristic behaviours of a superconducting regime, as described in [12]. In short, according to our ansatz, as a consequence of the closed orbits of the wave-function imposed by the CA evolution law, $\theta(\mathbf{x}, t)$ can only vary by discrete amounts running along the contour Σ in which the EM field is a pure gauge $A_\mu(\vec{x}, t) = \partial_\mu \theta(\vec{x}, t)$. The Stokes theorem gives a quantization of the magnetic flux through the area $\mathcal{S}_\Sigma$ limited by Σ,

$$\int_{\mathcal{S}_\Sigma} \mathbf{B}(\mathbf{x}, t) \cdot d\mathbf{S} = \oint_\Sigma \mathbf{A}(\mathbf{x}, t) \cdot d\mathbf{x} = \oint_\Sigma \nabla\theta(\mathbf{x}, t) \cdot d\mathbf{x} = n\frac{\varphi_0}{2}\,. \tag{11}$$

Since the magnetic flux is quantised, the electric current cannot smoothly decay while flowing around the torus, and there is not electric resistance, i.e. we have SC, [12] and [5, 6, 7]. The quantisation (10) means that the Goldstone field $\theta(\mathbf{x}, t)$ transforms as the phase of a condensate of a fermionic pair operator $\langle \epsilon_{\alpha\beta}\Phi^\alpha\Phi^\beta \rangle$ of charge $-2e$. That is, it plays the role of the usual Cooper pair of the BCS microscopic theory of SC. From (10) it is also manifest that the quantisation of the Goldstone field implies a breaking of the EM gauge $U_{EM}(1)$ to $\mathcal{Z}_2$.

Now we consider a EM gauge connection to the CA physical state $\Phi(x') = e^{i\frac{e}{\hbar c}\int_x^{x'} A_\mu dx^\mu}\Phi(x')$. Indeed, gauge interaction has a particular geometrodynamical meaning in CA, as revealed by using the close analogy with the EC theory [7]. This can be alternatively obtained by using the minimal substitution $\hbar k_\mu(x) \to \hbar k_\mu(x) - eA_\mu(x)$ in the solution of the interacting CA. By imposing the CA evolution law in space-time with orbifold topology $\mathbb{S}^1/\mathcal{Z}_2$, i.e. the PBCs in space-time, this yields the Dirac quantisation condition for magnetic monopoles (in analogy with the Bohr-Sommerfeld quantisation)

$$\oint A_\mu dx^\mu = n\frac{\varphi_0}{2}\,. \tag{12}$$

The spatial component yields the quantisation of the magnetic flux described above whereas the time component yields the Josephson effect. Let us consider a junction between superconductors with a voltage difference ΔV (the BCs in this case is determined by the isolating barrier). Since the Maxwell equation in this case implies that $\Delta V = -\partial_t\theta$, the Stokes' theorem yields $T_{junct}\Delta V/c = \varphi_0/2$ corresponding to the frequency $f_{junct} = 1/T_{junct}$ of the Josephson effect. Alternatively this can be derived by considering that Lagrangian at the junction can only depend on the phase difference of the Goldstone fields $\mathcal{L}_{junct} = \mathcal{F}(\Delta\theta)$. The Josephson effect follows from the fact that the evolution law implies that the Goldstone can only vary by finite steps, see eq.(10), so that $\mathcal{F}(\Delta\theta) = \mathcal{F}(\Delta\theta + n\varphi_0/2)$ [12].

Due to the quantum recurrence, which in our description originates from the CA evolution law or by the PBCs, the EM gauge field A_μ, similarly to the Goldstone field, is constrained to have a periodic behaviour. This means that it can be expanded in vibrational eigenmode which, for the sake of simplicity, we assume here to be harmonic $A_\mu(\vec{x}, t) = \sum_{n\in\mathbb{Z}} e^{-in\omega^\gamma t}A_{(n)\mu}(\vec{x})$. This implies that the EM Lagrangian can be expanded in vibrational eigenmodes as well. In the temporal (unitary) gauge

$$\mathcal{L}_{YM} = -\frac{1}{4}\int d^4x F_{\mu\nu}F^{\mu\nu} = \sum_{n\in\mathbb{Z}}\int d^3x \left[-\frac{1}{4}F_{(n)\mu\nu}F^{\mu\nu}_{(n)} - \frac{1}{2}\frac{(\omega^\gamma_n)^2}{c^2}A_{(n)\mu}A^\mu_{(n)}\right] \tag{13}$$

However we must consider that at very low temperature $T \ll \beta$ only the lower mode of the system, that here we will denote by the bar sign, is populated. This means that the summation of the above Lagrangian must be truncated to the fundamental mode. As well-known for gauge invariance in extra dimensional theories (see Higgsless, Composite Higgs or unified gauge-Higgs models), this implies that the low temperature is no more gauge invariance. So we find again an effective EM gauge breaking. Roughly speaking, at very low temperature we can imagine that, similarly to a Bose-Einstein condensation, all the vibrational modes of energy $\omega^\gamma_n = n\omega^\gamma$ condensate on the fundamental mode.

This symmetry breaking can also be seen from the effective matter Lagrangian $\mathcal{L}_S$ obtained by integrating out from the equations of motion the contributions of the higher modes as a function of the fundamental one. By expanding $\mathcal{L}_S$ for small perturbations around the fundamental mode, which can be effectively written as $\hat{\Phi} = \rho e^{i2e\theta}$, we get

$$\begin{aligned}\mathcal{L}_S &\simeq \int d^3x \left[-\frac{1}{2}\hat{\Phi}^* |(\vec{\nabla} - i2e\vec{A})|^2 \hat{\Phi} - \frac{1}{2}\alpha|\hat{\Phi}|^2 - \frac{1}{4}\beta|\hat{\Phi}|^4 \right] \\ &= \int d^3x \left[-2e^2\rho^2(\vec{\nabla} - i2e\vec{A})^2 - \frac{1}{2}(\vec{\nabla}\rho)^2 - \frac{1}{2}\alpha\rho^2 - \frac{1}{4}\beta\rho^4 \right] . \end{aligned} \quad (14)$$

This describes (with appropriate normalisation) the BCS theory of ordinary SC. The first term describe the fundamental vibrational mode without the perturbation of the higher modes. Due to gauge transformations it can be written as $\bar{\mathcal{L}}_S[A_\mu - \partial_\mu \phi] \sim \frac{V}{\Lambda^2}(A_\mu - \partial_\mu \theta)$, V is the volume and $\Lambda = \frac{1}{\sqrt{4e^2\langle\rho^2\rangle}}$ is the penetration length of the Meissner effect. Indeed, deep inside the superconductor where the EM field is pure gauge, we have $F_{\mu\nu}(\partial_\chi \theta) = 0$ and the magnetic field is vanishing $\vec{B} = 0$. $\bar{\mathcal{L}}_S$ describes the fact that the energy cost to expel the magnetic field is small with respect to the that of the classical configuration. By considering that the minimum of ρ is $\langle\rho^2\rangle = -\frac{\alpha}{\beta}$ we get $\Lambda = \frac{1}{2e}\sqrt{-\frac{\alpha}{\beta}}$, in agreement with ordinary SC. Finally, by expanding ρ around its minimum $\rho = \langle\rho\rangle + \rho'$ we get $\vec{\nabla}^2\rho' = -2\alpha\rho'$. This allows us to introduce the coherence length $\xi = \frac{1}{-2\alpha}$. This describes the stability of the vortices and of the flux-quantised magnetic field. Indeed the energy density of the superconducting state is lower than the energy density of the normal state. We have $\frac{\alpha^2}{4\beta} = \frac{1}{32e^2\lambda^2\xi^2}$. Thus, in agreement with the ordinary description, the case $\xi > \Lambda$ and $\xi < \Lambda$ describe Type I SC and Type II SC, respectively.

4. Conclusions

In this paper we have shown that the interest of the 't Hooft CA theory, or similarly of the EC theory, is not only limited to our knowledge of foundational aspects of QM: it also represents a powerful tool to derive and describe in a very simple way nontrivial quantum phenomena such as the behaviour of the ECCs in CNs and the essential phenomenology of SC. Indeed we have derived the CN electronic properties, and essential quantum effects associated to SC, by simply considering the relativistic modulations of the vibrational modes, and the role the temperature in the cyclic dynamics, characterising the evolution law of CA.

Acknowledgments

We acknowledge G. 't Hooft for useful comments and partial financial support from the project FAR "*Control and enhancement of superconductivity by engineering materials at the nanoscale*".

References

[1] 't Hooft G 2003 *Int. J. Theor. Phys.* **42** 355 (*Preprint* hep-th/0104080)
[2] 't Hooft G 2014 The Cellular Automaton Interpretation of Quantum Mechanics. (*Preprint* arXiv:1405.1548)
[3] Dolce D and Perali A 2014 *Found. Phys.* **44** 905 (*Preprint* arXiv:1307.5062)
[4] Dolce D and Perali A 2015 *EPJ Plus* **103** 41 (*Preprint* arXiv:1403.7037)
[5] Dolce D 2013 *Europhys. Lett.* **102** 31002 (*Preprint* arXiv:1305.2802)
[6] Dolce D 2011 *Found. Phys.* **41** 178 (*Preprint* arXiv:0903.3680v5)
[7] Dolce D 2012 *Annals Phys.* **327** 1562 (*Preprint* arXiv:1110.0315)
[8] Dolce D 2012 *Annals Phys.* **327** 2354 (*Preprint* arXiv:1110.0316)
[9] Dolce D 2015 Unified description of Relativity and Quantum Mechanics from 't Hooft CA *(In preparation)*
[10] Charlier J C, Blase X and Roche S 2007 *Rev. Mod. Phys.* **79**(2) 677
[11] de Woul J, Merle A and Ohlsson T 2012 *Phys.Lett.* **B714** 44 (*Preprint* arXiv:1203.3196)
[12] Weinberg S 1996 *The quantum theory of fields. Vol. 2: Modern applications*, p 489 (Cambridge Univ. Press)
[13] White C T and Mintmire J W 1998 *Nature* **394** 29

Emergence of four dimensional quantum mechanics from a deterministic theory in 11 dimensions

G Doyen[1] **and D Drakova**[2]

[1]Ludwig-Maximilians Universität, München, Germany
[2]Faculty of Chemistry, University of Sofia, Bulgaria

E-mail: nhdd@chem.uni-sofia.bg

Abstract. We develop a deterministic theory which accounts for the coupling of a high dimensional continuum of environmental excitations (called gravonons) to massive particle in a very localized and very weak fashion. For the model presented Schrödinger's equation can be solved practically exactly in 11 spacetime dimensions and the result demonstrates that as a function of time an incoming matter wave incident on a screen extinguishes, except at a single interaction center on the detection screen. This transition is reminiscent of the wave - particle duality arising from the "collapse" (also called "process one") postulated in the Copenhagen-von Neumann interpretation. In our theory it is replaced by a sticking process of the particle from the vacuum to the surface of the detection screen. This situation was verified in experiments by using massive molecules. In our theory this "wave-particle transition" is connected to the different dimensionalities of the space for particle motion and the gravonon dynamics, the latter propagating in the hidden dimensions of 11 dimensional spacetime. The fact that the particle is detected at apparently statistically determined points on the screen is traced back to the weakness and locality of the interaction with the gravonons which allows coupling on the energy shell alone. Although the theory exhibits a completely deterministic "chooser" mechanism for single site sticking, an apparent statistical character results, as it is found in the experiments, due to small heterogeneities in the atomic and gravonon structures.

1. Introduction

In the spirit of Feynman's statement [1] the double slit experiment contains the essence of quantum mechanics and its unresolved problems. In order to understand quantum mechanics we have to understand the double slit experiment with electrons, photons, matter waves. Modern experiments have clearly demonstrated that single massive particles (heavy rare gas atoms [2] and He clusters [3], fullerenes and their derivatives of high molecular weight [4]-[12], meso-tetraphenylporphyrin [6], phthalocyanine and its derivatives [5, 7, 8, 13]), when scattered from nano lattices and optical gratings, form diffraction patterns in quite the same way as particles of smaller mass and photons. The diffraction pattern can be understood as resulting from matter wave interfering with itself and is readily obtained as a solution of Schrödinger's equation. The formation of the interference pattern on the detection screen occurs, however, as a result of many repeated experiments. In each experiment a single particle is detected at a well defined local site on the screen. This transition from wave to particle behaviour is commonly claimed to be a "collapse" of the wave function, which itself cannot be understood by solving Schrödinger's equation or within any kind of well defined dynamics.

In this contribution we present an investigation of the diffraction of atoms and molecules from nano lattices from a novel theoretical point of view. From the experimental point of view diffraction has been observed for nearly any kind of atom or molecule, e.g. H [14], H_2 [15], He, Ne, Ar, Kr, Xe [2, 3, 16, 17], C_{60}, C_{70} [4]-[9], [11, 12], phthalocyanine [5, 7, 10, 13]. Diffraction of He, Ne, H_2 has also been observed from crystalline structured surfaces [2, 18, 19, 20]. From the experimental findings it has to be concluded that right in front of the screen (surface) we have a wave front (wave function) which contains the diffraction pattern verified in the experimental outcome. The experimental observations are then: the diffracting particle *sticks* at a definite site on the surface, the stuck particle *stays* at least as long as the diffraction pattern has formed and finally the particle *moves* on the surface with a telegraph-signal like dynamics. This is the experimental finding which is referred to as the collapse of the wave function and as "wave - particle duality". Decoherence theory does not explain any of these observations.

The experimental outcome is at the same time interpreted as the sticking process which is believed to be accessible by standard quantum mechanical methods. From the experimental point of view collapse and sticking are identical phenomena, but sticking is considered to be completely inside the framework of the quantum mechanical formalism, whereas collapse is considered to be completely outside (cf. e.g. ref. [21]).

Within the Copenhagen interpretation the described experiments are considered as a manifestation of the wave - particle duality. The matter wave diffracts from the nano lattice and then collapses as a particle localized at a definite site on the screen. The collapse is considered outside the realm of the time dependent Schrödinger equation. This is not completely in accordance with the experimental point of view described above. In the Copenhagen interpretation the time dependent Schrödinger equation takes over again immediately after localization (collapse) and on cold flat surfaces this usually means a plane wave like motion parallel to the surface which is, however, not observed experimentally.

From the view of the time dependent Schrödinger equation alone wave packet calculations for H_2 dissociation and sticking on closed packed single crystal metal surfaces are feasible on six dimensional potential energy surfaces [22] and, after appropriate statistical averaging over initial states, in accordance with experimental results [23]. Wave packet calculations, however, cannot reproduce the telegraph like motion (diffusion) of H atoms and CO molecules on the surface. This has not yet been recognized as a significant problem.

Decoherence theory suggests that the time dependent Schrödinger equation could solve the problem, if coupling to the environmental degrees provided by the screen (surface) has to be included. It would calculate the reduced density matrix ρ_{red} which just comprises the degrees of freedom of the diffracted field (particle) after summing over the environmental degrees of freedom. Decoherence theory then claims that ρ_{red} becomes diagonal, the diagonal matrix elements specifying the probabilities for different outcomes of the experiment: sticking at various sites all over the surface or reflection into various directions.

Decoherence theory often treats localization as an exemplification of the quantum Zeno effect due to scattering of particles belonging to the environment [24]. The assumptions for this approach are: preparation of an initial state with an incoming particle from the environment; measurement of the system with the environmental particle in a definite state; probability interpretation of the transition rate in a definite final state and rate equations like the generalized Ehrenfest theorem or Fermi's golden rule. If the system (e.g. a particle) is initially in a delocalized zero energy state (e.g. Bloch state on an ordered surface) at zero temperature, then excited environmental states cannot be mixed in, because the total energy would not remain constant. If the system is initially in a localized gaussian at zero temperature of the environment, then it is not in a zero energy state. Mixing in of excited environmental modes is then possible, but would force the system to lower its energy, because of energy transfer from the system, which would mean delocalization of the system.

Journal of Physics: Conference Series **626** (2015) 012063 doi:10.1088/1742-6596/626/1/012063

Within decoherence theory and especially the quantum Zeno effect, localization of a particle can only be obtained, if the environmental degrees are capable of localizing the particle. This is not the case for atoms or small molecules at cold surfaces, where experimentally localization is clearly observed. Hence decoherence theory cannot explain the experimental findings for localization of atoms and molecules on cold metal surfaces.

2. The Hamiltonian

The model Hamiltonian is:

$$\begin{aligned} H &= E_g n_g + \sum_p E_p n_p + \sum_w E_w n_w + \\ &+ \sum_p V_{gp}(c_g^+ c_p + c_p^+ c_g) + \sum_{p,w} V_{wp}(c_w^+ c_p + c_p^+ c_w) \\ &+ \varepsilon_0 n_0 + \sum_{\kappa_w, w} \varepsilon_{\kappa_w} n_{\kappa_w} + \sum_{\kappa_w, w} [V_{0,\kappa_w} n_w a_0^+ a_{\kappa_w} + V_{\kappa_w,0} n_w a_{\kappa_w}^+ a_0] \end{aligned} \tag{1}$$

The environmental modes are existing in high dimensional spacetime (11 spacetime dimensions), have a high density of states and are called gravonons in the following. The meaning of the symbols in eq. (1) is: $n_g, n_p, n_w, n_0, n_{\kappa_w}$: number operators for gas particle at source, slits and screen, and gravonons; $c_i^+, c_i, a_0^+, a_0, a_{\kappa_w}^+, a_{\kappa_w}$: creation and annihilation operators for gas particle and gravonons; $E_g, E_w, E_p, \varepsilon_0, \varepsilon_{\kappa_w}$: energy quanta of gas particle and gravonons; $V_{gp}, V_{wp}, V_{0,\kappa_w}$: matrix elements for hopping of the gas particle and gravonon excitations. The gravonons are massive bosons which emerge in the limit of weak and local gravitational interaction in high (11) dimensional spacetime [25]. The coupling to gravonons is effective only within spacetime deformations called warp resonances represented by n_w, c_w^+, c_w, etc. In these warp resonances the gas particle is localized perpendicular and parallel to the screen surface. V_{0,κ_w} is calculated from the 11 dimensional gravitational potential $V_{grav}^{(11)}(r) = -\frac{G^{(11)}M}{\pi^7 r^8}$. The physical nature of the gravonons as massive bosons, as well as the derivation of an effective Schrödinger equation in high dimensional spacetime yielding gravonons similar to the common quantum particles as its solution, are described in detail in ref. [25]. The gravonons are the only fields which reside not only in our 4 dimensional spacetime, but in the additional compactified hidden spacial dimensions.

The Hamiltonian is visualized in figure 1. The many-particle wave has the form: $\mid \Psi\rangle = C_0(t) \mid \phi\rangle \otimes \mid 0\rangle + \sum_{w,\kappa_w} C_{w\kappa_w}(t) \mid w\rangle \otimes \mid \kappa_w\rangle$. $\mid \phi\rangle \otimes \mid 0\rangle$ describes a tensor product of the particle in the wave $\mid \phi\rangle$ between source and screen and the gravonons in the initial configuration $\mid 0\rangle$. $\mid w\rangle \otimes \mid \kappa_w\rangle$ has the particle in $\mid w\rangle$ at a specific point on the screen with gravonon excitations $\mid \kappa_w\rangle$. The time dependent coefficients are obtained from Schrödinger's equation: $i\hbar \frac{\partial \Psi(t)}{\partial t} = \mathrm{H}\Psi(t)$.

3. Solution for a single site on the screen

The scattering of a matter field from a double slit and the "wave - particle duality" observed in these experiments is in our theory interpreted as follows. If there is a site on the screen which is energetically degenerate with the initial wave $\mid \Psi(t=0)\rangle$, the configuration $\mid K\kappa_{proj}\rangle$ with the gas particle field having significant strength on this site will get strongly entangled with the degenerate gravonon configurations on this site. As the excitations of the gravonons in the hidden dimensions depend on the matter field being extremely localized near the chosen site the matter field strength cannot drift away as long as the excited gravonons are moving around in the large hidden dimensions. If this takes time long enough for other physical-chemical processes to be initiated, it will result in an experimentally detectable event. Such an event would be called "collapse" in the framework of Copenhagen quantum mechanics.

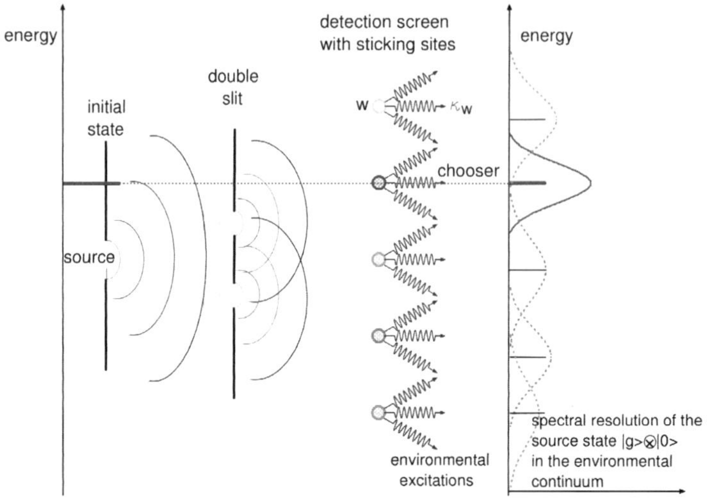

Figure 1. Model of adsorbate sticking in a double-slit diffraction experiment: a matter wave incident from a source interferes with itself between the slits and the screen. Hitting the detection screen allows entanglement with the gravonons. The initial matter wave determines the energy shell and, since the interaction with the gravonons is very weak, only on-shell coupling with a selected site on the screen, named chooser in the drawing, is effective, leading to particle sticking. Many experiments with different initial matter waves lead to the interference pattern due to adsorbed particles on the detection screen.

This situation has been treated theoretically in a simplified manner in ref. [25]. The Green function $G^+_{K\kappa_{proj}}$ has the form

$$G^+_{K\kappa_{proj}}(\epsilon_\kappa) = \frac{1}{\epsilon_\kappa - \alpha(\epsilon_\kappa) + i\Gamma(\epsilon_\kappa)} \tag{2}$$

where α and Γ are the real and imaginary parts of the self-energy, respectively, with

$$\Gamma(\epsilon_k) = \pi \sum_{\kappa_w} | V_{0,\kappa_w} |^2 \, \delta(E - \epsilon_\kappa) = \pi \frac{U^2}{\Delta} \ \text{ if } | \epsilon_k | < \Delta \tag{3}$$

which defines U. The last equation implies an approximation where the matter wave - gravonon interaction $V_{0,\kappa_w} = \frac{U}{\sqrt{N}}$ is non-zero and energy independent over an energy range of width Δ and zero otherwise. N is the number of gravonons in the energy interval Δ. Defining the density ρ_ϵ of gravonon levels at $\epsilon_\kappa = 0$

$$\frac{\rho_\epsilon}{N} = -\frac{1}{\pi}\mathrm{ImG}^+_{K\kappa_{proj}} = \frac{1}{\Delta} = \frac{1}{\pi\Gamma}, \tag{4}$$

yields $\Delta = \pi\Gamma$. In order to study the time dependence due to the coupling to the gravonons we start from the formal solution of the time dependent Schrödinger equation

$$| \Psi(t)\rangle = e^{-i\mathrm{H}t} \mid 00\rangle = \sum_{\kappa_w} \mathrm{e}^{-i\epsilon_\kappa t} \mid K\kappa_w+\rangle \frac{V_{0,\kappa_w}}{\epsilon_\kappa - \alpha(\epsilon_\kappa) - i\Gamma(\epsilon_\kappa)} \tag{5}$$

where we assumed V_{0,κ_w} to be real. $\{| K\kappa_w+\rangle\}$ describe gravonons modified by the incoming gas matter wave. For the projection of $| \Psi(t)\rangle$ on these configurations we obtain:

$$\sum_{\lambda_w} | \langle K\lambda_w+ | \Psi(t)\rangle |^2 \approx 1 - e^{-2\Gamma t}, \tag{6}$$

where from eq. (3) $\frac{U^2}{\Delta} = \Gamma/\pi$ and $\Delta = \pi\Gamma$ from eq. (4).

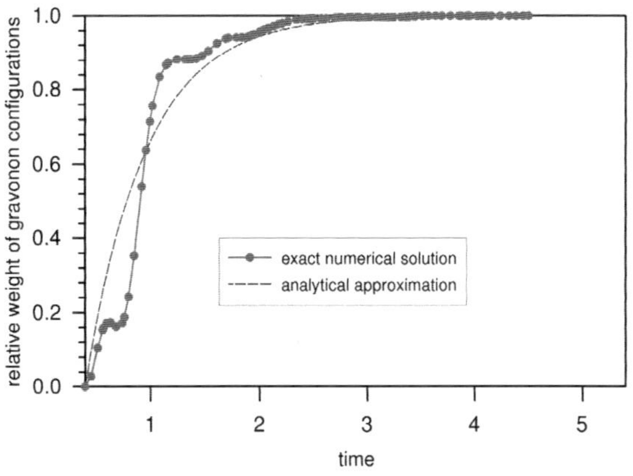

Figure 2. The chooser at work: the numerical solution (full curve) in comparison to the analytical approximation eq. (6) (dashed curve). The time is in units of $\hbar/2\Gamma$.

Equation (6) and fig. 2 show that we have an exponential increase with time of the weight of the incident wave packet in the warp resonance due to entanglement with the gravonon continuum, i.e. the incident wave becomes localized on the adsorption site.

4. The sticking time problem

The time of flight between the nano lattice and the screen can be reduced arbitrarily by reducing the distance between them. At one point the gravonon coupling becomes too weak to capture the particle during the short time of flight. The particle then moves back to the nano lattice where it is diffracted again or reflected. At some reduced distance the experimentally observed diffraction pattern should change and/or display a reduced intensity. If this does not occur, the coupling gravonon modes need to have a sufficiently large frequency.

This would be understandable within the framework of the $\gamma - \eta$ model [26]. In the limit where the frequency of the softest gravonon mode tends to zero one has for the coupling strength $\gamma = \frac{\omega_A^2}{\omega_S^2} \approx \frac{1}{1-\eta}$ where ω_A characterizes the motion of the gravonon image of the gas particle, i.e. the spacetime deformation of the gas particle, relative to the gravonon images of the atoms in the screen surface. ω_S is the unperturbed gravonon frequency. η is the parameter specifying the importance of forces between the local spacetime deformations in the gravonon space. If $\eta \rightarrow 1$ these forces are of little importance, γ becomes large and the softest gravonon mode which dominates the gravonon coupling can have a relatively large frequency ω_{soft}, provided it remains softer than any of the modes in the four dimensional environment.

In this way the sticking times might differ by eight orders of magnitude, explaining eight orders of magnitude different diffusion rates of hydrogen on Cu(001) and Ni(001) [27, 28]. In the experiment on the diffraction of phthalocyanine waves [13] we do not know the coupling strength, however, knowing the sticking time of approximately 10^{-4} seconds allows us to make an estimate of the coupling strengths V_{gp} and V_{wp} of the order of 10^{-11} eV, which is also the order of magnitude of ω_S and ω_A.

If the time of flight between the nano lattice and screen becomes smaller than $1/\omega_{soft}$, a standing wave between lattice and screen will develop. The change of the matter wave will change the diffraction pattern and this might offer a possibility for experimentally determining the sticking time. In this way it might be possible to demonstrate experimentally that collapse (which is equal to sticking in this case) is not infinitely fast but needs a finite time. This would give a clue, whether decoherence is at work (which is predicted to be extremely fast) or the process displayed in fig. 2.

5. Time dependent solution of the model Hamiltonian neglecting the coupling to the gravonons

The treatment of the problem with a "hopping hamiltonian" (also called tight binding hamiltonian in solid state physics) requires many empty lattice points to simulate plane wave motion. To understand the problem consider in the following a one dimensional situation. If for a distance L between source and screen there are N empty lattice points separated by a lattice constant a, then the largest k-vector is $k_{max} = \frac{\pi}{a} = N\pi/L$. As an example take $L = 10^{10}$ bohr (roughly 50 cm), the mass of the diffracting particle $M = 10^5$ (roughly 50 proton masses) and the velocity $k/M = 10^{-5}$ (roughly 50 m/sec). The k-vector is then $k \approx 1$. Then we need at least $N = kL/\pi \approx 3 \times 10^9$ empty lattice points. This is much too large for our numerical simulations because of lack of symmetry in the system source - slits - screen and because of the high dimensionality of the gravonon fields.

If we would have only two empty lattice points between the source and the warp resonances in the screen, then choosing the highest energy of such a hopping model equal to the kinetic energy of the diffracting particle (≈ 1 meV) would yield a "transfer time" (now this is the time for half a Rabi-like oscillation) of $t = 10^5$ a.u. $= 10^{-12}$ seconds, which differs by 10 orders of magnitude from the experimental transfer time $ML/k \approx 10^{15}$ a.u. $\approx 10^{-2}$ seconds. Because reasonable time scales are essential for our investigation, we choose the hopping parameter in such a way that the Rabi-oscillation time agrees with the experimental transport time. This means that for the discussed example the hopping matrix element should be of the order of 10^{-12} eV.

Figure 3 shows snapshots of the time development of a wave packet created in the source and interacting with the two slits and several sites on the screen. The area of the spots scales linearly with the amplitude squared of the wave packet at different places in the source, the slits and on five sites on the screen, as a function of time. At short time a big spot in the source, nothing in the slits and on the screen corresponds to the initial situation with the wave packet in the source. Gradually with time the wave packet is distributed everywhere, in the slits and on different sites on the screen. An interference pattern starts evolving at the position of the screen. The time dependent Schrödinger equation of the wave packet in interaction with the source, the slits and the screen shows the tendency to build the interference pattern. However, Rabi-like oscillations bring back the system to the initial situation and the cycle is repeated. This is not shown in fig. 3. Using the first two lines of the Hamiltonian eq. (1) leads to the interference pattern on the right hand side of fig. 3. However, this is not what is observed experimentally. The first two lines of the Hamiltonian do not lead to sticking of the particle on one site on the screen. There is no definete choice, but a distribution over many sites with different intensity. Furthermore permanent Rabi-like oscillations from the source to the screen and back occur.

The theoretical interference pattern is the same as in experiment. Within the Copenhagen interpretation of quantum mechanics solving the time dependent Schrödinger equation, using

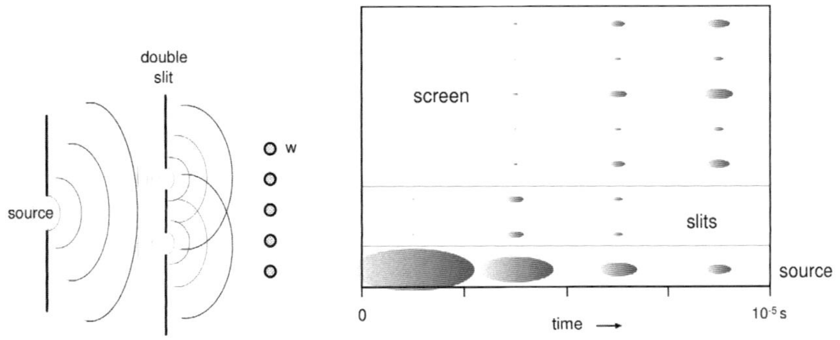

Figure 3. Left: geometric model of a double slit experiment with matter wave, without accounting for the interaction of the gas particle with the gravonons. Right: interference pattern created after 10^{-5} seconds.

the collapse and probabilistic concepts, the experimental interference pattern is reproduced and interpreted. However, the time dependent Schrödinger equation in four dimensional spacetime cannot explain the experiment: the collapse of the wave packet on a definite point on the screen cannot be predicted, with Copenhagen quantum mechanics we have no chooser, we have only the possible choices if we know what the final states are, however, there is no mechanism to select a single definite choice at a time.

6. Time dependent solution of the model Hamiltonian including the coupling to the gravonons

With the entanglement to the gravonons included (the last line in eq. 1) the result is displayed in fig. 4 for a single particle wave. In experiment the interference pattern on the screen develops as one particle after the other in successive emission of wave packets from the source sticks on the screen. The enhancement of the spot on the screen with time means that the incident wave becomes localized on a selected site transforming in a localized particle. This mechanism of sticking on a definite site on the screen we call the chooser. In each sticking event we have a definite site on the screen where the incident wave localizes. So instead of "God playing dice" we have the chooser mechanism defining the final site for sticking. There are no Rabi-like oscillations which bring the particle back to the source, the particle stays on this site on the screen for a long time, enabling the building up of the interference pattern in many successive sticking events, before it diffuses to a different site.

Then how does the chooser work? The condition to come to this result is a local and very weak interaction with the gravonons. Experimentally one can prepare a flat and ideal surface, however, the sites on the surface are never identical. Tiny differences between the sites matter because the interaction with the gravonons is very weak. The choice depends on the weakness of the interaction. If the interaction is zero we have the result with no chooser mechanism, hence no definite outcome. Turning on a very weak interaction with the gravonons means involving just those environmental configurations which are resonant or near-resonant with the initial configuration. (Perturbation theory shows that their involvement is inversely proportional to the energy difference with the initial configuration.) And, since the sites on the screen differ in energy, only those sites, whose energy differs from that of the initial configuration on the order of their interaction energy, will be involved. For a very weak interaction this can be just one site.

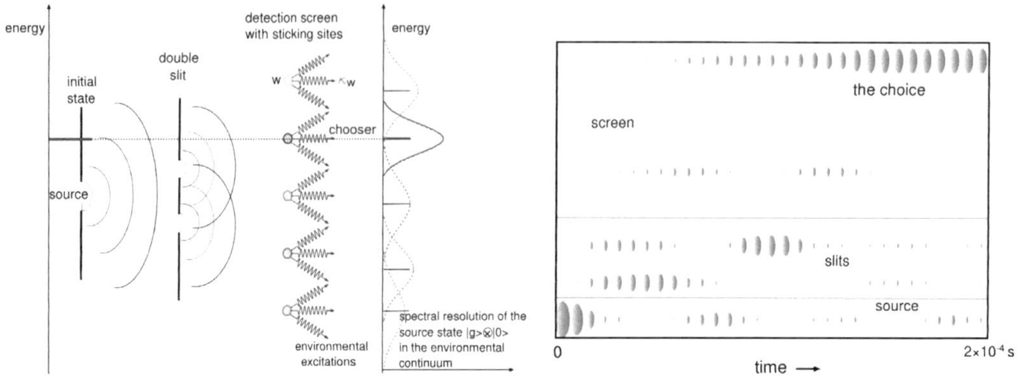

Figure 4. Left: geometric model of a double slit experiment with matter wave, taking into account the entanglement of the gas particle with the gravonons. Right: solving the time dependent Schrödinger equation a point in the diffraction pattern is created by a single sticking particle.

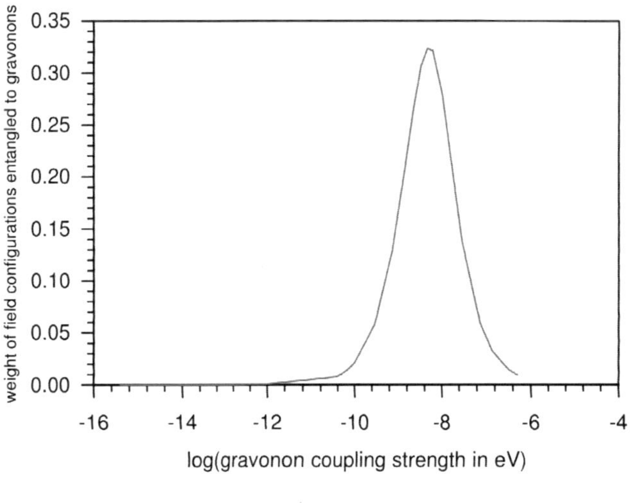

Figure 5. Summed weight of all field configurations entangled to gravonons versus the logarithm of the gravonon coupling strength $\sqrt{\sum_{\kappa_w} | V_{0,\kappa_w} |^2}$ in eV.

With the next incident wave packet prepared, the condition for degeneracy coupling will be satisfied on a different site on the screen. The interaction is so weak that it is weaker than the energy differences between the sites, hence just one site will satisfy the resonance condition.

Is this behaviour dependent on the coupling strength V_{0,κ_w}? If the interaction with the gravonons is localized and strong then environmental excitations of different nature over a broad energy interval will be involved, states describing particle reflection back into the vacuum included. So the particle wave will be preferentially reflected from the screen rather than to accomodate and stick as an adsorbate. The reflected state is the preferred final state rather than sticking because with sticking there is the need to excite many highly energetic modes

Journal of Physics: Conference Series **626** (2015) 012063 doi:10.1088/1742-6596/626/1/012063

(phonon modes, tomonagons) to dissipate the energy. However, particle wave reflection is not seen in experiment.

Very local and weak interaction V_{0,κ_w} with the gravonons warrants coupling of the initial wave packet with degenerate gravonons as shown in fig. 1 by the dashed horizontal line and the spectral distribution of the initial configuration $\mid g\rangle \otimes \mid 0\rangle$ in the continuum of environmental configurations $\{\mid w\rangle \otimes \mid \kappa_w\rangle\}$ shows. Since the entangled gravonons are degenerate the density of environmental configurations is high, the energy differences are small, hence the time for particle localization is long.

Even if the warp resonances are exactly on the energy shell, the coupling to the gravonons must not become arbitrarily small, if sticking is required to occur. The reason is that the gravonons represent a local spacetime structure exhibiting few modes of extremely small but non-zero frequency. If the matrix elements for gravonon coupling become smaller than the softest gravonon mode, gravonons will no longer entangle with the particle motion. We have studied this by calculating the degree of entanglement as a function of gravonon coupling strength. In fig. 5 the summed weight of all field configurations entangled to gravonons is plotted versus the logarithm of the gravonon coupling strength in eV. Entanglement to gravonons is effective over two orders of magnitude and becomes unimportant for weaker and stronger interaction. A variation of this interaction strength by 10^{-6} eV or even less from site to site can decide, whether the particle sticks here or somewhere else. The results of fig. 5 are, of course, only valid for the model structure and the parameters applied here. They depend on the nature of the particle (shape, mass, electronic structure) and on the nature of the screen.

7. Conclusions

A particle incident as a delocalized wave sticks at one point on the screen. A delocalized matter wave is transformed in a particle stuck on a selected site on the detection screen. Adsorbate sticking is thus the result of solving Schrödinger's time dependent equation, taking into account the entanglement with gravonons in high dimensional spacetime. The wave-to-particle transition is due to the local and weak interaction with gravonons. The interference pattern develops in the matter wave between the slits and the screen. Transition from wave to particle occurs where the amplitude of the interference wave is high because of enhanced entanglement with the gravonons. Minima in the amplitude of the interference wave mean attenuated entanglement with gravonons. Particle localization and sticking in the developed theory are apparently statistical. It has been demonstrate here, however, that they are deterministic. In this sense the problem of definite outcome has been solved. The apparent statistical behaviour is due to different initial conditions. (This is in contrast to refs. [29, 30, 31].) The incoming state, as prepared in experiment, is degenerate with warp resonances at a small number of sites on the screen. In the limiting situation just a single on-shell site is the only one where entanglement with gravonons can be efficient because the interaction energy with gravonons is weaker than the energy differences of the sites for adsorbate sticking. The theory is nonlocal, non-locality is implicit as it is implicit in the Schrödinger equation. The theory is quantum realistic. Whereas in the Copenhagen interpretation of quantum mechanics the wave function serves just to provide the probability for different outcomes of a measurement, in our theory the world wave function is the real system itself.

References

[1] Greene B 1999 in *The Elegant Universe: Superstrings, Hidden Dimensions and the Quest for the Ultimate Theory* (Vintage Ser. , W.W. Norton & Co) p 97 writes: "Feynman was fond of saying that all of quantum mechanics can be gleaned from carafully thinking through the implications of this single experiment [the double slit experiment] so it's worth discussing".

[2] Bruch L W, Schöllkopf W and Toennies J P 2001 in *Atomic and Molecular Beams The state of the Art 2000* ed R Campargue (Springer Verlag) p 615

[3] Brühl R, Guardiola R, Kalinin A, Kornilov O, Navarro J, Savas T and Toennies J P 2004 *Phys. Rev. Lett.* **92** 185301
[4] Arndt M, Nairz O, Vos-Andreae J, Keller C, van der Zouv G and Zeilinger A 1999 *Nature* **401**, 680
[5] Hornberger K, Gerlich S, Haslinger Ph, Nimmrichter S and Arndt M 2012 *Rev. Mod. Phys.* **84** 157
[6] Hackermüller L, Uttenthaler S, Hornberger K, Reiger E, Brezger B, Zeilinger A and Arndt M 2003 *Phys. Rev. Lett.* **91** 090408
[7] Gerlich S, Eibenberger S, Tomandl M, Nimmrichter S, Hornberger K, Fagan P J, Tüxen J, Mayor M and Arndt M 2011 *Nature Communications* **2** 263
[8] Nairz O, Brezger B, Arndt M and Zeilinger A 2001 *Phys. Rev. Lett.* **87** 160401
[9] Nairz O, Arndt M and Zeilinger A 2003 *Am. J. Phys.* **71** 319
[10] Juffmann Th, Truppe S, Geyer Ph, Major A G, Deachapunya S, Ulbricht H and Arndt M 2009 *Phys. Rev. Lett.* **103** 263601
[11] Gerlich S et al. 2007 *Nature Physics* **3** 711
[12] Hornberger K, Gerlich S, Ulbricht H, Hackermüller L, Nimmrichter S, Goldt I V, Boltalina O and Arndt M 2009 *New J. Phys.* **11** 043032
[13] Juffmann Th, Milic A, Müllneritsch M, Asenbaum P, Tsukernik A, Tüxen J, Mayor M, Cheshnovsky O and Arndt M 2012 *Nature Nanotech.* **7** 297
[14] Hoinkes H and Wilsch H 1992 in *Helium Atom Scattering from Surfaces* ed E Hilpke (Springer Verlag) chapter 7, p 113; Wilsch H and Rieder K H 1983 *J. Chem. Phys.* **78** 7491
[15] Díaz C, Martin F, Kroes G J, Minniti M, Farías D and Miranda R 2012 *J. Phys. Chem.* C **116** 13671
[16] Grisenti R E, Schöllkopf W, Toennies J P, Hegerfeld G C and Köhler T 1999 *Phys. Rev. Lett.* **83** 1755
[17] Schöllkopf W and Toennies J P 1994 *Science* **266** 1345
[18] Rieder K H 1994 *Surf. Rev. Lett.* **1**, 51; Farías D and Rieder K H 1998 *Rep. Prog. Phys.* **61** 1775
[19] Minniti M, Díaz C, Cuñado J L F, Politano A, Maccariallo D, Martin F, Farías D and Miranda R 2012 *J. Phys.: Condensed Matter* **24** 354002
[20] Drolshagen G, Kaufhold A and Toennies J P 1985 *J. Chem. Phys.* **83** 827; Toennies J P 1993 *J. Phys.: Condensed Matter* **5** A25
[21] Stapp H 1997 in *Quantum future - from Volta and Como to the present and beyond* eds Ph Blanchard and A Jadczyk (Springer Verlag) p 173
[22] Groß A 1998 *Surf. Sci. Rep.* **32** 291
[23] Bonn M, Kleyn A W and Kroes G J 2002 *Surf. Sci.* **500** 475
[24] Joos E 2007 in *Quantum Decoherence* Poincaré Seminar 2005, Ser. Progress in Math. Physics vol. 48, eds J M Raimond and V Rivasseau V (Basel: Birkhäuser) p 177
[25] Doyen G and Drakova D 2014 *Preprint* arXiv:1408.2716
[26] Drakova D and Doyen G 2013 *J. Physics: Conf. Series* **442** 012049
[27] Lin T S and Gomer R 1991 *Surf. Sci.* **255** 41
[28] Lauhon L J and Ho W 2000 *Phys. Rev. Lett.* **85** 4566; 2002 *Phys. Rev. Lett.* **89** 079901
[29] Bassi A and Ghirardi G 2000 *Phys. Lett.* A **275** 373
[30] Bassi A, Lochan K, Satin S, Singh T P and Ulbricht H 2013 *Rev. Mod. Phys.* **85** 471
[31] Grübl G 2002 *Preprint* arXiv: quant-ph/0202101v1

Exact analytical non-Hermitian formulation of the time evolution of decay of one and two identical quantum particles

Gastón García-Calderón

Instituto de Física, Universidad Nacional Autónoma de México, Apartado Postal 20 364, 01000 México, Distrito Federal, México

E-mail: gaston@fisica.unam.mx

Abstract. An analytical solution to the time evolution of decay of one and two identical noninteracting particles is presented using the formalism of resonant states. It is shown that the time-dependent wave function and hence the survival and nonescape probabilities for the initial state of a single particle and entangled symmetric and antisymmetric initial states of two identical particles evolve in a distinctive form along the exponential and long-time nonexponential decaying regimes. In particular, for the last regime, they exhibit different inverse power of time behaviors.

1. Introduction

Historically, the notion of quantum decay was developed to explain α-decay in radioactive nuclei. In 1928, Gamow derived the analytical expression for the exponential decay law $\exp(-\Gamma t/\hbar)$, with Γ the decay rate, by imposing purely outgoing boundary conditions to the solutions to the Schrödinger equation that describes the decaying system [1]. Typically this is modeled by a potential having a barrier where the particle is initially confined prior to decay by tunneling into open space. The approach by Gamow constituted one of the first theoretical treatments of an open quantum system. The outgoing boundary conditions, however, imply the vanishing of the coefficients of the incoming waves that appear in the general solution to the Schrödinger equation outside the interaction region and lead, due to time-reversal considerations, to discrete complex energy eigenvalues. One sees, therefore, that purely outgoing boundary conditions imply a non-Hermitian formulation of decay. Nevertheless, this approach has been very successful in describing particle decay as a tunneling process not only in nuclei [2], but also in other realms of physics as, for example, electronic decay in double-barrier semiconductor heterostructures [5] or atomic decay in ultracold traps [6]. In the energy domain, the vanishing of the incoming wave coefficients mentioned above corresponds to the complex poles of the S-matrix to the problem. This has led to formulations of nuclear reactions involving resonance expansions of the cross section [3, 4]. It is worth noticing, however, that this provides a link between the energy and the time domains, corresponding respectively, to scattering and decay, and to a definition of resonances as an intrinsic property of open quantum systems.

Near the end of the fifties of last century Khalfin demonstrated that if the energy spectra E of the system is bounded by below, *i.e.*, $E \in (0, \infty)$, the exponential decay law cannot hold at long times, behaving instead as an inverse power of time [7]. Studies on the short-time behavior of decay predicted also a departure from the exponential decay law. This is related, however, to the existence of the energy moments of the Hamiltonian H [8, 9]. Here short and long times refer to the time scale set by the lifetime of the decaying system. The short-time behavior has been the subject of much discussion, particularly in connection with the quantum Zeno effect [10, 11]. The experimental search of departures from the exponential decay law remained elusive for decades [12, 13]. A few years ago, however, it was verified in the short-time regime using ultracold atoms [14] and more recently, in the long-time regime using organic molecules in solution [16]. The quantum Zeno effect has also been observed in ultracold decaying systems [15]. These experimental results seem to contradict theoretical claims made in the seventies of last century, that argued that due to the influence of the measurement apparatus on the decay process, the exponential decay law should hold at all times [17, 18].

One sees, therefore, that the theoretical description of the dynamics of particle decay by tunneling has become more complex than was expected in older times. In general, it seems to consist of three regimes: Exponential and non-exponential at short and long times. Clearly the approach initiated by Gamow requires substantial modification to deal with the non-exponential contributions. At the end of the 1960s and along the 1970s there were some relevant developments on the properties of resonance states in the energy domain that involved consideration of the analytical properties of the outgoing Green's function to the full problem [19]. This function provided a framework to study the issues of normalization and eigenfunction expansions involving resonant states. Here it is adopted the normalization procedure that arises from the residue at a complex pole of the outgoing Green's function [20] and resonance expansions using the Cauchy integral theorem [21, 22]. The above considerations form the basis for the time-dependent treatment that I shall review below to describe the time evolution of decay of one and two particles.

It might be worthwhile to point out that the formulation considered here, involving the full hamiltonian H to the problem, should be contrasted with approaches where the Hamiltonian is separated into a part H_0 corresponding to a closed system and a part H_1 responsible for the decay which is usually treated to some order of perturbation theory, as in the work by Weisskopf and Wigner to describe the decay (also exponential) of an excited atom interacting with a quantized radiation field, published as well in the old days of quantum mechanics [23]. This approximate approach has become a standard procedure for treating the class of decay problems where perturbation theory can be justified.

The paper is organized as follows. Section 2 discusses the time evolution of decay of a single particle. Section 3 refers to the decay of symmetric and anti-symmetric identical particles and, finally, section 4 presents some concluding remarks.

2. Decay of a single particle

Consider a single particle confined at $t = 0$ along the internal region of a real spherically symmetrical finite-range potential, *i.e.*, $V(r) = 0$ for $r > a$. For simplicity we consider s waves and choose as units $\hbar = 2m = 1$. The solution to the time-dependent Schrödinger equation in the radial variable r, as an initial value problem, may be written at time $t > 0$ in terms of the retarded Green's function $g(r, r'; t)$ of the problem as

$$\Psi(r,t) = \int_0^a g(r,r',t)\Psi(r',0)\,\mathrm{d}r', \tag{1}$$

where $\Psi(r, 0)$ stands for the arbitrary state initially confined within the internal interaction region. Since the decay refers to tunneling into the continuum, for the sake of simplicity it is assumed that the potential does not possess bound nor antibound states. It is convenient to express the retarded time-dependent Green's function in terms of the outgoing Green's function

$G^+(r,r';k)$ of the problem. Both quantities are related by a Laplace transformation. The Bromwich contour in the k complex plane corresponds to a hyperbolic contour along the first quadrant that may be deformed to a contour that goes from $-\infty$ to ∞ along the real k axis,

$$g(r,r';t) = \frac{i}{2\pi}\int_{-\infty}^{\infty} G^+(r,r';k)\mathrm{e}^{-ik^2t}\,2kdk. \tag{2}$$

This allows to make use of the resonant expansion of the outgoing Green's function [26]

$$G^+(r,r';k) = \frac{1}{2k}\sum_{n=-\infty}^{\infty}\frac{u_n(r)u_n(r')}{k-\kappa_n}, \quad (r,r')^\dagger \le a \tag{3}$$

where the notation $(r,r')^\dagger$ means that the point $r = r' = a$ is excluded in the above expansion (otherwise it diverges) and the set of functions $\{u_n(r)\}$ correspond to the resonant states (also known as quasinormal modes) of the problem. They follow from the residues at the complex poles $\{\kappa_n\}$ which also provide its normalization condition [20, 26]

$$\int_0^a u_n^2(r)dr + i\frac{u_n^2(a)}{2\kappa_n} = 1. \tag{4}$$

The resonant states satisfy the Schrödinger equation of the problem $[\kappa_n^2 - H]u_n(r) = 0$ with outgoing boundary conditions $u_n(0) = 0$, $[du_n(r)/dr]_{r=a} = i\kappa_n u_n(a)$,. The complex energy eigenvalues are $\kappa_n^2 = E_n = \mathcal{E}_n - i\Gamma_n/2$, where $\mathcal{E}_n$ yields the resonance energy of the decaying fragment and Γ_n stands for the resonance width, which yields the lifetime (recalling that $\hbar = 1$) $\tau_n = 1/\Gamma_n$ of a given resonance level. The lifetime of the system is defined by the longest lifetime, *i.e.*, the shortest width. The complex poles $\kappa_n = a_n - ib_n$ are distributed along the third and fourth quadrants of the complex k plane in a well known manner [19]. Notice that writing $G^+(r,r';k) = G^+(r',a;k)\exp[ik(r-a]$ one may get a resonant expansion for $r' < a$ with $r > a$. The above representation for $G^+(r,r';k)$ satisfies the closure relation for resonant states [26],

$$\mathrm{Re}\left\{\sum_{n=-\infty}^{\infty} u_n(r)u_n(r')\right\} = \delta(r-r'), \quad (r,r')^\dagger \le a \tag{5}$$

and the sum rules,

$$\mathrm{Im}\left\{\sum_{n=-\infty}^{\infty} \frac{u_n(r)u_n(r')}{\kappa_n}\right\} = 0, \quad (r,r')^\dagger \le a \tag{6}$$

and

$$\mathrm{Im}\left\{\sum_{n=-\infty}^{\infty} u_n(r)u_n(r')\kappa_n\right\} = 0, \quad (r,r')^\dagger \le a \tag{7}$$

The above results permit to write the retarded Green's function as [26]

$$g(r,r';t) = \sum_{n=-\infty}^{\infty}\begin{cases} u_n(r)u_n(r')M(y_n^\circ), & (r,r')^\dagger \le a \\ u_n(r')u_n(a)M(y_n), & r' < a,\ r > a, \end{cases} \tag{8}$$

where the functions $M(y_n)$, the so called Moshinsky functions, are defined as [26]

$$M(y_n) = \frac{i}{2\pi}\int_{-\infty}^{\infty}\frac{\mathrm{e}^{ikr}\mathrm{e}^{-ik^2t}}{k-\kappa_n}dk = \frac{1}{2}\mathrm{e}^{(imr^2/2t)}w(iy_n), \tag{9}$$

Journal of Physics: Conference Series **626** (2015) 012064 doi:10.1088/1742-6596/626/1/012064

$y_n = e^{-i\pi/4}(1/2t)^{1/2}[(r-a) - 2\kappa_n t]$, and the function $w(z) = \exp(-z^2)\mathrm{erfc}(-\mathrm{i}z)$ stands for the Faddeyeva or complex error function [24] for which there exist efficient computational tools [25]. The argument y_n° of the functions $M(y_n^0)$ in (8) is that of y_n above with $r = a$.

In what follows the discussion will be concerned with two quantities of interest in decaying problems: The survival probability $S(t)$ that yields the probability that at time t the system remains in the initial state and the nonescape probability $P(t)$ that provides the probability that at time t the particle still remains within the confining region of the potential. The survival probability follows immediately from the expression for the survival amplitude

$$A(t) = \int_0^a \Psi^*(r,0)\Psi(r,t)\,dr, \quad S(t) = |A(t)|^2, \tag{10}$$

and the nonescape probability reads

$$P(t) = \int_0^a \Psi^*(r,t)\Psi(r,t)\,dr. \tag{11}$$

One see that both quantities require only of the solution along $r \leq a$. Hence inserting the first equation of (8) into equation (1) yields for the time-dependent decaying solution the exact expression

$$\Psi(r,t) = \sum_{n=-\infty}^{\infty} C_n(r)u_n(r)M(y_n^{\circ}), \qquad r < a, \tag{12}$$

where the coefficients C_n are defined as

$$C_n = \int_0^a \Psi(r,0)u_n(r)dr. \tag{13}$$

Equation (15) may be used to study the short-time behavior of the survival probability. This will no be discussed here and refer the interested reader to ref. [27], where it is shown that in general the short-time behavior of the survival probability S(t) has a dependence on the initial state and may behave as either $S(t) = 1 - \mathcal{O}(t^{3/2})$ or $1 - \mathcal{O}(t^2)$. This might be of some interest because there is the widespread opinion that the short-time behavior of $S(t)$ must be quadratic.

Since the potential is real, it follows from time-reversal invariance that $u_{-n}(r) = u_n^*(r)$ and $\kappa_{-n} = -\kappa_n^*$ [4] that allows to express equation (13) as a sum running from $n = 1$ to ∞. Also, using some properties of the Faddeyeva function allows to write the Moshinsky function for the poles lying on the fourth quadrant as [24, 26],

$$M(y_n^{\circ}) = e^{-i\kappa_n^2 t} - M(-y_n^{\circ}). \tag{14}$$

Substitution of equation (14) into equation (15) leads to the expression

$$\Psi(r,t) = \sum_{n=1}^{\infty} C_n(r)u_n(r)e^{-i\kappa_n^2 t} - \sum_{n=1}^{\infty} I_n(r,t), \quad r < a, \tag{15}$$

where $I_n(r,t)$ stands for the non-exponential contribution,

$$I_n(r,t) = [C_n u_n(r)M(-y_n^{\circ}) - C_{-n}u_{-n}(r)M(y_{-n}^{\circ})], \tag{16}$$

where the argument y_{-n}° is similar to that of y_n°, defined above, with κ_n substituted by $\kappa_{-n} = -\kappa_n^*$ and C_{-n} follows from equation (13) by recalling that $u_{-n}(r) = u_n^*(r)$. At long times these two

Journal of Physics: Conference Series **626** (2015) 012064 doi:10.1088/1742-6596/626/1/012064

functions exhibit an inverse power of time behavior. In particular, for a real initial state, the decaying solution behaves along the exponential and long-time regimes as [26],

$$\Psi(r,t) \approx \sum_{n=1}^{\infty} C_n u_n(r) \mathrm{e}^{-i\mathcal{E}_n t} \mathrm{e}^{-\Gamma_n t} - i\eta \,\mathrm{Im} \left\{ \sum_{n=1}^{\infty} \frac{C_n u_n(r)}{\kappa_n^3} \right\} \frac{1}{t^{3/2}}; \; r \leq a, \tag{17}$$

where $\eta = 1/(4\pi i)^{1/2}$.

Assuming that the initial state $\Psi(r,0)$ is normalized to unity, there is an interesting relationship that follows from the closure relation (5),

$$\mathrm{Re} \sum_{n=1}^{\infty} \{C_n \bar{C}_n\} = 1, \tag{18}$$

where

$$\bar{C}_n = \int_0^a \Psi^*(r,0) u_n(r) dr. \tag{19}$$

Equation (18) indicates that although $\mathrm{Re}\{C_n \bar{C}_n\}$ cannot be interpreted as a probability, since in general is not a positive quantity, nevertheless it represents the 'strength' or 'weight' of the initial state in the corresponding resonant state. If this has a value close to unity one may ignore the rest of the coefficients in the expansion for $\Psi(r,t)$. One may see the coefficients $\mathrm{Re}\{C_n \bar{C}_n\}$ as some sort of quasi-probabilities [28]. This deserves further study.

It is worth mentioning that there is another route to analyze the long-time behavior of $g(r,r';t)$ which follows by closing the Bromwich contour mentioned above along a straight line C_l that is 45° off the real axis and goes through the origin as discussed in ref. [29, 26] that will not be considered here.

3. Decay of two identical noninteracting particles

In the case of a system of identical non interacting particles, it is known that the Hamiltonian H must be symmetric under the permutation of the indices of the particles so the exchange operator and H necessarily commute. Thus, it is enough to impose the appropriate symmetry/antisymmetry on the initial state $\Psi(y_1, y_2, 0)$ since symmetry is conserved as time evolves. Hence, the time evolution for decay of two identical particles may be written as

$$\Psi(\mathbf{r},t) = \int_0^a \int_0^a g(r_1, y_1, t) g(r_2, y_2, t) \Psi(\mathbf{y}, 0) \, \mathrm{d}y_1 \, \mathrm{d}y_2, \tag{20}$$

where $\mathbf{r}$ and $\mathbf{y}$ denote, respectively, (r_1, r_2) and (y_1, y_2).

A simple choice, which corresponds to a symmetric state, is given by the product of single particle states $\psi_\alpha(y_1)$ and $\psi_\alpha(y_2)$, with α denoting the state,

$$\Psi(\mathbf{y},0) = \psi_\alpha(y_1)\psi_\alpha(y_2). \tag{21}$$

Substitution of (21) into (20), using the first equation in (8), yields the *factorized symmetric* state

$$\Psi(\mathbf{r},t) = \left(\sum_{p=-\infty}^{\infty} C_{p,\alpha} u_p(r_1) M(z_p) \right) \left(\sum_{q=-\infty}^{\infty} C_{q,\alpha} u_q(r_2) M(z_q) \right) \tag{22}$$

where $C_{n,\alpha}$, with $n = p, q$, is given by

$$C_{n,\alpha} = \int_0^a u_n(y) \psi_\alpha(y) \, \mathrm{d}y. \tag{23}$$

Another choice for the initial state consists of the linear combination of single-particle states $\psi_s(y_1)$ and $\psi_s(y_2)$. Here, $s = \alpha, \beta$ refers to the possible states of the two particles, and hence we may write

$$\Psi(\mathbf{y}, 0) = \frac{1}{\sqrt{2}}(\psi_\alpha(y_1)\psi_\beta(y_2) \pm \psi_\beta(y_1)\psi_\alpha(y_2)), \tag{24}$$

where respectively, the plus sign refers to *entangled symmetric* and the minus sign to *entangled antisymmetric* states. Then, substitution of (24) into (20), using the first equation in (8), yields

$$\Psi(\mathbf{r}, t) = \frac{1}{\sqrt{2}} \sum_{p,q=-\infty}^{\infty} (C_{p,\alpha}C_{q,\beta} \pm C_{p,\beta}C_{q,\alpha}) u_p(r_1) u_q(r_2) M(y_p) M(y_q), \tag{25}$$

where the coefficients $C_{n,\beta}$ follow by replacing α for β in (23). It is worth recalling that the coefficients $\{C_{n,s}\}$, which involve only single-particle states, fulfill the relationship [26]

$$\mathrm{Re}\left(\sum_{n=1}^{\infty} C_{n,s}\bar{C}_{n,s}\right) = 1, \tag{26}$$

where $\bar{C}_{n,s}$ is defined as (23) with $\psi_s(y)$ substituted by $\psi_s^*(y)$. Hence for real initial states, $\bar{C}_{n,s} = C_{n,s}$. Although the $C_{n,s}$ are complex and its real part may be negative, they play a most relevant role in time-dependent expansions.

In a similar fashion as for the single-particle case, one may write the two-particle states as a sum of exponential decaying contributions plus a long-time inverse power term. For the *entangled symmetric* state one may obtain from equation (25)

$$\Psi(\mathbf{r}, t) \approx \frac{1}{\sqrt{2}} \sum_{p,q=1}^{\infty} (C_{p,\alpha}C_{q,\beta} + C_{p,\beta}C_{q,\alpha}) u_p(r_1) u_q(r_2) \mathrm{e}^{-i(\mathcal{E}_p+\mathcal{E}_q)t} \mathrm{e}^{-\frac{1}{2}(\Gamma_p+\Gamma_q)t} - A\frac{1}{t^3}. \tag{27}$$

and, similarly, for the *entangled antisymmetric* state,

$$\Psi(\mathbf{r}, t) \approx \frac{1}{\sqrt{2}} \sum_{p,q=1}^{\infty} (C_{p,\alpha}C_{q,\beta} - C_{p,\beta}C_{q,\alpha}) u_p(r_1) u_q(r_2) \mathrm{e}^{-i(\mathcal{E}_p+\mathcal{E}_q)t} \mathrm{e}^{-\frac{1}{2}(\Gamma_p+\Gamma_q)t} + B\frac{1}{t^5}. \tag{28}$$

where in the above expressions A and B are constants. These quantities have been obtained explicitly for the model of a *delta*-shell interaction [30]. The relevant point is the distinct long-time inverse power behavior of *symmetric* and *antisymmetric* decaying states, which seems to possess a general character [31].

The survival amplitude of a two-particle system is defined as

$$A(t) = \int_0^a \int_0^a \Psi^*(r_1, r_2, 0)\Psi(r_1, r_2, t)\, \mathrm{d}r_1\, \mathrm{d}r_2; \tag{29}$$

hence, the survival probability is given simply by $S(t) = |A(t)|^2$. The nonescape probability of a two-particle system is defined as

$$P(t) = \int_0^a \int_0^a |\Psi(\mathbf{r}, t)|^2\, \mathrm{d}r_1\, \mathrm{d}r_2. \tag{30}$$

Once $\Psi(\mathbf{r}, t)$ is known, the calculation of $S(t)$ and $P(t)$ follows using the above expressions.

Lack of space prevents to illustrate the analytical expressions derived above for calculations of the survival and nonescape probabilities for the single and two-particle cases. I refer the interested readers to go to refs. [29] and [30].

4. Concluding remarks

I hope I have been able to discuss the essential aspects of the formalism of resonant states for the time evolution of decay. It is surprising to see that over the years most of the attention has been for single particle decay. However, recent developments in the design and control of the number of atoms in ultracold traps and its interactions [6, 32, 33] might lead to new and interesting findings involving the decay by tunneling of several particles.

Acknowledgments

I acknowledge the partial financial support of UNAM-DGAPA-PAPIIT IN111814.

References

[1] Gamow G 1928 *Z. Phys.* **51** 204
[2] Gamow G and Critchfield C L 1949 *Theory of Atomic Nucleus and Nuclear Energy-Sources* (Oxford: Clarendon Press)
[3] Siegert A. F. J. 1939 *Phys. Rev.* **56** 750
[4] Humblet J and Rosenfeld L 1961 *Nucl. Phys.* **26** 594
[5] Tsuchiya M, T. Matsusue T and Sakaki H 1987 *Phys. Rev. Lett.* **59** 2356
[6] Serwane F, Zürn G , Lompe T, Ottenstein T B, Wenz A N and Jochim S 2011 *Science* **332** 336
[7] Khalfin L A 1958 *Sov. Phys. JETP* **6** 1053
[8] Khalfin L A 1968 *JETP Lett.* **8** 65
[9] Muga J G, Wei G W and Snider R F 1996 *Europhys. Lett.* **35** 247
[10] Misra B and Sudarshan E C G 1977 *J. Math. Phys.* **18** 756
[11] Koshino K and Shimizu A 2005 *Phys. Reports* **412** 191
[12] Norman E B, Gazes S B, Crane S G and Bennett D A 1988 *Phys. Rev. Lett.* **60** 224; Norman E B, Sur B, Lesko K T, Lammer R-M, DePaolo D J and Owens T L 1995 *Phys. Lett.* **B 357** 521
[13] OPAL Collaboration, Alexander G *et.al.* 1966 *Phys. Lett. B* **368** 244
[14] Wilkinson S R, Bharucha C F, Fischer M C, Madison K W, Morrow P K, Niu Q, Sundaram B and Raizen M G 1997 Nature **387** 575
[15] Fischer M C, Gutiérrez-Medina B, and Raizen M G 2001 *Phys. Rev. Lett.* **87** 040402
[16] Rothe C, Hintschich S I and Monkman A P 2006 *Phys. Rev. Lett.* **96** 163601
[17] Ekstein H and Siegert A J F 1971 *Ann. Phys. (N. Y.)* **68** 509
[18] Fonda L, Ghirardi G C and Rimini A 1978 *Rep. Prog. Phys.* **41** 587
[19] Newton R G 2002 *Scattering Theory of Waves and Particles* (New York: Dover Publications) chapter 12
[20] García-Calderón G and Peierls R 1976 *Nucl.Phys. A* **265** 443
[21] García-Calderón G 1976 *Nucl. Phys. A* **261** 130
[22] García-Calderón G and Berrondo M 1979 *Lett. Nuovo Cimento* **26** 562
[23] Weisskopf V F and Wigner E P 1930 *Z. Phys.* **65** 54
[24] Abramowitz M and Stegun A I 1964 *Handbook of Mathematical Functions* (New York: Dover Publications)
[25] Poppe G M P and Wijers C M J 1990 *ACM Trans. Math. Softw.* **16** 38
[26] García-Calderón G 2010 *Adv. in Quant. Chem.* **60** 407
[27] Cordero S and García-Calderón G 2012 *Phys. Rev. A* **86** 062116
[28] Halliwell JJ and Yearsley JM 2013 *Physical Review A* **87** 022114
[29] García-Calderón G, Maldonado I and Villavicencio J 2007 *Phys. Rev. A* **76** 012103
[30] García-Calderón G and Mendoza-Luna L G 2011 *Phys. Rev. A* **84** 032106
[31] del Campo A 2011 *Phys. Rev. A.* **84** 012113
[32] Pons M, Sokolovski D and del Campo A 2012 *Phys. Rev. A* **85** 022107
[33] Rontani M 2013 *Phy. Rev. A* **88** 043633

Journal of Physics: Conference Series **626** (2015) 012065 doi:10.1088/1742-6596/626/1/012065

Chemical and mechanical instabilities in high energy heavy-ion collisions

G Gervino[1,3], A Lavagno[2,3] and D Pigato[2,3]

[1]Dipartimento di Fisica, Università di Torino, I-10126 Torino, Italy
[2]Department of Applied Science and Technology, Politecnico di Torino, I-10129 Torino, Italy
[3]INFN, Sezione di Torino, I-10125 Torino, Italy

Abstract. We investigate the possible thermodynamic instability in a warm and dense nuclear medium where a phase transition from nucleonic matter to resonance-dominated Δ-matter can take place. Such a phase transition is characterized by both mechanical instability (fluctuations on the baryon density) and by chemical-diffusive instability (fluctuations on the isospin concentration) in asymmetric nuclear matter. Similarly to the liquid-gas phase transition, the nucleonic and the Δ-matter phase have a different isospin density in the mixed phase. In the liquid-gas phase transition, the process of producing a larger neutron excess in the gas phase is referred to as isospin fractionation. A similar effects can occur in the nucleon-Δ matter phase transition due essentially to a Δ^- excess in the Δ-matter phase in asymmetric nuclear matter. In this context, we study the hadronic equation of state by means of an effective quantum relativistic mean field model with the inclusion of the full octet of baryons, the Δ-isobar degrees of freedom, and the lightest pseudoscalar and vector mesons. Finally, we will investigate the presence of thermodynamic instabilities in a hot and dense nuclear medium where phases with different values of antibaryon-baryon ratios and strangeness content may coexist. Such a physical regime could be in principle investigated in the future high-energy compressed nuclear matter experiments where will make it possible to create compressed baryonic matter with a high net baryon density.

1. Introduction

One of the very interesting aspects in nuclear astrophysics and in the heavy-ion collisions experiments is a detailed study of the thermodynamical properties of strongly interacting nuclear matter away from the nuclear ground state.

The new accumulating data from x-ray satellites provide important information on the structure and formation of compact stellar objects. Concerning the structure, these data are at first sight difficult to interpret in a unique and selfconsistent theoretical scenario, since some of the observations indicate rather small radii and other observations indicate large values for the mass of the star.

On the other hand, the information coming from experiments with heavy ions in intermediate- and high-energy collisions is that the EOS depends on the energy beam but also sensibly on the electric charge fraction Z/A of the colliding nuclei, especially at not too high temperature [1, 2, 3, 4]. Moreover, the study of nuclear matter with arbitrary electric charge fraction results to be important in radioactive beam experiments and in the physics of compact stars.

In this article, we to study a hadronic equation of state (EOS) at finite temperature and density by means of a relativistic mean-field model with the inclusion $\Delta(1232)$-isobars [5, 6, 7] and by requiring the Gibbs conditions on the global conservation of baryon number and net electric charge. Transport model calculations and experimental results indicate that an excited state of baryonic matter is dominated by the Δ resonance at the energies from the BNL Alternating Gradien Synchrotron (AGS) to RHIC [8]. Moreover, in the framework of the nonlinear Walecka model, it has been predicted that a phase transition from nucleonic matter to Δ-excited nuclear matter can take place and the occurrence of this transition sensibly depends on the Δ-meson coupling constants [9, 10].

The main goal of this paper is to show that, for asymmetric warm and dense nuclear medium, the possible Δ-matter phase transition is characterized by mechanical and chemical-diffusive instabilities. Similarly to the liquid-gas phase transition [11], chemical instabilities play a crucial role in the characterization of the phase transition and can imply a very different electric charge fraction Z/A in the coexisting phases during the phase transition. In this context we show that a hadron phase transition of the baryon-antibaryon plasma can take place in the physical region of finite net baryon density ($\rho_0 \leq \rho_B < 2\rho_0$, where ρ_0 is the nuclear saturation density) and temperature ($T \simeq 120 \div 150$ MeV).

2. Hadronic equation of state

Let us briefly to introduce the adopted nuclear equation of state (EOS) to describe the interaction in the many-particles system involved in high energy heavy ion collisions.

The relativistic mean-field model is widely successful used for describing the properties of finite nuclei as well as hot and dense nuclear matter [4, 5, 12]. The Lagrangian for the self-interacting octet of baryons (p, n, Λ, Σ^+, Σ^0, Σ^-, Ξ^0, Ξ^-) can be written as [5]

$$\begin{aligned}\mathcal{L}_{\rm octet} &= \sum_k \overline{\psi}_k\,[i\,\gamma_\mu\,\partial^\mu - (M_k - g_{\sigma k}\,\sigma) - g_{\omega k}\,\gamma_\mu\,\omega^\mu - g_{\rho k}\,\gamma_\mu\,\vec{t}\cdot\vec{\rho}^{\,\mu}]\,\psi_k \\ &+\frac{1}{2}(\partial_\mu\sigma\partial^\mu\sigma - m_\sigma^2\sigma^2) - \frac{1}{3}a\,(g_{\sigma N}\,\sigma)^3 - \frac{1}{4}\,b\,(g_{\sigma N}\,\sigma^4) \\ &+\frac{1}{2}\,m_\omega^2\,\omega_\mu\omega^\mu + \frac{1}{4}\,c\,(g_{\omega N}^2\,\omega_\mu\omega^\mu)^2 + \frac{1}{2}\,m_\rho^2\,\vec{\rho}_\mu\cdot\vec{\rho}^{\,\mu} \\ &-\frac{1}{4}F_{\mu\nu}F^{\mu\nu} - \frac{1}{4}\vec{G}_{\mu\nu}\vec{G}^{\mu\nu}\,, \end{aligned} \tag{1}$$

where the sum runs over the full octet of baryons, M_k is the vacuum baryon mass of index k, the quantity $\vec{t}$ denotes the isospin operator that acts on the baryon.

In regime of finite values of temperature and density, a state of high density resonance matter may be formed and the $\Delta(1232)$-isobar degrees of freedom are expected to play a central role [6, 8]. In particular, the formation of resonances matter contributes essentially to baryon stopping, hadronic flow effects and enhanced strangeness. The Lagrangian density concerning the Δ-isobars (Δ^{++}, Δ^+, Δ^0, Δ^-) can be expressed as [13]

$$\mathcal{L}_\Delta = \overline{\psi}_{\Delta\,\nu}\,[i\gamma_\mu\partial^\mu - (M_\Delta - g_{\sigma\Delta}\sigma) - g_{\omega\Delta}\gamma_\mu\omega^\mu]\psi_\Delta^\nu\,, \tag{2}$$

where ψ_Δ^ν is the Rarita-Schwinger spinor for the Δ-isobars (Δ^{++}, Δ^+, Δ^0, Δ^-). Due to the uncertainty on the meson-Δ coupling constants, we limit ourselves to consider only the coupling with the σ and ω meson fields ($x_{\sigma\Delta} = g_{\sigma\Delta}/g_{\sigma N}$, $x_{\omega\Delta} = g_{\omega\Delta}/g_{\omega N}$), more of which are explored in the literature. The adopted coupling constants are the same of Ref. [6].

The ρ_i^B and ρ_i^S are the baryon density and the baryon scalar density, respectively. They are given by

$$\rho_i^B = \gamma_i \int \frac{{\rm d}^3k}{(2\pi)^3}[n_i(k) - \overline{n}_i(k)]\,, \tag{3}$$

$$\rho_i^S = \gamma_i \int \frac{{\rm d}^3k}{(2\pi)^3}\,\frac{M_i^*}{E_i^*}\,[n_i(k) + \overline{n}_i(k)]\,, \tag{4}$$

where $n_i(k)$ and $\overline{n}_i(k)$ are the fermion particle, antiparticle distributions function, given by

$$n_i(k) = \frac{1}{\exp(E_i^*(k) - \mu_i^*)/T + 1}, \tag{5}$$

$$\overline{n}_i(k) = \frac{1}{\exp(E_i^*(k) + \mu_i^*)/T + 1}. \tag{6}$$

The effective chemical potentials μ_i^* are given in terms of the chemical potentials μ_i by means of the following relation

$$\mu_i^* = \mu_i - g_{\omega i}\,\omega - g_{\rho i}\,t_{3i}\,\rho\,, \tag{7}$$

where t_{3i} is the third component of the isospin of i-th baryon. The baryon effective energy is defined as $E_i^*(k) = \sqrt{k^2 + {M_i^*}^2}$, where the effective mass of the ith baryon is defined as

$$M_i^* = M_i - g_{\sigma i}\sigma\,. \tag{8}$$

The thermodynamical quantities can be obtained from the grand potential Ω_B in the standard way. More explicitly, the baryon pressure $P_B = -\Omega_B/V$ and the energy density ϵ_B can be written as

$$\begin{aligned}
P_B &= \frac{1}{3}\sum_i \gamma_i \int \frac{\mathrm{d}^3k}{(2\pi)^3}\,\frac{k^2}{E_i^*(k)}\,[f_i(k) + \overline{f}_i(k)] - \frac{1}{2}\,m_\sigma^2\,\sigma^2 \\
&- \frac{1}{3}a\,(g_{\sigma N}\,\sigma)^3 - \frac{1}{4}\,b\,(g_{\sigma N}\,\sigma^4) + \frac{1}{2}\,m_\omega^2\,\omega^2 \\
&+ \frac{1}{4}\,c\,(g_{\omega N}\,\omega)^4 + \frac{1}{2}\,m_\rho^2\,\rho^2\,, \qquad (9)\\
\epsilon_B &= \sum_i \gamma_i \int \frac{\mathrm{d}^3k}{(2\pi)^3}\,E_i^*(k)\,[f_i(k) + \overline{f}_i(k)] + \frac{1}{2}\,m_\sigma^2\,\sigma^2 \\
&+ \frac{1}{3}a\,(g_{\sigma N}\,\sigma)^3 + \frac{1}{4}\,b\,(g_{\sigma N}\,\sigma^4) + \frac{1}{2}\,m_\omega^2\,\omega^2 \\
&+ \frac{3}{4}\,c\,(g_{\omega N}\,\omega)^4 + \frac{1}{2}\,m_\rho^2\,\rho^2\,. \qquad (10)
\end{aligned}$$

Finally, from a phenomenological point of view, we can take into account the meson (boson) particle degrees of freedom by adding their one-body contribution to the thermodynamical potential, that is, the contribution of a quasi-particle Bose gas with an effective chemical potential μ_j^* and an effective mass m_j^* for the j-meson, which contain the self-consistent interaction of the meson fields [4]. Following this working hypothesis, we can evaluate the pressure P_M, the energy density ϵ_M and the particle density ρ_j^M of mesons as

$$P_M = \frac{1}{3}\sum_j \gamma_j \int \frac{\mathrm{d}^3k}{(2\pi)^3}\,\frac{k^2}{E_j(k)}\,g_j(k)\,, \tag{11}$$

$$\epsilon_M = \sum_j \gamma_j \int \frac{\mathrm{d}^3k}{(2\pi)^3}\,E_j(k)\,g_j(k)\,, \tag{12}$$

$$\rho_j^M = \gamma_j \int \frac{\mathrm{d}^3k}{(2\pi)^3}\,g_j(k)\,, \tag{13}$$

where $\gamma_j = 2J_j + 1$ is the degeneracy spin factor of the j-th meson ($\gamma = 1$ for pseudoscalar mesons and $\gamma = 3$ for vector mesons), the sum runs over the lightest pseudoscalar mesons (π,

K, η, η') and the lightest vector mesons (ρ, ω, K^*, ϕ), considering the contribution of particle and antiparticle separately. In Eqs.(11)-(13) the function $g_j(k)$ is the boson particle distribution given by

$$g_j(k) = \frac{1}{\exp\{(E_j^*(k) - \mu_j^*)/T\} - 1}\,, \tag{14}$$

where $E_j^*(k) = \sqrt{k^2 + {m_j}^{*2}}$. The corresponding antiparticle distribution will be obtained with the substitution $\mu_j^* \to -\mu_j^*$. Moreover, the boson integrals are subjected to the constraint $|\mu_j^*| \leq m_j^*$, otherwise Bose condensation becomes possible. We have verified that such a condition is never realized in the considered range of temperature and density.

Finally, the total pressure and energy density will be

$$P = P_B + P_M\,, \tag{15}$$

$$\epsilon = \epsilon_B + \epsilon_M\,. \tag{16}$$

3. Stability conditions and phase transition

We are dealing with the study of a multi-component system at finite temperature and density with two conserved charges: baryon (B) number and zero net strangeness (S) number ($r_S = \rho_S/\rho_B = 0$). For what concern the electric charge (Q), we work in symmetric nuclear matter with a fixed value of $Z/A = 0.5$ and we do not consider fluctuation in the electric charge fraction, due to the high temperature regime. Therefore, the electric charge results to be separately conserved in each phase during the phase transition.

The chemical potential of particle of index i can be written as

$$\mu_i = b_i\,\mu_B + s_i\,\mu_S\,, \tag{17}$$

where b_i and s_i are, respectively, the baryon and the strangeness quantum numbers of i-th hadronic species.

For such a system, the Helmholtz free energy density F can be written as

$$F(T, \rho_B, \rho_S) = -P(T, \mu_B, \mu_S) + \mu_B\rho_B + \mu_S\rho_S\,, \tag{18}$$

with

$$\mu_B = \left(\frac{\partial F}{\partial \rho_B}\right)_{T,\rho_S}\,, \qquad \mu_S = \left(\frac{\partial F}{\partial \rho_S}\right)_{T,\rho_B}\,. \tag{19}$$

In a system with N different particles, the particle chemical potentials are expressed as the linear combination of the two independent chemical potentials μ_B and μ_S and, as a consequence, $\sum_{i=1}^{N} \mu_i\rho_i = \mu_B\rho_B + \mu_S\rho_S$.

Assuming the presence of two phases (denoted as I and II, respectively), the system is stable against the separation in two phases if the free energy of a single phase is lower than the free energy in all two phases configuration. The phase coexistence is given by the Gibbs conditions

$$\mu_B^I = \mu_B^{II}\,, \qquad \mu_S^I = \mu_S^{II}\,, \tag{20}$$

$$P^I(T, \mu_B, \mu_S) = P^{II}(T, \mu_B, \mu_S)\,. \tag{21}$$

At a given baryon density ρ_B and at a given zero net strangeness density $r_S = \rho_S/\rho_B = 0$, the chemical potentials μ_B are μ_S are univocally determined by the following equations

$$\rho_B = (1-\chi)\,\rho_B^I(T, \mu_B, \mu_S) + \chi\,\rho_B^{II}(T, \mu_B, \mu_S)\,, \tag{22}$$

$$\rho_S = (1-\chi)\,\rho_S^I(T, \mu_B, \mu_S) + \chi\,\rho_S^{II}(T, \mu_B, \mu_S)\,, \tag{23}$$

where $\rho_B^{I(II)}$ and $\rho_S^{I(II)}$ are, respectively, the baryon and strangeness densities in the low density (I) and in the higher density (II) phase and χ is the volume fraction of the phase II in the mixed phase ($0 \le \chi \le 1$).

An important feature of this conditions is that, unlike the case of a single conserved charge, baryon and strangeness densities can be different in the two phases, although the total ρ_B and ρ_S are fixed.

For such a system in thermal equilibrium, the possible phase transition can be characterized by mechanical (fluctuations in the baryon density) and chemical instabilities (fluctuations in the strangeness number). As usual the condition of the mechanical stability implies

$$\rho_B \left(\frac{\partial P}{\partial \rho_B}\right)_{T,\rho_S} > 0\,. \tag{24}$$

By introducing the notation $\mu_{i,j} = (\partial\mu_i/\partial\rho_j)_{T,P}$ (with $i,j = B,S$), the chemical stability can be expressed with the following conditions [6]

$$\mu_{B,B} > 0\,, \quad \mu_{S,S} > 0\,. \tag{25}$$

In addition to the above conditions, for a process at constant P and T, it is always satisfied that

$$\rho_B\,\mu_{B,B} + \rho_S\,\mu_{S,B} = 0\,, \tag{26}$$

$$\rho_B\,\mu_{B,S} + \rho_S\,\mu_{S,S} = 0\,. \tag{27}$$

Whenever the above stability conditions are not respected, the system becomes unstable and the phase transition take place. The coexistence line of a system with one conserved charge becomes in this case a two dimensional surface in (T, P, r_S) space, enclosing the region where mechanical and diffusive instabilities occur.

4. Antibaryon-baryon and strangeness phase transition

By increasing the temperature and the baryon density during the high energy heavy ion collisions, a multi-particle system may take place and the formation antiparticles become much more relevant [7, 14, 15].

In analogy with the liquid-gas case, we are going to investigate the existence of a possible phase transition in the nuclear medium by studying the presence of instabilities (mechanical and/or chemical) in the system.

As already observed, during a phase transition with two conserved charges, the strangeness fraction $r_S = \rho_S/\rho_B$ is not locally conserved in the single phase but only globally conserved. Therefore, during the compression of the system, the appearance of particles with strangeness could, in principle, shift the diffusive instability region to negative (or positive) values of r_S, even if the system is prepared with a zero net strangeness. Such a feature has no counterpart in the liquid-gas phase transition and it turns out to be very relevant in order to properly determine the instability region through the binodal phase diagram.

Taking into account that Eq. (27) becomes in this case

$$\left(\frac{\partial \mu_B}{\partial r_S}\right)_{T,P} + r_S \left(\frac{\partial \mu_S}{\partial r_S}\right)_{T,P} = 0\,, \tag{28}$$

the chemical stability condition is satisfied if

$$\left(\frac{\partial \mu_S}{\partial r_S}\right)_{T,P} > 0 \ \text{ or } \ \begin{cases} \left(\frac{\partial \mu_B}{\partial r_S}\right)_{T,P} < 0\,, & \text{if } r_S > 0\,, \\[2ex] \left(\frac{\partial \mu_B}{\partial r_S}\right)_{T,P} > 0\,, & \text{if } r_S < 0\,. \end{cases} \tag{29}$$

Journal of Physics: Conference Series **626** (2015) 012065 doi:10.1088/1742-6596/626/1/012065

Let us report some numerical results by considering the hadronic EOS discussed here. First of all, we observe that the condition of mechanical and chemical stability is not satisfied in a large range of temperature and density, therefore, the system goes to a phase transition. In this context, let us remark that the Δ-isobar degrees of freedom play a crucial role in the presence of the instability conditions of the EOS and in the nature of the phase transition from nucleonic matter to a resonance-dominated Δ-matter.

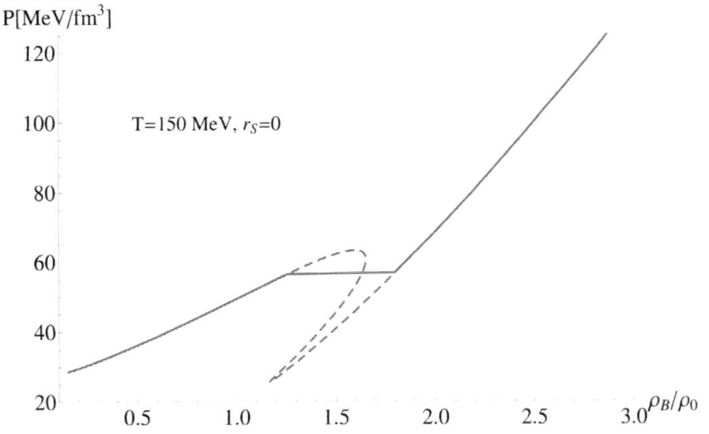

Figure 1. Pressure as a function of baryon density at fixed temperature. The continuous (dashed) lines correspond to the solution obtained with (without) the Gibbs construction.

To better clarify this matter of fact, in Fig. 1, we report the pressure as a function of baryon density (normalized to the nuclear saturation density ρ_0) at fixed temperature and zero net strangeness. In this case the condition of mechanical instability (29) is not satisfied. It is possible to verify that also chemical-diffusion instability takes place and the system goes to a phase transition. Let us remark that, although the system has globally zero net strangeness, the mixed phase is realized with two phases with different values of baryon-antibaryon and strangeness content.

In Fig. 2, we report the antibaryon-barion ratio R (left panel) and the strangeness fraction Y_S (right panel) for particle and antiparticle as a function of baryon density. The continuous (dashed) lines correspond to the solution obtained with (without) the Gibbs construction. During the phase transition (between the two vertical dashed lines reported in the left panel) a strong variation in the antibaryon-barion ratio takes place with a significant increase of antibaryon ($\bar{B}$) particles. This feature is reflected in the strangeness fraction with a stronger increase in positive values of Y_S due to antibaryon $\bar{B}$ (antihyperon) particles with respect to negative values due to baryons B (hyperons). Also meson M and antimeson ($\bar{M}$) particles undergo to fluctuation in the strangeness fraction but this feature appears to be less pronounced.

This matter of fact appears more explicitly evident in Fig. 3, where the antiproton-proton ratio (left panel) and kaon-antikaon ratio (right panel) are reported as a function of baryon density at $T = 150$ MeV. Also in this particular case, let us point out the abrupt variation of the ratios during the phase transition region.

In conclusion, similarly to the liquid-gas phase transition in a warm and low density nuclear matter, a pure hadronic phase transition also can occur at higher temperatures and densities due to the presence of both mechanical and chemical instabilities. During the phase transition the two hadronic phases have a different baryon density and different strangeness fraction in

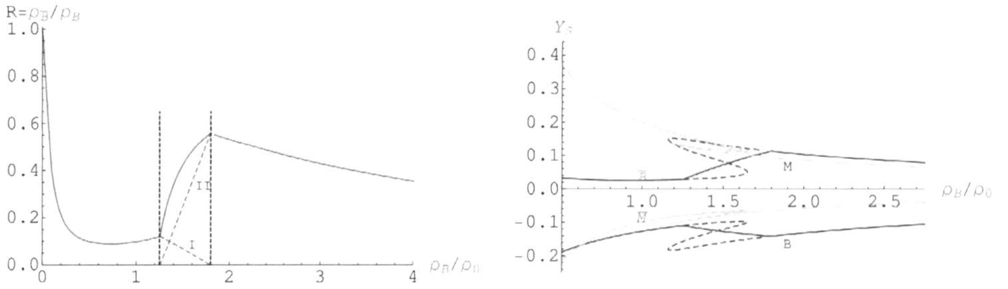

Figure 2. Left panel: Antibaryon-baryon ratio (R) versus baryon density at fixed temperature $T = 150$ MeV. The vertical dashed lines indicate the beginning and the end of the phase transition. The dashed lines with the symbols I and II give the values of R in the two coexisting phases. Right panel: Strangeness fractions (Y_S) of baryons (B), mesons (M) and their antiparticles as a function of baryon density in the pure hadronic phase and in the mixed phase at fixed temperature and zero net strangeness. The continuous (dashed) lines correspond to the solution obtained with (without) the Gibbs construction.

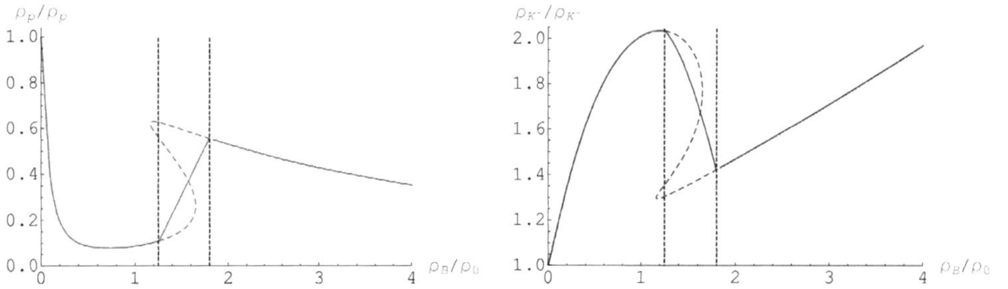

Figure 3. Antiproton-proton $\overline{p}/p$ (left panel) and kaon to antikaon K^+/K^- (right panel) ratios as a function of baryon density. The continuous (dashed) lines correspond to the solution obtained with (without) the Gibbs construction.

the mixed phase, although the global strangeness is fixed to zero. This matter of fact appears to be strictly connected to a strong variation of the antibaryon to baryon ratio during the phase transition. As a consequence, in the mixed phase, an increase of antihyperons (positive strangeness) with respect to hyperons (negative strangeness) is compensated by a suppression of strange mesons (kaon particles with positive strangeness) with respect to stange antimesons (antikaons particle with negative strangenss). This feature could be phenomenological very relevant in order to identify such phase transition in the future compressed baryonic matter experiments like CBM-FAIR experiment at GSI Darmstadt and NICA at JINR Dubna

References

[1] Di Toro M *et al.* 2006 *Nucl. Phys.* A **775** 102
[2] Bonanno L, Drago A and Lavagno A 2007 *Phys. Rev. Lett.* **99** 242301
[3] Alberico W M and Lavagno A 2009 *Eur. Phys. J.* A **40** 313
[4] Lavagno A 2013 *Eur. Phys. J.* A **49** 102

Journal of Physics: Conference Series **626** (2015) 012065 doi:10.1088/1742-6596/626/1/012065

[5] Glendenning N K 1992 *Phys. Rev.* D **46** 1274
[6] Lavagno A 2010 *Phys. Rev.* C **81** 044909
[7] Lavagno A and Pigato D 2012 *J. Phys. G: Nucl. Part. Phys.* **39** 125106
[8] Hofmann M, Mattiello R, Sorge H, Stocker H and Greiner W 1995 *Phys. Rev.* C **51** 2095
[9] Li Z, Mao G, Zhuo Y and Greiner W 1997 *Phys. Rev.* C **56** 1570
[10] Lavagno A and Pigato D 2012 *Phys. Rev.* C **86** 024917
[11] Müller H and Serot B D 1995 *Phys. Rev.* C **52** 2072
[12] Drago A, Lavagno A and Pagliara G 2014 *Phys. Rev.* D **89** 043014
[13] Li Z, Mao G, Zhuo Y and Greiner W 1997 *Phys. Rev.* C **56** 1570
[14] Gervino G, Lavagno A, Pigato D 2012 *Cent. Eur. J. Phys.* **10(3)** 594
[15] Alberico W M *et al.* 2008 *Physica A* **387** 467

Journal of Physics: Conference Series **626** (2015) 012066 doi:10.1088/1742-6596/626/1/012066

Representations of the ultrahyperbolic BMS group $\mathcal{UHB}(2,2)$. I. General Results

Evangelos Melas

Technological Educational Institution of Lamia. Department of Informatics. 35100 3nd National Road Athens–Lamia

E-mail: evangelosmelas@yahoo.co.uk

Abstract. The ordinary Bondi–Metzner–Sachs (BMS) group B is the common asymptotic symmetry group of all radiating, asymptotically flat, Lorentzian space–times. As such, B is the best candidate for the universal symmetry group of General Relativity (G.R.). However, in studying quantum gravity, space–times with signatures other than the usual Lorentzian one, and complex space-times, are frequently considered. Generalisations of B appropriate to these other signatures have been defined earlier. In particular, the generalisation $B(2,2)$ appropriate to the ultrahyperbolic signature $(+,+,-,-)$ has been described in detail, and the study of its irreducible unitary representations (IRs) of $B(2,2)$ has been initiated. We continue this programme by introducing a new group $\mathcal{UHB}(2,2)$ in the group theoretical study of ultrahyperbolic G.R. which happens to be a proper subgroup of $B(2,2)$. In this paper we report on the first general results on the representation theory of $\mathcal{UHB}(2,2)$. In particular the main general results are that the all little groups of $\mathcal{UHB}(2,2)$ are compact and that the Wigner–Mackey's inducing construction is exhaustive despite the fact that $\mathcal{UHB}(2,2)$ is not locally compact in the employed Hilbert topology.

1. Introduction

In 1939 Wigner laid the foundations of special relativistic quantum mechanics and relativistic quantum field theory by constructing the Hilbert space strongly continuous unitary irreducible representations (IRs) of the (universal cover) of the Poincare group P. The Bondi–Metzner–Sachs (BMS) group B is the common asymptotic group of all curved real lorentzian space–times which are asymptotically flat in future null directions [1, 2], and is the best candidate for the universal symmetry group of G.R.. In a quantum setting the universal property of B for G.R. make it reasonable to attempt to lay a similarly firm foundation for quantum gravity by following through the analogue of Wigner's programme with B replacing P. Some years ago McCarthy constructed explicitly [3, 4, 5, 6, 7, 8, 9, 10] the IRs of B for exactly this purpose. This work was based on G.W.Mackey's pioneering work on group representations [11, 12, 13, 14, 15]; in particular an extension to the relevant infinite–dimensional case of his semi–direct product theory.

It is difficult to overemphasize the importance of Piard's results [16, 17] who soon afterwards proved that *all* the IRs of B, when this is equipped with the Hibert topology, are derivable by the inducing construction. This proves the exhaustivity of McCarthy's list of representations and renders his results even more important.

However, in quantum gravity, complexified or euclidean versions of G.R. are frequently considered and the question arises: Are there similar symmetry groups for these versions of the theory? McCarthy constructed [18], in abstract form, all possible analogues of B, both real and in any signature, or complex, with all possible notions of asymptotic flatness 'near infinity'. There are, in fact, forty−one such groups. These abstract constructions were given in a quantum setting; the paper was concerned with finding the IRs of the groups $\mathcal{G}$ in Hilbert spaces (especially for the complexification $\mathcal{CB}$ of B itself). It was argued that these Hilbert space representations were related to elementary particles and quantum gravity (via gravitational instantons).

Let $B(2,2)$ be the BMS group $\mathcal{G}$ appropriate to the 'ultrahyperbolic signature' and asymptotic flatness in null directions. $B(2,2)$, is like B itself, based on a null cone [18], and it is given by

$$B(2,2) = C_e^{\infty}(T^2, R) Ⓢ_T G^2 \tag{1}$$

,i.e., it is the semi−direct product of the group G^2, where $G = SL(2,R)$, times the abelian normal subgroup $C_e^{\infty}(T^2, R)$ of so called supertranslations; $C_e^{\infty}(T^2, R)$ is the set of *even* real−valued infinitely−differentiable functions defined on the 2−Torus $T^2 = S^1 \times S^1$, S^1 being the set of vectors of unit length in $R^2 - \{0\}$. That is the functions $\alpha(m,n) \in C_e^{\infty}(T^2, R)$ satisfy the even−ness condition

$$\alpha(m,n) = \alpha(-m,-n),$$

where, $m = \frac{x}{|x|}, x = (x_1, x_2) \in R^2 - \{0\}$, and, similarly, $n = \frac{y}{|y|}, y = (y_1, y_2) \in R^2 - \{0\}$. The representation theory of $B(2,2)$ has been initiated elsewhere [19, 20].

The present paper reports the first general results on the representation theory of

$$\mathcal{UHB}(2,2) = C^{\infty}(P_1(R) \times P_1(R), R) Ⓢ_T G^2, \tag{2}$$

$P_1(R) = S^1/Z_2$ being the one−dimensional real projective space (the circle quotient the antipodal map). $\mathcal{UHB}(2,2)$ arises naturally in the construction of the generalizations of B given in [18] but it remained unnoticed in [18]. The crucial difference between $\mathcal{UHB}(2,2)$ and $B(2,2)$ is that for $\mathcal{UHB}(2,2)$ the supertranslations are *completely unconstrained,* whereas, for $B(2,2)$ they are described by *even* functions on the torus T^2. The representation theory of $\mathcal{UHB}(2,2)$ has been initiated in [21, 22, 23].

2. The group $\mathcal{UHB}(2,2)$

Recall that the ultrahyperbolic version of Minkowski space is the vector space R^4 of row vectors with 4 real components, with scalar product defined as follows. Let $x, y \in R^4$ have components x^{μ} and y^{μ} respectively, where $\mu = 0,1,2,3$. Define the scalar product $x.y$ between x and y by

$$x.y = x^0y^0 + x^2y^2 - x^1y^1 - x^3y^3. \tag{3}$$

Then the ultrahyperbolic version of Minkowski space, sometimes written $R^{2,2}$, is just R^4 with this scalar product.

In [21] it was shown that

Theorem 1 *The group $\mathcal{UHB}(2,2)$ can be realised as*

$$\mathcal{UHB}(2,2) = L^2(\mathcal{P}, \lambda, R) Ⓢ_T G^2 \tag{4}$$

with semi−direct product specified by

$$(T(g,h)\alpha)(x,y) = k_g(x)s_g(x)k_h(w)s_h(w)\alpha(xg, yh), \tag{5}$$

Journal of Physics: Conference Series **626** (2015) 012066 doi:10.1088/1742-6596/626/1/012066

where $\alpha \in L^2(\mathcal{P}, \lambda, R)$ and $(x, y) \in \mathcal{P}$. For ease of notation, we write $\mathcal{P}$ for the torus $\mathrm{T} \simeq P_1(R) \times P_1(R)$, $P_1(R)$ is the one–dimensional real projective space, and $\mathcal{G}$ for $G \times G$, $G = SL(2, R)$. In analogy to B, it is natural to choose a measure λ on $\mathcal{P}$ which is invariant under the maximal compact subgroup $\mathrm{SO}(2) \times \mathrm{SO}(2)$ of $\mathcal{G}$. $L^2(\mathcal{P}, \lambda, R)$ is the separable Hilbert space of real–valued functions defined on $\mathcal{P}$.

Moreover, if $g \in G$ is

$$\begin{bmatrix} a & b \\ c & d \end{bmatrix}, \tag{6}$$

then the components x_1, x_2 of $\mathbf{x} \in R^2$ transform linearly, so that the ratio $x = x_1/x_2$ transforms fraction linearly. Writing xg for the transformed ratio,

$$xg = \frac{(\mathbf{x}g)_1}{(\mathbf{x}g)_2} = \frac{x_1 a + x_2 c}{x_1 b + x_2 d} = \frac{xa + c}{xb + d}. \tag{7}$$

The factors $k_g(x)$ and $s_g(x)$ on the right hand side of (5) are defined by

$$k_g(x) = \left\{ \frac{(xb + d)^2 + (xa + c)^2}{1 + x^2} \right\}^{\frac{1}{2}}, \tag{8}$$

$$s_g(x) = \frac{xb + d}{|xb + d|}, \tag{9}$$

with similar formulae for yh, $k_h(y)$ and $s_h(y)$.

It is well known that the topological dual of a Hilbert space can be identified with the Hilbert space itself, so that we have $L^{2'}(\mathcal{P}, \lambda, R) \simeq L^2(\mathcal{P}, \lambda, R)$. In fact, given a continuous linear functional $\phi \in L^{2'}(\mathcal{P}, \lambda, R)$, we can write, for $\alpha \in L^2(\mathcal{P}, \lambda, R)$

$$(\phi, \alpha) = <\phi, \alpha> \tag{10}$$

where the function $\phi \in L^2(\mathcal{P}, \lambda, R)$ on the right is uniquely determined by (and denoted by the same symbol as) the linear functional $\phi \in L^{2'}(\mathcal{P}, \lambda, R)$ on the left. The representation theory of $\mathcal{UHB}(2, 2)$ is governed by the dual action T' of $\mathcal{G}$ on the topological dual $L^{2'}(\mathcal{P}, \lambda, R)$ of $L^2(\mathcal{P}, \lambda, R)$. The dual action T' is defined by:

$$< T'(g, h)\phi, \alpha > = < \phi, T(g^{-1}, h^{-1})\alpha > \cdot \tag{11}$$

A short calculation gives

$$(T'(g, h)\phi)(x, y) = k_g^{-3}(x) s_g(x) k_h^{-3}(y) s_h(y) \phi(xg, yh). \tag{12}$$

Now, this action T' of $\mathcal{G}$ on $L^{2'}(\mathcal{P}, \lambda, R)$, given explicitly above, is like the action T of $\mathcal{G}$ on $L^2(\mathcal{P}, \lambda, R)$, continuous. The 'little group' L_ϕ of any $\phi \in L^{2'}(\mathcal{P}, \lambda, R)$ is the stabilizer

$$L_\phi = \{(g, h) \in \mathcal{G} \mid T'(g, h)\phi = \phi\}. \tag{13}$$

By continuity, $L_\phi \subset \mathcal{G}$ is a closed subgroup.

3. Representation theory

Let A and $\mathcal{G}$ be topological groups, and let T be a given homomorphism from $\mathcal{G}$ into the group of automorphisms $\mathrm{Aut}(A)$ of A. Suppose A is abelian and $\mathcal{H} = A \circledS_T \mathcal{G}$ is the semi–direct product of A and $\mathcal{G}$, specified by the continuous action $T : \mathcal{G} \longrightarrow \mathrm{Aut}(A)$. In the product topology of $A \times \mathcal{G}$, $\mathcal{H}$ then becomes a topological group. It is assumed that it becomes a separable locally compact topological group.

In order to give the operators of the induced representations explicitly it is necessary ([11], [12], [13], [14], [15] and references therein) to give the following information

(i) An irreducible unitary representation U of L_{ϕ_o} on a Hilbert space D for each L_{ϕ_o}.

(ii) A $\mathcal{G}$ –quasi–invariant measure μ on each orbit $\mathcal{G}\phi \approx \mathcal{G}/L_{\phi_o}$; where L_{ϕ_o} denotes the little group of the base point $\phi_o \in A'$ of the orbit $\mathcal{G}\phi_o$; A' is the topological dual of A.

Let D_μ be the space of functions $\psi : \mathcal{G} \to D$ which satisfy the conditions

$$(a) \qquad \psi(gl) = U(l^{-1})\psi(g) \qquad (g \in \mathcal{G},\ l \in L_\phi)$$

$$(b) \quad \int_{\mathcal{G}\phi_o} < \psi(q), \psi(q) > \mathrm{d}\mu(q) < \infty,$$

where the scalar product under the integral sign is that of D. Note, that the constraint (a) implies that $< \psi(gl), \psi(gl) >=< \psi(g), \psi(g) >$, and therefore the inner product $< \psi(g), \psi(g) >$, $g \in \mathcal{G}$, is constant along every element q of the coset space $\mathcal{G}/L_{\phi_o} \approx \mathcal{G}\phi_o$. This allows to assign a meaning to $< \psi(q), \psi(q) >$, where $q = gL_{\phi_o}$, by defining $< \psi(q), \psi(q) >:=< \psi(g), \psi(g) >$. Thus the integrand in (b) becomes meaningful due to the condition (a). A pre–Hilbert space structure can now be given to D_μ by defining the scalar product

$$< \psi_1, \psi_2 >= \int_{\mathcal{G}\phi_o} < \psi_1(q), \psi_2(q) > \mathrm{d}\mu(q), \tag{14}$$

where $\psi_1, \psi_2 \in D_\mu$. It is convenient to complete the space D_μ with respect to the norm defined by the scalar product (14). In the resulting Hilbert space, functions are identified whenever they differ, at most, on a set of μ–measure zero. Thus our Hilbert space is

$$D_\mu = L^2(\mathcal{G}\phi_o, \mu, D). \tag{15}$$

Define now an action of $\mathcal{H} = A \circledS_T \mathcal{G}$ on D_μ by

$$(g_o\psi)(q) = \sqrt{\frac{\mathrm{d}\mu_{g_o}}{\mathrm{d}\mu}(q)}\psi(g_o^{-1}q), \tag{16}$$

$$\alpha\psi(q) = e^{i<g_o\phi_o,\alpha>}\psi(q) \tag{17}$$

where, $g_o \in \mathcal{G}$, $q \in \mathcal{G}\phi_o$, and $\alpha \in A$. Eqs. (16) and (17) define the IRs of $\mathcal{HB}$ induced for *each* $\phi_o \in A'$ and *each* irreducible representation U of L_{ϕ_o}. The 'Jacobian' $\frac{\mathrm{d}\mu_{g_o}}{\mathrm{d}\mu}$ of the group transformation is known as the *Radon–Nikodym* derivative of μ_{g_o} with respect to μ and ensures that the resulting IRs of $\mathcal{HB}$ are unitary.

The central results of induced representation theory ([11], [12], [13], [14], [15] and references therein) are the following

(i) Given the topological restrictions on $\mathcal{H} = A \circledS_T \mathcal{G}$ (separability and local compactness), any representation of $\mathcal{H}$, constructed by the method above, is irreducible if the representation U of L_{ϕ_o} on D is irreducible. Thus an irreducible representation of $\mathcal{H}$ is obtained for *each* $\phi_o \in A'$ and *each* irreducible representation U of L_{ϕ_o}.

(ii) If $\mathcal{H} = A \circledS_T \mathcal{G}$ is a regular semi–direct product (i.e., $A^{'}$ contains a Borel subset which meets each orbit in $A^{'}$ under $\mathcal{H}$ in just one point) then *all* of its irreducible representations can be obtained in this way.

4. Obstructions and resolutions

Two remarks are in order regarding the representations of $\mathcal{UHB}(2,2)$ obtained by the above construction

(i) As it is explained in [21] the subgroup $L^2(\mathcal{P},\lambda,R)$ of $\mathcal{UHB}(2,2) = L^2(\mathcal{P},\lambda,R) \circledS_T \mathcal{G}$ is topologised as a (pre) Hilbert space by using a natural measure on $\mathcal{P} = P_1(R) \times P_1(R)$ and by introducing a scalar product into $L^2(\mathcal{P},\lambda,R)$. If R^8 is endowed with the natural metric topology then the group $\mathcal{G} = SL(2,R) \times SL(2,R)$, considered as a subset of R^8, inherits the induced topology on $\mathcal{G}$. In the product topology of $L^2(\mathcal{P},\lambda,R) \times \mathcal{G}$ $\mathcal{UHB}(2,2)$ is a non–locally compact group (the proof follows without substantial change Cantoni's proof [24], see also [3]). (In fact the subgroup $L^2(\mathcal{P},\lambda,R)$, and therefore the group $\mathcal{UHB}(2,2)$ can be employed with many different topologies. The Hilbert type topology employed here appears to describe quantum mechanical systems in asymptotically flat space–times [9]). Since in the Hilbert type topology $\mathcal{UHB}(2,2) = L^2(\mathcal{P},\lambda,R) \circledS_T \mathcal{G}$ is not locally compact the theorems dealing with the irreducibility of the representations obtained by the above construction no longer apply (see e.g. [13]). However, it can be proved that the induced representations obtained above *are* irreducible. The proof follows very closely the one given in [6] for the case of the original BMS group B.

(ii) Here it is assumed that $\mathcal{UHB}(2,2)$ is equipped with the Hilbert topology. It is of outmost significance that it can be proved [21] that in this topology $\mathcal{UHB}(2,2)$ is a regular semi–direct–product. The proof follows the corresponding proof [16, 17] for the group B. Regularity amounts to the fact that [12] $L^{2^{'}}(\mathcal{P},\lambda,R)$ can have no equivalent classes of quasi–invariant measures μ such that the action of $\mathcal{G}$ is strictly ergodic with respect to μ. When such measures μ do exist it can be proved [12] that an irreducible representation of the group, with the semi–direct–product structure at hand, may be associated with each that is not equivalent to any of the IRs constructed by the Wigner–Mackey's inducing method. In a different topology it is not known if $\mathcal{UHB}(2,2)$ is a regular or irregular semi–direct–product. Irregularity of $\mathcal{UHB}(2,2)$ in a topology different from the Hilbert topology would imply that there are IRs of $\mathcal{UHB}(2,2)$ that are not not equivalent to any of the IRs obtained above by the inducing construction. Strictly ergodic actions are notoriously hard to deal with even in the locally compact case. Indeed, for locally compact non–regular semi–direct products, there is no known example for which all inequivalent irreducibles arising from strictly ergodic actions have been found. For the other 41 groups defined in [18] regularity has only been proved for B [16, 17] when B is equipped with the Hilbert topology. Similar remarks apply to all of them regarding IRs arising from strictly ergodic actions in a given topology.

5. Results

The new results are: A new group $\mathcal{UHB}(2,2)$ is introduced for the group theoretical study of ultrahyperbolic G.R.. $\mathcal{UHB}(2,2)$ is a proper subgroup of the group $B(2,2)$ initially proposed in [18] as appropriate to the 'ultrahyperbolic signature' and asymptotic flatness in null directions. Both $\mathcal{UHB}(2,2)$ and $B(2,2)$ are based on the null cone $\mathcal{N}$ of $R^{2,2}$. The crucial difference between $\mathcal{UHB}(2,2)$ and $B(2,2)$ is that the supertranslations in the case of $\mathcal{UHB}(2,2)$ are *free* functions defined on $P_1(R) \times P_1(R)$, whereas in the case of $B(2,2)$ the supertranslations are *even* functions defined on $S^1 \times S^1$. Remarkably, it is the nature of the supertranslations of $\mathcal{UHB}(2,2)$ which allows to establish contact [21] of the representation theory of $\mathcal{UHB}(2,2)$ with

Journal of Physics: Conference Series **626** (2015) 012066 doi:10.1088/1742-6596/626/1/012066

standard representation theory; something which is not feasible for the representation theory of $B(2,2)$. $\mathcal{UHB}(2,2)$ captures more efficiently, via its subgroup $L^2(\mathcal{P},\lambda,R)$, the fundamental characteristic of $R^{2,2}$, namely, that there is no clear–cut distinction of past and future in $R^{2,2}$. In [21] it is proved that when $\mathcal{UHB}(2,2)$ is employed with the Hilbert topology *all* little groups of $\mathcal{UHB}(2,2)$ are compact. Moreover in [21] it is shown that the Wigner–Mackey's inducing construction is exhaustive despite the fact that $\mathcal{UHB}(2,2)$ is not locally compact in the employed Hilbert topology. This result is rather important because other group theoretical approaches to quantum gravity which invoke Wigner–Mackey's inducing construction (see e.g. [25]) are typically plagued by the non–exhaustiveness of the inducing construction which results precisely from the fact that the group in question is not locally compact in the prescribed topology. Exhaustiveness is not just a mathematical nicety: If the inducing construction is not exhaustive one cannot simply know if the most interesting information or part of it is coded in the irreducibles which cannot be found by the Wigner–Mackey's inducing procedure. These results, compactness of the little groups and exhaustiveness of the inducing construction, not only are they significant for the group theoretical approach to quantum gravity advocated here, but also they have repercussions [21] for the other approaches to quantum gravity.

In comparing the representation theory of $\mathcal{UHB}(2,2)$ [21, 22, 23] with the representation theory of $B(2,2)$ [19, 20], we find both similarities and differences. The key difference between $\mathcal{UHB}(2,2)$ and $B(2,2)$, the supertranslations in the case of $\mathcal{UHB}(2,2)$ are *free* functions defined on $P_1(R)\times P_1(R)$, whereas in the case of $B(2,2)$ the supertranslations are *even* functions defined on $S^1\times S^1$, leaves its trace on the representation theory: the proof [21] of compactness for little groups of $\mathcal{UHB}(2,2)$ is similar to, but subtly different from, the corresponding proof [19] for $B(2,2)$. On the other hand, an interesting similarity between $\mathcal{UHB}(2,2)$ and $B(2,2)$ lies in the structure of their little groups: Their one–dimensional little groups form an unexpected family of continuous/discrete groups (with many connected components). Also, their finite little groups involve subgroups of direct products of the symmetry groups of the regular polygons only; the regular polyhedra do not appear at all here. The regular polyhedra appear [5] in the representation theory of the ordinary Bondi–Metzner–Sachs group.

References

[1] Bondi H, Van Der Berg M G J and Metzner A W K 1962 *Proc. R. Soc. Lond. A* **269** 21
[2] Sachs R K 1962 *Proc. R. Soc. Lond. A* **270** 103
[3] McCarthy P J 1972 *Proc. R. Soc. Lond. A.* **330** 517
[4] McCarthy P J 1972 *J. Math. Phys.* **13** 1837
[5] McCarthy P J 1973 *Proc. R. Soc. Lond. A* **333** 317
[6] McCarthy P J and Crampin M 1973 *Proc. R. Soc. Lond. A.* **335** 301
[7] McCarthy P J 1975 *Proc. R. Soc. Lond. A* **343** 489
[8] McCarthy P J 1978 *Proc. R. Soc. Lond. A* **358** 141
[9] McCarthy P J and Crampin M 1974 *Phys. Rev. Lett.* **33** 547
[10] McCarthy P J 1972 *Phys. Rev. Lett.* **29** 817
[11] Wigner E 1939 *Annals of Mathematics* **40** 149
[12] Mackey G W 1968 *Induced representations of groups and quantum mechanics* (Benjamin)
[13] Mackey G W 1955 *The theory of group representations* (The University of Chicago Press)
[14] Simms D J 1968 *Lie groups and quantum mechanics* (Bonn notes) (Berlin: Springer)
[15] Isham C J 1984 Topological and quantum aspects of quantum theory *Relativity groups and topology* B S DeWitt and R Stora eds
[16] Piard A 1977 *Rep. Math. Phys.* **11** 259
[17] Piard A 1977 *Rep. Math. Phys.* **11** 279
[18] McCarthy P J 1992 *Soc. Lond. A.* **338** 271
[19] McCarthy P J and Melas E 2003 *Nucl. Phys.* B **653** 369
[20] Melas E 2006 *J. Phys. A : Math. Gen.* **39** 3341

[21] Melas E Representations of the Ultrahyperbolic BMS group $\mathcal{UHB}(2,2)$. I. General Results (unpublished)
[22] Melas E Representations of the ultrahyperbolic BMS group $\mathcal{HB}$. II. Determination of the representations induced from infinite little groups *Preprint* arXiv:1312.0532v1
[23] Melas E Representations of the ultrahyperbolic BMS group $\mathcal{HB}$. II. Determination of the representations induced from finite little groups *Preprint* arXiv:1402.1428v1
[24] Cantoni V 1967 On the representations of the Bondi–Metzner–Sachs group Ph.D. Thesis, University of London
[25] Isham C J and Kakas A C 1984 *Class. Quantum Grav.* **1** 633

Journal of Physics: Conference Series **626** (2015) 012067 doi:10.1088/1742-6596/626/1/012067

Quantum mechanics from invariance principles

Florin Moldoveanu

Logic and Philosophy of Science Research Group, University of Maryland, College Park, MD 20742, USA

E-mail: fmoldove@gmail.com

Abstract. Quantum mechanics is an extremely successful theory of nature and yet it lacks an intuitive axiomatization. In contrast, the special theory of relativity is well understood and is rooted into natural or experimentally justified postulates. Here we introduce an axiomatization approach to quantum mechanics which is very similar to special theory of relativity derivation. The core idea is that a composed system obeys the same laws of nature as its components. This leads to a Jordan-Lie algebraic formulation of quantum mechanics. The starting assumptions are minimal: the laws of nature are invariant under time evolution, the laws of nature are invariant under tensor composition, the laws of nature are relational, together with the ability to define a physical state (positivity). Quantum mechanics is singled out by a fifth experimentally justified postulate: nature violates Bell's inequalities.

Quantum mechanics is a very unintuitive theory: it predicts only probabilistic outcomes but it supposes to be the "whole story" [1], it exhibits correlations between separable systems which cannot be explained by "local realistic" means [2], it is based on an abstract formalism involving hermitean operators and complex vector spaces. Also it is not even clear what we mean by "quantum" [3].

In contrast, the special theory of relativity is well understood and is rooted in two simple postulates: the laws of nature are invariant under changes in inertial frames of reference and the principle of invariant light speed. The first postulate of the special theory of relativity is easy to understand. The second postulate is ultimately justified by experimental evidence (starting with the Michelson - Morley experiment [4]).

It is the aim of this paper to start deriving quantum mechanics in a very similar way to how the special theory of relativity is obtained.

The special theory of relativity has a kinematical foundation, but emphasizing this fact obscures a larger point that it is based on a specific invariance of the laws of nature. In particular, the special theory of relativity uses only one kind of invariance, related of inertial reference frames. Figure 1 presents one possible line of argument for deriving the special theory of relativity.

Similarly, quantum mechanics can be derived from several other invariances and from natural or experimentally justified postulates (see Figure 2).

When talking about dynamics, we need to allow interaction between any two physical systems. The critical element is the invariance of dynamics under tensor composition of two subsystems. Composability was originally discoverd in the 1970s by Emile Grgin and Aage Petersen [5] as the necessary ingredient of understanding classical and quantum mechanics in a single mathematical formalism. The original motivation was a belief by Bohr (as reported by his personal assistant

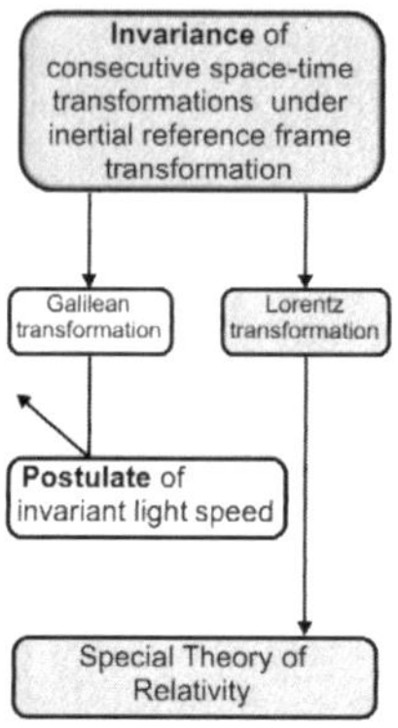

Figure 1. Deriving the special theory of relativity line of argument.

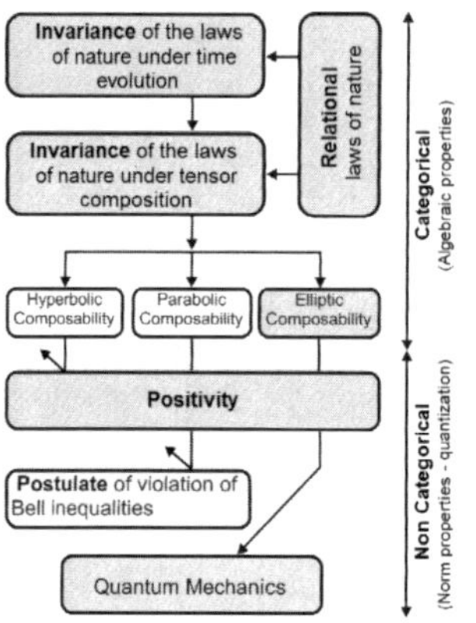

Figure 2. Deriving quantum mechanics line of argument.

Aage Petersen) that the correspondence principle has more to reveal. This grounbreaking work resulted in what is now called a Jordan-Lie algebra [6].

The Jordan-Lie algebra requires a norm and an important observation is that this norm is unique [7] given the spectral distance. Therefore to derive quantum mechanics we have to derive the necessity of the Jordan algebra of observables and the Lie algebra of generators together with their compatibility condition. From the algebraic properties of the Jordan-Lie structure and positivity, the usual Hilbert space formulation is recovered in special cases using Berezin deformation quantization [8].

In a Jordan-Lie approach to quantum mechanics, the starting point is the existence of two products, one describing the generators and the other one the observables, which form a duality known as "the equivalence of observables and generators" [9] or "dynamic correspondence" [10]. However, demanding the two products is too strong a requirement and we can start from much milder assumptions arriving at them.

The starting point is requiring the existence of time and a configuration space manifold Q endowed with a product (a tensor product) with unit.

At a point $p \in Q$ one can define a tangent space T_pQ and a cotangent space $T_p{}^*Q$. We will also assume that there are some C^∞ functions over the cotangent bundle M for which there is

Journal of Physics: Conference Series **626** (2015) 012067 doi:10.1088/1742-6596/626/1/012067

a way to generate a vector field out of them, and from now on we will restrict the domain of discussion only for those functions.

Let time evolution be represented by a one parameter group of transformations ϕ defined as follows:

$$\phi : M \times \mathbb{R} \to M, \tag{1}$$

with $\phi(x, 0) = x$ and $\phi(\phi(x, t), s) = \phi(x, t + s)$.

Suppose that there is an unspecified family of local operations $\{\circ\}$ which describe the laws of nature (for example Poisson bracket, Jordan product, commutator, etc). Introducing the notation: $\phi_t(x) \equiv \phi(x, t)$, the invariance of the laws of nature under time evolution reads:

$$(g \circ h)(\phi_{\Delta t}(p)) = g(\phi_{\Delta t}(p)) \circ h(\phi_{\Delta t}(p)), \tag{2}$$

with $p \in M$ a point in the cotangent bundle M and g, h, functions defined in the neighborhood of p. In other words, we demand the existence of a universal local morphism which preserves all algebraic relationships under time translation.

If X_f^i is a vector field arising out of a function f (corresponding to a particular time evolution), define $\mathcal{T}_{f_\epsilon}$:

$$\mathcal{T}_{f_\epsilon} = I + \epsilon X_f^i \frac{\partial}{\partial u^i}, \tag{3}$$

with I the identity operator and u^i the coordinate set in a local $\mathbb{R}^{2n}$ chart covering the point $p \in M$.

Because f is in one-to-one correspondence with X_f^i we can introduce a time translation transformation T_{f_ϵ} and a product $\alpha \in \{\circ\}$ between a distinguished f and any g as follows:

$$f\alpha g = \lim_{\epsilon \to 0} \frac{g - T_{f_\epsilon} g}{\epsilon}, \tag{4}$$

which is the Lie derivative of g along the vector field generated by f corresponding to a particular time evolution. Equivalently $T_{f_\epsilon} = (I - \epsilon f\alpha\cdot)$.

We generalize the product α for all f's and g's by repeating the argument for all conceivable dynamics. To make sure the domains of f and g are identical and well behaved, in case of pathologies, we can restrict the set of g to the span of all possible f.

Then the invariance of the laws of nature under time evolution can be expressed as:

$$T_{f_\epsilon}(g \circ h) = [T_{f_\epsilon} g] \circ [T_{f_\epsilon} h], \tag{5}$$

which to first order in ϵ implies a left Leibniz identity:

$$f\alpha(g \circ h) = (f\alpha g) \circ h + g \circ (f\alpha h). \tag{6}$$

From the relational property of the laws of nature, we demand for any f that $1\alpha f = 0$. Also we have $f\alpha 1 = 0$ from the definition of the Lie derivative. α will be shown later to be the usual commutator in quantum mechanics (or the Poisson bracket in classical mechanics) but it has no skew-symmetry property yet.

When talking about dynamics, it is usual to consider the interaction of two systems. The key property of a configuration space of many systems is the ability to concatenate the sub-systems using a tensor product.

Inspired by Grgin's groundbreaking work [5] we can introduce a composability category $\mathcal{U} = \mathcal{U}(\otimes, \mathbb{R}, \alpha, \cdots)$ which respects the existence of a unit element, the real numbers field $\mathbb{R}$ (understood as the set of constant functions), meaning $\mathcal{U} \otimes \mathbb{R} \simeq \mathcal{U} \simeq \mathbb{R} \otimes \mathcal{U}$.

First it can be shown that the product α is not enough and this demands the existence of a secondary product $\sigma \in \{\circ\}$. The invariance of the laws of nature under composability demands that a bipartite product α_{12} should be built out of the products listed in the composability category. If only a product α exists in the composability category, it must be trivial due to Leibniz identity. Only by adding another product σ we can construct non-trivial mathematical structures.

The most general way to construct the products α and σ in a bipartite system is as follows:

$$\begin{aligned}(f_1 \otimes f_2)\alpha_{12}(g_1 \otimes g_2) = a(f_1\alpha g_1) \otimes (f_2\alpha g_2) + \\ b(f_1\alpha g_1) \otimes (f_2\sigma g_2) + c(f_1\sigma g_1) \otimes (f_2\alpha g_2) + \\ d(f_1\sigma g_1) \otimes (f_2\sigma g_2)\end{aligned} \tag{7}$$

$$\begin{aligned}(f_1 \otimes f_2)\sigma_{12}(g_1 \otimes g_2) = x(f_1\alpha g_1) \otimes (f_2\alpha g_2) + \\ y(f_1\alpha g_1) \otimes (f_2\sigma g_2) + z(f_1\sigma g_1) \otimes (f_2\alpha g_2) + \\ w(f_1\sigma g_1) \otimes (f_2\sigma g_2)\end{aligned} \tag{8}$$

with f_1, f_2, g_1, g_2 arbitrary functions over the manifold M at a point p.

Then we can use the existence of the unit of the composability category to determine the coefficients b, c, d, y, z, w. If we normalize the definition of product σ such that $1\sigma 1 = 1$, we can show that $d = y = z = 0$ and $b = c = w = 1$. Applying the Leibniz identity on the bipartite products demands $a = 0$. Hence the bipartite products in the shorthand notation are:

$$\begin{aligned}\alpha_{12} &= \alpha_1\sigma_2 + \sigma_1\alpha_2 \\ \sigma_{12} &= \sigma_1\sigma_2 + x\alpha_1\alpha_2\end{aligned} \tag{9}$$

We can now observe that if α is a skew-symmetric product and σ is a symmetric product the symmetry and skew-symmetry is maintained under composition.

So far the product α is undefined. We know it is a derivation because it obeys the Leibniz identity, meaning it forms a Loday algebra [11], but it can be further proved that the product α is skew-symmetric. In turn this shows that the product σ is symmetric.

Let us summarize what we have up to this point. First, there is a skew-symmetric bilinear product α which obeys both a Leibniz identity and a Jacobi identity. As such it forms a Lie algebra. Then there is a symmetric bilinear product σ and a universal parameter x which is a constant of nature. The normalized parameter x can be -1, 0, or $+1$ corresponding to elliptic, parabolic, or hyperbolic composability classes. In quantum mechanics case $x = -\hbar^2/4$ and the fact that the Planck constant is invariant is a non-trivial fact due to composability [13]. A side-effect of the invariance of the Planck constant as a consequence of composability is the impossibility to have a consistent theory of mixed classical-quantum world because classical and quantum mechanics belong in disjoint composability classes [13].

From $\sigma_{12} = \sigma_1\sigma_2 + x\alpha_1\alpha_2$ it is easy to see that there is a linear transformation J from the space of observables to the space of generators called 'dynamic correspondence" [10] or "the equivalence of observables and generators" [9] such that $J\sigma J = \sqrt{x}I$.

This map is the root cause of the existence of complex numbers in all quantum mechanics formulations even when quantum mechanics is represented over real numbers [14].

We can now investigate the relationship between the product α and the product σ.

In general the products α and σ are not necessarily associative. For an arbitrary product $*$, a measure of non-associativity is the associator:

$$[f,g,h]_* = (f*g)*h - f*(g*h) \tag{10}$$

Using the Jordan and Leibniz identities along with the skew-symmetry of the product α, one can show [5] that there is a relationship between the α associator and σ associator, called the Petersen identity:

$$[f,g,h]_\sigma + J[f,g,h]_\alpha = 0 \tag{11}$$

In turn this means that an associative product β can be introduced as: $\beta = \sigma + J\alpha$.

When $J \neq 0$, in the quantum case, by choosing $f = h$ and $f = h\sigma h$ in Eq. (11) and using the Leibniz and skew-symmetry property of the product α we obtain: $[h,g,h]_\sigma = 0$ and $[h\sigma h, g, h]_\sigma = 0$. This means that the product σ obeys the flexible law and Jordan identity [5].

For the classical case, the product σ is always an associative product in addition to being commutative (symmetric).

A direct consequence of those results implies that quantum mechanics cannot be formulated over octonions because the product β has to be associative and therefore the Jordan algebra of observables σ cannot be special. The classification of real Jordan algebras restricts the allowed number systems for quantum mechanics.

We can also confirm that the tensor product is associative by verifying that: $\alpha_{(12)3} = \alpha_{1(23)}$ and $\sigma_{(12)3} = \sigma_{1(23)}$ by direct application of Eq. (9). Similarly we can see that the tensor product is also commutative: $\alpha_{(12)} = \alpha_{(21)}$ and $\sigma_{(12)} = \sigma_{(21)}$, thus making the composability category a commutative monoid.

If in quantum mechanics one changes the imaginary unit of complex numbers from $\sqrt{-1}$ to $\sqrt{+1} \neq \pm 1$, one obtains the so called "hyperbolic quantum mechanics" over split-complex numbers [15]. This corresponds to the hyperbolic composability case, and this case violates the Stone-von Neumann theorem because it has non-equivalent representations even in the finite dimensional case.

More important, it can be proved that in the hyperbolic case one encounters negative probabilities and the hyperbolic case can be eliminated by requiring positivity.

Therefore the only physical cases remaining are classical mechanics with the product α the Poisson bracket, and the product σ the point-wise function multiplication, and quantum mechanics with the product α the commutator and the product σ the anti-commutator. By composability, nature can only be in one composability class, and the way in which we can distinguish in nature between the elliptic and parabolic composability classes is by experimental evidence.

It is well known that classical mechanics obeys Bell inequalities [2] and quantum mechanics can achieve higher correlations. The strong experimental evidence in favor of violations of Bell inequalities starts with the Aspect experiment [16].

Collecting all the results above we introduce the following definition:

Definition 1 *A composability two-product algebra is a real vector space $\mathfrak{A}_\mathbb{R}$ equipped with two bilinear maps σ and α such that the following conditions apply:*

$$\begin{gathered}\alpha \text{ is a Lie algebra,}\\ \sigma \text{ is a Jordan algebra,}\\ \alpha \text{ is a derivation for } \sigma \text{ and } \alpha,\\ [A,B,C]_\sigma + \frac{J^2\hbar^2}{4}[A,B,C]_\alpha = 0,\end{gathered}$$

where $J \to (-J)$ is an involution, $1\alpha A = A\alpha 1 = 0$, $1\sigma A = A\sigma 1 = A$, and $J^2 = -1, 0, +1$.

Journal of Physics: Conference Series **626** (2015) 012067 doi:10.1088/1742-6596/626/1/012067

Quantum mechanics corresponds to $J^2 = -1$ (elliptic composability), classical mechanics corresponds to $J^2 = 0$ (parabolic composability), and the unphysical hyperbolic quantum mechanics corresponds to $J^2 = +1$ (hyperbolic composability).

The composability two-product algebra is a consequence of the algebraic constraints due to the invariance of the laws of nature under tensor composition. We can reconstruct the usual quantum mechanics formulation from it in a particular case using deformation quantization. Quantum mechanics can be formulated over real numbers, complex numbers, quaternions, or $SL(2,\mathbb{C})$. In the case of real numbers, complex numbers, or quaternions, quantum mechanics predicts probabilities by using Born rule. In the less known case of $SL(2,\mathbb{C})$ quantum mechanics predictions are 4-current probability densities. This formulation leads to spinors and Dirac's equation. For the sake of simplicity we will only consider the usual formulation of quantum mechanics over complex numbers. In the classical mechanics case we also restrict ourselves to symplectic manifolds.

In the classical mechanics case the the product σ is commutative and associative. Hence it is isomorphic with regular function multiplication $f\sigma g = fg$. From Darboux theorem [12] we obtain that the product α is a bracket:

$$f\alpha g = \{f,g\} = f\overleftrightarrow{\nabla} g = \sum_{i=1}^{n} \frac{\partial f}{\partial q^i}\frac{\partial g}{\partial p_i} - \frac{\partial f}{\partial p_i}\frac{\partial g}{\partial q^i}. \tag{12}$$

From classical mechanics we can now use deformation quantization to construct the Hilbert space formulation. However it is intructive to see how the this is achieved in flat $\mathbb{R}^{2n}$ space.

First we can build a Kähler manifold as follows. In elliptic composability we have a parameter J satisfying $J^2 = -1$. Because the Hamiltonian formalism is defined over the real numbers, J cannot be a scalar and must have a matrix representation.

To simplify the problem we consider a one-dimensional physical system. In this case the maximum matrix dimension for the representation of J is two. The only possibility is for J to have the same representation as the representation of the complex numbers imaginary unit:

$$i = J = \begin{pmatrix} 0 & -1 \\ 1 & 0 \end{pmatrix}. \tag{13}$$

In general, J is not the imaginary complex number unit but a tensor of rank $(1,1)$: $J = J^I{}_J$ with the following matrix representation:

$$J^I{}_J = \begin{bmatrix} 0 & -1_n \\ 1_n & 0 \end{bmatrix}. \tag{14}$$

In turn J allows to build a metric tensor and an indefinite inner product. We arive at a special case of a Kähler manifold: a complex projective space.

On $\mathbb{R}^{2n}$ the Berezin quantization [8] is the following prescription to construct compact operators from continuous functions on phase space:

$$Q_\hbar(f) = \int_{\mathbb{R}^{2n}} \frac{dpdq}{2\pi\hbar} f(p,q) |\Phi_\hbar^{(p,q)}\rangle\langle\Phi_\hbar^{(p,q)}|, \tag{15}$$

where $\Phi_\hbar^{(p,q)}$ are coherent states defined as:

$$\Phi_\hbar^{(p,q)} = (\pi\hbar)^{-1/4} e^{-ipq/2\hbar} e^{ipx/\hbar} e^{-(x-q)^2/2\hbar}. \tag{16}$$

At this point we recovered the Hilbert space formulation of complex quantum mechanics. This can be double checked by using the Gelfand-Naimark-Segal (GNS) construction [17] after extracting the C*-algebra condition for any bounded operators T as follows:

$$
\begin{aligned}
||T\Phi||^2 = \langle T\Phi, T\Phi\rangle = \langle T^*T\Phi, \Phi\rangle \leq \\
||T^*T\Phi||\ ||\Phi|| \leq ||T^*T||\ ||\Phi||^2 \\
||T||^2 \leq ||T^*T|| \leq ||T^*||||T|| = ||T||^2 \\
||T^*T|| = ||T||^2.
\end{aligned}
\tag{17}
$$

The $\mathbb{R}^{2n}$ case is special and there are other quantization methods available. The approach can be generalized if we start from Poisson manifolds. Any Poisson manifold admits a deformation quantization [18] but in this general case we may not obtain a Kähler manifold.

Next we ask how to handle the case of quantum systems which do not have a classical analog like the case of spin. Spin quantum systems are finite dimensional and in this case a recent result showed that the composability two-product algebra is enough to uniquely recover the Hilbert space [19]. The full reconstruction problem of quantum mechanics is still open and in general we can obtain either C*-algebras or C*-Hilbert modules. The complete number system classification for quantum mechanics is another open problem.

In the reconstruction cases obtained so far both composition and information considerations are essential in deriving quantum mechanics. The role of composition arguments is unsurprising given Bell's theorem because for a single particle there are classical models which reproduce exactly all quantum mechanics predictions. Pure composition arguments are unable to eliminate non-physical theories like hyperbolic quantum mechanics.

Composition arguments are special cases of categorical considerations which we proved that determine the algebraic relationships. Information theoretical considerations demand positivity as a pre-requisite, and positivity can recover the norm properties in Hilbert space formulation in certain cases.

We have seen that the distinction between quantum and classical mechanics has a composition (algebraic), not an information origin. Why does nature prefer elliptic composition over parabolic composition? This is no different from asking why nature prefers having a maximum speed limit over unlimited speeds. For quantum mechanics the answer cannot be given by either categorical or positivity arguments and we have to appeal to experimental evidence.

References

[1] Einstein A, Rosen N and Podolsky B 1935 Can quantum-mechanical description of physical reality be considered complete? *Phys. Rev.* **47** (10) 777
[2] Bell J S 1964 On the Einstein–Poldolsky–Rosen paradox *Phys.* **1** 195
[3] Popescu S and Rohrlich D 1994 Nonlocality as an axiom *Found. Phys* **24** 379
[4] Michelson A A and Morley E W 1887 On the relative motion of the Earth and the luminiferous ether *Am. J. Sc.* **34** 333
[5] Grgin E and Petersen A 1976 Algebraic implications of composability of physical systems *Comm. Math. Phys.* **50** (2) 177
[6] Landsman N P 1998 *Mathematical Topics Between Classical and Quantum Mechanics* (New York: Springer)
[7] Connes A 1994 *Noncommutative Geometry* (San Diego: Acad. Press)
[8] Berezin F A 1975 General concept of quantization *Commun. Math. Phys.* **40** 153
[9] Grgin E and Petersen A 1974 Duality of observables and generators in classical and quantum mechanics *J. Math. Phys.* **15** (6) 764
[10] Alfsen E M and Shultz F W 2003 *Geometry of State Spaces of Operator Algebras* (Boston: Birkhäuser)
[11] Loday J–L 1992 *Cyclic homology* (Berlin: Springer)
[12] Laurent-Gengoux C, Pichereau A and Vanhaecke P 2013 *Poisson Structures* (Heidelberg: Springer)
[13] Sahoo D 2004 Mixing quantum and classical mechanics and uniqueness of Planck's constant *J. Phys.* A **37** (3) 997
[14] Adler S L 1995 *Quaternionic Quantum Mechanics and Quantum Fields* (New York: Oxford Univ. Press)
[15] Khrennikov A and Segre G 2006 Von Neumann uniqueness theorem doesn't hold in hyperbolic quantum mechanics *Int. J. Theor. Phys.* **45** (10) 1869

[16] Aspect A, Grangier P and Roger G 1982 Experimental realization of Einstein–Podolsky–Rosen–Bohm gedankenexperiment: a new violation of Bell's inequalities *Phys. Rev. Lett.* **49** (2) 91
[17] Arveson W 1981 *An Invitation to C^*-Algebra* (New York: Springer-Verlag)
[18] Kontsevich, M. 2003 Deformation quantization of Poisson manifolds. *Lett. Math. Phys.* **66** (3), 157
[19] Kapustin, A. 2013 Is quantum mechanics exact? *J. Math. Phys.* **54**, 062,107

Journal of Physics: Conference Series **626** (2015) 012068 doi:10.1088/1742-6596/626/1/012068

Light propagated distance and redshift of a distant star

Shigeto Nagao

Ishibashi, Ikeda, Osaka, 563-0032 Japan

E-mail: snagao@lilac.plala.or.jp

Abstract. According to the formerly reported 4-D spherical model of the time and universe, factors affecting the redshift are discussed. In addition to the factor from the space expansion, two other factors derived from the light speed variation are proposed. One is the energy density factor of the wave medium, which was formerly reported to determine the light speed. The second is a newly proposed factor caused by the electromagnetic interaction of light with substances in the 3-D space. Subsequently the direct correlation of the light propagated distance with the redshift is given. Superimposed graphs of it on the real observed data from the Supernova Cosmology Project exhibited an excellent fit for a case that the current radius (equal to our observed time) of the universe is between 0.7 and 0.8 of its maximum. This could be an important ground for a possibility of the 4-D spherical model, which implies that the universe has been expanding at a constant speed by our observed time.

1. Introduction

The author of this article formerly proposed a model of the time and universe, in which the space energy ("Spacia") is spread with expansion in a 3-dimensional surface of a 4-dimensional sphere ("4-D Spherical Model" or "4DS Model") [1, 2]. What we detect as energy in the 3-D space is a vibration of the intrinsic space energy in the 3-D space. What we observe as the time passing at a constant speed ("Time") is the radius dimension of the 4-D sphere for the distribution of the space energy. Space expansion is proportional to the Time as shown by $|\boldsymbol{r}| = |\boldsymbol{x}| * |\boldsymbol{\theta}|$, where $\boldsymbol{\theta}$ denotes the angle from the center of universe for a 3-D space vector $\boldsymbol{r}$ and $\boldsymbol{x}$ is the radius of the 4D sphere of universe equal to the Time. The model insists on the changing speed of light propagation in accordance with the expansion of the universe. Due to the existence of the wave medium for energy in the 3-D space there is a special frame stationary to the intrinsic space energy Spacia. According to the model, I further proposed previously to abandon the Special Relativity and newly introduce the acceleration factor (f_a) shown by eqs. (1) and (2) as a feature of wave [3].

$$\alpha = \frac{F}{m} f_a \,, \quad F = \frac{m}{f_a}\alpha \tag{1}$$

$$f_a = \left(1 - \frac{v^2}{c^2}\right)^n \quad (n \geq 1) \tag{2}$$

In the former article of ref. [2], it was shown that the light propagated distance from a distant star is not proportional to the time passed since it was emitted due to the changing speed of light.

Journal of Physics: Conference Series **626** (2015) 012068 doi:10.1088/1742-6596/626/1/012068

The formulas and graphs shown there were not direct relations of the light propagated distance with the redshift but with the time passed since emission. In this article, factors affecting the redshift are discussed and the graph of light propagated distance versus redshift, which are the real observables in the Supernova Cosmology Project, is given.

In the section 4 for discussion, the reason why the Michelson-Morley type experiments are incapable to detect a shift of interference fringe is also discussed because the presence of a wave medium for light and other energy in 3D space is a prerequisite for the proposed 4DS Model.

2. Redshift

2.1. Definition of terms related to Time

At the beginning, define the following terms and abbreviations related to Time.

x (Radius of universe): Radius of the 4D sphere of universe, equal to the Time (T).
CU (Cosmic Unit): Unit of x and T, being 1 at its maximum.
(Time-related variables are expressed in the CU in this article.)
T_E (Time of Emission): Time when the light was emitted.
T_P (Present Time): Present Time of universe, when the light reaches us.
T_{ER} (Relative Time of Emission): Relative ratio of T_E to T_P. $T_{ER} \equiv T_E/T_P$
T_B (Back in Time): Back in Time from present for T_E. $T_B \equiv T_P - T_E$
T_{BR} (Relative Back in Time): Relative ratio of T_B to T_P. $T_{BR} \equiv T_B/T_P$
T_C (Time Clear): Time when the space became transparent to light.
T_{CR} (Relative Time Clear): Relative ratio of T_C to T_P. $T_{CR} \equiv T_C/T_P$

2.2. Wavelength and frequency

As potential factors affecting the redshift there would be three types; Doppler effect, the space expansion and the speed of light. In case of inter-galaxy observation, Doppler effect due to the relative velocity to the wave medium can be neglected. Both the emitter and the observer can be regarded stationary to the wave medium Spacia at their respective positions. In the current cosmology, only the space expansion is considered as a factor for redshift along with adjustment from the theory of relativity because the light speed is regarded constant. However, the factor by light speed variation should be taken into consideration according to the 4DS Model.

In general for wave, the wave propagation speed is proportional to the wavelength whereas the frequency remains constant only depending on the emitter, shown as

$$v = f * \lambda \ . \tag{3}$$

This is a case for light propagation in a substance like water or glass as well as for sound propagation. From the factor of light speed variation, the frequency remains constant whereas the wavelength changes. On the other hand, from another factor corresponding to the space expansion itself, the light speed is not altered but the wavelength is expanded and the frequency is decreased. Combined the two factors, the change of frequency and that of wavelength since the light was emitted are shown by the following equations

$$\frac{f\,(T_P)}{f\,(T_E)} = \frac{1}{n} \tag{4}$$

$$\frac{\lambda\,(T_P)}{\lambda\,(T_E)} = n \times \frac{C\,(T_P)}{C\,(T_E)} \tag{5}$$

in which n denotes the "Space expansion ratio" defined by $n \equiv T_P/T_E$.

2.3. Light speed

The light propagation speed should be affected by the energy density of the medium ("Energy density factor" or "f_D") and also by electromagnetic interaction of light with substances in 3-D space ("Electromagnetic interaction factor" or "f_{EM}"). I propose that the light speed by our observed Time is expressed by the following formula

$$C(T) = C(x) = \frac{dL}{dx} = K * f_D * f_{EM} \,, \tag{6}$$

where K is a constant. The Energy density factor f_D was discussed in my former articles [1] and [2], and is given by the following formula using the Cosmic Unit (CU) for x. When the universe stops expansion due to gravity, x takes its maximum value 1 in the CU.

$$f_D = \frac{1}{x\sqrt{1-x}} \qquad (T_C \leq x < 1) \tag{7}$$

An example of the Electromagnetic interaction factor f_{EM} is the fact that light propagates slower in the water or glass than in vacuum. Light interacts with substances and receives scattering, which results in slowing propagation speed. Before the formation of hydrogen atoms, the space of universe was not transparent to light propagation due to interaction of light with electrons and protons in plasma. After completion of atom formation, the space became cleared up for light. It is generally said that the Cosmic Microwave Background (CMB) radiation started to go straight around 370,000 years after the Big Bang. The temperature of the universe was around 3,000 K at the time, corresponding to an energy of about 0.25 eV, which is much less than the 13.6 eV ionization energy of hydrogen [4]. At the early time known as "the time of last scattering", light could not go straight even after the completion of plasma conversion to atoms. This can be interpreted due to high density of atoms. In accordance with the space expansion, the incidence of light to interact with atoms should decrease corresponding to the descending density of atoms. We could expect the incidence is reversibly proportional to the cube of the radius x. Furthermore, when the CMB radiation started ($x = T_C$), the Electromagnetic interaction factor would be zero. I propose the following formula for f_{EM}.

$$f_{EM} = 1 - \frac{T_C^3}{x^3} \qquad (T_C \leq x < 1) \tag{8}$$

If CMBR started 370 thousand years after the Big Bang and the current universe is 13.8 billion years old [4], T_{CR} would be 2.68×10^{-5}. However, it would be likely that current universe is regarded older than the real one due to the lack of adjustment of changing speed of light. Let's take a second scenario for a larger T_{CR} value corresponding to one million years for the initial CMBR and 10 billion years for the age of current universe, which gives 1×10^4 for T_{CR}.

$$T_{CR}(1) = \frac{3.7 \times 10^{-5} years}{1.38 \times 10^{-10} years} = 2.68 \times 10^{-5} \tag{9}$$

$$T_{CR}(2) = \frac{1 \times 10^{-6} years}{1 \times 10^{-10} years} = 1 \times 10^{-4} \tag{10}$$

From eqs. (6), (7) and (8), the following formula is given for the light speed.

$$C(x) = K * \frac{1}{x\sqrt{1-x}} * \left(1 - \frac{T_C^3}{x^3}\right) \tag{11}$$

We don't know yet either T_C or T_P. If T_{CR} is 2.68×10^{-5} and T_P is 0.75, then T_C is 2×10^{-5}. In Fig. 1, a graph of light speed versus the radius $x(= T)$ in semi-logarithmic scale is given.

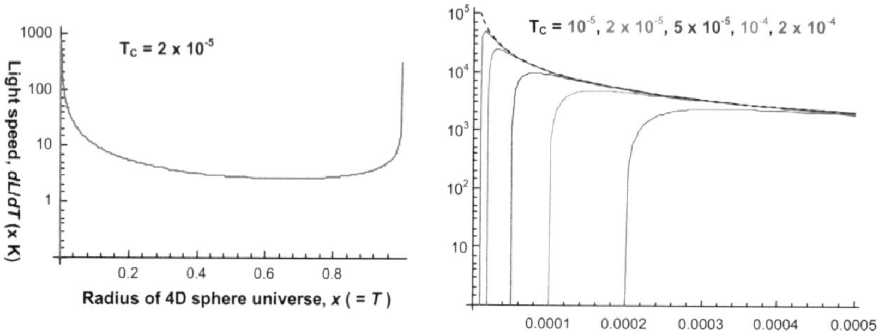

Figure 1. Time-course of light speed by the Time. **Left**: Whole range graph of light speed for the Time Clear $T_C = 2 \times 10^{-5}$. The maximum in small $x(=T)$ value area is not correctly shown due to limitation of sampling points. **Right**: Enlarged graphs in small $x(=T)$ value area for $T_C = 10^{-5}$ to 2×10^{-4}. The dotted line in black is for the case of $f_{EM} = 1$.

The right part is an enlarged graph in very small area of x for plural T_C values from 10^{-5} to 2×10^{-4}. The dotted line in black represents the light speed graph for $f_{EM} = 1$. As shown in the graph, for the area of x greater than 5×10^{-4}, the effect of f_{EM} can be ignored and treated as $f_{EM} = 1$ subject to $T_C \le 2 \times 10^{-4}$. For $x \ge 5 \times 10^{-4}$, light speed can be treated as

$$C(x) = \frac{K}{x\sqrt{1-x}} \; . \tag{12}$$

2.4. Redshift

The prolongation ratio of wavelength at T_P from T_E is given as follows from eqs. (5) and (11).

$$\frac{\lambda(T_P)}{\lambda(T_E)} = n \times \frac{C(T_P)}{C(T_E)} = \frac{T_P}{T_E} \times \frac{T_E\sqrt{1-T_E}}{T_P\sqrt{1-T_P}} \times \frac{1-\left(\frac{T_C}{T_P}\right)^3}{1-\left(\frac{T_C}{T_E}\right)^3} = \frac{\sqrt{1-T_E}}{\sqrt{1-T_P}} \times \frac{1-T_{CR}^3}{1-\left(\frac{T_{CR}}{T_{ER}}\right)^3} \tag{13}$$

Because T_{CR} is expected smaller than at least 10^{-3}, T_{CR}^3 in $1 - T_{CR}^3$ of eq. (13) can be ignored. The redshift z is given as follows.

$$z + 1 = \frac{\lambda(T_P)}{\lambda(T_E)} = \frac{\sqrt{1 - T_P T_{ER}}}{\sqrt{1-T_P}} \times \frac{1}{1-\left(\frac{T_{CR}}{T_{ER}}\right)^3} \tag{14}$$

$$z = \sqrt{1 + \frac{T_P(1-T_{ER})}{1-T_P}} \times \frac{1}{1-\left(\frac{T_{CR}}{T_{ER}}\right)^3} - 1 \tag{15}$$

Replacing T_{ER} by T_{BR}, we get the following equation.

$$z = \sqrt{1 + \frac{T_P T_{BR}}{1-T_P}} \times \frac{1}{1-\left(\frac{T_{CR}}{1-T_{BR}}\right)^3} - 1 \tag{16}$$

Graphs of redshift z versus T_{BR} for various T_P values are given in Fig. 2. The left graph is for $T_{CR} = 2.68 \times 10^{-5}$, but it is almost same as that for $T_{CR} = 1 \times 10^{-4}$ in this scale. The right graph is in very large T_{BR} area greater than 0.999 for $T_{CR} = 2.68 \times 10^{-5}$ and $T_{CR} = 1 \times 10^{-4}$

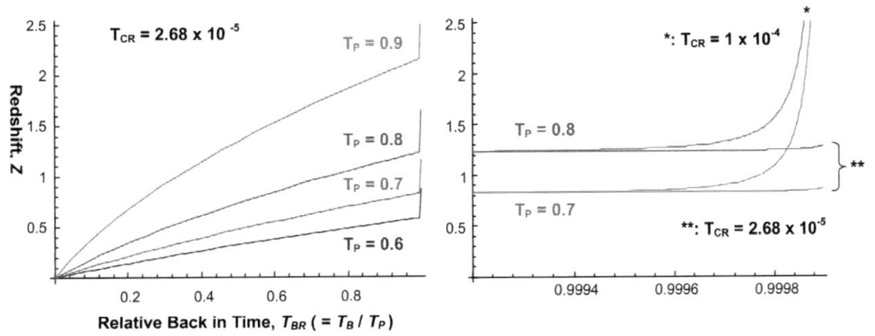

Figure 2. Redshift versus the Relative Back in Time (T_{BR}). **Left**: Whole range graph of redshift for the Relative Time Clear $T_{CR} = 2.68 \times 10^{-5}$. **Right**: Enlarged graph in high T_{BR} area for $T_{CR} = 2.68 \times 10^{-5}$ and $T_{CR} = 1 \times 10^{-4}$.

with $T_P = 0.7$ or 0.8. Redshift values for some representing T_{ER} values are given in Table 1. As we see in the figure and the table, for the area of T_{ER} greater than 0.001 (or $T_{BR} \leq 0.999$), we can ignore the influence by f_{EM}, and z can be approximated to be

$$z \approx \sqrt{1 + \frac{T_P T_{BR}}{1 - T_P}} - 1 \,. \tag{17}$$

Table 1. Redshift z.

$T_{CR} = 2.68 \times 10^{-5}$ (corresponding to 3.7×10^5year/1.38×10^{10}year)							
	T_{ER}						
T_P	0.1 (1380MY)	0.01 (138MY)	0.001 (13.8MY)	1×10^{-4} (1.38MY)	5×10^{-5} (0.69MY)	3×10^{-5} (0.414MY)	2.7×10^{-5} (0.373MY)
0.6	0.533	0.576	0.581	0.612	0.869	4.508	70.680
0.7	0.761	0.819	0.825	0.862	1.158	5.360	81.769
0.8	1.145	1.227	1.235	1.280	1.643	6.789	100.371
$T_{CR} = 1 \times 10^{-4}$ (corresponding to 1×10^6year/1×10^{10}year)							
	T_{ER}						
T_P	0.1 (1000MY)	0.01 (100MY)	0.001 (10MY)	2×10^{-4} (2MY)	1.1×10^{-4} (1.1MY)	1.01×10^{-4} (1.01MY)	
0.6	0.533	0.576	0.582	0.807	5.358	52.761	
0.7	0.761	0.819	0.827	1.086	6.341	61.077	
0.8	1.145	1.227	1.237	1.555	7.991	75.028	

3. Light propagated distance

Light propagated distance of a star, which we detect now, is given as

$$L(T_E) = \int_{T_E}^{T_P} C(x)dx = \int_{T_E}^{T_P} \frac{K}{x\sqrt{1-x}} \left(1 - \frac{T_C^3}{x^3}\right) dx \,. \tag{18}$$

Journal of Physics: Conference Series **626** (2015) 012068 doi:10.1088/1742-6596/626/1/012068

If $T_{ER} \geq 10^{-3}$ or $T_{BR} \leq 0.999$, we can approximate it as

$$L(T_E) \approx \int_{T_E}^{T_P} \frac{K}{x\sqrt{1-x}} dx \ . \tag{19}$$

As shown in the reference [2], eq. (19) is written by T_{BR} as follows.

$$\int \frac{1}{x\sqrt{1-x}} dx = \log\left(\frac{1-\sqrt{1-x}}{1+\sqrt{1-x}}\right) + k \tag{20}$$

$$L(T_{BR}) = K \cdot \log\left(\frac{1-\sqrt{1-T_P}}{1+\sqrt{1-T_P}} \cdot \frac{1+\sqrt{1-T_P+T_PT_{BR}}}{1-\sqrt{1-T_P+T_PT_{BR}}}\right) \tag{21}$$

From eq. (17), T_{BR} is replaced by z as

$$(z+1)^2 = 1 + \frac{T_PT_{BR}}{1-T_P} \tag{22}$$

$$T_{BR} = \frac{(1-T_P)z(z+2)}{T_P} \ . \tag{23}$$

Putting eq. (23) into eq. (21), we get the following formula of light propagated distance versus the redshift for the range of $T_{ER} \geq 10^{-3}$ or $T_{BR} \leq 0.999$.

$$\sqrt{1-T_P+T_PT_{BR}} = \sqrt{1-T_P+(1-T_P)z(z+2))} \tag{24}$$

$$L(z) = K \cdot \log\left(\frac{1-\sqrt{1-T_P}}{1+\sqrt{1-T_P}} \cdot \frac{1+\sqrt{1-T_P+(1-T_P)z(z+2)}}{1-\sqrt{1-T_P+(1-T_P)z(z+2)}}\right) \tag{25}$$

Because we don't know yet the value of K, let's take the relative light propagated distance to that at $z = 0.05$ ("RLD") in order to superimpose its graph on the observed data from the Supernova Cosmology Project [5].

$$\begin{aligned} RLD \equiv \frac{L(z)}{L(0.05)} &= \log\left(\frac{1-\sqrt{1-T_P}}{1+\sqrt{1-T_P}} \cdot \frac{1+\sqrt{1-T_P+(1-T_P)z(z+2)}}{1-\sqrt{1-T_P+(1-T_P)z(z+2)}}\right) \\ &\times \left(\log\left(\frac{1-\sqrt{1-T_P}}{1+\sqrt{1-T_P}} \cdot \frac{1+\sqrt{1.1025(1-T_P)}}{1-\sqrt{1.1025(1-T_P)}}\right)\right)^{-1} \end{aligned} \tag{26}$$

Its graphs for $T_P = 0.6, 0.7, 0.8$ and 0.9 are given in Fig. 3. Respective T_{ER} values corresponding to $z = 0.05$ for $T_P = 0.6$, 0.7 and 0.8 are shown in Table 2. RLD values respectively for $T_{ER} = 0.1$, 0.01 and 0.001 are also shown in Table 2 for $T_P = 0.6$, 0.7 and 0.8 . As we see there, if the RLD value is smaller than about 65, T_{ER} should be greater than 0.001 subject to $T_P \geq 0.6$. For the area, we can ignore the effect by f_{EM}, and the approximation expressed by eq. (26) can be applied. In Fig. 3, RLD graphs for $RLD \leq 50$ and $z \leq 1.17$ are shown in dual logarithmic scale.

For the small z range $z \leq 0.05$, not only f_{EM} but also the influence of f_D to light speed variation can be ignored. Light speed can be regarded constant and the RLD is given as follows.

$$L = c(T_P - T_E) = cT_PT_{BR} \tag{27}$$

$$T_{BR} = \frac{T_B}{T_P} = \frac{z}{z+1} \tag{28}$$

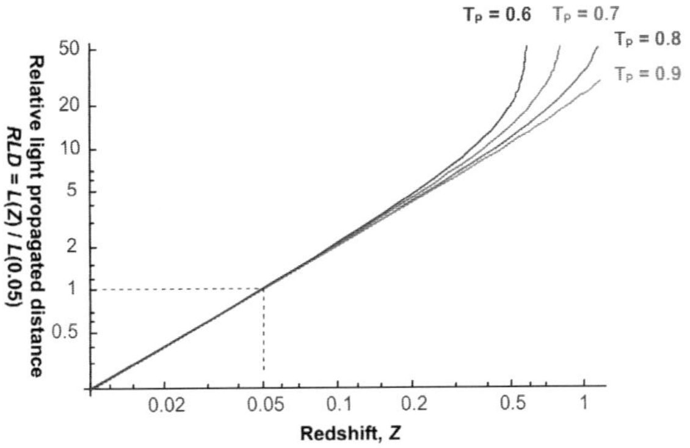

Figure 3. Relative light propagated distance (RLD) versus redshift.

Table 2. Relative light propagated distance $RLD = L(z)/L(0.05)$.

T_P	T_{ER} at $z = 0.05$	T_{ER}				
		0.1	0.01	0.001	$T_{CR}(2) = 10^{-4}$	$T_{CR}(1) = 2.68 \times 10^{-5}$
0.6	0.93167	24.5	45.9	67.0	< 88.1	< 100.2
0.7	0.95607	34.7	63.9	92.7	< 121.5	< 138.0
0.8	0.97438	51.4	92.7	133.4	< 174.1	< 197.3

$$RLD = \frac{cT_PT_{BR}}{cT_P \cdot 0.05/1.05} = 21 \cdot T_{BR} = 21 \cdot \frac{z}{z+1} \tag{29}$$

Superimpose RLD versus z graphs in a way that they match to the observed data for various T_P values in the area $z \leq 0.05$ because eq. (29) is independent of T_P. The results are shown in Fig. 4. The observed data give an excellent fit to the expectation for a case that T_P is between 0.7 and 0.8 in the range $RLD \leq 50$. The dotted line in black in Fig. 4 is the graph of eq. (29), which would be applicable if light speed had been constant.

Provided that the current Time T_P is roughly 0.75, the K value for light speed is given as follows from eq. (12), where c denotes the current light speed.

$$K = \left(0.75 \times \sqrt{0.25}\right) \cdot c \approx 0.375 \times c \approx 1.125 \times 10^8 m \cdot s^{-1} \tag{30}$$

How would be the RLD in a case $0.001 > T_{ER} \geq T_{CR}$? The redshift shown by eq. (15) diverges to infinity when T_E approaches to T_C. On the other hand the light propagated distance shown by eq. (18) has at least the following upper limit.

$$L(T_E) = \int_{T_E}^{T_P} \frac{K}{x\sqrt{1-x}} \left(1 - \frac{T_C^3}{x^3}\right) dx < \int_{T_E}^{T_P} \frac{K}{x\sqrt{1-x}} dx \leq \int_{T_C}^{T_P} \frac{K}{x\sqrt{1-x}} dx \tag{31}$$

The RLD also has an upper limit when T_E approaches to T_C.

$$RLD = \frac{L(x = T_E)}{L(z = 0.05)} < \frac{1}{L(z = 0.05)} \int_{T_E}^{T_P} \frac{K}{x\sqrt{1-x}} dx \leq \frac{1}{L(z = 0.05)} \int_{T_C}^{T_P} \frac{K}{x\sqrt{1-x}} dx \tag{32}$$

Values of the right side of eq. (32) are shown in Table 2 for T_{CR} is 10^{-4} or 2.68×10^{-5}.

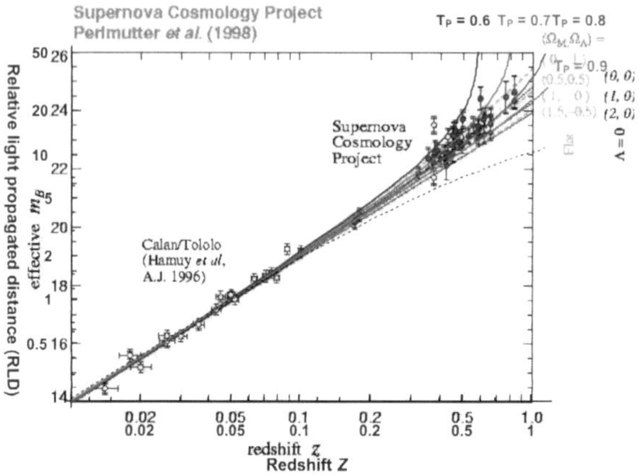

Figure 4. Superimposed graph of the Relative light propagated distance (RLD) versus redshift on the Supernova Cosmology Project data.

4. Discussion and conclusion

The presence of the medium for light and other energy in the 3-D space is a prerequisite for the proposed 4DS Model. If the Michelson-Morley experiment has duly proved the absence of the light medium without any doubt as repeatedly confirmed by many scientists over more than 100 years [6]-[8], the 4DS Model is denied. Let's check key points of the experiment, which were partly discussed previously in the reference [3], in a more comprehensive manner.

In the experiment, a light beam is split by a half-mirror into a straight beam and a reflected right angle beam. Both beams are reflected at the respective ends of arms, return to the half-mirror and are combined for detection [7, 8]. Firstly let's see the experiment by the reference frame stationary to the medium and then by another frame fixed to the apparatus.

1) By the reference frame stationary to the medium

- Phase propagation speed of a wave is constant independent of the speed of the emitter to the medium. Therefore, the light speed is constant by the medium's frame.
- The route of the beam A parallel to the apparatus movement is longer than the route of the perpendicular beam B for the round trip between the half-mirror and the end mirror. For details, please refer to the reference [3]. The beam A takes Δt longer time than the beam B for the round trip.
- The displacement of the beam A is released at the splitter Δt earlier than that of the beam B is released so that the two displacements reach the same point at the same instance for interference detection. There is a phase difference of $\omega\Delta t$ between the two beams at the detector, where ω denotes the angular frequency of the beams [3].
- The reflecting point at the end mirror varies by the time if the mirror moves toward the medium. Even if the reflecting point is varied, the reflected wave is shown by the same wave equation and the frequency remains constant. We can confirm it as follow. Take a wave shown by $A = \sin(x - vt)$. At $t = t_1$ the wave is reflected. The reflected wave is shown by $A = \sin(x + vt_1 + v(t - t_1)) = \sin(x + vt)$, which is independent of t_1, because reflected wave phase speed is $-v$ independent of the mirror speed to the medium.
- In the case of Michelson-Morley experiment, the respective reflected beams have the same light speed and same frequency with the phase difference of $\omega\Delta t$. At the point where the

Journal of Physics: Conference Series **626** (2015) 012068 doi:10.1088/1742-6596/626/1/012068

two beams are combined to interfere, the sum of amplitudes can be expressed as follows, where A is the maximum amplitude of beam A and beam B.

$$\begin{aligned} U_a + U_b &= A\sin\omega t + A\sin\omega(t+\Delta t) \\ &= A\sin\left(\omega t + \frac{\omega\Delta t}{2} - \frac{\omega\Delta t}{2}\right) + A\sin\left(\omega t + \frac{\omega\Delta t}{2} + \frac{\omega\Delta t}{2}\right) \\ &= 2A\cos\frac{\omega\Delta t}{2}\sin\left(\omega t + \frac{\omega\Delta t}{2}\right) \end{aligned} \tag{33}$$

- The combined wave can be shown as follows, where x denotes the distance from the combined point in the direction to the interferometer and k denotes the angular wave number given as $k = \omega/c$, in which c is the light speed toward the medium. By the variation of Δt there is no change of the frequency while the maximum amplitude and the phase vary.

$$U_a + U_b = 2A\cos\frac{\omega\Delta t}{2}\sin\left(kx - \omega t - \frac{\omega\Delta t}{2}\right) \tag{34}$$

2) By the frame fixed to the apparatus

- The light traveled distance is same for both the beams A and B by the apparatus' frame.
- The time is same as that by the medium's frame.
- The average light speed during the round trip between the half-mirror and the end mirror is different among the beams A and B shown by $c_A = 2L/(t+\Delta t)$ and $c_B = 2L/t$, respectively.
- However, the direction of light propagation at the detector is identical for the beam A and the beam B. Therefore, light speed at the detector is common for the both beams.
- Therefore, the combined wave for detection can be also expressed similarly to eq. (34) as follows, where U_a', U_b', k' and x' are values measured by the apparatus' frame. k' is given as $k' = \omega/c'$, where c' is the light speed by the apparatus' frame in the direction of x'.

$$U_a' + U_b' = 2A\cos\frac{\omega\Delta t}{2}\sin\left(k'x' - \omega t - \frac{\omega\Delta t}{2}\right) \tag{35}$$

3) Interference fringe
Either by the medium's frame or the apparatus' frame, there are following aspects for the interference fringe:

- The combined wave goes into the interferometer, and passes through the two slits with a very small distance.
- On the detection screen, interference fringe is shown.
- The location and pattern of the fringe depends on the slit pitch and the distance from the slits to the screen.
- The fringe pattern varies by the light frequency or wavelength, but is not altered by the variation of the phase of light.

As a conclusion, by the variation of Δt no change of position of the interference fringe should be detected while the brightness of the fringe would be altered.

The latest modern Michelson-Morley type experiment would be the Herrmann et al's experiment [9]. It concluded that the anisotropy of the speed of light $\Delta c/c$ is less than 1×10^{-17}. The experiment uses two laser emitters with two crossed orthogonal optical resonators that are implemented in a single block of fused silica instead of a single emitter and a splitter. For frequency comparison, split-off fractions of the two beams are overlapped on a photodiode to generate a beat note at a difference frequency [9]. In order to generate a beat note, the two beams should be combined in the same direction. Therefore, by the apparatus' frame the light speed at the beat note generator is common for the both beams even if respective light speeds in the two resonators are different. The reason why the Michelson-Morley like experiments should be incapable to detect the interference fringe shift is because the two beams are finally combined in a common direction for measurement of interaction. Because the direction is common for both beams, the respective light speeds by the apparatus' frame are equal to each other at the detector. Difference in the route of beam before being combined causes only difference in phase.

It was striking to be shown from the Supernova Cosmology Project (SCP) by Perlmutter et al that the expansion of the universe was accelerating in spite of the gravitational attraction of the universe [10]- [13]. The dark energy resulting from an unknown repulsive force is proposed. Many attempts to rationale the new force are under investigation but no one has succeeded in a perfect form. There is another possibility to interpret the observed data from the SCP, that is, the light speed may be varied by the time passing since the light was emitted. According to the 4DS Model, light speed varies with space expansion due to the existence of the wave medium.

Light speed variation by the time was discussed in the reference [2]. In this article, the direct correlation between the light propagated distance and the redshift, which are the real observables in the SCP, was given. As a factor affecting the light speed, the electromagnetic interaction factor causing scattering by substances (f_{EM}) is newly introduced in addition to the formerly proposed energy density factor of the wave medium (f_D) for determining light speed. f_D and f_{EM} are expressed by eqs. (7) and (8), respectively. The light speed by our observed Time is expressed by eq. (11). The redshift is accordingly shown as eq. (15) or (16). Except for the very early time since the Big Bang, that is, in the range $T_{ER} \geq 0.001$ or $T_{BR} \leq 0.999$ f_{EM} can be ignored. The redshift value can be approximated to be eq. (17).

If $T_{BR} \leq 0.999$ for a supernova, the light propagated distance can be expressed by redshift as eq. (25). The relative light propagated distance to that of $z = 0.05$ (RLD) is finally derivatized to eq. (26) shown in Fig. 3. The RLD graph in a dual logarithmic scale has shown an excellent fit to the observed data from the SCP for the Present Time of universe T_P being between 0.7 and 0.8 as shown in Fig. 4. The SCP data should be a supportive evidence for a possibility that the universe has been expanding at a constant speed according to the 4DS Model.

I will highly appreciate other investigators to verify the relation of light propagated distance to redshift in more detail from observed data as well as any discussions and arguments.

References

[1] Nagao S 2009 Nature of tracing dimension, imaginary order of freedom and our observed time passing at constant speed *J. Phys.: Conf. Ser.* **174** 012069 doi: 10.1088/1742-6596/174/1/012069

[2] Nagao S 2011 Light speed and the expansion of the universe *J. Phys.: Conf. Ser.* **306** 012073 doi: 10.1088/1742-6596/306/1/012073

[3] Nagao S 2013 Acceleration factor for propagation of a stationary wave in its wave medium: movement of energy in the 3-D space *J. Phys.: Conf. Ser.* **442** 012067 doi: 10.1088/1742-6596/442/1/012067

[4] Cosmic microwave background radiation *Wikipedia* (accessed 20 July 2013) http://en.wikipedia.org/wiki/Cosmic_microwave_background_radiation

[5] Cosmology from Type Ia Supernovae *Supernova Cosmology Project* (accessed 10 January 2008) http://www-supernova.lbl.gov/public/papers/aasposter198dir/wwwposter2b.jpg

[6] Michelson A A 1881 *American Journal of Science* **22** 120

[7] Michelson A A and Morley E W 1887 *American Journal of Science* **34** 333

[8] Michelson-Morley experiment *Wikipedia* (accessed 28 August 2013) http://en.wikipedia.org/wiki/Michelson-Morley_experiment

[9] Herrmann S *et al* 2009 *Physical Review D* **80** 105011 (arXiv:1002.1284)

[10] Perlmutter S *et al* 1997 *Astrophys. J.* **483** 565

[11] Perlmutter S *et al* 1998 *Nature* **391** 51

[12] Perlmutter S *et al* 1999 *Astrophys. J.* **517** 565

[13] Schmidt B *et al* 1998 *Astrophys. J.* **507** 46

Journal of Physics: Conference Series **626** (2015) 012069 doi:10.1088/1742-6596/626/1/012069

Brain Dynamics, Chaos and Bessel Functions

W J Freeman[1], A Capolupo[2], R Kozma[3], A Olivares del Campo[4] and G Vitiello[2]

[1] Department of Molecular and Cell Biology, University of California Berkeley, CA 94720-3206 USA

[2] Dipartimento di Fisica "E.R.Caianiello" Universitá di Salerno and INFN Gruppo collegato di Salerno, Fisciano (SA) - 84084, Italy

[3] Department of Mathematics, University of Memphis, Memphis TN 38152, USA

[4] The Blackett Laboratory, Imperial College London, Prince Consort Road, London SW7 2BZ, UK

E-mail: vitiello@sa.infn.it

Abstract. By resorting to Freeman's observations showing that the distribution functions of impulse responses of cortex to sensory stimuli resemble Bessel functions, we study brain dynamics by considering the equivalence of spherical Bessel equation, in a given parametrization, to two oscillator equations, one damped and one amplified oscillator. The study of such a couple of equations, which are at the basis of the formulation of the dissipative many-body model, reveals the structure of the root loci of poles and zeros of solutions of Bessel equations, which are consistent with results obtained using ordinary differential equation techniques. We analyze stable and unstable limit cycles and consider thermodynamic features of brain functioning, which in this way may be described in terms of transitions between chaotic gas-like and ordered liquid-like behaviors. Nonlinearity dominates the dynamical critical transition regimes. Linear behavior, on the other hand, characterizes superpositions within self-organized neuronal domains in each dynamical phase. The formalism is consistent with the observed coexistence in circular causality of pulse density fields and wave density fields.

1. Introduction

The distribution functions of impulse responses of cortical evoked potentials has been analyzed in terms of ordinary differential equations (ODE) [1]. The present study starts with the observation that impulse responses resemble Bessel functions (see e.g. Figure 2.24a, Section 2, page 111 of ref. [1]). In addition, it is commonly assumed that pulse and wave modes coexist simultaneously in circular causality. The resemblance of functions representing neurological processes with Bessel functions is a motivation for us to investigate the equivalence of the spherical Bessel equation and a system of two equations, one for a damped harmonic oscillator, the other one for an amplified harmonic oscillator, provided a specific parametrization is used [2]. The analysis of the system of such equations, which are at the basis of the formulation of the dissipative many-body model [3]-[8], reveals the structure of the root loci of poles and zeros of solutions of Bessel equations. The results we find are consistent with the results obtained using ordinary differential equation techniques. Moreover, we find that consideration of wave modes cannot be omitted in favor of pulse mode, as often done in the study of the individual neurons. Although necessary, the knowledge of single neuron properties is not sufficient for the comprehension of global functional activity of the brain.

We study stable and unstable limit cycles and consider thermodynamic features of brain functions, which may be described in terms of transitions between chaotic gas-like and ordered liquid-like behaviors [7, 8, 9]. Nonlinearity dominates the dynamical critical transition regimes. Linear behavior instead characterizes superpositions within self-organized neuronal domains in each dynamical phase. The dissipative character of the brain system dynamics leads us, following refs. [3, 4, 5, 10], to consider collective properties in the many-body physics framework, focusing on the quantum degrees of freedom associated with the electrical dipole vibrational modes of water molecules and other molecules of biological interest. We stress, however, that neurons and glia cells are considered to be classical objects in the dissipative many-body model. Remarkably, Bessel equations are classical equations, which we thus understand as the macroscopic manifestation of the underlying quantum field theory dynamics [11].

2. Brain dynamics and Bessel equations

As already mentioned, laboratory observations in brief time windows of brain activity show [1] that pulses and wave modes coexist simultaneously in circular causality determining the dynamics of neuronal populations. Much attention is given in current neuroscience to the study of pulse modes. In this paper our attempt is to combine in one unified dynamical frame the continuum of the wave mode and the action potential point processes of pulse modes. Our analysis is based on the fact that impulse responses of cortex to electric shocks (evoked potentials) exhibit distribution functions resembling to Bessel functions. We will see that pulse/wave-modes combined in a dynamical circular causality emerge. Motivated thus by the experimental observations (see e.g. Figure 2.24a, Section 2, page 111 in [1]), we consider the the spherical Bessel equation of order n:

$$z^2\frac{d^2}{dz^2}M_n + 2z\frac{d}{dz}M_n + [z^2 - n(n+1)]M_n = 0 \, , \tag{1}$$

with integer $n = 0, 1, 2, ...$. Solutions of Eq. (1) are the first and second kind Bessel functions and their superposition called Hankel function [12].

Eq. (1) is invariant under the transformation $n \to -(n+1)$, so that M_n and $M_{-(n+1)}$ are solutions of the same equation. By following ref. [13], this degeneracy is removed by introducing the parametrization: $q_{n,l} = M_n \cdot (x_n)^{-l}$, $z = \epsilon_n x_n$ and $x_n = e^{-\frac{t}{\alpha_n}}$. Here ϵ_n and α_n are arbitrary parameters and t can be taken to be the time parameter. Then, by setting $n(n+1) = l(l+1)$, two different equations for $l = -(n+1)$ and $l = n$, are obtained: the symmetry under the transformation $n \to -(n+1)$ is broken [13]. These differential equations can be reduced to:

$$\ddot{\xi} + L\dot{\xi} + \omega_n^2(t)\,\xi \;=\; 0 \, , \tag{2}$$

$$\ddot{\zeta} - L\dot{\zeta} + \omega_n^2(t)\,\zeta \;=\; 0 \, . \tag{3}$$

The shorthand notation: $\xi \equiv q_{n,-(n+1)}$, $\zeta \equiv q_{n,n}$, $L \equiv \frac{2n+1}{\alpha_n}$ has been used. Dots represent derivatives with respect to time, L describes properties intrinsic to the system, such as biological, physiological or behavioral properties. The time-dependent frequencies appears to be $\omega_n(t) = \frac{\epsilon_n}{\alpha_n} e^{-\frac{L\,t}{2n+1}} = \omega_0\, e^{-\frac{t}{\alpha_n}}$, with $\omega_0 \equiv \frac{\epsilon_n}{\alpha_n} = k\,c \equiv \omega_{0,k}$; k is the wave-number $k = 2\,\pi/\lambda$. L can be expressed as $L_k = \frac{(2n+1)}{\epsilon_n} k\,c$. This is done in such a way that, by fixing ϵ_n, L_k and $\omega_{0,k}$ are independent of n (and of time). The transformation $n \to -(n+1)$ leads to solutions (corresponding to $M_{-(n+1)}$) which can be obtained by time-reversal $t \to -t$ in $\omega_n(t)$. The k index will be omitted for simplicity whenever no misunderstanding arises.

We remark that Eqs. (2) and (3) are nothing else than the couple of equations at the basis of the dissipative model of brain [3] with time-dependent frequency $\omega(t)$. One can reduce them to the single parametric oscillator equation [14]

$$\ddot{r}_n + \Omega_n^2(t) r_n = 0 \, , \tag{4}$$

where $\xi \equiv q_{n,-(n+1)} = \frac{1}{\sqrt{2}} r_n e^{-\frac{Lt}{2}}$, $\zeta \equiv q_{n,n} = \frac{1}{\sqrt{2}} r_n e^{\frac{Lt}{2}}$ and

$$\Omega_n(t) = \left[\omega_n^2(t) - \frac{L^2}{4}\right]^{\frac{1}{2}} \geq 0 \,, \tag{5}$$

is the common frequency of oscillation. Proof of the mechanism is the demonstration that the signals are in quadrature; the frequency and decay rate are always identical, and the phase of the inhibitory signal lags a quarter cycle behind the excitatory signal. Note that $\xi = q_{n,-(n+1)}$ and $\zeta = q_{n,n}$ are "harmonically conjugated". $\Omega(t)$, assumed to be real in order to avoid the overdamping regime, becomes time-independent in the limit $t \to \infty$ and it is fully specified by characteristic parameters of the system. We remark that temporal frequency modulation (smooth decrease of 10% of the mean, but rarely an increase) is observed in gamma bursts of oscillation in the ECoG lasting long enough (> 3 to 5 cycles). The impulse responses (averaged evoked potentials) has fixed time-invariant frequencies. This is equivalent to time t going to infinity.

Quantization of Eqs. (2) and (3) leads to the introduction of modes a_k and $\tilde{a}_k$, respectively. In the dissipative quantum model of brain [3], they represent molecular electric dipole vibrational modes called dipole wave quanta (DWQ). They are identified with the Nambu-Godstone (NG) boson quanta generated in the dynamical mechanism of the breakdown of the rotational symmetry of the electrical dipoles of the molecules of water and other biomolecules [3, 4, 6]. Through sensorial perception, the brain is permanently coupled to its environment (called its *Double* [3, 4, 6]). External stimuli induce the symmetry breakdown with the consequent dynamical process of condensation of NG boson modes in the system ground state. Since these are (quasi-)massless modes, they propagate through the whole system volume and are thus responsible for long range correlations among its molecular components (formation of ordered coherent patterns) [3, 4, 10, 11]. These correlations trigger and promote the ephaptic interaction among neurons implying the formation of the observed coherently oscillating neuronal amplitude modulated (AM) assemblies [1, 5, 6, 15].

In the dissipative model, the constrain $\mathcal{N}_{a_k} - \mathcal{N}_{\tilde{a}_k} = 0$, with $\mathcal{N}_{a_k}$ and $\mathcal{N}_{\tilde{a}_k}$ denoting the number of a_k and $\tilde{a}_k$ modes, respectively, is imposed on the NG boson condensation. It guaranties that the ground states has the lowest energy. However, the number of k-modes, for each k, in the ground state is not fixed. Infinitely many lowest energy states (called the memory states in the dissipative model) with distinct $\mathcal{N} \equiv \{\mathcal{N}_{a_k} = \mathcal{N}_{\tilde{a}_k}, \forall\, k\}$ are then possible. $\mathcal{N}$ is then referred to as the "order parameter".

3. Linearity and nonlinearity in brain activity

Laboratory observations (see in particular Section 2.5.3 in ref. [1]) have shown that linearity plays a crucial role in brain activity, so that linearity cannot be neglected in favor of nonlinearity, which also plays a crucial role, as we will see. Brain functional activity presents an intriguing interplay between linear and nonlinear dynamical regimes. Such an interplay depicts a scenario neither new nor surprising if compared to what normally observed in condensed matter physics. Unfortunately, in recent years neuroscience studies have mostly focused on fashionable nonlinear dynamical models, which, we repeat play a most important role in brain activity, but, nevertheless, are not sufficient for the comprehension of the brain functional activity. One aspect of the relevance of linearity has to do with the fact that the fitting functions describing oscillatory processes of assemblies of neurons are linear superpositions of the form [1])

$$v(t) = \sum_j^n A_j\, e^{-p_j\, t} \,. \tag{6}$$

Journal of Physics: Conference Series **626** (2015) 012069 doi:10.1088/1742-6596/626/1/012069

A_j, $p_j \in \mathbb{C}$. They have positive coefficients, i.e. $Re(p_j) \geq 0$ and $Im(p_j) \geq 0$. When p_j is real, A_j is also real. Complex p_j and A_j always come in conjugate pairs, yielding damping *sin* or *cos* terms, such as $\sin\big(Im(p_j)\,t + \varphi_j\big)e^{-Re(p_j)\,t}$, with φ_j a constant phase term in A_j. For example, $\sin(\Gamma_j\,t + \varphi_j)\,e^{-\gamma_j\,t}$ can be rewritten as $A_j e^{-p_j\,t} + A_j^*\,e^{-p_j^*\,t}$, where $A_j = e^{-i(\varphi_j - \frac{\pi}{2})}/2$, $p_j = \gamma_j + i\Gamma_j$ and similarly for a damping *cos* term. Note that $A_j \equiv A_{k_j}$ and $p_j \equiv p_{k_j}$, i.e. A_j and p_j depend on the wave-number k_j, for each j, which in our field theory frame varies over a continuum set of values. Whenever no misunderstanding arises, the simplified notation A_j and p_j will be used. Eq. (6) has the Laplace transform

$$V(s) = \frac{\prod_i^m (s + z_i)}{\prod_j^n (s + p_j)}\,, \tag{7}$$

The zeros are at z_i and the poles are lying in the left-hand side of the complex s-plane.

Each term in Eq. (6) satisfies Eq. (2) and corresponds to a DWQ a_{k_j} mode. The modeled output potential is thus described by a linear superposition of a_{k_j} modes, representing solutions of the n-order Bessel equation. The meaning of this is that a particular stimulus will excite an $\mathcal{N}$-set, to which different $\mathcal{N}_{a_k}$ components belong [3].

We remark that, due to mathematical completeness, we need to "double" the degrees of freedom of the system by introducing the basis functions $e^{+p_j\,t}$, so to complete the basis $\{e^{-p_j\,t},\, e^{+p_j t}\}$. Thus, one cannot avoid to consider also the basis element $e^{+p_j t}$.

Use of $A_j(t) = A_j\,e^{-p_j\,t}$, $A_j \neq 0$, in Eq. (2) eventually gives:

$$\gamma_{k_j} = \frac{L_{k_j}}{2}\,, \qquad \Gamma_{k_j} = \Big[\omega^2_{n,k_j}(t) - \gamma^2_{k_j}\Big]^{\frac{1}{2}} \equiv \Omega_{k_j,n}(t)\,, \tag{8}$$

where the k subscript notation has been restored. The requirement of doubling the degrees of freedom, i.e. introducing terms with $e^{+p_j t}$ basis as solutions of Eq. (3), the same equations as in (8) are obtained. The same common frequency $\Omega_{k_j,n}(t)$ thus establishes for each k-mode and a particular n the resonating link between the damped/amplified systems.

We observe that in each of the damped/amplified oscillators in Eq. (2) and Eq. (3) time-reversal symmetry is broken due to their dissipative/amplified character. The "arrow of time", in each of the damped and amplified oscillators, thus emerges as a distinctive dynamical feature of the dissipative quantum model. It can be related to the "thermodynamical" and the "psychological" arrows of time. Details on this point may be found in refs. [16, 17].

In closing this Section, we remark that the linearity of Eqs. (2) and (3) allows us to consider the summations in the fitting functions with $\{e^{-p_j\,t},\, e^{+p_j t}\}$ basis. Linearity has originally motivated the analysis in terms of coupled ODEs [1, 19], which results are confirmed by the present analysis in terms of Bessel functions. However, our modeling also implies nonlinearity since the damped/amplified oscillators are in fact nonlinearly coupled in the coherent state they generate at quantum level, where they appear as an entangled two modes system (see [3, 6] for details). On the other hand, the root loci studied in [1, 18] can be also analyzed in the present treatment (cf. next Section). The study of the root loci reveals that the biophysical nonlinearity at the microscopic level allows that phase transition resembling a Hopf bifurcation are energetically possible.

Moreover, we observe that the frequency $\omega_n(t)$ reaches the time independent value ω_0 for higher order n of the Bessel equation, which is the limit of energy conservation, corresponding in turn to the maximal coupling between the damped and the amplified oscillators, forming in that limit a closed system. This is consistent with observations in the limit of stationarity, where most cortical impulse responses (evoked potentials from single shock electrical stimulation) last only 1.5 to 2.5 cycles, those few lasting longer have fixed stationary frequency. Ergodicity has to be also assumed in order to use time ensemble averaging to estimate pulse density in the linear range, which can be done also in the present case. We thus see that linearity, stationarity and ergodicity are compatible with Bessel functions in the dissipative quantum model.

In conclusion, the complex interplay of linearity and nonlinearity and the coexistence of pulse modes and wave modes are features of brain functional activity which manifest in a very rich phenomenology.

4. Limit cycle

In this Section we consider the limit cycle occurrence as one of the examples discussed in [2], which, together with [1], we will closely follow in our presentation. We reproduce the results obtained in [1] thus obtaining explicit connection with the analysis of the present paper.

From Section 6, Page 380 in ref. [1] one can derive that the dynamics is determined by the rules:

(1) If the roots all lie in the left half of the s-plane, the nonlinear KII set has a stable equilibrium to which the active states return when input is terminated.

(2) If complex conjugate poles lie on root loci in the right of the s-plane, the poles move on the root loci in the direction of decreasing K to the $i\omega$ axis and stay there, and the KII set has a stable limit cycle.

(3) If two conjugate root loci cross the $i\omega$ axis from the left half into the right half of the s-plane with decreasing K_n, the KII set has an unstable limit cycle.

Time evolution carrying through left and right hand side of the locus plots, i.e. damped and amplified solutions, appears to be deeply related with the chaotic dynamics that is observed in the brain [1, 20].

Let $D(s)$ denote the transfer function for a KIII set output from a KII set with negative feedback from a KI set. It is given by

$$D(s) = \frac{C(s)}{1 + K_e C(s) P(s)} \,. \tag{9}$$

$C(s) = (6.25 \times 10^6)/(s + i250)(s - i250)(s + 100)$ and $P(s) = 100/s(s + 100)$ are, respectively, the KII and KI set transfer functions in a stable limit cycle state.

The poles of Eq. (9) change as K_e changes:

$$D(s) = \frac{6.25 \times 10^6 s(s + 100)}{(s + p_1)(s + p_2)(s + p_3)(s + p_4)(s + p_5)} \,. \tag{10}$$

The p_n poles can be real or complex. Figure (1) shows the root locus plot of $D(s)$. The larger poles cross the imaginary axis from the left to the right as K_e decreases, which signals an unstable limit cycle. The limit cycle will be stable at frequencies where the poles are purely imaginary (≈ 160 rad/sec).

As observed in Section 2, decaying solutions and amplifying ones arise in our framework when one removes the degeneracy between the n and $-(n+1)$ solutions of the Bessel equation. The crossing of the imaginary axis thus means the transition between these two sets of Bessel equation solutions as the gain increases (from decaying solutions to amplifying ones).

When the rate of oscillation of the pulses ($\Gamma_{k_j,n}$) changes with time at a fixed frequency, the stable limit cycle is obtained. This is consistent with our analysis, since identifying the limit cycle frequency with $\Omega(t)$, which, as observed above, links the damped and amplified solutions, the stable equilibrium state (constant pulse frequency) corresponds to fully coupled solutions (obtained for $n \to \infty$, with $\Omega(t)$ becoming indeed time independent in such a limit).

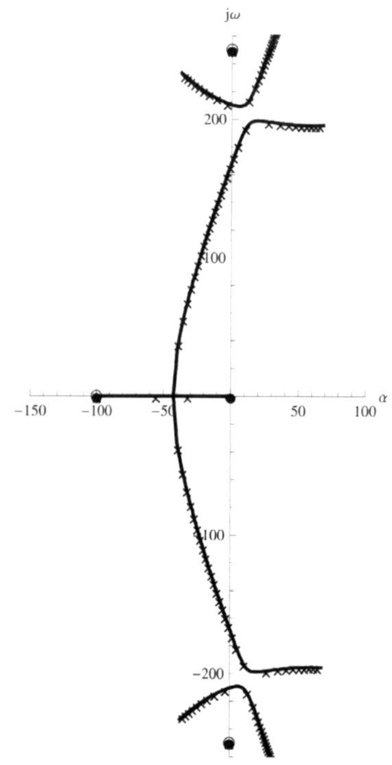

Figure 1. Root locus plot of Equation (9) with poles (×) and zeros (•).

5. Conclusion

We have discussed the laboratory observations showing that impulse responses of cortex to sensory stimuli exhibit distribution functions resembling to Bessel functions and that, in the brief time windows of brain activity accessible to observations, pulses and wave modes coexist simultaneously in circular causality determining the dynamics of neuronal populations. The small-signal linear dynamic range allows the possibility to decompose cortical impulse responses by fitting to them the sums of linear basis functions. These have been found to be solutions of ODE [1, 19] and in the present paper have been shown to be solutions of the spherical Bessel equation of order n. This equation turns out to be equivalent to the couple of damped/amplified oscillator equations at the basis of the formulation of the dissipative quantum model of brain [3, 4, 6, 13]. The harmonic conjugated solutions, represented in terms of the quantum modes a_k and their conjugate $\tilde{a}_k$, are expressed in terms of the common frequency of oscillation $\Omega(t)$ and create a mirror image on the imaginary axis in the s-space in the Laplace transform representation. The dissipative many-body model offers thus a possibility in the study of pulse-wave duality and in combining the complexity of nonlinear interactions with linear, stationary, ergodic solutions at the observational level. Our results are consistent with the ODE treatment [1, 18, 19] and the connection with such a formalism is made explicit by considering specific applications in [2]. As an example, the occurrence of the limit cycles has been reported in the present paper.

In ref. [2] it has been discussed how the coefficients that can be determined from the plot of the poles of the fitting functions in s-space may be related to the lifetime of a specific neuronal

Journal of Physics: Conference Series **626** (2015) 012069 doi:10.1088/1742-6596/626/1/012069

("memory") process. Here we do not discuss such issues. Rather, we observe that couples of damped/amplified oscillator solutions, $\{n,\, k_j\}$ and $\{n,\, \tilde{k}_j\}$ (the notation $\tilde{k}_j$ refers to the $\tilde{a}_{k_j}$), may be linked through the values of the parameters defining the system $\{L_{k_j}, \omega_{k_j,n}\}$. With infinitely many k modes, one can always pair them up adding modes in pairs in s-space. If one k_j mode, for a given j, has a shorter lifetime than the other one in the couple, the linked system becomes unpaired ("widowed"). However, since k_j and $\tilde{k}_j$, for each j, belong to continuum sets, new links (new couples) can be created, thus recovering the lost links. Due to the continuum, we may also have couples with different i and j, for any i and j, with continuously changeable i and j. In the observed neuronal assemblies [5, 6, 15], due to the continuous interchange of partners in the couples, no neuron is "wired" to someone else. The neuronal dynamical assembly of correlated neurons is made of constantly changing, unwired neuronal connections, described by coherent states in the dissipative quantum model of brain [3]-[6].

References

[1] Freeman W J 1975 (Reprinted 2004) *Mass Action in the Nervous System* (New York: Academic Press)
[2] Freeman W J, Capolupo A, Kozma R, Olivares del Campo A and Vitiello G 2015 Bessel functions in mass action modeling of memories and remembrances *Preprint*
[3] Vitiello G 1995 *Int. J. Mod. Phys.* **B9** 973
[4] Vitiello G 2001 *My Double Unveiled* (Amsterdam: John Benjamins Publ. Co.)
[5] Freeman W J and Vitiello G 2006 *Physics of Life Reviews* **3** 93
[6] Freeman W J and Vitiello G 2008*J. Phys. A: Math. Theor.* **41** 304042
[7] Freeman W J 2015 *Current Opinion in Neurobiology: Brain Rhythms and Neurodynamics* **31** G Buzsáki and W J Freeman (Eds.) (Amsterdam, NL: Elsevier) scheduled for April 2015
[8] Vitiello G 2014 *Current Opinion in Neurobiology* CONEUR1426 PII: S0959-4388(14)00145-7 Doi: 10.1016/j.conb.2014.07.017
[9] Kozma R, Puljic M and Freeman, W J 2014 in D Plenz (Ed.) *Criticality in Neural Systems* (N.Y: John Wiley & Sons) p.153 (*Preprint* arXiv:1206.1108)
[10] Ricciardi L M and Umezawa H 1967 *Kibernetik* **4**, 44 Reprint in: Globus G G, Pribram K H and Vitiello G (Eds.) *Brain and being* (Amsterdam: John Benjamins, 2004) p 255
[11] Blasone M, Jizba P and Vitiello G 2011 *Quantum Field theory and its macroscopic manifestations* (London: Imperial College Press)
[12] Arfken G B and Weber H J 2013 *Mathematical Methods for Physicists* (Waltham, MA: Academic Press)
[13] Alfinito E and Vitiello G 2000 *Int. J. Mod. Phys.* **14** 853
[14] Celeghini E, Rasetti M and Vitiello G 1992 *Ann. Phys. (N.Y.)* **215** 156
[15] Freeman W J and Quian Quiroga R 2013 *Imaging brain function with EEG* (New York: Springer)
[16] Alfinito E and Vitiello G 1999 *Phys. Lett.* **A252** 5
[17] Alfinito E, Mańka R and Vitiello G 2000 *Class. Quant. Grav.* **17** 93
[18] Ilin R and Kozma R 2006 *Phys. Lett.* A **360** 66
[19] Kozma R and Freeman W J 2001 *Int. J. Bifurcation and Chaos* **11(6)** 1607
[20] Vitiello G 2004 *Int. J. Mod. Phys.* **18** 785

Journal of Physics: Conference Series **626** (2015) 012070 doi:10.1088/1742-6596/626/1/012070

Ellipsoidal nested sampling, expression of the model uncertainty and measurement

C Palmisano[1], G Mana[2] and G Gervino[1,3]

[1] Dipartimento di Fisica, Università di Torino, Via P. Giuria, 1 - 10125 Torino, Italy
[2] INRIM - Istituto Nazionale di Ricerca Metrologica, Str. delle Cacce, 91 - 10135 Torino, Italy
[3] Istituto Nazionale di Fisica Nucleare (INFN), Sezione di Torino, Via P. Giuria, 1 - 10125 Torino, Italy

E-mail: palmisan@to.infn.it, g.mana@inrim.it, gervino@to.infn.it

Abstract. The measurand value, the conclusions, and the decisions inferred from measurements may depend on the models used to explain and to analyze the results. In this paper, the problems of identifying the most appropriate model and of assessing the model contribution to the uncertainty are formulated and solved in terms of Bayesian model selection and model averaging. As computational cost of this approach increases with the dimensionality of the problem, a numerical strategy, based on multimodal ellipsoidal nested sampling, to integrate over the nuisance parameters and to compute the measurand post-data distribution is outlined. In order to illustrate the numerical strategy, by use of *MATHEMATICA* an elementary example concerning a bimodal, two-dimensional distribution has also been studied.

1. Introduction

Use of Bayesian methods as ways to incorporate the model uncertainty into the data analysis and the uncertainty budget is often computationally expensive, but advances in computing technology have allowed to take these methods into account. Consequently, the choice between two or more hypotheses, for example when we have to the fit the shape of an experimental curve, can be made by considering different models, each indexed by one or more parameter, where the Bayesian model selection and model averaging provides the probabilistic framework to simultaneously treat both the model and data uncertainties.

Let $\boldsymbol{x}$ the list measurement data, $(A, B, \ldots, M, \ldots)$ the list of different possible models, θ the measurand, and $(\theta_A, \theta_B, \ldots, \theta_M \ldots)$ the relevant nuisance parameters. The joint distribution of the data, measurand, parameters, and model can be written in terms of conditional probabilities as

$$P(\boldsymbol{x}, \theta, \theta_M, M) = L(\theta, \theta_M, M|\boldsymbol{x})\pi(\theta, \theta_M|M)\Pi(M), \tag{1}$$

where we have introduced the likelihood $L(\theta, \theta_M, M|\boldsymbol{x}) = P(\boldsymbol{x}|\theta, \theta_M, M)$, the conditional prior probability distribution of the measurand and parameters $\pi(\theta, \theta_M|M)$, and the prior probability of the model $\Pi(M)$. By conditioning the left hand side of Eq. (1) on the data and the model, and applying the chain rule, we can write (1) as

$$P(\theta, \theta_M|\boldsymbol{x}, M) = \frac{L(\theta, \theta_M, M|\boldsymbol{x})\pi(\theta, \theta_M|M)}{Z(\boldsymbol{x}|M)}, \tag{2}$$

where we introduced the so-called "evidence" $Z(\boldsymbol{x}|M) = P(\boldsymbol{x}|M)$. It is the normalizing factor of the numerator in (2), i.e. it satisfies the relationship

$$Z(\boldsymbol{x}|M) = \int_{\Theta \oplus \Theta_M} L(\theta, \theta_M, M|\boldsymbol{x}) \pi(\theta, \theta_M|M) \, \mathrm{d}\theta \, \mathrm{d}\theta_M \,. \tag{3}$$

By applying the Bayes theorem to $P(\boldsymbol{x}, M)$ we obtain the discrete probability $\mathrm{Prob}(M|\boldsymbol{x})$ for the model M provided by the data

$$\mathrm{Prob}(M|\boldsymbol{x}) = \frac{Z(\boldsymbol{x}|M)\Pi(M)}{P(\boldsymbol{x})}, \tag{4}$$

which, by use of the law of total probability, can be rewritten in terms of evidence $Z(\boldsymbol{x}|M)$ of each model as

$$\mathrm{Prob}(M|\boldsymbol{x}) = \frac{Z(\boldsymbol{x}|M)\Pi(M)}{\sum_M Z(\boldsymbol{x}|M)\Pi(M)} \,. \tag{5}$$

Eq. (5) gives the probability of a model given the observed data, and a simple way to select a model is to choose the most probable. However, when the relevant probabilities are approximately equal and no single model stands out, it is necessary to average over the considered models.

2. Multimodal, ellipsoidal, nested sampling

Since the evaluation of evidence Z in (3) becomes impractical when the integration space has $20 \div 30$ dimensions, and, as in the most practical case, the likelihood is different from zero in a small fraction of the integration domain, we have investigated a nested sampling technique relating the likelihood values to the prior volume [1, 2].

Firstly, p likelihood samples $L_{\theta_1}, \ldots, L_{\theta_p}$ are sampled in $\Theta \oplus \Theta_M$ according to $\pi(\theta, \theta_M|M)$. Next, the smallest, indicated as L_1, is removed and replaced by a new sample, L_{new}, subject to the constraint $L_{\mathrm{new}} > L_1$. The prior volume enclosed by the iso-likelihood surface $L = L_1$ is estimated as $p/(p+1)$, where $p/(p+1)$ is the mean value of the largest of p uniform samples in $[0, 1]$, the total volume of $\Theta \oplus \Theta_M$ being normalised to 1. The discharge of the lowest likelihood L_n, sampling of a replacement constrained to $L_{\mathrm{new}} > L_n$, and shrinking of the prior volume of the associated iso-likelihood surface to $V_n = p^n/(p+1)^n$ are repeated until some stopping criterion is satisfied, for example when the contribution to (3) of the surviving likelihood samples, i.e. $L_{\mathrm{max}} p^n/(p+1)^n$, where L_{max} is the maximum sample, is less than some pre-defined value.

By using the sequence of the discarded likelihoods $0 < L_1 < L_2 \ldots < L_N$, and the differences $\Delta V_n = V_{n-1} - V_n$ of the associated prior volumes $V_0 = 1 > V_1 > V_2 > \ldots V_N > 0$, Eq. (3) can be approximated as (rectangle method)

$$Z(\boldsymbol{x}|M) \approx \sum_{n=1}^{N} L_n \Delta V_n + L_{\mathrm{max}} \left(\frac{p}{p+1} \right)^N . \tag{6}$$

The challenge in implementing (6) is sampling within the iso-likelihood surfaces $L > L_n$. As the Monte Carlo Markov Chain algorithm may be inefficient, improvements have been proposed in [3]-[6], which we are presently investigating for application to metrology.

Ellipsoidal sampling [3] replaces the iso-likelihood surface $L = L_n$ by a hyper-ellipsoid given by the covariance matrix of the living samples and centered in their mean value, and the $L_{\mathrm{new}} > L_n$ is sampled within the intersection of the domain of integration and this hyper-ellipsoid [7].

When the integrand in (3) is a multimodal function, the hyper-ellipsoid will be often bonded to the maxima of the integrand and the sampling will result in an unacceptable decrease in the acceptance rate of L_{new}.

Journal of Physics: Conference Series **626** (2015) 012070 doi:10.1088/1742-6596/626/1/012070

In this case, a possible strategy is partitioning the set of all living points in clusters, and then enclose each cluster in a "small" hyper-ellipsoid, centered in the mean value of the cluster points and defined by the covariance matrix calculated with the cluster points.

In [4] is described a method to avoid hyper-ellipsoids to overlap, but we will consider a method to manage the overlapping according to [5, 6, 8]. At each step of the method the set of all living points is partitioned in clusters. If the number of hyper-ellipsoids is greater than 1, a single hyper-ellipsoid is sampled with probability proportional to its volume; then a $L_{\text{new}} > L_n$ is sampled and accepted with probability $1/(\text{number of hyper-ellipsoids containing } L_{\text{new}})$. If the point is rejected, a new sampling of a hyper-ellipsoid takes place.

2.1. Bimodal example in $\mathbb{R}^2$

By use of *MATHEMATICA* (Wolfram Inc.), we applied the ellipsoidal nested sampling to this elementary integral

$$I = V_{\text{D}} \int_{30}^{45} \left(\int_{31}^{44} f(x,y)\, \mathrm{d}y \right) \frac{1}{V_{\text{D}}}\, \mathrm{d}x, \tag{7}$$

where $V_{\text{D}} = (45-30) \times (31-44) = 195$ is the volume of the integration domain and (see Fig. 1)

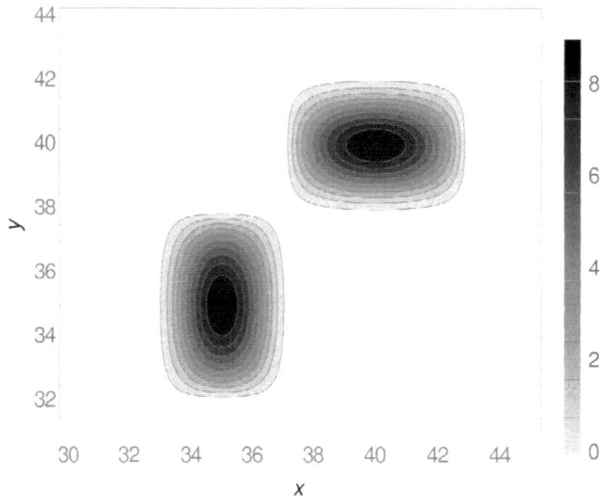

Figure 1. Contour plot of the function $f(x, y)$ defined in Eq. 8.

$$\begin{aligned} f(x,y) = {} & \cos^2\Big(\frac{2\pi}{10}(x-35)\Big)\texttt{UnitBox}\Big(\frac{x-35}{5}\Big)\big(-(y-35)^2+9\big)\texttt{UnitBox}\Big(\frac{y-35}{6}\Big) \\ & + \big(-(x-40)^2+9\big)\texttt{UnitBox}\Big(\frac{x-40}{6}\Big)\cos^2\Big(\frac{2\pi}{10}(y-40)\Big)\texttt{UnitBox}\Big(\frac{y-40}{5}\Big). \end{aligned} \tag{8}$$

The bimodal function $f(x, y)$, which is centered in $(x, y) = (35, 35)$ and $(x, y) = (40, 40)$, plays the role of a likelihood; $1/V_{\text{D}}$ is the prior distribution and $\texttt{UnitBox}(x)$ is by definition equal to 1 for $|x| \leqslant 1/2$ and 0 elsewhere. Our goal is to reobtain the value of the analytical calculation, that is $I = 180$. Fig. 2 shows an ellipsoid (red dashed curve) calculated after sampling uniformly 200 points from the domain of integration. The external ellipsoid, red continuous curve corresponding to $h\, k_1$, where $h = 1.1$, enlarge the sampling volume and is useful to better study the contour lines (iso-likelihood surfaces) of the integrand. It is also shown the minimum point of the integrand (green large disk). Strictly speaking, the integrand

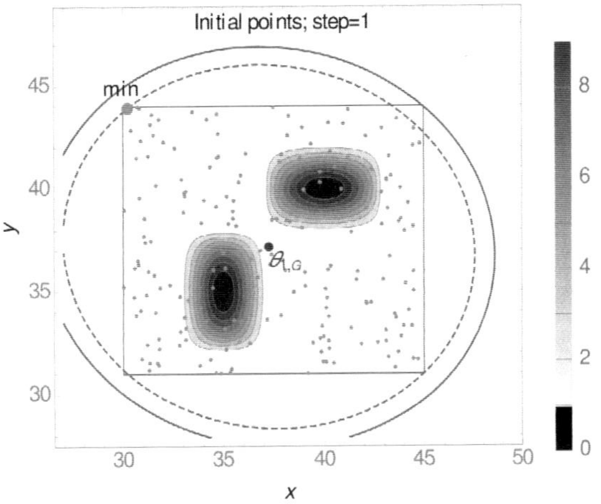

Figure 2. Contour plot of the function $f(x, y) + l(x, y)$; are also shown 200 sampled points (green), a couple of ellipsoids (red curves, corresponding to k_1 and $h\, k_1$, with $h = 1.1$) containing the sampled points, the mean value $\theta_{1,G}$ (blue) and the minimum point (green disk).

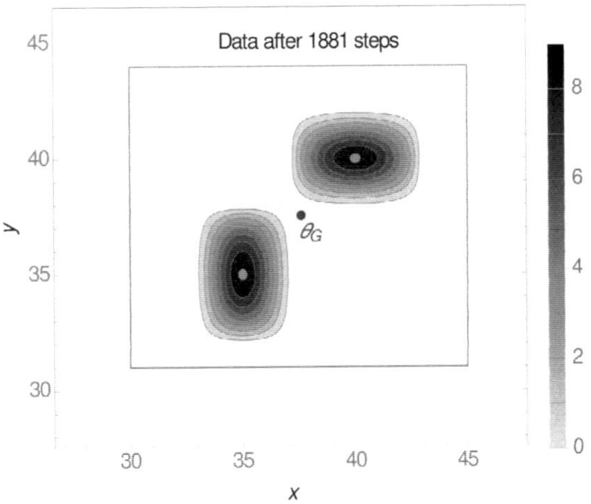

Figure 3. Contour plot of the function $f(x, y) + l(x, y)$; after 1881 steps the 200 living points (green) shrunk around the maximum points of $f(x, y)$. The internal rectangle indicates the integration domain $[30, 45] \times [31, 44]$.

should be equal to zero outside the support of $f(x, y)$, but in order to implement the nested sampling it is necessary to add to $f(x, y)$ a little appropriate function $l(x, y)$. In our example with function (8), after 1881 steps, the set of the 200 living points is totally shrunk in the maximum points of the integrand $f(x, y)$, see Figs. 3; moreover, a file containing a list of 1881 "minimal likelihood" values L_n is available to calculate the evidence Z according to Eq. (6). With the trapezoid rule, the final result is integral= 148.7 with uncertainty= 14.3. The uncertainty has been assessed by use of one of the equivalent formulae suggested in [9]. The set of the lowest likelihood points is shown in Fig. 4 and the list of the number of clusters at each step is shown in Fig. 5 A realization of 100 independent calculations of the integral with 200 points, that is 100

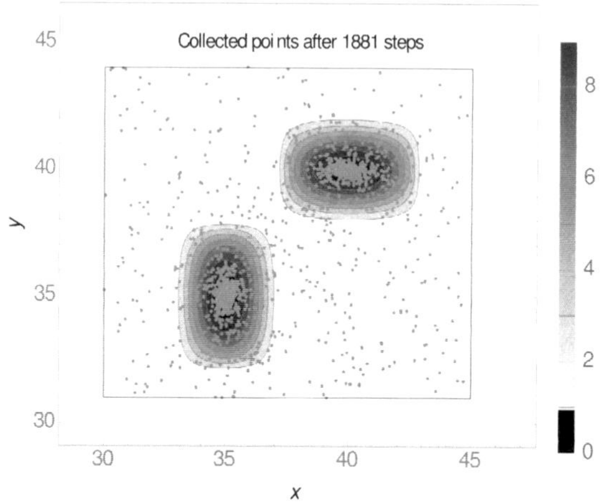

Figure 4. Set of the lowest likelihood points (green), corresponding to the discharged likelihood values L_n by which to calculate the integral in Eq. (6). The internal rectangle indicates the integration domain $[30, 45] \times [31, 44]$.

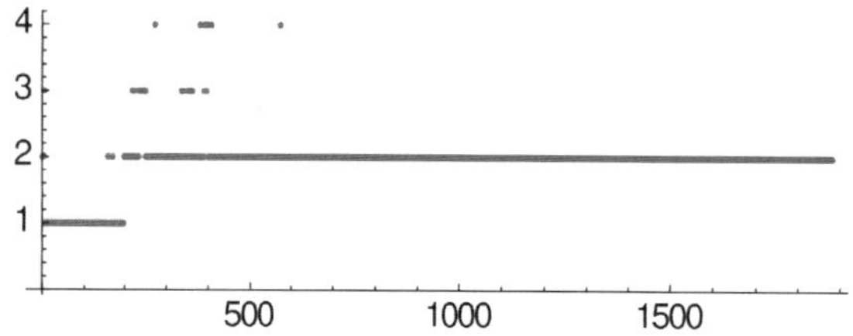

Figure 5. Number of clusters at each step.

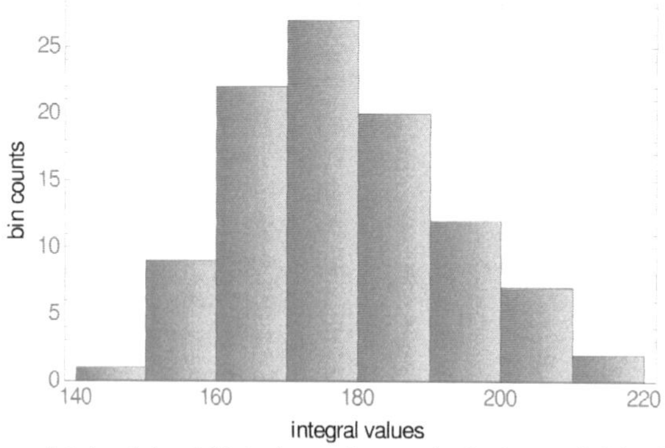

Figure 6. Histogram obtained by 100 independent calculations of (7). Mean value = 178.1, standard deviation of the mean = 1.5, exact value = 180.

different sets of 200 initial points, has generated the histogram in Fig. 6 having a mean value = 178.1 and a standard deviation of the mean = 1.5, the trapezoid rule having been exploited again. Obviously, the mean value of the histogram is very close to the analytical calculation, while the value obtained by a single calculation can be very different from the real value.

3. Conclusion

The role of the Bayesian approach to the expression of the model uncertainty in measurements consists of the quantification of both the sampling and model contributions to the post-data uncertainty. The multimodal nested sampling has been fruitful in astrophysics, but at authors knowledge has not yet been exploited in metrology. In any case, to develop an algorithm and the relevant software to tackle numerical integration of multimodal functions over high-dimensional spaces can be fruitful from a more general point of view.

Acknowledgments

This work was jointly funded by the European Metrology Research Programme (EMRP) participating countries within the European Association of National Metrology Institutes (EURAMET) and the European Union. (JRP: NEW04-Uncertainty).

References

[1] Skilling J 2004 *AIP Conf. Proc.* vol 735, ed R Fischer, R Preuss and U von Toussaint pp 395-405
[2] Skilling J 2006 *Bayesian Analysis* **1** 833
[3] Mukherjee P, Parkinson D and Liddle A R 2006 *The Astr. J.* **638** L51
[4] Shaw R, Bridges M and Hobson M P 2007 *Mon. Not. R. Astron. Soc.* **378** 1365
[5] Feroz F and Hobson M P 2008 *Mon. Not. R. Astron. Soc.* **384** 449
[6] Feroz F, Hobson M P and Bridges M 2009 *Mon. Not. R. Astron. Soc.* **398** 1601
[7] Allison R and Dunkley J 2014 *Mon. Not. R. Astron. Soc.* **437** 3918
[8] Feroz F, Hobson M P, Cameron E and Pettitt A N 2013 Importance Nested Sampling and the MultiNest Algorithm *Preprint* astro-ph/1306.2144v2
[9] Keeton C R 2011 *Mon. Not. R. Astron. Soc.* **414** 1418

7th International Workshop DICE2014 Spacetime – Matter – Quantum Mechanics IOP Publishing
Journal of Physics: Conference Series **626** (2015) 012071 doi:10.1088/1742-6596/626/1/012071

General relativistic quantum theories

Foundations. Expanding Universes

Claudio Parmeggiani

University of Milano-Bicocca - Italy (UNIMIB)

E-mail: c.parmeggiani@campus.unimib.it

Abstract. Here the space-time is represented by the usual, four-dimensional manifold and at every space-time point is assigned an infinite-dimensional Hilbert space, seat of a (local) quantum description: states, probabilities and expectations. On the space-time manifold is assigned a metric tensor and it is assumed that the quantum fields commutations relations do not only depend on the metric tensor but also on its Ricci tensor: this is a fundamental postulate. This assumption has many relevant consequences: the theory is regularized; the commutators and the propagators are well defined functions and, applying the theory to electroweak interactions, we can obtain a finite and discrete specter of leptons masses.

1. Introduction and summary

We shall discuss (the foundations of) a General Relativistic Quantum Theory: now the Hilbert space of the quantum description is replaced by a *fiber bundle* (here called *quantum bundle*) based on a four-dimensional Riemannian manifold (the space-time); the fibers are complex, infinite-dimensional, separable Hilbert spaces; on every fiber (that is at every space-time point) are defined quantum states (vectors or rays), physical quantities (operators), probabilities and expectation-values (Sections 2, 3).

At every space-time point it is also possible to define, in the usual way, *creation* and *annihilation* operators; they are assumed to depend on the value of the *metric tensor* (of the space-time Riemannian manifold) *and* of the value of its *Ricci tensor*: this is the *main hypothesis*. Now the *quantum fields* and the *quantum Lagrangian* are, fixed a bundle frame, well defined operator-valued functions; the Pauli-Jordan distributions and the Feynman propagators are true, smooth functions (metric and Ricci tensors dependent): the theory is "physically" *regularized*, there are no more divergences. The standard quantum field theories are formally recovered for a "null" Ricci tensor, in particular for a "constant", special relativistic metric tensor; but, in fact, for an isotropic, expanding or contracting metric (Friedmann, Lemaitre), the theory assumes a very different shape. Quite interesting is the case of an "exponentially" expanding space-time metric. (Sections 4, 5).

The metric-dependent modifications of the commutation relations not only regularize the theory but, as a consequence, the values of the *leptons masses* (and the number of *leptons generations*) are, by some supplementary assumptions, theoretically predictable. (In principle also the masses of strong interacting particles are predictable). In fact, imposing the equality of the space-time dependent *self-mass* (logarithmically diverging, when the Ricci tensor goes to zero or when it is simply proportional to the metric tensor) and of a space-time dependent

Journal of Physics: Conference Series **626** (2015) 012071 doi:10.1088/1742-6596/626/1/012071

counter term, we arrive at a second order partial differential equation; this equation links the d'Alembertian of some classical *scalar field* to the self-mass term; the *finite* set of self-values of this equation are related to the leptons masses. The counter term is, eventually a remnant of some quantum field which "breaks the symmetries"; note that here the vacuum expectation-values are space-time dependents (Sections 6, 7).

2. Quantum bundles

A *quantum bundle* $(\mathcal{H}, X, \pi)$ (or $\mathcal{H}_X$) is a vector bundle (see, for example, Bourbaki [2], Dieudonné [3], Chapters XVI-XX, Lang [8]) where the base space X is a 4-dimensional Riemannian manifold (the space-time) and the fibers at $x \in X$, $\mathcal{H}_x$ are infinite-dimensional, complex, separable Hilbert spaces, all isomorphic to a typical fiber $\mathcal{H}_0$, consistently to the requirement of *space-time homogeneity*. Alternatively we can assign on the total space $\mathcal{H}$ an equivalence relation $\mathcal{R}$; then $X := \mathcal{H}/\mathcal{R}$ and $h\mathcal{R}k$ if and only if $\pi(h) = \pi(k)$ $(h, k \in \mathcal{H})$; the relation $\mathcal{R}$ can be thought as "at the same place, at the same time". A bundle (moving) *frame* is an isomorphism, F_x, $\mathcal{H}_x \mapsto \mathcal{H}_0$; if s is a section of $\mathcal{H}_X$, $F_x \circ s$ is an ordinary map from X to $\mathcal{H}_0$.

The quantum bundle is here built starting from a (nuclear) vector bundle $\mathcal{E}_X$ and assigning, for every $x \in X$, a bilinear, separately continuos, symmetric and positive form k_x on $\mathcal{E}_x$ (a metric along the fibers). The map:

$$(f, g) \in \mathcal{E}_x \times \mathcal{E}_x \mapsto k_x(\overline{f}, g) = \overline{k_x(f, \overline{g})} \tag{1}$$

define, on $\mathcal{E}_x$, a pre-Hilbert structure ($f \mapsto \overline{f}$ is a conjugate-linear involution); its completion can be identified to $\mathcal{H}_x$ (see Schwartz [11] or Gelfand [5], Chapter I, for all mathematical details).

If the vectors Φ and Ψ belong to $\mathcal{H}_x$, we shall denote by $\langle\Phi, \Psi\rangle_{k,x}$ (or simply by $\langle\Phi, \Psi\rangle_x$) their *inner product*, at $x \in X$; $\|\Phi\|_x$ $(:= \sqrt{\langle\Phi, \Phi\rangle_x})$ is the *norm* of Φ. An explicit expression of the inner product, dependent on the space-time metric g and of its Ricci tensor $R^{(g)}$, will be introduced in the following Sections.

The *physical states* at x are now represented by (not null) vectors of $\mathcal{H}_x$. The (bounded) *physical quantities* are represented by elements of $\mathcal{A}_x := Lin(\mathcal{H}_x)$. The *average value* of the physical quantity represented by $\xi \in \mathcal{A}_x$, the system being in the state Φ, at $x \in X$, can consequently be written as:

$$\frac{\langle\Phi, \xi.\Phi\rangle_x}{\langle\Phi, \Phi\rangle_x}, \tag{2}$$

a generalization of the Born formula.

3. The base space (space-time)

We shall assume that the space-time (the base space of the quantum bundle) is a 4-dimensional, globally- and time-oriented, *Riemannian manifold* (X, g, τ, o). $\mathcal{T}_x$ and $\mathcal{T}'_x$ (its dual) are the tangent and the cotangent vector spaces, at $x \in X$; g_x is a $(+1, -1, -1, -1)$ quadratic form on $\mathcal{T}_x$; τ_x is a time-orientation function (that is a $\{+1, -1\}$ valued function, constant on each one of the two connected g_x-cones of $\mathcal{T}_x$); o_x is an orientation of $\mathcal{T}_x$ (and of $\mathcal{T}'_x$). Therefore g is a section of $\mathcal{T}'_X \otimes \mathcal{T}'_X$ and $\widehat{g}$ (the dual of g) is a section of $\mathcal{T}_X \otimes \mathcal{T}_X$; identifying quadratic forms to linear maps, $\widehat{g}.g = I_{\mathcal{T}}$, $g.\widehat{g} = I_{\mathcal{T}'}$.

$R^{(g)}$ is the Ricci tensor of g and $\widehat{R^{(g)}}$ is its dual, that is, if $p \in \mathcal{T}'_x$, $\widehat{R}_x(p,p) = R_x(\widehat{g}_x.p,\ \widehat{g}_x.p) = \sum_{0\leq\alpha\beta\gamma\delta\leq 3} R_{\gamma\delta}g^{\alpha\gamma}g^{\beta\delta}p_\alpha p_\beta$. In the following we shall need the Ricci tensor of the Friedmann-Lemaitre *metric* $g_x = dt_x^2 - \frac{a^2(t(x))}{b^2(r(x))}d\mathbf{r}_x^2$ (expressed in a non-static chart $(t, r_1, r_2, r_3, U) = (t, \mathbf{r}, U)$, $U \in X$):

$$R^{(g)} = 3\left(h'(t) + h^2(t)\right).dt^2 - \frac{a^2(t)}{b^2(r)}\left(h'(t) + 3h^2(t) + \frac{2K}{a^2(t)}\right).d\mathbf{r}^2. \tag{3}$$

Here $dt_x^2 = dt_x \otimes dt_x$, $d\mathbf{r}_x^2 = \sum_{1\leq j\leq 3} dr_x^j \otimes dr_x^j$; a is a differentiable, positive function and $a(0) = 1$; $h(t) = a'(t)/a(t)$; $r = \sqrt{(r^1)^2 + (r^2)^2 + (r^3)^2}$, $b(r) = 1 + \frac{K}{4}r^2$, K $(-\infty < K < +\infty)$ is the spatial, constant, sectional curvature. Fixed $s \in \mathbb{R}$, $K/a^2(s)$ is the curvature of the space-like submanifold $\Sigma_s = \{x \in X : t(x) = s\}$. See, for example, Einstein [4], Tolman [12] or Weinberg [13].

For a "generalized" de Sitter metric (the expansion parameter L is > 0 but, typically, $\neq K$)

$$a(t) = \cosh(\sqrt{L}t), \quad h(t) = \sqrt{L}\tanh(\sqrt{L}t), \quad h'(t) = L/\cosh^2(\sqrt{L}t); \tag{4}$$

in this case $R^{(g)} = 3L.g + 2(K - L)\frac{a^2(t)}{b^2(r)}.d\mathbf{r}^2$.

4. Creation and annihilation operators

The choice of the typical fiber $\mathcal{H}_0$ (of the quantum bundle) and of the corresponding vector space $\mathcal{E}_0$ (Section 2) are quite arbitrary and are done in dependence of what we want to describe. If the main objective is to build a theory of elementary particles and of their interactions, the obvious candidate for the role is a Fock space, direct sum of tensor product (symmetric or antisymmetric) of the *one-particle* Hilbert spaces, $\mathcal{H}_x^{(ONE)}$, invariant under the action of the (local) g_x–orthogonal group (Wigner [17], Schwartz [11], Weinberg [14], Chapter II).

Here $\mathcal{E}_x^{(ONE)}$ $(\subset \mathcal{H}_x^{(ONE)})$ is chosen to be the nuclear space $\mathcal{S}(\mathcal{T}'_x, \mathbb{C}^N)$ (a Schwartz space of indefinitely differentiable functions of rapid decrease), $1 \leq N < +\infty$; the vectors of $\mathcal{T}'_x$ represent the particles energy-impulse. Then, the positive and $\mathbb{C}$–bilinear *kernel* (which is a tempered distribution of $\mathcal{S}'(\mathcal{T}'_x \times \mathcal{T}'_x, \mathbb{C}^N \otimes \mathbb{C}^N)$)

$$K_x : f, g \mapsto K_x(f, g); \quad f,\, g \in \mathcal{S}(\mathcal{T}'_x, \mathbb{C}^N),\ x \in X \tag{5}$$

assigns a pre-Hilbert structure on $\mathcal{S}(\mathcal{T}'_x, \mathbb{C}^N)$: $\langle f, g\rangle_x = K_x(\,\overline{f}, g)$; the completion of this pre-Hilbert space can be identified to the one-particle Hilbert space, as in [11] and [17]. Afterwards we can define the local *Jordan-Wigner* creation and annihilation operators (Fermi statistics), $a_x^*(f) = \sum_n a_{x,n}^*(f_n)$, $a(f) = (a^*(\overline{f}))^*$, such that

$$[a_x(f),\ a_x^*(g)]_+ = K_x(f, g).I_x = \sum\nolimits_{mn} K_{x,mn}(f_m, g_n).I_x \tag{6}$$

and similarly for the local *Heisenberg* creation and annihilation operators (Bose statistics). Here $m, n = 1, 2, \ldots, N$, $x \in X$, $f,\ g \in \mathcal{S}(\mathcal{T}'_x, \mathbb{C}^N)$ and I_x is the identity of $\mathcal{H}_x$; the maps $f \mapsto a(f)$, $f \mapsto b(f)$ are $\mathbb{C}$*–linear* and continuous.

Then let be $\mu_{x,M}^{\pm}$ $(x \in X,\ M \in \mathbb{R}^+)$ a Lebesgue measure, concentrated on the local mass hyperboloid $\Omega_{x,M}^{\pm} = \{p \in \mathcal{T}'_x : \widehat{g}_x(p, p) = M^2,\ \tau_x(p) = \pm 1\}$; M is the *effective* mass of the particle we want to describe and we shall assume (this is the *fundamental postulate*) that:

$$\langle f, g\rangle_x = K_x(\overline{f}, g) = \sum\nolimits_n \int_{\Omega_{x,M}^+} \overline{f_n(p)} g_n(p).d\mu_{x,M}^{(R)}(p) \tag{7}$$

where $d\mu_{x,M}^{(R)} := \rho_x^{(R)}.d\mu_{x,M}^+$ and $\rho_x^{(R)}$ is a real, continuos function of rapid decrease, *dependent on the Ricci tensor* of g, $R^{(g)}$; in fact, for invariance reasons, we really do not have much more choices. Subsequently it is assumed that:

$$\rho_x^{(R)}(p) = \mathcal{N}(x, p)\exp\left(-\frac{\lambda}{2M_P^2}\widehat{R}_x(p, p)\right), \tag{8}$$

M_P is the Planck mass (or some other universal parameter), λ is a real parameter, a priori > 0 or < 0. $\mathcal{N}(x,p)$ is a *normalization factor* chosen in such a way that on some fixed spatial submanifold $\Sigma_0 = \{x \in X : t(x) = t_0\}$ (Section 3) the theory "reduces" to the standard one, that is $\rho_x^{(R)}(p) = 1$, on Σ_0. In the following we shall always set $t_0 = 0$.

For the de Sitter metric described at the end of Section 3, if $p = \varepsilon.dt + \frac{a}{b}\mathbf{p}.d\mathbf{r}$, $\mathbf{p} \in \mathbb{R}^3$ (so $M^2 = \widehat{g}(p,p) = \varepsilon^2 - \mathbf{p}^2$),

$$\widehat{R}(p,p) = 3M^2L + 2\frac{L-K}{\cosh^2(\sqrt{L}t)}\mathbf{p}^2, \tag{9}$$

when $\varepsilon^2 = M^2 + \mathbf{p}^2$; therefore, posing $\lambda_a(t) := \lambda\frac{K-L}{M_P^2}\tanh^2(\sqrt{L}t)$,

$$\rho_x^{(R)}(p) = \exp\left(-\lambda_a(t(x))\mathbf{p}^2\right) \tag{10}$$

which play the role of a "smooth" cut-off, physically generated. Evidently $\lambda(K-L)$ has to be > 0.

5. Quantum fields and interactions

If the $w(p)$ are, for every $p \in \mathcal{T}'_x$, linear maps $\mathbb{C}^{2N} \mapsto \mathbb{C}^N$and F, G are functions of $\mathcal{S}(\mathcal{T}'_x, \mathbb{C}^{2N})$ we can define the bounded, *self-adjoint*, anti-commuting operators ($n = 1, 2, \ldots, N$, $i, j = 1,\ 2,\ \ldots,\ 2N$)

$$\begin{aligned}\xi_x(F) &= \sum\nolimits_j \xi_{x,j}(F_j) = a_x^*(\overline{w.F}) + a_x(w.F) \\ &= \sum\nolimits_{nj}\left(a_{x,n}^*\left(\overline{w_{nj}F_j}\right) + a_{x,n}\left(w_{nj}F_j\right)\right)\end{aligned} \tag{11}$$

hence

$$[\xi_x(F),\ \xi_x(G)]_+ = \left(\Delta_x(\overline{F},G) + \Delta_x(\overline{G},F)\right).I_x. \tag{12}$$

Δ_x is a kernel of $\mathcal{S}(\mathcal{T}'_x, \mathbb{C}^{2N})$, $\Delta_x(\overline{F},G) = \sum_{ij}\int\left(\overline{F_i}m_{ij}G_j\right)(p).d\mu_{x,M}^{(R)}(p)$; $m_{ij}(p) = \sum_n \overline{w_{n,i}(p)}w_{n,j}(p) = \overline{m_{ji}(p)}$; if $\Delta_x(\overline{F},F) = 1$, the $\xi_x(F)$ are unitary. The ξ are essentially a local version of the anti-commuting Majorana "real variables" ([9], see also Weyl [16], Chapter IV); they describe, if $N = 2$, Majorana neutrinos. A "charged" particle, like the electron, can be described by a couple of such operators, (ξ,η); usually, the corresponding Dirac fields are defined as $\psi = \frac{1}{\sqrt{2}}(\xi + i.\eta)$ so $2[\psi^*,\ \psi]_+ = [\xi,\ \xi]_+ + [\eta,\ \eta]_+$.

It is now possible to define a metric dependent version of the anti-commuting *quantum fields*, as ($u \in \mathcal{T}_x$, $\chi_u(p) = e^{-i(p.u)}$):

$$\xi_j(x,u) := \xi_{x,j}(\chi_u) = \sum\nolimits_n\left(a_{x,n}^*\left(\overline{\chi_u w_{nj}}\right) + a_{x,n}\left(\chi_u w_{nj}\right)\right) \tag{13}$$

which are *well defined* operator-valued functions, assigned on $\mathcal{T}_X$. Note that it is alternatively possible to build the theory starting from the quantum fields, operator-valued functions defined on $\mathcal{T}_X$. For a flat metric, the measures μ_M do not depends on x, $\rho^{(R)} = 1$ and the tangent space can be (fixed a space-time origin) identified to X: we shall obtain the usual, distributional, quantum fields.

We can also build a metric dependent version of the *Pauli-Jordan functions*:

$$\Delta_{ij,M}(x,u) = \int_{\Omega_{x,M}^+} e^{ip.u}m_{ij}(p).d\mu_{x,M}^{(R)}(p) \tag{14}$$

so $[\xi_i(x,u),\ \xi_j(x,v)]_+ = \left(\Delta_{ij,M} + \overline{\Delta_{ij,M}}\right)(x;u-v).I_x$. The $u \in \mathcal{T}_x \mapsto \Delta_{ij,M}(x,u)$ are now true, differentiable and bounded functions. In fact the functions m_{ij} are tempered, for every

Journal of Physics: Conference Series **626** (2015) 012071 doi:10.1088/1742-6596/626/1/012071

representation of the $g-$orthogonal group, so $|m_{ij}(p)| \leq c\left(\mathbf{p}^2/M^2\right)^d$ (from some $c > 0, d \geq 0$) and

$$|\Delta_{ij,M}(x,u)| \leq c\int_{\mathbb{R}^3}\left(\frac{\mathbf{p}^2}{M^2}\right)^d \frac{e^{-\lambda_a(t(x))\mathbf{p}^2}}{2\sqrt{M^2+\mathbf{p}^2}}d\mathbf{p} \tag{15}$$
$$= c\pi M^2\int_{\mathbb{R}+} e^{-\lambda_a(t(x))M^2 q}.q^{d+\frac{1}{2}}.(1+q)^{-\frac{1}{2}}.dq;$$

the last integral is proportional to a confluent hypergeometric function $\lambda_a \mapsto \mathbf{\Psi}\left(d+\frac{3}{2}, d+2, \lambda_a M^2\right)$ (see Gradshteyn-Ryzhik [6], Section 9.2) which is analytic, for $\lambda_a > 0$, but diverge, as $(1/\lambda_a)^{d+1}$, for $\lambda_a \to 0^+$(that is for $\lambda \to 0$ or $M_P \to +\infty$; or when $t \to 0$).

The (quantum) *Lagrangian densities* (the free and the interaction terms) are here built in the usual way (see, for example, [1], [14] and, particularly, Jauch and Rohrlich, [7]) starting from a multilinear, self-adjoint expression, $\mathcal{L}(x,u)$, of the quantum fields; $x \in X$, $u \in \mathcal{T}_x$. The *action integral*, at $x \in X$, is obtained formally "summing" over all the vectors of $\mathcal{T}_x$:

$$\mathcal{A}_x = \int_{\mathcal{T}_x}\mathcal{L}(x,u).d\nu(u) \tag{16}$$

where ν is some Lebesgue measure assigned on $\mathcal{T}_x$. Note that the measure ν provides eventually a space-time cut-off and it is physically irrelevant, the relevant *ultraviolet divergences* (vacuum fluctuations, self masses, self charges) are controlled by the function $\rho^{(R)}$, of rapid decrease.

Hence, when $\lambda_a \approx 0$, we shall *almost* recover every standard, special relativistic theory. Almost, because some terms, in a perturbation expansion, became very large, when $\lambda_a \approx 0$. For example, when we try to calculate the electron "self-energies", in quantum electrodynamics, the Weisskopf-Furry term (Weisskopf [15]), depending on λ_a and now logarithmic diverging when $\lambda_a \to 0^+$ (from integrals like to $\int_{\mathbb{R}+} e^{-\lambda_a(t(x))M^2 q}.q^{\frac{1}{2}}.(1+q)^{-\frac{3}{2}}.dq$) is always there; we have to face this problem (Sections 6, 7).

6. Self-masses and counter terms

The Electro-Weak one-loop calculations of the leptons self-energies (self-masses) proceed now in the standard way, the factor $\exp\left(-\lambda_a(t)\mathbf{p}^2\right)$ provides only a "smooth" cutoff. The calculations can be done directly, as in [15] or by the propagators machinery ([1], [7], [14]). In the latter case, for every loop, we have to add to M (the effective mass of the lepton we are considering) a term like (t is always the time coordinate):

$$M_L.\alpha_X^*.I(t, M_L, M_X); \tag{17}$$

α_X^* is a constant, proportional to the coupling parameter α_X; M_L is the mass of the lepton of the loop; M_X is the mass of the particle which mediate the interaction (photons, $Z-$particles or $W-$particles). The functions I can be computed by standard techniques (Feynman parametrization, Wick rotation); for $\lambda_a \to 0$ and $M_X = 0$

$$I(t, M_L, 0) \approx -\log\left(\lambda_a(t)M_L^2\right) + \tfrac{5}{3} - c$$
$$= -\left(\log\tanh^2(\sqrt{L}t) + \log\left(M_L/M_P\right)^2\right) + c_1 \tag{18}$$

while if $M_L << M_X$:

$$I(t, M_L, M_X) \approx -\log\left(\lambda_a(t)M_X^2\right) + \tfrac{7}{6} - c$$
$$= -\left(\log\tanh^2(\sqrt{L}t) + \log\left(M_X/M_P\right)^2\right) + c_2. \tag{19}$$

c $(= \psi(\frac{3}{2}) + 2\gamma)$, c_1 and c_2 are constant, not depending on t or on M_L and M_X.

Ignoring the $W-$term, proportional to the neutral lepton mass, the charged leptons self-masses, M_{SELF}, are obtained adding the photon term, $M.\alpha^*_A.I(t, M, 0)$ and the $Z-$term, $M.\alpha^*_Z.I(t, M, M_Z)$:

$$M_{SELF}(t)/M = -\left(\alpha^*_A + \alpha^*_Z\right)\log\tanh^2(\sqrt{L}t) - \alpha^*_A \log\left(M_L/M_*\right)^2, \tag{20}$$

the constants are absorbed by M_*. M_{SELF} now *depends* on x, we are forced to assume that the interaction Lagrangian contains a mass-like *counter-term* such that we have to add to M_{SELF} another real function of x, M_Θ, proportional to M and such that $M_{SELF}(x) + M_\Theta(x) = 0$, for every $x \in X$. Alternatively (and perhaps more perspicuously) it is possible to start from a "bare" mass dependent on x, $M(x)$ $(= M + M_\Theta(x))$ and to impose that $M(x) + M_{SELF}(x) = M$ where M is the effective mass, not dependent on x. In any case we shall assume that the M_Θ term is associated to a *scalar* ("classical") field Θ by a relation like to:

$$\Delta_g \Theta(x) = -\frac{\theta}{M} M_\Theta(x).\Theta(x); \tag{21}$$

here Δ_g is the Laplace-Beltrami differential operator of the metric tensor g and θ $(\neq 0)$ is some real parameter.

When $K > 0$, X has the topology of $\mathbb{R} \times \mathbb{S}^3$; therefore, if $(t, \mathbf{n})$ is a diffeomorphism $X \mapsto \mathbb{R} \times \mathbb{S}^3$ $(\mathbf{n} \in \mathbb{R}^4, \mathbf{n}^2 = 1)$, posing $\Theta(x) = T(t(x)).S_\ell(\mathbf{n}(x))$

$$T''(t) + 3h(t).T'(t) + \frac{K\ell(\ell+2)}{a^2(t)}T(t) = \frac{\theta}{M}M_{SELF}(t).T(t); \tag{22}$$

the S_ℓ are the $\mathbb{S}^3-$ spherical harmonics, $\ell = 0, 1, 2, \ldots$; note that there are $(\ell+1)^2$ independent $\mathbb{S}^3-$ spherical harmonics, for every ℓ. Setting $a(t)^{3/2}T(t) = U(\sqrt{L}t)$ $(\sqrt{L}t = s)$, $a(t) = \cosh(\sqrt{L}t)$ (Section 3)

$$U''(s) = \left(w - w_A \log\tanh^2 s - \frac{w_B}{\cosh^2 s}\right)U(s) \tag{23}$$

where $w_A = \frac{\theta}{L}\left(\alpha^*_A + \alpha^*_Z\right)$, $w_B = \frac{3}{4} + \frac{K}{L}\ell(\ell+2)$ and $w = -\frac{\theta}{L}\alpha^*_A \log\left(M/M_*\right)^2$. For $w < 0$, U oscillate (at infinity) while, for $w > 0$ (that is $M < M_*$), U generally diverge (at infinity), except for a *finite* set of value of w: w_n, $n = 0, 1, 2, \ldots, n_0$; in this case U goes to zero exponentially, at infinity; besides, if n is even (odd), U is even (odd).

Approximating $\log\tanh^2 s$ by $-1/\cosh^2 s$, $U''(s) = \left(w - \frac{(w_B - w_A)}{\cosh^2 s}\right)U(s)$ and this differential equation can be integrate analytically in term of Jacobi polynomials ([6], Section 8.9), $z = \tanh s$, $n = 0, 1, 2, \ldots$

$$\mathbf{P}^{(\alpha,\beta)}_n(z) = \frac{1}{2^n}\sum_{\substack{a,b\in\mathbb{N}\\ a+b=n}} \binom{n+\alpha}{a}\binom{n+\beta}{b}(z+1)^a (z-1)^b. \tag{24}$$

Then $w_n = \left(\sqrt{w_B - w_A + \frac{1}{4}} - \frac{1}{2} - n\right)^2$, $n < \sqrt{w_B - w_A + \frac{1}{4}} - \frac{1}{2}$, $\alpha = \beta = \sqrt{w_B - w_A + \frac{1}{4}} - \frac{1}{2} - n$.

7. Lepton masses

Now we shall assume that the lepton case correspond to $\ell = 1$; if $w_A = 0.40$ and $w_B = 8.32$ (that is $\frac{K}{L} = 2.52$) a numerical integration of the U differential equation leads to $n_0 = 2$ (so there are exactly *three generations* of leptons) and to

$$w_0 = 4.798, \ \ w_1 = 1.708, \ \ w_2 = 0.0681. \tag{25}$$

Inserting the experimental value of α_Z/α_A ($\Gamma_Z\,(el, el) \approx \frac{1}{3}M_Z\alpha_Z$, but $\Gamma_Z = 0.084\ GeV$, $M_Z = 91.2\ GeV$ (PDG [10]) hence $\alpha_Z = 2.76 \cdot 10^{-3}$ and $\alpha_Z/\alpha_A = 0.378$) we get $\frac{\theta}{L}\alpha_A^* = w_A\frac{\alpha_A}{\alpha_A+\alpha_Z} = 0.29$ (and $M_* = 1.99\ GeV$). Then

$$\log\frac{M_n}{M_0} = \frac{w_0 - w_n}{0.58}; \tag{26}$$

$n = 0$ for the electron, $n = 1$ for the $\mu-$particle, $n = 2$ for the $\tau-$particle, therefore

$$\log\frac{M_\mu}{M_{el}} = 5.33, \quad \log\frac{M_\tau}{M_{el}} = 8.15 \tag{27}$$

compatibly to experimental data.

Finally observe that the real function $s \mapsto -w_A \log\tanh^2 s - w_B/\cosh^2 s$ (w_A, $w_B > 0$) diverge positively at the origin, has a negative minimum (if $w_B > w_A$) and goes to zero at infinity; on the contrary if, for example, $a(t) = e^{\frac{L}{2}t^2}$ we have to consider a function like $s \mapsto w_A \log\left(1 - e^{-s^2}\right) + w_B e^{-s^2} - w_C s^2$; consequently there are typically a denumerable infinity of self values: the metric expansion is, so to speak, too fast.

References

[1] Bogoliubov N N and Shirkov D V 1980 *Introduction to the Theory of Quantized Fields* (New York: John Wiley and Sons)
[2] Bourbaki N 2007 *Eléments de Mathématique. Varietes Differentielles et Analytiques: Fascicule de Resultats* (Berlin: Springer-Verlag)
[3] Dieudonné J 1970 *Eléments d'Analyse* (Paris: Gauthier-Villars Editeur)
[4] Einstein A 1955 *The Meaning of Relativity* (Princeton University Press)
[5] Gelfand I M and Vilenkin N Ya 1964 *Generalized Functions. Applications of Harmonic Analysis* vol 4 (New York: Academic Press)
[6] Gradshteyn I S and Ryzhik I M 1965 *Table of Integrals, Series and Products* (New York: Academic Press)
[7] Jauch J M and Rohrlich F 1955 *The theory of Photons and Electrons* (Cambridge MA: Addison-Wesley Publishing Company)
[8] Lang S 1985 *Differential Manifolds* (Berlin: Springer-Verlag)
[9] Majorana E 1937 Teoria simmetrica dell'Elettrone e del Positrone *Il Nuovo Cimento* **5** 171-184
[10] Olive K A et al. (Particle Data Group) 2014 *Chinese Physics* C **38** 090001
[11] Schwartz L 1968 *Application of Distributions to the Theory of Elementary Particles in Quantum Mechanics* (London: Gordon & Breach)
[12] Tolman R C 1987 *Relativity, Thermodynamics and Cosmology* (New York: Dover Publications)
[13] Weinberg S 1972 *Gravitation and Cosmology: Principles and Applications of the General Theory of Relativity* (New York: John Wiley and Sons)
[14] Weinberg S 1995 *The Quantum Theory of Fields* (Cambridge University Press)
[15] Weisskopf V F 1939 On the Self-Energy and the Electromagnetic Field of the Electron *Physical Review* **56** 72
[16] Weyl H 1950 *The Theory of Groups and Quantum Mechanics* (New York: Dover Publications)
[17] Wigner E P 1939 On Unitary Representations of the Inhomogeneous Lorentz Group *Annals of Mathematics* **40** 149-208

Spacetime representation of electromagnetic radiation in a (2+1)-dimensional universe

Thomas Reynolds

Department of Computer Science, University at Albany, 1400 Washington Ave., Albany, NY 12222 USA

E-mail: tracm2@acm.org

Abstract. Electromagnetic radiation's nature embodying both classical and quantum phenomena suggests the fundamental interest of understanding its status within any view of spacetime. Maxwell's equations have well proven their correctness as a formal representation of electromagnetic radiation. A speculative physical model of the light quantum developed within a (2+1)-dimensional view of spacetime is compared to the two mutual induction laws of Maxwell's Equations.

1. Introduction

Questioning the essential nature of spacetime is not straightforward. Whatever technical theories anyone may follow, everyone nevertheless seems personally and socially to operate habitually within the long shared human presumption of events manifesting themselves within a Newtonian arena. Einstein and Minkowski revised the technical scientific understanding of how spacetime bookeeping works, and quantum theory added an element of uncertainty to the bookkeeping, but all have mostly left spacetime at three spatial and one temporal dimensions. More recently, highly technical and speculative string theories have suggested as alternatives some or other increased dimensionality.

However, theories which consider a picture based on reduced spacetime dimensionality are much less common, and seem to be based on the theorists adopting a separable objective viewpoint from which they can see and describe spacetime.[1] Perhaps Archimedes could have moved the Earth if he had had a lever long enough and a place to stand (and a fulcrum). But the ability of the human mind to objectify the Universe as a whole may lead to error if there is not really a place to stand outside of the Universe to view it.

I have for some time been developing an idea not dependent on having a "place to stand", and have found it useful heuristically if not yet a fully integrated theory. In its emphasis on physical motion and change it may be categorized as Machian,[2] and can be simplistically stated as being fundamental spacetime strictly founded on no more than three dimensions, two spatial plus an additional one spatial (or spatiotemporal). This view has been the subject of a few colloquia presentations and various informal discussions but has been so far unpublished. In private discussion the question has been raised, given the foundational importance and success of Maxwell's equations, how to think of them in a way consistent with both their usual interpretation and this new view. The *DICE2014 Spacetime - Matter - Quantum Mechanics* conference presents an appropriate opportunity briefly to consider such an explication.

2. A thought experiment

Reading Maxwell's original writings, one can find him puzzling through the data and abstracted relationships from which he synthesized the now well-established "Equations".[3] Despite the widespread existence now of physical proofs—such as fantastically complex smartphones, for example—of the correctness of the ideas of Maxwell, of relativity, of quantum mechanics, and so on, it is still possible to puzzle about the physical world. One interesting puzzle, philosophical as well as physical, is the status of space and time. A phenomenon which Maxwell's equations formalize, electromagnetic radiation of energy ("EMR"), seems a good choice for special attention in this puzzle. Particularly, the quanta of that fundamental phenomenon must have each in itself an intimate relationship with the essential nature of the spatial, temporal world, or it would not count as fundamental.

This is a more than philosophical question. For example the important cultural enterprise of teaching physics, especially as physics incorporates the attoworld, may be more or less efficient or successful depending on the conceptual understanding of EMR.[4][5] And the work of engineers, whose thinking may be more concrete than physicists', may be similarly affected. Consider one very common representation used to express conceptual understanding, the simple transverse waveform shown in Figure 1. Some version of this probably is found in every introductory physics textbook.

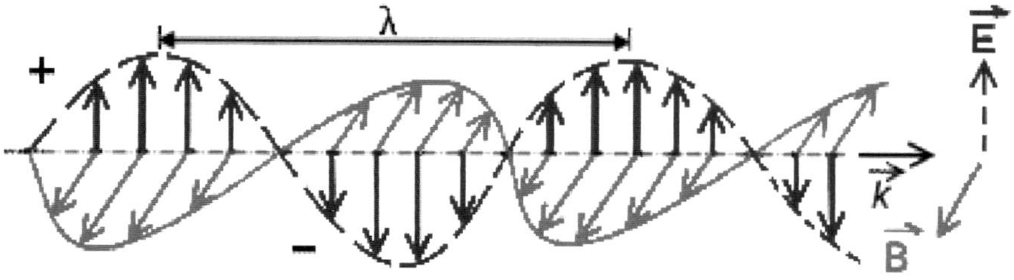

Figure 1. Electromagnetic radiation waveform

This is a formal graphical representation of varying correlated values with no explicit reference to space or to time. It is not especially a representation of a physical entity in spacetime without bringing in additional, unseen concepts. One can use a thought experiment to examine exactly what ideas might be needed.

Einstein famously cited a youthful thought experiment involving chasing a light beam as the beginning of what became the special relativity world view.[6] As a variation on that thought experiment, rather than chase a light beam imaginatively looking *at* it, ride *on* a single light quantum in free space and see as it "sees". In such case there is, for example, no waveform to be seen. If the light quantum is alone in free space, there is nothing to experience except the present moment's electromagnetic field, *ie* the raw quantum itself, without even any way to distinguish any moment from any other. Lacking data storage, a history to which to refer, one cannot say more. (One might want to say that one would be experiencing also the present moment's energy, but I omit direct consideration of energy in this paper.)

It is an important fact that an EMR packet carries no past with it. But maybe oneself is carrying data storage of some kind. If so, one can have a list of field strength values in sequence as one has "seen" them. One can relate the entries to each other, and note that they consist of a variety of values which repeat periodically. But one is not required to treat the data sequence as temporal, as a history. It can alternatively, and sufficiently, be treated solely as spatial, as values "here" and "behind" rather than "now" and "past". Do the data values go with clicks of a surveyor's wheel, or with ticks of a clock? It is merely an issue of adding place stamps rather than time stamps to the raw data items. There is no need to use both, and given that the idea of physical existence presumes spatiality that may be chosen as more basic.

In any case, riding on the light quantum, since there is a behind one might allow oneself to think of there being a "forward", at least in the sense of continuing the periodic sequence. However, there is no "ahead" nor "future". In this thought experiment one is not allowed to bring in the whole complicated extra apparatus of an usual space and time arena. So nothing exists until the quantum gets there, which consists merely of field strength values being added in sequence to the list. Whether space or time, the light quantum seems to be creating its own.

Before leaving the thought experiment, consider the light quantum itself, taken as an object separate from an observer. What might be said about what it is like? In the attempted view of an observer in a rest frame, a photon moving by at c is like a passing instantaneous presence, it cannot be "seen" coming or going since it is itself the criterion of visibility. In one instant, it need not have any more than a two-dimensional form. The person riding on the photon has no reason to suggest it is any other form, and any alternative would not be meaningful without an external reference. The two-dimensionality is like a physical realization of length contraction at the extreme of relative velocities.[7]

Can this understanding be justified? Is there a simple physical model which reasonably well coheres both with the understanding produced by the thought experiment and with all the usual scientific ideas about EMR?

3. A visualizable simple model

An heuristic image, a visualized interpretation of equation or graph, can be useful. After all, even scientists spend most of their time processing information using their non-technical mental habits. A good model works because it offers imagery involving recognizable physical things in manifest spacetime which preserve important technical scientific relationships. The Bohr model of the atom, for example, was quite productive despite its flaws. And its derivative vocabulary, of "orbits", energy "jumps", and "spin", is still, for better or worse, widely used in teaching.

What picture to aid thought might suit the EMR quantum discussed above, knowing only what has been seen so far? What are the formal features of the Maxwell equations (below) with which the model should correlate? First of all, how to understand the "∂t" has already been indicated. It is no more than the infinitesimal of the third dimensional forward spatial (or spatiotemporal) movement, correlating with the instantaneity of the light quantum.

Part of the beauty of Maxwell's equations is their simplicity, and their remaining features are references to electric and magnetic fields. What model, what shape, might realize the characteristic phenomena of electromagnetism to match with those remaining mathematical features? Following the principle that simplicity is a virtue, and trying not to think anything extra, allow the electromagnetic fields to have spatial existence, which they certainly do as ordinarily understood. This suggests a picture of the instantaneous EMR event as embodied by a two dimensional form of some kind, spatially oscillating radially as successive field strength states periodically occur. Consistent with a principle of spatial isotropy, and with orthogonality of $\vec{E}$ and $\vec{B}$, this would be two rather than one-dimensional, and discoid as its default shape. The sequence of periodic variations corresponds to the accumulating of successively ∂t-forward positions. Note that, at least in the case of a single light quantum, this need not be thought of as occupying successive positions in some spatial arena, which would assume an unnecessary separate concept. Rather, the discoid is constructing the third dimension of its own existence.

Merely by reducing the instantaneous state to a two dimensional form, and allowing oscillation of that form to proceed as (not *in*) a third dimension, one finds one has already a realization of a particle-wave hybrid: orderly variation of a two dimensional particle constituting three (but not four) dimensional wavelike behavior. Figure 1 represents just such a phenomenon.

The issue of course is how well a minimal two-plus-one dimensional understanding of fundamental spacetime could cohere with the great corpus of physics, both theory and experiment. Simplicity may be elegant without being correct. That Maxwell's equations legitimately include the case of a single quantum is ok, see for example [8] for one analysis. And so consideration continues with a speculative rough sketch.

4. Trying the model

Only two of Maxwell's equations will be considered here, the Faraday and Ampere-Maxwell laws.[9]

$$\text{Faraday Law: } \vec{\nabla} \times \vec{E} = -\partial \vec{B}/\partial t \qquad \text{Ampere-Maxwell law(simplified): } \vec{\nabla} \times \vec{B} = \partial \vec{E}/\partial t$$

Conceptually, Maxwell's equations for EMR, shown in derivative form here, include only electric and magnetic fields and characteristics of their activity and interrelationship. Considering the basic understanding of the fields as developed from the 19th Century to now, how might one optimize correlation of that knowledge with the so-far uncharacterized 2-D discoidal entity?

The first, obvious, correlation is of field strengths with spatial extent. Then the field partial differentials in the wave equations may be taken as describing characteristics of changing spatial features. From the result of the thought experiment and this one simple conceptual correlation, everything else including the cross products follows. That $\vec{E}$ by itself is more associated with localizable linear phenomena than with areal, and $\vec{B}$ vice versa, and that a 2-D discoid's essential physical features are just circumference and area (radius being a mathematical summary), together suggest correlating $\vec{E}$ with circumference and $\vec{B}$ with area. (I will use "Θ" for circumference and "A_m" for area.)

As always seeking a minimum interpretation, find $\vec{B}$ to be the usual supposed magnetic field lines perpendicular to and (by simplest hypothesis) filling the discoid surface's area "A_m". Suitable to the two-dimensionality, the $\vec{B}$ field lines, though having direction, are constituted actually as instantaneous 2-D transverse sections of field lines, which is not the same as points. An interesting question which will not be considered in this short paper is whether the $\vec{B}$ field lines might be found actually curving back outside of Θ and closing themselves, perhaps giving some evanescent or virtual $\vec{B}$ presence outside of Θ (in contextual circumstances when "outside" might make a difference). But here it is sufficient to leave them as merely 2-D slices with a direction + or − sense associated with them. The obvious choice for relevant spatial change of $\vec{B}$, distinct from sense direction changes, is greater or lesser area of A_m as field strength varies periodically. But keep in mind that for induction effects, the change that counts is relative motion of the field lines of $\vec{B}$ versus an electric charge or analogue to such.

Looking at Θ, its relevant spatial change can be as periodic increase and decrease of circumferential length, its own expansion and contraction occurring with A_m's. This seems an appropriate phenomenon to count as $\partial\vec{E}$, the displacement current which according to the Ampere-Maxwell law varies directly with $\vec{B}$ and has the same direction as its curl. Similar to the $\vec{B}$ slices, one must allow Θ to have no transverse thickness, leaving radial "size" to be as it may be. Thereby they may be said to have physical characteristics matching the formal representations "$\vec{B}$" and "$\vec{E}$".

Considering further, one can continue to see this configuration agreeing with Maxwell's equations. The displacement current has just been identified. Using the right hand rule and calling the thumb direction along Θ the + direction, curling one's fingers transversely around the circumference Θ matches the $\vec{B}$ lines slices within A_m and gives their + direction. The outmost $\vec{B}$ slices are effectively co-extensive with the circumferential displacement current and thereby provide loci of non-zero circulation of $\vec{B}$ to justify the curl $\nabla \times \vec{B}$ in the Ampere-Maxwell law left side which, by the same right hand rule, has as it should the same direction as $\partial\vec{E}$.

As for the Faraday law, there is not much to choose from to be circulation of $\vec{E}$ except the displacement current, the curvature of which varies from tighter at Θ_{min} to slighter at Θ_{max}. Curving one's right hand fingers along Θ finds one's thumb, the + direction of the $\vec{E}$ curl, pointing in the + direction of the $\vec{B}$ which is of course the direction of the $\vec{B}$ lines whose $\partial\vec{B}$ would be lines being effectively cut as Θ expands and contracts.. Since the Faraday law includes the "−" changing the sign of the right side, one uses the left hand rule to find the direction of the current induced by the cut lines, and one finds one's thumb pointing in the − direction of the displacement current, opposing it consistent with the usual understanding. So the model's basic geometry seems to be acceptable.

There may be problems related to how well numerical calculations, presumably referencing two rather than three dimensions, might match data. Given that present three-dimensional calculations do work, how the (2+1)-D model may produce results properly agreeing with empirical observations will need to be considered. Of course the rough model reported here does not suggest that the usual three-dimensional world does not exist or that science is not done in it. The project is merely to try to figure out how it may be possible to systematize seeing the world without reifying the time construct. This has long been suggested as appropriate, but as far as I know only in general without many specifics. Presumably a great number of adjustments may be needed to coordinate the differences between a (2+1)-D world view and the usual manifest one. In the "Comments" section I mention a couple of the many intriguing ideas about how usual three-dimensional materials may arise from 2-D energy packets. But this is indeed not a project for one single person to contemplate.

A different issue here is that so far + and − have been arbitrary as I limited my discussion to the general interrelationships found in Maxwell's equations. Explaining the decision among options how the + and − senses might correlate specifically with discoidal features or activities would require another discussion as long and detailed as I have just given. So I will merely mention that I use the obvious simple idea that field magnitudes correspond directly to area of A_m and length of Θ, and the less simple idea that there are two discoid oscillation periods per one light quantum wavelength period. The rest may be seen sketchily in the following brief discussion of light quantum activity using Figure 2. Note that given continuously mutual effects it is necessary to pick some one point and begin to describe the process from there.

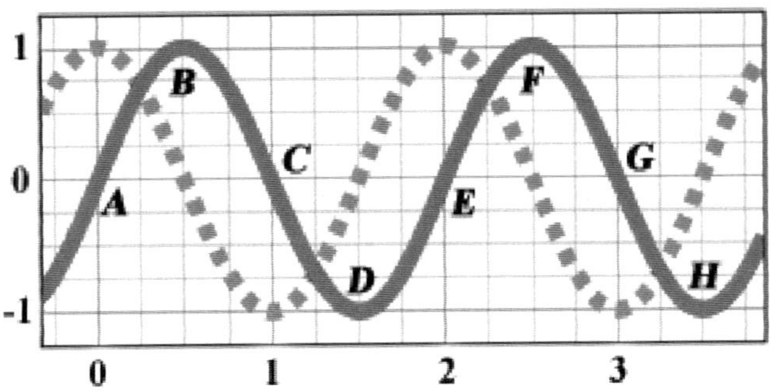

Figure 2. Simplified and annotated form of Figure 1. Solid red line is sine; dotted green line is cosine, the first derivative of sine. In arbitrary units.

At E in Figure 2, the fields have magnitude nominally 0, at which polarity flips within a continuous process the rate of change of which is here at maximum. This could correspond to the induced counter-field change in the positive direction, which began at D and is now at E,

(i) having nullified the negative fields the absolute value of which had maximized at D, now (ii) beginning to increase the discoid field again but of positive polarity. As the positive fields increase, the induced counter field begins to oppose them and will bring the positive-direction increase to a halt at F, where the lack of change leaves the fields to collapse and to begin to decrease toward G. This is the beginning of a change in the negative direction the rate of which will maximize at G and then taper off at H, at which point the process described just above from D repeats.

5. Comments

In this discussion I used only the concept "circumference" not *eg* "boundary". Electromagnetic energy in the real, well-filled Universe as we ordinarily understand it certainly is affected by surrounding objects which themselves, ex hypothesis, are ultimately constituted of electromagnetic quanta in various configurations. But it should be an empirical question whether a discoidal form such as this paper postulates would be best understood as an oscillating self-delimited object in a void or conversely as a pulsating void delimited by something like an aetherial substance. An aether theorist—which it is heuristically very tempting, but I think not necessary, to become with this view—might think of this raw electromagnetic energy as moving through a universal, uniform, immobile, and isotropic elastic medium.

Of course the exact form might not be discoid, or not always, as the phenomenon of polarization brings up. The representation of quanta as line-like and rotating in helical orientations seems inconsistent with spatial isotropy, nor is it easy to think of paired $\vec{E}$ and $\vec{B}$ as forming no more than a two dimensional stick. Spin is another puzzle which I have not covered in this paper, nor the reduction to this simple geometry of the complex imagery of string theory.

In any case, combination of individual light quanta with other entities would be likely to require some spatial distortion of the quanta. Whether completely losing their identity in merging with something, retaining an individuation while combining in a complex dance of trajectories, or splitting into parts, the transitions between isolated two dimensionality to functional three (but not four) dimensionality are an interesting spacetime puzzle. Heuristic possibilities are not hard to imagine, such as something like "feelers" of $\vec{B}$ lines extending from contiguous quanta as their geometries intertwined, then breaking under specific stresses. In all of these cases, though, the "arrow of time", of essential third spatiotemporal dimension progression, would remain as part of the fundamental nature of spacetime.

6. Conclusion

Since the physical models of light developed during the heyday of the wave–particle debate withered in the face of quantum theory, models of light quanta have been mostly nonphysical. However, as part of activity related to understanding the place of time in scientific thinking, a question was raised whether it is possible to model Maxwell's equations as describing a physical, electromagnetic entity in a spacetime of no more than three dimensions, two spatial plus an additional one spatial (or spatiotemporal). The belief is that given the increasing geometrization of all of the sciences, contributions should be made toward the development of a shared scientific culture in which thinking spatially has at least as much status as thinking temporally A model has been found which is at least heuristically successful and worth refining in future work.

References

[1] Barbour J 1999 *The End of Time* (Oxford UP: Oxford, UK)
Barbour J 2009 The nature of time *Preprint* arXiv gr-qc/0903:3489v1.
[2] Mach E 1960 *The Science of Mechanics* (LaSalle, IL USA: Open Court) "It is utterly beyond our power to *measure* the changes of things by *time*. Quite the contrary, time is an abstraction, at which we arrive by

means of the changes of things; made because we are not restricted to any one *definite* measure, all being interconnected." p.273, italics in original

[3] Simpson T K 1997 *Maxwell on the Magnetic Field* (New Brunswick, NJ USA: Rutgers UP)

[4] www.attoworld.de

[5] Byun T and Lee G 2014 Why students still can't solve physics problems after solving over 2000 problems *Am. J. Phys.* **82** (9) 906–13

[6] Einstein A 1951 Autobiographical notes *Albert Einstein: Philosopher-Scientist* ed P A Schlip (New York, NY USA: Tudor Publishing) pp. 2–95

Norton J D 2013 Chasing the light: Einstein's most famous thought experiment *Thought Experiments in Philosophy, Science, and the Arts* ed M Frappier *et al* (New York, NY USA: Routledge) pp. 123–140; Norton covers three separate accounts of the story, two written by Einstein and one reported by Gestalt psychologist Max Wertheimer who interviewed Einstein in 1916

[7] Rowe E G P 2001 *Geometrical Physics in Minkowski Spacetime* (London, UK: Springer) pp. 62–63

[8] Raymer M G and Smith B J 2005 The Maxwell wave function of the photon *Proc. SPIE 5866 The Nature of Light: What is a Photon?* (Aug. 4, 2005 San Diego, CA USA) pp. 293–7

[9] Fleisch D 2008 *A Student's Guide to Maxwell's Equations* (Cambridge, UK: Cambridge UP)

mergence of classical gravit and t e objective reduction of t e quantum state in deterministic models of quantum mec anics

icardo Galle o Torromé

[1]Departamento de Matemática, Universidade Federal de São Carlos, Rodovia Washington Luís, km 235, SP, Brazil

[2]The Abdus Salam International Center for Theoretical Physics, Strada Costiera 11, Trieste, Italy

E-mail: rigato39@gmail.com

Abstract. Models for deterministic quantum mechanics of Cartan-Randers type are introduced, together with the fundamental notions of the concentration of measure theory. We explain how the application of the concentration of measure to Cartan-Randers models provides a framework from which it emerge 1. The invariance under infinitesimal diffeomorphisms of the macroscopic dynamics 2. A mechanism for reduction of the quantum state and 3. The Weak Equivalence Principle.

1. Introduction

Deterministic emergent quantum mechanics denotes several new approaches to the foundations of quantum mechanics based on deterministic descriptions of an underlying level of physical reality [1, 2, 4, 7, 10, 11, 12, 19]. In deterministic emergent quantum mechanics there are two physical scales. First, there is a *fundamental scale*, which is usually associated with the Planck scale. The degrees of freedom at this scale are deterministic. The second scale is associated with quantum scales (Standard Model scale, or atomic or even molecular scales). The main objectives of such frameworks to reproduce the mathematical formalism of quantum mechanics as an emergent description from an underlying fundamental deterministic framework.

It is in this context of deterministic models for quantum mechanics that the author proposed a particular type of geometric micro-statistical models [8]. This paper describes a theoretical mechanism for quantum state reduction in the framework of such deterministic models. It turns out that in the dynamical regime where the quantum state reduction of the quantum state happens, the interaction driving the reduction has a strong resemblance with the gravitational interaction. This provides a newer look to the possibility that the dynamical reduction of the quantum state is related with the gravitational interaction.

The structure of this paper is the following. We first describe the dynamical systems that we will consider. After this, a succinct introduction to the *concentration of measure phenomena* in measure metric spaces is provided. This is a mathematical property that appears in several areas of geometry, functional analysis and probability theory. Then we apply concentration of measure to show how the reduction of the wave packet happens spontaneously in our deterministic models for quantum systems, independently of the presence of a measurement device or process. The concentration of measure shows how the weak equivalence principle emerges as well.

. Deterministic Cartan- anders Models

In the deterministic models that we consider the physical degrees of freedom describe point particles, whose evolution determines the worldsheets that are parameterized by a *two dimensional time parameter* (t, τ), where τ is a time coordinate associated to a macroscopic observer and t corresponds to a new parameter that enters in the dynamics. We called t the *internal time*. These two parameters are logically independent. These worldsheets are submanifolds of a configuration manifold which has a product structure

$$TM \cong \prod_{k=1}^{N} TM_4^k, \tag{1}$$

where $\{M_4^k,\ k = 1, ..., N\}$ are 4-manifolds diffeomorphic to a fixed 4-manifold M_4 and $N >> 1$. Then we have a collection of diffeomorphisms $\{\phi_k : M_4^k \to M_4,\ k = 1, ..., N\}$. The algebra of functions $\mathcal{F}_D(T^*TM)$ on T^*TM that we consider are the *diagonal functions*, obtained by an algebra embedding

$$\theta : \mathcal{F}(T^*TM_4) \to \mathcal{F}(T^*TM), \quad f \mapsto (f_1, ..., f_N),\ f_k = f,\ k = 1, ..., N. \tag{2}$$

that are elements of $\mathcal{F}(T^*TM)$ of the form

$$\big((u_1, p_1), ..., (u_N, p_N)\big) \mapsto \big(f(u_1, p_1), ..., f(u_N, p_N)\big).$$

It is relevance for our considerations that the average values of the functions $f \in \mathcal{F}_D(T^*TM)$ are well defined, that is, independent of the diffeomorphisms $\{\phi_k : M_4^k \to M_4,\ k = 1, ..., N\}$. This can be achieved if the measure μ_P in T^*TM is of the form

$$\mu_P = \prod_{k=1}^{N} \mu(k)_P, \tag{3}$$

where $\mu(k)_P$ is a probability measure in $T^*TM_4^k$. Then one defines

$$\widetilde{\mathcal{O}}_{CM}(\mathcal{S}, t, \tau) = \int_{T^*TM} \widetilde{\mathcal{O}}(k, t, \tau)\, \mu_P. \tag{4}$$

The classical Hamiltonian function for our deterministic systems is of the form

$$H(t, \tau, u, p) = \sum_{n=1}^{8N} \beta^n(u, t, \tau) p_n, \tag{5}$$

where $u \in TM$, $p \in T^*TM$, TM is the *configuration manifold* with $N >> 1$ and constrained by the requirement that, under an underlying metric structure $\hat{\eta}$ on T^*TM,

$$\|\beta\|_{\hat{\eta}} < 1. \tag{6}$$

These conditions are equivalent to have bounded acceleration and speed for each fundamental degrees of freedom. The theory that we propose goes beyond these requirements and assume the existence of universal maximal acceleration and maximal speed. We have called these dynamical systems deterministic Cartan-Randers models (in short, DCRM). These models were called Finslerian models in [8], but the new denomination is more adequate.

The two time parameters (t, τ) are used to describe a *double dynamics* U_t and U_τ. The dynamical equations for this Hamiltonian for the U_τ dynamics are

$$\frac{du^i}{d\tau} = \beta^i(t, \tau, u), \quad \frac{dp_i}{d\tau} = -\sum_{k=1}^{8N} \frac{\partial \beta^k(t, \tau, u)}{\partial u^i} p_k, \quad i, j, k = 1, ..., 8N. \tag{7}$$

The U_t should hold have several properties. There must consists of cycles. Each cycle start with an *ergodic regime* follow by a *concentration regime*, followed by a *expanding regime*, followed by the next *ergodic regime*, etc... The concentration regime is described by geometric flow of the geometric structures on T^*TM defining the Cartan-Randers system (that is, a dual metric $g \in \Gamma T^{(2,0)}TM$ and a vector field $\beta \in \Gamma TM$) and must allow for equilibrium states, characterized by the 6-dimensional hypersurface $\Sigma = \prod_{k=1} \Sigma_4 \times S_3$, where Σ_4 is the unit hyperboloid and S_3 is the 3-sphere. Although we do not propose a particular form of the flow here, neither in [8], we will be able to extract falsifiable consequences for our theory. The research of several possible U_t dynamics is under current investigation.

One of the techniques highlighted by G.'t Hooft is to describe classical systems quantum mechanically, that is, using Hilbert space theory (such ideas are closely related with Koopman's approach to dynamical systems [13, 16]). In Hooft's theory the standard canonical quantization relations are adopted and used. Thus, commutators are taken for operators at equal time τ. The quantum operators that we will consider are quantization of the diagonal algebra $\mathcal{F}_D(T^*TM)$. However, in DCRM time is associated with a 2-dimensional parameter and we need to specify the commutation relations at each fixed value of the pair (t, τ). Therefore, we adopt the following quantization rules defined as follow,

- The values of the position coordinates $\{x^\mu_k,\ k = 1, ..., N,\ \mu = 1, ..., 4\}$ and the velocity coordinates $\{y^\mu_k,\ k = 1, ..., N,\ \mu = 1, ..., 4\}$ of the fundamental degrees of freedom appear as eigenvalues of certain hermitian operators $\{\hat{x}^\mu_k, \hat{y}^\mu_k,\ k = 1, ..., N,\ \mu = 1, ..., 4\}$,

$$\hat{x}^\mu_k | x^\mu_l, y^\nu_l, \mu, \nu = 1, ..., 4\rangle = \sum_l \delta_{kl}\, x^\mu_l\, | x^\mu_l, y^\nu_l, \mu, \nu = 1, ..., 4\rangle,$$
$$\hat{y}^\nu_k | x^\mu_l, y^\nu_l, \mu, \nu = 1, ..., 4\rangle = \sum_l \delta_{kl}\, y^\nu_l\, | x^\mu_l, y^\nu_l, \mu, \nu = 1, ..., 4\rangle. \tag{8}$$

- There are a set of Hermitian operators $\{\hat{p}_{\mu x k}, \hat{p}_{\mu y k},\ k = 1, ..., N,\ \mu = 1, ..., 4\}$ that generates local diffeomorphism on TM along the integral curves of the local vector fields $\{\frac{\partial}{\partial x^\mu_k}, \frac{\partial}{\partial x^\nu_k},\quad \mu, \nu = 1, 2, 3, 4,\ k = 1, ..., N\}$.
- The canonical commutation relations at fixed 2-time (t, τ) hold good,

$$[\hat{x}^\mu_k, \hat{p}_{\nu x l}] = \imath\, \hbar\, \delta^\mu_\nu\, \delta_{kl}, \quad [\hat{y}^\mu_k, \hat{p}_{\nu y l}] = \imath\, \hbar\, \delta^\mu_\nu\, \delta_{kl} \tag{9}$$

and any other canonical commutator is equal to zero.

The Hamiltonian (5) can be quantized to obtain the following Hermitian operator,

$$\widehat{H}(t, \tau, \hat{u}, \hat{p}) = \frac{1}{2} \sum_{k=1}^{8N} \left(\beta^k(t, \tau, \hat{u})\, \hat{p}_k + \hat{p}_k\, \beta^k(t, \tau, \hat{u})\right) \tag{10}$$

When the quantum conditions (9) are applied to the Heisenberg's equation for the Hamiltonian (10), the equations (7) for the eigenvalues $\{x^\mu_k, y^\mu_k\ k = 1, ..., N,\ \mu = 1, ..., 4\}$ in the eigenbasis

(8) are obtained. Moreover, in order to be consistent between the Hilbert formalism and the geometric formalism, we need to impose the constraints

$$\hat{y}_k^\mu = \frac{d\hat{x}_k^\mu}{d\tau}, \quad \forall\, k = 1, ..., N, \quad \mu = 1, ..., 4. \tag{11}$$

These relations are consistent, despite the fact that in quantum mechanics the coordinate operators $\{\hat{X}^a, a = 2,3,4\}$ and the velocity operators $\{\dot{\hat{X}}^a,\ a = 2,3,4\}$ do not commute for each quantum degree of freedom a. However, let us remark that the operators $\{(\hat{x}_k^a, \hat{y}_k^a)\}_{k=1,a=2}^{N,4}$ do not coincide with the quantum operators $\{X^a, \dot{\hat{X}}^a,\ a = 2,3,4\}$. The operators $\{\hat{X}^a, \dot{\hat{X}}^a,\ a = 2,3,4\}$ have as spectrum the possible outcomes of measurements of position and velocity for each a. The operators $\{\hat{X}^a, \dot{\hat{X}}^a,\ a = 2,3,4\}$ should emerge in DCRM together with the wave function for the quantum state, that appears as phenomenological description of the ergodic regime when one considers the projection $(t,\tau) \mapsto \tau$. It is currently under investigation to construct systematically the operators $\{\hat{X}^a, \dot{\hat{X}}^a\, a = 2,3,4\}$ from the canonical operators $\{x^\mu, y^\mu, \hat{p}_{\mu xk}, \hat{p}_{\mu,yk},\ \mu = 1,2,3,4, = 1,...,N\}$.

As a consequence of this interpretation, the quantum states that are obtained from DCRM are generically non-localized in both position and speed (or canonical momentum) and both $\hat{X}^a$ and $\dot{\hat{X}}^a$ have generally non-zero dispersion. Therefore, the quantum states must hold a representation of a non-commutative algebra,

$$[\hat{X}_a^\mu, \hat{X}_b^\nu] = A^{\mu\nu}\,\delta_{ab}, \quad [\dot{\hat{X}}_a^\mu, \dot{\hat{X}}_b^\nu] = B^{\mu\nu}\,\delta_{ab}, \quad [\hat{X}^\mu, \dot{\hat{X}}^\nu] = C^{\mu\nu}\,\delta_{ab}, \tag{12}$$

with $\mu, \nu = 1,2,3,4, a, b = 1, ..., N$. The requirement of local invariance under the Lorentz group of this algebra implies that the spacetime must be a quantum spacetime compatible with an universal maximal acceleration and a universal maximal speed. A geometric realization of quantum spacetime with maximal acceleration and maximal speed is under current investigation.

A fundamental problem with the quantum Hermitian Hamiltonian operator (10) is that, being linear on the momentum operators, it is not direct that the quantum Hamiltonian $\widehat{H}(t,\tau,\hat{u},\hat{p})$ has a stable vacuum eigenstate. This problem can be solved if there is a dynamical mechanism that drastically reduces the dimensionality of the Hilbert space. In Hooft's proposal a gravitational type interaction originates the loss of information required for such reduction [4, 11]. However, there is not logical need for gravity from a formal point of view.

In DCRM there is a natural mechanism to bound from below the spectra of the quantum Hamiltonian $\widehat{H}(t,\tau,\hat{u},\hat{p})$ without introducing the gravitational interaction from the beginning [8]. In particular, in the U_t dynamics the classical Hamiltonian (5) evolves towards an identically zero classical Hamiltonian. Thus, we impose on the U_t operator the quantum constraint

$$\lim_{t\to T} \widehat{H}(t,\tau,\hat{u},\hat{p})|\psi\rangle_t = 0 \tag{13}$$

This constraint shows the emergent character of the local diffeomorphism invariance for the U_t and U_τ dynamics when acting on an generic quantum state $|\psi\rangle \in \mathcal{H}$ in the domain $t \to T$.

. Concentration o measure

Under the hypothesis that in the domain $t \to T$ the U_t is a 1-Lipschitz operator, we can find several interesting consequences. In particular, under this hypothesis, *concentration of measure* as it appears in asymptotic theory of finite normed spaces [14], metric geometry [9] and probability theory [18] can be applied in DCRM. Let us describe very briefly what is concentration of measure. A measure metric space is a triplet $(\mathcal{T}, \eta_P, \varepsilon)$ where $\mathcal{T}$ is a topological space, η_P is a measure on $\mathcal{T}$ and ε is a metric function on $\mathcal{T}$. Then the general property of concentration is usually quoted as follows [14],

In a measure, metric space, real 1-Lipschitz functions of a large number of real variables are almost constant almost everywhere.

The metric structure ϵ is necessary to define the 1-Lipschitz condition; the measure structure η_P is necessary to have a notion of almost everywhere.

The intuitive reason behind this phenomenon can be explained as follows in the case that $\mathcal{T}$ is a topological manifold (therefore, the notion of dimension of $\mathcal{T}$ is defined). When the function $f : \mathcal{T} \to$ is under strict control, as it is the case of a 1-Lipschitz function, the possibility that the difference between two values on the image is significant when they are found as a translation in a given direction is sharply cut off with the distance in the images between them. This is because when the number of variables is large, since similar differences will appear in other directions, the 1-Lipschitz condition will be spoiled. They can be exceptions to these bounds, but the probability that this happens is zero. The principle of concentration is a bast generalization of the central limit theorem in probability theory.

To illustrate how the concentration of measure arises in DCRM we assume that the internal dynamics operator U_t is 1-Lipschitz for values of the parameter t close enough to T. This assumption is compatible with the fact that there is an equilibrium limit is 1-Lipschitz (by assumption). Then let us consider a 1-Lipschitz function $f \in \mathcal{F}_D(T^*TM)$ of the form

$$f : \prod_{k=1}^{N} T^*TM_4^k \to R$$

with $N >> 1$. Then one has that almost everywhere

$$f(u_1, ..., u_N) = f(v_1, ..., v_N) + \mathcal{E}(N),$$

where $\mathcal{E}(N)$ is a small error depending on the natural number N. This error $\mathcal{E}(N)$ quantified in function of $\epsilon(u, v)$ in the form of *concentration maps* [14, 18]. These concentration maps are usually exponential maps. The power of the concentration phenomenon will be exemplified in the following two sections.

. Exam le o concentration o measure and a lication to dynamical reduction o quantum states

Let us consider an application of the concentration of measure in R^q with $q >> 1$. η_P is a measure and $f : R^q \to R$ a real 1-Lipschitz function. Then there is concentration given by (see for instance [18] , pg. 8)

$$\eta_P(|f - M_f| > \rho) \leq \frac{1}{2}\exp\Big(-\frac{\rho^2}{2\rho_P^2(f)}\Big), \tag{14}$$

where we have adapted the example from [18] to a Gaussian distribution η_P with mean M_f and standard deviation $\rho_P(f)$. $\rho_P(f)$ has the physical interpretation of being the minimal resolution attainable when measuring the observable associated with the classical function $f : R^q \to R$.

We apply this *example* of concentration to the function $f \in \mathcal{F}_D(T^*TM)$. For f there is a maximal resolution $\rho_p(f)$ in their possible measurement outcome values. In the 1-Lipschitz dynamical regime of the evolution operator U_t, the function f must be constant almost everywhere, since f is 1-Lipschitz in (u, p) and t. Moreover, for macroscopic observable effects one expect a relation of the type

$$\frac{\rho^2}{\rho_P^2(\tilde{f})} \simeq N^2, \quad 1 << N. \tag{15}$$

Let us restrict our considerations to the case when the quantum system corresponds to a fundamental particle which quantum fields appear in the Standard Model. The natural number N provides a measure of the *complexity* of the fundamental system compared with the complexity of the associated quantum system. The degree of complexity of a quantum state is of order 1, since there is one quantum particle involved compared with the degree of complexity of a DCRM, which is of order N. This order of complexity 1 is of the same order than the dimension of the model space-time manifold M_4, the number of spin degrees of freedom, and other quantum numbers associated with the quantum mechanical description of a fundamental quantum particle. Let us consider the case $TM \cong R^{8N}$. If we make use of (15), the concentration relation (14) applied to the function f in the 1-Lipschitz dominated regime of U_t becomes

$$\eta_P\Big(|f - M_f| > \rho_P(f)\Big) \leq \frac{1}{2}\exp\Big(-32\,N^2\Big). \tag{16}$$

Note that M_f depends on the initial conditions $(u_1(0),...,u_{8N}(0),p_1(0),...,p_{8N}(0))$. Since by assumption $\rho_P(f)$ is small compared with $|f - M_f|$ and $N_0 >> 1$, there is concentration of measure around the mean M_f. Thus, if a measurement of an observable associated with f is performed, the result M_f will be obtained with high certainty. This corresponds with a reduction of the quantum state. These reduction of the state not only happens when the system is being *measured*, but it is an *spontaneous process*. Such spontaneous processes happen even if there is not measurement. The driven force for this to happen is purely classical, since it appears in the limit $t \to T$ only (in the ergodic regime $\hat{H} \neq 0$ and therefore, it is not present). A measurement involves an additional process which is not classical and mediated by effective quantum gauge interactions between the system and a external quantum particle (for example, electromagnetic interaction). Such quantum process is amplified by the detector device.

. Concentration o measure and emer ence o the ea equivalence rinci le in DC M

We denote by *observable coordinate* $X^\mu(\mathcal{S}_i(\tau)$ associated to the system $\mathcal{S}_i$, $i \equiv \mathcal{S}, A, B$ the coordinate of the system $\mathcal{S}_i$ measured by a macroscopic observer. In the concentration(1-Lipschitz dynamical domain), such observable are 1-Lipschitz. The observable coordinate does not depend on t, since emerge in the effective description of DCRM in the projection $(t,\tau) \to \tau$. Therefore, they correspond to measurements of position observable performed by a macroscopic observer.

The configurations of the subsystems A and B can be represented in some special local coordinates on TM by the coordinate representation

$$A \equiv (u_1(\tau),...,u_{N_A}(\tau),0,...,0) \quad \text{and} \quad B \equiv (0,...,0,v_1(\tau),...,v_{N_B}(\tau)),$$

with $N = N_A + N_B, N_A, N_B >> 1$ and $u(\tau) = u(t = T, \tau)$. Locally, the full system $\mathcal{S}$ can be represented as

$$\mathcal{S} \equiv (u_1(\tau),...,u_{N_A}(\tau),v_1(\tau),...,v_{N_B}(\tau)).$$

By the concentration property (14), the evolution under the same initial conditions for the observable coordinates $X^\mu(\mathcal{S}(\tau))$, the configuration $X^\mu(A(\tau))$, the configuration $X^\mu(B(\tau))$ will differ after the evolution along the time τ such that

$$\eta_P\big(|X^\mu(\mathcal{S}_i(\tau)) - M^\mu(\tau)| > \rho\big)_{t\to T} \sim C_1(i)\exp\big(-C_2(i)\frac{\rho^2}{2L_p^2}\big).$$

Journal of Physics: Conference Series **626** (2015) 012073 doi:10.1088/1742-6596/626/1/012073

$C_2(i)$ is of order 32. For the U_t dynamics it can happens that there is an interchange of fundamental degrees of freedom in the form of fundamental interactions with the environment. This can affect the motion of the center of mass $M^\mu(\mathcal{S}_i(t,\tau))$ in a rather intricate way, depending on the system $\mathcal{S}_i(\tau)$. However, if the system is in *free fall* in the sense of absence of interchange of matter with the environment, the center of mass coordinates $M^\mu(\mathcal{S}_i(\tau))$ follow a well defined ordinary differential equation,

$$\frac{d}{d\tau}M^\mu(\tau) = h^\mu(\tau) \tag{17}$$

where the function h^μ is fixed by the equations of motion of the k-degrees of freedom by the properties of the measure μ_k and it does not depend on the system. We can assume that $\frac{d}{d\tau}\mu_k(t,\tau) = 0$. Then the center of mass coordinates $M^\mu(\tau)$ will not depend on the system $\mathcal{S}$, A or B. In this case, let us consider geometric configurations for the systems $\mathcal{S}_i$, $i = 1,2,3$ such that the mean value functions $\{M^\mu(\tau) : R \to R, \tau \mapsto M^\mu(\tau)\}_{\mu=1}^4$ do not depend on the system A, B or $\mathcal{S}$ at any time τ. Since for macroscopic or quantum systems the quotient $\frac{\rho^2}{2L_p^2} \geq 32$, there is concentration of the functions $\{X^\mu(\tau)\}_{\mu=1}^4$ around the same mean $\{M^\mu(\tau)\}_{\mu=1}^4$.

In the limit $t \to T$ the Hamiltonian is constrained by the condition (13). Such condition suggests that in the dynamical evolution described by the operator U_τ, the gravitational interaction must be included, since the dynamics is invariant under diffeomorphisms. Therefore, in the limit $t \to T$ we can decompose

$$\hat{H}_t = \hat{H}_{matter} + \hat{H}_{gravity}, \tag{18}$$

where $\hat{H}_{gravity}$ is the Hamiltonian responsible for the U_t dynamics and must be introduced in order that the constraint (13) holds. In this interpretation, gravity is associated with the purely internal dynamics U_t in the $t \to T$ limit. Moreover, if we assume the value N, we can make a falsifiable prediction: in DCRM framework, the weak equivalence principle applied to the observable coordinates $\{X^\mu\}_{\mu=1}^4$ is exact up to a precision $O(\exp(-32N^2)$ or higher. Thus, for $N = 1$ the error should be of order $\exp(-32)$, while for $N = 2$ it should be of order $\exp(-128)$.

. Gravity as emer ent interaction

If we collect all the characteristics of the 1-Lipschitz interaction U_t in the regime $t \to T$ we have that,

(i) It is described by a theory invariant under infinitesimal diffeomorphism transformations, since the constraint (13) holds good,

(ii) The weak equivalence principle for the center of mass functions $M^\mu(\mathcal{S}(\tau))$ holds good,

(iii) There is a local maximal speed for the fundamental degrees of freedom of a DCRM and local Lorentz invariance holds,

(iv) It is a classical, macroscopic and universal interaction,

(v) It must be compatible with the existence of a maximal acceleration.

Then the identification of a part of the 1-Lipschitz interaction U_t in the regime $t \to T$ and the gravitational contribution to the external interaction itself U_τ is a natural step, since the constraint (13) must hold. Also, let us remark that the acceleration is universally bounded, in contrast with general relativity. Maximal acceleration appears because the local character of interactions, the existence of an universal minimal universal length, and the existence of an universal maximal speed for any dynamical degree of freedom. All these properties are present in DCRM.

The suggestion that the gravitational interaction and objective reduction of the wave packet are related is not new (see for instance [3, 5, 15, 17]). However, we have argued that in the DRCM framework classical gravity and the reduction of the wave packet are two aspects of the same concentration of measure phenomena. Moreover, our argument supports the classical and emergent nature of the gravitational interaction. From what we have said, the emergent nature is clear. The classical nature is irremediably associated to the emergence, since it is an interaction that appears significatively only in the regime where physical quantum and macroscopic systems are localized in position and speed variables.

c no led ments

This work was supported by PNPD-CAPES n. 2265/2011, Brazil and by ICTP, Italy.

e erences

[1] Acosta D, Fernández de Córdoba P, Isidro J M and Santander J L G 2013 *Int. J. Geom. Methods Mod. Phys.* **10** 1350007
[2] Adler S L 2001 *Chance in Physics Lect. Notes Phys.* **574** ed Bricmont J*et al* (Berlin: Springer) pp 103-114
Adler S L 2004 *Quantum Theory as an Emergent Phenomenon: The Statistical Mechanics of Matrix Models as the Precursor of Quantum Field Theory* (New York: Cambridge University Press)
[3] Adler S L Gravitation and the noise needed in objective reduction models Preprint arXiv:1401.0353 [gr-qc]
[4] Blasone M, Jizba P and Vitiello G 2004 *Decoherence and Entropy in Complex Systems Lect. Notes Phys.* **633** Ed. Elze H T (Berlin: Springer) pp 151-163
[5] Diósi L 1989 *Phys. Rev. A* **40** 1165
[6] Diósi L 2009 *J. Phys.: Conf. Series* **174** 012002
[7] Elze H T 2006 *Trends in General Relativity and Quantum Cosmology* **253** Ed. Benton C V (New York: Nova Science Publ.) 79-101
Elze H T 2009 *Int. J. Qu. Info.* **7** 83
[8] Gallego Torromé R 2006 *J. Math. Phys.* **47** 072101
[9] Gromov M 1999 *Riemannian structures for Riemannian and non-Riemannian spaces* PM**152** Ed. LaFontanine J and Pansu P(Boston: Birkhäuser)
[10] Grössing G 2014 *Int. J. Mod. Phys. B*, **28** 1450179
[11] 't Hooft G 2000 Determinism and Dissipation in Quantum gravity Preprint hep-th/0003005
't Hooft G 2007 *J. Phys.: Conf. Series* **67** 012015
[12] 't Hooft G 2014 The Cellular Automaton Interpretation of Quantum Mechanics: a View on the Quantum Nature of our Universe, Compulsory or Impossible? Preprint arXiv:1405.1548 [quant-ph]
[13] Koopman B O 1931 *Proc. Nat. Acad. Sci. (USA)* **17** 315
[14] Milman V D and Schechtman G 2001 *Asymptotic Theory of Finite Dimensional Normed Spaces* Lecture notes in Mathematics (Berlin: Springer)
[15] Penrose R 1996 *Gen. Rel. and Grav.* **8** 581
Penrose R 2005 *The Road to Reality* (London: Vintage)
[16] Reed M and Simon B 1980 *Methods of Modern Mathematical Physics I: Functional Analysis, Revised and Enlarged Edition* (New York: Academic Press)
[17] Smolin L 1986 *Class. Quant. Grav.* **3** 347
[18] Talagrand M 1986 *Ann. Probab.* **24** 1-34
[19] Wetterich C 2009 *J. Phys.: Conf. Series* **174** 012008
[20] Will C M 2006 *Living Rev. Relativity* **9** 3

Symmetry as a foundational concept in Quantum Mechanics

Houri Ziaeepour

Institut UTINAM-CNRS, Université de Franche Comté, Besançon, France

E-mail: houriziaeepour@gmail.com

Abstract. Symmetries are widely used in modeling quantum systems but they do not contribute in postulates of quantum mechanics. Here we argue that logical, mathematical, and observational evidence require that symmetry should be considered as a fundamental concept in the construction of physical systems. Based on this idea, we propose a series of postulates for describing quantum systems, and establish their relation and correspondence with axioms of standard quantum mechanics. Through some examples we show that this reformulation helps better understand some of ambiguities of standard description. Nonetheless its application is not limited to explaining confusing concept and it may be a necessary step toward a consistent model of quantum cosmology and gravity.

1. Introduction

Looking at the history of science and technology of 20th century and these early years of 21st century, we find that most of them are directly or indirectly the fruit of understanding and application of quantum mechanics. From electronic components in smart phones which have become an indispensable gadget in human life, to electricity generated by nuclear or solar energy facilities, to understanding of photosynthesis phenomenon and maybe its artificial imitation in future, couldn't be discovered, understood, controlled, and applied without quantum mechanics.

However, despite owing so much to this most fundamental and profound law of physics, and despite its success in harshest and most stringiest tests that present science and technology allow, it is yet considered by many, including some of participants of this conference[1], to be incomplete and only the state of knowledge rather than complete reality. For this reason since its formulation in 1930 by Dirac and von Neumann, many alternative interpretations of quantum phenomena are suggested. Some of these alternatives such as deterministic interpretations and local hidden variables are ruled out [1, 2] - although there are claims that not all loop holes are closed and deterministic probably testable extension of quantum mechanics exist [3]. Furthermore, many of alternative interpretations do not present a discriminating test and in this regard are equivalent to quantum mechanics but less natural and more obscure. Example of these models are Bohmian quantum mechanics [4] and Everett *many worlds* interpretation [5].

In this proceedings we review a new proposal [6] for reformulating axioms of the standard quantum mechanics - rather than a new interpretation - using the logical and physical concept of symmetry. We show that some of rules such as Born law for determination of expectation values

[1] 7th *International Workshop DICE2014, Spacetime - Matter - Quantum Mechanics, Sep. 15-19, 2014, Castiglioncello, Tuscany, Italy.*

Journal of Physics: Conference Series **626** (2015) 012074 doi:10.1088/1742-6596/626/1/012074

can be obtained in place of being postulated and introduce by hand in the theory. Moreover, in this new description confusing concepts such as nonlocality and measurement problem can be better explained and understood. Indeed, in modeling quantum systems and phenomena, in particular in particle physics and condensed matter, symmetries are extensively used and play a central role in mathematical formulation and physical interpretation. But they do not explicitly appear in quantum foundation and postulates as suggested by Dirac and von Neumann. In mathematical logic, symmetry can be considered as an extension of fundamental logical concept of equality. It describes and quantifies degeneracies and maximum amount of information carried by a system in an abstract manner and independent of details.

In the following sections we first present axioms of quantum mechanics according to this new description. Then, through mathematical demonstration and case examples we show that the theory obtained from these axioms is equivalent to the standard formulation of quantum axioms. Nonetheless, confusing quantum observations acquire straightforward logical and mathematical interpretation directly related to symmetries.

2. Quantum mechanics postulates in symmetry language

(i) A quantum system is defined by its symmetries. Its state is a vector belonging to a projective vector space -called state space - which represents symmetry groups of the system. Observables are associated to self-adjoint operators and a set of independent observables is isomorphic to a subspace of commuting elements of the space of self-adjoint operators acting on the state space.

(ii) The state space of a composite system is homomorphic to the direct product of state spaces of its components. For non-interacting (separable) components, this homomorphism becomes an isomorphism.

(iii) Evolution of a system is unitary and ruled by conservation laws imposed by its symmetries and their representation by state space.

(iv) Decomposition coefficients of a state to eigen vectors[2] of an observable presents the symmetry/degeneracy of the system with respect to its environment according to that observable. Its measurement is by definition the operation of breaking this symmetry/degeneracy. The outcome of the measurement is the eigen value of the eigen state to which the symmetry is broken. This spontaneous[3] symmetry breaking reduces the state space (the representation) to subspace generated by other independent observables.

(v) A probability independent of measurement details is associated to eigen values of an observable as the outcome of a measurement. It presents the amount of symmetry/degeneracy of the state before its breaking by the measurement process.

3. Weirdnesses and unsettled issues of quantum mechanics

Before describing how the new description of quantum mechanics helps understand microscopic world and quantum phenomena, we briefly review observations and concepts which seem confusing and weird:

Complete but random The concept of randomness in physics existed before the discovery of quantum mechanics. But it was considered that it is an apparent rather than inherent and basic rule induced by inability of observing and adding to (classical) mechanical formulation of all

[2] To be more precise we should use the term *rays* in place of vectors because vectors differing by a constant are equivalent. Thus, we assume that states are normalized.

[3] We explain in Sec. 7.2 why the breaking of degeneracy between states should be classified as spontaneous.

degrees of freedom. Therefore, the fact that quantum mechanics only predicts a probability for outcomes of measurements and observations has been considered to be due to incompleteness of our information about physical world and its rules rather than fundamental. Consequently, *Hidden variables* were introduced to *complete* the information, and remove the randomness. However, observations, notably validation [1] of Bell inequalities [7], have ruled out the presence of a local hidden physics. There yet claims of nonlocal hidden physics [3, 5] but there is no proposition how their existence can be proved.

Nonlocality Classical world is local and and causal. In any process there is a well distinction between cause and effect. Even in Einstein relativity where order of events can be different for spacelike observers, causality is preserved for timelike observers who can exchange a signal with the limited and constant speed of light in vacuum. In quantum mechanics there are cases where cause and effect cannot be disentangled. This means that physical systems can be entangled and *communicate* with each other seemingly instantaneously. Einstein called this phenomenon *the spooky action at a distance.*

Contextual In classical mechanics a random process stays random for ever. A coin bouncing back by an elastic surface has an observable and definitive state at the time of touching the surface, i.e. its initial state before bouncing can be known. Nonetheless its state in the next touch-down is again random and observation or non-observation of its initial state does not change the probability of outcomes in the next observation/measurement. The same is not true in quantum mechanics in which once an observable is measured, repetition of the operation gives exactly the same value. In addition, in the case of vector quantities, in some situations it is not clear which component is observed. This contextuality of quantum mechanics raises a number of questions:

- Is a quantum state a physical reality ? If yes, why does it seem to have a non-unitary evolution when it is measured ?
- Why do measured quantities behave classically ?
- Does the wave function collapse?

In the following sections we describe answer to these questions in the framework of the reformulation of axioms of quantum mechanics based on symmetries.

4. State space and its properties

Symmetries are our only tool for abstraction of the Universe, and their breaking is the only logical and physical mean for distinguishing between objects. This observation is abstracted in representation of symmetry groups by vector spaces. Therefore, in contrast to the standard description of quantum mechanics which introduces an abstract state (Hilbert) space without specifying its relation with the system, postulate (i) explicitly associates it to the representation of symmetries of the physical system. This space presents, in a unique manner, all the properties of a system and what can be learned about them. Of course since the discovery of quantum mechanics vector spaces representing symmetries of systems were used as the abstract Hilbert space mentioned in the quantum mechanics axioms à la Dirac and von Neumann, but the axioms do not clarify and impose this choice. Consequently, it stays unclear whether another space - *a more complete space* - can exist which explains properties of the system in a deterministic rather than random manner. The reformulated axiom (i) emphasizes on the fact that only symmetries are meaningful and clarifies the abstract and unspecified nature of the state space. We remind that this postulate does not assume that the vector space of symmetry representation is a Hilbert space. We will prove this property in section (8).

Like any vector space the state space is generated by eigen vectors of its commuting subgroup, which according to axiom (i) defines independent observables. Decomposition of an arbitrary state vector depends on the selected basis, but a change of basis only modifies the decomposition of states not their identity or nature as a distinguishable member of the maximal abelian subalgebra of the associated algebra of the symmetry group. Therefore, in this description observables have an identity independent of the chosen basis. This fact is important, e.g. for interpretation of double Stern-Gerlach experiment, see the next subsection.

4.1. Active manifestation of symmetries in quantum mechanics and contextuality

In contrast to classical mechanics, in quantum systems symmetries have an operative role and induce degeneracies which reflect history and relation of a system with its environment. This solves preferred basis issue of quantum mechanics. A transformation by $U \in G$ where G is the symmetry group leads to:

$$|\psi\rangle = \sum_{\alpha} a_\alpha |\alpha\rangle, \; |\alpha\rangle \rightarrow |\alpha'\rangle = U|\alpha\rangle, \quad \Longrightarrow \quad |\psi\rangle \rightarrow |\psi'\rangle = \sum_{\alpha'} a_\alpha |\alpha'\rangle$$

Therefore the maximal abelian subspace is projected to a new subspace. In general $|\alpha'\rangle$ are no longer eigen vectors of the new abelian subspace and coefficients $a_{\alpha'}$ do not have the same value and physical interpretation as a_α. For instance, in a double Stern-Gerlach experiment electrons are polarized in the first detector and only one of polarizations, e.g. Z^+ is kept. We identify $\{|\alpha\rangle\} \equiv \{|Z^+\rangle, |Z^-\rangle\}$, $a_{Z^+} = 1$, $a_{Z^-} = 0$. The passage of electrons through the second detector with a field in X direction is equivalent to application of a symmetry transformation which projects the abelian subspace to X direction. According to postulate (i) and in agreement with observations, after passage to the second detector, observables belong to the new abelian subspace, and both polarizations in X direction become possible. Therefore, the origin and naturalness of contextuality is the fact that physical systems actively present symmetries.

Noncommuting position and momentum which both represent translation symmetry are considered to be example of preferred basis because it seems that collapsed/projected wave function of macroscopic objects prefers position basis. This misinterpretation has even led to construction of quantum gravity models based on position as preferred observable [8]. This apparent preference is evidently a misinterpretation and caused by the fact that most objects are non-relativistic with respect to us, and $\langle \hat{X}(t) - \hat{X}(t_0)\rangle \approx 0$. Thus, they look to be in a position eigen state. On the contrary, photons look like to prefer momentum basis. Therefore, preference of a basis is just a misconception.

5. State and symmetries in quantum and classical systems

Degeneracies of a system determine the representation realized by its state space and induce a symmetry that we call *state symmetry*. For instance, in a hydrogen atom the Lagrangian has $O(3) \cong SU(2)$ symmetry, but electron state space is generated by $(r, \phi, \theta) \in O(3)\mathbb{T} \times \mathbb{T} \times \mathbb{T} \cong U(1) \times U(1) \times U(1) \subset \mathbb{R}^3 \supset O(3)$. Once an observable is *measured*, according to axiom (iv) the state symmetry or degeneracy between eigen vectors breaks, and thereby the representation is reduced to trivial. Therefore, state symmetry is related to systems history, initial condition, and preparation, presumably it is ephemeral and may be total i.e. all states have similar probability, or partial i.e. some are preferred to others. In the former case coefficients of decomposition of the state to eigen vectors are equal, otherwise the symmetry is partial, meaning that the system's history or environment somehow discriminates between eigen states but cannot single out one of them. The symmetry of states is by nature associated to system-environment symmetries and is inseparable from it. Therefore, breaking of degeneracies is comparable and analogue to spontaneous symmetry breaking in which the symmetry is broken in states but not (or softly)

Journal of Physics: Conference Series **626** (2015) 012074 doi:10.1088/1742-6596/626/1/012074

in the Lagrangian. Moreover, if state space is decomposed to orthogonal subspaces, breaking of symmetry in one block does not affect others. Notably, if a symmetry orthogonal to spacetime is broken, the effect will be global and in contrast to classical physics, spacetime is not treated in a special manner. This explains the origin of nonlocality in quantum mechanics and why there is no classical analogue for it.

Only in vector spaces defined over complex numbers $\mathbb{C}$, operators with non-zero determinant are always diagonalizable[4]. This property is necessary to ensure the existence of a basis composed of eigen vectors for every observable. Because, properties of a system are by definition conserved under application of its symmetry group G, and observables are self-adjoint (hermitian) operators, representation of symmetry groups by state space must be unitary. This requirement enforces the necessity of definition of state space over $\mathbb{C}$. Although many groups are unitarily represented by vector spaces defined over real numbers, complete set of representations are realized on a complex field.

Symmetry and similarity make sense only if there are at least two objects to be compared. Thus, a nontrivial Universe U must have at least two non-empty nontrivial subsets S (system), and A (apparatus/observer). This fact explains why it seems that quantum mechanics needs the presence of an observer.

5.1. Comparison with classical physics

In Classical physics symmetries are passive, only reduce the number of degrees of freedom, and in many-body systems studied statistically, it makes distribution of some observables similar. A classical system with n observables has a configuration space isomorphic to $\bigotimes^n U(1) \cong \mathbb{R}^n$ irrespective of symmetries, except for conservation relations which impose constraintes. Consequently, at each moment the state of a system is a point in the configuration/phase space rather than a vector, and probabilities (observables) are $n-1$ simplexes in $\mathbb{R}^n$ and in general do not present a representation of symmetries of the system.

6. Inseparability of composite systems

Properties of state space explained in section 5 along with axiom (ii), symmetries in particle physics, and universality of gravity, show that decomposition of system to components is never completely orthogonal and it is not possible to find two quantum subsystems S_1 and S_2 such that $S_1 \cap S_2 = \oslash$. This rule also applies to the Universe U as a whole and means that it is an ensemble of intertwined vector spaces and is topologically open. Thus, for any system $S \subset U$ there is a system S' such that $S \subset S' \subset U$. Consequently, locality and separability are approximations and Universe is inherently nonlocal, composite, and inseparable.

As an application of this conclusion, we apply compositeness axiom to proposition of D.N. Page and W.K. Wootters [9] to use conditional rather than absolute probabilities in quantum mechanics. Their aim is to provide a consistent description of evolution of quantum systems with diffeomorphism symmetry which have total Hamiltonian $\mathcal{H} = 0$. This proposition has been criticized [10] mainly due to presence of an unobservable clock in the model. In the framework of symmetry description, the problem of this model can be recognized easily. One of their main assumptions is division of the Universe in two non-interacting parts which one is considered as system and the other as clock. According to postulate (ii) such a division is impossible, otherwise each part can be considered as a separate universe. Indeed modification of Page and Wootters proposal to solve the problem of unobservable clock [11] and its experimental realizations [12] include at least 3 subsystems which is consistent with minimum request of symmetry description. On the other hand, in the framework of symmetry description $\mathcal{H} = 0$ due to global diffeomorphism does not mean that Universe must be static, because it is by

[4] Strictly speaking this is true for finite representations.

Journal of Physics: Conference Series **626** (2015) 012074 doi:10.1088/1742-6596/626/1/012074

definition a composition of its subsystems which vary with respect to each others. One of these subsystems can be considered as a quantum clock and its entanglement with at least one other subsystem transfers the time to others. Such a clock is observable despite the absence of its direct interaction with all subsystems except one. Details of the operation of such clock will be discussed elsewhere.

7. Measurement

In standard description of quantum mechanics there are experiments in which it is not clear which quantity is measured. For example measurement of photon polarization with an interaction $\propto S_1 \times S_2$ (Dzyaloshinskii-Moriya interaction [13, 14]). Considering S_1 and S_2 as of system and apparatus, respectively, the transverse plane to $\mathbf{S}_2$ can be considered as preferred plane for measurement. Then, if the direction of the spin to be measured is already known to be in the plane, for instance if system consist of photons propagating in vacuum parallel to $\mathbf{S}_2$, there is no preferred direction on the plane. In standard description of quantum mechanics this seems problematic because it is not clear which component of spin vector is measured. On the other hand, in symmetry description there is no ambiguity, and in all cases the measured observable is the unique generator of abelian subalgebra of $SU(2)$ which is independent of arbitrary definition of coordinates on the plane.

7.1. Ambiguity of measurement basis

Modern literature on foundation of quantum mechanics present measurement as entanglement between system and apparatus or environment [15, 16]:

$$\sum_{\alpha} s_\alpha |\alpha\rangle \otimes |a_0\rangle \to \sum_{\alpha} s_\alpha |\alpha, a_\alpha\rangle \tag{1}$$

where $|a_0\rangle$ is the initial state of the apparatus or environment and $|\alpha\rangle$ is an arbitrary basis for the system. In the framework of symmetry description, system and apparatus must have an interaction related to the symmetry represented by states $|\alpha\rangle$. Assuming for simplicity finite dimensional representations, dimension M of the apparatus/environment initial representation must be larger than system dimension N. The entanglement by interaction breaks the $[N] \otimes [M]$ dimensional representation of system-apparatus to an $[N']$ dimensional representation. If $N' > N$ some outcomes would be equivalent. If $N' < N$ the apparatus will not be able to single out some states and they may create an interference. This reduction of state space, and thereby the symmetry, is usually explicit and related to the design of the apparatus, but there is no guarantee that $|\alpha, a_\alpha\rangle$ states be orthogonal to each others. In symmetry description this is not a problem because the right hand side of (1) presents the only physically meaningful entity, namely observable(s) associated to the symmetry in the representation realized by $|\alpha, a_\alpha\rangle$, even if these states are not eigen vectors.

7.2. Decoherence and symmetry breaking

Entanglement approach does not clarify the final breaking of degeneracies and decoherence. In symmetry description, the large number of degrees of freedom of environment present an infinite dimensional representation of symmetries, in general a degenerate representation containing many copies or equivalent subspaces. Entanglement reduces the state to:

$$|\psi\rangle \otimes |E\rangle = |\psi, E\rangle = \sum_{\alpha,\ \epsilon_i^\alpha, i=0}^{i\to\infty} f(\alpha, \{\epsilon_i^\alpha\})|\alpha, \epsilon_1^\alpha, \epsilon_2^\alpha, \cdots, \epsilon_i^\alpha, \cdots\rangle \tag{2}$$

where α index characterizes a basis for symmetry representation of the system and ϵ_i presents equivalence class of representations of the system induced to environment/apparatus by interaction with system. Random selection of one member of these equivalence classes breaks the symmetry and its realization by the system becomes trivial. Thus, in contrast to classical systems immediate repetition of the measurement on the same system gives the same outcome. Because equivalence classes realize the same representation of symmetry as the system, they play the role of a grand canonical ensemble and one can define a probability for outcomes which would be determined by the symmetry/degeneracy of the state of system and completely independent of the environment. This shows the crucial role of symmetry and its breaking for measurement and decoherence.

7.3. Symmetry breaking and phase transition

The process of entanglement of system with environment is similar to a phase transition. The state of system-environment can be projected to an arbitrary basis similar to (2) and projection coefficients can be treated as a random field. The Sinai theorem [17] for phase transition in classical systems proves that for perturbations - due to noise or temperature - lower than a critical value, the probability distribution function of a random field to which a Hamiltonian is associated, is a limit Gibbs distribution for a single phase, and under suitable conditions the system has a probability close to 1 to be in one of the stable ground states. This theorem can be also applied to quantum systems, specially because the presence of equivalence classes of symmetry representations in the environment/apparatus is mathematically analogue to the classical case. Therefore if the setup of a measurement is suitable, a symmetry breaking does occur and it is a direct consequence of quantum entanglement presented in the r.h.s. of (2) which makes the system to approach a ground state and stays there. This theorem also predicts the existence of surfaces in coupling space in which multiple phases can coexist. For quantum systems they correspond to cases where interaction does not completely break the symmetry and produce an interference.

For instance, in a Stern-Gelach experiment, the environment is composed of photons of the magnetic field and spins of atoms in the detector. Although all these components contribute in system-environment state $|\psi, E\rangle$, only a limited number of photons get the chance to interact with the passing electron. The interaction Hamiltonian between two spins is proportional to $\hat{\vec{S}}_1.\hat{\vec{S}}_2$. Therefore, Hamiltonian of this ensemble depends on $s_i s_j$ where s is the eigen value of projection of spins. Considering the large number of photons, $\{s_i\}$ behave as a random field s interacting with electron's spin. Although the state of the electron is initially a superposition of two polarizations, according to Sinai theorem, the photon spin field - the environment - pushes the electron spin - the system - to approach randomly to one of two minima of the Hamiltonian, corresponding to eigen value to the Cartan subalgebra of $\mathbb{Z}_2$ symmetry of polarized electrons along or opposite to detector's magnetic field, with a probability very close to 1 which breaks polarization superposition.

8. Randomness, probability, and Born law

In symmetry description of quantum mechanics Born law for probabilities associated to possible outcomes of measurements can be proved rather than postulated. Following axiom (i), state of a system contains all obtainable information about it. As probabilities f_i are by definition positive or zero real values, they must depend on absolute value of projection coefficients of the state on the eigen vectors of the observable operator $|a_i|^2$ with $\sum_i^n f_i(|a_i|^2) = 1$. In addition, without loss of generality and due to projective nature of state space, states can be normalized such that $\sum_i^n |a_i|^2 = 1$. Considering the state space of a system composed of two similar part:

$$|\psi\rangle = \frac{1}{2}(|\psi_1\rangle \otimes |\psi_2\rangle + |\psi_2\rangle \otimes |\psi_1\rangle) = \sum_i a_i^2 |i\rangle \otimes |i\rangle + \sum_{i \neq j} a_i a_j (|i\rangle \otimes |j\rangle + |j\rangle \otimes |i\rangle) \quad (3)$$

Journal of Physics: Conference Series **626** (2015) 012074 doi:10.1088/1742-6596/626/1/012074

Symmetrization over two subsystems means that due to similar preparation they are indistinguishable. From this expression we conclude that the probability for subsystems to be in (i, j) state is $f(|a_i a_j|^2)$. On the other hand, because these subsystems are prepared independently, the mutual probability for the system to be in (i, j) state is $f(|a_i|^2)f(|a_j|^2)$. Therefore:

$$f(|a_i|^2)f(|a_j|^2) = f(|a_i|^2|a_j|^2) \tag{4}$$

It has been proved [18] that the only function with such property is a positive power-law. Finally, by applying (4) to summation properties, one finds the Born law:

$$f(|a_i|^2) = |a_i|^2 \tag{5}$$

This relation also proves that state space must be a Hilbert space and show that the reformulation of quantum axioms based on symmetry is equivalent to standard description. Nonetheless, it gives a more logical structure to the model and its predictions

9. Outline

Introduction of symmetry in axioms and fundamental structure of quantum mechanics is not just for clarification of some subtleties and confusing predictions. Symmetry is a fundamental concept in construction of physical Universe and its inclusion in foundation of quantum mechanics is most probably crucial and necessary for building an abstract model of the Universe and its content. An evidence for this is the central role of gauge symmetry in particle physics and diffeomorphism symmetry of spacetime. We highlighted a couple of consequences of symmetries for global view of cosmology in this proceedings, but many others must be explored and understood, notably in the path to discovery of a unified model of quantum process which includes gravity.

References

[1] Rowe M A, Kielpinski D, Meyer V *et al.*2001 *Nature* **409** 791
[2] Colbeck R and Renner R 2011 *Nature Commu.* **2** 411 (*Preprint* arXiv:1005.5173)
[3] Ghirardi G and Raffaele R 2013 *Phys. Rev. Lett.* **110** 170404
[4] Bohm D 1952 *Phys. Rev.* **85** 166; *do.* 1952 *Phys. Rev.* **85** 180
[5] Everett H I 1957 *Rev. Mod. Phys.* **29** 454
[6] Ziaeepour H 2013 *Preprint* arXiv:1305.4349
[7] Bell J S 1964 *Physics* **1** 195
[8] Dowker F, Henson J and Sorkin R 2004 *Mod. Phys.* A **19** 1829
[9] Page D N and Wooters W K 1983 *Phys. Rev.* D **27** 2885
[10] Kuchař K 1992 in *Proceedings of the* 4th *Canadian Conference on General Relativity and Relativistic Astrophysics* ed G Kunstatter, D Vincent and J Williams (Singapore: World Scientific)
[11] Gambini R, Porto R A, Pullin J and Torterolo S 2009 *Phys. Rev.* D **79** 041501 (*Preprint* arXiv:0809.4235)
[12] Moreva E, Brida G, Gramegna M Giovannetti V, Maccone L and Genvese M 2013 *Preprint* arXiv:1310.4691
[13] Dzyaloshinskii I E 1958 *J. Phys. Chem. Solids* **4** 241
[14] Moriya T 1960 *Phys. Rev.* **120** 91
[15] Schlosshauer M 2004 *Rev. Mod. Phys.* **76** 1267 (*Preprint* quant-ph/0312059)
[16] Schlosshauer M and Fine A 2006 in *Quantum Mechanics at the Crossroads: New Perspectives from History, Philosophy and Physics* ed J Evans and A S Thorndike (Berlin: Springer)
[17] Sinai Ya G 1982 *Theory of phase transition: Rigorous results* international series in natural philosophy (New York: Pergamon Press)
[18] Hardy L 2001 *Preprint* arXiv:quant-ph/0101012